Transnational Accounting

Second Edition

TRANSACC

Transnational Accounting

Second Edition

Edited by
Dieter Ordelheide
and
KPMG

VOLUME
3

palgrave

Published 2001 by
PALGRAVE
Houndmills, Basingstoke, Hampshire RG21 6XS and
175 Fifth Avenue, New York, N.Y. 10010
Companies and representatives throughout the world

PALGRAVE is the new global academic imprint of St. Martin's Press LLC Scholarly and Reference Division and Palgrave Publishers Ltd (formerly Macmillan Press Ltd).

Editors
Dieter Ordelheide
KPMG

Outside North America
ISBN 0–333–71020–7

In North America
ISBN 1–56159–246–3

First published by Macmillan Press and Stockton Press, 1995
Reprinted 1996

This book is printed on paper suitable for recycling and made from fully managed and sustained forest sources.

A catalogue record for this book is available from the British Library.

10 9 8 7 6 5 4 3 2 1
10 09 08 07 06 05 04 03 02 01

Printed in Great Britain by
Bath Press, Bath

TRANSNATIONAL ACCOUNTING

Volume 3

CONTENTS

TRANSNATIONAL ACCOUNTING

Volume 1

CONTENTS

TRANSNATIONAL ACCOUNTING

Volume 2

CONTENTS

List of Abbreviations (Volume 3: Portugal–USA)

Portugal

ACE	*Agrupamento Complementar de Empresas* (Complementary Company Groupings)
ATOC	*Associação dos Técnicos Oficiais de Contas* (Chamber of Official Accountants)
CCom	*Código Comercial* (Commercial Code)
CIRC	*Código do Imposto sobre o Rendimento das Pessoas Colectivas* (Companies Income Tax Code)
CMVM	*Comissão do Mercado de Valores Mobiliários* (Portuguese Securities Market Commission)
CODMVM	*Código do Mercado de Valores Mobiliários* (Securities Market Code)
CNC	*Comissão de Normalização Contabilística* (Accounting Standardization Commission)
CRC	*Código do Registo Comercial* (Commercial Registry Code)
CROC	*Câmara dos Revisores Oficiais de Contas* (Chamber of Registered Statutory Auditors)
CSC	*Código das Sociedades Comerciais* (Companies Business Code)
CTOC	*Câmara dos Técnicos Oficiais de Contas* (Chamber of Official Accountants)
DGCI	*Direcção-Geral das Contribuições e Impostos* (Tax authority)
DR	*Diário da República* (Official government journal)
GAAP	Generally accepted accounting principles
IASC	International Accounting Standards Committee
ISP	*Instituto de Seguros de Portugal* (Insurance Institute of Portugal)
LDA	*Sociedade por Quotas* (Private Limited (liability) company)
OROC	*Ordem dos Revisores Oficiais de Contas* (Chamber of registered Statutory Auditors)
POC	*Plano Oficial de Contabilidade* (Official Accounting Plan)
PTE	*Escudos* (Portuguese Monetary Unit (Escudos))
ROC	*Revisor Oficial de Contas* (Registered Statutory Auditors)
SA	*Sociedade Anónima* (Public limited company or corporation)
TOC	*Técnicos Oficial de Contas* (Official Accountant)

Spain

AECA	*Asociación Española de Contabilidad y Administración de Empresas* (Accounting and Business Administration Association)
BOE	*Boletín Oficial del Estado* (Official Gazette)
BOICAC	*Boletín Oficial del Instituto de Contabilidad y Auditoria de Cuentas* (Official Gazette of the Accounting and Audit Institute)
CCom	*Código de Comercio* (Commercial Code)
CNMV	*Comisión Nacional del Mercado de Valores* (National Securities and Exchange Commission)
ESP	Spanish Peseta
ET	Entity Theory
FIAMM	*Fonds de inversión en activos del mercado monetario* (Money market asset investment funds)
IACJCE	*Instituto de Auditores-Censores Jurados de Cuentas de España* (Institute of Chartered Accountants of Spain)

ICAC	*Instituto de Contabilidad y Auditoria de Cuentas* (Accounting and Audit Institute)
IVA	*Impuesto sobre el valor añadido* (Value added tax)
LSA	*Ley de Sociedades Anónimas* (Corporations Law)
LSRL	*Ley de Sociedades de Responsabilidad Limitada* (Limited Liability Companies Law)
NC	*Normas para la formulación de las cuentas anuales consolidadas/Normas de consolidación* (Standards for the preparation of consolidated annual accounts/Consolidation standards)
PCT	Parent Company Theory
PGC	*Plan General de Contabilidad* (National Chart of Accounts)
RD	*Real Decreto* (Royal Decree)
RDL	*Real Decreto Legislativo* (Legislative Royal Decree)
REA	*Registro de Economistas Auditores* (Register of Economic Auditors)
SA	*Sociedad anónima* (Corporations)
SC	*Sociedad Colectiva* (General partnership)
SCom	*Sociedad comanditaria simple* (Ordinary limited partnership)
SComPA	*Sociedad comanditaria por acciones* (Limited partnership with shares)
SRL or SL	*Sociedad de responsabilidad limitada* (Limited liability company)
UTES	*Uniones temporales de empresas* (Joint ventures)

Sweden

ABL	*Aktiebolagslagen* (Companies Act)
BFL	*Bokföringslagen* (Accounting Act)
BFN	*Bokföringsnämnden* (Accounting Standards Board)
EES	*Europeiska ekonomiska samarbetsområdet*, EFTA countries, except Switzerland (European Economic Cooperation Area)
EFTA	European Free Trade Association
FAR	*Föreningen Auktoriserade Revisorer* (Association of Authorized Public Accountants)
FPG	*Frivillig pensionsgaranti* (Compulsory Insurance Scheme)
NBK	*Näringslivets börskommitté* (Industry Committee on Stock Exchange Affairs)
PRV	*Patent och registreringsverket* (Registration Authority)
RNFS	*Revisorsnämndens föreskrifter* (Supervisory Board of Public Accountants Act)
RR	*Redovisningsrådet* (Financial Accounting Standards Council)
SEK	Swedish Krone
SFF	*Sveriges Finansanalytikers Förening* (Society of Swedish Financial Analysts)
SI	*Sveriges Industriförbund* (Federation of Swedish Industries)
SURV	*Skatteutjämmingsreserv* (Tax Equalization Reserve)
ÅRFL	*Lag om årsredovisning i försäkringsföretag* (Annual Report of Insurance Companies Act)
ÅRKL	*Lag om årsredovisning i kreditinstitut och värdepappersbolag* (Annual Report of Financial Institutions Act)
ÅRL	*Årsredovisningslagen* (Annual Report Act)

Switzerland

ABV	*Verordnung über die ausländischen Banken in der Schweiz* (Ordinance regulating Foreign Banks)
AFG	*Bundesgesetz über die Anlagefonds* (Federal Investment Funds Act)
AFV	*Verordnung über die Anlagefonds* (Ordinance regulating Investment Funds)

AG	*Aktiengesellschaft* (Stock corporation)
AHVG	*Bundesgestz über die Alters- und Hinterlassenen-Versicherung* (Federal Old Age, Widows' and Widowers' Insurance Act)
AKAD	*Aktivische Konsolidierungsdifferenz/Kapitalaufrechnungsdifferenz* (Positive consolidation difference)
ALVG	*Bundesgesetz über die Arbeitslosenversicherung* (Federal Unemployment Insurance Act)
ARR	Accounting and Reporting Recommendations in Switzerland
AVO	*Verordnung über die Beaufsichtigung von privaten Versicherungseinrichtungen* (Ordinance governing the Regulation of Private Insurance Companies)
BSV	*Bundesamt für Sozialversicherung* (Federal Social Insurance Agency)
BVG	*Bundesgesetz für berufliche Vorsorge* (Federal Pension Fund Law)
CHF	Swiss Franken
EEA	European Economic Area
EBK	*Eidgenössiche Bankenkommission* (Swiss Federal Banking Commission)
EStV	*Eidgenössische Steuerverwaltung* (Swiss Federal Tax Authorities)
FER	*Fachempfehlungen zur Rechnungslegung* (Swiss Accounting and Reporting Recommendations)
GmbH	*Gesellschaft mit beschränkter Haftung* (Limited liability company)
GoR	*Grundsätze ordnungsmäßiger Rechnungslegung* (Commercially accepted accounting principles in Switzerland)
GV	*Generalversammlung* (General meeting of shareholders)
HB I	*Handelsbilanz I* (Legal statutory accounts)
HB II	*Handelsbilanz II* (Accounts for consolidation purposes)
HGB	*Handelsgesetzbuch* (German Commercial Code)
HR	*Handelsregister* (Commercial Register)
HWV	*Höhere Wirtschafts- und Verkehrsschulen* (Commercial junior colleges)
IAS	International Accounting Standards
IASC	International Accounting Standards Committee
KG	*Kommanditgesellschaft* (Limited partnership)
KUVG	*Bundesgesetz über die Kranken- und Unfallversicherung* (Federal Health and Accident Insurance Act)
OECD	Organization for Economic Co-operation and Development
OR	*Obligationenrecht* (Swiss Code of Obligations)
PC	Swiss Penal Code
p.m.	pro memoria
PS	*Partizipationsschein* (participation certificate or non-voting share)
RHB	*Revisionshandbuch der Schweiz* (Swiss Auditing Handbook)
SA	*Société Anonyme* (French name for Corporation)
SBB	*Schweizerische Bundesbahnen* (Swiss Federal Railways)
SEC	Securities and Exchange Commission in the USA
SHAB	*Schweizerisches Handelsamtsblatt* (Swiss Trade Journal)
SWX	Swiss Stock Exchange
US GAAP	Generally accepted accounting principles in the USA
VAG	*Bundesgesetz betreffend die Aufsicht über die privaten Versicherungseinrichtungen* (Federal Act on the Regulation of Private Insurance Companies)
VR	*Verwaltungsrat* (Board of directors)
WPH	*Wirtschaftsprüferhandbuch* (Auditing Handbook)

UK

APB	Auditing Practices Board
ASB	Accounting Standards Board
ASC	Accounting Standards Committee
APC	Auditing Practice Committee
ASSC	Accounting Standards Steering Committee
BBA	British Bankers' Association
BCCI	Bank for Credit and Commerce International
CA	Companies Act
CCA	Current cost accounting
CCAB	Consultative Committee of Accounting Bodies
CICA	Canadian Institute of Chartered Accountants
CIPFA	Chartered Institute of Public Finance and Accountancy
COSA	Cost of sales adjustment
CPP	Current purchasing power
DCF	Discounted cash flow
DP	Discussion Paper
DTI	Department of Trade and Industry
ED	Exposure Draft
EPS	Earnings per share
EU	European Union
EUV	Existing use value
FASB	Financial Accounting Standards Board in the USA
FRC	Financial Reporting Council
FRED	Financial Reporting Exposure Draft
FRRP	Financial Reporting Review Panel
FRS	Financial Reporting Standard
GAAP	Generally accepted accounting principles
GBP	British Pound Sterling
IBF	Irish Bankers' Federation
IASC	International Accounting Standards Committee
IAS	International Accounting Standard
ICAEW	Institute of Chartered Accountants in England and Wales
ICAS	Institute of Chartered Accountants in Scotland
IFAC	International Federation of Accountants
IIMR	Institute of Investment Management and Research
IMRO	Investment Management Regulatory Organization
IOSCO	International Organization of Securities Commissions
ISE	International Stock Exchange
MWC	Monetary working capital
MWCA	Monetary working capital adjustment
OBSF	Off-balance sheet financing
OFR	Operating and Finance Review
OIAC	Oil Industry Accounting Committee
OMV	Open market value
PLC	Public limited company
RAA	Reasonable accounting analogy
RSB	Recognized supervisory bodies
S	Section

SAS	Statements of Auditing Standards
Sch	Schedule
SEC	Securities and Exchange Commission
SFAS	Statement of Financial Accounting Standards
SI	Statutory Instrument
SOP	Statement of Principles
SOPED	Statement of Principles Exposure Draft
SORP	Statement of Recommended Practice
SP	Statement of Principles
SSAP	Statement of Standard Accounting Practice
STRGL	Statement of Total Recognised Gains and Losses
TR	Technical Release
UITF	Urgent Issues Task Force
USM	Unlisted Stock Markets

USA

AAA	American Accounting Association
ABA	American Bar Association
AcSEC	Accounting Standards Executive Committee
AICPA	American Institute of Certified Public Accountants
APB	Accounting Principles Board
ABO	Accumulated Benefit Obligation
ARB	Accounting Research Bulletin
ASB	Auditing Standards Board
ASR	Accounting Series Release
CAP	Committee on Accounting Procedures
CPA	Certified Public Accountant
CPE	Continuing Professional Education
EDGAR	Electronic Data Gathering, Analysis, and Retrieval
EITF	Emerging Issues Task Force
EPS	Earnings per share
FAF	Financial Accounting Foundation
FARS	Financial Accounting Research System
FASAC	Financial Accounting Standards Advisory Council
FASB	Financial Accounting Standards Board
FIN	FASB Interpretation
FRR	Financial Reporting Release
GAAP	Generally Accepted Accounting Principles
GASB	Government Accounting Standards Board
GAAS	Generally Accepted Auditing Standards
IASC	International Accounting Standards Committee
IAS	International Accounting Standard
IDS	Integrated Disclosure System
IFAC	International Financial Accounting Committee
IPO	Initial Public Offering
IPR&D	In-process Research & Development
IRC	Internal Revenue Code
IRS	Internal Revenue Service

ISA	International Standards on Auditing
LLC	Limited Liability Company
LLP	Limited Liability Partnerships
MD&A	Management Discussion & Analysis
NAARS	National Automated Accounting Retrieval System
NASDAQ	National Association of Securities Dealers' Automated Quotation
PBO	Projected Benefit Obligation
SAB	Staff Accounting Bulletin
SAS	Statements on Auditing Standards
SEC	Securities and Exchange Commission
SFAC	Statement of Financial Accounting Concepts
SFAS	Statement of Financial Accounting Standards
SOP	Statement of Position
SPE	Special-purpose Entity
USD	US Dollar

PORTUGAL
INDIVIDUAL ACCOUNTS

António C. Pires Caiado

Acknowledgements

The author wishes to thank his colleagues Ana Pinto and Ana Marques for their help and contributions.

Note

While writing this book, two official Portuguese bodies of accountants have changed their names. The association of official accounting technicians, Associação dos Técnicos Oficiais de Contas (ATOC) is now called Câmara dos Técnicos Oficiais de Contas (CTOC). Also, the auditors' chamber has changed its name from Câmara dos Revisores Oficiais de Contas (CROC) to Ordem dos Revisores Oficiais de Contas (OROC).

CONTENTS

I Introduction and background

1 Standards and standard setting

Portugal's accounting regulation is eminently public, since financial statements are drawn up and presented essentially in line with statutory accounting standards which are set forth in legislative instruments of varying status – Law (*Lei*), Decree-Law (*Decreto-Lei*), Circular (*Circular*). The standards are drawn up by various government bodies. They are published in the Official Journal (*Diário da República*, DR).

The following bodies are currently empowered to issue accounting regulations:

- The Parliament and Council of Ministers (in the form of Laws and Decree-Laws).
- The Ministry of Financial Affairs.
- The Accounting Standards Committee (*Comissão de Normalização Contabilística*, CNC).

Furthermore, the Bank of Portugal (*Banco de Portugal*, BoP), the National Securities and Exchange Commission (*Comissão do Mercado de Valores Mobiliários*, CMVM) and the Insurance Institute of Portugal (*Instituto de Seguros de Portugal*, ISP) may regulate the accounting systems of entities and groups under their surveillance.

Members of the accounting profession (at the academic, private professional association, individual professional accountant and auditor level) participate significantly in the preparation of the mandatory accounting standards and in the ongoing development, criticism and updating thereof.

2 Historical development of standard setting

Normalization of companies' accounts in Portugal followed the development of other countries. Dispositions of Commercial Code (*Código Comercial*, CCom) of 1888, (articles 29° to 44°) set rules about the obligation of a merchant to undertake bookkeeping and maintain the books in order according to certain procedures. Since 1943, official entities were trying to put an end to extravagances, abuses and frauds that, in accounting matters, some companies were involved in – the law nº 1995, of 27 May 1943 assigned the Verifiers Chamber with competencies for the establishment of rules of normalization.

The Fiscal Reform of the 1960s brought about the necessity for companies to use a uniform terminology in their accountancy, identical meanings for the accounts used, equal forms of presentation of the financial statements, etc. Thus, Decree-Law nº 45103, of 1 July 1963, approved the Code of Industrial Contribution (*Código da Contribuição Industrial*) and introduced important innovations in the accounting of large companies (group A). Some of the innovations were:

- The results of companies would, for fiscal purposes, be determined by the company's accountancy.
- The taxable profit would have to be disclosed in the profit accounts of the year, elaborated according to sound accounting principles.
- The fiscal declaration model 2 would have to be accompanied by certain elements that were compulsory and should only consider the effective results.
- The companies would have to centralize, in a proper establishment, located in a specific department, the bookkeeping of their several departments.

In order to enforce the Fiscal Council's (*Conselho Fiscal*) actions, Decree-Law nº 49381, of 15 November 1969, introduced changes to the fiscal regime of public limited companies (*Sociedades Anónimas*, SA). Articles 30º to 38º presented dispositions relative to the elaboration of the balance sheet and income statement, manager's report (*relatório do Conselho de Administração*), Fiscal Council's reports and accounts disclosure.

The pressure that the economists and the accountants had been making on the Government to revise the above-mentioned articles from the Commercial Code, led to the appointment of a Commission, supported by the Fiscal Centre of Studies of the Financial Ministry (*Centro de Estudos Fiscais do Ministério das Finanças*). This Commission prepared a draft of the Accounting Act (*Plano Oficial de Contabilidade*, POC), that was submitted to evaluation and criticisms of several entities.

Meanwhile, other works were being developed by diverse entities, but these were not accepted by the interested parties on the accounting information:

- Accounting Act-project.
- Contribution for the Portuguese Accounting Act of the National Syndicate of Employers – Centre of Studies – (1965).
- Accounting Act for Companies, of the working group of Technical Accounts Syndicate of Professional Workers of Lisbon (1970).
- Portuguese Accounting Act of Portuguese Society of Accounting (1974).

In this context, the Budget State Department created a Commission in November 1974, confirmed by dispatch published in the Official Journal, II series, of 18 March 1975, with the purpose of establishing uniformity of concepts and designations of the different accounts that integrate the basic accounting documents and imposing the adoption of sound accounting policies, so that the disclosed elements may provide confidence and transparency, which are indispensable concepts to a generalized appreciation, not only by the stockholders and workers, but by the public in general, interested in the development of the economic and financial development of the country.

The Commission work involved two phases: first, there was the production of a work named 'Accounting Normalization-I phase', which included:

- Commission Report.
- Uniform Complete Balance Sheet.
- Synthetic Balance Sheet.
- General notes about the balance sheet.
- Specific notes about the balance sheet.
- Income statement (by nature).
- General notes about the income statement.
- Appendix to the balance sheet and income statement.
- Chart of Accounts and list of accounts utilization.
- List of accounts of the revenue's components.
- Map of origins and applications of funds.

The subsequent phase of the Commission's work consisted of the revision of the preceding study and the elaboration of new documents – income statement by functions, complete chart of accounts and definitions and annotations on most of the accounts.

Initially, the Commission admitted that it would only oblige the presentation of the final documents: balance sheet, income statement and appendix. However, the Government's opinion was that the work should be more complete and it approved and published Decree-Law nº 47/77, that established that the Accounting Act (POC) was of compulsory application for all companies, with the exception of credit institutions and insurance companies.

In 1986, as Portugal joined the European Community, there was the necessity to adapt the Accounting Act to the Fourth and Seventh Directives. This led to the approval of Decree-Laws nº 410/89 and nº 238/91, of 21 November and 2 July, respectively.

3 The Accounting Standards Committee

The Dispatch of the Budget's Secretary of State, of 27 February 1975, created the Commission for the Study of Accounting Standardization that functioned until the publication of Decree-Law nº 47/77 – the Accounting Act.

Article 4º of this diploma envisaged the creation of an Accounting Standards Committee, an independent organ that would function, administratively and financially, under the scope of the Finance Ministry. The entities integrated in the Committee were established by Decree (*Portaria*) and represented the official and private institutions directly interested and technically more valid.

With the responsibility of ensuring the efficient continuation of the national accounting standardization, the Committee's regulations were approved by the publication of Decree nº 262/87, of 3 April, whose article nº 1 defined their attributions: 'without prejudice of the general objectives' in nº 3 of article 4º of Decree-Law nº 47/77, of 7 February, these are the specific attributions of Accounting Standards Committee (CNC):

(a) Promote the necessary studies for the adoption of principles, concepts and accounting procedures that we should consider of general acceptance.
(b) Elaborate the projects that involve alterations, complements (*aditamentos*) and interpretative rules of the POC.
(c) Determine and elaborate specific accounting acts or to pronounce about them, when elaborated by other entities.
(d) Elaborate opinions (*pareceres*) about legislative decrees whose dispositions may influence the accountancy of private companies and public sector companies.
(e) Answer to questions asked by private companies and by public sector companies related to the applications or interpretation of the POC, taking into account the disposed on the end of nº 2 of article 5º of Decree-Law nº 47/77.
(f) Participate on international meetings where matters related to accounting standardization are treated, with the purpose of giving technical opinions.

For the prosecution of these attributions the Accounting Standards Committee has three organs: President (*Presidente*), General Council (*Conselho Geral*) and Executive Commission (*Comissão Executiva*).

The General Council has, besides the President (designated freely by the Finance Ministry), 41 members representing: the general interests of the Government (10 members), the professional associations of technicians (10 members), the teaching and scientific institutions (9 members), the public companies sector with the exception of the banks and insurance companies (4 members) and the private sector of the economy (8 members).

The Executive Commission has 13 members. The President is elected by private election between the members of the General Council and the other members which represent: the departments of the Finance Ministry (2 members), the Chamber of Official Auditors of Accounts (*Câmara dos Revisores Oficiais de Contas*, CROC) (1 member), the Portuguese Accountants Association (1 member), the professional associations of technicians (2 members), the teaching and scientific institutions (2 members), the public companies sector (1 member), the private sector of the economy (1 member), the representative of the National Securities and Exchange Commission (CMVM) and one representative of the Official Accounting Technicians Association (*Associação dos Técnicos Oficiais de Contas*, ATOC).

The attributions of the Executive Commission are, according to nº 11 of the Decree, to:

(a) Promote the execution of works determined by the General Council.
(b) Create working groups and co-ordinate their actions, through the nomination of some members for the effect.
(c) Deliberate, under the presentation of General Council, studies and projects that are to be made by their own initiative, by work groups or by third parties.
(d) Prepare general programmes of activities, that are submitted annually to the General Council, and respective budgets.
(e) Promote the publication of a periodic bulletin, whose execution was committed to a work group.

The General Council meets ordinarily every three months and extraordinarily when summoned by the President. In order to function it needs the simple majority of its members or, half an hour after the beginning of the meeting, the presence of one-third of its members.

The Executive Commission meets ordinarily four times a month and extraordinarily when summoned by the President, and needs the presence of at least six elements for the deliberations.

Between the various opinions that the Executive Commission presents we can distinguish answers concerning doubts from entities of public and private sectors and concerning projects for specific accounting acts elaborated by other entities (credit institutions, state services, universities, etc.).

The adaptation of the POC to the EU Directives occupied a substantial part of the work done by the Committee. In order to clarify some aspects of the POC the Committee elaborated and approved the following Accounting Directives (*Directrizes Contabilística*), which were approved by the Finance Minister:

- Accounting Directive nº 1 – Accounting treatment of concentration of companies' activities.
- Accounting Directive nº 2 – Accounting, by the donor, of assets donated.
- Accounting Directive nº 3 – Accounting treatment of construction contracts.
- Accounting Directive nº 4 – Accounting for contractual obligations of concessionaire companies.
- Accounting Directive nº 5 – Accounting treatment of rights and obligations inherent to the concession of the bingo game.
- Accounting Directive nº 6 – Elimination of the results not achieved in transactions between subsidiaries.
- Accounting Directive nº 7 – Accounting of research and development expenses.
- Accounting Directive nº 8 – Clarification of the expression 'Regularization of infrequent and of substantial significance', relatively to account 59 – 'Retained results'.
- Accounting Directive nº 9 – Accounting in the individual accounts of the property of subsidiaries and associates.
- Accounting Directive nº 10 – Transitory regime of financial leases.
- Accounting Directive nº 11 – Inter-community VAT (*Imposto sobre o Valor Acrescentado*, IVA).
- Accounting Directive nº 12 – Accounting concept of goodwill.
- Accounting Directive nº 13 – Concept of fair value.
- Accounting Directive nº 14 – Cash flow statements.
- Accounting Directive nº 15 – Amortization and remission of stocks.
- Accounting Directive nº 16 – Revaluation of assets.
- Accounting Directive nº 17 – Accounting treatment of future contracts.
- Accounting Directive nº 18 – Objectives of financial statements and generally accepted accounting principles.
- Accounting Directive nº 19 – Retirement benefits.
- Accounting Directive nº 20 – Income statement by functions.
- Accounting Directive nº 21 – Accounting of the effects of the introduction of the euro.
- Accounting Directive nº 22 – Transactions subjected to special taxes on consumption.
- Accounting Directive nº 23 – Relations between accounting entities of the same juridical entity.
- Accounting Directive nº 24 – Joint ventures accounting.
- Accounting Directive nº 25 – Leases.
- Accounting Directive nº 26 – Revenue.

4 Other regulatory bodies

4.1 Bank of Portugal

According to Article 115º of the General Regime of Credit Institutions and Financial Companies (approved by Decree-Law 298/92) it is up to the Bank of Portugal (BoP) to establish accounting rules for the institutions subjected to its supervision, as long as they do not go against the National Securities and Exchange Code (código do MVM), and define the elements these institutions must submit and publish. Accordingly, Instruction nº 4/96 of the BoP establishes the Accounting Act for the Banking System and sets up specific rules for the accounting of those institutions.

4.2 Insurance Institute of Portugal

The Insurance Institute of Portugal (ISP) published the Insurance Companies Accounting Act on 1 June 1994. This follows the Fourth European Directive. Its structure is similar to that of the general Accounting Act. Furthermore, some adaptations are made by circular or norm addressed to each company.

4.3 National Securities and Exchange Commission (CMVM)

The National Securities and Exchange Commission's role is to protect funds invested in the capital markets, regulate the information provided to investors and ensure the correct working of the stock and futures markets.

Among its statutory powers we must point out the power to regulate the information, which is fulfilled by:

- The issue of an approval on prospectus made available to investors as soon as there is a public call for investment.

- The verification of legally required publication and disclosures of companies.
- The periodic publications which are available from quoted companies.
- Publication of information on events that could affect the quote price.
- Specific accounting regulations related to investment funds.

On this last item the CMVM published the following regulations: nº 95/14 – Securities Investment Funds Accounting; nº 96/6 – Real State Investment Funds Accounting.

II Forms of business organization

1 Commonly used legal forms of business organization

Article 1º, nº 2 of the Companies Code (*Código das Sociedades Comerciais*, CSC), considers that we are in the presence of companies when they:

- Have as their object the practice of commercial acts.
- Adopt the type of general partnership (*sociedade em nome colectivo*), private limited company (*sociedade por quotas*), public limited company (*sociedade anónimas*) or ordinary limited partnership (*sociedade em comandita*).

Therefore, a company is characterized by two fundamentals features:

- The object: commercial exercise.
- The type: predicted in law.

These two conditions (object and type) define the regular company, but we cannot forget to refer to irregular enterprises (with a commercial objective, but without commercial type), the civil enterprises of commercial type (without a commercial objective, but with commercial type) and the civil enterprises (without objective or commercial type).

As stated above, in legal terms companies may assume several types:

- General partnerships (*Sociedades em Nome Colectivo*).
- Private limited companies (*Sociedades por Quotas*, SQ).
- Public limited companies (*Sociedades Anónimas*, SA).
- Ordinary limited partnership (*Sociedades em Comandita*).

Nowadays, just public limited companies (SA) and private limited companies (SQ) are relevant in economical and juridical practice, as the number of and capital invested in general partnerships and ordinary limited partnerships has been decreasing. In fact, private limited companies (SQ) are the most common form in Portugal, representing more than 85 per cent of the total number of companies. But public limited companies (SA) have more money invested, representing more than 70 per cent of the total investment. Together, both types of companies represent more than 90 per cent of the total (in quantity) and 99 per cent of the money invested.

Legally, what distinguishes most of the different types of company is the responsibility of the partners, which is:

- Unlimited in general partnerships.
- Limited in public limited companies (SA) and private limited companies (SQ).
- Both unlimited and limited in ordinary limited partnerships.

Economically, these companies are different mainly because of the way they perform the following activities:

- Financing.
- Attributions of the partners.
- Distribution of profits.

Each type of company has specific merits and faults that must be taken into consideration when creating a company and choosing the adequate legal form.

There are big differences between enterprises of people (whose perfect type is the general partnership) and capital enterprises (whose model is the public limited company). The limited liability company can be seen as an in-between. It all depends on the number of partners and their role in the administration. The administrators must prepare annual accounts, as required generally of all businesses by the Commercial Code

(CCom), Article 34°, n° 1, within three months of the financial year end.

2 General partnerships

General partnerships are based on the mutual trust of their partners, whose responsibility is not only unlimited, but also shared. All private assets of the partners are a guarantee to the obligations of the society and all partners have total responsibility.

The main characteristics of this type of company are:

- The partner not only answers individually for his or her share but also answers, subsidiarily, for the social obligations of the company, with the other partners. The partner that supports more than his or her share of the obligations of the society has the right to receive the excess from the other partners afterwards.
- Partners of industry are admitted.
- The contract of a general partnership has to refer to:
  - (i) The type and characterization of the entry of each partner, in industry or assets, as well as the value attributed to the assets.
  - (ii) The value attributed to the industry contribution of the partners, in order to establish the distribution of profits and losses.
  - (iii) The part of the capital which corresponds to the assets entry of each partner.
- No titles representing social parts can be emitted.
- The firm of the general partnership should, when it doesn't individualize all partners, have the name of at least one partner, with the addition, condensed or in full, 'e Companhia' (and partners) or other addition that proves the existence of other partners.
- The value of the industry contribution is not taken into consideration on the common equity.
- The industry partners are not responsible, in internal relations, for the social losses, except if the contract of the society says anything otherwise. In case the industry partner answers for the social losses and contributes with capital, he or she will be given a part of the capital correspondent to this contribution.

Taking into consideration these characteristics, it is understandable that these societies are formed between a small number of persons of a family or friends, that all partners have great concern about the company and that the share of one partner can only be transmitted with the acceptance of all other partners. This also applies in the case of the admission of a new partner.

3 Ordinary limited partnerships

There are two possible types of ordinary limited partnership: the simple one and the one with shares. In both types the responsibility is unlimited for some partners (collective partners, managers) and limited for others (ordinary limited partners). This is why the ordinary limited partnership has an 'in-between' position to general partnerships and corporations. The ordinary limited partnerships facilitate the association of active persons who do not have enough resources or who have only their industry with others who have capital and want to invest it without great risk.

This is the less-known legal type of company. One of the inherent problems is the difficulty to protect the market from unreliable managers or to sell a share without loss. These companies, that were few and far between, lost all their importance with the appearance of limited liability companies and, actually, tended to disappear.

4 Private limited companies

These companies have the advantage of attracting capital that would not be invested in other legal types of company: investors not willing to assume unlimited responsibility and investors not willing to have limited responsibilities in ordinary limited partnerships (where they can only be managers or collective partners).

According to the Companies Code (CSC), the main characteristics of this type of company are the following:

- The capital is divided into quotas, and the partners stand by each other as defined in the social contract. They are only obliged to put more money into the company when such is established by law or in the contract.

- Only the social property of the company is responsible for all payables, unless an exception is made in the contract.
- The capital cannot be lower than 1,000,000 PTE or 5,000 euro and there must be at least two partners.
- The contract of the company must mention:
  - (i) The value of each quota of capital and the identification of the respective owner.
  - (ii) The amount of money put in by each partner in the contract and the amount of money of the deferred entries.
- The firm must be formed, with or without acronym, by the name of all, some or one of the partners, or by a particular denomination, or by the union of both of these elements, but in all cases it must end with the word '*Limitada*' (limited) or by the abbreviation '*Lda*'.
- Entries can be made in money or other assets – which must be previously audited – but no industry contributions are admitted. Furthermore, the deferred entries cannot be more than half of the money entries.

Most partners' resolutions are taken at a general meeting. This is the case for the call and payment of supplementary entries, the depreciation of quotas, the acquisition, alienation and financial burdening of own quotas and the agreement to divide or end quotas, the exclusion of partners, the destitution of managers and of the members of the inspection organ, the approval of the manager's report, etc.

There must be an annual general meeting in the first three months of each civil year to deliberate on the manager's report, the accounts of the year and the proposal for the application of profits, and in order to proceed with a general appreciation of the management and inspection and election of these organs.

If the contract of the company institutes a Fiscal Council, the rules by which it must work will be the same as the Fiscal Council of public limited companies (SA). The private limited companies that do not have a Fiscal Council must designate an Official Auditor of Accounts (*Revisor Oficial de Contas*, ROC) to proceed with the legal revision/auditing if, during two consecutive years, two of the three following limits are crossed:

(a) Total of balance: 350,000,000 PTE
(b) Total of liquid sales and other profits: 600,000,000 PTE
(c) Average number of employees during the year: 50.

Private limited companies are similar to public limited companies (SA) in some ways, and to general partnerships in others. They are similar to public limited companies (SA) because the partners' responsibility is limited, because they can have a Fiscal Council, etc. They are similar to general partnerships because:

- They can adopt a firm for a social reason.
- The number of partners is usually small.
- All the partners can be directly involved in the management.
- The partner substitution is not easy and is not so frequent.

5 Public limited companies

In Portugal the introduction of this type of company – which is based on the representation of capital in shares and on the representation of the partners' rights by financial titles – was very important and had enormous economic and social consequences. Even though this type of company was designed to allow the dispersion of capital, it is frequent to find public limited companies whose shares belong to people of the same family, with a great concentration of capital.

The main characteristics of public limited companies are the following:

- The capital is divided into shares and each shareholder's responsibility is limited to the value of his or her shares.
- The company must have at least five shareholders, except when the law states otherwise.
- The contract of the company must mention:
  - (i) The nominal value and the number of shares.
  - (ii) The particular conditions, if they exist, in which share transmissions are submitted.
  - (iii) The category of shares that are created,

with express indication of the number of shares and rights attributed to each category.

(iv) If the shares are nominative or bearer and the rules for eventual conversions.

(v) The capital realized and the term of the realization of capital signed.

(vi) The authorization, if granted, for the emission of bonds (*obrigações*).

(vii) The structure adopted for the administration and inspection of the company.

- The name of the company will always end with the phrase *'Sociedade anónima'* (anonymous company) or the abbreviation 'SA', independently of the way it was formed.
- The common equity and the shares must be expressed in nominal value.
- The minimum common equity is 5,000,000 PTE. The shares are indivisible and all have the same nominal value, that cannot be lower than 1,000 PTE.
- Entries of money can only be deferred by 70 per cent of nominal value of the shares. The maximum time for deferment is five years.
- Entries of assets different from money cannot be deferred and are subject to a report compiled by an Official Auditor of Accounts (ROC).
- The money entries should be deposited in a credit institution before the completion of the contract, in an account opened in the name of the future company, and the deposit bill must be shown in the notary.
- The public limited companies can be formed with an appeal to public subscription, according to article 279º of CSC. In these circumstances the National Securities and Exchange Code (*Código do MVM*) must also be followed.
- The administration and inspection of the company can be structured by one of two ways:
  (i) Administration Council and Fiscal Council.
  (ii) Direction, General Council and Official Auditor of Accounts (ROC).

The limitation of liability and the small value of shares allow these to be acquired even by people who have small savings. Because of that, this is the legal form adopted by companies that intend to make big economic undertakings. Most partners do not interfere directly in the company's business. The limited liability and the great importance of capital justify certain measures of inspection by the shareholders, creditors and state: publication of manager accounts, inspection by special control bodies, etc.

6 Co-operatives

Co-operatives are collective persons, of free constitution, of variable capital and composition, that aim, through co-operation and mutual help of their members and according to co-operative principles, for the satisfaction, without profits, of economical, social and cultural needs. Complementary, they can realize transactions with third parties. They are regulated by the Co-operative Code, approved by Decree-Law nº 454/80, of 9 October.

The main characteristics of co-operatives are the following:

- Admission or exit constitute a free and voluntary act, being the number of members and the capital variable, although submitted to legal established rules.
- The admission or exclusion of co-operatives cannot be object of restrictions or discriminations.
- The social organs are elected by democratic methods, according to statutes and subordinated to the principle of equality of all members.
- The right to vote in co-operatives of first degree is based on the attribution of one singular vote to each member, independently of his or her participation in the capital, although complementary legislation can predict other forms of the right to vote.
- The attribution of the right to vote in co-operatives of superior degree should be defined in a democratic base and in a more adequate form.
- The remuneration given to the members of co-operatives for their participation in common equity, in compulsory deposits and for the realization of investment titles should be limited, and the respective rate fixed by the General Assembly.
- The surplus (profit, in companies) can be distributed by co-operators according to the eco-

nomic operations realized with the co-operative or to the work and services rendered by them.
- Relations with other co-operatives should be privileged.

III Objectives, concepts and general principles

1 Objectives of financial reporting

Information about the economic resources controlled by the company and its capacity in the past to modify these resources, is useful to predict the capacity of the company to generate liquidity in the future. Information about the financial structure is useful to predict the financial needs. Information about liquidity and solvability is useful to predict the company's capacity to honour its financial obligations. All the information taken together permits future predictions of profits and dividends.

The Accounting Act states that the financial statements should provide information on the financial position, the changes it suffers and the results from operations, so that they can be useful to investors, creditors and other users, enabling them to invest rationally, grant credit and take other decisions. Furthermore, the information provided must contribute to the efficient functioning of the national securities market.

Still, according to the POC, the information must be comprehensible to those that wish to analyze and evaluate it, helping them to distinguish the uses of economical resources that are efficient from the ones that are not and disclosing the net profit of the year. The responsibility of preparing the financial information that is going to be disclosed lies with the administration.

The most important characteristic of the information disclosed by the financial statements is comprehensibility. This means that financial information included in the financial statements must be quickly understood by users, assuming they have reasonable knowledge about managerial, economic and accounting aspects, as well as the will to study the financial information. However, no information about complex matters should be excluded from financial demonstrations on the grounds that it would not be understood.

The utility of the information is determined by the following characteristics:

- Relevance – information must have the quality to influence the decisions taken by its users, by helping them evaluate past, present or future events and by confirming or correcting their evaluations. Relevance is highly related to materiality, as no information that is not material can be relevant. The relevance of the information can be lost if there are delays in disclosing it.
- Reliability – information must not contain material errors or prejudice, representing correctly operations and other events that are expected to arise. Most financial information is exposed to the risk of being less than a faithful representation of reality. This is due to inherent difficulties either in identifying the transactions and other events to be measured, or in devising and applying measurement and presentation techniques that can convey messages that correspond with those transactions and events. Furthermore, there is always uncertainty about the collectability of accounts, the lifetime of fixed assets, the assurances, etc. Therefore, reliability is related to prudence.
- Comparability – the disclosure and quantification of the financial effects of operations and other events must be registered in a consistent way by the company throughout its life, in order for it to be possible to identify trends in its financial position and in the profits from its operations. Moreover, companies must adopt normalized ways of registry in order to enable users of financial information to make comparisons between them. Users must be informed about accountancy policies used in the preparation of financial statements and of any changes in those policies. The necessity of comparability should not be confused with uniformity and it must not be allowed to become an impediment to the introduction of better accounting policies.

2 Concepts and general principles

The conceptual base that supports the financial statements is the Accounting Directive nº 18, which defines a hierarchy of norms:

- Accounting Act (POC).
- Accounting Directives formulated by the Accounting Standards Committee (CNC).
- International accounting standards formulated by the IASC – International Accounting Standards Committee.

The fourth chapter of the Accounting Act presents the accounting principles used in financial statements (see VI):

- Going concern.
- Consistency.
- Accrual.
- Historical cost.
- Prudence.
- Substance over form.
- Materiality.

3 Interrelationship of financial reporting and tax accounting

Tax law is set out in the Company Taxation Code (*Código do Imposto sobre o Rendimento das Pessos Colectivas*, CIRC) and is distinct from the accounting law. CIRC sets out (Article 17º, (a)) that accounting should be organized according to accounting normalization.

Tax and accounting results are closely linked (although they may be very different due to temporary and permanent differences) since numerous tax-generated items are included in the accounting records and accounts. In addition, the tax return actually includes the balance sheet and income statement approved by the company's directors. Sometimes fiscal law has specific rules. Companies that wish to take advantage of certain tax benefits are obliged to make some purely fiscal entries in their accounting records and hence to include them in their financial statements. Some of these entries may even distort an accounting evaluation of the financial statements, e.g. uncollectable provisions.

Accounting Directives issued by the Accounting Standards Committee (CNC) over the past years have sought to restrain the profound impact of these fiscal adjustments on the adequate presentation of the financial statements.

The tax law sets out separate rules for the deductibility of certain expenses and the taxation of certain revenues, in the form of Decree-Laws and Official Pronouncements (*Ofícios Circulares*) by the fiscal administration, but the general principle in Portugal is that accounting records must give a true and fair view of the financial position and all transaction records must be supported by documentary evidence.

In the case of a dispute it is up to the tax authorities to prove that there has been a miscalculation. The right to determine a separated assessment has to be related to incorrect profits chargeable to corporation tax.

Article 18º, nº 1 of CIRC states that revenues and expenses, and other positive or negative components of taxable profits, are charged to their exercise, according to the principle of accrual. Revenue comprises the total amount received (or to be received) in return for the goods sold and services rendered. It also includes governmental subsidies and compensations, among others. Furthermore, article 98º determines the accounting obligations of companies.

IV Bookkeeping and preparation of financial statements

1 Duty to prepare financial statements

1.1 Preparation of annual accounts

Companies are compelled to present and disclose their balance sheet (*dar balanço e prestar contas*) – CCom (Article 18º, nº4) – namely, present their assets and liabilities and account them for in the 'book of inventory and balances' (*livro de inventário e balanços*) – CCom, Article 62º.

The economic year is the same as the civil year, and it is compulsory for all businesses to close their accounts having the 31 December as a reference (Decree-Law nº 16731, 13 April from 1929, arº137º).

Nevertheless, the CIRC allows a different fiscal period for:

- Companies and other entities subject to company tax whose headquarters are outside Portugal.
- Other entities who for economic reasons ask for permission at the Ministry of Finance.

The fiscal period can be less than a year (in the case of beginning or ending a business) or over a year (liquidation purposes).

There are some disadvantages in having a different fiscal period because there are disclosures at the year end (civil year) connected with the balance sheet date that demand two or more balance sheets a year (value-added tax disclosures, legal social security disclosures, etc.).

Accounting information must be made worthwhile in the eyes of partners and third parties for internal users. The documents considered to be financial statements (balance sheet, income statement, appendix and cash flow statement) must show the company's financial position, its changing and the results of operations, so they can be useful to investors, creditors and other users, and help them invest wisely and make decisions.

The Portuguese Accounting Act, the Company Code and Accounting Directives issued by CNC consider that the financial statements for companies, as a whole, include:

- Manager's report (including suggestions on how to apply profit/loss).
- Balance sheet.
- Income statement by nature and functions.
- Appendix.

Certified companies (the companies that are compelled to have an Official Auditor – ROC) must also present a cash flow statement, the Official Audit Report (*Certificação Legal de Contas*), the Annual Auditing Report, and the Fiscal Council Report (should there be one).

These are the documents to be presented to the General Assembly of stockholders and deposited with the Commercial Registration Body, accompanied by a written document that provides proof of the statements' approval (the minute book).

The Accounting Act presents two models of balance sheet following the guidelines of the Fourth Directive. Assets include fixed assets (plant, land, equipment, tools, long-term financial investments), current assets (inventory, receivables, short-term financial investments, liquidities) and accruals and deferrals (income accrual and expenses deferral). Equity, reserves, results, liabilities (payables, provisions for risks) and accruals and deferrals (accrued expenses and deferred income) are presented on the other side.

The Accounting Act standards were adapted to the Fourth Directive and are characterized by their thoroughness – necessary for inter-company comparisons – and flexibility – due to the current variety of accounting systems in several Member States.

We find thoroughness in the following aspects:

- The inclusion of compulsory financial statements.
- Obligation to present financial statements according to specific models.
- Apart from exceptional circumstances, consistency is compulsory. Otherwise, a special disclosure must be made in the Appendix.
- Each item on the balance sheet or income statement must include amounts related to the past economic year in order to enforce comparability.
- It is forbidden to compensate assets with liabilities or income with expenses.

We find flexibility in the following aspects:

- Exclusion in scope of certain businesses' activities (e.g. banks, insurance companies).
- Possibility to adapt the chart of accounts to the company's needs and resources. Items belonging to the balance sheet or income statement can be:
  - Divided, as long as they respect the actual structure.
  - Omitted, if it does not include any amount from the current or preceding year.
  - New items may be adjusted if their content isn't covered by any other account.
  - Items preceded by numbers can be regrouped.
- It is possible to classify an operation in terms of a priority procedure, as long as it is disclosed in the Appendix its relation to other items.

- The structure, definition and wording of the balance sheet and income statement must be adapted if the company needs it.
- Small and medium-sized companies (CSC, article 262º) may present abridged financial statements.

1.2 Interim financial statements

The article 342º of the Exchange Market Code (Decree-Law nº 142-A/91) establishes the following criteria for half-yearly reporting:

- Companies with shares on the Stock Exchange must publicly disclose, until the 30 September of each year, information about their activities and results of the first semester. The elements contained must provide sufficient information to investors, in order to permit them to form a rational judgement about the company's activities and results, in comparison with the previous year. It should also aid them to predict future economic or financial behaviour.
- The above-mentioned information must include, among other optional disclosures:
  (i) Company's name and headquarters.
  (ii) Balance sheet at the closing date (30 June of the current year).
  (iii) Income statement at the same closing date, reporting anticipated dividends paid or to be paid.
  (iv) Amount of investments made, total sales and expenses description of some significance.
  (v) Any other information stated in the Code.

The growing market globalization – especially in the case of shares in the Stock Exchange – demands further disclosure to information users. In Portugal, Decree nº 1222/97 of 12 December demands the presentation and public disclosure – within a period of 30 days – of quarterly reporting including operations information, profit/loss and economic and financial situation of companies with shares in the Stock Exchange. Such a measure is in fact positive, as it allows information users the knowledge and regular activity update.

2 Bookkeeping and Accounting Act

2.1 Formal aspects of bookkeeping

The Commercial Code states that bookkeeping is compulsory in its article 18º, nº2. The formal aspects of bookkeeping are governed by article 31º. Every company must keep orderly accounts appropriate to its business which enable all its transactions to be monitored in chronological sequence and enable balance sheets and inventories to be drawn up periodically.

Companies must also keep a minute book or books, showing all the resolutions adopted by the general and special shareholders' meetings and other corporate bodies. These resolutions must be filed by the directors in the commercial registry within 30 days of approval of the minutes.

The books must be kept by an accountant (registered at the Official Accountants Association). Companies must retain the books, correspondence, documentation and supporting paperwork relating to their business, in due order, for 10 years from the date of the last entry in the books.

Regardless of the procedure used, all books of account and accounting documents must be kept clearly, in chronological order, without blanks, spaces, interpolations, crossings out or deletions.

The CCom indicates the books that are compulsory (article 31º):

- Inventory and balance sheets: initial inventory, summarized balance sheet and annual balance sheets.
- Diary: includes operations that change assets and liabilities in chronological order.
- Ledger: transcription of all operations registered in the Diary, ordered by accounts.
- Copy book: serves to maintain a file of all correspondence issued in chronological dates.

The books may have loose leaves to enable printing.

There are some external obligations: a stamp tax in all books and legalization of the 'Inventory and Balance Sheet Book' and 'Diary Book' (in a competent registry).

2.2 Accounting Act

Based upon the Decree-Law n° 47/77, of 7 February, and amended by Decrees-Law n° 410/89 and 238/91, the Accounting Act is structured in 14 chapters:

1 Introduction.
2 Technical considerations.
3 Financial information characteristics.
4 Accounting principles.
5 Valuation procedures.
6 Balance sheets.
7 Income statements.
8 Appendix to the balance sheet and income statement.
9 Resources and application statements.
10 Table of accounts.
11 Chart of accounts.
12 Explanatory notes.
13 Consolidation rules.
14 Consolidated financial statements.

The Accounting Act's main characteristics are:

- The accounts classification is decimal with subdivisions.
- The table of accounts (see Table IV.1) considers classes of accounts from 1 to 8, being classes from 1 to 5 for the balance sheet preparation, classes 6, 7 and 8 for the income statement, class 9 for cost accounting and class 0 for other considerations (e.g. cash flows accounting).
- The balance sheet models were conceived according to the Fourth Directive format. The assets accounts are developed in growing liquidity and the liabilities accounts in descending order of demand.
- The depreciation amounts diminish the corresponding assets accounts and provisions are inserted as corrections of assets or liabilities (depending on their source).
- Classes 6 and 7 are designated Costs and losses and Income and gains, respectively, with grouped accounts to accommodate operational, financing and extraordinary results, class 8. This class also contains an account for taxes on income purposes.
- The balance sheet and income statement also include amounts from the preceding year.
- The income statement by functions is now compulsory.
- The appendix demands several areas of information in order to enlighten some accounts and operations.
- There is no explicit reference to costing systems and class 9 is still reserved to cost accounting, without any suggested development.

The Accounting Act isn't applicable to banks, insurance companies and other financial entities (Decree-Law n° 410/89, Article 2°, n° 2), as well as other entities mentioned in n° 1 which, though practising commercial, industrial and agricultural activities, do not exceed 30,000,000 PTE in business volume (Article 3°, n° 5).

Article 3°, n° 1 allows individual companies under the Commercial Code, individual limited responsibility establishments, partnerships, companies with limited responsibility and co-operatives which haven't exceeded two of the three limits established in Article 262° of the Company Code, to present the less developed models of the balance sheet, income statement and appendix.

2.3 Inventory

The Commercial Code (1888) requires that all sole traders and companies make and value an annual inventory of their assets and liabilities by physical count or list and record them in the Inventory and Balance Book.

In the absence of permanent inventory records this must be done at the year end or within a few days on either side, provided that it is possible to determine the movements in the inventory period and calculate actual year end quantities.

In order to have better financial information about the inventory, the recent Decree-Law n° 4/99, of 12 February, requires all companies with audited accounts by an ROC to have a permanent inventory by the year 2000 (some exceptions are made).

Table IV.1 Table of accounts

1 Liquidity	2 Third parties	3 Inventory	4 Fixed assets	5 Equity	6 Costs and losses	7 Income and gains	8 Results	9 Cost accounting
11 Cash	21 Customers	31 Purchases	41 Financial investments	51 Common stock	61 Raw materials consumed and merchandise sold costs	71 Sales	81 Operational results	
12 Deposits	22 Suppliers	32 Merchandise	42 Tangible fixed assets	52 Own shares	62 External supplies and services	72 Services	82 Financial results	
13 Long-term deposits	23 Loans	33 Work in progress	43 Intangible fixed assets	53 Supplementary capital	63 Taxes	73 Supplementary income	83 (Current results)	
14 Other deposits	24 State and other public entities		44 Fixed assets in process	54 Premium	64 Personnel costs	74 Subsidies	84 Extraordinary results	
15 Marketable securities	25 Shareholders (partners)			55 Adjustments in investments	65 Other operational costs	75 Self-constructed assets	85 (Results before taxes)	
	26 Other debtors and creditors	36 Raw materials and others		56 Revaluation reserve	66 Depreciation	76 Other operational income	86 Taxes on income	
	27 Accruals and deferrals	37 Advances on purchases		57 Reserves	67 Provisions			
18 Other temporary investments	28 Doubtful receivables provision	38 Inventory corrections	48 Accumulated depreciation		68 Financial costs	78 Financial income	88 Net results	
19 Short-term investment provisions	29 Risks and contingencies provisions	39 Inventory depreciation provisions	49 Financial investments provisions	59 Retained earnings	69 Extraordinary costs	79 Extraordinary income	89 Anticipated dividends	

V Balance sheet and profit and loss account formats

1 General classification principles

The Accounting Act establishes that the annual accounts comprise the balance sheet, the income statement and the appendix, which together constitute a reporting unit. The information contained in each of these statements must, in every aspect, match the information contained in the others. The annual account formats in the Accounting Act (POC) are mandatory for all Portuguese companies, regardless of their legal form. The information presented must be relevant (material and timely) and reliable. Comparability, which includes consistency, is a secondary quality that interacts with relevance and reliability to contribute to the usefulness of information. Furthermore, the annual accounts must, at all events, be drawn up clearly and must present a true and fair view of the companies' net worth, financial position and results. According to the Eighteenth Accounting Directive, the annual accounts must provide information on the current financial position, the changes it suffered and the profits, so that they can be useful to investors, creditors and other users. This way users can invest rationally, give credit and take similar decisions, contributing to an efficient functioning of the capital's market. When, in exceptional circumstances, the structure of the balance sheet or income statement is modified, this fact must be indicated and justified in note nº 1 of the appendix (see IX). Where it is necessary to adjust the previous year's figures to make them comparable, this fact must be indicated and explained in note nº 2 (see IX).

The National Securities and Exchange Commission's Code states, in its article 161º, that financial information must respect the principles of sufficiency, veracity, objectivity and actuality, and defines each one of these characteristics.

Not all norms concerning the annual accounts are purely private. Many of them are inspired in public interests and cannot be broken. It is the case of all dispositions that defend the integrity of the capital.

2 Balance sheet

Portuguese regulations prescribe a single balance sheet layout in the form of an account. With some minor variations, this format coincides with that envisaged in Article 9º of the Fourth Directive and is the format traditionally used. Balance sheet items are ordered in terms of the liquidity of the assets (from lower to higher) and of the claimability of the liabilities (from lower to higher). Thus the structure of the assets and liabilities is based on a classification in terms of maturity (current and short-term versus fixed and long-term). The sections into which the assets and liabilities are grouped are further divided into subsections based on the nature of the items. There are separate sections for financial investments, accounts receivable, long-term debt and current liabilities.

In order to include in the balance sheet adjustments to the value of asset items, the assets are shown in the balance sheet at full acquisition or production cost, and the accumulated depreciation and related provisions for the whole class of items are recorded separately on the assets side of the balance sheet as negative items.

The format of the balance sheet, according to the Accounting Act, and taking into consideration the changes made in 1991, is shown in Table V.1.

Some practical aspects on the elaboration of the balance sheet:

- Financial investments maturing at not more than one year must be recorded under the heading 'short-term financial investments'.
- Borrowings maturing at not more than one year must be recorded under the heading 'Accounts payable – short-term liabilities'.
- If the retained earnings are negative (showing losses from previous years) the value must be presented with a minus sign in retained earnings, in order to show the actual net amount of equity.
- The net profit (or loss) for the financial year must be recorded as a separate equity item, with a plus or minus sign, depending on whether it is an increase (income) or decrease (loss) of net worth.

Table V.1 Full format of the balance sheet in English and in Portuguese

Assets	**Activo**
C Fixed assets:	**C Imobilizado:**
I. Intangible fixed assets:	**I. Imobilizações incorpóreas:**
1 Implementation expenses	1 Despesas de instalação
1 Research and development expenses	1 Despesas de investigação e desenvolvimento
2 Industrial property and other rights	2 Propriedade industrial e outros direitos
3 Goodwill	3 Trespasses
4 Work in process	4 Imobilizações em curso
4 Prepayments	4 Adiantamentos por conta de imob.s incorpóreas
II. Tangible fixed assets:	**II. Imobilizações corpóreas:**
1 Land and natural resources	1 Terrenos e recursos naturais
1 Buildings and other constructions	1 Edifícios e outras construções
2 Basic equipment	2 Equipamento básico
2 Transportation equipment	2 Equipamento de transporte
3 Tools and utensils	3 Ferramentas e utensílios
3 Administrative equipment	3 Equipamento administrativo
3 Containers	3 Taras e vasilhame
3 Other tangible fixed assets	3 Outras imobilizações corpóreas
4 Work in process	4 Imobilizações em curso
4 Prepayments	4 Adiantamentos por conta de imob.s corpóreas
III. Financial investments:	**III. Investimentos financeiros:**
1 Parts of capital in group companies	1 Partes de capital em empresas do grupo
2 Loans to group companies	2 Empréstimos a empresas do grupo
3 Parts of capital in associated companies	3 Partes de capital em empresas associadas
4 Loans to associated companies	4 Empréstimos a empresas associadas
5 Titles and other financial applications	5 Títulos e outras aplicações financeiras
6 Other loans granted	6 Outros empréstimos concedidos
6 Work in process	6 Imobilizações em curso
6 Prepayments	6 Adiantamentos por conta de invest. financeiros
D Current assets:	**D Circulante:**
I. Inventories:	**I. Existencias:**
1 Raw, subsidiary and consuming materials	1 Matérias-primas, subsidiárias e de consumo
2 Goods and work in process	2 Produtos e trabalhos em curso
3 By-products, waste and residue	3 Subprod.s, desperdícios, resíduos e refugos
3 Finished and intermediate goods	3 Produtos acabados e intermédios
3 Merchandise	3 Mercadorias
4 Prepayments	4 Adiantamentos por conta de compras
II. Accounts receivable – medium and long term	**II. Dívidas de terceiros – médio e longo prazo**
II. Accounts receivable – short-term assests:	**II. Dívidas de terceiros – curto prazo:**
1 Trade accounts receivable – current account	1 Clientes, c/c
1 Trade accounts receivable – bills receivable	1 Clientes – títulos a receber
1 Doubtful debtors	1 Clientes de cobrança duvidosa
2 Group companies	2 Empresas do grupo
3 Participated and participant companies	3 Empresas participades e participantes

Table V.1 (contd.)

4 Other stockholders (partners)	4 Outros accionistas (sócios)
4 Prepayments to suppliers	4 Adiantamentos a fornecedores
4 Prepayments to fixed assets suppliers	4 Adiantamentos a fornecedores de imobilizado
4 State and other public entities	4 Estado e outros entes públicos
4 Other debtors	4 Outros devedores
5 Underwriters of capital	5 Subscritores de capital
III. Marketable securities:	**III. Títulos negociáveis:**
1 Shares in group companies	1 Acções em empresas do grupo
3 Bonds and participation titles in group companies	3 Obrigações e títulos part. em emp.s do grupo
3 Shares in associated companies	3 Acções em empresas associadas
3 Bonds and particip. titles in associated companies	3 Obrigações e títulos part. em emp.s associadas
3 Other tradable titles	3 Outros títulos negociáveis
3 Short-term financial investments	3 Outras aplicações de tesouraria
IV. Bank deposits and cash:	**IV. Depósitos bancários e caixa:**
Bank deposits	Depósitos bancários
Cash	Caixa
E Accruals and deferrals:	**E Acréscimos e diferimentos:**
Income accrual	Acréscimo de proveitos
Deferred costs	Custos diferidos
Total assests	*Total do activo*
Equity and liabilities	**Capital próprio e passivo**
A Equity:	**A Capital próprio:**
I. Common stock:	**I. Capital:**
Own shares – nominal value	Acções (quotas) próprias – valor nominal
Own shares – discounts and premiums	Acções (quotas) próprias – descontos e prémios
Supplementary entries	Prestações suplementares
II. Share premiums	**II. Prémios de emissão de acções (quotas)**
III. Adjustments in parts of capital in associated	**III. Ajust.s de partes de capital em filiais e associadas**
Revaluation reserves	Reservas de reavaliação
IV. Reserves:	**IV. Reservas:**
1/2 Legal reserves	1/2 Reservas legais
3 Statutory reserves	3 Reserves estatuárias
4 Contractual reserves	4 Reservas contratuais
4 Other reserves	4 Outras reservas
V. Retained earnings	**V. Resultados transitados**
VI. Net profit;	**VI. Resultado líquido do exercício;**
Anticipated dividends	Dividendos antecipados
Total equity	*Total do capital próprio*

Table V.1 (contd.)

Liabilities:	Passivo:
B Provisions for risks and contingencies	**B Provisões para riscos e encargos:**
1 Provisions for pensions	1 Provisóes para pensões
2 Provisions for taxes	2 Provisões para impostos
3 Other provisions for risks and charges	3 Outras provisões para riscos e encargos
C Accounts payable – long-term liabilities	**C Dívidas a terceiros – médio e longo prazo**
C Accounts payable – short-term liabilities:	**C Dívidas a terceiros – curto prazo:**
1 Issued bonds:	1 Empréstimos por obrigações:
Convertible	Convertíveis
Non-convertible	Não convertíveis
1 Loans by participation titles	1 Empréstimos por títulos de participação
2 Debts to credit institutions	2 Dívidas a instituições de crédito
3 Prepayments	3 Adiantamentos por conta de vendas
4 Suppliers, current account	4 Fornecedores, c/c
4 Suppliers – unregistered invoices	4 Fornecedores – facturas em recepção/conferência
5 Suppliers – titles to pay	5 Fornecedores – títulos a pagar
5 Suppliers of fixed assets	5 Fornecedores de imobilizado-títulos a pagar
6 Group companies	6 Empresas do grupo
7 Participated and participant companies	7 Empresas participadas e participantes
8 Other stockholders (partners)	8 Outros accionistas (sócios)
8 Prepayments of clients	8 Adiantamentos de cleintes
8 Other obtained loans	8 Outros empréstimos obtidos
8 Suppliers of fixed assets, current account	8 Fornecedores de imobilizado, c/c
8 State and other public entities	8 Estado e outros entes públicos
8 Other creditors	8 Outros credores
D Accruals and deferrals:	**D Acréscimos e diferimentos:**
Cost accrual	Acréscimos de custos
Deferred income	Proveitos diferidos
Total liabilities	*Total de passive*
Total equity and liabilities	*Total do capital próprio e do passivo*

Table V.2 Full format of the income statement in English and in Portuguese

A Costs and losses	**A Custos e perdas**
2.a) Cost of merchandise sold and consumed materials:	2.a) Custo das mercadorias vendidas e das matérias consumidas:
Merchandise	Mercadorias
Materials	Matérias
2.b) External supplies and services	2.b) Fornecimentos e serviços externos
3. Personnel costs:	3. Custos com o pessoal:
3.a) Remuneration	3.a) Remunerações
3.b) Social expenses:	3.b) Encargos sociais:
Pensions	Pensões
Others	Outros
4.a) Depreciation of tangible and intangible fixed assets	4.a) Amortizações do imobilizado corpóreo e incorpóreo
4.b) Provisions	4.b) Provisões
5. Taxes	5. Impostos
5. Other operating costs and losses	5. Outros custos e perdas operacionais
6. Losses in group and associated companies	6. Perdas em empresas do grupo e associadas
6. Depreciation and provisions of financial investments	6. Amortizações e provisões de aplicações e investimentos financeiros
7. Interests and similar costs:	7. Juros e custos similares:
Concerning group companies	Relativos a empresas do grupo
Others	Outros
10. Extraordinary costs and losses	10. Custos e perdas extraordinários
8+11. Corporate tax of the financial year	8+11. Imposto sobre o rendimento do exerício
13. Net profit	13. Resultado líquido do exercício
B Income and earnings	**B Proveitos e ganhos**
1. Sales:	1. Vendas:
Merchandise	Mercadorias
Goods	Produtos
1. Services rendered	1. Prestações de serviços
2. Variation of production	2. Variação da produção
3. Self-constructed assets	3. Trabalhos para a própria empresa
4. Supplementary income	4. Proveitos suplementares
4. Subsidies to exploration	4. Subsídios à exploração
5. Earnings in group and associated companies	5. Ganhos em empresas do grupo e associadas
5. Income from parts of capital	5. Rendimentos de participações de capital
6. Earnings from tradable titles and other financial applications:	6. Rendimentos de títulos negociáveis e de outras aplicações financeiras:
Concerning group companies	Relativos a empresas do grupo
Others	Outros
7. Others interests and similar income:	7. Outros juros e proveitos similares:
Concerning group companies	Relativos a empresas do grupo
Others	Outros
9. Extraordinary income and earnings	9. Proveitos e ganhos extraordinários

3 Income statement

Portugal opted for only one of the four income statement formats offered by the Fourth Directive: the one where expenses are classified by nature (see Table V.2). This is the format envisaged in Article 24º of the Fourth Directive and is essentially the one traditionally used. The income statement by nature is laid out as an account in which expenses and revenues are classified according to the nature of the transactions. Furthermore, since 1999 all companies, except small and medium companies, must present an income statement by functions (see IX).

The format of the income statement by nature is shown in Table V.2.

For the purposes of presenting the annual accounts inventories are accrued for the net differences:

- By showing the simple variation during the period between opening and closing inventories (increase or decrease) of the products handled by the company: finished goods, semi-finished goods, work in progress and other products of the company.
- By directly reflecting consumption: this method is used for inventories of goods acquired from outside for resale or for inclusion in the production process (merchandise)

Extraordinary items (costs and losses, income and earnings) are those that are, unusual by nature or not frequent. Some examples are earnings or losses due to natural catastrophes, political convulsions, expropriations or prohibitions imposed by new regulations. Also included in these extraordinary items are losses in goods and investments, prior year's corrections, etc. Extraordinary items in the income statement by nature have a different reading from those considered in the income statement by functions. In the income statement by functions, to be considered extraordinary one item has to be material and cumulatively unusual by nature and not frequent. In these items are also included the materially relevant values that are a result of the extinction of liabilities.

4 Common aspects

The Accounting Act of 1989 establishes two levels of detail for the formats:

- The standards format, which must be used for the annual accounts of corporations, employee-owned companies, limited liability companies and limited partnerships with shares.
- The abridged formats, in which the items are more aggregated and the reporting requirements less stringent (Article 3º of the POC). The abridged annual accounts formats may be used by sole traders, general partnerships and ordinary limited partnerships, as well as small and medium-sized companies.

The POC emphasizes that items may be added for which no allowance is made in the standard or abridged forms of the balance sheet and income statement, provided their content is not already covered by existing items. Additionally, items can be further subdivided, so long as the structure is not affected. Balance sheet and income statement items denoted by Arabic numerals may be grouped if their amount is not material to a true and fair view or if the aim is to improve clarity.

5 Special legislation

Decree-law nº 298/92 regulates credit institutions and financial companies. This includes banks and other financial institutions. This decree-law states that these institutions are supervised by the Bank of Portugal.

VI Recognition criteria and valuation

1 Accounting principles

The fundamental accounting principles are laid down in the first part of the Accounting Act of 1989. The application of these principles is mandatory and does not follow any particular hierarchy:

(a) Going concern (*princípio da continuidade*): it is assumed that a company will continue in business for the foreseeable future. Accordingly, it is generally understood that there are no intentions to liquidate or significantly reduce the operations volume.

(b) Consistency (*princípio da consistência*): it is assumed that a company will not change its accounting policies from one economic year to another. If it does and it is material, it should be stated accordingly in the appendix (note nº 1).

(c) Accrual (*princípio da especialização ou do acréscimo*): income and expenses should be recognized when incurred, despite cash flows, and included in the financial statements.

(d) Historical cost (*princípio do custo histórico*): accounting registers should be based upon acquisition or production costs, either using constant or current prices.

(e) Prudence (*princípio da prudência*): it is possible to include some degree of caution in the financial statements in situations where predictions are uncertain. Nevertheless, this should not allow for hidden reserves or excessive provisions or the deliberate imbalance of assets and liabilities.

(f) Substance over form (*princípio da substância sobre a forma*): operations should be accounted for according to substance and financial situation and not just their legal form.

(g) Materiality (*princípio da materialidade*): financial statements must disclose all the relevant items that may affect evaluations or decision-making by users.

These principles can be found explicitly or implicitly in the Fourth Directive. In both legislative standards annual accounts are disclosed with the purpose of presenting a true and fair view. Nevertheless, the Fourth Directive considers some general principles which weren't translated to the Accounting Act such as:

- Identity (*princípio da correspondência entre o balanço de abertura de um exercício e o balanço de encerramento do exercício anterior*): the opening balances of the current economic year must match the ending balance of the previous year.
- No offsetting (*princípio da não compensação*): net amounts are not admitted.

The first two principles are commonly used so there was no need to translate them.

These generally accepted accounting principles (GAAP) are laid down in the Accounting Directive nº 18 'Financial statement objectives and generally accepted accounting principles'. This Directive states that the GAAP are considered in the Accounting Act, Accounting Directives and, in issues not studied in the former, the International Accounting Standards issued by IASC, as previously stated.

2 Value concepts

The value concepts are included in Chapter 5 of the Accounting Act. The Accounting Act adopts the historical cost principle – acquisition or production cost – for inventories, fixed tangible assets, fixed intangible assets and some financial investments. The Fourth Directive proposes this general valuation standard in Article 32º. There are exceptions to this rule which will be discussed in due course.

Acquisition cost is the amount of cash or cash equivalents paid given to acquire an asset at the time of its acquisition or construction. The items generally included in the initial cost of an acquired asset are: purchase price, duties and non-refundable taxes paid, costs directly attributable to bringing the asset to working condition for its intended use (e.g. site preparation, freight, installation costs, professional costs).

Conversion or production costs which are capitalized as part of inventory include costs directly related to the units of production such as: direct materials, direct labour (including taxes and benefits) and an allocation of the fixed and variable production overheads incurred in converting materials into finished goods (e.g. depreciation, rent and rates, maintenance and factory management and administration).

Borrowing costs may be capitalized if they could have been avoided if the qualifying asset had not taken place. When borrowings are made specifically to finance a qualifying asset, it is relatively easy to identify the borrowing assets which would have been avoided. The capitalization of borrowing costs should cease when all the activities necessary to prepare the qualifying asset for its intended use or sale are complete.

The research and development costs should be expensed, unless underlying conditions prove otherwise (see VII).

With regard to the possibility of capitalizing exchange differences, the POC stipulates that the potential gain or loss arising from long-term foreign currency loans aimed at specific asset financing can be recorded as a reduction of or an addition to the related asset value.

3 Valuation rules for fixed assets

The Accounting Act defines fixed assets as those items retained in the company for use in a continuum and not destined to be sold or transformed in the course of business. The legal property may belong to the company or they may have a financial lease. Nowadays, accounting professors defend the IASC definition in order to introduce the 'economic benefits' definition.

3.1 Fixed tangible assets

3.1.1 Definition

Includes tangible items, mobile or not, which a company uses in its operational activity, not destined to be sold or transformed, for over a year. It also includes subsequent expenditures to be capitalized.

Items under a financial lease are also included and follow a specific accounting (see VII).

3.1.2 Measurement

As a general rule, when an item of property, plant and equipment qualifies for recognition as an asset, it is initially measured at its cost. This applies even to assets acquired using funds obtained from government grants. The assets acquired under finance leases require special treatment (see VII).

The initial cost generally includes the same items normally set in an acquisition cost.

If an asset is constructed for use by the company, the determination of the initial cost of the asset is made using the same general principle of the initial measurement at cost in relation to acquired assets. Therefore, costs which are directly attributable to the construction of the asset may be capitalized including:

- Work subcontracted.
- Materials.
- Labour.
- Overheads directly attributable to the construction of the assets.
- Fees and permits.

The capitalization of these items as fixed assets must be recognized as revenue in the income statement.

3.1.3 Subsequent expenditure

Judgement is required in determining whether costs incurred subsequent to the initial purchase of an item of fixed tangible assets qualify for capitalization. Expenditure that merely restores or maintains the originally assessed standard of performance of the existing asset is expensed when incurred.

3.1.4 Revaluation

There are two accepted revaluation procedures:

- Legal revaluation: under a decree-law that defines the actualization of the historical cost for current prices (inflation).
- Free revaluation: undertaken by an independent, professionally qualified valuer.

The accounting treatment of revaluation surplus, deficit or realisation is further explained in VIII.

3.1.5 Depreciation

The depreciation is shown in the balance sheet by its accumulated value associated with the asset depreciated. The appropriate value adjustments are made taking into consideration the diminution in the original value of fixed assets.

Only one type of fixed tangible assets are exempt from the requirement to depreciate: land. Investment in property is considered a financial investment and the building component, if in use, is also subject to systematic depreciation. The difference is that depreciation in fixed tangible assets is considered operational and in investment in property (financial investment) it is considered financial.

The general rules for depreciation of all assets, tangible or intangible, are in Regulatory Decree nº 2/90, of 12 January. The depreciable amount of the asset is the cost of an asset, less its residual value. If the item was revalued, the depreciable amount is the revalued amount less any residual value.

The useful life of the asset is the period of time over which an asset is expected to be used, or the period calculated according to the fiscal rates proposed by the Regulatory Decree.

There are several depreciation methods but the ones recommended and accepted by the Regulatory Decree are just two: straight-line and diminishing balance. The straight-line method is commonly used by Portuguese companies. There is the possibility to increase or decrease the rate of depreciation when using the straight-line method, when proved that an asset is under- or over-used, respectively. Not all items can be subject to the diminishing balances method, there are exceptions (e.g. vehicles).

3.1.6 Impairment

The Accounting Act defines that if, at the balance sheet date, any of the items of the fixed tangible or intangible assets, with limited or unlimited economic life, has a market value of less than the one in the accounting registers, then the surplus of the carrying amount should be expensed. This reduction shall cease as soon as there's a reversal of the situation.

3.1.7 Retirements and disposals

There is no specific rule concerning assets when no future economic benefits are expected to flow to the company in association with them. The common procedure is to remove them from the balance sheet once they're no longer useful. The gain or loss is calculated as the difference between the estimated net disposal proceeds and the carrying amount of the asset and is recognized in the income statement. The revaluation surplus, should there be one, is transferred to retained earnings when surplus is realized.

3.2 Intangible fixed assets

Intangible fixed assets include implementation expenses, expansion expenses, research and development expenses (see VII), royalties and goodwill. Intangible fixed assets are considered to have an economic life of no more than five years.

The word goodwill is translated into Portuguese as *'trespasse'* which has been considered a poor definition as it has legal implications. The goodwill arises as the result of a business combination accounted for as an acquisition and is the excess of the cost of the acquisition over fair value (Accounting Directive nº 13 'Fair value'). Goodwill is capitalized and amortized over a period not exceeding five years unless a longer period not exceeding 20 years can be justified.

3.3 Financial investments

Apart from investments in subsidiaries and associates, long-term and current investments, financial assets must be carried in the balance sheet at their acquisition cost. The acquisition cost includes charges such as brokerage, fees, duties and bank fees. Should the acquisition cost exceed the market value, the surplus amount must be provisioned.

For individual accounts the Accounting Act proposes two types of treatment for associates: the acquisition cost and the equity method. Accounting Directive nº 9 states that the equity method is always mandatory, apart from some limited exceptions. In terms of subsidiaries, the ownership of a company in the individuals accounts also follows the equity method. Table VI.1 summarizes the procedure.

4 Valuation rules for current assets

4.1 Inventories

4.1.1 Definition

Inventories are assets held for sale in the course of business, in the process of production for such sale, in the form of materials or supplies to be consumed in the production process or rendering of services or, by-products or waste with commercial value.

4.1.2 Measurement

Inventories should be measured at the acquisition cost or production cost, regardless of the exceptions considered for by-products and in cases that the market price is lower than the historical cost (see Table VI.2).

Replacement: the cost that a company would have to incur if it had to replace the product under the same conditions.

Net realizable value: expected selling price deducted of selling expenses.

Cost may be assigned to individual items of inventory either by specific identification method or by use of a cost formula. When the use of a specific identification method is inappropriate, the Accounting Act presents several optional cost formulas in: the weighted average formula, 'first-in, first-out' formula (FIFO), 'last-in, last-out' formula (LIFO), which are denoted as benchmark treatments and standard costs, which is the allowed alternative treatment (in very specific cases). However, the same cost formula must be used for inventories having the same characteristics.

The specific identification method entails assigning the specific cost of an item of inventory to a particular item of inventory and is only feasible when each item of inventory can be separately identified.

The weighted average cost formula assigns a value to each item of inventory based on a weighted average of items in inventory at the beginning of the period and the weighted average of items acquired or produced during the period. The FIFO cost formula assumes that the items of inventory which were purchased or produced first were first used. Therefore, at the end of the

Table VI.1 Accounting methods in financial investments

Classification	Position	Ownership	Method
Other investments	No significant influence	(less than 20%)	Acquisition cost
Associates	Some influence	(less than 50% and more than 20%)	Acquisition cost Equity method
Subsidiaries	Significant influence	(more than 50%)	Equity (individual) Consolidation (group)

Table VI.2 Classification of inventories

Type	Method	Market price
Raw materials and merchandise	Acquisition cost	Net realizable value
Product	Conversion or production cost	Replacement cost
By-products, wastes and residues	Net realizable value	(N/A)

Table VI.3 Formulas applicable to certain activities

Activity	Formula	Alternative formula
Agriculture, cattle-raising, forestry	Standard costs	Net realizable value after margin deduction
Mining	Production costs	Net realizable value after margin deduction
Fishing industries	Production costs	Net realizable value after margin deduction
Retail establishments	Selling price after margin deduction	

period, the items are valued at the most recent purchase prices. The LIFO cost formula assumes the contrary, that the items in the inventory which were acquired or produced last were used first.

Some activities may vary its formulas when a cost–benefit analysis finds that is impossible to apply the general formulas (see Table VI.3).

When the net realizable value or replacement cost of an item in inventory is less than its cost, the excess is not written off immediately in the income statement, but a provision for depreciation is created. This provision influences the value of the liquid asset and the income statement.

4.2 Receivables and other debtors, cash and cash equivalents and short-term investments

In practice, receivables and other amounts due from third parties are recognized when an amount becomes receivable by the company (e.g. goods shipped to a customer, or services rendered). If the receivables are expected to be realized in the normal course of the operating cycle of the company, they are classified as current.

If by any chance, the company is aware that some of the receivables are uncollectable then they should be written off. If they are just doubtful, a provision must be created for the amount considered to be doubtful. This provision will adjust the gross amount in the balance sheet.

Portuguese short-term investments include marketable securities and other types of investments with a maturity of less than one year. So, some short-term investments may be considered cash equivalents (if less than three months) and others not. They're accounted for at the acquisition cost. If this acquisition cost is in excess of the market value, than a provision has to be created to correct the surplus.

5 Valuation rules of liabilities and provisions

5.1 Liabilities

Liabilities include all debts that are certain at the balance sheet date and which will be fulfilled by payment and the liabilities due to fixed assets and stocks capitalized in the balance sheet of which the entity has economic (substance over form) ownership.

In general, the classification of liabilities depends on when they fall due for repayment. If they are due in less than a year they're considered short-term liabilities (current), if due in more than a year, long-term liabilities.

Liabilities are generally stated at face value. An exception is made for finance lease liabilities which are discounted. Unless we're in the presence of a liquidation balance sheet, it is not permissible to value debts at discounted value.

There may be liabilities not expressed in the balance sheet and their amount and nature must be disclosed in the Appendix.

5.2 Provisions

A provision is a present obligation arising from past events, the settlement of which is expected to result in an outflow of resources from the company, or, the possibility that some assets may suffer write-offs.

There are two types of provision:

- Value adjustments to assets (*provisões activas*), such as provisions for obsolescence, doubtful accounts, inventory depreciation, marketable securities and financial investments. These provisions are deducted from the gross value of the related asset. The Accounting Act asks provisions to cease if the related situation isn't realistic any more.
- Contingent provisions (*provisões passivas*): classified separately on the liability side, as they are potential outflows. The provision is an estimation.

VII Special accounting areas

1 Deferred taxation

There are no specific rules concerning deferred taxes. The Accounting Act demands a disclosure of situations that might affect future taxes in note n°6 of the Appendix.

In terms of basis or treatment nothing else is

mentioned. Accounting Directive nº 18 places the International Accounting Standards (IAS) as one of the accounting procedures that should be taken into consideration after POC and Accounting Directives issued by CNC.

Technical Interpretation nº 4 (*Interpretação Técnica nº 4*) from CROC, issued in December 1993, advises auditors to verify the disclosure in Note nº 6 of the Appendix in terms of future taxes arising from the reinvestment of assets selling surpluses. For fiscal purposes (through Law nº 71/93, 26 November) the surplus is deferred during the economic life of the new investment.

2 Post-balance sheet events

There are no specific accounting standards concerning events after the balance sheet date. According to Accounting Directive nº 18, when no specific Portuguese accounting standards exist, IAS should be followed, which means the 1999-revised IAS nº 10 'Events after the balance sheet date'.

Only one of the 26 Accounting Directives refers to subsequent events: Accounting Directive nº 17 'Futures contracts'. In its fifth item in Note nº 5 it asks for 'disclosure of significant variations occurred in the fair value of future contracts and covered positions (should they exist) between the end of an economic period and the issue date of financial statements'.

Article 66º of the CSC demands that the manager's report includes references to facts occurring after the balance sheet date and before the financial statements issue.

Because of this, there is a common procedure among certified companies (the companies that are compelled to have a Certified Accountant – *Revisor Oficial de Contas*) to inform auditors of subsequent events, meaning those that arise after the balance sheet date and before the date of the Official Audit Report (*Certificação Legal de Contas*).

Subsequent events concerning conditions that existed at the balance sheet and that are material should lead to an adjustment of the balance sheet. Subsequent events concerning conditions that arose after the balance sheet date require disclosure in Note nº 48 of the Appendix.

3 Long-term contracts

In Chapter 5 'Valuation procedures' of the Portuguese Accounting Act, item 3.17 indicates that products in process resulting from pluri-annual activities should be accounted for using the percentage of completion method. An alternative method is also allowed, the completed contract.

Accounting Directive nº 3 'Accounting treatment of construction contracts' applies to construction contracts that relate to unique construction projects (e.g. bridges, ships, buildings). There is no reference to the length of a contract in its definition, but there is a requirement that the contract should begin in one economic year and end in another economic year. The Accounting Directive assumes that the contract includes a pre-established price or a margin over costs supported by the company.

When the outcome of a contract can be estimated reliably, revenues and costs should be estimated by the percentage of completion method, meaning as the work progresses rather than when the work is completed. Expected losses should be recognized immediately. When the construction is near its end, costs should be estimated and accrued. The conditions for reliable estimation are:

- Revenue can be reliably estimated.
- Future costs and stages of completion can be measured reliably.
- Costs can be identified and measured reliably.

In summary, under this method, contract revenue is matched with the contract cost incurred in reaching the stage of completion at the balance sheet date, resulting in the reporting of revenue, expense and profit which can be attributed to the proportion of work completed.

If the outcome cannot be measured reliably, costs should be expensed and revenues should be recognized in line with the completed contract method.

A company should disclose in note nº 3 of the Appendix the method used to account for construction contracts and in note nº 48 of the Appendix:

- The amount of costs and profits of the contracts in progress that contributed to the income statement.

- The amount received and the amounts that will be received.

Additional Note: for fiscal purposes only the above-mentioned methods are accepted. Number 2 of Article 21° of the CIRC demands the percentage of completion method for some situations:

- Public or private construction projects with contracts when there is a pre-established price, partial invoices and the work in progress matches invoicing.
- Construction projects of a company sold in separate parts as they are concluded and delivered, although the total amount of costs may be unknown.

4 Service contracts

Accounting Directive n° 26 'Revenue' presents the method by which revenue from services rendered can be recognized. When the outcome of a transaction can be estimated reliably, the revenue associated with the completion phase should be accrued at the balance sheet date. The Accounting Directive considers that an outcome can be estimated reliably when:

- The revenue amount can be measured reliably.
- The costs incurred and to be incurred until service completion can be measured reliably.
- It is probable that the economic benefits will flow to the entity.
- The completion phase of the transaction can be measured reliably at the balance sheet date.

When outcome can be estimated reliably the reference method to recognize revenue is the percentage of completion method described in point number 3.

5 Accounting for leases and sale and lease back agreements

Finance leases are treated in three different sources: Accounting Act, Accounting Directive n° 10 'Transitory regime of finance leases' and Accounting Directive n° 25 'Leases'.

Although Accounting Directive n° 25 was the last to be issued, it is there that the two types of lease are differentiated: finance leases, where the lessee acquires rights and obligations that are similar to those acquired by an outright purchase of the legal title to an asset; and operating leases, where such rights and obligations are not acquired by the lessee.

For finance leases, the substance and legal form of the transaction are not the same. Accounting Directive n° 25 requires finance leases to be accounted for on the basis of substance, regardless of ownership, as in a finance lease there's a transfer of all risks and advantages – this Accounting Directive was based on IAS 17 'Leases', but IAS states '...substantially all risks and rewards...' and here we find '...all risks and advantages...' – incident to ownership of an asset. The Accounting Directive states that it is enough to verify one of the following situations to regard a lease as financial:

- The lease transfers ownership of the asset to the lessee by the end of the lease term.
- The lessee has the option to purchase the asset at a price which is expected to be sufficiently lower than the fair value at the date the option becomes exercisable such that, at the inception of the lease, it is reasonably certain that the option will be exercised.
- The lease term is for the major part of the economic life of the asset even if the title is not transferred.
- At the inception of the lease the present value of the minimum lease payments amounts to or is superior to the fair value of the leased asset.
- The leased assets are of specialized nature such that only the lessee can use them without modifications being made.
- The lessee may cancel the lease but will bear any losses arising from it.
- Gains or losses from fluctuations in the fair value of the residual fall to the lessee.
- The lessee has the ability to continue the lease for a secondary period at a rent substantially lower than the market rent.

The explanatory note to account 42 'Fixed Assets' (*Imobilizações corpóreas*), in conjunction with Accounting Directive n° 10, explains how to recognize finance leases as both assets and liabilities in the balance sheet:

- The lessee records an asset at the date of the contract, being the lower of capitalized fair value of rights to use the asset over the lease

term, which is subject to depreciation. The discount rate is the one inherent to the lease or the market interest rate.
- It also records a liability to make payments to the lessor; rentals paid are allocated between a capital element which reduces the outstanding liability and interest expense on the liability.
- Depreciation on leased assets should be calculated using useful life, unless there is no reasonable certainty of eventual ownership. In the latter case, the shorter of useful life and lease term should be used.

The lessor does not record a physical asset, but rather a rental receivable; rentals received are allocated between a capital element, which reduces the receivable and interest income.

For operational leases, accounting can be made on the basis of legal form. The lessee records rentals payable as an expense and the lessor records ownership interest in the physical asset, which is subject to depreciation. Rentals received and receivable are recorded as income.

Accounting Directive nº 10 enforces that all lease contracts outstanding that fall under the definition of financial leases should form part of the balance sheet as if they had always been there. The loss or profit that arises from the transfer should be considered an extraordinary item and should be taken to the income statement.

Should the amount be of such a significance that it distorts results, it can be taken directly to retained earnings (account 59) in the balance sheet according to Accounting Directive nº 8.

Note nº 15 of the Appendix demands a disclosure of all assets used through finance leases. Accounting Directive nº 25 also states that for sale and leaseback with results in a finance lease, any excess of proceeds over the carrying amount should be deferred and amortized over the lease term. In the case of an operational lease the excess should be recognized at once.

6 Research and development costs

According to Accounting Directive nº 7 'Accounting of research and development expenses', research costs – the Accounting Directive considers that research expenses must have to be original and planned with the purpose of obtaining new scientific or technical knowledge – should be expensed, unless it can be proved that economic benefits will arise in the future.

Development costs should also be expensed, unless they meet the following criteria, in which case they must be capitalized:

- Clearly defined product with costs separately identified and reliably measured.
- Technically feasible project.
- Enterprise intends to proceed.
- Market can be demonstrated.
- Adequate resources can be demonstrated.
- Capitalization should not exceed recoverable amount.

Capitalized development costs should be amortized on a systematic basis reflecting the pattern of recognition of benefits but, according to item 4.8 of chapter 5 'Valuation procedures' of the Portuguese Accounting Act, for no more than five years.

Note nº 8 of the Appendix demands comments about the capitalization of research and development costs.

7 New financial instruments

Accounting Directive nº 17 'Futures Contracts' does not deal with financial instruments in general. It includes rules concerning recognition, measurement, presentation and disclosure of a particular type of derivatives, futures contracts.

The scope of this Directive is rather narrow, as international standards on the subject tend to include all financial instruments. The purpose established in the Accounting Directive is 'accounting treatment of future contracts', although it is not clear whether or not it includes commodities futures contracts.

In a futures contract there is an established right to an asset exercisable on maturity, by the same amount, there is an initial margin (*margem inicial*) and profits and losses (*ajustes diários de ganhos e perdas*) are computed on a day-to-day basis, meaning 'marking to the market'. Financial futures, in particular, are traded on organized exchanges that set standard terms for the contracts. In Portugal there is the Derivatives Exchange in Oporto (*Bolsa de Derivados do Porto*).

Accounting Directive nº 17 does not recognize the contracts in the balance sheet, as there is a compensation of rights and obligations, many contracts finish before its maturity and there is no assurance that contracts will materialize. Nevertheless, disclosure is compulsory in note nº 48 of the Appendix.

The initial margin is considered a short-term investment and is accounted in 15 'Tradable titles'. This means that the initial margin is not considered a receivable but a financial asset, contrary to what can be found in IAS.

The costs related to fees and taxes and interest related to the initial margin (should there be interest) are taken to financial income.

The daily gains and losses are treated differently. It all depends on the configuration of the contract. If the contract is intended to speculate, gains and losses are taken to financial income, but if the contract has hedging purposes, gains and losses are deferred and accounted in 275 'Accruals and deferrals – Deferred daily adjustments in futures contracts' (*Acréscimos e diferimentos – Ajustes diários diferidos em contratos de futuros*).

At the hedging contract inception, there may be gains or losses. They can be accounted to operational income or financial income, depending on the nature of the future contract. In the case of speculating futures contracts there are no deferred gains or losses to be recognized.

The Accounting Directive demands disclosure on note nº 48 of the following items:

- Future contracts: types and maturity, type of position, markets, purpose, type of risk covered.
- Hedging position: position, coverage indicators, fair value of covered positions.
- Global risk evaluation method in speculating and hedging operations.
- Amounts due to initial margins.
- Significant variations occurred in the fair values of futures contracts between the balance sheet date and the issuing date of the financial statements.
- Gains and losses from speculating operations.
- Initial margins realized with financial instruments.

8 Foreign currency translation

As far as the Portuguese Accounting Act is concerned, transactions should be translated on the date of the transaction. Subsequently, monetary balances should be translated at the closing and non-monetary balances at the rate that relates to the valuation basis (e.g. historical cost). Differences on monetary items should be taken to income. There are two major exceptions, the case of long and medium receivables or payables and loans concerning assets in progress.

In the first case, favourable foreign currency translation should be deferred if there are reasonable expectations of a change (reversal) in the currency rate. As monetary inflows or outflows take place, the deferred amount must be taken into income.

In the second case, it is possible (but not mandatory) to charge assets in progress with the positive or negative difference.

The closing rate used for currency translation at the balance sheet date must be disclosed in Note nº 4 in the Appendix.

9 Employee benefits

The Accounting Directive nº 19 'Retirement Benefits' addresses formal and informal plans, funded and unfunded plans and national, state, industry plans. It does not deal with other forms of employee remuneration or post-employment benefits (e.g. deferred compensation, health and welfare, bonus plan, long-service leave benefits). For the latter, revised IAS 19 can be followed according to Accounting Directive nº 18.

For defined contribution plans, the contributions of a period should be recognized as expenses. For defined benefit plans there are three different accounting treatments depending on the existence of funds and their management type (within the entity or by a third party), nevertheless, the global treatment is to recognize the current service cost as an expense.

Past service costs, experience adjustments and

changes in actuarial assumptions in relation to existing employees should be taken to income over expected remaining working lives. When termination, curtailment or settlement is probable, cost increases should be charged immediately but any gains should be recognized when the events occur. For retired employees, plan amendments should be recognized immediately at the present value of the effects.

The benchmark actuarial valuation method is the projected benefit valuation method. Actuarial assumptions should include projected salary levels.

Note nº 31 of the Appendix demands disclosure on the type of plan (defined contribution or defined plan) and a complete description.

VIII Revaluation accounting

The Accounting Directive nº 16 approved by the Accounting Standards Committee (CNC) in 1995 concerns the revaluation of the elements of tangible fixed assets (*activos imobilizados tangíveis*), stating that it can be done based on the variation of the acquisitive power of the currency or on the fair value (referred to in Accounting Directive nº 13). If a free revaluation is made (therefore, not following one of the two methods referred to) it must be stated in Note nº 1 of the appendix.

The companies that revalue their tangible fixed assets should divulge, in Note nº 39 of the appendix, when the case is materially relevant, three aspects: the class the revalued element belongs to, the surplus obtained (and the method used to calculate it) and the identification of the entity that carried out the revaluation. Furthermore, in the additional statements, the purpose of the revaluation should be disclosed.

More recently, Decree-Law nº 31/98 has authorized the revaluation of the elements of tangible fixed assets which are used in commercial, industrial or agricultural activities, as long as they exist and are being used by the companies on the date of the revaluation. This decree-law complies with the rules established by the Sixteenth Directive. This authorization allows companies to face the devaluation of the currency by actualizing the historical costs of the elements of the tangible fixed assets, but it only applies to those elements that have an expected life of five or more years. The coefficients that must be used in the revaluation are also defined.

IX Notes and additional statements

1 Notes

The Accounting Act of 1989 requires the following notes on the appendix of the financial statements:

1 Disclosure and justification of the rules of the Accounting Act that, in exceptional cases, have not been followed and the effect that has on the financial statements, as they must give a true and fair view of the assets, the liabilities and the profits of the company.
2 Disclosure and comments on the accounts of the balance and the income statement whose contents can not be compared with the ones from the previous year.
3 Valuation criteria used as well as the methods used in the adjustments of values (depreciation and provisions).
4 Foreign exchange rates used in the conversion of the values that were in foreign currency.
5 Extent to which the net profit was affected in order to obtain fiscal advantages.
6 Disclosure of situations that affect significantly the future taxes.
7 Average number of people working in the company, divided into employees and wage-earners.
8 Commentary on the accounts 431 'Installation expenses' (*despesas de instalação*) and 432 'Research and development expenses' (*despesas de investigação e desenvolvimento*).
9 Justification of the depreciation of the trespasses on a period longer than five years.
10 Changes in the assets accounts that are on the balance sheet. The respective depreciation and provisions must be disclosed in a specified table, that divides the assets into intangible assets, fixed assets and financial investments.
11 Disclosure of the costs incurred in the exer-

cise that are related to the financing of assets, during the construction, that have been capitalized in this period.

12 Disclosure of the legal diplomas on which the revaluation of the fixed assets and the financial investments were made.

13 Elaboration of a specified table discriminating the revaluation values.

14 Disclosure of several aspects concerning the fixed and the ongoing assets.

15 Disclosure of the assets used on the basis of a financial lease, as well as their accounting values.

16 Firms and headquarters of the group companies, the associated companies and the subsidiaries, as well as the part of the capital the company holds, their equity and net profit of the previous year.

17 Firms of other companies where the company holds shares that have an accounting value of more than 5 per cent of its circulating assets and that are accounted for in the 'negotiable titles' account.

18 Discrimination of the account 4154 'Funds' (*fundos*) and indication of the respective affectations.

19 Global indication, by assets category, of the materially relevant differences between the costs of the elements of the circulating assets, calculated according to the valuation criteria adopted, and the respective market values.

20 Explanation of the special circumstances that led to the situation described in note n° 19.

21 Indication and justification of the extraordinary provisions related to elements of the circulating assets for which stable descents are predictable, as a result of value fluctuations.

22 Global values of the inventories that are not with the company (dispatched, on transit, to the guard of others).

23 Global value of potential uncollectable receivables included in each of the different receivables accounts.

24 Indication, by organ, of the money advanced or lent to each one of the administration, direction and inspection organs, as well as the interest rate, the main conditions, the amounts already disbursed and the responsibilities taken by guaranties.

25 Global value of the debts that concern the personnel of the company.

26 Global value of the debts that are titled, by items of the balance, when not evidenced on it.

27 Quantity and nominal value of convertible bonds, of participation titles and other similar titles, issued by the company, with indication of the rights they confer.

28 Discrimination of the debts included in the 'state and other public entities' (*Estado e outros entes públicos*) account that are overdue.

29 Value of the liabilities that have a term longer than five years.

30 Value of the liabilities that are covered by a guaranty of the company, with details.

31 Global value of the financial responsibilities that are not included in the balance, as long as their indication is useful to the appreciation of the financial condition of the company.

32 Description of the responsibilities of the company by given guaranties, divided into categories.

33 Indication of the difference, when taken to the assets, between the values of the liabilities and the correspondent amounts already received.

34 Division of the provisions into categories, according to a defined table.

35 Form of realization of the equity and its increases and decreases, only in the year in which they took place, as well as indication of the equity still not realized.

36 Number of shares of each category that the capital divides into and their nominal value.

37 Participation in the subscribed capital of each of the collective entities that hold at least 20 per cent of it.

38 Number and nominal value of the shares and quotas subscribed in the capital, during the year, below the limits of the authorized equity.

39 Indication of the variations of the revaluation reserves occurred in the year, with details.

40 Disclosure and justification of the movements occurred in the year in each of the items of equity that are referred to in the balance sheet, that have not been disclosed yet.

41 Statement of the cost of inventory sold and materials consumed (see additional statements).

42 Statement of the variation of production (see additional statements).

43 Disclosure, global to each of the organs, of the remuneration attributed to the members of the social organs that are related to the exercise of the respective functions.

44 Separation of the net value of sales and services rendered, calculated on the accounts 71 'Sales' (*vendas*) and 72 'Services rendered' (*prestação de serviços*), by activity and by market, whenever those activities and markets are considerably different.

45 Financial profit statement (see additional statements).

46 Extraordinary profit statement (see additional statements).

47 Information demanded by legal diplomas.

48 Other information considered relevant for a better understanding of the financial position and the profits.

2 Additional statements

2.1 Statement of changes in financial position

According to the Accounting Act the companies must also disclose the statement of changes in financial position (*demonstração da origem e da aplicação de fundos*), which has the format shown in Table IX.1.

2.2 Statement of variation in working capital

According to the Accounting Act the companies must also disclose the statement of variation in working capital (*demonstração das variações dos fundos circulantes*), which completes the information given by the previous statement and has the format shown in Table IX.2.

2.3 Cash flow statement

The Accounting Directive nº 14 approved by the CNC introduces the Cash Flow Statement (*Demonstração de Fluxos de Caixa*). Its main purpose is to present information about the cash receipts and cash payments of a company in a certain period. The statement has to act as a complement to the documents that show the financial performance of a company. This demand is applicable to companies in general, including financial

Table IX.1 Format of the statement of changes in financial position

Sources of funds			**Application of funds**		
Internal:	x		Distributions:	x	
Net profit of the year	x		By profit application	x	
Depreciation	x		By reserve application		x
Variation of provisions		x			
			Decreases in equity:		
External:			Decrease in common stock and supplements		x
Increase in equity:					
Increase in common stock and supplements	x				
Increase in emission premiums and special reserves	x		Long term financial operations:		
			Increase in financial investments	x	
Loss offset	x		Decrease of long-term payables	x	
		x	Increase in long-term receivables	x	
Long-term financial operations:					x
Decrease in financial investments	x		Increase in assets:		
Decrease in long-term receivables	x		Work for the company	x	
Increase in long-term payables	x		Acquisition of fixed assets	x	
		x			x
Decrease in assets:					
Transfer of assets		x			
Decrease in working capital		x	Increase in working capital		x
Total sources of funds		x	*Total applications of funds*		x

Table IX.2 Format of the statement of variation in working capital

Increase in inventory	x	Decrease in inventory	x
Increase in short-term receivables	x	Decrease in short-term receivables	x
Decrease in short-term payables	x	Increase in short-term payables	x
Increase in liquidity	x	Decrease in liquidity	x
Decrease in working capital	x	Increase in working capital	x
	x		x

companies, insurance companies and even small or medium-sized companies.

It should be noted that the cash flows in the statement are classified in three categories: operating activities, financing activities and investing activities. In its preparation one can follow the direct or the indirect method (see Tables IX.3 and IX.4). Under the direct method, activities are reported separately disclosing major classes of gross cash receipts and gross cash payments. This method allows stakeholders to understand the way in which the company uses and generates cash at her disposal. Under the indirect method, the activities are reported adjusting net profit (or loss) for the effects of transactions of a non-cash nature, accruals related to receipts or payments due in the future or costs related to cash flows connected with investing or financing. Companies ought to use the direct method as it gives more thorough and accurate information.

2.4 Income statement by functions

All companies that use the POC must disclose, at the end of the year, an income statement by nature. Decree-Law 44/99 states that they must also disclose an income statement by functions, in a format that is defined by the Accounting Act. The accounting directive nº 20 'Income statement by functions' approved by the Accounting Standards Committee (CNC) defines the income statement as being compulsory for the companies who must have official audit reports. Companies that must elaborate an income statement by functions must also present, in note nº 42 of the appendix, a cost of sales and services rendered statement.

The format this accounting directive defines for the statement is different from the one that is optional to every company, according to the Accounting Act of 1989. The main differences relate to the financial component of the statement and to the introduction of two new items: other operating costs and losses (*outros custos e perdas operacionais*) and profit per share (*resultados por acção*).

As with the cash flow statement, there are three categories: operating activities, financing activities and investing activities. The format defined by the accounting directive is shown in Table IX.5.

2.5 Statement of the cost of inventory sold and materials consumed

This statement must be disclosed in note nº 41 of the appendix (see Table IX.6).

2.6 Cost of sales and services rendered statement

This statement must be disclosed by all the companies that are required to present an income statement by functions, and must be in note nº 42 of the appendix (see Table IX.7).

2.7 Statement of the variation of production

This statement must be disclosed in note nº 42 of the appendix (see Table IX.8).

2.8 Financial profit statement

This statement must be disclosed in note nº 45 of the appendix (see Table IX.9).

2.9 Extraordinary profit statement

This statement must be disclosed in note nº 46 of the appendix (see Table IX.10).

Table IX.3 The cash flow statement (direct method)

OPERATING ACTIVITIES

Receipts from clients
Payments to suppliers
Payments to personnel

Flows generated by operations

Payment/receipt of profit tax
Other receipts/payments due to the operating activities
Receipts related with extraordinary items
Payments related with extraordinary items

Flows generated by extraordinary items

Flows generated by operations
Flows generated by extraordinary items

Flows from operating activities [1]

INVESTING ACTIVITIES

Receipts from:
- Financial investments
- Fixed assets
- Intangible assets
- Subsidies to investments
- Interest and similar income
- Dividends
- … … …

Payments due to:
- Financial investments
- Fixed assets
- Intangible assets
- … … …

Flows from investing activities [2]

FINANCING ACTIVITIES

Receipts from:
- Loans borrowed
- Increase in capital shares and advance on equity
- Subsidies and donations
- Disposal of company's shares
- Loss coverage
- … … …

Payments of:
- Loans borrowed
- Leasing obligations
- Interest and similar costs
- Dividends
- Decrease in capital and advance on equity
- Acquisition of the company's own shares
- … … …

Flows from financing activities [3]

Changes in cash and cash equivalents
[4] = [1] + [2] + [3]
Effect of changes in foreign currency
Cash and cash equivalents in the opening period
Cash and cash equivalents at the closing period

Table IX.3(b) The cash flow statement (indirect method)

OPERATING ACTIVITIES

Net Result

Adjustments:
Depreciation
Provisions
Financial results
Increase/decrease in receivables
Increase/decrease in inventories
Increase/decrease in payables
Decrease in deferred credits
Decrease in deferred charges
Profit on disposals
Losses on disposals
… … …

Flows from operating activities [1]

INVESTING ACTIVITIES

Receipts from:
- Financial investments
- Fixed assets
- Intangible assets
- Subsidies to investments
- Interest and similar income
- Dividends
- … … …

Payments due to:
- Financial investments
- Fixed assets
- Intangible assets
- … … …

Flows from investing activities [2]

FINANCING ACTIVITIES

Receipts from:
- Loans borrowed
- Increase in capital shares and advance on equity
- Subsidies and donations
- Disposal of company's shares
- Loss coverage
- … … …

Payments of:
- Loans borrowed
- Leasing obligations
- Interest and similar costs
- Dividends
- Decrease in capital and advance on equity
- Acquisition of the company's own shares
- … … …

Flows from financing activities [3]

Changes in cash and cash equivalents
[4] = [1] + [2] + [3]
Effect of changes in foreign currency
Cash and cash equivalents in the opening period
Cash and cash equivalents at the closing period

Table IX.5 Format of the income statement by functions

	N	N-1
Sales and services rendered	X	X
Cost of sales and services rendered	–X	–X
Gross profit	+/– X	+/– X
Other operating profits and earnings	X	X
Distribution costs	–X	–X
Administrative expenses	–X	–X
Other operating costs and losses	–X	–X
Operating profit	+/– X	+/– X
Net financing costs	–X	–X
Earnings (losses) in subsidiaries and associated companies	+/– X	+/– X
Earnings (losses) in other investments	+/– X	+/– X
Current profit	+/– X	+/– X
Tax on current profit	–X	–X
Current profit after tax	+/– X	+/– X
Extraordinary profit	+/– X	+/– X
Tax on extraordinary profit	–X	–X
Net profit	+/– X	+/– X
Profit per share	+/– X	+/– X

Table IX.6 Format of the statement of cost of inventory sold and materials consumed

Movements	**Inventory**	**Raw, subsidiary and consuming materials**
Beginning inventory		
Acquisitions		
Inventory regularization		
Ending inventory		
Costs of the year		

2.10 Management report

According to article 65º of the CSC, the members of the administration must elaborate and submit to the competent organs of the company the management report (among other things) that must be elaborated according to the law and signed by all members of the administration. The same code defines, in its article 66º, that this report must include, at least, a faithful and clear exposition on the evolution of the business and the situation of the company. This article defines that the management report must disclose, specially:

- The evolution of the management in the different sectors where the company has undergone activity, in what concerns the conditions of the market, the investments, the costs, the

Table IX.7 Format of the statement of cost of sales and services rendered statement

	Finished and intermediate goods	By-products, waste and residue	Services rendered
Beginning inventory			
Entries from production			
Inventory regularization			
Exists to production and fixed assets			
Ending inventory			
Cost of sales			

Table IX.8 Format of the statement of the variation of production

	Finished and intermediate goods	By-products, waste and residue	Work in process
Ending inventory			
Inventory regularization			
Beginning inventory			
Increase/decrease of the year			

profits and the research and development activities.

- The relevant facts that occurred after the end of the year.
- The predictable evolution of the company.
- The number and the nominal value of quotes or shares of the company bought or sold during the year, the reasons and the respective prices, as well as the number and the nominal value of all the own quotes and shares the company holds at the end of the year.
- The authorizations granted to businesses between the company and its administrators.
- A proposal for the application of profits, duly substantiated.
- The existence of branches of the company.

The proposal for the application of the profits must be made according to the law. The Companies Code states, in its article 33º, that first of all the profits must cover any accumulated losses. Article 295º states that after all the losses are covered, then the company must put in a legal reserve at least 5 per cent of the profits. After this, the company should pay dividends of at least 50 per cent of the profits still available (article 294º). The rest of the profits can be put in free reserves.

Table IX.9 Format of financial profit statement

Costs and losses	Years N	Years N – 1	Profits and earnings	Years N	Years N – 1
681 – Supported interests			781 – Interests obtained		
682 – Remuneration to participation titles			782 – Income from participation titles		
683 – Depreciation of investments in real estate			783 – Income from real estate		
684 – Provisions for financial applications			784 – Income from capital participation		
685 – Unfavourable exchange rate differences			785 – Favourable exchange rate differences		
686 – Granted ready payment discounts			786 – Obtained ready payment discounts		
687 – Losses in the sale of treasury applications			787 – Earnings in the sale of treasury applications		
688 – Other financial costs and losses			787 – Other financial profits and earnings		
Financial profits					

Table IX.10 Format of the extraordinary profit statement

Costs and losses	Years N	Years N – 1	Profits and earnings	Years N	Years N – 1
691 – Donations			791 – Restitution of taxes		
692 – Uncollectable debts			792 – Recuperation of debts		
693 – Losses in inventory			793 – Earnings in inventory		
694 – Losses in assets			794 – Earnings in assets		
695 – Fines and penalties			795 – Benefits of contractual penalties		
696 – Increases in depreciation and provision			796 – Decreases in depreciation and provision		
697 – Corrections from previous years			797 – Corrections from previous years		
698 – Other extraordinary costs and losses			797 – Other extraordinary profits and earnings		
Extraordinary profits					

X Auditing

1 Appointment of the auditor

In Portugal, the recognized auditor is called 'Revisor Oficial de Contas' (ROC); this chapter is dedicated exclusively to this type of auditor.

An audit is required for businesses of the following types:

- Public companies.
- Public limited companies.
- Private limited companies with a Fiscal Council.
- Private limited companies which for more than two years present two of the three following criteria:
  - i) Total liquid assets above 350 million PTE.
  - ii) Total income above 600 million PTE.
  - iii) Average number of employees above 50.

This means that only small limited liability companies are excluded from a mandatory audit. This is in conformity with the Fourth Directive, art. 51. Some hybrid companies such as Complementary Company Groupings (*Agrupamento Complementar de Empresas*, ACE) are compelled to have Official Audits.

The audit covers the financial statements (which include balance sheet, income statement by nature and functions, cash flow statements and the appendix) and the management report. The financial statements are audited to see if they are in accordance with generally accepted accounting principles. An opinion on the quality of management decision-making in business terms is not part of the auditor's duty. However, one of the accounting principles analyzed is the going concern principle (*continuidade*). The audit encompasses the accounting records, official books, mail, third party's letters of confirmation, etc.

In practice, management and the Official Accounting Technician (*Técnico Oficial de Contas*, TOC) sign a declaration (*Declaração de Responsabilidade*) affirming that all records and information were presented, that there are no intentions to reduce or endanger the business, etc.

There are no legal regulations as to how the audit should be performed but several recommendations have been issued by the body of the audit profession (CROC) named Auditing Technical Standards (*Normas Técnicas de Auditoria/Revisão*). Although they are not legally enforced if not performed that can lead to several sanctions within the professional body.

The auditor is chosen and approved by the shareholders' representatives usually for a mandate of three years. That information has to be deposited in the Commercial Registration Body (*Conservatória do Registo Comercial*, CRC).

The Official Audit Report is delivered to management so that it can join the financial statements and manager's report. These documents are later approved in an Assembly of Shareholders and deposited in the Commercial Registration Body (CRC), accompanied by a written document that states approval of the accounts. The fiscal model (*Modelo* 22) also demands the Official Audit Report.

2 Auditor requirements

Decree-Law n° 422-A/93 regulates the auditing activity in terms of responsibilities, qualifications, incompatibilities, minimum fees and sanctions. A recognized auditor (ROC) can only go into practice after inscription in a list (*Lista dos ROC*).

To enter the profession in Portugal one has to:

- Have Portuguese nationality (but foreigners can be admitted under some conditions).
- Be morally apt.
- Be aged over 25 and less than 65.
- Have no civic or political limitations.
- Never have been convicted of any crime.
- Have a degree in law, economics, management, auditing or equivalent courses in accounting.
- Gain approval in a traineeship.
- Pass an examination.

The auditor is subject to an annual quality control with the purpose of verifying if the auditor's working methods and working papers are correct in terms of standards and the company's complexity. The selection is random and listed every year.

The Exchange Market Code (Decree-Law n° 142-A/91, article 103° and 104°) created the *Auditor Externo* which is a recognized auditor (ROC) inscribed in the Exchange Market Committee and with the purpose of giving a second 'official' opinion about the financial statements of entities supervised by this body.

The auditor's main obligation, considered of public interest, is to audit companies. The Company Code has broadened the commitments of auditors to include, among others:

- Approve capital rise with reserves integration (article 91º).
- Co-operate with any shareholder that demands his or her right to information (article 214º, nº 4).
- Intervene in the court processes of account approval (article 67º, nº 4).

3 Auditor's report

The auditor's report is the Official Audit Report (*Certificação Legal de Contas*) and an Annual Report (*Relatório Anual de Fiscalização*) – this report includes the tests performed and major recommendations and comments. The Official Audit Report also called 'Opinion' is a document where auditors express their professional and independent opinion about the company's financial position and the result of its operations and cash flows.

There are four main paragraphs. The first identifies the documents examined and the amount of total liquid assets, equity and profit/loss. The second explains the company's and the auditor's responsibilities. The third gives a summarized description of the audit's scope. The fourth and last reflects the auditor's opinion. There are six major types of opinion:

- Unqualified opinion: financial statements give a true and fair view.
- Unqualified opinion with explanatory notes (*ênfases*): financial statements give a true and fair view but there are fundamental uncertainties.
- Qualified opinion except for limitation (*reserva por limitação de âmbito*): not all the evidence reasonably expected was obtained but the effect on the financial statements is not misleading.
- Qualified opinion except for disagreement (*reserva por desacordo*): the financial statements were not prepared according to the generally accepted accounting principles but the effect on the financial statements is not misleading.
- Qualified disclaimer: not all the evidence reasonably expected was obtained but the material effect on the financial statements can be misleading.
- Qualified opinion – adverse opinion: the financial statements were not prepared according to the generally accepted accounting principles but the effect on the financial statements is misleading.

The last two types of opinion are not usually used.

The auditor must report immediately if facts come to light that seriously endanger the company. That legal obligation is in the Decree-Law nº 422-A/93, in the Companies Code (article 420ºA and 262ºA) and in the Penal Process Code (*Código do Processo Penal* – article 242º and 386º). As a matter of fact, the Company Code expects auditors to perform duties of prevention and vigilance.

The Bank of Portugal also demands an oral or written report in the audits of financial societies in the General Regime of Credit Institutions and Financial Societies (*Regime General das Instituições de Crédito e Sociedades Financeiras* – article 121º).

This legal obligation places the auditor in a difficult position as he performs his duty, because the technical standards issued by CROC, to which he is obliged, expressly state that fraud detection is not the purpose of the official audit.

XI Filing and publication

1 Company Code

Law defends the right to information. In partnerships, the individual partner is entitled to full accessibility of information and does not need to justify himself in any way (Companies Code, article 214º, nº 1, 2, 3 and 5). In companies whose capital is composed of shares, that right is somewhat limited in terms of who (can access information) and what (can be accessed).

The financial statements from the last three years, including Fiscal Council's or Auditor's Reports are subject to a special publicity, so they

Table XI.1 Responsibilities for the presentation and preparation of financial information

	Accountancy	Remaining departments	Managers	Fiscal Council	Auditor
Preparation					
Manager's report		X	X		
Balance sheet	X				
Income statement	X				
Appendix	X				
MOAF	X				
Fiscal Council Report				X	
Official Auditor's Report					X
Other information		X	X		
Presentation					
Manager's report			X		
Balance sheet			X		
Income statement			X		
Appendix			X		
MOAF			X		
Fiscal Council Report				X	
Official Auditor's Report					X
Other information			X		

are accessible to all users (Companies Code – CSC, article 288º, nº 1, al.a)).

The manager's report and the financial statements have two major recipients: the company itself (General Assembly) and the State (fiscal).

The managers, administrators or directors are obliged to annually present (for the General Assembly approval) within three months (CSC, article 65º and article 376º) from the balance sheet date (the year ends on 31 December, so the deadline is 31 March) the manager's report and the financial statements (CSC, article 451º, nº 1).

All the members of the administration still in duty must sign them (manager's report and financial statements) before their presentation. Nevertheless, should there be any doubts, former administrators are compelled to disclose the requested information.

However, to allow partners/shareholders to correctly evaluate the financial situation of the company whose accounts they are going to approve, managers must present the statements at least, 15 days in advance (15 March).

If the company is obliged to present an Official Auditor's Report, and the law gives him 15 days to issue a report (CSC, article 452º, nº 4), then all the information to be presented to the General Assembly (plus the Official Auditor's Report and the Annual Report) must first be presented to the Auditor by 1 March.

The financial statements and the manager's report, after approval, must be deposited at the Commercial Registry Body (Commercial Registry Code, article 15º) and published in the Official Journal (DR). This rule nowadays is compulsory for all companies. The deposit has to be made within the period of 30 days from the approval date (30 April).

For fiscal purposes, there is not much difference between the contents of information presented to the General Assembly – only the format or reading changes. The deadline for the presentation of Modelo 22 (the fiscal model) and its appendixes is 31 May.

2 Stock Exchange requirements

Article 341º of the Code of MVM requires approval and publication of the annual reports with its publication within 30 days after the approval of: manager's report, balance sheet, income statement, appendix and the other compulsory elements such as Official Auditor's Report, annual report, Fiscal Council's Report fiscal, (article 100°) and minutes.

Article 339° demands the publication of information in one journal of great circulation, on the quotation bulletins sent to the National Securities and Exchange Commission (CMVM).

XII Sanctions

The shareholder to whom information may have been denied (or wronged), when he was entitled to it, may require a court inquiry. The court may order company information to be released or, as it is stated in the civil process code, dismiss the people responsible for the omission.

The manager, administrator or director who blatantly refuses to provide – illegally – information required by partners or shareholders, fails to present it on time, or presents false information, may be punished with imprisonment for three months and a penalty for 60 days (each day represents a pre-established amount).

The lack of presentation of the signed manager's report and financial statements of the year end within two months from deadline (31 May), invites partners to require a court inquiry in order to figure out the company's financial situation and possible responsibilities. The fees may vary from 10,000 PTE to 300,000 PTE (CSC, article 528°, n° 1).

If the deposit of the financial statements and manager's report is forgotten, the fee can vary from 1,000 PTE to 10,000 PTE (which is absurd) for companies with equity of less than 100,000 PTE – or from 10,000 PTE to 100,000 PTE for companies with equities exceeding the above amount. Most companies choose not to deposit the financial elements.

The fiscal sanctions are much heavier than the ones presented previously. The failure to present adequate, or on time, relevant fiscal information can have consequences from mere fees to criminal responsibility. According to the current fiscal infraction regime, managers, administrators or directors are considered primary agents of several crimes and fiscal infractions.

On the other hand, companies must respond along with their representatives for the payment of fines and penalties should they be considered guilty as charged. Moreover, the company's representatives may be responsible along with the company for the payment of fines and penalties, if the company's assets are inadequate. The auditors themselves may have to pay along with the company if it is proved that their work was inefficient and they could have reported the irregularities (CSC, article 81° and 82°).

Decree-Law n° 20-A/90, of 15 January, gave legal course to several fiscal infractions, for example:

- Refusal to present accounting documents – from 10,000 PTE to 10,000,000 PTE – article 28°.
- Forgery, destruction, hiding or damaging relevant fiscal elements (if fraud is not proved) – fines ranging from 50,000 PTE to three times the unpaid tax up to 5,000,000 PTE – article 33°.
- Lack of compulsory books of accounting – from 10,000 PTE to 1,500,000 PTE – article 35°.

XIII Financial reporting institutions

Câmara dos Técnicos Oficiais de Contas (CTOC)
Avenida 24 de Julho, 58, 1200 Lisboa
Tel: ++351 21 393 93 00
Fax: ++351 21 397 33 53
Email: atoc@atoc.mailcom.pt
http://www.ctoc.pt/

Ordem dos Revisores Oficiais de Contas (OROC)
Rua do Salitre, 51, 1250 Lisboa
Tel: ++351 21 353 60 76
Fax: ++351 21 353 61 49
Email: croc@mail.telepac.pt
http://www.cidadevirtual.pt/croc/

Comissão de Normalização Contabilística (CNC)
Rua Angelina Vidal, 41, 1196 Lisboa
Tel: ++351 21 814 78 93
Fax: ++351 21 813 87 42

Comissão do Mercado de Valores Mobiliários (CMVM)
Av. Fontes Pereira de Melo, 21, 1050 Lisboa
Tel: ++351 21 353 70 77
Fax: ++351 21 353 70 78
Email: cmvm@cmvm.pt
http://www.cmvm.pt

Associação Portuguesa de Técnicos de Contabilidade (APOTEC)
Rua Rodrigues Sampaio, nº 50-3º Esq, 1150 Lisboa
Tel: ++351 21 314 03 91
Fax: ++351 21 315 60 68
Email: apotec@mail.telepac.pt
http://www.apotec.pt

For national databases in Portugal, see Portugal – Group Accounts, XIV.

Bibliography

Borges, A. and Ferrão, M. (1995). *A Contabilidade e a Prestação de Contas* (Rei dos Livros, Lisbon).

Borges, António, Rodrigues, Azevedo and Rodrigues, Rogério (1998). *Elementos de Contabilidade* (Rie dos Livros, Lisbon).

Caiado, António C. Pires and Madeira, Paulo Jorge (1999). *Aspectos Contabilísticos e Fiscais da Prestação de Contas* (Vislis, Lisboa).

Correia, Fernando Augusto (1998). A Informação Financeira Intermédia. *Jornal de Contabilidade*, 252, Março, pp. 73–82.

Costa, Carlos Baptista da and Alves, Gabrile Correia (1997). *A Contabilidade Financeira* (Vislis, Lisboa).

Ferreira, Rogério Fernandes (1998). Reavaliação do Activo Imobilizado Corpóreo (ou Tangível) – Decreto-Lei nº 31/98, de 11 de Fevereiro. *Jornal de Contabilidade*, 253, Abril, pp. 105–117.

Pascoal, Telmo (1999). Justo Valor das Participações Sociais em Processo de Fusão. *Jornal de Contabilidade*, 263, Fevereiro, pp. 37–41.

Pinto, José Alberto Pinheiro (1994). Problemática Contabilística do Imobilizado. *Jornal de Contabilidade*, 204, Março, pp. 54–61.

Reis, José Vieira dos (1998). Revisor Oficial de Contas: A Profissão em Portugal. *Revisores & Empresas*, Julho/Setembro, pp. 12–26.

Reis, José Vieira dos (1994). Determinação Fiscal e Contabilística dos Resultados das Obras de Carácter Plurianual. *Jornal de Contabilidade*, 206, pp. 116–230.

Ribeiro, Carlos Manuel Marques and Lagos, José Manuel Bruno (1998). Impostos Diferidos. *Jornal de Contabilidade*, 259, pp. 298–309.

Rodrigues, Lúcia Maria Portela de Lima (1997). Contabilização dos Contratos de Futuros: uma Análise Comparativa. *Jornal do Técnico de Contas e da Empresa*, 383/384, pp. 212–214.

Silva, F. V. Gonçalves da and Pereira, J. M. Esteves (1996). *Contabilidade das Sociedades* (Plátano, Lisboa).

Vaz, Eduardo José Escaleira (1998). Contabilização de Liquidações e Cortes em Planos de Benefícios Definidos, e dos Benefícios de Terminação do Emprego, pela Entidade Empregadora. *Jornal de Contabilidade*, 254, pp. 143–148.

PORTUGAL
GROUP ACCOUNTS

Leonor Fernandes Ferreira

Note

While printing this book, two official Portuguese bodies of accountants have changed their name. The association of official accounting technicians, Associação dos Técnicos Oficiais de Contas (ATOC) is now called Câmara dos Técnicos Oficiais de Contas (CTOC). Also, the auditors' chamber has changed its name from Câmara dos Revisores Oficiais de Contas (CROC) to Ordem dos Revisores Oficiais de Contas (OROC).

CONTENTS

I Introduction and background information

1 Legal background on Portuguese accounting

Table I.1 summarizes the major events in the development of Portuguese accounting regulations during the last 25 years.

The process of modernizing Portuguese accounting regulations took a step forward in 1976 with the law establishing the Accounting Standards Commission (*Comissão de Normalização Contabilística* or CNC) and with the issuing of the first Official Accounting Plan (*Plano Oficial de Contabilidade* or POC) approved by the Decree Law No. 47/77 of 7 February 1977. This signified the official adoption of the institutional model for accounting regulation suggested by a prior Finance Ministry study group, namely the French model. This model involves a standard setting body attached administratively to the Ministry of Finance, with powers to lay down a national accounting plan, the provisions of which are given legal force in the form of decree or decree law.

Turning towards Europe, Portugal has, since January 1986, been a member state of the European Union, with all that such membership entails for the country's economic orientation and its accounting and financial reporting practices. In November 1989, a revised version of the POC was issued to give effect to the Fourth EU Directive (Decree Law No. 410/89, dated 21 November 1989). Later, the 1989 version of the POC was modified and amended in July 1991 to attend to the requirements of the EU Seventh Directive on consolidated accounts (Decree Law No. 238/91 of 2 July 1991). According to this law, group accounting is compulsory to Portuguese parent companies in line with international rules for the years of 1991 and after. It is perhaps noteworthy that the publication took the form of a document annexed to a Decree Law approved by the Council of Ministers, and not simply by the

Table I.1 Major events in the development of Portuguese accounting regulation

1977	Creation of the Accounting Standardization Commission (CNC) Approval of the first Official Accounting Plan (POC)
1986	Approval of the Business Companies Code Approval of the Commercial Registry Code
1987	Powers given to the Accounting Standardization Commission
1989	Revision of the Official Accounting Plan to comply with EU Fourth Directive
1990	Reform of the Accounting Plan for the banking system
1991	Reform of the stock market regulation Creation of the Securities Market Commission (CMVM) Amendment of the Official Accounting Plan to comply with EU Seventh Directive First year of compulsory presentation of consolidated accounts by Portuguese groups
1992	Regulation of group accounting for financial institutions
1993	Accounting plan for financial institutions in line with the 1986 EU Directive Regulation of cash flow disclosures New regulation for the auditing profession in line with EU Eighth Directive
1994	Accounting plan for insurance companies in line with the 1991 EU Directive Regulation of group accounts of insurance companies
1995	Regulation of the accounting profession Creation of the official body for the accounting profession (ATOC)
1997	Approval of the first accounting plan for the public sector

Finance Ministry, a fact which may indicate the national importance attributed to such a document.

The CNC which is presently the main Portuguese regulatory body in accounting was responsible for both revisions of the POC. The influences of French ideas and practices, which have in the past been particularly significant in the field of accounting, have begun to lose their importance, as the international standards, mainly from English-speaking countries are gaining more and more followers. Such influence has backing in the 23 opinions *(directrizes contabilísticas)* approved by the general council of the CNC. These opinions, published in the official journal between 1987 and January 1999, are intended to round out the POC. Contrary to what happens with the official accounting plan, however, these opinions are not laws, either being only approved by the general council of the CNC or under the instruction of the Secretary of State and not by any minister.

Public filing requirements on individual and consolidated accounts are generally stated in the Companies Business Code (*Código das Sociedades Comercais*, or CSC, Decree Law No. 262/86, dated 2 September 1986) and in the Commercial Register Law (*Código de Registo Comercial*, or CRC, Decree Law No. 403/86, dated December 3 1986). In 1991, a securities market reform justified the creation of the Portuguese Securities Market Commission (*Comissão do Mercados de Valores Mobiliários*, or CMVM) along with the approval of the Securities Market Regulation (*Código do Mercado de Valores Mobiliários*, or CODMVM Decree Law No. 142-A/91, dated 10 April 1991). This law has influenced the disclosure and publication of accounts, with additional requirements for listed companies.

Statutory audit requirements are established in the CSC and in the legislation which regulates the auditing profession (Decree Law No. 422-A/93, dated 30 December 1993) that is in line with the EU Eighth Directive. The auditing profession has issued some important accounting-related guidelines (see XII). On the contrary, the accounting profession has not been an important standard setting body. This is perhaps due to the fact that the accounting profession recently assisted in the creation of its official body (*Associação dos Técnicos Oficiais de Contas* or ATOC), after the publication, in 1995, of a Decree which regulates the profession, as well as undertakes disciplinary measures (Decree Law No. 265/95 of 17 October 1995). Membership in ATOC is now required in order to exercise the official accounting profession. This means to be responsible for and authorized to sign the financial accounts and income tax returns of a company or a group of companies.

2 Historical development of group accounting in Portugal

The growth of large public companies in the 1960s and the emerging separation of control from ownership led eventually to the need for consolidated accounts. Some Portuguese groups used to prepare them for purposes of internal control of investment and financing decisions. So consolidated accounts began to be disclosed on a voluntary basis by few groups in Portugal, from the early 1960s onwards, before a legal requirement existed.

The first reference to consolidated accounts in the Portuguese legislation dates back to 1968, for the banking sector.

Following the political revolution of 1974 the main sectors of activity in the Portuguese economy have been nationalized and the former private groups have become state-owned new groups. The second legal reference to consolidated accounts occurred in 1976 with the approval of the statutes of the State Owned Undertakings Institute (Decree Law No. 496/76 of 26 June 1976, art. 41, No. 3). Under certain conditions specified in that law, state-owned parent companies should prepare a consolidated balance sheet. Later on, an accounting plan for the banking system stated the obligation to prepare consolidated accounts for credit institutions with foreign subsidiaries (Decree Law No. 455/78, dated 30 December 1978). These laws did not contain, however, references to methods and procedures of consolidation.

Thus, the political and economic vicissitudes of the 1970s and early 1980s have resulted in the temporary demise of industrial and financial groups, so that consolidated accounts were hardly called for until new industrial groups emerged. By that time, topics about consolidated accounts could only be found in the curricula of intermediate and university courses to a minor extent. This helps to explain the lack of consolidated accounts by Portuguese groups and the absence of legislation.

From 1987 onwards, Portuguese companies with listed shares became legally obliged to prepare and disclose the consolidated balance sheet as part of the interim information reporting to 30 June (Decree Law No. 235/87, dated 12 June 1987). But the law did not define the scope of companies to include in the consolidation.

The presentation of consolidated financial statements was not regulated in detail before the implementation of the EU Seventh Directive and the general obligation to prepare consolidated accounts had not applied to Portuguese companies till then (Decree Law No. 238/91, of 2 July 1991). Thus, until 1991, companies in Portugal had not been required to publish consolidated accounts, although they have been permitted, under certain restricted circumstances to file tax returns on a group basis since 1988. In most cases, however, taxes are not levied on the basis of the consolidated accounts.

3 Current legal framework of Portuguese group accounting

The main legal sources of consolidated reporting requirements are the Official Accounting Plan, the Companies Business Code, the Commercial Registry Code and the Securities Market Code. Some CNC opinions, recommendations by the Chamber of Registered Statutory Auditors (*Câmara dos Revisores Oficiais de Contas*, or CROC) and instructions by the CMVM are also relevant for group accounting reporting. When preparing their rules, these bodies have been strongly influenced by the IAS.

The main standard setters for different group accounting rules are the CNC, the Bank of Portugal and the Insurance Institute of Portugal. Current regulation of group accounting in Portugal is summarized in Table I.2.

The content and layout of the consolidated accounts are prescribed by the POC (Decree Law No. 238/91, dated July 2 1991, Annexe II). The POC was drawn up in the light of the EU Fourth and Seventh Directives, which apply neither to banks and other financial institutions nor to insurance companies. The former are subject to a separate set of accounting requirements in line with the 1986 Directive on the financial statements of credit and financial institutions (implemented through Decree Law No. 36/92 of 28 March 1992). Group accounting of insurance companies is regulated by special legislation which was drawn up in light of the 1991 EU Directive (implemented by Decree Law No. 147/94 of 25 May 1994).

These decrees made it mandatory for all groups controlled by a parent company to consolidate. The objective of consolidated accounts is to give a true and fair view of the group activity, its financial position, income and equity (see II). Some groups are exempted, however, for reasons of size or because they belong to consolidated larger groups with parent companies in EU member countries (see III) and some subsidiaries must or may be excluded, under certain conditions, from consolidation (see IV).

Regulation on consolidated accounts was added as chapters 13 and 14 to the POC, at the revision of July 1991 and completed by opinions and guidelines issued by some official bodies. The contents of these chapters and titles of the CNC opinions on topics of group accounting are presented in Tables I.3 and I.4.

The requirements of the Portuguese accounting regulations are based on those of the EU Seventh Directive and the wordings in the POC and in the Directive are in general coincident. The Portuguese legislator decided to keep the obligation to consolidate to a minimum and to permit departures if they are justified in the notes.

The layout of the consolidated accounts of certain kinds of companies differs from those in the POC. This happens with banks and other financial institutions whose accounting rules are issued by the Bank of Portugal (*Banco de Portugal, Instruções técnicas sobre consolidação de contas das instituições bancárias, anexa à circular Série A*, No. 235, dated April 21 1992). Consolidated financial statements of insurance companies are regulated by a specific plan prepared by the Insurance Institute of Portugal (*Instituto de Seguros de*

Table I.2 Current regulation of group accounting in Portugal

COMPANIES SUBJECT TO THE POC	
Scope of consolidation	Decree-Law No. 238/91 Decree-Law No. 127/95
Basic accounting principles	POC, sections 3,4,5 CNC, Opinions 12,18,21
Methods and procedures of consolidation	POC, section 13 CNC, Opinions 1,6
Consolidated financial statements formats	POC, section 14 CNC, Opinion 14 CMVM, Instruction 11/98
FINANCIAL INSTITUTIONS	
Scope of consolidation	Decree Law No. 36/92 BP, Circular A-235
INSURANCE COMPANIES	
Scope of consolidation	Decree Law No. 147/94 Decree Law No. 86/94 ISP, *Norma Reg.* No. 31/95-R ISP, *Norma Reg.* No. 6/96
GENERAL	
General principles	CSC, art. 508-B
Consolidated director's report	CSC, art. 508-C
Disclosure	CSC, arts. 167, 508-A, 508-E CRC, art. 70 CODMVM, arts. 339–344
Audit	CSC, art. 508-D CROC, Recommendations (see Table XII.1) CROC, Interpretations (see Table XII.1) CODMVM, art. 100
Sanctions	
To the board of directors	CSC, title VII CODMVM, art. 351
To the auditors	Decree-Law No. 422-A/93, arts. 71–84
To the accountants	Decree-Law No. 265/95, arts. 26–41

BP – *Banco de Portugal*
ISP – *Instituto de Seguros de Portugal*

Portugal, *Norma* No. 31/95, of 28 December 1995 and *Norma* No. 6/96, of 5 March 1996).

The consolidated accounts (which encompass the consolidated balance sheet, the consolidated profit and loss statement and notes to the consolidated statements), together with the consolidated directors' report, the group auditor's report and, if appropriate, the consolidated cash flow statements, must be made public by being deposited within the Commercial Registry (see X and XI). Provisions implementing the requirements of the EU Directives regarding the audit and publication of consolidated accounts and laying down requirements for a directors' report on group activities, are also included in the legislation already published (see XII), as well as sanctions for non-compliance with the consolidating reporting requirements (see XIII).

Table I.3 Contents of consolidated accounts in the POC

METHODS AND PROCEDURES OF GROUP ACCOUNTING (POC, Section 13)
- Introduction
- Group accounts statements:
  - General rules
  - Objective of group accounts
  - Date of elaboration
  - Changes in the scope of the consolidation
- Methods of consolidation
- Full consolidation:
  - General rules
  - Valuation criteria
  - Deferred taxation
  - Elimination of intragroup transactions
  - Minority rights
- Proportional consolidation
- Valuation criteria for associated companies:
  - Equity method
  - Exclusions from the equity method
- Provisional rules

CONSOLIDATED FINANCIAL STATEMENTS (POC, section 14)
- Consolidated balance sheet format
- Consolidated profit and loss account (horizontal format)
- Consolidated profit and loss account (functional layout)
- Contents of the notes to consolidated accounts

4 Relationship between group accounts and tax accounts

Financial accounting in Portugal has traditionally been tax-orientated. Indeed, the 1977 POC was quite related to the tax authorities and was designed more to facilitate tax inspections and to justify a company's income tax (because the basis for the calculation of the tax was the individual income, that is the profit from accounting books) more than to disclose information to the shareholders or members and to the public on the company's financial situation and operations report. In practice, this orientation continues today, although user groups specifically cited in the POC are, in the order there, investors, lenders, workers, suppliers and other creditors, government and other official authorities, and the public, in general (POC, 3).

Portuguese accounting practice in the sphere of valuation and income measurement has been affected by the body of decrees and decree laws dealing with the calculation of taxable profit and with tax treatments which are mandatory if certain tax allowances are to be obtained. The most important fiscal law related to accounting is the Companies Income Tax Code (*Código do IRC* or CIRC) which provides a set of rules for the valuation and recording that affect the financial disclosures.

Table I.4 CNC opinions relating to group accounting

Year Published	Opinion Number	Description
1991	1	Accounting for business combinations
1992	6	Elimination of profits and losses resulting from transactions between undertakings included in a group
1993	9	Accounting for investments in subsidiaries and associated companies in the individual accounts of the undertaker company
1993	12	The accounting concept of goodwill
1993	13	The concept of fair value
1997	21	Accounting for the consequences of the adoption of the euro
1998	24	Joint ventures accounting (draft, waiting for approval)

In general, tax accounts are based on individual accounts. From 1988 on, consolidation for tax purposes is possible in certain restricted conditions stated in the law (Decree Law No. 414/87, dated 31 December and later on, CIRC, arts. 59, 59-A and 60). Taxation based on consolidated accounts is optional and depends upon an authorization required by the parent company of the group and given by the Ministry of Finance for a period of five years (see VIII).

According to the regulation of the accounting profession recently approved, individual and group accounts, as well as tax returns of Portuguese companies and groups, must be signed by an accountant *(técnico oficial de contas)* registered with ATOC.

5 Relationship between individual and group accounts

Portuguese legislation on group accounting is an extension of that on individual accounts, being part of the same documents. Consolidated accounts contain information which complements the individual accounts of the parent company. The obligation to prepare consolidated financial statements and a consolidated directors' report does not liberate group companies from the obligation to prepare their individual accounts and directors' report, but groups with listed securities may ask CMVM for an authorization not to publish both sets of accounts (CSC, arts. 65 and 508-A).

Consolidated accounts are not usually the basis for the definition of the distributive profit. Dividend payout requirements are based on the income presented in the individual accounts and dividends may only be paid out of distributive reserves (CSC, arts. 32 and 33). The net annual profit and the accumulated profits from previous years and other reserves which are not locked up according to the law or the articles of association (revaluation reserves and legal reserves may not be distributed) may be used for profit distribution purposes (CSC, arts. 218 and 225). However, the dividend cannot exceed what is realistic, according to the financial situation of the company.

Changes in the net equity accounts must be disclosed under the notes to individual accounts. The individual profit and loss accounts do not disclose the income distribution which is part of a proposal included in the directors' report of the company to be approved by the general meeting.

6 Empirical data

Most business in Portugal is conducted on the basis of sole proprietorship or limited companies, either private (*sociedades por quotas,* or LDA) or public (*sociedades anónimas*, or SA), some of the latter having a stock exchange listing. These forms of business organization are governed primarily by the CSC and the CRC, approved in 1986. In general, Portuguese companies were not obliged to prepare consolidated accounts prior to the enactment of the EU Directives on consolidated accounts into the national law (for companies to which the POC applies in 1991, for banks and other financial institutions in 1992 and for insurance companies in 1994). Until that time, Portuguese groups did not usually disclose consolidated accounts. Although it is now possible to assess the impact of the EU Accounting Directives on the business community, as far as we know, there are no general surveys on the importance of consolidated accounts presented by Portuguese groups and few empirical studies have been carried out on the application of the legislation governing consolidated accounts.

Some official and private entities have their own databases on financial reporting information and treat these data for internal purposes (see Portugal – Individual Accounts, XIII). This is the case of the CMVM, who publish an annual report on the general situation of the securities markets. Such publication contains information about periodical and non-periodical information disclosed by companies with listed securities since 1995. CMVM focuses not only on the quality of information disclosed but also on the moment in which the information is disclosed.

Statistics on the Portuguese groups preparing consolidated accounts are not available. Evidence, however, reveals the following:

- Strong influence of the legislation on consolidated accounts due to the fact that most of the companies did not publish consolidated accounts before the legal requirement existed.

- Majority of voting rights as being the most commonly used criteria to consolidate.
- Extensive use of the clauses of subsidiaries exclusion from consolidation.
- Limited application of proportional consolidation method.
- Formal compliance with the presentation formats in the POC and other accounting regulation issued by the CNC, the CROC and the CMVM.
- Omission of certain recommended disclosures under the notes to consolidated accounts.
- Delay regarding publication and deposit requirements imposed by the CSC, the CRC and the CODMVM.

II Objectives, concepts and general principles

1 Objectives of group accounting

The main objective of group accounts is to give a true and fair view (*imagem verdadeira e apropriada*) of the assets, liabilities, financial position and profit or loss of the undertakings included in the consolidation taken as a whole (POC, 13.2.2.a) and CNC, Opinion No. 18). The notes should disclose information related to this objective. The rules for consolidated accounts may not be used if contrary or insufficient to the objective unless in exceptional circumstances and provided that the notes disclose the causes and effects of such departures in the balance sheet and in the profit and loss account (POC, 13.2.2 b, 14.4.4.8 and 14.4.4.9).

The POC includes a brief summary of the objectives and qualitative characteristics of financial statements expressed in the FASB's Statements of Financial Accounting Concepts 1 and 2 (POC, 3). The usefulness of information in the financial statement depends on relevance, reliability and comparability. These three characteristics, together with concepts, principles and accounting rules, make it possible to achieve the true and fair view objective.

Group accounts should be suitable for making rational economic decisions and hence contributing to the functioning of efficient capital markets and the accountability of management. The POC states that users of financial statements will be better able to analyze the capacity of the firm, in terms of the timing and certainty of the cash flows it may generate, if they are provided with information that focuses on its financial position, results of operations and changes in financial position (POC, 3).

Group accounts in Portugal are used mainly to inform shareholders or members but also to protect minorities. They are not used for profit distribution and only in a few cases are they a basis for tax accounts. Consolidated income is only relevant for tax purposes when the group has previously been authorized by the tax authorities to opt for the consolidated income tax system. Such option requires that the controlling company owns more than 90 per cent of the capital stock of the other companies in the fiscal group. Thus, some groups which are taxed under the consolidated income tax system are part of larger groups which include companies that do not meet the conditions to enter in the taxation based on the consolidated income (see VIII).

2 Underlying concepts

It is not clear which theory dominates the legal framework of Portuguese group accounting because elements of all can be found in the Portuguese accounting regulation and different conceptual approaches are taken at a time. Sometimes consolidated accounts are conceived as an extension to the accounts of the parent company while in others the group is viewed as an economic unit. Portuguese group accounting is based on a mix of elements from the parent theory, the entity theory and the proprietorship theory. Details about the underlying conceptual focus adopted in the Portuguese regulation of group accounting is described below. Table II.1 summarizes the most relevant issues in order to understand those concepts.

The scope of consolidation includes related companies in which the parent company has a majority interest and also joint ventures and other associated companies. The concept of group adopted by the legal Portuguese rules relies on *de jure* and *de facto* criteria. The parent theory prevails in the wording used to define group-related companies (such as mother companies and affiliated companies) and in the condition that having a participation in capital is necessary for the obligation to consolidate.

Furthermore, methods other than the full method are used for consolidation of joint ventures and associated companies. Elements of the parent's theory can be found in the optional use of proportional consolidation for joint ventures and elements of the proprietorship theory can be found when the equity method is used.

Another example of the parent theory relates to the different status of group shareholders or members of affiliated companies and minority interests. This means that minority interests are not included under the equity heading in the balance sheet and that their amount is computed on the basis of the book value of the net worth of the affiliated companies before elimination of non-realized intragroup result.

The concept of consolidated accounts as an expression of an economic unit (entity theory) influenced the way of computing elimination of intragroup profits and losses, which is made in full, without taking into account the percentage of

Table II.1 Underlying concepts in Portuguese group accounting

Objectives of group accounts	Consolidated statements must give a true and fair view of the assets, liabilities, equity and net result of the group companies as if they were one single company, and Consolidated accounts are an extension of the individual statements of the parent company prepared from the perspective of its main users
Main users of consolidated financial statements	Shareholders or members of the parent company
Definition of subsidiary	The obligation to prepare consolidated accounts requires being a shareholder or member, sometimes together with elements of *de facto* control
Group net result	Result to the shareholders or members of the parent company after deducting minority interests
Goodwill	Positive goodwill is an intangible fixed asset and should be depreciated Negative goodwill is usually included in the consolidated equity and exceptionally charged to the profit and loss account
Minority interests	Minority interests are presented in the consolidated balance sheet as a separate item between consolidated equity and liabilities; they are included at book value Minority interest are presented in the profit and loss statement as a separate item before reaching net group result; they are not part of the consolidated result
Consolidation of assets and liabilities (full method)	Assets (and liabilities) of subsidiaries are consolidated at book value plus the excess value of identifiable assets (and liabilities) if allocation is possible
Elimination of intragroup gains and losses	Non-realized gains are 100 per cent eliminated after deducting the share of the parent company result which is allocated to the minority interests. Elimination of sales and costs are not allocated to the minority interests

ownership of the subsidiaries. Also in line with the entity theory, the consolidated results for fiscal purposes incorporate both the parent company's income (or loss) and the part allocated to the external shareholders of dependent companies. On the contrary, both the consolidated balance sheet and the consolidated profit and loss account in the POC reflect in separate accounts the amounts of income relating to the different types of shareholders or members.

3 General principles

In Portugal, there is no tradition of accounting principles gaining general acceptance through recognition by the accounting profession. This is perhaps because, until 1996, there were many associations representing the accounting profession but none of them was the official body for the profession.

Fundamental accounting principles which are promulgated in the POC are those of continuity, consistency, accruals, and prudence, similar to those stated in the EU Fourth Directive (POC, 4). The POC requires the use of 'costs of acquisition or production, either in nominal or in constant *escudos*'. This suggests that current purchasing power accounting would be legally acceptable in Portugal, but the acceptability of current cost accounting is more doubtful.

The Portuguese regulatory accounting system has been approved by a structured hierarchy of laws. One may be tempted to conclude that the Official Accounting Plan comes at the top in this hierarchy, followed by the *directrizes contabilísticas*, and the IAS, given the fact that the POC was approved by the Government and was published in the Annexe to a Decree Law while the *directrizes contabilísticas* are approved by the General Council of the Accounting Standardization Commission and signed by its president. Thus, the set of accounting rules that parent companies must apply by descending order of importance are the POC, the CNC opinions and the IAS (CNC, Opinion No. 18).

A few Portuguese groups have been preparing their consolidated financial statements according to international standards in order to present them to the Securities Exchange Commission of the United States and to the NYSE. These financial statements and those prepared according to the national regulation disclose the same amounts of assets, equity and group income but in different formats, which makes it difficult to undertake a comparison (e.g. Portugal TELECOM, SA).

In order to accomplish the true and fair view objective, the methods and procedures of consolidation must be maintained from one financial year to another (POC, 13.2.1.a)).

If the effects of non-compliance with the true and fair view are not material, departures from general principles are allowed, provided that the notes state the circumstances under which that has happened. This may happen in the following cases, based on grounds of absence of materiality:

- Exclusion from consolidation of subsidiaries being non-material for the true and fair view objective (Decree Law No. 238/91, art. 4, Nos. 1–2).
- Non-use of the equity method to account for an immaterial investment in an associated company (POC, 13.6.2).
- Non-disclosure in the notes to consolidated financial statements of information related to undertakings indirectly held up to 10 per cent of the share capital (POC, 14.4.4.6).
- Application of non-uniform accounting valuation criteria between consolidated accounts and individual accounts of the parent company if the difference in results is immaterial (POC, 13.4.2.c).
- Elimination of transactions or profits between the undertakings included in the scope of the group (POC, 13.4.4.c).
- Differences between the date of the consolidated balance sheet and the date of the individual balance sheet of the parent company (POC, 13.2.3.c).

III Group concepts and preparation of group accounts

1 Group concept

1.1 Overview

In the same way as in the Seventh EU Directive, the word *group* is avoided in Portuguese accounting regulations. Instead, elements for a definition are included in the POC and in Decree Law No. 238/91. The CSC and the CIRC also present their group concepts, which are different. It is mandatory for all groups controlled by a parent company to consolidate. Some groups are exempted, however, for reasons of size or because they belong to consolidated larger groups with parent companies in EU member countries. In any case, parent companies of listed groups must always present their consolidated accounts.

1.2 The accounting law

The POC provides definitions of parent company (*empresa-mãe*), associated companies (*empresas associadas*), and other undertakings (*outras participadas*). A group is defined as including the parent company and its subsidiaries but not associated companies and other undertakings (POC, 2.7).

The definition of a subsidiary company is given in the POC and in the legislation which implemented the Seventh Directive using the same wording in both laws. Subsidiary company (*empresa filial*) is defined as the companies over which a parent company has the power to dominate or control (POC, 2.7 and Decree Law No. 238/91, art. 1). Both *de jure* and *de facto* control are relevant for the accounting definition of group leading therefore to a legal and economic concept of group. Although participation in share capital must exist in order to exercise control, it is not necessary to own over 50 per cent of the voting rights to control a company. A minority of voting rights together with other mechanisms of control may be sufficient. Portuguese legislation did not include the options in the Seventh Directive in which the controlling company need not be a shareholder.

The concept of associated company includes direct and indirect participation. Holding a participating interest of at least 20 per cent of the capital of another company may result in exercising a significant influence over the operating and financial policy of the undertaking and, thus, permits the presumption that those companies are associated undertakings. (POC, 13.6.1.b)).

The item financial investments (*investimentos financeiros*) includes undertakings in group companies, associated companies, and other undertakings which are those undertakings held at less than 20 per cent of the capital. It also includes bonds and other securities held for more than one year, as well as land and property (real estate) held as financial investment (POC, 4.2.3). But this item does not include own shares, which are recorded (deducted) in the equity.

1.3 The companies law

The Companies Business Code refers to group relationships (*relação de grupo*) and participation relationships (*relação de participação*) (CSC, title VI, arts. 483 to 493). Distinction is made between totally controlled subsidiaries (*sociedades em relação de domínio total*) and simply controlled subsidiaries (*sociedades em relação de domínio simples*). The former is a company that holds 90 per cent or more of the capital of another, either directly or indirectly. Companies dominated through a contract or agreement specified in the law (*contrato de grupo paritário* and *contrato de subordinação*) together with totally controlled subsidiaries, are parts of a group. Simply controlled subsidiaries which are not included in the group concept are those companies that hold a majority of the shares or the majority of the voting capital of another, either directly or indirectly, or have the right to appoint a majority of the members of the management or supervisory bodies.

1.4 The fiscal law

The entity theory dominates the fiscal concept of group (CIRC, art. 59, No. 1). Concepts of group for accounting and taxation purposes differ. While legal control in the POC is based on the voting rights, for tax purposes, the concept of subsidiary company (*empresa dependente*) depends on the percentage of participation in the share capital. Consolidation for income tax purposes requires that the parent company (*empresa dominante*) holds at least 90 per cent of the share capital, directly or indirectly. Additionally, all the subsidiaries must be located in Portugal, while the accounting concept extends consolidation to subsidiaries on a worldwide basis. So far, it is usual that a group authorized to be taxed on the basis of the consolidated income is part of a larger group of companies which is obliged to prepare and disclose consolidated accounts.

2 The obligation to consolidate

2.1 The obligation to consolidate

Accounting consolidation has been compulsory in Portugal since 1991 (since 1992 for banks and other financial institutions and from 1995 on in the case of insurance companies). The obligation applies to companies having a participation in capital (vertical groups) and includes direct and indirect participated companies. Own shares and own parts of capital are out of the scope of consolidation. Horizontal groups are not obliged to present consolidated accounts in Portugal.

Although participation in the share capital is a necessary requirement to consolidate, the control depends also upon *de facto* elements, unless in the case of a majority of voting rights.

The definition of a parent company in the POC envisages a simple majority of voting rights. In practice, however, differences exist in the case of non-voting shares or parts of capital and plural voting rights.

2.2 Legal forms of the consolidating company

The scope of application of Portuguese consolidated accounts is the same as that laid down in the Seventh Directive. Corporations, limited liability companies and limited partnerships with shares must prepare consolidated accounts (Decree Law No. 238/91, art. 2, No. 2).

From 1 January 1995 partnerships and limited partnerships could also qualify as parent companies if all their partners and limited partners are corporations or private limited companies or limited partnerships which have a public limited company as the general partner (Decree Law 127/95 of 1 June 1995). A practical consequence of this rule is that having a group structure involving partnerships or limited partnerships is no longer a way of avoiding the presentation of consolidated financial statements by dividing the entire group into subgroups.

Parent companies which are not organized under one of the legal forms mentioned above are not obliged to present consolidated accounts. That is the case of co-operatives and sole entrepreneurships, which are both common forms of business organization in Portugal.

3 Exemptions from consolidation

3.1 Small groups

An entity is exempted from the requirement to consolidate provided it has not exceeded two out of the three following limits: total assets: 1,500 million *escudos* (PTE); total turnover: PTE 3,000 million; number of employees: 250 (Decree Law No. 238/91, art. 3, Nos. 1–2). The thresholds are calculated based on the individual accounts of the last financial year and before the elimination of intragroup transactions, and so the parent company does not have the possibility to find itself exempted from the obligation to consolidate after the consolidation due to elimination carried out during the process of consolidation. These size exemptions do not apply to groups, however, if one or more of their member companies are listed on any EU stock exchange (Decree Law No. 238/91, art. 3, No. 3.) The notes to the individual accounts of the exempted parent company must mention the reason for the exemption (POC, 8.16).

3.2 Subgroups

The Seventh Directive's exemption of subgroups, when the ultimate parent company prepares consolidated accounts under the said Directive, also applies. The exemption is conditional upon compliance with a number of cumulative conditions (Decree Law No. 238/91, art. 3, Nos. 4–6).

Firstly, exemption requires that the parent company is a subsidiary undertaking and its own parent company is situated in the EU and holds at least 90 per cent of the capital of the exempted Portuguese parent company (note that this percentage can not correspond to the percentage of voting rights).

Another condition for the exemption is that the minority shareholders or members in a Portuguese exempted parent company, who may hold up to 10 per cent of the capital of this company, expressly authorize the exemption. The Portuguese law omits regarding deadlines for the minority shareholders demand of consolidated accounts and regarding the exclusion of undertakers with no voting rights from the process of deciding this exemption. Consolidation request by the employees, their representatives and administrative or judicial authority is not mentioned under the Portuguese legislation.

A third condition for the exemption is that the ultimate parent company of the group is presenting group accounts within the EU. Portugal does not exercise the option in article 11 of the Seventh Directive if the parent company is not located in the EU, perhaps intending to facilitate comparability.

Finally, the exemption to consolidate is not allowed unless none of the companies in the group has listed securities in any EU stock exchange.

The exempted parent company has the obligation to comply with a cumulative number of preparing, auditing and disclosing requirements in order to assure protection of the minorities (Decree Law No. 238/91, art. 3, No. 5, d)). The exempted company must disclose in the notes to its individual accounts information such as the consolidated accounts of the ultimate parent company where it is included as a subsidiary. The balance between the advantages of non-consolidation and the costs of additional disclosures imposed to an exempted company has meant that some Portuguese groups prefer not to make use of the exemption, in particular those which are part of international groups.

IV Consolidated group

1 Group relevant definitions

The Portuguese accounting regulations contain a number of legal definitions regarding related companies. They are indicated in the POC, in Decree Law No. 238/91 and in the group accounting laws specific for banks, other financial institutions and insurance companies.

According to the POC a group includes the parent company and its subsidiaries (*filiais*). An associated company (*empresa associada*) is one company that holds at least 20 per cent and no more than 50 per cent of the voting capital of another and cannot be considered either as a parent company or a subsidiary.

Other undertakings (*outras participadas*), those companies where participation is less than 20 per cent, are out of the scope of consolidation and hence they are valued both in individual and consolidated accounts at historic cost with a provision for depreciation, when necessary.

The obligation to consolidate is based on the legal definition of subsidiary in the POC and in the Decree Law No. 238/91, which are consistent. Decree Law No. 36/92 and Decree Law No. 147/94 include similar definitions, respectively for banks and other financial institutions and insurance companies.

The general rule is that subsidiaries are to be consolidated using full consolidation and associated companies included by using the equity method. Proportional consolidation may be used for joint ventures as an alternative to the equity method. The main method of full consolidation mentioned in the POC is the purchase method described in the Directive, the pooling-of-interests method being exceptional (see VI).

2 Subsidiaries

A number of conditions for the preparation of consolidated accounts in terms of relationships between a parent and a subsidiary undertaking are specified in the law.

Portuguese parent companies must consolidate all their direct and indirect subsidiaries, regardless of where their registered offices are located

(Decree Law No. 238/91, art. 2, No. 2; Decree Law No. 36/92, art. 3, No. 1; Decree Law No. 147/94, art. 3, No. 1). The worldwide consolidation principle can justify the accounting revision of foreign subsidiaries of Portuguese parent companies.

In Portugal, consolidation was made compulsory under legal power and economic control (Decree Law No. 238/91, art. 1; Decree Law No. 36/92, art. 2; Decree Law No. 147/94, art. 2).

With respect to *de jure* control, only the majority of voting rights was implemented. Capital held by any person acting on the behalf of a group undertaking must be added to that of the parent company in order to calculate the voting rights. The holding of the direct or indirect majority of the voting rights may happen with a minority of participating capital due to the existence of non-voting shares or as a consequence of the dispersion of capital voting rights.

The Portuguese Law also adopted two *de facto* criteria of control in conjunction with additional *de jure* presumptions. These two criteria are the appointment or removal of the majority of board members and dominant influence. Reference to *de facto* control in terms of unified management has not been stated.

The Portuguese law requires consolidation when a parent company has the power to control not necessarily dependent exclusively on legal rights but also evidenced by the right to appoint or remove the majority of the members of the administrative, management or supervisory body and is at the same time a shareholder or a member of it. Those rights of appointment and removal may exist in the absence of a majority of voting rights because they are attached to preferential shares. In legal terms, the presumption may be removed wherever it is proved that in fact it does not determine a dominant influence. In practice the action of the supervisory bodies are usually not determinant for the future of a company in order to set alone a dominant influence.

When the parent undertaking has the right to exercise a dominant influence over a subsidiary, pursuant to a contract (*contratos de grupo paritário* and *contratos de subordinação*) or to a provision in its articles of association, consolidation is also compulsory. Portugal chose to limit it to those cases where the parent is a shareholder, by implementing the option to make the obligation to consolidate dependent on the ownership of 20 per cent of the voting rights (for banks and insurance companies the percentage is increased up to 40 per cent). Although the Seventh Directive has granted to member states the option to provide details concerning the form and content of such agreements, the Portuguese legislator did not take it up. In practice, the participation comes usually before the contract, but this is not obligatory.

Controlling alone based on an agreement with other shareholders or members of the undertaking and holding a majority of the voting rights pursuant to an agreement with other shareholders (*acordo parasocial*) are also under the conditions for the obligation to consolidate in Portugal.

Different combinations of the *de jure* and *de facto* control criteria explain the possible variability of situations under which Portuguese companies are obliged to prepare consolidated accounts. The current definition of a subsidiary adopted in Portuguese reporting practice has been most of the times the majority of voting rights. However, some groups do not disclose the criteria used to define a subsidiary or the notion of control.

3 Exclusions of subsidiaries from the consolidation

3.1 Optional exclusions

Optional exclusions are on the grounds of materiality, severe long-term restrictions and temporary holding. Subsidiaries of banks and insurance consolidating companies may also be excluded from consolidation if it is impossible to obtain any accounts from them without disproportionate expense or undue delays (Decree Law No. 36/92, art. 5, No. 5, b and Decree Law No. 147/94, art. 5, No. 5, b). These optional exclusions depend upon the judgement of the managers of the parent company. The auditors will also judge them in terms of reasonability.

The materiality of companies excluded must be

considered as a whole, not individually (Decree Law No. 238/91, art. 4, Nos. 1 and 2).

The exclusion based on severe long-term restrictions exists in subsidiaries located in foreign countries over which the group's control is restricted by the policies of the governments of those countries or when there are difficulties in transferring funds (Decree Law No. 238/91, art. 4, No. 3, a). This happened to some Portuguese companies having subsidiaries in the former Portuguese territories in Africa.

Subsidiaries may also be excluded if they are held on a short-term basis (Decree Law No. 238/91, art. 4, No. 3, b). This is not the situation, however, when it is expected that a subsidiary which has been held for a long time will be sold in the near future.

In order to prevent abuses, the POC requires a number of disclosures under the notes about the excluded companies, such as the names, registered offices, the proportion of capital held by the parent company or any person acting on its behalf and a justification for the exclusion but not necessarily the percentage of the voting rights (POC, 14.4.4.2).

3.2 Compulsory exclusions

As required by the Seventh Directive, subsidiaries must be excluded from consolidation if their activities are of such a different nature from those of the rest of the group that their inclusion would militate against the true and fair view (Decree Law No. 238/91, art. 4, Nos. 4 and 6). In such cases, the equity method should be used and information must be given under the notes to consolidated accounts (POC, 14.4.4.2).

The law states that the exclusion rule cannot be applied merely because companies in the scope of consolidation reflect different activities, such as industrial, commercial or services unless divergence in activities is contrary to the true and fair view objective (Decree Law No. 238/91, art. 4, No. 5). Similar rules of compulsory exclusion apply to consolidated accounts of banks and other financial institutions (Decree Law No. 36/92, art. 5) and to insurance companies (Decree Law No. 147/94, art. 5).

On the question of whether a bank or an insurance company may be included in the consolidated accounts when being subsidiaries of an industrial group, the Portuguese law is not clear but the practice and prevailing doctrine have been in favour of exclusion. In the case of a consolidating bank or other financial institution, the exclusion of industrial, commercial, services and insurance subsidiaries is compulsory.

4 Changes in the scope of consolidation

If the composition of the undertakings included in the consolidation has changed significantly in the course of a financial year, additional information must be given in order to provide the necessary comparability. Such information must be disclosed either by preparing adjusted consolidated financial statements or in the notes (POC, 13.2.4 and 14.4.4.14).

V Uniformity of accounts included

1 Formats

The models provided in the POC assure the uniformity of formats used for Portuguese group accounts (see IX).

2 Balance sheet dates

Companies incorporated in Portugal adopt in general a financial year ending on 31 December. However, some companies make use of the authorization given by the Companies Business Code to adopt a different financial year end, for the years after 1995, provided that the financial period used for accounting disclosures is a period of 12 complete months (CSC, art. 65-A). This facility is in accordance with a fiscal permission in force since 1989 (CIRC, art. 17).

In general, the consolidated balance sheet should be drawn up as of the same date as the individual balance sheet of the parent undertaking (POC, 13.2.3 a). However, another date may be considered in order to take account of the largest number or the most important undertakings included in the consolidation, but the reasons therefore shall be disclosed in the notes to the consolidated accounts together (POC, 13.2.3.b and 14.4.13.a). This happens when the Portuguese consolidating company is also a subsidiary included in the consolidating scope of a foreign group.

Usually, the parent company and its subsidiaries adopt the same financial year end. However, in some cases the financial year end of a consolidated subsidiary is different from that of its parent company. When the difference in time is no longer than three months, the individual financial statements of the subsidiary may be used, but disclosure must be made of important events occurring in the period between the two dates which may have influenced the assets, liabilities, financial position and net income (POC, 14.4.13.b). Where the balance sheet date of a consolidated subsidiary precedes the consolidated balance sheet date by more than three months, that subsidiary shall be consolidated on the basis of interim accounts drawn up as at the consolidated balance sheet date (POC, 13.2.3.c). Auditing of such special financial statements is required to be part of the general audit of the group.

3 Recognition criteria and valuation rules

The basic group accounting rules include the accounting principles and the valuation criteria laid down by the POC (POC, 4 and 5). Recognition criteria and valuation rules should conform to those criteria and be the same as those used in the individual accounts of the parent undertaking (POC, 13.4.2.a and 13.4.2.b).

In order to assure the uniformity of group accounts, assets, liabilities and equity included in the consolidated accounts must be valued according to uniform methods. If necessary (material), consolidating adjustments should be made to the figures of companies being consolidated in order to meet these criteria (POC, 13.4.2c). But it is possible to avoid the adjustments when they are not material, provided that the notes give additional information (POC, 13.4.2.d and 14.4.14).

Portuguese accounting rules for consolidated accounts are very flexible. The fact that many options and derogations in the EU Directive were implemented explains why the notes to the consolidated accounts are so numerous. Disclosure of accounting policies must be done under the notes in order to facilitate comparisons. Examples of such disclosures are the valuation criteria used in the consolidated accounts, methods for adjustments of depreciation and provisions of assets and the exchange rate used to translate to Portuguese currency the items first expressed in other currencies (POC, 14.4.4.23 and 14.4.4). Disclosure of other information relevant to a better understanding of the financial position and the income of the companies included in the consolidation is also recommended (POC, 14.4.4.50). Examples of other information reported are segmental information, accruals and deferrals and contingencies.

4 Translation of foreign currencies

According to the POC, Portuguese group accounts must be showed in *escudos* (PTE, the *escudo* has been the monetary unit of Portugal for about 100 years), in thousands of *escudos* or in ecus (POC, 2.11). Reporting in euros also has been possible since the country joined the *euro* in January 1999 and it will be compulsory after 2001 (CNC, Opinion, No. 21).

The POC states that foreign currency operations are registered in the individual accounts at the exchange rate in force at the date of the operation. For balance sheet purposes, receivables, liabilities and foreign currency deposits are stated in PTE at the exchange rate in force at the end of the year (POC, 5.1.1 and 5.2.1). The exchange differences on non-hedged foreign currency monetary items are to be recognized in the profit and loss statement in the period they arise. In certain circumstances, however, unrealized exchange gains may be deferred. This is the case when there is an expectation of reversal of the exchange gains (POC, 5.2.2). Disclosure of the exchange rate used to express items included in the individual and consolidated accounts which are, or were, originally expressed in foreign currencies should be done under the notes (POC, 8.4). Usually, companies adopt the exchange rates published by the Bank of Portugal. Where foreign currency balances are covered by forward exchange contracts, the contract rate is used for currency conversion.

An opinion recently issued by the CNC refers to methods for the translation of accounts from *escudos* to *euros*: the temporal method or net investment method (CNC, Opinion No. 21). There are no specific requirements in the Portuguese legislation regarding the translation of foreign currencies in the consolidated accounts, besides the recommendation to disclose in the notes the exchange rates used to convert the financial statements of foreign subsidiaries (POC, 14.4.4.24). It is noteworthy that this is less than the basis of conversion required by the EU Seventh Directive.

Few Portuguese groups have foreign subsidiaries with the obligation to consolidate but it is common to find Portuguese subsidiaries of foreign groups. In the cases reported, transactions denominated in a currency other than the reporting currency are usually to be accounted for in the consolidated accounts as they are in the individual accounts. The financial statements of the foreign subsidiaries of Portuguese parent companies have been translated generally into *escudos* at the following exchange rates: equity at the historical exchange rate, profits at the average exchange rate, and the remaining accounts at the exchange rate in force at the end of the year. Assets and liabilities denominated in foreign currencies are translated into *escudos* at the average market rates of exchange at 31 December. The financial statements of foreign subsidiaries are also translated at the year end average market rates. Exchange differences arising on the opening net investments in foreign subsidiaries are taken to reserves.

VI Full consolidation

1 General principles

The full method is the rule in the case of subsidiaries included in the consolidation. This is the line-by-line method, with assets, liabilities, income and expense being included in full and with minority interests being shown where appropriate.

Two methods are allowed for the consolidation of capital: purchase method and pooling-of-interests, the former being the rule. This section describes the purchase method and criteria for the application of the pooling-of-interests method as set out in the accounting regulation (POC, 13.4 and CNC, Opinion No. 1).

2 Consolidation of capital

The consolidation of capital consists of offsetting the book value of direct or indirect participation in the capital stock of the subsidiary company against the proportional part of its equity (excluding the results of the current year) at the date of its inclusion in the scope of consolidation. This date is either the beginning of the year in which the group first presents consolidated accounts or the year of acquisition of the participation (POC, 13.4.1.c). Rights given to own shares and quotas are suspended and accordingly they are not considered for the purpose of calculating the percentage of control (POC, 13.4.1.f).

Where there is a difference and it is possible to identify the undervalued or overvalued items in the assets and liabilities of the subsidiary company, such difference must be allocated directly to the item in the balance sheet of the subsidiary company to which it is related. The remainder difference between the acquisition cost and the restated value of assets and liabilities of the participated company, either positive or negative, is shown as 'consolidation difference' in the first consolidation (POC, 13.4.1.d).

2.1 Consolidation differences

The positive difference in the first consolidation is allocated to the assets or liabilities of the subsidiary company, as far as possible. If the net asset value of the acquired assets exceeds the acquisition cost, the remaining difference (goodwill) is recorded in the assets side of the consolidated balance sheet as intangible fixed asset (POC, 14.3.4.1.e and 14.3.4.1.g1). The POC requires goodwill to be written off normally over five years. When the depreciation exceeds five years, the notes to the consolidated accounts must justify the fact (POC, 14.4.4.17). In practice, amortization is usually written off on a straight-line basis over periods up to the pay back on the investment.

The negative consolidation difference in the first consolidation is allocated to the assets or liabilities of the subsidiary company, as far as possible. The remaining difference is recorded under the equity of the consolidated balance sheet. This difference may be brought to the consolidated profit and loss account only in the case of low profitability of the subsidiary company or when there is a realized capital gain as consequence of disposal of the participation (POC, 13.3.4.1.e and 14.3.4.1.g2).

According to provisional rules in the POC, when a former participated undertaking is first included in the scope of consolidation some years after belonging to the group, the consolidated balance sheet should include an amount equal to the percentage of the capital and reserves of the undertaking, and the difference between that amount and the book value of the participating interest in the parent's accounts (either positive or negative difference) should be disclosed separately in the equity capital of the consolidated balance sheet at the relevant date (POC, 13.7).

The consolidated difference will not be re-evaluated, unless in case of modification of the percentage of control, by subsequent acquisitions or disposal of participation. In the case of expected future losses, the consideration of a provision is not allowed.

2.2 Minority interests

Minority interests include the proportion of capital, reserves and result of the year in the equity of the subsidiary company which corresponds to their percentage of participation in the equity capital of the subsidiary. Calculation involves both direct and indirect participation, excludes own shares and is made before the elimination of intragroup transactions. Most groups use the historical cost principle (book values) and do not charge consolidation differences to minority interests.

When the loss of the subsidiary company exceeds the equity attributed to the minority interests, excluding own shares, that excess will be allocated to the consolidating company provided that the minority shareholders or members confine their responsibility to the amounts contributed and that there are no agreements relating to additional contributions. If the minority has agreed or is obliged to refund this part of the loss or to deduct the loss, the minority interests may be disclosed as a negative item in the liability side of the consolidated balance sheet (CNC, Opinion No. 6).

2.3 Multi-level groups

The principles used in preparing the consolidating accounts of a subsidiary that is indirectly participated by a Portuguese parent company do not differ from those which apply in the case of direct participation. The accounting law omits the techniques to consolidate multi-level groups. In practice, some groups use a step-by-step approach, with intermediate levels of consolidation at each subgroup, considering the direct percentages of participation, while other groups prefer to consolidate all the subsidiaries directly in the ultimate parent company of the group.

2.4 Changes in the percentage of control

When there is an increase in the percentage of capital held in a subsidiary, the initial difference of consolidation is affected. The sale of a participation originates a reduction in the percentage of control and ownership in the subsidiary and, consequently, there will be a reduction in the consolidation difference calculated in the first year of consolidation. Although there are no Portuguese legal rules about accounting treatment of disposals of participations, Portuguese groups usually recognize the reduction in the consolidated profit and loss account.

3 Consolidation of debt

Assets and liabilities of companies included in the consolidation are considered in full (POC, 13.4.1.a). Credits and debits between companies included in the consolidation must be eliminated in full, but only after adjustments for uniformity of the reporting standards are made (POC, 13.4.1.a).

For the rules governing the consolidation of debits and credits in foreign currency, see V.

4 Consolidation of profit and loss account

Revenues and expenses of companies included in the scope of consolidation are considered in full (POC, 13.4.1.b). The portion of the result attributed to minority interests must be shown in a separate heading in the consolidated profit and loss account. Elimination of sales, services, costs, income and expenses relating to transactions between group undertakings included in the consolidation are usually the following:

- Gains and losses from intragroup transactions included in the book value of stocks in the consolidated balance sheet.
- Gains and losses arising from intragroup fixed assets transactions.
- Gains and losses arising from services rendered intragroup.
- Dividends attributed between companies included in the consolidation.
- Gains and losses on disposal of participating interests in enterprises included in the consolidation.
- Deferred taxes.

The prudence principle applies, so that derogation from the need to eliminate may be accepted if elimination would involve undue expense or if the profit or loss is also verified by similar transactions outside the group. Profits and losses are eliminated in full regardless of the percentage of ownership held in the subsidiary companies (POC, 13.4.4 and CNC, Opinion No. 6).

5 Pooling-of-interests method

Most of the business combinations are acquisitions and, thus, the purchase method is used for consolidation. The pooling-of-interests method is reserved for mergers (CNC, Opinion No.1).

A merger happens when two (or more) companies give rise to a new company or when out of two companies one of them disappears while the other issues shares which are paid essentially with the net assets of the former. In 1997, four mergers were registered within the CMVM and in 1996 there were six.

In order to apply the pooling-of-interests method, the following conditions are necessary:

- Independence of both companies before the merger: none of the merged companies is allowed to have been a subsidiary of the other or has participated in its capital with 10 or more per cent during the two years preceding the merger.
- The merger does not involve substantial payments in cash: if the equality of share values requires a cash payment between merged companies, this may not exceed 10 per cent of the value of the shares issued.
- At least 90 per cent of the shares in the acquiring company must be acquired in a unique transaction or in a series of transactions in accordance with a plan and during a period shorter than one year.
- Absence of certain planned transactions after the merger: both groups of shareholders continue after the merger to exercise influence over the business combination and there must be no plans to divest an important part of the merged companies during the two years after the merger has occurred.

VII Proportional consolidation and equity method

1 Overview

The Portuguese legislation contains requirements for evaluating undertakings which do not meet the requirements of full consolidation. This section deals with the scope of and simplifications under the equity method and proportional consolidation. The circumstances under which the former applies for joint ventures, associated companies and some excluded undertakings are described in detail. Proportional consolidation is reserved for joint ventures, as an option.

2 Proportional consolidation

According to the POC, the method of proportional consolidation (line-by-line) is optional and may be used if an undertaking included in a consolidation manages another undertaking jointly with one or more undertakings not included in that consolidation. In that case, the inclusion of that other undertaking in the consolidated accounts may be in proportion to the rights in its capital held by the undertaking included in the consolidation (POC, 13.5.a).

The same rules apply as those described for full consolidation, however with the necessary adaptation that assets, liabilities, income and expenditure items are included according to the percentage of the capital held by the group undertaking. Goodwill and negative consolidation differences are also calculated using the same rules as for the full method of consolidation (POC, 13.5.b).

The notes to the consolidated accounts must disclose the names, addresses, the factors on which joint management is based and the proportion of their capital held by the company included or by persons acting on their behalf (POC, 14.4.4.5).

An undertaking managed jointly is usually an associated company and so, when the proportional consolidation is not used, the equity method should apply (POC, 15.5.c). Because the former requires more information about the undertaking companies than the equity method, the latter has been more frequently used by Portuguese joint ventures.

3 Equity method

The Portuguese legislator balanced between classifying the equity method as a method of consolidation (POC, 13.3) and as a valuation method (POC, 13.6).

The equity method should be applied in the following three circumstances:

- Subsidiaries excluded from the scope of consolidation due to differences in the activity (Decree Law No. 238/91, art. 4, No.4).
- Associated companies (POC, 13.6.1).
- Joint ventures not proportionately consolidated (POC, 13.5, c).

If the associated undertaking is not material for the purposes of providing the true and fair view objective, the equity method need not be applied, but the notes to consolidated accounts must mention the fact (POC, 13.6.2 and 14.4.4.4).

In the same way as in the EU Seventh Directive, the Portuguese accounting law allows for two variants of the equity method. At the first time of use of the equity method the participating interest shall be shown in the consolidated balance sheet (POC, 13.6.1), either:

- At its book value calculated in accordance with the valuation rules laid down for individual accounts (cost acquisition). The difference between that value and the amount corresponding to the proportion of capital and reserves represented by that participating interest shall be disclosed separately or in the notes to the accounts (POC, 14.4.4.19) and shall be calculated either at the date at which that method is used for the first time or date of acquisition or, where they were acquired in two or more stages, as at the date on which the undertakings became an associated undertaking; or
- At an amount corresponding to the proportion of the associated undertaking's capital and reserves represented by that participating interest. The difference between that amount and the value shall be disclosed separately in the consolidated balance sheet. That difference shall be calculated as at the date at which that method is used for the first time.

Where an associated undertaking's assets or liabilities have been valued by methods other than those used for consolidation (full method), a revaluation by the methods used for consolidation is required for the purpose of calculating the differences referred above. Where such revaluation has not been carried out, the fact must be disclosed in the notes to the consolidated accounts (POC, 14.4.4.20).

The book value in the first method above, or the amount corresponding to the proportion of the associated undertaking's capital and reserves referred to in the second method above, shall be increased or reduced by the amount of any variation which has taken place during the financial year in the proportion of the associated undertaking's capital and reserves represented by that participating interest. It shall also be reduced by the amount of the dividends relating to that participating interest (POC, 13.6.1.h). In so far as the positive difference referred to above cannot be related to any category of assets or liabilities it shall be dealt with in accordance with the method described under VI.2 (POC, 13.6.1.g).

The proportion of the profit or loss of the associated undertakings attributable to such a participation must be shown in the consolidated profit and loss statement as a separate item (POC, 13.6.1.i).

Under the equity method, elimination of profits included in the assets should normally be done proportionally as far as the availability of information allows it and if they are material. However, in practice, the necessary information is not available and thus frequently the consolidating company does not make that elimination (POC 13.6.1.j).

4 Equity method and individual accounts

The equity method is also valid, as an option, for the valuation in the individual accounts of group and associated companies undertakings, according to the most recent version of POC, dating from July 1991 (POC, 5.4.3). CNC states, however, that the equity method is compulsory and not optional (CNC, Opinion No. 9) and some auditors have qualified their reports if the method is not used.

The topic has been a matter of debate considering that sometimes the latter may be opposite to the POC and that the auditing profession (and in a minor extent, some accountants) are pushing companies to follow the CNC opinions where they are not in accordance with the letter of the POC. Defenders of the use of the equity method argue that the profit of the parent company could be manipulated by its directors while opposers of the method refer to an existing conflict between two accounting rules of different position in the Portuguese hierarchy of laws.

VIII Deferred taxation

1 Legislation and practice of deferred taxation

According to the POC, account shall be taken in the consolidated financial statements of any difference arising on consolidation between the tax chargeable for the financial year and for the preceding financial years, and the amount of tax paid or payable in respect of those years (POC, 13.4.3 and 14.4.4.38).

The POC recommends disclosure of deferred taxes in the consolidated balance sheet, in the consolidated profit and losses statement and in the notes. However, the accounting regulation does not specify how to calculate those tax differences and which tax rate to apply. So far, calculation and disclosure of deferred taxes has not been very common among Portuguese companies, who usually apply the liability method. A few larger companies, however, account for deferred taxes in line with IAS 12 and FAS 109. This is the case with Portuguese subsidiaries of international major groups and companies with listed securities or those audited by international firms. Auditors have been requiring, although still in a limited extent, the use of the deferral method as a condition to avoid qualification in their audits. As an example, deferred taxes related to unrealized capital gains resulting from disposals of tangible fixed assets under certain conditions of future reinvestment are required and non-compliance with this justifies a qualification in the audit report (CROC, *Interpretação Técnica* No. 4).

Portuguese companies that account for deferred taxes in the consolidated accounts have adopted the following procedures:

- Deferred taxes which arise in the current year are usually shown as income tax for the year in the consolidated profit and loss account and as accruals or tax debt in the consolidated balance sheet.
- Deferred taxes relating to prior years are written back as an extraordinary gain as a result of the reversal of timing differences in the year.
- Provision is made for deferred taxation, using the liability method, on material timing differ-

ences to the extent that it is probable that a liability will crystallize.
- The charge for taxation is based on the adjusted taxable profit for the year of each company or group of companies which consolidate for income tax purposes.
- The amount of income tax included in the consolidated profit and loss account as an expense is assessed at a standard rate of 34 per cent (according to CIRC) and on the top of this local authorities levy a local tax (*derrama*), of up to 10 per cent.

The computation of the expense incurred varies whether the group is being taxed in a group or in the basis of individual income of companies included in the consolidation.

2 Individual taxable income and sources of deferred taxes

In general, corporate income is levied on the companies individual worldwide earnings. Taxable income is defined as comprehensive income and, by reference to CIRC, it is based broadly on the net equity change during the tax period. Taxable income is the net result for the year shown in the individual financial statements, excluding capital increases, revaluation reserves, some equity increases, dividends and capital decreases. Adjustments between the taxable and accounting income of Portuguese companies are resumed in the corporate income tax return (*Mod. 22 de IRC, Q17*). Examples of permanent and timing tax differences (only the latter are relevant for deferred taxation) which are common to individual and consolidated accounts, as indicated in the income tax return, include the following:

- Accounting charges not accepted if above certain limits (such as excessive depreciation and provisions for doubtful debts and obsolete inventories).
- Fiscal charges accepted above the accounting costs (such as certain donations for cultural, scientific and humanitarian purposes).
- Excess of depreciation due to revaluation of tangible fixed assets.
- Capital gains resulting from disposals of tangible fixed assets if reinvested under certain conditions.
- Charges not accepted at all (such as fines related to infringement of laws, confidential expenses and provisions for depreciation of financial assets).
- Capital gains adjustments (such as 95 per cent of dividends from investments in subsidiaries and associated companies where the participation represents no less than 25 per cent of the share capital).
- Reduction of the adjusted accounting result by the previous 6 years' losses carried forward.
- Tax incentives and tax credits from international bilateral tax agreements.

According to the Chamber of Auditors, deferred taxes to account for may have three origins: individual accounts, adaptation of individual accounts to consolidated accounts and adjustments to accounting in order to assess taxation based on the consolidated income (CROC, *Recomendação Técnica,* No. 9). Timing differences arising from consolidation are due to adjustments from the elimination of intragroup results, use of uniform rules, translation of foreign currencies and differences in the consolidation methods.

3 Taxation based on the consolidated income

Taxation on the basis of the consolidated income is optional.

Consolidated income is only relevant for tax purposes if the group had obtained an authorization for being taxed under the rules of the group income tax (CIRC, art. Nos. 59, 59-A and 60 and *Circular da DGCI* No. 15/90 of 4 May). The authorization is requested by the parent company and is valid for a period of five years. The three following cumulative conditions must be met in order to obtain the said authorization:

- The parent company must hold, either directly or indirectly, at least 90 per cent of the nominal capital of its subsidiaries.
- Headquarters of every company in the fiscal group must be located in Portugal.
- None of the companies in the group may benefit from any special regime of income tax.

State-owned companies, co-operatives and sole-proprietorships are not allowed to be taxed on the basis of the consolidated profit. When the composition of the group changes, a renewal of the request in the year following the modification is necessary.

It is noteworthy to mention that under the group taxation regime the income tax is calculated over the whole income of the subsidiaries, including the proportion of the result of the minority shareholders or members. Accounting depreciation of goodwill is not allowed for tax purposes and so it results in a deferred tax position, to be included as a separated item in the consolidated financial statements of companies under the consolidated income tax system.

The parent company of the fiscal group must disclose, together with the group tax returns, information about the individual income tax of each company in the fiscal group and the tax losses arising from the application of the consolidated income tax system. The notes to consolidated accounts must include tax differences arising as a result of the consolidated income tax system. However, companies do not usually indicate the year in which the timing differences arise nor the reversals in each year. And companies usually do not disclose breakdown of credits and debits between group companies due to the tax effect which resulted from the application of the consolidated income tax system.

The number of Portuguese groups under this regime has been increasing since 1987. For the fiscal year of 1988 there were about 100 groups authorized to be taxed on group income extending to more than 1,000 companies.

IX Formats

1 Overview

Consolidated financial statements should comprise a balance sheet, a profit and loss statement and notes to the accounts (annexe). These documents should make up a composite whole.

Formats according to the models prescribed in the accounting law and the level of detail of them are essentially the same as those for the parent company (legal entity) accounts, except that specific items arising from consolidation may appear in the consolidated accounts.

The formats of the balance sheet and profit and loss account include the requirement to present comparative figures for the previous year and the possibility to omit headings where the amount is zero. The principle of consistency of presentation must be followed from year to year, otherwise departures should be reported under the notes.

No specific consolidated formats are required with respect to the consolidated cash flow and consolidated sources and application of funds.

2 Consolidated balance sheet

The layout required for the consolidated balance sheet set out in the POC (POC, 14.1) is presented in Table IX.1.

With regard to content, the presumed format is very detailed in which most lines of the balance sheet correspond to a single two- or three-digit account code within the national chart of accounts. It may be noted that three columns are provided for the current year's asset figures. The gross values of consolidated assets, provisions for depreciation or amortization and net values, respectively, provide the complete format in the POC.

Compared to individual accounts, differences in the assets side of the consolidated balance sheet are positive consolidation difference and the omission of investments in affiliated undertakings (which appear under the financial fixed assets in the individual accounts). The item 'associated companies' in the assets side of the consolidated balance sheet includes the value of the affiliated companies excluded from consolidation due to differences in activity and other undertakings which are valued by the equity method (see VII). Under the equity heading there is the negative consolidation difference and the consolidated result. Minority rights in the capital of the subsidiaries included in the consolidation appear between equity and liabilities. Capital includes only the capital of the parent company. The same applies to the share premium account. Reserves are the consolidated reserves of the previous year plus the consolidated result for the current year.

Table IX.1 Consolidated balance sheet format

	Years n GA	DP	NA	n−1 NA
ASSETS				
Fixed Assets				
Intangible Fixed Assets				
Formation expenses	x	x	x	x
Research and development expenses	x	x	x	x
Industrial property and other rig	x	x	x	x
Goodwill	x	x	x	x
Expenses in progress	x		x	x
Advances to suppliers	x		x	x
Consolidation differences	x	x	x	x
	x	x	x	x
Tangible Fixed Assets				
Land and natural resources	x	x	x	x
Buildings and other constructions	x	x	x	x
Machinery and equipment	x	x	x	x
Transport equipment	x	x	x	x
Tools and utensils	x	x	x	x
Furniture and office equipment	x	x	x	x
Containers	x	x	x	x
Other fixed assets	x	x	x	x
Construction in progress	x		x	x
Advances to suppliers of fixed assets	x	x	x	x
	x	x	x	x
Investments				
Investments in associated companies	x	x	x	x
Loans to associated companies	x	x	x	x
Investments in others participate	x	x	x	x
Loans to others participated companies	x	x	x	x
Securities and other investments	x	x	x	x
Other loans	x	x	x	x
Investments in progress	x		x	x
Advances	x		x	x
	x	x	x	x
Current Assets				
Inventories				
Raw materials and consumables	x	x	x	x
Work in progress	x	x	x	x
Spoilage, waste and scraps	x	x	x	x
Finished and semi-finished products	x	x	x	x
Merchandises	x	x	x	x
Advances to suppliers	x		x	x
	x	x	x	x
Medium- and Long-term Receivables (a)	x	x	x	x
Short-term receivables				
Trade accounts receivable – current	x	x	x	x
Trade accounts receivable – bills	x	x	x	x
Doubtful debtors	x	x	x	x
Group companies	x	x	x	x
Other participated companies	x	x	x	x
Other shareholders and partners	x	x	x	x
Advances to suppliers	x		x	x
Advances to suppliers of fixed assets	x		x	x
State and other public entities	x	x	x	x
Other debtors	x	x	x	x
Shares' and *quotas*' subscribers	x		x	x
	x	x	x	x
Negotiable Securities				
Shares in associated companies	x	x	x	x
Bonds in associated companies	x	x	x	x
Other negotiable securities	x	x	x	x
Other short-term investments	x	x	x	x
	x	x	x	x
Cash and Bank Balance				
Bank deposits	x		x	x
Cash	x		x	x
	x		x	x
Accruals and Deferrals				
Accrued income	x		x	x
Deferred expenses	x		x	x
	x		x	x
Total accumulated depr		x		
Total provisions		x		
TOTAL ASSETS	x	x	x	x

(a) To discriminate as the Short Term Receivables Item NA – Net Assets

	Years n	n − 1
CAPITAL AND LIABILITIES		
Capital	x	x
Own shares (*quotas*) – nominal value	−x	−x
Own shares (*quotas*) – premiums and discounts	−+x	−+x
Supplementary capital	x	x
Share issue premiums (*quotas*)	x	x
Consolidation differences	−+x	−+x
Net equity adjustments in group and associated companies	−+x	−+x
Revaluation reserves	x	x
Reserves		
Legal reserves	x	x
Statutory reserves	x	x
Contractual reserves	x	x
Other reserves		
Past years results	−+x	−+x
Subtotal	−+x	−+x
After tax result of the year	−+x	−+x
Anticipated dividends	−+x	−+x
TOTAL NET EQUITY	−+x	−+x
Minority rights	−+x	−+x
Liabilities		
Provisions for Risks and Charges		
Provisions for pensions	x	x
Provision for taxes	x	x
Provisions for other risks and charges	x	x
	x	x
Medium- and Long-term Liabilities (a)	x	x
Short-term Liabilities		
Bond borrowing		
Convertible bonds	x	x
Non-convertible bonds	x	x
Participating bonds borrowing	x	x
Debits to credit institutions	x	x
Advances from customers	x	x
Suppliers – current accounts	x	x
Suppliers – invoices outstanding	x	x
Suppliers – bills payable	x	x
Accounts payable to suppliers of fixed assets – bills	x	x
Group companies	x	x
Other participated and participating companies	x	x
Other shareholders and quotaholders	x	x
Advances from customers	x	x
Other borrowing	x	x
Suppliers of fixed assets suppliers – current account	x	x
State and other public entities	x	x
Other creditors	x	x
	x	x
Cost accruals	x	x
Deferred income	x	x
	x	x
TOTAL LIABILITIES	x	x
TOTAL OF NET EQUITY, MINORITY RIGHTS AND LIABILITIES	x	x

GA – Gross Assets DP – Depreciation and provisions

Table IX.1 (contd.)

	Exercícios n AB	AP	AL	n – 1 AL
ACTIVO				
Imobilizado				
Imobilizações incorpóreas				
Despesas de instalação	x	x	x	x
Despesas de investigação e desenvolvinmento	x	x	x	x
Propriedade industrial e outros diseitos	x	x	x	x
Trespasses	x	x	x	x
Imobilizações em curso	x		x	x
Adiantamentos por conta de imobilizacoés incorpoíeas	x		x	x
Diferenças de consolidação	x	x	x	x
	x	x	x	x
Imobilizações incorpóreas				
Terrenos e recursos naturais	x	x	x	x
Edifícios e outras construções	x	x	x	x
Equipamento básico	x	x	x	x
Equipamento de transporte	x	x	x	x
Ferramentas e utensílios	x	x	x	x
Equipamento administrativo	x	x	x	x
Taras e vasilhame	x	x	x	x
Outras imobilizações corpóreas	x	x	x	x
Imobilizações em curso	x		x	x
Adiantamentos por conta de imobilizacoés corpoíeas	x		x	x
	x	x	x	x
Investimentos financeiros:				
Partes de capital em empresas ass	x	x	x	x
Empréstimos a empresas associadas	x	x	x	x
Partes de capital em outras empre	x	x	x	x
Empréstimos a outras empresas par	x	x	x	x
Títulos e outras aplicações finan	x	x	x	x
Outros empréstimos concedidos	x	x	x	x
Imobilizações em curso	x		x	x
Adiantamentos por conta de investimento financeiros	x		x	x
	x	x	x	x
Circulante				
Existências				
Matérias-primas, subsidiárias e de comsumó	x	x	x	x
Produtos e trabalhos em curso	x	x	x	x
Subprodutos, desperdícios, Resíduos e refugos	x	x	x	x
Produtos acabados e intermédios	x	x	x	x
Mercadorias	x	x	x	x
Adiantamentos por conta de compras	x		x	x
	x	x	x	x
Dívidas de terceiros – Médio e longo	x	x	x	x
Dívidas de terceiros – Curto prazo				
Clientes, c/c	x	x	x	x
Clientes – Títulos a receber	x	x	x	x
Clientes de cobrança duvidosa	x	x	x	x
Empresas associadas	x	x	x	x
Empresas participadas e participantes	x	x	x	x
Outros accionistas (sócios)	x	x	x	x
Adiantamentos a fornecedores	x		x	x
Adiantamentos a fornecedores de-imobilizado	x		x	x
Estado e outros entes públicos	x	x	x	x
Outros devedores	x	x	x	x
Subscritores de capital	x		x	x
	x	x	x	x
Títulos negociáveis:				
Acções em empresas associadas	x	x	x	x
Obrigações em empresas associadas	x	x	x	x
Outros títulos negociáveis	x	x	x	x
Outras aplicações de tesouraria	x	x	x	x
	x	x	x	x
Depósitos bancários e caixa:				
Depósitos bancários	x		x	x
Caixa	x		x	x
	x		x	x
Acréscimos e diferimentos:				
Acréscimos de proveitos	x		x	x
Custos diferidos	x		x	x
	x		x	x
Total de amortizações		x		
Total de provisões		x		
	x	x	x	x

(a) A discriminar como as dívidas a curto prazo

	Exercícios n	n – 1
CAPITAL PRÓPRIO E PASSIVO		
Capital próprio		
Capital	x	x
Acções (quotas) próprias – Valor nominal	−x	−x
Acções (quotas) próprias – Decontos e prémios	− +x	− +x
Prestações suplementares	x	x
Prémios de emissão de acções (quotas)	x	x
Diferenças de consolidação	− +x	− +x
Ajustamentos de partes de capital em filiais e associadas	− +x	− +x
Reservas de reavaliação	x	x
Reservas		
Reservas legais	x	x
Reservas estatutárias	x	x
Reservas contratuais	x	x
Outras reservas	x	x
Resultados transitados	− +x	− +x
	− +x	− +x
Resultado líquido do exercício	−x	− +x
Dividendos antecipados	− +x	− +x
Total do capital próprio	− +x	− +x
Interesses minoritários	− +x	− +x
Passivo		
Provisões para riscos e encargos		
Provisões para pensões	x	x
Provisões para impostos	x	x
Outras provisões para riscos e encargos	x	x
	x	x
Dívidas a terceiros – Médio e longo prazo (a)	x	x
Empréstimos por obrigações		
Convertíveis	x	x
Não convertíveis	x	x
Empréstimos por títulos de participação	x	x
Dívidas a instituições de crédito	x	x
Adiantamentos por conta de vendas	x	x
Fornecedores, c/c	x	x
Forcecedores – Facturas em recepçõus e conferência	x	x
Forcecedores – Títulos a pagar	x	x
Forcecedores de imobilizado – Títulos a pagar	x	x
Empresas associadas	x	x
Empresas participadas e participantes	x	x
Outros accionistas (sócios)	x	x
Adiantamentos de clientes	x	x
Outros empréstimos obtidos	x	x
Fornecedores de imobilizado, c/c	x	x
Estado e outros entes públicos	x	x
Outros credores	x	x
	x	x
Acréscimos e diferimentos	x	x
Acréscimos de custos	x	x
Proveitos diferidos	x	x
	x	x
TOTAL DO PASSIVO	x	x
TOTAL DO CAPITAL PRÓPRIO, DOS INTERESSES MINORITÁRIOS E DO PAS	x	x

AB – ACTIVO BRUTO AP – AMORTIZAÇÕES E PROVISÕES

3 Consolidated profit and loss statements

Table IX.2 provides the format for the consolidated profit and loss statement by nature in the POC (POC, 14.4.2).

Table IX.2 Consolidated profit and loss account format

	Exercícios N		N-1	
Custo e perdas				
Custo das mercadioras vendidas e das matérias consumidas:				
Mercadorias	x		x	
Matérias	x	x	x	x
Fornecimentos e serviços externos		x		x
Custos com o pessoal:				
Remunerações	x		x	
Encargos sociais:				
Pensões	x		x	
Outros	x	x	x	x
Amortizações do imobilizado corpóreo e incorpóreo	x		x	
Provisões	x	x	x	x
	x		x	
Outros custos e perdas operacionais	x	x	x	x
(A)		x		x
Amortizações e provisões de aplicações e investimentos financeiros		x		x
Juros e custos similares:				
Relativos e empresas associadas	x		x	
Outros	x	x	x	x
(C)		x		x
Perdas relativas a empresas associadas		x	x	
Custos e perdas extraordinários		x	x	
(E)		x		x
Imposto sobre o rendimento do exercicio		x		x
(G)		x		x
Interesses minoritários		(−+)x		(−+)x
Resultado consolidado liquido do exercicio		(−+)x		(−+)x
		x		x
Proveitos e ganhos				
Vendas:				
Mercadorias	x		x	
Produtos	x		x	
Prestações de serviços	x	x	x	x
Variação da produção		(−+)x		(−+)x
Trabalhos para a própria empresa		x		x
Proveitos suplementares	x		x	
Subsidios à exploração	x		x	
Outros proveitos e ganhos operacionais	x	x	x	x
(B)		x		x
Ganhos de participações de capital:				
Relativos e empresas associadas	x		x	
Relativos a outras associadas	x		x	
Rendimentos de titulos negociáveis e de outras aplicações financeiras:				
Relativos a empresas associadas	x		x	
Outros	x		x	
Outros juros e proveitos similares:				
Relativos a empresas associadas	x		x	
Outros	x	x	x	x
(D)		x		x
Proveitos e ganhos extraordinários		x		x
(F)		x		x
Resumo:				
Resultados operacionais: (B) − (A) =		x		x
Resultados financeiros: (D − B) − (C − A) =		x		x
Resultados correntes: (D) − (C) =		x		x
Resultados antes de impostos: (F) − (E) =		x		x
Resultados consolidado com os interesses minoritários do exercicio: (F) − (G) =		x		x

	Year N		N-1	
Expenses and losses				
Costs of goods sold and materials consumed:				
Goods	x		x	
Materials	x	x	x	x
External supplies and services		x		x
Personnel expenses:				
Wages and salaries	x		x	
Social security costs:				
Pension costs	x		x	
Others	x	x	x	x
Depreciation of intangible and tangible fixed assets	x		x	
Provisions	x	x	x	x
Taxes	x		x	
Other operating costs	x	x	x	x
(A)		x		x
Losses in associated companies		x		x
Depreciation and provisions for investments	x		x	
Interest expenses and similar costs:				
Associated companies	x		x	
Others	x	x	x	x
(C)		x		x
Losses in associated companies		x		x
Extraordinary costs and losses		x		x
(E)		x		x
Income tax of the year		x		x
(G)		x		x
Minority interests (result for the year)		(−+)x		(−+)x
Group net result for the year		(−+)x		(−+)x
		x		x
Revenues and gains				
Sales:				
Goods	x		x	
Products	x		x	
Services	x	x	x	x
Increase or decrease in stocks of finished products and work in progress		(−+)x		(−+)x
Work performed by the undertaking for its own purposes	x		x	
Supplementary revenues	x		x	
Subventions received	x		x	
Other operating revenues	x	x	x	x
(B)		x		x
Dividends on shares and profits from undertakings:				
Gains in associated companies	x		x	
Gains in other related companies	x		x	
Income from negotiable securities and other short-term investments:				
From associated companies	x		x	
Other	x		x	
Other financial revenues:				
From associated companies	x		x	
Other	x	x	x	x
(D)		x		x
Extraordinary revenues and gains		x		x
(F)		x		x
Resume				
Operating Result: (B) − (A) =		x		x
Financial Result: (D − B) − (C − A) =		x		x
Current Result: (D) − (C) =		x		x
Result Before Taxes: (F) − (E) =		x		x
Group Result including minority rights for the year: (F) − (G) =		x		

The POC opted for the horizontal format classified by nature, with expenses and losses on the left-hand side and revenues and income for the year on the right. This traditional format provides a number of subtotals which are of interest for analytical purposes in a *résumé* at the bottom of the statement and the notes to the consolidated accounts complete the information on that statement. Disclosures such as breakdown for sales, cost of sales, financial expenses and extraordinary items are required in the notes.

Specific items in the consolidated profit and loss statement are the shares of the group shareholders or members and minority interests in the consolidated results. The share of income in associated companies is not separately mentioned.

A vertical format of consolidated profit and loss statement is provided in the accounting law. Some groups have voluntarily disclosed this format, in addition to the full version of the statement by nature, but could not replace it (POC, 14.4.3). According to a recently approved law, disclosure of a vertical format of consolidated profit and loss statement, which includes earnings per share (EPS) disclosure, should be obligatory for the years after 2000 (Decree Law No. 44/99 of 12 February 1999).

4 Group specific items

4.1 Consolidation differences

Positive consolidation difference is presented in the assets side of the consolidated balance sheet and classified as an intangible fixed asset. It must be amortized in no more than five years, unless the reasons for a longer useful life are explained under the notes.

Negative consolidation difference is included under the equity. Its inclusion in the consolidated profit and loss statement is allowed only in exceptional circumstances (see VI).

4.2 Minority interests

Where the parent company does not hold 100 per cent of the capital of a subsidiary included in the consolidation, the consolidated balance sheet and profit and loss account must show the minority interests in appropriate headings. Minority interests in the balance sheet will appear between the equity (out of it) and liabilities and in the profit and loss account before the group result. Separation and analysis of the minority rights in the balance sheet into capital, reserves and income by each included subsidiary undertaking is not required. Own shares in affiliated companies are deducted before calculation of the minority rights in the consolidated balance sheet and must be disclosed under the equity. Minority rights in a company with negative equity will appear with zero amount in the consolidated balance sheet, unless there is any agreement or clause in the articles of association imposing to that minority the obligation to cover the accumulated losses.

4.3 Consolidated result for the year

The consolidated result for the year should be shown in both the consolidated balance sheet and profit and loss account, with separated line for the shares of the group and the minorities. The latter is not broken down by companies and is calculated before elimination. The accounts present the result before the appropriation which is approved by the annual general meeting. The relationship between the consolidated appropriation of profit is not mentioned in the consolidated accounts, but is included in the individual director's report.

X Notes and additional information

1 Notes

The notes to consolidated accounts required by the POC have been designed to meet the requirements of the EU Seventh Directive. Together with consolidated balance sheet and profit and loss accounts they form a whole and should give a true and fair view of the group. They include a set of information intended to amplify and explain items in the consolidated financial statements. The POC emphasizes that the quality of the financial information given by companies is very dependent on the content of the notes which cover the following topics (number of disclosures in brackets):

- Scope of consolidated accounts (7).
- Information required to report a true and fair view (2).
- Consolidation procedures adopted (11).
- Financial commitments (2).
- Accounting policies (2).
- Specific items in the consolidated financial statements (24).
- Sundry information (2).

The notes to consolidated accounts must indicate the names, addresses, percentage of capital held of companies included in the consolidation, companies excluded, companies accounted for as associated undertakings (using the equity method) and companies accounted for by the proportional method (POC, 14.4.4.1 to 14.4.4.7). The notes to the individual accounts of the parent company must also contain information relating to companies included in or excluded from the consolidation, such as a list of subsidiaries, associated companies and undertakings in which the parent company holds at least 10 per cent of the capital. Associated companies and undertakings which are not required to publish their financial statements may be omitted (POC, 8.16).

Departures from the requirement of the POC in order to show a true and fair view are allowed but the reasons for such exceptional departure must be disclosed in the notes (POC, 14.4.4.8 and 14.4.4.9).

Disclosures regarding group specific information are presented in Table X.1.

Information which is similar in the consolidated and individual accounts includes at least the following items:

- Financial commitments not included in the consolidated balance sheet and responsibilities for guarantees given.
- Valuation methods and exchange rates used to convert operations and to translate amounts originally expressed in foreign currency for inclusion in the consolidated financial statements.
- Revaluation and other details of tangible fixed assets.
- Interest payable capitalized during the year and start-up, research and development costs.
- Amounts owing repayable after more than five years, amounts covered by guarantees and differences between amounts owing and the corresponding amounts received.
- Accounting values of leased assets.
- Details of the equity capital and changes occurred during the year.
- Breakdown of financial charges, extraordinary items and movements of provision during the year and turnover by activity and market.
- Information about employment such as average number of employees, employment costs, advances and loans guaranteed to boards members of the parent company.
- Situations having an impact on future taxation and accounting methods used in order to obtain tax benefits:
  - (i) The use of exceptional asset valuation principles.
  - (ii) Depreciation in excess of what is economically justified.
  - (iii) Extraordinary provisions (write-downs) against assets.

Table X.1 List of specific disclosures for consolidated accounts

- Analysis of goodwill from consolidation and comment on its depreciation over more than five years
- Exceptions to general accounting principles applied to consolidated financial statements
- Consolidation adjustments not made as considered non material
- Material events occurred after the date of the consolidated balance sheet
- Implication to comparability of financial statements of changes to the scope of consolidation
- Differences between accounting policies and valuation methods used in individual and consolidated accounts
- Accounting choices for associated undertakings
- Analysis of minority interests

Although the notes are too numerous (50 entries), they are intended to be the minimum information required by law. Other information that could be useful for the reader or be relevant to inform about the financial position of the group should be disclosed. So far, voluntary disclosures have been exceptional.

2 Consolidated directors' report

The contents of the directors' report for the individual and group accounts are in general identical (CSC, art. 508-C).

The consolidated directors' report must include a fair view of the development of the business and position of the consolidated undertakings. Furthermore, the report must comment on important events which have occurred subsequent to the end of the financial year and exceptional circumstances which have influenced the financial statements disclosure of relevant amounts, if possible.

The consolidated directors' report must also mention the number and nominal value (or in their absence, the nominal value) or the according par value of the consolidating company shares held by the parent company, other group undertakings or their nominees or by a person acting in his own name but on behalf of those undertaking and information concerning acquisitions of own shares. As an alternative, this information may be shown in the notes.

The consolidated accounts together with the consolidated directors' report and the annual report of the parent company may be included in a single document (*relatório e contas*). But the disclosure of such information is not permitted in the notes.

3 Other reports

Parent companies of groups with listed securities must disclose two additional statements: a consolidated cash flow and a functional profit and loss account statement. The law does not provide a format for the former.

Most of the groups adopt for cash flow disclosures the format prescribed for individual accounts which is in line with IAS requirements. This format separates the annual cash flows from operating, investing and financing activities, the former using either the direct or the indirect method. A set of notes completes the disclosures about consolidated cash flow of the year (CNC, Opinion No. 14).

Disclosure of consolidated profit and loss statement by function (and earnings per share) has been made voluntarily, but it will become compulsory after 2000 (POC, 14.4.3).

The preparation of a consolidated statement of sources and applications of funds is also recommended, although no specific format is provided for it. The model prescribed to individual accounts may be used on a voluntary basis (POC, 2.6 and 9). Most of the Portuguese groups do not present such a statement but comparative figures together with information disclosed in the notes allow an estimation of the consolidated funds flows for the year.

There are no requirements for disclosure of human resource accounting and green accounting in a group basis. Few companies, however, disclose at group level other information, in so far as there are peculiarities.

Groups with securities listed in the official stock market have additionally the obligation to disclose mid-year consolidated financial statements, as an alternative to mid-year individual accounts, and since 1999 quarterly consolidated information, the latter according to specific abridged formats in the securities markets law (CMVM, Instruction No. 11/98).

XI Auditing

1 Audit requirements

The auditing profession consists of statutory auditors (*revisores oficiais de contas*, or ROCs) registered with the Chamber of Registered Statutory Auditors (*Câmara dos Revisores Oficiais de Contas* or CROC). This is the officially recognized professional body for auditors and it is responsible for issuing standards and undertaking disciplinary measures. The auditing profession has been regulated in line with the Eighth EU Directive since January 1994 (Decree Law No. 422-A/93, of 30 December 1993).

The consolidated financial statements and the consolidated directors' report must be audited before they are submitted to the supervisory board of the parent company (CSC, art. 508-D).

The group's auditor verifies if the consolidated accounts comply with the true and fair view objective. The audit of the consolidated financial statements has to state whether the legal requirements and the company's articles have been observed and has to express an independent and professional opinion on concordance between the information given in the consolidated financial statements and the consolidated director's report (CSC, art. 508-D). Auditors must also audit the consolidated intermediate information.

The group's auditor must examine the work of the auditors of the undertakings included in the consolidation and also audit the individual financial statements of the companies consolidated when they have not been audited with similar standards (CROC, *Recomendação Técnica* No 9). The aim of such work is to verify the conformity of the individual financial statements with generally accepted accounting and to confirm if the consolidation rules have been adhered to.

Finally, auditors are obliged to provide the information requested by shareholders or members at the annual general meeting.

2 Appointment of the legal auditors

As for auditing, Portuguese corporations (SA) and private limited companies (LDA) must appoint a statutory audit board (*conselho fiscal*), the size and composition of which depend on the size and legal form of the company. Corporations usually have a three-member supervisory board, one of whom is the company's ROC who acts as auditor. For private limited companies above a certain size as specified in the law implementing the EU Fourth Directive, the same applies as in the case of a corporation (50 employees, net assets of 1,500,000 euros, and sales turnover and other revenues of 3,000,000 euros). For smaller LDAs, no statutory audit is required.

Auditors are appointed by shareholders meeting for a period of between two and four years. Although only one member of the statutory board of auditors need be professionally qualified, the following restrictions apply to all members. No member of the audit board may:

- Be a director or employee of the company, or receive any special benefits from it.
- Be a director, employee or audit board member of the company's parent or affiliated companies or of any company which is in a position to control or to be controlled by the company because of any special contractual obligations.
- Be a shareholder, owner or employee of any company carrying out in its own name any of the functions, or subject to the restrictions, described above.
- Perform any functions in a competitor company.
- Have a close family relationship with persons in any of the preceding categories.
- Be legally incapacitated from carrying out public duties.

Normally the auditor of the parent company's individual accounts is appointed to audit the group accounts. If a different auditor is to be appointed he must be separately elected by the shareholders of the parent company.

3 Audit report

Portuguese corporations and those private limited companies over a certain size are subject to an annual audit which is to be presented to the shareholders or members at the annual meeting.

At least one auditor who has qualified as a member of the professional auditing body must undertake such work (CSC, art. 36). The appointed auditor takes part in the activities of the statutory audit board and issues, independently and outside the board, an annual statutory audit (*Certificação legal de contas*). The audit report must be issued pursuant to CROC auditing standards and presented in a standard form (CSC, art. 452).

This means that the auditor confirms if the consolidated accounts comply with the objective of giving a true and fair view and examines whether the consolidated financial statements are free of material mis-statement. The examination includes the following:

- Verification, on a test basis, of evidence supporting the amounts in the consolidated financial statements and assessment of significant estimates based on judgements and criteria defined by the directors and adopted in the consolidated financial statements.
- Assessment of the adequacy of accounting principles adopted, their disclosure and the

appropriateness of the going concern principle.

- Evaluation of the presentation of the consolidated financial statements.

The ROC also monitors the activity of the group and their management and issues a report as a result of that work (*Relatório e parecer de Conselho fiscal*).

Listed companies are compelled to present an additional audit report by an auditor registered under the Portuguese Securities Market Commission (*auditor externo*). This external audit extends to semi-annual financial reports. It provides a reasonable basis for the auditor's opinion which satisfies the principles of completeness, correctness, objectivity and timeliness required by the Securities Market Code (CODMVM, art. 100). In 1996, about 30 per cent of the external audits report include qualifications and more than 65 per cent of them include emphasis. In their comments, the auditors refer mainly to insufficient provisions, valuation of tangible fixed assets and non-use of the equity method.

In addition, a number of the largest Portuguese companies employ international accounting firms to carry out voluntary independent audit in accordance with international auditing standards. In practice, increasing competition along with growing relationships with international firms have contributed to the quality and independence of the Portuguese auditor's work.

4 Audit guidelines

Auditing standards are specified by the CROC in its periodically updated *Manual* which embraces international auditing principles. It also includes a set of technical recommendations (*recomendações técnicas*) and opinions (*interpretações técnicas*) that guide the auditing practice. Table XI.1 lists the guidelines issued by the CROC which relate to the audit of consolidated financial statements.

5 Co-ordination between auditors

Co-ordination between auditors of the parent company, subsidiaries and jointly-controlled companies is necessary in order to issue a reliable consolidated audit report.

In Portugal, the audit of subsidiary companies has to rely on the individual accounts auditors' work. The auditors of undertakings included in the consolidation must give to the auditors of the parent company the information they needed, which is sometimes confidential. The auditors of the consolidated accounts will take into account the quality of the audit of the individual accounts according to the relative importance of the consolidated company or companies in the group. In some circumstances additional audit testing of an affiliated company may be necessary. The auditor of the consolidated accounts will judge whether additional controls are necessary, based on the grounds of the materiality of this company or companies. If a company in the group refuses to allow the auditor to do such testing, the fact should be mentioned in the auditors' report of the group. In the case of foreign subsidiaries where there is not an acceptable external audit, the auditors of the consolidated accounts can require the non-audited company's directors to appoint an auditor who follows professional rules equivalent to the Portuguese ones. It can require specific controls from the internal audit department or introduce certain controls on his or her own behalf.

Report methods and responsibilities of issuing reports when an auditor uses the work or report of other independent auditor are defined by professional standards (CROC, *Recomendação Técnica* No. 19).

Table XI.1 Guidelines on consolidated accounts issued by the CROC

RECOMMENDATIONS	
9	Statutory audit of consolidated accounts
10	Information and qualifications in the statutory audit report
19	Use of others auditors working papers
INTERPRETATIONS	
1	Comparative figures in the consolidated financial statements
3	Audit of the first consolidated accounts disclosed by certain financial institutions
5	Applicability of the equity method
11	Wording and quantification of qualifications in an audit report

XII Filing and publication

1 Rules of disclosure

The rules of disclosure for consolidated accounts are in general similar to those for individual accounts. The type of information that companies must disclose depends on size criteria, their legal forms and whether they are listed companies.

The required consolidated accounts (the balance sheet, the profit and loss account and the annexe), together with the consolidated statutory audit and directors' report, must be submitted to the group's auditor in order to issue the legal certification of accounts. These documents should be available for inspection by the shareholders or partners at the offices of the parent company at least 15 days before the calling of the meeting where the consolidated accounts are to be approved. This meeting shall occur no later than five months after the end of the financial year (CSC, arts. 65 and 508-A). Then, the consolidated financial reporting documents shall be public by depositing them with the Commercial Registry before the end of the sixth month after the balance sheet date (CSC, art. 508-E and CRC, arts. 3.n and 42, No. 1).

Special disclosure rules apply to small groups, for which consolidated accounts are not mandatory as their parent companies may be exempted. The consolidated accounts where the exempted parent company appears as a subsidiary must be included in the notes to individual accounts of the exempted company.

If the consolidating company is not established as a corporation, as private limited company or as a limited partnership, it must at least make the consolidated statements available to the public at its registered office. Copies of part or the whole of the documents must be provided on request and at no more than cost (CSC, art. 508-E, No. 2).

Groups with listed companies have some additional disclosure requirements, as follows:

- Obligation to send the consolidated financial reports to the CMVM and the Lisbon Stock Exchange Association no later than the date when they are delivered for publishing (CODMVM, art. 339, No.2)
- Obligation to send to the CMVM and to the Lisbon Stock Exchange Association semi-annual information (individual or consolidated) not later than three months after the end of the semester (CODMVM, art. 342).
- Obligation to disclose quarterly consolidated information, only in the case of companies listed on the securities market with official quotations, since September 1998 (CODMVM, art. 343 and Ministerial Order No. 1222/97, of 12 December 1997).
- Submission of an additional independent audit report (optional for companies with securities admitted to trading for an unlimited period on the market without quotations of the Lisbon Stock Exchange and in some circumstances of semi-annual statements) (CODMVM, art. 100).
- Obligation to comply with publication requirements in the stock market gazette (*Boletim oficial de cotações*) and in a newspaper of the region where its headquarters are located no later than 30 days after the approval by the annual general meeting (CSC, art. 167 and CODMVM, art. 339).
- Obligation to disclose information about cash flows (this obligation only applies for companies with listed shares on the market with official quotations and on the second market in the Lisbon Stock Exchange and it is recommended in other cases of listed companies) (CMVM, Instruction, No. 11/93).

Listed groups may request from CMVM an exemption from publishing both the individual and consolidated statements reported to the end of the financial year (CODMVM, art. 341, No. 4).

2 The practice

Most of the Portuguese parent companies have been complying with disclosure obligations of consolidating statements.

In its Annual Report (1997), the CMVM informs that, for the year of 1996, out of the 228 global issuers with listed securities in the Lisbon Stock Exchange (companies with listed shares in the market with official quotations and on the second market), the number of companies disclosing annual information as at 31 December 1997 was

88. The number of companies disclosing half-year information reports as at 30 June 1997 was 86 (71 for the market with official quotations, 15 for the second market and two companies voluntarily).

A few Portuguese groups do not publish the individual and the consolidated financial statements reported to the end of the financial year, although the approval of both is mentioned in the directors' and audit reports. In 1996, out of the 40 companies which submitted exemption applications to the CMVM, 37 companies were exempted from publication of one set of statements. Thirty-four companies opted to publish the consolidated accounts and the three remaining exempted companies did not express any preference but the CMVM granted to them exemption from publication of the individual statements.

Some groups have delayed sending the consolidated accounts to the Portuguese Securities Market Commission and to the Commercial Registry. In 1997, five companies sent their annual financial reports to CMVM with a delay of more than 30 days, against 26 companies in the previous year, but none of the faulty companies was listed in the market with official quotations. In the same year, about 96 per cent of companies published their semi-annual information in due time in a newspaper of wide circulation and in the Lisbon Stock Exchange Official Journal.

An increasing number of listed companies have published quarterly reports before this obligation entered in force, mainly through press releases.

Consolidated statements of Portuguese groups are always published in Portuguese but an increasing number of companies – primarily subsidiaries of foreign concerns – present bilingual group financial reports.

XIII Sanctions

The Companies Business Code and the Securities Market Code include a threat of sanction to the directors of a company who have in particular the responsibility for preparing and presenting consolidated financial statements. The organization of the process and the decision to impose penalties to the board of directors is of the responsibility of the Commercial Registry and the Portuguese Securities Market Commission. Severe cases of non-compliance may be brought to court as criminal acts and if serious effects are proved imprisonment and fines should apply. Those who violate the rules relating to group accounting objectives, principles and valuation criteria, disclosure requirements, the rules relating to the audit of consolidated accounts or the filing and publication requirements, are subject to the following sanctions:

- Opposition to audit and supervision should justify the payment of a fine and prison up to six months (CSC, art. 522).
- Non-submission or impediment to submitting to the general meeting the consolidated accounts in due time will impose the payment of a fine between PTE 10,000 and PTE 300,000 (CSC, art. 528).
- Penalties for non-approval of accounts vary according to the capital, turnover, economic situation of the wrongdoer and the values of the information.
- Violation of the rules about filing and publication of group accounts implicates a fine, according to the gravity of the fault.
- Any non-compliance with the obligation to deposit the accounts within the Commercial Registry, the refusal to disclose information and the disclosure of false or incomplete information will originate the imposition of fines (CSC, arts. 518 and 519).
- Distribution of locked up profits or reserves of the company are also punished (CSC, art. 514).

The CMVM issues mandatory injunctions, admonitions and recommendations, the former in order to prevent infringements of the rules regulating the markets and the latter to improve the functioning of the markets. Non-compliance with the obligation to publish periodic information may cause the imposition of cumulative penalties by the Portuguese Securities Market Commission to listed companies. Fines may vary between PTE 50,000 and PTE 300,000 depending on the seriousness of the infringements (CODMVM, art. 351).

Professional organizations of auditors and

accountants impose sanctions on their members for non-compliance with professional rules. Penalties range from warnings and admonitions to temporary or permanent suspension of the activity.

The CROC issues penalties, once a complaint has been processed, to an auditor by its board of directors (Decree Law No. 422-A/93, arts. 71 to 84). Sanctions to auditors are according to the grade of their non-compliance, verbal warning, notice, censure, fines (ranging from PTE 50,000 to PTE 500,000), temporary suspension for up to five years and withdrawal from the professional association.

Accountants have to ensure that the tax statements are in accordance with the accounting law and professional rules. Sanctions imposed by the professional association ATOC depend on the accountant's error and in certain circumstances are the same as for non-compliance with some fiscal laws. They range from warnings and penalties (payment of a fine up to five times the minimum national, in case of bad performance of the function) to temporary suspension (when the accountant gives up their work without a valid reason or reveals any secret) and withdrawal from the association (in the case of destruction, omission or falsification of data). Sanctions may either be maximized under proof of guilt or reduced if the accountant confesses the failure (Decree Law No. 265/95, arts. 26 to 41).

The refusal to sign the accounts of a company is punished with the suspension of the authorization to practise or even the withdrawal of membership from the association, which means the prohibition indefinitely to practise.

XIV National databases

Databases on Portuguese group accounts are not available to the public and Internet connections about consolidated accounts in Portugal are not relevant yet. Therefore, access to complete and updated data may still be a problem for those wanting to carry out empirical studies on Portuguese group accounting.

On the contrary, much information on individual accounts is available from public and private sources and may be consulted either free of charge or by the payment of a fee. The most generally known databases on Portuguese accounting have been those of *Banco de Portugal (Central de Balanços)*, *Banco Português do Atlântico* and the Lisbon Stock Exchange Association (*Sistema de Informação e Serviços de Bolsa*). The latter contained data about the individual accounts of listed companies. Information in these databases includes the balance sheet, the profit and loss accounts and some financial ratios, but not the notes. The SIIB database is no longer available. A new improved database, called DATHIS, is being constructed and is expected to be available in the near future. It is at its experimental stage and consultation is still restricted, mainly for research and academic purposes.

International companies operating in Portugal have provided accounting database services for a long time. The most well-known is Dun & Bradstreet, which annually edits a booklet with 14 financial ratios based on the individual accounts data. Banks, other financial institutions and insurance companies have also prepared their own accounting databases as a support for investment consulting activity.

Accounting data necessary to construct consolidated accounting databases has been collected by various official bodies, such as the Commercial Registry, the Portuguese Securities Market Commission, the Lisbon Stock Exchange Association, the Ministry of Finance (Tax Authority), The Bank of Portugal and the Insurance Institute of Portugal.

The obligation to deliver the consolidated accounts to the Portuguese Securities Market Commission justifies a current project by CMVM to have a database of consolidated accounts of companies with listed securities. As a consequence of the obligation to deposit them within the Commercial Registry, it is expected that a general database on Portuguese group accounts will soon be available.

The Tax Authority (under the Ministry of Finance) has been receiving annually the returns related to income taxation based on group accounts. The information covers the period of application of this special taxation system regime and all the groups under the regime. It consists of a sample with information on about 100 national groups and more than 1000 companies.

The Bank of Portugal and the Insurance Institute of Portugal receive periodically the consolidated accounts from the companies operating in Portugal in the banking and insurance industries. As supervisory bodies for these specific sectors both bodies treat the financial data of groups of companies operating in those sectors but the results of the analysis are not available to the public.

Professional associations may also have a helpful role in promoting databases of consolidated financial reporting. The Portuguese Association of Financial Analysts (*Associação Portuguesa de Analistas Financeiros*) annually awards the Portuguese company with the best financial report and thus this professional association has been collecting the consolidated accounts of major Portuguese parent companies.

During the past 10 years developments in Portuguese financial accounting have included a number of major changes resulting from Portugal's membership of the European Union and its implementation of the Fourth, Seventh and Eighth Directives. As far as these changes are being digested, consolidated accounts will take on a much higher profile and further significant developments can be expected, relating to national databases and Internet connections to make possible empirical studies on Portuguese accounting. Thus, one may anticipate the development of financial reporting in Portugal as an aid to financial analysis.

Bibliography

Azevedo Rodrigues, J. (1998). *Práticas de Consolidação de Contas – Empresas Subordinadas ao POC e Instituições Financeiras*, (Áreas Editora, Lisboa).

Belo, L. (1994). As novas regras de tributação pelo lucro consolidado. *Fisco*, **67**, 3–14.

Braz Machado, J. (1993). *Consolidação de Contas* (Editorial Notícias, Lisboa).

CMVM (1997). *Relatório anual* (CMVM, Lisboa).

Fernandes Ferreira, R. (1994a). Valorimetria, contabilização e/ou registo, na empresa-mãe, das participações (partes de capital em filiais e associadas). *Jornal do Técnico de Contas e da Empresa*, **342**, 65–8.

Fernandes Ferreira, R. (1994b). Novas anotações sobre consolidação de contas. Uma questão passível de controvérsia (!). *Jornal do Técnico de Contas e da Empresa*, **343**, 94–5.

Ferreira, D. (1991). Normas sobre procedimentos contabilístico-fiscais no âmbito da consolidação fiscal. *Fisco*, **28**, 3–9.

Ferreira, D. (1992). Consolidação fiscal e economias de imposto – aspectos fiscais, financeiros e contabilísticos. *Fisco*, **40**, 7–17.

Ferreira, D. (1993). A consolidação das contas dos grupos de sociedades. *Fisco*, **57**, 3–16.

Ferreira, L. (1993). Group accounting in Portugal: In Gray, S., Coenenberg, A. and Gordon, P., *International Group Accounting – Issues in European Harmonization*, 191–201 (Routledge, London and New York).

Ferreira, L. (1994). *European Financial Reporting: Portugal* (Routledge, London and New York).

Ferreira, L. (1998). *Accounting in Portugal European Accounting Guide*, 3rd ed., edited by Alexander, D., Archer, S., 798–877 (Harcourt Brace & Company Professional Publishing, San Diego).

Gonçalves da Silva, F. and Pereira, J. (1998). *Contabilidade das Sociedades* (Plátano Editora, Lisboa).

Jorge, J. (1992). Consolidação fiscal e consolidação contabilística – críticas e sugestões. *Fisco*, **43/44**, 62–8.

Lopes, C. (1997). *Aspectos Contabilísticos e Fiscais da Consolidação de Contas*, (Rei dos Livros, Lisboa).

Lousa, M. (1994). O regime de tributação pelo lucro consolidado. *Fiscália*, **11**, 4–11. See also Lousa, M. (1995). *Eurocontas*, **6**, 1–9.

Pinheiro Pinto, J. (1996). Contabilização de investimentos financeiros pelo método da equivalência patrimonial. *Revista de Contabilidade e Comércio*, **211**, Vol. LIII, 301–18.

Pinto, J. (1997). Transposição para escudos das demonstrações financeiras expressas em outra moeda. *Revista de Contabilidade e Finanças*, **6**, II série, 5–7.
Vieira dos Reis, J. (1991). *A Consolidação de Contas* (Rei dos Livros, Lisboa).
Vieira dos Reis, J. (1994a). O trespasse. *Jornal do Técnico de Contas e da Empresa*, **342**, 69–72.
Vieira dos Reis, J. (1994b). Tratamento contabilístico da diferença de aquisição (remanescente). *Fisco*, **63**, 27–32.
Xavier, C. (1993). Coligação de sociedades comerciais. *Revista da Ordem dos Advogados*, **III**, 575–607.

Legal texts for group accounting

Decreto-Lei n° 496/76, de 26 de Junho, D.R. n.°148, I Série, p.p 1420(1)–(10), approval of the statute of the holding company of State owned companies (Instituto das Participações do Estado).

Decreto-Lei n.° 47/77, de 7 de Fevereiro, D.R. n.° 31, I Série, 2° Suplemento, pp. 200(6)–(53), approval of the first official accounting plan and creation of the accounting standardization committee.

Decreto-Lei n.° 455/78, de 30 de Dezembro, D.R. n.° 299, I Série, 13° Suplemento, pp. 2798(241)–(276), approval of the accounting plan to the banking system.

Decreto-Lei n.° 262/86, de 2 de Setembro, D.R. n.° 201, I Série-A, pp. 2293–2385, approval of the companies business code.

Decreto-Lei n.° 403/86, de 3 de Dezembro, D.R. n.° 278, I Série, pp. 3623–3638, approval of the commercial registry code.

Decreto-Lei n.° 235/87, de 12 de Junho, D.R. n.° 134, I Série, pp. 2299–2301, enactment of Directive 82/121/CEE, of February, 15, about information to be published by companies with listed securities.

Decreto-Lei n.° 414/87, de 31 de Dezembro, D.R. n.° 300, I Série-A, 5° Suplemento, pp. 4440(244)–(245), first regulation about taxation based on group income.

Decreto-Lei n.° 410/89, de 21 de Novembro, D.R. n.° 268, I Série, 1° Suplemento, pp. 5112(2)–(32), approval of the official accounting plan in line with the EU Fourth Directive.

Decreto-Lei n.° 142-A/91, de 10 de Abril, D.R. n.° 83, I[a] Série-A, 1° Suplemento, pp. 1918(2)-(118), approval of the stock market code and creation of the stock market supervisory body.

Decreto-Lei n.° 238/91, de 02 de Julho, D.R. n.° 149, I[a] Série-A, pp. 3364–89, amendment to the official accounting plan and enactment of the EU Seventh Directive.

Declaração de rectificação n.° 236-A/91, de 31 de Outubro, D.R. n.° 251, I Serie-A, pp. 5610(10), correction of *Decreto-Lei n.° 238/91.*

Decreto-Lei n.° 36/92, de 28 de Março, D.R. n.° 74, I[a] Série-A, pp. 1482–84, definition of the scope of consolidation for banks and other financial institutions.

Instruções Técnicas sobre consolidação de contas das instituições financeiras, Anexo à Circular do Banco de Portugal, 21 de Abril de 1992, procedures for consolidated accounting of financial institutions.

Decreto-Lei n.° 422-A/93, de 30 de Dezembro, D.R. n.° 303, I Série-A, 2° Suplemento, pp. 7240(6)–7240(32), regulation of the auditing profession in line with the EU Eighth Directive.

Decreto-Lei n.° 147/94, 25 de Maio, D.R., n.° 121, I[a] Série-A pp. 2766–69, definition of the scope of consolidation for insurance companies.

Rectificação 86/94, de 30 de Junho, D.R. n.° 149, I[a] Série-A, 3° Suplemento, pp. 3442(5), correction of *Decreto-Lei n.*° 147/94.

Decreto-Lei n.° 127/95, de 01 de Junho, D.R. n.° 127, I[a] Série-A, pp. 3456–57, enactment of EEC/90/604 and EEC/90/605, Directives of November 8.

Decreto-Lei n.° 265/95, de 17 de Outubro, D.R. n.° 240, I Série-A, pp. 6442–6450, creation of the official body of the accounting profession.

Decreto-Lei n.° 328/95, de 09 de Dezembro, D.R. n.° 283, I[a] Série-A, pp. 7692–96, amendments to companies business code and commercial registry code

Norma Regulamentar n.º 31/95-R, de 28 de Dezembro de 1995, procedures for group accounts of insurance companies.

Decreto-Lei n.º 257/96, de 31 de Dezembro, D.R. n.º 302, Iª Série-A, pp. 4702–4710, amendment to companies business code and to commercial registry code.

Norma 6/96, de 5 de Março de 1996, Instituto de Seguros de Portugal, rules for group accounting of insurance companies.

Portaria n.º 1222/97, de 12 de Dezembro de 1997, D.R. n.º286, Iº Série-B, p. 6578, obligation of companies with listed securities to inform quarterly about their activity, results and economic and financial situation.

Decreto-Lei n.º 44/99, de 12 de Fevereiro, D.R. n.º 36/99, Iª Série-A. pp. 4702–4710, amendment of the official accounting plan.

Directrizes contabilísticas, CNC
Recomendações técnicas and *Interpretações técnicas*, CROC
Instructions, CMVM

Taxation based on the income of the group

Decreto-Lei n.º 414/87, de 31 de Dezembro, D.R. n.º 300, I Série-A, 5º Suplemento, pp. 4440(244)–(245), first regulation of taxation based on group income (revoked).

Decreto-Lei n.º 442-B/88, de 30 de Novembro, D.R. n.º 277, I Série, 2º Suplemento, pp. 4754(38)–(71), approval of companies income tax code.

Circular n.º 4/90, de 9 de Janeiro, da DGCI regulation of taxation based on group income (revoked).

Decreto-Lei n.º 377/90, de 30 de Novembro, D.R. n.º 277, I Série, pp. 4948(2)–(4), amendment to the companies income tax.

Decreto-Lei n.º 251-A/91, de 16 de Julho, D.R. n.º 161, I Série-A, 1º Suplemento, pp. 3628(2)–(6), amendment to the companies income tax code.

Lei n.º 71/93, de 26 de Novembro, D.R. n.º 277, I Série-A, 1º Suplemento, pp. 6664(2)–(4), contains amendments to the companies income tax code.

Circular n.º 15/94, de 6 de Maio, da DGCI, regulation of taxation based on group income.

Lei n.º 71/98, de 26 de Novembro, D.R. n.º 277, I Série-A, 1º Suplemento, pp. 6664(2)–(4), contains amendments to the companies income tax code.

SPAIN
INDIVIDUAL ACCOUNTS

Antonio López Díaz
Pedro Rivero Torre

Acknowledgements

We gratefully acknowledge the help of Elena Fernández, Javier De Andrés, Ignacio Martínez del Barrio and Mariano Cabellos Velasco.

CONTENTS

I Introduction and background

1 Standards and standard setting

Spanish accounting regulation is eminently public, since financial statements are drawn up and presented essentially in line with statutory accounting standards which are set forth in legislative instruments of varying status – Law (*Ley*), Royal Decree (*Real Decreto,* RD), Ministerial Order (*Orden Ministerial*), Resolution (*Resolución*) or Circular (*Circular*). The standards are drawn up by various government bodies. They are published in the Official Gazette (*Boletín Oficial del Estado,* BOE) and in the Official Gazette of the Accounting and Audit Institute (*Boletín Oficial del Instituto de Contabilidad y Auditoria de Cuentas,* BOICAC) in the case of resolutions issued by the institute. The following bodies are currently empowered to issue accounting regulations:

- The Parliament and Council of Ministers (in the form of laws and Royal Decrees).
- The Ministry of Economic and Financial Affairs.
- The Accounting and Audit Institute (*Instituto de Contabilidad y Auditoria de Cuentas*, ICAC).

Additionally, in association with the ICAC, the Bank of Spain (*Banco de España*), the National Securities and Exchange Commission (*Comisión Nacional del Mercado de Valores,* CNMV) and the Directorate General of Insurance (*Dirección General de Seguros*) may regulate the accounting systems of the entities and groups under their surveillance.

Members of the accounting profession (at the academic, private professional association, individual professional accountant and auditor level) participate significantly in the preparation of the mandatory accounting standards and in the ongoing development, criticism and updating thereof.

2 Historical development of standard setting

2.1 Prior to 1990

Financial accounting standard setting by the government was inaugurated in 1973 with the approval under Decree 530/1973 of the National Chart of Accounts (*Plan General de Contabilidad,* PGC 1973). Whilst the chart is modelled on the French chart of accounts, the commission which drafted it (*Comisión Central de Planificación Contable*) took particular note of the work of the EU's Group of Accounting Experts.

The 1973 chart of accounts was the basic instrument for standardizing accounting in Spain and made a sweeping contribution to bringing accounting practices up to date. Initially optional, its use was gradually but progressively made mandatory for companies which decided to avail themselves of later laws allowing the revaluation of accounts. However, the chart of accounts was adopted voluntarily by most businesses. Arising out of the 1973 chart, in 1974 a chart of accounts was approved for small and medium-sized enterprises, but it was not as well received as had been expected.

The PGC 1973 was not the only effort to set accounting standards. Special mention should be made of the work of the Spanish Accounting Planning Institute (*Instituto de Planificación Contable*) and of professional bodies, notably the Accounting and Business Administration Association (*Asociación Española de Contabilidad y Administración de Empresas*, AECA).

The Accounting Planning Institute was established by Royal Decree in 1976 in order to update, refine and encourage accounting planning and to secure Spain's full integration into the worldwide standard setting movement. The Institute's numerous achievements include the adaptation of the chart of accounts to various sectors of industry, and the issuing of recommendations for the consolidation of the annual accounts of corporate groups as well as the accounting treatment of value-added tax (*impuesto sobre el valor añadido*, IVA).

In parallel with the accounting standards issued by the government, private-sector standard setting arrived with the formation in 1979 of the

AECA. This professional body set about filling the gaps left by the government's mandatory accounting standards, issuing a number of principles and standards which have been widely accepted by the professional associations of auditors.

2.2 Since 1 January 1990

Spain's accession to the EU on 1 January 1986 made it essential to adapt the system of commercial regulation to Union requirements on company law. This process affected the legal status of companies and their economic and accounting framework, substantially modifying the legal regime governing businesses' accounting practice. The Reform Law (Law 19/1989) implemented into national legislation the First, Second, Third, Fourth, Sixth and Seventh EC Directives. The Eighth Directive was enacted through the Audit Law (Law 19/1988), which was subsequently implemented by the related regulations, approved on 20 December 1990 by Royal Decree 1636/1990 (see Table I.1).

But the process of internationalizing accounting in Spain has not stopped at mere adaptation of the directives, rather it is currently continuing. From 1995 onwards, by means of the additional second disposition in the Law of Limited Liability Companies (Law 2/95) which introduced changes into the Corporations Law, annual and consolidated accounts can be published in ECUs as well as in pesetas indicating the exchange rate in the notes to the accounts, which would be that prevailing on the day of closing the balance. The introduction of the euro is also going to have beneficial effects on the international comparability of financial statements drawn up in Spain. On 15 September 1997 the Accounting and Audit Institute (ICAC) resolved to create a working party to draw up a report about aspects of accounting derived from the introduction of the euro. This working party has drawn up a report which will be used in drawing up the final rules to facilitate the introduction of the euro in accounting from 1 January 1999. These rules regulate, among other things, the valuation of certain asset items as a result of the introduction of the euro (exchange rate differences in states participating in the euro, exchange rate contracts or transactions of participating states' currencies, investment and expenses derived from the introduction of the euro) and the information that the enterprises should include in their annual accounts in relation to this subject matter.

Spain is currently considering allowing organizations that are quoted in secondary markets to present their accounts in line with International Accounting Standards, making an exception to

Table I.1 Harmonization of Spanish law with EU directives and further regulatory instruments

EU Directive	Spanish harmonizing legislation	Regulatory implementation
Eighth (84/253)	Audit Law 19/1988	Regulations implementing Law 19/1988 (Royal Decree 1636/1990)
First (69/151) Second (77/91) Third (78/855) Fourth (78/660) Sixth (82/891) Seventh (83/349)	Law 19/1989 on the partial reform and adaptation of commercial law to EU company law directives[1]	Corporations law, as amended (Legislative Royal Decree 1564/1989) Commercial Register Regulations (Royal Decree 1784/1996) National Chart of Accounts (PGC) (Royal Decree 1643/1990) Regulations for the preparation of the consolidated annual accounts (Royal Decree 1815/1991)

[1] Legislation amended by Law 19/1989:
- Commercial Code (*Código de Comercio*), and the regulations therein relating to limited partnerships with shares
- Corporations Law (*Ley de Sociedades Anónimas*)
- Limited Liability Companies Law (*Ley de Sociedades de Responsabilidad Limitada*)
- Law relating to Employee-owned Companies (*Ley de Sociedades Anónimas Laborales*)

the application of national rules. Even though the rules have yet to be drawn up, work is currently being done on their preparation.

3 The Accounting and Audit Institute (ICAC)

The Accounting and Audit Institute (*Instituto de Contabilidad y Auditoria de Cuentas*, ICAC) was established by Law 19/1988 as an independent arm of the government reporting to the Ministry of Economic and Financial Affairs; into it was merged the now defunct Accounting Planning Institute (*Instituto de Planificación Contable*). Accordingly, the new institute assumed the standard setting and planning powers of the old one. Besides being responsible for accounting matters it is also the regulatory body for auditing.

The accounting-related functions assigned to the new body were set forth systematically in Royal Decree 302/1989, which approved the institute's articles and functions:

- To undertake technical studies and draw up proposals for a chart of accounts conforming to European directives and the law relating to these matters, and to approve modified versions of the chart of accounts designed to meet the needs of particular sectors of trade and industry.
- Establishing criteria for the inclusion of items in the chart of accounts to that end and for adapting it to such sectors as may be deemed necessary to facilitate the working of the regulations. Such criteria will be published in the institute's official gazette.
- Continuing fine-tuning and updating of accounting strategy and auditing, for which purpose the institute will advise the Ministry of Economic and Financial Affairs of changes needed in laws or regulations to harmonize them with EU legislation or with advances in accounting and auditing standards.
- To undertake or encourage such research, investigation, documentation and publication activities as may be necessary to implement and fine-tune accounting and auditing standards.
- Co-ordination and technical co-operation in the area of accounting and auditing with international bodies, particularly the EU, and with Spanish public law corporations or associations dedicated to research. In discharging these functions, with the approval of the appropriate agency of the Ministry of Foreign Affairs, representatives of the institute will attend the meetings held by commissions or working parties specializing in these matters of which Spain is a member and which represent international governmental organizations.

The Accounting and Audit Institute first assumed its regulatory responsibilities with the implementation of the new legal framework established as a result of the accounting reforms, preparing and drafting the new version of the chart of accounts, approved on 20 December by Royal Decree 1643/1990 (PGC 1990; see IV.3) and the regulations for the preparation of consolidated annual accounts, approved on 20 December by Royal Decree 1815/1991 (*Normas para la formulacíon de las cuentas anuales consolidadas*; see Spain – Group Accounts).

In addition, Royal Decree 1643/1990 in its final provisions empowered the Ministry of Economic and Financial Affairs and the ICAC:

> to issue obligatory regulations implementing the Spanish national chart of accounts and adaptations thereof to business sectors in terms of valuation methods and the standards governing the preparation of annual accounts. These regulations will be issued in the form of Resolutions by the President of the Institute and will be published in the Official Gazette (BOE) and in the Official Gazette of the Institute itself (BOICAC).

Since the Royal Decree came into force the Institute has made use of the powers conferred on it to issue a number of new rules – most of them extending the valuation methods in the chart of accounts (PGC 1990) – and other sundry implementations, particularly the adaptation of the chart of accounts to certain business sectors (see IV.3.3).

Additionally, upon the recommendation of the institute, the Ministry of Economic and Financial Affairs may, by issuing a Ministerial Order:

- Approve versions of the chart of accounts adapted to the needs of particular business sectors.

- Amend the method of amortizing research and development expenses prescribed in the chart of accounts.
- Adapt the valuation methods and the standards governing the preparation of the annual accounts to the specific situation of the reporting entity.

4 Private sector standard setting

Standard setting by the accounting profession is closely linked with the formation and growth of the Spanish Accounting and Business Administration Association (AECA), established in 1979 with the aim of contributing to the development of accounting and business administration. To date it has established four commissions to deal with the following core areas of research:

- Accounting principles and standards.
- Company valuation.
- Management accounting.
- Organization and methods.

The association follows a policy of involving in the drafting of its documents not merely the issuers but also the auditors and the users of financial information.

To date the association has issued 20 documents on accounting principles and standards (see Table I.2). The first, Accounting Principles and Standards in Spain (*Principios y normas de contabilidad en España*), defines a conceptual mini-framework which includes notes on the economic climate, and the characteristics of and requirements governing accounting information. The other documents deal with accounting standards, applicable to the preparation of annual accounts, on specific questions arising from the general principles.

In short, the association made up for the shortcomings of the government-issued accounting regulations prior to the recent reforms by issuing standards which are generally accepted by the profession and, therefore, introduced a dynamism which was lacking in the legal framework of accounting. These professional standards are now

Table I.2 Acounting principles and standards issued by the Spanish Accounting and Business Administration Association (AECA)

Statement No.	Title
1	Accounting principles and standards in Spain
2	Tangible fixed assets
3	Intangibles and deferred charges
4	Exchange differences
5	Suppliers, creditors and other Accounts payable
6	Trade accounts receivable, debtors and other accounts receivable
7	Accrual accounts
8	Inventories
9	Income tax
10	Shareholders' equity
11	Provisions, contingencies and events subsequent to the date of the financial statements
12	Deferred income
13	Revenues
14	Provision for the return of assets to the State
15	Financial investments
16	Pension funds
17	Expenses
18	Financial liabilities
19	Futures and options on inventory
20	Cash flow statement

All titles are published in Spanish only.

the essential complement of the legislation governing accounting and are a reference framework for recording transactions or situations not expressly envisaged in the legislation (see III.1.2.1, III.1.2.3).

5 Other government regulatory bodies

In addition to the Accounting and Audit Institute, other agencies have specific functions with regard to the accounting and policing of certain types of business. Chief among these are the Directorate General of Insurance (*Dirección General de Seguros*), the Bank of Spain (*Banco de España*) and the National Securities and Exchange Commission (*Comisión Nacional del Mercado de Valores*, CNMV).

5.1 Directorate General of Insurance

Through the Directorate General of Insurance the Ministry of Economic and Financial Affairs supervises the insurance industry, insurance advertising, and the financial position and solvency of insurance entities. The Directorate General of Insurance has decision-making powers, conferred upon by RD 1348/1985, except in matters reserved to the Minister or the government.

Law 13/1992, regulating the equity and supervision on a consolidated basis of finance entities (which amended various articles of the Private Insurance Law, Law 33/1984), authorized the Minister of Economic and Financial Affairs, on the recommendation of the Directorate General of Insurance and after consultation with the Accounting and Audit Institute (ICAC), to:

- Prescribe or amend the accounting standards, valuation methods and format required for the annual accounts of insurance entities.
- Exercise identical powers with regard to 'consolidable groups of insurance entities', as defined in Law 13/1992, Art. 25.4.

Additionally, at the instance of the Directorate General of Insurance, the Ministry determines the frequency with which the information is to be supplied and the form in which it is to be presented (see Table I.3).

5.2 Bank of Spain

By Order dated 31 March 1989 of the Ministry of Economic and Financial Affairs, the Bank of Spain was granted the power to:

- Issue and modify in line with relevant EC Directives (86/635 and 89/117) the accounting standards and the formats for the balance sheet,

Table I.3 Accounting provisions regarding insurance entities

Ministry of Economic and Financial Affairs Order of 28 December 1992 regarding the valuation of investments by insurance entities in fixed-income marketable securities.
Directorate General of Insurance resolution of 31 December 1992, publishing the internal rate of return referred to in valuation method 4 of the Order of 28 December 1992, regarding the valuation of investments by insurance entities in fixed-income marketable securities.
Resolution of 1 December 1994, of the Accounting and Audit Institute, for which the technical auditing rules of drawing up the special and complementary report to the annual account audit of insurance companies, requested by the General Directorate of Insurance.
Resolution of 25 June 1997, of the General Intervention of the Social Security, which establishes the accounting procedures for the registration of mutual funds for accidents at work and professional illnesses, of transactions relating to debts with the Social Security which have been the object of payment postponement or fractionation, or for those whose payment collection is under executive procedural action.
Royal Decree 2014/1997, of 26 December, which approves the Accounting Plan for insurance organizations and the rules for formulating the accounts of groups of insurance organizations.

Published in Spanish only.

- profit and loss account and other financial statements of banks and other financial institutions, and for the consolidated financial statements envisaged in Law 13/1985.
- Establish the frequency and level of detail with which the related data must be supplied to the regulatory authorities and be generally made public by the credit entities themselves.

However, before prescribing or modifying the format of the published balance sheet and profit and loss account or prescribing or modifying valuation methods, the Bank must first consult the Accounting and Audit Institute (ICAC).

The Bank of Spain discharges this duty by issuing circulars with which banks and other financial institutions subject to its regulation are required to comply (see Table I.4).

5.3 National Securities and Exchange Commission (CNMV)

One of the cornerstones of the reform of the securities market undertaken in 1988 (Law 24/1988) was the establishment of the Securities and Exchange Commission (*Comisión Nacional del Mercado de Valores*, CNMV) as the agency responsible for supervising and monitoring the securities markets.

The commission's numerous functions include, in the accounting sphere, the power (with the prior express authority of the Minister of Economic and Financial Affairs) to regulate the accounting regime of stock exchange management companies, securities companies and agencies, the securities settlement service, members of the stock exchanges, collective investment institutions, portfolio management companies, management companies of official futures and options markets and collective investment institution management companies.

These powers were expressly conferred upon the commission by the Ministry of Economic and Financial Affairs through various ministerial orders which authorize the commission to:

- Prescribe or amend the accounting standards and the format of the annual accounts of the aforementioned companies.
- Establish the frequency and level of detail with which the relevant data must be supplied to the commission and generally made public.

Table I.4 Circulars issued by the Bank of Spain

4/1991	Regarding accounting standards and financial statement formats for credit entities
7/1991	Regarding accounting standards for credit entities
4/1992	Amending circular 4/1991
7/1992	Regarding accounting standards and financial statement formats for credit entities
9/1992	Regarding the published balance sheets of credit entities
14/1992	Regulating the market in inter-bank deposits
18/1992	Regarding credit entities' fixed income security portfolio write-downs
4/1993	Amending the accounting standards for credit entities in circular 4/1991
5/1993	Setting and controlling the minimum equity requirements for credit entities
3/1994	Amending circular 4/1991
6/1994	Amending circular 4/1991
2/1995	Amending circular 4/1991
2/1996	Amending circular 4/1991
7/1996	Amending circular 4/1991
1/1997	Deposit Guarantee Fund provision accounting
3/1997	Regarding minimum shareholders' equity
5/1997	Amending circular 4/1991
3/1998	Regarding information to be provided to the Bank of Spain by approved valuation companies and services
7/1998	Amending circular 4/1991

Published in Spanish only.

Table I.5 Circulars issued by the National Securities and Exchange Commission

1/1990	Regarding the accounting standards, formats of confidential and published financial statements, published annual accounts and the audit of securities market management companies.
3/1990	Regarding the accounting standards, formats of confidential and published financial statements, formats of supplementary financial statements, published annual accounts and the audit of securities companies.
5/1990	Regarding the accounting standards, formats of confidential and published financial statements and published annual accounts of securities companies and agencies.
7/1990	Regarding the accounting standards and formats of the confidential financial statements of collective investment institutions.
8/1990	Regarding the determination of the liquidation value of investment funds and the operating ratios and investment ceilings of collective investment institutions.
1/1992	Amending the formats for periodical reporting by credit entities whose securities are listed.
2/1992	Regarding the accounting standards, formats of confidential and published financial statements, and the published statistics and financial reporting of official futures and option market management companies.
4/1992	Regarding the accounting standards, formats of confidential and published financial statements, formats of supplementary financial statements and published annual accounts of the securities settlement service.
5/1992	Regarding the accounting standards and confidential financial statements of collective investment institution management companies and portfolio management companies.
6/1992	Regarding the equity requirements for securities companies and agencies and their consolidable groups.
1/1993	Regarding the accounting standards for consolidation applicable to groups and subgroups of securities companies and agencies.
2/1993	On the information regarding foreign collective investment institutions which have registered the prospectus offering their participations or shares with the National Securities and Exchange Commission.
4/1993	Regarding the accounting standards for investment funds.
3/1994	On the information that listed companies must submit to the National Securities and Exchange Commission.
4/1994	Regarding the accounting standards for stock investment funds and collective investment companies.
1/1995	Regarding the accounting standards for brokerage houses.
2/1996	On the information that listed financial institutions must submit to the National Securities and Exchange Commission.
1/1997	On stockholders' equity of the brokerage houses.
9/1997	On relevant facts and their disclosure.
2/1998	On the valuation of the underwritings and reimbursements of the shares in investment funds and daily publishing liquidating value.
4/1998	On the agreement between the Stock Exchange Councils and the National Securities and Exchange Commission to simplify the disclosure of relevant equity holdings, relevant facts, quotation on the Stock Exchange, and submission of the annual accounts and audit report of listed companies.
7/1998	On the advertising of collective investment institutions.

Published in Spanish only.

- Exercise similar powers with regard to the content and public dissemination of the audits of the said companies.

The commission must first consult the Accounting and Audit Institute (ICAC) before prescribing or amending the format of the published financial statements, prescribing or amending valuation methods and generally issuing any other provision specifically affecting the audit of financial statements. Thus the institute participates indirectly in the preparation of the standards governing companies involved in the securities markets.

Since its formation the commission has exercised these powers by issuing circulars whose requirements are binding upon the companies subject to its regulation (see Table I.5).

II Forms of business organization

1 Most commonly encountered legal forms of business organization

1.1 General partnership

General partnerships are governed by the Commercial Code (CCom Arts. 125–44). A general partnership (*sociedad colectiva*, SC) is a firm in which all the partners, unless it is agreed otherwise, participate collectively in the management of the business and contribute personally to achieving its ends. The rights and obligations of partners include the following in particular:

- Participation in the profits, which are divided in proportion to the partner's contribution, unless a different arrangement is built into the partnership agreement. Losses are distributed in the same proportion. Moreover, partners are entitled to receive a share of the assets of the firm upon liquidation.
- Partners may examine the administration statement and the accounts.
- Partners may not conduct outside the partnership transactions which properly pertain to it.
- Partners must indemnify the partnership for any loss due to malice, abuse of powers or gross negligence on the individual's part, if so required by the other partners.
- Partners, whether or not they are managing partners, are subsidiarily jointly and severally liable on a personal basis, with their entire assets, for any liabilities arising from the firm's activities.

The administrators must prepare annual accounts, as required generally of all businesses by CCom Art. 34.1, within three months of the financial year end.

Pursuant to the Commercial Code, once the annual accounts have been approved, the profit will be distributed as set forth by the partners in the partnership agreement or, in the absence of any such provision, in proportion to each partner's participation in the capital. Any surplus will be taken to reserves (either mandatory reserves, if so provided in the partnership agreement, or voluntary reserves, if all the partners so resolve).

1.2 Limited partnership

Limited partnerships are governed by CCom Arts. 145–57. They may be ordinary limited partnerships (*sociedad comanditaria simple*, SCom), if the capital is not divided into shares, or limited partnerships with shares (*sociedad comanditaria por acciones*, SComPA) where the capital is so divided. The partnership will consist of general partners, who contribute assets and/or are active in the business, and limited partners, who only contribute assets. The general partners are liable for the firm's debts under the same terms as in a general partnership, whereas the limited partners are liable only to the extent of their contribution to the capital.

In an ordinary limited partnership, unless otherwise agreed, limited partners have an equal claim with the general partners to participate in the firm's profits and to a share of the assets on liquidation. However, unlike general partners, limited partners may not take part in the management of the firm nor examine the administration statement or the accounts unless the partnership agreement provides otherwise.

In limited partnerships with shares, limited partners have the same right to information as shareholders of corporations (*sociedad anónima*, SA), since the law relating to corporations is applicable to such firms except where it clashes with CCom Arts. 151–7, governing limited partnerships with shares.

In ordinary limited partnerships the result is determined and distributed as in a general partnership. In the case of limited partnerships with shares, the result is determined and distributed as set forth in the legislation governing corporations.

1.3 Limited liability company

Limited liability companies (*sociedad de responsabilidad limitada*, SRL or SL) are governed by the Law of 23 March 1995, which has replaced the law of 17 July 1953. The company is constituted with a set amount of capital, which may not be less than ESP 500,000, made up of the shareholders' contributions and divided into equal indivisible participations which may not be included in marketable securities nor called shares, but which confer on their holders the same rights as in the case of a corporation (SA). The capital must be fully subscribed and paid up from the beginning. Participants are liable for the company's debts only to the extent of their contribution.

The conduct of the company's business is decided by a majority of the capital. However, the articles may waive the requirement of such a majority for the adoption of certain resolutions. A meeting of participants is required only if the latter number more than 15 or if the articles so require. Otherwise, resolutions may be adopted by any means which guarantees that the declared will of the participants is accurately reflected.

The management of the company will be able to be left to a sole administrator, various administrators who act jointly, or to a management board. In the case of a management board, either the company statutes or the partners meeting at a General Meeting, will set the minimum and maximum number of management board members without them being in any such case less than 3 or more than 12.

Any participant is entitled to examine the company's financial statements pursuant to the public deed of incorporation. Participants are entitled to share in the distributable income in proportion to their holding in the company, any agreement to the contrary being null and void.

1.4 Corporation

A corporation (*sociedad anónima*, SA) is governed by the Corporations Law (LSA), approved by Royal Decree 1564/1989. Its capital is divided into shares, the possession of which legitimates the bearer as a shareholder and confers the corresponding rights recognized in the law and in the articles. There may be ordinary shares, preference shares and non-voting shares. The capital may not be less than ESP 10 million, which must be fully subscribed, with at least 25 per cent of the par value of every share paid up at the date of incorporation. Shareholders are not personally liable for the corporation's debts, since their liability is limited to the amount of their contribution to the capital.

The shareholders' meeting is the sovereign decision-making body in a corporation, and its resolutions express the will of the company. Resolutions are adopted by majority vote, and the meeting's sovereignty is limited by the law, the articles and the interests of the company. The shareholders' meeting has the power to:

- Approve the conduct of business.
- Approve the annual accounts.
- Resolve the distribution of profit.
- Resolve modifications of the capital stock, the issue of debentures, transformations, mergers, spin-offs and any other amendment of the articles.
- Appoint or remove directors.
- Appoint auditors if the company is required to have its annual accounts audited.
- In the resolution to distribute dividends, determine the time and form of payment.

The directors are the persons who manage and represent the company at all times. When the administration of the company is assigned collectively to two or more persons, the latter constitute the board of directors. The functions of the board of directors include the following:

- Administer the company and represent it in and outside court.
- Call and attend the shareholders' meeting.
- Prepare the annual accounts, the management report, the proposed distribution of income and, where appropriate, the consolidated annual accounts and management report.
- Register corporate resolutions in the Commercial Register.

As stated above, the directors are responsible for preparing the annual accounts, the management report and the proposed distribution of income and, where appropriate, the consolidated annual accounts and management report, within three months of the financial year end.

The shareholders' meeting must take place

during the first six months of each year, to pass judgement on the conduct of business, approve, where appropriate, the accounts of the previous year, and resolve the distribution of the profit (see V.5.2.3).

2 Relative importance of the different legal forms of business organization

The new legislative framework established following the commercial reform favoured the creation of limited liability companies at the expense of corporations, since, as indicated above, the requirements for the constitution of a corporation are much tougher than those demanded for the constitution of a limited company, especially with respect to minimum capital. From 1990 onwards the limited company is therefore much more frequent than the corporation, especially among small and medium-sized businesses which predominate in the Spanish economy. This situation accounts for the figures shown in Table II.1 and gives an idea of the dominance of small and medium-sized companies. (Note that limited liability companies must have at least ESP 500,000 of capital, with no maximum limit, as compared with the ESP 50 million maximum stipulated by the former legislation). Table II.2 gives further data with respect to commercial companies classified by legal form respectively.

The Bank of Spain's Central Balance Sheet Office (*Central de Balances*) began publication in 1982, and its data is drawn from non-financial companies which voluntarily supply information to the Bank of Spain at the latter's request (see Table II.3). The Central Balance Sheet Office participates in the BACH (Bank for the Accounts of Companies Harmonized) project, which also involves several European countries, Japan, and the United States.

The sample of companies is a unique set, with certain biases:

- There is a preponderance of large companies. Thus, in 1996, over 75 per cent of the gross added value reflected in the database was earned by 505 companies with over 500 employees.
- There is a preponderance of state-owned companies, which represented 34.6 per cent of the gross added value in 1996.
- There is a preponderance of permanent employees, accounting for 79.5 per cent of all employees in 1991.
- Business activities are not equally represented. Only the following industries are well represented: manufacture of transport material (66.4 per cent); transport, storage and communications (58.9 per cent); energy and water (98.7 per cent); chemical products (51 per cent); transport and communications (58.9 per cent).

3 Stock corporations

It is noteworthy that the economy as a whole is not very well represented on the various stock exchanges. The Spanish stock market is characterized by high sectoral concentration, banks, electricity utilities and *telefónica* (communications) having a disproportionate presence by comparison with the other sectors, as evidenced in Tables II.4 and II.5.

Most share trading in Spain is currently carried out through the Spanish Stock Exchange Interconnection System (SIBE), a real-time electronic transaction system which allows the four Spanish stock exchanges (Madrid, Barcelona, Bilbao and Valencia) to send their orders via computer terminals to the same central computer. The introduction of this system has led to a single price for each share on the four Spanish markets since 1995. Of the four stock exchanges, the most important is Madrid, which handles most of the trading, followed by Barcelona, Bilbao, and Valencia. Shares have the most important weighting. The market capitalization of shares on the Madrid Stock Exchange in 1997 was over ESP 44,260,000 million, whereas fixed-income securities amounted to ESP 3,369,000 million (Bolsa de Madrid, 1998).

Table II.6 shows those companies which were listed on one or more of the main foreign stock exchanges in 1998.

Table II.1 Total number of commercial companies formed, 1994–96, classified by business activity

Sector		1994 No.	1995 No.	1995 Capital	1996 No.	1996 Capital
(a)	Agriculture, livestock, hunting, forestry and fishing	2,110	2,919	25,671	2,387	21,580
(b)	Energy and water	118	167	1,702	142	848
(c)	Extraction industry	294	305	1,379	223	1,734
(d)	Manufacturing industry	16,611	18,563	118,848	13,495	103,541
(e)	Energy and water	139	190	4,512	121	2,227
(f)	Construction	16,111	19,473	99,610	13,919	83,879
(g)	Shops, repairs	27,629	25,698	84,988	23,992	83,331
(h)	Restaurants and hotels	6,136	6,171	15,791	6,404	14,633
(i)	Transport, storing and communications	4,462	4,181	17,580	3,954	14,353
(j)	Financial institutions and insurance	840	704	22,703	788	42,997
(k)	Services to companies and rentals	21,321	21,947	163,468	22,831	230,992
(l)	Public administration, defence and obligatory social security	2	0	0	0	0
(m)	Education	1,093	1,731	2,919	1,293	2,340
(n)	Health and veterinary activities, social work	1,075	1,155	3,225	1,167	3,733
(o)	Others	3,480	3,453	10,072	3,106	6,970
Total		**101,421**	**106,657**	**572,468**	**93,811**	**613,160**

Note: Capital in ESP million.
Source: Statistics on commercial companies prepared by the National Statistics Institute from information in the mercantile registers. *Instituto Nacional de Estadistica, Monthly Statistics Bulletin*, July 1998.

Table II.2 Total number of commercial companies established in the period 1993–97, classified by their legal form

Year	Total No.	Total Capital	Stock corporations No.	Stock corporations Capital	Limited liability companies No.	Limited liability companies Capital	Other companies No.	Other companies Capital
1993	83,077	602,600	3,919	287,200	79,053	314,526	105	874
1994	101,421	554,489	4,009	238,526	97,365	315,018	47	945
1995	106,657	572,468	3,655	213,552	102,967	358,852	35	64
1996	93,811	613,160	3,357	265,737	90,434	347,137	20	285
1997	91,248	842,981	3,685	376,466	87,531	465,119	32	1,396

Note: Capital in ESP million.
Source: Statistics on commercial companies prepared by the National Statistics Institute from information in the mercantile registers. *Instituto Nacional de Estadistica, Monthly Statistics Bulletin*, July 1998.

Table II.3 Classification by size and nature of enterprises that co-operate with the Bank of Spain's Central Balance Sheet Data Office

	Enterprises		Gross added value		Turnover	
SIZE	Number	%	ESP billion	%	ESP billion	%
Small	5,873	75.7%	905.8	7.7%	3,706.0	9.6%
Medium	1,381	17.8%	1,865.1	15.8%	7,024.4	18.3%
Large	505	6.5%	9,059.8	76.6%	27,707.5	72.1%
TOTAL	7,759	100%	11,830.7	100%	38,438.0	100%
NATURE						
Public	420	5.4%	4,091.7	34.6%	8,196.0	21.3%
Private	7,339	94.6%	7,739.0	65.4%	30,242.0	78.7%

Source: Bank of Spain's Central Balance Sheet Data Office. *Resultados Anuales de las Empresas no Financieras. Noviembre 1997.*

Table II.4 Total number of companies included in the Madrid Stock Exchange index

	1992	1993	1994	1995	1996	1997	1998
Banking and finance	13	14	19	19	17	16	15
Electricity, gas and water	10	11	11	11	11	11	11
Food, beverages and tobacco	9	10	10	12	11	10	13
Construction	14	14	13	16	16	13	14
Portfolio/investment	6	5	5	6	6	5	6
Communications	7	9	6	6	7	7	7
Engineering	14	14	13	12	11	13	16
Petroleum and chemical	11	10	10	10	11	13	16
Other industries and services	10	9	11	15	16	22	24
TOTAL	94	96	98	107	106	110	122

Source: *Bolsa de Madrid,* http://www.bolsamadrid.es

Table II.5 Relative importance of the various sectors included in the Madrid Stock Exchange index

	1992	1993	1994	1995	1996	1997	1998
Banking and finance	29.83	29.08	30.29	31.65	31.06	28.99	34.19
Electricity, gas and water	20.54	23.64	24.64	21.55	26.48	28.87	22.56
Food, beverages and tobacco	5.28	3.75	2.90	3.12	2.94	2.80	3.10
Construction	7.86	6.62	7.03	7.52	5.25	4.37	6.11
Portfolio/investment	5.41	3.85	3.03	2.91	2.74	2.48	2.37
Communications	12.15	13.05	13.69	12.07	11.83	14.33	14.01
Engineering	3.47	3.12	3.15	3.64	2.93	2.93	3.61
Petroleum and chemical	10.25	11.42	10.53	10.19	9.68	8.61	7.98
Other industries and services	6.11	5.47	4.74	7.35	7.09	6.62	6.07

Source: *Bolsa de Madrid,* http://www.bolsamadrid.es

Table II.6 Companies listed on the main foreign stock exchanges in 1998

Company	New York	Frankfurt	London	Paris	Tokyo
Argentaria	X		X		
Banco Bilbao Vizcaya	X	X	X		X
Banco Central Hispano	X	X		X	X
Banco Popular				X	
Banco de Santander	X	X		X	X
Cepsa		X			
Dragados		X			
Endesa	X	X	X		
Ercros		X			
Fasa Renault		X			
Iberdrola		X	X		
Inburesa		X			
Pryca			X		
Repsol	X	X	X		
Tabacalera		X			
Telefónica	X	X	X	X	X
Unión Fenosa			X		

Source: *Expansión* (daily newspaper), 28 September 1998.

III Objectives, concepts and general principles

1 Objectives and principles of financial reporting

1.1 Objectives of financial reporting

The main users of accounting information on Spanish companies have traditionally been the tax authorities and, to a lesser degree, banks, creditors and shareholders. The significant influence of tax regulations on accounting in Spain and the high dependence of Spanish companies on bank financing as compared with the scant influence of the stock exchanges, helped to encourage an equity view of accounting, focusing on shareholder and creditor equity protection, to the detriment of the information content of financial reports. This situation particularly hindered the adoption by Spanish companies of the principles and standards that were beginning to be applied in Spain, either through the requirements of the 1973 chart of accounts (PGC 1973), or by those who sought to standardize Spanish practice in line with generally accepted international accounting principles.

However, the pace of economic growth in recent years, the increasingly responsible attitude of Spanish companies towards accounting and financial information reporting, and the advent of legal regulation in 1989–90 on such important aspects as disclosure and, in certain cases, the audit of annual accounts, enable us today to refer to genuine accounting regulations which are fully independent of tax regulations, and which are aimed at making the economic and financial information supplied by companies useful for economic decision-making by the various groups or echelons with an interest in the progress of a company's business.

It should be borne in mind that the regulations are legalistic and, accordingly, only governmental provisions command the authority to require compliance. However, in the formulation of the standards considerable attention was paid to the views and suggestions of academics, practitioners, representative auditors, the business sectors involved, and in general all users of the financial information.

The Commercial Code, the cornerstone of accounting law, expressly states that 'the annual accounts must be drafted clearly and give a true and fair view of the company's net worth, financial position and results, pursuant to the applicable provisions' (CCom Art. 34.2). The concept of a true and fair view, previously unknown in Spain, was introduced into Spanish law through the Fourth EC Directive. The goals deriving from this ambiguous term are contained principally in the Reform Law (Law 19/1989) and, therefore, incorporated in the current regulatory framework.

The introduction to the 1990 chart of accounts (PGC 1990), which is the basic instrument regulating financial reporting in Spain (see IV.3), states that companies' economic and financial reporting must be:

- Comprehensible. Within the complexity of the economic world, the information must be easily understood by the user.
- Relevant. It should include the information which is of genuine relevance to the user without adding excess information, to the detriment of the preceding characteristic.
- Reliable. So that it can meet the desired objective, the information must not contain serious errors.
- Comparable. The information must be consistent and uniform over time and among different companies.
- Up to date. The information must be produced at a time when it is useful to users and not after a significant delay.

Although these characteristics may conflict, they must be applied in order to achieve the desired balance.

1.2 Accounting principles

1.2.1 Generally accepted accounting principles

It is considered that, as a corollary of the systematic and regular application of accounting principles (understood to be the mechanism for representing the reality of the transactions recorded), the annual accounts will give a true and fair view of a company's net worth, financial position and results.

In Spain, there are basically two sources of accounting standards applicable today: legal (mandatory) and professional (generally accepted) standards.

The PGC 1990 sets out the statutory accounting principles which all companies must observe in their financial reporting, expanding on and supplementing the accounting principles initially established in the Commercial Code (CCom). The list of accounting principles in the PGC is as follows (see VI.1):

- Prudence.
- Going concern.
- Recording.
- Acquisition cost.
- Accruals.
- Matching of revenues and expenses.
- No setting off.
- Consistency.
- Materiality.

In addition, the Accounting Principles and Standards Commission of the AECA (*Comisión de Principios y Normas de Contabilidad de AECA*) is currently the only source of accounting rules accepted by nearly all audit corporations, in addition to the support it is accorded by all state-owned and private entities and organizations. Its statements were fundamental in the formulation of the PGC 1990, and were duly reviewed to ensure that their content was consistent with the new legal framework established by the accounting reforms (see I.4).

1.2.2 Hierarchy of accounting principles

The PGC 1990 also establishes a ranking of accounting principles and a set of rules to apply in cases of conflict between them:

- In the event of a conflict between mandatory accounting principles, the one which best

leads the annual accounts to give a true and fair view of the company's net worth, financial position and results should prevail.
- Without prejudice to the provision in the preceding paragraph, the principle of prudence must predominate above all others.

Thus the application of the various principles should lead to a true and fair view. However, all other factors being equal *vis-à-vis* the true and fair view, the prudence principle must predominate.

1.2.3 Flexibility in the application of accounting principles

Although the commercial law expressly sets out statutory accounting principles, it sanctions a degree of flexibility in their application and allows accounting experts to apply alternative solutions which depart from the required principles when necessary in exceptional cases to ensure a true and fair view. Part 1 of the PGC 1990 which discusses the accounting principles, states:

> When the application of the accounting principles in the said regulation is not sufficient in order for the annual accounts to express the aforementioned true and fair view, the necessary explanations must be given in the notes on the financial statements regarding the accounting principles applied.
>
> In exceptional cases where the application of an accounting principle or of any accounting standard prevents the annual accounts from presenting a true and fair view, it shall not be applied. This shall be mentioned in the notes on the financial statements, with an explanation of the reasons and and indication of the effect on the company's net worth, financial position and results.

In this connection, note 2 on the financial statements must include an explanation of the exceptional reasons for which, in order to convey a true and fair view, the legal rules on accounting have not been applied, as well as information on the optional principles which may have been applied, indicating the effect on the company's net worth, financial position and results.

1.3 Valuation methods

The commercial law relating to accounting regulates in detail the process of preparing annual accounts, both in its quantitative aspects, with a broad enumeration of accounting principles and standards and valuation methods, and in its qualitative aspects, establishing general rules for preparing the annual accounts and prescribing formats for them. The valuation methods formulated in the legal framework for accounting are regarded as the implementation of generally accepted accounting principles, and embody the criteria or rules to be applied to particular transactions or events and to the various items comprising the annual accounts (assets and liabilities, expenses and revenues).

The philosophy for applying valuation methods must necessarily be the same as that prescribed by the accounting regulations for accounting principles, since the valuation methods are just an implementation of the accounting principles, i.e. the application of the accounting principles and valuation standards is subordinate to the achievement of a true and fair view.

The valuation methods set out in the accounting regulations are based particularly on the principle of prudence. It should also be noted that the acquisition cost principle predominates in practice and that only write-downs (and not write-ups) are permissible.

2 Relationship between financial reporting and tax accounting

2.1 Traditional linkage between financial and tax accounts

Spanish legislation has traditionally aimed to harmonize accounting and tax criteria, although this does not mean that valuations for accounting purposes must be accepted indiscriminately for tax purposes. Art. 9.2 of the General Taxation Law (Law 25/1995) establishes that tax law is independent of other laws, and that, in the tax area, non-tax legislation is merely supplementary. An identical interpretation is derived from Law 43/1995 Art. 10.2 (Regulations implementing the

Corporate Income Tax). The reason for this autonomy lies in the disparity between the goals pursued by financial accounting and tax legislation, in the aims of tax policy, in the flexibility of the accounting regulations and, fundamentally, in the primacy in the accounting sphere of the principle of prudence in valuation. Accordingly, there is a selective division between commercial and tax regulations, and commercial regulations are effective taxwise but are subordinate to the tax precepts and principles.

Accounting is the basic means by which the profit taxable under corporate income tax is calculated. In this connection, the Corporate Income Tax Law (Law 43/1995) requires that, in calculating the profit for corporate income tax purposes, revenues and expenses are to be reckoned at their financial accounting values, provided the accounts at all times reflect the company's true net worth (Law 43/1995 Art. 10.3). However, profit per books and taxable profit may not coincide, owing to divergences in the criteria for the classification, valuation and allocation in time of revenues and expenses established by commercial and tax legislation. Any differences between the profit per books and the taxable profit for corporate income tax purposes must therefore be rectified through non-accounting adjustments to adapt the said profit to the taxable profit. Accordingly:

	Profit per books
±	Non-accounting adjustments
=	Taxable profit for corporate income tax purposes

However, the aforementioned principle is largely ignored in the Corporate Income Tax Regulations, which list an excessively detailed series of accounting requirements, and the reality is that tax rules on accounting have traditionally dominated business accounting in Spain. Even after the approval of the PGC 1973, companies continued to apply mainly accounting criteria established for tax purposes. This situation is due basically to the tendency to establish accounting regulations from a tax viewpoint, as indicated above, and, on the other hand, the absence for many years of commercial legislation substantially focused on the preparation of companies' external accounting information. Consequently the business attitude to establish accounting information has traditionally been dominated by criteria established by the tax authorities, in the firm belief that the accounts were drawn up solely in order to comply with tax regulations.

2.2 Interrelationships arising from the 1989–90 reforms

The Reform Law (Law 19/1989) introduced a significant change in the status of accounting regulations with regard to tax regulations and established a clear separation between the two, with the emphasis on co-ordination and harmonization between them but maintaining their mutual independence. The characteristics of the relationship between the tax and accounting spheres can be summarized in the following points:

- Non-interference of tax regulations in the accounting sphere. The seventh Final Provision of Royal Decree 1643/1990, which gave the royal assent to the national chart of accounts, expressly states that accounting law is independent of tax law, which had traditionally taken precedence in Spain. Consequently, companies' annual accounts must always furnish purely accounting information or strictly business information, free of outside influences (including tax), so as to present a true and fair view of the company's net worth, financial position and results.
- Priority of tax legislation in determining the taxable profits. Tax law is autonomous and, consequently, taxable profit for the purpose of the various taxes is determined pursuant to tax legislation. As a result, the aforementioned revocation by the chart of accounts affects only the tax provisions regulating accounting matters, not those regulating tax matters. That is, it does not in any way affect the application of the tax regulations regarding classification, valuation and recognition established for the various taxes and, in particular, those for determining taxable income for corporate income tax purposes.
- Harmonization of accounting and tax criteria. The accounting sphere is governed by accounting regulations, free from interference

from tax regulations, and the tax sphere is governed by tax regulations without prejudice to the supplementary nature of the accounting regulations. As a result, taxable income for corporate income tax purposes, which is a tax concept calculated pursuant to tax regulations regarding classification, valuation and recognition of revenues and expenses, may not be the same as the balance of the profit and loss account, which is an accounting concept calculated pursuant to the criteria established in the commercial legislation contained in the national chart of accounts. Thus the reconciliation of the profit per books with the taxable profit requires the appropriate non-accounting adjustments to be made and, where applicable, the related recording in the financial statements of the pre-payment or deferral of taxes.

However, corporate income tax does not involve a radical separation between the taxable income and book income. Rather, the tax is linked with the accounting, as manifested fundamentally in the principle of recording in the accounts (*principio de inscripción contable*). This principle establishes that an essential requirement for expenses to be tax deductible is that they must have been recorded previously for accounting purposes; with regard to revenues, it establishes that they may not be computed for tax purposes at a date subsequent to that on which they were recorded for accounting purposes (Law 43/1995 Art. 19), without prejudice to the exceptions which the law may allow.

Another nexus linking the book income and taxable income lies in the notes on the financial statements. Note 15 must state any difference between the calculation of the income of the period and that which would have arisen had items been valued pursuant to tax criteria owing to differences between the latter and the statutory accounting principles.

IV Bookkeeping and preparation of financial statements

1 Duty to prepare financial statements

1.1 Preparation of annual accounts

The Commercial Code (CCom) makes it a requirement for all businesses, without distinction, to prepare annual accounts, comprising the balance sheet, profit and loss account and the notes. These documents constitute a reporting unit and must be prepared by the entrepreneur or the directors within three months of the financial year end.

Additionally, the Corporations Law (LSA) obliges companies which limit the liability of their shareholders (i.e. corporations, employee-owned companies, limited liability companies and limited partnerships with shares) to file within the aforementioned period a management report, a proposed distribution of income and, where appropriate, the consolidated annual accounts and management report.

As regards the structure of the annual accounts, two formats are available:

- The standard format is applicable to capital companies, i.e. corporations (*sociedad anónima*, SA) including employee-owned companies (*sociedad anónima laboral*, SAL), limited liability companies (*sociedad de responsabilidad limitada*, SRL) and limited partnerships with shares (*sociedad comanditaria por acciones*, SComPA), except small and medium-sized capital companies as defined in LSA Arts. 181, 190 and 201 (see Table IV.1), which may use the abridged format for the balance sheet and notes and, where appropriate, the profit and loss account.
- Companies with other legal forms and sole traders may use the abridged formats.

Only those companies obliged to prepare the balance sheet and notes using the standard format must include a statement of changes in financial position in the notes.

1.2 Interim financial statements

Companies which have issued securities admitted to listing on the stock exchange are required to file interim financial reports.

- Half-yearly reporting. A balance sheet and a profit and loss account, as well as information on the evolution of business, on net turnover and on other items, and, where appropriate, the corresponding consolidated information, must be filed for each half of the calendar year within two months of the end of that period (see IX.5.1).
- Quarterly reporting. Additionally, net turnover, the pre-tax result and other items (both individual and, where appropriate, consolidated) must be filed for the first and third quarters of the calendar year, within 45 days of the close of each period (see IX.5.2).

2 Bookkeeping

2.1 Statutory books

The formal aspects of bookkeeping are governed by CCom Arts. 25–33. Every enterprise must keep orderly accounts appropriate to its business which enable all its transactions to be monitored in chronological sequence and enable balance sheets and inventories to be drawn up periodically. Without prejudice to what may be required by law or in special provisions, every enterprise must maintain an inventories and annual accounts book and a journal.

Commercial companies must also keep a minute book or books, showing all the resolutions adopted by the general and special shareholders' meetings and other corporate bodies. These resolutions must be filed by the directors in the Commercial Register within eight days of approval of the minutes. Limited liability companies must keep a partner register book, and corporations and limited partnerships with shares must keep a registered shares book pursuant to the requirements of the Commercial Code for the other books. These books can be prepared using EDP applications. The books must be kept by the entrepreneur personally or by persons duly authorized for the purpose, without prejudice to the former's liability. Authorization will be deemed to have been given unless otherwise proved.

Regardless of the procedure used, all books of account and accounting documents must be kept clearly, in chronological order, without blank spaces, interpolations, crossing out or erasures. Any errors or omissions in the accounting entries must be remedied as soon as they are detected. Abbreviations or symbols whose meaning is not clear pursuant to the law, the regulations or generally accepted commercial practice may not be used.

The books which enterprises keep voluntarily are implicitly regulated under the freedom of accounting procedures indicated above. Such books may likewise be legalized by the commercial registrar.

Under Corporate Income Tax Regulations Law (Law 43/1995) and the related Regulations (RD

Table IV.1 Obligatory books and accounts for commercial companies

Commercial legislation	Corporate income tax legislation	VAT (IVA) legislation
Journal	Journal	Register book of invoices issued
Inventories and annual accounts book	Inventories and annual accounts book	Register book of invoices received
Minute book	Ledger	Capital investments register book
Register of participants	Register book of purchases and expenses	Register book of certain intra-EU transactions
Share register	Register book of sales and revenues Cash book	

537/1997), companies must keep the general books required by the Commercial Code, together with a ledger and the ancillary accounting records shown in Table IV.1. The various ancillary accounting records for each year must be kept separately, and the opening and closing balances must coincide with those shown in the related annual accounts.

As Art. 54 LIS indicates, permanent establishments will be obliged to keep separate accounts that refer to the transactions they carry out and the asset items affected. They will also be obliged to comply with other obligations of an accounting nature, registration or formalities applicable to tax-payers as personal obligation.

Also, value-added tax (*Impuesto sobre el valor añadido*, IVA) regulations (Law 37/1992 and the related regulations approved by RD 1624/1992) require tax payers to keep the accounting records listed in Table IV.1. The register book of invoices issued must contain the separate transactions subject to value-added tax, including tax-exempt and internal transactions. The register book of invoices received is to contain all invoices and customs documents relating to goods and services acquired or imported, including those for the company's own use. The capital investments register book must make accurate identification possible of all the invoices and customs documents relating to capital investments registered in the said register book for the purpose of calculating capital investment value-added tax deductions. Finally, a register book of certain intra-EU transactions is to be kept, where appropriate, pursuant to value-added tax regulations.

Both corporate income tax and value-added tax regulations permit these records to be kept on EDP applications, provided that the print-out sheets are subsequently numbered consecutively and bound for the purpose of presenting the required register books.

2.2 Formal requirements for the books and annual accounts

The annual accounts must indicate the date as of which they were prepared, and they must be signed by the sole trader, by all the partners, in the case of a partnership, or by the directors, in the case of a corporation or limited liability company. If any of the required signatures is missing, the reason must be indicated on all documents lacking the signature.

The balance sheet, profit and loss account and notes must be identified, each clearly bearing its title, the name of the company to which it refers and the period to which it relates. The annual accounts must be expressed in pesetas. However, figures may be expressed in thousands or millions of pesetas when their magnitude warrants it, in which case the fact must be indicated in the annual accounts.

Additionally, the opening balance of one year must coincide with the closing balance of the previous year.

Undertakings must legalize the books they are obliged to keep pursuant to the Commercial Code. This may be done:

- *A priori*. The undertaking presents the books which it is legally required to keep to its local commercial registry so that, before they are used, the first page of each book is endorsed with a certificate indicating the number of pages and the remaining pages are marked with the registry's stamp.
- *A posteriori*. Entries or postings may be made by any suitable method on loose sheets (e.g. EDP applications), which must later be bound in order to make up the legal books, and these must be legalized within four months of the financial year end.

Enterprises must retain the books, correspondence, documentation and supporting paperwork relating to their business, in due order, for six years from the date of the last entry in the books, unless otherwise specified by general or special provisions. Pursuant to the general rules of law, enterprises' books and other accounting documents have probative value in court.

3 The national chart of accounts

3.1 Guidelines and structure

The new chart of accounts (PGC 1990), approved by Royal Decree 1643/1990, established the basic technical framework for accounting standard setting in Spain. The PGC is structured as follows:

- Accounting principles.
- Chart of accounts.
- Definitions and accounting equations.
- Annual accounts.
- Valuation methods.

The first part expands on, systematizes and supplements the generally accepted accounting principles incorporated in the Commercial Code. It also establishes an order of priority for accounting principles and gives some guidelines for resolving situations of conflict between principles (see III.1.2).

The second part is the chart of accounts proper, which consists of seven groups of accounts, numbered 1–7, subclassified using decimal notation. The first five groups are balance sheet accounts and the remaining two are expense and revenue accounts.

The third part deals with definitions and accounting equations. Each of the groups, subgroups and accounts into which the chart of accounts is broken down is defined, describing its content and the salient characteristics of the transactions comprising it.

The fourth part of the chart of accounts discusses the annual accounts, giving the statutory rules for preparing these documents and the required formats (standard and abridged) which all businesses must use (see V).

The fifth part comprises the valuation methods (see VI and VII), which are ordered in the same sequences as the asset and liability items, revenues and expenses to which they refer. The valuation methods are viewed as an implementation of the accounting principles contained in the first section (see III.1.3).

The chart of accounts (PGC) is mandatory for all enterprises, regardless of their legal form and size. Specifically, all its contents relating to the rules for preparing and presenting financial reporting – namely parts 1 (accounting principles), 4 (annual accounts) and 5 (valuation methods) – are mandatory.

Another feature of the PGC is that it must be applied with a reasonable margin of flexibility, and not as a rigid formulation, to be followed to the letter. In short, it is a flexible instrument in the hands of companies, and in this connection businesses may use numbers, names and variations in the accounts other than those set forth in PGC parts 2–3, provided the final result of their accounting entries is the preparation of annual accounts in the formats prescribed by the chart of accounts and that the various items are valued in conformity with the valuation methods contained therein. If the terminology, etc. of parts 2–3 are not followed, similar account names should be used in order to facilitate the preparation of the annual accounts, for which there are prescribed rules governing structure, content and presentation.

The chart of accounts is conceived of as purely an accounting document, free of interferences from other spheres (including taxation).

3.2 Adaptations of the chart of accounts to different sectors of business

The first Final Provision of Royal Decree 1643/1990, which approved the PGC 1990, authorized the Ministry of Economic and Financial Affairs, at the instance of the Accounting and Audit Institute (ICAC), to approve, by ministerial order, versions of the PGC 1990 adjusted to the needs of individual business sectors.

In recent years the ICAC has been working hard at adapting the PGC to the characteristics of different sectors of the economy. Up to now, diverse sector plans have been approved and these are indicated in Table IV.2. In addition, various commissions have been set up and are currently working on drafts of plans that will be approved in the future.

3.3 Resolutions of the Accounting and Audit Institute implementing the chart of accounts

The fifth Final Provision of Royal Decree 1643/1990 authorized the Accounting and Audit Institute (ICAC) to approve, by resolution, mandatory regulations implementing the PGC and its sectoral versions in terms of valuation methods and rules for preparing annual accounts. The ICAC has issued the resolutions listed in Table IV.3 which implement, clarify or expand the contents of the chart of accounts.

In its bulletin (BOICAC) the Institute has also published rulings in response to queries which it

Table IV.2 Sectoral versions of the 1990 chart of accounts (PGC)

Construction companies, by means of Ministerial Order from the Economics and Treasury Ministry (MO of 27 January 1993).
Sporting federations (MO of 2 February 1994).
Real estate agencies (MO of 28 December 1994).
Sporting corporations (MO of 23 June 1995).
Health centres (MO of 23 December 1996).
Insurance companies (Royal Decree 2014/1997, of 26 December).
Electric Utility Companies (Royal Decree 437/1998, of 20 March).
Charitable organizations (Royal Decree 776/1998, of 30 April).

Published in Spanish only.

Table IV.3 Resolutions of the Accounting and Audit Institute (ICAC)

Resolution of 15 January 1991, establishing a working party to formulate regulations on the recording of mergers and spin-offs.
Resolution of 16 May 1991, setting general criteria for determining net turnover.
Resolution of 30 July 1991, issuing valuation methods for tangible fixed assets.
Resolution of 25 September 1991, setting criteria for recording pre-paid taxes in respect of provision for pensions and similar obligations.
Resolution of 21 January 1992, issuing valuation methods for intangibles.
Resolution of 30 April 1992, regarding certain aspects of valuation method No. 16 of the Spanish national chart of accounts.
Resolution of 27 July 1992, issuing valuation methods for participations in capital arising from non-monetary contributions to the constitution or increase of capital of companies.
Resolution of 21 January 1992, regarding the criteria for recording participations in money market asset investment funds (*fondos de inversión en activos del mercado monetario*, FIAMM)
Resolution of 16 December 1992, implementing certain criteria to be applied in valuing and recording the Canary Islands General Indirect Tax.
Resolution of 8 February 1992, establishing a working party on the recording of financial options and futures.
Resolution of 20 December 1996 regarding the setting of general criteria to determine the concept of net worth with respect to capital reductions and dissolution of companies.
Resolution of 20 January 1997 which deals with the accounting treatment of the special regimes established for VAT and the Canary Islands General Indirect Tax.
Resolution of 9 October 1997 regarding some aspects of the sixteenth valuation rule in the General Accounting Plan (PGC 1990) (accounting for Corporate Tax).
Report on accounting aspects of adapting to the euro (conclusions of the working party appointed in resolution of 15 September 1997).

Published in Spanish only.

considers to be of general interest. Although such rulings do not have administrative status, i.e. are not binding, they are guidelines for businesses which encounter the situations to which the rulings refer.

V Balance sheet and profit and loss account formats

1 Regulations governing the annual accounts

CCom Art. 34.1 and LSA Art. 172 establish that the annual accounts comprise the balance sheet, the profit and loss account and the notes, which together constitute a reporting unit. The information each of these contains must in every respect match that contained in the others. Whereas the Commercial Code (CCom) merely lays down very general rules as to the composition of the annual accounts, the Corporations Law (LSA), which applies also to limited liability companies (SRL), employee-owned companies (SAL) and limited partnerships with shares (SComPA) gives a detailed structure of accounts, in line with that envisaged and accepted by the Fourth Directive. At all events, CCom Art. 35.7 and the first Final Provision of the Corporations Law (LSA) both refer to a future regulatory enactment to determine the final content of such accounts.

The aforementioned regulatory enactment took place through Royal Decree 1643/1990, which approved the chart of accounts. PGC part 4 entitled 'Annual Accounts' implements and expands the rules in commercial law relating to the preparation of annual accounts. It also prescribes specific formats for the balance sheet and profit and loss account which are applicable to Spanish businesses in general, and details the minimum contents on the notes on the financial statements.

2 General classification principles

The annual account formats in the PGC are mandatory for all Spanish enterprises, regardless of their legal form, without prejudice to the provisions of sectoral versions of the chart of accounts or to legislation specifically relating to certain enterprises. The annual accounts must, at all events, be drawn up clearly and must present a true and fair view of the enterprise's net worth, financial position and results (CCom Art. 34.2).

CCom Art. 35.6 includes a list of the general rules for preparing the annual accounts which are mandatory for all enterprises:

- Prior year figures and comparability. Each item of the balance sheet, profit and loss account and statement of changes in financial position must show, in addition to the current year's figures, those for the previous year. For this purpose, when the figures are not comparable, either because the layout of the document has changed, or there has been a change in recognition methods, the previous year's figures must be adjusted for presentation alongside those of the current year.
- The balance sheet and profit and loss account must not contain items which have no balance, unless they had a balance in the prior year.
- No set-offs. No setting off is allowed between asset and liability items or between revenue and expense items.
- The structure of the balance sheet and the profit and loss account may not be changed from year to year, save in exceptional circumstances, which must be duly described and justified in the notes.
- Additionally, the chart of accounts states expressly that, following the principle of consistency, the accounting criteria (i.e. valuation rules) may not be changed from one year to the next save in exceptional circumstances, which must be duly described and justified in the notes, and must meet the criteria authorized in the chart of accounts.

3 Format and legal form, size, business sector, stock exchange listing

3.1 General formats and sectoral versions

Spanish regulations prescribe a single balance sheet layout in the form of an account. With some minor variations, this format coincides with that envisaged in the Fourth EU Directive, Art. 9 and is the format traditionally used in Spain.

Spain opted for only one of the four profit and loss account formats offered by the Fourth Directive and chose the account format with expenses classified by type. This is the format envisaged in the Fourth EU Directive, Art. 24, and is essentially the one traditionally used in Spain.

Current commercial legislation also makes it mandatory to prepare notes on the annual accounts in order to complete, expand upon and discuss the information contained in the balance sheet and profit and loss account. The format envisaged for the notes in the chart of accounts indicates the minimum content, which should be supplemented with any other information required to facilitate comprehension of the annual accounts.

PGC 1990 establishes two levels of detail for the formats:

- The standard formats, which must be used for the annual accounts of corporations, employee-owned companies, limited liability companies and limited partnerships with shares.
- The abridged formats, in which the items are more aggregated and the reporting requirements less stringent. The abridged annual account formats may be used by sole traders, general partnerships and ordinary limited partnerships, as well as small and medium-sized capital companies as defined in LSA Arts. 181, 190 and 201, i.e. corporations, employee-owned companies, limited liability companies and limited partnerships with shares, if in two consecutive years the latter do not exceed at least two of the three thresholds shown in Table V.1.

Under the current commercial legislation the structure, nomenclature and terminology of the annual accounts may be adapted to the characteristics and nature of the business of certain sectors, provided the Ministry of Economic and Financial Affairs sanctions such modification at the recommendation of the Accounting and Audit Institute (ICAC) (see IV.3.3).

3.2 Special legislation

Banks and other financial institutions – Bank of Spain circular No. 4/1991 sets out the accounting standards and formats which in line with relevant EC Directives (86/635 and 89/117) must be used for the annual accounts of banks and other financial institutions, the type and frequency of reporting, and the confidential and public documents which the following entities must file:

- Deposit-taking entities (banks, state-owned financial institutions and credit co-operatives).
- Finance entities.
- Leasing companies.
- Mortgage loan companies.
- Money market intermediary companies.

Institutions in the securities markets – institutions in the securities market which are regulated by the National Securities and Exchange Commission (CNMV) must prepare their financial statements in accordance with the specific accounting standards and forms of presentation established by that regulatory body in its circulars (see Table I.5).

Interim reporting by listed companies – a Finance Ministry order of 18 January 1991 specified the nature of the information which entities whose securities are listed on the stock exchange must disclose every quarter and half-year, and distinguished between different types of entity on the basis of their business activity. Such information must be prepared, in line with the instructions and formats in annexes I and II of the ministerial order, according to the business activity carried out by the company (see IX.5).

4 Basis of presentation of the annual accounts

Comparability – in order to harmonize corporate reporting, the general rules for preparing annual accounts, expressed in the Commercial Code and other mercantile legislation, must be observed (see V.2).

The notes must include (note 2) the following information regarding the presentation of the annual accounts:

- Where, in exceptional circumstances, the structure of the balance sheet or profit and loss account is modified pursuant to CCom Art. 36, this fact must be indicated and justified in the notes.

- Where it is necessary to adjust the previous year's figures to make them comparable (see V.2), or if they cannot be made comparable, the fact must be indicated and explained in the notes.

Distinction between short-term and long-term items – items are classified as current or non-current, and short or long-term, depending on whether the period envisaged for maturity, disposal or cancellation is less than or more than 12 months, regardless of the initially agreed conditions.

Inter-company transactions – for the purposes of presenting annual accounts, groups of companies, associated companies and multigroup companies are defined (see also Spain – Group Accounts) as follows:

- Groups of companies. For the purposes of presenting the annual accounts of a company, any other company will be deemed to form part of the group if it is linked by a direct or indirect relationship of control, similar to that envisaged in CCom Art. 42.1 as regards groups of companies or companies directly or indirectly controlled by the same entity or individual.
- Associated company. Though not a group company as defined above, one or more group companies or the controlling entity or individual may exercise a strong influence over a company. For these purposes, such a strong influence will be deemed to exist where the associated company is 20 per cent owned (or 3 per cent owned if it is listed).
- Multigroup companies. The items relating to associated companies must also include those relating to multigroup companies, defined as companies which are managed jointly by a group company or the controlling entity or individual and by one or more third parties outside the group.

Addition, subdivision and grouping of items – items may be added for which no allowance is made in the standard or abridged forms of the balance sheet and profit and loss account established in the chart of accounts, provided their content is not already covered by existing items. Additionally, items may be further subdivided, so long as the structure of the chart of accounts is not affected. Balance sheet and profit and loss account items denoted by arabic numerals may be grouped if their amount is not material to a true and fair view or if the aim is to improve clarity. Items which are grouped in the balance sheet and profit and loss account must be broken down in the notes (note 2). Likewise, asset and liability items which appear in two or more balance sheet items must be identified, together with the amount shown under each item.

5 Standard balance sheet format

5.1 Balance sheet structure

Balance sheet items are ordered in terms of the liquidity of the assets (from lower to higher) and of the claimability of the liabilities (from lower to higher). Thus the structure of the assets and liabilities is based on a classification in terms of maturity (current and short-term versus fixed and long-term).

The areas or sections into which the assets and liabilities are grouped as outlined above are

Table V.1 Size criteria for applying the abridged formats

Criterion	Advanced balance sheet and notes	Abridged profit and loss account
Total assets (million ESP)	395	1,580
Turnover (million ESP)	790	3,160
Average number of employees[1]	50	250

[1] In accordance with the chart of accounts, all the persons that have or have had an employment relationship with the company during the year should be taken into account and given a weighting on the basis of hours worked for the purpose of calculating the average head account.

further divided into subsections based on the nature of the items. There are separate sections for 'Financial investments', 'Accounts receivable', 'Long-term debt' and 'Current liabilities', broken down in terms of group companies, associated companies and others (see Tables V.2(a) and V.2(b)).

Lastly, in order to include in the balance sheet adjustments to the value of asset items, the assets are shown in the balance sheet at full acquisition or production cost, and the accumulated amortizations (systematic depreciation) and related provisions (reversible write-downs) for the whole class of items (e.g. tangible fixed assets) are recorded separately on the assets side of the balance sheet as negative items (the net values technique).

The total amount of rights to assets acquired under financial lease transactions must be recorded on the asset side under a separate item. For this purpose an item must be created in section B II of the asset side, called 'Rights on leased assets'. The debts relating to these transactions must be recorded under a separate head, for which purpose the items 'Long-term lease payments payable' and 'Short-term lease payments payable' must be created in sections D II and E II, respectively, on the liabilities side of the balance sheet (see VI.3.5.5–6).

If the enterprise holds own shares for amortization which were acquired pursuant to a resolution to reduce capital adopted by the shareholders' meeting, section A VIII, entitled 'Own shares for capital reduction', must be created on the liability side. This item, whose balance will always be negative, will reduce the balance of equity.

Financial investments maturing at not more than one year must be recorded under heading D IV, 'Short-term financial investments'.

Unpaid capital on shares constituting permanent financial investments which has not yet been called but which, pursuant to LSA Art. 42 of the Corporations Law, is claimable at short notice must be recorded under heading E V 3 on the liability side of the balance sheet.

Borrowings maturing at not more than one year must be recorded in section E of the liability side under 'Current liabilities'.

If there are provisions for liabilities and expenses with maturities of not more than one year, section F of the liability side, entitled 'Short-term provisions for liabilities and expenses', must be created.

To record trade accounts receivable which mature at more than one year, item BVI, 'Long-term trade accounts receivable' must be added on the asset side, and must be suitably broken down.

To record trade accounts payable which mature at more than one year, item D VI 'Long-term trade accounts payable' must be added on the liability side, and must be suitably broken down.

5.2 Equity items

5.2.1 Presentation of equity items in the balance sheet

Shareholders' investment includes the following items:

- Subscribed capital stock.
- Share premium account.
- Revaluation reserve.
- Reserves (from income).
- Profit and loss brought forward.
- Income or loss for the year.
- Interim dividend paid during the year.

The prior year losses and interim dividend paid during the year must be recorded with a minus sign under shareholders' investment in order to show the actual net amount of equity. Likewise, the amount of own shares acquired by the company pursuant to a resolution to reduce capital adopted by the Shareholders' Meeting (LSA Art. 170) must be recorded under liabilities in the balance sheet as a reduction in equity.

The profit or loss for the financial year must be recorded as a separate equity item, with a plus or minus sign, depending on whether it is an increase (income) or decrease (loss) of net worth.

However, capital subscribed but not paid and own shares acquired by the company for subsequent sale are not recorded as negative items in the net worth but are explicitly recorded in sections A and B V, respectively (or D I and D V, as appropriate) on the asset side of the balance sheet.

5.2.2 Rules according to which equity items may be set up and released

The capital stock item arises when a commercial company is formed and its first balance is always the amount appearing in the deed of incorporation. With regard to the materialization of these

Table V.2(a) Layout of the assets side of the standard balance sheet

Assets	Activo
A Subscribed capital unpaid and uncalled	A Accionistas (socios) por desembolsos no exigidos
B Fixed and other non-current assests	B Inmovilizado
I Formation expenses	I Gastos de establecimiento
II Intangibles	II Inmovilizaciones inmateriales
1 Research and development expenses	1 Gastos de investigación y desarrollo
2 Concessions, patents, licenses, trademarks, etc.	2 Concesiones, patentes, licencias, marcas y similares
3 Goodwill	3 Fondo de comercio
4 Transfer rights	4 Derechos de traspaso
5 EDP applications	5 Aplicaciones informáticas
6 Advances	6 Anticipos
7 Provisions	7 Provisiones
8 Amortization	8 Amortizaciones
III Tangible fixed assets	III Inmovilizaciones materiales
1 Land and structures	1 Terrenos y construcciones
2 Technical installations and machinery	2 Instalaciones técnicas y maquinaria
3 Other installations, tools and furniture	3 Otras instalaciones, utillaje y mobiliario
4 Advances and construction in progress	4 Anticipos e inmovilizaciones materiales en curso
5 Other tangible fixed assets	5 Otro inmovilizado
6 Provisions	6 Provisiones
7 Amortization	7 Amortizaciones
IV Financial fixed assets	IV Inmovilizaciones financieras
1 Holdings in group companies	1 Participaciones en empresas del grupo
2 Loans to group companies	2 Créditos a empresas del grupo
3 Holdings in associated companies	3 Participaciones en empresas asociadas
4 Loans to associated companies	4 Créditos a empresas asociadas
5 Long-term investment securities	5 Cartera de valores a largo plazo
6 Other loans	6 Otros créditos
7 Long-term deposits and guarantees	7 Depósitos y fianzas constituidos a largo plazo
8 Provisions	8 Provisiones
V Own shares	V Acciones propias
C Deferred charges	C Gastos a distribuir en varios ejercicios
D Current assets	D Activo circulante
I Subscribed capital called but not paid	I Accionistas por desembolsos exigidos
II Inventories	II Existencias
1 Commercial inventories	1 Comerciales
2 Raw materials and other supplies	2 Materias primas y otros aprovisionamientos
3 Work in progress and semi-finished goods	3 Productos en curso y semiterminados
4 Finished goods	4 Productos terminados
5 By-products, wastage and recovered materials	5 Subproductos, residuos y materiales recuperados

Table V.2(a) (contd.)

Assets	Activo
6 Advances	6 Anticipos
7 Provisions	7 Provisiones
III Accounts receivable	III Deudores
1 Customer receivables for sales and services	1 Clientes por ventas y prestaciones de servicios
2 Receivables from group companies	2 Empresas del grupo, deudores
3 Receivables from associated companies	3 Empresas asociadas, deudores
4 Sundry accounts receivable	4 Deudores varios
5 Employee receivables	5 Personal
6 Tax receivables	6 Administraciones públicas
7 Provisions	7 Provisiones
IV Short-term financial investments	IV Inversiones financieras temporales
1 Holdings in group companies	1 Participaciones en empresas del grupo
2 Loans to group companies	2 Créditos a empresas del grupo
3 Holdings in associated companies	3 Participaciones en empresas asociadas
4 Loans to associated companies	4 Créditos a empresas asociadas
5 Short-term investment securities	5 Cartera de valores a corto plazo
6 Other loans	6 Otros créditos
7 Short-term deposits and guarantees	7 Depósitos y fianzas constituidos a corto plazo
8 Provisions	8 Provisiones
V Short-term own shares	V Acciones propias a corto plazo
VI Cash at bank and in hand	VI Tesorería
VII Pre-payments and accrued income	VII Ajustes por periodificación
Total (A+B+C+D)	Total general (A+B+C+D)

funds under assets, LSA Art. 36 refers only to monetary and non-monetary contributions, and does not envisage the possibility of the labour being treated as a contribution to equity.

Capital stock may be represented by ordinary shares, preference shares, non-voting shares or shares with restricted rights. Non-voting shares merit special mention, since their par value may not exceed 50 per cent of the disbursed capital. However, they grant their holders alternative rights, including the preferential right, over all other shares, to a 5 per cent minimum dividend. Capital increases must be resolved by the shareholders' meeting (LSA Art. 152) and may take place through new contributions by shareholders, the offsetting of loans against the company, the transformation of reserves or profits into capital, the conversion of debentures, the absorption of another company or the capitalization of capital gains. Capital may be reduced to refund shareholders' contributions, remit uncalled capital, record voluntary reserves, increase legal reserves or offset losses (LSA Art. 163). In a capital reduction it is necessary to analyze how it will affect the guarantees of creditors and third parties.

The share premium is a contribution of funds by the subscription of shares at above their par value. LSA Art. 47.3 states that it must be fully disbursed on subscription. This is an unrestricted reserve, that is, it can be released in the case either of capital increases and offset of losses or of distribution of dividends.

The revaluation reserve is the contra item of the restated value of assets. The values are restated pursuant to revaluation laws aimed at compensating for the effects of inflation on companies' net

Table V.2(b) Layout of the liabilities side of the standard balance sheet

Shareholders' investment and liabilities	Pasivo
A Shareholders' investment	A Fondos propios
I Subscribed capital stock	I Capital suscrito
II Share premium account	II Prima de emisión
III Revaluation reserve	III Reserva de revalorización
IV Reserves	IV Reservas
1 Legal reserve	1 Reserva legal
2 Reserves for own shares	2 Reservas para acciones propias
3 Reserves for shares of the controlling company	3 Reserva para acciones de la sociedad dominante
4 Reserves required by the articles	4 Reservas estatutarias
5 Other reserves	5 Otras reservas
V Profit and loss brought forward	V Resultados de ejercicios anteriores
1 Unallocated income	1 Remanente
2 Prior year losses	2 Resultados negativos de ejercicios anteriores
3 Shareholders' contributions to offset losses	3 Aportaciones de socios para compensación de pérdidas
VI Profit or loss for the year	VI Pérdidas y ganancias (beneficio o pérdida)
VII Interim dividend paid during the year	VII Dividendo a cuenta entregado en el ejercicio
B Deferred revenues	B Ingresos a distribuir en varios ejercicios
1 Capital subsidies	1 Subvenciones de capital
2 Exchange gains	2 Diferencias positivas de cambio
3 Other deferred revenues	3 Otros ingresos a distribuir en varios ejercicios
C Provisions for liabilities and charges	C Provisiones para riesgos y gastos
1 Provisions for pensions and similar obligations	1 Provisiones para pensiones y obligaciones similares
2 Provisions for taxes	2 Provisiones para impuestos
3 Other provisions	3 Otras provisiones
4 Reversion reserve	4 Fondos de reversión
D Long-term debt	D Acreedores a largo plazo
I Debentures and other marketable debt securities	I Emisiones de obligaciones y otros valores negociables
1 Non-convertible debentures	1 Obligaciones no convertibles
2 Convertible debentures	2 Obligaciones convertibles
3 Other marketable debt securities	3 Otras deudas representadas en valores negociables
II Payable to credit entities	II Deudas con entidades de crédito
III Payable to group and associated companies	III Deudas con empresas del grupo y asociadas
1 Payable to group companies	1 Deudas con empresas del grupo
2 Payable to associated companies	2 Deudas con empresas asociadas

Table V.2(b) (contd.)

Shareholders' investment and liabilities	Pasivo
IV Other accounts payable	IV Otros acreedores
1 Notes payable	1 Deudas representadas por efectos a pagar
2 Other accounts payable	2 Otras deudas
3 Long-term guarantees and deposits received	3 Fianzas y depósitos recibidos a largo plazo
V Uncalled capital payments payable	V Desembolsos pendientes sobre acciones no exigidos
1 Group companies	1 De empresas del grupo
2 Associated companies	2 De empresas asociadas
3 Other companies	3 De otras empresas
E Current liabilities	E Acreedores a corto plazo
I Debentures and other marketable debt securities	I Emisiones de obligaciones y otros valores negociables
1 Non-convertible debentures	1 Obligaciones no convertibles
2 Convertible debentures	2 Deudas con empresas convertibles
3 Other marketable debt securities	3 Otras deudas representadas en valores negociables
4 Interest on debentures and other securities	4 Intereses de obligaciones y otros valores
II Payable to credit entities	II Deudas con entidades de crédito
1 Loans and other accounts payable	1 Préstamos y otras deudas
2 Accrued interest payable	2 Deudas por intereses
III Short-term payables to group and associated companies	III Deudas con empresas del grupo y asociadas a corto plazo
1 Payable to group companies	1 Deudas con empresas del grupo
2 Payable to associated companies	2 Deudas con empresas asociadas
IV Trade accounts payable	IV Acreedores comerciales
1 Advances received on orders	1 Anticipos recibidos por pedidos
2 Payables for purchases and services	2 Deudas por compras o prestaciones de servicios
3 Notes payable	3 Deudas representadas por efectos a pagar
V Other non-trade payables	V Otras deudas no comerciales
1 Accrued tax payable	1 Administraciones públicas
2 Notes payable	2 Deudas representadas por efectos a pagar
3 Compensation payable	3 Otras deudas
4 Other accounts payable	4 Remuneraciones pendientes de pago
5 Short-term guarantees and deposits received	5 Fianzas y depósitos recibidos a corto plazo
VI Operating liabilities provisions	VI Provisiones para operaciones de tráfico
VII Accruals and deferred income	VII Ajustes por periodificación
Total (A+B+C+D)	Total general (A+B+C+D)

worth. They are applied as provided by the revaluation law under which the reserve was recorded (see VIII).

The legal reserve is a permanent item in corporations and is a guarantee similar to the capital stock (see V.5.2). Its availability and the reasons for which it can be released are the same as for the capital stock item.

The special reserve may be subject to the Corporations Law (it includes the Reserve for Reciprocal Holdings specified in LSA Art. 84) or to special laws affecting companies involved in particular lines of business. It is entered out of undistributed income and its availability depends on the regulations under which this reserve was made obligatory.

The reserve for shares of the parent company and the reserve for own shares are governed by LSA Arts. 75 and 79, under which these reserves must be entered at the amount at which the acquired shares are carried on the asset side. The reserves become available when the shares are written off or disposed of.

The article reserve and voluntary reserves are entered out of undistributed income by order of the company's articles or by resolution of the shareholders' meeting. The availability of the article reserve depends on the rules that govern it in the articles, whereas voluntary reserves are unrestricted, i.e. they can be released for any purpose.

The reserve for redeemed capital stock is to be entered out of unrestricted reserves or undistributed income, and cannot be released, since it constitutes a guarantee to the company's creditors which was previously represented by capital stock (LSA Art. 167).

The retained earnings constitute an unrestricted reserve generated out of income not appropriated for any other purpose and, accordingly, companies may use them for the same purpose as voluntary reserves.

The item 'shareholders' contributions to offset losses' shows contributions by the shareholders to restore the company's net worth. They may take the form of any type of asset which offsets the losses.

5.2.3 Approval of the annual accounts and distribution of income

The annual accounts of capital companies (LSA Art. 95) must be submitted to the shareholders' meeting for approval. The meeting must be held within the first six months of each year to examine the conduct of business, approve, where appropriate, the accounts of the previous year and resolve on the distribution of the profit.

Once the balance sheet has been approved, the shareholders' meeting decides on the application of the profit (LSA Art. 213), taking into account the following restrictions imposed by that law:

- Before all else, 10 per cent of income must be appropriated to the legal reserve until its balance represents at least 20 per cent of the capital stock. Until the legal reserve exceeds 20 per cent of the capital stock it can be used only to offset losses, provided that no other available reserves are sufficient for the purpose (LSA Art. 214).
- Once the appropriations required by the Law or the company's articles have been made, dividends can be distributed out of the income of the period or out of unrestricted reserves only if distribution does not leave the net worth per books less than the capital stock amount (LSA Art. 213).
- If, owing to losses in prior years, the company's net worth would be less than the capital stock amount, income of the period must be used to offset losses (LSA Art. 213).
- Until the balances of formation expenses, research and development expenses and goodwill have been fully amortized, no income may be distributed unless the unrestricted reserves are at least equal to the amount of the unamortized balances under these headings (LSA Art. 194).
- Where the capital has been reduced, a company cannot declare a dividend until its legal reserve has reached at least 10 per cent of the new capital stock amount (LSA Art. 168.4).

An interim dividend may be paid to shareholders (if so decided by the shareholders' meeting or by the directors), in which case the directors must draw up a financial statement evidencing sufficient liquidity to cater for the distribution. The amount to be distributed may not exceed the amount of income derived from the end of the previous year net of losses carried forward, appropriations to reserves required by law or by the articles and the estimated tax to be paid on such income (LSA Art. 216). The amount of the interim dividend must be disclosed in the notes (note 3) and the projection evidencing sufficient liquidity required under LSA Art. 216 must be included. This statement must cover a period of one year from the date on which the distribution of the interim dividend is approved.

Additionally, the proposed distribution of income must also be disclosed in note 3, according to the scheme shown in Table V.3.

6 Standard profit and loss account format

6.1 Structure of the profit and loss account

The profit and loss account format envisaged in the national chart of accounts is laid out as an account in which expenses and revenues are classified according to the nature of the transactions. Additionally, pursuant to CCom Art. 35.2, the result is segmented into various levels by the grouping of individual balances to yield the subtotals shown in Table V.4. The standard profit and loss account format is shown in Tables V.5(a) and V.5(b).

6.2 The analytical statement of income

The chart of accounts provides the possibility of voluntarily presenting an analytical statement of income, which must be included, where appropriate, in note 21 of the notes on the annual accounts in accordance with the format prescribed by the chart of accounts (see Table V.6(a)). However, the format adopted in the chart of accounts (PGC) is very similar to that used by the Bank of Spain's Central Balance Sheet Office, and is adapted to the format periodically required by the National Securities and Exchange Commission (CNMV) of listed companies as part of their interim financial information, and to the analytical BACH project model.

The purpose of the analytical model is to derive a series of intermediate management aggregates, on which is based the allocation of income to production factors. Certain partial income levels are also expressed. The intermediate management balances are calculated as follows:

Table V.3 Appropriation account

Basis of distribution
Profit or loss for the year
Retained earnings
Voluntary reserves
Reserves...
Total
Distribution
To legal reserve
To special reserves
To voluntary reserves
To...
To dividends
To...
To offset losses carried forward
Total

Table V.4 Segments of income

	Operating income/loss
±	Financial income/loss
=	Ordinary operating income/loss
±	Extraordinary income/loss
=	Income/loss before tax
±	Corporate income tax
=	Income/loss for the year

- Production value – the production figure is calculated by adding to net turnover other operating revenues arising from a company's secondary and ordinary activities. The unsold production or unfinished products in stock must also be included, together with production for in-house use and operating subsidies. However, it should be noted that the production figure is a heterogeneous aggregate, since it measures sales at market prices and stocks and in-house work at production cost.
- Value added – this is calculated as the difference between production value and outside

consumption. This value, together with dividend and interest income and extraordinary income, is distributed between the parties contributing to its formation: employees (staff costs), lenders (financial expenses), self-financing, the State and shareholders (provisions, corporate income tax and profit and loss account balance).

This document also makes it possible to distinguish between the analytical result levels shown in Table V.6(b).

Finally, it should be noted that the preparation of the analytical statement of income does not pose additional calculation difficulties, since the transfer of data from the statutory profit and loss account to the analytical statement of income requires no prior calculations or transactions.

6.3 Definition of major items

6.3.1 Net turnover

A resolution of the Accounting and Audit Institute (ICAC) dated 16 May 1991, establishes criteria for determining net turnover. This concept is related to the company's ordinary business, which is taken to mean that which is regularly conducted

Table V.5(a) Standard profit and loss account format (expenses)

Debit	Debe
A EXPENSES	**A GASTOS**
1 Decrease in finished product and work in progress inventories	1 Reducción de existencias de productos terminados y en curso de fabricación
2 Supplies	2 Aprovisionamientos
(a) Merchandise consumed	(a) Consumo de mercaderías
(b) Raw materials and other consumables consumed	(b) Consumo de materias primas y otras materias consumibles
(c) Other external expenses	(c) Otros gastos externos
3 Personnel expenses	3 Gastos de personal
(a) Wages, salaries, etc.	(a) Sueldos, salarios y asimilados
(b) Employee welfare expenses	(b) Cargas sociales
4 Provision for depreciation and amortization	4 Dotaciones para amortizaciones de inmovilizado
5 Variation in operating provisions	5 Variación de las provisiones de tráfico
(a) Variation in inventory provisions	(a) Variación de provisiones de existencias
(b) Variation in provisions for, and losses on, non-collectible receivables	(b) Variación de provisiones y pérdidas de créditos incobrables
(c) Variation in other operating provisions	(c) Variación de otras provisiones de tráfico
6 Other operating expenses	6 Otros gastos de explotación
(a) Outside services	(a) Servicios exteriores
(b) Taxes other than income tax	(b) Tributos
(c) Other current operating expenses	(c) Otros gastos de gestión corriente
(d) Provision to the reversion reserve	(d) Dotación al fondo de reversión
I Operating income (B1+B2+B3+B4−A1−A2−A3−A4−A5−A6)	**I Beneficios de explotación** (B1+B2+B3+B4−A1−A2−A3−A4−A5−A6)
7 Financial expenses and other similar charges	7 Gastos financieros y gastos asimilados
(a) Group companies	(a) Por deudas con empresas del grupo
(b) Associated companies	(b) Por deudas con empresas asociadas
(c) Other	(c) Por deudas con terceros y gastos asimilados
(d) Losses on financial investments	(d) Pérdidas de inversiones financieras

by the company from which it obtains periodic revenues. Where several types of business are conducted simultaneously, the revenues from the various businesses are included in ordinary income if they are obtained regularly and periodically and arise from the economic cycle of production and marketing or the rendering of services pertaining to the company, i.e. of the production of goods and rendering of services which comprise its activity. Net turnover is obtained by aggregating sales and services rendered, and it is reduced by the amount of returns and volume discounts.

The following are component parts of turnover:

- The balance of sales and services rendered. Sales are valued at the invoiced amount, in the case of collections maturing at less than one year. If they mature at more than one year, the amount is recorded net of the presumed interest accruing on the transaction. Taxes on sales, and the expenses arising from such sales, are not included here but are recorded under the appropriate expense items. For this purpose, prompt payment discounts are treated as financial expenses, whether or not they are recorded on the invoice. Discounts for other reasons which are included in the invoice are treated as a lower amount of sales.

Table V.5(a) (contd.)

Debit	**Debe**
8 Variation in the financial investment provisions	8 Variación de las provisiones de inversiones financieras
9 Exchange losses	9 Diferencias negativas de cambio
II Net financial income (B5 + B6 + B7 + B8 − A7 − A8 − A9)	**II Resultados financieros positivos** (B5 + B6 + B7 + B8 − A7 − A8 − A9)
III Ordinary operating income (AI + AII − BI − BII)	**III Beneficios de las actividades ordinarias** (AI + AII − BI − BII)
10 Variation in intangible and tangible fixed asset and control portfolio provisions	10 Variación de las provisiones de inmovilizado inmaterial, material y cartera de control
11 Losses on intangible and tangible fixed assets and control portfolio	11 Pérdidas procedentes del inmovilizado inmaterial, material v cartera de control
12 Losses on transactions with own shares and own debentures	12 Pérdidas por operaciones con acciones y obligaciones propias
13 Extraordinary expenses	13 Gastos extraordinarios
14 Prior years' expenses and losses	14 Gastos y pérdidas de otros ejercicios
IV Net extraordinary income (B9+B10+B11+B12+B13−A10−A11−A12−A13−A14)	**IV Resultados extraordinarios positivos** (B9+B10+B11+B12+B13−A10−A11−A12−A13−A14)
V Income before tax (AIII + AIV − BIII − BIV)	**V Beneficios antes de impuestos** (AIII + AIV − BIII − BIV)
15 Corporate income tax	15 Impuesto sobre sociedades
16 Other taxes	16 Otros impuestos
VI Income for the year (AV − A15 − A16)	**VI Resultado del ejercicio (beneficios)** (AV − A15 − A16)

Table V.5(b) Standard profit and loss account (revenues)

Credit	Haber
B REVENUES	**B INGRESOS**
1 Net turnover	1 Importe neto de la cifra de negocios
(a) Sales	(a) Ventas
(b) Services rendered	(b) Prestaciones de servicios
(c) Sales returns and volume discounts	(c) Devoluciones y 'rappels' sobre ventas
2 Increase in finished goods and work in progress inventories	2 Aumento de las existencias de productos terminados y en curso de fabricación
3 Capitalized expenses of in-house work on fixed assets	3 Trabajos efectuados por la empresa para el inmovilizado
4 Other operating revenues	4 Otros ingresos de explotación
(a) Other sundry and current operating revenues	(a) Ingresos accesorios y otros de gestión corriente
(b) Subsidies	(b) Subvenciones
(c) Overprovision for liabilities and charges	(c) Exceso de provisiones para riesgos y gastos
I Operating loss (A1+A2+A3+A4+A5+A6–BI–B2–B3–B4)	**I Pérdidas de explotación** (Al +A2+A3+A4+A5+A6–Bl–B2–B3–B4)
5 Income from shareholdings	5 Ingresos de participaciones en capital
(a) Group companies	(a) En empresas del grupo
(b) Associated cornpanies	(b) En empresas asociadas
(c) Other	(c) En empresas fuera del grupo
6 Income from other marketable securities and investment revenues	6 Ingresos de otros valores negociables y de créditos del activo inmovilizado
(a) Group companies	(a) De empresas del grupo
(b) Associated companies	(b) De empresas asociadas
(c) Other	(c) De empresas fuera del grupo
7 Other interest and similar revenues	7 Otros intereses e ingresos asimilados
(a) Group companies	(a) De empresas del grupo
(b) Associated companies	(b) De empresas asociadas
(c) Other interest	(c) Otros intereses
(d) Income from sales of financial investments	(d) Beneficios en inversiones financieras
8 Exchange gains	8 Diferencias positivas de cambio
II Net financial loss (A7+A8+A9–BS–B6–B7–B8)	**II Resultados financieros negativos** (A7+A8+A9–B5–B6–B7–B8)
III Ordinary operating loss (BI+BII–AI–AII)	**III Pérdidas de las actividades ordinarias** (BI+B II–AI–AII)
9 Gains on intangible and tangible fixed asset and control portfolio disposals	9 Beneficios en enajenación de inmovilizado inmaterial, material y cartera de control
10 Gains on transactions with own shares and own debentures	10 Beneficios por operaciones con acciones y obligaciones propias
11 Capital subsidies transferred to income for the year	11 Subvenciones de capital transferidas al resultado del ejercicio
12 Extraordinary revenues	12 Ingresos extraordinarios
13 Prior years' revenues and income	13 Ingresos y beneficios de otros ejercicios

Table V.5(b) (contd.)

Credit	Haber
IV Net extraordinary loss (A10+A11I+A12+A13+A14−B9−B10−B11−B12−B13)	**IV Resultados extraordinarios negativos** (A10+A11+A12+A13+A14−B9−B10−B11−B12−B13)
V Loss before tax (BIII+BIV−AIII−AIV)	**V Pérdidas antes de impuestos** (BIII+BIV−AIII−AIV)
VI Loss for the year (BV+A15+A16)	**VI Resultado del ejercicio (Pérdidas)** (BV+A15+Al6)

- Deliveries of merchandise or goods for sale and the rendering of services by the company in exchange for non-monetary assets or in consideration of services which represent an expense to the company are included in annual turnover and are valued at the acquisition cost or production cost of the goods or services delivered or at the market value of the consideration received, whichever is lower. They must be shown under sales and services rendered.

The following components are not included in the amount for annual turnover:

- Units of product for sale which are consumed in-house, and in-house work for the company. The latter must be recorded under capitalized expenses of in-house work on fixed assets on the revenue side of the profit and loss account (item 3 in the standard format, item 1(b) in the abridged format).
- The item 'Other operating revenues' (revenue item 4 in the profit and loss account, item 1(b) in the abridged format), since it refers to revenues which are not received periodically.
- Value-added tax (IVA).
- Subsidies, save in exceptional cases where they are itemized on the basis of units of product sold, in which case they would be shown under the related 'sales' or 'services rendered' headings and would be computed in the annual turnover.
- In all cases, financial revenues, including those deriving from instalment sales of goods and services and those obtained for prompt payment, must not be included in net turnover unless the undertaking is a credit entity.
- The amount of special tax on the manufacture or importation of certain goods must be excluded from the manufacturer or importer's net turnover, since they are single phase taxes which are passed on to the acquiror.
- Because of the special conditions under which they are performed, certain transactions are not considered as sales, e.g. sales with a commitment to buy back are excluded, since they are really secured credit transactions.

The following items are deducted to obtain the net annual turnover:

- The amount of sales returns.
- Volume discounts on sales and services rendered.
- Trade discounts given on the revenues which form part of the annual turnover.

The annual turnover of undertakings in which other undertakings have an interest through participation accounts (regulated by CCom Arts. 239 etc.) will comprise the total sales or services rendered, before deduction of the portions due to participants other than the manager. Undertakings which participate in others through such accounts must not include in their annual turnover the amount accruing from such participation.

Undertakings conducting a business which is jointly managed with others must include in their

Table V.6(a) The analytical statement of income

	Item
	Net sales, service and other operating revenues
±	Variation in finished goods and work in progress inventories
±	Capitalized expenses of in-house work on fixed assets
+	Operating subsidies
=	**Value of production**
–	Net purchases
±	Variation in merchandise, raw materials and other consumable inventories
–	External and operating charges
=	**Value added by the company**
–	Other expenses
+	Other revenues
–	Staff costs
=	**Gross operating income/loss**
–	Provisions for depreciation and amortization
–	Provision to reversion reserve
–	Bad debts and variation in operating provisions
=	**Net operating income/loss**
+	Financial revenues
–	Financial expenses
–	Provisions for depreciation and amortization and financial provisions
=	**Ordinary operating income/loss**
+	Income from fixed assets and extraordinary revenues
–	Loss on fixed assets and extraordinary expenses
–	Variation in intangible and tangible fixed asset and control portfolio provisions
=	**Income before tax**
±	Income tax
=	**Income/loss after tax**

Table V.6(b) Result levels in the analytical statement of income

	Value added
–	Staff costs
=	**Gross operating income/loss**
–	Depreciation and amortization provisions
=	**Net operating income/loss**
±	Financial income/loss
=	**Ordinary operating income/loss**
±	Extraordinary income/loss
=	**Income/loss before taxes**
±	Corporate income tax
=	**Income/loss for the year**

own turnover the proportion of the other business's turnover which is attributable to them by virtue of their participation in the joint venture agreement. Such undertakings must not include in their turnover their dealings with the joint venture.

Undertakings whose ordinary revenue consists of commission must include in their net annual turnover the amount of commission accrued in the financial year. Undertakings arranging sales to third parties which also act as depositaries of the merchandise without assuming the risks of the merchandise sold must include in their net turnover the compensation collected as mediators in such transactions.

If the financial year is shorter than a calendar year, the net annual turnover will be the amount earned in such financial year.

6.3.2 Production and administrative expenses

For the purposes of presenting the annual accounts, inventories are accrued for only in respect of net differences:

- By showing the simple variation during the period between opening and closing inventories (increase or decrease) of the products handled by the company: finished goods, semi-finished goods, work in progress, and other products of the company.
- By directly reflecting consumption. This method is used for inventories of goods acquired from outside for resale or for inclusion in the production process – merchandise, raw materials and other supplies. Such consumption must be quantified using the valuation methods established in the chart of accounts (PGC Valuation Standard No. 17), which are shown in Table V.7.

Table V.7 Concept of purchases

	Purchases[1]
–	Purchase returns and similar transactions[2]
–	Purchase volume discounts[3]
±	Variation in inventories[4]
=	Consumption during the year

[1] Amount of purchase expenses, including transport and tax on acquisitions, excluding deductible value-added tax (IVA) borne discounts and suchlike which are included in the invoice and are not for prompt payment are treated as a lower amount of purchases. Discounts, etc. granted to the undertaking for prompt payment, regardless of whether or not they are included in the invoice, are treated as financial revenues in all cases.

[2] The amount of discounts, etc. subsequent to receipt of the invoice which arise owing to quality defects, failure to meet delivery deadlines or other similar causes.

[3] The amount of discounts, etc. granted to the undertaking for attaining a certain volume of orders.

[4] The variation between opening and closing inventories.

6.3.3 Financial expenses and revenues

Financial expenses and revenues are conceived as results connected with the obtaining of finance or derived from the undertaking's financial investments in the broad sense of the term. Thus financial expenses include payments of interest on the undertaking's short- and long-term debts, non-extraordinary losses on the sale of financial investments other than holdings in group and associated companies, definitive defaults on non-trade loans, and foreign currency exchange losses. Financial expenses also include the net variation in the provision for financial investments, excluding the control portfolio.

Conversely, financial revenues include interest and dividends, non-extraordinary gains on the sale of financial investments other than holdings in group and associated companies, and foreign currency exchange gains, when the chart of accounts allows them to be recorded as such.

6.3.4 Extraordinary expenses and revenues

The Corporations Law (LSA) defines extraordinary expenses and revenues vaguely as those not arising from the undertaking's ordinary operations. In this connection the PGC 1990 states in general that a loss or expense (and a gain or revenue) is classified as an extraordinary item only if it arises from events or transactions which, in view of the sector in which the undertaking operates, meet two conditions:

- They fall outside the undertaking's ordinary habitual activities.
- They can reasonably be expected not to recur frequently.

It is necessary therefore to take into consideration the undertaking's particular commercial activities in order to classify a given event or transaction as extraordinary or ordinary.

The chart of accounts also gives a broad list of items which may be classified as extraordinary, including, for example, expenses arising from flood, fire or other accidents, tax or criminal penalties and fines, recoveries of receivables written off as bad debts, etc. and the following:

- Net variation in provisions for intangibles, tangible fixed assets and control portfolio.
- Gain/loss on the sale or retirement of intangibles, tangible fixed assets or holdings in group and associated companies (control portfolio).
- Gain/loss on transactions involving the company's own securities (shares and debentures).
- Capital subsidies taken to income of the period.
- Prior year items. This item must include the results of transactions accrued in prior years but which are evidenced or which arise in the current year. Only expenses/losses/revenues/gains which are relatively non-material may be classified by type.
- Changes in accounting methods. In exceptional cases where accounting methods change from one year to the next, the change

will be considered to arise at the beginning of the year, and the extraordinary results will include the accumulated effect of the asset and liability variations as of that date which are due to the change in methods (see VII.4).

6.3.5 Tax on profits

An important feature of the new commercial legislation is that the tax on profits is shown as an expense, and is calculated on an accruals basis, regardless of the amount payable or paid in the year (the tax effect method; see VII.1.1). However, the portion of the tax relating to ordinary profits is not differentiated from that relating to extraordinary profits.

If the undertaking reports a loss for tax purposes in the year, and if certain requirements are met, the accounting regulations allow the resultant tax credit for tax loss carry-forwards to be recognized (see VII.1.2). Accordingly, the profit and loss account will show this item with a minus sign in expenses under item 15 (item 14 in the abridged format).

7 Abridged annual accounts formats

The fourth section of the chart of accounts presents the abridged formats that can be used by sole traders, partnerships, and small or medium-sized capital companies pursuant to the criteria set out in Table V.1. These versions permit greater item aggregation and are less demanding as regards data collection than the standard formats, in line with the lower administrative capacity of small companies (see Tables V.8(a) (b) and V.9(a) (b)).

Table V.8(a) Layout of the assets side of the abridged balance sheet

Assets
A Subscribed capital unpaid and uncalled
B Fixed and other non-current assets
I Formation expenses
II Intangibles
III Tangible fixed assets
IV Financial investments
V Own Shares
C Deferred charges
D Current assets
I Subscribed capital called but not paid
II Inventories
III Accounts receivable
IV Short-term financial investments
V Short-term own shares
VI Cash at bank and in hand
VII Pre-payments and accrued income
Total (A+B+C+D)

Table V.8(b) Layout of the liabilities side of the abridged balance sheet

Shareholders' investment and liabilities
A Capital and reserves
I Subscribed capital stock
II Share premium account
III Revaluation reserve
IV Reserves
V Profit and loss brought forward
VI Profit and loss for the year
VII Interim dividend paid during the year
B Deferred income
C Provisions for liabilities and charges
D Long-term debt
E Current liabilities
Total (A+B+C+D+E)

Table V.9(a) Layout of the abridged profit and loss account (expenses)

Debit
A EXPENSES
1 Materials consumed in operations
2 Personnel expenses
(a) Wages, salaries, etc.
(b) Employee welfare expenses
3 Provision for depreciation and amortization
4 Variation in operating provisions and losses on non-collectible receivables
5 Other operating expenses
I Operating income (B1–A1–A2–A3–A4–A5)
6 Financial expenses and other similar charges
(a) Group companies
(b) Associated companies
(c) Other parties
(d) Losses on financial investments
7 Variation in financial investment provisions
8 Exchange losses
II Net financial income (B2+B3–A6–A7–A8)
III Ordinary operating income (AI+AII–BI–BII)
9 Variation in intangible and tangible fixed asset and control portfolio provisions
10 Losses on intangible and tangible fixed asset and control portfolio
11 Losses on transactions with own shares and own debentures
12 Extraordinary expenses
13 Prior years' expenses and losses
IV Net extraordinary income (B4+B5+B6+B7+B8–A9–A10–A11–A12–A13)
V Income before tax (AIII+AIV–BIII–BIV)
14 Corporate income tax
15 Other tax
VI Income for the year (AV–A14–A15)

Table V.9(b) Layout of the abridged profit and loss account (revenues)

Credit
B REVENUES
1 Operating revenues
(a) Net turnover
(b) Other operating revenues
I Operating loss (A1+A2+A3+A4+A5–B1)
2 Financial revenues
(a) Group companies
(b) Associated companies
(c) Other
(d) Income from sales of financial investments
3 Exchange gains
II Net financial loss (A6+A7+A8–B2–B3)
III Ordinary operating loss (BI+BII–AI–AII)
4 Gains on intangible and tangible fixed asset and control portfolio disposals
5 Gains on transactions with own shares and own debentures
6 Capital subsidies transferred to income for the year
7 Extraordinary revenues
8 Prior years' revenues and income
IV Net extraordinary loss (A9+A10+A11+A12+A13–B4–B5–B6–B7–B8)
V Loss before tax (BIII+BIV–AIII–AIV)
VI Loss for the year (BV+A14+A15)

VI Recognition criteria and valuation

1 Accounting principles

The accounting principles are laid down in the first part of the PGC 1990. The application of these principles is mandatory.

Prudence principle (*principio de prudencia*) – only realized profits are shown at year end. In contrast, projected risks and possible losses, including prior year losses, are recorded as soon as they become known. To this effect, it will be necessary to draw a distinction between reversible or possible risks and losses and realized or irreversible losses.

Consequently, at year end, all foreseeable risks and losses are shown, regardless of the source thereof. Where such risks and losses become known between the accounting close and the date of presentation of the annual accounts, the information should be provided in the notes on the annual accounts, regardless of whether the said risks and losses are reflected in the balance sheet and profit and loss account.

Also, all types of diminution in value should be recorded, irrespective of whether a profit or loss has been reported for the year.

With regard to the importance of the prudence principle within the accounting system, the chart of accounts supports the stipulation in the Commercial Code that this principle should prevail over all others. An instance may be cited which is very common in companies to which the principle is most relevant (Cañibano, 1991). When the acquisition of intangible fixed assets is recorded, it is uncertain whether the related costs should be capitalized or shown as a period expense. The acquisition expenses may be clearly linked to a future source of revenues and, as such, would be capitalized per the matching principle. However, uncertainty about when the revenues will be received or about their effective value, suggests, on a strict application of the prudence principle, that these expenses should be recorded as a period expense.

Going concern principle (*principio de empresa en funcionamiento*) – it is assumed that a company will continue in business for the foreseeable future. Accordingly, accounting valuations are not aimed at calculating the value of equity for the purposes of the sale or liquidation of a company.

Realization principle (*principio de registro*) – this principle is stated in the chart of accounts as follows: 'Business transactions must be recognized as soon as they give rise to rights and obligations.' This principle delineates a precise rule for deciding when a transaction should be recorded. The correct application of the principle requires a great many transactions to be shown in the annual accounts that, for various reasons, were never previously included. An example of a liability that once formed no part of the balance sheet might be pension commitments to employees. Pension commitments are now regarded as deferred salaries which are accrued on a daily basis as with any other type of remuneration. This type of obligation is nowadays shown in the annual accounts. The accounting treatment of trade bill discounting also stems from the application of the recording principle. The financing obtained from trade bill discounting at finance entities is considered as a debt that must be reflected in the balance sheet until the bills are paid by customers.

Cost principle (*principio del precio de adquisición*) – as a general rule, all assets and rights are recorded at their acquisition price or production cost. As statement 1 of the Accounting and Business Administration Association explains, this criterion applies both to assets and to liabilities. In the case of liabilities, acquisition price means the amount to be reimbursed; in the case of equity items, the amount of shareholders' investments and retained profits. The value resulting from the application of this principle has to be maintained in the balance sheet for as long as the relevant asset or liability appears. The value must not be subject to any adjustments possible on the basis of other principles or to any adjustments to correct for the effect of inflation.

The Fourth EC Directive introduces the cost principle as the basis of accounting valuation, but permits member States to incorporate revaluation principles into their domestic law. Apart from balance sheet regularizations in the light of various legal provisions, the last being in 1996, Spain has not availed itself of this right and, to date, does not permit any other valuation system (see VIII).

Accruals principle (*principio del devengo*) – revenues and expenses are recognized when the actual flow of the related goods and services occurs, regardless of when the resulting monetary or financial flow arises. The need to calculate a periodic profit or loss requires comparison of two contrary flows: revenues and expenses. Under the accruals principle, of the three separate occasions on which profits per books could be recognized (after production, on the sale of the goods or services or on collection of the revenues), the second is the one preferred. Where there is a significant delay between the date of the sale and that moment of collection, a contra item should be recorded as a result of the accruals principle not having been replaced by the cash principle in such a way that recognized profits are fully guaranteed. In other words, the application of the accruals principle requires provisions to be set up at the same time. The provisions are intended to highlight the risk attached to these collection rights.

Matching principle (*principio de correlación de ingresos y gastos*) – the result for the year comprises the revenues for that period, net of the expenses incurred in attaining it, and profits and losses not directly connected with the company's normal line of business. Accordingly, the expenses incurred in any given period must offset the revenues arising in the same period. However, certain expenses may be clearly linked to future revenues, such as R&D costs and, in general, all costs aimed at improving a company's business performance. Such costs may be shown as assets. Similarly, revenues arising in a given year may relate to future costs. In that case, such revenues should be recorded on the liabilities side of the balance sheet as revenues deferred.

No offsetting principle (*principio de no compensación*) – under no circumstances may asset and liability items be used to offset each other, nor may expenses and revenues shown in the profit and loss account, as the model annual accounts make clear. This principle stems from the need for clarity in the presentation of the annual accounts. A company should show the effective amounts of its rights and obligations separately in its balance sheet, never the net effect of the two, irrespective of whether they relate to the same parties. The same applies to revenues and expenses.

Consistency principle (*principio de uniformidad*) – once a rule for applying an accounting principle has been adopted, it should be retained over time and applied to all like items, provided that the antecedent circumstances remain the same. If the circumstances change, the rule may be modified. In that event, however, the change must be explained in the notes on the accounts, along with its quantitative and qualitative effects. For comparative purposes the rules adopted must be invariable, and changes should be made only in order to render them more acceptable, and compatible with the ultimate object of giving a true and fair view of a company's economic and financial position.

Materiality principle (*principio de importancia relativa*) – certain accounting principles may be applied less strictly, provided the effect of doing so is slight in quantitative terms and does not work against the accounts as a true and fair view of the company's financial position.

This principle is related to the requirement that accounting information should be relevant. An obvious example would be to show as a period expense the acquisition of fixed asset items of little economic value.

2 Value concepts: acquisition cost or production cost

LSA Arts. 195.1 and 196.1, as amended, enacting the provisions of the Commercial Code, and specifically in connection with the acquisition cost principle, stipulate that fixed and current assets must be valued at their acquisition price or production cost. The chart of accounts supports this general valuation standard in its chapter on tangible fixed assets. However, it should be appreciated that the rule is applicable to other fixed asset items too.

2.1 Acquisition cost

Acquisition cost includes the amount invoiced by the supplier and all the additional expenses arising until the item enters into service: site preparation, transport, customs tariffs, insurance, installation, assembly and similar expenses.

Financial expenses may be included in the acquisition cost provided they were incurred before the entry of the asset into service and provided they were invoiced by the supplier, or relate to loans or any other type of external financing of the asset. In that event the inclusion of these expenses in the acquisition cost of the asset should be mentioned in the notes.

Indirect taxes to which tangible fixed items are subject may be included in the acquisition cost thereof only when they are not directly recoverable from the tax authorities.

2.2 Production cost

The production cost of goods manufactured or built by the company itself is calculated by adding to the acquisition cost of raw materials or other consumables the other expenses directly allocable to the assets. A reasonable proportion of the costs indirectly allocable to the related assets may be added, provided that such costs relate to the asset's period of manufacture or construction.

Financial expenses may be included in the production cost provided they were incurred before the entry of the asset into service and provided they were billed by the supplier or relate to loans or any other type of external financing of the manufacture or construction of the asset. In that event the inclusion of such expenses in the production cost should be indicated in the notes.

2.3 Capitalization of financial expenses

The Accounting and Audit Institute (ICAC) established norms for tangible fixed asset valuation in a Resolution of 30 July 1991. According to these rules, the only financial expenses deemed to be capitalizable as tangible fixed assets are interest and fees incurred through the use of outside sources of financing, i.e. debts to fixed asset suppliers and debts arranged specifically for asset acquisitions or construction.

Financial expenses may be capitalized up to the market value of the related assets but only prior to the entry into service thereof, deemed to be the moment at which the asset can regularly produce revenues after its trial period. If an asset comprises several parts that may be used separately, the capitalization of financial expenses will cease at different moments for each portion of the asset.

If no outside sources of finance have been used specifically for the acquisition or construction of the asset, the financial expenses incurred in general-purpose external financing (except for trade debts) may be capitalized. These capitalizable financial expenses are generally calculated on the basis of the effective average interest rate on the debts, which is applied to the investment after discounting the proportion financed with specific borrowed funds and equity. This discounted amount of investment is limited to the amount of borrowed funds not aimed specifically at the acquisition of the asset, after discounting trade debts. Accounting regulations do not permit equity-related financial charges to be capitalized.

2.4 Capitalization of exchange differences

With regard to the possibility of capitalizing exchange differences, the chart of accounts stipulates that the potential gain or loss arising on long-term foreign currency loans aimed at specific asset financing can be recorded as a reduction of, or an addition to, the related asset value, provided that all the following conditions are met:

- The loan giving rise to the exchange differences must be used solely for specific and clearly identifiable fixed asset additions.
- The installation term for such fixed assets must exceed one year.
- The exchange rate variation must arise before the fixed assets enter service.
- The amount resulting from the transfer of the exchange difference to the cost of the assets may in no circumstances exceed the market or replacement value of the assets.

The chart of accounts states that special regulations may exist which are applicable to certain industries where high levels of long-term foreign currency borrowing are usual. These specific cases are described in the relevant industry regulations or other related standards.

2.5 Capitalization of indirect costs

The only rules about what should be included in the acquisition cost and production cost have already been mentioned. Neither in the chart of accounts nor in the ICAC Resolution on tangible fixed asset valuation is there a clear definition of what indirect costs should be incorporated into acquisition cost and production cost. Only the Accounting and Business Administration Association (AECA) has issued further guidelines in the second publication in its series of Accounting Principles and Standards, devoted to tangible fixed assets.

The association recommends the inclusion in

acquisition cost of the expenses outlined above (see VI.2.1). Moreover, concerning goods manufactured or built by the company itself, it recommends taking into account the whole effective production cost. This means all costs resulting from the rules usually applied by the company in calculating the cost of its production process. It totally discourages the inclusion of concepts such as structure charges and others not objectively attributable to the related assets.

2.6 Acquisitions free of charge

Assets acquired for nothing are governed by the ICAC Resolution of 30 July 1991 on rules for tangible fixed asset valuation. Assets acquired free of charge are shown at the resale value, that is, the price a purchaser would be prepared to pay, given the condition and location of the asset. This value is reckoned in the context of the company's position and on the general assumption that the asset will continue to be used.

The subgroup 'Deferred revenues' is used as contra item for this special type of acquisition and the valuation standards relating to capital subsidies applied. Consequently, the amounts included under this heading will be credited to the profit and loss account in direct proportion to the depreciation rate of the asset acquired free.

2.7 Asset swap transactions

The direct exchange of tangible fixed assets is also covered by the ICAC Resolution of 30 July 1991 on norms for tangible fixed asset valuation. In this special type of transaction the fixed assets received are valued at the net realizable value of the asset relinquished (provided that the latter value does not exceed the market value of the asset received, and that any loss arising from the transaction is recognized). The expenses arising, from receipt of the assets up to their entry into service, are capitalized, provided that the value of the assets is not thereby led to exceed their market value.

2.8 Acquisition costs in foreign currencies

Tangible and intangible fixed assets denominated in foreign currencies are generally translated into pesetas at the exchange rate prevailing at the date on which the items were included in assets. Depreciation, amortization and the related provisions should be calculated on the basis of the amount, in pesetas, resulting from the application of the aforementioned rule. (For the valuation of foreign currency receivables and liabilities see VII.5).

3 Valuation rules for fixed assets

3.1 Systematic depreciation

The way in which depreciation is shown in the balance sheet is unusual in international accounting terms but not difficult to comprehend. Instead of showing items net, the standard balance sheet shows the accumulated depreciation fund and the amount of the acquisition cost of the tangible fixed assets.

The appropriate value adjustments have to be made when taking into consideration the diminution in the original value of fixed assets. Pursuant to the ICAC Resolution of 30 July 1991, establishing fixed asset valuation standards, the annual provision for these should be based on a combination of certain factors, as follows:

- The depreciable value of the asset, defined as the amount resulting from applying the standards in the chart of accounts and in the aforementioned resolution establishing fixed asset valuation standards.
- The projected net residual value, whether it be negative or positive (if negative, a provision must be set up), of the amount expected to be recouped, net of the expenses associated with the sale of the assets at the end of the estimated period of service.
- Useful life, which is the estimated period, calculated on the basis of various factors such as use and physical wear and tear, technical or commercial obsolescence and legal limits on asset use, over which the fixed assets usually generate returns.
- The depreciation method applied, which will not depend on tax considerations, or a company's profitability, but which will be chosen as the most appropriate for distributing the cost of the fixed assets over the years of

their useful life. The most commonly used method is straight-line depreciation, though any method that conforms with the aforementioned provision would be acceptable.

The switch from one method to another would be acceptable only on the grounds of a change in the technical or economic utilization of the asset. Such a switch tends to be infrequent, since it may seriously jeopardize the consistency principle.

Depreciation methods which distribute the costs of the fixed assets over the years of their useful life in accordance with a technical and economic yardstick may be applied. The depreciation method must be chosen independently of tax considerations or of the profitability standards the company sets.

Ceasing systematic depreciation when the current value of an asset is higher than its accounting value is not permissible.

- The date at which the depreciation process commences, which should coincide with the moment the relevant fixed asset is available for regular use.

Although the above-mentioned rules are to be found in regulations relating to tangible fixed assets, they also apply to intangible assets. Another ICAC Resolution, dated 21 January 1992, relating specifically to intangible assets, includes one or two interesting points in this connection:

- Useful life is the estimated period of returns from the assets and is updated annually. Modifications to the estimated useful life of an intangible asset give rise to prospective variations in the related amortization. Errors in estimation must be modified retrospectively with a charge to extraordinary profits as a result of the impact on accumulated amortization.
- Amortization methods. Any systematic method that takes the technical and economic characteristics of the depreciated asset into consideration may be used, including the constant rate method and the increasing or declining variable rate methods.

3.2 Value reductions

Under CCom Art. 39, neither the acquisition cost nor the production cost can exceed the market value of the assets. If, at the closing date of the balance sheet, the market value of the asset items is lower than the result of applying acquisition or production cost, the appropriate valuation adjustments are made. Accordingly, the ICAC resolution of 30 July 1991 states that where there is a non-definitive diminution in value to below the net book value of an asset, which is not recoverable through the revenue generation cycle, a reversible provision should be set up. If the reasons underlying the value adjustment cease to exist, the lower value should not be maintained.

3.3 Write-backs

As the ICAC Resolution of 30 July 1991 indicates, assets must be recorded at the lower of cost or market value. If the accounting value falls below the market value the allocation of a provision is mandatory in every case. If the diminution in value is irreversible, the related losses should be shown directly.

3.4 Formation expenses

These expenses are valued at the acquisition price or production cost of the related goods and services. Formation and capital increase expenses include lawyers' fees, public-deeding and registration expenses, the printing of bulletins and securities, advertising, commission, other security placement expenses, etc. arising from corporate formation or capital increases.

Pre-opening expenses should include fees, travel expenses and expenditure on preparatory technical and economic studies, on corporate launch advertising, the hiring, training and distribution of personnel, etc. arising from the setting up of a company.

Formation expenses should be amortized systematically over a maximum of five years. No profits may be distributed to shareholders until such expenses have been fully amortized, except where the amount of uncommitted reserves is at least equal to the aforementioned unamortized expenses (LSA Art. 194).

3.5 Intangible assets

The Accounting and Audit Institute (ICAC) approved a Resolution dated 21 January 1992, enacting the provisions of the chart of accounts in connection with valuation rules for intangible assets. This resolution follows the rules laid down by the Resolution of 30 July 1991. In particular, the following provisions are to be applied in connection with the assets and rights indicated.

3.5.1 Research and development expenses

The second valuation rule of the ICAC Resolution dated 21 January 1992 refers to R&D expenses. Research expenses are defined as original and planned in-depth studies aimed at acquiring new knowledge and a deeper understanding of scientific and technical matters. Development expenses are defined as the specific application of the results obtained from research in a phase prior to the commencement of corporate production.

R&D expenses, including those relating to activities charged to other companies, are expensed currently, given the difficulty of excluding similar expenses that cannot be identified as R&D expenses, the element of uncertainty regarding the success of the projects concerned, and the need to sustain a continuing R&D effort. However, R&D expenses may be capitalized provided the following conditions are met from the moment they become known:

- Existence of specific and individual aims for each R&D activity.
- Clear cost allocation, recording and distribution criteria for each project.
- Reasonable expectations of technical success with regard to operations or sales, and reasonably assured profitability, which must be evaluated jointly in the case of interrelated research activities with a common goal.
- Guaranteed funding sufficient for the completion of projects in hand.

Projects are valued at acquisition or production cost, which, in the case of in-house development, includes personnel, materials and service costs and the depreciation and amortization of the fixed assets directly used for the project, plus the proportion of indirect costs which can reasonably be allocated thereto, but in no case may include underactivity, financial or general structural costs. The amount of development costs capitalized may in no case include research expenses. Where the R&D activity gives rise to amounts allocable to patents and rights, they should be treated in line with the regulations governing such assets and should be amortized accordingly.

There are no separate items in the balance sheet for research and development expenses. Nevertheless, it is possible to break the overall figure down into individual projects, differentiating between research projects and development projects.

The amounts capitalized are amortized systematically over a maximum period of five years, in the case of research expenses, from the year in which they are capitalized and, in that of development expenses, from the date on which the project is concluded. The conditions that give rise to the capitalization of development expenses are updated each year over which the project runs in order to verify the advisability of the practice. If it can no longer be justified, the residual value has to be allocated to period results.

R&D expenses are a good example of the balance that has to be struck between the prudence principle and the principle that revenues and expenses should match up. If there is a large measure of uncertainty about the profits which the project is expected to generate, the prudence principle would suggest that R&D expenditure should be charged to the year's result. On the other hand, if there is a strong conviction that the project will generate revenue (and the conviction rests on the assumption that the conditions in the PGC and the ICAC Resolution of 21 January 1992 will be met), R&D expenditure should be capitalized as an application of the matching principle.

3.5.2 Goodwill

Goodwill is stated only where it has been acquired for a consideration. As it stands the legislation does not envisage internally produced goodwill being shown. Statement No. 3 of the Accounting and Business Administration Association (AECA), on intangibles, points out that most companies invest considerable sums with the aim of increasing goodwill. But it is far from easy to determine whether such investment has actually generated the goodwill intended. It must therefore be charged to the profit and loss account.

Goodwill is defined as a set of intangible assets which add to the value of a company, such as its customers, corporate name, market share, competitiveness, and its human and other resources. It is calculated as the difference between the amount paid to acquire the assets/liabilities and their actual value, up to the market value of the assets and the present value of the liabilities, in line with market interest rates, without prejudice to their being shown at redeemable value in the balance sheet.

When the difference is negative, the reasons have to be analyzed and the difference recorded as a provision for contingencies and expenses or as a reduction in the value of the assets.

As AECA Statement No. 3 observes, goodwill can be shown only where it arises as a result of an acquisition, that is to say, where it arises from the purchase of a going concern. As the same AECA statement indicates, goodwill must relate to the acquisition of a company as a whole and not to a group of assets or liabilities. In the latter case, should there be a difference between the price paid for the acquisition and the true value of the assets or liabilities, the difference must be attributed to the assets and liabilities acquired rather than to goodwill.

Consequently, it should be understood that an acquisition has taken place in cases of merger or absorption. In the case of a partnership purchase the rules to be applied are those that govern financial investments.

Goodwill is amortized systematically over the period in which revenues are generated, up to a maximum of 5 years, with extension to 10 years permissible, provided it is explained and justified in the notes.

3.5.3 Other intangible fixed assets: administrative concessions, patents and similar rights, transfer rights

Administrative concessions are defined as the transfer by a State agency to an individual of the right to a public asset or the management of a public service. They are included in a company's assets at the total amount of the expenses incurred in securing such rights, and the cost is amortized systematically with a charge to results over a period not exceeding the term of the concession. If the administrative concession involves tangible fixed assets, to be returned to the granting entity, a reversion allowance should be recorded, with systematic provisions over the term of the concession, calculated on the basis of foreseeable revenues, so that the allowance amounts to the net book value of the assets plus the expenses incurred in the return thereof at the date when it takes place.

In the event of failure to comply with the terms of the concession giving rise to the loss of the rights thereto, the total residual value of the assets at that date must be charged to losses and the necessary provisions set up to cover potential liabilities.

Patents and similar rights are valued at the cost incurred to acquire the right to their use. Patents and intellectual industrial property are both valued on the basis of acquisition or production cost, and if they arise as a result of a development project the unamortized cost allocable to the appropriate project may be included. Valuation adjustments are made pursuant to fixed asset regulations.

Transfer rights are deemed to be the letting of premises to a third party, who assumes all the rights and obligations conveyed by the original lease agreement in exchange for payment. In order for such rights to be recorded as assets, the transfer must be in respect of a consideration. The amounts capitalized in this connection must be amortized over the period in which revenues are expected to arise. That period may in no case exceed the period over which goodwill is amortized.

3.5.4 Computer applications

The amounts paid to acquire ownership of or access to computer software projected to be used over several years are capitalized, including the cost of programs developed in-house which meet with the conditions specified above (see VI.3.5.1). Under no circumstances is computer software to be regarded as a tangible fixed asset. The amounts are amortized systematically over the period in which revenues are projected, which may not exceed five years.

Always pursuant to the ICAC Resolution of 21 January 1992, on intangible assets, existing program modification and modernization expenses, personnel training expenses, program maintenance expenses and expenses arising from consultation services or global computer system reviews may not be capitalized, even when the services are performed and invoiced by other companies.

3.5.5 Financial leased assets

Under the valuation standards of the PGC, the lessee records the leased asset and the related liability for the debt assumed in its balance sheet. Leased assets must be shown as intangibles as recognition of the right to the use thereof, until, in the event of the lessee exercising the purchase option, the assets are transferred to tangible fixed assets. The aforementioned ICAC Resolution of 21 January 1992, prescribing intangible fixed asset valuation standards, governs the fine print of these transactions. The accounting record would be as shown in Table VI.1.

Financial leasing is defined as an agreement permitting the use of an asset in exchange for periodic instalments and including a purchase option on the asset concerned. Where, during the period of the validity of the lease, the exercise of the purchase option is formally agreed or otherwise guaranteed, the transaction is recorded as an asset acquisition with payment deferred. If the purchase option has not been exercised but the terms of the contract give rise to no reasonable doubt that it will be, the resulting rights are shown, as an intangible asset at its nominal value, and the total debt from the instalments plus the

Table VI.1 Disclosure of financial leased assets

On signature of the financial leasing contract:		
Intangible fixed assets		
Deferred charges		
	to	Other accounts payable
On the payment of each instalment:		
Other accounts payable		
	to	Cash at bank and in hand
Financial expenses		
	to	Deferred charges
For the depreciation of the asset received:		
Provision for depreciation		
	to	Accumulated depreciation

amount of the purchase option must be reflected in liabilities. The difference between both amounts, deriving from the financial expenses of the transaction, is recorded as deferred charges, to be charged to the profit and loss account in accordance with a suitable financial criterion.

Accounting regulations do not specify the financial criterion to be used. However, the most sensible solution is to charge deferred expenses to results at the same rate as the financial expenses accrue. Since the financial expenses must be disclosed separately in the leasing contract, the accounting procedure becomes much clearer. The example in Table VI.2 will help to elucidate matters. The following points should be noted in connection with the foregoing:

- No reasonable doubt regarding the exercise of the purchase option is deemed to exist if on signature of the contract the option price is immaterial or nominal in relation to the total contract price, or lower than the estimated residual value of the asset at the date when the purchase option is to be exercised.
- Unaccrued value-added tax (IVA) charged on future instalments is not shown under liabilities in the balance sheet. If the instalments are not tax-deductible, they are deemed to be period expenses and do not affect the initial value of the fixed assets.
- When the purchase option is exercised, the value of the rights recorded and the related accumulated amortization are transferred from intangible assets to the relevant tangible fixed asset accounts.

The ICAC Resolution of 21 January 1992, enacting intangible fixed asset valuation rules, applies only to financial leases. An operating lease contract must be wholly regarded as a conventional rent, with the periodic payments expensed currently. The accounting record of the operation would be as shown in Table VI.3.

3.5.6 Sale and lease-back agreements

This type of contract is a financial procedure based on the sale of an asset, which provides financing, combined with its subsequent lease to the vendor with a purchase option on the same asset. The instalments comprise both repayment

Table VI.2 Accounting treatment of financial leased assets (*arrendamiento financiero)*

Price of the asset acquired: 1,500 units

Contract term: 2 years

Asset useful life: 10 years

Residual value: 500 units

The split of the lease rates would be as follows:

Year	Depreciation charge	Financial charge	Total lease rate
1	460	140	600
2	540	60	600

And the accounting record in the first year would be:

On signature of the financial leasing contract:

1,500	Intangible fixed assets		
200	Deferred charges	to	Other accounts payable 1,700

On the payment of the instalment:

600	Other accounts payable	to	Cash at bank and in hand 600
140	Financial expenses	to	Deferred charges 140

For the depreciation of the asset received:

150	Provision for depreciation	to	Accumulated depreciation 150

Table VI.3 Operating lease

At the moment when rental payments are due:

Rates and rentals
to Cash at bank and in hand

In the case where the lessee opts to purchase:

Tangible fixed assets
to Cash at bank and in hand

of the financing received and payment of the associated financial charges. This type of transaction is recorded as follows. In the sale of an asset to be subsequently leased, where the terms of the agreement show that the transaction is essentially a method of financing, the book value of the assets is transferred to intangible assets and no profit or loss is recognized on the sale. Also, the values of the instalments and the purchase option are recorded as debt. The difference between the sum of these values and the financing received is shown as deferred charges.

3.6 Tangible fixed assets

3.6.1 Additional standards

With regard to tangible fixed assets, in addition to the general standards referred to above, the third valuation rule of the chart of accounts sets out a series of specific standards applicable to the following assets and rights:

- Unbuilt land – refurbishment and enclosure expenses, soil removal expenses, clearing and drainage expenses and demolition expenses preparatory to new construction work, together with inspection and planning expenses prior to the acquisition of the land, are included in the acquisition cost.
- Construction – all permanent installations and items, construction levies and project and works management fees form part of the acquisition or production cost. The value of the land, and of the buildings or other structures, should be shown separately.
- Technical installations, machinery and tools – the value of these items must include all

acquisition or manufacturing and construction expenses, up to the time when the items are brought into service.

- Utensils and hand tools – utensils and hand tools incorporated into mechanical items are subject to the valuation and depreciation rules applicable to machinery.

  In general, utensils and hand tools which do not form part of machinery, and whose estimated period of use does not exceed one year, must be expensed currently. If the estimated period of use does exceed a year it is advisable, for simplicity's sake, to tackle annual restatements by way of a physical count.
- Spare parts – the recording of spare parts is governed by the Fifth Standard of the ICAC Resolution of 30 July 1991. They are defined as components to be fitted to productive tangible fixed assets in replacement of similar parts.

  These items are valued under general valuation rules and are shown as inventories if their storage period is less than a year, or as tangible fixed assets if the storage period exceeds one year. Spare parts are depreciated at the same rate as the parts they are to replace.
- Construction in progress – expenses incurred on in-house work on fixed assets are charged to the related expense accounts. The balances of the construction in progress accounts are charged at year end, with a credit to the 'in-house work on fixed assets' account on the revenue side of the profit and loss account.

3.6.2 Renewal, extension and improvement of tangible fixed assets

The third valuation standard of the PGC stipulates that these costs should be included among assets, since they are an augmentation of the value of the asset concerned in so far as they increase its capacity, productivity or useful life, and provided that the net book value of the items replaced is known or can be reasonably estimated so that they may be written off. However, the second and third standards of the ICAC Resolution dated 30 July 1991 on tangible fixed assets set down more specific rules for these costs.

Renewal of tangible fixed assets is deemed to comprise the recovery of the initial characteristics of the renewed asset, the cost of which is capitalized as an addition to the value of tangible fixed assets. The items replaced or surrendered in exchange for the new items and the related valuation adjustments are eliminated from tangible fixed asset accounts and the difference (profit or loss) between the net cost and the recovered product is recognized. If the renewal affects a part of an asset whose depreciation or valuation adjustment is not identifiable because it was not recorded separately, the provisions of the rule on fixed asset repairs come into play.

Extension is defined as the incorporation of new elements into fixed assets, thus achieving greater production capacity, whilst an improvement increases the efficiency of the fixed asset. To qualify for capitalization, the cost relating to the extension or improvement of an asset, including the demolition or removal of the items replaced, must give rise to an increase in production capacity, substantially increased productivity, or an extension of the estimated useful life of the asset. In any case, the amount to be capitalized will be limited to the market value of the tangible fixed asset.

3.6.3 Repair and maintenance of tangible fixed assets

These costs are governed by the Fourth Standard of the ICAC Resolution dated 30 July 1991 on tangible fixed assets. Repair is deemed to be the process of returning tangible fixed assets to full working order, whilst maintenance comprises maintaining an asset's production capacity. Expenses arising as a result are expensed currently. However, if in the course of the repair a replacement is needed, it is governed by the standard on renewals. Where an extraordinary repair is required at the end of an asset's use cycle, a provision must be recorded each year for the proportion of the estimated amount of the repairs to be performed.

3.7 Marketable securities

3.7.1 Valuation at acquisition cost

Both fixed-interest and equity securities are generally valued at their acquisition cost on subscription or purchase. This cost comprises the total amount paid or to be paid for the purchase of the securities, including the expenses inherent in the transaction. Security valuation based on the equity method even for holdings in group and associated companies is not permissible in individual accounts, as stipulated in the Fourth Directive, Art. 59.

Uncalled payments payable on shares form part of the acquisition cost and are recorded as debt. Uncalled payments payable on shares which constitute long-term financial investments are shown as long-term liabilities on the liabilities side of the balance sheet. On the other hand, uncalled payments payable on shares which constitute a short-term financial investment should be shown as a reduction in current assets. In both cases, when said payments are called, they are transferred to current liabilities in the balance sheet.

The amounts of preferential subscription rights are considered to be included in the acquisition cost. Accordingly, when new rights are acquired, the amounts are recorded as additions to the subscription cost. When subscription rights are sold, the acquisition cost of the remaining marketable securities is reduced by the amount of the rights sold. This amount is calculated by applying a generally accepted valuation formula in line with the accounting principle of prudence. At the same time, the amount of the valuation adjustments recorded is progressively diminished.

Accrued dividends and unmatured and accrued explicit interest at the date of the purchase are not included in the acquisition cost. Such dividends and interest are shown separately in the relevant accounts receivable on the basis of their maturity. Explicit interest is defined as the returns not included in the refundable value.

In any event, the average price or average weighted cost by homogeneous group methods should be applied. (Homogeneous groups are deemed to be securities with the same rights.)

The regulations do not permit the capitalization of interest on borrowing intended for the acquisition of securities.

3.7.2 Special marketable security acquisition methods

As in the case of tangible assets, acquisitions of marketable securities free of charge should initially be recorded at their saleable value, which is the marketable value in the case of quoted securities. When the acquisition of shares free of charge arises from the allocation of shares with a charge to reserves, the increase in the cost of the holding is not shown, since there is a reduction in the value of the shares.

Share swap transactions are governed by the ICAC Resolution dated 30 July 1991 on tangible fixed assets (see VI.2).

3.7.3 Valuation adjustment because of lower current values

Reversible losses may arise on marketable securities, in which case a provision should be created. If irreversible losses arise, the diminution in value of the related asset should be shown directly.

Marketable securities quoted on an organized secondary market are recorded at year end at the lower of cost or market value. If the market value prevails, provision should be made to reflect the depreciation of the securities. However, where there is clear and well-founded evidence that the value of the securities is less than market value, it should be written down accordingly. Market value is taken to be the lower of the average market price in the last quarter or the market price at the balance sheet close (or, failing that, on the previous day).

If there is accrued and unmatured implicit and explicit interest at year end, which should be included among assets, the valuation adjustment is to be calculated by comparing the market price with the sum of the acquisition costs of the securities and the unmatured and accrued interest at year end.

Unquoted marketable securities are stated at acquisition cost. However, where the acquisition cost is higher than the amount resulting from the application of rational valuation standards accepted in normal practice, a provision must be recorded for the difference. In this regard, share holdings are recorded at their underlying book value, adjusted by any unrealized gains existing at the moment of acquisition and still existing at the date of subsequent valuation.

These criteria also apply to holdings in the capital stock of group and associated companies. The amount of the provisions recorded is based on variations in the equity of the investee company, even if the securities are quoted on an organized secondary market.

3.7.4 Own shares and debentures

Own shares are valued in line with the valuation rules applicable to all marketable securities. However, it should be noted that, when own shares or those of the controlling company are purchased, a restricted reserve to the same amount should be recorded (LSA Art. 79). When the shares are written off, capital is reduced by the par value of the said shares. The result of the transaction is determined by taking the difference between the acquisition cost of the shares and their par value. This amount should be charged or credited, as appropriate, to reserves. When the shares are sold, the result obtained is recorded under the 'Losses on transactions in own shares and debentures' or 'Gains on transactions in own shares and debentures' heading in the profit and loss account.

Where a company acquires its own debentures and bonds for redemption, differences may arise between the acquisition cost (excluding unmatured accrued interest) and reimbursable amounts (excluding implicit interest recognized as deferred charges) which are charged or credited, as appropriate, to the profit and loss account.

3.7.5 Valuation of financial investments denominated in foreign currencies

Equity securities denominated in foreign currencies are translated into pesetas at the exchange rates prevailing at the date on which the relevant items are included in assets. Their value should not exceed the value that would result from applying the year end exchange rate to the market value of the assets.

Exchange differences relating to fixed-interest securities and credits denominated in foreign currencies are dealt with below (see VII.5).

3.8 Other financial investments

Loans which mature in over one year are included in fixed assets (both trade and non-trade loans).

Non-trade loans are recorded at the amount paid. Implicit interest on such loans, i.e. the difference between the amount paid and the par value of the credits, should be progressively taken to income as interest revenues in the year in which they are accrued, following a financial criterion and recognizing the related interest credit on the assets side of the balance sheet.

Loans for fixed asset disposals are valued at the sale price, excluding the interest included in the par value of the loans, which is recorded as per the criteria outlined in the preceding paragraph.

Unmatured and matured accrued interest is recorded in long- or short-term credit interest accounts in accordance with a financial criterion on the basis of maturity. This criterion differs from that used for long-term trade receivables in accordance with Spanish regulations, and therefore has a different effect on the balance sheet (see VI.4.2).

The appropriate valuation adjustments should be made at the balance sheet closing date and, if appropriate, the related provisions should be recorded on the basis of the risk of doubtful debts involved in the collection of such assets.

3.9 Deferred charges

The seventh valuation standard of the PGC states in connection with this heading:

> Debt arrangement expenses are valued at their acquisition price or production cost. In principle, these expenses should be recognized in the year in which they arise but may, in exceptional cases, be deferred. In that case, they must be charged to the profit and loss account following a financial criterion and within the term of the related debts. These amounts must be fully allocated to the profit and loss account when the related debts mature.
>
> Deferred interest charges are valued at the difference between the redeemable value and the issue value of the related debts. These expenses are charged to the profit and loss account following a financial criterion within the term of the related debts.

4 Valuation of current assets

4.1 Stocks

Assets included in stocks are valued at their acquisition or production cost. The acquisition cost includes the amount billed plus the additional expenses arising until the goods are stored, such as transport expenses, customs duties, insurance expenses, etc. The amount of the indirect taxes to be levied on stock acquisitions is included in the acquisition cost only when it is not directly refundable by the tax authorities. Production cost is calculated by adding the acquisition cost of raw materials and other consumables to those costs which are directly allocable to the products. On top of these must be added also an amount that can reasonably be assigned to costs indirectly attributable to the relevant products, provided that such costs relate to the manufacturing period (see VI.2.2).

In the case of assets whose acquisition cost is not individually identifiable, the average price or average weighted cost will be used on a general basis. FIFO, LIFO or other similar methods are acceptable and may be used if the company sees fit.

In exceptional cases, and for some lines of business, certain raw materials and consumables may be valued at a fixed quantity and value, provided the following conditions are met:

- They are renewed constantly.
- Their overall value and nature do not materially vary.
- This overall value is not significant to the company.

The application of this system should be described in the notes on the accounts, evidencing the application and the amount of the fixed value.

Valuation adjustments – when the market value of an asset or any other relevant value is lower than its acquisition or production cost, the asset value is adjusted and the appropriate provision recorded if the diminution in value is reversible. If the loss in value is irreversible, this fact will be taken directly into account in the valuation of the stocks. The market value in such a case is deemed to be as follows:

- Raw materials – the lower of replacement value or net realizable value.
- Trade stocks and finished goods – realizable value, net of the related marketing expenses.
- Work in progress – realizable value of the finished goods, net of manufacturing costs and marketing expenses not incurred.

Valuation of stocks denominated in foreign currencies – where necessary, the value of stock denominated in foreign currencies is translated into pesetas at the exchange rate prevailing on the date of each acquisition. This procedure should be followed regardless of whether weighted average price, FIFO, LIFO or other similar methods are used.

A provision must be recorded when the value thus obtained exceeds the market value of the stocks at the closing date of the annual accounts. If the said market value is denominated in a foreign currency, it should be translated into pesetas at the exchange rate prevailing on the date.

4.2 Other current assets

Other current assets include the following:

- Customer and operating receivables – trade receivables are stated in the balance sheet at their face value. Long-term interest included in this value appears under the 'Deferred revenues' heading on the liabilities side of the balance sheet, and is transferred annually to results after the application of a financial criterion.

  Doubtful debtors' accounts are shown in a specific account, the value of which is restated through the recording of a provision.

  There are two possible ways of stating this provision: by overall calculation at year end of debtor and customer accounts, or by an individualized system of monitoring debtor and customer account balances.
- Other debtors: short-term non-trade receivables – the same valuation methods are used for other debtors as for non-trade loans or financial investments (see VI.3.8). Non-trade loans are stated at the amount paid. The difference between this amount and the par value of the loans should be taken to income as interest revenues in the year in which such implicit interest is accrued, following a financial criterion and recognizing the interest credit on the assets side of the balance sheet.

  The related adjustments must be made in the case of doubtful debts.
- Short-term investments in securities – the accounts included in this subgroup are the same as those in financial investments but in this case are short-term accounts. Accordingly, the appropriate valuation methods are

those used for marketable securities which constitute long-term investments (see VI.3.7).

- Other non-bank accounts – these are current cash accounts held with companies of all kinds (groups, associated or other), with shareholders, administrators or any other individual or company, but excluding banks, bankers, credit entities, customers and suppliers.

  PGC part 2 includes in this subgroup current accounts with group and associated companies, shareholders and administrators, i.e. current cash accounts with the individuals or companies that the heading title indicates.
- Cash – liquid amounts in cash and immediately available current accounts at banks and other finance entities. Wherever an overdraft arises in these accounts the appropriate liability account should be shown. Cash accounts denominated in foreign currencies are translated into pesetas at the exchange rates prevailing at the date of their inclusion among assets. At year end these amounts are shown in the balance sheet at the exchange rates then prevailing. If, as a result of this valuation, exchange gains or losses arise, they are charged or credited, as appropriate, to the profit or loss for the year.

4.3 Accrual accounts

The 'pre-paid expense' account includes, by application of the accruals principle, the expenses recorded in the year which relate to subsequent years. Similarly, the pre-paid interest account reflects interest paid in the current year but which relates to the following year.

5 Valuation of shareholders' investments and liabilities

The equity items are discussed in detail above (see V.5.2).

5.1 Deferred revenues

Deferred revenues are those that are stated currently but which must be allocated to future years as they are earned pursuant to the accruals principle (e.g. deferred interest revenues) or as the costs relating to them are incurred (e.g. capital subsidies) or in the year in which they are realized or effectively settled (e.g. exchange gains). These headings reflect long-term revenue flows, unlike accrual accounts, which reflect short-term revenues, spanning portions of two calendar years. The accounts included under this heading in the PGC are as follows:

- Capital subsidies – non-refundable amounts received as a contribution to the financing of fixed assets. They must be allocated to the profit and loss account in proportion to the depreciation of the assets that have been financed with the subsidies, or in the year the asset is sold or written off if it is not amortizable.
- Deferred interest revenues – interest passed on to customers or debtors which, since it refers to long-term receivables, cannot yet be considered as revenue in the year in which the transaction is recorded, but will be progressively in subsequent years as the implicit interest is accrued.
- Exchange rate gains – where deferral is required by accounting regulations (see VII.5).

5.2 Provisions for liabilities and expenses

These provisions are defined by the chart of accounts as funds aimed at covering current or prior years' expenses and clearly identified losses or debts which at year end are probable or certain but cannot be exactly quantified, nor can the exact date on which they will arise be predicted with any accuracy. In order to value these provisions correctly, it is usually necessary to form reasonable estimates. The estimates must be based on an examination of each individual case, since general-purpose provisions can affect the quality of financial information. Where there are no grounds on which a sensible estimate can be based, the event in question may be treated as a contingency. As complete an explanation as possible should be included in the notes.

A clear distinction therefore needs to be drawn between provisions and contingencies. Provisions are an accounting reflection of events or situations giving rise to losses in the year, but which involve a degree of uncertainty, either because not all the relevant details or circumstances are yet known, or because, although the event in question has not yet taken place, it is highly likely that it will. A contingency relates to an event that may happen, even though there is no reason to suppose that it is imminent or even likely. In short, the difference lies in the degree of uncertainty. In line with the quantity and quality of the information to be recorded about such events, it should be mentioned in the notes on the accounts with an indication of the nature of the contingency, foreseeable variations and factors on which the contingency is dependent, plus as detailed an assessment as possible of the effects on the equity and profit and loss account. A highly probable contingency should be recorded as a provision.

The PGC provisions subgroup includes the following accounts:

- Provisions for pensions and similar obligations – funds aimed at covering legal or contractual obligations to company employees for retirement or other welfare benefits (to widows, orphans, etc.). Provisions for pensions and similar obligations include the estimated amounts required according to actuarial calculations in order to furnish the necessary internal funds to meet the legal or contractual obligations, without prejudice to the statement in the provision of interest in favour of the fund.
- Provision for taxes – this provision should include tax whose amount and date of payment are not known with certainty. Accordingly, it can be used under the following circumstances:
  - (i) Where an estimate of corporation tax is required during the year, when all the factors determining the final amount to be paid in this connection are not yet known.
  - (ii) Where a company wishes to record tax contingencies, since it is deemed that certain circumstances have made the contingencies probable but the exact amount thereof must be estimated.
  - (iii) Where appeals against tax assessments are highly likely to fail.
- Provisions for liabilities – an amount aimed at covering probable or certain liabilities arising from litigation in progress or from outstanding indemnity payments or obligations of an undetermined amount, such as collateral and other similar guarantees provided by a company.
- Provisions for major repairs – this provision is recorded for extraordinary overhauls or repairs of tangible fixed assets. Applying the accruals principle, the costs incurred in such repairs must be accrued annually, usually on the basis of wear and tear.
- Reversion reserve – this reserve relates to the recoupment of the value of the revertible assets on the basis of the reversion conditions stipulated in the concession. Accordingly, the chart of accounts does not include this reserve in equity. The licensor recovers the value of the assets over the term of the concession through the creation of a provision, the amount of which must be estimated and recorded on a systematic basis, regardless of the favourable or unfavourable results arising from the management of the company's activities.
- Provisions for operating liabilities – amounts aimed at covering sales return expenses, repairs, after-sale inspection and similar liabilities under warranty. The chart of accounts places these provisions on the liabilities side of the balance sheet under the heading 'Current liabilities'.

5.3 Payables

- Non-trade payables – non-trade debt is recorded at its redeemable value. The difference between this amount and the amount received is shown separately as deferred charges on the asset side of the balance sheet. The difference is progressively charged to results over the accounting years to which the transaction relates by reference to a financial criterion. Debts on fixed asset purchases are valued at face value. The interest included in the face value, excluding that which is included in the value of the fixed assets, is stated separately on the asset side of the balance sheet, and progressively charged to results according to a financial criterion. Credit accounts are recorded at the drawn amount and must be explained in the notes on the accounts. Debts

Table VI.4 Information to be disclosed in the notes with regard to valuation rules

Items	Information to be included in the notes
Formation expenses	The criteria used for capitalization, depreciation and, where appropriate, write-off.
Intangible fixed assets	The criteria used for capitalization, depreciation, provision allocation and, where appropriate, write-off. If the goodwill depreciation period is extended beyond five years, it must be explained and justified. The criteria applied to leasing transactions.
Tangible fixed assets	The criteria applied to the following items must be fully detailed: • Depreciation and provision allocation. • Capitalization of financial expenses and exchange losses. • Accounting for the renewal, extension and enhancement of assets. • Valuation of the costs incurred on in-house work on fixed assets. • Those tangible fixed assets which are included on the asset side of the balance sheet in fixed amounts. • Regularization of asset value in conformity with legislation.
Marketable securities and similar financial investments, both long- and short-term	The valuation criteria, and in particular the criteria applied to valuation adjustments.
Non-trade loans, both long- and short-term	The valuation criteria, and in particular the criteria applied to valuation adjustments and, where appropriate, the interest accrual.
Stocks	The valuation criteria, and in particular the criteria applied to valuation adjustments. In addition, the valuation criteria for those items which are included on the asset side of the balance sheet in fixed amounts must also be detailed.
Own shares purchased by the company	The valuation criteria applied.
Capital subsidies	The criteria by which they are credited to the profit and loss account.
Provisions for pensions and similar obligations	The accounting criterion and the method of assessing the risks covered by the related provisions.
Other provisions for liabilities and expenses	The accounting criterion and the method of assessing the risks covered by the related provisions.
Debts, both long- and short-term	The valuation criteria and the financial criterion used for charging both implicit and explicit interest to results.
Corporate tax	The accounting criterion
Foreign currency transactions	It will be necessary to indicate: • Valuation criteria of foreign currency balances. • Exchange rates used for translating into pesetas the items originally denominated in foreign currencies. • Accounting for exchange rate differences.
Revenues and expenses	The accounting criteria applied.

denominated in foreign currencies are translated into pesetas at the exchange rates prevailing at the transaction date; at year end they are valued at the exchange rates ruling then. In exchange rate hedging transactions (exchange rate insurance and similar coverage) only the unhedged element of the transaction is taken into consideration.

The accounting treatment of exchange rate gains and losses arising as a result of applying the valuation method outlined above is explained in greater detail below (see VII.5).

- Trade accounts payable – trade loans are stated in the balance sheet at their face value. Long-term interest included in this value is recorded in the deferred charges account on the asset side of the balance sheet, and is transferred annually to results by reference to a financial criterion.

5.4 Accrual accounts

These accounts, by applying the accruals basis of accounting, under the heading 'Revenues collected early', comprise any revenues recorded in the current year which relate to the following year. Also, the 'Interest collected early' account shows interest collected during the year which relates to subsequent years.

6 Information to be disclosed in the notes with regard to valuation rules

The accounting rules followed in connection with certain items must be indicated in the notes (see Table VI.4).

VII Special accounting areas

1 Accounting treatment of corporate income tax

1.1 Calculation of the tax incurred

The accounting treatment of Corporate Tax is regulated by the PGC Valuation standard 16 which has been developed through three ICAC resolutions. The first of these was on 25 September 1991, the second on 30 April 1992 and the third which brings the second up to date and substitutes it was on 9 October 1997.

The accounting of Corporate Tax is based on two basic premises:

- Corporate Tax is an expense in calculating the business profit and should be accounted as such.
- Like any expense, it should be calculated following generally accepted principles, especially with respect to the amount due, the correlation of income and expenses and prudence in valuation.

In accounting for the Corporate Tax expense, Spanish mercantile norms have opted for the Tax Effect method, so that on the one hand the expense to be accounted should be calculated as a tax on profits in line with accounting practices, and on the other a tax effect assessment is carried out in order to calculate the amount to be paid to the public purse, taking account of existing regulations governing this tax in the fiscal area. The two calculations will generally give two different results since accounting and fiscal norms fail to coincide on many points. See Tables VII.1 and VII.2.

As can be seen, the differences that exist between the accrued Corporate Tax expense and the differential tax payable can only be due to two reasons:

- Timing differences between fiscal and accounting areas which arise from the exist-

Table VII.1 Calculation of accrued corporate income tax

	Profit per books before tax
±	Permanent differences
=	Adjusted profit per books
×	Tax rate
=	Gross tax
±	Tax relief and tax credits
=	**Accrued corporate income tax**

Table VII.2 Calculation of the tax liability

	Economic result before taxes
±	Permanent differences
±	Timing differences
=	Preliminary tax basis
–	Negative tax basis from previous years
=	Tax basis
×	Rate of levy
=	Gross payable tax
–	Deductions and allowances
=	Net payable tax
–	Tax deducted and payments on account
=	**Tax liability**

ence of different methods of allocating expenses and revenues over time.

- The tax basis from previous years, given that fiscal regulations allow a business that makes a loss in one year to compensate for them over the following seven years, as set forth in Art. 23 of the Corporate Tax Law (*Ley del Impuesto sobre Sociedades)* (Law 43/1995).

If the timing difference implies that the business is going to pay a sum that is less than the accrued corporate income tax expense, then the difference arising is of negative sign, that is, an obligation to the tax authorities is created which is accounted for as deferred tax. If the contrary is the case with a sum payable being greater than the accrued corporate income tax expense, then a positive entry is made and implies the creation of right which is expressed in the accounts as anticipated tax. The differences, be they negative or positive, will be eliminated in future financial years when they revert with contrary sign to their creation, something that will imply payment of a greater amount in the first case, and in the second, a lesser amount to the tax authorities in these later years.

The possibility of compensating for losses in future financial years that arise in the year in which a negative result is achieved gives rise to a credit with the tax authorities, something that would mean a reduction in payable Corporate Tax in future financial years (a maximum of seven as commented above).

What emerges from all of the above is that in Spain the basis for calculating the amount due in Corporate Tax is taken from the profit and loss account. Nevertheless, international regulations (FAS-109 and NIC-12) suggest a new method for calculating Corporate Tax based on the balance sheet. This means the method used in Spain would give rise to 'Timing Differences', whilst international norms give rise to 'Temporary Differences'. Nevertheless, it must be stated that at the moment Spain is beginning to consider this new proposal based on the balance sheet even though there is no regulation in Spain which refers to it for the time being.

Finally, it should be pointed out that there are certain special tax regimes which exist because of territorial questions by virtue of what are called Economic Arrangements *(Conciertos Económicos)*, which are agreed between the State and an autonomous community (regional government). Specifically, there are two arrangements of this kind, one with the Basque Country by virtue of Law 38/1997, of 4 August 1997, and another with Navarra in accordance with Law 19/1998, of 15 June 1998. We should highlight the fact that determination of the applicable regulation and the tax regime will basically depend on the place where the operations are carried out so that the same business may pay tax to various administrations.

The Autonomous Regulation of Corporate Tax in Navarra is governed by Autonomous Community Law *(Ley Foral)* 24/1996, of 30 December; by Autonomous Community Law 22/1997, of 30 December, which modifies some of the articles in the previous law; and by an Autonomous Community Decree, whilst in the Basque Country there are different regulations for each of the historically chartered provinces *(fueros)* that make up this autonomous community, these being Vizcaya, Álava and Guipúzcoa.

Comparatively speaking, the Autonomous Regulation of Corporate Tax includes an incentive scheme which is substantially more favourable than State regulations, the following being the most relevant: greater deductions (for both investment and employment creation); more benevolent tax rates; special regimes for small and medium-sized enterprises; co-ordination, management and financial centres; companies to promote enterprise; measures to assist shipping and fishing enterprises.

1.2 Accounting treatment of timing differences

PGC Valuation Rule no. 16 and the ICAC Resolution of 9 October 1997 will have to be taken into account as this lays down that the liability items arising from deferred taxes should be accounted for in every case, whilst the assets relating to anticipated tax payments and credits for losses to be compensated will only be accounted for when future realization is reasonably certain, all of this being by virtue of the principle of prudence. In addition, other assets already included in the accounts will have to be removed along with those about which logical doubts arise in relation to their future recovery. In this way the inclusion of doubtful assets on the asset side of the balance sheet is avoided, since the annual accounts should show a true and fair view of the business's assets, financial situation and results.

In the case of credits derived from negative tax bases, these will only be included in the accounts if they meet the following conditions:

- That the loss is the consequence of something that is not habitual in the management of the business.
- That sufficient future profits are foreseen over a period not exceeding seven years in order to compensate for said negative bases.

Concerning anticipated taxes, it is presumed that the future realization is not sufficiently guaranteed and therefore cannot be accounted for:

- When reversion is going to take place in a period of time exceeding 10 years.
- When the business habitually runs up losses.

Nevertheless, anticipated taxes may be included in the accounts in the case of recovery in more than 10 years if there are deferred taxes of an equal or greater amount and which have the same reversion term.

According to the Final Disposition 7.1 of RD 1643/1990 which approves the PGC, it should be noted that fiscal assets and liabilities that arise as a result of the differences commented on above cannot be compensated for, not even those that derive from positive and negative differences that come about in the same year.

Account must be taken of what is set out in PGC valuation rule no. 16, which indicates that when modification of tax legislation or the evolution of the economic situation of the enterprise give rise to a variation in the amount of anticipated taxes, tax credits or deferred taxes, then the balance of the aforementioned accounts will be adjusted and included in the profit and loss as income or expense as is the case.

In this regard, the Eighth rule of the ICAC Resolution of 9 October 1997 establishes that any change known before the formulation of the annual accounts which affects the accounting of the tax effect will have to be taken into account in quantifying the accrued corporate income tax expense. In particular, if the variation in the tax rate is known before formulation of the annual accounts then the amount of anticipated and deferred tax will be adjusted, as will credits derived from negative tax bases.

1.3 Accrual for permanent differences and for tax credits and tax relief

The ICAC Resolution of 9 October 1997 establishes that permanent differences, as well as for tax credits and tax relief may be the object of periodification in the following cases:

- When the permanent differences mean an abatement of the accrued corporate income tax expense.
- Any deduction and tax rebate on Corporate Tax, excluding deductions and payments on account, even so, only those deductions and tax rebates fiscally applied in the tax returns of the year in question may be considered deferrable.

The deferrals mentioned above are effected by correlating the reduction in the expense item of Corporate Tax with the depreciation of the asset which caused the permanent difference, deduction or tax rebate. Moreover, information will have to be provided in section 4 of the PGC notes to the accounts forms devoted to 'Valuation rules' about the criterion used in deferring over all of the financial years it affects and until it is completed.

1.4 Foreign taxes

The ICAC Resolution of 9 October 1997 makes reference to foreign taxes that are similar in nature to the Spanish Corporate Tax, establishing that the accrued expenses of said taxes are recorded in the same way as accrued corporate income tax expense, even so, 'agreements on double taxation' should always be taken into account. Moreover, companies that are subject to tax payment abroad should provide full information in the report about all those aspects that affect their annual accounts.

1.5 Permanent differences

Permanent differences arise as a consequence of different criteria when defining the expenses and income from an accounting and fiscal point of view. The main permanent differences are those which are commented on below.

The Corporate Tax Law (Law 43/1995, Art. 14) establishes that the following cannot be considered tax-deductible expense and therefore do not give rise to non-accounting adjustments which come about from the creation of permanent positive differences:

- Those which represent a payment from own funds. This point should highlight the case of management share of profits, since when legal and statutory reserves are endowed and shareholders are paid a dividend of 4 per cent or more, if this is what is set forth in the statutes, said amount will be considered a tax deductible expense.
- Those derived from accounting for Corporate Tax.
- Fines and administrative and penal sanctions, penalty payments for late presentation of returns/settlements and self-assessments.
- Gambling losses.
- Donations and charitable subscriptions. On this point it should be noted that the law indicates that deductions are allowed on neither expenses relating to public relations with customers or suppliers or those affecting the uses and customs that are effected with respect to the business's personnel, nor those carried out to directly or indirectly promote the sale of goods and the provision of services, nor those that are found to be correlated with income.
- Expenses for services corresponding to operations carried out directly or indirectly with persons or entities resident in countries or territories that are defined as tax havens, or payments that are made through persons or entities resident in said countries, except when it can be shown that the accrued expense is in response to an operation or transaction which has effectively been carried out. It should be made clear that the rules governing international fiscal transparency will not be taken into account in the case of tax deductible expenses.

The accounting result derived from the transfer of tangible and intangible fixed assets is calculated by the difference between the sale price and its net accounting value. Nevertheless, the Corporate Tax Law (Law 43/1995, Art. 15.11) indicates a different treatment when incorporating said result into the tax basis when dealing with assets that were acquired after 1 January 1983, given that it is not desirable to encumber the purely monetary aspect of the result and hence allow the accounting profit to be reduced to the estimated extent corresponding to the effects of inflation. The calculation of monetary depreciation is carried out through the use of correcting indices that are approved annually and are applied to both the acquisition value as well as to periodic amortization; this always means a permanent difference between the accounting and fiscal results.

Account should also be taken of the fact that the monetary correction will be the same as in the case of updated asset items being transferred as laid down in Art. 5 of Royal Decree-Law 7/1996, since this is permitted under the provisions of Law 12/1996. But what is more, and must be made clear on this point, is that Royal Decree-Law 7/1996 contains a rule devoted to protecting the interests of the tax authorities for cases where losses are incurred in the transfer of elements whose value has been updated, minimizing the losses incurred, to the effects of integration within the tax basis, on the amount corresponding to the element in the 'Revaluation reserve' account, so that to this end it will be necessary to make a positive non-accountable adjustment due to the permanent difference which is made apparent (see VIII).

Finally, it is necessary to take account of Law 43/1995, Art. 15, which sets out fiscal valuation rules, and note that accounting values are generally accepted, that is, the acquisition price or production cost is taken up as its value. Nevertheless, the same article sets out some precautions to be taken in valuing certain operations where a certain distortion can occur with respect to market values. Of all of the cases that are set forth in fiscal norms, the following two are worth highlighting: entailed transactions and exchange transactions. The following contemplates both of these concepts.

According to Law 43/1995, Art. 16 and the Regulation governing Corporate Tax (Royal Decree 537/1997, Arts. 15 to 28), it can be seen that permanent differences with either a positive or negative sign may appear when there are transactions between entailed persons, specifically when the transfer price is less than the market price. This has a negative effect on the tax authorities either because of the reduction in levy or because of its deferral. The Administration has the prerogative to modify the transfer price to bring it in line with market prices, so that the adjustment carried out should be bilateral and tax payments of the adjustment will affect the period in which the transactions were carried out. Since the taxpayer is not obliged to make any adjustment, when the administration decides to carry out an adjustment on the valuation carried out by the enterprise, the corresponding permanent positive or negative differences arise as is the case.

With regard to the exchange transactions, these are regulated under Law 43/1995, Arts. 15, 16 and 17, where it is made clear that the discrepancy between fiscal and accounting interpretations arise when the market price of what is received is greater than the net accounting value of what is ceded. Since, fiscally speaking, the former prevails and accounting values come second, this will lead to the creation of future permanent differences which will become apparent when amortization is carried out or when the item is transferred.

1.6 Timing differences

The different timing allocation criteria in accounting and taxation when allocating income and expenses are going to mean the creation of timing differences. The passage of accounting profit recorded on the books, to the fiscal profit will take place once positive and negative non-accounting adjustments are considered and those that are accounted for having been derived from possible timing differences which will lead to the creation of assets and liabilities.

1.6.1 Interaction between accounting and tax methods of allocation

Art. 19 of Law 43/1995 contemplates the timing of income and expense allocation, setting out the general criterion of allocation during the tax period in which it is accrued, that is, the principle of accrual is accepted as it is used in accounting. Nevertheless, Corporate Tax Law contemplates the possibility of an enterprise exceptionally using different criteria of income and expense allocation, as long as this is to achieve a true and fair view of its net worth, financial situation and results, in which case the taxpayer should request approval of the criteria used from the Revenue Authority. To do this, the procedures laid down in Royal Decree 537/1997, Art. 29. have to be followed. Even so, in cases like this, the following have to be taken into account:

- Expenses that have not been accounted for in the profit and loss account or in a reserve account, if such is in line with a legal norm or regulation, will not be tax-deductible, with the exception of those asset items which can be freely amortized.
- Expenses accounted for in the profit and loss account in a tax period later than the one in which it should have been recorded, or the income allocated in said account in an earlier period, will each be allocated to the tax period in which they have been assigned in the accounts, as long as this does not mean less tax payment than would have been payable by applying general timing allocation rules.

1.6.2 Specific types of timing differences

The following sections will consider the main timing differences that can crop up in the Corporate Tax returns that companies have to file.

1.6.2.1 Instalment or deferred payment transactions

In analyzing deferred transactions, Law 43/1995, Art. 19 states that revenue will be understood to be proportionally obtained as payments are made, except when the company decides to use the accrual principle. For this reason, as long as the company opts for the former, that is, fiscal allocation along the lines of cash criterion, since in accounting the principle of accrual is followed, a timing difference will arise with a negative sign and this will give rise to a deferred tax which will revert when payments for corresponding sales come about. A deferral in the payment of tax therefore comes about. In the second case, that is, when the company also follows the accrual principle for tax purposes, no difference will arise.

Law 43/1995 establishes what should be understood as deferred transactions or with deferred payment, indicating that these will be those sales and work carried out in which the price is wholly or partially received by means of successive payments or a sole payment, as long as the time taken between delivery and the maturity of the last or sole payment is greater than a year.

1.6.2.2 Financial leasing

When we consider fiscal regulation of financial lease transactions we have to go to Law 43/1995, Art. 128, which firstly sets forth the type of contracts to which said article refers. They are those which make reference to the seventh additional clause of Law 26/1988 with regard to Lending Agencies' Discipline and Intervention. Specifically, it considers those contracts that have a minimum duration of 2 years when they are the object of movable assets, and 10 years when they are the object of fixed assets or industrial establishments.

It should be remembered in relation to these contracts that the financial charges paid by the leaseholder to the leasing company will be considered a tax-deductible expense. In the same way, the leaseholder will also account for said amount as an expense, that is, no differences between accounting and tax treatment will become apparent.

On the other hand, the financial leasing instalments paid that correspond to the recovery of the cost of the good will also be a tax-deductible expense in relation to Corporate Tax as long as it is depreciable, but there is a limit for this fiscal expense item whose amount is the following: two times the linear amortization coefficient according to the amortization tables applied to the asset in question. If the return cost of the asset exceeds this limit then the excess will be deductible in successive tax periods in which the limit will also have to be respected. This fiscal expense will have to be compared with the accounting expense, which will be the amount that the company has accounted for amortization of the leased item and which will always be of a lower amount. This means a timing difference with a negative sign will arise, that is, a deferred tax appears and this will revert once the purchase option is taken up by the leaseholder so that by then there will be no tax-deductible expense while the accounts will show the expense as amortization of the acquired asset.

1.6.2.3 Provisions for contingencies and expenses

Various provisions are included in the PGC in this section, even though in every case its funding will imply accounting for it as an expense. Nevertheless, it is very different from a fiscal point of view. Analysis of the first section of Law 43/1995, Art. 13 specifically establishes that as a general rule, funding of provisions to cover foreseeable risks, eventual losses, expenses or likely debts will not be deductible. Nevertheless, Corporate Tax Law later states a series of exceptions to the aforementioned general rule which are commented upon below.

It should be remembered that funding not specifically dealt with by law will be considered non-deductible and will therefore give rise to a positive timing difference with the creation of the subsequent anticipated tax that will revert when the risk, loss or expense cease to be foreseeable and therefore really become apparent.

The following will be tax-deductible:

- Provisions relating to legal responsibility underway or derived from indemnity payments or duly documented payments pending whose amount is not finally established.
- Provisions for the recuperation of returnable

assets, attending to the return conditions established in the concession, without prejudice to the amortization of items susceptible to the same, in such a way that the balance of the return fund be equal to the accounting value of the asset at the time of the return, including the amount for damages demanded by the assignor entity for reception of the asset.

- Provisions that sea-fishing and marine and air navigation enterprises set aside for essential large-scale repairs because of the general inspections to which sea-going vessels and aircraft are obliged to be submitted. Permission must be sought from the Tax Authority for provision funding to be tax-deductible for other types of extraordinary repairs of asset items, as laid down in Royal Decree 537/1997.
- Provisions for covering repair and service guarantees and to cover sundry expenses arising from returned goods up to the following amount: (A×C)/(A+B)
  Where:
  A: The sales total in the tax period whose provision allowance we are calculating.
  B: The sales total in immediately preceding two tax periods.
  C: The expense total incurred in providing coverage over the three periods referred to in points A and B.

The treatment applicable to pensions and similar obligations should, in turn, also be highlighted. The Corporate Tax Law lays down that funding of external funds that are subject to Law 8/1987 governing Pension Funds and Plans, will be deductible. Even so, it should be remembered that said Law has been modified by Law 30/1995, covering Private Insurance Supervision and Regulation. This law lays down that the contribution to the fund cannot exceed one million pesetas per participant and this is therefore the amount that can be deducted. So, if the enterprise follows the same criteria in its accounting then no kind of differences ought to arise. If the limit is not respected in the accounts, then a timing difference will come about and will need a positive non-accounting adjustment.

Law 43/1995 also mentions coverage of contingencies similar to those of pension plans, which can equally be a tax-deductible expense as long as they comply with the following requirements:

- That contributions be fiscally imputed to those persons associated with the benefits.
- That the right to receive future benefits be irrevocably transferred.
- That the ownership and management of the resources that make up said contributions be transferred.

Finally, it is necessary to look at Art. 14 of Law 43/1995 which establishes that provisions for internal funds will not be considered tax-deductible. In this case an anticipated tax payment will come about because when the enterprise records the endowment it will be collected with an accountable but not fiscal expense. But when the time comes to make the payments to the employee from the fund, then it will not be an accountable expense but it will be a fiscal one. This means that what arises in this case is a positive timing difference.

Following discussion of the above, all that remains is to take account of the ICAC Resolution of 25 September 1991, which establishes that only anticipated taxes derived from the endowment of pension provision and similar obligations will be able to be included on the asset side of the balance sheet, up to the amount corresponding to payments for benefits that are going to be realized using said provision in the 10 years following the closing date of the financial year and as long as there are no reasonable doubts as to whether sufficient profits are obtained to meet payments. Nevertheless, the ICAC permits fiscal credits which correspond to payments over periods greater than 10 years, when they are covered by deferred taxes whose reversion year is the same as that of the payment of benefits. This said, it may be noted that it is possible for the aforementioned differences to be permanent rather than timing differences.

1.6.2.4 Depreciation and amortization

In this section the first thing to observe is Law 43/1995, Art. 11, since this is where the general requirements are established so that amortizations are a tax-deductible expense. Specifically, the first section of said article establishes the tax-deductible status of amounts that, as amortization of tangible or intangible fixed assets, correspond to the effective depreciation which different items suffer due to functioning, use, enjoyment or obsolescence. It also indicates that depreciation is effective when:

- It is the result of applying linear amortization coefficients established in the Official Amortization Tables, which figure in the Annexe to Royal Decree 537/1997.
- It is the result of applying a constant percentage on the outstanding value of depreciation.
- It is the result of applying the digit number method.
- It falls within a plan formulated by the taxpayer and accepted by the Tax Authority.
- The taxpayer justifies the amount.

In addition to all of the above, it is also necessary to analyze Royal Decree 537/1997, Art. 1, which establishes more general rules:

- The acquisition price or cost of production will be depreciable, with residual value being excluded.
- Depreciation will be practised item by item. When dealing with asset items of a similar nature or subject to a similar degree of use, depreciation will be able to be carried out on them as a group, as long as the accumulated depreciation corresponding to each asset item can be known at all times.
- In dealing with an asset element, different amortization methods will not be able to be used either simultaneously or successively. Nevertheless, in exceptional cases which are indicated and justified in the annual accounts report, the provisions of this Royal Decree allow for the application of a method of amortization that differs from that which had been previously employed.

After looking at the general criteria which affect both tangible and intangible fixed assets, the following section considers specific aspects to take into account.

Tangible fixed assets – asset items of this kind start to depreciate as soon as they are ready for use and throughout the whole of its useful life. From an accounting point of view, depreciation is practised taking account of the effective depreciation; nevertheless, differences sometimes appear during the financial years in which the amortization is carried out and these give rise to timing differences which can be either positive or negative.

Worth a special mention is the freedom to amortize referred to in 43/1995, Art. 11.2, since this allows tax payment to be delayed because the amortization is carried out much more quickly than in the accounts, hence giving rise to timing differences with a negative sign.

Intangible fixed assets – these items also have to be depreciated taking account of their useful life. This is understood as the period in which it is reasonably hoped that they will produce income. Up to this point, fiscal and accounting rules stay fairly close together, although in order to see the possible differences that can arise, specific aspects of both have to be considered. Accounting practice indicates the following:

- Research and development expenses and computer software have to depreciated over a maximum period of five years.
- The trade fund and transfer rights are systematically depreciated within a period that cannot exceed the period during which they contribute towards obtaining income for the company, with a maximum limit of 10 years, even though from the fifth year the annual report should include the appropriate justification.
- Industrial property and administrative concessions are, in general, depreciated within the period of years during which it is hoped that the company's operations can benefit.

The following have to be taken into account in considering the fiscal implications:

- Research and development expenses can take advantage of freedom to depreciate, in which case a negative timing differences will arise.
- Computer software is not specifically included in Corporate Tax legislation and should be included in Art. 11.5.c of Law 43/1995, which indicates that all intangible fixed assets not expressly mentioned and without a specific expiry date are depreciated over 10 years. In view of this being five years in accounting practice, a negative timing difference will arise.

- For the trade fund a maximum period of 10 years is granted so that in general no differences arise in relation to this concept.
- Transfer rights also have to depreciated in 10 years, except when the contract has a shorter duration, in which case the time limit will be less and will be set by the contract itself. This means that no differences will arise from this item
- Industrial property and administrative concession also have the same 10-year limit and since accounting depreciation is carried out over a longer period there will be negative timing differences.

1.6.2.5 Provision for depreciation of investment securities

A distinction has to be made on this point between investment securities that are quoted on a secondary organized market from those that are not quoted, by virtue of Law 43/1995, Art. 12. Provisions relating to shares in the capital of enterprises that are quoted, which are not part of group or associated companies are treated differently, and are treated equally for tax and accounting purposes so that no differences arise. When dealing with shares that are not quoted, there is a fiscal limit to the amount that can be deducted for provisions and this is calculated as the difference between the theoretical accounting value at the beginning and end of the financial year. In this case differences can appear between accounting and fiscal areas because from an accounting point of view the provision which is funded is calculated by comparing the theoretical accounting value with the historical entry price, which should be corrected for possible tacit capital gains existing at the time of acquisition and which still exist at the later valuation. On the other hand, when analyzing what happens with provisions that affect fixed-interest securities, it should be noted that differences can arise both if they are quoted and if they are not quoted in an organized secondary market. In the first case, there is a fiscal limit on the amount to deduct on the basis of depreciation and this consists of the sum obtained from global depreciation in the tax period in the taxpayer's entire portfolio of quoted fixed-interest securities. Nevertheless the analysis for accounting purposes is carried out on each investment and the calculated amounts may exceed the previous limit, so positive timing differences would arise. Finally, considering the case of non-quoted fixed-interest securities, tax legislation forbids the deduction of endowment for depreciation of security that has a certain repayment value and which are not quoted on organised secondary markets. This means that if a endowment provision is included in the accounts for these stocks then a timing difference will arise and will, in the same way as the previous case, also be positive.

1.6.2.6 Allowance for bad debts

This aspect is regulated by Law 43/1995, Art. 12, which first establishes the circumstances which allow the deduction of endowment and these are as follows:

- A year has passed since the maturity of the right to receive payment from the debtor.
- The debtor is declared bankrupt, in insolvency proceedings, temporary receivership or involved in similar proceedings or situations.
- The debtor is being prosecuted for fraud.
- The rights have been claimed in court or are the object of legal proceedings or arbitration proceedings whose outcome will determine payment.

At the same time, the circumstances which block the deductibility of endowment are established:

- The credits are owed or guaranteed by public corporation.
- The credits are guaranteed by financial institutions or by reciprocal guarantee companies.
- The credits are guaranteed by means of pledge, possession reserve agreement and retainer rights, except in cases of loss or degradation of the guarantee.
- Those guaranteed by means of a credit or risk insurance contract.
- Those that have been the object of renewal or extended.

In addition, it also establishes that endowment based on global estimations of the risk of customer and debtor insolvency will not be deductible, something which is allowed in accounting. Due to the principle of prudence, accountancy legislation indicates the need to endow this type of provision as long as there is the possibility of non-payment so that timing differences will arise in many cases with a positive sign causing the appearance of the corresponding anticipated tax.

1.7 Tax incentives for small businesses

Law 43/1995 devotes the whole of Chapter XII to tax incentives for small businesses, as do Arts. 40 to 45 of Royal Decree 537/1997. In defining a business as small, tax regulations take account of its business turnover so that when it is less than 250 million pesetas in the tax period prior to the one under consideration, the enterprise in question can take advantage of tax incentives which are analyzed below:

- Depreciation freedom: in order to enjoy this incentive, an additional requirement is to reach an increase in the enterprise's total average workforce over the 24 months following the start of the tax period in relation to the previous 12 months, and, moreover, that said increase is maintained for the following 24 months. Compliance with this means that the amount that can be taken advantage of from this incentive will be what results from multiplying 15,000,000 by the average increase in the workforce. The depreciation freedom indicated here can be applied to new tangible fixed assets as well as those acquired through leasing, as long as the purchase option is exercised. In addition, there is also freedom of depreciation without complying with the requirement relating to the workforce for items of tangible fixed assets whose unitary value does not exceed 100,000 pesetas, with the limit for this kind of deduction being 2 million pesetas.
- Depreciation of new tangible fixed assets: new tangible fixed assets may be amortized on the basis of the coefficient resulting from multiplying the maximum linear amortization coefficient set forth in official tables by 1.5.
- Endowment for debtor insolvency: the limit to tax deductions is increased for small businesses so that the maximum tax-deductible limit established for these is one per cent of the balance of payments due at the end of the tax period.
- Exemption due to reinvestment: this incentive will mean non-inclusion in the tax basis of asset increases obtained from the sale of tangible fixed assets, as long as the amount obtained is reinvested in assets of a similar nature. Nevertheless, there is a limit of 50 million pesetas so that any sum exceeding said amount can be used for the reinvestment of non-recurring profits. In addition, the taxpayer will have the chance to present special reinvestment plans, as long as special circumstances prevail.
- Leasing: in this case the enterprise may deduct three times the maximum linear amortization coefficient because, on the one hand, on acquiring new intangible assets the depreciation is tax-deductible, as stated above, and can be calculated applying the percentage resulting from multiplying the maximum linear coefficient by 1.5. On the other hand, that acquired through leasing, as also already seen in studying this kind of operation, tax-deduction is allowed in relation to double depreciation of what would be obtained by applying maximum linear coefficient. From all that is commented upon it can be deduced that the permitted deduction is three times the normal maximum depreciation, the calculation of said deduction being the following: the maximum linear amortization coefficient in official tables multiplied by 2 and by 1.5, that is, multiplied by 3.

1.8 Presentation of deferred taxation in the annual accounts

The ICAC Resolution of 30 October 1997 establishes the content that should be included in the annual accounts with respect to everything relating to Corporate Tax.

The effects of timing differences must be included in both the balance sheet and the profit and loss account, as well as in the report (see Table VII.3).

In addition, conciliation of the result with the tax base for Corporate Tax should be included in point 15 of the report (see Table VII.4).

Table VII.3 Presentation of deferred tax in the annual accounts

Information to be presented on the Balance Sheet

Anticipated taxes shall be presented as an asset in section D.III 'Circulating assets', entry, 'Public Administration'.

Anticipated taxes shall be presented as a liability in section E.V. 'Short term creditors', entry, 'Public Administration'.

If there is credit for losses to be compensated in future financial years, since these represent an asset similar in nature to the anticipated tax, these are included in the same asset section on the balance sheet.

When accounting for long-term anticipated taxes or credits for losses to be compensated for in a future year in the long term, they should be included in these accounts on the asset side of the balance sheet in section B.IV. 'Fixed financial assets', entry 'Long term Public Administration' which is created to such effect.

When accounting for long-term deferred taxes or credits, they should be included on the liability side of the balance sheet in section D.IV. 'Other creditors', entry 'Long-term Public Administration' which is created to such effect.

Information to be presented in the profit and loss account

The amount due in Corporate Tax for the year in which the annual accounts are drawn up should be reflected. The account that includes this amount is called 'Tax on profits' and is an expense.

Nevertheless, there is still the possibility of a credit balance cropping up and this implies less expense for the enterprise. It will figure in the results account after calculating the 'Pre-tax result' for arrival to the 'Year result', which may be positive or negative, indicating respectively the profit or loss that the enterprise has obtained.

Information to be presented in the company report

Note 15 of the report is concerned with collating everything relating to the enterprise's 'Tax situation', where in the first place presentation of the accounting result conciliated with the tax basis for Corporate Tax which is shown in Table VII.3. But the following information should also be presented:

- An explanation of existing anticipated and deferred taxes, whether they have been generated in the current year or if they come from previous years, indicating the variations in the current year and those that will come about in later years.
- An explanation of the accounting and tax valuation differences arising in the year.
- Negative tax bases pending compensation and the terms for carrying this out.
- An explanation of the deductions and allowances generated in the current year and those that are pending application, independent of the year from which they come.
- Accepted promises in relation to tax incentives.
- Treatment applied to anticipated taxes and credits for compensation of negative tax bases in Corporate Tax and regarding the circumstances which led to them being recorded or not on the asset side of he balance sheet.
- The amounts of anticipated tax and the credits for compensation of negative tax bases not recorded on the asset side of the balance sheet, indicating as is the case, the term and conditions for being able to do it.
- When anticipated tax or credits for negative tax bases arise that come from a previous year and had not been recorded, information will be provided about the circumstance that caused them to arise with respect to those existing at the time at which said balance sheet assets were not recorded.
- The expense item of tax arising in the year and that whch is derived from previous years. Adequate details, as long as these are significant, will be provided about the amounts of deferred and anticipated tax which revert in the current year and those which revert in later years, as well as credits for negative tax bases which are compensated in the current year.
- The criteria used in periodifying permanent differences and deductions from the payable amount.
- Contingencies derived from Corporate Tax and in particular on the expense that derives from previous years.

In addition, note 4 of the report devoted to 'Valuation rules' should also include information relating to tax on profits, showing the criteria used in accounting for it.

If the company is able to formulate an abbreviated report it only has to provide the information for note 4 devoted to 'Valuation rules' since note 15 is not required in these cases.

Table VII.4 Tax reconciliation statement

Net profit			X
	Increases	*Decreases*	
Income tax	X	X	X
Permanent differences	X	X	X
Timing differences			
– originated in this period	X	X	X
– originated in prior periods	X	X	X
Prior years' tax loss carry-forwards			X
Taxable net profit			X

1.9 Rules governing the rate to be applied as of 1996

The standard tax rate in Spain is 35%. The following specific rates may apply:

- Insurance companies, social work institutions and the accident at work and professional illness insurance funds of the Social Security which meet the requirements set forth in the regulations will pay a tax rate of 25 per cent; reciprocal guarantee companies and re-finance companies regulated by Law 1/1994 which governs the Legal Status of Reciprocal Guarantee Companies, registered with the Bank of Spain; co-operative credit agencies and rural savings banks, except in relation to non-co-operative results which are taxed at the flat rate; professional associations, business associations, official chambers, trade unions and political parties; foundations, establishments, institutions and non-profit-making institutions that do not meet the requirements to enjoy the tax regime established in Law 30/1994 regarding Foundations and Fiscal Incentives for Private Participation in Activities of General Interest; employment promotion funds constituted under the *aegis* of Art. 22 of Law 27/1984 regarding Reconversion and Re-industrialization; and the unions, federations and confederations of co-operatives.
- Fiscally protected co-operative companies will pay 20 per cent, except for those aspects of the results that are not co-operative and on which the flat rate is levied.
- Entities that meet the requirements to enjoy the tax regime established in Law 30/1994 affecting Foundations and Fiscal Incentives for Private Participation in Activities of General Interest will pay 10 per cent.
- Property investment companies and property investment funds regulated by Law 46/1984, of Collective Investment Institutions shall pay 7 per cent. These have the nature of non-financial collective investment institutions and the sole objective of investing in urban property to rent it out. Residential property must represent at least 50 per cent of total assets.
- One per cent will be levied on investment companies regulated by Law 46/1984 affecting Collective Investment Institutions, whose shares capital is quoted on the Stock Exchange; investment companies and investment funds regulated by Law 46/1984, which have the nature of non-financial investment institutions and whose sole purpose is investment in residential property for renting it out; the fund for regulating the public nature of the mortgage market set forth in Art. 25 of Law 2/1981, of Regulation of the Mortgage Market.
- 40 per cent will be levied on entities devoted to the research and exploitation of hydro-carbons in the terms laid down in Law 21/1974, regarding Research and Exploitation of Hydrocarbons.

- 0 per cent will be levied on pension funds regulated by Law 8/1987, of Pension Plans and Funds.

2 Long-term contracts in construction companies

The adaptation of the chart of accounts to construction companies, enacted by a Ministry of Economic and Financial Affairs Order of 27 January 1993, established the regulations applicable to companies engaged on construction work spanning more than one business year. The ICAC regulation refers to companies which undertake the following activities:

- Preparatory work: demolition, earthmoving and drilling.
- General construction of buildings and other civil engineering work: erection of buildings, specialized civil engineering works (bridges, tunnels, etc.), surfacing and enclosure, motorways, highways, landing fields, railway lines, sports centres and other special-purpose structures.
- Installations in buildings and on work sites: electrical installations, thermal, acoustic and anti-vibratory insulation, plumbing, air conditioning and heating, etc.
- Completion of buildings and work sites.
- Hiring out construction and demolition equipment with operator.

The modification of the chart of accounts to suit long-term construction industry contracts takes into account the essential features of the industry, which is a major contributor to GDP, performs work to order, and can never be sure how the price quoted on a long-term project will be affected during the course of it. This situation has lead increasingly to the formation of joint ventures (*uniones temporales de empresas*, UTES), which are now a familiar feature of the industry.

The legislation opted for the general use of the percentage of completion method as the best suited to afford a true and fair view of annual results. Where technical and administrative reasons preclude the use of this method, because the conditions for recognizing revenues before completion of the work are not met, the completed contract method is applied instead. This method does not allow the allocation of revenues by periods on the basis of the percentage completed at the date of the financial statements. All the revenues are recognized in the financial year in which the completed work is handed over, or subsequently, when the construction contract has been completely fulfilled and all the requirements for recognizing the revenues have been met.

Order of 27 January 1993, point 18 prescribes the method of determining sales, revenues from completed work, and other revenues, for which it distinguishes 'commissioned work performed under contract' from 'non-commissioned or non-contractual work performed for subsequent sale'.

In the first type of work, revenues are recognized by the percentage of completion method and are determined according to one of the following two procedures:

- Through valuation of units of work at prices established in the contract, for which it is necessary to define a unit of measure representing the consumption of factors, or units of the end product.
- Based on the percentage of total revenues established with regard to the costs incurred.

Point 18 of the Order opts for neither of the two procedures, but merely reiterates the principle of consistency, according to which the same procedure must be used for all work, whether it involves short- or long-term projects. Nevertheless, if the aim of the two procedures is to attribute the revenues and expenses of each period appropriately, this goal is better achieved with the second, particularly if suitable cost accounting procedures are set up.

The Order confines the application of the percentage of completion method to fulfilment of two conditions:

- The contract estimates, as well as the revenues, costs and percentage of completion, can be reasonably accurately determined at any given moment.
- There are no unusual or extraordinary risks to hinder the development of the project, or doubts as to the customer's acceptance of the order.

Failure to meet either of these conditions must be explained in the notes on the annual accounts and, in accordance with the principle of prudence, the completed contract method must be applied.

With regard to 'non-commissioned or non-contractual work performed for subsequent sale', revenues are recognized when what the regulation refers to as the 'effective transfer' of the assets under construction has taken place, in accordance with the sales conditions, and the foreseeable costs, if any, incurred in completing the project already sold are known.

3 Post-balance sheet events

The chart of accounts does not lay down guidelines for distinguishing between events subsequent to the year end which are to be regarded as affecting the annual accounts at that date and those which are not. All it does is require information to be given in note 19 on the annual accounts about subsequent events which do not affect the accounts at that date but which outside users will find it helpful to know about, and subsequent events which could have an impact on the going concern principle.

4 Changes of accounting methods

Changes in accounting methods are permitted only in exceptional circumstances, since they constitute a departure from the principle of consistency, and must be duly reported in the notes on the accounts (note 2). In this connection, PGC Valuation Standard No. 21 explicitly states that the change must comply with the principles established in such standards and will be deemed to have taken effect at the beginning of the year. The cumulative effect of the variations in assets and liabilities calculated at that date must be shown as extraordinary results in the profit and loss account.

In this connection, changes of accounting method affecting items valued on the basis of estimates need not be regarded as such where they are due simply to additional information having become available or new circumstances having been ascertained.

5 Foreign currency translation (receivables and payables: foreign exchange gains and losses)

The accounting treatment of foreign exchange differences is subject to PGC Valuation Standard No. 14. The method adopted by the chart of accounts in connection with monetary items whose value is conditioned by foreign exchange fluctuations is conservative, based on the principle of prudence. It is inverse in that it requires unrealized exchange losses to be charged to profit, whereas exchange gains may be taken to profit only in very specific, well evidenced cases.

In accordance with this standard, fixed-income securities and receivables and payables denominated in foreign currencies are translated into pesetas at the exchange rates ruling on the transaction date and are adjusted at year end to the exchange rates then prevailing. In cases of exchange rate hedging, only the unhedged portion of risk should be taken into account.

The application of this translation method will give rise to positive or negative exchange differences, i.e. unrealized gains or losses, which will be realized in the year in which the underlying transactions are settled. The accounting treatment of such differences is based on the following rules:

- Positive or negative exchange differences, whether receivable or payable, must be classified by due date and currency, and for this purpose currencies which, although different, are officially convertible in Spain are grouped together.
- The unrealized negative differences in each group are generally charged to profit.
- The unrealized positive differences in each group are not usually credited to profit for the year but deferred until the maturity or early cancellation of the relevant fixed-income securities, receivables or payables, and recorded under the heading of revenues deferred on the liability side of the balance sheet.
- Notwithstanding the foregoing, provided that exchange losses in each homogeneous group have been charged to prior years' profit or to period profit, unrealized positive differences

may be credited to profit up to the amount of such negative differences.

- In addition, positive differences deferred in prior years may be credited to income as negative exchange differences to the same or a higher amount are recognized in each homogeneous group, or in the year in which the underlying transaction is settled.

At the instance of the Accounting and Audit Institute (ICAC) the Ministry of Economic and Financial Affairs recognized the specific accounting problems arising in the regulated sectors (electricity utilities, motorway toll concessionaires and Telefónica) and issued an Order dated 12 March 1993 covering the accounting treatment of foreign currency exchange differences by companies in those sectors. These new rules may be summarized as follows:

- Negative exchange differences must be distributed over the life of the foreign currency receivables, payables and fixed-income securities to which they refer.
- Negative differences attributable to prior years and to the current year must be charged to profit, in line with the principle of prudence.
- Negative exchange differences attributable to subsequent years, provided there is reasonable confidence that they will be recovered through future rate increases, must be recorded as deferred charges to achieve a due correlation between revenues and expenses.
- As a general rule, in view of the principle of prudence, positive differences are deferred until the due date of the underlying transaction.
- Provided that in prior years negative differences have been charged to profit, positive differences may be allocated to profit for the year up to the total of the negative differences charged to profit in prior years.

After this date, the Economics Ministry dictated two orders (of 18 March 1994 and 23 March 1994, respectively), which indicated the accounting treatment of foreign exchange gains and losses created in specific enterprises ('Canal de Isabel II', 'Hispasat', 'Ferrocarrils de la Generalitat de Catalunya', 'Iberia', 'Vuelos Internacionales de Vacaciones' y 'Aviación y Conmercio'. The accounting treatment specified in these rules, in much the same way as the previous ministerial order, also implies a less strict application of the principle of prudence in accounting for exchange rate differences.

6 Derivative instruments

Transactions relating to these kinds of instruments are not expressly mentioned in either the Commercial Code or in the Corporations Law. The PGC 1990 specifies that companies should provide information in the notes to the accounts about firm promises to purchase and sell inventory, as well as about future contracts relating to inventory (note 8 of the standard notes). The notes should also contain information about firm promises to buy or sell negotiable shares and other financial investments of a similar nature (note 8 of the standard notes).

As far as specific regulation of options and futures is concerned, we would highlight AECA document no. 19 which proposes a possible accounting treatment for options and futures on inventory, and no. 15 on financial investments, which also contains some dispositions relating to this kind of instruments. Nevertheless, because of the private nature of these documents, like any others issued by AECA, they are not obligatory. At the same time, in April 1995 ICAC published a draft rule for the accounting treatment of future contracts, even though the rule has still not been approved (July 2000). This draft regulates the accounting criteria for recording the main effects of futures transactions, the information to include that relates to this type of contract in the annual accounts and the specific accounts where the transactions should be recorded.

As far as sector rules are concerned, we should underline the fact that the Bank of Spain's circulars, that regulate aspects relating to the accounting and to the preparation of credit-granting institutions' financial statements, and the Securities and Exchange Commission's (CNMV) circulars that regulate the same aspects of Stock Brokers and Companies and the Collective Investment Institutions, contain rules relating to accounting for options and futures.

VIII Revaluation accounting

1 General aspects

There are two types of accountable revalorization:

(a) Free revalorization comes about when a company records an increase in the market value of any of its assets and proceeds to reflect this situation in its accounts.
(b) Revalorization authorized by law, in which the enterprises increase the value of their assets on the basis of a legal rule enacted to such effect, generally with the objective of eliminating distortions that inflation causes in accounting figures.

As far as the former are concerned, they have been abolished following mercantile reform so that the current rules (Commercial Code, or *Código de Comercio*, LSA, PGC 1990) establish that asset items should be registered in the accounts at the acquisition price or the production cost, without prejudice to the opportune value corrections that reflect effective or potential losses, the latter being application of the principle of prudence in valuation.

This means a change with respect to the situation that existed before the accounting reform, since firms were permitted to increase the value of their assets within the normative framework of the old General Accounting Plan, as long as the revalorization was beyond doubt.

In the second case, company asset revalorization has been permitted eight times on the basis of legally established rules. These readjustments have taken place in the years 1961, 1964, 1973, 1977, 1979, 1981, 1983 and 1996. In addition, companies that held shares abroad were able to regulate them in 1980 and those that were considered Basque, in relation to the Economic Arrangement (*Concierto Económico)* with the Basque Country, had the opportunity to carry out another update in 1990.

With the exception of the 1996 readjustment and the Basque one in 1990, there is no sign in the accounting of Spanish enterprises of other readjustments since transitory disposition no. 18 of Law 43/1995 (Corporate Tax Law) established the transfer to the legal reserve or to freely disposable reserves of the balances of accounts which were used as a counter-entry to previous revalorizations. The section below provides detailed analysis of the characteristics of the 1996 readjustment, regulated by Royal Decree Law 7/1996, of 7 June and by Royal Decree 2607/1996, of 20 December, because their effects on the balance sheets of a significant percentage of Spanish companies have been outstanding.

2 The 1996 revaluation

2.1 Conditions for taking advantage of the law

The revaluation could be taken up on a voluntary basis by:

(a) Those subject to payment of Corporate Tax because of personal obligation to contribute or because of the real obligation to contribute by setting up permanently on Spanish soil, as long as they carried out their accounting in accordance with that established in the Commercial Code. Companies that made up part of a group of companies had to carry out the readjustment operations on an individual basis.
(b) Those taxpayers liable for Personal Income Tax because of personal obligation to contribute and who carry out business or professional activities and have a real obligation to contribute by being permanently established in Spain, as long as they comply with the following requirements:
   - Taxpayers carrying out business activities who do their accounting in accordance with the Commercial Code.
   - Physical persons carrying out professional activities who use the record books (income, expenses, capital equipment and provisions for funds and supplies) with due diligence.
(c) Civil Companies, trust funds, co-ownerships and other tax-paying entities that carry out professional or business activities, as long as their accounts are in accordance with that laid down for the two previous sections in the appropriate case.

In spite of the above, it should be pointed out that those entities that pay tax in accordance with the laws prevailing in the historically chartered territories of Guipuzcoa, Vizcaya and Navarra, even though they could have taken advantage of the opportunity to readjust their balance sheets in 1996, to do so they were not subjected to the state rules commented on here, but rather to specific rules affecting the Autonomous Community *(Norma Foral)* 6/96 of revaluation in Vizcaya, Autonomous Community Rule 236/96 of revaluation in Navarra and Autonomous Community Decree 9/97 of revaluation in Navarra.

As far as revaluable assets are concerned, the following could be revalorized:

(a) Tangible fixed asset items situated in both Spain and abroad, whether or not they had any effect on economic transactions. Dealing with taxpayers with a real obligation to contribute because of being permanently established in Spain, the asset items should be also be affected. Asset items under construction, adaptation or assembly could also be adjusted.
(b) Leased fixed assets, whether or not the purchase option had been taken up. In the latter case the adjustment effects were conditional on subsequent exercise of the purchase option.
(c) Land and building sites belonging to property development enterprises, including those inside constructed buildings or those under construction.

It is important to underline the fact that revaluation was necessarily to refer to all asset items that could be adjusted, and that in the case of buildings the readjustment should be made distinguishing between the value of the ground and that of the construction. At the same time, in the case of depreciable assets, these were to be effectively serviceable and not be fiscally amortized.

2.2 Revaluation procedure

2.2.1 Revaluation operation

Value readjustment was carried out with respect to asset items that could be readjusted which figured on the first balance sheet closed on 9 June 1996 or later.

The revaluation coefficients applied are shown in Table VIII.1. The coefficients indicated were

Table VIII.1 1996 revaluation coefficients

Year of acquisition or production of the asset item	Coefficients
1983 and previous years	1,810
1984	1,640
1985	1,520
1986	1,430
1987	1,360
1988	1,300
1989	1,240
1990	1,190
1991	1,150
1992	1,130
1993	1,110
1994	1,090
1995	1,050
1996	1,000

maximum in nature and could be applied in proportion. The resulting coefficients of the chosen proportion were to be applied with respect to all asset items, with the exception of those listed below, for which other proportionally different ones could be used:

- These affecting savings banks' charitable-social work.
- These affecting co-operative societies' education and promotion funds.
- These affecting activities that make up part of the statutes or specific aim of non profit-making institutions and foundations.
- These affecting historical and artistic patrimony.

The established coefficients were applied in the following manner:

- On acquisition price or production cost, taking account of the acquisition or production year of the asset item. The applicable coefficient for extensions and improvements was the one corresponding to the year in which they had been carried out.
- On depreciation corresponding to the tax-deductible acquisition price or production cost, taking account of the year in which they were deducted.

The coefficients relating to fixed asset acquired through leasing were applied in the following manner:

- On the accounting value of the asset item at the time of contract maturity, taking account of the year in which it had come about.
- On the financial leasing rate, on the part corresponding to tax-deductible cost recovery, taking account of the year in which they were deducted.
- When the purchase option has been exercised, on tax-deductible depreciation corresponding to said option, taking account of the year in which they were deducted.

In the case of asset items included in the first balance sheet closed on or after 31 December 1983, the coefficients were applied in the following manner:

- On the value of the asset items in said balance sheet, taking account of the year in which it was closed.
- On the accumulated depreciation in the aforementioned balance sheet and on those carried out later which were not tax-deductible, taking respective account of the year in which the balance was closed and the years in which the depreciation was deducted.

The difference between the amounts determined by applying the appropriate coefficients and the values taken into considerations for applying the aforementioned coefficients is the amount of capital gain arising from monetary depreciation or the value increase in the revaluated asset item.

A reduction had to be applied to this capital gain, something established with the aim of taking account of the way in which the firms were financed in determining value increases. To do this, the interested party could choose between one of the two following options:

- Reduce the item value increase and the corresponding amortizations by 40 per cent.
- Reduce the item value increase and the corresponding amortizations by the percentage that results from subtracting from 1 the following ratio:

$$\frac{\text{Owner' equity}}{\text{Total liabilities} - \text{Credit rights} - \text{Cash at bank and in hand}}$$

The determining magnitudes of this ratio were those that existed during the time of ownership of each revaluable asset item or in the year corresponding to the revaluated balance sheet and the five previous financial years, whichever the interested party may choose. The magnitudes taken into account were those of the closing balances for each of the corresponding financial years.

In addition, those firms that opted to apply what is laid down in the latter option would not have to carry out any reduction whatsoever when the aforementioned ratio corresponding to the year of the readjusted balance sheet and the five previous financial years were superior to 0.4.

Once the reduction procedure was chosen it was applied to all of the asset items that could be revaluated.

2.2.2 Revaluation accounts

Once the capital gain was calculated using the procedure described above, it was entered into the accounts under 'Revaluation reserve Royal Decree-Law 7/1996, of 7 June', using the accounts corresponding to the revaluated asset items as a counter-entry without altering the accumulated depreciation accounted for, and taking into account that the new resulting value of the readjustment could not exceed the market value of the readjusted asset item, taking account of its readiness for use as a function of wear and tear and the use made of it by the interested party.

A flat rate of 3 per cent was levied on the balance of the account 'Revaluation reserve Royal Decree-Law 7/1996, of 7 June'. This sole levy was not considered to be either Corporate Tax nor Personal Income Tax. Its amount was charged to the account 'Revaluation reserve Royal Decree-Law 7/1996, of 7 June' and was not considered a tax-deductible expense in determining the tax basis for the taxes mentioned above, rather it was considered directly a debt.

As far as the information to include in the annual accounts relating to the revaluation is concerned, the following items of information should be included:

- The balance sheet should show the amount corresponding to the revaluation carried out, net of the sole readjustment levy, under the item 'Revalorization reserve' in the grouping corresponding to owners' equity. The levy will figure in an entry in the grouping corresponding to short term debts (only in the 1996 annual accounts).

- In the notes to the annual accounts corresponding to the financial years in which the revaluated items are still part of the organization's assets, information relating to the following aspects:
  - i) Criteria used in making revaluation, with an indication of the entries affected in the annual accounts.
  - ii) Movement over the year in the account 'Revaluation reserve Royal Decree-Law 7/1996, of 7 June', indicating opening balance, increases, decreases and transfers to capital or other entries in the year, with an indication of the nature of this transfer and the closing balance.
  - iii) Tax treatment of the entry 'Revaluation reserve'.
  - iv) Information should be provided about the items that have been the object of revaluation including: the law which authorizes it, the revaluation amount for each entry on the balance sheet, an indication of the most significant items, the effect of revaluation on provision for amortization, and, therefore, on the following year's result and amount of net accumulated revaluations at the end of the financial year carried out under the aegis of Royal Decree-Law 7/1996, of 7 June, and the effect of revaluations on amortizations and provisions for the year.
- In addition, the consolidated annual accounts should contain the same information as that required in the individual annual accounts, adding information relating to the part corresponding to the revalorization reserve which affects the following entries:
  - i) Reserves in companies consolidated by global or proportional integration.
  - ii) Reserves offset in companies.
  - iii) External partners.

Concerning the fate of the 'Revaluation reserve Royal Decree-Law 7/1996, of 7 June' account, we should comment that from the date on which Tax Inspection has approved and accepted its balance, or in the case of three years having passed without approval then said balance will be used to:

- Eliminate negative results. Losses to be offset may be accumulated from years prior to that in which approval is given as well as those of the same year in which it is approved or following years.
- Increases in share capital.
- Freely disposable reserves, once 10 years have passed from the date of the balance sheet which reflects the readjustment operations.

In the case of revaluation reserves coming from leased asset items, the purchase option must have been exercised in addition to the requirements indicated above.

The balance of the account 'Revalorization reserve Royal Decree-Law 7/1996, of 7 June' cannot be distributed either directly or indirectly unless the capital gain has been realized. This will be understood to be the case when:

- Readjusted asset items have been amortized in the part corresponding to said amortization.
- Revaluated asset items have been transferred or eliminated from the books.

Finally, we should comment that the distribution of this revalorization reserve gives rise to deduction for twin levy on dividends, as laid down in article 28 of Law 43/1995 (Corporate Tax Law) and in article 78 of Law 18/1991 (Personal Income Tax Law).

IX Notes and additional statements

1 Notes

The notes are conceived as an indispensable component of the reporting package which constitutes the annual accounts. The purpose of the notes is to clarify and supplement the information in the balance sheet and profit and loss account (CCom Art. 35.3 and LSA Art. 199).

The content of the notes is left open, and they are not restricted merely to what the law prescribes (basically set forth in the Commercial Code and the Corporations Law, specifically LSA Art. 200). Rather, the notes should include all the data and clarification needed to convey a truer and fairer view of the undertaking.

PGC part 4, dealing with annual accounts, stipulates the minimum disclosure required in the notes by implementing and enlarging upon the requirements set out in commercial legislation. When the legally required information is not significant, however, the related sections need not be completed. In this connection, two formats are envisaged for the notes: the standard format and an abridged format. The undertakings entitled to use the abridged notes format are those entitled to use the abridged balance sheet format (see V.3.1).

No specific layout is imposed, although the information must be presented concisely, in an orderly fashion, in compliance with the principle of clarity, which must govern the preparation of the annual accounts (CCom Art. 34.2).

The contents of the standard notes required by the chart of accounts are set out in Table IX.1. Sole traders, partnerships and small capital companies (as defined in LSA Art. 181; see Table V.1) may use the abridged notes format, the minimum contents of which, pursuant to the chart of accounts, are also shown in Table IX.1.

Two parts of the content of the notes are especially interesting:

- Financial statements that do not have sufficient substance for inclusion as independent documents, but which Spanish legislation has opted to include as one of the points in the notes. These include:
  i) The income distribution statement (note 3, see Table V.3) which is included in the standard and the abridged notes.
  ii) The conciliation of the accounting result with the Corporate Tax basis (note 15 of the standard notes, see Table VII.4). This statement is only obligatory for enterprises that formulate the standard notes.
  iii) Statement of changes in the financial position (note 20 of the standard notes), which is dealt with in the section below and which is also only obligatory for companies that formulate the standard notes.
  iv) The analytical profit and loss account (note 21 of the standard notes, see Table V.6) which is drawn up on a voluntary basis. This account consists of a reorganization of the entries in the profit and loss account and provides the most significant magnitudes such as production value, added value generated by the firm, the gross operating result and the net operating result.
- Information that completes that provided in some of the balance sheet or the profit and loss account entries. A summary of the main content of the standard notes in relation to this point can be seen in Tables IX.2 (information which completes the standard balance sheet content), and IX.3 (information which completes the standard profit and loss account content).

2 Statement of changes in the financial position

The Reform Law (Law 19/1989) requires the notes to include a statement of changes in the financial position of the business when the law so demands. The Corporations Law (LSA) availed itself of the opportunity to include the statement of changes in the financial position as an integral part of the notes for all companies, excluding only those which, because of their size, are entitled to present the abridged notes format (see Table V.1). Thus, Law 19/1989 gave the notes a broader content than is required by the Fourth EC Directive.

The statement of changes in the financial position must describe the funds acquired in the year, their various sources, and their application in fixed assets or working capital (LSA, Art. 200). In point 8 of the section on annual accounts, the chart of accounts gives precise rules for preparing the statement of changes in the financial position and, as stated above in connection with the notes, point 20 contains a specific format for this statement in which all the items are detailed with the exception of 'Funds obtained from operations' and 'Funds applied in operations', which are the first items in the 'Source of funds' and 'Application of funds' columns, respectively (see Table IX.4).

In this connection point 8(d) sets out the main corrections to be made to the result in order to arrive at the 'Funds obtained from operations' if the adjusted period result is a profit, or the 'Funds applied in operations' if the adjusted period result is a loss. Such information on the reconciliation of

Table IX.1 Supplementary notes required by the PGC

Standard[1]	Abridged
1 Line of business	1 Line of business[2].
2 Basis of presentation of the annual accounts. The accounting principles and presentation criteria used to prepare the annual accounts must be disclosed: true and fair view, accounting principles, comparability, grouping of items and items recorded under various headings (see III. 1.2.3, V.4).	2 Basis of presentation of the annual accounts[3].
3 Distribution of income. Information must be included about the limitations on the distribution of income, together with the proposed distribution of income schedule and liquidity projection statement, if interim dividends were distributed (see V.5.2.3).	3 Distribution of income[2].
4 Valuation methods. The valuation criteria applied to the various items of the annual accounts must be explained (see VI.3–5).	4 Valuation methods[2].
Notes 5–16 must disclose detailed information on the various balance sheet items, including analysis of variations under the main headings, details of certain items and reconciliation of the book income with the corporate income tax profit (see Tables V.3 and VII.2)	
5 Formation expenses	
6 Intangibles	
7 Tangible fixed assets	5 Fixed assets. Analysis of the variations in each item and in the relevant accumulated depreciation and amortization and provisions.
8 Marketable securities and suchlike financial investments	
9 Stocks	
10 Shareholders' equity	6 Capital stock. Where there are several classes of shares, indicate the number and par value of those in each class.
11 Subsidies	
12 Provision for pensions and similar obligations	
13 Other provisions for liabilities and charges	
14 Non-trade debts	7 Debts. Total amount of debts with a residual maturity of over five years, and all debts secured by guarantees *in rem*, indicating form and nature.
15 Tax matters	8 Group and associated companies. Name and registered offices of the undertakings directly or indirectly more than 3% owned (if listed) or 20% owned (if unlisted) by the reporting undertaking, indicating: fraction of capital held and amount of capital, reserves and last year's results.
16 Guarantee commitments to third parties and other contingent liabilities	
17 Revenues and expenses. This section must include information on revenues and expenses, average number of employees, and accrual accounts, detailed on the basis of transactions with group, associated and multigroup companies, foreign currency transactions and distribution of the turnover by segment (see Table V.5).	9 Expenses. Breakdown of the two items: 'Employee welfare expenses', distinguishing between pension contributions and other welfare expenses, and 'Variation in operating provisions and losses on non-collectible receivables', distinguishing between bad debts and the variation in the allowance for bad debts.

Table IX.1 (contd.)

Standard[1]	Abridged
18 Other information concerning the undertaking's governing body must be disclosed. (a) Overall amount, detailed by type, of salaries, travel allowances and other compensation of any kind. (b) Advances and loans, indicating the conditions and basic characteristics. (c) Pension and life insurance obligations. (d) Transactions covered by some type of guarantee, and the assets covering such guarantees.	10. Other information regarding the members of the governing body: amount of salaries, *per diems* and other compensation, advances and loans, pension and life insurance obligations.
19 Events subsequent to year end (see VII.3).	
20 Statement of changes in financial position (see IX.2).	
21 Analytical statement of income (on a voluntary basis) (see V.6.2)	

[1] Certain information may be omitted from the notes if its inclusion might be seriously detrimental to the company (e.g. information on group and associated companies or the distribution of the turnover by segment). Such omission must be justified (LSA Art. 200).
[2] Same content as in the standard format.
[3] Same content as in the standard format except for the grouping of items heading.

the result per the books to the actual sources or applications of funds from operations must be disclosed in a note on the statement of changes in the financial position.

The statement of changes in the financial position must show the figures for the current year and those for the immediately preceding year. Should the previous year's figures not be comparable, they must be adjusted accordingly.

When a transaction leads to a simultaneous source and application of funds for the same amount (e.g. conversion of debentures to shares) both the source and the application must be shown. This increases the information content of the statement of changes in the financial position, even if the information is not relevant for calculating the variation in working capital during the year, which is the basic aim of the statement.

The statement of changes in the financial position uses the same criteria for differentiating short- and long-term items as are used in drawing up the annual accounts (i.e. maturity at under or over one year). Surveys have revealed a general feeling in Spain that the statement of changes in the financial position is not all it might be, and that a cash flow statement would be a more useful instrument of financial analysis.

3 Segmental Reporting

In Spain the most relevant standard relating to segmental reporting is that included in note 17 of the standard notes. This indicates that the companies should show the distribution of the net turnover corresponding to the firm's ordinary activities by type of activity as well as by geographical markets. Nevertheless, the rule allows the required information on this point to be omitted when because of its character its provision may seriously prejudice the firm. This means that in practice the application of this rule is voluntary

Table IX.2 Information to be included in the standard notes with regard to the balance sheet

Note	Further information to be disclosed
5 Formation expenses[1]	• Information on significant items: nature and amount
6 Intangibles[1]	• Detailed information on significant items
7 Tangible fixed assets[1]	• Revaluations: law authorizing them, amount, effect on results • Depreciation rates • Assets which are fully written down, obsolete or no longer used • Investments abroad • Items subject to guarantees and reversion • Firm purchase and sale commitments • Items not directly assigned to operations • Capitalized interest and exchange differences • Subsidies and donations received in connection with tangible fixed assets
8 Marketable securities and other financial investments[1]	• Participating interest • Marketable securities and other such financial investments • Non-trade loans
9 Stocks	• Firm purchase and sale commitments, and information on future contracts
10 Shareholders' equity[1]	• Detailed information on the number of shares and their par value, amounts pending disbursement and claimability date • Detailed information on capital increases under way • Rights attached to founders' shares, reimbursed stock, convertible debentures and similar • Availability of reserves • Information on own shares
11 Subsidies	• Amount and nature of subsidies received • Compilance or otherwise with the conditions attached thereto
12 Provision for pensions[1] and similar obligations	• Contingencies and expenses which are covered • Type of capitalization used in provisions for pensions and similar obligations
13 Other provisions for liabilities and charges	
14 Non-trade debts	Details of the following items: • Section D IV 2 of the liabilities side of the balance sheet, 'Other accounts payable', distinguishing between debts which can be transformed into subsidies, payables to suppliers of fixed assets and others • Sections E III 1 and E III 2 of the liabilities side of the balance sheet, 'Payable to group companies' and 'Payable to associated companies', distinguishing between loans and other debts and interest payable Details of the following items: • Average interest rate on long-term non-trade payables • Debts secured by guarantees *in rem* • Amounts not drawn in discount and credit lines • Accrued unpaid financial expenses • Detail of outstanding debentures and bonds
15 Tax matters	See Table VII.3

Table IX.2 (contd.)

Note	Further information to be disclosed
16 Guarantee commitments to third parties and other contingent liabilities	• Detail by guarantee type of the total amount of guarantees and of those recorded in the balance sheet • Nature of the contingent liabilities; system for estimating them and factors on which they depend, indicating the effects on net worth and results in the event of their materializing

[1] The movements between the opening and the closing positions have to be analyzed.

Table IX.3 Further information in the standard notes relating to the profit and loss account

Note	Further information to be disclosed
17 Revenues and expenses	• Details of items A 2(a) 'Merchandise consumed' and 2(b) 'Raw materials and other consumables consumed', distinguishing between purchases and movements in inventories • Details of item A 3(b) 'Employee welfare expenses', distinguishing between contributions to and provisions for pensions and other welfare expenses • Transactions with group and associated companies: purchases, sales, services received and rendered, interest paid and charged, dividends and other profits distributed • Foreign currency transactions: purchases, sales, services received and rendered • Breakdown of turnover by line of business and by geographical market (this information may be omitted if its disclosure might be seriously detrimental to the undertaking, in which case such omission must be justified, LSA Art. 200) • Average number of employees, by category • Extraordinary expenses and revenues, including those relating to prior years • Expenses and revenues recorded during the year which relate to a subsequent year • Expenses and revenues recognized in the year which will be settled in a subsequent year

and the majority of companies choose not to include this information in their annual accounts (for example, more than 60 per cent of the companies that prepare standard notes analyzed by the Principality of Asturias' Central Balance Sheet Data Office (http://www.caef1.derecho.uniovi.es) do not provide this information).

4 Management report

This report should be drawn up by all companies that have to prepare the standard balance sheet and notes (LSA Art. 202). The report does not form part of the annual accounts but is subject to the same requirements as regards disclosure and, where appropriate, auditing. If the company is subject to a statutory audit, the auditors must check that the management report is consistent with the annual accounts (LSA Arts. 218 and 208–9). The content of the management report is governed by LSA Art. 202. It must contain a truthful summary of the progress of the business and the company's situation, and should report any significant events since year end, the outlook for

Table IX.4 Statement of changes in financial position

Sources of funds	Year N	Year N–1
1 Funds obtained from operations		
2 Shareholders' contributions		
(a) Capital increases		
(b) Loss offset		
3 Capital subsidies		
4 Long-term debt		
(a) Debt securities and similar liabilities		
(b) Group companies		
(c) Associated companies		
(d) Other companies		
(e) Fixed asset and other suppliers		
5 Fixed asset disposals		
(a) Intangibles		
(b) Tangible fixed assets		
(c) Financial investments		
(i) Group companies		
(ii) Associated companies		
(iii) Other financial investments		
6 Disposals of own shares		
7 Early amortization or short-term transfer of financial investments		
(a) Group companies		
(b) Associated companies		
(c) Other financial investments		
Total funds obtained		
Funds applied in excess of funds obtained (decrease in working capital)		
Application of funds		
1 Funds applied in operations		
2 Formation and debt arrangement expenses		
3 Fixed asset additions		
(a) Intangibles		
(b) Tangible fixed assets		
(c) Financial investments		
(i) Group companies		
(ii) Associated companies		
(iii) Other financial investments		
4 Acquisition of own shares		
5 Capital stock reductions		
6 Dividends		

Table IX.4 (contd.)

Sources of funds	Year N		Year N–1	
7 Repayment or transfer to short-term of long-term debt				
(a) Debt securities and similar liabilities				
(b) Group companies				
(c) Associated companies				
(d) Other debt				
(e) Fixed asset and other suppliers				
8 Provisions for liabilities and charges				
Total funds applied				
Funds obtained in excess of funds applied (increase in working capital)				
Variation in working capital	**Year N**		**Year N–1**	
	Increase	**Decrease**	**Increase**	**Decrease**
1 Subscribed capital called but not paid				
2 Inventories				
3 Accounts receivable				
4 Accounts payable				
5 Short-term financial investments				
6 Own shares				
7 Cash				
8 Accrual accounts				
Total				
Variation in working capital				

the future, research and development activities and acquisitions of own shares. Companies that by preparing abridged balance sheet and notes do not have to draw up the management report should provide information, where appropriate, in the notes about own or holding company share acquisition (see article 79 of the LSA).

5 Interim reports

Under Spanish regulations, issuers of securities which are listed on formal secondary markets are obliged to file interim reports. Such reports, containing the quarterly and half-yearly information prescribed in the Finance Ministry Order of 18 January 1991, depending on the reporting entity's business, must be filed with the Spanish Securities and Exchange Commission and with the management companies of the stock markets on which the securities are listed.

In this connection, Law 24/1988, which reformed the securities market, charged the Ministry of Economic and Financial affairs with responsibility for deciding the information which such entities must publish every quarter and half-year. By virtue of these powers the Ministry issued a ministerial Order of 18 January 1991, specifying the requirements for listed entities.

This ministerial order specified the information that listed entities must publish periodically, and differentiated between various types of organization on the basis of their business activities. In addition to a general-purpose reporting format it established specific formats for credit entities, financial leasing companies, insurance companies and portfolio investment companies. The individual financial information and, where appropriate, the consolidated information, must be released six-monthly and quarterly, giving the information for the current and preceeding periods. All this information must be issued in the formats and according to the requirements of the aforementioned ministerial order.

This information must be sent to the National Securities and Exchange Commission (*Comision Nacional del Mercado de Valores*, CNMV) and to the management companies of the relevant stock exchanges in the terms set out in the ministerial order. Within 15 days of the filing deadline the stock exchange management companies will publish the six-monthly and quarterly information they have received in their quotation bulletins (*boletín de cotización*). Additionally, the CNMV will hold the information for public inspection at any time and it is possible to examine the information provided at the CNMV web site (http://www.cnmv.es).

5.1 Half-yearly reporting

The requisite half-yearly filing comprises a balance sheet and profit and loss account, as basic financial statements. In addition, each half-year the following information must be filed:

- Presentation basis and valuation rules.
- Survey of business developments.
- Distribution of net turnover by line of business.
- Average number of employees.
- Dividends distributed.
- Significant events (including a detailed list of the relevant or significant events which have occurred during the relevant period).
- Changes in the companies comprising the consolidated group.
- Annexe explaining the significant events.

The information on the second half of the year must coincide exactly with the annual accounts and management report, if they have already been prepared by the directors. If they have not yet been prepared, any discrepancies with the half-yearly information already filed must be resolved by reporting to the National Securities and Exchange Commission and to the stock exchanges concerned within 10 days. This procedure also applies to any qualifications expressed in the auditors' report on the annual accounts and management report with regard to the half-yearly information already filed.

Additionally, a ministerial Order of 30 September 1992, amending that of 18 January 1991, provides that, where the auditors' report on the annual accounts includes a qualified opinion, an adverse opinion or a disclaimer of opinion, the entity must furnish a special auditors' report with the information for the subsequent half-year, disclosing whether the reasons for the qualification have been remedied or whether they still persist. If they have been remedied, the impact of the corrections on the information for the current half-year must be indicated. If they have not been remedied, the effects of including such qualifications on the results and equity appearing in the current period's interim reporting must be indicated.

5.2 Quarterly reporting

The quarterly filing is less complex and involves only the following:

- Net turnover.
- Result before taxes.
- Subscribed capital stock.
- Average number of employees.
- Survey of business developments.
- Dividends distributed.
- Significant events.
- Annexe explaining the significant events.

In its rule 12, on the preparation of annual accounts, the PGC states that interim financial statements must be prepared in the same form and using the same criteria as the annual accounts. Interim statements must be prepared separately from the annual accounts, which must not be affected by the presentation of the former, and any entries made for the purpose of presenting such statements must be cancelled. However, the current regulations offer no specific method of preparing or auditing such interim financial information.

X Auditing

1 The audit and the audit requirement

The Audit Law (Law 19/1988 Art. 1) defines an audit as 'the process of reviewing and verifying accounting documents for the purpose of issuing a report which may affect the interests of third parties'. This same law also states, with regard to the auditing of annual accounts, that 'it shall consist of verifying and opining whether such financial statements present a true and fair view of the audited company or other entity's financial position and of the results of its operations and of the funds that have come into its possession and been put to use during the period examined . . .; it shall also comprise verification of the consistency of the management report with such statements'.

LSA Art. 203, as amended, states that 'the annual accounts and management report must be reviewed by auditors. Companies which are eligible to present an abridged balance sheet are exempt from this requirement'. It should be recalled that under the same law companies which meet at least two of the following conditions at year end for two years running are eligible to present an abridged balance sheet:

- Their assets do not exceed ESP 395 million.
- Their net annual turnover is less than ESP 790 million.
- Their average workforce during the year does not exceed 50 employees.

Additionally, LSA Art. 205.2 stipulates that, where a company is not required to be audited, shareholders representing 5 per cent of the capital may request the commercial registrar of the district in which the company's registered offices are situated to appoint an auditor to review the annual accounts for a given year, provided that no more than three months have elapsed since the financial year end. The expenses attendant upon such an audit are to be borne by the company. Furthermore, pursuant to the first Additional Provision of the Audit Law (Law 19/1988), the annual accounts must be audited of companies active in financial markets. Those which are listed or which issue debentures through public offers or which are financial brokers must submit their accounts to the scrutiny of auditors. This requirement also applies to companies engaging in insurance activities or which receive subsidies from, undertake work for, provide services or supply goods to the State and its agencies.

The regulations set out in the above-mentioned first Final Provision do not apply to organizations which are State agencies or State-owned companies, except in the case of companies which issue marketable securities and are required to publish a prospectus. Such companies are also generally required to be audited.

2 Appointment of auditors

Auditors may be appointed by three procedures:

- Generally speaking, it is the shareholders' meeting that appoints the auditors. The appointment must be finalized before the end of the financial year to be audited. The auditors will be contracted for an initial fixed period of time which may not be less than three years and not more than nine years from the starting date of the first financial year to be audited. Once the initial period is completed they may be contracted annually.
- If the shareholders' meeting fails to appoint auditors before the end of the business year to be audited, or the auditors nominated decline the appointment, or cannot carry out their duties, the directors, the commissioner of the bondholders' syndicate, or any other shareholder, may request that auditors should be appointed by the commercial registrar within whose jurisdiction the company's registered offices are situated. This applies even where there is no legal requirement for an audit, if shareholders representing five per cent of the share capital so request.
- Should there be just cause (the only grounds on which appointed auditors can be dismissed), the directors and individuals empowered to move the appointment of the auditors can lodge a plea with the court of first instance for the appointment approved by the shareholders' meeting or by the commercial registrar to be set aside and another nominee appointed.

3 Auditors

Individuals or legal entities registered as practising auditors with the official registrar of auditors of the Institute of Accounting and Auditing (ICAC) may perform audit functions. Individuals may register under one of three categories in the official register of auditors:

- Practising.
- Employed (by a practising auditor or by an audit firm).
- Not practising.

Audit firms must meet the following requirements:

- All the partners must be individuals.
- A majority (at least) of the partners must be registered practising auditors and most of the share capital and voting rights must belong to them.
- A majority of the administrators and directors must be practising auditor partners. Whatever the circumstances, the sole administrator of firms of this kind must be an auditor.
- An audit firm must be registered with the official registrar of auditors.

Each year the ICAC publishes an updated list of the auditors on the official register. If auditors or audit firms belong to more than one professional body of auditors, they must choose a single one for the purpose of registration.

4 Organization of the audit profession

As indicated earlier, individuals or legal entities whose names are entered in the official register of auditors of the Institute of Accounting and Auditing (ICAC) are qualified to carry out audits. Registration is open to anyone who meets the following conditions:

- University graduate.
- Completion of courses in, *inter alia*, the theory of auditing, financial analysis, consolidation of accounts and cost accounting. The courses must have been taught or accredited by corporate bodies representing firms engaging in audit activities: the Institute of Chartered Accountants (*Instituto de Auditores-Censores Jurados de Cuentas*), the Register of Economist Auditors (*Registro de Economistas-Auditores*, set up by the General Council of Spanish Economists' Associations) and the Register of Commercial Graduates (*Registro de Titulares Mercantiles*).
- Practical training over a minimum period of three years in financial and accounting work, at least two of them in auditing.
- Passing a professional aptitude exam.

The corporate bodies representing auditors are public law entities comprising auditors and audit firms. They are responsible, among other things, for the following activities:

- Formulation, adaptation and revision of audit standards, upon their own initiative or at the request of the Accounting and Audit Institute (ICAC).
- Holding professional aptitude exams.
- Organizing and, where appropriate, participating in, theoretical training courses.
- Organizing of on-going training and professional refresher courses.
- Encouraging members' co-operation in the necessary practical training, monitoring their compliance.
- Quality control on professional practice, etc.

5 Auditors' liability

Auditors are directly, jointly and severally held accountable to companies or entities audited by them and to third parties for any detriment arising from failure to fulfil their obligations. To cover the risk, auditors must provide security in the form of a cash deposit, government debt stock, the surety of a financial entity or third-party liability insurance. The sum needed for the first year of business (the minimum amount in subsequent years) is ESP 50 million for individuals. This figure is multiplied by the number of partners (whether or not they are auditors) in the case of firms. The minimum security must be increased by 30 per cent of the total billings exceeding the preceding year's minimum guarantee in subsequent years.

If an audit is performed by a member of an audit firm, both the individual auditor and the firm will be held liable. The other partners who did not sign the report will be jointly and severally liable for any damages. Infringements, minor or major, of professional standards are punishable, without prejudice to the auditors' liability to third parties, by the imposition of sanctions by the Accounting and Audit Institute (ICAC).

Auditors are required to maintain strict confidentiality about any information that may pass through their hands. They may not use such information for any purpose other than the audit of their client's accounts. Auditors are also required to retain and safeguard for a period of five years from the date of the report, all documentation relating to each audit performed, including working papers, which constitutes the evidence and supporting paperwork on the matters which are the subject of the report.

6 Auditors' independence

Audit Law 19/1988 requires auditors to be independent in the discharge of their duties with regard to the companies or entities they audit. No auditor is eligible to audit a company or entity who:

- Holds a management position, is a member of the board or is an employee of the company or entity.
- Is a shareholder or partner in the company or entity with an ownership interest in it exceeding 0.5 per cent of the face value of the share capital, or whose ownership interest accounts for more than 10 per cent of his or her personal net worth.
- Is related, up to the second remove of kinship, to an executive, manager or director of the company or entity.
- Is deemed unsuitable by virtue of other legal requirements.

These rules apply to the three financial years preceding that in which the work is actually done. Thus if the 1998 figures were being audited in March 1999 the rules would cover the financial years 1995–98.

It is also a legal requirement that for three years after the term of his or her appointment no auditor may enter the employ of the business he or she has audited or serve on its governing or management bodies.

7 Sanctions

In auditing practice the following are regarded as serious transgressions against professional standards of conduct:

- Failure to undertake an audit after having entered into a firm agreement to do so.
- Discrepancies between the report submitted and the evidence on which it is purportedly based.
- Breach of confidentiality.
- Deriving personal advantage from information encountered in the course of the audit, or enabling a third party to do so.
- Failure to heed the grounds of disqualification from auditing a particular company.
- Failure to observe the regulations governing the audit, resulting in financial loss to the client company or third parties.
- Taking on too much work.
- Cumulative minor transgressions.

The Accounting and Audit Institute (ICAC) is responsible for imposing penalties. Once a complaint has been processed, the facts are put before the chairman of the ICAC, who adjudicates on the merits. The chairman will take into consideration the advice of the Consultative Committee of the ICAC, comprising four appointees of the Finance Ministry and six from the profession. The penalties may range from a warning to temporary or permanent suspension.

8 Content and availability of the audit results

Article 4 of the regulation implementing Audit Law 19/1988 states that the auditors' report is a commercial document which must contain at least the following data:

- Identity of the company audited.
- Identity of the individuals or legal entities that requested the audit and to whom it is addressed.
- Identity of the annual accounts audited.
- Reference to the methods adopted in carrying out the audit.
- An express statement that the notes on the annual accounts include the requisite informa-

tion for a proper interpretation and understanding of the financial position and net worth of the company or entity audited, and of its results for the year.

- A technical opinion as to whether the annual accounts present a true and fair view of the net worth and financial position of the company, of the results of its operations and of the funds obtained and applied by it in the period examined. Reference must be made to the conformity of the presentation of the annual accounts with generally accepted accounting principles, to events subsequent to year-end and to possible infringements of legal or internal regulations which might affect the true and fair view.
- Signature of the auditor or auditors who prepared the report.

Also subject to audit regulations are reviews or audits by an auditor of other accounting statements and documents for the purpose of reporting to third parties on whether they genuinely record the business transactions they are supposed to.

It is the responsibility of the audited company's directors to furnish third parties with copies of the annual accounts and auditors' report. The auditor may do so only with the express authority of the audited company, except in certain cases provided for in the Audit Law. When a company's annual accounts are filed in the commercial register they must be accompanied by the auditors' report. Any quotation from the auditors' report must adhere strictly to the wording of the original.

If the annual accounts are published in unabridged form, the auditors' report must also be reproduced in full. If the annual accounts are published in abridged form, the auditors' report may be omitted, but the opinion expressed therein must be indicated. In no case may the auditors' report be published in part or in excerpt form.

XI Filing and publication

LSA Art. 218, which also applies to limited liability companies (*sociedades de responsabilidad limitada*, SRL), employee-owned companies (*sociedades anónimas laborales*, SAL) and limited partnerships with shares (*sociedades comanditarias por acciones*, SComPA), makes it obligatory for all these forms of capital companies to lodge copies of the following documents with the commercial registrar within a month of the approval of the annual accounts by the shareholders' meeting (see II.1):

- Certificate of the resolutions of the shareholders' meeting approving the annual accounts and the distribution of income. If one or more of the elements comprising the annual accounts has been presented in abridged form, the fact must be documented, with an explanation of the reason.
- The annual accounts.
- The management report, if the company is obliged to draw it up.
- The auditors' report if the company is subject to statutory audit or if an audit was performed at the request of a minority of the shareholders.

  A copy of the document relating to owners equity, when the company is obliged to draw it up in compliance with what is laid down in article 97 of the Stock Market Law (the official form or this document was established in the ministerial order of 14 June 1995).
- Certificate to the effect that the annual accounts submitted are the audited annual accounts.

Article 365 of the Mercantile Register Rules (RD 1784/1986) extends the previous obligation to reciprocal guarantee companies and pension funds. At the same time the additional disposition to Law 7/1996 of Retail Trade Regulation establishes that the deposit obligation is extended to organisations of any legal nature, except sole traders, whose sales exceed ESP 100 million. The balance sheet and profit and loss accounts to be deposited should be drawn up in accordance with the official form, established by order of the Ministry of Justice on 14 January 1994. Companies that have to draw up specific account models because of sector adaptation are excepted from the use of compulsory forms. It is possible to file the accounts in Spain's official languages other than Spanish (Basque, Galician, Catalan and Valencian).

The registrar must declare whether the annual accounts meet the legal requirements within 15 days of their presentation. The official gazette of the commercial registrar must include a list of the companies that have duly deposited the requisite documents in the preceding month. Anyone may inspect or obtain a copy of the documents deposited with the commercial registrar. There is an agency of the commercial register in the capital of every province. The register is under the authority of the Ministry of Justice.

Apart from the publication requirements for official documents, the stock market requires specific information on quoted companies. As mentioned above (see IX.5), the Securities Market Law (Law 24/1988) allotted to the Ministry of Economic and Financial Affairs the task of defining the information which such entities have to make public. A subsequent ministerial Order of 18 January 1991 laid down specific requirements for companies to submit quarterly and half-yearly reports on their financial position to the National Securities and Exchange Commission (CNMV) and to the management companies of the relevant stock exchanges, as detailed above (see IX.5). The management companies publish the reports in their bulletins and the CNMV makes the information available to the public (http://www.cnmv.es).

Spanish companies listed on stock exchanges in other countries are equally subject to their regulatory regimes.

XII Sanctions

In the first month of each year the commercial registrars send to the Directorate General of Registries and Notaries Public a list in alphabetical order of companies which failed to comply with the requirement to deposit their annual accounts the previous year. In the second month of each year the Directorate General forwards the lists to the ICAC so that proceedings can be initiated.

Non-compliance by the organization's administrators with the obligation to file the required documents within the established period will mean that no document referring to the company will be registered in the Mercantile Register while such non-compliance persists, with the exception of those relating to the dismissal or resignation of administrators, managing directors, general managers or receivers, and the revoking or relinquishment of powers of attorney, as well as the dissolution of the company and the appointment of receivers and the entries (noted in the Mercantile Register) ordered by the judicial or administrative authority.

Non-compliance will also mean a fine being imposed on the company for an amount of between ESP 200,000 and 10 million by the ICAC. The fine to be imposed is determined in line with the size of the company (LSA Art. 221).

Apart from the sanctions consequent upon failure to publish, the Penal Code will penalize certain behaviour relating to not keeping accounts or to keeping them in an irregular manner. Article 290 establishes that company administrators who falsify the annual accounts or other documents, which should reflect the legal or economic situation of the organization, in a manner designed to cause economic prejudice to the company, to one of its partners, or to a third party, will be punished with a fine and a prison sentence of one to three years. In addition, article 310 establishes that those obliged by tax law to keep accounts, books or tax ledgers shall be subject to a sentence of 7 to 15 weekends arrest and a fine if they have:

(a) Completely failed to meet the requirement.
(b) Kept different accounting books relating to the same activity or financial year, concealing or misrepresenting the true financial situation of the business.
(c) Failed to record business transactions in the books of account or falsified the record.
(d) Fraudulently stated non-existent items.

The Penal Code also establishes in article 261 that if in bankruptcy proceedings, tender or suspension of payments, false information is given relating to the account statement, with the aim of wrongfully achieving their filing, will be punished with a prison term of one to two years and a fine.

There are basically two kinds of infringement with regard to the financial information that listed companies must submit to the regulatory bodies of the securities market:

- Failure by companies or other entities whose securities are publicly quoted to submit quarterly and six-monthly reports to the regulatory

bodies of the official secondary markets or to the National Securities and Exchange Commission (Law 24/1988 Art. 100) is considered serious misconduct.
- Failure by companies or other entities whose securities are publicly quoted to comply with the requirement to have their annual accounts independently audited (Law 24/1988 Art. 99) is considered very serious misconduct.

Serious misconduct may entail a public warning, the imposition of a fine, limitation of the volume of transactions, the suspension of quotation or suspension from membership of the market for a period not exceeding one year. Very serious misconduct may result in a fine, limitation of the volume of transactions, suspension of quotation or suspension from membership for a period not exceeding five years.

XIII Financial reporting institutions and national databases

Accounting and Audit Institute (ICAC)
Huertas 26, 28014 Madrid
Tel: 34 1 4290960
Fax: 34 1 4299486
E-mail: icac@icac.meh.es
Internet: http://www.icac.meh.es

Spanish Accounting and Business Administration Association (AECA)
Alberto Aguilera 31–5°, 28015 Madrid
Tel: 34 1 5474465
Fax: 34 1 5413484
E-mail: info@aeca.es
Internet: http://www.aeca.es

National Securities and Exchange Commission
Comisión Nacional del Mercado de Valores (CNMV)
P° Castellana 19, 28046 Madrid
Tel: 34 1 5851500
Fax: 34 1 3193373
E-mail: dap@cnmv.es
Internet: http://www.cnmv.es

Bank of Spain
Alcalá 50, 28014 Madrid
Tel: 34 1 3385180
Fax: 34 1 3385320
E-mail: webmaster@bde.es
Internet: http://www.bde.es

Directorate General of Insurance
P° de la Castellana 44, 28071 Madrid
Tel: 34 1 3397000
Fax: 34 1 3397113
E-mail: webmaster@meh.es
Internet: http://www.meh.es

Institute of Chartered Accountants of Spain (IACJCE)
General Arrando 9, 28010 Madrid
Tel: 34 1 4460354
Fax: 34 1 4471162
E-mail: cmartin@iacjce.es
Internet: http://www.iacjce.es

Register of Economist Auditors (REA)
Claudio Coello, 18 1°, 28001 Madrid
Tel: 34 1 4310311
Fax: 34 1 5750698
E-mail: rea@mad.servicom.es
Internet: http://www.rea.es

Register of Commercial Graduates
Zurbano 76, 28010 Madrid
Tel: 34 1 4412067
Fax: 34 1 4417002
E-mail: arestru@eucmax.sim.ucm.es
Internet: http://www.ucm.es/info/evce/euee/

ARDAN database
http://www.ardan.es

CABSA database
http://www.cabsa.es

Dun & Bradstreet database
http://www.dun.es

Centre for Economic and Financial Analysis
(Centro de Análisis Económico y Financiero del Principado de Asturias)
http://caef1.derecho.uniovi.es/

Bibliography

Almela Díez, B. (1997). Evolución legislativa de las relaciones entre fiscalidad y contabilidad. *Técnica Contable,* **577**, enero, 33–44.

Asociación Española de Contabilidad y Administración de Empresas (1992). Serie sobre Principios contables 1–20 (AECA, Madrid).

Alvarez Melcón, S. (1992). Accounting and tax regulations. In J. A. Gonzalo (ed.) *Accounting in Spain, Fifteenth Annual Congress of the European Accounting Association*, 245–254 (Asociación Española de Contabilidad y Administración de Empresas, Madrid).

Amerigo Cruz, E. *et al.* (1996). *Curso práctico de Contabilidad* (Instituto de Estudios Fiscales, Madrid).

Anglá, J. *et al.* (1996). *Curso de Contabilidad. Una aplicación práctica* (Media, Barcelona).

Antolinez, S. (1990). Imagen fiel y principios contables. *Nuevo Plan General de Contabilidad* I, *Revista Española de Financiación y Contabilidad,* **63**, 351–361.

Arthur Andersen (1992). *Análisis práctico de la contabilidad en España* (Arthur Andersen, Madrid).

Banco de España (1991). Circular 4 de 14 de junio. a las entidades de credito sobre normas de contabilidad y modelos de estados financieros. *BOE,* **153**, 27 June.

Banco de España (1997). *Central de Balances. Resultados anuales de las empresas no financieras. 1996* (Banco de España, Madrid).

Blanco Richart, E. (1998). *Manual práctico. Principios y normas contables* (Fundación Universitaria San Pablo CEU, Valencia).

Blasco Lang, J. J. (1990). Análisis de las cuentas anuales: la memoria. *Partida Doble,* **1**, 41–47.

Bolsa de Madrid (1998). *Informe Anual* (Bolsa de Madrid, Madrid).

Bolufer Nieto, R. (1990). El PGC: instrumento del nuevo marco de la información económico-financiera. *Partida Doble,* **1**, 11–15.

Bolufer Nieto, R. (1992). The Accounting and Audit Institute and the standard national chart of accounts. In J. A. Gonzalo (ed.) *Accounting in Spain, Fifteenth Annual Congress of the European Accounting Association,* 75–82 (Asociación Española de Contabilidad y Administración de Empresas, Madrid).

Broto, J. and Condor, V. (1989). El principio de prudencia versus principio de correlación de ingresos y gastos. *Lecturas sobre principios contables* 2, 261–296 (Asociación Española de Contabilidad y Administración de Empresas, Madrid).

Bueno Campos, E. (1990). Leasing y lease-back. *Cómo aplicar el Plan de Contabilidad* 2, 81–92 (Asociación Española de Contabilidad y Administración de Empresas, Madrid).

Cañibano, L. (1989a). Los principios contables de AECA y la IV Directriz de la Comunidad Económica Europea. *Lecturas sobre principios contables,* 177–208 (Asociación Española de Contabilidad y Administración de Empresas, Madrid).

Cañibano, L. (1989b). *The new Legal Framework of Accounting in Spain to harmonize with the EEC Directives,* Instituto Universitario de Administración de Empresas 16 (Universidad Autónoma de Madrid, Madrid).

Cañibano, L. (1991). Principios contables. *Cómo aplicar el Plan de Contabilidad,* 5–20 (Asociación Española de Contabilidad y Administración de Empresas, Madrid).

Cañibano, L. (1992a). Professional standards: AECA Accounting Principles. In J. A. Gonzalo (ed.) *Accounting in Spain, Fifteenth Annual Congress of the European Accounting Association,* 85–99 (Asociación Española de Contabilidad y Administración de Empresas, Madrid).

Cañibano, L. (1992b). *La reforma contable en España.* Instituto Universitario de Administración de Empresas 26 (Universidad Autónoma de Madrid, Madrid).

Cañibano, L. (1992c). *Contabilización de los planes de pensiones,* Instituto Universitario de Administración de Empresas 27 (Universidad Autónoma de Madrid, Madrid).

Cañibano, L. (1996). *Contabilidad: análisis contable de la realidad económica,* 7th edition (Pirámide, Madrid).

Carmona Moreno, I. M. (1996). *Contabilidad Financiera. Curso teórico-práctico acelerado* (Paraninfo, Madrid).

Casanovas Parella, I. (1992a). *La empresa y la contabilidad* (Praxis, Barcelona).

Casanovas Parella, I. (1992b). Spain and the process of harmonization with EU accounting regulations. In J. A. Gonzalo (ed.) *Accounting in Spain, Fifteenth Annual Congress of the European Accounting Association*, 169–184 (Asociación Española de Contabilidad y Administración de Empresas, Madrid).

Cea García, J. L. (1990). Revisión panorámica de los modelos de cuentas anuales en el Plan General de Contabilidad de España (PGCE). *Nuevo Plan General de Contabilidad* I, *Revista Española de Financiación y Contabilidad*, 401–423.

Cea García, J. L. (1992). Official auditors' register and the professional bodies. In J. A. Gonzalo (ed.) *Accounting in Spain, Fifteenth Annual Congress of the European Accounting Association*, 349–360 (Asociación Española de Contabilidad y Administración de Empresas, Madrid).

Clavijo Hernández, F. *et al.* (1997). *Impuesto sobre sociedades: aspectos fundamentales* (Lex nova, Madrid).

Corona Romero, E. (1996). *Curso práctico de Contabilidad* (Instituto de Estudios Fiscales, Madrid).

Cubillo, C. (1990a). La reforma contable en España. *Nuevo Plan General de Contabilidad* I, Revista Española de Financiación y Contabilidad, **63**, 301–315.

Cubillo, C. (1990b). Acciones propias y acciones de la sociedad dominante. *Cómo aplicar el Plan de Contabilidad* 3, 49–59 (Asociación Española de Contabilidad y Administración de Empresas, Madrid).

Cuervo, A., Rodríguez, S. L. and Parejo, J. A. (1993). *Manual de sistema financiero español* (Ariel Economía, Madrid).

Esteban Marina, A. (1997). *Contabilidad y base imponible en el nuevo impuesto sobre sociedades* (Marcial Pons, Madrid).

Fernández López, J. A. (1996). *Contabilidad de tributos e impuesto sobre sociedades* (Centro de Estudios Financieros, Madrid).

Fernández Peña, E. (1990). Valoración del circulante. *Nuevo Plan General de Contabilidad 2, Revista Española de Financiación y Contabilidad*, **64**, 587–592.

Gallizo, J. L. (1993). *Los estados financieros complementarios* (Pirámide, Madrid).

García Arthus, E. (1996). Resultado contable y base imponible en el nuevo impuesto sobre sociedades. *Partida Doble*, **65**, marzo, 5–11.

García-Olmedo Domínguez, R. (1997). Las diferencias temporarias: Otro enfoque en el tratamiento contable del impuesto sobre beneficios. *Técnica Contable*, **586**, octubre, 665–680.

Gómez Artime, A. G. (1995). *Cuentas Anuales del Plan General de Contabilidad y Registro Mercantil* (Paraninfo, Madrid).

Gómez Valls, F. (1998). *Tratamiento contable del impuesto sobre sociedades* (Pirámide, Madrid).

Gonzalo, J. A. and Gallizo, J. L. (1992). *European Financial Reporting: Spain* (Routledge, London).

Herranz, F. (1990). Gastos e ingresos a distribuir en varios ejercicios. *Nuevo Plan General de Contabilidad* 2, *Revista Española de Financiación y Contabilidad*, **64**, 607–623.

Instituto de Contabilidad y Auditoría de Cuentas (1991). Resolución de 16 de mayo de 1991, del ICAC, por la que se fijan criterios generales para determinar el 'importe neto de la cifra de negocios'.

Instituto de Contabilidad y Auditoría de Cuentas (1991). Resolución de 30 de julio de 1991, del ICAC, por la que se dictan normas de valoracion del inmovilizado material.

Instituto de Contabilidad y Auditoría de Cuentas (1991). Resolución de 25 de setiembre de 1991, del ICAC, por la que se fijan criterios para la contabilización de los impuestos anticipados en relacion con la provisión para pensiones y obligaciones similares.

Instituto de Contabilidad y Auditoría de Cuentas (1992). Resolución de 21 de enero de 1992, del ICAC, por la que se dictan normas de valoración del inmovilizado immaterial.

Instituto de Contabilidad y Auditoría de Cuentas (1992). Resolución de 30 de abril de 1992, del ICAC, sobre algunos aspectos de la norma de valoración numero dieciseis del Plan General de Contabilidad.

Instituto de Contabilidad y Auditoría de Cuentas (1992). *Normativa sobre contabilidad en España* (Instituto de Contabilidad y Auditora de Cuentas, Madrid).

Instituto de Contabilidad y Auditoría de Cuentas (1994). Resolución de 1 de diciembre de 1994, del ICAC, por la que se publica la norma técnica de auditoría de elaboración del informe especial y complementario al de auditoría de las cuentas anuales de las entidades de seguros, solicitado por la Dirección General de Seguros.

Instituto de Contabilidad y Auditoría de Cuentas (1996). Resolución de 20 de diciembre de 1996 por la que se fijan criterios generales para determinar el concepto de patrimonio contable a efectos de los supuestos de reducción de capital y disolución de sociedades regulados en la legislación mercantil.

Instituto de Contabilidad y Auditoría de Cuentas (1997). Resolución de 20 de enero de 1997 por la que se desarrolla el tratamiento contable de los regimenes especiales establecidos en el Impuesto sobre el Valor Añadido y en el Impuesto General Indirecto Canario.

Instituto de Contabilidad y Auditoría de Cuentas (1997). Resolución de 9 de octubre de 1997 sobre algunos aspectos de la norma de valoración decimosexta del Plan General de Contabilidad (contabilización del impuesto de sociedades).

Instituto de Contabilidad y Auditoría de Cuentas (1997). Resolución de 15 de septiembre de 1997, del ICAC, por la que se constituye un grupo de trabajo para la elaboración de un informe sobre aspectos contables derivados de la introducción del Euro.

Instituto de Contabilidad y Auditoría de Cuentas (1997). Resolución de 9 de octubre de 1997, del ICAC, sobre la contabilización del Impuesto de Sociedades.

Instituto de Contabilidad y Auditoría de Cuentas (1998). Informe sobre los aspectos contables de la adaptación al euro (conclusiones del grupo de trabajo nombrado por resolución del 15 de septiembre de 1997).

Kirchner Colom, C. (1996). *Manual de Plan General de Contabilidad* (Praxis, Barcelona).

Labatut Serer, G. and Llombart Fuertes, M. (1996). Diferencias permanentes y temporales por aplicación del método del efecto impositivo según la nueva ley sobre el impuesto de sociedades. *Técnica Contable,* **566**, febrero, 87–104.

Laínez, J. A. (1993). *Comparabilidad internacional de la información financiera* (Instituto de Contabilidad y Auditoría de Cuentas, Madrid).

Larriba, A. (1990). Diferencias de cambio en moneda extranjera. *Nuevo Plan General de Contabilidad* 2, *Revista Española de Financiación y Contabilidad*, **64**, 625–663.

Larriba, A. (1992). Contabilidad de entidades de crédito en España. *Revista Española de Financiación y Contabilidad*, **73**, 807–853.

Larriba, A. (1993). *Fusiones y escisiones de sociedades* (Instituto de Auditores–Censores Jurados de Cuentas de España, Madrid).

Lizcano, J. (1990). Cuentas anuales: balance y cuenta de pérdidas y ganancias. *Nuevo Plan General de Contabilidad* I, *Revista Española de Financiación y Contabilidad*, **63**, 425–450.

Llorente Sanz, M. S. (1997). *Aspectos contables del impuesto sobre sociedades* (Instituto de Contabilidad y Auditoría de Cuentas, Madrid).

López Combarros, J. L. (1992). Auditing standards. In J. A. Gonzalo (ed.) *Accounting in Spain, 1992, Fifteenth Annual Congress of the European Accounting Association*, 363–373 (Asociación Española de Contabilidad y Administración de Empresas, Madrid).

López Díaz, A. (1991). La financiación básica en el Plan General de Contabilidad. *Cuadernos Aragoneses de Economía*, **1**, 213–234.

López Diaz, A. and Menéndez, M. (1991). *Contabilidad financiera* (Editorial AC, Madrid).

López Díaz, A. (dir.) (1998). *Análisis económico-Financiero de las empresas de Asturias por sectores de actividad. 1993–1994* (Principado de Asturias, Oviedo) (abstract in http://caef1.derecho.uniovi.es).

Mallado Rodríguez, J. A. and Correa Ruiz, M. C. (1997). Efecto real del Capítulo XII de la Ley 43/1995, del Impuesto de sociedades para las pequeñas empresas. *Técnica Contable,* **577**, enero, 45–54.

Marmolejo Oña, M. (1997). *Actualización de balances 1996* (Centro de Estudios Financieros, Madrid).

Martínez Arias, A. and García Díez, J. (1992). *Supuestos de Contabilidad Financiera* (Pirámide, Madrid).

Martínez Arias, A. and Prado Lorenzo, J. M. (1991). *Nuevo Plan General de Contabilidad* (Pirámide, Madrid).

Martínez Churiaque, J. (1990). Amortización y saneamiento del inmovilizado. *Nuevo Plan General de Contabilidad 2, Revista Española de Financiación y Contabilidad,* **64**, 67–79.

Martínez García, F. J., Garay González, J. A. and García de la Iglesia, M. I. (1995). *Contabilidad. Comentarios, esquemas, ejercicios y cuestiones* (Pirámide, Madrid).

Menéndez Menéndez, M. (dir.) (1996). *Contabilidad Financiera Superior* (Civitas, Madrid).

Menéndez Menéndez, M. (dir.) (1997). *Contabilidad General,* second edition (Civitas, Madrid).

Montesinos Julve, V. (1990a). Correcciones valorativas. *Cómo aplicar el Plan de Contabilidad,* 44–64 (Asociación Española de Contabilidad y Administración de Empresas, Madrid).

Montesinos Julve, V. (1990b). La valoración del inmovilizado en el nuevo Plan de Contabilidad. *Nuevo Plan General de Contabilidad 2, Revista Española de Financiación y Contabilidad,* **64**, 559–586.

Montesinos Julve, V. (1991). La memoria en el nuevo Plan General de Contabilidad. Plan General de Contabilidad 90, *Partida Doble,* **10**, 51–57.

Montesinos Julve, V. (1992). Financial statements of Spanish companies. In J. A. Gonzalo (ed.) *Accounting in Spain, Fifteenth Annual Congress of the European Accounting Association,* 187–204 (Asociación Española de Contabilidad y Administración de Empresas, Madrid).

Montesinos Julve, V., García, M. A. and Vela, J. M. (1989). El principio del devengo: algunas consideraciones en torno a su concepto y aplicación en Contabilidad', *Lecturas sobre principios contables,* 209–233 (Asociación Española de Contabilidad y Administración de Empresas, Madrid).

Moreno Rojas, J. (1997). *Contabilidad y fiscalidad. Diferencias entre resultado contable y base imponible en el nuevo impuesto sobre sociedades* (Universidad de Sevilla, Sevilla).

Muñoz Merchante, A. (1997). *Fundamentos de Contabilidad* (UNED, Madrid).

Noguero-Salinas, A. (1990). El descuento de los efectos comerciales a cobrar. *Cómo aplicar el Plan de Contabilidad,* **3**, 61–72 (Asociación Española de Contabilidad y Administración de Empresas, Madrid).

Norverto, L. M. (1990). El tratamiento de las nuevas reservas en la reforma del Plan General de Contabilidad. *Nuevo Plan General de Contabilidad* **3**, *Revista Española de Financiación y Contabilidad,* **65**, 839–863.

Omeñaca García, J. (1996). *Amortización del inmovilizado. Tratamiento contable y fiscal* (Deusto, Bilbao).

Omeñaca García, J. (1998). *Plan General de Contabilidad comentado* (Deusto, Bilbao).

Pascual Pedreño, E. (1996). *Impuesto sobre sociedades y contabilidad* (Lex nova, Madrid).

Pulido, A. (1990). Provisiones para riesgos y gastos. *Cómo aplicar el Plan de Contabilidad* 3, 105–119 (Asociación Española de Contabilidad y Administración de Empresas, Madrid).

Rivero Torre, P. (1989). Los principios contables y las directrices de la Comunidad Europea', in *Lecturas sobre principios contables,* 113–152 (Asociación Española de Contabilidad y Administración de Empresas, Madrid).

Rivero Torre, P. (1992). Financial reporting by regulated enterprises. In J. A. Gonzalo (ed.) *Accounting in Spain, Fifteenth Annual Congress of the European Accounting Association,* 117–128 (Asociación Española de Contabilidad y Administración de Empresas, Madrid).

Rivero Torre, P. (1995). *Análisis de balances y de estados complementarios,* 7th edition (Pirámide, Madrid).

Rivero Romero, J. (1993). *Contabilidad financiera* (Trivium, Madrid).

Rivero Romero, J. and Rivero Menéndez, M. R. (1996). *Contabilidad* (Campomanes libros, Guadalajara).

Saez Torrecilla, A. (1990). Los principios contables y la imagen fiel en el nuevo PGC. *Partida Doble,* **1**, 16–23.

Saez Torrecilla, A. and Corona, E. (1991). *Análisis sistemático y operativo del Plan General de Contabilidad* (McGraw-Hill, Madrid).

Saez Torrecilla, A. and Gómez Aparicio, J. M. (1995). *Contabilidad General* (McGraw-Hill, Madrid).

Saez Torrecilla *et al.* (1995). *Contabilidad de empresas* (McGraw-Hill, Madrid).

Saez Torrecilla, A. *et al.* (1997). *Contabilidad* (McGraw-Hill, Madrid).

Salvador Cifre, C. and Pla Vall, A. (1998). *Impuesto sobre sociedades: régimen general y empresas de reducida dimensión* (Tirant lo blanch, Valencia).

Trujillano Olazarri, J. (1997). Las diferencias permanentes y temporales en la nueva ley del impuesto sobre sociedades. *Técnica Contable*, **579**, marzo, 185–200.

Trujillano Olazarri, J. (1998). *Problemática contable y fiscal del impuesto sobre sociedades* (Centro de estudios financieros, Madrid).

Tua Pereda, J. (1990). El Plan General de Contabilidad y el derecho contable. *Nuevo Plan General de Contabilidad* 3, *Revista Española de Financiación y Contabilidad*, **65**, 823–837.

Tua Pereda, J. (1992). The legal framework of accounting. In J. A. Gonzalo (ed.) *Accounting in Spain, Fifteenth Annual Congress of the European Accounting Association*, 59–72 (Asociación Española de Contabilidad y Administración de Empresas, Madrid).

Urias Valiente, J. (1997). *Introducción a la Contabilidad. Teoría y supuestos*, (Pirámide, Madrid).

Vela, M., Montesinos, V. and Serra, V. (1992), *Manual de contabilidad*, 2nd edition (Ariel, Barcelona).

Vilardell Riera, I. *et al.* (1997). *Introducción a la Contabilidad General* (McGraw-Hill, Madrid).

Yebra, O. (1990). Análisis de las cuentas anuales: el balance y la cuenta de pérdidas y ganancias. *Partida Doble*, **1**, 33–40.

Yebra, O. (1990). Contabilización de los planes de pensiones. *Cómo aplicar el Plan de Contabilidad*, 4, 3–24 (Asociación Española de Contabilidad y Administración de Empresas, Madrid).

Legal texts

Law of 17 July, 1953: Ley de 17 de julio de 1953 de Régimen Jurídico de las sociedades de Responsabilidad Limitada.

Law 21/1974: Ley 21/1974, de 27 de junio, sobre Investigación y Explotación de Hidrocarburos.

Law 61/1978 (Corporate Income Tax Law): Ley 61/1978, de 27 de diciembre, del Impuesto sobre Sociedades.

Law 2/1981: Ley 2/1981, de 25 de marzo, de Regulación del Mercado Hipotecario.

Law 5/1983: Ley 5/1983, de 29 de junio, sobre medidas urgentes en materia presupuestaria, financiera y tributaria.

Law 9/1983 (General State Budget Law): Ley 9/1983, de 13 de julio, de Presupuestos Generales del Estado.

Law 27/1984: Ley 27/1984, de 26 de julio, sobre Reconversión y Reindustrialización.

Law 33/1984 (Private Insurance Law): Ley 33/1984, de 2 de agosto, sobre Ordenación del Seguro Privado.

Law 46/1984: Ley 46/1984, de 26 de diciembre, de Instituciones de Inversión Colectiva.

Law 15/1986: Ley 15/1986, de 25 de abril, de Sociedades Anónimas Laborales.

Law 8/1987: Ley 8/1987, de 8 de junio, de Planes y Fondos de Pensiones.

Law 19/1988 (Audit Law): Ley 19/1988, de 12 de julio, de Auditoría de Cuentas.

Law 24/1988 (Securities Market Law): Ley 24/1988, de 28 de julio, del Mercado de Valores.

Law 26/1988: Ley 26/1988, de 29 de julio, sobre disciplina e intervencion de las entidades de crédito.

Law 19/1989 (Reform Law): Ley 19/1989, de 25 de julio, de reforma parcial y adaptación de la legislación mercantil a las Directivas de la Comunidad Económica Europea (C.E.E.) en materia de Sociedades.

Law 13/1992: Ley 13/1992, de 1 de junio, de recursos propios y supervision en base consolidada de las entidades financieras.

Law 37/1992 (Value-Added Tax Law): Ley 37/1992, de 28 de diciembre del Impuesto sobre el Valor Añadido.

Law 1/1994: Ley 1/1994, de 11 de marzo, sobre Régimen Jurídico de las Sociedades de Garantía Recíproca.

Law 30/1994: Ley 30/1994, de 24 de noviembre, de Fundaciones y de Incentivos Fiscales a la Participación Privada en Actividades de Interés General.

Law 2/1995: Ley 2/1995, de 23 de marzo, de Sociedades de Responsabilidad Limitada.

Law 10/1995: Ley Orgánica 10/1995, de 23 de noviembre, por la que se aprueba el Código Penal.

Law 25/1995 (General Taxation Law): Ley 230/1963, de 28 de diciembre, General Tributaria.
Law 30/1995: Ley 30/1995, de 8 de noviembre, de Ordenación y Supervisión de Seguros Privados.
Law 43/1995: Ley 43/1995, de 27 de diciembre, del Impuesto sobre Sociedades.
Law 7/1996: Ley 7/1996, de 15 de enero, sobre ordenación del comercio minorista.
Law 37/1998: Ley 37/1998, de 16 de noviembre, de reforma de la Ley 24/1988, de 28 de julio, del Mercado de Valores.
Order 31 March 1989: Orden, de 31 de marzo de 1989, por la que se faculta al Banco de España para establecer y modificar las normas contables de las Entidades de Credito.
Order 18 January 1991: Orden, de 18 de enero de 1991, sobre informacíon en Bolsas de valores.
Order 30 September 1992: Orden, de 30 de septiembre de 1992, por la que se modifica la de 18 de enero de 1991 sobre informatión pública periódica de las entidades emisoras de valores admitidos a negociación en Bolsa.
Order 27 January 1993: Orden del Ministerio de Economía y Hacienda, de 27 de Enero de 1993, por la cual se aprueban las Normas de Adaptación del Plan General de Contabilidad a las empresas constructoras.
Order 12 March 1993: Orden del Ministerio de Economía y Hacienda, de 12 de Marzo de 1993, sobre tratamiento contable de las diferencias de cambio en moneda extranjera en empresas reguladas.
Order 14 January 1994: Orden de 14 de enero de 1994, por la que se aprueban los modelos obligatorios de cuentas anuales a presentar en los Registros Mercantiles.
Order 2 February 1994: Orden de 2 de febrero de 1994, por la que se aprueban las normas de adaptación del Plan General de Contabilidad a las federaciones deportivas.
Order 18 March 1994: Orden de 18 de marzo de 1994, sobre diferencias de cambio en empresas reguladas.
Order 23 March 1994: Orden de 23 de marzo de 1994, sobre diferencias de cambio en empresas del transporte aereo.
Order 28 December 1994: Orden de 28 de diciembre de 1994, por la que se aprueban las normas de adaptación del Plan General de Contabilidad a las empresas inmobiliarias.
Order 14 June 1995: Orden de 14 de junio de 1995, por la que se regula la información sobre autocartera a remitir al Registro Mercantil.
Order 23 June 1995: Orden de 23 de junio de 1995, por la que se aprueban las normas de adaptación del Plan General de Contabilidad a las sociedades anónimas deportivas.
Order 23 December 1996: Orden de 23 de diciembre de 1996, por la que se aprueban las normas de adaptación del Plan General de Contabilidad a las empresas de asistencia sanitaria.
Order 14 January 1997: Orden de 14 de enero de 1997 por la que se modifica la de 14 de enero de 1994, por la que se aprobaban los modelos de presentación de cuentas anuales para su depósito en el Registro Mercantil correspondiente.
Order 12 February 1998: Orden de 12 de febrero de 1998 sobre normas especiales para la elaboración, documentación y presentación de la información contable de las sociedades de garantía recíproca.
Royal Decree of 22 August 1885: Real Decreto de 22 de agosto de 1885 por el que se publica el Código de Comercio.
Royal Decree 2631/1982 (Corporate Income Tax Regulations): Real Decreto 2631/1982, de 15 de octubre, por el que se aprueba el Reglamento del Impuesto sobre Sociedades.
Royal Decree 1348/1985: Real Decreto 1348/1985, de 1 de agosto, por el que se aprueba el Reglamento de Ordenación del Seguro Privado.
Royal Decree 302/1989: Real Decreto 302/1989, de 17 de marzo, por el que se aprueba el Estatuto y la estructura orgánica del Instituto de Constabilidad y Auditoría de Cuentas.
Royal Decree 1564/1989: Real Decreto 1564/1989, de 22 de diciembre, por el que se aprueba el texto refundido de la Ley de Sociedades Anónimas.
Royal Decree 1636/1990: Real Decreto 1636/1990, de 20 de diciembre, por el que se aprueba el Reglamento de desarrollo de la Ley de Auditoría de Cuentas.
Royal Decree 1643/1990: Real Decreto 1643/1990, de 20 de diciembre, por el que se aprueba el Plan General de Contabilidad.

Royal Decree 1815/1991: Real Decreto 1815/1991, de 20 de diciembre, por el que se aprueban las normas para la formulación de las cuentas anuales consolidadas.
Royal Decree 1624/1992: Real Decreto 1624/1992, de 29 de diciembre, por el que se aprueba el Reglamento del Impuesto sobre el Valor Añadido.
Royal Decree 1784/1996: Real Decreto 1784/1996, de 19 de julio,por el que se aprueba el Reglamento del Registro Mercantil.
Real-Decreto Ley 7/1996, de 7 de junio, sobre Medidas urgentes de carácter fiscal y de fomento y liberalización de la actividad económica.
Royal Decree 2607/1996: Real Decreto 2607/1996, por el que se aprueba el Reglamento de la Actualización de Balances.
Royal Decree 572/1997: Real decreto 572/1997, por el que se revisan los límites contables de los artículos 181 y 190 del texto refundido de la Ley de Sociedades Anónimas, aprobado por el Real Decreto Legislativo 1564/1989, de 22 de diciembre.
Royal Decree 537/1997: Real Decreto 537/1997, de 14 de abril, por el que se aprueba el Reglamento del Impuesto sobre Sociedades.
Royal Decree 2014/1997: Real Decreto 2014/1997, de 26 de diciembre, por el que se aprueba el Plan de Contabilidad de las entidades aseguradoras y las normas para la formulación de las cuentas de las entidades aseguradoras.
Royal Decree 437/1998: Real Decreto 437/1998, de 20 de marzo, por el que se aprueban las normas de adaptación del Plan General de Contabilidad a las empresas del sector eléctrico.
Royal Decree 776/1998: Real Decreto 776/1998, de 30 de abril, por el que se aprueban las normas de adaptación del Plan General de Contabilidad a las entidades sin fines lucrativos y las normas de información presupuestaria de estas entidades.
Resolution of 8 May 1996 of the General Directorate of Registers and Notaries, which approves the translation into the other official languages of the obligatory forms for annual accounts to be presented at the Mercantile Register for filing.
Resolution of 25 September 1996 of the General Directorate of Registers and Notaries, which approves the translation into an official language of the obligatory forms for annual accounts to be presented at the Mercantile Register for filing.
Resolution of 25 June 1997 of the General Intervention of the Social Security which establishes the accounting procedures for the registration of mutual funds for accidents at work and professional illnesses, of transactions relating to debts with the Social Security which have been the object of payment postponement or fractionation, or for those whose payment collection is under executive procedural action.
Navarra Charter Law 23/1996: Ley Foral 23/1996, de 30 de diciembre, de Actualización de Valores.
Navarra Charter Law 24/1996: Ley Foral 24/1996, de 30 de diciembre, sobre el Impuesto de Sociedades en Navarra.
Navarra Charter Law 22/1997: Ley Foral 22/1997, de 30 de diciembre, por la que se modifican parcialmente los impuestos de la renta de las personas físicas, sobre patrimonio, sobre sociedades y sobre sucesiones.
Vizcaya Charter Norm 6/1996, of 27 November, of Revaluation Accounting.
Guipuzcoa Charter Norm 11/1996, of 5 December, of Revaluation Accounting.
Navarra Charter Decree 9/1997, of 20 January, which is used to develop Navarra Charter Law 23/1996, of 30 December, of Revaluation Accounting.

SPAIN
GROUP ACCOUNTS

Antonio López Díaz
Pedro Rivero Torre

Acknowledgements

We gratefully acknowledge the help of Elena Fernández, Javier De Andrés, Ignacio Martínez del Barrior and Mariano Cabellos Velasco.

CONTENTS

I Introduction and background

1 Historical background

Until the publication of Royal Decree (*Real Decreto*, RD) 1815/1991 approving the 'Standards for the Preparation of Consolidated Annual Accounts' (*Normas para la formulación de las cuentas anuales consolidadas*), hereafter referred to as 'Consolidation Standards' (*Normas de Consolidación*, NC) the preparation of consolidated annual accounts was more the exception than the rule for groups of companies. This was due mainly to the absence of a legal obligation to consolidate. Royal Decree 1815/1991 enacts the provisions of the Commercial Code (*Código de Comercio*, CCom) which, for the purpose of adapting commercial law to EU regulations, requires the presentation of consolidated annual accounts.

However, certain regulations did exist which were a precursor to Royal Decree 1815/1991. In the tax area Royal Decree-Law (*Real Decreto-Ley*) 15/1977, relating to tax, financial and public investment measures, was the first regulation to set down the method of taxation of the consolidated profits of groups of companies. In addition, specific regulation of the consolidated income tax system is contained in Royal Decree 1414/1977. Law (*Ley*) 18/1982 also touched on this area, regulating the tax system applicable to joint ventures and companies developing regional industry. This law introduced a restricted concept of a group of companies for the purpose of a consolidated taxation system, requiring dependent companies to have holdings exceeding 90 per cent for them to be considered part of a tax group.

In the strictly accounting field, the predecessor of the Spanish Accounting and Audit Institute (*Institute de Contabilidad y Auditoria de Cuentes*, ICAC), the Institute of Accounting Planning (*Instituto de Planificatión Contable*), proposed an order dated 15 July 1982, which established the first standards for the preparation of consolidated accounts. This order, which was strictly an optional regulation, was solely aimed at promoting the consolidation of accounts for general information purposes. The order is of considerable importance, despite its non-binding nature, since it was based on the draft Seventh Directive which was approved the following year on 13 June 1983. Its clear association with current legislation is evidenced by its content: the main objective of preparing consolidated annual accounts is to reflect a true and fair view of the business situation of the group.

Two other events, each relating to a specific sector of the Spanish economy, affected the consolidation of accounts:

- The regulatory body of the Spanish electricity industry initiated a substantial financial improvement plan, for which it required, among other reporting procedures, the preparation, presentation and external audit of accounts of electricity utilities and their subsidiaries. This included companies in which electricity utilities had a minority holding. The regulation of these areas is contained in an order dated 26 April 1984 and in Royal Decree 441/1986.
- As a result of the grave crisis suffered at the end of the seventies and in the early eighties and of the high cost involved in cleansing the affected entities, the banking sector was forced to establish a much stricter accounting reporting system in order to strengthen the control of the Bank of Spain (*Banco de España*). Thus, Law 13/1985 governing investment coefficients, equity and reporting obligations of financial intermediaries, followed by Royal Decree 1371/1985, required

Table I.1 Historical development of consolidated accounts

1977–82	Regulation of the Consolidated Income Tax System
1982	Standards for the preparation of consolidated accounts (non-binding)
1984–86	Regulation of consolidated annual accounts for the electricity industry
1985	Regulation of consolidated annual accounts for finance entities
1989–91	Standards for the preparation of consolidated annual accounts (RD 1815/1991).

deposit-taking entities such as private banks, savings banks and credit co-operatives, to consolidate their balance sheets and profit and loss accounts with those of other deposit-taking and finance entities which together comprised a decision-making unit. The historical development of consolidated accounts is shown in Table I.1.

2 Harmonization with European Directives

In order to harmonize Spanish legislation with the European directives governing commercial matters, Law 19/1989 was passed which adapted commercial legislation to EU corporate regulations. These regulations introduced the provisions of EU corporate law, including the Seventh Directive governing consolidated accounts into Spanish law. The reform affected the major commercial regulations, i.e. the Commercial Code, the Corporations Law (*Ley de Sociedades Anónimas*, LSA), and the Limited Liability Companies Law (*Ley de Sociedades de Responsabilidad Limitada*, LSRL).

Consequently, Title III of Book I of the Commercial Code was reworded (Arts. 42 to 49, § 3) to incorporate the provisions of the Seventh Directive, and now constitutes the basic legal framework for the preparation of consolidated annual accounts.

The Commercial Code governs the following areas:

- The compulsory preparation of consolidated annual accounts in certain circumstances.
- The parameters for deciding whether a group of companies is required to prepare consolidated accounts.
- The basic consolidation standards to be taken into account.
- The obligations of controlling companies with regard to audits and disclosure of their consolidated accounts.
- A set of technical rules referring to such matters as the structure of the consolidated accounts, the eliminations to be carried out and the valuation of assets and liabilities.

Despite the provisions of the Commercial Code, the numerous variations that might occur in terms of the structure of a group of companies, in the application of the various consolidation methods and procedures contained in the law, and in the transactions performed by group companies, required further development of the regulations. To take these variations into account, Royal Decree 1815/1991 approving the 'Consolidation Standards' (NC) was implemented.

Thus current Spanish legislation on the consolidation of annual accounts arose as a consequence of the need to harmonize Spanish commercial legislation with EU regulations. However, as well as the need to comply with the Seventh Directive, Spanish law has also been influenced by the rulings of other regulatory bodies, such as IAS 27 and 28 of the International Accounting Standards Committee (IASC).

With this regard, it is necessary to bear in mind the communication launched by the European Commission in November, 1995. Instead of either extensive modification of the existing Accounting Directives or the creation of a European Accounting Standard Setting Body, the Commission agreed to put the Union's weight behind the international harmonization process which is already well under way in IASC, allowing the presentation of consolidated accounts in accordance with the IASC standards as long as those standards do not dissent with European Accounting Directives.

So far, neither the National Securities and Exchange Commission (*Comisión Nacional del Mercado de Valores*, CNMV) nor the ICAC have taken any initiative aimed at allowing the utilization of international standards for the preparation of consolidated accounts to be presented in the Spanish Stock Exchange.

3 Regulations for financial institutions

As far as banks and other financial institutions are concerned, the Bank of Spain's Circular 4/1991 (Banco de España, 1991), regarding accounting standards and financial statement formats for credit entities, stipulates specific rules and presentation models for the published consolidated financial statements of the groups of credit entities requiring consolidation listed in Law 13/1985 and its related regulations.

The Ministry of Economy and Finance Order

(*Orden Ministerial*) dated 18 January 1991 specifies the information which must be disclosed every quarter and half-year by companies listed on the stock exchange which are required to present consolidated annual accounts. Depending on the nature and business activity of the group, this information must be released in such a way that it conforms to the instructions laid down in the order. Several presentation models are envisaged for this purpose; in addition to a general reporting format, specific formats are prescribed for credit entities, financial leasing companies, insurance companies and portfolio companies.

II Objectives, concepts and general principles

1 Relevant objectives

The objective of consolidated accounts is to reflect the true and fair view of the group of companies included in consolidation in the same terms as those laid down for individual companies (CCom Art. 44.3).

The consolidation standards supplement the contents of the national chart of accounts (*Plan General de Contabilidad*), broadening the wording of the commercial accounting regulations to cover the accounting of groups of companies.

If, within a group of companies with their own individual nature, one company exercises effective control over the others so that that company directly or indirectly takes decisions about the whole group, the individual annual accounts of each company do not reflect a true and fair view of the economic reality of the whole group, and relevant information may be excluded. Reciprocal credits and debits and inter-group commercial or financial transactions may cloud the economic situation of the companies implicated. Only the joint accounts give a true picture of the financial or net worth situation and of the results of the economic entity formed by the group.

Despite the fact that the true and fair view of the controlling company can only be ascertained from the consolidated accounts, these accounts contain information which complements the individual accounts of the parent company. Accordingly, CCom Art. 42.2 establishes that the obligation to prepare consolidated annual accounts and a consolidated management report does not liberate group companies from the obligation to prepare their related individual annual accounts and management report in accordance with the regulations governing their specific case.

The objective of the consolidation standards is to ensure that the users of the resulting accounting information can obtain a picture of the group's financial and economic potential. For this purpose, the consolidation standards leave room for interpretation with regard to the concept of a group of companies. Not only is a group of companies defined by the fact that the controlling company owns a certain percentage of the dependent companies, but a company also forms part of a group when it is subject to the effective control of a superior decision-making unit (see III.1).

However, in order for a group of companies to opt for the consolidated income tax system, a substantially more restrictive condition is imposed: the controlling company must own more than 90 per cent of the capital stock of the other group companies. Furthermore, such a consolidated tax system requires the authorization of the tax authorities.

In summary, consolidated results are only relevant for tax purposes if the group qualifies for, and is authorized to opt for, the consolidated income tax system. A common case may be that of a mixed system under which part of the group is taxed under the consolidated income tax system and other companies in the group, which do not reach the minimum degree of control to accede to the consolidated tax system, file individual tax returns. The rules governing the accounting treatment of corporate income tax for groups and their relationship with the rules for filing individual tax returns are dealt with in full detail below (see VIII).

1991 was the first year that Spanish groups were required to present consolidated annual accounts. The studies that have been carried out since that date on the application of the legislation governing these consolidated annual accounts reveal the following:

Firstly, the study by Benitez *et al.*, (1993, pp. 31–55):

- The strong influence of commercial legislation on the preparation of consolidated accounts.

Most of the companies presenting consolidated annual accounts did so for the first time in 1992 (referring to the year 1991), in accordance with Royal Decree 1815/1991.

- The importance of the controlling company in the consolidated group, with a high proportion of ownership of around 90–100 per cent, the most common link to ownership being through the majority vote of the board of directors.
- The similar nature of the activities of the group companies.
- The small number of multigroup companies.
- The scant use of clauses excluding dependent companies.
- Formal compliance with the standards on presentation formats included in the Consolidation Standards (NC) approved by Royal Decree 1815/1991.
- The omission in certain cases of data on companies requiring consolidation (companies owning the holding, their registered offices, reasons for exclusion, activity and their actual percentage of ownership).

Secondly, the study by Archel *et al.*, (1995, pp. 59–79), based on the 1991 and 1992 annual accounts presented on the stock exchange:

- A significant improvement in the quality of the financial information available due to the publication of Royal Decree 1815/1991. The degree of compliance with the obligation of presenting consolidated annual accounts is regarded as 'more than acceptable'.
- However, a certain lack of information is detected with respect to several significant matters, such as date of the first consolidation or the goodwill amortization period.
- The existing options with regard to the election of the first date of consolidation and its eventual allocation to income make the degree of comparability achieved rather unsatisfactory.
- The election of the date for calculating the first consolidation or the period of amortization of goodwill do not lead to significant statistical differences in those ratios regarded as indicative of the financial and economic position of the groups of companies.

Finally, it is worth mentioning the study by Ansón et al., (1997, pp. 917–934), that underlines the usefulness of consolidated information vs. the parent company individual accounts in order to analyze the prospects of stock returns of the parent company of a group.

2 Underlying concepts and general principles

The introduction to the consolidation standards states that consolidated accounts are conceived as an extension to the accounts of the parent or controlling company (the parent company theory, PCT). However, despite the statement included in the introduction, certain regulations suggest a different conceptual approach to consolidation according to which the consolidated annual accounts are seen as an expression of a group, as a tangible economic unit (the entity theory, ET). Several examples of the solutions adopted in the consolidation regulations and their underlying conceptual focus are described below, so as to give an impression of the hybrid nature of the approach chosen:

- This mixed nature is apparent in the treatment given to the external shareholders of controlled companies. In accordance with the perception of the consolidated accounts as an extension of the accounts of the parent company (PCT), the external shareholders of controlled companies are considered to have a different status to those of the related controlling company. Consequently, the interests of these external shareholders are not included within equity in the balance sheet. However, the financing provided by external shareholders is not considered as external financing either. It seems to be considered as a particular kind of external financing for the shareholders of the parent company, which is neither external nor internal.
- The influence of the concept of consolidated accounts as an expression of a differentiated economic unit (ET) is also seen on the computation of minority interests. These are calculated after making the appropriate eliminations from the profits or losses arising from intercompany transactions rather than on the basis of the book value of the net worth of these companies before making the eliminations.

- In line with the entity theory of consolidation (ET), the elimination of inter-group profits and losses is made in full, without taking into account the percentage of ownership of the dependent companies, or whether the controlling or the dependent companies play the role of seller.
- The concept of consolidated accounts as an expression of a tangible economic unit (ET) also seems to impact on the way consolidated results are determined. Under Spanish legislation, consolidated results incorporate the controlling company's income (loss), as well as that allocated to the external shareholders of dependent companies. However, both the consolidated balance sheet and the consolidated profit and loss account reflect in a separate account the amounts relating to the different types of shareholder.
- The 'consolidation scope' not only includes related companies in which the parent company has a majority interest, but also, in accordance with the concept of consolidated annual accounts as an expression of a tangible economic unit (ET), multigroup and associated companies. A consolidation method or procedure other than the global integration method is used for multigroup and associated companies (see VII).
- The valuation of differences in consolidation (goodwill and negative differences and the portion of these differences allocated to specific asset and liability items) is limited to the percentage of the capital stock of the related companies, owned by the controlling company. This is in accordance with the concept of consolidated accounts as an extension of the accounts of the controlling company (PCT).

III Group concept and preparation of group accounts

1 Group accounts

The only commercial regulation of groups of companies is that of the consolidation of annual accounts. Until the approval of Law 19/1989 partially amending and adapting commercial law to European directives, the concept of a group was unheard of in Spanish commercial law. This law, which requires the consolidation of the annual accounts of groups of companies, was the first legislation to regulate the accounting of corporate groups. Prior to Law 19/1989, the concept of a group only appeared in certain regulations governing specific areas such as the consolidated income tax system, the establishment of limits to insurance companies' investments using reserves, the quantification of credit entity equity and with respect to the regulation of the stock market.

The Consolidation Standards (NC) approved by Royal Decree 1815/1991 define a group of companies, for the sole purpose of the consolidation of accounts, as that comprising a controlling company and one or several controlled companies (NC Art. 1).

The controlling/controlled relationship is defined by NC Art. 2, which describes a controlling company as a commercial company which, being a shareholder of a commercial or non-commercial company, has one of the following characteristics with regard to the investee (controlled) company:

- It holds the majority voting rights: directly, through another dependent company, or through agreement with other shareholders.
- It has the power to name or remove from office the majority of the members of the board of directors.
- It has appointed with its votes the majority of the members of the board of directors who hold office in the year in which the consolidated annual accounts are being prepared and in the two preceding years. The consolidation of accounts is not required if the company, whose directors have been appointed, is related to another company in either of the ways mentioned above (holds majority voting rights or has the power to name the majority of directors).

For the purpose of computing the voting rights, to those held directly by the controlling company itself are added those relating to the companies controlled by it or to other persons acting in their own name but on behalf of another group company.

As described above, the criterion used to determine whether a group of companies exists and, consequently, whether there is an obligation to present consolidated annual accounts, depends on whether one company can exercise direct or indirect control over others. There are two viewpoints from which the analysis of the concept of a group of companies can be seen:

- The legal viewpoint, under which the obligation to consolidate arises when the controlling company has a holding of at least 50 per cent in the capital stock of the dependent company.
- The economic viewpoint, for which the relevant criterion defining the existence of a group is the management unit.

Once again the mixed nature of the Spanish consolidation rules comes into focus. Although the legal viewpoint has not been selected as the basis for Spanish consolidation regulations, it is difficult to hold that the economic viewpoint is prevailing, since Spanish regulations do not take into account certain situations which arise frequently in practice and which are dealt with by the Seventh Directive (Art. 12). These are situations in which there is no capital stock between the group companies, despite the fact that there is a clear co-ordination between them. The Seventh Directive describes two possible situations of this type:

- Where there is a contract establishing one management unit.
- When the managing bodies of the implicated companies mainly comprise the same people in the year in which the consolidated annual accounts are being prepared.

The existence of a co-ordination group does not give rise to the obligation to consolidate. All the cases dealt with by the Consolidation Standards require a relationship of dependence, that is to say, there must be a controlling company and a controlled company (as defined by NC Art. 2). The holding of a securities portfolio denoting the existence of associated companies or multigroup companies does not give rise to the obligation to consolidate unless there are other controlled companies.

The definition of the controlling company applied by Spanish regulations restricts the scope of the consolidation regulations to the field of commercial companies, i.e. general partnerships, limited partnerships, limited liability companies and corporations including employee-owned companies. However, even though the controlling company must always be a commercial company, the dependent companies need not necessarily be so.

2 Subgroup accounts

The obligation to present consolidated annual accounts applies to any commercial company which complies with the controlling/controlled relationship with respect to other companies (see III.1). Accordingly, controlling companies dependent on others might exist, and this is usually called a subgroup. This is one of the two cases (the other being small groups – see III.3) in which Spanish accounting regulations do not require the presentation of consolidated annual accounts.

Certain requirements must be met in order for a controlling company dependent on a superior level group to be exempt from consolidation. Since the external shareholders of the exempted controlling company might be adversely affected by such exemption, Spanish law gives them the right to require the preparation of consolidated annual accounts by the subgroup in which they have a direct ownership interest, as long as this participation is at least 10 per cent of the capital stock.

Therefore, in accordance with NC Art. 9, the presentation of consolidated annual accounts is not required where the controlling company is at least 50 per cent owned by another company, provided that the minority shareholders holding 10 per cent of the capital stock do not request such consolidation of the subgroup's annual accounts at least six months prior to the year end.

The following additional requirements must be met:

- None of the subgroup companies can be a listed company.
- The controlling company must be an EU-resident company.
- The exempted subgroup must be consolidated with a superior level group by the global integration method.
- The exempted company must describe this exemption from consolidation in the notes to its annual accounts, as must the group to

which it belongs and the registered offices of the controlling company.

- The consolidated accounts of the controlling company, its management report and the auditors' report must be filed with the commercial register at the location of the registered offices of the exempted Spanish company.

The obligation to consolidate arises when there is a group of companies, i.e. when there is at least one controlling company and one or several controlled companies in accordance with NC Art. 2. The existence of investment securities denoting that there are associated or multigroup companies does not necessarily give rise to the obligation to consolidate unless there are other dependent companies which are controlled; neither does the existence of a co-ordination group.

According to CCom Art. 47.3, associated companies are defined, for the sole purpose of consolidation, as companies in which one or several group companies exercise a significant influence over their management. Multigroup companies are defined, for the sole purpose of consolidation, as those which are jointly managed by a group company and one or more non-group companies (see VII.1 and VII.2.1).

3 Exemptions for small groups

Small groups are exempted from preparing group accounts. The objective of this exemption is to reduce the reporting obligations of groups which, due to their size, do not have a significant economic materiality. In accordance with the provisions of the Seventh EC Directive, the Commercial Code bases the size of small group on three thresholds: ESP 1,200 million of assets, ESP 2,400 million of turnover and an average number of employees of 250 (CCom Art. 43. 1). Subsequently, Royal Decree 1815/1991 maintained the values contained in the Commercial Code; however in its sole temporary provision, it states that up to the last year which closes before 1 January 2000 the limits will be as follows: ESP 2,300 million of assets, ESP 4,800 million of turnover and 500 employees.

In accordance with NC Art. 8, several additional requirements must be met:

- None of the group companies can be a listed company.
- At the accounting date of the controlling company, the aggregates of the group companies must not exceed two of the three thresholds mentioned above for two consecutive years.
- The calculation of the thresholds must take into account the figures of all the group companies, even if certain dependent companies are excluded from consolidation for any of the reasons described below (see IV). The prior adjustments and eliminations relating to the consolidation must also be taken into account. If these adjustments and eliminations are not included, the limits shall be the total assets and turnover magnified by 20 per cent (the limit relating to the average number of employees is not affected). The average number of employees is determined by considering all the people who have worked for the group companies during the year and the duration of their service and averaging this out over the year.

IV Consolidated group

1 The consolidated set

The consolidated set (*conjunto consolidable*) comprises companies to which the full or proportional integration method is applied (NC Art. 13). The full integration method is applied to group companies, i.e. the controlling company and the controlled companies forming the group. Associated companies are included in the consolidated accounts by the equity method. The proportional integration method is optional for multigroup companies. If the proportional integration method is not applied, the controlling company must include the multigroup company under the equity method.

In its definition of the controlling/dependent relationship determining the existence of a group of companies, NC Art. 2 establishes the criteria which define the presence of a management unit (see III.1).

As the Consolidation Standards state, the criterion for determining the existence of a group of companies and, accordingly, the obligation to present consolidated annual accounts, is the effective control exercised by one company over another. For example, one company may hold less than 50 per cent of the capital stock of another but exercise effective control over it; those two companies would form a group of companies and, consequently, would be obliged to present consolidated annual accounts.

In addition, from the description of the controlling/dependent relationship described in NC Art. 2, it can be inferred that the mere potential capacity (not the formal or effective capacity) to exercise control of the majority voting rights in a dependent company determines the existence of a group of companies. For instance, if a company has the potential power to name or remove the majority of members of the board of directors the existence of a group of companies will be assumed, even though in practice this option is never exercised.

However, according to the available empirical evidence (see II), it is the presence of a clear control relationship, defined by a high ownership interest in the dependent company, that determines in practice the existence of a group of companies and the need to prepare consolidated annual accounts. The possibility of preventing the inclusion of a company in a group by means of a contractual agreement transferring the capacity to exercise voting rights is not envisaged by Spanish consolidation regulations.

The Consolidation Standards define the consolidation scope (*perímetro de la consolidación*) including companies acting under the sole management of a controlling company, together with associated companies, i.e. companies over which the controlling company exercises a significant degree of direct or indirect influence, and multigroup companies, i.e. non-dependent companies jointly managed by the controlling company and/or other group companies and other nongroup companies (NC Art. 15).

2 Exemptions from full consolidation for dependent companies

NC Art. 11 of the consolidation standards envisages the possibility of excluding from the full consolidation method those companies with one or more of the following characteristics:

- When a dependent company has a scant material interest with respect to the true and fair view that should be expressed by the consolidated accounts. If several group companies have a scant materiality individually, they are only excluded from full integration if they collectively have a scant materiality with regard to expressing a true and fair view.
- When there are significant long-term restrictions hindering the controlling company from exercising its rights over the equity or management of the dependent company, e.g. legally declared insolvency or legal or governmental intervention, etc. Other circumstances, such as limitations to the remittance of funds or governmental intervention over certain net worth items, are not necessarily reasons for excluding companies. However, these circumstances must be described in the notes to the consolidated annual accounts with reference to their possible effect on the group's net worth, financial situation and results.
- When the information required to prepare the consolidated annual accounts can only be obtained by incurring clearly disproportionate expenses or by giving rise to delays preventing the preparation of such accounts on schedule, i.e. at the accounting close and for the same period as for the individual annual accounts of the controlling company.
- When the holdings in the dependent companies were acquired and are owned exclusively for the purpose of their subsequent transfer within one year from the acquisition date.
- When the dependent companies have such differing business activities that their inclusion would contravene the objective of the consolidated accounts. However, the Consolidation Standards restrict this exception, so it generally does not apply when the companies included in consolidation are partially industrial or commercial and partially engaged in the rendering of services, or when they perform differing industrial or commercial activities or render differing services. Specifically, the Consolidation Standards state that

when the companies involved include finance companies and insurance entities, together with others whose corporate purpose comprises commercial or industrial activities or the rendering of services, their inclusion would contravene the purpose of consolidation.

Mere asset-holding and portfolio companies are not deemed to perform differing business activities. Asset-holding companies are defined as follows: more than one half of their real assets during more than six months out of twelve are assigned to activities which are not business, professional or artistic. Portfolio companies are defined as follows: more than one half of their real assets during more than six months out of twelve comprise marketable securities, provided that these securities are not allocated to any other activity originally envisaged in the company's articles but are merely held by those companies.

If any company is excluded from consolidation for any of these reasons and is integrated in the consolidated accounts by the equity method, NC Art. 11 stipulates that this must be mentioned in the consolidated notes.

NC Art. 11 also states that when the excluded company is not domiciled in Spain, its individual annual accounts must be attached to the consolidated annual accounts, or alternatively these documents must be made available to the public.

Several authors have analyzed the possible effects and implications of the exemptions from full consolidation:

- Based on the available empirical evidence, Benitez *et al.*, (1993, pp. 31–55) suggest that in practice the only reason for exclusion is the scant materiality of the excluded dependent companies with respect to the true and fair view which should be expressed by the consolidated accounts.
- Robleda (1996, p. 18) asserts that, very often the reason argued for excluding numerous dependent companies from the full consolidation method is their scant material interest with respect to the true and fair view, while, really, the true reason is the severe financial problems of some of these dependent companies.
- Gonzalo (1994, pp. 19–20) explains that the exemption based on the differing business activity of the dependent company cannot be justified from the accounting point of view. The reason why this exemption is maintained in Spain and several other countries is because of the interest of the insurance and banking regulatory authorities in taking their decisions by using genuine insurance or banking information, without any sort of interference caused by the existence of companies not belonging to these sectors.
- Blasco (1997, pp. 87–9 and 146–147) is also very critical concerning the exemption from full consolidation of dissimilar dependent companies. First of all, because of its optional character. Secondly, because the different nature of some activities cannot be a valid reason for exemption in a world where the appearance and proliferation of big diversificated conglomerates has been a relatively frequent event. Finally, because the exemption from full consolidation and the application of the equity method can lead to an entirely different picture of the profitability, liquidity, credit standing and financial structure of the group. However, the empirical analysis that the author carries out shows that, in spite of the option, the full consolidation is preferred by the groups included in the sample.

V Uniformity of accounts included

1 Uniformity of formats

Before aggregating all the information on the group companies, their financial statements must first be sufficiently standardized so that the information to be aggregated meets the criteria used by the controlling company to prepare its individual annual accounts.

Spanish regulations require the presentation of both the individual annual accounts of the controlling company and the consolidated annual accounts of the group companies (NC Art. 6.3) (see II.1).

2 Balance sheet date of the companies included

The consolidated annual accounts must refer to the same accounting close and encompass the same period as the annual accounts of the controlling company (NC Arts. 17 and 62). However, if a dependent company closes its accounting year less than three months before the close of the consolidated annual accounts, it can be included in consolidation at the book values relating to its annual accounts at that accounting close date provided that the duration of its accounting year is the same (12 months) as that of the consolidated annual accounts.

When significant transactions take place between the accounting close of the dependent company and that of the consolidated annual accounts (unusually high increase in production, high gains or losses on the disposal of fixed assets, the acquisition of a significant stake in another company, the obtainment of a significant subsidy, etc.) these transactions must be included in consolidation. If the transaction involves another company of the group, the appropriate eliminations must be made and a description of all the circumstances must be given in the notes to the consolidated annual accounts.

If a dependent company closes its accounts more than three months before the close of the consolidated accounts, or if the duration of the period to which they refer does not coincide with that of the consolidated annual accounts, specific annual accounts must be prepared for the same length of time and with the same closing date as the consolidated annual accounts.

If a company joins a group, or is excluded from a group, the individual profit and loss account to be included in consolidation must refer solely to the portion of the year in which the dependent company was part of the group (NC Art. 17).

3 Recognition criteria and valuation rules

The standardization of valuation rules is defined by the Consolidation Standards. This states that the asset and liability items and the revenues and expenses of consolidated companies must be valued according to uniform criteria laid down in the valuation principles and standards established by the Commercial Code, the Corporations Law (LSA), as amended, the Chart of Accounts, and other legislation governing this topic.

If any asset or liability item or any revenue or expense is valued for a dependent company's accounts using different criteria from those applied in consolidation, this item must be revalued for the sole purpose of consolidation, and in accordance with the criteria applied by the controlling company. Unless the result of the new valuation is scantily material to the true and fair view of the group, the required adjustments must be made.

The controlling company must apply the same valuation criteria to the consolidated annual accounts as to its own individual accounts (NC Art. 18.3).

In exceptional cases, different valuation criteria may be applied by a dependent company, provided that these criteria are more appropriate in the circumstances than those applied by the controlling company. This situation must be duly described and justified in the consolidated notes.

If the controlling company and a dependent company apply different criteria for the depreciation of similar items, the principle of uniformity can be interpreted with flexibility. It would be possible to claim, for instance, that although the asset items are technically similar, they might have different useful lives because of different working hours, different technological evolution in the case of different countries, or a more or less hostile environment. If any of these reasons can be given for not applying the principle of uniformity, this fact must be described in detail in the notes to the consolidated annual accounts.

4 Foreign currency translation

The Consolidation Standards stipulate that the balance sheet and profit and loss account items of foreign companies consolidated by the full or the proportional consolidation method must be translated to pesetas under one of the following methods:

- Year end rate method.
- Monetary/non-monetary method.

In general, the year end rate method will be applied except when the activities of the foreign company are so closely linked to those of a Spanish group company that it can be considered as an extension of the activities of the Spanish company. In this case, the translation is made under the 'monetary/non-monetary method'.

When the year end rate method is used, it is considered that since the dependent company engages in an activity which is different from that of the controlling company, it also has accounting autonomy. At year end the annual accounts are received for the purpose of consolidation and a single translation rate is applied. Consequently, the whole business of the foreign company is subject to currency risk rather than simply its individually considered assets and liabilities. The investor is only interested in his initial investment in pesetas and in the flow of dividends received.

When the foreign company's business is closely linked to that of the Spanish controlling company, the 'monetary/non-monetary method' is applied. There is no one basic characteristic (it may be that the company is marketing or producing the raw materials to be used by the group, or is attracting outside financing) which defines this link. Factors or indicators other than those defined under the standards (weighted by the company under its own criteria) must be valued. Under the 'monetary/non-monetary method' transactions made by the foreign company are subject to the same currency risk as when the parent company makes transactions directly.

Since the activities of foreign companies may be linked in different ways to those of the controlling company, the two translation methods may be used by companies within the same group. Accordingly, part of the translation differences may be reflected as a separate item within equity and part may directly affect the profit and loss account.

4.1 Year end rate method

This method is described in NC Art. 55 which prescribes the following rules:

- All assets, rights and obligations are translated to pesetas at the exchange rate prevailing at the close of the annual accounts of the foreign company to be included in consolidation.
- The profit and loss account items are translated at the exchange rates prevailing at the transaction dates. An average exchange rate may be used provided that it is weighted on the basis of the volume of transactions made in each period (month, quarter, etc.).
- The difference between the equity of the foreign company, which includes the balance of the profit and loss account, translated at the historic exchange rate, and the net worth position resulting from the translation of the assets, rights and obligations will be included as a positive or negative balance in equity under the Translation Differences heading. This balance will be net of the portion of the difference relating to external shareholders, which must be recorded separately under the Minority Interests account.

In short, Spanish standards favour the exchange rates prevailing on the transaction dates as a general criterion for account items, even if the year end rate has been applied to balance sheet items. The average exchange rate is only considered as a possibility (not a preference) in cases where the historic identification of the period items is difficult provided that a weighting is made on the basis of the volume of transactions made each month, quarter or year. The application of this method gives rise to translation differences which must be recorded as additions to or subtractions from the group's equity, but with no effect on the consolidated profit and loss account.

When shares or participations in foreign companies are sold, the translation differences are considered to be reserves of the company which has sold them but are not included in the profit and loss account.

4.2 The monetary/non-monetary method

The following rules included in NC Art. 56 must be applied:

- Non-monetary balance sheet items are translated at the historic exchange rate prevailing at the date the item is included in the company's net worth or, where appropriate, at the first consolidation date. The valuation adjustments made to non-monetary items are translated by using the exchange rate applicable to the items.

- Monetary items in the foreign company's balance sheet are translated at the exchange rate current at its accounting year end. Monetary items are deemed to be cash as well as all items relating to collection rights and payment obligations. Accrual accounts are translated at the exchange rate relating to the item concerned.
- Profit and loss account items are translated at the exchange rates prevailing at the transaction dates, except for those relating to non-monetary items. The average exchange rate for the year can be applied, provided that it is weighted on the basis of the volume of transactions made in each period (month, quarter, etc.).
- Non-monetary revenues and expenses, such as depreciation, amortization and provisions, the allocation to results of deferred expenses and revenues, and the cost to be considered in determining gains or losses on the disposal of such items, are translated at the historic exchange rate applied to these non-monetary items.
- Translation differences arising from the application of this translation method are allocated to results and are disclosed separately under the Translation Gains (Losses) heading as appropriate. Alternatively, they may be treated in line with the controlling company's criteria for exchange differences arising from foreign currency transactions. These criteria are defined by Valuation Standard No. 14 of the PCG (see Spain – Individual Accounts, VII.5).

4.3 Elimination of results arising from inter-company transactions

The elimination of results from inter-company transactions is performed at the exchange rate prevailing at the transaction date. However, when the monetary/non-monetary method is applied and, consequently, the historic exchange rate is used for non-monetary items, inter-company results arising from the disposal of the non-monetary assets of foreign companies will be eliminated so that these items are reflected at the book value of the disposing company. The exchange rate prevailing at the date of the inclusion of the item in the company's net worth or, as appropriate, at the date of the first consolidation is applied to this book value (NC Art. 58).

4.4 Foreign companies subject to high inflation rates

When dependent foreign companies are subject to high inflation rates, the balance sheet and profit and loss account items must be adjusted before their translation to pesetas is made by the year end rate method, because of the effects of price changes. Alternatively, they may be translated to pesetas under the monetary/non-monetary method, provided that the application of this method leads the annual accounts to reflect a true and fair view.

Adjustments for inflation are made by following the standards set down in the country where the foreign company is located. Additionally, information on these adjustments and on the criteria applied must be disclosed in the consolidated notes (NC Art. 57).

The translation of annual accounts denominated in foreign currencies in the case of companies carried by the equity method is discussed below (see VII. 2.4).

4.5 Accounting for the introduction of the euro

Once the fixed conversion rates between the national currency units of the EMU participating Member States are in effect, the translation differences on consolidation of foreign subsidiaries operations denominated in one of the other participating currencies will become fixed amounts.

The Spanish accounting authority (ICAC) considered the possibility of using for this purpose a conversion method different from the one normally used when foreign dependent companies of a group are consolidated, treating the translation differences as a reserve. The treatment of these differences as a reserve could be justified on account of their irrevocable nature and the exceptionality of the situation created as a result of the introduction of the euro.

The final decision has been to recommend the groups to use the same method that has been applied in the past. Accordingly, if the year end method has been used, the translation differences will be treated as a reserve and included within equity, whereas if the monetary/non-monetary method has been used, translation differences will be allocated to results.

VI Full consolidation

1 Consolidation of capital

Under NC Art. 2 the full integration method is applied to group companies, i.e. the nucleus comprising the controlling company and one or more controlled companies.

Consolidation Standards (NC Art. 16) describe the full integration method as follows:

> The application of the full integration method requires the inclusion in the balance sheet of the controlling company of all the assets, rights and obligations comprising the net worth of the dependent companies, and the inclusion in the former's profit and loss account of all the revenues and expenses with an impact on the latter's period results; all this without prejudice to prior standardizations and the related eliminations.

The first elimination that must be made is the investment/equity elimination, defined by the Consolidation Standards (NC Art. 22) as:

> Offsetting the book value of the direct or indirect holding of the controlling company in the capital stock of the dependent company against the proportional part of the equity of the consolidated (dependent) company relating to the holding at the date of the first consolidation.

The first consolidation date for each dependent company is that at which it is included in the group, i.e. when at least one of the following conditions established by NC Art. 2 is met:

- The controlling company holds the majority voting rights.
- The controlling company has the power to appoint the majority of the members of the board of directors.

However, it may be considered that the inclusion of a dependent company in the group occurs at the beginning of the year in which the group was first required to present consolidated annual accounts, or in which they were prepared voluntarily, provided that this date is later than that of the effective incorporation into the group.

Under the fourth final provision of Royal Decree 1815/1991, the obligation of dominant companies to prepare consolidated annual accounts and a management report came into force for annual accounts closing after 31 December 1990.

1.1 Differences in the first consolidation

Under NC Art. 23, the positive or negative difference in the consolidation is deemed to be the difference between the book value of the direct or indirect holding in the capital stock of the dependent company and the value of the proportion of the equity of the dependent company which is allocable to that holding at the date of the first consolidation.

The book value of this holding consists of the acquisition price determined under the valuation standards set down by the chart of accounts (PGC Valuation Standard No. 8 relating to marketable securities), i.e. the price includes the full amount paid, or that must be paid for the acquisition, including preferential share subscription rights and the expenses inherent in the acquisition. Unmatured accrued dividends at the acquisition date are not included in the acquisition price.

This amount is reduced by valuation adjustments, provisions or losses arising prior to the first consolidation and after the appropriate valuation standardization has been made if different accounting methods have been used from those applied by the controlling company.

Equity is regarded as defined by the chart of accounts, net of own shares, without prejudice to the recalculation of the related percentage of ownership. Own shares as defined by the Corporations Law have their rights suspended (LSA Art. 79) and, accordingly, are not taken into account for the purpose of determining the degree of control.

1.2 Goodwill in consolidation

When there are positive differences in the first consolidation, as far as possible they are directly allocated, for the sole purpose of preparing consolidated accounts, to the dependent company's items. This increases the values of the assets, which then exceed their book values, or decreases the values of the liabilities up to the limit that is allocable to the controlling company on the basis of its percentage of ownership of the capital stock of the dependent company of the difference between the book value of the related net worth item and its market value at the first consolidation date. The amounts resulting for the balance sheet item based on the allocation are amortized, if appropriate, by the same methods as those applied before the allocation.

The positive differences remaining to be allocated are recorded under the 'Goodwill in consolidation' heading in the consolidated balance sheet. Goodwill in consolidation is amortized systematically over the period in which it will contribute to the generation of income for the group up to a maximum of 10 years. When the amortization exceeds five years, the appropriate justification must be included in the notes to the consolidated annual accounts (NC Art. 24). This goodwill means that the dependent company has a capacity to obtain income above the average of its sector, as it has certain differing factors with respect to its competitors. The concept thus resembles the goodwill registered in individual accounts when it has been acquired for a consideration.

1.3 Negative differences in consolidation

Where there is a negative difference in the first consolidation, as far as possible it is allocated directly, for the sole purpose of preparing consolidated accounts, to reduce the value of the assets or to increase the value of the liabilities up to the limit that is allocable to the controlling company under the criterion for positive differences (see VI.1.2).

The portion of the negative difference which cannot be directly allocated by this method is recorded under the 'Negative differences in consolidation' heading on the liabilities side of the consolidated balance sheet.

Negative differences in consolidation can only be taken to income in the following cases:

- When the difference is based on the poor performance of the related company or on the reasonable projection of the expenses related to this performance, provided that this projection actually comes about.
- When the differences relate to a realized capital gain. The capital gain is considered to have been realized when the related asset is sold or when it is released from inventory. It is also considered to have been realized in the related portion when the holding in the capital stock of the dependent company is fully or partially sold.

In the first case, the negative goodwill in consolidation acts as a provision: when the projected losses or the projected expenses actually arise, the provision is applied to the projected losses and expenses and is charged to the consolidated profit and loss account. In the second case, the negative goodwill in consolidation acts as a deferred revenue since it is justified by an advantageous acquisition, and the negative goodwill from the first consolidation remains on the liabilities side of the consolidated balance sheet until the advantageous situation becomes a realized gain (NC Art. 25).

As it has already been mentioned (see VI.1), there are two alternatives for the first consolidation date, either the date when the dependent company is first included in the group or the date when consolidated annual accounts are presented for the first time. Since there is not unified temporary criterion for the book value of the investment in the controlled company and the proportional part of its equity, it becomes necessary to analyze the evolution of the equity in the controlled company from the investment date until the first consolidation date in order to understand the meaning of the negative difference in consolidation (Blasco, 1997, pp. 44–49). Two different situations may arise:

- If the controlled company has gone through a negative evolution in that period, the difference between the book value of the controlled company and the proportional part of its equity will correspond to the provision which the controlling company would have to have allocated for compensating the negative evolution of the controlled company equity. Therefore if the difference arisen between book value and equity is positive and the perpetuation of this negative evolution in the controlled company is anticipated, the positive difference should be allocated to the year income, though Spanish Consolidation standards do not impose any obligation of this sort.

- If the evolution in the controlled company since the investment date has been positive, as a revaluation of the investment book value is not allowed, an offsetting between two disparate amounts will have to be done and the negative difference arisen will have to be regarded as a reserve under NC Art. 25.4. The problem is that the further the distance between the investment date and the first consolidation date, the less meaningful the consolidation difference becomes and it will be increasingly difficult for the analyst to determine the original sources of the difference, and to account for which part of the difference responds to projected losses at the investment date and which part responds to a realized capital gain.

1.4 Subsequent consolidations

In the second and subsequent consolidations, the investment/equity elimination is made under the same conditions as those established for the first consolidation.

Once the group has been formed and effective control begins to be exercised, the controlling company is allocated a portion, equal to the percentage of capital stock owned, of the variations which arise in the dependent company's equity between the first consolidation date and the current consolidation date. NC. Art. 27 states that:

> Reserves generated by dependent companies since the date of the first consolidation, including those which have not passed through the profit and loss account, shall be recorded in the liabilities side of the consolidated balance sheet under the 'Reserves at consolidated companies' heading, net of the portion of the reserves relating to external shareholders.

The 'Reserves at consolidated companies' shall be positive and included in equity in the balance sheet when there is an increase in the equity of the dependent company between the first consolidation date and the current consolidation date. Otherwise they will be negative and included with a negative sign in equity in the consolidated balance sheet.

Equity is calculated by excluding the results for the year, and taking into account the adjustments made in prior years' results relating to intragroup transactions. In any event, the valuation adjustments relating to the holding investments in the dependent company made after the inclusion of the dependent company will already have been eliminated. If the assets revalued in the first consolidation continue to increase their market value, these assets are not revalued in subsequent consolidated balance sheets as this would be inconsistent with the acquisition cost principle (Martínez, 1993, pp. 12–16).

1.5 Minority interests

Minority interests comprise the portion of equity including the period result of the dependent companies which does not relate to the group, once the adjustments and eliminations relating to intra group transactions have been made.

Under the full integration method, all asset and liability items must be included in the consolidated balance sheet, even if the holding of the controlling company in the dependent company is less than 100 per cent. In this case the difference between the total amount of equity and the group's share in the equity is allocated to external shareholders. These shareholders are not strictly minority shareholders, but external shareholders, since the group might constitute a minority and yet be required to prepare consolidated annual accounts, not because of their percentage of ownership, but on the basis of their control over the majority voting rights or their power to appoint the directors.

NC Art. 26 refers to minority interests as follows:

> The proportion of equity at the date of the first consolidation relating to non-group third

parties shall be recorded under the 'Minority interests' heading on the liabilities side of the consolidated balance sheet. This amount is not reduced by the portion of unpaid capital on the shares relating to the third parties, which will be recorded separately on the assets side of the balance sheet.

The allocation of the profits of the dependent companies to external shareholders is governed by NC Art. 43 as follows:

- The share in the consolidated income or loss for the year relating to external shareholders must be recorded in a separated heading in the consolidated profit and loss account. This share is calculated on the basis of the percentage of ownership by the external shareholders in the capital stock of the dependent company, excluding own shares, and taking into account the profits of the company after the consolidation adjustments and eliminations and having considered the related corporate income tax expense.
- Where the losses allocable to external shareholders of a dependent company exceed the portion of the company's equity relating to the external shareholders prior to the required consolidation adjustments and eliminations, the excess will be allocated to the controlling company provided that the external shareholders confine their liability to the amounts contributed and that there are no pacts or agreements relating to additional contributions.

1.6 Multi-level groups

When a company directly dependent on a controlling company is in turn the controlling company of another company directly dependent on it, and in turn, the latter company controls a fourth company, and so on, the methods and principles applied in the preparation of consolidated accounts does not differ from those used in the dependence relationship in the first degree. Under NC Art. 6.2, dependent companies which are also controlling companies must prepare consolidated annual accounts and a management report in accordance with the standards envisaged in this text.

However, in accordance with NC Art. 9, Spanish controlling companies which are at least 50 per cent-owned by a superior-level EU-resident company are exempted from the obligation to prepare consolidated annual accounts, as long as the minority shareholders owning 10 per cent of the capital stock of the exempted company do not request the consolidation of the subgroup's annual accounts and that the additional requirements stipulated in NC Art. 9 are complied with (see III. 2).

It must be borne in mind that Spanish regulations favour the effective dependence/control method over the method of taking the percentage of ownership that results from multiplying the respective partial percentages for the purpose of determining the control relationship and, consequently, the obligation to consolidate. Accordingly, NC Art. 3 stipulates that:

> In order to determine the voting rights, to the rights held directly by the controlling company must be added those held by the related dependent companies or to other persons acting in their own name but for the account of other group companies. In these cases, the number of votes held by the controlling company with respect to its indirectly dependent companies is that which is held by the dependent company that has a direct holding in the capital stock of these indirectly dependent companies.

The Consolidation Standards opt for consolidation by successive stages rather than the alternative method of simultaneous consolidation in order to make the investment/equity elimination in the case of indirect holdings. In accordance with NC Art. 30, the stages over which the investment/equity elimination is made in cases of indirect holdings are as follows:

- The investment/equity elimination relating to the dependent companies with no direct holding in any other dependent company is made.
- The successive investment/equity eliminations are made in the following order: the consolidation of the dependent companies controlling an already consolidated subgroup, followed by the consolidation of the controlling company of the higher level group.

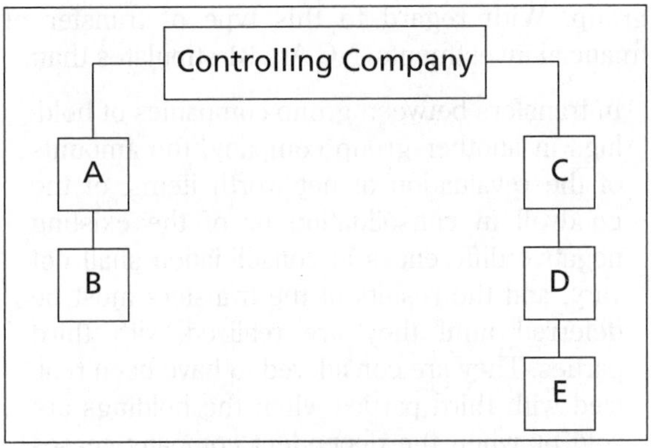

Figure VI.1 Successive investment/equity eliminations

In the case shown in Figure VI.1, the order of the successive consolidations would be as follows:

- B in A and E in D.
- D in C.
- A and C in the controlling company.

Reserves at consolidated companies arising in prior stages must be taken into account for the purpose of determining the equity of those consolidated companies. Minority interests in every stage are regarded as liabilities arising in prior stages and, accordingly, will not be eliminated.

1.7 Special problems

1.7.1 Holdings in the capital stock of the controlling company

NC Art. 32 states that holdings in the capital stock of the controlling company will be recorded under the Shares of the Controlling Company heading on the assets side of the consolidated balance sheet. These holdings are recorded at their book value without prejudice to the related valuation standardization adjustments.

1.7.2 Problems arising from non-instantaneous acquisitions

If the percentage of ownership required to exercise effective control has been obtained through the successive acquisition of shares of the investee company, the difference in the first consolidation is calculated by comparing the book value of the holding (obtained by adding the acquisition prices of the successive acquisitions) with the portion (based on the percentage of ownership giving the effective control) of the equity of the dependent company. The calculation of this difference shall refer to the date on which the number of shares acquired gives the possibility of exercising effective control.

When the percentage of ownership in the capital stock of the dependent company increases, the rules of the general method are applied (NC Art. 28). In this way, at each new acquisition of shares giving rise to an increase in the percentage of ownership, a new difference in consolidation must be calculated by comparing the acquisition price of the new investment with the proportional part of the equity of the dependent company relating to it at that acquisition date. The consolidation differences arising as a result of successive investments in one dependent company can be offset for the purpose of presenting the consolidated balance sheet, and this must be reported in the notes to the consolidated annual accounts (NC Art. 23.7).

Between the first consolidation and the new investment, funds may be generated by the investee companies. These funds are recorded under the Reserves at Consolidated Companies heading. After the second or subsequent investments, the total percentage of ownership is taken into account for the computation of the amount of these reserves or losses at consolidated companies.

When the new investment does not give rise to an increase in the holding in the dependent company, there will be no new differences with respect to the first investment/equity elimination in the first consolidation. When the increase in the percentage of ownership in the dependent company arises with no additional investment, the goodwill or negative differences in consolidation are not modified and, if appropriate, the reserves at consolidated companies must be altered, with an express description of this circumstance in the notes to the consolidated annual accounts (NC Art. 28.3).

When the dependent company acquires own shares in the market from external shareholders, there is an increase in the percentage of ownership by the controlling company in this dependent company. The rules stipulated by NC Art. 28.3 concerning increases in the percentage of ownership with no additional investment are applicable in this connection.

1.7.3 Reciprocal holdings in dependent companies

According to LSA Art. 82, reciprocal holdings exceeding 10 per cent of the capital stock of the investee companies cannot be established. This prohibition also applies to circular holdings in subsidiaries.

When there is a reciprocal relationship between dependent companies, the difference in consolidation for each holding is calculated in the same way as in cases of direct control and the calculation of equity is not affected by that reciprocity. However, for the calculation of the reserves at consolidated companies and for the distribution of period results among the controlling company and external shareholders, it is necessary to take this reciprocity into account. The timing of the holdings must also be taken into consideration; if the reciprocal holdings arose prior to the acquisition of the holding by the controlling company, the required calculations deriving from this interrelation which were not considered in the preparation of prior consolidated accounts must be made (NC Art. 33).

In turn, NC Art. 43.3 states that in order to calculate the portion of the results of dependent companies with a reciprocal ownership interest which are allocable to minority interests, the appropriate adjustments and eliminations for inter-group transactions and the reciprocity arising from the holdings must be considered for the purpose of calculating the result of each of these companies.

1.7.4 Transfer of holdings between group companies

When a group company sells a holding to another company in the same group, there is a gain or loss for the selling company that must be deferred for the purpose of preparing the consolidated accounts until the year in which it is realized. As this is an intragroup transaction, the result is not considered to have been realized until the acquiring company sells the holding to a third party or until it ceases to form part of the group. With regard to this type of transfer of financial investments, NC Art. 31 stipulates that:

> In transfers between group companies of holdings in another group company, the amounts of the revaluation of net worth items, of the goodwill in consolidation or of the existing negative differences in consolidation shall not vary, and the results of the transfers must be deferred until they are realized with third parties. They are considered to have been realized with third parties when the holdings are sold or when the dependent company ceases to form part of the group. In this connection, the result that must be deferred is based on the value of the holding taking into account the reserves generated at consolidated companies, including those generated in the current year, without prejudice to whether they are reflected in consolidated results until the transfer date or the date on which the dependent company ceases to form part of the group.

1.7.5 Reductions in the percentage of control and of ownership

The sale by the controlling company of the shares of the dependent company gives rise to a reduction in the percentage of ownership with the related diminution in the first consolidation difference and adjustment of the reserves at consolidated companies. It is necessary to verify whether the number of shares sold leads to the forfeit of effective control over the dependent company, which would give rise to a change in the consolidation method applied, and also to verify the date of the sale of the shares, so that the related percentages of ownership might be properly applied.

In order to determine accurately the results of this sale, NC Art. 29.1 states that three factors must be taken into account:

- The diminution in value of the reserves at consolidated companies will be considered for consolidation purposes to be reserves of the company whose ownership interest has been reduced and not a result from the sale.
- The results in the dependent company until the transaction date.
- The proportion of goodwill and negative differences in consolidation allocated to results.

This analysis intends to separate two different kind of results (see Cervera, 1997, pp. 9–13). On the one hand, the gains or losses generated by the investment as a result of the evolution of the dependent company's equity. On the other hand, the gains and losses generated as a result of the action of any factors inducing a difference between share market value and share book value.

In cases where the reduction in the investment does not give rise to a decrease in the percentage of ownership, neither the goodwill in consolidation nor the negative differences in consolidation are modified. However this does not affect the adjustments that must be made to the reserves at consolidated companies, which must be disclosed in the notes to the consolidated annual accounts.

2 Consolidation of debt

Under NC Arts. 34 and 35, the reciprocal credits and debits and revenues and expenses between group companies must be eliminated in full after the appropriate adjustments have been made for the purpose of attaining the necessary uniformity in reporting standards.

The elimination of reciprocal credits and debits is considered necessary for the reflection of the true and fair view of the consolidated accounts, since if the elimination is not made, the net worth and economic situation of the group would not be adequately portrayed to third parties. Inter-group credits and debits might arise as a result of commercial or financial transactions, and their respective balances must be equal. If the criteria followed by the debtor and the creditor companies differ or their balances do not coincide, the required prior adjustments must be made (NC Art. 19).

The consolidation of reciprocal credits and debits denominated in foreign currencies is governed by the regulations described above (see V.4).

3 Elimination of intragroup profits

3.1 General

Consolidation Standards define intragroup transactions as those made between two group companies after the date of the inclusion of both companies in the group. The full amount of profits or losses arising from intragroup transactions are eliminated and deferred until they are realized against third parties. Profits or losses are considered to be realized against third parties when one of the parties participating in the transaction leaves the group or when different provisions relating to the various operations dealt with by the Consolidation Standards have been met (see VI.3.2. to VI.3.6).

The profits or losses on intragroup transactions made the year of the first consolidation must be deferred. The elimination of results arising from intragroup transactions made in the current year affects the consolidated profit or loss, whilst the eliminations involving prior years' transactions affect reserves. Adjustments to results and reserves affect the position of the company selling the related asset or rendering the related service.

When a controlling company becomes a dependent company, inter-group transactions in the new group are considered to be the same as those in the former group.

All these provisions are applied when a third party acts in its own name and for the account of a group company.

3.2 Elimination of gains or losses arising from intragroup stock transactions

NC Art. 37 defines intragroup stock transactions as those in which one group company purchases stocks from another group company, regardless of whether they are reflected in the seller company's accounts as stocks or fixed assets. Results are deferred until the year in which they are actually realized. The profit or loss to be deferred is equal to the difference between the acquisition price or production cost, net of provisions, and the selling price.

Results are considered to have been realized at the date of the sale to third parties of the goods acquired. If the purchasing company is engaged in manufacturing and the purchased stocks are included in the production process, the result is realized when the products formed from the raw materials purchased are sold. Losses are considered to have been realized when there is a diminution in value with respect to the acquisition price or the production cost of the stocks up to the limit of this diminution in value. In these cases the appropriate provision must be recorded.

When, as a result of an intragroup transaction, a fixed asset item is included in inventories (e.g. if the purchasing company engages in the marketing of used machinery), the related change must be recorded under the 'Fixed assets transformed into stocks' heading on the debit side of the consolidated profit and loss account at cost, net of intragroup results. If such transactions are material, this must be described in the notes to the consolidated annual accounts.

When the stocks acquired increase the cost of fixed assets, the standards contained in NC Art. 38 must be applied (see VI.3.3).

3.3 Elimination of gains or losses arising from intragroup fixed asset transactions

NC Art. 38 defines intragroup fixed asset transactions as those in which a group company purchases fixed asset items from another group company, regardless of whether these fixed assets are recorded as such or as stocks in the selling company's accounts.

The results arising from these transactions, computed as the difference between the acquisition price or production cost, net of the related amortizations and provisions, and the selling price, must be deferred until the year in which they are realized. Results are considered to have been realized:

- When the purchased fixed assets are sold to a third party.
- When the asset to which the related depreciation had been incorporated as cost is sold to third parties; or if this depreciation was not incorporated as cost, in proportion to the depreciation of the asset or its release from inventory each year.
- In the case of losses, when there is a diminution in value with respect to the acquisition price or production cost of the fixed assets (net of related amortizations and provisions), up to the limit of this diminution in value.

When the amortization of the cost of the purchased fixed assets is included as an addition to the cost of the related stocks, the regulations contained in NC Art. 37 must be applied (see VI.3.1).

When, as a consequence of an intragroup transaction, a current asset item is added to fixed assets (e.g. if the selling company engages in the production of manufactured machinery) this change must be recorded under the 'Capitalized expenses of group in-house work on fixed assets' heading on the credit side of the consolidated profit and loss account at cost, net of intragroup results. When these transactions are material, they must be recorded in the notes to the consolidated annual accounts.

3.4 Elimination of results arising from intracompany service transactions

NC Art. 39 defines intragroup service transactions as those in which one group company renders services to another group company, including financial services. The results arising from these transactions, calculated as the difference between the acquisition price or production cost and the selling price, must be deferred until the year in which they are realized.

When the services acquired are incorporated into the cost of stocks or fixed assets, these results are considered to have been realized when the requirements contained in NC Arts. 37 and 38 are met.

If these services are included as a fixed asset cost, they must be recorded under the 'Capitalized expenses of group in-house work on fixed assets' heading on the credit side of the consolidated profit and loss account at cost, net of intragroup results. When these results are material, reference must be explained in the notes to consolidated annual accounts.

3.5 Elimination of results arising from intragroup financial investments transactions and acquisitions of financial assets from third parties

Under NC Art. 40, intragroup financial investment transactions are defined as those in which a group company acquires financial assets from another group company, excluding holdings in group companies.

The gains or losses arising from these transactions, calculated as the difference between the acquisition price, net of provisions, and the selling price, must be deferred until the year in which they are realized. These gains or losses are considered to have been realized when the financial investments are sold to third parties. If losses are involved, they are considered to have been realized when there is a diminution in value of the financial investments with respect to the acquisition price of the assets up to the limit of this diminution. In these cases the related provision must be recorded.

Also, NC Art. 41 stipulates, for the sole purpose of the preparation of consolidated accounts, that gains or losses can arise from the acquisition from third parties of securities issued by group companies, excluding holdings in the capital stock of these companies.

The gain or loss on such transactions is computed as the difference between the repayment value, net of the related deferred financial expenses, and the acquisition price of such investments, less the uncollected accrued interest. The result is recorded as extraordinary results under the 'Gains/losses on transactions with shares of the controlling company and financial liabilities of the group' headings in the consolidated profit and loss account, as appropriate.

3.6 Elimination of intragroup dividends

NC Art. 42 defines intragroup dividends as those revenues recorded in the year by one group company which have been distributed by another group company. Intragroup dividends are considered to be mere transfers of liquidity from one company to another and, accordingly, they must be eliminated from consolidated results and included as reserves of the company receiving the dividends. If at year end there are agreed unpaid dividends, the reciprocal credits and debits at the companies receiving and distributing the dividends must be adjusted. Interim dividends are eliminated with a credit to the related debit account in the accounts of the distributing company.

VII Proportional consolidation and equity method

1 Proportional consolidation

The proportional integration method is regulated by CCom Arts. 47.1 and 47.2 and by Chapter IV, Section 1 of the Consolidated Standards approved by Royal Decree 1815/1991.

CCom Art. 47.1 stipulates that the proportional integration method may be applied to multigroup companies. In this connection, NC Art. 4 states that:

> Multigroup companies should be considered, for the sole purpose of the consolidation of accounts, to be those which are not included as dependent companies and which are managed by one or several group companies with holdings in their capital stock, together with one or more non-group companies. Specifically, this joint management situation is considered to arise when, besides there being an ownership interest, one or both of the following circumstances exist:
>
> - The company's articles establish that there is joint management.
> - There are pacts or agreements enabling the shareholders to veto corporate resolutions.

If the group's controlling company preparing the consolidated annual accounts decides not to apply the proportional integration method, these companies must be included in the consolidated reporting through the equity method of consolidation in accordance with NC Art. 14.1.

However, Spanish accounting standards do not specify the situations or circumstances in which it is preferable to apply either the proportional integration method or the equity method. Accordingly, it is the responsibility of the managing board of the controlled company entrusted with the preparation of the consolidated annual accounts to make this decision on the basis of the ability of each method to reflect a true and fair view of the specific contribution of each group to the management of a multigroup company.

The regulations adapting the chart of accounts to construction companies, which were approved by the Ministry of Economy and Finance Order dated 27 January 1993, state that each company participating in a joint venture must include in its balance sheet and profit and loss account the proportion of the balances of the joint venture which relate to it on the basis of its percentage of ownership (PGC Valuation Standard No. 21). The inclusion is made after the required timing standardizations, taking into account the accounting closing date and business year of the company, the pertinent valuation standardization if the joint venture applies valuation methods other than those used by the company, and the required reconciliations and reclassifications of the related items have been made. It is also necessary to eliminate unrealized gains on transactions between the company and the joint venture in the appropriate proportion, as well as the reciprocal assets and liabilities and revenues and expenses. This regulation is not considered to be a consolidation standard but a valuation method for the integration and recording of the joint venture's transactions in the participating companies. These criteria apply also to non-construction companies as stated expressly in the Ministry of Economy and Finance Order dated 27 January 1993.

2 Equity method

2.1 Applicability

The equity method of consolidation is regulated by the Commercial Code (CCom Arts. 47.3, 47.4 and 47.5), and by Chapter II, Section 3 and Chapter IV, Section 2 of the Consolidated Standards.

CCom Art. 47.3 stipulates that the equity method must be applied to the integration of associated companies in consolidated annual accounts. NC Art. 5 stipulates that:

> Associated companies should be considered for the sole purpose of the consolidation of accounts to be those in which one or several group companies exercise a significant influence on its management, under the following circumstances:
>
> - One or several group companies have holdings in the company's capital stock.
> - There is a lasting relationship with the company's business activities.

CCom Art. 47.3 also stipulates that these requirements are considered to have been met when one or several group companies have holdings of at least 20 per cent in an unlisted (3 per cent if listed) company's capital stock.

However, CCom Art. 47.5 does not require the application of the equity method when the holdings in an associated company are not material to the information that must be provided by the consolidated annual accounts.

NC Art. 14 states that the equity method must be used in the preparation of the consolidated annual accounts in the case of investments in:

- Multigroup companies to which the proportional integration method is not applied.
- Dependent companies which are not consolidated by the full integration method in accordance with NC Art. 11.2-e, i.e. when dependent companies have such diverse business activities that their inclusion would contravene the purpose of the preparation of consolidated annual accounts (see IV).

As it is the existence of a group of companies which determines the obligation to present consolidated annual accounts, the mere existence of associated or multigroup companies does not necessarily give rise to this obligation unless there are other dependent companies, regardless of whether these companies are excluded from consolidation for certain technical reasons. This situation is set forth explicitly in Section 3 of the Introduction to the Consolidation Standards.

Spanish accounting regulations do not permit the application of the equity method for the purpose of recording holdings in individual annual accounts. These holdings must be valued at their acquisition cost, and valuation reductions can only be made if there is a decrease in the value of the securities. However, the diminution in value of the holdings in the capital stock of group and associated companies must be calculated and the related provisions recorded, even in the case of listed securities on the basis of the variations in the equity of the investee (PGC Valuation Standard No. 8).

2.2 Criteria for the application of the equity method

NC Art. 46 states that the equity method of consolidation consists of 'replacing the book value of a holding in a group company's accounts by the proportion of the investee's equity relating to the group on the basis of its percentage of ownership of these companies'. This amount must be reflected on the assets side of the balance sheet under the 'Holdings carried by the equity method' heading.

The book value of the investment is the acquisition price calculated in accordance with the chart of accounts (PGC Valuation Standard No. 8 on marketable securities) net of valuation adjustments, provisions or losses recorded prior to the company being considered an associated company if the holding in the investee was acquired in several portions until it became an associated company. Equity is as defined by the chart of accounts, net of the amount of own shares.

When a group of companies exerts significant influence over a company which is the controlling company of another group, NC Art. 51 opts for consolidating this company on the basis of its consolidated accounts, i.e. it must be the controlling company's consolidated accounts which are taken for the application of the equity method.

Therefore the equity method consists of a simple restatement of the value of the holding on the basis of the equity of the associated company. It does not involve aggregating the investee's assets, liabilities, revenues and expenses with those of the group. Consequently, credits/debits and revenues/expenses arising from transactions between the companies consolidated by the equity method and the other group companies are not eliminated, since those relating to the company carried by the equity procedure are not included in the consolidated annual accounts. However, the results arising from transactions between the associated company and the other consolidated companies must be eliminated, provided that they have not been realized with third parties and for the amount corresponding to the related group company's holding in the associated company's capital stock. Under NC Arts. 64.3-f) and 65.3-f), the uneliminated portion of the credits and debits and revenues and expenses with companies consolidated using the proportional integration method, as well as the credits and debits and revenues and expenses with companies carried by the equity method, must be included separately under the related balance sheet and profit and loss account headings.

Consequently, certain commentators do not consider the equity method to be a consolidation method, but simply a 'valuation procedure'. Companies are divided into the 'consolidated set' (*conjunto consolidable*) (NC Art. 13), which includes companies consolidated by the global or proportional consolidation methods, and the 'consolidation scope' (*perímetro de la consolidación*) (NC Art. 15), comprising both the companies forming part of the 'consolidated set' and those to which the equity method is applicable.

2.2.1 Timing and valuation standardization

Under NC Art. 47, in order to carry out the equity method, the annual accounts of the investee must first be sufficiently standardized if they are prepared on the basis of different valuation criteria from those used by the group, provided that the required information is available and that the differences in valuation are material. Likewise, if appropriate, the necessary timing standardization of the annual accounts of the company carried by the equity procedure with the consolidated annual accounts must be made.

NC Arts. 17 and 18 on the full integration method are also applicable to the equity method for the purposes of timing and valuation standardization.

2.2.2 Calculation of differences in consolidation

NC Art. 48 sets forth the criteria that should be applied for the purpose of restating the value of the holding in the investee:

- The restatements shall be made by comparing the book value of the holding with the amount of the investee's equity corresponding to the holding. This comparison must be based on the book values at the date that the associated company is considered to be such, or at the beginning of the year in which the company is included in consolidation for the first time.

- The portion of the negative or positive difference resulting from this comparison, which is allocable to specific asset and liability items at the company carried by the equity method whose book value does not coincide with the market value, must be incorporated into or deducted from the value of the holding in the consolidated balance sheet. For this purpose, the limits established by NC Art. 23.4 for the 'first-time consolidation difference' in the global integration method should be taken into account, i.e. that the variation in the value of the holding is exclusively limited to the portion of the divergence between the market and book values of the related items which is allocable to the group on the basis of its percentage of ownership of the investee.
- The higher value of the holding which might arise from applying the above criterion must be reduced in subsequent years with a charge to consolidated results, and as the related asset and liability items diminish in value or as they are released from the accounts or sold to third parties.
- The difference remaining after the allocation is recorded in the consolidated balance sheet under the 'Goodwill in consolidation' or 'Negative differences in consolidation' headings, according to whether the difference is positive or negative and separately from those relating to companies consolidated by the global or proportional integration methods. The provisions of NC Arts. 24 and 25 on the global integration method are also applicable to the equity method in this connection; in particular, goodwill in consolidation must be amortized over the period in which it contributes to the obtainment of revenues up to a maximum of 10 years, and must be appropriately justified in the notes to the consolidated annual accounts when it is amortized over more than five years. These positive and negative differences can only be mutually offset when they relate to the same associated company, and this must be described in detail in the notes to the consolidated annual accounts.

2.2.3 Modification of the holding

When the group company owning the holding acquires new shares of the investee but does not have control over the latter, NC Art. 49 envisages that the process described above (see VII.2.2.2) must be repeated on the basis of the percentage of the additional holding with regard to the investee's capital stock. In the case of reductions in the holding, NC Art. 49 stipulates that the account representing the holding must be valued by the percentage of the investee's equity which remains after the reduction in accordance with NC Art. 29 relating to the global integration method (see VI.1.7.5).

2.2.4 Subsequent consolidations

The investor company must restate holdings carried by the equity method in each year subsequent to that in which it was consolidated for the first time by increasing or decreasing its value for the purpose of reflecting the variations in the equity of the associated company.

NC Art. 50 stipulates the following:

- The value of the holding must be restated on the basis of the percentage relating to the investor company of the period income obtained by the associated company. This amount must be reflected in the consolidated profit and loss account under the 'Equity in income of companies carried by the equity method' or 'Share in losses of companies carried by the equity method' heading, as appropriate.
- The amount is obtained after the elimination, where appropriate, of the results arising from intragroup transactions, provided that they have not been realized with third parties outside the consolidated group. These eliminations must only affect the percentage of the results of the associated company which relate to the group company and must only be made provided that the necessary information can be obtained. NC Arts. 36 to 41 relating to the elimination of the results arising from intragroup transactions for the purpose of the global integration method are also applicable in this case.
- If the investee incurs losses, the reduction of the balance of the account representing the holding is limited to the book value of the holding. However, provisions must be recorded for additional losses when the group company with the ownership interest has

undertaken to absorb the losses, and this situation must be described in the notes to the annual accounts.

- The value of the holding must also be corrected by the portion of the reserves generated by the associated company from its first consolidation which corresponds to the investment. All the adjustments and eliminations in prior years' results must also be taken into account. The increase in reserves is included in the equity in the consolidated balance sheet under the 'Reserves at companies carried by the equity method' heading.
- The profits distributed by the company carried by the equity method are considered to be reserves at the investor company, and the value of the holding must be reduced accordingly. In the case of interim dividends, the book value of the holding is reduced with a charge to the results of the recipient company.

2.3 Rules for modifying the valuation of investments from the equity method to the proportional or global integration method

NC Art. 52 envisages that when a company carried by the equity method in prior years is consolidated by the full or the proportional integration method, the adjustments and eliminations that had been made under the equity method must be taken into account.

2.4 Foreign currency translation

Under NC Art. 59, the annual accounts of foreign companies carried by the equity method must be translated into pesetas in accordance with the 'year end exchange rate method'. NC Arts. 55, 57 and 58 governing the translation of annual accounts denominated in foreign currencies with regard to the 'year end exchange rate method', 'foreign companies subject to high inflation rates' and the 'elimination of results arising from intragroup transactions' are applicable in this connection (see V.4). However, NC Art. 59 states that in cases where companies subject to high inflation do not present adjusted accounts due to the effect of changes in prices and because the information required for the calculation of these adjustments or to apply the 'monetary/non-monetary method' is not available, these companies will be excluded from consolidation, and the necessary information included in the notes to the consolidated annual accounts.

The translation differences arising from the application of this method which relate to the holding of the controlling company must be recorded in equity in the consolidated balance sheet under the 'Translation differences' heading. The value of the holding carried by the equity method is increased or reduced, as appropriate, by the translation differences.

Consequently, Spanish regulations reserve the 'year end rate method' for foreign associated companies. The 'monetary/non-monetary method' is applied to foreign companies whose activities are an extension of the controlling company's activity. This is not the case for associated companies, since the link will generally be much weaker in this type of company.

VIII Corporate income tax in consolidated accounts

The amount of the corporate income tax expense that must be reflected in the consolidated profit and loss account must be calculated in accordance with PGC Valuation Standard No. 16 (see Spain – Individual Accounts, VII.1). Those criteria specified in this valuation standard are entirely applicable to the accounting of corporate income tax in consolidated accounts:

- The annual corporate income tax expense is recorded on the basis of the annual accounting profit, irrespective of the tax payable by the company (tax effect method).
- The recognition of any tax credits and tax relief in the accrued corporate income tax.
- The necessary application of the prudence accounting principle with regard to recognizing the tax effect of losses which can be offset fiscally.

The computation of the expense incurred depends on the group's tax system, that is to say, whether the consolidated companies file individual tax returns or they are taxed under the consolidated income tax system, which is optional.

1 Groups of companies filing individual tax returns

If the consolidated companies file individual tax returns, the adjustments and eliminations made as a result of the consolidation process may give rise to timing and permanent differences between the accounting and tax recognition of certain expenses and revenues, and the accounting implications of the tax effect will be reflected in the consolidated annual accounts.

Specifically, timing differences may derive from the adjustments required to achieve the appropriate timing and valuation uniformity at the companies included in consolidation and from the elimination of intragroup results. All these tax effects will be eliminated in the future as the differences originating them reverse to future periods. The recording in the consolidated financial statements of the tax effect arising from such timing differences will be made in accordance with the following rules established for individual companies by Valuation Standard No. 16 of the chart of accounts and the related ICAC Resolution dated 9 October 1997:

- Prepaid taxes are only taken into account if they are material to the future tax charge and if there are no reasonable doubts regarding their future recovery. Accordingly, it is presumed that the future recovery of prepaid taxes is not sufficiently guaranteed when this is expected to take place over a period exceeding 10 years, or when companies habitually reporting losses are involved. Otherwise, prepaid taxes can only be recognized in the consolidated annual accounts when the same or a higher amount of deferred taxes are recorded and the reversal period of the deferred taxes is equal to that of the prepaid taxes.
- Credits arising from tax losses qualifying for offset can be recorded as long as they are derived from an unusual fact in the operation of the company which is unlikely to reappear in the future. The general recovery period is seven years although in some specific cases a recovery period that cannot exceed 10 years may also be admitted.
- Deferred taxes must always be recorded.
- The tax assets and liabilities arising from timing differences must never be offset, and must therefore be reflected, as appropriate, on the assets and liabilities side of the consolidated balance sheet.

2 Companies taxed under the consolidated income tax system

The definition of a group for the purpose of access to the tax consolidation system is more restricted than for the purpose of presenting consolidated financial reports. Corporate Tax Law 43/1995, dated 27 December 1995 regulates (*Ley del Impuesto solare Sociedades*, LIS, Title VIII, Chapter VII, arts. 78–96) this special regime of tax consolidation. It establishes the requirement to have a direct or indirect holding of more than 90 per cent in the capital stock of dependent companies in order to apply for access to the consolidated income tax system. This fact means that both companies filing individual tax returns and companies which are taxed under the consolidated income tax system may form part of the same consolidated group.

The ICAC Resolution dated 9 October 1997 contains precise rules that the companies forming the tax group must observe in order to determine the corporate income tax expense that each company must reflect in its profit and loss account. This Resolution favours the recording in the annual accounts of each company comprising the tax group of the corporate income tax expense accrued in accordance with the tax characteristics imposed by the consolidated income tax system. Accordingly, for the calculation of this expense, it is necessary to take into account, in addition to the parameters relevant to individual taxation, the timing and permanent differences arising from the elimination of results on transactions between group companies which is brought about by the process of determining the consolidated income tax base (see Table VIII.1).

Table VIII.1 Reconciliation of book income for the year with the taxable income for companies taxed under the consolidated tax system

	Profit per books for the year
±	Individual permanent differences
±	Individual timing differences
=	**Individual taxable income**
±	Permanent differences from tax consolidation
±	Timing differences from tax consolidation
=	**Taxable income**

Regulation 6 of the Resolution stipulates the following rules for calculating the corporate income tax expense and other aggregates relating to the tax effect of each company included in the consolidated tax system:

- If, as a result of the elimination for tax purposes of results arising from inter-company transactions, there is a deferral of the recognition of these results by the group until they have been realized with third parties outside the consolidated group, the company that has recorded these results for accounting purposes will record the related tax effect in its financial statements. Prepaid taxes and credits for tax loss carry-forwards, where appropriate, are only recorded when the requirements of the accounting principle of prudence for individual companies are met (see VIII.1), provided that they can actually be collected by the set of companies comprising the consolidated group for tax purposes.
- If a non-reversible difference arises from the elimination of dividends distributed among tax group companies and of other results arising from inter-company transactions, the company that had recorded the result will have a permanent difference with respect to the taxable profit that would have arisen had the companies filed individual tax returns. This difference will be recorded in accordance with the general rules established for individual companies (see Spain – Individual Accounts, VII.1.1).
- If a tax group company, taking into account the above, reports tax losses and the set of companies which comprise the group for tax purposes partially or fully offset them in their consolidated income tax return, the related tax effect must be recorded as follows:
  (i) For the portion of the tax loss offset, a reciprocal credit and debit will arise between the company reporting the loss and the companies offsetting it.
  (ii) For the portion of the tax loss not offset by the group companies, the company reporting the loss will record a credit only if the overall tax group expects to generate taxable profits within a maximum period of seven years, which is the period established by current tax regulations for carrying forward consolidated tax losses.

When the tax losses are incurred prior to consolidated taxation, the consolidated tax group must be able to offset the losses in order to be able to record the related tax effect.

The Resolution stipulates that the tax deductions and relief applied in the corporate income tax settlement for the year must modify the calculation of the tax expense accrued by each company by an amount that is determined by the consolidated income tax system and not by the higher or lower amount that would relate to each company had they filed individual tax returns. The tax deductions and relief applied under the consolidated income tax system are allocated to those companies that, according to the regulations governing the consolidated tax system, have performed the activity or obtained the earnings giving rise to them (fixed asset investments, formation of branches abroad, etc.). The amounts of prepayments, including tax withholdings, are allocated to the company that has effectively borne them.

If, as a consequence, the sum of the tax deductions and relief applied to a company gives rise to a negative amount of tax payable, this amount shall be taken into account when determining the reciprocal credit or debit between that company and the other companies belonging to the tax group.

If, as a result of tax regulations or legal private relationships between the tax group companies, the distribution of the tax charge does not coincide with the tax payable which has been calculated in accordance with the stipulations contained in this resolution, the difference for each company is treated as follows:

- If the differences arise as a result of the application of a distribution method derived from tax regulations, the company lowering its tax charge must record a credit to the corporate income tax expense account with a charge to a credit account before the company that increases its tax charge. In turn, the company increasing its tax charge must make a charge to the corporate income tax expense account with a credit to a debit account before the company lowering its tax charge.
- If the differences arise as a result of the application of a distribution method derived from a private agreement between the tax group companies, each company must make the adjustments described in the preceding paragraph and record the effect in the related extraordinary expense and revenue accounts, without modifying the corporate income tax expense account.

Also, each tax group company must include in the notes to its annual accounts the related chart of accounts indications, together with any circumstance relating to this special tax system, including most notably:

- Permanent and timing differences arising as a result of the consolidated income tax system, indicating in the case of timing differences the year in which they arise and the reversal made in each year.
- The offsetting of tax losses arising from the application of the consolidated income tax system.
- The breakdown of the credits and debits between group companies arising from the tax effect resulting from the application of the consolidated income tax system.

3 Applicable tax rate

The tax scale applicable to groups of companies that have requested and obtained the consolidated tax status is that applicable to companies that file individual tax returns.

Modifications to tax regulations altering the applicable tax rate will give rise to an adjustment to the balance of the recorded tax asset and liability accounts. The effect of this adjustment will be computed for the purpose of calculating the accrued corporate income tax expense.

IX Formats

1 The consolidated annual accounts

CCom Art. 44.1 refers to the documents to be incorporated into the consolidated annual accounts using the same criteria as those laid down by the chart of accounts for the individual accounts of any company. CCom Art. 44.1 states that 'the consolidated annual accounts shall comprise the consolidated balance sheet, the consolidated profit and loss account and the consolidated notes to the annual accounts as one unit'. These documents must 'be clearly presented and reflect a true and fair view of the net worth, financial situation and results of the group in accordance with the Commercial Code (CCom), the Corporations Law (LSA), as amended, the chart of accounts (PGC) and the Consolidation Standards (NC) approved by Royal Decree 1815/1991' (NC Art. 61). The dates and period to which the consolidated annual accounts refer are the same as those used for the individual annual accounts of the controlling company.

The consolidated annual accounts must be prepared by the directors of the controlling company within the period established for the preparation of the latter's individual annual accounts, i.e. within three months of the accounting close. As at the controlling company, the group's consolidated accounts must be signed by the controlling company's directors and reasons must be given for the absence of any signatures. The identity of the group and of the period to which the accounts refer is also essential.

2 The consolidated balance sheet

2.1 The structure of the consolidated balance sheet

The structure of the consolidated balance sheet (see Tables IX.1(a) and IX.1(b)) must respect as far as possible the format established by the chart of accounts for individual companies. Under NC Art. 64, the consolidated balance sheet

> shall comprise, duly separated, the assets, rights and obligations of the controlling company and of the dependent companies to which the global integration method has been applied, without prejudice to the appropriate eliminations and adjustments, and the group's equity and, under a separate heading, the portion of the equity relating to external shareholders.

The consolidated balance sheet must also include:

> the assets, rights and obligations of the multigroup companies to which the proportional integration method has been applied, on the basis of the group's percentage of ownership of these companies, and without prejudice to the related adjustments and eliminations.

NC Art. 64 sets down standards for the preparation of the consolidated balance sheet which, in part, reflect those for the preparation of individual accounts. However, they add certain specific requirements for consolidated accounts, which relate to consolidation techniques as follows:

- When the composition of the companies included in consolidation varies significantly from one year to the next, the notes to the consolidated annual accounts must contain the information required for the comparison of accounts. This regulation has a greater effect on the notes than on the balance sheet, although the balance sheet structure is actually altered.
- The portion of the uneliminated credits and debits with companies consolidated by the proportional integration method, as well as the credits and debits with companies carried by the equity method, must be recorded separately under the related asset and liability headings.
- Credits and debits with dependent, associated and multigroup companies, as defined by the chart of accounts, not included in consolidation must be recorded separately under the related asset and liability headings.

2.2 Analysis of specific consolidated balance sheet headings

Noteworthy consolidated balance sheet headings included in the format envisaged by the Consolidation Standards are as follows:

Subscribed capital unpaid and uncalled: this heading includes the uncalled capital amounts due at the consolidation date as a result of capital increases at the controlling company and at dependent companies, provided that they are subscribed by external shareholders.

The capital amounts due on dependent company shares acquired by the controlling company are eliminated in consolidation. However, the capital amounts receivable by external shareholders are recorded separately on the assets side of the balance sheet in accordance with NC Art. 26.

Financial investments: within this heading, the following accounts arising from the application of the different consolidation methods and procedures merit special consideration:

- Holdings carried by the equity method: this account was defined and its accounting treatment described above (see VII. 2.2).
- Loans to companies carried by the equity method: these loans, granted by group companies to companies carried by the equity method, form part of the group's finances and, accordingly, must be recorded separately from other loans granted to non-group companies. Another similar account must be added for the

Table IX.1(a) Assets side of the consolidated balance sheet

Assets
A Subscribed capital unpaid and uncalled
B Fixed and other non-current assets
I Formation expenses
II Intangibles
1 Intangible assets and rights
2 Provisions and amortization (–)
III Tangible fixed assets
1 Land and structures
2 Technical installations and machinery
3 Other tangible fixed assets
4 Advances and construction in progress
5 Provisions and amortization (–)
IV Financial investments
1 Holdings carried by the equity method
2 Loans to companies carried by the equity method
3 Long-term investment securities
4 Other loans
5 Provisions (–)
V Shares of the controlling company
C Goodwill in consolidation
1 Of companies consolidated by the global or proportional integration method
2 Of companies carried by the equity method
D Deferred charges
E Current assets
I Subscribed capital called but not paid
II Inventories
III Accounts receivable
1 Customer receivables for sales and services
2 Receivable from companies carried by the equity method
3 Other debtors
4 Provisions (–)
IV Short-term financial investments
1 Short-term investment securities
2 Loans to companies carried by the equity method
3 Other loans
4 Provisions (–)
V Short-term shares of the controlling company
VI Cash
VII Accrual accounts
TOTAL (A+B+C+D+E)

recording of the uneliminated portion of loans granted to multigroup companies consolidated by the proportional integration method, since these companies form a more integral part of the consolidated group even than associated companies (NC Art. 64.3.f)). In turn, these loans must be included in fixed or current assets, depending on their maturity.

- Long-term investment securities: this account includes financial investments in non-consolidated companies. These may be holdings in group companies excluded from consolidation for any of the reasons included in NC Art. 11, holdings in associated or multigroup companies which are of scant materiality to the true and fair view of the group, or they may

Activo

A Accionistas por desembolsos no exigidos
B Inmovilizado
  I Gastos de establecimiento
  II Inmovilizaciones inmateriales
    1 Bienes y derechos inmateriales
    2 Provisiones y amortizaciones (–)
  III Inmovilizaciones materiales
    1 Terrenos y construcciones
    2 Instalaciones técnicas y maquinaria
    3 Otro inmovilizado
    4 Anticipos e inmovilizaciones materiales en curso
    5 Provisiones y amortizaciones (–)
  IV Inmovilizaciones financieras
    1 Participaciones puestas en equivalencia
    2 Créditos a sociedades puestas en equivalencia
    3 Cartera de valores a largo plazo
    4 Otros créditos
    5 Provisiones (–)
  V Acciones de la sociedad dominante
C Fondo de comercio de consolidacion
    1 De sociedades consolidadas por integración global o proporcional
    2 De sociedades puestas en equivalencia
D Gastos a distribuir en varios ejercicios
E Activo circulante
  I Accionistas por desembolsos exigidos
  II Existencias
  III Deudores
    1 Clientes por ventas y prestaciones de servicios
    2 Empresas puestas en equivalencia
    3 Otros deudores
    4 Provisiones (–)
  IV Inversiones financieras temporales
    1 Cartera de valores a corto plazo
    2 Créditos a empresas puesta em equivalencia
    3 Otro créditos
    4 Provisiones (–)
  V Acciones de la sociedad dominante a corto plazo
  VI Tesorería
  VII Ajustes por periodificación

TOTAL (A+B+C+D+E)

simply be minority investments which do not give the investor a strong influence over decision-making at these companies. These holdings are valued under valuation standard 8 of the chart of accounts.

Goodwill in consolidation: this heading includes two entries, one for companies consolidated by the global or proportional integration method and the other for companies carried by the equity method. The definition and accounting treatment of this heading are described above (see VI.1.2). Despite its inclusion in the balance sheet as an independent heading, it might seem reasonable to include it under 'Intangibles' in a separate account as it is an intangible asset.

Table IX.1(b) Liabilities side of the consolidated balance sheet

Shareholders' investment and liabilities
A Shareholders' investment
I Subscribed capital stock
II Share premium account
III Revaluation reserve
IV Other reserves of the controlling company
1 Unrestricted reserves
2 Restricted reserves
3 Profit and loss brought forward
V Reserves at companies consolidated by the global or proportional integration method
VI Reserves at companies carried by the equity method
VII Translation differences
1 Of companies consolidated by the global or proportional integration method
2 Of companies carried by the equity procedure
VIII Income (loss) allocable to the controlling company
1 Consolidated income (loss)
2 Income (loss) allocated to minority interests
IX Interim dividend paid during the year (–)
B Minority interests
C Negative differences in consolidation
1 Of companies consolidated by the global or proportional integration method
2 Of companies carried by the equity method
D Deferred revenues
1 Capital subsidies
2 Other deferred revenues
E Provisions for liabilities and charges
F Long-term debt
I Debentures and other marketable debt securities
II Payable to credit entities
III Payable to companies carried by the equity method
IV Other accounts payable
G Current liabilities
I Debentures and other marketable debt securities
II Payable to credit entities
III Payable to companies carried by the equity method
IV Trade accounts payable
V Other non-trade payables
VI Operating liabilities provisions
VII Accrual accounts
TOTAL (A+B+C+D+E+F+G)

For the purpose of its presentation in the balance sheet, the controlling company can follow the direct amortization method, in which case the goodwill is valued at its net book value, or it may apply the indirect amortization method, in which case the goodwill is valued at the original amount, net of the related accumulated amortization.

Shares of the controlling company: This account includes the amount of the controlling company's capital stock which is held by the controlling company itself or by its dependent companies, together with the proportion of the controlling company's shares which is held by multigroup companies consolidated by the proportional integration method.

Pasivo

A Fondos propios
- I Capital suscrito
- II Prima de emisión
- III Reserva de revalorización
- IV Otras reservas de la sociedad dominante
  - 1 Reservas distribuibles
  - 2 Reservas no distribuibles
  - 3 Resultados de ejercicios anteriores
- V Reservas en sociedades consolidadas por integración global o proporcional
- VI Reservas en sociedades puestas en equivalencia
- VII Diferencias de conversión
  - 1 De sociedades consolidadas por integracion global o proporcional
  - 2 De sociedades puestas en equivalencia
- VIII Perdidas y ganancias atribuibles a la sociedad dominante (beneficio o pérdida)
  - 1 Pérdidas y ganancias consolidadas
  - 2 Pérdidas y ganancias atribuibles a socios externos
- IX Dividendo a cuenta entregado en el ejercicio

B Socios externos

C Diferencia negativa de consolidaci6n
- 1 De sociedades consolidadas por integración global o proporcional
- 2 De sociedades puestas en equivalencia

D Ingresos a distribuir en varios ejercicios
- 1 Subvenciones de capital
- 2 Otros ingresos a distribuir en varios ejercicios

E Provisiones para riesgos y gastos

F Acreedores a largo plazo
- I Emisiónes de obligaciones y otros valores negociables
- II Deudas con entidades de crédito
- III Deudas con sociedades puestas en equivalencia
- IV Otros acreedores

G Acreedores a corto plazo
- I Emision de obligaciones y otros valores negociables
- II Deudas con entidades de crédito
- III Deudas con sociedades puestas en equivalencia
- IV Acreedores comerciales
- V Otras deudas no comerciales
- VI Provisiones para operaciones de tráfico
- Vll Ajustes por periodificación

TOTAL (A+B+C+D+E+F+G)

When these shares were acquired to amortize capital, it would seem reasonable, for the purposes of treating them in the same way as own shares in the individual annual accounts as prescribed by the chart of accounts, to include these shareholdings in liabilities as a deduction from the group's equity. However, this possibility is not expressly mentioned by the Consolidation Standards.

Shareholders' investment: shareholders' investment in the consolidated balance sheet includes the following specific accounts:

- Reserves at companies consolidated by the global or proportional consolidation method: this account includes the reserves gained by the controlling company in dependent companies consolidated by the global consolidation method or in multigroup companies consolidated by the proportional integration method from the date of the first consolidation to the consolidated balance sheet date. Reserves are calculated taking into account the adjustments in and eliminations from the equity of dependent and multigroup companies arising from intragroup transactions, without including the result for the year. If these reserves have a negative sign, this account is recorded as a reduction in the group's equity.

- Reserves at companies carried by the equity method: this account includes the reserves gained by the controlling company in companies carried by the equity method from the date of the first consolidation to the consolidated balance sheet date. These reserves are calculated under the same criteria as those used for reserves at companies consolidated by the global or proportional integration methods and are recorded in the balance sheet as an addition to or reduction in equity, as appropriate.
- Translation differences: this account includes the positive and negative differences arising as a result of the application of the 'year end rate method' to the translation into pesetas of the individual annual accounts of foreign consolidated companies in accordance with NC Art. 55. The portion of the difference allocable to external shareholders is recorded under heading B 'Minority interests' on the liabilities side of the consolidated balance sheet. Under NC Art. 54, the 'year end rate method' must be generally applied to the companies consolidated by the global or proportional integration method, except when the activities of the foreign companies are so closely linked to those of a Spanish group company that they may be considered an extension of the latter, in which case the translation is made under the 'monetary/non-monetary method'. The annual accounts of foreign companies carried by the equity method must be translated into pesetas in accordance with to the 'year end exchange rate method' (NC Art. 59). When the 'monetary/non-monetary method' is applied, NC Art. 56 states that the company has two options for the treatment of translation differences. They may be computed as a profit or loss for the year, as appropriate; or they may be treated under the provisions of the chart of accounts for translation differences arising from foreign currency transactions (see Spain – Individual Accounts, VII.5). Accordingly, the controlling company can decide on the translation method to be applied and therefore has the possibility of exercising an influence on consolidated profits or losses through the translation differences. The Consolidation Standards include two accounts under the 'Translation differences' heading, one for companies consolidated by the global or proportional integration method, and another for companies carried by the equity method.
- Income (Loss) allocable to the controlling company: this account includes the share of the controlling company in the consolidated companies' results. This share is determined by subtracting the share of external shareholders in the consolidated results.

Minority interests: this account records the portion of the equity of the companies consolidated by the global integration method which relates to external interests. Its presentation in the consolidated balance sheet after equity would seem to indicate that Spanish regulations follow the theory of the extension of the parent company. However, the fact that minority interests are not included in liabilities (external shareholders' financing does not actually give rise to a repayment obligation for the group) suggests that this theory is not applied strictly, but that the regulations adopt in this case, as in others, an intermediate position, taking advantage of the theory of the extension without discarding the entity theory.

Negative differences in consolidation: this account includes the excess value of identifiable assets and liabilities over the price paid for the financial investments in dependent, associated and multigroup companies at the investment date. The nature of this heading and the method under which it is allocated to results must necessarily depend on the event giving rise to it (see VI.1.3).

The model consolidated balance sheet includes two entries in which to record the negative goodwill, one for companies consolidated by the global or proportional integration method, and the other for companies carried by the equity method.

Debt payable to companies carried by the equity method: the comments made about the 'Loans to companies carried by the equity method' also apply to this heading *mutatis mutandis*. Although the model consolidated balance sheet makes no express reference to this, the related accounts must be recorded separately for the uneliminated portion of debts with multigroup companies consolidated by the proportional integration method as well as for debts with non-consolidated dependent, associated or multigroup companies (as defined by the chart of accounts) (NC Art. 64.3-f,g).

3 The consolidated profit and loss account

3.1 Structure of the consolidated profit and loss account

Under NC Art. 65:

> The consolidated profit and loss account shall comprise, duly separated, the revenues and expenses of the controlling company and of the dependent companies consolidated by the global integration method, without prejudice to the appropriate adjustments and eliminations, and the consolidated profit or loss, with the inclusion in a separate account of the portion relating to minority interests. The consolidated profit and loss account shall also include the revenues and expenses of multigroup companies consolidated by the proportional integration method in the proportion represented by the group's stake in its capital stock, without prejudice to the appropriate adjustments and eliminations.

Tables IX.2(a) and IX.2(b) show the complete profit and loss account format. Unlike the chart of accounts (PGC), Spanish consolidation regulations do not envisage the possibility of presenting a consolidated analytical statement of income. Nevertheless, the controlling company must feel free to disclose a consolidated analytical statement of income as an extension to its own individual accounts (individual companies may disclose voluntarily an analytical statement of income in note 21 to the annual accounts).

The preparation of the consolidated profit and loss account is subject to the same requirements as that of the profit and loss accounts of individual companies, together with others arising from consolidation techniques, summarized below:

- When the composition of the companies included in consolidation varies significantly over the year, the information required for the comparison of the consolidated annual accounts must be recorded in the notes to the accounts.
- The uneliminated portion of the revenues and expenses arising from transactions with companies consolidated by the proportional integration method, as well as the revenues and expenses with companies carried by the equity method must be recorded separately.
- Likewise, the revenues and expenses arising from transactions with group, multigroup and associated companies, as defined by the chart of accounts, not included in consolidation, must be recorded separately in the profit and loss account.
- When, as a result of an inter-group transaction, a fixed asset item is converted to stocks, the 'Fixed assets converted to stocks' heading will be created to record this change. Similarly, when as a result of an inter-group transaction, inventories are converted into fixed assets, this change is reflected at the net cost of the conversion under the 'Capitalized expenses of group in-house work on fixed assets' account on the revenues side of the profit and loss account.

3.2 Analysis of specific profit and loss account headings

Noteworthy consolidated profit and loss account headings derived from the application of the different consolidation methods and procedures are as follows:

Translation losses/gains: these headings include the negative and positive translation differences arising from the application of NC Art. 56 relating to the 'monetary/non-monetary method' when these differences are taken to

Table IX.2(a) Consolidated profit and loss account (expenses)

Debit
A Expenses
1 Decrease in finished product and in progress inventories
2 Materials used and other external expenses
3 Personnel expenses
a) Salaries, wages and similar expenses
b) Employee welfare expenses
4 Provision for period depreciation and amortization
5 Variation in operating provisions
6 Other operating expenses
I *Operating income*
7 Financial expenses
8 Losses on short-term financial investments
9 Variation in financial investment provisions
10 Exchange losses
11 Translation losses
II *Net financial income*
12 Share in losses of companies carried by the equity method
13 Amortization of goodwill in consolidation
III *Ordinary operating income*
14 Losses on fixed assets
15 Variation in tangible and intangible fixed asset provisions
16 Losses on disposals of holdings in companies consolidated by the global or proportional integration method
17 Losses on disposals of holdings in companies carried by the equity method
18 Losses on transactions with shares of the controlling company and financial liabilities of the group
19 Extraordinary expenses and losses
IV *Net extraordinary income*
V *Consolidated income before tax*
20 Corporate income tax
VI *Consolidated income for the year*
21 Income allocated to minority interests
VII *Income for the year allocated to the controlling company*

results. Their inclusion in financial losses or profits is a result of their being similar to exchange differences.

Share in losses/income of companies carried by the equity method: these headings include the controlling company's share in the profits or losses of companies carried by the equity method. However, if the activities of these companies relate to extraordinary activities of the group, it would be reasonable to include these companies' results in a specific account under extraordinary results, with a due explanation of this circumstance in the notes to the consolidated annual accounts.

Amortization of goodwill/reversal of negative differences in consolidation: these accounts are included in the consolidated profit and loss account, together with the 'Share in losses/income of companies carried by the equity method' accounts, after financial profits (losses) and computed for the purpose of determining the 'Ordinary operating income (loss)'.

- Amortization of goodwill in consolidation: in accordance with the Consolidation Standards, goodwill is recorded as an asset in the balance sheet and it must be amortized systematically over the period in which it contributes to the obtainment of revenues by the group up to a

Debe

A Gastos

1 Reducción de existencias de productos terminados y en curso de fabricación
2 Consumos y otros gastos externos
3 Gastos de personal
a) Sueldos y salarios asimilados
b) Cargos sociales
4 Dotaciones para amortizaciones de inmovilizado
5 Variación de provisiones de tráfico
6 Otros gastos de explotación

I *Beneficios de explotación*

7 Gastos financieros
8 Pérdidas de inversiones financieras temporales
9 Variacion de provisiones de inversiones financieras
10 Diferencias negativas de cambio
11 Resultados negativos de conversión

II *Resultados financieros positivos*

12 Participación en pérdidas de sociedades puestas en equivalencia
13 Amortización del fondo de comercio de consolidación

III *Beneficios de las actividades ordinarias*

14 Pérdidas procedentes del inmovilizado
15 Variacion de provisiones de inmovilizado material e inmaterial
16 Pérdidas por enajenaciones de participaciones en sociedades consolidadas por integración proporcional
17 Pérdidas por enajenaciones de participaciones puestas en equivalencia
18 Pérdidas por operaciones con acciones de la sociedad dominante y con pasivos financieros del grupo
19 Gastos y pérdidas extraordinarias

IV *Resultados extraordinarios positivos*

V *Beneficios consolidados antes de impuestos sobre beneficios*

20 Impuesto sobre beneficios

VI *Resultado consolidado del ejercicio (beneficios)*

21 Resultado atribuido a socios externos (beneficio)

VII *Resultado del ejercicio atribuido a la sociedad dominante (beneficio)*

maximum of 10 years (NC Art. 24). Despite its presentation in the model profit and loss account established by the Consolidation Standards, it might seem reasonable to include this account under 'Operating income (loss)', since the goodwill arises from the acquisition of a controlling interest in dependent companies or in order to exercise direct influence over decision-making at associated companies, both of which suggest the existence of a consolidated group.

- Reversal of negative differences in consolidation: this credit account attempts either to offset the group's expenses incurred to reverse a loss-making trend at the investee or the payments necessary in order for the group to maintain its investment, or it serves to recognize as profits for the year the portion of the saving in the purchase price relating to a realized capital gain arising from the sale of an asset or its release from inventory in accordance with NC Art. 25.3 (see VI.1.3).

Losses/gains on disposals of holdings in consolidated companies: the extraordinary nature of the transactions generating these results is clear, since the objective of the holdings is to enable the investor to exercise control or a notable influence over the related investees.

Table IX.2(b) Consolidated profit and loss account (revenues)

Credit
B Revenues
1 Net turnover
2 Increase in finished product and work in progress inventories
3 Capitalized expenses of group in-house work on fixed assets
4 Other operating revenues
I *Operating loss*
5 Income from shareholdings
6 Other financial revenues
7 Income from short-term financial investments
8 Exchange gains
9 Translation gains
II *Net financial loss*
10 Share in income of companies carried by the equity method
11 Reversal of negative differences in consolidation
III *Ordinary operating loss*
12 Gains on fixed assets
13 Gains on disposals of holdings in companies consolidated by the global or proportional integration method
14 Gains on disposals of holdings in companies carried by the equity method
15 Gains on transactions involving the controlling company's shares and financial liabilities of the group
16 Capital subsidies transferred to income for the year
17 Extraordinary revenues or gains
IV *Net extraordinary loss*
V *Consolidated loss before tax*
Vll *Consolidated loss for the year*
18 Loss allocated to minority interests
Vll *Loss for the year allocated to the controlling company*

Losses/gains on transactions with shares of the controlling company and with financial liabilities of the group: these headings include the results arising from transactions with shares of the controlling company, either by the controlling company itself or by the dependent companies. They also include results of early amortization through the acquisition by a group company of the financial liabilities issued by another group company.

Consolidated income/loss for the year: this heading includes the results of the controlling company and of the consolidated dependent companies (including the portion allocable to minority shareholders), plus the share of the controlling company in the results of the multigroup and associated companies on the basis of the related percentages of ownership.

Income/loss allocated to minority interests: this heading includes the share of minority shareholders in the results for the year of partially-owned dependent companies after the appropriate adjustments and eliminations for intragroup transactions have been made (NC Art. 43).

Income/loss for the year allocated to the controlling company: this is calculated by subtracting the share of the external shareholders from the consolidated income or loss for the year. This heading includes the results of the controlling company, plus the share of the latter in the results obtained in the period by consolidated companies after the related adjustments and eliminations for intragroup transactions have been made.

Haber

B Ingresos
1 Importe neto de la cifra de negocios
2 Aumento de existencias de productos terminados y en curso de fabricación
3 Trabajos efectuados por el grupo para el inmovilizado
4 Otros ingresos de explotación
I *Pérdidas de explotación*
5 Ingresos por participaciones en capital
6 Otros ingresos financieros
7 Beneficios de inversiones financieras temporales
8 Diferencias positivas de cambio
9 Resultados positivos de conversión
II *Resultados financieros negativos*
10 Participación en beneficios de sociedades puestas en equivalencia
11 Reversión de diferencias negativas de consolidación
III *Pérdidas de las actividades ordinarias*
12 Beneficios procedentes del inmovilizado
13 Beneficios por enajenación de participaciones en sociedades consolidadas por integración global o proporcional
14 Beneficios por enajenación de participaciones puestas en equivalencia
15 Beneficios por operaciones con acciones de la sociedad dominante y con pasivos financieros del grupo
16 Subvenciones en capital transferidas al resultado del ejercicio
17 Ingresos o beneficios extraordinarios
IV *Resultados extraordinarios negativos*
V *Pérdidas consolidadas antes de impuestos*
VI *Resultado consolidado del ejercicio (pérdida)*
18 Resultado atribuido a socios externos (pérdidas)
VII *Resultados del ejercicio atribuidos a la sociedad dominante (pérdida)*

X Notes and additional statements

1 Notes to the consolidated annual accounts

The consolidated notes must contain all the information required to complete, amplify or further discuss the information included in the consolidated balance sheet and profit and loss account in order to show a true and fair view of the related group. The content of the consolidated notes is established by the Consolidation Standards (see Tables X.1 and X.2), which stipulate in line with CCom Art. 48 the minimum information to be included, unless this information is considered to be non-material (NC Art. 66).

2 The consolidated statement of changes in financial position

NC Art. 66-c expressly stipulates the optional nature of the consolidated statement of changes in financial position. However, groups opting to prepare this document must include it in note 25 to the consolidated annual accounts in the format established by the Consolidation Standards (NC) in line with the rules relating to individual companies.

Despite the optional nature of the consolidated statement of changes in financial position, it is apparent that the mere analysis of an individual statement of changes in financial position may lead to significant mistakes. For example, in the case of a controlled company being financed by its controlling company, the balance between the capacity to obtain funds from operations and the

Table X.1 Analysis of heading variations and details to be included in the consolidated notes

A breakdown, by company, of the following headings:

- Goodwill in consolidation
- Negative differences in consolidation
- Holdings in companies carried by the equity method
- Reserves at companies consolidated by the global and proportional integration methods and reserves at companies carried by the equity method
- Translation differences
- Minority interests
- Net turnover, by activity and geographical markets, relating to group activities (alternatively the net turnover of all the companies consolidated by the global and proportional integration methods can be described)
- Translation gains/losses
- Consolidated income/loss for the year

An analysis of variations in the year of the items contained in each of the following headings:

- Goodwill in consolidation
- Negative differences in consolidation
- Intangible assets
- Tangible fixed assets
- Long-term marketable securities
- Non-trade loans
- Equity
- Minority interests
- Provisions for liabilities and charges
- Reconciliation of the consolidated results to the taxable base

investment and funding decisions could only be analyzed considering the acting economic unit, no matter how many legal entities make up this economic unit (Ruiz, 1998, p. 130)

In addition to the regulations established by the chart of accounts, NC Art. 67 sets down the following specific rules for the preparation of the consolidated statement of changes in financial position:

- When the composition of the companies included in consolidation varies considerably from one year to the next, this should be described in the notes so that a reasonable comparison can be made of the consolidated annual accounts.
- The reciprocal sources and applications of funds resulting from inter-company transactions must be fully eliminated if the companies involved are consolidated by the global integration method, and in the proportion relating should the companies involved be consolidated by the proportional integration method.
- In addition to the adjustments set out by the chart of accounts relating to individual companies, the appropriate corrections must be made to the base of consolidated results in order to calculate the 'Funds obtained from operations' if the adjusted period result is a profit or the 'Funds applied in operations' if the adjusted period result is a loss. Specifically, the required corrections will be made to eliminate:
  (i) Amortization of goodwill in consolidation.
  (ii) Reversal of negative differences in consolidation.
  (iii) Profits or losses at companies carried by the equity method.
- The effect on the variation in the consolidated working capital, arising from the acquisition or disposal of holdings in companies to be included in or excluded from consolidation for the first time, must be shown in one account in the statement of changes in financial position as a source or application of funds, as appropriate, and the effect of the acquisitions and the disposals must be indicated separately. In these cases, an exhibit to the consolidated statement of changes in financial

Table X.2 Specific notes relating to consolidation

Note	Content
1 Dependent companies	Detailed description of the dependent companies with their activity and the percentage of controlled capital. If appropriate, the exclusion from global integration and the impact of the application of the equity method on the net worth, financial position and results of the group must also be expounded. The reason for the exclusion from consolidation and the breakdown equity of the companies excluded must be mentioned.
2 Associated and multigroup companies.	The same information as in note 1, but relating to associated and multigroup companies. If the controlling company opts to carry multigroup companies by the equity method, this decision must also be justified.
3 Basis of presentation of the consolidated annual accounts	Justification of the accounting principles applied for the purpose of showing a true and fair view of the group. The following information must also be included for comparison purposes: • Explanation of the adaptation of the prior year's figures or the inability to do so. • Information on changes in the 'consolidation scope' or 'consolidated set' and the effect of these changes on the group's equity, financial position and results. • Information on significant transactions between group companies when the business year of one ends up to three months before the consolidated accounting close.
4 Valuation standards	The accounting principles applied to consolidated account items are indicated in this note, particularly the valuation standards adopted with regard to accounts arising in the consolidation process: goodwill in consolidation, negative differences in consolidation, transactions between consolidated companies (indicating the methods applied for the elimination of inter-group results and reciprocal debts and credits), standardization of individual annual accounts, translation of accounts denominated in foreign currencies, shares of the controlling company, corporate income tax, revenues and expenses, etc.
5 Goodwill in consolidation	Analysis of variations in the consolidated balance sheet accounts included in this section and detail, by investee, of final balances.
6 Negative differences in consolidation	As for note 5.
7 Holdings in companies carried by the equity method	Breakdown of this heading by company, indicating variations in the year and the reasons for the variations.
8 Formation expenses	
9 Intangible assets	
10 Tangible fixed assets	
11 Marketable securities	This note must include the following: • Detail of marketable securities by maturity (short- and long-term), by type of security, and by issuer entity (companies carried by the equity method and other debtors). • Analysis of period variations in long-term securities and in the related provisions. • Information on holdings of companies included in the 'consolidated set' exceeding 5 per cent of the capital stock in companies not included in notes 1 and 2.
12 Non-trade loans	
13 Stocks	
14 Equity	Analysis of period variations in these accounts and breakdown, by consolidated company, of the 'Reserves at companies consolidated by the global and integration methods', 'Reserves at companies carried by the equity method' and 'Differences in consolidation' accounts. The following information must also be included:

Table X.2 (contd.)

Note	Content
	• Capital increases at group companies and the amount of capital authorized by the related shareholders' meeting, indicating the period of authorization. • Controlling company shares held by group companies, the proposed use and amount of the reserves for the acquisition of controlling company shares. • Indication of non-group and related companies owning at least 10 per cent of the capital stock of a group company.
15 Minority interests	Breakdown of this heading, indicating, for each dependent company, the variations in the year, the reasons for the variations, and the breakdown, by item, of the balance at year end (capital, revaluation reserves, results, etc.).
16 Subsidies	
17 Provisions for liabilities and charges	
18 Non-trade debts	
19 Tax matters	
20 Guarantees to third parties and other contingent liabilities	
21 Revenues and expenses	• Information on non-consolidated and associated companies: tangible fixed asset additions, operating purchases and sales, services received and rendered, interest paid and collected, dividends and other distributed results. • Foreign currency transactions with information on consolidated foreign companies. • Breakdown, by foreign company, of translation gains/losses. • Contribution of each consolidated company to consolidated results, indicating the portion relating to minority interests. • Distribution, by activity and geographical markets, of the turnover of the group's ordinary activities unless the publication of this information may cause serious harm to the company, in which case the omission must be justified and the turnover of all the companies consolidated by the global and proportional integration methods must be described. • Information on the disposal of shares in the capital stock of consolidated companies and of the debentures issued by the companies consolidated by the global and proportional integration methods.
22 Transactions with related companies	Information on period transactions of group companies and on debits and credits at year end with individuals and legal entities directly or indirectly related to the controlling company.
23 Other information on the board of directors	• Overall information, by type of remuneration, of all compensation earned by the directors of the controlling company and of the group, multigroup or associated company obliged to pay it. • Detailed information on advances or loans granted by any group company to the controlling company's directors, and on pension commitments to present or former directors.
24 Events subsequent to year end	
25 Consolidated statement of changes in financial position (on a voluntary basis)	If prepared, it must be presented in accordance with the model disclosed in Table X.3.

position must include the following information at the acquisition or disposal date for all companies acquired or sold:
(i) Amount of fixed assets.
(ii) Amount of long-term liabilities.
(iii) Amount of working capital.
(iv) Acquisition/selling price of the holdings.

- When a company carried by the equity method is subsequently consolidated by the global or proportional consolidation method or *vice versa*, the effect of this change on consolidated working capital is disclosed in one account as a source or application of funds, as appropriate. In this connection, the information relating to the acquisition or disposal of companies described in the preceding section must be included as an exhibit to the consolidated statement of changes in financial position.
- The translation into pesetas of the statement of changes in financial position of foreign companies consolidated by the global or proportional integration method is made at the exchange rates ruling at the transaction dates. An average exchange rate may be used provided that it is duly weighted on the basis of the volume of transactions made in each period (month, quarter, etc.). The effect on working capital of exchange rate variations must be shown as a source or application of funds, as appropriate.

The consolidated statement of changes in financial position (see Table X.3) may be prepared under either of the following procedures:

- On the basis of the statement of changes in financial position of each of the companies consolidated by the global or proportional integration methods.
- Under the method envisaged by the chart of accounts for individual companies, i.e. on the basis of the consolidated annual accounts of two consecutive years.

Regardless of the procedure used, the consolidated statement of changes in financial position must describe 'the funds obtained in the year, and the application or use of these funds in fixed or current assets' (NC Art. 67.1).

However, the method of preparing the consolidated statement of changes in financial position on the basis of two consecutive balance sheets presents fewer difficulties than the alternative method since the following complications are avoided:

- The obtainment of the statement of changes in financial position of all the companies included in the 'consolidated set', when certain of these companies may be exempt from this obligation.
- The performance of laborious adjustments and eliminations of inter-group transactions which are simultaneously included as a source of funds for one company and as applications for another.

The following comments intend to give an explanation of those special items in the consolidated statement of changes in financial position (see Table X.3), that is to say, of those sources and applications which are peculiar to this statement (see Ruiz, 1998, pp. 133–50).

- Funds obtained/applied from/in operations. Apart from the specific corrections which have to be made, the format of the consolidated statement of changes in financial position obliges the making of an allocation of funds obtained/applied to the controlled company and to external shareholders. This requirement necessitates distributing operating funds between the controlling company and external shareholders for all necessary corrections, since the adjustments to be made can affect the controlling company and the external shareholders in a different way.

  For example, the necessary adjustment for the elimination of the amortization of goodwill in consolidation can only affect the controlling company's shareholders and not external shareholders.
- Acquisitions of the controlling company's shares. The applications included in this item consist of the amounts used for the acquisition of the controlling company's shares from third parties, either by the controlling company or by any of its subsidiaries.
- Capital stock reductions at the controlling company. This item reflects the amounts paid or to be paid to the controlling company's shareholders as a result of a capital stock reduction. It also shows the pay back to be

Table X.3 Consolidated statement of changes in financial position

Source of funds	Year N Increase	Year N–1 Decrease
Funds obtained		
1 Funds obtained from operations		
(a) Allocated to the controlling company		
(b) Allocated to minority interests		
2 Contributions from external shareholders and from the controlling company's shareholders		
3 Capital subsidies		
4 Long-term debt		
(a) Debt securities and other similar liabilities		
(b) Other debt		
5 Fixed asset disposals		
(a) Intangible assets		
(b) Tangible fixed assets		
(c) Financial investments		
6 Disposals of the controlling company's shares		
7 Partial disposals of holdings in consolidated companies		
8 Funds used for disposal of consolidated companies		
9 Early amortization or transfer to short-term of financial investments		
Total funds obtained		
Funds applied in excess of funds obtained (Decrease in working capital)		
Funds applied		
1 Funds applied in operations		
(a) Allocated to the controlling company		
(b) Allocated to minority interests		
2 Formation and debt arrangement expenses		
3 Fixed asset additions		
(a) Intangible assets		
(b) Tangible fixed assets		
(c) Financial investments		
4 Acquisition of the controlling company's shares		
5 Capital stock reductions at the controlling company		
6 Dividends		
(a) Of the controlling company		
(b) Of group companies allocated to minority interests		
7 Acquisition of additional holdings in consolidated companies		
8 Funds applied for acquisition of consolidated companies		
9 Repayment or transfer to short-term of long-term debt		
10 Provisions for liabilities and charges		
Total funds applied		
Funds obtained in excess of funds applied (Increase in working capital)		

Table X.3 (contd.)

	Year N Increase	Year N Decrease	Year N–1 Increase	Year N–1 Decrease
Variation in working capital				
1 Subscribed capital called but not paid				
2 Inventories				
3 Accounts receivable				
4 Accounts payable				
5 Short-term financial investments				
6 Shares of the controlling company				
7 Cash				
8 Accrual accounts				
Total				
Variation in working capital				

made to external shareholders due to their participation in the capital reimbursement of the controlled companies.

- Application for dividends. It refers exclusively to dividends paid to the controlled company's shareholders, since the dividends attributable to the controlling company for its participation in the income of the controlled companies must be eliminated in the consolidation process.
- Acquisition of additional holdings in consolidated companies. This item reflects the acquisition price of new investments in companies already consolidated, either fully consolidated or consolidated by the proportional method.
- Funds applied for acquisition of consolidated companies. Instead of reflecting separately the sources and applications of funds derived from assets and liabilities acquired with a new consolidated company, an application is shown only in one account, reflecting the effect on the variation in the consolidated working capital, that is to say, the acquisition price of the participation minus the working capital of the company which enters into the group.
- Contribution from external shareholders and from the controlling company's shareholders. This item must reflect true contributions, without considering any unpaid capital stock increases.

Therefore, the participation that the controlling company subscribes in the capital increase of a controlled company must not be taken into account. A capital stock increase in a controlled company may lead to a modification in the capital distribution between the controlling company and external shareholders. In that case, there will be a source or application of funds from or to external shareholders.

- Partial disposals of holdings in consolidated companies. The sale price must be separately reflected as a source of funds.
- Funds obtained from disposal of consolidated companies. This item refers to participation sales which lead to the departure of some controlled companies from the consolidated group.

The preparation of a consolidated cash flow statement is neither required nor encouraged by Spanish Consolidation Standards (nor for individual companies), despite the fact that this statement explains the variations in cash as a variable more objectively than the statement of changes in financial position does with respect to variations in working capital.

3 The consolidated management report

The consolidated management report must be added to the consolidated annual accounts even though it does not form a part of the accounts (CCom Art. 44.1) (see Table XII.1). Both the consolidated annual accounts and the management report must be signed by all the directors of the controlling company, who are liable for their accuracy (CCom Art. 44.8).

CCom Art. 49 regulates the content of the consolidated management report and establishes a series of minimum requirements. However, although group management has a considerable degree of freedom in its preparation, it must be relatively consistent with the content of the consolidated annual accounts. According to CCom Art. 49, the management report must show 'a true and fair view of the evolution of the business and situation of the consolidated companies and their future expectations'.

Furthermore, the following verifiable information must also be included:

- Significant events subsequent to year end for the consolidated companies.
- Group research and development activities.
- Number and par value, or in their absence, the book value, of the controlling company's shares or participations held by it, by subsidiaries, or by third parties acting in their own name but for the account of a group company.

However, it would seem reasonable to consider that the above points 'should be included in the consolidated notes and not the management report, the latter being confined to secondary matters and projections'.

XI Auditing

1 Audit requirements

Under Spanish regulations, the obligation to present consolidated annual accounts carries with it the obligation to submit these documents to review and examination by independent auditors so that the latter may issue a qualified report on whether they reasonably reflect the financial situation and results of the group in accordance with generally accepted accounting and consolidation principles and standards.

The Commercial Code states that 'the controlling company's board of directors must appoint auditors to control the group's annual accounts and management report. The auditors shall verify that the management report is consistent with the consolidated annual accounts' (CCom Art. 42.5). This appointment and the related approval must be registered with the Commercial Register (*Registro Mercantil*) (Arts. 153 and 154 of the Commercial Register Regulations, *Reglamento del Registro Mercantil*).

The auditors' report on the consolidated annual accounts must be presented in accordance with the same deadlines as those for individual annual accounts in general, and for the controlling company in particular, although with the additional requirement that the auditors' reports on group companies must be available for the completion of the auditors' report on the group (see Table XII.1).

The auditors' report must be deposited alongside the consolidated annual accounts and management report with the commercial register in the month after their approval, so ensuring the disclosure of the information.

2 Co-ordination between auditors

The appointment of the group's auditors does not exclude the possibility of dependent, multigroup and associated companies appointing other auditors. In this case the actions between the auditors must be co-ordinated, taking into account that the overall responsibility lies with the group's auditors, who must therefore direct the actions of the other auditors.

The ICAC Resolution dated 16 December 1992, publishing the technical regulations for auditor relationships, defines the responsibilities and report methods of issuing reports when one auditor uses the work or reports of another independent auditor.

Specifically, when an auditor expressing an opinion on the consolidated annual accounts has to take into consideration the reports issued by other auditors authorized by Spanish law and referring to investees of the company whose accounts are under examination, the following rules must be observed:

- Auditors are always liable for the opinions expressed in their report and must therefore reach their own conclusions.
- Auditors of consolidated annual accounts must refer to the report of the other auditor when expressing their opinion on such accounts,

clearly indicating in their report the difference in liability between the portion of the annual accounts covered by their report and that covered by the work of the other auditor.

- The reference in the report on the consolidated accounts to the fact that part of the examination was performed by another auditor should not be considered to be a qualification, but a definition of the liabilities of the respective auditors.
- Auditors of consolidated annual accounts must take into account the qualifications contained in the other auditor's opinion and mention them in their report when these qualifications are material to the overall group accounts, unless they have been corrected by accounting corrections or through the consolidation process.

3 Presentation of the auditors' report

The general regulations governing auditors' reports described in Spain – Individual Accounts (see X) are also applicable to auditors' reports on consolidated accounts. However, the features exclusively affecting consolidation must also be taken into account.

There are two options for the presentation of the auditors' report on consolidated accounts:

- In a specific format with the structure established by the ICAC's technical auditing standards, referring to the consolidated balance sheet, profit and loss account and the related notes and consolidated management report. In this way, two consecutive complete reports in accordance with the standard auditors' report format must be issued: one relating to the annual accounts and management report of the controlling company, and the other relating to those of the consolidated group.
- Alternatively, one sole auditors' report may be issued for the controlling company and the group, accompanied by separate annual accounts and management report for the controlling company and the group.

An argument exists for abolishing the option of issuing a joint auditors' report with joint annual accounts of the controlling company and of its group, since this would contravene the principle of clarity in the presentation of consolidated annual accounts (NC Art. 61.2), because of the requirement to present the balance sheet and profit and loss account in a four-column format (two for the controlling company and two for the group, one for each year). This would become more complicated in the event of the presentation of joint notes (Condor and Gabás, 1991).

XII Filing and publication

The Commercial Code expressly establishes the obligation to deposit the consolidated annual accounts, management report and the related auditors' report with the Commercial Register in order to ensure that this information is adequately disclosed (CCom Art. 42.6). The deposit and disclosure of these documents are made in accordance with the regulations governing the annual accounts of corporations (see Table XII.1). Any person can obtain information from the documents deposited.

The Ministry of Economy and Finance Order dated 18 January 1991 that governs publication of information by companies issuing marketable securities listed on the stock exchange stipulates that those entities which are obliged to present consolidated annual accounts must furnish the National Securities and Exchange Commission (*Comisión Nacional del Mercado de Valores*, CNMV) and the related stock exchange management companies with their own six-monthly and quarterly individual financial reports as well as those referring to the group controlled by them in accordance with the formats and instructions contained in the order.

This information must be made public by the stock exchange management companies in their 'Quotation Bulletins' within 15 days from the end of the related deadline (two months for six-monthly information and 45 days for quarterly information). Also, the CNMV must keep the information received available to the public at all times.

Table XII.1 Deadlines for the preparation, approval and disclosure of financial statements

	Closing date (if 31 December)
Preparation of annual accounts and management report and proposed distribution of profits (and, where appropriate, of the consolidated annual accounts and management report).	31/3/19X (3 months)
Availability of the annual accounts to the shareholders.[1]	15/6/19X
Submission of the annual accounts (individual and consolidated) to the approval of the ordinary shareholders' meeting.[2]	30/6/19X (6 months)
Deposit of the annual accounts and management and auditors' reports (individual and consolidated) with the Commercial Register.[3]	31/7/19X (7 months)

[1] LSA Art. 212.2, as amended, states that any shareholder may obtain immediately and free of charge from the date that the shareholders' meeting has been summoned, the documents submitted for approval at the meeting and the auditors' report. Notice must be given of the ordinary shareholders' meeting at least 15 days prior to the date on which it is scheduled to be held (LSA Art. 97).

[2] The ordinary shareholders' meeting must be held in the first six months of each year to monitor corporate management, to approve, if appropriate, the prior year's financial statements, and to decide on the distribution of results (LSA Art. 95).

The ordinary shareholders' meeting of the controlling company will approve, if appropriate, the consolidated annual accounts and management report simultaneously with the financial statements of the controlling company itself (CCom Art. 42.6).

In turn, the auditors have at least one month from the delivery of the accounts signed by the directors to present their report. If, as a consequence of the auditors' report, the annual accounts are modified, the auditors will extend their report to include the changes introduced (LSA Art. 210)

[3] A certificate must be filed with the Commercial Register containing the resolutions approving the financial statements and the distribution of income, together with a copy of each of the statements (individual and consolidated), of the management report and of the auditors' report (LSA Art. 218 and CCom Art. 42.6).

XIII Sanctions

LSA Art. 221.1, as amended, establishes that any non-compliance by the directors with their obligation to deposit with the Commercial Register the accounting documents described in LSA Art. 218 (see Table XII.1) shall lead to the imposition of a fine ranging from ESP 200,000 to ESP 2 million for each year's delay in depositing the accounts, the statute of limitations for these infringements being three years. These sanctions are also applied if the controlling company's directors fail to deposit the group's accounting documents as stipulated in CCom Art. 42.6, i.e. the consolidated annual accounts and management report and the auditors' report.

Non-compliance with the obligation to furnish the periodic public information governed by the Ministry of Economy and Finance Order dated 18 January 1991 is considered a serious offence and may give rise to the imposition of any of the sanctions contained in Law 24/1988 Art. 103 (Securities Market Law of 28 July 1988): public warning, a fine, suspension of activities or suspension as a member of the market for up to one year.

In accordance with Law 26/1988 governing discipline and intervention in credit entities, the credit entities listed in Royal Legislative Decree 1298/1986 (Art. 1), and the persons holding administrative or managerial posts in these entities, shall be liable to be penalized if they contravene regulatory and disciplinary rules. These infringements are classified as very serious, serious and slight. Serious and very serious infringements are sanctionable at the transgressing credit entity according to the penalties listed in Table XIII.1. The statute of limitations period for these penalties is five years.

In addition to the penalty imposed on the credit entity, for very serious infringements one of the following penalties will be imposed on the persons holding administrative or managerial posts who have brought about the infringement by fraudulent or negligent behaviour:

Table XIII.1 Penalties system for credit entities

Very serious infringements	Serious infringements
• Not keeping the required accounting records or the existence of essential irregularities in the records that prevent the reflection of the entity's net worth and financial situation. • Non-compliance with the obligation to submit financial statements to audit in accordance with regulations. • Non-compliance with the duty of presenting a truthful situation to the shareholders or to the general public, provided that the number of persons affected or the materiality of the information is considered to be especially relevant; otherwise, this shall be considered to be a serious infringement.	• Non-compliance with current regulations governing the recording of transactions and the preparation of balance sheets, profit and loss accounts and other financial statements that must be notified to the competent administrative body.
Penalties	**Penalties**
• Fine of up to 1 per cent of equity or of ESP 5,000,000, if 1 per cent of equity is lower than the latter amount. • Revocation of the entity's authority.	• Private warning • Fine of up to 0.5 per cent of equity or of ESP 2,500,000, if 0.5 per cent of equity is lower than the latter amount.

- A fine to each of no more that ESP 10 million.
- Suspension from holding the related post for up to three years.
- Removal from office and disqualification from holding administrative or managerial posts at the same credit entity for up to five years.
- Removal from office and disqualification from holding administrative or managerial posts at any credit entity for up to ten years.

The third or fourth penalty may also be imposed together with the fine of ESP 10 million.

Bibliography

Almela Díez, B. (1991a). *Panorama normativo de la consolidatión fiscal*, VI Congreso de la Asociacíon Española de Contabilidad y Administracíon de Empresas (AECA), Vigo.

Alvarez Melcón, S. (1991b). Contabilidad y fiscalidad de los grupos de sociedades. *Proceedings of the Sixth Congress of the Asociación Española de Contabilidad y Administracin de Empresas* (AECA), Vigo.

Alvarez Melcón, S. (1989). *Análisis contable del régimen de declaración consolidada de los grupos de sociedades*, Instituto de Contabilidad y Auditoria de Cuentas (ICAC), Madrid.

Alvarez Melcón, S. (1991b). Las cuentas anuales consolidadas. *Cómo Consolidar las Cuentas Anuales de los Grupos de Sociedades,* **1**, Asociación Española de Contabilidad y Administración de Empresas.

Alvarez Melcón, S. (1993). *Consolidación de Estados Financieros* (McGraw Hill, Maidenhead).

Ansón, J. A., Blasco, M. P. and Brusca, M. I. (1997). Utilidad de la información contable para evaluar la rentabilidad de sociedades dominantes: papel de la información consolidada frente a la individual, *Revista Española de Financiación y Contabilidad,* **93**, October-December, 917–934.

Archel, P. (1991). *La conversión de partidas en las filiales de diferente nacionalidad,* VI Congreso de la Asociacíon Española de Contabilidad y Administracíon de Empresas (AECA), Vigo.

Archel, P. (1994). Son comparables las cuentas anuales consolidadas que se publican en España?, Evidencia Empírica. *Partida Doble,* **42**, February, 34–38.

Archel, P., Robleda, H. and Santamaría, R. (1995). Una aproximación empírica al estudio de las eliminaciones en las cuentas consolidadas. *Revista Española de Financiación y Contabilidad,* **82**, January-March, 59–79.

Banco de España (1991). *Circular 4 de 14 de junio, a las entidades de crédito sobre normas de contabilidad y modelos de estados financieros,* **153**, 27 June.

Benitez, J., Blasco, P., Condor, V. and Costa, A. (1993). La información consolidada publicada en el Registro Mercantil. Una aproximación. *Proceedings of the Seventh Congress of the Asociación Española de Contabilidad y Administración de Empresas* (AECA), 31–55.

Blanco, M. A. and Martinez, M. (1991). Responsabilidad contable y fiscal de la sociedad dominante en el grupo de sociedades. *Cómo Consolidar las Cuentas Anuales de los Grupos de Sociedades,* **3**, Asociación Española de Contabilidad y Administración de Empresas (AECA), 33–57.

Blasco, P (1997). *El Análisis de las Cuentas Anuales Consolidadas. Una Aproximación Conceptual y Empírica,* Asociación Española de Contabilidad y Administración de Empresas (AECA). Madrid.

Cañibano, L. (1991). Las cuentas consolidadas de los grupos de sociedades. *Cómo Consolidar las Cuentas Anuales de los Grupos de Sociedades,* **1**, Asociación Española de Contabilidad y Administración de Empresas, pp. 9–17.

Cea Garcia, J. L. (1991). *Comentarios sobre las partidas especificas de los modelos de balance y cuenta de pèrdidas y ganancias consolidadas según el borrador del ICAC.* VI Congreso de la Asociacíon Española de Contabilidad y Administracíon de Empresas (AECA), Vigo.

Cea Garcia, J. L. (1992). Algunas anotaciones sobre la imagen fiel y sobre el concepto de cuentas anuales consolidadas de los grupos de sociedades. *Revista de Estudios Financieros,* **108**, March, 23–40.

Cervera, N. (1997). La consolidación en el caso de venta de acciones. *Partida Doble,* **75**, February, 5–14.

Condor López, V. (1988). *Cuentas consolidadas, aspectos fundamentales en su elaboración.* Instituto de Contabilidad y Auditoría de Cuentas (ICAC), Madrid.

Condor López, V. (1992). *Methodologia de las cuentas consolidadas en Europa.* Instituto de Contabilidad y Auditoria de Cuentas (ICAC), Madrid.

Condor López, V. (1992). Las normas sobre formulación de cuentas anuales consolidadas, aspectos más relevantes. *Partida Doble,* **26**, September, 4–19.

Condor, V. and Gabás, F. (1991). Los grupos de sociedades y su auditoría. *Proceedings of the Ninth National Congress of the Auditores-Censores Jurados de Cuentas,* Valencia.

Confederación Española de Organizaciones Empresariales (CEOE) (1987). *Grupos de sociedades. Su adaptación a los normas de las Comunidades Europeas,* Madrid.

Corona Romero, E. (1991). *El cuadro de financiación consolidado en el borrador de normas para la formulación de cuentas anuales consolidadas,* VI Congreso de la Asociacíon Española de Contabilidad y Administracíon de Empresas (AECA), Vigo.

Corona Romero, E. (1992). Eliminaciones de resultados en el método de integración global según el Real Decreto 1815–1991. *Revista de Estudios Financieros,* **110**, May, 103–44.

Fernandez Fernandez, J. M. (1993). *Consolidación de Estados Contables* (Editorial AC, Madrid).

Gonzalo Angulo, J. A. (1993). Lectura e interpretación de las cuentas anuales consolidadas. *Revista de Estudios Financieros,* **122**, 73–138.

Gonzalo Angulo, J. A. (1994). *Lectura e interpretación de las cuentas anuales consolidadas* (Centro de Estudios Financieros, Madrid).

Heras Miguel, L. (1993). *Normas de Consolidación: Comentarios y Casos Prácticos* (Centro de Estudios Financieros, Madrid).

Iglesias Sánchez, J. (1992). Utilidad, fiabilidad y relevancia de los estados financieros consolidados. Partida Doble, **26**, September, 20–30.

Labatut Serer, G. (1991). *Efecios impositivos derivados de la tribuiación individual de las empresas que forman grupo consolidable: Referencia a las operciones vinculadas por elementos circulantes.* VI Congreso de la Asociacíon Española de Contabilidad y Administracíon de Empresas (AECA), Vigo.

Labatut Serer, G. (1992). *Contabilidad y Fiscalidad del Resultado Empresarial* (Instituto de Contabilidad y Auditoría de Cuentas, Madrid).

Lainez Gadea, J. A. (1991). *La conversión de cuentas anuales en moneda extranjera de sociedades dependientes y asociadas,* VI Congreso de la Asociacíon Española de Contabilidad y Administracíon de Empresas (AECA), Vigo.

Lara, L. and Martín, A. (1991). *La consolidación contable.* Consorcio de la Zona Franca de Vigo, Vigo.

Larriba, A. (1991). Las cuentas consolidadas en la banca. *Cómo Consolidar las Cuentas Anuales de los Grupos de Sociedades,* **3**, Asociación de Contabilidad y Administración de Empresas, 77–102.

López García, G. (1991). La contabilidad del impuesto sobre sociedades en los grupos. *Proceedings of the Sixth Congress of the Asociación Española de Contabilidad y Administración de Empresas (AECA)*, Vigo.

López Santacruz, J. A. (1992). Régimen tributario de los grupos de sociedades. *Revista de Estudios Financieros,* **108**, March, 3–23.

Marín, L. (1994). Aspectos contables de la consolidación. *Partida Doble,* **42**, February, 6–15.

Martínez, F. J. (1993). Revisión de los valores de los activos en consolidación. *Partida Doble,* **38**, October, 12–16.

Montesinos, V., Serra, V. and Gorgues, R. (1990). *Manual Práctico de Consolidación Contable* (Ariel Economía, Barcelona).

Robleda, H. (1996).Modificaciones en las normas sobre consolidación en la Ley S.R.L. *Partida Doble,* **64**, February, 15–18.

Rosa Vargas, G. (1992). La integración proporcional, método de consolidación para las sociedades multigrupo. *Partida Doble,* **26**, September, 31-36.

Ruiz Lamas, F. (1998). El análisis externo del cuadro de financiación consolidado. *Revista de Contabilidad y Tributación,* **187**, October, 123–192.

Tua Pereda, J. (1991). La obligación de consolidar en la legislación mercantil española. *Partida Doble,* **14**, July, 12–23.

Yebra, O. (1991). Homogeneización de la información contable. *Cómo Consolidar las Cuentas Anuales de los Grupos de Sociedades,* **1**, Asociación Española de Contabilidad y Administración de Empresas, 59–79.

Legal texts and professional pronouncements

ICAC Resolution dated 9 October 1997: Resolución de 9 Octubre 1997, del Instituto de Contabilidad y Auditoría de Cuentas, sobre algunos aspectos de la norma de valoración numero dieciseis del Plan General de Contabilidad.

Law 18/1982: Ley 18/1982, de 26 de mayo, de régimen fiscal de las agrupaciones y uniones temporales de empresas y sociedades de desarrolo industrial regional.

Law 24/1988: Ley 24/1988, de 28 de julio, del Mercado de Valores.

Law 13/1985: Ley 13/1985, de 25 de mayo, de coeficientes de inversión, recursos propios y obligaciones de información de los intermedianos financieros.

Law 26/1988: Ley 26/1988, de 29 de julio, sobre disciplina e intervención de las entidades de credito.

Law 19/1989: Ley 19/1989, de 25 de julio, de reforma parcial y adaptación de la legislación mercantil a las Directivas de la Comunidad Económica Europea (CEE) en materia de Sociedades.

Law 13/1992: Ley 13/1992, de 1 de junio, de Recursos propios y supervisión en base consolidada de las Entidades Financieras.

Law 43/1995: Ley 43/1995, de 27 de diciembre, del Impuesto sobre Sociedades.

Commercial Register Regulations: Reglamento del Registro Mercantil, aprobado por Real Decreto 1597/1989, de 29 de diciembre.

Ministry of Economy and Finance Order dated 18 January 1991: Orden de 18 de enero de 1991, sobre información en Bolsas de valores.

Ministry of Economy and Finance Order dated 29 December 1992: Orden de 29 de diciembre de 1992 sobre recursos propios y supervisión en base consolidada de las sociedades y Agencias de Valores y sus grupos.

Royal Decree-law 15/1977: Real Decreto-Ley 15/1977, de 25 de febrero, sobe medidas fiscales financieras y de inversión pública.

Royal Decree 1414/1977: Real Decreto 1414/1977, de 17 de junio, por el que se dictan las normas paras la formulación de la declaración del beneficio consolidado de los grupos de sociedades.

Royal Decree 1371/1985: Real Decreto 1371/1985 de l de agosto, que desarrolla la ley 13/1985, de 25 de mayo, de coeficientes de inversión, recursos propios y obligaciones de información de los intermediarios financieros.

Royal Decree 441/1986: Real Decreto 441/1986 de 28 de febrero.

Royal Decree Law 1298/1986: Real Decreto Legislativo 1298/1986 de 28 de junio, sobre adaptación del derecto vigente en maleria de Entidades de crédito al de las Comunidades Europeas.

Royal Decree 1815/1991: Real Decreto 1815/1991, de 20 de diciembre, por el que se aprueban las normas para la formulación de las cuentas anuales consolidadas.

Royal Decree 1343/1992: Real Decreto 1343/1992, de 6 de noviembre, por el que se desarrolla la Ley 13/1992, de 1 de junio, de recursos propios y supervisión en base consolidada de las entidades financieras.

Royal Decree 1345/1992: Real Decreto 1345/1992, de 6 de noviembre, por el que se dictan normas para la adaptación de las disposiciones que regulan la tributación sobre el beneficio consolidado a los grupos de sociedades cooperativas.

Royal Decree 2024/1995: Real Decreto 2024/1995, de 22 de diciembre, por el que se modifica parcialmente el Real Decreto 1343/1992, de 6 de noviembre, por el aue se desarrolla la Ley 13/1992, de 1 de junio, de Recursos Propios y supervisión en base consolidada de la Entidades Financieras, y se incluye un nuevo Título V sobre las reglas especiales de vigilancia aplicables a los grupos mixtos no consolidables de entidades financieras.

SWEDEN
INDIVIDUAL ACCOUNTS

Rolf Rundfelt

Acknowledgement

This description of Swedish financial reporting in individual accounts has been compiled by KPMG Sweden. Åke Näsman has made a thorough review of the entire manuscript for which I am greatly thankful. I remain, of course, responsible for any errors or omissions.

CONTENTS

I Introduction and background

1 Historical development of financial accounting and reporting

1.1 Introduction

In terms of accounting tradition Nobes (1993, p. 30) divides countries into two basic categories. Sweden falls into the Continental, as opposed to the so-called Anglo-Saxon, group. Continental accounting regimes are characterized as having the following general features:

- Legalism.
- Credit orientation.
- Secrecy.
- Tax domination.
- Form over substance.
- Government rules.

A superficial look at Swedish accounting regulation may give the impression that Sweden is, indeed, a country strongly influenced by the continental European tradition. This impression is strengthened when reading FEE's Comparative Study on Conceptual Accounting Frameworks in Europe (FEE, 1997). There, Sweden is included together with 'Other Continental European Countries', a group that includes France, Spain and Portugal. Swedish accounting is described as being

> based on the continental European approach where the government used to be the dominant force by introducing accounting rules based on national economic policies. The most important requirement in the Accounting Act (now replaced by the Annual Accounts Act) is that accounts should be based on good accounting practice. This has been interpreted as the accounting principles used in practice by a qualitatively representative circle of companies. The law does not include any formal references to recommendations issued by standard setting bodies. However, such recommendations play an important role in the interpretation of what is good accounting practice. (FEE 1997, p. 63)

While factually correct, the picture of Swedish financial reporting is much more mixed. This is manifest in the fact that several Swedish companies have received prizes for their annual reports from financial analysts and others. It could be that both Nobes and FEE base their conclusions on the majority of companies, i.e. unlisted rather than listed companies. For the large, listed companies it is more appropriate to classify Sweden as a member of what FEE calls the 'Anglo-Saxon Countries' including Denmark, Ireland, the Netherlands, Norway and the United Kingdom.

Anyhow, the intention of major companies to provide their shareholders with relevant information within the constraints of a Continental legalistic system makes the development of Swedish accounting an interesting study.

Also of interest is the fact that the Swedish Financial Accounting Standards Council in 1997 announced that it plans to issue some 30 new standards, based on International Accounting Standards (IAS), before the end of year 2000. If successful, Sweden will become the first European country to fully implement IAS in its regulatory framework.

1.2 Current legislation

In Sweden the regulation of accounting is based upon the Annual Accounts Act (*Årsredovisningslagen, ÅRL*). The act covers all types of business, large or small, incorporated or not. Thus it is rather general in nature, with only a few specific rules on valuation which will be dealt with later (see Table I.1). The act emphasizes the need to protect the creditors and stresses the principle of prudence.

One of the more important requirements in the Annual Accounts Act is that the accounts should be based on good accounting practice (*god redovisningssed*). This has been interpreted as the accounting principles used in practice by a qualitatively representative circle of companies. The law does not include any formal references to recommendations issued by standard setting bodies. Still, such recommendations have become increasingly more important, especially for the listed companies. Listed companies are committed to follow the recommendations issued by the

Table I.1 The Annual Accounts Act (1995)

Scope

Covers all kinds of business.

Important features

Based on the Fourth and Seventh EC Directives.

Accounts should be based on good accounting practice and give a true and fair view.

True and fair override not permitted.

The act mentions seven fundamental accounting principles of which the principle of prudence is the most important.

Almost no detailed valuation rules.

Detailed formats for balance sheet and profit and loss account but not for the funds statement.

Section on group accounting much expanded in relation to the previous Accounting Act.

Swedish Financial Accounting Standards Council subject to a true and fair override.

The annual report is publicly available. All limited liability companies, including family-owned companies, have to file their accounts with a registrar. Copies of the report can be obtained for a small fee.

1.3 Historical development

The Annual Accounts Act has recently been revised as a consequence of Sweden signing the EEA Agreement. The act is applicable to financial years starting on 1 January 1997.

The dates of the various Swedish Accounting and Companies Acts are given in Table I.2. As can be seen, there have been rather few revisions. Apart from Sweden becoming a member of the EEA, only in one case has a major event triggered a revision. This came in the aftermath of the Krueger crash in the early 1930s. The subsequent investigation revealed that, by forming a holding company with interests in many different industries, Krueger had been able to create profits by selling assets at inflated prices from one company to another. The remedy, embodied in the 1944 Companies Act, was to require that a consolidated balance sheet (*koncernredogörelse*), a calculation of the equity of the group as a whole, should be drawn up. It did not have to be published but there was a requirement to report the group's uncommitted equity and some details of transactions between group companies. The ability of the parent company to pay dividends now became dependent on the financial position and income of the whole group. The publication of a consolidated balance sheet became mandatory under the 1975 Companies Act (Kedner, 1989, p. 24). The 1975 Act also made it mandatory to publish a consolidated profit and loss account and, for larger companies, a funds statement.

1.4 International harmonization

As a small country with an open economy, it is natural that Sweden should have a number of international companies. Some 30 of the largest companies are listed on foreign stock exchanges, 15 of them in the United States (for details see Table I.3). Only two of the companies mentioned are actively traded, i.e. Astra and Ericsson, both in the United States. Consequently there is strong interest in the development of international accounting standards and the harmonization of financial reporting between countries.

For the larger companies, until the last few years, international accounting standards have been synonymous with US GAAP. The major reason is, of course, the need to reconcile Swedish GAAP with US GAAP in the case of those companies that have to file their reports with the Securities and Exchange Commission (SEC).

Recently the standards issued by the International Accounting Standards Committee (IASC) have gained increasing acceptance. In their 1997 annual reports some 20 companies announced that their accounting policies were more or less in line with International Accounting Standards (IAS). The most prominent example is SAS, the Scandinavian airline. The ownership of SAS is split between Denmark, Norway and Sweden. It has been agreed between the three countries that SAS, therefore, should base its accounting on IAS rather than on the national standards of any one of the parent countries. As a result, SAS financial statements are based in their entirety on IAS.

Another example is Stora, a major company in the paper and pulp industry. In 1998 Stora

Table I.2 Historical development of Swedish Accounting and Companies' Acts

Annual Accounts Act (Årsredovisningslagen, ÅRL)

Original Act introduced in 1929. The emphasis is on the protection of creditors. The only requirement is the publication of a balance sheet.

Revised Act introduced in 1976. Interests other than those of creditors are mentioned for the first time. The emphasis is still on the protection of creditors, i.e. the principle of prudence. A profit and loss account is also required.

Current Act introduced in 1995. Applies to financial years starting on 1 January 1997. The act is based on the Fourth and Seventh EC Directives.

Companies Act (Aktiebolagslagen, ABL)

Original Act introduced in 1848. The basic requirement of the act was that a limited liability company had to be approved by the government.

New acts passed in 1895 and 1910. They included detailed rules on how a limited liability company should be formed, the responsibilities of the management, audit requirements, etc.

The Act of 1944 was heavily influenced by the experience of the Krueger crash. The act extended the rules on management responsibilities and on reporting. A new requirement was for a group balance sheet to be prepared by all groups of companies. There was no requirement that the balance sheet should be published, however.

The Companies Act was revised in 1975. The major changes were that the group balance sheet had to be published and there were new requirements on the publication of a group profit and loss account. All larger companies were required to publish funds statements.

A further revision was made in 1995 when most of the accounting requirements were moved to the Accounting Act.

merged with Enso, a Finnish company. The new company is domiciled in Finland. It has been announced that Stora-Enso will base their accounts on IAS.

In most cases the effect on earnings by applying IAS is minor. There is nothing in the Annual Accounts Act that prevents a company from adopting IAS. Thus, the only case where an adjustment is necessary is where a company has taken advantage of an option in the act that is not in compliance with IAS. One such example is revaluations. In Sweden companies are allowed to revalue their fixed assets on an *ad hoc* basis that is not in conformity with IAS.

The effect on earnings is much greater for those companies reconciling their accounting principles with US GAAP. One of these companies is Ericsson, the telephone company. US GAAP require the capitalization of expenditure on software. In Swedish accounts this item is normally expensed. Owing to the company's rapid expansion into mobile telecommunications, its US earnings are significantly higher than earnings under Swedish GAAP.

Other significant items include accounting for goodwill, accounting for sale and lease-back transactions, accounting for deferred taxation pensions and marketable securities.

An unusual feature is that Swedish companies provide details of the reconciliation in their annual report. It is thus possible to compare earnings according to Swedish and US GAAP for several years running. The picture that emerges is mixed. In some years the difference is positive, in others negative. For most companies positive and negative differences appear in a seemingly random fashion. The same is true for equity. Table I.4 shows differences between equity according to US and Swedish GAAP for some Swedish companies providing reconciliation to US GAAP. For Ericsson the difference is mainly explained by capitalized software. Netcom has capitalized start-up expenses in a way which is not accepted under US GAAP which explains the big negative difference. In Volvo, finally, the positive difference is mainly caused by including unrealized gains on securities that are available for sale. A note detailing the kind of information given by companies that reconcile their Swedish GAAP financial statements to US GAAP is given in Table I.5.

Table I.3 Swedish equities listed abroad

March 1998 Company	Market: *Nordic countries* Copenhagen	Oslo	*Europe* Brussels	London	Paris	Switzerland	Germany	*Other* USA OTC (ADR)	NASDAQ (ADR)	NYSE New York	Tokyo
ABB	X			X			O		O		
AGA				X		X		O			X
Astra				X				O		X	
Atlas Copco				X			X	O			
Avesta-Sheffield				O							
Biacore Int.									X		
BTL	X										
Dahl	X										
Electrolux				X	X	X	O		X		
Ericsson				X	X	X	X		X		
Esselte				X							
Incentive				O							
Investor				O							
Lindab	X										
MoDo				O							
NetCom Systems									X		
Nordbanken Hold.				X				O			
Nordström & Thulin		X									
Perstorp				X							
S-E-Banken				O							
Sandvik				X				O			
SCANIA										X	
Securitas				O							
Skandia	X			X							
SKF				X	X	X			X		
Stora				X			X				
Swedish Match									X		
Sv. Cellulosa SCA				X			X	O			
Sv. Handelsbanken				O							
Trelleborg				O							
Volvo			X	X	X	X	X		X		X

X = official listing. O = unofficial listing.
Source: STOCKHOLM STOCK EXCHANGE – FACT BOOK 1998

Table I.4 Differences between equity according to US and Swedish

Company	1996	1997
Astra	−1	0
Atlas Copco	+4	+3
Electrolux	−3	−1
Ericsson	+11	+9
Netcom	−28	−17
Sandvik	+2	+3
SKF	−1	−1
Volvo	+14	+10

Note: Plus or minus signs indicate that equity under US GAAP is larger or smaller respectively.
Source: Rundfelt (1992), p. 60.

It is not meaningful to carry out a similar comparison with IAS, as there are only a few cases with a full reconciliation. However, the description of their accounting polices indicates that they are basically in line with the current IAS.

Another reason for the small stated differences between Swedish GAAP and IAS may be a lack of awareness of what is actually required by the IASC. Few people have a detailed enough knowledge of IAS to say with certainty whether the accounting principles used in a specific case are in line with IAS or not. This is likely to change when companies become aware of the requirement in IAS 1 according to which it is not permitted to state that the accounts are based on IAS unless they comply fully with all the requirements of each applicable standard.

1.5 Comparison between IAS and Swedish GAAP

Table I.6 gives an overview of how Swedish GAAP compare with IAS (August 2000). For each IAS, a reference is made to an applicable recommendation, if any, published by the Swedish Financial Accounting Standards Council (RR X) or the Accounting Standards Board (R X).

2 Standard setting

2.1 Swedish Financial Accounting Standards Council (SFASC)

The leading standard setting body in Sweden is the Swedish Financial Accounting Standards Council (*Redovisningsrådet*). The council was set up in 1989 by the government, the Federation of Swedish Industries *(Sveriges Industriförbund)* and the Swedish Institute of Authorized Public Accountants (*Föreningen Auktoriserade Revisorer*). Each of the three bodies nominate three persons to the board. The board then appoints nine members to the council for a term of six years at the most. At present three of the members of the council are public accountants, three are senior financial managers and three are academics. In 1997 the Stockholm Stock Exchange replaced the government in the council. At the same time some other business organizations became members. The change was made to underline that the council should be seen as a private standard setting body. The change was also made to facilitate funding of the council. For 1998, the council received SEK six million, half of which was raised by increasing the fee paid by listed companies to the Stock Exchange.

The secretariat is small, with only one full-time professional and four to five technical staff seconded by the large accounting firms. Much of the research, editing, etc. is done by members of the council. Working procedures include hearings, exposure drafts that are open to comment and a final recommendation. Meetings of the Council are not open to the public.

The council has translated the conceptual framework issued by the IASC. By doing that, it has declared its intention of adhering closely to the standards issued by the IASC (see *Redovisningsrådet*, 1991). Over time, 'adhering closely' has come to be interpreted as literal translations of the relevant IAS. The only differences accepted from 1997 onwards are those caused by the Annual Accounts Act. For example, the act does not permit reverse acquisitions.

An important decision, taken in 1997, is that the council shall 'catch up' with the IASC in the year 2000. That means that as of the year 2001 Sweden will have some 40 recommendations in place, virtually identical to the standards issued by the IASC (so far, the SFASC has published 20 recommendations, see Table I.7). The decision to translate the relevant IAS shall be seen in the light of this ambitious undertaking. The only way it can be achieved is by refraining from additions to or

Table I.5 A reconciliation between Swedish and US GAAP for Ericsson 1997.

NOTE 24 ACCOUNTING PRINCIPLES GENERALLY ACCEPTED IN THE UNITED STATES

Elements of the Company's accounting principles which differ significantly from generally accepted accounting principles in the United States (US GAAP) are described below:

(A) REVALUATION OF ASSETS

Certain tangible assets have been revalued at amounts in excess of cost. Under certain conditions, this procedure is allowed according to Swedish accounting practice. Revaluation of assets in the primary financial statements is not permitted under US GAAP, why depreciation charges of such items are reversed to income.

(B) CAPITALIZATION OF SOFTWARE DEVELOPMENT COSTS

In accordance with Swedish accounting principles, software development costs are charged against income when incurred. Under US GAAP, FAS No. 86 "Accounting for the Cost of Computer Software to be Sold, Leased or Otherwise Marketed", these costs are capitalized after the product involved has reached a certain degree of technological feasibility. Capitalization ceases and amortization begins when the product becomes available to customers. The Group has adopted an amortization period for capitalized software of three years. Capitalization amounting to SEK 5,232 m. (SEK 4,282 m. in 1996) has increased income and amortization amounting to SEK 3,934 m. (SEK 3,341 m. in 1996) was charged against income for the period when calculating income in accordance with US accounting principles.

(C) PENSIONS

The Company participates in several pension plans, which in principle cover all employees of its Swedish operations as well as certain employees in foreign subsidiaries. The Swedish plans are administered by an institution jointly established for Swedish industry (PRI) in which most companies in Sweden participate. The level of benefits and actuarial assumptions are established by this institution and, accordingly, the Company many not change these.

Effective 1989, the Company has adopted FAS 87, Employer's Accounting for Pensions, when calculating income according to US GAAP.

The effects for the Company of using this recommendation principally relate to the actuarial assumptions, and that the calculation of the obligation should reflect future compensation levels. The difference relative to pension liabilities already booked at the introduction in 1989 is distributed over the estimated remaining service period.

(D) CAPITALIZATION OF INTEREST EXPENSES

In accordance with Swedish accounting practice, the Company has expensed interest costs incurred in connection with the financing of expenditures for construction of tangible assets. Such costs are to be capitalized in accordance with US GAAP, and depreciated as the assets concerned.

(E) CAPITAL DISCOUNT ON CONVERTIBLE DEBENTURES

According to Swedish accounting principles, the 1997/2003 convertible debenture loan and its nominal interest payments are valued at present value, based on the market interest rate at the time of the issue. The difference to the nominal amount, the capital discount, is credited directly to equity. According to US GAAP, convertible debenture loans shall be reported as liabilities at nominal value. When calculating income and equity according to US GAAP, the effects of the capital discount are reversed.

(F) OTHER

Stock issue costs

The costs incurred by the Company relating to the stock issue in 1995 have been charged to income in accordance with accounting principles generally accepted in Sweden. In accordance with US GAAP such costs are charged directly to stockholders' equity.

Sale-leaseback of property

In 1987, group companies sold properties which were leased to subsidiaries under contracts expiring in 1997.

Under US GAAP, the sales during 1987 are considered financing arrangements and the gains are deferred and the proceeds are therefore treated as a liability. In accordance with Swedish accounting practice at the time, no deferral of profit had to be made if the sale price did not exceed the market value and if leasing costs did not exceed normal market leasing rates.

During 1995 and 1997, Ericsson waived its options to repurchase and canceled the rental contract on the properties. Consequently, the income portion of the sales proceeds were recognized in income for the periods in accordance with US GAAP.

Hedge accounting

Ericsson has currency forward exchange contracts and options regarding firm commitments as well as budgeted cash flows regarding sales and purchases. According to Swedish accounting practice both kinds are considered hedges and are not valued at market. According to US GAAP, contracts and options not related to firm commitments must be valued at market.

Tax on undistributed earnings in associated companies

In accordance with Swedish accounting practice, no accrual is made for withholding taxes on undistributed profits of companies that are consolidated applying the equity accounting method. Under US GAAP, the company holding shares should accrue for withholding taxes on possible dividends.

Business combination adjustments

When applying Swedish accounting practice, the Company shows negative goodwill as a deferred credit which is released to income over on average 5–10 years (see also Notes to the financial statements, Accounting Principles (B) and Note 6). In accordance with US GAAP, negative goodwill should be applied as a reduction of non-current assets acquired and be amortized over the economic life of each asset.

(G) DEFERRED INCOME TAXES

Deferred tax is calculated on all US GAAP adjustments to income.

SUMMARY

Application of US GAAP as described above would have had the following approximate effects on consolidated net income and stockholders' equity. It should be noted that, in arriving at the individual items increasing or decreasing reported net income, consideration has been given to the effect of minority interests.

Table I.5 (contd.)

ADJUSTMENT OF NET INCOME

	1997	1996	1995
ITEMS INCREASING REPORTED NET INCOME			
Depreciation on revaluation of assets	**22**	24	23
Capitalization of software	**1,298**	942	1,242
Capital discount on convertible debentures	**13**	–	–
	1,333	966	1,265
ITEMS DECREASING REPORTED INCOME			
Pensions	**72**	−158	−182
Capitalization of interest expenses, net after depreciation	**10**	−24	−15
Other	**207**	−97	−122
Deferred income taxes	**137**	379	376
	426	100	57
Net increase in net income	**907**	866	1,208
Net income as reported in the consolidated income statements	**11,941**	7,110	5,439
Approximate net income in accordance with US GAAP	**12,848**	7,976	6,647
Reported income per share	**12.15**	7.27	5.83
Approximate income per share in accordance with US GAAP	**13.06**	8.15	7.11

ADJUSTMENT OF EQUITY

	1997	1996	1995
INCREASES			
Capitalization of software	**7,398**	6,100	5,158
Pensions	**674**	746	588
Capitalization of interest, net after cumulative depreciation	**339**	349	325
	8,411	7,195	6,071
REDUCTIONS			
Revaluation of assets	**425**	403	414
Capital discount on convertible debentures	**803**	–	–
Other	**305**	97	194
Deferred income taxes	**2,138**	2,230	1,848
	3,671	2,730	2,456
Adjustment of stockholders' equity, net	**4,740**	4,465	3,615
Reported Stockholders' equity	**52,624**	40,456	34,263
Approximate equity according to U.S. GAAP	**57,364**	44,921	37,878

ADJUSTMENT OF CERTAIN BALANCE SHEET ITEMS ACCORDING TO US GAAP

SEK m.	As per reported balance sheet Dec 31 1997	Dec 31 1996	As per US GAAP Dec 31 1997	Dec 31 1996
Intangible assets	**748**	919	**8,146**	7,019
Tangible assets	**19,225**	17,754	**19,048**	17,607
Other long-term receivables	**3,365**	2,208	**3,414**	2,108
Other receivables	**19,133**	10,514	**20,176**	11,412
Minority interest in equity	**4,395**	3,410	**4,427**	3,440
Provisions	**21,095**	17,938	**23,606**	20,171
Convertible debentures	**6,034**	1,772	**6,837**	1,772
Other long-term liabilities	**510**	1,687	**510**	1,711
Other current liabilities	**26,577**	18,481	**26,804**	18,481

Provisions include the tax effect of undistributed earnings in associated companies.

The company in principle follows FAS 95 when preparing the statement of cash flows.

According to FAS 95, however, only cash, bank and short-term investments with due dates within 3 months shall be considered cash and cash equivalents, rather than within 12 months. Applying this definition would mean following adjustments of reported cash:

CONSOLIDATED, SEK m.	1997	1996	1995	1994
Short-term cash investments, cash and bank, as reported	**29,127**	19,060	15,385	11,892
Adjustment for items with maturity of 4–12 months	**−15,004**	−8,396	−7,902	−5,166
Cash and cash equivalents as per US GAAP	**14,123**	10,664	7,483	6,726

Table I.6 Swedish GAAP and IAS

IAS 1	Financial statements	A recommendation based on IAS 1 is expected in 2001.
IAS 2	Inventories	Corresponds to RR 2. The LIFO method is not allowed in Sweden.
IAS 7	Cash flow statements	Corresponds to RR 7.
IAS 8	Extraordinary items	The recommendation on accounting for extraordinary items is RR 4. Accounting for changes in accounting is dealt with in RR 5. The two recommendations correspond to IAS 8.
IAS 9	R&D	Corresponds to R 1.
IAS 10	Contingencies	There is no similar Swedish recommendation. Contingencies will be part of a new recommendation based on IAS 37, Provisions, effective 1 January, 2001.
IAS 11	Construction contracts	A recommendation based on IAS 11 takes effect on 1 January, 2001.
IAS 12	Income taxes	A recommendation based on IAS 12 takes effect on 1 January, 2001.
IAS 14	Segments	Corresponds to R 9.
IAS 16	PPE	A recommendation based on IAS 16 takes effect on 1 January, 2001.
IAS 17	Leases	Corresponds to RR 6.
IAS 18	Revenue	A recommendation based on IAS 18 takes effect on 1 January, 2001.
IAS 19	Retirement benefits	A recommendation based on IAS 19 is expected during 2001.
IAS 20	Government grants	Corresponds to R 5.
IAS 21	Foreign exchange	Corresponds to RR8.
IAS 22	Business combinations	Corresponds to RR 1.
IAS 23	Borrowing costs	A recommendation based on IAS 23 is expected 2001.
IAS 24	Related parties	A recommendation based on IAS 24 is expected in 2001.
IAS 25	Investments	There is no similar recommendation.
IAS 26	Benefit plans	There is no similar recommendation.
IAS 27	Consolidated statements	Part of IAS 27 is covered by RR 1.
IAS 28	Associated companies	A recommendation based on IAS 28 takes effect on 1 January, 2001.
IAS 29	Hyperinflation	There is no similar recommendation.
IAS 30	Disclosures in banks	There is no similar recommendation. Most of the requirements are, however, dealt with in regulations from the Swedish Financial Supervisory Authority.
IAS 31	Joint ventures	A recommendation based on IAS 31 takes effect on 1 January, 2001.
IAS 32	Financial instruments	A recommendation based on IAS 32 is expected in 2001.
IAS 33	Earnings per share	A recommendation based on IAS 33 takes effect on 1 January, 2001.
IAS 34	Interim reporting	A recommendation based on IAS 34 takes effect on 1 January, 2001.
IAS 35	Discontinuing operations	A recommendation based on IAS 35 takes effect on 1 January, 2001.
IAS 36	Impairment	A recommendation based on IAS 36 takes effect on 1 January, 2001.
IAS 37	Provisions	A recommendation based on IAS 37 takes effect on 1 January, 2001.
IAS 38	Intangibles	A recommendation based on IAS 38 takes effect on 1 January, 2001.
IAS 39	Financial instruments	There is no similar recommendation.

deletions from the original texts. Still, it remains to be seen whether the council will succeed. An even greater question is whether Swedish companies are prepared for the fundamental change caused by introducing a very strict accounting regulation in such a short time.

The statutes of the Swedish Financial Accounting Standards Council state that its recommendations apply to public companies, i.e. basically listed companies. In a foreword to the recommendations, the view is expressed that the general principles of recognition and valuation should be applied also by non-public companies. The Swedish Institute of Authorized Public Accountants has expressed the opinion that, for example, RR 1, the first recommendation issued by the council, should be applied by all companies.

An interesting problem arises from the fact that the council's recommendations are not mandatory. The council can bring no sanctions to bear. So far, most public companies seem to have adopted the recommendations, including one that requires goodwill to be amortized over a period not exceeding 20 years. This is partly explained by the listing agreement signed by all listed companies. In the listing agreement companies declare that they will follow all recommendations published by the council unless there are compelling reasons for not doing so. Thus, one can say that the listing agreement contains a true and fair override in relation to recommendations issued by the council but not to requirements in the Annual Accounts Act.

2.2 Emerging Issues Task Force (EITF)

The Emerging Issues Task Force, EITF, was set up in 1995. The EITF has five members, four of which are members of the SFASC. The objective of

Table I.7 Recommendations and ongoing projects of the Swedish Financial Accounting Standards Council

Recommendations	
RR 1	Business combinations and group accounts
RR 2	Accounting for inventories
RR 3	Accounting for receivables and liabilities with special interest rate conditions
RR 4	Extraordinary items
RR 5	Changes in accounting principles
RR 6	Leases
RR 7	Cash flow statements
RR 8	Foreign currency translation
RR 9	Income taxes
RR 10	Construction contracts
RR 11	Revenue
RR 12	Property, plant and equipment
RR 13	Associated companies
RR 14	Joint ventures
RR 15	Intangible assets
RR 16	Provisions
RR 17	Impairment
RR 18	Discontinuing operations
RR 19	Earnings per share
RR 20	Interim reporting
Ongoing projects	
• Pensions	
• Presentation of financial statements	
• Associated companies	
• Financial instruments – disclosure and presentation	

the EITF is to help promote good accounting practice by giving interpretations of the recommendations issued by the SFASC and by issuing statements on accounting issues which are not covered by any recommendation. The objective is also to work closely with the Standing Interpretations Committees set up by the IASC. Table I.8 lists some of the statements that have been issued.

A question which has not been resolved is the status of the EITF statements. It is expected that, when the SFASC publish a recommendation based on IAS 1, if a company states that it complies with recommendations published by the SFASC, compliance would have also to include statements from the EITF.

2.3 Accounting Committee of the Swedish Institute of Authorized Public Accountants

The main standard setting body until the advent of the council was the Accounting Committee

Table I.8 Statements issued by the Emerging Issues Task Force

- Accounting for zero-coupon bonds
- Stock appreciation rights
- Accounting for changes of the accounting year
- Expenses in relation to issues of shares
- Change in pension liability due to a change in actuarial assumptions
- Provisions for the year 2000
- Capitalization of interest costs
- Group contributions
- Classification of a lease contract

(*Redovisningskommittén*, AC) set up by the Swedish Institute of Authorized Public Accountants (*Föreningen Auktoriserade Revisorer*, FAR). The committee was formed in the late 1940s. Over the years some 20 recommendations (see Table I.9) have been published, together with a number of interpretations and pronouncements.

The Accounting Committee consists of a small number of leading public accountants. It used to publish what was called an exposure draft that was sent out to members of the Institute. After taking into consideration the comments received, the committee would publish a proposed recommendation (*förslag till rekommendation*) and the members were again asked to respond. The wording could be changed again before the board finally confirmed the recommendation. The idea of publishing both draft recommendations and proposals for recommendations was to give companies adequate time to change their accounting policies in line with the recommendations.

The recommendations were highly regarded throughout the 1970s. In the 1980s, however, partly as a result of a revival of interest in the stock market, the standing of the recommendations was gradually eroded. An early instance concerned accounting for sale and lease-back transactions. In the early 1980s the Accounting Committee published an exposure draft suggesting that such transactions should be accounted for as a loan if that was the substance of the transaction. Very few companies were in favour of this, and the exposure draft was later withdrawn.

In the late 1980s, the Institute looked for different

Table I.9 Recommendations, proposals and drafts by the Swedish Institute of Authorized Public Accountants

FAR 1	The Annual Reports of Limited Companies
FAR 2	Valuation of Inventories (replaced by RR 2)
FAR 3	Accounting for Tangible Assets
FAR 4	Accounting for Pension Provisions
FAR 5	Accounting for Credit Agreements (Bank Overdrafts and Construction Credits)
FAR 6	Accounting for Factoring
FAR 7	Accounting for Leasing Agreements
FAR 8	Accounting in Limited Companies for Credit Granting and Commitment Obligations to Related Parties
FAR 9	Accounting for Pledged Assets
FAR 10	Cash Flow Statements
FAR 11	Business Combinations (withdrawn and replaced by RR 1)
FAR 12	Accounting for Shares and Participation
FAR 13	Accounting for Extraordinary Items

- Translation of Foreign Subsidiaries Financial Statements for inclusion in Consolidated Financial Statements (proposal)
- Accounting for Investments in Associates (draft)
- Accounting for Standardized Options, Forward Agreements, Interest Rate Swaps and Currency Swaps (draft)
- Current Cost Accounting (draft)
- Balance Sheet for Liquidation Purposes (draft)

ways in which to rebuild confidence in the accounting recommendations. What finally emerged was a decision to work with representatives of government and industry and set up the Swedish Financial Accounting Standards Council (*Redovisningsrådet*). It was thought that enlarging the range of interests that had an active voice in standard setting would lead to more widespread compliance.

The Accounting Committee ceased to publish recommendations in 1987 although it is still in existence. Among other things, it is responsible for drawing up comments on exposure drafts from the SFASC and from the IASC. The Accounting Committee has also recently published guidelines for companies drawing up financial statements according to the new Annual Accounts Act (FAR, 1996).

2.4 Accounting Standards Board (ASB)

The Accounting Standards Board (*Bokföringsnämnden*) is a government body, set up in 1977. Its task was to stimulate the development of 'good accounting practice' in companies' bookkeeping and in their published accounts. An interesting aspect of its membership in the early years was a strong trade union presence. It was thought that union representation on the boards of the larger companies would lead to demands for more accounting information. As it turned out, apart from a few individuals, the trade unions showed no great interest in the work of the board. Their representation has since been reduced.

At the request of companies and the tax authorities the board has published a large number of statements and recommendations on bookkeeping and published accounts. During the 20 or so years of its existence, the board has worked in tandem with the Accounting Committee and later the SFASC. How have the different organizations determined what should be on their agenda?

From the outset, the board made it clear that it did not see itself as competing with the Accounting Committee. The board kept a rather low profile, concentrating mostly on bookkeeping issues. There has been one major exception. In the late 1970s the Accounting Committee adopted an exposure draft dealing with the valuation of receivables and payables in foreign currencies. The exposure draft was based on the traditional principle of prudence but with one departure. Unrealized gains were to be recognized to the extent that they could be netted against unrealized losses. The board thought that the netting proposed by the Accounting Committee was unnecessarily complex. Influenced, *inter alia*, by the US FAS 8 on foreign currency translation, the board put out an exposure draft of its own, according to which receivables and payables in foreign currency would be translated at the balance sheet rate. During the exposure process, however, it became evident that the approach was too radical to win the necessary support. Many companies were against it, fearing that unrealized gains would be included in taxable profit. Ten years later the time was more propitious. The board issued R 7, Translation of Receivables and Payables in Foreign Currency (for details see

VII.2.8). R 7 has since been superseded by RR 8 issued by the SFASC. Most of the ASB's other recommendations deal with less contentious issues. Table I.10 provides a list of the recommendations issued by the ASB. Apart from R 7, three more recommendations are likely to be superseded by recommendations issued by the SFASC. The three are R 1 on R&D (to be covered by a recommendation on Intangibles), R 5 on Government grants and R 9 on Segment reporting.

Probably more important than the recommendations have been the pronouncements issued by the board. As a governmental body, one of the ASB's prime responsibilities is to explain how 'good accounting practice' should be interpreted in particular cases, especially tax cases before the courts.

The future of the board is still under discussion. It is expected that the board will take prime responsibility for ascertaining that accounting regulation functions properly. Thus, the ASB will be given an oversight role. The ASB is also expected to be responsible for taking actions against non-listed companies whose accounts do not follow good accounting practice. (For listed companies the responsibility falls primarily on the Stock Exchange.) Finally, an important task will be to determine to which extent recommendations issued by the SFASC shall apply to non-public companies. For example, it is widely believed that not all of the disclosure requirements that are of interest to the capital markets are relevant for non-public companies. The ASB may also be of the opinion that some recommendations are too complex to be applied by smaller companies. The recommendations on tax accounting and leasing are obvious examples.

According to a recently published statement (January 1999) the Board has stated that, in the future, it will make public to what extent recommendations published by the Swedish Financial Accounting Standards Council apply to other than listed companies.

Table I.10 Recommendations issued by the ASB

R 1	Accounting for Research and Development Costs
R 2	Joint Verifications
R 3	Accounting Verifications
R 4	Disclosure of the Number of Employees and their Remuneration
R 5	Accounting for Government Grants
R 6	Accounting for Invoices
R 7	Translation of Receivables and Payables in Foreign Currency
R 8	On the Legal Requirement for Annual Reports
R 9	Segment Reporting
R 10	On the Documentation of Accounting Systems
R 11	Bookkeeping Requirements for Sole Proprietorships and Individuals

2.5 Industry Committee on Stock Exchange Affairs

An early contribution to the development of financial reporting was made by the Industry Committee on Stock Exchange Affairs (*Näringslivets Börskommitté*, NBK). In the 1960s the committee was instrumental in improving disclosure of what had previously been hidden reserves. The layout of the financial statements that was eventually adopted in the Accounting Act of 1976 was first proposed by the committee.

In 1983–84 the committee made an attempt to introduce a new method of accounting for income tax, based on US principles. A new format was also suggested for the income statement. Despite strong support from leading figures in trade and industry, the attempt failed. It proved to be the end of the committee's work in the field of accounting.

2.6 Stock exchange

The Stockholm Stock Exchange has taken an active interest in promoting IAS for public companies. To the Stock Exchange it is important that Swedish companies follow international accounting principles in order to encourage international investors to trade in the Swedish market. Therefore, the Stock Exchange has become a member of the board of the SFASC. The Stock Exchange is also a major contributor of funds to the council.

The Stock Exchange monitors compliance with the recommendations issued by the SFASC. If the Stock Exchange becomes aware of a possible case of non-compliance it refers the case to a small group of accounting experts. If this group of experts agrees that there is a problem, the case is brought to a disciplinary committee set up by the Stock Exchange. If the committee reaches the conclusion that the financial statements are misleading the company is fined. In extremely rare cases the company may be de-listed.

II Forms of business organization

1 Commonly used legal forms of business organization

The dominant means of conducting business is through a limited liability company (*aktiebolag*). The minimum share capital is SEK 100,000.

There is no legal distinction between private and public companies except that only the latter may offer its shares to the general public. Thus there is only one form of limited liability company. All limited liability companies have to file their accounts with a registrar. They are all also required to be audited by an authorized or approved public accountant. However, for accounting purposes there is a difference between private and public companies in so far as the recommendations issued by the Swedish Financial Accounting Standards Council, at least when it concerns disclosure, apply only to public companies, i.e. primarily listed companies.

Limited liability companies have a one-tier board. Employees have the right to appoint members to the board. Where the average number of employees exceeds 25, employees can appoint two board members. If the number of employees exceeds 1,000 they can appoint three. There is no limit on the number of board members the shareholders may appoint.

The board puts to the annual general meeting a proposal for the dividends to be paid out. The amount it proposes may be reduced, but, except for some rare occasions, the general meeting cannot vote for a higher dividend than the board has recommended. Dividends are paid once a year, after the annual general meeting.

The second most important form of business organization is the partnership. There are two types of partnerships: general commercial partnerships (*handelsbolag*) and limited partnerships (*kommanditbolag*). A general commercial partnership is a legal entity but the partners are jointly and severally liable towards third parties for the partnership's obligations. A limited partnership is characterized by the limitation of liability in respect of one or more, but not all, of the partners. Companies, as well as individuals, may be partners in both types of partnership. Partnerships are entered in a local commercial register kept at the county administration office (*länsstyrelsen*). The register is public.

Another form of business entity is the co-operative (*ekonomisk förening*). These are quite important, especially in agriculture, forestry and the retail sector. The consumers' co-operatives form one of the largest industries in Sweden.

A co-operative is an economic society formed to further the interests of the members by economic activities in which the members take part as customers or suppliers. Co-operatives are legal entities and the members are not personally liable for the co-operative's obligations. The county administration office keeps a register of all co-operatives. The register is public.

A business entity could also be run as a sole proprietorship. A sole proprietorship is not a legal entity.

According to the *Statistical Yearbook of Sweden 1997*, in 1995 approximately 2.1 million people were employed in the private sector in one or other of the business forms mentioned above. In addition, 1.3 million people were employed in the public sector (see Table II.1).

2 The Swedish securities market

The Swedish securities market consisted primarily of the Stockholm Stock Exchange with its main activity in equity trading and OM Stockholm which organizes trading in derivatives. Early in 1998 the two exchanges merged, creating a joint equities and derivatives exchange called Stockholm Stock Exchange. In addition there are a few very small markets where local shares are traded.

Table II.1 Legal form of business in Sweden

Legal form of business	No. of enterprises (thousands)	No. of employees (millions)
Sole proprietorship and individuals	226	40
Co-operatives	15	66
General commercial partnerships	77	35
Limited liability companies	210	1,908
Other forms (e.g. foundations)	29	100
Private sector total	557	2,149
Public sector	2	1,251

Source: Statistika Centralbyrån (1997).

The shares of 100 companies were listed on the Stockholm Exchange at 31 December 1997, with a market value of SEK 1.838 billion. In addition, 59 companies were listed on the over-the-counter market and shares in a further 102 were traded on an unofficial basis. The latter group comprises companies that may apply for a full listing when their record permits them to.

Turnover on the Stockholm Stock Exchange increased dramatically in the 1990s following the elimination of a turnover tax in December 1991. Measured by turnover, the Stockholm Stock Exchange is currently around the tenth largest in the world.

Not a great many foreign companies are listed on the Stockholm market. The largest, measured by trading volume is the Finnish company Nokia. Autoliv and Pharmacia & Upjohn, two companies that used to be Swedish but that are now domiciled in the United States, are also heavily traded.

Thirty-one Swedish companies' shares are traded on foreign stock exchanges, most of them in London. The securities most actively traded abroad are those of Ericsson, Astra, SKF and Volvo.

III Objectives, concepts and general principles

1 Objectives of financial reporting

As stated above (see I), the main objective of accounting legislation is to safeguard creditors. This is accomplished through prudent valuation of assets and liabilities and by restricting the amount of profit that is available for the payment of dividends. In addition, financial reporting should provide information to other interested parties, e.g. shareholders, employees and the fiscal authorities.

1.1 Distributable profits

A central section in the Companies Act (ABL) deals with the definition of the amount that may be distributed to the shareholders (distributable profits). The amount that can be distributed is made up of two parts:

- The reported net profit, less obligatory transfers to reserves, if any.
- The non-restricted reserves or retained profits.

Thus a company can make a distribution out of retained profits even if there is a net loss for the year.

For a group, how much may be distributed is determined as the lesser of the amounts that are available according to the parent company's accounts and the group accounts (ABL chapter 12 § 2).

Most companies find that the amount that can be distributed is less than the true profit. The reason is the existence of tax-based reserves. As will be explained below (see III.3), a company can allocate part of its profits to what are generally known as untaxed reserves. Any such allocation has to be accounted for as a charge in the income statement of the legal entity if it is to be tax-deductible. Consequently the reported net profit, and hence distributable profit, is significantly reduced.

In bad years the opposite is true. A loss-making company would, in theory, be able to pay a dividend even if there were no retained earnings. What the company would have to do would be to release some of its untaxed reserves. Because of the open nature of these untaxed reserves in the balance sheet and the appropriations in the income statement there is little risk that a user of the financial statements would be misled. The fact that they are unconcealed means that there is a big difference from the hidden reserves that are generally believed to exist in, for example, some German companies.

The requirement that distributable profits should be the lesser of what is available according to the parent company and the group accounts has another important implication. There is an increasing tendency to adopt different, i.e. less prudent, accounting principles in the group accounts. One example is equity accounting, which is not permitted in the individual accounts. As a result, if the share of the profits of the associated company exceeds the amount paid out in dividends, this difference could not normally be distributed and is therefore included in the group's restricted equity.

1.2 Financial reporting for shareholders

Information for the capital markets became very important during the 1980s as a result of the stock market boom. Most of the developments regarding information about the true profit took place without benefit of guidance from standard setting bodies. The listing requirements did not impose specific measurement rules, either. As a result a wide variety of accounting principles came to be used by quoted companies. In the 1990s, however, the trend was to harmonize accounting principles. This trend is expected to continue as the Financial Accounting Standards Council continues issuing new recommendations.

Another important factor in explaining the interest from companies in providing the capital markets with relevant information is the effort by the Stock Exchange to encourage companies to be more open. The Stock Exchange is one of the sponsors of a yearly competition for the best annual report. The award is presented on a day when invited speakers comment on trends in financial reporting and on examples of good (and not so good) accounting practices.

An important factor is also that leading Swedish companies, Astra, Ericsson, Volvo, etc. have outgrown the relatively small Swedish capital market. Thus, they have had to raise capital in foreign markets. In order to compete for capital these companies have had to provide information that is at least as good as that provided by their major competitors. By doing that, they have come to be seen as good examples inspiring other companies to provide better information.

1.3 Financial reporting for employees

Financial reporting for employees is not a major issue. However, in the 1970s, when the Companies Act (ABL) was passed, employees were given the right to appoint two or more board members in most companies. At the time it was thought that employees and the trade unions would take a much more active interest in the disclosure of financial information. This proved not to be the case.

There have been a few instances of companies printing a condensed version of their annual report especially for their employees. Such cut-down versions tend to use simpler language, with more illustrations and fewer technical notes. They are thus more of a simplified annual report than a true employee report.

The information in a typical annual report normally includes some information about employees. In addition to standard information such as the number of employees and levels of remuneration, companies may disclose staff turnover, absenteeism rates, costs of training, etc. A few companies also publish a value-added statement, which can be seen as a means of stressing the importance of the employees in the creation of wealth (this is dealt with further in section IX.2.3).

2 Concepts and general principles

The Annual Accounts Act lists a number of principles to be followed in drawing up the financial statements. Most of them are too general to have a direct effect on the accounting principles a company adopts. Even the concept of prudence, which most people would argue is the cornerstone of Swedish accounting, is at times being questioned (Artsberg, 1992). One reason is, of course, the growing importance of standards issued by the International Accounting Standards Committee.

Artsberg argues that, when examining the recommendations of the now defunct Accounting Committee between 1960 and 1987, a clear trend can be discerned away from prudence towards matching. If one were to examine the accounting principles actually followed by companies, the trend would probably appear even more pronounced.

Unlike the Accounting Act from 1976, the new act requires financial statements to show a true and fair view. They shall also be based on good accounting practice. Wherever there is no guidance in the act, accounts should be based on good accounting practice. In the preparatory work on the Annual Accounts Act, the concept of good accounting practice is described as the accounting principles used in practice by a qualitatively representative circle of companies. Even if recommendations issued by standard setting bodies are said to be of great importance to the interpretation of good accounting practice, companies still have a great deal of freedom in this respect.

There has been some discussion on the relationship between true and fair and good accounting practice (e.g. see Thorell, 1996). One view is that good accounting practice is more general in that it relates to legislation, accounting recommendation and established practices. The true and fair view, on the other hand, relates more to the circumstances in the individual company.

The difference should, however, not be exaggerated. In most cases, what is true and fair is determined by good accounting practice, in particular as the true and fair override only permits companies to depart from recommendations issued by standard setting bodies. It is not permitted to depart from requirements in the Annual Accounts Act.

3 Interrelationship of financial reporting and tax accounting

There is a strong link in Sweden between financial reporting (in individual accounts) and tax accounting. According to the major tax code, tax should be assessed on reported profits unless there are specific provisions in the Tax Code that state otherwise. A normal tax return from a company thus starts with the reported net profit. To this figure are added back those items in the income statement which are not taxable/deductible (see VII.1). Finally, an adjustment is made for items that are taxable/deductible even though they do not appear in the income statement for the purposes of arriving at taxable income.

The reason for having a strong link between financial reporting and tax accounting seems to be mostly pragmatic. It is widely assumed that, if there is only one set of rules, bookkeeping will be simplified. According to the head of the tax department of KPMG in Sweden, there was also a trend in the 1970s and 1980s to narrow the gap between taxable profit and the true profit. The most obvious example is that most untaxed reserves have been eliminated.

Because of the link between financial reporting and tax accounting, companies that want to claim a deduction for tax purposes normally also have to charge it in the income statement. One common example is accelerated depreciation. For tax purposes, any item classified as machinery may be depreciated over five years even though the economic life may be estimated at 20 years or more.

Assume a company has bought an aeroplane for

SEK 100 million. The company uses straight-line depreciation over 20 years. Normal depreciation will then be SEK 5 million each year. For tax purposes the deduction for depreciation is SEK 20 million. In the income statement both the normal and the extra depreciation will be shown, as follows:

Income before normal depreciation	SEK 75
Normal depreciation	−5
Income after normal depreciation	70
Extra depreciation for tax purposes	−15
Profit before tax	SEK 55

The extra depreciation for tax purposes in the profit and loss account is normally shown under a heading 'Appropriations' together with other items of a similar nature. In the balance sheet there will be a corresponding set of accounts under the heading 'Untaxed reserves'. This means that only the normal depreciation reduces the asset value in the balance sheet.

Of course, appropriations do not always reduce profit. Profit will sometimes be increased, perhaps to compensate for a poor trading result. In the example above, a company could reverse the extra depreciation in order to show a higher profit before tax. Appropriations therefore help the company to smooth reported net profits. This is stated as one of the purposes of the untaxed reserves. Being able to smooth reported profits, it was thought, would give companies a better chance of surviving an economic downturn. Another important argument is that untaxed reserves could be seen as an alternative to a loss carry-back system – a system that at present is not permitted in Sweden. This argument has less force today than it used to. Before 1990 Swedish companies had to claim losses within a six-year period in order to qualify for tax relief. Untaxed reserves could be used to extend this period by releasing reserves built up in profitable years. Nowadays companies are allowed to carry tax losses indefinitely. Partly because of this change, the ability of Swedish companies to create untaxed reserves has been drastically reduced.

From 1994 and onwards only two kinds of untaxed reserves are allowed that are material. The first and by far the most important is excess depreciation. For fiscal purposes most assets, except buildings, are depreciated over five years. As the economic life may be 20–30 years, as in the case of ships, printing equipment, etc. capital-intensive companies have access to extremely generous interest-free tax credits. The other kind of untaxed reserve that will be found in almost all companies is the allocation reserve (*periodiseringsreserv*). This reserve was introduced through legislation in December 1993. All companies are allowed to deduct, each year, 20 per cent of their net profit before tax and to allocate that amount to this reserve. The deduction is tax deductible. After five years the reserve has to be released. The amount released from the reserve increases net profit before tax and consequently taxable profits. It has been estimated that the economic benefit of the allocation reserve approximately corresponds to a reduction in the corporate tax rate from 28 to 25 per cent.

Anybody familiar with Swedish financial statements will know that the appropriations have to be disregarded in order to arrive at a meaningful profit figure. As the appropriations are normally shown clearly this does not give rise to any problems. The advantage of this method of reporting is that, despite the need to use the financial reports for tax accounting, it is still possible to arrive at a profit figure as if the link did not exist.

In spite of the appropriations, some differences can still exist between the reported profit and taxable income. The differences can be permanent or temporary. Permanent differences are those where some items are not taxable/deductible. One of the more important examples is dividend income, which is not taxable if the company receiving the dividend owns more than 25 per cent of the shares of the company paying it. The reasoning behind this is probably similar to the rationale for requiring the equity method for associated companies. If a company holds more than 25 per cent of the shares of another company, the fiscal authorities assume that the holding is part of the normal business rather than a financial investment. As the company paying the dividend has already paid tax on it, the exemption for the company receiving the dividend eliminates the double taxation that would otherwise occur.

One other example of a temporary difference is a write-down, for example, of shares. For tax purposes a loss often has to be realized before it can be recognized for tax purposes. A write-down in the accounts would thus have to be added back when calculating taxable income. This is typically the case with investments in shares. If a company buys shares that subsequently fall in value, the book value has to be written down. The write-down is not, however, recognized for tax purposes. Only when the shares are sold and the loss is realized will the company be able to claim a tax deduction. Provision for restructuring costs is another example of an item which is tax-deductible only in special circumstances. These and other examples are examined below (see VII).

As will be seen when we come to consider consolidated accounts, the last few years have seen a major change in group accounting. According to a recommendation of the SFASC, public companies should abandon the use of appropriations and untaxed reserves in their group accounts. Instead, reported tax expense in the group accounts should be based on comprehensive tax accounting.

An interesting question is the extent to which financial reporting is tax-driven. One would assume that because of the strong link between financial reporting and tax accounting the answer would be that there is a strong tax bias. In practice it is not so simple. Firstly, the separate disclosure of tax-based appropriations eliminates some of the distortions. Secondly, one has to distinguish between individual and group accounts. Since only the former are used for tax purposes, there is some scope for reducing the tax bias in the latter by using different accounting principles in the group accounts compared with the individual accounts. This is discussed further under Sweden – Group Accounts.

Among accountants, opinion is split as regards the desirability of maintaining the strong link between financial reporting and tax accounting. Some argue that the link is beneficial (Tidström and Hesselman, 1991). This view stresses the advantages of being able to work within one set of accounts. Severing the link might create a double set of accounts. This could be a problem, especially for the larger companies. Thus one of the proponents of retaining a single set of rules is the chief finance officer of one of the larger companies in Sweden, Saab-Scania, at the same time a member of the SFASC. Where smaller companies are concerned it seems reasonable to assume that there is only need for one set of accounts, i.e. tax-based, regardless of whether a strong link exists or not.

Other accountants believe that the present link makes it difficult to introduce new accounting standards that would increase net profit and thus taxable profit (Johansson and Östman, 1985). Examples include accounting for construction contracts, capitalization of interest and lease accounting. The SFASC has come to realize that for a standard-setting body with no sanctions to apply, it would be unwise to ask for a change in accounting principles that would cost companies (tax) money.

Therefore, the tax link is a matter of great concern to the council. The council would like to issue standards without having to worry about the tax consequences, and has lobbied in an attempt to change the act. The result, so far, is negative. An association for smaller companies have argued against such a change, believing that it would put an extra burden on their accounting. As a consequence, the council has found it necessary to issue recommendations that differ between group and individual accounts – reluctantly, because there seems to be general agreement that it is preferable for the same accounting principles to apply both to group accounts and to the accounts of individual companies. If both sets of accounts are claimed to reflect good accounting principles, it is difficult to see how different accounting principles can be equally good in the same annual report.

In Sweden, the taxable entity is the individual company. Through group contributions (*koncernbidrag*) it is possible for a group to achieve almost the same result as if the group was the taxable entity. This is further described in VII.1.

IV Bookkeeping and preparation of financial statements

1 Duty to prepare financial statements

Virtually all companies have to prepare financial statements. The only exceptions are partnerships and sole proprietorships where:

- The average number of employees is less than 10.
- If the company is the parent company of a group, the average number of employees in the group must be less than 10.
- The assets are less than 1,000 basic amounts. One basic amount in 1998 was SEK 36,400.

Financial statements are normally public. A copy has to be filed in the register (normally the Patent and Registration Office, which is located in Sundsvall) no later than seven months after the balance sheet day. Filing is not a requirement for partnerships and sole proprietorships with:

- Fewer than 200 employees.
- Assets less than 1,000 basic amounts.

If the company is a parent company, both limits must also be true for the group, otherwise the exemption ceases.

1.1 Companies limited by shares and co-operatives

The annual report of a limited liability company and of a co-operative consists of a balance sheet, an income statement and a management report. The balance sheet and the income statement should cover at least two years. In larger companies the annual report should include a cash flow statement. A company is defined as large if:

- The number of employees exceeds 200.
- The assets exceed 1,000 basic amounts.
- The company's shares or promissory notes are listed on the stock exchange or an authorized market.
- The company is the parent company of a group to which the first two points apply.

Large companies also have to publish at least one interim report. The report should cover not less than six months and not more than eight. Starting in 1999 listed companies have to publish quarterly reports.

All parent companies (except parent companies in subgroups) must, in addition, prepare a balance sheet and an income statement for the group and – depending on their size – a cash flow statement as well.

1.2 Partnerships and sole proprietorships

The annual report consists of a balance sheet and an income statement. Larger companies, i.e. companies which have to make their financial statements public (see IV.1.1), also have to include a management report and a cash flow statement. Companies of this type also have to prepare an interim report just like limited liability companies.

If a partnership or a sole proprietorship is a parent company and the average number of employees in the group is at least 10, then the same rules apply to the group accounts.

2 Bookkeeping and charts of accounts

2.1 Formal aspects of bookkeeping

The rules that govern bookkeeping are to be found in BFL §§ 4–10. The act requires continuous recording of transactions based on supporting documentation and a retention period for all such records of 10 years. The government is working on a revision of the act. One of the more important changes that have been proposed is that Swedish companies will be allowed to do the bookkeeping outside Sweden. At present, this is not allowed.

More detailed guidance on some of the requirements has been issued by the ASB. The board has published statements on what is meant by supporting documentation, how the archiving of paperwork can be organized, etc. (see Table IV.1).

Table IV.1 Recommendations and standards by the ASB on formal aspects of bookkeeping

Recommendations	
R 2	*Joint Verification.* Classifies the circumstances in which joint verification may be used.
R 3	*Accounting Verification.* Defines accounting verification and explains how documentation should be retained.
Statements	
U 89:5	*When should a transaction be accounted for in the book of first entry?* Normally, the next day. Only in exceptional cases is a delay of more than a month acceptable.
U 89:13	*How should purchases from a cash-and-carry outlet be verified?* A fairly detailed specification is required.
U 90:2	Categorizes when a company should account for receivables and payables.

2.2 Charts of accounts

Swedish companies have a long tradition of using charts of accounts. The first major industry-wide chart was the so-called M-Plan (*Mekankontoplanen*), published in 1945. The M-Plan was designed for companies in the engineering industry. Its purpose was to help companies develop a reporting system that could be used for both internal and external purposes.

The initiative was taken by the Federation of Engineering Companies. Hence, there was no governmental interference. Use of the plan was voluntary. As it was widely regarded as a significant improvement over earlier accounting systems, the M-Plan gained acceptance among almost all companies of any size and was even adopted in other industries.

With the introduction of computers, enthusiasm for the M-Plan subsided. Instead there was a demand for a more flexible chart of accounts that would better suit the needs of smaller companies. In 1976 the BAS-Plan was introduced to meet these demands (BAS 76, 1976). The BAS-Plan was developed jointly by the Federation of Swedish Industries and the Swedish Employers' Confederation. It is based on the following general principles:

- Transactions are recorded according to the rules of double-entry bookkeeping in order to facilitate reconciliation.
- The BAS-Plan concentrates on financial reporting for external purposes but can also be used for internal purposes.
- The revised layout of the balance sheet and the income statement complies with the Annual Accounts Act of 1995.
- Accounts are included to facilitate the production of mandatory statistical records.
- The plan is easy to use either with computers or manually.

BAS 76 uses nine classes of accounts:

1 Assets.
2 Liabilities and owners' equity.
3 Revenues/income.
4–7 Expenses/expenditure.
8 Financial items, extraordinary items, appropriations, tax.
9 Other (to be used for internal reconciliation, for example).

Within each class there is a list of all the possible accounts that a company might need to use. Companies may select the accounts they think useful. No specific valuation rules are included in the plan. Companies are assumed to follow good accounting practice.

The BAS-Plan has been slightly modified to fit in with the requirements of different industries. Most accounting systems supplied by software companies are now based on the plan. As a result the BAS-Plan is used by nearly all companies.

The tax authorities also have a vested interest in standardizing the way the accounts are organized so as to simplify the tax audit. In 1990 it was decided that all companies should provide the tax authorities with a detailed schedule of their assets and liabilities, revenue and costs in the form of a standardized excerpt from the accounts (*standardiserat räkenskapsutdrag*) following the BAS-Plan. The system became compulsory in 1992.

In all, 140 items have to be disclosed. The information may be submitted on a specially designed form or on floppy disk. Companies that submit the standardized excerpt from the accounts are no longer required to file a copy of their annual report with the tax authorities.

3 Inventory

The basic rules for taking inventory are set out in the law on stocktaking for tax purposes (1955: 257). The law requires each item of inventory to be listed and valued at the lower of cost or market. The items that are kept in the inventory are assumed to be the most recently acquired (the FIFO principle). If the inventory is not taken at the balance sheet date there has to be a system for continuous updating of the inventory records.

The list of inventory has to be signed and assurance is required that no items have been omitted.

Some of the problems with stocktaking have to do with how an item is defined and whether a certain item belongs to the inventory or not. As to the definition of an item some guidance is given in a recommendation from the SFASC. According to the recommendation different items may not be grouped except for homogenous goods such as raw materials (RR 2, para. 31). In the terms of the recommendation an item has to be seen as very narrowly defined. Since the lower of cost or market principle has to be applied to each and every item, the effect can be significant.

Sometimes it is not easy to say whether an item should be included in the inventory, e.g. when it is being moved to or from a warehouse. The criteria proposed in the recommendation of the SFASC (RR 2, para. 32) are based on risk and reward. Any given item should be included in the inventory of whoever bears the risks and reaps the rewards attached to it.

The question of which party is entitled to carry any particular item in its inventory was of much greater concern before 1990 than it is today. Before 1990 companies were allowed to write down the value of the inventory by 40 per cent for tax purposes. Thus the higher the inventory the higher the tax credit associated with the write-down. As was mentioned earlier (see III), there have been several changes in the Tax Code, reducing most of the tax-based appropriations. One of the changes was the abolition of the inventory write-down. Under the present rules, the main principle is the lower of cost or market. However, there is an alternative rule. Companies may, if they prefer, value their inventory at 97 per cent of cost. For a company with a low rate of obsolescence the alternative rule will yield a lower inventory value and, consequently, a lower tax charge.

A problem shared by many companies is how to value surplus inventory. An example may be an inventory of spare parts for machinery which is no longer being produced. Still, the company may decide to keep the inventory in order to be able to service its customers. Often, the inventory could last for many years. In spite of this, there is no obsolescence in the traditional sense. The spare parts are in good condition and there is no price deduction. Therefore, the tax authorities have been reluctant to accept a reduction of the cost price. In their recommendation on valuation of inventories, RR 2, the SFASC has, however, ruled that in cases like these, a company should value its inventory at the lower of cost and the present value of the estimated sales revenue. It is also acceptable to use simplified methods for calculating the value of inventory provided these are applied consistently and that they result in a value approximately equal to what would be achieved using a more refined method of calculation. An example of a simplified method may be a write-down of 10 per cent for one-year-old items, 30 per cent for two-year-old items, etc.

V Balance sheet and profit and loss account formats

1 General classification principles

Formats for the balance sheet and the profit and loss account are prescribed in the Annual Accounts Act, Chapter 3 and in Appendices 2–4. As to the balance sheet, the act provides for only one of the two formats that are allowed in the Fourth EU Directive, i.e. the one in Article 9 (horizontal format). Thus the act does not permit companies to use the format prescribed in Article 10 (vertical format) where current liabilities are shown as a deduction from total assets. For the profit and loss statement the act prescribes the two vertical formats taken from Articles 23 and 25 of the Fourth EU Directive. Companies with fewer than 10 employees may use a simplified layout provided that it conveys adequate information on revenues, costs, assets and liabilities. According to a recent proposal from the Ministry of Justice, a simplified layout is proposed also for companies with a turnover not exceeding SEK 100 million or assets not exceeding SEK 50 million.

In the annual report the balance sheet and the profit and loss account should include figures for at least two years. A few companies show figures for three years. Most of the latter companies are ones which have to file their accounts with the SEC in the United States.

A change in the classification of items in the financial statements must be applied retroactively in order to facilitate comparability. Changes in accounting principles may be applied retroactively (for a further discussion see VII.2.9).

In addition to the financial statements nearly all companies publish a five- or ten-year summary of the more important figures. These summaries are not mentioned in the act. It seems to be the exception rather than the rule for prior year figures to be restated, in spite of the frequency of changes in accounting principles. Consequently, there is a need for caution when analyzing the development of a company over time.

As for set-offs, the general rule is that they are not permitted.

2 Balance sheet

2.1 Regulation format

The format of the balance sheet is prescribed in the Annual Accounts Act (see Table V.I). In addition, the act lists a number of requirements that are mandatory for limited liability companies.

In certain industries it may be necessary to add other items. Construction companies, for example, normally include a heading for construction contracts.

In the balance sheet the least liquid assets come first, in line with the format used in the EC Directives. The principle is the same for equity and liabilities. In the 1976 Annual Accounts Act (BFL § 19) the order was the other way round. Thus, in Swedish financial statements before 1997 the balance sheet starts with the most liquid assets and with the most short-term liabilities.

2.2 Untaxed reserves

There is one group of items on the liabilities side in the individual accounts which is not normally encountered in other countries. It is headed 'Untaxed reserves' (see III.3). These reserves correspond to the appropriations which are shown in the profit and loss account.

For public companies, starting with 1992, untaxed reserves are eliminated in the group accounts. Seventy-two per cent of the reserves are included under equity, with the remainder being accounted for as a deferred tax liability. As will be seen when we come to group accounts, this change was a result of a recommendation from the Swedish Financial Accounting Standards Council (RR 1, Business Combinations and Group Accounts). One of the reasons for the change was to help foreign readers of Swedish financial statements to appreciate the true profit and financial position of companies.

In the individual accounts, untaxed reserves will be retained for the foreseeable future. In the notes, information should be given on the amount of deferred taxes related to the untaxed reserves. (See also VII.1.)

2.3 Asset movement schedules

According to Chapter 5, § 3 in the Annual Accounts Act, limited liability companies have to disclose acquisition cost, additions, disposals, transfers, accumulated depreciation, revaluations, write-downs, etc. The information should be presented for each fixed asset heading shown in the balance sheet. According to a statement by the Accounting Standards Board, all companies also must maintain a register of fixed assets (*anläggningsregister*) to enable the company to determine the residual value of any asset that is sold or scrapped.

Table V.2 illustrates a schedule of property, plant and equipment from the annual report of Volvo, 1997. In addition to the information provided in the group accounts, Volvo also gives information on the tax assessment value of land and buildings in Sweden in their parent-only accounts. This is an example of a rather outdated requirement. The intention was to provide readers with a figure that was supposed to reflect the true value of buildings, etc., better than the

Table V.1 Suggested format for the balance sheet

Called up company capital not paid	**Tecknat men ej inbetalt kapital**
Fixed assets	**Anläggningstillgångar**
Intangible fixed assets	**Immateriella anläggningstillgångar**
Research and development costs and similar items	Balanserade utgifter för forsknings- och utvecklingsarbeten och liknande arbeten
Concessions, patents, licences, trademarks and similar rights	Koncessioner, patent, licenser, varumärken samt liknande rättigheter
Rights of tenancies and similar rights	Hyresrätter och liknande rättigheter
Payments on account for intangible assets	Förskott avseende immateriella anläggningstillgångar
Tangible assets	**Materiella anläggningstillgångar**
Land and buildings	Byggnad och mark
Technical plant and machinery	Maskiner och andra tekniska anläggningar
Other fixtures and fittings, tools and equipment	Inventarier, verktyg och installationer
Payment on account for tangible fixed assets and tangible fixed assets in course of construction	Pågående nyanläggningar och förskott avseende materialla anläggningstillgångar
Financial fixed assets	**Finansiella anläggningstillgångar**
Participating interests in affiliated undertakings	Andelar i koncernföretag
Amounts owed by affiliated undertakings	Fordringar på koncernföretag
Participating interests in associated undertakings	Andelar i intresseföretag
Amount owned by associated undertakings	Fordringar på intresseföretag
Other investments and participating interests	Andra långfristiga värdepappersinnehav
Loans to partners and to persons related to partners	Lån till delägare eller till delägare närstående
Other amounts receivable	Andra långfristiga fordringar
Current assets	**Omsättningstillgångar**
Stocks	**Varulager m m**
Raw materials and consumables	Råvaror och förnödenheter
Work in progress	Varor under tillverkning
Finished goods and goods for resale	Färdiga varor och handelsvaror
Work in progress on behalf of third parties	Pågående arbete för annans räkning
Payments on account for goods	Förskott till leverantörer
Debtors	**Kortfristiga fordringar**
Trade debtors	Kundfordringar
Amounts owned by affiliated undertakings	Fordringar på koncernföretag
Amounts owned by associated undertakings	Fordringar på intresseföretag
Other amounts receivable	Övriga fordringar
Prepayments and accrued income	Förutbetalda kostnader och upplupna intäkter
Investments	**Kortfristiga placeringar**
Participating interests in affiliated undertakings	Andelar i koncernföretag
Own shares	Egna aktier
Other investments	Övriga kortfristiga placeringar
Cash at bank and in hand	**Kassa och bank**

book value. The tax assessment values are supposed to equal approximately 75 per cent of the market value at the time of the assessment. In practice, however, the relationship between tax-assessed values and fair values is so weak that tax-assessed values are seldom used by analysts.

More useful is the information given by Volvo on replacement costs and on depreciation based on replacement costs. This type of information is now rare among Swedish companies as is further discussed in section VIII.2.

Table V.1 (contd.)

Shareholders' equity	**Bundet eget kapital**
Restricted equity	**Bundet eget kapital**
Share capital	Aktiekapital
Share premium account	Överkursfond
Revaluation reserve	Uppskrivningsfond
Legal reserve	Reservfond
Non-restricted equity	**Fritt eget kapital**
Retained profit or accumulated losses	Balanserad vinst eller förlust
Net profit for the year	Årets resultat
Untaxed reserves	**Obeskattade reserver**
Liabilities	**Skulder**
Provisions	**Avsättningar**
Provisions for pensions and similar obligations	Avsättningar för pensioner och liknande förpliktelser
Provision for deferred taxation	Avsättningar för skatter
Other provisions	Övriga avsättningar
Long-term debt	**Långfristiga skulder**
Bond loan	Obligationslån
Amounts owed to credit institutions	Skulder till kreditinstitut
Amounts owed to affiliated undertakings	Skulder till koncernföretag
Amounts owed to associated undertakings	Skulder till intresseföretag
Other loans	Övriga skulder
Short-term debt	**Kortfristiga skulder**
Amounts owed to credit institutions	Skulder till kreditinstitut
Payments received on account from customers (may be accounted for as a deduction from stocks)	Förskott från kunder (får även redovisas som avdragspost under Varulager m m)
Trade creditors	Leverantörsskulder
Bills of exchange payable	Växelskulder
Amounts owed to affiliated undertakings	Skulder till koncernföretag
Amounts owed to associated undertakings	Skulder till intresseföretag
Corporation tax	Skatteskulder
Other loans	**Övriga skulder**
Accruals and deferred income	Upplupna kostnader och förutbetalda intäkter
Memorandum items	**Poster inom linjen**
Pledged items and contingencies	**Ställda säkerheter och ansvarsförbindelser**
A specification of assets pledged for liabilities or provisions	Panter och därmed jämförliga säkerheter som ställts för egna skulder och för förpliktelser som redovisas som avsättningar, varje slag för sig
A specification of other assets pledged	Övriga ställda panter och därmed jämförliga säkerheter, varje slag för sig
Contingencies	**Ansvarsförbindelser**
Pension liabilities not included among provisions or covered by assets in a pension fund	Pensionsförpliktelser som inte har upptagits bland skulderna eller avsättningarna och som inte heller har täckning i pensionsstiftelses förmögenhet
Other contingencies	Övriga ansvarsförbindelser

Table V.2 Property, plant and equipment: note taken from Volvo's annual report 1997.

Intangible and tangible assets

Acquisition cost	Value in balance sheet 1995	Value in balance sheet 1996	Invest-ments	Sales/ scrapping	Subsidiaries acquired and divested	Translation differences	Reclassi-fications	Value in balance sheet 1997
Goodwill	6,251	2,974	–	–	1,083	36	–	4,093
Patents	161	102	13	(2)	3	(5)		111
Aircraft engine costs	1,108	1,115	88		42	16	–	1,261
Total intangible assets	**7,520**	**4,191**	**101**	**(2)**	**1,128**	**47**	**–**	**5,465**
Buildings	14,065	13,166	1,173	(84)	29	175	516	14,975
Land and land improvements	2,391	2,116	117	(40)	3	54	49	2,299
Machinery and equipment[1]	37,713	38,782	6,453	(2,350)	178	330	3,104	46,497
Construction in progress including advance payments	2,840	4,860	2,019	(2)	–	43	(3,847)	3,073
Total property, plant and equipment	**57,009**	**58,924**	**9,762**	**(2,476)**	**210**	**602**	**(178)**	**66,844**
Assets under operating leases	3,917	6,698	9,773	(1,896)	1,303	665	(56)	16,487
Total tangible assets	**60,926**	**65,622**	**19,535**	**(4,372)**	**1,513**	**1,267**	**(234)**	**83,331**

Table V.2 (contd.)

Accumulated depreciation	Value in balance sheet 1995	Value in balance sheet 1996	Depre-ciation[2]	Sales/ scrapping	Subsidiaries acquired and divested	Translation differences	Reclassi-fications	Value in balance sheet 1997	Book value in balance sheet 1997[3]
Goodwill	820	811	196	–	–	11	–	1,018	3,075
Patents	84	50	14	(2)	–	2	–	64	47
Aircraft engine costs	990	1,053	43	–	–	3	–	1,099	162
Total intangible assets	**1,894**	**1,914**	**253**	**(2)**	**–**	**16**	**–**	**2,181**	**3,284**
Buildings	5,776	5,728	506	(57)	31	67	203	6,478	8,497
Land and land improvements	535	516	39	(3)	1	9	(8)	554	1,745
Machinery and equipment[1]	25,604	26,222	4,186	(1,493)	142	160	(198)	29,019	17,478
Construction in progress, including advance payments	–	–	–	–	–	–	–	–	3,073
Total property, plant and equipment	**31,915**	**32,466**	**4,731**	**(1,553)**	**174**	**236**	**(3)**	**36,051**	**30,793**
Assets under operating leases	1,070	1,730	1,812	(964)	284	143	(19)	2,986	13,501
Total tangible assets	**32,985**	**34,196**	**6,543**	**(2,517)**	**458**	**379**	**(22)**	**39,037**	**44,294**

[1] Machinery and equipment pertains mainly to production equipment.
[2] Includes accumulated write-downs.
[3] Acquisition value less accumulated depreciation.

Investments in intangible and tangible assets amounted to 9,863 (8,200; 6,491). Investments in assets under operating leases amounted to 9,773 (3,851; 2,585).

Investments approved but not yet implemented at the end of 1997 amounted to SEK 20.0 billion (16.8; 17.7).

Replacement cost (unaudited information)

At year end 1997, the replacement cost of buildings, machinery and equipment, based on methods of calculation applied by Volvo and which in certain cases involve the use of indexes, was estimated at SEK 47.3 billion after calculated depreciation. The corresponding value shown in the Volvo Group balance sheet was SEK 30.8 billion. Calculated depreciation based on the present replacement cost amounted to SEK 7.4 billion in 1997. The corresponding depreciation in the consolidated income statement, which is based on historical cost, was SEK 4.7 billion.

2.4 Classification of receivables and liabilities

If a receivable is classified as current, companies have to disclose amounts that are due after more than 12 months after the balance sheet date. For liabilities, companies have to disclose amounts that are due after more than 12 months and after more than five years after the balance sheet date.

2.5 Equity items in limited liability companies

Within equity, the main distinction is between restricted and unrestricted equity. Restricted equity is not available for the payment of dividends. In the individual accounts restricted equity is itself divided into share capital (*aktiekapital*), share premium account (*överkursfond)*, revaluation reserve *(uppskrivningsfond)* and other statutory reserves including the legal reserve *(reservfond)*. Unrestricted equity is divided into non-restricted reserves and net profit for the period. Non-restricted reserves may be further divided into different components, but this is more a hangover from the past when management tried to 'appropriate' earnings so that they would not be available for the payment of dividends. In the group accounts it is not unusual to find only share capital and restricted reserves within restricted equity, and non-restricted equity and net profit for the period within unrestricted equity as in the model balance sheet (see Table V.1). RR 1, the recommendation on business combinations and group accounts, states that a further division is possible but that it is normally of little value.

The legal reserve and the share premium account together should amount to 20 per cent of the share capital. If the amount is lower, 10 per cent of the net profit should be allocated to the legal reserve.

An allocation to the revaluation reserve is one of two alternatives available to a company that undertakes a revaluation of its fixed assets (see VI.5.6, VIII):

- The first alternative is to issue bonus shares. The revaluation amount would then be credited to share capital.
- The second alternative would be to credit the revaluation reserve.

In an economic sense these alternatives are equal; in both cases the amount would increase the restricted equity.

It is not usual in Sweden to account for translation gains and losses related to foreign subsidiaries separately within equity.

Changes in equity have to be disclosed. Normally the changes are explained in the notes. A table would give details of the changes in restricted and unrestricted equity respectively. This table could include changes due to an issue of new shares, the payment of dividends, translation differences taken directly to equity, etc. An example from the annual report of Volvo is shown in Table V.3.

The Swedish Financial Accounting Standards Council has started work on a recommendation similar to IAS 1. One of the requirements is a new statement that will detail the non-owner movements in equity on the face of the income statement. The new recommendation is likely to take effect from the year 2000 or 2001.

3 Profit and loss account

3.1 Regulation format

The format of the profit and loss account is prescribed in the Annual Accounts Act, appendices 3 and 4. Companies have to use a vertical format but can choose between splitting the costs according to function or to types of costs. For the largest companies it is somewhat more common to split their costs according to function. The profit and loss account shown in Tables V.4 and V.5 is taken from the guidelines for financial statements published by the Swedish Institute of Authorized Public Accountants (FAR 1996).

The same basic rules are applicable to both the profit and loss account and to the balance sheet. The format must be followed. The items listed in the format are to be regarded as a minimum requirement. As a rule, therefore, companies can include more headings but may exclude headings only in exceptional cases.

According to the current act, companies need to apply for exemption from the Patent and Registra-

Table V.3 An example of how changes in shareholders' equity can be reported

Shareholders' equity

The share capital of the Parent Company is divided into two classes of shares: A and B. Both classes carry the same rights, except that each Series A share carries the right to one vote and each Series B share carries the right to one tenth of a vote.

	Number of shares and par value			
	A (no.)	B (no.)	Total (no.)	Par
Dec 31, 1996	142,151,130	321,407,122	463,558,252	2,318
Redemption	(3,546,185)	(19,050,682)	(22,596,867)	(113)
New issue	–	+559,500	+559,500	+3
Bonus issue	–	–	–	+441[1]
Dec 31, 1997	**138,604,945**	**302,915,940**	**441,520,885**	**2,649**

[1] Par value per share rose from SEK 5 to SEK 6.

In accordance with the Swedish Companies Act, distribution of dividends is limited to the lesser of the unrestricted equity shown in the consolidated or Parent Company balance sheets after proposed appropriations to restricted equity. Unrestricted equity in the Parent Company at December 31, 1997 amounted to 28,160.

As of December 31, 1997 foundations connected to Volvo and the Volvo employee pension foundation's holdings in Volvo were 0.66% and 0.05% of the share capital and 1.44% and 0.10% of the voting rights, respectively.

As shown in the consolidated balance sheet as of December 31, 1997, unrestricted equity amounted to 41,309 (40,652; 34,618). It is estimated that 44 of this amount will be allocated to restricted reserves.

Change in shareholders' equity	Share capital	Restricted reserves	Unrestricted equity	Total equity
Balance, December 31 1994	2,220	14,545	26,567	43,332
Cash dividend	–	–	(1,512)	(1,512)
Net income	–	–	9,262	9,262
Conversion of debenture loans	98	1,510	–	1,608
Effect of equity method of accounting[1]	–	818	(818)	–
Transfer between unrestricted and restricted equity	–	(1,766)	1,766	–
Translation differences	–	(849)	(995)	(1,844)
Exchange differences on loans and futures contracts[2]	–	–	366	366
Other changes	–	6	(18)	(12)
Balance, December 31, 1995	**2,318**	**14,264)**	**34,618**	**51,200**
Cash dividend	–	–	(1,854)	(1,854)
Distribution of shareholding in Swedish Match	–	–	(4,117)	(4,117)
Net income	–	–	12,477	12,477
Effect of equity method of accounting[1]	–	373	(373)	–
Transfer between unrestricted and restricted equity	–	439	(439)	–
Translation differences	–	(222)	87	(135)
Exchange differences on loans and futures contracts[2]	–	–	40	40
Other changes	–	52	213	265
Balance December 31, 1996	**2,318**	**14,906**	**40,652**	**57,876**
Cash dividend	–	–	(1,993)	(1,993)
Redemption of shares	(113)	–	(5,694)	(5,807)
Bonus issue of shares	441	(113)	(328)	–
Net income		–	10,359	10,359
New issue of shares	3	113	–	116
Effect of equity method of accounting[1]	–	(34)	34	–
Transfer between unrestricted and restricted equity	–	92	(92)	–
Translation differences	–	1,396	(528)	868
Exchange differences on loans and futures contracts[2]	–	–	(665)	(665)
Accumulated translation difference on the Renault holding[3]	–	–	(552)	(552)
Other changes	–	113	116	229
Balance, December 31, 1997	**2,649**	**16,473**	**41,309**	**60,431**

[1] Mainly associated companies' contributions to net Group income, reduced by dividends received.
[2] Hedge net assets in foreign subsidiaries and associated companies.
[3] Difference pertains to Renault shares sold and, in connection with the sale, has affected consolidated capital gains.
Source: Volvo's annual report 1997

Table V.4 Format of the profit and loss account, classified by type of expenditure

Net turnover	**Nettoomsättning**
Change in stocks of finished goods, work in progress and work for others	Förändring av lager av produkter i arbete, färdiga varor och pågående arbete för annans räkning
Own work capitalized	Aktiverat arbete för egen räkning
Other operating income	Övriga rörelseintäkter
Operating costs:	*Rörelsens kostnader:*
Cost of raw materials and consumables	Råvaror och förnödenheter
Merchandise	Handelsvaror
Other external charges	Övriga externa kostnader
Staff costs	Personalkostnader
Depreciation, write-downs and reversals of write-downs on tangible and intangible assets	Avskrivningar och nedskrivningar (samt återföringar därav) av materiella och immateriella anläggningstillgångar
Write-downs of current assets over and above normal write-downs	Nedskrivningar av omsättningstillgångar utöver normala nedskrivningar
Exceptional items	Jämförelsestörande poster
Other operating charges	Övriga rörelsekostnader
Operating result	**Rörelseresultat**
Income from financial investments	***Resultat från finansiella investeringar:***
Income from participating interests in affiliated undertakings	Resultat från andelar i koncernföretag
Income from participating interests in associated undertakings	Resultat från andelar i intresseföretag
Income from other participating interests, investments and amounts receivable which are classified as fixed assets	Resultat från övriga värdepapper och fordringar som är anläggningstillgångar (med särskild uppgift om intäkter från koncernföretag)
Other interest receivable and similar income (detailing how much relates to affiliated undertakings)	Övriga ränteintäkter och liknande resultatposter (med särskild uppgift om intäkter från koncernföretag)
Interest payable and similar charges income (detailing how much relates to affiliated undertakings)	Räntekostnader och liknande resultatposter (med särskild uppgift om kostnader avseende koncernföretag)
Income after financial items	**Resultat efter finansiella poster**
Extraordinary income	Extraordinära intäkter
Extraordinary charges	Extraordinära kostnader
Appropriations	Bokslutsdispositioner
Tax on profit or loss for the year (income taxes, current and deferred)	Skatt på årets resultat (inkomstskatter, betalda och latenta)
Other taxes	Övriga skatter
Net profit or loss for the year	**Årets resultat**

Table V.5 Format of the profit and loss account, classified by function

Net turnover	**Nettoomsättning**
Cost of goods sold	Kostnad för sålda varor
Gross profit or loss	**Bruttoresultat**
Selling costs	Försäljningskostnader
Administrative expenses	Administrationskostnader
Research and development expenses	Forsknings- och utvecklingskostnader
Exceptional items	Jämförelsestörande poster
Other operating income	Övriga rörelseintäkter
Other operating charges	Övriga rörelsekostnader
Operating result	**Rörelseresultat**
Income from financial investments:	*Resultat från finansiella investeringar:*
Income from participating interests in affiliated undertakings	Resultat från andelar i koncernföretag
Income from participating interests in associated undertakings	Resultat från andelar i intresseföretag
Income from other participating interests, investments and amounts receivable which are classified as fixed assets	Resultat från övriga värdepapper och fordringar som är anläggningstillgångar (med särskild uppgift om intäkter från koncernföretag)
Other interest receivable and similar income (detailing how much relates to affiliated undertakings)	Övriga ränteintäkter och liknande resultatposter (med särskild uppgift om intäkter från koncernföretag)
Interest payable and similar charges income (detailing how much relates to affiliated undertakings)	Räntekostnader och liknande resultatposter (med särskild uppgift om kostnader avseende koncernföretag)
Income after financial items	**Resultat efter finansiella poster**
Extraordinary income	Extraordinära intäkter
Extraordinary charges	Extraordinära kostnader
Appropriations	Bokslutsdispositioner
Tax on profit or loss for the year (income taxes, current and deferred)	Skatt på årets resultat (inkomstskatter, betalda och latenta)
Other taxes	Övriga skatter
Net profit or loss for the year	**Årets resultat**

tion Office *(Patent och Registreringsverket)* in order to be allowed to omit items in the profit and loss statement. An exemption is given only rarely and only if the company can claim competitive disadvantage. According to a recent proposal from the government, non-listed companies with a turnover of less than SEK 100 million and with assets less than SEK 50 million may lump together change in inventories, own work capitalized, other operating income cost of raw materials and consumables, and other operating charges if the profit and loss account is classified by type of expenditure. For those companies using a statement classified by function, a somewhat more limited easing is proposed.

Turnover has to be reported by all companies unless a company can claim a competitive disadvantage and an exemption is granted by the Patent and Registration Office.

3.2 Definition of major items

Revenues should not include value-added tax or other sales taxes. Revenues from the sale of goods, etc. should be shown separately from other revenue such as rents, provisions, royalties, etc. If activities and/or markets differ significantly from one another, companies have to disclose net turnover for each activity and market separately.

Research and development costs (R&D) have to be disclosed in the notes if they are not shown in the profit and loss statement. R&D shall include depreciation on fixed assets that can be allocated to R&D work in a rational and consistent way.

Virtually all listed companies distinguish between normal or planned depreciation and accelerated depreciation for tax purposes. This distinction is not required in law although it is recommended by the Accounting Committee (FAR 1, The Annual Report of Limited Liability Companies). The reason for it is to enable readers to form an opinion of a company's true profits without the distortions introduced by the tax legislation.

It is permitted to revalue fixed assets (see VI.5.6, VIII). Depreciation has to be applied to the written-up value. This higher depreciation has to be charged as normal depreciation. When a revalued asset is sold, the gain is calculated as the difference between the selling price and the book value including revaluation.

Currency gains and losses are normally reported both as an operating item and as a financial item. The former would include gains and losses on trade-related items.

Gains and losses on the sale of marketable securities would normally be accounted for as a financial item.

Extraordinary items are rare following a recommendation issued by the SFASC on how to account for extraordinary items (RR 4). Examples given in the recommendation are costs following acts of war or earthquakes. Extraordinary items have to be detailed in the notes if they are not shown in the profit and loss account. Taxes related to extraordinary items also have to be separately disclosed.

Many Swedish companies include a line for exceptional items *(jämförelsestörande poster)* among operating costs even though the SFASC has rejected this idea. The council believes that were it acceptable to report exceptional items while at the same time extraordinary items were disallowed, the likely effect would be that items which used to be classified as extraordinary would simply be recategorized. Typical examples of exceptional items are gains/losses on the sale of fixed assets and reorganization cost. The Accounting Committee has recommended in a guide on financial reporting (*Om årsredovisning i aktiebolag)* that the reporting of exceptional items should be limited to cases where the number of such items is so large (which could mean four or more items) as to make the profit and loss statement unbalanced. If the number of items is smaller, companies are recommended to report each item separately. This recommendation is, however, not widely followed as most companies seem to use a line for exceptional items even when they report only one or two items. In addition, exceptional items often include both gains and losses which would seem to be counter to the prohibition on netting in the Annual Accounts Act.

One of the companies reporting exceptional items is Electrolux. Electrolux' consolidated income statement and footnote 5 is shown in Table V.5. As can be seen in the table, Electrolux uses the term 'items affecting comparability' rather than exceptional items. To report operating income before and after items affecting comparability is clearly not in line with RR 4.

Table V.6 An example of how exceptional items can be reported

Consolidated income statement (SEKm)		1997	1996
Net sales	(Note 2)	**113,000**	110,000
Cost of goods sold		**−81,916**	−80,057
Gross operating income		**31,084**	29,943
Selling expenses		**−21,449**	−20,025
Administrative expenses		**−4,830**	−5,269
Other operating income	(Note 3)	**149**	123
Other operating expense	(Note 4)	**−404**	−324
Operating income before items affecting comparability		**4,550**	4,448
Items affecting comparability	(Note 5)	**−1,896**	–
Operating income	(Notes 2, 6, 24)	**2,654**	4,448
Interest income	(Note 7)	**1,285**	1,453
Interest expense	(Note 7)	**−2,707**	−2,651
Income after financial items		**1,232**	3,250
Minority interests in income before taxes		**51**	−218
Income before taxes		**1,283**	3,032
Taxes	(Note 8)	**−931**	−1,182
Net income		**352**	1,850
Net income per share, SEK	(Note 9)	**4.80**	25.30

Note 5. ITEMS AFFECTING COMPARABILITY (SEKm)	**1997**
Costs of restructuring	**−2,500**
Capital gain	**604**
Total	**−1,896**

Source: Electrolux Annual Report, 1997

Appropriations were mentioned earlier (see III.3). The aim is to eliminate distortions created by the need to account for all deductions claimed for tax purposes in the profit and loss account.

Sometimes other items may be found among appropriations which do not fit in with the ordinary profit and loss account. The most common example are the so-called group contributions. A group contribution is a transfer of profits between companies in a group. The companies have to be Swedish and at least 90 per cent owned by the parent company. For the company that gives a contribution, the contribution is tax-deductible. For the receiving company the contribution is taxable. Thus, the main purpose is to minimize taxes for a group even though a group is not permitted to file tax returns for the group as a whole. In both cases it has been regarded as good accounting practice to account for such contributions in the profit and loss account. To highlight the fact that they are not ordinary revenue or expense items they are often classified as appropriations. According to a statement from the Emerging Issues Task Force in 1998, companies are now recommended, in the normal case, to report such items as increases or decreases directly in equity.

Prior year items normally arise in connection with a change in accounting policy. Accounting for changes in accounting policy is discussed below (see VII.2.9). According to a recommendation of the Swedish Financial Accounting Standards Council (RR 5) adopted in December 1993, changes in accounting principles should normally be accounted for retrospectively.

Tax reported in the profit and loss account of an individual company normally equals tax payable. The reason is that most timing differences are accounted for under appropriations. Thus, in principle, the profits shown in the profit and loss account before tax equals income that is reported for the assessment of tax.

In the group accounts, of public companies at least, the appropriations have been eliminated, as was mentioned above (see III). In consequence the tax charge in the profit and loss account of the group would include deferred tax based on comprehensive tax accounting.

3.3 Distributable profits

The amount of profits that can be distributed to shareholders is based on the balance sheet. Distributable profits include net profit for the period and any unrestricted reserves, less required allocations to legal reserves.

For a parent company distributable profits are identified as the lesser of the amounts that are available according to the parent company accounts and the group accounts.

Distributable profit will in most cases be less than the true profit because of the existence of untaxed reserves. Appropriations to untaxed reserves reduce net profit and hence profits available for distribution. The elimination of untaxed reserves in the group accounts does not solve the problem, as the board has to decide the dividend pay-out on the basis of the lesser of the unrestricted equity available in the group and in the individual accounts. For the same reason, distributable profits cannot normally be increased by using different – i.e. less prudent – accounting principles in the group accounts. This is becoming more common but has no effect on distributable profits.

The determination of distributable profits in a group is discussed in the recommendation 'Business Combinations' issued by the Swedish Financial Accounting Standards Council (RR 1). The recommendation is to start with the distributable profits available to the parent company. To this amount should be added distributable profits of the subsidiaries (excluding minority interests, if any) to the extent that they can be distributed without any need for the parent to write down the book value of its shares in the subsidiary. Intragroup profits would also have to be eliminated. Other adjustments are made in special circumstances. These are further explained under Sweden – Group Accounts.

VI Recognition criteria and valuation

1 General recognition criteria

Recognition is defined in the IASC's Conceptual Framework as the process of incorporating an item in the balance sheet or income statement that meets the definition of an element and satisfies the criteria for recognition which are:

(a) It is probable that any future economic benefit associated with the item will flow to or from the enterprise.
(b) The item has a cost or value that can be measured with reliability (paragraphs 82 and 83).

The Annual Accounts Act is not so specific. According to the act, Chapter 2, § 4:

(a) Only profits made at the balance sheet date may be included.
(b) Account must be taken of all foreseeable liabilities and potential losses.
(c) Account must be taken of all depreciation.
(d) Account must only be taken of income and charges relating to the financial year.

In addition, the Annual Accounts Act, Chapter 3, § 1, says that a company has to include all assets, provisions and liabilities in the balance sheet. However, the act does not explain what is meant by 'profits made' nor how assets, provisions and liabilities should be defined.

BFL 4 contains the following paragraph on recognition which is somewhat more specific:

> 'Any business transaction that changes the composition or the size of the capital shall be recorded continuously. Examples of business transactions are cash payments or amounts payable or receivable'.

The act does not specify when a payable or receivable arises. According to the preparatory work a payable or receivable normally arises when an invoice has been received or delivered. In order to avoid companies being able to influence the timing of the accounting there is an additional requirement according to which a payable or receivable should be recorded when recording is due on the basis of good accounting practice.

This paragraph has traditionally been given a legalistic interpretation. As an example, lease obligations have not normally been included among a company's liabilities. Following the issue of RR 6 from the SFASC on lease accounting this has changed (see VII.5). Another example was a case in 1989 where the National Board of Trade (*Kommerskollegium*) criticized two auditors for not having recognized invoices received. A construction company had subcontracted some work. The contract was terminated, however, for alleged failure to achieve the quality standards agreed. The company refused to accept the invoices sent in by the subcontractor. The invoices were sent instead to an arbitrator. According to the National Board of Trade a company could not decline to record an invoice unless it was patently false.

As the Swedish Financial Accounting Standards Council has stated that their recommendations in the future will be based on IAS, it can be assumed that Swedish companies, more and more, will adopt the same recognition criteria as are required in IAS.

2 General valuation principles

In general, fixed assets are valued at cost and current assets at the lower of cost or market. There are some exceptions. The most important exception is that revaluation of fixed assets is permitted in certain circumstances.

3 Basic principles

The Annual Accounts Act mentions seven basic accounting principles: going concern, consistency, prudence, accrual and matching principle, individual valuation, prohibition against set-offs and continuity.

3.1 Going concern principle

Financial statements are based upon the going concern principle. The going concern principle is also used when a company has to draw up a balance sheet for liquidation purposes (*likvidationsbalansräkning*). Such a balance sheet is required where more than half the stated capital has been lost (ABL, Chapter 13 § 2). The aim is to establish whether the company has sufficient remaining assets to avoid filing for bankruptcy. Thus the company is not required to write its fixed assets down to a liquidation value. The only difference compared with a normal balance sheet is that some assets could be recorded at market value if it exceeds the book value.

In a case that was brought before the Supreme Court (Blomberg and Oppenheimer, 1993), a company which had gone into bankruptcy in 1984 was sued by a supplier for not having done so earlier. According to the supplier it was already obvious a year earlier that the company's net worth was negative. The company argued that if its R&D had been capitalized the negative net worth would have disappeared. The question before the court was whether it would have been appropriate to capitalize R&D in a company which had obvious liquidity problems. The verdict came out in favour of the defending company. According to the Supreme Court, R&D is an asset that can be included in a balance sheet for liquidation purposes so long as there is a possibility that the company may conclude its R&D activities and market the new products.

The only time when the going concern principle does not apply is when a company decides to liquidate itself.

3.2 Consistency principle

Consistency is one of the principles that is often mentioned in connection with the preparation of financial statements. The requirement for consistency is normally taken to imply that the accounting principles upon which the financial statements are based should be the same from year to year.

In practice this is rarely the case. Companies change their accounting principles frequently. In most cases the changes are made in order to adopt more international accounting principles. A recent example is that many companies have started to account for deferred tax assets after having reported net losses.

It is often pointed out that a change in accounting principles will increase reported profits or improve the debt to equity ratio. The inference is that the change is designed to achieve this effect rather than to adopt more international accounting principles. It is difficult to discriminate between these two possible reasons why a company might change its accounting principles. International accounting principles are normally taken to mean IASC principles or US principles, both of which are clearly capital market-oriented. Either will often give rise to higher reported profits or an improved debt to equity ratio as compared with accounting principles conforming to the prudence principle.

According to a recommendation of the Swedish Financial Accounting Standards Council on accounting for changes in accounting principles, companies may change an accounting principle only after a recommendation has been issued or when the change clearly leads to improved accounting (RR 5, Changes in Accounting Principles). The latter is taken to mean that the new accounting principle is more in line with international practice.

3.3 Prudence principle and the realization principle

In Swedish accounting legislation and practice the prudence principle is generally considered to be more important than any other principle. This is manifested both in the general valuation rules for assets and liabilities and in the requirement that profits have to be 'made' in order to be recognized.

It is interesting to note that, despite their oft-repeated adherence to the prudence principle, companies sometimes account for transactions in a way that is definitely not prudent. Perhaps the best example are sale and lease-back transactions. Particularly in the late 1980s several companies sold property to insurance companies and others. The seller agreed to rent the property for a period of 10–25 years. Typically the rate would equal the buyer's borrowing costs. At the end of the term the seller had the right (but no obligation) to buy the property back at the same price it had been sold at.

Many people would argue that such a transaction is borrowing and not a sale. Yet, without exception, companies that decided to enter into sale and lease-back agreements were able to show large gains. The resulting lease-back obligation did not have to be recorded because it was treated as an operating lease.

Another example of 'imprudent' accounting relates to exchanges of similar assets. Two companies may agree upon an exchange of assets, e.g. property, woodland or ships. Most transactions of this nature are accounted for at market values, although it could be argued that, so long as the exchange involves like assets, no gain should be recognized.

A third example is revaluations. The act permits the revaluation of fixed tangible and financial items in certain circumstances, if the increase in value is deemed to be of a permanent nature. See V.6 for a further discussion.

The fact that the Annual Accounts Act uses the words 'profits made' rather than the more common 'realized profits' is of no practical importance (see Thorell, 1996). According to Thorell the meaning is the same. When looking for an interpretation, guidance can be found in, *inter alia*, IAS 18 on Revenues. It has been accepted in practice that both the closing rate method for valuation of receivables and payables in foreign currency and the percentage-of-completion method may be used without having a conflict with the realization principle.

In the past some companies valued their holdings of marketable securities at market prices. This was permitted under the old Accounting Act. However, the new act is interpreted as requiring companies to use the lower of cost or market for marketable securities. SFASC, among others, have argued that it makes little sense in requiring lower of cost or market for marketable securities and at the same time permitting the percentage-of-completion method for construction contracts. A Royal Commission on Accounting obviously agreed. In their final report they proposed to make a change in the 1995 Annual Accounts Act so that marketable securities should be allowed to be valued at market. As such a rule would be counter to the Fourth Directive, the Commission suggested to invoke the true and fair override principle. Although a final decision has not been taken at the time of this writing, it seems most unlikely that the proposal will be enacted. The Fourth Directive only permits the true and fair override to be applied to individual companies.

3.4 Accrual and matching principle

The accrual principle is well accepted in practice as is the matching principle. An interesting observation is, though, that some analysts do not seem to believe in the accruals as they think that the principle gives the management too much discretion in determining when costs should be recognized in the income statement. Therefore, there has been a revival of interest in the cash flow statement. As to the matching principle, recent IASs seem to focus more on defining what assets and liabilities shall be included in the balance sheet than on the effects on the income statement. One example is IAS 22 (revised 1998) on Business Combinations. IAS 22 now makes it difficult to set up a provision for restructuring costs even though many companies would argue that such a provision is necessary in order to achieve a correct matching of income and expenses. A Swedish standard based on IAS 22 (revised 1998) takes effect on 1 January 2001.

3.5 Individual valuation

The Annual Accounts Act provides that the components of assets and liabilities must be valued individually. Thus, it is not allowed to use the portfolio method where one asset could be carried at cost exceeding the market value on the grounds that at the same time other assets had a market value much higher than their book value.

In RR 2 on inventories, the problem is discussed in some detail. Paragraph 31 states that the lower of cost or market should normally be applied to each item in the inventory. An exception can be made for like goods or where valuation item by item would be too expensive. Raw materials are an example of like goods. A decline in the value of one type of goods may not be offset against an increase in the value of another.

There can be little doubt that the same principle should apply to companies owning investment properties. However, it is clear from reading the accounting principles adopted by some of these companies that the portfolio method is sometimes used. A plausible explanation is that such companies look upon their property holdings as a portfolio of assets and that a write-down of some assets to be required when the market value of the whole portfolio exceeds book value does not make economic sense to them.

A draft statement from the Emerging Issues Task Force does make it clear, however, that the principle of individual valuation should be applied also by companies owning investment properties. The only exception that is accepted is when some properties are managed as a unit.

Companies owning marketable securities also sometimes use the portfolio method in spite of the legal requirements. For investment companies this may make sense. The use of the method in such cases is supported by a recommendation of the Accounting Committee (FAR 12). According to the recommendation, the portfolio method may be used for investments in shares when the investments are intended to create a diversified portfolio. However, the method is sometimes used by other companies whose holdings are not sufficiently diversified to be called a portfolio.

4 Value concepts

4.1 Acquisition costs

The basic rule is that an acquired asset should be valued at cost. For inventories cost is defined in RR 2. Paragraph 12 states that by cost is meant purchase price, import duties, transport costs and other costs directly attributable to the goods in question. Any rebates, bonuses, etc. should be deducted from the purchase price.

Where the price of an item is stated in foreign currency the purchase price is derived by translating the stated price at the rate prevailing when the acquisition was recorded. If the price has been hedged, the rate at the time the company entered into the hedge is normally used. However, for short-term hedging, e.g. hedges for less than three months, the contracted rate may be used instead.

Sometimes an acquisition is made where the supplier makes long-term finance available at an artificially low rate of interest. In that case RR 3, a recommendation on how to account for loans bearing an abnormal rate of interest, requires the value of the cheap finance to be deducted from the purchase price. 'Long-term' is taken to mean a period exceeding one year. The same principle applies to a sale of goods.

Accounting for barter or an exchange of assets is discussed by the Swedish Financial Accounting Standards Council in an exposure draft on property, plant and equipment. The accounting treatment follows IAS 16 closely. For an exchange of unlike assets markets values are used. A company can use the market value of the asset acquired or given up, depending on which value is the more visible. For like assets book value should be used unless the market value is lower, in which case the market value is the one to use.

The recommendation on accounting for an exchange of assets take effect on January 1 2000. At present no recommendations provide guidance in this area. As a result, several companies that have on occasion exchanged like assets have used market values, thus showing a profit on the transaction.

4.2 Production costs

For a manufacturing company, acquisition cost comprises the acquisition cost of raw materials, etc., other costs directly attributable to the goods, e.g. direct labor costs, including social security costs, and a fair portion of the fixed and variable production overheads that relate to bringing the inventories to their present location and condition. Production overheads include depreciation of fixed assets used in production. Interest on those assets, however, is not among the costs that may be capitalized.

Sales costs, administrative costs and costs of research and development are not production costs unless they are directly attributable to a specific order.

Interest costs are not normally part of those costs that are necessary to bring the goods to their present location and condition. An exception is made for goods where time is an essential element in the production process, as with cheese or wine.

The allocation of fixed production overheads should be based on the normal capacity of the production facilities. Normal capacity is the production expected to be achieved on average over a number of periods. Overheads which are not allocated are recognized as an expense in the period in which they are incurred. In periods of abnormally high production the proportion of overhead allocated to a unit of production should be reduced so that inventories are not measured above historical cost.

Costs of internally produced fixed assets are dealt with in a recommendation from the Accounting Committee (FAR 3). According to this recommendation borrowing costs can be capitalized as part of the production cost. In an exposure draft on property, plant and equipment from the Swedish Financial Accounting Standards Council the relevant paragraph was changed as to require capitalization. Because of comments received on the exposure draft the council has decided to defer the question on whether or not to require capitalization of interest cost to a separate recommendation based on IAS 23.

4.3 Lower attributable values

If the value of a fixed asset has fallen permanently below its carrying amount, the Annual Accounts Act Chapter 4, § 5 requires a write-down. The act does not prescribe how the extent of the write-down should be calculated. There is merely a reference to good accounting practice. Write-downs are further dealt with below (see VI.5.5).

For current assets the Annual Accounts Act 4:9 prescribes the lower of cost or market. The market price is normally defined as an estimated selling price less selling costs. This concept is discussed in more detail in RR 2, paras 22 to 25. Selling costs should include the cost of warehousing goods not expected to be sold within one year. In that case interest costs should also be taken into account. This means that the expected future selling price should be discounted.

It may be difficult to estimate a selling price for work in progress. Replacement cost may be the only practical alternative. If replacement cost is used, RR 2 states that an effort should be made to ascertain the likelihood of the cost being recovered.

5 Valuation rules for fixed assets

5.1 R&D expenses, formation expenses and other intangible items

According to the Annual Accounts Act, Chapter 4, § 2, R&D expenses may be capitalized if it is likely that the expenses will be of material value to the company in coming years. The same rule applies to other intangible items like patents, licenses, trademarks and goodwill. Intangible assets shall be amortized over a period not exceeding five years unless a longer time can be justified. Expenses incurred in connection with the formation of a company or an increase in share capital must not be entered as an asset but should be expensed when incurred.

R&D expenses are also the subject of a recommendation by the Accounting Standards Board, R 1. This recommendation specifies the criteria that have to be met before R&D expenses can be capitalized. Among the criteria are:

- The expenses have to be clearly delineated.
- There must be a certain application for the R&D expenses.
- It must be probable that the R&D will lead to revenues sufficient to recover the capitalized amount.
- There must be sufficient resources for the company to complete the project and to undertake the necessary marketing effort.

According to R 1 it is good accounting practice to give information about the costs of R&D. Companies should also disclose the amount of R&D that has been capitalized. A survey of listed companies' accounts indicates that these requirements are indeed met.

Capitalization of R&D expenses is rare. One reason is that in this area tax accounting is based on what is recorded in the financial accounts. That means that capitalization reduces the tax advantage that could otherwise be obtained. Thus most companies which do capitalize R&D are in a no-tax position because of (for example) large loss carry-forwards. An exception is to be found in the defence industry, where the capitalization of R&D expenses is sometimes required in government contracts. It could be argued that a company could capitalize R&D expenses in the group accounts while at the same time expensing them in the individual accounts in order to benefit from the tax advantages. So far, this has not been done. Therefore, it is likely that also other factors explain why companies are reluctant to use the option provided for in the Annual Accounts Act.

When calculating distributable profits no adjustment needs to be made if a company has capitalized R&D or other intangible items.

Depreciated amounts may not be written back.

5.2 Goodwill

Goodwill shall be accounted for in the same way as other intangible items. This implies that it should be amortized over a period of five years unless a longer period can be justified. The act relates to individual company accounts. The goodwill it mentions is consequently directly acquired goodwill and not the more common kind that arises only in group accounts. Goodwill in the individual accounts results from the acquisition of the individual assets of another company when the acquiring company pays more than can be allocated to specific assets.

Obviously there is little difference between the kind of goodwill that arises in individual accounts and that which arises in group accounts. Both are due to the acquisition cost being higher than the fair value of the particular assets taken over. Nevertheless, in practice, goodwill arising on consolidation is often treated differently from goodwill in the individual accounts.

The dominant practice as regards goodwill in the individual accounts is to amortize the capitalized amount over five years (Rundfelt, 1991). The reason is that this type of goodwill is tax-deductible, and the Tax Code accepts a five-year amortization period.

For goodwill arising on consolidation practice has been varied. After the SFASC published RR 1 on Business Combinations most companies have used an amortization period of between 10 and 20 years. Direct write-offs have all but disappeared.

The calculation of goodwill is governed by para. 12 in RR 1. Goodwill is defined as the difference between the purchase price and the fair value of the assets and liabilities acquired. Intangible assets, such as brands, may be separately identified. If so, they would be required to be amortized over the same period as that used for goodwill.

Goodwill that has been amortized cannot be written back.

The equity method is allowed only in the group accounts. Thus goodwill related to an acquisition of an associated company can arise only in the group accounts. Such goodwill should be accounted for in the same way as other goodwill, according to RR 1.

5.3 Other intangible fixed assets

The most common examples of other intangible fixed assets are patents and leasehold rights. Brands, on the other hand, are only rarely to be found in Swedish company accounts. One reason is that, as the amortization rules for goodwill also cover such intangibles as brands, there is no incentive for companies to distinguish between different types of intangibles.

Expenditure on software is discussed in two pronouncements from the Accounting Standards Board (U 88:15 and U 88:16). According to these statements expenditure may be capitalized if there is reason to believe that the software being developed for sale will be commercially viable. Expenditure on the development of software intended for use by the company itself should normally be expensed as it occurs.

There has been some discussion whether it is possible for a company to revalue brands once they have been expensed. The general revaluation rules do not exclude the possibility for externally acquired brands that are separately identified to be revalued. A necessary condition is that the brand should demonstrably have a lasting value which significantly exceeds the current book value. Few would expect this condition to be met in practice. In the case of internally generated brands the Annual Accounts Act does not allow reinstatement.

The SFASC has recently issued a standard dealing with intangibles. It will take effect on 1 January 2001.

5.4 Tangible fixed assets

A fixed asset is defined in the Annual Accounts Act 4:1 as an asset that is intended to be held or used for a longer period, i.e. for more than a year. Equipment of lesser value and equipment with an expected useful life of less than three years may, however, be expensed for tax purposes on acquisition (Municipal Income Tax Code § 23).

Major spare parts and special tools are accounted for as fixed assets (Accounting Standards Board, U 88:3). If the value is low they would normally be accounted for as supplies and included among inventory.

Fixed assets with a finite life have to be depreciated over the economic life. This is normally referred to as planned depreciation or depreciation according to plan. The reason is that for tax purposes a much shorter depreciation period, normally five years for plant and equipment, is allowed. A necessary condition, however, is that depreciation for tax purposes should coincide with depreciation in the accounts. In most companies this is achieved by charging depreciation according to plan against operating revenues and treating the difference between this kind of depreciation and that allowed for tax purposes as a fiscally required appropriation (see example in III.3 above).

No special method for the calculation of depreciation is prescribed. The dominant method is straight-line depreciation without taking residual values into account. In one or two industries – for example, in the energy sector – there are common standards for estimating the economic life of assets.

There are no special rules for investment properties. Therefore, investment properties shall be depreciated over their useful life.

No special requirements have to be met before the depreciation method can be changed. If a change is implemented, including a revision of the economic life of the asset, it should be accounted for as a change of estimate according to RR 5.

5.5 Impairment write-downs

If the value of a fixed asset has been permanently diminished, the Annual Accounts Act, Chapter 4, § 5, requires a write-down by an amount which is in line with good accounting practice.

Some guidance in this gray area is given by FAR 3, a recommendation on tangible fixed assets, and FAR 12, on financial assets, issued by the Accounting Committee.

As regards tangible fixed assets, the Accounting Committee states that a write-down may be necessary if there has been a sudden fall in value, e.g. because of a decision to close a factory. A write-down may also be necessary in cases where the rate of return is unacceptably low for other than temporary reasons.

According to the Accounting Committee the correct value that should be calculated in such cases is the discounted value of future cash flows. For assets that may be sold in the near future an estimated selling price, less selling costs, may be used instead.

It is not known to what extent this recommendation has been followed in practice. It is unlikely to have been widely adopted. The usual case where a write-down occurs is in connection with restructuring. If such restructuring involves a write-down of assets the fact is not normally disclosed. Still, it seems likely that most write-downs involve the closure of plant where there is no doubt that the residual value is almost zero.

The reference to a permanent decline in value has made it difficult to require a write-down if the company argues that the decline is only temporary. This was the case in the early 1990s with real estate. One company claimed that there was no need for a write-down as it expected the market value to exceed book value in 10 years' time.

A Swedish recommendation takes effect on 1 January 2001. It is likely that this recommendation will be controversial as it requires companies to test their assets for impairment every time that, for instance, profitability has fallen below the cost of capital. The test requires management to assess future cash flows and to discount them in order to verify that the book value of an asset does not exceed its recoverable amount.

It could be argued that IAS 36, in principle, is not very different from FAR 3, i.e. the present Swedish recommendation. However, FAR 3 is based on the idea that a write-down is only necessary when the fall in value is deemed to be permanent. Therefore, an introduction of the concepts in IAS 36 is likely to have a substantial impact on Swedish company accounts.

For financial assets, some guidance is given by the Accounting Committee in FAR 12. FAR 12 requires companies to write down the value of shares if the value has fallen and the fall is other than temporary. It is specifically stated that a write-down is not required if the fall in value is due to a downturn in the economy that can be looked upon as temporary. In consequence there are some examples of companies holding marketable securities with a market value below book value. According to FAR 12 companies must give reasons for not writing down the value. The reason normally given is that the decline in value is expected to be only temporary. Presumably this could go on for years, provided the company has the ability to hold on to its assets.

If the value of an impaired asset has been written down, the write-down should be reversed if the reasons for the write-down are no longer valid.

5.6 Revaluations

Revaluations of fixed assets are covered by the Annual Accounts Act, Chapter 4, § 6. The revaluation of fixed assets is permitted provided the value of the assets is substantially in excess of book value and the excess can be expected to last for a considerable time. This last condition has been interpreted in practice as meaning not less than 10 years.

The amount at which an asset has been revalued may not be taken to income. Instead, the company can choose to increase either capital (through a bonus share issue) or non-distributable reserves (see V.2.5).

The Swedish Financial Accounting Standards Council has been working for several years on a recommendation on Property, Plant and Equipment. One of the controversial areas has been whether to allow revaluations or not. The main argument in favour is that revaluations are permitted in the Annual Accounts Act. As the council has no power to enforce its recommendations it could be unwise to try to restrict something that is permitted by the law. On the other hand there are other cases where the SFASC has decided to issue a standard that narrows down the options contained in the Annual Accounts Act. A second argument is that from a user perspective it is better to account for the assets at a value that is closer to their market value rather than an outdated cost. This is particularly true for long-lived assets such as forest land. Most of the forest land of the larger Swedish forest-based companies was acquired more than 50 years ago when prices were a fraction of what they are today. A third, closely-related argument, is that by revaluing assets, a company can improve its observable debt to equity ratio. It has been claimed that lenders pay more attention to the figures actually recorded in the balance sheet.

There are also strong arguments against revaluations. Most important is that revaluations along the lines in the Annual Accounts Act do not comply with IAS 16. It could also be put in question if they conform with the requirements in the Fourth Directive. Thus, a Swedish company could revalue their property on an *ad hoc* basis. There is no need to revalue a whole class of assets, nor to do it at regularly intervals. Finally, the wording of the act has been interpreted as not allowing for revaluations all the way up to the market value. The requirement that the market value must be expected to remain over book value for a considerable time implies that there has to be a margin between the revalued amount and the market value, although the size of the margin has never been spelt out.

Another argument against revaluations that has been raised is that IASC is currently working on standards on agriculture and investment properties. It has been proposed that assets covered by these standards should be valued at market. Taken together with IAS 39 on Financial Instruments, presumably all assets where it could be said that there is a reliable market value would be valued at market, thus making revaluations unnecessary.

The SFASC voted on a final recommendation on Property, Plant and Equipment in early 2000. It will take effect on 1 January 2001.

If an asset has been revalued, depreciation should be based on the revalued amount.

Information about revaluations should be given in the notes to the accounts.

5.7 Special problem areas

Expenditure on fixed assets of minor value may be expensed as it is incurred. What is meant by minor value has been defined by the National Tax Board. Items which cost less than SEK 2,000 are normally considered to be of minor value. Larger companies, i.e. those with more than 200 employees, can expense items costing up to SEK 10,000.

Expenditure on repairs and maintenance is normally expensed as incurred. The definition of repairs and maintenance is based upon such action as is taken within the expected life of the asset in order to maintain it in original condition. If the repair involves a part of the asset which represents most of its value, a new investment is normally assumed to have taken place.

5.8 Participating interests

The Annual Accounts Act, Chapter 7, §§ 24–29, prescribes that the equity method must be used when accounting for associated companies in group accounts. As to joint ventures, companies can choose between the equity method and proportional consolidation. An associated company is defined as a company in which the investor has a significant influence, other than temporary. If the investor holds more than 20 per cent of the voting power of the investee, it is presumed that the investor does have significant influence, unless it can be clearly demonstrated that this is not the case. A joint venture is defined as an operation which is subject to joint control by two or more parties.

Accounting for associated companies and joint ventures is also on the agenda for the SFASC. One of the problems being discussed is how to account for investments in listed companies by closed-end investment companies that exceed the 20 per cent threshold. Formally, these investments meet the criteria for being accounted for by the equity method. The investment companies remain, however, critical. They argue that the equity method is costly and unnecessary as their holdings are typically evaluated based on current share prices. Also, as the investment companies normally produce their accounts early, before the investee has produced their accounts, the share of the results of the investee refer to a different accounting period.

The Annual Accounts Act provides for a choice between the equity method and proportional consolidation for joint ventures. In practice both methods are used. Even though the equity method is the more common, proportional accounting is used by some large companies like SCA and companies investing in real estate. The SFASC has discussed whether to require proportional accounting for all joint ventures. The most likely outcome of the discussion is, however, to retain both of the options in the act.

The equity method is not allowed in the individual accounts except for investments in partnerships. Thus, in the individual accounts, shares in associated companies are accounted for at cost. Shareholdings in associated companies shall be shown separately from other shareholdings.

6 Valuation of current assets

6.1 Raw materials and consumables

The basic assumption for inventories is that the goods that were acquired first are those which are sold first (FIFO). The LIFO method is not permitted. If there are practical difficulties in applying the FIFO method, the average cost method may be used instead, provided it can be presumed to give a reasonable approximation to the FIFO method (see RR 2). It is not known to what extent the average cost method is used in practice. Normally companies refer to the FIFO method when explaining their accounting policies.

A comparison has to be made between the FIFO cost and market value (for finished products). A market value cannot normally be estimated for raw materials, as they make up only a small part of the finished product. Replacement cost is therefore the practical alternative.

A write-down can sometimes be avoided even though the replacement cost has fallen. This is the case where there is good reason to assume that the selling price will not be affected.

Guidelines on accounting for consumables are given in a pronouncement by the Accounting Standards Board (U 88:3). Consumables which are essential to a company and are kept in a central warehouse should be accounted for as part of the inventory. The cost of consumables may be expensed if it is immaterial or if the consumables have been delivered to different departments (end users).

6.2 Work in progress and finished goods

Cost flow assumptions are the same as for raw materials (see above).

The basic valuation rule is the lower of cost or market, where market is normally defined as the net selling price. The same rules apply for tax purposes. According to the Tax Code, companies are allowed to reduce the actual cost by 3 per cent in order to take minor obsolescence into account. That is the reason why most companies describe their accounting policy for inventories as the lower of cost, determined in accordance with FIFO, less obsolescence, and market.

Obsolescence could, of course, exceed 3 per cent, in which case the company will have to estimate a net selling price to see whether it yields a lower value. If stocks are so high that it is expected to take more than a year to sell them, sales costs should include interest cost. That means that the selling price has to be discounted (RR 2, para. 24).

Production costs have to include a share of production overheads, based on normal capacity. It has been common practice to include among overheads a capital cost for plant and equipment that is made up of depreciation and an imputed interest cost. The new recommendation on inventories, RR 2, para. 18, states, however, that interest is not a cost that can be inventoried except in the case of goods where time is an essential element in the production process.

The disclosure section in RR 2 is fairly brief. Paragraph 33 states that a company should give information about its valuation principles and the extent to which a portfolio method has been used rather than individual valuation, with a breakdown of the inventory in a way that is appropriate to the company.

In the unlikely case where inventories have been written down to net realizable value and prices have thereafter risen, the write-down shall be reversed. This is a consequence of requiring inventories to be valued at the lower of cost or market.

6.3 Receivables and marketable securities

Receivables are accounted for at nominal value less a provision for bad debts. The provision for bad debts has to be based on an individual examination of each customer where there is any risk of insolvency. Companies with instalment sales to a very large number of customers may reckon their provision for bad debts as a percentage based on previous experience.

Marketable securities are covered by a recommendation of the Accounting Committee (FAR 12, Accounting for Shares and Participation). According to this recommendation a company that invests liquid assets in marketable securities should normally account for them as current assets even though they may be held for more than a year. The basic valuation rule is the lower of cost or market. If companies invest in marketable securities on a portfolio basis the recommendation accepts the portfolio method.

7 Valuation of liabilities and provisions

7.1 Liabilities

The distinction between liabilities and equity is based mainly on the legal form of the security. Thus preference shares are accounted for as equity even though the holder may have an option to redeem the shares at par.

Convertible securities have traditionally been accounted for as a liability. According to a recommendation (RR 3) of the Swedish Financial Accounting Standards Council, convertibles issued after 1 July 1993 should be accounted for using split accounting. This means that a convertible has to be split into two parts, the conversion right and the straight debt. The estimated value of the conversion right is taken to restricted equity. Thus, the value of the conversion right is treated in the same way as if the company had issued new shares at a price exceeding par value. The debt, based on the discounting of contractual payments at the market rate of interest at the time the convertible is issued, is accounted for as (long-term) debt.

The Annual Accounts Act lays down no explicit valuation rules on liabilities. It has, however, long been taken for granted that liabilities denominated in a foreign currency should be valued at the higher of cost or market (see VII.2.8).

Liabilities are normally accounted for at the repayable amount. For loans with a stated interest that differs from the current market rate, RR 3 prescribes that the loan should be accounted for at an amount that is equal to the present value of future payments. A typical example is zero-coupon bonds. Another example is convertible securities (see above).

RR 3 does not permit the use of the gross method, where the difference between the amounts received and repayable is accounted for as an asset.

The recommendation also requires the borrower to increase the liability each period with the imputed interest, based on the market rate when the loan was taken with the increase treated as interest expense in the profit and loss statement – so that the value equals the repayable amount when the loan matures.

As has already been mentioned, RR 3 requires

discounting of receivables and payables where the stated interest rate differs from the market rate. There are some exceptions to this rule. For example, if a company grants an employee a loan at a low rate of interest it is not required to adjust the interest. The reason is that such an adjustment would lead only to a corresponding adjustment of the cost of wages. As net profit will not be affected, it is felt that such an adjustment would provide no useful information. Likewise, an interest-free loan to a customer or from a client will affect only the allocation between operating profit and the financial net. As the amounts are likely to be insignificant, no adjustment is required. However, if the credit extends beyond a year, net profit will be affected. Hence in that case an adjustment has to be made.

Other exceptions include subsidized loans from the government. These should be accounted for according to recommendation R 5 of the Accounting Standards Board, Government Grants. In addition, tax assets and liabilities are excluded from the scope of RR 3.

Some liabilities are non-interest-bearing, e.g. tax liabilities, social security charges, trade creditors, etc. A special feature, at least among the majority of listed companies, is that such liabilities are often kept separate in the balance sheet from the interest-bearing ones. The separation is useful for financial analysis, as it enables the user to compare the interest charges with the amount of liabilities.

7.2 Provisions

A provision has to be set up for specific losses, liabilities or costs which are attributable to the current or previous financial years and which, at the date of the balance sheet, are likely or certain to be incurred but uncertain as to the exact amount or the time of settlement (Annual Accounts Act 4:16).

Provisions are accounted for as a separate category of items between equity and liabilities. This has been criticized as it may be seen that provisions are not liabilities. Two Swedish companies (Esselte and Svedala) have taken note of this criticism and reported their provisions as a subset of Liabilities.

The most common examples of provisions are pension provisions, provisions for deferred taxes (primarily in the group accounts), provisions for warranties and provisions for restructuring. The latter is the more controversial. In the last few years several Swedish companies including Electrolux, Investor, S-E-Banken and Skandia, have set up large provisions for restructuring. In some cases the provisions have been made at the time of a management change. This has led to a debate on 'big-bath accounting' and a call for stricter rules on what may be included in a provision.

The SFASC has issued a standard on IAS 37 on Provisions. It is almost identical to the international standard and takes effect on 1 January 2001.

Provisions for restructuring costs are normally not tax deductible. Some, but not all, companies therefore recognize a deferred tax asset at the same time they set up the provision. (See VII.1 for a further discussion of how to account for deferred taxes.)

The largest provision typically relates to pensions. Part of the pension payments comes from the state. The state pension is financed by social security contributions equal to approximately 13 per cent of wages in 1998. In addition, most employees will receive an additional pension paid for by their employer. Thus, practically every company has a pension liability. The liability can be covered in a number of ways. Most small companies pay regular premiums to an insurance company, which assumes the pension liability. Larger companies normally prefer to make a provision for their pension liabilities in their own accounts. The liability is calculated on an actuarial basis. It does not take future salary increases into account. On the other hand the discount rate does not take account of inflation; it has remained in the 3–4 per cent range for a long time now. The pension liability is calculated by a central institute (the *Pensionsregistrerings-Institutet*, PRI), where all employees are registered. At year end each company is notified of its pension liability. The increase in the liability is accounted for as a pension cost. This cost is tax-deductible. For accounting purposes the pension cost is divided in two parts, at least for larger companies. The first part is equal to an estimated interest cost on the pension liability at the beginning of the year. This part is reported as a financial expense. The remainder is accounted for as part of the social security charges among operating costs.

Companies that make provision for their pension liabilities are administered by a specialized insurance company (*Försäkringsbolaget Pensionsgaranti*, FPG) that guarantees the employees their pensions even if the company goes bankrupt. The insurance is administered by a special company to which all businesses contributing to the scheme pay a small premium.

As the system covers all salaried employees, amendments of pension plans are rare. An amendment that would increase provisions would probably be taken in the year of the change. There are no rules permitting amendments to be allocated over some arbitrary period of time.

A few companies, mostly banks, have independent pension funds. The fund is an independent entity and may not be consolidated. Their investment policies are strict. A fund may not invest in the shares of the company sponsoring the fund.

If the pension liability exceeds the fund's assets, the difference constitutes a pension cost to the company. If the fund builds up surplus assets the company is not allowed to claw them back. Therefore, the surplus is not accounted for as an asset of the sponsoring company. However, the surplus can be used to reduce future pension costs.

Accounting for pension costs is one more of the topics that is currently under discussion by the SFASC. The council has committed itself to introduce a recommendation based on IAS 19. This is a controversial decision as the pension liability under IAS 19 is estimated taking future wage increases into account. Also, in order for pension costs to be tax-deductible, companies have to account for the costs as calculated by the PRI. Therefore, a recommendation based on IAS 19 can only be applied in the group accounts. In the individual accounts, the pension liability would have to be based on the old rules. This means that companies would have to make two estimates of their liabilities which could be costly.

The council has also debated the accounting for pension funds. The main question is whether a Swedish pension fund can be looked upon as 'plan assets' as defined by IAS 19. The problem is that, in Sweden, pensions are paid out by the company. The company can then be reimbursed from the fund for what it has paid out. In IAS 19 it is assumed that it is the fund that makes the payments to the employees. This is looked upon as a prerequisite for allowing companies to net their pension liability with the assets in the pension fund. Whether Swedish companies will be allowed to make such a netting and still assert that they comply with IAS 19 is still an open question.

Under IAS 19 a surplus of funds over pension liabilities will normally be accounted for as an asset in the sponsoring company. The asset will be classified as a receivable, i.e. it represents the value of not having to pay pension costs in coming years. It is likely that the council will adopt this recommendation. It is also expected that actuarial gains and losses will be amortized over the expected average working lives of the employees participating in the plan.

The Annual Accounts Act permits a company to show its pension liability as a contingent liability. This exception is normal for pension obligations that are not tax-deductible. An example would be pension commitments to management above the ceiling permitted by the Tax Code. As the pension obligation is binding on the company it is unquestionably a liability and should be accounted for as such. This is almost certainly what the SFASC will recommend as well when it issues its recommendation on accounting for pension costs.

Provisions for deferred income tax are also common. They are dealt with below (see VII).

7.3 Value adjustments and untaxed reserves

Value adjustments occur only in rare cases. Normally, they are accounted for net, i.e. as a deduction from the asset to which the adjustments relate.

Untaxed reserves figure prominently in Swedish accounting. As explained above (see III.3), untaxed reserves make it possible to give the true and fair view the capital markets need and at the same time comply with the tax rules. The most important of the untaxed reserves since the tax reforms of the late 1980s is connected with excess depreciation. Most plant and equipment may be depreciated over five years for tax purposes. The economic life could be 10 or 20 years. The difference between tax-based depreciation and planned depreciation would be accounted for as an appropriation in the income statement and as an untaxed reserve in the balance sheet. The latter would be found as a separate category between owners' equity and provisions (but only in the individual accounts). For the purposes of financial analysis it would normally be allocated to deferred tax liability and to owners' equity. The percentages would be 28 and 72 respectively from 1994 onwards.

It was expected that untaxed reserves would disappear with the new Annual Accounts Act. As mentioned earlier, they have already been eliminated from the group accounts of public companies following the publication of RR 1. The stated reason for keeping untaxed reserves is that it helps the tax authorities when they check company tax returns. It is difficult to believe, however, that the same result could not be achieved by simply adding a few lines to the tax return form. Still, severing the link between financial reporting and tax accounting is not likely in the foreseeable future.

VII Special accounting areas

1 Deferred taxation

According to the Tax Code, taxable profit should be equal to the accounting profit. Thus Sweden has a tax-conformity regime. There are, however, a number of exceptions. Some income, e.g. most dividend income, is not taxable. Some costs are not tax-deductible, e.g. donations, certain costs of entertainment, the incorporation costs of a company, etc. From time to time there have also existed some items which appeared only in the tax assessment. The most important of these was known as the Annell deduction (*Annell-avdrag*). The Annell deduction could be claimed by companies when they issued new shares. Dividends on the shares were deductible up to a certain amount, the exact amount depending on the date of the issue and on who was the recipient of the dividend. Total deductions could not exceed the amount raised through the issue of shares. The dividend deduction has now been repealed. Some form of relaxation of the double taxation of corporate profits is, however, likely to reappear in the near future.

All the instances above are examples of permanent differences. The amount in such cases tends to be rather small. Temporary differences in the individual accounts are also on the small side. This is because of the peculiar Swedish requirement that all tax-based appropriations must be shown in the income statement. (As was pointed out earlier, RR 1 eliminates these appropriations in the group accounts. This is explained more fully under Sweden – Group Accounts.)

For the purposes of analysis the tax-based appropriations can be regarded as temporary differences. The most important is excess depreciation. Companies that are capital-intensive, such as shipping, steel, woodpulp, etc. benefit particularly from the generous rules on fiscal depreciation. Another appropriation common in the early 1990s was the tax equalization reserve (*Skatteutjämningsreserv*, SURV). All companies were entitled to a deduction of almost 30 per cent of that part of profit that is retained. In 1993 it was decided to replace SURV with an allocation reserve. At the same time nominal corporate income tax was cut from 30 per cent to 28 per cent. Companies can allocate up to 20 per cent of their profits before tax to the allocation reserve. After five years the amount allocated to the reserve has to be brought back as a negative allocation with a corresponding increase in taxable profits. Thus, the real benefit arises from the fact that the money set aside is interest free. The value of this benefit has been estimated to equal around three percentage points which means that the effective tax rate in Sweden is approximately 25 per cent.

There are a few exceptions to the rule that there are no temporary differences in the individual accounts. Buildings may be depreciated for tax purposes even though there is no depreciation in the financial statements. Particularly, companies with investment properties have taken advantage of this opportunity by claiming a maximum depreciation for tax purposes but only a minimum in the financial statements. Normally, these companies have not set up a provision for deferred taxes. This will have to change now that the SFASC has published their new recommendation on tax accounting, which takes effect on 1 January 2001.

Another example of a temporary difference relates to revaluations. Companies which have revalued their assets have normally not provided for any deferred tax on the revalued amount. This will also have to change when the new recommendation becomes effective. A company which has made a revaluation in the past will have to transfer an amount equal to 28 per cent of the difference between book value and the tax base from retained earnings to the deferred tax liability.

Even though few companies report deferred taxes separately in the balance sheet, disclosure has to be made according to the Annual Accounts Act. Information has to be given on the difference between taxes payable and the tax that relates to the year, unless insignificant. Although the act does not specify what is meant by taxes relating to a certain year, it has been interpreted as those taxes that would have been reported under a system a comprehensive tax accounting.

Tax losses can be carried forward without any restrictions. Up to now, there is no Swedish recommendation on how to account for tax losses. During the 1990s it has, however, become increasingly common to set up a deferred tax asset equal to 28 per cent of the tax loss carried forward provided it is probable that the company will have taxable profits within the not-too-distant future. Normally deferred tax assets are netted against deferred tax liabilities. Netting is expected to be allowed in the forthcoming recommendation dealing with tax accounting, on the condition that both the tax assets and the tax liabilities relate to the same tax authority.

In Sweden, the taxable entity is the individual company. Thus, groups are not allowed to file a tax return for the group as a whole. This could lead to excessive taxes having to be paid if one company in a group makes losses while other companies report profits. In order to mitigate such consequences, companies in a group are allowed to make group contributions (*koncernbidrag*). Group contributions mean that one group company can transfer its profits to a loss-making company in order to minimize taxes. A precondition is that the companies are at least 90 per cent-owned by the parent.

To illustrate, assume there are two wholly-owned companies in a group: company A is reporting profits (before tax) of 100 and company B is reporting a loss of 100. Company A decides to make a contribution to B of 100. According to a statement from the Emerging Issues Task Force, the contribution should be accounted for through equity. Thus, company A's distributable reserves are reduced by 100 and company B's increased by the same amount.

The group contribution is deductible for tax purposes for company A. Following the principles in IAS 12, if the tax relates to an item accounted for in equity, the related tax should also be accounted for in equity. Consequently, the net group contribution made by company A will be 72 (provided the tax rate is 28 per cent). Likewise, company B will report an increase in equity of 72.

The contra item is the reported tax in the income statement. Company A therefore reports a profit before tax of 100 minus a notional tax charge of 28. For company B the loss is reduced from 100 to 72 by re-posting a tax benefit of 28.

2 Post-balance sheet events

Accounting for post-balance sheet events is dealt with in the Annual Accounts Act, Chapter 6, § 1. The act requires that information about transactions or events of special importance to the company which have occurred during the year or after the balance sheet date has to be included in the management report.

Examples of transactions or events of special importance instanced in commentaries on the act are agreements concluded with other parties, major changes in capacity utilization, major changes in the prices of essential raw materials, foreign currency changes and major acquisitions or disposals.

3 Long-term contracts

Traditionally, long-term contracts have been accounted for using the completed contract method. It was generally assumed that the old Accounting Act did not allow the percentage of completion method other than in special cases. One such case would be a long-term contract which is almost completed, where revenue and costs can be estimated with almost total certainty. The Annual Accounts Act is less restrictive and opens up the possibility to use the percentage of completion method in many more cases. This has had the effect that, starting in 1997, all of the leading companies in the building industry changed their accounting principle to the percentage of completion method. Also, many of the large engineering companies (ABB, Electrolux, Ericsson, etc.) use the percentage of completion method.

Guidelines on accounting for construction contracts are set out in FAR 2, the recommendation on inventories published by the Accounting Committee. FAR 2 permits the percentage of completion method in exceptional circumstances. One condition is that work on the contract should be almost completed, so that revenues and expenses can be estimated with near certainty.

This recommendation has been overtaken by a recommendation based on IAS 11, Construction Contracts, that takes effect on 1 January 2001. The recommendation requires the use of the percentage of completion method for all construction contracts. The completed contract method can, however, be used in individual accounts in order to minimize taxes payable.

4 Service contracts

The SFASC has also published a recommendation on accounting for service contracts that takes effect on 1 January 2001. There was some discussion whether to combine this standard with the one on Construction contracts as both are based on the percentage of completion method. In the end, the council opted for two recommendations as IASC has done. As with construction contracts, companies will be permitted to use the completed contract method in their individual accounts.

5 Accounting for leases

Up to 1997 companies did not have to capitalize their finance leases. Thus, except in the accounts of a few large multinationals, leases were an off-balance-sheet item. This was generally regarded as the biggest deficiency in Swedish financial reporting in the mid-1990s. Therefore, the SFASC published a recommendation on lease accounting, based on IAS 17 in 1995 with an effective date of 1 January 1997.

The recommendation makes a distinction between operating and financial leases. Finance leases have to be capitalized. The criteria for a finance lease are the same as in IAS 17. All companies have to disclose their lease commitments, both operating and financing, with a breakdown of the contractual payments on short (less than one year), medium (one to five years) and long-term (more than five years).

The leasing industry has, not surprisingly, been very critical of the recommendation. They have argued that the recommendation is complex and that it will ruin a valuable source of finance, particularly for smaller companies. They also criticized the criteria used to distinguish finance leases, claiming that the criteria were not operational.

It is still too early to say anything about whether the recommendation has had any effect on the propensity to use lease financing. What we can see in the 1997 annual accounts is that the amount of finance leases being capitalized is rather small. One reason that may explain this is that the recommendation only applies to lease contracts signed in 1997. Thus, the recommendation need not be applied retroactively. More disturbing is that the information given on contractual lease payments is lacking in many companies. As the payments should include rents for offices, warehouses, etc. one would expect most companies to disclose at least something. A company which provides above-average disclosure of their leasing transactions, is Ericsson (see Table VII.1).

On the issue of whether the criteria for a finance lease are operational or not, a question was put to the Emerging Issues Task Force. A trucker leased a lorry from a finance company. The finance company bought the lorry from a manufacturer. The lease term was five years with no automatic option for renewal. Estimated useful life is 10–15 years. At the end of the lease term the unamortized amount was 65 per cent of the acquisition cost. The manufacturer had agreed to buy back the truck for 40 per cent of the acquisition cost, if asked – 40 per cent was assumed to be close to the fair value of the used truck. The trucker had agreed to guarantee the residual

Table VII.1 Example of footnote with information on lease transactions

NOTE 23 LEASING

LEASING OBLIGATIONS

Equipment under Financial leases, recorded as tangible assets, consists of:

FINANCIAL LEASES	1997	1996
ACCUMULATED ACQUISITIONS		
Land and buildings	**310**	315
Machinery	**48**	52
Other equipment	**38**	48
	396	415
ACCUMULATED DEPRECIATION		
Land and buildings	**44**	37
Machinery	**38**	39
Other equipment	**22**	20
	104	96
Net value	**292**	319

At December 31, 1997, future payment obligations for leases were distributed as follows:

	Financial leases	Operating leases
1998	51	1,469
1999	44	1,164
2000	39	940
2001	33	763
2002	33	694
2003 and later	181	1,530
	381	6,560

Expenses for the year for leasing of assets were SEK 1,659 m. (SEK 1,358 m. in 1996 and SEK 801 m. in 1995).

LEASING INCOME

Some consolidated companies lease equipment, mainly telephone exchanges, to customers. These leasing contracts vary in length from 1 to 12 years.

The acquisition value of assets leased to others under Operating leases amounted to SEK 473 m. at December 31, 1997 (December 31, 1996: SEK 494 m.). Accumulated depreciation amounted to SEK 390 m. and net investments to SEK 83 m. at December 31, 1997 (December 31, 1996: SEK 392 m. and SEK 102 m., respectively).

Net investment in Sales-type leases and Financial leases amounted to SEK 155 m. at December 31, 1997 (December 31, 1996: SEK 187 m.).

Future payments receivable for leased equipment are distributed as follows:

	Sales-type and Financial leases	Operating leases
1998	32	54
1999	23	38
2000	16	33
2001	5	34
2002	1	24
2003 and later	127	6
	204	189
Less: interest	49	–
Net investment	155	189

Source: Ericsson's annual report, 1997

value if lower than 65 per cent of the acquisition cost. He did not guarantee the residual value that fell below 40 per cent of the acquisition cost as that was effectively guaranteed by the manufacturer. The question asked was: based on the facts above shall the trucker account for the lease as a finance lease?

The EITF answered no. Even though it could be argued that the only party exposed to significant risks related to the truck was the trucker, the fact that the unguaranteed value of the truck (unguaranteed by the trucker) was 40 per cent led the EITF to conclude that the lease contract did not transfer all of the significant risks to the trucker.

In 1999 the SFASC revised its recommendation on lease accounting, RR 6. The revision was based on the revised IAS 17. The new recommendation is effective starting 1 January 2001. The most significant change is that the revised recommendation expands the disclosure requirements.

6 Sale and lease-back agreements

Sale and lease-back arrangements have been very common in Sweden, especially in the late 1980s. Companies would sell their head offices, etc. to what were generally known as 'renting companies'. The renting company would then let the property back to the seller. The rent would in most cases be based on interest rates guaranteeing the renting company a margin to cover its capital costs. The renting company would also write a call option that gave the seller the right to buy the property back after 10–25 years, normally at the same price as the asset was sold for.

It is fairly clear that these sale and lease-back transactions were nothing more than loans, especially when the call option is taken into account. An old draft recommendation on leasing published by the Accounting Committee, FAR 7, included a paragraph requiring sale and lease-back agreements that in substance were loans to be accounted for as such. This recommendation was in effect ignored. The Accounting Committee therefore decided to issue a revised recommendation which said, *inter alia*, that if the selling price did not exceed an estimated market value the sale could be accounted for as a sale and not as a loan. Needless to say, this recommendation has been universally followed. It has afforded many companies a welcome opportunity to report a substantial gain and to improve their debt to equity ratios.

In the 1990s the volume of sale and lease-back transactions has all but disappeared. The most important economic reason was the collapse of the real estate market in the early 1990s. In addition, the new recommendation on leasing, RR 6, requires capitalization when the lease-back meets the criteria for being a finance lease.

Some of the larger Swedish companies have entered into 'big-ticket' sale and lease-back transactions as buyers and lessors. As an example, an engineering company could have bought a jumbo jet which then was leased out. The advantage to the company is that all machinery, including aeroplanes, are deductible for tax purposes over five years even though the economic life is significantly longer. Therefore, the transaction would, in effect, create an interest-free loan from the government. The highest tax court has, however, recently made a decision to reject the arrangement based on the argument that the transaction was outside the normal operations of the company. Thus, several companies stand to lose hundreds of millions of Swedish kronor.

7 Government grants

Accounting for government grants is discussed in recommendation R 5 of the Accounting Standards Board. The recommendation follows IAS 20 closely. According to R 5, a grant that does not have to be repaid should be accounted for as income. If there is a conditional obligation to repay, revenue recognition is permitted only if there is a high degree of probability that repayment will not be required.

Investment grants for fixed assets should be amortized over the life of the asset. This is achieved by reducing the acquisition cost by the amount of the grant. Investment grants relating to fixed assets may not be taken into income when received.

Information on the amount of government grants received and the terms of the grants should be disclosed.

8 New financial instruments

The Accounting Committee has issued a draft recommendation on options, futures and swaps. The draft recommendation is based on the principle of prudence. The premium received for a call or put option is not to be shown as revenue until the option has expired. A loss should be recognized on futures and forwards if the market price is lower than the price in the contract at the balance sheet date. In a swap agreement the company should account for the position it has received through the swap. Information should be given in the notes about swap agreements and the possible risks.

The EITF has also made a pronouncement on stock-appreciation rights (*syntetiska optioner*). A stock-appreciation right gives an employee, normally a member of the management, the right to obtain an amount of cash equal to the difference between the share price and a pre-determined price, normally the share price at the time when the right is awarded. The value of the right at the time when the right is awarded, is estimated using, for example, the Black & Scholes model. The employee normally pays this amount to the company to avoid negative tax consequences.

Any money received by the company is accounted for as income. The difference between the share price and the predetermined price at the end of each accounting period is accounted for as a liability and as a cost. The cost should be classified as a wage cost based on the assumption that the transaction is part of the pay-packet to the management.

The SFASC has discussed whether to start work on a more general recommendation on financial instruments based on IAS 32 or IAS 39. A major obstacle for the council is the fact that market values are not admissible under the Annual Accounts Act. Therefore, the council has decided to wait for work on a revision of the act, following the expected change to the Fourth Directive, to be completed before taking any major decisions in this area.

9 Foreign currency translation

Foreign currency translation is dealt with in a recently published recommendation from the Financial Accounting Standards Council, RR 8, effective 1 January 1999 and based on IAS 21. The new recommendation replaces a recommendation from the Accounting Standards Board on the translation of receivables and liabilities (R7), and a draft recommendation from the Accounting Committee on the translation of the financial statements of foreign subsidiaries.

All receivables and liabilities denominated in a foreign currency should be translated using the rates ruling at the balance sheet date. Unrealized gains are taken to income.

The recommendation is silent on accounting for hedging transactions with the exception of hedges of net investments in foreign subsidiaries. (These are dealt with in the section on group accounting.) The old recommendation from the ASB did, however, include a few paragraphs on hedging in individual companies. These paragraphs have been included in an annexe to RR 8 in order to give some guidance to companies. Hedge accounting is thus permitted when a receivable or liability in foreign currency is hedged by another item. It is not quite clear what can be counted as 'another item'. Examples suggested in the paragraph are shares in foreign subsidiaries or foreign property financed locally.

In practice, hedge accounting is widespread, especially in group accounts. It is less common in individual accounts partly because hedge accounting normally means that losses on foreign currency are deferred for tax purposes. Even so, there are several companies that state among their accounting principles that hedge accounting is used, e.g. for shares in foreign subsidiaries.

Most companies seem to have interpreted the recommendation of the ASB as permitting the hedging of anticipated transactions. Certainly the hedging of such transactions is widespread. Most companies state that they have a policy of hedging their anticipated transactions, normally for a period of 3–12 months. In a few companies the period is several years. Volvo, for example, hedges their anticipated sales for a period of up to three years. In its financial statements for 1997, Volvo disclosed a loss before tax on hedging anticipated transactions of SEK 5 billion. This information is tucked away in the footnote where Swedish GAAP is reconciled with US GAAP.

Volvo, like most other public companies, include an extensive discussion in the annual report of their exposure to currency risks and their hedging policy. Thus, one often finds information on inflows and outflows in major currencies and sometimes also information on net assets (see Table VII.2 for information provided by Volvo).

Table VII.2 Example of information given for foreign currency translation and financial instruments

Board of Directors' report

Impact of exchange rates on Volvo's operating income

Average spot rates in 1997 were significantly more favorable for Volvo than in the preceding year. As shown in the table below, the net effect of changed spot rates was positive in the amount of SEK 4,830 M. The favorable effects of higher average spot rates for "inflow" currencies, mainly U.S. dollars and British pounds, and Japanese yen, were reinforced by lower averaged spot rates for "outflow" currencies: mainly the Dutch guilder but also German mark and Belgian franc.

However, since Volvo hedges large portions of payment flows in foreign currencies, the changes in spot rates do not have an immediate impact on earnings. In 1997 the effect of forward contracts and option contracts on earnings amounted to a loss of 1,180 (1996: gain 1,100), resulting in a negative impact of 2,280 on operating earnings in 1997, compared with 1996.

Changes in spot exchange rates in connection with the translation of foreign subsidiaries' earnings, as well as the revaluation of balance sheet items in foreign currencies, also had an impact.

The total effect on Group operating income of changes in foreign exchange rates in 1997, compared with 1996, amounted to 2,600.

Total income effect due to changes in foreign exchange rates

	Net flow 1997	Income effect
Effect of changes in spot rates in each currency		
USD	2,600	2,390
GBP	650	1,310
CAD	430	250
ITL	1,195,400	170
NLG	(2,230)	140
AUD	240	100
Other		470
Effect of changed spot rates, net		4,830
Effect of forward contracts and options contracts[1]		(2,280)
Translation of foreign subsidiaries' operating income		80
Revaluation of balance sheet items in foreign currency		(30)
Total effect		**2,600**

[1] Group sales are reported at average spot rates and the effect of hedging is included among other operating income/expenses.

The Volvo Group's net commercial flows in various currencies, 1997

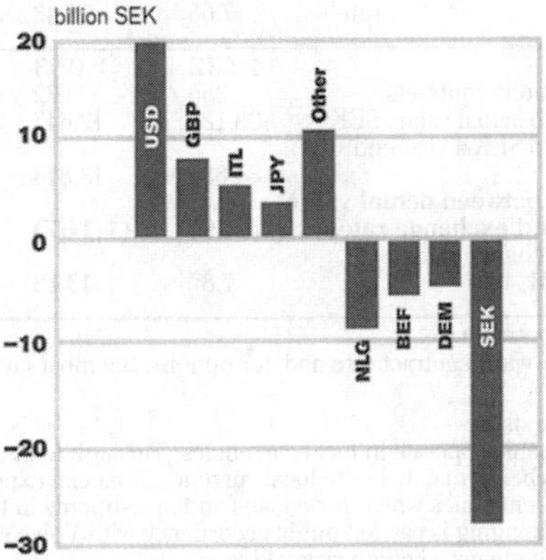

Volvo operates in an international environment with 11% of all sales in Sweden, 47% in the rest of Europe, 28% in North America, 4% in South America and 8% in Asia. Production and assembly are carried out on all continents, but mainly in Northern Europe.

Table VII.2 (cont.)

Note 30: Financial risks

In the course of its operations the Volvo Group is exposed to various types of financial risks. Group-wide policies form the basis for each Group company's action program. Monitoring and control is conducted continuously in each company as well as centrally. Most of the Volvo Group's financial transactions are carried out through Volvo's in-house bank, Volvo Group Finance, which conducts its operations within established risk mandates and limits.

Foreign exchange risk

Volvo's currency risk is related to changes in contracted and projected flows of payments (commercial exposure), to payment flows related to loans and investments (financial exposure), and to the revaluation of assets and liabilities in foreign subsidiaries (equity exposure).

The objective of Volvo's foreign exchange risk management is to reduce the impact of foreign exchange movements on the Group's income and financial position.

Commercial exposure

Volvo uses forward exchange contracts and currency options to hedge the value of future payment flows. Contracts related to hedging of anticipated sales and purchases of foreign currency normally do not exceed 36 months. In accordance with the Group's currency policy, between 40% and 80% of the net flow in each currency is hedged for the coming 12 months, 20% to 60% for months 13 through 24 and 0% to 40% for months 25 through 36. The value of all forward and options contracts as of December 31, 1997 was SEK 51.9 billion (40.2; 41.0).

Volvo Group outstanding currency contracts pertaining to commercial exposure, December 31, 1997

The table shows contracts hedging future flows of commercial payments.

		Inflow currencies				Outflow currencies		Other currencies	Total
		USD	GBP	ITL	JPY	DEM	BEF	Net expressed SEK M	
Due date 1998	amount	2,281	487	933,459	41,741	(1,212)	(19,011)	4,654	
	rate[1]	7.38	11.64	0.0043	0.0838	4.64	0.2186		
Due date 1999	amount	1,526	351	368,888	34,489	(768)	(7,823)	3,036	
	rate[1]	7.53	12.20	0.0044	0.0697	4.53	0.2167		
Due date 2000	amount	625	215	24,500	17,230	(147)	(1,705)	952	
	rate[1]	7.68	12.43	0.0045	0.0694	4.40	0.2155		
Total		**4,432**	**1,053**	**1,326,847**	**93,460**	**(2,127)**	**(28,539)**	**8,642**	
of which options contracts		355	72	4,341	15,974	(273)	0	260	
Translated to actual value, SEK[2]		33,128	12,647	5,792	7,253	(9,884)	(6,222)	9,234	**51,948**
Translated to SEK a year end 1997 rates		34,887	13,814	5,904	5,664	(9,351)	(6,082)	8,642	**53,478**
Difference between actual value and year-end exchange rates		**(1,759)**	**(1,167)**	**(112)**	**1,589**	**(533)**	**(140)**	**592**	**(1,530)**
Year-end exchange rates, December 31, 1997		7.87	13.12	0.0045	0.0606	4.40	0.2131		

[1] Average contract rate.
[2] Average forward contract rate and, for options, the most favorable of the year end rate and contract rate.

Financial exposure

Group companies operate in local currencies. Through loans and investments being mainly in the local currency, financial exposure is avoided. In companies which have loans and investments in foreign currencies, hedging is carried out in accordance with Volvo's financial policy, which means a limited risk-taking.

Equity exposure

In conjunction with translation of the Group's assets and liabilities in foreign subsidiaries to Swedish kronor, a risk arises that the currency rate will have an effect on the consolidated balance sheet. In accordance with the Group's currency policy, net assets (shareholders' equity) in foreign subsidiaries and associated companies are hedged up to 50%. Hedging is mainly done through borrowing in the same currency as the net assets. At year end 1997, net assets in subsidiaries and associated companies outside Sweden amounted SEK 25 billion, of which 17% was hedged.

	1995			1996			1997		
Total outstanding currency contracts, December 31	**Notional amount**	**Carrying amount**	**Estimated fair value**	**Notional amount**	**Carrying amount**	**Estimated fair value**	**Notional amount**	**Carrying amount**	**Estimated fair value**
Foreign exchange contracts									
– receivable position	127,290	5,771	8,320	56,238	3,388	3,492	40,349	1,951	2,006
– payable position	261,736	(5,669)	(6,956)	50,303	(3,377)	(1,477)	84,591	(2,150)	(3,762)
Foreign exchange swaps									
– receivable position	4,652	–	49	83,301	4	2,623	27,268	–	1,416
– payable position	123	–	(1)	81,565	(1)	(2,517)	26,045	–	(1,604)
Options – purchased									
– receivable position	13,558	21	1,255	11,163	33	1,075	5,135	–	379
– payable position	802	(23)	(23)	34	–	–	4,156	–	(147)
Options – written									
– receivable position	27	–	–	–	–	–	2,080		57
– payable position	7,235	(4)	(36)	5,806	–	(17)	4,274	–	(61)
Total		96	2,608		47	3,179		(199)	(1,716)

The notional amount of the derivative contracts represents the gross contract amount outstanding. To determine the estimated fair value, the major part of the outstanding contracts have been marked-to-market. Discounted cash flows has been used in some cases.

Table VII.2 (cont.)

Interest-rate risks

Interest-rate risk relates to the risk that change in interest-rate levels affect the Group's earnings. By matching fixed-interest periods of financial assets and liabilities, Volvo reduces the effects of interest-rate changes. Interest-rate swaps are used to shorten the interest-rate periods of the Group's long-term loans. Exchange-rate swaps make it possible to borrow in foreign currencies in different markets without incurring currency risk.

Interest futures contracts and futures contracts are commonly used by Volvo to offset fluctuations in variable interest rates on short-term loans. Volvo has liquid contracts, such as standardized futures contracts in Eurodollar with maturity terms of up to three years. Most of these contracts are hedging transactions for short-term borrowing.

	1995			1996			1997		
Total outstanding interest relatedcontracts, December 31	**Notional amount**	**Carrying amount**	**Estimated fair value**	**Notional amount**	**Carrying amount**	**Estimated fair value**	**Notional amount**	**Carrying amount**	**Estimated fair value**
Interest-rate swaps									
– receivable position	84,354	11	2,905	10,349	–	335	20,322	9	875
– payable position	69,746	(167)	(963)	15,416	(34)	(462)	28,142	(25)	(1,760)
Forwards and futures									
– receivable position	83,621	–	52	153,553	–	303	165,186	–	435
– payable position	169,882	–	(378)	170,670	–	(550)	190,866	(17)	(465)
Options – purchased									
– receivable position	–	–	–	426	–	2	145	–	1
– payable position	–	–	–	–	–	–	–	–	–
Options – written									
– receivable position	–	–	–	–	–	–	–	–	–
– payable position	400	(1)	(1)	–	–	–	–	–	–
Interest-rate caps and floors purchased									
– receivable position	–	–	–	301	1	1	376	–	1
– payable position	–	–	–	8	(8)	(8)	159	–	–
Interest-rate caps and floors written									
– receivable position	–	–	–	300	–	–	–	–	–
– payable position	–	–	–	88	(44)	(44)	88	(44)	(44)
Total		(157)	1,615		(85)	(423)		(77)	(957)

Credit risks in financial instruments

Counterparty risks

The derivative instruments used by Volvo to reduce its foreign-exchange and interest-rate risk in turn give rise to a counterparty risk, the risk that a counterparty will not fulfill its part of a forward or options contract, and that a potential gain will not be realized. Risks are limited through careful credit checking and the establishment of maximum levels of exposure. Where appropriate, the Volvo Group arranges master netting agreements with the counterparty to reduce exposure. The credit exposure in interest-rate and foreign exchange contracts is represented by the positive fair value – the potential gain on these contracts – as of the reporting date. The risk exposure is calculated daily. The credit risk in futures contracts is limited through daily or monthly cash settlements of the net change in value of open contracts. Since contracts off the balance sheet involve substantial credit risks, agreements on future contracts are made only with those banks for which limits are assigned based on careful credit checking. The estimated exposure in currency interest-rate swaps, forward exchange contracts and futures and options purchased amounted to 2,291, 2,441 and 380 as of December 31, 1997.

Volvo does not have any significant exposure to an individual customer or counterparty.

An increase of credit risk – on or off the balance sheet – occurs in financial instruments when counterparties have similar economic characteristics that could cause their ability to meet contractual obligations to be impaired by unfavorable developments in the business community.

Credit risk in financial investments

The liquidity in the Group is invested mainly in local cash pools or directly with Volvo Group Finance. This concentrates the credit risk to the Group's in-house bank. Volvo Group Finance invests the liquid funds in the money and capital markets.

All investments must meet criteria for low credit risk and high liquidity.

Calculation of fair value of financial instruments

Volvo has used various methods and assumptions, which were based on evaluations of market conditions and risks existing at the time, to estimate the fair value of the Group's financial instruments as of December 1995, 1996 and 1997, respectively. In the case of certain instruments, including cash and cash equivalents, non-trade accounts payable and accruals and short-term debt, it was assumed that the carrying amount approximated fair value for the majority of these marketable instruments, due to their short maturities.

Quoted market prices or dealer quotes for the same or similar instruments were used for the majority of marketable securities, long-term investments, long-term debt, forward exchange contracts and options contracts. Other techniques – such as estimating the discounted value of future cash flows, replacement cost and termination cost – have been used to determine fair value for the remaining financial instruments. These values represent approximate valuations and may not be realized.

Unrealized exchange losses are charged to income for hedging transactions related to booked items, while unrealized gains are not. Unrealized gains are taken into account when calculating market value.

Source: Volvo's annual report 1997.

10 Prior year adjustments

Prior year adjustments are dealt with in two recommendations of the Swedish Financial Accounting Standards Council published in 1993, RR 4 and RR 5. RR 4 deals with exceptional and extraordinary items. One example of an exceptional item relates to prior year items due to the discovery of a fundamental error. In practice, if a fundamental error were to be discovered in a previous financial statement it would be accounted for in the year when the error is discovered. As such cases are very rare, the council decided that there was no need to require fundamental errors to be accounted for retrospectively along the lines of IAS 8. Instead the recommendation is that the correction of fundamental errors should be accounted for as part of the normal profit, possibly as an exceptional item, with adequate disclosure of the facts in the notes on the accounts.

The second recommendation, RR 5, is concerned with changes in accounting principles. This recommendation requires most such changes to be accounted for retrospectively. This is controversial, as it could be argued that changing a balance sheet retrospectively is not in accordance with the Annual Accounts Act. It means that in many cases the book value of an asset will be increased. One example could be a requirement to capitalize interest costs. Since revaluations are permitted only in special circumstances this could be regarded as against the law.

The SFASC spent much time debating the problem before coming to its conclusion. One decisive argument for retrospective amendment was that otherwise, in most cases, it would be possible to change an accounting principle only with forward effect. As the Annual Accounts Act also requires information in the financial statements to be comparable, the council concluded that the preparatory work to the Annual Accounts Act could not have taken changes in accounting principles into consideration when revaluation was discussed and hence that retrospective application would be the best solution.

11 Accounting for the euro

A Royal Commission has recently published a report on how to account for the euro. The main proposals are that companies may use the euro in their accounts but that tax returns still have to be filed in Swedish kronor. A decision to change the reporting currency requires that a decision has been taken at an annual general meeting. Therefore, no change was possible before the year 2000. A company that opts for euro has to state its share capital in euro. Swedish shares have to have a par value which is normally an even amount, e.g. SEK 100. When translating this into Euros, the figure may become EUR 10,5263, which is obviously impractical. It was therefore discussed to abolish the system of par value shares, replacing them with quota shares. No proposal for such a change has, however, been made at this moment.

A few companies already provide information on their financial statements in euro. A change to euro is not expected to bring any other accounting changes.

VIII Revaluation accounting

1 Revaluation accounting

The Annual Accounts Act gives companies a limited right to revalue fixed assets. The rules were outlined above (see VI.5.6). It is not known to what extent companies actually use this right. In some industries, notably the forest-based industry, revaluations are common. The assets revalued are usually timberland and hydroelectric plants. One reason for undertaking revaluations is to increase share capital through a bonus issue in order to improve the visible debt to equity ratio. Also, it could be argued that revaluing the assets means that the balance sheet gives a better picture of the net worth of the company.

2 Current cost accounting

In 1980 the Accounting Committee put out an exposure draft on current cost accounting. The draft was based on the seminal work of Edwards and Bell (1965). The draft included several alternatives. One of them meant asking companies to revalue their assets to current values and calculate an operating profit based on the revalued amounts. The amount at which the assets were revalued was included in the income statement as a holding gain. This was in marked contrast to the British current cost proposal, for example, where holding gains were excluded. In the Swedish proposal the holding gains were added to the operating profit in order to arrive at a true nominal profit. From this nominal profit an amount was to be deducted equal to the reduction in general purchasing power of the equity. The result was an inflation-adjusted or real profit.

At about the same time as the Accounting Committee issued its exposure draft a Royal Commission started work on a proposed new corporate tax system based on real profit. The commission published its findings in 1982, at a time when business profits had fallen substantially. Thus interest in adjustment for inflation had all but disappeared.

A few companies have published supplementary accounts based on the exposure draft from the Accounting Committee. However, these accounts received little attention. One reason was the inclusion of the holding gains, which, although correct in theory, was regarded as permitting phantom profits to be shown in the income statement.

No listed company continues to publish current cost accounts in their annual report. A few companies provide, however, information on current costs for property, plant and equipment. One of these is Volvo, see Table V.2. The exposure draft on current cost accounting has not been officially withdrawn but it is no longer included in the annual volume of accounting recommendations published by the Swedish Institute of Authorized Public Accountants (FAR).

IX Notes and additional statements

1 Notes

The Annual Accounts Act, Chapter 5, requires the following information to be disclosed in the notes on the financial statements if the information is not already given in the statements:

- The principles adopted for valuing assets and liabilities.
- The methods used for depreciating fixed assets, including estimates of the economic life of various assets.
- Information on fixed assets, including acquisition cost, additions, disposals and transfers during the financial year, current and accumulated depreciation and write-downs.
- The assessed value of properties (applies only to properties in Sweden).
- Net turnover of lines by business or geographical segments that are materially different from each other.
- Information on intragroup purchases and sales.
- Information about revaluations.
- Information on the revaluation reserve.
- Details of shares held in subsidiaries and in other companies where the holding exceeds 20 per cent, including the names of the companies, the number of shares owned, the book value of the shares and the net result. Information does not have to be published if deemed to be insignificant or if it could give rise to competitive harm.
- Information on receivables due after more than 12 months and on loans due after more than 12 months and after more than five years.
- Information on assets pledged.
- Loans to shareholders, directors and employees.
- Details of convertible loans and warrants.
- The number and value of shares issued. A breakdown of equity detailing funds which are restricted and which are available for distribution. An explanation of changes in the equity from the previous year.
- Details of the different kinds of shares that have been issued.
- Information on deferred taxes if not given in the income statement.
- Average number of employees including a breakdown on males and females.
- Wages including social costs. Amounts paid to board members and to the chief executive director shall be disclosed separately.
- Retirement benefits to board members and to the chief executive director.
- Agreements on severance pay to board members and to the chief executive director.
- Subsidiaries shall give information on the name and location of the parent company.

- Information on the amount of remuneration paid to the auditor detailing the amount which is paid for other services than auditing. This requirement became effective from 1 January 1999.

The requirement to provide information on severance payments was taken from a recommendation published by the Stock Exchange and the Industry Committee on Stock Exchange Matters (*Näringslivets Börskommitté*) in 1993 (NBK, 1993).

At that time it was realized that a number of directors had received very generous compensation when they resigned. In some cases it came as a surprise, since there had been no mention of the terms in the annual report. The NBK recommendation applies to all listed Swedish companies. For each of the following:

- The chairman of the board.
- Any member of the board who receives special compensation over and above what was approved by the AGM.
- The chief executive officer.

Information should be given on:

- The total amount of compensation and other benefits.
- The essential terms of agreements on future pensions.
- The essential terms of severance payments when the person leaves the company.

For other people in the management team, summary information should be given on the last two items above.

This recommendation is still in effect.

According to the act, companies have to disclose all shareholdings exceeding 20 per cent, even indirect. For a large group this could mean that they have to include several pages in the annual reports with details of shareholdings of little public interest. In late 1998 it was decided that a parent company may limit the information to its direct holdings and that holdings that are not important for providing a true and fair view need not be included in the printed annual report. However, all direct holdings must be included in the copy of the report that is sent to the Patent and Registration Office.

2 Additional statements

2.1 Funds statement/cash flow statement

The Annual Accounts Act requires larger limited liability companies to publish a funds statement. By 'larger' is meant companies with, *inter alia*, over 200 employees (see IV.1.1 for details).

The act does not lay down a format for the funds statement. Instead, a format is to be found in a recently published recommendation, RR 7, from the SFASC, effective from 1 January 1999. The Swedish recommendation is a translation of IAS 7. Thus, the recommendation gives companies an option to choose between the direct and the indirect method. It is expected that almost all companies will choose the latter starting with net profits.

When the exposure draft was published, several companies criticized the draft for not permitting the reporting of an operating cash flow. Operating cash flow is defined as cash flow from operations before financial cash flows and tax but after capital investments. Operating cash flows are of interest to both management and users as it can be calculated for segments. In spite of this, the SFASC decided not to depart from the format prescribed in IAS 7. Instead, companies are recommended to include as an extra item, information on operating cash flows.

The SFASC also recommended companies to provide supplementary information on changes in net debt. Several companies argued that the change in liquid assets, as defined in IAS 7 and RR 7, is not very relevant to users. The change in net debt is a more meaningful figure. The SFASC agreed recommending that this information should be added.

An example of a cash flow statement is given in Table IX.1. It is taken from SCA, one of the largest forest-based companies in Sweden. The forest industry has been active in promoting the cash flow statement as a management tool and SCA has been one of the companies criticizing the format laid out by the SFASC for being too rigid. As can be seen, the SCA format differs from the format in RR 7. Still, according to many users it provides much better information on the cash flows of the company.

Table IX.1 SCA's cash flow statement 1993

		1993	1992
Source of liquid funds			
Funds provided from operations	Income before taxes and minority interests	2,616	4,865
	Income taxes	−1,043	−783
	Other items not affecting liquidity	574	1,675
	Withdrawals from blocked accounts	5	24
	Total funds provided from operations (a)	2,152	5,781
Funds used in operations	Change in inventories	−648	161
	Change in operating receivables	−1,152	−1,471
	Change in operating liabilities	1,586	1,696
	Total funds used in operations (b)	−214	386
Investments	Investments in FICE, net effect on liquidity	−2,915	–
	Investments in PPE, etc.	−2,648	−2,488
	Investment in shares and participations	−635	−236
	Investment in intangible assets	−113	−1,582
	Sales of fixed assets	1,444	1,953
	Change in group composition	−398	−1,701
	Total investments, net (c)	−5,265	−4,054
Total liquid funds provided (d)=(a)+(b)+(c)		**−3,327**	**2,113**
Financial borrowing and lending	Change in financial lending	−73	297
	Change in financial borrowing	2,045	1,002
	Change in pension provision	81	−226
	Change in other long-term liabilities	−23	−37
	Change in minority interests	−50	46
	Translation difference	−167	−207
Total financial borrowing and lending (e)		**1,813**	**875**
Dividends		−900	−799
Total transactions with owners (f)		**−900**	**−799**
Total change in liquid assets during the period (d) + (e) + (f)		**−2,414**	**2,189**
Opening balance		8,532	6,343
Changes	Changes in cash and bank deposits	−35	−471
	Change in short-term investments	−2,379	2,660
Closing balance in liquid assets		6,118	8,532

Source: SLA's annual report, 1993

2.2 Segment reporting

Information on net turnover per segment is required by the Annual Accounts Act where companies have more than one line of business or geographical segments that are substantially different from each other. Guidance on the legal requirements is given by the Accounting Standards Board in its recommendation R 9. The requirement in the Annual Accounts Act applies to all companies, even small ones. The Accounting Standards Board recommendation focuses primarily on public companies, as the need for information on individual lines of business is greater in their case.

The board's R 9 follows IAS 14 (before it was revised) closely. Companies are recommended to disclose, *inter alia*, turnover, operating profit and net assets for each line of business. A line of business should normally comprise at least 10 per cent of turnover or assets.

For geographical segments the recommendation is to disclose turnover, operating profit, the number of employees and capital investments.

In practice, the quality of segmental reporting among listed companies is very high. Most companies provide the information recommended by the Accounting Standards Board for five or more lines of business. Moreover, information is often added about competition, capital investment and the prospects for the coming year. The information about geographical segments, however, tends to be less detailed.

Some of the information on SCA's business areas is shown in Table IX.2. In addition, SCA provides an extensive verbal discussion on each business area in the annual report.

2.3 Value-added statements and social accounting

A few companies include a value-added statement in the annual report. This is probably a legacy of the 1970s when there was lively debate about social accounting. Interest has since subsided. There are no recommendations either on social accounting or on value-added statements. The formats of the value-added statements often differ from each other. The value of the information they convey is dubious. An example of a value-added statement taken from Electrolux is given in Table IX.3.

2.4 Earnings per share

Almost all listed companies publish data on earnings per share. Often earnings per share are calculated both before and after dilution and before and after extraordinary items.

The most authoritative guidelines on calculating earnings per share are those published by the Society of Swedish Financial Analysts (*Sveriges Finansanalytikers Förening*, SFF, 1992). In the society's recommendation, earnings are defined as, in normal cases, the net profit after tax divided by the average number of shares.

Unlike in some other countries, it is not recommended that goodwill should be excluded. Extraordinary items should be included in the earnings figure but enough detail should be given to make it possible for analysts to take out the items they consider non-recurring. Tax should include deferred tax.

The main difference between the Swedish recommendation and IAS 33 relates to the calculation of diluted earnings per share when companies have issued warrants. IAS 33 assumes that the dilutive effect is equal to the difference between the average price for the share during the year and the exercise price. The Swedish recommendation, on the other hand, calculates dilution as the difference between the share price at the end of the year and the exercise price. Furthermore, the Swedish recommendation recommends that the exercise price is discounted. IAS 33 is silent on discounting.

There is also a difference between the Swedish recommendation and IAS 33 on presentation. IAS 33 requires that information on earnings per share is given on the face of the income statement. There is no similar Swedish requirement.

3 Management report

The information that should be given in the management report (*förvaltningsberättelse*) is detailed in The Annual Accounts Act, Chapter 6, §§ 1 and 2. By the terms of the act, information should be provided on such facts as are essential to form an opinion on the profitability and the financial position of a company and which are not given in the financial statements. Information should also be

Table IX.2 Operations in brief 1997

Group total[1]

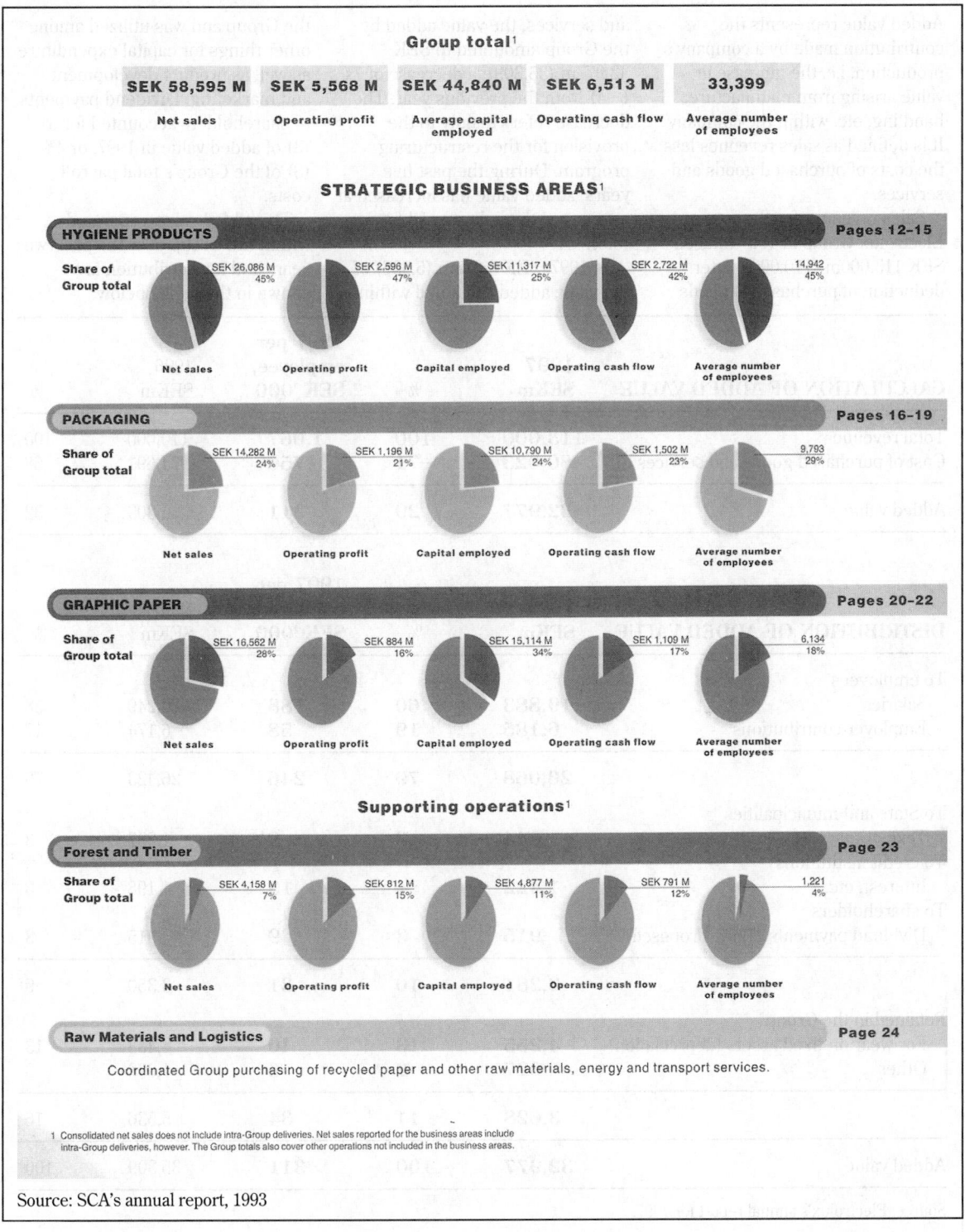

SEK 58,595 M	SEK 5,568 M	SEK 44,840 M	SEK 6,513 M	33,399
Net sales	Operating profit	Average capital employed	Operating cash flow	Average number of employees

STRATEGIC BUSINESS AREAS[1]

HYGIENE PRODUCTS — Pages 12–15

Share of Group total

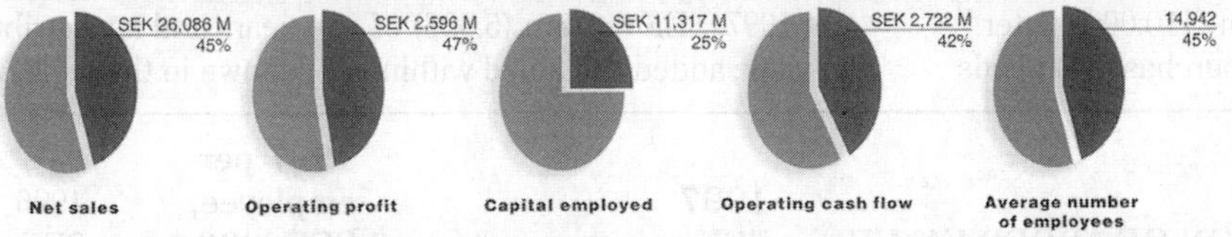

PACKAGING — Pages 16–19

Share of Group total

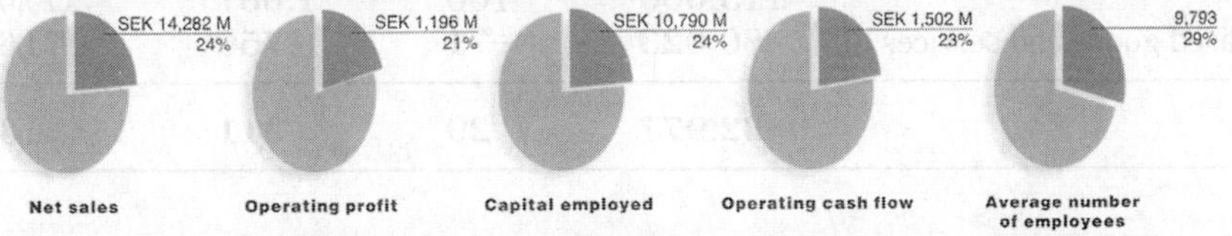

GRAPHIC PAPER — Pages 20–22

Share of Group total

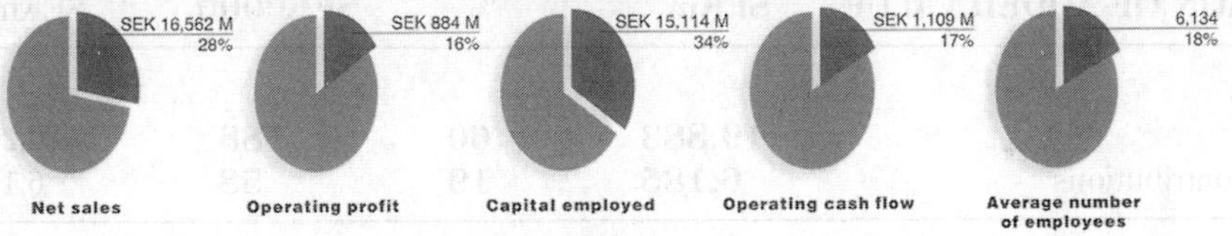

Supporting operations[1]

Forest and Timber — Page 23

Share of Group total

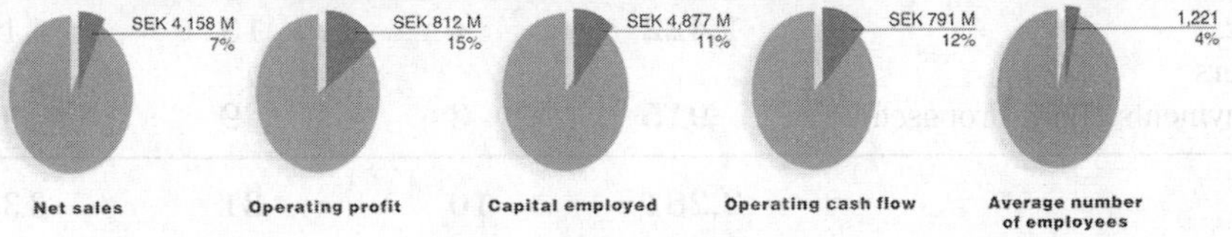

Raw Materials and Logistics — Page 24

Coordinated Group purchasing of recycled paper and other raw materials, energy and transport services.

1 Consolidated net sales does not include intra-Group deliveries. Net sales reported for the business areas include intra-Group deliveries, however. The Group totals also cover other operations not included in the business areas.

Source: SCA's annual report, 1993

Table IX.3 Example of statement of added value

Added value represents the contribution made by a company's production, i.e. the increase in value arising from manufacture, handling, etc. within the company. It is defined as sales revenues less the costs of purchased goods and services.

Sales revenues for the Electrolux Group in 1997 totalled SEK 113,000m (110,000). After deduction of purchases of goods and services, the value added by the Group amounted to SEK 32,977m (35,309), a decrease of 7% (−4) from the previous year. The decrease refers mainly to the provision for the restructuring program. During the past five years, added value has increased at an average annual rate of 4.5% (5.8).

In 1997, SEK 3,628m (5,536) of the value added remained within the Group and was utilized among other things for capital expenditure as well as product development and marketing. Dividend payments to shareholders accounted for 3% (3) of added value in 1997, or 4% (3) of the Group's total payroll costs.

The added value generated within the Group over the past two years and its distribution are shown in the tables below.

CALCULATION OF ADDED VALUE	1997 SEKm	%	1997 per employee, SEK '000	1996 SEKm	%
Total revenues	**113,000**	**100**	**1,067**	110,000	100
Cost of purchased goods and services	**−80,023**	**−71**	**−756**	−74,691	−68
Added value	**32,977**	**29**	**311**	35,309	32

DISTRIBUTION OF ADDED VALUE	1997 SEKm	%	1997 per employee, SEK '000	1996 SEKm	%
To employees					
Salaries	**19,883**	**60**	**188**	20,249	58
Employer contributions	**6,185**	**19**	**58**	6,174	17
	26,068	**79**	**246**	26,423	75
To State and municipalities					
Taxes	**944**	**3**	**9**	1,237	3
To credit institutions					
Interest, etc.	**1,422**	**4**	**13**	1,198	3
To shareholders					
Dividend payments (1997: Proposed)	**915**	**3**	**9**	915	3
	3,281	**10**	**31**	3,350	9
Retained in the Group					
For wear on fixed assets (depreciation)	**4,255**	**13**	**40**	4,438	13
Other	**−627**	**−2**	**−6**	1,098	3
	3,628	**11**	**34**	5,536	16
Added value	**32,977**	**100**	**311**	35,309	100

Source: Electrolux's annual report for 1997.

supplied on important transactions or events that have occurred during the financial year or since the end of the reporting period.

In addition, companies have to provide information on expectations for future years, on ongoing research and development work and on foreign branches.

Finally, the management report should give information on the appropriation of profits for the year.

In practice the management report has become a formal document with little information value. Instead, most companies publish an overview explaining what has happened during the year under review. It resembles the North American management discussion and analysis but it is not a mandatory part of the annual report.

4 Information on the environment

A new requirement, effective 1 January 1999, is that some companies have to provide information in the management report on how their operations affect the environment. Companies that are included are those who require a special licence from the authorities because their operations may be harmful to the environment. An annual report has to be sent in to the authorities before 31 March. The report should specify what conditions on environment that apply to the company and what activities the company has undertaken in order to meet the conditions. The report only covers operations in Sweden.

Example of how a company may report according to a statement from the Accounting Standards Board:

Company X runs a business which requires a special licence from the authorities. The licence relates to a laundry which uses different kinds of organic solvents. The organic solvents affect the environment when dissolved in the air. During year 20x1 the licence will have to be renewed because the volume handled by the laundry is expected to increase. The part of operations that is effected by the licence accounts for 90 per cent of the company's net turnover.

5 Interim reports

Under the Annual Accounts Act, interim reports must be published by all companies that are required to have an authorized public accountant. One of the criteria for the requirement is a payroll of over 200 employees. At least one interim report has to be published. It should cover at least six but not more than eight months. (For details of the different corporate forms and the requirement to publish interim reports, see IV above.) The report should give information about the company's operations, including turnover, pre-tax profit, capital investments and changes in liquidity, during the period.

According to a new listing agreement, listed companies have undertaken to publish quarterly interim reports, starting from 1999. The report should be published within two months from the end of the reporting period. The interim report should contain:

- A condensed income statement with comparative figures for the last year. Net income should be reported on an after-tax basis.
- Diluted and undiluted earnings per share.
- A condensed balance sheet with comparative figures for the last year.
- A condensed cash flow statement with comparative figures for the last year.
- Turnover and/or operating income per segment if reported in the annual report.
- A comment on important factors necessary to understand the company's performance, including the effect of significant exceptional items.
- Information on capital investments.
- Information on when the next interim report is expected to be published.

The SFASC is working on a recommendation, based on IAS 34, detailing what information should be provided.

There is no requirement for interim reports to be audited. However, the stock exchange recommends a review. If there has been no review by the auditors, the fact has to be disclosed.

X Auditing

The auditing profession consists of two main groups, authorized public accountants (*auktoriserade revisorer*) and approved accountants (*godkända revisorer*). There are approximately 2,000 members in each of the respective professional organizations, the Institute of Authorized Public Accountants (*Föreningen Auktoriserade Revisorer*, FAR) and the Swedish Society of Registered Accountants (*Svenska Revisorssamfundet*). The institute is a member of the IFAC, IASC and FEE.

Members have to be actively in practice. If an auditor moves into industry, the qualification must be surrendered.

The auditing profession has been regulated by the Supervisory Board of Public Accountants (*Revisorsnämnden*) since 1995. The board issues authorizations, publishes instructions and hears complaints against authorized or approved accountants.

The requirements for authorization are broadly an MA plus a degree in business, economics, information technology, company law, tax law, etc. The applicant also has to demonstrate five years of practice under the supervision of an authorized accountant. Formerly there was no need to pass a formal test before qualifying. Since 1994, however, all applicants have had to pass an examination.

The academic requirements for becoming an approved accountant are less comprehensive.

All companies that have to prepare financial statements must also be audited. (For the definition of such companies see IV.1.) All limited liability companies have to engage at least one auditor who is either authorized or approved. In larger companies the auditor has to be authorized. A larger company is one that has:

- Assets that exceed 1,000 basic amounts as determined under the National Insurance Act.
- Over 200 employees.
- A stock exchange listing.

Large economic associations also have to have either an authorized or an approved auditor. A large association is defined using the same criteria as in the Annual Accounts Act.

The audit covers the annual report. The annual report consists of an income statement, a balance sheet and the management report, plus a funds statement in the case of larger companies. The annual report should be handed over to the auditors no later than six weeks before the annual general meeting of the shareholders. The annual general meeting has to be held no later than six months after the end of the financial year.

The auditors should render an audit report to the meeting. The report should be submitted to the board of directors no later than three weeks beforehand. The audit report should include a statement as to whether the annual report has been prepared in accordance with the Annual Accounts Act (or the Economic Associations Act, see Table X.1).

The audit report should also include a statement discharging the board members and the managing directors from liability. There should be a qualification in the report if the company has not fulfilled any of its obligations to:

- Make deductions for preliminary tax in accordance with the Tax Collection Act.
- Apply for registration for the purposes of value-added tax.
- Submit tax returns.
- Pay tax on time.
- Register for VAT.
- Withhold employee taxes on income.
- Pay sales taxes and charges on time.
- Submit returns for employee taxes, social charges and VAT.

As from 1999, the auditor must also report on violations of a number of specified laws.

In practice it is very rare for an auditor to qualify the annual report. Nevertheless, the auditor may have some critical comments to add. These should be communicated to the board of directors or the managing director.

The auditor may not pass to an individual shareholder or third party any information about the company which he or she has obtained in the course of his or her duties.

XI Filing and publication

The publication of financial statements is governed by ABL and the Annual Accounts Act (*Årsredovisningslagen*) 8:3. Under the Annual Accounts Act all companies with limited liability

Table X.1 Auditors' report

To the general meeting of the shareholders of ________

Registered Number ________

We have audited the parent company and the consolidated financial statements, the accounts and the administration of the board of directors and the managing director of ________ for (the XX month period ending on month XX) 20__. These accounts and the administration of the Company are the responsibility of the board of directors and the managing director. Our responsibility is to express an opinion on the financial statements and the administration based on our audit.

We conducted our audit in accordance with Generally Accepted Auditing Standards in Sweden. Those Standards require that we plan and perform the audit to obtain reasonable assurance that the financial statements are free of material misstatement. An audit includes examining, on a test basis, evidence supporting the amounts and disclosures in the financial statements. An audit also includes assessing the accounting principles used and their application by the board of directors and the managing director, as well as evaluating the overall presentation of information in the financial statements. We examined significant decisions, actions taken and circumstances of the Company in order to be able to determine the possible liability to the Company of any board member or the managing director or whether they have in some other way acted in contravention of the Companies Act, the Annual Accounts Act or the Articles of Association. We believe that our audit provides a reasonable basis for our opinion set out below.

In our opinion, the parent company and the consolidated financial statements have been prepared in accordance with the Annual Accounts Act, and, consequently we recommend

that the income statements and the balance sheets of the Parent Company and the Group be adopted, and

that the profit (loss) of the Parent Company be dealt with in accordance with the proposal in the Administration Report.

(A separate list of loans, assets pledged and contingent liabilities has been prepared in accordance with the stipulations in the Companies Act.)

In our opinion, the board members and the managing director have not committed any act or been guilty of any omission, which could give rise to any liability to the Company. We therefore recommend

that the members of the board of directors and the managing director be discharged from liability for the financial year.

Place and date

Name
Authorised Public Accountant

have to publish their annual accounts. No later than one month after the annual general meeting the accounts have to be sent to the central registry (which for limited companies is the Patent and Registration Office).

For other companies there are three different size tests. Very small, unregistered firms are not required to prepare annual accounts. Such firms are those with a turnover that does not normally exceed 20 basic amounts according to the National Insurance Act.

The annual accounts of larger firms – partnerships, economic associations, etc. – have to be available on the company's premises for inspection by those who are interested. These accounts should contain an income statement and a balance sheet.

The largest companies, those with assets in excess of 1,000 basic amounts, or over 200 employees, must file their annual accounts with the county administrative board. The accounts should contain the same information as that which is required of large limited liability companies.

In practice, copies of annual accounts are readily obtainable. The Patent and Registration Office has a special department that distributes copies of the annual accounts of limited liability companies for a small fee. Naturally the annual accounts of very small companies and of those companies in financial difficulty may be missing or incomplete. However, the Patent and Registration Office has threatened to delist companies that do not file their accounts within two years of the end of the financial year.

XII Sanctions

Infringement on the rules of the Annual Accounts Act can lead to directors and the managing director being fined. At the extreme they might be sentenced to imprisonment. This would probably happen in a case of serious fraud.

Failure to follow an accounting recommendation could lead to fines imposed by the disciplinary committee of the Stock Exchange (*Fondbörsens disciplinkommitté*). According to the listing agreement companies have undertaken to observe good accounting practice and to follow recommendations published by the Financial Accounting Standards Council. A departure from the recommendations may, however, be made provided an explanation is given. So far, few cases of failure following an accounting recommendation have been brought to the disciplinary committee. One of the few relates to Frontec, a company in the IT industry.

In their annual report for 1994, Frontec had stated that it had a policy of not capitalizing expenditure on R&D. Still, in their 1995 accounts, some expenditure had been capitalized. The amount was significant as the capitalization enabled Frontec to report a profit equal to what had been forecasted. Without capitalization, profits would have remained unchanged from 1994. In addition, Frontec provided no information that it had changed its accounting policy related to R&D.

Before the disciplinary committee, Frontec explained that the expenditure that had been capitalized related to a new operation being set up in the United States. Thus, it was argued that Frontec had not changed any accounting policy but that the company instead had adopted a new accounting policy for a different kind of R&D. Somewhat surprisingly, the committee accepted the argument. Still, Frontec was fined because the information given about the adoption of the new accounting policy was lacking.

In addition it should be mentioned that the auditing profession requires its members to encourage the use of the recommendations issued by the Accounting Standards Board, the Financial Accounting Standards Council and the Accounting Committee. There is no requirement, however, to report on whether a company has opted to follow the recommendations or not.

Despite the lack of sanctions, the recommendations published by the Swedish Financial Accounting Standards Council seem, so far, to have been adopted by almost all listed companies. But it needs to be borne in mind that the big test will come in the years ahead when the SFASC has completed its programme of 'catching up' with the IASC.

XIII Financial Reporting institutions and national databases

Institution	Address	Telephone	Fax No.	E-mail/Web page
FAR	Norrtullsgatan 6 Box 6417 113 82 Stockholm	46 823 41 30	46 834 14 61	sekr@far.se http://www.far.se
Redovisningsrådet	Odeng.65 Box 6417 113 82 Stockholm	46 815 18 95	46 832 12 50	http://www. redovisningsradet.se
Bokföringsnämnden	Regeringsg. 48 Box 7831 103 98 Stockholm	46 8787 80 28	46 821 97 88	–
Finansinspektionen	Box 7831 103 98 Stockholm Regeringsgatan 48, 7 tr (Salén-huset)	46 8787 80 00	46 824 13 35	finansinspektionen@fi.se http://www.fi.se
Revisorsnämnden	Karlavägen 104 Box 24014 104 50 Stockholm	46 8783 18 70	46 8783 18 71	rn@revisorsnamnden.se

Databases

BolagsFakta
Gives information about companies in the Nordic region
http://www.bolagsfakta.se/international/

Six
Gives information about Financial markets
http://www.six.se/en_index.html

Bibliography

Artsberg, K. (1992). *Normbildning och redovisningsförändring* (Lund University Press, Lund).
BAS (1976). *Kontoplan med anvisningar* (Näringslivets, Stockholm).
Blomberg, G. and Oppenheimer, B. (1993). Nytt utslag i HD gör det lättare att aktivera utvecklingskostnader. *Balans*, **11**, pp. 16–20.
Edwards, E. and Bell, P. (1965). *The Theory and Measurement of Business Income* (University of California Press, Berkeley and Los Angeles).
FEE. (1997). *FEE Comparative Study on Conceptual Accounting Frameworks in Europe* (FEE, Brussels).
Financial Accounting Standards Board (1993). *Accounting for the Impairment of Long-lived Assets* (FASB, Norwalk, Conn.).
Föreningen Auktoriserade Revisorer (1982). *Swedish Accounting and Auditing* (FAR, Stockholm).
Föreningen Auktoriserade Revisorer (1996). *Om årsredovisning i aktiebolag* (FAR, Stockholm).
Fridman, B. and Hanner, P. V A. (1987). Överväganden kring FARs rekommendation om koncernredovisning. *Balans*, **12**, pp. 36–47.
Johansson, S-E. and Östman L. (1985). Återvändsgränd för nya redovisningsstandards. *Balans*, **4**, pp. 12–26.
Kedner, G. (1989). *Svensk anpassning till EGs normer för årsredovisning*, Report 1989:2 (Bokföringsnämnden, Stockholm).
KPMG Bohlins (1993). *Bohlins Skattehandbok* (KPMG Bohlins, Stockholm).
Malmström, L. (1993). Stockholmsbörsen har ingen tradition att styra på redovisningsområdet. *Balans*, **6–7**, pp. 44–46.
Näringslivets Börskommitté (1993). *Näringslivets Börskommitté rekommendation om information angående ledande befattningshavares förmåner* (NBK, Stockholm).
Nobes, C. (1993). International accounting: differences and how to remove them. *Professional Investor*, **2**, pp. 28–30.
Redovisningsrådet (1991). *Information om Redovisningsrådet* (Redovisningsrådet, Stockholm).
Rundfelt, R. (1991). *Tendenser i börsföretagens årsredovisningar 1991* (Stockholms Fondbörs, Stockholm).
Rundfelt, R. (1992). *Tendenser i börsföretagens årsredovisningar 1992* (Stockholms Fondbörs, Stockholm).
Rundfelt, R. (1993). *Tendenser i börsföretagens årsredovisningar 1993* (Stockholms Fondbörs, Stockholm).
SOU 1998:136, *Accounting and share capital in euro, Stockholm 1998.*
Stockholm Stock Exchange (1998). *Fact Book 1998* (Stockholms Fondbörs, Stockholm).
Statistiska Centralbyrån (1997). *Statistical Yearbook of Sweden 1992* (Statistiska Centralbyrån, Stockholm).
Sveriges Finansanalytikers Förening (1992). *Finansanalytikernas rekommendationer* (SFF, Stockholm).
Tidström, G. and Hesselman, A. (1991). Beskattningen – skall den utgå från god redovisningssed? *Balans*, **6–7**, pp. 13–17.
Thorell, P. (1996) *ÅRL. Årsredovisningslagen* (Justus förlag, Uppsala).

SWEDEN
GROUP ACCOUNTS

Walter Schuster

Acknowledgement

The author wishes to thank Professor Lars Ostman, who actually co-wrote the original version of this text, and has acted as an advisor on this edition. He also wishes to thank Professor Emeritus Sven–Erik Johansson for his helpful comments on the manuscript to the first edition, and Anja Lagerstrom and Thomas Hjelstrom for great practical assistance.

CONTENTS

I Introduction and background

The idea of group accounting was introduced in Sweden at the beginning of the 1930s, both in practice, although to a limited extent, and as a subject for discussion in circles interested in the development of financial accounting. A major event which sparked off these ideas occurred in 1932. Ivar Kreuger was the leading industrialist in Sweden at the time who also had a remarkable reputation among financial institutions abroad. He had, for instance, restructured much of the match industry of the world as well as shaping the Swedish Match Company and its subsidiaries. Another of his core companies, Kreuger & Toll, was a construction company. His empire came under heavy financial stress around 1930 and Kreuger took his life. Afterwards, it became apparent that he had used transfers between his companies to conceal the true financial position of the group. Shareholders and creditors alike lost money. Politicians, journalists and ordinary people reacted emotionally and legal action was taken, in particular against a leading bank manager close to Kreuger. The manager was convicted; although whether this was the right verdict has since been disputed (af Trolle, 1992).

The fact that the financial problems of the Kreuger Group had been concealed was attributed to the lack of group accounting. Ideas already introduced thus gained further support. The accounting laws did not require group accounting and such requirements obviously became a major issue for new legislation. A new Companies Act (*Aktiebolagslagen*, ABL) was proposed in 1941 and came into force in 1944 (see Table I.1). According to this Act, parent companies were obliged to prepare either a complete balance sheet for the group or a special report on parent company and group equity with regard to distributable group equity. The auditors of the parent company had to review these documents. As a minimum in terms of publication, a report on distributable group equity had to be published by the parent company.

After the introduction of this Act, there was a voluntary evolution towards complete group accounting. This tendency was especially strong during the 1960s. Gradually more and more companies which were registered on the Stockholm Stock Exchange started publishing consolidated income statements and balance sheets. Pioneer companies, like Asea, Electrolux and SKF, took the lead. Standards introduced by private standard-setting institutions played an important role in this development. In 1968 the Swedish Industry and Commerce Stock Exchange Committee (*Näringslivets börskommitté*, NBK), with the Federation of Swedish Industries (*Sveriges Industriförbund*, SI) and the Stockholm Chamber of Commerce (*Stockholms Handelskammare*) as principals, published an influential recommenda-

Table I.1 Historical development of Swedish group accounting

1944	New Swedish Companies Act. Parent companies obliged to prepare a group balance sheet or a special report on parent company and group equity with regard to distributable dividends.
1968	NBK published a recommendation on income statements emphasizing groups as the most important accounting entities.
1971	Accounting Committee of the FAR published a draft recommendation on methodological issues. Still no legal requirement nor obligation to prepare group accounts.
1975	New Swedish Companies Act. Parent companies required to publish consolidated income statements, balance sheets and statements of source and application of funds.
1976	Swedish Accounting Act (Bokföringslagen) issued.
	Accounting Committee of the FAR published a new recommendation, relying heavily on the draft from 1971.
1991	First recommendation from the RR published.
1995	Annual Report Act (Årsredovisningslagen) issued.
1996	New recommendation from the RR on group accounting published.
1998	Recommendation from the RR on accounting for effects of changes in the exchange rates.

tion on income statements emphasizing groups as the most important accounting entities. The Accounting Committee of the Swedish Institute of Authorized Public Accountants (*Föreningen Auktoriserade Revisorer*, FAR), the leading standard-setting body at the time, published a draft recommendation on methodological issues in 1971. Still there existed neither a legal requirement, nor any obligation from the Stock Exchange as such, to prepare group accounts.

The new Companies Act of 1975 more or less validated the standard of financial accounting of leading Swedish companies at the time (see Table I.2 for the current regulatory system). According to the new Act, parent companies had to publish consolidated income statements, balance sheets and statements of source and application of funds. In response to the new Act, the Accounting Committee of the FAR published a proposal for a recommendation on group accounting. The proposal relied heavily on the draft of 1971. This text, which was issued as a recommendation in 1976, was influential in Sweden for about a decade. However, when important mergers and acquisitions with considerable goodwill values became increasingly common, deviations from the guidelines also became common. Parts of the business community no longer regarded the recommendation as sufficiently relevant to actual business life. In the late 1980s the rate of compliance was therefore low (Johansson/Östman, 1995).

Table I.2 The current regulatory system

- The Swedish Companies Act, 1975
- The Annual Report Act, 1995
- The Annual Report in Financial Institutions Act, 1995
- The Annual Report in Insurance Companies Act, 1995
- The proposed recommendation on accounting for associated companies by FAR
- The RR's recommendation on consolidated statements (RR 1:96)
- The RR's recommendation on accounting for effects of changes in foreign exchange rates (8:98)
- The recommendation for banks and other financial institutions issued by the Financial Supervisory Authority

During the late 1980s, Swedish financial accounting showed an increase in divergent practices which provoked a discussion about the standard-setting system (Johansson/Östman, 1995). A new standard-setting body was founded in 1989, the Swedish Financial Accounting Standards Council (*Redovisningsrådet*, RR), with the aim of becoming standard setter for public companies. The registration contract at the Stockholm Stock Exchange stipulates that companies which do not follow the recommendations of the RR have to state this explicitly in their annual reports. There were three founders: the government – represented by the standard-setting body, the Swedish Accounting Standards Board (*Bokföringsnämnden*, BFN) – the Institute of Authorized Public Accountants (FAR) and the Federation of Swedish Industries (SI). Like many other countries, Sweden chose to tackle the compliance problems by including representatives of companies which were obliged to furnish information in the standard-setting body.

The first recommendation of the new standard-setting body was published in 1991 on the subject of group accounting. The new recommendation on consolidated statements applied to accounts for financial years beginning on 1 January 1992 or later. Many companies were already complying with the new rules in the accounting year 1991, while others followed in 1992. When evaluating compliance with the new recommendation, it should be noted that, because of general business development at the time, acquisitions with considerable goodwill values were a minor phenomenon.

Since then the RR has been reorganized. Formally an association is the principal of the council. The Swedish government is no longer a formal member. Instead there are more representatives from the business community and the Stockholm Stock Exchange. The explicit intention of the RR is that its recommendations should, as far as possible, comply with those of the International Accounting Standards Committee (IASC). RR has to date (August 2000) published recommendations.

A new act, the Annual Report Act (*Årsredovisningslagen*, ÅRL), came into force in 1997. This act was passed with the purpose of complying with the EU Directives, since Sweden had become a member in the EU.

It should be noted that the intention of the legislator is that accounting methods should develop from accounting practice and therefore Swedish accounting laws and recommendations sometime lack detailed rules.

In some specific areas comments are made on common practice in Swedish companies. Some empirical findings are also presented. Unless otherwise mentioned, these are based on Rundfelt (1997). Rundfelt has investigated trends in the financial accounting of companies registered at the Stockholm Stock Exchange which are presented in an annual review published by the Stockholm Stock Exchange and the Accounting Standards Board (BFN). This contains a survey of and commentary on published financial information and accounting principles used by the one hundred largest companies registered on the most important list of the Stockholm Stock Exchange.

II Objectives, concepts and general principles

By tradition, Swedish financial accounting has focused on the protection of creditors. Furthermore, the nature of legislation on income distribution has given accounting the function of determining distributable equity. The accounts for groups, as well as for individual companies, have therefore been prepared with the explicit aim of identifying the limits of distributable equity. According to the Swedish Companies Act (ABL), a company may not distribute dividends exceeding the amount of unrestricted shareholders' equity reported in the balance sheet. In addition, a parent company may not distribute dividends exceeding the amount of unrestricted shareholders' equity reported in the consolidated balance sheet, reduced by the amount that shall be transferred to restricted shareholders' equity according to the annual reports of the individual group companies. Unrestricted shareholders' equity must therefore be shown separately in the consolidated balance sheet. Partially as a consequence of the focus on the protection of creditors, conservatism as an ideal has had, and still has, a dominant role in the minds of Swedish accountants and business people.

Another important influence on Swedish accounting has been the close connection between taxation and financial accounting. In principle, the income shown in the financial accounts determines the taxable income. Explicit exceptions to this rule are not common. Groups as such are not subject to taxation; group accounting thus has no direct effect on taxation.

Over the years, especially since the mid 1980s, views on the objectives of financial accounting have started to change, at least in leading transnational firms. As mentioned above, several of these are also registered on foreign stock exchanges. A general emphasis on owner and stock market perspectives has been reinforced. Mass media coverage on the financial development of companies has increased dramatically. Not surprisingly, certain aspects of the reactions of financial markets are being increasingly considered by companies which are obliged to furnish information. These aspects have also become important considerations for Swedish standard setters.

According to current regulations, a parent company is required to publish financial statements for itself and for the group. Basically, items have to be valued in the same way at both levels (see V), but certain exceptions may be acceptable. Sometimes deviations are used as a tool for handling more or less unsolvable problems, as for example the dilemma of combining compliance with the basic valuation principles of the parent company with openness to international accounting developments, without provoking unnecessarily unfavourable tax effects.

The recommendation on group accounting of 1991 has been replaced by a new recommendation (RR 1:96). Basically, the recommendation follows IAS 22 on business combinations (see XV concerning differences). The following issues are included:

- Scope.
- Definitions.
- General issues (a brief interpretation of the legal rules about reporting obligations; the meaning of consistency between company and

group levels; minority interests; elimination of transactions between group companies; equity in consolidated financial statements).
- Nature of a business combination.
- Accounting for acquisitions.
- Accounting for unitings of interest.
- A comparison with IAS 22.
- Disclosure.

Accounting for foreign subsidiaries is covered in another recommendation (RR 8:98). This recommendation closely follows IAS 21 on the effects of changes in foreign exchange rates (see XV concerning differences).

The ideas behind these recommendations are based on the developments discussed above and are consistent with the views of leading practitioners and academics. Regarding group accounting, a group as an accounting entity is similar to a legal entity. Assets and liabilities of subsidiaries should be accounted for rather than the shares, as if they had been bought by the parent company. Under the purchase method, which is the main consolidation method used in Sweden, the acquisition of subsidiary shares is basically regarded as analogous to an event that obviously has not occurred: the acquisition by the parent company, and therefore the group, of the subsidiary's assets and liabilities. This view was introduced in Sweden by Johansson (1954).

Conflicting goals in standard setting, of the kind mentioned above, were important considerations when the recommendation on group accounting was developed. A reasonable standard for the information needs of international and national stock markets was an important objective. Consequently, an implicit view is to look at group accounting in terms of valuations which are meaningful for such users, rather than simply as a formal set of computation rules to follow regardless of intrinsic meaning. With this underlying idea, legal rules from another time, issued in another context, tend to represent restrictions rather than objectives for development. An important example of such a restriction is that an allocation of 'untaxed reserves' between deferred taxes and equity is not allowed in individual accounts. However, such an allocation is now made in group accounts (see V.3) in order to make them more comprehensible to foreign readers.

III Group concept and preparation of group accounts

Regulations relating to the preparation and presentation of consolidated accounts are set out in the Annual Report Act (ÅRL). Each parent company is required to prepare consolidated accounts for each financial year (ÅRL 7:1). The Annual Report Act applies to limited companies. Financial institutions (*kreditinstitut och värdepappersbolag*) and insurance companies (*försäkringsföretag*) are regulated in two special annual report acts (*Lag om årsredovisning i kreditinstitut och värdepappersbolag*, ÅRKL, and *Lag om årsredovisning i försäkringsföretag*, ÅRFL, respectively). These special acts often refer to the general Annual Report Act. Regarding financial holding companies (*finansiella holdingbolag*), these should follow the Annual Report Act. They should prepare group accounts according to ÅRKL if they primarily are holding companies for subsidiaries that are financial institutions. The same applies for insurance companies. For banks and other financial institutions special recommendations have been issued by the Financial Supervisory Authority (*Finansinspektionen*). These special recommendations are not discussed in this text.

However, existing regulation governing the financial reporting of other types of legal entity corresponds closely to the requirements of the Annual Report Act. Such similar requirements can be found in the Act governing the financial reporting, etc. of special legal entities. According to a recommendation issued by the Accounting Standards Board (BFN R 8) which interprets this law for special legal entities, most entities, regardless of legal form, can be the parent of a group, for example a sole trader proprietorship (i.e. a registered firm managed by one individual and not employing any others). However, such companies cannot be subsidiaries of other companies. Most other kinds of legal entity can be parent as well as subsidiary.

The Annual Report Act defines the parent/subsidiary relationship (see IV.1). The obligation to prepare group accounts is thus based on a legal definition of a parent company.

A parent company that itself is a subsidiary does not have to prepare group accounts if the parent company and all of its subsidiaries are included in the group accounts prepared by another parent company, this other parent company falls under the jurisdiction of a country within the *Europeiska ekonomiska samarbetsområdet* (EES, Area of European Economic Cooperation, which is the EFTA except Switzerland), and the group accounts are prepared and audited in accordance with the Seventh EU Directive. In that case the group accounts of the parent company at the higher level should be sent in to the registration authority (ÅRL 7:2).

In practice, one way for a parent company to avoid the obligation to consolidate one of its subsidiaries is to alter the relationship to this subsidiary, so that it no longer qualifies as a parent/subsidiary relationship from a legal point of view. The parent company may, for example, transfer some of its holdings in a subsidiary to an independent company that is managed on a unified basis. This type of transaction is not unusual.

The annual report of an individual company should include an income statement, a balance sheet and an administrative report. Larger companies should also attach a statement of changes in financial position (*finansieringsanalys*). The annual report of a parent company should in addition include a consolidated income statement and a consolidated balance sheet. If the parent company submits a statement of changes in financial position, a consolidated statement should also be included. Notes to the financial statements are considered to be an integral part of the financial statements and thus also of the annual report (see X).

The consolidated accounts should be prepared for the same period of time as the parent company's accounts (ÅRL 7:9). The consolidated accounts should be available for the auditors at least one month before the annual general meeting of shareholders of the parent company (ÅRL 8:2); the general meeting should take place within six months of the balance sheet date (ABL 9:5). No later than one month after the consolidated income statement and balance sheet have been adopted by the annual general meeting of shareholders, copies of the accounts and the audit report should be sent to the registration authority (*Patent och registreringsverket*, PRV, ÅRL 8:3). All public companies in Sweden publish group accounts (see XII).

IV Consolidated group

1 Subsidiaries

The obligation to prepare consolidated accounts is based on the legal definition of a parent/subsidiary relationship as defined by the Annual Report Act (ÅRL 1:4). A company is the parent of another company (its subsidiary) if it either directly, or jointly with or through subsidiaries, holds more than half of the voting rights in the legal entity, or holds shares in the legal entity and by virtue of an agreement with other owners controls more than half of the voting rights in the legal entity, or holds shares in the legal entity and has the right to select or dismiss more than half of the members of its board of directors, or holds shares in the legal entity and has the exclusive right to a controlling interest by virtue of an agreement with the legal entity or through an instruction in the articles of association of the legal entity. There are no detailed rules about how these prerequisites relate to each other. It has been debated whether this means that a legal entity can have more than one parent and be included in more than one group (Rutberg and Skog, 1997).

When calculating the majority of voting rights, indirect votes are included in so far as the parent company controls these rights. Consider, for example, that company X holds 60 per cent of the voting rights in company A, which holds 60 per cent of the voting rights in company B. In this case X holds only 36 per cent of the voting rights in B, but since it has a controlling interest in A, it controls 60 per cent of the voting rights in B. B is therefore a subsidiary to X. Consider another example in which company X holds 40 per cent of the voting rights in each of companies A and B. Furthermore, company A holds 40 per cent of the voting rights in B. In this case X indirectly holds 56 per cent of the voting rights in B. However, since X does not control the majority of the votes in A, B is not considered to be a subsidiary to X.

Companies in which the group controls less than 50 per cent of the voting power, but in which the group nevertheless has a controlling influence because the remaining shares are spread among many other small shareholders, should normally not be included as subsidiaries in the consolidated accounts.

2 Exemptions from inclusion in the group accounts

Consolidated financial statements should include the parent and all its subsidiaries, with a few exceptions (ÅRL 7:4). Subsidiaries do not have to be included if they are insignificant, or the control of the parent company is severely limited by substantial and permanent obstacles, or necessary information could not be obtained without unreasonable expense or within reasonable time, or the holding is temporary.

The first exception applies, for example, to dormant companies. The second exception applies, for example, to warlike conditions or long-term political unrest. The third exception might in very rare cases be applied to acquisitions of foreign subsidiaries where the accounts have not yet been co-ordinated with those of the parent. The fourth exception applies to subsidiaries that have been acquired solely with the purpose of being divested, normally within one year from the time of purchase, and where the holding thus is temporary. A holding must not be reclassified as temporary once it has been included in the group accounts (RR 1:96).

Subsidiaries which are not included for the above reasons should be included in the accounts as investments in shares and information on their exclusion is to be provided in the annual report.

There is an explicit exception to the rule for consolidation for subsidiaries whose activities are very different from the rest of the group. This is supposed to be interpreted restrictively. Such subsidiaries should be consolidated according to the equity method (see VII.1).

According to regulations concerning special legal entities, very small groups do not have to prepare group accounts. Small, in this case, is defined by the number of employees, 10 being the critical figure. The smallest among those groups which are obliged to prepare group accounts do not have to fulfil all requirements on supplementary information.

V Uniformity of accounts included

1 Accounting period

The Annual Report Act (ÅRL 7:9) states that the group accounts should refer to the balance sheet date of the parent company. Normally companies belonging to the same group have the same balance sheet date. The Accounting Act (BFL § 12) requires that companies belonging to the same group should have the same financial year. Exemptions from this requirement are only allowed in extraordinary circumstances. One exemption is, for example, foreign subsidiaries which are required to have a different financial year by local legislation. These cases are of course unusual. The standard for group accounts (RR 1:96) requires that the accounts of the parent and subsidiary, that are used as a base for the group accounts, should be for the same financial period. If the financial periods are different, the subsidiaries should prepare accounts for the same financial period as the parent, as a base for the group accounts. If this is unfeasible, accounts for different periods may be used if the difference is not more than three months and the same from period to period. However, important transactions between the respective balance sheet dates should affect the group accounts.

2 Uniformity of recognition criteria, valuation rules and format

Basically, a requirement of the Annual Report Act (ÅRL 7:5, 7:7) is that the consolidated income statement and balance sheet should take the form of a combination of the income statements and the balance sheets of the parent company and its subsidiaries, prepared in a way which conforms with generally accepted accounting principles (*god redovisningssed*). The Financial Accounting Standards Council recommendation on consolidated statements (RR 1:96, § 8) further requires that the income statements and balance sheets, on which the consolidated accounts are based, are prepared in accordance with uniform principles both in terms of valuation and classification. The accounting methods to be used for this purpose should be those applied by the parent company.

It is recognized that it is not always possible to co-ordinate the accounting methods used in the different group companies. If the accounting methods applied by the subsidiary in its original accounts are different from those of the parent company, it may be necessary to adjust its income statement and balance sheet before the accounts are used as a basis for the consolidated statements.

It is thus a general rule that the accounting methods (recognition criteria, valuation rules and format) used to prepare consolidated financial statements must be the same as those used to prepare the accounts of the separate group companies. According to the Annual Report Act (ÅRL 7:11) different accounting principles may be used in the group accounts if there are specific reasons for this. This makes it possible to better suit the accounting information to the requirements of the users, primarily investors. The Financial Accounting Standards Council observes that in certain cases the accounting principles that are applied by the parent could be influenced by tax considerations. However, in those cases where there are no relevant Swedish standards, the accounting principles applied in the group accounts must not be in conflict with standards published by the International Accounting Standards Committee. Furthermore, information should be given in the notes about differences between accounting principles in the parent and in the group.

3 Untaxed reserves

Swedish tax laws allow companies to defer taxation by making certain provisions for that purpose only, as long as these provisions and the changes in them are reflected in the balance sheet of the individual companies as untaxed reserves and in the income statement as changes in the untaxed reserves. The Annual Report Act (ÅRL) further prescribes that untaxed reserves and changes in these should be separately disclosed in the individual balance sheet and in the income statement. However, according to the Financial Accounting Standards Council's recommendation on group accounts (RR 1:96 § 10), the consolidated untaxed reserves must be allocated to restricted shareholders' equity and deferred tax liability in the consolidated balance sheet. The consolidated changes in untaxed reserves must be allocated in a similar manner (see VIII concerning deferred taxes).

An important characteristic of the Swedish Companies Act is that a parent may not distribute dividends exceeding the amount of unrestricted shareholders' equity in the consolidated balance sheet. It requires that the amount of the unrestricted equity (that is distributable equity), or the accumulated deficit of the group after deduction of internal profits, is disclosed in the consolidated balance sheet (see also VIII). This will have consequences for group accounts in several respects, for example when accounting for foreign subsidiaries and for associated companies (see VII).

4 Foreign currency translation

4.1 Overview

Swedish companies should (ÅRL 2:6) prepare their reports in SEK. The amounts may, in addition to this, be expressed in another currency too. The foreign operations have to be translated for inclusion in the group accounts. Formerly, the accounting standards in Sweden have differentiated between translation of foreign subsidiaries for inclusion into the group accounts and accounting for exchange rate effects in a legal entity. There have been two different recommendations. This is no longer the case. According to the Financial Accounting Standards Council (RR 1:96 Enclosure 1), the reason for this is that fundamentally it is just two different aspects of the same problem area, even if the effects occur in two different settings (accounts for a legal entity and group accounts respectively).

The choice of method for translating the financial statements of foreign operations should, according to the Financial Accounting Standards Council (RR), be determined by an assessment of the operational and financial characteristics of those subsidiaries. Foreign subsidiaries are regarded as belonging to one of two categories: independent or integrated. In a recommendation (Accounting for Effects From Changed Exchange

Rates, RR 8:98) the RR sets out guidelines on how to identify these two categories and how to translate the financial statements for inclusion in the consolidated statements. These guidelines for translation are explained below (see V.4.2 and V.4.3).

The identification of these two categories follows closely IAS 21. An integrated subsidiary carries on its business as if it were an extension of the reporting enterprise's operations. For example, such a foreign operation might only sell goods imported from the reporting enterprise and remit the proceeds to the reporting enterprise. On the other hand, an independent subsidiary accumulates cash and other monetary items, incurs expenses, generates income and perhaps arranges borrowings, all substantially in its local currency. It may also enter into transactions in foreign currencies, including transactions in the reporting currency. There are a number of indications that a foreign operation is independent:

- While the reporting enterprise may control the subsidiary, its activities are carried out with a significant degree of autonomy.
- Transactions with the reporting enterprise are not a high proportion of the subsidiary's activities.
- The activities of the subsidiary are financed mainly from its own operations or local borrowings rather than from the reporting enterprise.
- Costs of labour, material and other components of the subsidiary's products are primarily paid or settled in the local currency rather than in the reporting currency.
- The subsidiary's sales are mainly in currencies other than the reporting currency.
- Cash flows of the reporting enterprise are not affected directly from day-to-day activities of the subsidiary.

A foreign branch that is not a separate legal entity, but could be classified as independent, should be translated at the closing rate method. A transfer from one category to the other category may become necessary if, for example, significant changes occur in the nature of its operations.

If a group consists of a number of subgroups, the group accounts could either be prepared in steps by the respective subgroups, or for all group companies simultaneously. According to the Financial Accounting Standards Council, the classification of the nature of each subsidiary in both those cases should be made from the respective group level.

4.2 Independent subsidiaries

The financial statements of an independent subsidiary should be translated using the closing rate method. Under this method, assets and liabilities are translated using the rate at the balance sheet date. Items in the income statement are translated using the exchange rates at the dates of the transactions. For practical reasons a rate that approximates the actual exchange rates could be used at the translation. The subsidiary is in effect treated as a separate company in which the parent has a net investment. The book value of the investment changes as the exchange rate of the foreign currency changes. Exchange differences are seen as expressions of unrealized changes in value and should therefore not be included in the consolidated income. Instead, such exchange differences are taken directly to shareholders' equity. On the disposal of a foreign entity, the cumulative amount of the exchange differences, which have been deferred and which relate to that foreign entity, should be recognized as income or as expense in the same period in which the gain or loss on disposal is recognized.

4.3 Integrated subsidiaries

In the case of an integrated subsidiary, non-monetary assets and liabilities are translated at the rate corresponding to the basis for the subsidiary's valuation of the assets. When assets are stated at acquisition costs (less depreciation), the investment rate should be used. Non-monetary items may in some cases be stated at the market value on the balance sheet date (if this is lower than cost). These items should be translated using the closing rate. Monetary assets and liabilities are translated using the rate at the balance sheet date.

Items in the income statement are translated using the average rate, with the exception of depreciation and write-downs of fixed assets and the cost of goods sold. In the case of depreciation and write-downs, the rate applicable to the asset is used. This is normally the investment rate. The cost of goods sold is calculated as the sum of purchases during the period translated at the average rate and the difference between the opening and the closing inventory, translated as explained in the preceding paragraph. If the inventory turnover rate is high and the changes in the translation rate are insignificant during the period, the cost of goods sold may be translated using the average rate.

Since exchange differences under this method relate exclusively to monetary items in the subsidiary's accounts, they are included in the reported income for the group as other exchange gains or losses.

The financial statements of foreign subsidiaries operating in highly inflationary economies (the example given corresponds to the definitions in IAS 29) are also translated using this method. Alternatively, inflation-adjusted financial statements of independent subsidiaries may be translated using the closing rate method. However, when choosing a method in this respect, it should be recognized that existing price indices may be irrelevant for this purpose. In those cases, it is inappropriate to include foreign subsidiaries into the group accounts on the basis of inflation-adjusted financial statements translated at the closing rate method, and the method used for integrated subsidiaries should be used. Thus in this respect the Swedish recommendation differs from IAS, where only the translation of inflation-adjusted financial statements at the closing rate is allowed. However, the Financial Accounting Standards Council observes that both methods produce the same outcome assuming that the exchange rates move in relation to inflation differences and used price indices are relevant.

Certain exchange differences relating to the translation of the financial statements of independent subsidiaries may, as a consequence of the underlying logic of the closing rate method, be treated in a special manner. For example, exchange differences arising on a non-current inter-company monetary item, that is, in effect, an extension to or deduction from a parent's net investment in an independent foreign subsidiary, may be taken directly to consolidated equity.

Furthermore, if a parent company has taken up foreign currency loans to provide a hedge against the net investment in an independent subsidiary, the exchange differences arising on such loans should be taken to consolidated equity. According to former practice in Sweden (FAR's proposed recommendation on Translation of Foreign Subsidiaries' Financial Statements for Inclusion in Consolidated Financial Statements), this applies only to the extent that the exchange differences were covered by exchange differences for the year arising on the net investment, taking tax effects into consideration. The RR has not decided on this issue, but awaits the IASC recommendation in this area. In the meantime the RR observes that it is the net investment that is hedged, and consequently it is only the amount in the group accounts that could be hedged. Schuster (1996) analyzes the tax effects.

A hedge could be located either in the parent company or in another subsidiary, and the hedge could be effected, not only by borrowing an amount in foreign currency, but also through other means, for example by selling a currency forward.

Information should be provided on the amount of exchange rate differences taken directly to equity (if these have been reduced by hedging, the effect from this should be disclosed too) and the total amount of accumulated exchange differences taken directly to equity at the beginning and the end of the year respectively, and an explanation of the difference.

At the disposal of the net investment such exchange differences should be recognized as income or expense.

4.4 Empirical results

According to Rundfelt (1997), the choice of translation method has significant accounting consequences. Many companies seem to prefer taking exchange differences in independent companies directly to equity, because it probably means a lower volatility in income. What makes the classification hard to make is that few subsidiaries in reality are completely independent. The Swedish car manufacturer Volvo, for example, is generally regarded as an integrated group. To speak of 'independent' subsidiaries seems misleading, according to Rundfelt. Regarding Swedish practice, Rundfelt's opinion is that more operations are classified as independent, than would be warranted by the recommendation. The reason for this, says Rundfelt, is that there is a rule, stating that if the major part of the foreign operations can be classified as independent, all subsidiaries may be translated at the closing rate (it should be noted that since 1998 this rule is no longer applicable).

Our view is consistent with Rundfelt's observation. It seems as if companies tend to classify their foreign subsidiaries as independent, even if the subsidiaries from a fundamental business perspective, and by a strict application of the indications mentioned in V.4.1 above, could be characterized as integrated.

Furthermore, according to Rundfelt (1997) practice regarding translation of goodwill and assets that are recorded at a higher amount in the group accounts than in the individual subsidiaries is diverse (in the new recommendation on accounting for the effects from changed exchange rates, it is stated that those items should be translated at the closing rate – Table V.1). Nineteen companies have provided information about their principles for translation of foreign subsidiaries in highly inflationary countries. The monetary–non monetary method is dominating completely (it is common in Sweden to refer to the temporal method as the monetary–non monetary method). Three companies translate inflation-adjusted financial statements.

4.5 Consistency

When there is a change in the classification of a foreign operation, the translation procedures applicable to the revised classification should be applied from the date of the change in the classification. When a foreign operation is reclassified from integrated to independent, there is an exchange difference in the non-monetary assets, calculated as the difference between the exchange rate at the time of the reclassification and the exchange rates that previously have been used at translation. This exchange difference should be taken directly to equity. When a foreign operation is reclassified from independent to integrated, the translated amounts for non-monetary items at the date of the change are treated as the historical cost for those items in subsequent periods. Exchange differences which have been deferred are not recognized as income or expense until the disposal of the operation.

If a significant foreign subsidiary has been reclassified from independent to integrated or vice versa, information should be provided about the nature of the reclassification, the reason for the reclassification, how the reclassification has affected the equity of the group, and how the profits of each period presented in the annual report should have been affected had the reclassification been made at the earliest period presented.

4.6 Presentation under equity

As mentioned above (see V.2), the Swedish Companies Act (ABL) requires that the amount of the unrestricted equity, or the accumulated deficit, of the group after deduction of internal profits is disclosed in the consolidated balance sheet.

The unrestricted shareholders' equity of a group includes both the unrestricted shareholders' equity in the parent company and the contribution of each subsidiary, positive or negative, to consolidated unrestricted shareholders' equity (see IX.1). This applies both to domestic and foreign subsidiaries. However, there is a restriction that the contribution of a subsidiary may not exceed the group's share of the unrestricted shareholders' equity of the subsidiary. The unrestricted shareholders' equity according to the accounts of the subsidiary is for a foreign subsidiary interpreted as the unrestricted shareholders' equity translated at the closing rate. This applies regardless of which method is used in the translation of the foreign subsidiary for inclusion in the group accounts.

Table V.1 Accounting for goodwill in foreign currency units (FCU) – example

Before acquisition

Parent (SEK million)			
Assets (cash)	300	Equity	300

Subsidiary (FCU million)			
Assets	100	Equity	40
		Liabilities	60

Purchase price	FCU 60 million
Currency rate at purchase date	FCU 1 = SEK 5

Purchase analysis

Purchase price	SEK 300 million	
Market value of assets less market value of liabilities	SEK 200 million	(in this case market values are assumed to equal book values)
Goodwill	SEK 100 million	

After acquisition

Parent (SEK million)			
Shares in subsidiary	300	Equity	300

One year after acquisition

Assume that one year later the accounts of the parent and the subsidiary look the same.

Current currency rate FCU 1 = SEK 6

The group's balance sheet, assuming no amortization of goodwill, will look as follows:

Group (SEK million)			
Assets	600	Equity	300
		Translation difference	40 *
Goodwill	100	Liabilities	360
	700		700

* Net assets: FCU 40 million × (SEK 6 – SEK 5) = SEK 40 million

If however, the accounting for goodwill is in FCUs the balance sheet will look as follows:

Group (SEK million)			
Assets	600	Equity	300
		Translation difference	60 *
Goodwill	120	Liabilities	360
	720		720

* Net assets: FCU 60 million (incl. goodwill) × (SEK 6 – SEK 5) = SEK 60 million

VI Full consolidation

1 Overview

According to the Annual Report Act (ÅRL 7:17), a parent company and its subsidiaries should be combined using the purchase method, the pooling-of-interests method, or the equity method. The act defines the special circumstances when the pooling method or the equity method may be used. The pooling method (ÅRL 7:22) may be used if the parent company's shares represent more than 90 per cent of the nominal value of the total number of shares in the subsidiary, and the shares are acquired primarily through an issue of shares (a cash part must not exceed 10 per cent of the nominal value of the shares issued), and it is consistent with generally accepted accounting principles (*god redovisningssed*) and a true and fair view. The equity method (ÅRL 7:23) should be used if the operations of the subsidiary is so different from those of the rest of the group that a consolidation according to the purchase or pooling method would not give a true and fair view.

According to the Financial Accounting Standards Council's recommendation on consolidated statements (RR 1:96, §§ 19–20), a business combination can be designated either as an actual acquisition of a subsidiary on the part of the parent company, or as a merger of the financial interests of two (or more) companies. The choice of consolidation method depends on how the combination is classified. In most cases, it is an acquisition. Acquisitions must be accounted for in accordance with the purchase method. In rare cases, when the combination meets the requirements of a merger, the pooling-of-interests method should be used. This should be the case when it is impossible to identify one company as the acquirer, or the dominating owner in the new entity (see VI.3).

This section describes the criteria for the application of the purchase method and the pooling-of-interests method (the equity method is described in section VII) and the peculiarities of the two methods as set out in the recommendation on consolidated statements. The description of the purchase method encompasses guidelines for determining the group's acquisition cost of the shares of a subsidiary, of the acquired assets and liabilities taken over and hence also of the consolidated goodwill. The treatment of minority interests in subsidiaries in which the group does not hold all of the voting power is also described. The section on the pooling-of-interests method emphasizes the criteria for its application. Lastly, consolidation of intragroup debts, profits, revenues and expenses is explained. These techniques generally apply regardless of which method of consolidation is used (minority interests are not relevant to the pooling-of interests method).

When reorganizations occur within groups, sometimes operations are transferred from one subsidiary to another at an amount that differs from the fair value. The book value of the shares in the subsidiary in the parent company should then be adjusted for goodwill and other differences between the valuation according to the purchase analysis and the amounts relating to the transferred operations that are reported in the subsidiary.

In the case of a merger of a parent company and one or more of its subsidiaries into one legal entity, the indirect ownership of the subsidiary's assets and liabilities is transformed into a direct ownership. This type of transaction raises many valuation issues in the parent company's financial statements, but the group accounts should normally remain unaffected. The obligation for the subsidiary to keep separate accounts ceases, when the merger is registered by the registration authority (*Patent och Registreringsverket*, PRV).

2 Purchase method

2.1 Concept

The purchase method, according to the Financial Accounting Standards Council (RR 1:96, § 27), accounts for a business combination as the acquisition of one company by another. The acquisition is seen as a transaction whereby the acquiring company indirectly acquires the assets and takes over the liabilities held by the subsidiary company. In the consolidated balance sheet, the identifiable assets and liabilities must therefore be valued at the acquisition cost for the group, that is at fair value on the date of acquisition. These values, as well as the consolidated acquisition cost of the shares and any resultant goodwill on consolidation, are determined at the acquisition. This is called a purchase analysis (*förvärvsanalys*). In the case of non-instantaneous acquisitions, when the shares in the subsidiary are purchased on different occasions with long intervals between each acquisition, a purchase analysis is required for each acquisition.

The special problems relating to the purchase analysis when not all the shares in the subsidiary are acquired are discussed below (see VI.2.8).

As a general rule, the original purchase analysis should remain unchanged (RR 1:96, § 50); it may not be changed to accommodate subsequent events. However, if it becomes apparent that certain of the premises on which the original purchase analysis was based were either incomplete or erroneous, the purchase analysis may be adjusted. The use of this exception is limited by a time restriction: adjustments to the purchase analysis may not be made later than at the end of the financial year starting after the time of acquisition. Adjustments are also limited to conditions that were known at the date of acquisition, but which could not be quantified at that date, and to purely erroneous data.

2.2 Acquisition cost

The payment for the shares in the subsidiary could be made in cash or by other means. The acquisition cost consists of the cash outlay or the fair value of any other means of payment, such as an issue of securities in the parent company (RR 1:96, § 30). If the payment is made through an issue of securities that are traded at an exchange, the fair value should be calculated on the basis of the average price during the last 10 trading days before the acquisition became publicly known.

The purchase agreement may contain contingent conditions. The purchase price may, for example, depend on the income of the company during an interim period. If the final settlement of the purchase price is to be made by means of an additional payment or a repayment at a later date and if, at the date of acquisition, it is likely that an additional payment will be made and it is possible to estimate the amount with a reasonable degree of accuracy, the initial purchase analysis should be based on the estimated final acquisition cost (RR 1:96, § 33). When the final acquisition cost is known, an adjustment should be made to the consolidated acquisition cost and the purchase analysis.

2.3 Valuation of 'acquired' assets and liabilities

For some types of assets and liabilities there exist certain guidelines for determining the estimated fair value to be used in the purchase analysis (RR 1:96, § 42). Inventories and work in progress should be valued at the amount that should have been paid at a direct acquisition. Inventories of raw material should be valued at replacement cost. Receivables and liabilities with an interest rate lower than the normal market rate should be valued at net present value. Tangible fixed assets may be valued using an estimated fair value, or utilization value. There is no specific rule governing how this utilization value should be determined.

The financial statements of the acquired subsidiary may have been prepared using different accounting principles from those used by the parent and the group. In this situation one has to establish the amount at which the items in the subsidiary's accounts would have been recorded, had the parent company's accounting principles been used.

The identifiable assets and liabilities may include assets and liabilities which are not included in the subsidiary's balance sheet, such as patents, tenancy rights and tax loss carry-forwards. These may be assigned separate values in the consolidated balance sheet if they represent future economic benefits, and could be valued in a reliable way (RR 1:96, §§ 34–35). This rule also applies to trademarks in acquired subsidiaries. If these assets are fixed assets, they should be depreciated over estimated useful life in the same manner as consolidated goodwill (see VI.2.4).

It is not unusual for a subsidiary on acquisition to own assets which the purchasing company intends to divest in the near future. If this is the case, the acquisition of the subsidiary and the divestment of parts should be treated as one transaction (RR 1:96, § 44). Accordingly, the assets sold should be included in the purchase analysis at the amount received for them. If they have not been sold at the balance sheet date, they should be included at the amount expected to be received for them. The advantage of this method is that no capital gains or losses are reported on sales which are integral parts of the purchase transaction.

Groups are not subject to taxation in Sweden. Therefore, write-ups in connection with consolidation will not have any consequences on tax payments. However, it is required (RR 1:96, Enclosure 3) that deferred taxes are accounted for on the difference between the estimated fair value and the residual taxable value in the subsidiary for each balance item, to the extent that this difference is not included among untaxed reserves. This is because the tax implications of an indirect acquisition of assets and liabilities are different from those of a direct acquisition. In a situation where a company acquires fixed assets, the acquisition cost of these assets constitutes the basis for the future tax-related depreciations. If the same assets are acquired indirectly by means of the purchase of shares, the tax implications are not dependent on the consolidated values, but on the residual values for tax purposes in the subsidiary at the date of acquisition. The deferred tax liability, or tax receivable, that is required to be recorded, is thus seen as a measure of the additional future taxes, or future reduction in taxes, which are caused by the indirect acquisition compared to a direct acquisition.

The fair values of the indirectly acquired assets are shown in full on the left-hand side of the balance sheet and the deferred taxes are shown on the right-hand side (see Table VI.1).

There has been some criticism of this view of deferred taxes. Eriksson (1991) states that there

Table VI.1 Deferred taxes in purchased assets – example

Before acquisition

Parent (SEK million)				**Subsidiary (SEK million)**			
Assets	100	Equity	40	Assets	50	Equity	20
		Liabilities	60			Liabilities	30

Purchase price amounts to SEK 40 million, and is entirely financed by new liabilities.

Purchase analysis

Purchase price		SEK 40 million
Market value of assets if brought directly	SEK 70 million	
– Market value of subsidiary's liabilities	– SEK 30 million	
– Deferred taxes in purchased assets	– SEK 6 million*	– SEK 34 million
Goodwill		SEK 6 million

*Market value less book value = 70 –50 = SEK 20 million
(Book value is in this case assumed to equal the residual value for tax purposes.)
Assuming a tax rate of 30% deferred taxes are: 30% × SEK 20 million = SEK 6 million.

After acquisition

Parent (SEK million)				**Group (SEK million)**			
Assets	100	Equity	40	Assets	170	Equity	40
Shares in subsidiary	40	Liabilities	100	Goodwill	6	Deferred tax liability	6
						Liabilities	130

is a risk of increased complexity when deferred taxes in purchased assets are identified. Furthermore, constructing a purchase cost based on a direct purchase is in effect constructing a cost that has never existed. He also says that there is an inconsistency in treating assets and goodwill differently in this aspect. However, the RR has stated that these points have been considered but rejected.

2.4 Goodwill from consolidation

The difference between the group's acquisition cost of the shares in a subsidiary and the net amount of the fair value of the identified assets acquired and the liabilities taken over, including deferred taxes, constitutes the consolidated goodwill (RR 1:96, § 39). However, in many cases a difference is due to reasons other than goodwill. It is therefore important that a thorough review of the purchase precedes the purchase analysis in order to identify all assets acquired and liabilities taken over, including deferred tax liabilities and receivables. The purchase price may, for example, include an amount for tax loss carry-forwards that the group expects to be able to utilize or which have a quantifiable market value. In this case an estimated tax claim should be reported as an asset in the consolidated balance sheet, even though it may not be recognized as such in the subsidiary's accounts.

Goodwill is recognized as a fixed asset in the consolidated balance sheet and must therefore be amortized over its estimated useful economic life. The amortization method should be straight-line unless another method is more appropriate. General rules on how to determine the amortization period do not exist. Normally, however, it should not exceed five years and it must never exceed 20 years. An amortization period of more than five years may be used for situations such as long-term strategic acquisitions, or market or technological conditions that are considered to be particularly stable. If the amortization period is set at more than five years, a description of the reasons for this must be included in the annual report immediately following the acquisition. An analysis of consolidated goodwill is to be presented in a note setting out the acquisition cost and accumulated amortization (see X.1).

The amortization period for goodwill has been much debated in Sweden as well as in other countries. A main argument in the Swedish debate has been that Swedish groups are at a disadvantage relative to foreign groups in the competition for acquisitions of other companies if and when the amortization period in Sweden is shorter than in many other countries. The reasonability of the argument is dependent on the view of how the investors on the capital markets process and interpret accounting information. The starting point for the Financial Accounting Standards Council is that the accounting rules should be designed using the assumption that the choice of accounting method is important for the external view of the company. Then it is bad when amortization periods differ between countries. Against this background the council had decided to use International Accounting Standards as a guideline for Swedish standards. In this particular case the Council closely follows the development in IAS concerning a maximum period for amortization of goodwill.

The value of reported goodwill must be continually assessed (RR 1:96, § 47). If, for example, after acquisition a subsidiary (group), which carries a goodwill value in the consolidated financial statements, develops in an unsatisfactory manner in relation to the forecasts established at the acquisition, this goodwill value should be further amortized or written down.

According to the Annual Report Act (ÅRL 4:5), a write-down of a fixed asset should be reversed if there are no longer any reasons for it. The reversal should affect the income statement. There is no exception for consolidated goodwill. The cases where a reversal is appropriate are rare. In addition to this, the measurement problems are severe. Against this background reversals of write-downs of goodwill must be treated very restrictively (RR 1:96, § 47).

If an investment in a consolidated subsidiary is written down in the income statement of the parent company, this should have no direct implications for the consolidated statements. The write-down in the parent company is accordingly treated as an intragroup item which is eliminated in the consolidated statements. The reasons for write-down in the parent may, however, also have implications on the appraisal of the consolidated goodwill.

2.5 Negative consolidation difference

If the difference between the group's acquisition cost of the subsidiary and the net market value of the identified assets acquired and the liabilities taken over is negative, non-monetary fixed assets should be written down in proportion to the negative difference. Negative goodwill remaining after non-monetary fixed assets in the subsidiary have been written down to zero should be treated as deferred income (RR 1:96, § 49).

When the difference is treated as deferred income, a plan for reversals must be established at the date of the acquisition, reflecting the expectation at that date. The reversals are reported in the consolidated income statement as part of operating income. The amount in question must either be disclosed in the income statement or in a note to the income statement.

2.6 Provisions for future expenses

The acquisition of a new entity could make it necessary to take measures to integrate the new entity into the acquiring company's operations. Two examples of this type of actions are reductions in the workforce due to a reorganization and the closure of factories. If these actions have been given a concrete form, so that the costs relating to them could be estimated at the time of acquisition, provisions should be made for those costs when establishing the acquisition balance (RR 1:96, § 36). The actions have been given a concrete form if information is available about the operations affected by the actions, the amount of the outlays (estimated with a reasonable degree of accuracy), and the time period during which the actions will be undertaken (and that material changes are not probable during this period).

Furthermore, the financial statements of the group should contain information on the size of the provision and the effect its reversals will have on reported income for the group (RR 1:96, § 69).

2.7 Divestment of a subsidiary

In the event of a divestment of a subsidiary, the difference between the amount received for the shares and the consolidated value of the subsidiary's net assets at the date of divestment should be reported in the consolidated income statement as a capital gain or loss (RR 1:96, § 55). This amount is usually different from the amount reported in the accounts of the parent company.

Instead of divesting a percentage of the shares in a subsidiary, a third party may be invited to subscribe to new shares in the subsidiary. A resultant change in the group's shareholders' equity, including the equity portion of untaxed reserves, can be viewed as a profit/loss attributable to the group. The change should be reported as a profit/loss in the consolidated income statement (RR 1:96, § 56).

There are no specific rules for the situation in which holdings in a previously consolidated subsidiary are reduced so that it in effect only qualifies as an associate company. That is, there are no explicit rules governing a change from full consolidation to equity consolidation.

2.8 Minority interests

If a group does not hold all the shares in a subsidiary, there is a minority interest in the subsidiary's net income and shareholder's equity from the group's point of view. Total revenues, expenses, assets and liabilities of a subsidiary are included in the consolidated financial statements, even if the group does not hold all the shares in the subsidiary. The group is therefore seen as a single unit with no account taken of the ownership question. The shareholders of the parent company are, however, assumed to be interested mainly in their portion of the group's net income and shareholders' equity (including the equity portion of untaxed reserves). The minority interests in the subsidiary's income and shareholders' equity must therefore be shown separately in the consolidated financial statements. These minority items include both direct and indirect minorities (see below) (RR 1:96, § 11).

Minority interests in shareholders' equity and in the equity portion of untaxed reserves are reported as a separate line item between liabilities and consolidated shareholders' equity in the consolidated balance sheet. The minority share in the tax portion of untaxed reserves is reported together with the majority interest in the deferred tax liability in untaxed reserves.

Minority interests in the consolidated net result

for the period, including the equity portion of changes in untaxed reserves, is shown as a separate item in the consolidated income statement immediately before net income. The reported net income for the group thus only includes the portion of the group's net income attributable to the shareholders of the parent company.

A minority interest in a subsidiary, which due to losses has negative equity, should be reported as an asset in the group only in the case where the minority has a legal obligation to cover the losses and will be able to do that (RR 1:96, § 12).

When calculating the minority interests, both direct and indirect interests are included. Consider the following example: company X holds 70 per cent of the voting rights in company A and 35 per cent in company B. A also holds a further 30 per cent of the voting rights in B. Both A and B are subsidiaries to X. Minority interest in A amounts to 30 per cent. Minority interest in B amounts to 44 per cent, consisting of a direct minority of 35 per cent and an indirect minority of 9 per cent. The latter represents the interests in B pertaining to the minority interest in A.

In the event of remaining minority interests after the acquisition of a subsidiary, the purchase analysis should only relate to the group's percentage ownership in the subsidiary (RR 1:96, § 11). Since the minority interests in the acquired assets and liabilities taken over have in fact not been acquired, they are considered to remain unaffected by the purchase. Accordingly, minority interests in shareholders' equity (and equity portion of untaxed reserves) and in income are based on the subsidiary's pre-acquisition book values. No account is taken of the group's acquisition cost of the acquired net assets, consolidated depreciations or reversals of negative goodwill or special provisions.

There has been some criticism of this view. Eriksson (1974) has argued that the use of this method means that a valuation of an asset in a partly-owned subsidiary in the group accounts can be made on mixed grounds, since the share of the majority is valued at the group's investment cost and the share of the minority is valued at the subsidiary's purchase cost. However, the Financial Accounting Standards Council has stated that this point too has been considered but rejected.

In the case of intragroup profits or losses accruing from transactions with subsidiaries in which a minority interest exists, all of these should be eliminated (RR 1:96 § 13; see also the general discussion of the elimination of intragroup items in VI.4).

3 Pooling-of-interests method

The pooling-of-interests method accounts for a business combination in a way which unites the financial and ownership interests of two or more companies. This method of consolidation may only be applied in the rare cases when the combination meets the requirements of a merger. A merger is defined as a transaction whereby the owners of the merging companies together get the control over the combined net assets and operations in such a way that they will share the risks and rewards of the new entity in such a way in the future that none of the owner groups could be identified as the acquirer (RR 1:96, § 4).

It is recognized that mergers are normally arranged so that one of the companies formally acquires the other. If the requirements of a merger are satisfied, such a business combination should be reported in accordance with the pooling-of-interest method, although from a legal point of view it is an acquisition.

To ensure that the risks and rewards connected to the new entity is equally shared, all or almost all of the voting shares should be affected by the transaction, and the fair value of one company should not be significantly different from the value of the other company. The possibility to identify an acquirer increases, the larger the difference between the fair values of the companies, the smaller the share of the voting rights exchanged in the transaction compared to the total number of votes in each company, and when a certain group of owners is favoured relative to other groups, or the share in the new entity of the owners of one of the companies is made dependent on how the contributed part of the new entity will develop after the merger. If an acquirer could be identified, there is no equal sharing of the risks and rewards, and consequently there is not a merger. According to the Annual Report Act (ÅRL 7:22), additional conditions should be satisfied for a business combination to qualify as a merger (see VI.2.1). There could be both acquisitions and mergers in a group, and thus different accounting methods could be applied in the same group accounts.

The aim of the standard (RR 1:96, Enclosure 1) is to limit the use of the pooling method to the exceptional cases when there is an actual merger. There are not any formal criteria for when the pooling method is allowed in the standard, except from what is stated in the Annual Report Act. In the original standard published in 1991, there were such criteria modelled on US GAAP. According to the Council this means that after the change the pooling method will be less common.

Under the pooling-of-interests method, the values of the assets and liabilities in the balance sheets of the combined entities are, as a general rule, retained in the consolidated balance sheet. If the accounting principles of the combined companies are different, these must be co-ordinated and adjustments to the reported values may thus be necessary (RR 1:96, § 59). The group accounts should be prepared as if the companies had been combined from the beginning of the earliest period presented (RR 1:96, § 60).

4 Elimination of effects from intragroup transactions

The basic logic of the consolidated accounts is to show how the accounts would have appeared had the group been housed in a single legal entity. Regardless of which consolidation method is applied, intragroup items, such as debt/claims, revenues/expenses and profits/losses, must therefore be eliminated. There are no further rules or explicit guidance in the Swedish regulatory system on how to eliminate intragroup debt/claims or revenues/expenses.

Profits or losses on intragroup transactions should be eliminated to the extent that they are reflected in the book values of assets to be included in the consolidated balance sheet. When adjusting the consolidated financial statements, the effect on tax costs and liabilities must be taken into account.

In the consolidated income report, operating costs are increased by the amount of the intragroup profit before tax. Pre-tax profit for the group is thus reduced, so producing a positive effect on tax costs. The after tax profit for the group is thereby reduced with a net amount consisting of intragroup profit less tax effect. In the consolidated balance sheet, the pre-tax amount of the intragroup profit is deducted from the book value of the assets. On the liability side of the consolidated balance sheet, reported profit is reduced by the net amount of intragroup profits and tax liabilities are reduced by the decrease in reported tax cost.

Intragroup profits and losses accruing from transactions with companies in which a minority share exists, should be eliminated in their entirety (RR 1:96, § 13, Enclosure 1).

VII Proportional consolidation and equity method

Guidelines for the use of the proportional consolidation method and the equity method in the group accounts can be found in the Annual Report Act (ÅRL) and in an exposure draft on the accounting for associated companies issued by the Institute of Authorized Public Accountants (FAR).

1 Applicability

The Annual Report Act (ÅRL 1:5) states that if a company holds shares in a legal entity that is not a subsidiary, and has a significant influence over the operational and financial control of that entity, and the holding is a part of a permanent relation between the company and the legal entity, the latter is an associated company. Such significant influence is presumed to exist when a company directly or indirectly holds more than 20 per cent of the voting rights of another legal entity, if there is no indication to the contrary. Such indications to the contrary could be the existence of another owner, which holds all other voting rights and thus makes all important decisions, or the fact that the holding is of a temporary nature. An associated company should be accounted for according to the equity method in the group accounts.

Furthermore, if the operations of a subsidiary

are so different from the rest of the group, that a full consolidation into the group accounts would not be consistent with a true and fair view, the subsidiary should be consolidated according to the equity method (ÅRL 7:23). This would be applicable only when the operations of the subsidiary are very different from the rest of the group, for example in the case where a bank or an insurance company is a subsidiary in a group, where the parent company is an industrial enterprise. If the problems that are associated with full consolidation could be solved through segment reporting, this alternative should be used instead.

The Annual Report Act (ÅRL 7:29) states that if a company, that is included in the group accounts, together with other companies that are not included, manages another company, that is not a subsidiary, the latter may be accounted for according to the proportional consolidation method. The proportional consolidation method is recommended in the exposure draft on the accounting for associated companies for accounting for interests in joint ventures, in the form of jointly controlled entities, in the consolidated financial statements. The equity method is also presented as a permitted alternative.

A joint venture in the form of a jointly controlled entity is defined as a separate entity, in which each party has an interest. Major decisions are made jointly by the parties. It is often the existence of a joint venture agreement that distinguishes a joint venture from, for example, an associate company.

2 Equity method

Under the equity method, the investment is initially recorded at acquisition cost. The book value of the investment in the consolidated balance sheet is thereafter increased or decreased in order to recognize the investor's share of the associate's reported profits and losses after the date of acquisition. From a general standpoint, the use of this method can be considered to be in violation of basic principles of Swedish law. However, it is explicitly stated that it should be used in the consolidated accounts.

The calculations of the investor's share of profits and losses in an associate should generally be based on the (consolidated) financial statements for the associate. If the accounting principles of an associate are substantially different from those of the group, adjustments should, if possible, be made to the income statement and balance sheet prior to these accounts being used as a basis for the consolidated statements.

If an associate has a different financial year from the rest of the group, the previous year end accounts for the associate may be used as a basis for the consolidated accounts.

An important requirement when using the equity method is that the unrestricted (distributable) reserves in the consolidated statements are not increased through the use of this method. The parent company's share of the undistributed earnings of an associate must therefore be classified as restricted reserves in the consolidated financial statements. If the book value of the investment at the end of the year is higher than at the beginning of the year, the difference should be classified as an equity fund (***kapitalandelsfond***).

At acquisition the group's cost of the investment is compared with the fair values of the net identifiable assets and accounted for in the same manner as under the purchase method (see VI.2). The investor's share of the profit or loss after the acquisition is adjusted for extra depreciation of assets based on their fair values on acquisition and goodwill amortizations.

If the consolidation difference is negative, non-monetary fixed assets may be written down in proportion to the negative difference. This means that, later, the investor's share of the associate's profit is increased by the difference between the depreciation in the associate's accounts and the depreciation based on the lower acquisition cost for the group. However, negative goodwill may not be reported as long-term liability, although this is allowed under the purchase method in full consolidation (see VI.2.5). If the investor is unable to relate the difference to identifiable assets, the carrying amount of the investment should not be adjusted. When it can be shown that the reason for the negative difference (normally a low profitability) is no longer relevant, the book value of the investment may be revalued if the equivalent amount of the appreciation is taken directly to the restricted reserves in the consolidated financial statements. If the shares of the associate are quoted on a stock exchange, the book value after revaluation may not exceed its market value.

Significant profits or losses on transactions between the investor and the associate should be eliminated in the consolidated accounts. This is achieved by reducing the book value of the investment by the group's portion of these profits or, in the case of losses, by adding the group's portion of these losses to the book value of the investment.

A loss in value of an investment that is other than a temporary decline should be recognized in the same manner as a loss in value of other fixed assets. This means that the book value of the investment should be reduced to recognize the decline in value.

The reduction of an investment in the associated company due to losses does not have to go below a zero value, unless the investor has guaranteed obligations of the investee.

Sale of an investment in an associate is reported as a (capital) gain or loss equal to the difference at the time of sale between selling price and book value of the investment in the consolidated accounts. Thus the capital gain in the consolidated accounts may be different from that reported for the individual company.

There are no specific rules concerning a change in the valuation of investments from the equity method to proportional or full consolidation. Nor is there any specific guidance on how to account for changes in the investor's proportionate interest in the associate.

Financial statements for associates reporting in a currency which is different from that of the investor are accounted for as if the investee were a subsidiary. According to the Institute of Authorized Public Accountants (FAR), associated companies should normally be classified as independent because of the limited voting power of the investor.

VIII Deferred taxation

The Financial Accounting Standards Council is currently working on a recommendation on the accounting for income taxes. The aim for this recommendation is that it should present a coherent view on the accounting for taxes, and that it should be based on International Accounting Standard 12. Awaiting this, the tax-related issues are presented in an enclosure to the standard on consolidation (RR 1:96 Enclosure 3).

As already mentioned (see VI.4), the tax effects must be taken into account when eliminating intragroup transactions.

Tax liabilities/receivables in the difference between estimated market values and pre-acquisition book values should be taken into account when establishing the purchase analysis (see VI.2.3). A deferred tax liability/receivable is amortized over the estimated useful life of the relevant asset and thus affects the tax cost of the group in future periods. Since there are no tax liabilities associated with goodwill in the purchase analysis, the amortization of goodwill does not have any effect on the consolidated tax costs.

The effect on tax liabilities and tax cost must also be taken into account when adjusting balance sheet and income statement items to attain uniform accounting principles. These effects are considered to arise due to the fact that often the tax costs of a subsidiary would have been different had the subsidiary based its accounts on the adjusted principles.

The effects of deferred taxation also have to be considered in the treatment of untaxed reserves in the consolidated balance sheet. The nature and background of untaxed reserves has already been mentioned above (see V.3). Basically, the Swedish tax laws allow companies to make provisions for tax purposes in their individual accounts if these are reflected in the income statement. The Annual Report Act (ÅRL Enclosure 2) further states that such provisions must be reported separately in the financial statements. In the income statement they should be reported under a separate heading, appropriations (*bokslutsdispositioner*). In the balance sheet, appropriations should be reflected as increases in untaxed reserves. One exception to the general rule that the consolidated financial statements should be a summary of the parent company's and the subsidiaries' income statements and balance sheets is the treatment of untaxed reserves in the consolidated financial statements.

In the preparation of the consolidated balance sheet, all untaxed reserves are divided into two parts: deferred tax liability and equity. The deferred tax liability is reported separately as a long-term liability, while the balancing portion is reported under shareholders' equity within non-distributable reserves. Changes in untaxed reserves in the consolidated income statement are eliminated in a similar manner. The tax portion is reported as a tax charge for the year, either as a separate item in the income statement or in a special note. The equity portion is included in the net income for the year. As a consequence, untaxed reserves disappear completely from the consolidated financial statements. This allows the consolidated unrestricted shareholders' equity to be calculated in a true and fair manner (see IX.1).

Foreign subsidiaries and associate companies are not dealt with explicitly in the recommendation on group accounting (RR 1:96). There is no evidence, however, that the purchase analysis in these cases should differ as far as deferred taxes are concerned, although the situation is not clear.

Deferred tax is calculated using the current income tax rates prevailing in the various countries where subsidiaries are located. Normally the rate for the next fiscal year is used. In the event of a change in the applicable tax rate, the resultant change in the tax liability is reported as an increase or decrease in the tax charge for the year. If the effect of the change is significant, it should preferably be reported as a separate line item.

Deferred tax liabilities should not be discounted. An exception to this rule is considered permissible in a purchase analysis if the value of a tax liability is an essential part of an acquisition and if the relationship between the purchase price and the valuation of the tax liability is documented. This may be the case, for example, in the acquisition of dormant companies.

No provisions should be made for potential future tax costs related to the distribution of unrestricted equity from a subsidiary to a parent.

IX Formats

1 Formats to be used

As has already been mentioned (see V.2), the consolidated financial statements should be a combination (summary) of the parent company's financial accounts and those of its subsidiaries. The income statement and balance sheet of the separate entities should adopt the formats specified in the Annual Report Act (ÅRL). Departures from the required format are allowed if there are specific reasons for that, especially when tax reasons have influenced the accounting of the parent company.

As regards the consolidated statements, additional items are needed to account for those items which are specific to group accounting. The balance sheet includes 'Deferred tax liability' and 'Minority interest'. Equity is reported using two subheadings: 'Restricted shareholders' equity' and 'Unrestricted shareholders' equity (or accumulated losses)' (for more information see IX.2). 'Minority interest in subsidiaries' income' is an additional item in the income statement.

The share of equity and profits in a subsidiary accruing to shares held by a party outside the group should be accounted for separately in the group's balance sheet and income statement (ÅRL 7:8).

As regards the consolidated statement of changes in financial position, the same principles should be used as in the parent company (ÅRL 7:30). The Financial Accounting Standards Council has published an exposure draft to a recommendation on cash flow statements.

2 Presentation of equity

According to the Financial Accounting Standards Council (RR 1:96), consolidated shareholders' equity should be reported using two subheadings: 'Restricted shareholders' equity' and 'Unrestricted shareholders' equity (or accumulated losses)'. Restricted shareholders' equity consists of share capital (in the parent company) and any restricted reserves. Unrestricted shareholders' equity consists of unrestricted reserves (or loss carried forward) and consolidated income/loss for the year.

In the Annual Report Act it is required that restricted shareholders' equity is specified in more items than mentioned above. Generally the same principles as in a legal entity should be applied in the group accounts. However, it is also recognized that an adjustment should be made when that is necessary due to the differences between the accounts of a group and those of a legal entity. This is especially applicable to the accounting for equity components. Certain concepts that are used in the specification of restricted shareholders' equity are meaningful only for a legal entity. Such a specification is not called for in the group accounts. However, there should be a separate accounting for the accumulated increase in the value of investment in associated companies after the acquisition (*kapitalandelsfond*).

Restricted equity is defined as equity that is not classified as unrestricted (see below), for example share capital, non-distributable funds (such as legal reserves and their international equivalents) and the shareholders' equity portion of untaxed reserves.

The consolidated unrestricted shareholders' equity is, according to the Company Act, one of the restrictions of the distribution of profits from a parent company. The size of unrestricted shareholders' equity must be evident in the balance sheet of the group. It is calculated from the assumption that unrestricted shareholders' equity in a subsidiary's own accounts could be accounted for as unrestricted equity in the group to the extent that it could be distributed to the parent without the shares in the subsidiary having to be written down.

The consolidated unrestricted shareholders' equity in the group consists of:

- The parent company's reported unrestricted shareholders' equity, after deduction of anticipated dividends from subsidiaries and less the differences between the accounting principles of the parent company and the group, which reduce the equity of the group.
- The contribution of each subsidiary, positive or negative, to the unrestricted shareholders' equity in the group, calculated individually for each subsidiary.
- After deduction of the inter-company profit eliminated from the consolidated balance sheet.

The administrative report of the parent company should specify the amount of consolidated unrestricted shareholders' equity which, according to the annual reports of the consolidated subsidiaries, is to be transferred to restricted shareholders' equity. Since the information refers to transfers from unrestricted shareholders' equity as reported in the consolidated balance sheet, it does not include those transfers that utilize the acquired unrestricted shareholders' equity which is eliminated in the preparation of the consolidated balance sheet.

According to the Annual Report Act changes in the components of consolidated shareholders' equity should be specified. This is best done in a note. Transfers between restricted and unrestricted shareholders' equity could be summarized in one item.

When the purchase method is used, a subsidiary's contribution to unrestricted shareholders' equity is the lowest of the following (see Table IX.1):

- The group's share of the subsidiary's unrestricted shareholders' equity. It should be noted that when accumulated losses in subsidiaries have been acquired or taken into account by the parent company through the write-down of shares in the subsidiary, these are not included.
- The difference between the value of the subsidiary's assets and liabilities as reported in the consolidated balance sheet, thus including for example, goodwill, and the book value of the shares in the subsidiary in the parent company.

When the pooling method is used, a subsidiary's contribution to the consolidated unrestricted equity consists of the unrestricted equity in the subsidiary after deduction of amounts that have been eliminated at the time of acquisition. At that time the book value of the shares in the subsidiary is eliminated, firstly against the subsidiary's shareholders' equity, secondly against the consolidated shareholders' equity.

Table IX.1 Calculation of a subsidiary's contribution to consolidated unrestricted shareholders' equity when the unrestricted shareholders' equity in the subsidiary is positive (example)

+	Total equity in the subsidiary according to its own accounts, adjusted to the accounting principles of the group	2000
–	minority interest	–400
+	Unamortized goodwill relating to the subsidiary	100
=	The subsidiary's net assets in the consolidated balance sheet	1700
–	The book value of the shares of the subsidiary in the parent company	–1200
=	The group's share of the net assets in the consolidated balance sheet less the book value of the shares in the subsidiary (1)	**500**
	Unrestricted shareholders' equity in the accounts of the subsidiary	1800
–	minority interest	–360
=	The group's share of the unrestricted shareholders' equity in the subsidiary's own accounts (2)	**1440**
	The lowest of the amounts (1) and (2) above	**500**

X Notes and additional statements

1 Notes

Notes to the financial statements are considered to be an integral part of these statements. If information is not required by law to be presented in the consolidated income statement or the consolidated balance sheet, it is recommended that notes are used so that the principal financial statements do not become too complicated. Notes may also be used to simplify the format of the statement of changes in financial position, although this is less frequent in practice.

As has already been mentioned (see V.2), the provisions governing the consolidated accounts are to a large extent similar to those governing the accounts of a single entity. This applies also to the disclosure requirements. The notes to the consolidated financial statements therefore mainly correspond to those of the parent company. Although certain group specific information is also required (see Table X.1), notes for the parent company and the group are often integrated. Some of the information required can be burdensome to compile on a consolidated basis. Exemption from the regulations on the grounds of exceptional difficulties may be applicable in this instance although restrictiveness should be observed. Furthermore, it should be noted that there is no obligation to disclose the total of untaxed reserves on a consolidated basis. This means that tax driven depreciations (in excess of depreciation according to an appropriate depreciation plan) need not be specified on a consolidated basis.

One example of group specific information to be included in the notes is the description of the accounting principles which are used in the consolidated accounts. Accounting principles which are normally described include the method of consolidation of subsidiaries (including the definition of subsidiaries), and the accounting method used for associated companies and joint ventures, and the translation of the financial statements of subsidiaries operating in foreign countries (and changes in these principles).

The notes should include information in a number of other respects too. If a parent company does not prepare group accounts because it is a subsidiary to a parent on a higher level, information should be provided about the name, organizational number, and location (*säte*) of the parent on the higher level. Any differences in accounting principles used for the group and the individual companies should be disclosed. The name, organizational number, and address should be provided for each subsidiary, as well as information about the share of equity capital that the parent holds. If the share of the voting rights is different than the

Table X.1 Specific additional disclosures regarding the group

- Information pertinent to an evaluation of the result of the group's operation and/or financial position that is not disclosed on the balance sheet or the income statement.
- Events of material significance to the group.
- Accounting principles for consolidation, translation of foreign subsidiaries, valuation principles, depreciation periods used for consolidated goodwill and reserves for restructuring costs.
- Differences in accounting principles used by the group and the individual companies.
- The share of the parent company's purchases and sales for the year that relate to group companies.
- The name, organizational number, and location of the parent on the higher level if a parent company does not prepare group accounts because it is a subsidiary to a parent on a higher level.
- The name, organizational number, and location of each subsidiary, the share of equity capital that the parent holds, and the share of the voting rights, if that is different than the share of equity capital.
- The reasons why one or more of the subsidiaries are not included in the group, and the name of the companies that are not included because of the fact that the parent does not have a controlling interest, even though it holds more than 50 per cent of the voting rights, and the reasons why a subsidiary is included even though the parents company does not hold more than 50 per cent of the voting rights.
- Information that makes it possible to compare the group accounts of this year with the year before that, if the composition of the group has changed significantly during the financial year.
- Regarding each acquisition and merger during the year: the name of the added company and its operations, which method of consolidation that has been applied, the date of the acquisition or merger, and the operations of the subsidiary that is to be divested according to a plan.
- Regarding each large acquisition during the financial period: the acquired share of equity and voting rights, the price, the conditions of payment, and the provisions for restructuring (in following periods: actions taken, outlays incurred and the amount of the remaining provisions).
- The accounting treatment of goodwill (and negative goodwill) including the amortization period, the underlying reasons in those cases where the economic life of goodwill is estimated to be more than five years, and the reasons for a non-linear amortization of goodwill.
- A reconciliation of the goodwill values at the beginning and at the end of the period showing:
  i) Accumulated acquisition cost and accumulated depreciation at the beginning of the period.
  ii) Goodwill added during the period.
  iii) Depreciation during the period.
  iv) Adjustments and reductions during the period.
  vi) Accumulated acquisition cost and accumulated depreciation at the end of the period.
- Reasons why the acquirer has been unable to value certain assets or liabilities in the purchase analysis made at the acquisition.
- Reasons why an acquisition price has not been finally established (information about adjustments of these amounts should be given in the period they are accounted for).
- Regarding each merger during the year: the number of shares that have been issued, the merging companies' shares of equity and voting affected by the merger, the value of the assets and liabilities that each of the merging companies have contributed, and the part of sales, other revenues, and profits for each company that was generated before the merger.
- Changes arising during the year in the amounts of the shareholders' equity components in the consolidated balance sheet.
- The classification of foreign subsidiaries as independent or integrated, and which translation method is used for subsidiaries in highly inflationary economies.
- The amount of exchange differences taken directly to equity (if these have been reduced by hedging, the effects from this should be disclosed too).
- If a significant foreign subsidiary has been reclassified: the nature of the change in classification, the reason for the change, the impact of the change in classification on shareholders' equity, and the impact on net profit or loss for each period presented had the change in classification occured at the beginning of the earliest period presented.
- The effect on the financial statements of a foreign operation of a change in the exchange rate occurring after the balance sheet date (if the change is of such importance that it could be assumed that non-disclosure would affect the ability of users to make proper evaluations and decisions).
- Currency exposure, e.g. regarding net investments in independent subsidiaries in foreign currencies, and the company's policy concerning hedging against such exposure (desirable information).

share of equity capital, that should be disclosed too. There should be information about the reasons why one or more of the subsidiaries are not included in the group, and the name of the companies that are not included because of the fact that the parent does not have a controlling interest, even though it holds more than 50 per cent of the voting rights, and the reasons why a subsidiary is included even though the parent company does not hold more than 50 per cent of the voting rights.

If the composition of the group has changed significantly during the financial year, information should be provided that makes it possible to compare the group accounts of this year with the year before that.

Regarding each acquisition and merger during the year, information should be provided about the name of the added company and its operations, which method of consolidation that has been applied, the date of the acquisition or merger, and the operations of the subsidiary that is to be divested according to a plan.

Regarding each large acquisition during the financial period, information should be provided about the acquired share of equity and voting rights, the price, the conditions of payment, and the provisions for restructuring (in following periods information should be given about which actions that have been taken, which outlays that have been incurred and the amount of the remaining provisions).

Information should be provided about the accounting treatment of goodwill (and negative goodwill) including the amortization period, the underlying reasons in those cases where the economic life of goodwill is estimated to be more than five years, and the reasons for a non-linear amortization of goodwill. A reconciliation of the goodwill values at the beginning and at the end of the period should be presented, showing:

- Accumulated acquisition cost and accumulated depreciation at the beginning of the period.
- Goodwill added during the period.
- Depreciation during the period.
- Adjustments and reductions during the period.
- Accumulated acquisition cost and accumulated depreciation at the end of the period.

If the acquirer has been unable to value certain assets or liabilities in the purchase analysis made at the acquisition, the reasons for that should be given. The same applies if the acquisition price has not been finally established. Information about adjustments of these amounts should be given in the period they are accounted for.

Regarding each merger during the year, information should be provided about the number of shares that have been issued, the merging companies shares of equity and voting affected by the merger, the value of the assets and liabilities that each of the merging companies have contributed, and the part of sales, other revenues, and profits for each company that was generated before the merger.

Changes arising during the year in the amounts of the shareholders' equity components in the consolidated balance sheet should be specified in accordance with the Annual Report Act (ÅRL). This is conveniently done in the form of a note to the consolidated balance sheet. While there is no problem in clearly describing movements between restricted and unrestricted shareholders' equity (for example, 'transfer to legal reserve') in the corresponding specification for an individual company, such changes in the consolidated statements depend on several different underlying factors (which include write-down of shares in subsidiaries, depreciation of the consolidated acquisition cost of acquired assets, transfers to restricted reserves, allocation to or reversal of untaxed reserves). A specification of these components, particularly for larger groups, is cumbersome and is unlikely to interest readers. These types of change may be summarized under the heading 'Movements between restricted and unrestricted shareholders' equity'. However, changes during the year which have affected the total group's shareholders' equity should of course be specified (for example, a new share issue in the parent company, dividends from the parent company as well as exchange differences arising from the use of the current rate method in translating the accounts of foreign subsidiaries).

Information should be provided on the classification of foreign subsidiaries as independent or integrated and on which translation method that is used for subsidiaries in highly inflationary economies. Furthermore, information should be given about the amount of exchange differences taken directly to equity. If these have been reduced by hedging, the effects from this should be disclosed too. If a significant foreign subsidiary has been reclassified, a company should disclose the nature of the change in classification, the reason for the change, the impact of the change in classification on shareholders' equity, and the impact on net profit or loss for each period presented had the change in classification occured at the beginning of the earliest period presented. Information should also be provided on the effect on the financial statements of a foreign operation of a change in the exchange rate occurring after the balance sheet date if the change is of such importance that it could be assumed that non-disclosure would affect the ability of users to make proper evaluations and decisions. It is desirable that information is given about currency exposure, e.g. net investments in independent subsidiaries in foreign currencies, and the company's policy concerning hedging against such exposure.

2 Additional statements

In the case of a parent company, the administrative report should contain certain additional information regarding the group. This information is of two types:

- Information required for the separate entity should also be reported on a consolidated basis. This means, for example, that information should be included that is pertinent to an evaluation of the result of the group's operations and its financial position, but which is not required to be disclosed in the consolidated income statement or in the consolidated balance sheet. Also, information on events of material significance to the group which occurred during the financial year or thereafter should be disclosed.
- Specific additional disclosures are required such as the methods and valuation principles which have been used in the preparation of the consolidated accounts, proposed transfers of unrestricted equity to restricted equity, how large a share of the parent company's purchases and sales for the year relate to group companies and the reason for any departure from the legal regulations concerning the consolidated financial statements. The Annual Report Act permits such departures when there are exceptional difficulties involved in applying the regulations (see III). Valuation principles which have been used in the preparation of the consolidated accounts are in practice normally disclosed separately in a statement of valuation principles attached to the notes to the financial statements.

XI Auditing

According to the Swedish Companies Act (ABL), an auditor of a parent company should also give a report on the group. This report should be delivered two weeks before the annual general meeting of shareholders. Three kinds of information are mandatory:

- Whether the annual report of the group is in accordance with the Companies Act.
- Whether the auditors recommend that the annual general meeting discharges the board members and top executive from liability.
- Whether the auditors recommend that the annual general meeting adopts the consolidated balance sheets and income statements, including the dividend proposal.

A written report with this content should be published. At the annual general meeting of shareholders, the auditors are obliged to provide any information that is requested, unless this would be prejudicial to the company or group. They may not give special information to individual shareholders if this could hurt the company or group.

There are no formal requirements that the audits of the parent company and the group must be co-ordinated with those of the subsidiaries, joint ventures and associates.

Except for those auditing the accounts of very small companies, at least one auditor should be authorized by the Supervisory Board of Public Accountants *(Revisorsnämnden)*. Criteria for this authorization are found in two acts *(RNFS 1997:1* and *RNFS 1996:1)*: Authorized auditors should have auditing as their profession, live in the EES area (see III), have a degree from a university or business school in Sweden or any other Nordic country and have five years of professional experience. In addition, an exam at the Supervisory Board of Public Accountants should be taken.

Almost all authorized auditors are members of the Institute of Authorized Public Accountants (FAR). As in other countries, the auditing industry became more concentrated during the 1980s which led to a small number of firms dominating the market, each with some kind of relationship to an international firm. For a number of years, competition between these firms has visibly increased.

It remains to be seen whether there are any long-term implications of this increased competition on auditing quality in a broad sense and on independence. The questions that have arisen concerning the independence of auditing firms are not different from those raised in many other countries. The profession has been discussing this issue for many years and, as in many other countries, the discussion has become more intense, not least during periods of defaults in well-known companies (Johansson/Östman, 1995).

XII Filing and publication

No later than one month after the consolidated income statement and balance sheet have been adopted by the annual general meeting of shareholders, the parent company must send a copy of the annual report, including financial statements for the group, to the registration authority (PRV). As has already been mentioned (see III), the annual report of an individual company should include an income statement, a balance sheet and an administrative report. Larger companies should also attach a statement of changes in financial position (*finansieringsanalys*). The annual report of a parent company should, in addition, include a consolidated income statement and a consolidated balance sheet. If the parent company submits a statement of changes in financial position, a consolidated statement should also be included. Notes to the financial statements are considered to be an integral part of the financial statements and thus also the annual report (see X). Accounts for subsidiaries should follow the same rules as those for individual companies. Anybody may request data held by the registration authority, which may make a nominal charge.

Annual reports including financial statements for larger groups are printed and distributed by the parent company. The information in these reports is often focused at group level. There are cases where this focus is so strong that the question arises as to whether the accounting for the legal entity is actually in accordance with the law.

Almost all companies follow these rules. There is, however, a tendency among smaller companies and groups to delay sending their reports to the PRV.

XIII Sanctions

There is no clear general system of sanctions in Sweden to deal with companies that do not comply with accounting laws, but a government committee has made some proposals of which the outcome is still not clear. At the moment, very severe cases of non-compliance may be brought to court as criminal acts. Basically, it is perceived that serious effects must be proved, i.e. that it has not been possible to get a reasonably fair picture of the company or group from the accounting information, or that the accounts are deceptive or fraudulent. If this is proved, the legal system may impose imprisonment and/or fines on the persons responsible.

For companies registered at the Stockholm Stock Exchange there are special conditions. There is a law regulating stock exchanges. The stock exchange is required to have a review system including a disciplinary committee. This committee has to make decisions on all kinds of violations of the contract between the firm and the stock exchange. That also includes accounting principles. In practice, this seems to be restricted to a small number of cases.

For auditors there are two types of sanctions:

- The Supervisory Board of Public Accountants can investigate their actions. The sanction may be a warning or a withdrawal of authorization.
- The damage regulations in the Swedish Companies Act also apply to auditors.

XIV Differences between Swedish standards and the corresponding International Accounting Standards

As we have previously mentioned, the explicit intention of the RR is that its recommendations should, as far as possible, comply with those of the International Accounting Standards Committee. However, as the discussion above shows, there are differences in a number of respects. These differences are discussed in this section.

One difference is that, according to IAS 22, a write-down of goodwill must not be reversed. However, according to the Annual Report Act (ÅRL 4:5) a write-down of a fixed asset should be reversed if there are no longer any reasons for it. The reversal should affect the income statement. There is no exception for consolidated goodwill in the Annual Report Act. The cases where a reversal is appropriate are rare though and in addition to this, the measurement problems are severe. Against this background reversals of write-downs of goodwill must be treated very restrictively (RR 1:96, § 47).

Furthermore, according to IAS 22, negative goodwill should be recognized as income on a systematic basis over a period not exceeding five years unless a longer period, not exceeding 20 years from the date of acquisition, can be justified. According to RR 1:96 negative goodwill should be recognized as income in a systematic way that relates to the expected profitability.

IAS 22 states that, when using the pooling method, any difference between the amount recorded as share capital issued plus any additional consideration in the form of cash or other assets, and the amount recorded for the share capital acquired should be adjusted against equity. According to RR 1:96, the adjustment should be made against consolidated equity, firstly against the subsidiary's equity, and secondly against the restricted and unrestricted equity of the group.

According to IAS 27, a parent that is a wholly-owned subsidiary, or is virtually wholly owned, need not present consolidated financial statements provided, in the case of one that is virtually wholly owned, the parent obtains the approval of the minority interest. RR 1:96 refers to the Annual Report Act (ÅRL 1:4), where there are more exceptions.

RR 1:96 requires further information about provisions for restructuring expenses than IAS 22. IAS 27 (p. 32a) requires certain information about significant subsidiaries. RR 1:96 requires the same information for all subsidiaries.

The definitions of a group and an associated company is rather different in the Swedish Annual Report Act, and consequently in RR 1:96, compared to IAS 22. Furthermore, there are no recommendations for the accounting treatment of reverse acquisitions in RR 1:96, since it is not clear whether the method recommended in IAS 22 is consistent with the Swedish Annual Report Act.

According to IAS 22, in determining the cost of acquisition, marketable securities issued by the acquirer are measured at their fair value that is their market price at the date of the exchange transaction. According to RR 1:96, the fair value should be calculated as the average market value during the last 10 trading days before the acquisition became publicly known.

IAS 22 regulates the case where the acquirer is required to make subsequent payment to the seller as compensation for a reduction in the value of the purchase consideration. This is not regulated in RR 1:96, since this is not relevant to Swedish companies.

Another difference is that IAS 27 does not allow for acquired subsidiaries to be consolidated according to any other method than the purchase method. RR 1:96 has been adapted to the Annual Report Act which requires that certain acquired subsidiaries be consolidated according to the equity method.

Regarding the accounting for foreign subsidiaries, IAS 21.17 concerns exchange differences arising on a monetary item that, in substance, forms part of an enterprise's net investment in a foreign entity. The paragraph assumes that a parent company accounts for a subsidiary according to the equity method. That is not allowed according to Swedish law. RR 8:98 p. 22 assumes that the parent accounts for the subsidiary according to the cost method.

Furthermore, IAS 21.19 concerns hedging of a net investment in foreign entity (independent subsidiary). Only the case when the hedge is in the parent company, and in the form of a loan in foreign currency is discussed. According to RR 8:98, other ways of hedging, and cases where the hedge is in another company in the group, are treated in the same manner.

According to RR 8:98, the financial statements of foreign subsidiaries operating in highly inflationary economies are translated using the same method as for integrated subsidiaries. Alternatively, inflation-adjusted financial statements of independent subsidiaries may be translated using the closing rate method. However, when choosing a method in this respect, it should be recognized that existing price indices may be irrelevant for this purpose. In those cases, it is inappropriate to include foreign subsidiaries in the group accounts on the basis of inflation-adjusted financial statements translated at the closing rate method, and the method used for integrated subsidiaries should be used. Thus in this respect. the Swedish recommendation differs from IAS 21, where only the translation of inflation-adjusted financial statements at the closing rate is allowed. However, the Financial Accounting Standards Council observes that both methods produce the same outcome assuming that the exchange rates move in relation to inflation differences and used price indices are relevant.

Finally IAS 21.42b requires a disclosure of a reconciliation of the amount of net exchange differences classified as equity at the beginning and end of the period. RR 8:98 requires additional information about the amount by which those differences has been reduced by hedging activities.

Bibliography

Eriksson, L. (1974). *Koncernredovisningens informationsinnehåll* (Studentlitteratur, Hermods).

Eriksson, L. (1991). Förändringar–men knappast till det bättre. *Balans,* **12**, 12–18.

Johansson, S-E. (1954). Koncernbalansräkningens värderingsproblem. *Handelshögskolans i Göteborgs skriftserie*, **2**.

Johansson, S-E. and Östman, L. (1995). *Accounting Theory: Integrating Behaviour and Measurement* (Pitman Publishing, London).

Rundfelt, R. (1991a). Svensk rekommendation och internationell praxis. *Balans,* **12**, 10–11.

Rundfelt, R. (1997). *Tendenser i börsbolagens årsredovisningar* (Stockholms Fondbörs skriftserie).

Rutberg, A. and Skog, R. (1997). Det nya koncernbegreppet. *Svensk Skatterättslig Tidskrift*, **6–7**, 571–84.

Schuster, W. (1996). Foreign Exchange Exposure from an Accounting Perspective – an Analysis of Foreign Currency Hedging Loans of a Larger Amount than the Net Investment. EFI Research Paper 6556, Stockholm School of Economics.

af Trolle, U. (1992). Bröderna Kreuger, Torsten och Ivar. *Svenska Dagbladet.*

SWITZERLAND
INDIVIDUAL ACCOUNTS

Peter Bertschinger

Acknowledgements

Dedicated to Günther Schultz, retired Senior Partner of KPMG Fides, Switzerland, in recognition of his contribution to improve financial reporting in Switzerland.

CONTENTS

I Introduction and background

1 Introduction

As in most of the continental European countries, Swiss accounting regulation derives from the Napoleonic Code. This codified system of law imposes on all corporations certain requirements geared mainly to the protection of creditors. They include, but are not confined to, the following:

- Assets may normally be valued only at acquisition cost (the historical cost convention).
- Legal reserves have to be created before dividends can be paid, favouring internal financing through retained earnings.

The country's physical asset base suffered no destruction in the two World Wars. Additionally the currency, the Swiss franc, has remained quite stable for over 100 years. Switzerland has seen none of the hyperinflation, currency crises or currency reforms that have destroyed nominal values in other European countries. Many companies have been in existence for 50 years or more. The requirement of historical cost accounting means that often significant hidden reserves have been created.

The tax authorities use basically the same values as those stated in the statutory accounts as approved by the general meeting of the shareholders. Companies' boards and management have therefore always sought to reduce net income, i.e. taxable profit, by understating assets in the individual accounts and overstating liabilities. Minimal disclosure requirements have helped them to conceal profits. The attitude of the fiscal authorities has always been very flexible towards these practices in the past, as indirect taxation (social, value-added (VAT), gas, withholding taxes, etc.) are a more important source of revenue than direct taxation. Extensive income tax audits are rare.

Reliance on bank financing tends to reinforce such attitudes. Banks favour conservative accounting because it helps to reduce the risk their lending is exposed to. It also strengthens the hand of the board and the management as against minority shareholders and employees. Employees enjoy no right of representation through trade unions on the governing bodies of the organizations they work for.

Given these nowadays rather abnormal circumstances, the standards of accounting and financial reporting required in the individual accounts are somewhat relaxed, notably in that the creation of hidden reserves is perfectly legal. Under the new Corporations Act of 1 July 1992 their release has to be disclosed in the notes on the accounts. The Act also requires consolidated financial statements to be prepared and audited. In 1984 a Foundation for Accounting and Reporting Recommendations (*Stiftung für Fachempfehlungen zur Rechnungslegung*) was set up to improve the standards of financial reporting, especially in group accounts. The standards promulgated by the foundation follow internationally accepted accounting standards such as those of the IASC and the European Directives. They are binding upon the approximately 500–600 public companies with shares and/or debentures quoted at the Swiss Exchange (SWX).

Multinational corporations provide substantially more than the statutory minimum accounting and reporting information. They normally issue consolidated financial statements fully audited by one of the Big Five international accounting firms. Most such statements conform to the International Accounting Standards (IAS) of the IASC. However, the individual accounts of the parent company are drawn up under the terms of the Companies Act and do not normally comply with IASs.

In the past the more European-oriented multi national companies followed the Fourth and Seventh Directives. The industrial groups have now switched to International Accounting Standards. The big Swiss banks have been following the European Banking Directive, which is an adaptation of the Fourth and Seventh Directives to the circumstances of the banking industry. The big insurance companies have been following the European Insurance Directive, which embodies the requirements of the Fourth and Seventh Directives as they affect the insurance industry. The big Swiss banks and insurance groups have announced that they will adopt either IAS or US GAAP in the future to be more in line with their international competitors.

2 Historical development of Swiss financial accounting

Switzerland has a system of codified law typical of continental European countries. It has been influenced by the Napoleonic Code and by German law. Corporate Law is based upon the Code of Obligations (*Obligationenrecht*, OR) (Choi and Mueller, 1978, p. 92).

The Code of Obligations is set out in the fifth volume of the Swiss Civil Code. It governs all legal matters relating to commerce, finance and industry (business law), e.g. contracts, forms of business organization, accounting and auditing, etc. It is the equivalent of company law as found in other countries.

The present version of the Code of Obligations was enacted in 1911, with minor revisions in 1936. That the law had become seriously out of date became obvious during the economic boom years after World War II, and major efforts had been under way for more than three decades to change the code. Not until the end of 1991 was there a majority in parliament ready to adopt the changes, which also affect accounting and auditing. After years of debate the revised law became effective as of 1 July 1992.

Banking confidentiality remains important in law. One reflection of this is the absence of any requirement upon privately-owned companies to publish their financial statements.

The most extensive and authoritative history of Swiss accounting under the previous legislation is Käfer (1991).

In 1995 the Swiss government established a committee of experts in accounting and reporting (Groupe Mengiardi) to revise the Swiss corporation and company law and make it more compatible with European Directives. Their report was submitted in June 1998 and was published as a draft in October 1998. If adapted it would substantially expand the scope of accounting and reporting to all kinds of organizations and the related audit requirements. However, the draft law is subject to change by parlimentary discussions. This will most likely take several years until the law will become effective.

3 Standards and standard setting

3.1 The law relating to financial accounting

The legal requirements for accounting are very general in nature. All business organizations have to comply with the general rules embodied in articles 957–64 of the Code of Obligations. More stringent rules apply to corporations (OR Arts. 662–77). The relevant articles are listed in Table I.1. In general, however, the legal requirements in themselves are too broad to offer substantive guidance for effective corporate financial reporting.

The most unusual feature of the existing statutory regime is the unrestricted freedom to use undisclosed (i.e. secret or hidden) reserves. Only when they are released does the existence of such reserves have to be disclosed. To summarize, the major thrust of proposals for reform of company law (the Code of Obligations (OR)) is aimed at:

- Significantly greater disclosure requirements, e.g. in the notes.
- Preparation of consolidated financial statements.
- Revised accounting valuation rules, especially as regards the utilization of hidden reserves.

Politically it proved impossible to introduce the concept of 'true and fair view'. Small businesses in particular feared that the abolition of hidden reserve accounting would mean increased demands on them from the tax authorities, employees and outside shareholders. Additionally it was felt that more sophisticated accounting systems were what was really needed. The compromise was therefore arrived at that, while hidden reserves could still be created and held secretly, only their release would need to be shown.

The leading commentaries on the revised corporation law are Böckli (1992) and Forstmoser (1996). The Swiss Institute of Certified Accountants (*Schweizerische Treuhand-Kammer*) has interpreted the legal requirements extensively in its *Auditing Handbook* (*Revisionshandbuch* I, RHB, 1992, pp. 41–316), which is discussed below (see

Table I.1 Legal requirements on accounting matters

OR Art.	Requirement
General bookkeeping requirements affecting all companies	
957	Bookkeeping requirements
958	Duty to prepare inventory, balance sheet and income statement
959	Generally accepted bookkeeping standards
960	Currency requirement (Swiss francs) and valuation
961	Duty to sign the accounts and inventory
962	Duty to retain records for 10 years
963	Duty to publish accounts before court
964	Penalties for failure to comply with general bookkeeping requirements
Additional requirements upon corporations	
662	Duty to prepare directors' report and financial statements
662a	Generally accepted accounting principles
663	Minimum format of the income statement
663a	Minimum format of the balance sheet
663b	Minimum format of the notes on the accounts
663c	Disclosure of shareholders of public companies
663d	Annual report
663e	Duty to prepare consolidated financial statements
663f	Exemption for subholding companies
663g	Preparation of consolidated financial statements
663h	Protection in cases of confidentiality
664	Incorporation costs
665	General valuation of fixed assets
665a	Valuation of participations
666	Valuation of inventories
667	Valuation of securities
668	[Deleted]
669	Depreciation, value adjustments and provisions
670	Revaluation of participations and real estate
671	Allocation of reserves
671a	Reserve for own shares
671b	Revaluation reserve
672	Statutory reserve
673	Welfare reserve
674	Hidden reserves
675	Dividends
676	Construction dividends (*Bauzinsen*)
677	Payments to directors (*tantièmes*)

I.3.3). The RHB was revised in 1999 and renamed the *Handbuch der Wirtschaftsprüfung* (HWP).

3.2 Accounting and reporting recommendations

As the reform of company law took so long, an initiative was launched by the Swiss Institute of Certified Accountants. In consequence, the Swiss Foundation for Accounting and Reporting Recommendations (*Schweizerische Stiftung für Fachempfehlungen zur Rechnungslegung*) was set up in 1984. The foundation is the legal body of the Technical Committee on Financial Accounting and Reporting Recommendations (*Fachkommission für Empfehlungen zur Rechnungslegung*) which, as an independent private institution, issues the Recommendations on Financial Accounting and Reporting (*Fachempfehlungen zur Rechnungslegung*, FER). Various economic interests are represented, such as employer/employee organizations, banks, insurance companies, industry, financial analysts, the media, academics, the tax authorities, accountants and auditors. The standard setting body is organized as a foundation under federal supervision but funded by private parties with an interest in financial reporting. The funding is modest compared with that of Anglo-American standard setting bodies. As the members receive no remuneration, the only expenses are those of printing, publishing, meetings, etc. About half the income derives from the sale of the standards and other publications. Big companies normally contribute a modest sum each year to ensure the foundation's independence. The most active groups in the Technical Committee are those who actually have to prepare financial statements (banks, insurance companies, industry). They tend to water the recommendations down, although the larger public companies are voluntarily adopting IASC standards that go well beyond the recommendations.

The topic at the head of the foundation's agenda has been consolidated financial statements, as company law confines itself to the basic principles and lays down no detailed rules.

Table I.2 lists the recommendations that have been issued so far or are in preparation. The Recommendations on Financial Accounting and Reporting (FER) are in two parts:

- The recommendation itself.
- Explanations.

They are published in the three official languages (German, French and Italian) and also in English. Although the recommendations are not binding, they are the outcome of a broad consensus and widespread application is usual.

The first drafts are prepared by a project group of specialists in financial reporting. The results are refined by an executive committee and then discussed and approved by a commission of some 30 business people, as mentioned above. Once the proposals have been approved, an exposure draft is published and the views of interested parties are sought. After due consideration of the comments a final version is adopted and translations are prepared.

The recommendations take into consideration the guidelines of the OECD, European Directives and the International Accounting Standards of the IASC. Switzerland is a member of the OECD and the Swiss Institute a member of the IASC.

3.3 Swiss Institute of Certified Accountants

Formed in 1925, the Swiss Institute of Certified Accountants and Tax Consultants (*Treuhand-Kammer*) deals on a private basis with accounting and auditing issues. The institute comprises not only public accountants (independent auditors, *dipl. Wirtschaftsprüfer*, until end of 1997 called *dipl. Bücherexperte*) but also internal auditors and tax experts (*dipl. Steuerexperte*). It has both individual members and institutional members. There are some 3,000 individual members, about a quarter of them working as individual sole practitioners.

As mentioned above, tax consultants have also joined the institute. Similar training and certification programmes are available to those seeking to attain the qualification of Certified Tax Expert (*dipl. Steuerexperte*).

The third group within the institute is an association of internal audit departments from around 100 major Swiss firms numbering over 1,000 internal auditors. Most of the heads of the departments concerned are Swiss Certified Public Accountants.

Table I.2 FERs (year of issue) and proposals

0	*Objectives, Subject, Procedures of the Recommendations* (1985, revised 1993)
1	*Components of Individual Company Accounts and Consolidated Financial Statements* (1985, revised 1993)
2	*Consolidated Financial Statements* (1986, revised 1993)
3	*Generally Accepted Accounting Principles* (1990, revised 1993)
4	*Translation of Financial Statements Expressed in Foreign Currencies for Consolidation Purposes* (1990, revised 1993)
5	*Valuation in the Consolidated Financial Statements* (1990)
6	*Funds Flow Statement* (1992)
7	*Presentation and Format of the Consolidated Balance Sheet and Income Statement* (1992)
8	*Notes to the Consolidated Financial Statements* (1992)
9	*Intangible Assets* (1995)
10	*Off-balance Sheet Transactions* (1997)
11	*Taxes in the Consolidated Financial Statements* (1995)
12	*Presentation of Interim Statements* (1995)
13	*Accounting for Leases by the Lessee* (1997)
14	*Consolidated Financial Statements of Insurance Companies* (1995)
15	*Related Party Transactions* (1997)
16	*Employee Benefit Obligations* (1998)
	Under consideration are the following topics:
17	*Inventories*
18	*Tangible Fixed Assets*
19	*Individual Company Accounts*
20	*Impairment of Assets*
21	*Accounting for Non-Profit Organizations*
22	*Construction Contracts (PoC-Method)*

The institute organizes both basic education and continuing education programmes for professionals, sets the professional examinations, issues auditing and accounting statements and promotes the interests of the profession as a whole.

The institute's publications programme is quite extensive. The institute publishes the leading monthly professional journal *Der Schweizer Treuhänder*. It also issues a well-known series of professional and research monographs. Selected doctoral dissertations on accounting subjects from throughout Switzerland are included in this series. So far over 150 volumes have appeared.

An important product is the extensive four-volume *Auditing Handbook* (revised 1999), which contains accounting and auditing standards and interpretations. Although not legally binding, these find wide acceptance among accounting, legal and auditing professionals. Swiss accounting professionals are increasingly observing and implementing handbook recommendations, especially as regards the interpretation of the company law. Worries about exposure to professional liability claims have, however, acted as something of a brake on progress in this direction. In any case, it is imperative to ascertain the extent to which handbook recommendations have been followed in the work of preparation for the purposes of financial analysis, business negotiations or the evaluation of business opportunities (Mueller, 1980).

In 1999 an expanded four-volume handbook (*Handbuch der Wirtschaftsprüfung*, HWP) was published. It is also available on CD-ROM. It runs to approximately 2,000 pages and is the leading reference source on accounting and auditing in Switzerland. In its entirety the handbook is available only in German and French (*Manuel Suisse de révision 1998*, ed. Treuhand-Kammer, Zurich).

While the *Auditing Handbook* is concerned primarily with auditing procedures, its chapter 2 covers generally accepted bookkeeping and accounting principles. It covers in detail international accounting standards for the preparation of consolidated financial statements (group accounts). As with other Swiss financial accounting practices, the recommendations in the handbook go well beyond the requirements of commercial law.

Since the handbook recommendations are close to internationally accepted accounting and financial reporting practice, there is broad agreement with its contents, in theory at least. At the same time, however, the handbook's recommendations are in no way binding upon management or the board, and the accounting profession often has a difficult time persuading management to adopt them. Yet many accounting professionals feel that merely observing the requirements of company law could leave the auditor open to legal action for not meeting the handbook's more stringent standards. The fear of increased professional liability exposure has made the publication of the handbook somewhat controversial.

While the handbook annotates and explains all applicable provisions of the Code of Obligations (OR) in detail, it goes significantly beyond legal stipulations and sets forth underlying financial accounting conventions and professional accounting recommendations, so far as they exist.

3.4 National securities exchanges

In 1997 a revised Stock Exchange Act was introduced. There had never before been a federal law on dealing in securities. This has been the domain of the cantonal stock exchanges, the largest being Zurich, Basle and Geneva. These were consolidated into the new computerized Swiss Stock Exchange (SWX) in 1996. The Admission Committee of the SWX (*Zulassungsstelle*) has adopted that the Accounting and Reporting Recommendations (FER) should become a compulsory standard for all publicly quoted companies (see I.3.2).

As an important international financial centre Zurich has one of the more active securities exchanges on the European continent. Yet this is a relative matter. In comparison with the volume of trading on the New York, Tokyo or London exchanges, Zurich's turnover is quite small. Continental European companies simply employ much less equity capital than their Anglo-American counterparts. Most of the funding of industry comes, as in Germany, from the major 'universal' (non-specialist) banks and not from equity issues. Consequently, continental European equity markets are considerably thinner than those in the United Kingdom and the United States (Mueller, 1980). However, there is an increasing trend for securization of debts into marketable instruments.

Most underwriting and secondary trading business in Zurich is conducted by the brokerage departments of the large commercial banks. This fact must be borne in mind when the Swiss Stock Exchange listing regulations are considered.

Despite the modest capacity for portfolio investments in corporate securities in the Zurich market, a number of large non-Swiss multinational corporations (especially Japanese and US) have arranged for their securities to be quoted on the Swiss exchange for reasons of convenience. Also, many non-Swiss corporations are keen to issue bonds in Switzerland denominated in Swiss francs. Often these are arranged for institutional investors by way of private placements.

The regulations are fairly standard by continental European measures of comparison. London and especially New York listing regulations in respect of accounting, auditing and financial reporting requirements are far more comprehensive, as well as more stringent.

The listing requirements are set by the SWX. They have always been quite modest compared, for instance, with the regulations of the Securities and Exchange Commission (SEC) in the United States. The same applies to the continuous reporting required of listed companies (see Listing Rules of SWX).

The shares of only the very largest multinational companies (such as Nestlé, Novartis, ABB, CS Group, Swiss Re etc.) are listed on foreign stock exchanges (Frankfurt, London, etc.). Some of these companies are contemplating quotation in the United States and are therefore preparing to meet the stringent filing requirements of the SEC.

In connection with the introduction in 1997 of federal stock exchange regulations, the quality of accounting has also been improved. As already mentioned, the Accounting and Reporting Recommendations (FER), which were voluntary, have now become a minimum requirement for quoted companies. The revised Companies Act also meant that consolidated accounts have to be prepared, and all companies whose shares are quoted, or which have debentures outstanding, have to publish consolidated financial statements, audited by suitably qualified auditors. Additionally, such companies have to publish full interim financial statements for the first half-year from 1997 on, thus have been significantly increasing and improving interim information.

4 Regulated industries

4.1 Banking and insurance

The banking industry is subject to quite extensive accounting, auditing and financial reporting regulation. Obviously, the requirements of Swiss banking law are much more comprehensive and specific than those of the Code of Obligations (OR). One big difference from international accounting regulation is that banks and their auditors have to report to the Swiss Banking Commission (*Eidgenössische Bankenkommission*, EBK) in Bern, which is responsible for enforcing banking law. The comprehensive reports are confidential and are therefore not published. Mutual investment funds are regulated in much the same way as banks. The most important sources of law are:

- Federal Law on Banks and Savings Banks (*Bundesgesetz über die Banken und Sparkassen*, BankG).
- Ordinance regulating the Federal Law on Banks and Savings Banks (*Verordnung zum Bundesgesetz über die Banken und Sparkassen*, BankV).
- Ordinance regulating Foreign Banks (*Verordnung über die ausländischen Banken in der Schweiz*, ABV).
- Bulletins of the Federal Banking Commission (*Bulletin der Eidgenössischen Bankenkommission*).
- Circulars of the Federal Banking Commission (*Rundschreiben der Eidgenössischen Bankenkommission*).

The leading commentaries are: Auditing Handbook, Volume 3 (1999); Bodmer *et al.* (1991); Albisetti *et al.* (1987).

The new Stock Exchange Act introduced in 1997 has subjected the securities industry to further regulation. This includes a mandatory takeover code.

As mentioned above, mutual funds are subject to a regime similar to that of the banks. They are governed by special legislation:

- Federal Investment Funds Act (*Bundesgesetz über die Anlagefonds*, AFG).
- Ordinance regulating Investment Funds (*Verordnung über die Anlagefonds*, AFV).

The leading commentaries are: Auditing Handbook, Volume 4, 1999; Schuster, 1975.

Swiss insurance companies and Swiss branches of foreign insurers are closely supervised by the Swiss Insurance Agency (*Eidgenössisches Versicherungsamt*), which prescribes, for instance, investment and reserve policies.

The most important laws are:

- Federal Health and Accident Insurance Act (*Bundesgesetz über die Kranken- und Unfallversicherung*, KUVG).
- Ordinance No. 1 implementing the Federal Health and Accident Insurance Act in Respect of Accounting and Regulation (*Verordnung 1 über das Bundesgestz über die Kranken- und Unfallversicherung betreffend Rechnungswesen und Kontrolle*).
- Federal Act on the Regulation of Private Insurance Companies (*Bundesgesetz betreffend die Aufsicht über die privaten Versicherungseinrichtungen*, VAG).
- Ordinance governing the Regulation of Private Insurance Companies (*Verordnung über die*

Beaufsichtigung von privaten Versicherungseinrichtungen, AVO).

The leading commentaries are: Auditing Handbook, Volume 3, 1999; Kuhn, 1989; Erb, 1986.

The State social security system is governed by many laws which include accounting and auditing provisions. The main Acts and regulations, which are enforced by the Federal Social Insurance Agency (*Bundesamt für Sozialversicherung*, BSV), are:

- Federal Old Age, Widows' and Widowers' Insurance Act (*Bundesgestz über die Alters- und Hinterlassenen-Versicherung*, AHVG).
- Federal Unemployment Insurance Act (*Bundesgesetz über die Arbeitslosenversicherung*, ALVG).

4.2 Pension funds

Employee pension funds have also been heavily regulated since 1986 under the Federal Pension Fund Act (BVG). The Act lays down a chart of accounts, valuation and investment policies including notes in an annexe to the financial statements, requirements for audit by professional auditors and periodic review by professional actuaries. The requisite reporting to cantonal regulatory bodies and to beneficiaries is also quite stringent and formalized. The federal Acts and regulations are enforced by the Federal Social Insurance Agency (*Bundesamt für Sozialversicherung*, BSV).

The main laws are as follows:

- Federal Pension Fund Act (*Bundesgesetz über die berufliche Alters-, Hinterlassenen- und Invalidtätsvorsorge*, BVG).
- Ordinance governing the Regulation and Registration of Pension Funds (*Verordnung über die Beaufsichtigung und die Registrierung der Vorsorgeeinrichtungen*, BVV 1).
- Ordinance governing Pension Funds (*Verordnung über die berufliche Alters-, Hinterlassenen- und Invaliditätsvorsorge*, BVV 2).

The more important commentaries are: *Handbuch der Personalvorsorge*, Amt für berufliche Vorsorge des Kantons Zürich (ed.), 1991; Helbling, 1990; Auditing Handbook, Volume 4, 1999.

4.3 Nationalized industries and governmental bodies

Most countries in continental Europe, including Switzerland, have created state-owned enterprises to:

- Run public transport systems (railways and bus services).
- Provide postal services.
- Supply domestic and industrial users with gas, water and electric power.

Governments often make substantial financial subsidies available (if and when necessary) to private enterprises whose survival is in the national interest or to companies that face unfair international competition (for example, national airlines). The *quid pro quo* usually entails a degree of government influence, if not control. Other regulated enterprises include nationalized industries and governmental bodies.

These mostly nationalized industries have to follow accounting and reporting rules established by governmental agencies. These include, for example, the Swiss Federal Railways (*Schweizerische Bundesbahnen*, SBB), postal and telecommunications services (*Die Post*, the phone services went public in an IPO in 1998 and were renamed *Swisscom*), pipelines, power stations (a deregulation will become effective after the year 2000) and other utilities.

The relevant legislation is the Railways Act (*Eisenbahngesetz*), the Shipping Act (*Schiffahrtsgesetz*), the Pipelines Act (*Rohrleitungsgesetz*) and the Emergency Public Services Act (*Landesversorgungsgesetz*).

Various regulations exist concerning governmental accounting for:

- The federal government and its agencies.
- The 26 cantonal governments, including their offshoots, for example cantonal banks.
- Municipalities (approximately 3,020). Many cantons have a standard chart of accounts which is laid down in detailed accounting and reporting manuals.

II Forms of business organization

1 Overview

There are two main types of company in Switzerland, the stock corporation (*Aktiengesellschaft*, AG; *société anonyme*, SA) and the limited liability company (*Gesellschaft mit beschränkter Haftung*, GmbH). There are some 170,000 stock corporations. They constitute a substantial majority of Swiss companies, since that form of business organization has a number of advantages over the limited liability company, but the rules and regulations governing them can be taken as applying equally to limited liability companies unless specifically stated otherwise. According to estimates by the Federal Statistical Agency (1983) the number of corporations can be broken down approximately as follows:

- 12,000 holding companies.
- 10 corporations with more than 5,000 employees (e.g. UBS Bank).
- 40 corporations with more than 2,000 employees.
- 300 corporations with more than 500 employees.
- 900 corporations with more than 200 employees.

Hence most of the registered corporations have only one employee or none at all. Only about 3,000 corporations have sales exceeding CHF 50 million; many of them are only trading, not manufacturing, concerns. In the French-speaking region of Switzerland many corporations have been established for the sole purpose of holding a piece of real estate (*société immobilière*, SI).

The following types of business organization are to be found in Switzerland (Guhl, 1991):

- Stock corporations (*Aktiengesellschaft*, AG).
- Sole proprietorships (*Einzelunternehmung*).
- General partnerships (*Kollektivgesellschaft*, & Co.).
- Limited partnerships (*Kommanditgesellschaft*, KG).
- Limited liability companies (*Gesellschaft mit beschränkter Haftung*, GmbH).
- Co-operatives (*Genossenschaft*).
- Ordinary partnerships (*einfache Gesellschaft*).
- Foundations (*Stiftung*).
- Branches (*Zweigniederlassung, Sitz*).

The legal form known as a trust in the Anglo-American world does not exist in Swiss law.

2 Stock corporation (AG)

As mentioned earlier, the stock corporation (AG) is by far the most important form of business organization. It is also popular for small businesses and for a variety of business activities. A stock corporation has a fixed share capital and limited liability. The share capital has to be at least CHF 100,000, 20 per cent or CHF 50,000 of which has to be paid up or contributed in kind. The voting shares have a nominal value of CHF 10 or more. Both registered (*Namenaktien*) and bearer shares (*Inhaberaktien*) are common. Often the registered shares carry greater voting rights (for example, CHF 100 per share) than bearer shares (for example, CHF 1,000 per share). There are also non-voting shares called participation certificates (*Partizipationsscheine*, PS). They enjoy the same rights to dividend and liquidation payments but carry no voting rights. They are sometimes used by public companies and therefore are mostly quoted. Often no share certificates are issued for private companies.

The main governing bodies of the stock corporation are as follows (German terminology in brackets):

- Shareholders' meeting (*Generalversammlung*, GV).
- Board of directors (*Verwaltungsrat*, VR).
- Officers:
  - (i) Delegate of the board (*Delegierter des Verwaltungsrates*, Del.VR).
  - (ii) Management (*Geschäftsleitung*, GL).
  - (iii) Managers (*Direktoren*).
  - (iv) Proxies with limited powers of signature (*Prokuristen*, p.p.a.).
- Auditors (*Revisionsstelle, Revisoren*; see X).

The delegate of the board is both a member of the board and head of the management team. He or she therefore has significant power in running the company. The role is similar to that of the chief executive officer (CEO) in the United States or the managing director in the United Kingdom. Switzerland does not have the German two-tier supervisory board system where the executive management team (the *Vorstand*) is supervised by a board (the *Aufsichtsrat*) which includes outside directors and worker representatives (trade unions). However, more and more Swiss public companies are establishing an audit committee of outside board members.

At least three shareholders are required at the date of foundation. Subsequently it is permissible for the number to be reduced to a single shareholder (when the corporation is known as a one-person corporation (*Einmannaktiengesellschaft*).

The articles of incorporation have to include the firm's name (often in more than one language), its legal domicile (residence), the object of the company, and its capital stock (number of shares and nominal value per share category).

The annual general meeting of shareholders has to be held not later than six months after the balance sheet date as stated in the articles. In 90 per cent of all companies the balance sheet date will be the end of the calendar year, 31 December. The financial statements, including the auditors' report and the report of the directors, have to be sent to shareholders or made available at least 20 days before the general meeting.

Among other things the annual general meeting has the following obligations and rights (Ernst & Young, 1991, p. 13):

- Approval of the financial statements.
- Approval of the company's annual report.
- Power of decision on the profit distribution (dividends) as proposed by the board of directors.
- Election of the board of directors and the auditors.
- Discharge of the board of directors.
- Changes in the articles.
- Liquidation and winding up of the company.

The board of directors has to meet certain nationality requirements. The members have to be individuals (no business organizations). The majority of the members have to be Swiss citizens and have to live in Switzerland.

The board of directors can entrust executive functions to one (or more) delegates of the board. Outside directors are common in public companies. Their duties and rights are normally set out in writing in a document known as the organization rules (*Organisationsreglement*).

3 Other forms of business organization

3.1 Sole proprietorship

The business is carried on by one person alone, as in the case of a tradesman, farmer or a member of one of the liberal professions. If it reaches a certain size it has to be entered in the commercial register; the threshold is sales exceeding CHF 100,000. Liability is unlimited. This type of establishment is normally used only for starting up in business and liberal professions. Later there will be a switch to stock corporation status. As a rule the sole proprietor will follow the same accounting rules as stock corporations. The accounts are mainly tax driven. About 130,000 sole proprietorships are listed in the commercial register.

3.2 General partnership (& Co.)

A general partnership is an association of two or more persons for a commercial purpose. The disadvantage is that the partners' liability is unlimited, i.e. all the partners are liable to the full extent of their personal fortune. All the partners have to be natural persons: stock corporations may not become partners. The partnership has to be recorded in the commercial register. General partnerships normally use accounting rules similar to those of stock corporations. There are about 17,000 such partnerships on the commercial register.

3.3 Limited partnership (KG)

This is a similar form of business to the general partnership, except that the liability of one or more partners is limited to their equity share. This limit has to be stated in the commercial register. The general partners, whose liability is unlimited, have to be private persons, whereas the limited partners may be either individuals or corporate investors. There are only about 4,000 limited partnerships in the commercial register. Limited partnerships normally follow much the same accounting rules as stock corporations.

3.4 Limited liability company (GmbH)

Most of the requirements of company law also apply here. The share capital must not exceed CHF 2 million and the names of the principal shareholders have to be stated in the commercial register. This form has therefore never been attractive, as it is suitable only for a small company whose size would be signalled by the abbreviation GmbH after the name. The need to declare the main shareholders makes it doubly unattractive. Hence this form of business organization is not very often found. Even small businesses prefer the stock corporation alternative. Normally the same accounting and reporting rules apply as for stock corporations. The lack of enthusiasm for this corporate form is reflected in the fact that only about 20,000 are to be found in the commercial register, although there has been an increasing trend in the last couple of years for the GmbH for small businesses.

3.5 Co-operative

There is no restriction on the number of members. The main object of a co-operative is to promote or secure its members' interests. This form of organization is often used in agriculture, housing, insurance or the retail trade (e.g. the Co-op retail chain). Unless the accounting rules for the relevant regulated industry apply, co-operatives follow the same accounting rules as stock corporations. There are about 14,000 co-operatives in the commercial register. They normally enjoy some form of tax relief.

3.6 Ordinary partnership

This is a very loose form of organization or association between individuals and/or companies. It has no trading name and is established solely to carry out a specific project or joint venture (e.g. a construction contract). The accounting is normally done in the books of the joint venture partners. It cannot be entered in the commercial register.

3.7 Foundation

This is an amount of money or other assets (a fund) dedicated to a specific purpose. Normally the funds cannot be returned to the founder or sponsor. All foundations are supervised by cantonal or federal authorities to ensure that the assets and returns are properly used for the good of the beneficiaries. Virtually all pension funds in Switzerland are organized as foundations and are also subject to audit.

Special regulations govern the accounting, investment policies and financial reporting of pension funds (see I.4.2). There are about 21,000 foundations on the commercial register, most of them pension funds. They are normally tax-exempt.

3.8 Branch

Those can be branches of foreign or domestic corporations, usually banks, insurance companies, airlines, etc. They are mostly taxed in the same way as stock corporations if they are owned by a corporation with a foreign head office. Normally the accounting of the stock corporation as applied at the head office is adapted. There are no audit requirements for branches of foreign companies. The audit requirements apply only to the head office, which has to integrate the branch into its accounts. About 500 branches are listed in the commercial register. Many branches are not registered but are still treated as secondary tax domicile as each canton and municipality tries to tax all business establishments in their jurisdiction.

4 Commercial register

The commercial register (*Handelsregister*, HR) plays an important role in the commercial and business law. There are 26 cantonal commercial registers, which are co-ordinated by a federal agency to ensure uniformity. All changes in the data in the decentralized registers have to be published in the official Swiss Trade Journal (*Schweizerisches Handelsamtsblatt*, SHAB) which appears daily.

Excerpts are available on any company for a modest fee, to foreigners as well as Swiss nationals. More and more of these services are provided via the internet by private credit agencies. The information is restricted to the following, however, as financial statements and audit reports do not have to be filed:

- The company's name, domicile and legal business form.
- Its incorporation date and details of any changes in the information recorded in the register.
- The board members and other signatories, with their name and place of residence.
- The auditors of corporations (normally an audit firm) and their domicile. (This is a new requirement of company law.)
- Total amount of capital; the number, nominal value and types of shares.
- Authorized and contingent share capital (in the case of corporations).

5 Foreign listings of Swiss companies

There are about 200–300 companies that have their shares quoted on the Swiss Exchange (SWX). The following companies have multiple listings, i.e. they are also listed on foreign stock exchanges:

Name *Industry*	**Stock exchanges in addition to Swiss Exchange** *Accounting Standards, reporting currency if not CHF*
ABB *Electrical engineering*	Frankfurt, Vienna, Munich, London (Seaq), USA-OTC (ADR) *IAS, IAS-US GAAP reconciliation, consolidated financial statements in US$*
Adecco *Temporary work*	Paris, London (Seaq), New York (Nasdaq) *IAS and US GAAP*
algroup (Alusuisse-Lonza) *Aluminium, chemicals, packaging*	Frankfurt, London (Seaq) *IAS*
BB Biotech *Biotech investments*	Frankfurt *IAS*
BK Vision *Investments financial institutions*	Stuttgart *IAS*
Castle Private Equity *Investments unlisted companies*	Luxembourg *IAS, consolidated financial statements in US $*
Ciba SC *Special chemicals*	London (Seaq), USA-OTC (ADR) *US GAAP*
Clariant *Special chemicals*	London (Seaq) *IAS and FER*
Credit Suisse Group *Banking*	Frankfurt, Tokyo, London (Seaq), Paris (OTC), USA-OTC (ADR) *EU Banking Directives, US GAAP project*
Holderbank *Cement*	London (Seaq), USA-OTC (ADR) *IAS and FER*
Julius Baer *Banking*	Frankfurt *IAS*
Kühne & Nagel *Transportation*	Frankfurt *IAS and EU Directives*
LGT, Liechtenstein *Banking*	Frankfurt *IAS*

Logitech *Computer mice*	New York (Nasdaq) *US GAAP, consolidated financial statements in US$*
Mövenpick *Catering, hotels, food*	Frankfurt *IAS*
Nestlé *Food and beverage*	Amsterdam, Brussels, Frankfurt, Paris, Vienna, Tokyo, London (Seaq), USA-OTC (ADR) *IAS*
Novartis *Pharmaceuticals*	London (Seaq), USA-OTC (ADR) *IAS*
Pharma Vision *Investments in pharmaceuticals*	London (Seaq) *IAS*
Richemont *Tobacco, luxury goods*	USA-OTC (ADR), London (Seaq), Johannesburg (JDR) *IAS and FER, consolidated financial statements in £*
Roche *Pharmaceuticals*	London (Seaq), USA-OTC (ADR) *IAS*
Schindler *Elevators*	Frankfurt, Berlin *IAS and FER*
Selecta *Automatic food dispensers*	London (Seaq) *IAS*
Stillhalter Vision *Investments derivatives*	Stuttgart *IAS*
Sulzer *Technology, machinery*	London (Seaq) *IAS*
Sulzer Medica *Medical technology*	New York (ADS), London (Seaq) *IAS, IAS-US GAAP reconciliation*
Swatch Group *Watches*	London (Seaq) *EU-Directives and FER*
Swisscom *Telecommunications*	New York (ADS) *IAS, IAS-US GAAP reconciliations*
Swiss Re *Insurance*	London (Seaq), USA-OTC (ADR) *EU-Directives and FER, US GAAP project*
UBS *Banking*	Tokyo, London (Seaq), USA-OTC (ADR) *IAS*
TAG Heuer *Watches*	New York (ADS) *US GAAP*
VP Bank, Liechtenstein *Banking*	Frankfurt, Munich *FER*
Zurich Allied *Insurance*	London (Seaq) *FER, US GAAP project*

(Sources: mainly Swiss Stock Guide 98/99, Verlag Finanz und Wirtschaft, Zurich (Status of 30 June 1998))

Abbreviations:

ADR	=	American Depositary Receipt
ADS	=	American Depositary Share
EU Directives	=	Fourth and Seventh EU Directives of Company Law
FER	=	Swiss Accounting and Reporting Recommendations
IAS	=	International Accounting Standards
NASDAQ	=	National Association of Securities Dealers Automated Quotation System
New York	=	New York Stock Exchange (NYSE)
OTC	=	Over the Counter
Seaq	=	London Stock Exchange Automated Quotation Systems
SWX	=	Swiss Stock Exchange (electronic)
US GAAP	=	US Generally Accepted Accounting Standards

III Objectives, concepts and general principles

1 Principal users of accounts

As mentioned, most business activity in Switzerland is carried on by stock corporations. Some 170,000 figure on the commercial register. Only about 200–300 of them have their shares listed on a stock exchange. There are about another 300 with listed debenture bonds (*Obligationenanleihen*). Some stock corporations have both shares and bonds listed. It follows that over 99 per cent of all corporations are privately-owned, most of them by a single shareholder. They are therefore called one-man corporations (*Einmannaktiengesellschaften*). These companies are under no requirement to publish financial statements of any kind. The accounts are prepared mainly for the purpose of enabling the shareholder(s) to determine the dividend distribution. As mentioned, profits are often retained. Shareholders in the company can often realize a tax-free capital gain upon the sale of their holding. Where dividends are paid, both the company and the shareholder are liable for income tax (double taxation).

The statutory accounts are used only to assess income tax and determine the distribution of profit. For expenses to be tax-deductible they have to be included in the statutory accounts, i.e. booked in the general ledger. As a result, the amounts stated in the statutory accounts are usually taken directly into the tax return. To small companies this is a significant administrative relief.

For outside shareholders with minority interest the statutory accounts provide only minimal information about the stewardship of management. The limited disclosure and freedom to create hidden reserves mean that management can usually cover up its misjudgements without difficulty. Dissident shareholders have the power to call for a special investigation (*Sonderprüfung*) if they suspect irregularities. Since the introduction of the 1 July 1992 corporation law very few such investigations have been demanded. Most of those that have been set in train have been blocked by the courts.

Statutory accounts, including a report by a recognized audit firm, are often prepared at the request of the banks, even though banks tend to ensure they get better information so as to monitor the degree of risk their lending is exposed to (e.g. budgets, cash flow forecasts). All Swiss companies use these same statutory accounts also for declaring their tax liability. The tax return will normally state the same net income and equity as the statutory accounts adopted by the general meeting of shareholders.

Generally, the financial statements of Swiss companies will include only a balance sheet, profit and loss account and a brief section of notes with minimum disclosure. However, the trend is increasingly among larger companies, especially those operating internationally, to present more comprehensive information.

The (voluntary) Accounting and Reporting Recommendation FER 1 outlines the components of individual and consolidated financial statements which should preferably be included. Among the recommendations are the following:

- The individual and consolidated financial statements should comprise a balance sheet, an income statement (profit and loss account) and a funds flow statement as well as the notes.
- The funds flow statement should show the funds flow from operations in the individual and consolidated financial statements, the funds flow from investing/divesting activities and the funds flow from financing activities.
- The notes should disclose:
  - (i) The accounting policies applied in the individual and consolidated financial statements.
  - (ii) Further details on other parts of the financial statements.
  - (iii) Additional information not included elsewhere in the financial statements.
- Alongside the amounts for the reporting year the financial statements must show the comparative figures for the previous year. If they are not comparable, the fact must be explained.
- Where the company is part of a group, its financial statements may be presented in simplified form if consolidated financial statements are prepared and if the consolidated statements are made available to the user of the company's individual statements. In that case, the company's individual statements may omit the funds flow statement and the additional information in the notes except those required by law.

In general, users have only a limited degree of access to the accounts of Swiss companies, since there is no requirement to publish accounts or file them in the commercial register. The only exceptions are quoted companies, banks, insurance companies and mutual investment funds. Financial institutions are also subject to strict supervision. They must file accounts with the relevant regulatory authorities (e.g. the Swiss Banking Commission, the Swiss Insurance Agency, the Swiss Stock Exchange). In addition, banks and insurance companies are required to publish their balance sheets in the Swiss Trade Journal (SHAB) and further newspapers, as specified in their articles of association. These must be published annually, half-yearly or quarterly, depending on the size of the company.

All companies, even those not required to publish accounts, have to make the following documents available for inspection by shareholders:

- Balance sheet and auditors' report.
- Profit and loss account.
- Notes on the accounts.
- Annual report of the directors.
- The board's proposal for profit appropriation.

They must be available at head offices and branches at least 20 days before the ordinary general meeting of shareholders, which must be held within six months after the balance sheet date.

The primary users of published accounts are consequently shareholders. As mentioned, most businesses are privately-owned and therefore have few shareholders. Only about 200–300 companies are quoted on a Swiss stock exchange, i.e. have publicly traded shares. As in other continental European countries, only about 10 per cent of the population hold quoted shares. More significant are the diversified holdings of institutional investors. Such investors, Swiss and foreign, include insurance companies, mutual funds, and the pension funds of private and public employers. A not insignificant role is played by foreign private investors whose portfolios are managed by Swiss banks.

Banks normally try to avoid acquiring a big stake in quoted industrial companies. Where it happens it will usually be for historical reasons or because the company had to be bailed out, e.g. by the bank exchanging debt for equity. When the company is restored to financial health the bank will aim to sell the shares as soon as possible.

Banks are the source of most external finance, in the form of short- and long-term lending, frequently have seats on the board and access to detailed management accounts. Theoretically any creditor may apply directly to the company to see the balance sheets, the profit and loss account, the notes and the auditors' report. However, the company may forestall access by settling the debt. Other outsiders are not regarded as being entitled to information concerning the performance and activities of the business.

The effect of the banks' position in the user profile is to incline accounting, access and disclosure towards their banks' interests, i.e. conservatism, the protection of the security of their investments and the servicing of debt. Big companies, responding to pressure from international capital markets, have been publishing more complete consolidation accounts. Hence the group accounts are normally of much better quality than the individual accounts where disclosure is concerned. Most individual accounts understate equity because they are tax-driven.

2 Accounting concepts and standards

Accounting and reporting are governed by the legal requirements of the company law which is part of the Code of Obligations (OR). The law was amended as of 1 July 1992 to bring it more into line with the European Directives. It is expected that a new draft law on Accounting and Auditing in Switzerland (RRG) will considerably expand the accounting, reporting and auditing requirements in Switzerland and make it more compatible with European requirements.

Under the existing code, financial statements must be prepared in line with generally recognized commercial principles. According to OR Art. 662a the annual accounts in the individual financial statements should be prepared in accordance with recognized accounting principles in such a way as to offer the most reliable picture of the income and financial situation of the company (not a true and fair view). They should also include the previous year's figures. These recognized accounting principles (see also V.1), in particular, should follow the principles of:

- Completeness of the annual financial statements.
- Clarity and materiality of the financial statements.
- Prudence (conservatism).
- Continuity of the company's activities (going concern).
- Consistency in presentation and valuation.
- Prohibition of setting off assets and liabilities, as well as of setting off expenses and income (gross principle).

A regulation format is specified under the new law for minimum disclosure in the balance sheet and the income statement. The minimum level of disclosure required in the notes is also prescribed. This ensures a limited degree of standardization.

In general, accounts must be prepared in accordance with 'recognized accounting principles', a term which is not defined in any detail. In practice, such principles are determined primarily by the accounting profession and laid down by the Swiss Institute of Certified Accountants, which publishes a comprehensive *Auditing Handbook* (HWP) (see I.3.3).

IV Bookkeeping and preparation of financial statements

1 Bookkeeping

Under OR Art. 957 every firm entered in the cantonal commercial register is required to keep proper books of account. As a stock corporation can be created only by way of entry in the register, it is automatically subject to the provisions of general accounting law as set out in OR Arts. 957–64. The most important rules, which are also applicable to stock corporations, are presented below. Every company is required to keep such books of account as are necessary, given the nature and extent of the business, to reflect its financial state properly and to determine liabilities and claims in connection therewith, as well as the operating results of each year.

Some requirements are procedural only. For instance, OR Art. 957 requires every company to keep such records and books of accounts as are necessary, according to the nature of its business, and to prepare periodic (i.e. annual) balance sheets and income statements. Juridical interpretation of this article has expanded its applicability to include the maintenance of proper cost accounting systems where a manufacturing company is required to value work in progress and finished goods periodically. The accounting system therefore has elements of both financial accounting and cost accounting. The accounting system may necessitate the following, depending on the size and nature of the business:

- Vouchers: original accounting documents.
- Inventory: list of assets (for example, stocks or fixed assets) at a period end, showing quantity on hand, description and value/cost.
- Journal: book with transactions entered in chronological order.
- General ledger: summary book of accounting.
- Subsidiary ledgers: ancillary books of accounting, for example payroll, debtors, etc.
- Balance sheet: assets and liabilities as at a period end and the resulting net assets (shareholders' equity).
- Income statement: profit and loss account for a period, showing income and expenses by categories and the resulting profit/loss of the period.
- Cost accounting and calculation (to value work in progress and finished products and fixed assets manufactured internally).

2 Uniform system of accounts

As of 1 July 1992 a standard format was introduced for the accounts of stock corporations. The Code of Obligations (OR) requires additionally that accounts should be prepared in Swiss francs, that they should be complete and thorough and that they should give a clear picture of the trading position of the company, subject to the explicit acceptability of the creation and release of hidden reserves. However, under the revised corporation law the net amount released has to be disclosed in the notes.

The accounts of companies are prepared on a basis consistent with that required by the fiscal authorities. If deductions are to be claimed they must be included in the accounts, which leads to a form of presentation which may not reflect the company's true commercial position.

Switzerland has no general chart of accounts such as those of other continental European countries (e.g. France). However, a recommended chart (*Kontenplan*) has been published on a voluntary basis (Käfer, 1987). It is procedural only, as it is intended to provide bookkeepers with guidance on establishing a financial accounting system of a company. However, this chart of accounts was overtaken by the new company law which took effect on 1 July 1993. In 1996 a common chart of accounts was published for small and medium-sized businesses (Sterchi, 1996). It is the update of the standard chart of accounts originally developed by Karl Käfer, retired accounting professor of the University of Zurich. Many small businesses use this chart of accounts on their computerized PC bookkeeping systems. It takes into consideration the requirements of the Fourth and Seventh EU Company Directives.

3 Duty to prepare an inventory

The inventory is the list of assets (not of inventories or stocks only) showing the items, their value and the totals. (See also the special rules for corporations.) Other provisions of the Code of Obligations (OR) govern the running of internal control systems. For instance, OR Art. 958 requires companies to take annual inventories to ensure that the physical assets on the books are still in existence and that they are properly accounted for. For small companies inventories are determined on the basis of a stock-taking at year end. This exercise lists the identity of the articles, the quantity and the pricing, which are multiplied and added to the total which is taken into the balance sheet. Under Swiss law this listing has to be signed. The stock-taking at year end can include counting, measuring, weighing, etc., the articles. If a (computerized) perpetual inventory accounting system is in use, each article should be counted at least once a year during the year. At the year end a listing is printed out.

All articles have to be inventoried at least once a year, normally at year end. The inventory sheets, including quantities and valuation per article, have to be signed by management. However, it is accepted practice to include items, or groups of items, at a *pro memoria* CHF 1. This is also permissible for tax purposes. It could lead to the accumulation of substantial hidden reserves in the individual accounts.

4 Duty to prepare financial statements

According to OR Art. 958 every company in the commercial register is required to prepare an inventory of assets and liabilities, a balance sheet as of the end of each financial year and an income statement for the year then ended. These records must be completed within a reasonable time, having due regard to the normal business practice of the enterprise. For stock corporations this means within less than five months after the balance sheet date.

Under OR Art. 662 the board of directors of a corporation has to prepare a business report for each year. This consists of the audited annual financial statements, the annual report and the consolidated financial statements where such statements are required by law. The annual financial statements consist of the profit and loss account, the balance sheet and the notes.

Recorded transactions must be properly documented, and the documents underlying the preparation of financial statements must be preserved for at least 10 years. Also, the internal accounting records of Swiss companies must be physically located in Switzerland and must use (at least in their primary form) the national currency (Swiss francs). The records should be in one of the three main languages (German, French or Italian). English is widely used as well. In that case, the tax authorities may ask the company to provide them with translation into a Swiss national language. Of course, supplementary or subsidiary records may be in other currencies. The tax authorities will only accept financial statements expressed in Swiss francs.

Under OR Art. 960 the inventory sheets, the income statement and the balance sheet have to be prepared in the national currency, Swiss francs. This is also required by tax law. Rounding amounts to the nearest franc is possible. In group accounts, amounts are often rounded to the nearest thousand francs or to a million Swiss francs.

Under OR Art. 961 the inventory sheets, the income statement and the balance sheet must be signed by the persons responsible for the running of the company.

Under OR Art. 962 the books of account, the business correspondence and the accounting documents and vouchers have to be retained for a period of 10 years. The balance sheet and the income statement must be preserved in their (signed) original form; other records may be preserved in some other form, provided they can be accessed at any time, i.e. for the 10-year retention period. Ancillary ledgers can therefore be maintained on microfilm or COM films (Computer Output on Microfilm). Vouchers and accounting documents can even be maintained on magnetic information storage and retrieval systems (e.g. tape).

V Balance sheet and profit and loss account formats

1 Classification principles

1.1 Completeness

The financial statements should be complete and must not omit material information. All transactions should be fully entered in the bookkeeping system. Completeness includes the balances of the previous year being entered as the opening balances of the new financial year. The company is assumed to be a going concern. Either transactions are continuously booked from the beginning of the year to the end, or if this is not possible, e.g. with a permanent inventory system, a complete inventory-taking has to be performed at year end to identify all existing assets and liabilities. Even if items are fully written off (e.g. for tax purposes) they should still be recorded at a *pro memoria* CHF 1.

1.2 Clarity and materiality of the financial statements

Clarity is a concept which also figures in the general accounting conventions mentioned in OR Art. 958, which applies to all business organizations. It is doubly emphasized for corporations. The items should be clearly and precisely stated. Also, the additional information disclosed in the notes on the accounts should be self-explanatory and clear. Clarity is also achieved where a holding company prepares group accounts and therefore provides an insight into the economic substance of the holding company and its subsidiaries (substance over form).

According to FER 3, 'Fundamental accounting principles', issued in November 1990 (revised

March 1993), the annual financial statements fulfil the principle of clarity (para. 13) if:

- the accounts are transparently and properly classified (taking appropriate account of the specific nature and activities of the business);
- only similar accounts are grouped together;
- only in special cases, where the practice can be justified by objective criteria, may assets and liabilities, as well as income and expenses, be netted, and then only provided the financial statements are not rendered in any way misleading;
- the view of the financial position of the business is not prejudiced, either by the way in which the individual data are presented or through the omission of essential information, and the individual financial statement items are correctly labelled.

Within reason, rounding of the figures is acceptable.

The materiality concept was introduced only with the revised company law of 1992. Previously there were no such guidelines. Lawyers argued before that only financial statements detailed right down to centimes could be accurate. Now there is no question about rounding the figures in financial statements to francs, ignoring centimes. In group accounts it is in any case usual to round to the nearest thousand francs or, in the case of large groups, even to the nearest million.

1.3 Consistency of presentation and valuation

Consistency is important, so that prior year figures will be comparable. However, hidden reserve accounting makes the principle difficult to achieve. The presentation, the classification of headings, and the valuation and disclosure principles should be consistently applied. Any change should be explained and possibly quantified in the notes.

FER 3 defines consistency in the following way (para. 15):

> The annual financial statements are considered to have been prepared in accordance with the principle of consistency if in the current year the same accounting principles are applied as in the comparative period. Consistency relates to the presentation of the financial statements as a whole as well as to the content and the valuation of the individual account balances.

1.4 Prohibition of setting off assets and liabilities or expenses and income (the gross principle)

This principle is important for the sake of clarity in the financial statements. The structure of the balance sheet and income statement itself provides some guidance. The following have to be shown gross, for example:

- Accounts receivable and payable.
- Non-operating income and expense.
- Financial income and expenses.
- Extraordinary income and expenses.
- Gains on the sale of fixed assets (gross).
- Currency gains and losses.

However, it is accepted that the following may be shown in netted-out form:

- Net sales (gross sales less sales deductions).
- Net fixed assets (cost less accumulated depreciation).
- Net receivables (gross receivables less allowance for doubtful accounts).
- Release of hidden reserves net (to be disclosed in the notes).

Any departure from consistency must be justified in the notes.

2 Balance sheet

2.1 Minimum content

The basic format of the balance sheet required under OR Art. 663a is shown in Table V.1. The most important asset and liability headings are examined in detail overleaf (see VI). The equity accounts are discussed next.

2.2 Equity

Shareholders' equity normally includes the following items:

- Share capital: ordinary shares, registered or bearer.

Table V.1 Basic minimum format of the balance sheet

ASSETS	**AKTIVEN**
Cash	Flüssige Mittel
Accounts receivable: trade[1]	Forderungen aus Lieferungen und Leistungen
Accounts receivable: other[1]	Andere Forderungen
Inventories	Vorräte
Pre-paid expenses	Rechnungsabgrenzungsposten
Total current assets	**Umlaufvermögen**
Financial assets	Finanzanlagen
of which participations in other companies	*davon* Beteiligungen
Property, plant and equipment	Sachanlagen
Intangible assets	Immaterielle Anlagen
Incorporation, capital increase and organization costs	Gründungs-, Kapitalerhöhungs- und Organisationskosten
Own shares	Eigene Aktien
Unpaid share capital	Nicht einbezahltes Aktienkapital
Total fixed assets	**Anlagevermögen**
LIABILITIES AND SHAREHOLDERS' EQUITY	**PASSIVEN**
Accounts payable: trade[1]	Schulden aus Lieferungen und Leistungen
Accounts payable: other [1]	Andere kurzfristige Verbindlichkeiten
Accrued liabilities	Rechnungsabgrenzungsposten
Long-term liabilities[1]	Langfristige Verbindlichkeiten
Provisions	Rückstellungen
Total liabilities	**Fremdkapital**
Share capital	Aktienkapital
Participation capital	Partizipationskapital
Legal reserves (general reserves)	Gesetzliche Reserven (allgemeine Reserven)
Statutory reserves	Statutarische Reserven
Other reserves (free reserves)	Uebrige Reserven (freie Reserven)
Revaluation reserves	Aufwertungsreserven
Reserve for own shares	Reserve für eigene Aktien
Retained profits (accumulated losses)	Bilanzgewinn (Bilanzverlust)
Total shareholders' equity	**Eigenkapital**

[1] *Of which* due to and due from group companies and major shareholders.

- Participation capital: participation certificates are non-voting shares.
- General reserves: legal reserves from the assignment of profits, and share premiums or additional paid-in capital in excess of par.
- Free or special reserves.
- Undistributed earnings carried forward: retained earnings.
- Net profit/loss for the year.

Preference shares are seldom encountered in Swiss corporations.

Treasury stock (own shares) have to be capitalized and a corresponding reserve must be set up (see below).

2.2.1 Share capital and participation capital

Under OR Art. 621 the minimum share capital of a stock corporation is CHF 100,000. The shares may be bearer or registered. Both may be issued at the same time. They can be converted from one type into the other provided the articles of the company permit it. The minimum nominal value is set at CHF 10. Thus shares of no par value may not be issued. On the incorporation of the company a minimum of 20 per cent of the nominal value must be paid-up. In all cases the minimum paid-up capital must not be less than CHF 50,000. The unpaid amount has to be capitalized in the assets and termed 'unpaid share capital'.

It is also possible to contribute assets (contribution in kind) instead of paying the share capital in cash so long as a special audit report on the valuation is prepared.

Participation capital does not carry any voting rights. Otherwise it is equivalent to normal shares. However, it may not exceed double the amount of the normal voting share capital.

2.2.2 Reserves

2.2.2.1 Legal reserves – general reserves

Art. 671, para. 1 of the Code of Obligations (OR) requires certain allocations of profit to non-distributable reserves. These are normally called legal reserves or general reserves. The allocation must be as follows:

- Five per cent of the annual profit must be allocated to the general reserve until it reaches 20 per cent of the paid-up share capital.
- Also, once the 20 per cent level has been reached, the following must be allocated to this reserve (in brackets below: section of article):
  - (1) Any surplus over par value on the issue of new shares (additional paid-up capital in excess of par value, sometimes called a share premium), after deduction of the cost of the issue, to the extent that such surplus is not used for depreciation or welfare purposes.
  - (2) The excess of the amount which was paid up on cancelled shares over any reduction in the issue price of replacement shares.
  - (3) 10 per cent of the amounts which are distributed as a share of (annual) profits after payment of a dividend of 5 per cent.
- To the extent that it does not exceed half the share capital, the general reserve should be used only to cover losses or for measures which are sure to tide the company over times when business is poor, to counteract unemployment, or to soften its consequences (OR Art. 671, para. 3).
- The provisions of OR Art. 671, para. 2, section 3, and para. 3 do not apply to companies whose principal object is to hold a participating interest in other companies (holding companies) (OR Art. 671, para. 4).

Or Art. 671, para. 2, section 1 means in practice that value adjustments can be made directly to reserves. For example, overvalued participations are often directly value-adjusted by charging the amount to additional paid-up capital (share premium). The amount therefore does not affect the income statement as an (unusual) expense. In principle, the annual general meeting of shareholders has to approve this treatment. Shareholders and the board tend to favour it because no accumulated losses are created which would prevent the payment of dividends in future years. These direct charges to reserves (reserve accounting) should, however, be explained in the notes.

Similarly, all financing costs relating to a capital increase (e.g. emission stamp duty or bank commission) are normally also charged to the share premium, thus reducing the amount shown as additional paid-up capital.

Companies often stop making additional contributions to legal reserves out of profits when they reach 50 per cent of share capital, because the general meeting of shareholders has the power to distribute such amounts as dividends.

Holding companies normally cease allocations at 20 per cent. This rule has been introduced because it was argued that the subsidiaries already have to accrue legal reserves. On a consolidated basis the allocation would be doubled if the holding company allocated on the same basis as the operating company.

2.2.2.2 Reserves for own shares

Under the new company law companies are allowed to acquire their own shares if freely disposable equity (i.e. reserves) to the amount necessary for the purpose is available and if the total par value of the shares does not exceed 10 per cent of the capital (OR Art. 659).

Under OR Art. 659a the company must at the same time create a separate reserve to an amount corresponding to the acquisition value of its own shares held. This should ensure that no dividends can be paid in the respective amounts. This also applies to the parent if a subsidiary acquires the shares of its parent. In the event of disposal or cancellation of own shares, the reserve for own shares may be dissolved to the full amount of their acquisition value (OR Art. 671a).

If own shares are not disposed of within one year of acquisition the federal tax authorities might consider the price paid in excess of par value as a hidden dividend distribution to shareholders, subject to the withholding tax of 35 per cent. Hence any acquisition of own shares should be discussed beforehand with tax consultants.

2.2.2.3 Revaluation reserve

The revaluation reserve required by OR Art. 670 may be released only by conversion into share capital or by the depreciation or disposal of the revalued assets (OR Art. 671b; see VIII).

2.2.2.4 Reserves provided for by the articles of incorporation

The articles of incorporation may provide that higher amounts than 5 per cent of the annual profit shall be allocated to the reserve and that the reserve shall amount to more than 20 per cent of the paid-up share capital required by law (OR Art. 672, para. 1).

They may provide for the creation of further reserves and determine their purpose and use (OR Art. 672, para. 2).

2.2.2.5 Reserves for employee welfare purposes

The articles of incorporation may, in particular, provide that reserves may be set aside for the founding and support of institutions intended to promote the welfare of employees of the company (OR Arts. 673, 674, para. 3). This is, however, rare in practice, because in Switzerland pension funds are separate from and independent of the employer and are fully funded. Contributions are therefore normally expensed and paid over to the pension fund (in cash). This is tax-deductible.

2.2.2.6 Emergency reserves (Arbeitsbeschaffungsreserve)

These are tax-driven reserves that are created in boom times. They can be drawn upon during an economic downturn. Normally a corresponding (restricted) cash deposit is capitalized. It can be used to prevent unemployment. The government will also reimburse the company for the income tax it originally paid.

2.2.2.7 Relation between dividends and reserves

No dividend may be declared until the allocations to the legal reserve (OR Arts. 671–671b) and to the reserves required by the company's articles (OR Arts. 672–673) have been made in accordance with the law and the articles of incorporation (OR Art. 674, para. 1). The board of directors puts to the general meeting of shareholders the amount it proposes to pay out in dividends and the allocations it intends to make to reserves. The auditors have to examine the proposals and state their opinion as to whether they are in line with the law and the company's articles. This should also include an assessment of whether the company is able to finance the dividend, i.e. whether it has the necessary liquidity at its disposal.

Only when an affirmative audit report is available can the general meeting decide upon the distribution of profit. Thirty-five per cent of the dividend has to be sent to the Department of Withholding Tax at the federal tax office in Bern. Swiss shareholders can reclaim this tax in full if they declare the dividend income in their tax return. Foreign shareholders can claim most of the tax back under the terms of his country's double taxation treaty with Switzerland. If no treaty has been concluded or the dividend income has not been declared properly the tax is lost. Such unclaimed withholding tax generates considerable income for the Swiss government, indicating that it is not reclaimed.

The general meeting of shareholders may create other reserves which are required neither by law nor by the articles of incorporation or which go beyond their requirements to the extent that it is:

- Necessary for replacement purposes.

- Justified having regard to the continuing prosperity of the company or the distribution of a dividend as evenhandedly as possible, taking into account the interests of all shareholders.

Even where there is no such provision in the articles of incorporation, the general meeting of shareholders may also create reserves from the balance sheet profit for the establishment and support of institutions intended to promote the welfare of employees of the company, or for other welfare purposes.

Minority shareholders are not very well protected, as it is difficult to prove that the reserves are not in the best interests of the company. The difficulty is increased by the limited nature of the disclosure in the individual accounts. Dissident shareholders have the right to demand a special investigation (*Sonderprüfung*) under the new company law.

2.2.2.8 *Dividends, interest during construction*

Under OR Art. 676 the shareholders may receive a fixed rate of interest debited to the fixed asset account during the period of time which is required for preparation and construction until the company attains full operating capacity. Within these limits, the articles of incorporation (OR Art. 627, section 3) must stipulate a date for the termination of interest payments. This means that not only interest costs on debt but also dividends to shareholders can be capitalized as fixed assets. This is common procedure with power-generating corporations. Construction may take several years. No income statement is prepared until electricity is being produced. All costs incurred are capitalized during the period of construction. Swiss law therefore also allows dividends to be paid during the period of construction as long as the articles permit it. The dividends are also capitalized and later depreciated over the useful life of the plant.

3 Profit and loss account

3.1 Minimum content

Either the account format or the reporting format may be used for the profit and loss account.

The minimum content of the profit and loss account under OR Art. 663 is set (in account format) in Table V.2.

Under OR Art. 663 the minimum content of the profit and loss account (in the reporting format) is as shown in Table V.3.

Although not mentioned in the law, the Anglo-American type of profit and loss account, i.e. the cost of sales method, can also be used in the format shown in Table V.4.

This format is also permissible under FER 7, 'Presentation and format of the consolidated balance sheet and profit and loss account', issued in May 1992, for group accounts. However, the following types of cost required by OR Art. 663 have to be shown in the notes:

- Materials expenses.
- Personnel expenses.
- Depreciation expenses.

Table V.2 Minimum account format of the profit and loss account (OR Art. 663)

Expenses	**Revenues**
Materials use	Net sales
Depreciation	
Personnel expense	
Financial expenses	Financial income
Non-operating income	Non-operating expenses
	Gain from sale of fixed assets
Extraordinary income	Extraordinary expenses
Annual net profit	Annual net loss
Total expenses (and annual net profit)	**Total income (and annual net loss)**

Table V.3 Minimum reporting format of the profit and loss account (OR Art. 663)

	Net sales		**Erlös aus Lieferungen und Leistungen**
–	Materials use	–	Material- und Warenaufwand
–	Personnel expenses	–	Personalaufwand
–	Depreciation	–	Aufwand für Abschreibungen
+	Non-operating revenues	+	betriebsfremde Erträge
–	Non-operating expenses	–	betriebsfremde Aufwendungen
+	Financial revenues	+	Finanzertrag
–	Financial expenses	–	Finanzaufwand
+	Gain from the sale of fixed assets	+	Gewinne aus Veräußerungen von Anlagevermögen
+	Extraordinary income	+	außerordentliche Erträge
–	Extraordinary expenses	–	außerordentliche Aufwendungen
=	**Annual net profit/loss**	=	**Jahresgewinn/Jahresverlust**

Table V.4 Format of the profit and loss account: cost of sales method (FER 7)

	Net sales	**Nettoumsatz**
less	Cost of goods sold	Kosten der verkauften Produkte
	Gross margin	**Bruttomarge**
less	Selling and marketing expenses	Verkaufskosten
	General administrative expenses	Verwaltungskosten
	Research and development expenses	Forschung- und Entwicklungskosten
	Operating profit	**Betriebsgewinn**
less	Financial expenses net [1]	Finanzaufwand netto
less	Extraordinary expenses net [1]	Ausserordentlicher Aufwand netto
	Profit before tax	Gewinn vor Steuern
less	Income tax	Ertragssteuern
	Annual net income/loss	**Jahresgewinn/Jahresverlust**

[1] Gross amounts shown in the notes.

3.2 Net sales

This item comprises:

- The gross sales from the sale of merchandise, own products and services.
- Operating ancillary sales (e.g. sales of scrap metal, licence income).
- Sales deductions (returns, discounts, sales discounts, value-added tax, losses incurred on doubtful accounts, etc.).
- Currency gains and losses closely connected with sales transactions.

It should not, however, include the increase/decrease in work in progress and finished goods inventories and capitalized fixed assets manufactured by the company. Although this item is not mentioned in the law, it should be disclosed separately if it is significant. These items are valued at manufacturing cost and not at sales price, i.e. they do not include a profit margin. If these items are not material they may be included under other operating income.

3.3 Materials expenses

This item includes:

- Purchased merchandise, raw and auxiliary materials, etc. and changes in the position as compared with the previous year (increases and decreases).
- Purchases of components.
- Differences in inventory of purchased materials.
- Inventory value adjustments because of: obsolescence, losses, damage, price decline, etc.
- Direct freight costs, customs duties, etc.
- Purchase price deductions such as: returns, discounts, sales discounts, etc.
- Currency gains and losses closely connected with purchase transactions.
- General value adjustments for tax purposes (hidden reserves).

3.4 Personnel expenses

This item comprises:

- Salaries and wages, provisions and other employee remuneration.
- Legal and voluntary social expenses, including old age insurance pension costs, etc.
- Cost of hiring temporary staff.
- Creation of accrued liabilities for unpaid salaries, bonuses, holiday entitlement, overtime, etc.

3.5 Depreciation

This item comprises:

- Systematic and unscheduled depreciation on tangible fixed assets such as buildings, equipment, machinery, etc.
- Amortization and write-downs of intangible fixed assets such as patents, trademarks, copyrights, licences, goodwill, etc.
- Value adjustments of financial fixed assets, e.g. owing to a decline in value or losses on investments, loans, etc.

Value adjustments of current assets should not be included here.

3.6 Non-operating revenues and expenses

Differentiating between operating, non-operating and extraordinary revenues and expenses is not easy. Revenues and expenses may not be netted but have to be shown gross. 'Non-operating' means that the expense is not related to the objectives (goal) of the company as stated in its articles of incorporation. Examples are:

- Rents from non-operating property holdings by a manufacturing company.
- Royalty income and expenses.

3.7 Financial revenues and expenditure

Financial revenues comprise:

- Interest from bank accounts, notes receivable, loans receivable, etc.
- Interest and dividends on marketable securities.
- Gains from the sale of securities.
- Gains from reversal of downward value adjustments (write-downs) of financial assets.
- Realized and unrealized gains on financial assets in foreign currency.

Unrealized gains should be stated only if there is no doubt as to their realizability.

Financial expenses comprise:

- Interest paid/accrued on bank account overdrafts, notes payable, loans payable, etc.
- Interest on debenture loans.
- Losses on the sale of securities.
- Losses from value adjustments to financial assets.
- Realized and unrealized losses on financial assets in foreign currency.

Unrealized losses should be recognized as soon as the loss becomes evident, i.e. when it appears probable.

This item should not, however, include rent revenues/expenses or lease revenues/expenses, etc.

These should be shown separately under separate headings if material or, if not, under other operating revenues/expenses or possibly under non-operating revenues/expenses.

3.8 Gains from the sale of fixed assets

This item is peculiar to Swiss law, as only gains from such transactions have to be disclosed, not losses. It is understood that hidden reserves are often realized by selling property or other fixed assets (for example from the employer company, to pension funds). This would be disclosed separately under this heading. It should be borne in mind that significant tax effects may result. For example, many cantons levy special property turnover taxes (*Handänderungssteuern*) and capital gains taxes (*Grundstückgewinnsteuern*) on such transactions. These taxes should be included in tax expenses. They are not normally deducted directly from the gain (net of tax) but, rather, shown gross.

3.9 Extraordinary revenues and expenditure

Extraordinary revenues are defined as a gain from non-recurring transactions which are not related to the normal course of business. Examples of extraordinary revenue might include:

- A large payment received following a patent infringement case.
- An extraordinary gain resulting from an insurance pay-out for the loss of a building which was not rebuilt. Such gains are sometimes deferred for tax purposes as 'provisions for replacement' (*Ersatzbeschaffung*)

Extraordinary expenses are defined as losses on non-recurring transactions which are not related to the normal course of business. Examples of extraordinary expenses might include:

- A fire that destroyed inventories which were not insured.
- A tax penalty or fine that was imposed on the company.
- A large product liability case that was lost.

It is very difficult to distinguish between non-operating and extraordinary items. The use of extraordinary expenses should therefore be confined to items which are:

- Rare.
- Non-recurring.
- Material in amount.
- Unusual for the company, given its size and business purpose.

So that such items can be assessed it is recommended that their nature should be disclosed in the notes or in the income statement, especially in group accounts (see FER 8, 'Notes to the consolidated financial statements', para. 3).

3.10 Annual net profit/loss

This is the annual result of all operating revenues minus operating expenses, all non-operating revenues minus non-operating expenses, all financial revenues minus financial expenses, all extraordinary revenues minus extraordinary expenses, minus tax.

3.11 Revenue items which need not be disclosed

International accounting standards require many more income statement items to be disclosed on the face of the income statement or in the notes. These include, for example:

- Net operating result.
- Net financial result.
- Net extraordinary result.
- Changes in work in progress and finished goods.
- Capitalized fixed assets manufactured internally.
- Income before tax.
- Income tax.
- Details of personnel expenses.
- Board and management compensation, etc.

The above items need not be disclosed under Swiss law but are often presented voluntarily, above all in consolidated accounts. The first two can of course easily be calculated from the figures disclosed.

VI Recognition criteria and valuation

1 Principles of the recognition of assets, liabilities and provisions

1.1 Materiality

Materiality is a qualitative aspect influenced by subjective judgements (*Revisionshandbuch* I, p. 84). The question is what amount and/or omission will influence the decision to buy the investment, keep it or sell it. It may also involve the shareholders' approval of the financial statements or other decisions. Often the amounts concerned are taken into consideration alongside other balance sheet items or heads of the income statement, i.e.:

- Total assets.
- Shareholders' equity.
- Net sales.
- The net result.

It also depends whether net income is affected by the item or not (net of tax).

FER 3 (see V.I.2) defines materiality as follows (paras 9–11):

> All items which influence the valuation and presentation of the annual financial statements or their individual items to such an extent that the decision of a user of the financial statements might be influenced are to be considered material. Where the said criteria are an inadequate basis on which to judge whether an item is material or not, it may be useful to form a conclusion as to materiality by viewing the item in question in relation to appropriate key figures. If the accumulation of immaterial items amounts to a significant influence on the reliability of the financial statements, this factor is to be taken into consideration.

1.2 Prudence (conservatism)

Continental European countries, Switzerland included, have always stressed prudence as a key accounting concept. If quantitative decisions have to be taken the more conservative alternative is chosen. The key concept is that assets should not be overstated and liabilities should not be understated, so that equity is not overstated. This approach suits creditors such as the banks, since it reduces the risk their lending or investment is exposed to. It also minimizes the tax bill, employee demands and the accountability of management and board. Bad decisions are not as visible as they might otherwise be; losses can silently be made good from hidden reserves. Even for the government there are perceptible advantages. The tax flow is more consistent. It is lower when the company is in a phase of expansion. Sooner or later, however, the accumulation of hidden reserves will lead to tax liabilities, if only in the end upon liquidation.

It is obvious that the measurement of assets and risks is very much a matter of judgement. Continental European managers and auditors tend to be more conservative in their outlook than their Anglo-American counterparts.

Swiss statute and tax law has always favoured the creation of hidden reserves in the individual financial statements. Only in group accounts does the concept of a true and fair view enter into consideration.

It will not have escaped notice that the concept of conservatism is markedly at odds with the requirement of clarity. However, it is clearly the overriding principle.

FER 3 defines conservatism as follows (para. 14):

> The financial statements are prepared in accordance with the principle of prudence if:
>
> - Income is accounted for only when the goods have been delivered or, correspondingly, the service has been performed, and, when other types of income are recognized, only when the supplier's credit note has been received or an irrevocable claim exists, or, in general, where the circumstances support the recognition (the realization principle).
> - Losses and risks which are recognizable but not yet incurred are already provided for if the cause relates to the period under review or a prior period. This is so, even if they become known after the balance sheet date but before the financial statements

have been issued (the principle of conservatism).

- Independently of the valuation principles applied in the financial statements, the market value of inventories and work in process at the balance sheet date is taken into account if it is lower than the value determined according to the valuation rules (the lower of cost or market principle).

1.3 Continuance of the company's activities (going concern)

The going concern principle is also a very important concept. If the company is unable to continue (e.g. owing to bankruptcy), or is voluntarily wound up (e.g. upon liquidation), then the valuation basis has to be changed. Whereas strict application of the historical cost convention is necessary with a going concern, so giving rise to hidden reserves (owing to the effect of inflation), in a winding up or discontinuance of operations the assets are normally valued at fair market prices, less deferred tax, liquidation costs, etc.

FER 3, 'Generally accepted accounting principles', issued in November 1990 (revised March 1993), states in para. 1 and para. 8:

> The individual company accounts and the consolidated financial statements are based on the assumption that it is possible for the company to continue as a going concern for the foreseeable future. The continuation of the company as a going concern cannot be assumed if the dissolution of the company is anticipated or apparent, or if curtailments of a significant portion of the company's operations are planned or appear to be necessary.

Valuation problems in the event of liquidation or bankruptcy are treated in more detail below (see VIII).

1.4 Consistency of presentation and valuation

Consistency is important if prior year figures are to be comparable. However, hidden reserve accounting makes the principle difficult to implement fully. The basis of presentation, classification of items, valuation and disclosure should be consistent. Any change should be explained and possibly quantified in the notes. However, the principle is overridden by the company's ability to create hidden reserves. Therefore, in the individual accounts departures are highlighted only when hidden reserves are released. Examples would be:

- Netting income and expenditure.
- Failure to apply depreciation in a year of losses.
- A change of depreciation method.
- A change in the method of inventory valuation.

FER 3 defines consistency as follows (para. 15):

> The annual financial statements are considered to have been prepared in accordance with the principle of consistency if in the current year the same accounting principles are applied as in the period under comparison. Consistency relates to the presentation of the financial statements as a whole and not just to the content and the valuation of the individual account balances.

2 Valuation principles

There are few of the restrictive valuation principles found in other countries. Depreciation must be applied, value adjustments made and provisions set aside to the extent required by recognized accounting principles in business (OR Art. 669).

According to OR Art. 669, para. 1, sent. 2, 'Provisions are to be created particularly to cover uncertain contingent liabilities and potential losses from business transactions pending.' The code continues as follows:

> For purposes of replacement, the board of directors may take additional depreciation, make value adjustments and provisions and refrain from dissolving provisions which are no longer justified. Hidden reserves exceeding the above are permitted to the extent justified with regard to the continuing prosperity of the company or the distribution of a dividend as equal as possible, taking into account the interests of the shareholders (OR Art. 669, paras. 2–3).

This is the well known 'hidden reserve statute' (*stille Reserven*). Since the Code of Obligations stipulates only maximum accounting values (*Höchstwertprinzip*), undervaluation of assets and/or overstatement of liabilities can be implemented almost completely at management's discretion.

The auditors must be notified in detail of the creation and/or release of replacement reserves and any additionally created hidden reserves.

Such hidden reserves are sometimes also called latent or undisclosed reserves. To a large extent they are also accepted by the tax authorities. For example, one-third of inventories are deductible. In the canton of Zurich 80 per cent depreciation of tangible fixed assets with a limited useful life is allowed in the year of acquisition, and the remaining 20 per cent after four years. This single-stage depreciation is admissible in the tax accounts only if it is accounted for in the individual commercial accounts as well. The same canton permits an allowance for doubtful debt of 10 per cent on domestic trade receivables and 20 per cent on foreign receivables. Normally the tax return is identical to the statutory financial statements.

3 Incorporation, capital increase and organization costs

Under OR Art. 664 the costs of incorporation, capital increase and organization resulting from the establishment, expansion or reorganization of the business may be included in the balance sheet as an asset. They have to be shown separately and amortized within five years at most. Such costs would include the stamp duty on the issued capital (one per cent at present). Because of tax considerations companies normally capitalize the incorporation cost in the year of incorporation instead of amortizing. The following year the whole amount is written off, as this procedure is acceptable to the tax authorities.

4 Fixed assets

4.1 General principles

Under OR Art. 663a the following items are included in fixed assets:

- Financial assets (including investments in other companies).
- Property.
- Plant and equipment (tangible fixed assets).
- Intangible assets.

Included in tangible fixed assets are normally:

- Land.
- Buildings.
- Machinery.
- Equipment.
- Vehicles.
- Office furniture, computers.
- Construction in progress.
- Advances to suppliers of tangible fixed assets.

Under OR Art. 665 the maximum value of fixed assets is the acquisition or production cost less the requisite depreciation. This clearly reflects the historical cost convention. It means that an asset can be capitalized only at the amount which has been paid for it. It also implies a very conservative approach, indicating that 'depreciation, value adjustments and provisions must be effected to the extent required by generally accepted accounting principles'. Acquisition cost is determined on the basis of suppliers' invoices. Depreciation is determined by the method of depreciation and the estimated useful life of the assets. Tax considerations mean that often steep depreciation (from book values) is applied over a short period of time.

For certain categories of assets a portfolio approach to valuation is adopted which emphasizes the reduced importance, by comparison with other European countries, of the principle of separate valuation (see VI.5.5) in the individual accounts.

4.2 Land

This category includes purchased land at acquisition cost on the basis of a contract which needs registration in the land register of the local authority in whose area the land is situated. In addition, the following costs can also be capitalized:

- Estate agents' commission.
- Conveyancing fees.
- Valuation costs.

- Relevant property transfer taxes.
- The cost of bringing the land to condition required.

At acquisition, land is stated in the balance sheet at purchase price. In later years it has to be valued at the lower of acquisition cost or net realizable value. In the past years downward value adjustments have not been required for land. Inflation has meant that property prices had only one way to go: up. Depreciation is not normally applied to land, as its value is not used up.

Currently depressed property prices mean that devaluation may eventually become necessary if the property market continues to deteriorate (permanent impairment of value). A portfolio approach is often used in the individual accounts, where losses on certain pieces of land are compensated by gains in other items.

The value of land does not need to be disclosed separately. It is invariably included with that of the related buildings, as separation is often impracticable.

4.3 Buildings

Buildings are normally shown at acquisition or production cost less depreciation. The following costs can be included in production cost:

- Expenses for architects and engineers.
- Cost of excavation, etc.

Interest arising during the construction period can be capitalized.

Depreciation is determined mainly by tax law. The rate may vary from 2 per cent to 4 per cent of acquisition cost per annum. It can be doubled if the depreciation is calculated from the book value. The canton of Zurich allows a particularly favourable depreciation rate (see VI.2). To qualify as tax-deductible these amounts have to be booked through the statutory accounts.

A portfolio approach is often used in the individual accounts, where losses on certain buildings are compensated by gains in other items.

The book value of the building does not need to be shown separately. Usually it is included under a heading 'Land and buildings' or 'Property' as no separation is feasible.

However, the fire insurance value has to be disclosed in the notes (see IX). The value is derived from the (often semi-governmental) insurance company's annual fire insurance premium invoice, based on an inflation index for rebuilding costs. It more or less reflects the current cost of a new building that would have to be reimbursed in the event of destruction by fire. It therefore does not necessarily reflect the fair market value of the building (the depreciated value) because the insurance is geared to the new, replacement building. Additionally, the land does not need to be insured.

Mortgaged property also has to be disclosed in the notes, but only at its book value (see IX). As property normally is mortgaged the book value is visible.

4.4 Machinery, furniture and fixtures

These are valued at acquisition cost according to the suppliers' invoices. The following can also be capitalized:

- Freight, customs duties, etc.
- Freight insurance.
- The cost of installation.

Interest during the construction period can be capitalized. However, the above costs are seldom capitalized (mainly for tax reasons). Sometimes a fixed percentage is added to the purchase price, e.g. 2 per cent, based on experience. Discounts received are deducted from the purchase price.

Tools and other machines with a relatively low value are expensed directly, as they may have a short useful life. Normally only items higher than CHF 1,000 to CHF 10,000 are capitalized. Additionally, no tax requirements would limit direct write-offs.

Depreciation is often applied using the straight-line method based on acquisition cost. Normally the tax rates are used, which are rather generous. The tax depreciation rate can be doubled if the depreciation is calculated on the book value. For example, say the useful life is five years, straight-line depreciation of the acquisition cost would be 20 per cent, the rate from the book value would be 40 per cent. This second method does not require an indirect means of recording cumulative depreciation or a fixed assets register. More conservative depreciation methods are allowed in the canton of Zurich (see VI.2).

Additional depreciation is allowed for statutory accounting purposes (as also for quoted stock corporations). Some companies even value their fixed assets at a *pro memoria* CHF 1. However, the tax authorities may not allow the full depreciation amounts as tax-deductible.

The insurance value for all tangible fixed assets and the mortgaged amounts have to be disclosed in the notes (see IX).

4.5 Intangible assets

Intangible assets have to be shown separately in the balance sheet. They could include purchased intangible assets such as:

- Concessions, cartel quotas and similar rights.
- Patents, recipes, copyrights.
- Franchises, trademarks.
- Goodwill (if the assets and liabilities of a partnership or of a sole proprietor are acquired at more than the fair market value of the assets and assumed liabilities).
- Electronic data processing software (if not capitalized with the related hardware).

The differentiation from organization cost according to OR Art. 664 is vague. Internally generated intangible values (e.g. development costs) can therefore also be capitalized. However, the costs claimed have to be substantiated (hours spent, expenses, etc.). The determination of acquisition cost is based on suppliers' invoices or contracts. Amortization is determined by the method of depreciation and the estimated useful life of the assets. It is normally quite short (three to five years). Because of tax considerations intangibles are often directly expensed. Amortization is normally on a linear basis over five years. If the value of the assets diminish (permanent impairment), a complete write-off is recorded.

Research and development costs are rarely capitalized in practice. Sometimes project-related development costs are. They may also be included in inventories, work in progress, fixed assets, organization costs or deferred charges. However, their value should be monitored and, in the event of impairment, immediately written off. As instances of research and development costs being capitalized are virtually unknown, the subject has barely entered the Swiss accounting literature (see also VII.3).

4.6 Long-term loans

Long-term loans are stated at nominal value. If repayment is uncertain the nominal value is adjusted directly. Interest-free loans are normally discounted to their present value.

4.7 Participations

Participations are long-term equity stakes in other companies that give a significant degree of control. Voting rights of 20 per cent or more will normally ensure such control (OR Art. 664a, para. 3). Any participation that is essential to an assessment of the company's financial situation and income should be shown in the notes according to OR Art. 663b, section 7 (see IX).

Participations are normally an investment in the capital of:

- Stock corporations (shares).
- Limited liability companies.
- Co-operatives.

Acquisition cost is normally derived from purchase agreements or bank statements. If the intrinsic (fair market) value is less than acquisition cost, value adjustments become compulsory to reflect the long-term value impairment. In the case of quoted companies the intrinsic value is calculated from stock exchange prices. Where privately-held companies are concerned the value is more difficult to pin down. It is based on a business valuation. This is derived from capitalized future net profits (price/earnings ratio) and the fair market value of the non-operating assets less deferred taxation. Estimated future distributable net profits are capitalized at an interest rate which takes into consideration the risk-free interest rates of federal bonds and additionally the long-term nature or risk of the investment (e.g. 8–12 per cent).

A portfolio approach is often used in the valuation of the individual participations, where losses on certain items are compensated by gains in others.

The equity method may not be used in the individual accounts if the value exceeds acquisition cost. The goodwill in a participation can be written off for tax purposes only when the participation is in a bad way financially (continuing losses, little or no equity left, etc.).

Long-term investments in the shares of companies constituting less than 20 per cent of the capital and/or votes are stated at the lower of cost or intrinsic value.

5 Current assets

5.1 Inventories

Inventories include raw materials, work in progress and finished goods, as well as merchandise. Under OR Art. 666 they are all valued at no more than the acquisition cost or manufacturing cost. If the cost is higher than the market value generally at the date of the balance sheet, then such market value would be used (lower of cost or market, or net realizable value).

Value adjustments for slow-moving items (obsolescence) should be based on a formula which is applied consistently. On tax grounds the net value can additionally be reduced by one-third. This has to be recorded in the statutory accounts in order to be tax-deductible. Inventories normally also include: goods out on consignment or held by agents.

Instead of the specific identification method, the following cost flow assumptions can be used:

- (Weighted) average cost.
- 'First in, first out' (FIFO).
- 'Last in, first out' (LIFO).

The average cost method is the most commonly used in practice.

No definition of capitalizable acquisition or manufacturing cost is offered in the legislation. Many companies include only a few of the elements of cost, in order to reduce profits for tax purposes (e.g. only variable material cost). The Swiss Institute of Certified Accountants has defined the upper limit of cost which can be included in inventories (see Auditing Handbook, Volume 1, 1999, p. 147). It includes all acquisition costs that accrue until the inventories are in the company's warehouse (or have been brought to the intended location) such as:

- Invoice price of purchased merchandise, raw and ancillary materials, etc., and invoice price of components purchased.

  Plus purchasing costs such as:
- Direct freight.
- Customs duties.

  Less purchase price deductions such as:
- Returns.
- Discounts.
- Sales discounts.
- Volume bonuses, etc.

Manufacturing costs of work in progress and finished goods include all production costs that accrue until the inventories are ready for sale (or brought to the intended condition), such as:

- Acquisition cost of raw, ancillary and packaging materials, supplies and purchased components, as mentioned above.
- Materials overheads.
- Direct labour costs
- Labour overheads.
- Manufacturing overheads.
- Product-related development, testing and construction costs.
- Patent and manufacturing-related royalty costs.

Manufacturing costs can include both variable and fixed costs, such as:

- Fuel and power.
- Rents.
- Insurance premiums.
- Repairs.
- Small tools.
- Salaries of manufacturing management.
- Planned (systematic) depreciation (but not extraordinary depreciation).
- Interest on debt (but not calculated interest on equity).
- Scrap.

The above costs should be charged on the basis of normal production volume. Costs on idle capacity must not be included. Not capitalizable are:

- General and administrative costs.
- Selling costs.
- The cost of sales warehouses or stores.
- Sales provisions to agents.
- Other acquisition costs.

However, provisions and other acquisition costs may be capitalized as pre-paid expenses (for example, in the case of large construction contracts).

Storage costs are not capitalizable unless the inventories have to mature in the course of time, as in the case of cheese, whisky, wine and timber. Similarly, property companies can capitalize interest on land under development and/or construction in progress.

As mentioned above, the tax authorities do not prescribe any particular method of valuation. They are often very flexible if only a few of the above components are capitalized. To qualify as tax-deductible the statutory accounts have to show the same valuations as the tax return. From the stated value (lower of cost or market) the tax authorities grant additional tax relief of one-third (*Warenreservendrittel*). Inventories are therefore normally understated by comparison with international standards. This means that there are often significant hidden reserves in the individual accounts. In the group accounts these allowances are normally reversed after calculating deferred tax on the item.

Valuation at selling price is only rarely used, e.g. for readily marketable goods, such as gold ingots. This selling price may be above the acquisition or production cost. Where it is lower a value adjustment is necessary (lower of cost or market).

In certain industries other methods are used, e.g. retail companies are allowed to value inventories at the selling price less profit margin.

Only one amount of total inventories net of deductions has to be disclosed in the balance sheet. Pledged inventories have to be disclosed in the notes at their book value (see IX).

5.2 Long-term construction contracts

No specific accounting requirements prevail. For tax purposes the completed contract method is normally applied. This method defers the realization of the profit and revenues to the date when the project is completed. The same method is normally used for the statutory accounts as well.

The percentage of completion (PoC) method is admissible but rarely used. The following conditions have to be met for the PoC-Method:

- The contract must be clearly defined, the accounting system sophisticated and the profit secure. For a speculative contract the method is unacceptable.
- The company must be able to estimate the percentage of completion of the contract in a reliable manner.
- At the balance sheet date a claim for compensation must exist.
- Care should be taken with fixed price contracts or if cost-plus contracts have been agreed.

However, under both the completed contract method and the percentage of completion method recognized losses have to be provided for immediately (Bertschinger, 1991b).

5.3 Pre-paid expenses

Pre-paid expenses are expenses paid normally in advance for a future period. At the balance sheet date the item has not been used or consumed (e.g. insurance or rent paid in advance). Pre-paid expenses have to be disclosed separately in the balance sheet.

Non-material items are commonly expensed but not capitalized. This is a way of understating assets (hidden reserves) which is often accepted for both statutory and tax accounting purposes.

5.4 Cash

According to OR Art. 663a liquid funds (cash) must be disclosed as a separate item under current assets. They comprise petty cash (coin, bank notes in national and foreign currencies), cheques and money orders, postal and bank current accounts (cheque or savings) in Swiss francs and foreign currency that are valued at year end exchange rates. The valuation is usually at nominal values. Often the rates of the federal tax authorities are used. They are the average rates for the month of December, published in January the following year. Often they also include short-term deposits with banks (call money, 24-hour deposits, etc.). They can also be in the form of fiduciary deposits with banks.

Normally all amounts which are due within six months are included here. Under US GAAP only maturities of less than three months are included as cash.

Tied-up cash (e.g. blocked funds in a bank escrow account) should be disclosed. Under OR Art. 663b, para. 2, pledged funds should also be disclosed in the notes. If a company has a loan from its bank, or an overdraft, all assets held at the bank are automatically pledged in favour of the bank under the terms of the general contractual arrangements with most banks (see IX).

Although most banks provide cheque facilities with current accounts, cheques are seldom used in making payments. Even the banks themselves discourage it, since often substantial fees are charged for cashing and depositing cheques. Normally bank and postal transfers are made direct via the sophisticated giro clearing system. It can be done manually by sending the bank a payment order which indicates the payee and their bank or postal account number. Companies normally send a magnetic tape or diskette to the bank with electronically-stored payee information on it. The quality of these automatic payment systems is very high.

5.5 Securities

Quoted securities which form current assets may be valued at cost or their average stock exchange price during the month preceding the date of the balance sheet. They may not be valued at more than the highest of these two values. The company can therefore choose whether to value the securities at market or at cost. The market valuation is often used where portfolios are managed by outside investment managers (banks) who after some time are no longer able to define cost because the portfolio changes so frequently. Income is then booked as the change in the market value compared with the previous year. A portfolio approach is often used in the individual accounts, where losses on some items are compensated by gains on others.

Securities include quoted shares and bonds as well as precious metals, in the form of ingots or coin, with commodity prices (gold, platinum, silver). Switzerland is one of the countries where it is permissible to state marketable securities at values exceeding acquisition cost.

It is argued that these securities could easily be sold before the year end and then bought back in the new year to realize the same effect. Often the rates of the federal tax authorities are used. They are the average rates of the month of December, published in January the following year in a separate booklet. However, the tax authorities also recognize acquisition costs of securities which are lower than market value. Unrealized gains are therefore not liable to tax if they do not appear in the statutory accounts.

Unquoted securities are to be valued no higher than acquisition cost less the necessary value adjustments (OR Art. 667, para. 2). If permanent value impairment becomes obvious, write-downs are necessary. Sometimes it is necessary to undertake a separate evaluation of the company that has issued the securities.

Unrealized losses are taken to the income statement immediately and are tax deductible. Subsequent recoveries can be recognized for both accounting and tax purposes.

No disclosure is required by law under the heading 'Securities', although it is normally shown separately in the balance sheet.

5.6 Accounts receivable

Trade accounts receivable have to be shown separately under current assets under the terms of OR Art. 663a. These are invoices unpaid at year end sent to customers in the ordinary course of business (customer receivables for the delivery of goods or services). They are reduced by directly written-off debtors which are not collectible and a general allowance for doubtful debt (*Delkredere*). Tax law allows specific and general deductions for doubtful accounts. A full provision is accepted on specifically identified amounts where recovery is questionable or the amounts are long overdue. It is not necessary, however, for the debtor company to have entered into bankruptcy proceedings. After adjustments to specific items an additional provision for bad debt, normally 5 per cent of domestic debtors and 10 per cent on foreign accounts, is acceptable. The canton of Zurich even allows double these percentages for tax purposes if the amounts are also deducted in the statutory accounts.

Disclosure of the amounts deducted by way of allowance for bad debts is not required. Some companies nevertheless disclose the amount on the face of the balance sheet as a direct deduction from accounts receivable. Sometimes the allowance for doubtful debts is overstated (hidden reserves).

Inter-company receivables within the same group have to be disclosed separately in the balance sheet of parent and subsidiary companies or in the notes. Also, amounts due from major shareholders should be shown, under OR Art. 663a, para. 4 (see IX).

However, receivables from officers and employees do not need to be disclosed separately.

Other current receivables would normally include receivables from government authorities, especially for claims for domestic and foreign withholding tax refunds or VAT. They also have to be adjusted if the nominal value seems not fully collectable any more.

5.7 Notes receivable

Promissory notes are rarely used for the delivery of goods and services. They may be used for foreign customers who do not seem fully creditworthy. Notes are normally discounted immediately and deposited with banks. The resulting contingent liability in the event of default is often disclosed in the notes (see IX).

If the notes are held until maturity they may be reclassified as long-term. This might be the case if they were in substance a loan which would normally be rolled forward. Notes receivable from related parties have to be disclosed separately in the balance sheet or in the notes, under the terms of OR Art. 663a, para. 4 (see IX).

6 Liabilities

Current liabilities normally include:

- Trade accounts payable. This item would include unpaid invoices for purchased goods and services as well as purchases of fixed assets. Accounts payable to related parties have to be disclosed separately in the balance sheet or in the notes (OR Art. 663a, para. 4).
- Fiscal and social liabilities such as tax due and old age insurance taxes, pension contributions, salaries, etc.
- Accrued expenses.
- Deferred revenues.

Long-term liabilities normally include:

- Mortgages.
- Debenture bonds, etc.

Liabilities are stated at the nominal value which has to be repaid. Even if the market value is lower (e.g. the interest rate is well below the market rate) no adjustment is made. Liabilities to pension funds have to be disclosed in the notes.

7 Provisions

The relevant legislation is OR Art. 669, paras. 1–3. Long-term general provisions are created for all kinds of purposes and are often tax-driven:

- Product warranty and liability (normally 2 per cent of total sales can be accrued for tax purposes).
- Pending lawsuits, including tax contingencies.
- Repairs.
- Maintenance.
- Advertising and marketing costs (e.g. planned campaigns, catalogues, etc.).

Normally no special provision is made for staff dismissals. Swiss law provides only for very short periods of notice:

- One month for employees with less than one year of service.
- Two months for employees with two to nine years' service.
- Three months for employees with over 10 years' service.

Normal employment contracts provide for three months for rank-and-file employees and six months for middle management. Top management contracts sometimes specify longer periods.

Severance costs are accrued only after a decision has been taken to lay employees off. However, mostly the costs of overtime and holiday entitlement not taken up within the period are accrued. They would partly cover lay-off costs.

Provisions for tax are normally included in accrued (short-term) liabilities. Sometimes, however, no distinction is drawn between short- and long-term provisions.

VII Special accounting areas

1 Lease accounting

Leasing is less widespread in Switzerland than in other countries. The reason may be that favourable depreciation rates on fixed assets are available for tax purposes. Most leasing contracts are operating leases. Here lease fees (rent) are expensed.

Only rarely are finance (capital) leases encountered. Only sometimes are they capitalized as leased assets in the balance sheet of the lessee and depreciated over the lease period. At the same time a leasing obligation is set up under short and/or long-term liabilities. The monthly payment is regarded as partly interest expense and partly as a reduction in the liability. For the arrangement to qualify as a capital lease, one of the following has to apply (FER 13):

- At the end of the period of the lease ownership of the goods or property passes to the lessee.
- There is a purchase option under which the lessee can acquire the leased property for a nominal sum.
- The period of the lease is more than 75 per cent of the useful life.
- The discounted lease payments are at least 90 per cent of the fair market value.

These requirements are generally in line with international accounting standards.

Under Swiss law, leases that are not included in the balance sheet have to be shown in the notes (total of the unpaid leasing fees).

2 Foreign currency translation

There are no specific legal requirements regarding foreign currency translation. The federal tax authorities publish exchange rates that are to be used for tax purposes. These are the average of rates in the December before the balance sheet date. However, companies may use other rates (e.g. year end rates as published by the major banks) if they wish.

With regard to the translation of foreign currencies into Swiss francs in the individual accounts, Recommendation No. 2 (*Fachmitteilung* Nr. 2, *Fremdwährungen im Einzelabschluß*, ed. Treuhand-Kammer Zurich, Zurich 1990) of the Swiss Institute of Certified Accountants prefers all assets and liabilities stated in foreign currencies to be translated into Swiss francs at the current (i.e. balance sheet date) rate. The only exception are fixed assets and long-term liabilities, which are normally translated at the historical rate. As for the income statement treatment of the consequences of this method of translation, provisions are required with regard to any translation losses on long-term liabilities. Translation gains are often deferred until they are realized. This would apply, for example, to foreign branches with accounts in foreign currency. Such accounts must be translated into Swiss francs at the Swiss head office.

This recommendation also appears in the *Auditing Handbook*. It proposes four methods for the translation of foreign currency balances at year end:

- Current/non-current.
- Monetary/non-monetary.
- Modified monetary.
- Current rate.

In applying these methods it is necessary to ensure that, in general, the imparity principle is respected (i.e. the recognition of unrealized losses and deferral of unrealized gains) and that assets are stated at the lower of cost or market value (Coopers & Lybrand, 1991, p. S 109). The following guidelines are recommended for individual balance sheet items:

- Cash and sight deposits and overdrafts may be translated at the year end rate, since the gains and losses can be considered realized.
- Other short-term assets and liabilities, e.g. accounts receivable and payable: unrealized gains and losses can be offset to arrive at a net position which, if a gain, should be deferred as a provision.
- Inventories: care should be taken to ensure that the translation method applied does not result in a higher amount than historical cost in Swiss francs. Normally the valuation is done at the time the foreign currency invoice is received and booked to creditors at the actual exchange rate, and not adjusted at year end.
- Marketable securities: since such securities

can be adjusted to the year end market price (under OR Art. 667, para. 1), the related exchange rate gain or loss may be recorded as well, even if it is an unrealized gain.

- Long-term loans receivable and payable: unrealized gains and losses can be offset, provided they are in the same currency and of the same maturity. A net gain would be deferred as a provision and a net loss recognized in accordance with the imparity principle.
- Fixed assets that belong to foreign branches are translated into Swiss francs at historical rates. Translation at current rates would also be acceptable.
- Long-term investments are normally translated at historical rates, even if the value of the foreign currency has deteriorated permanently. Such translation at current rates is normally more conservative. The investment is stated at lower of cost or intrinsic value.

3 Research and development costs

Research and development costs normally include the following types of cost:

- Personnel expenses (payroll of the research and support staff).
- Materials expenses (materials used up for research purposes, e.g. drugs, chemicals, etc.).
- Depreciation expenses (depreciation of research computers, equipment and buildings).
- Overheads applicable to research and development departments.

Such costs are normally expensed. They are not normally shown on the face of the income statement because the income statement has to show the types of cost (materials, personnel, etc.) required by company law. They are therefore normally disclosed only in the notes on the (consolidated) financial statements (FER 8).

4 Government grants and subsidies

Government grants and subsidies are rare in Switzerland and normally take the form of favourable tax treatment. Some rural cantons grant other tax concessions or make land available at modest prices.

The construction of civil defence protection shelters used to be subsidized by central government. Such subvention is normally deducted from the related assets, thus reducing future depreciation charges. It is also possible to book a provision and release it over future periods.

Often no disclosure of government grants and subsidies is to be found in individual financial statements. Subsidies towards tangible fixed assets are usually booked direct to the respective assets, so reducing the asset book value and future depreciation. Sometimes the subsidies are shown as a provision and released to the income statement over the period required by the government contract. Amounts are seldom stated as extraordinary income.

After a period of indifference towards foreign companies, Switzerland is now making efforts to attract industrial investment from abroad (KPMG Fides Peat, 1989, p. 51). In the late 1970s federal and local governments started industry action programmes to create jobs in areas of high unemployment. Both Swiss and foreign firms are eligible for assistance. Federal aid to economically distressed areas is generally available only if new jobs are created. Assistance takes the form of federal government guarantees for bank credits and interest rate subsidies. The government will contribute up to a third, e.g. 3 per cent of a 9 per cent interest rate. Investment incentives vary from region to region but are most generous in districts such as Saint-Gall, Solothurn, Neuchâtel and Bern, which have troubled industrial sectors. More prosperous areas with a wide range of industries, such as Basle and Zurich, offer fewer concessions.

Municipalities interested in attracting investment often have land that they are willing to release below market rates for projects that aid industrial development. Local authorities are also willing to help retrain workers. For instance, watchmakers have been helped to enter the electronics industry. In Bern subsidies can amount to as much as 20 per cent of the wages paid.

5 Financial instruments

There is no specific guidance in the law or in the *Auditing Handbook* (HWP) on the treatment of new financial instruments, except in the banking industry. FER 8, 'Notes to the consolidated financial statements', mentions that disclosure in the notes is necessary of unusual business transactions pending and related risks. Such, for example, would be purchase or sales commitments for goods or commodities, including futures, financial instruments such as currency or interest rate swaps, forward foreign exchange transactions, etc. Recognizable losses from these transactions have to be accrued unless they represent hedging transactions.

It is not only banks and insurance companies that operate on the international financial markets. Large multinational companies (e.g. ABB) have professional treasury departments which habitually apply banking standards to their operations. They also follow international accounting standards (e.g. IAS 32 and 39).

The Swiss standard FER 10 is covering all kinds of off-balance sheet items and is therefore exceeding the scope of IAS 32. Mark-to-Market (MTM) is possible under FER 10.

6 Deferred tax

Deferred tax is very seldom applied in the individual financial statements. These normally show the same amounts as the tax return. Deferred tax accounting is therefore normally applied only in the group accounts, to achieve a true and fair view of the financial situation and results of operations. (See, for example, Tschopp, 1993, Helbling, 1988).

7 Factoring

Factoring means the transfer of receivables to a financial institution (a bank or specialist factoring company). The factor manages the accounting of the receivables. This includes also sending out reminders to the debtors. The receivables are thus booked out of the company. Sometimes the risk of doubtful accounts (*Delkredere*) is also transferred. No disclosure is necessary, even if the factoring company has a right of recourse against the seller of the receivables.

8 Loans and receivables to shareholders and other related parties

Under the new company law, loans and receivables to shareholders and related parties (individuals and companies) under common control have to be disclosed. This is important, as they could be granted without observing the arm's-length principle (see also FER 15).

They might in fact constitute a repayment of share capital. This would be in contravention of OR Art. 680, para. 2, which prohibits the shareholders' capital being returned. It might additionally be construed by the tax authorities as a concealed dividend, subject to 35 per cent withholding tax, especially if the shareholder is unwilling or unable to repay the loan from the company (construed dividend payment).

9 Trust agreements

A trust agreement is a fiduciary legal transaction by which a person gives assets (e.g. an investment, receivable, etc.) to a trustee. The risk remains with the trustor. The trustee, however, appears to third parties as the owner of the assets.

Interest and income from the assets are credited to the trustor. After the contract has expired the trustee returns the assets to the trustor.

These transactions are quite frequent in asset management. For a trust contract to be valid, the tax authorities must be satisfied that the following conditions apply:

- There is a written agreement between trustor and trustee dating from the commencement of the transaction.
- The risks and costs have to be borne by the trustor.
- The trustee receives a small commission which should not be less than two-thousandths of the fair market value of the asset.
- The assets and liabilities must be clearly identified as trust balances in the books of the trustee, on the face of the balance sheet, below the line or in the notes on the accounts.

It is recommended that the trust assets and liabilities should be shown in the notes.

10 Subsequent events

Events that take place after the balance sheet date but before the adoption of the balance sheet and/or the audit report have to be disclosed in the financial statements. They normally involve only losses incurred. If the reason for the loss relates to the old year, then a loss provision or value adjustment is necessary. Typical examples are:

- The bankruptcy of an account receivable which was included in the year end balance sheet.
- A tax dispute dating back to the previous year which is resolved by an adverse court decision in the new year.
- Raw materials ordered through a firm commitment to purchase but where there is a sharp fall in price the following year.

If an event bears no relation to the previous year it can be accounted for by disclosing it in the notes in the accounts. Examples might be:

- The uninsured portion of damage from a thunderstorm.
- A ship sunk in the new year.
- A sharp decline in a foreign currency to which the company is exposed.

Prudent accounting, however, would suggest, in these cases too, setting up provisions as of the last balance sheet date. This is normally done.

11 Pension accounting

Swiss companies are required by the Pension Fund Act (*Bundesgesetz für berufliche Vorsorge*, BVG) to establish independent legal entities in order to keep the assets due to the beneficiaries separate from those of the employer. These entities normally take the form of foundations or, in rare cases, co-operatives. If no such legal vehicle is created – for instance, if the size of the company does not warrant it – the company can join a mutual pension scheme run by a life insurance company or a bank. Companies normally treat the actual amounts paid or due to the pension fund during the year as period expense. For tax purposes generally there is no limit upon the employer contribution. Therefore, if the company's profits are on the high side, extra payments are made to reduce it, and it is the reduced profit that appears in the statutory accounts.

These pension funds may also contain hidden reserves in their financial statements. Normally they establish their balance sheets at historical cost. Often securities and real property are not adjusted to current cost. Excess funding is often used to increase benefits for the beneficiaries. The notes to the consolidated financial statements of the employer often reveal the excess funded status. Few companies capitalize such a surplus in the consolidated balance sheet.

If the pension fund is underfunded, i.e. if it shows a deficit of assets compared with the actuarial liabilities, the employer will in many cases make voluntary provision through a sense of moral obligation towards its former employees (constructive obligation).

The practice of allocating incurred costs to future periods is virtually unknown in Switzerland, given the conservative approach to all accounting in the statutory accounts (Coopers & Lybrand, 1991, p. 119). As mentioned above, it is not uncommon for large single payments to be made into pension funds over and above current requirements, usually to enable the company to avail itself of a tax deduction.

A few companies even make regular profit distributions by way of an appropriation of profits which has to be approved by the general meeting of shareholders. The amounts are deductible for tax purposes even though they are a distribution of profits (like dividends). In substance they are personnel expenses and normally reclassified as such in the consolidated financial statements.

Hidden reserves in pension funds are rarely regarded as deferred assets of the employer. Employers are not permitted to take them back into the balance sheet and income statement as foreign companies sometimes do. Pension funds are strictly regulated under cantonal supervision. Moreover, the board of trustees of pension foundations will consist of both employer and employee representatives. The latter could block such transactions. Only in extreme circumstances could an employer mobilize these excess funds.

Separate employer reserves (*Arbeitgeberbeitragsreserven*) have to be shown in the pension fund's balance sheet which can be used to pay the employer's contribution until they are exhausted. The employer would not capitalize the respective deferred assets (less deferred taxes). However, in merger and acquisition situations such 'hidden reserves' are regarded as assets of the target company (see Helbling, 1991). The issue is controversial, because it can be argued that such funds belong to the employees of the company taken over. Specialist advice and the approval of the government regulatory bodies are necessary before changes can be introduced in pension funds. A new FER Standard, FER 16 on pension liabilities, was issued in 1998 to be adopted in the year 2000. This standard was very controversial in its practical application.

As mentioned, surpluses cannot be repaid to the contributing company. Even the scope for investing such funds in the employer company is very limited. The following are possibilities:

- Current account payable of the employer.
- Short-term loan to the employer.
- Long-term loan to the employer.
- Investments in bonds of the employer or its subsidiaries.
- Investments in shares or non-voting shares of the employer or its subsidiaries.

All nominal investments (loans and current accounts) have to carry interest at market rates. The investment limits are very restrictive. Loans should normally be secured by a mortgage or other means. Securities should preferably be marketable and should not exceed certain rather low levels. If such investments exist, the auditors of the employer have to confirm to the cantonal authorities that the employer has recorded the same amount of debt and that the employer is a going concern (i.e. has more assets than liabilities).

Pension fund expenses do not have to be disclosed in the income statement of the employer. The employer's contributions are normally included in personnel expenses.

Under OR Art. 663b, para. 5 the total liability to the pension fund has to be disclosed separately in the notes on the accounts, with comparative figures for the previous year (see IX).

12 Prior period adjustments

It is very rare for Swiss companies to adjust prior years' retained earnings in the individual financial statements in order to correct prior year errors, or to restate for the effect of an accounting change. All such items are adjusted through the income statement. In exceptional circumstances, if the tax authorities will not allow depreciation or other deductions, then a credit is booked direct to retained earnings. This would, however, be disclosed under changes in the equity.

VIII Revaluation accounting

1 Revaluation to avoid bankruptcy

There is no requirement to provide information on the effects of changes in the price level in individual financial statements. In the individual financial statements this information is virtually never disclosed. Only write-backs of past depreciation can be booked as income. If the revaluation exceeds historical cost a revaluation reserve has to be set up, as explained below.

Where, owing to losses shown in the balance sheet, half the capital and statutory reserves are no longer covered, property or shareholdings whose real (intrinsic) value has risen above acquisition or production cost may be revalued to no more than the balance sheet deficit for the purpose of eliminating it (OR Art. 670).

Only property and equity investments in other companies (participations) may be revalued. Substantial hidden reserves may lurk under these headings because they have to be valued at historical acquisition cost. Many companies have been holding property for 50 years or more which has to be valued at the original purchase cost. Often it includes prime sites in the town centre which are thus considerably understated in the balance sheet. Some major companies acquired holdings in cities over 100 years ago, for example, at CHF 0.50 a square metre. The value of such land has increased to over CHF 5,000 a square metre nowadays, mainly thanks to inflation. Under the historical cost convention companies may not adjust the now meaningless acquisition cost to current values in the individual financial statements.

Only where the business is in difficulty, i.e. overindebted in the terms of OR Art. 725 where liabilities exceed assets, is an increase allowed. Of course, the company can always mobilize the hidden reserves by way of selling the assets.

In the event of revaluation the following must be disclosed in the notes:

- The subject, e.g. the piece of property or participation concerned.
- The amount of the revaluation (the extent of the increase).
- The date of the revaluation (normally the balance-sheet date).
- Reference to the special audit report required, which must be the work of specialist auditors.
- Reference to the valuation report, which must also be from experts.

In the case of property revaluation, a professional estate agent will be engaged to value the land and buildings on the basis of capitalized earnings, the substance (replacement cost), comparable transactions near by, land prices, etc.

Similar valuations are undertaken for equity investments in subsidiaries and/or affiliated companies, on the basis of capitalized estimated earnings, equity at fair market values, share prices in the case of public companies, etc. Obviously these valuations will be arrived at conservatively and provision for deferred taxation will be deducted.

An example of revaluation of property and equity investment is shown in Table VIII.1. In the case of marketable equity investments, revaluation to market value is permissible in accordance with OR Art. 667. The profit can be stated as an unrealized gain in the income statement.

2 Revaluation reserve

The amount of the revaluation should be shown separately as a revaluation reserve. The revaluation reserve is permissible only if the auditors confirm in writing to the general meeting of shareholders that the legal requirements have been met in terms of OR Art. 670. The revaluation reserve required by OR Art. 670 may be released only by conversion into capital, or by depreciation, or by the disposal of the revalued assets (OR Art. 671b; see V.2.2.2).

IX Notes and additional statements

1 The notes on the accounts

1.1 Overview

Under OR Art. 663b a certain minimum level of disclosure is required in the notes on the accounts. Under the old company law only two items of information had to be disclosed apart from the balance sheet and income statement: the insurance value of property and the total amount of contingent liabilities. These items were

Table VIII.1 Revaluation of property and equity investment

	Real estate	Investment
Acquisition cost	1,000	800
Book value	800	200
Current value	1,500	700
Possible revaluation with income effect	200	500
Possible revaluation without income effect under OR Art. 663b, para. 9	500	0

The revaluation of 500 has to be booked to a special equity account named 'Revaluation reserves'. It cannot be distributed until the revaluation is reversed or the assets are sold. Thus it is not possible to credit the amount to the income statement or to use it to reduce accumulated losses.

Source: Bertschinger (1992b).

normally disclosed on the face of the balance sheet as footnotes below the balance sheet total. The fire insurance value reflected the replacement cost of buildings, as land itself needs no fire insurance. This amount was derived from the company's fire insurance policy. All buildings in Switzerland must have compulsory insurance as regulated by the government. The details are the responsibility of the cantons (political subdivisions). Many cantons have a government-owned insurance company. Given the limited information and often understated book values disclosed in Swiss financial statements, analysts sometimes used the insurance figures to estimate the hidden reserves in real estate.

Alusuisse-Lonza Holding disclosed in its parent company financial statements for 1997 a *pro memoria* book value for its property holdings of CHF 1. In a footnote to the balance sheet, however, it mentions that the fire insurance value of the property of the holding company (a headquarters building on Lake Zurich) is CHF 94 million. This value would have been disclosed not to give an idea of the company's hidden reserves but more likely to demonstrate that the building was properly insured.

The footnotes in most Anglo-American countries grew to voluminous length, running to 10 or 20 pages of the annual report. Eventually they could no longer be placed at the foot of the balance sheet and had to appear as an appendix, or annexe, to the financial statements. They then became known as notes on the accounts.

The new company law (OR Art. 663b) requires 14 items of information to be disclosed in the notes. International accounting standards, including the Fourth and Seventh EU Directives and the International Accounting Standards promulgated by the International Accounting Standards Committee require much more disclosure than the Swiss requirements. FER No. 8, 'Notes to the consolidated financial statements', is fairly comprehensive and recommends additional disclosure in accordance with international standards.

The disclosures in the notes must be audited by the company's auditors. Comparative figures are required (OR Art. 662a, para. 1).

The mandatory information to be disclosed in the notes under OR Art. 663b is listed in Table IX.1.

1.2 Contingent liabilities

Under OR Art. 663b, para. 1 the 'total amount of guarantees and securities in favour of third parties' has to be disclosed in the notes on the financial statements. Normally only a single amount is disclosed. The following items are included as contingent liabilities:

- Guarantees
- Securities given in favour of third parties.
- Dividend guarantees to minority shareholders of subsidiaries.
- Obligations arising from the sale of receivables with recourse.
- Obligations arising from joint ventures in the form of partnerships where the partners' liability is unlimited.

The following need not normally be disclosed:

- Pending purchase or sales commitments.
- Obligations arising from endorsed promissory notes.
- Commitments arising from fixed assets on order.
- The portion of subsidiaries' share capital not yet paid up.
- Subordinated loans receivable.
- Comfort letters for the debt of subsidiaries (provided their substance is not a guarantee).
- Penalties in contracts.
- Trust agreements.

1.3 Pledged assets

Under OR Art. 663b, para. 2 the 'total amount of assets pledged against own debts and those where ownership is restricted' has to be disclosed in the notes. The book value of such assets has to be disclosed, not the current value, which would be more relevant. A single amount for all assets which serves as security is sufficient. Examples are:

- Property that serves as security for a loan at book value.
- Pledged inventories.
- Securities, term deposits, current accounts, etc. that are deposited or held with a bank if the bank has granted a loan to the company, or an overdraft. (According to the general agreement between banks and their cus-

Table IX.1 Minimum content of the notes to the accounts (annexe)

1	The total amount of guarantees, indemnity liabilities and pledges in favour of third parties.
2	The total amount of assets pledged or assigned for the securing of own liabilities, as well as of assets with retention of title. It is understood that only book values need be disclosed.
3	The total amount of liabilities from leasing not included in the balance sheet. According to the interpretation of the Swiss Institute in its *Auditing Handbook* this is required only for financial leases.
4	The fire insurance value of fixed assets (in a single amount only).
5	Liabilities to pension funds of the employees of the company.
6	The amounts, interest rates and maturities of debentures issued by the company.
7	Any participation which is essential to an accurate assessment of the company's financial situation and income. A participation is defined as an investment in another company commanding at least 20 per cent of the voting rights, according to OR Art. 665a, para. 3.
8	The total amount of hidden reserves released (under OR Art. 669, para. 2) and other hidden reserves (OR Art. 669, para. 3) to the extent that such total amount exceeds the total amount of the newly created reserves of that kind, if thereby the business result appears considerably more favourably.
9	Information on the subject and the amount of any revaluations (OR Art. 670).
10	Information on the acquisition or disposal and on the number of own shares held by the company, including its shares held by another company in which it has a majority participation. Equally, the circumstances in which the company has acquired or disposed of its own shares should be revealed (OR Arts. 659– 659b).
11	The amount of the authorized capital increase (OR Arts. 651–651a) and of the capital increase subject to conditions (OR Arts. 653–653i).
12	Departures from generally accepted accounting principles (OR Art. 662a, para. 3).
13	Disclosure of important shareholders (OR Art. 663a, para. 1).
14	Consolidation and valuation principles followed in the group accounts (OR Art. 663g).

tomers all assets with the bank serve automatically as collateral.)

- Receivables pledged to banks or financial institutions.
- Fixed assets that have been acquired under a hire-purchase agreement, where the asset is entered in a government public register, until it has been paid for in full. (In Switzerland there is an official register in which the ownership of goods bought on hire-purchase can be recorded.)
- Capitalized financial leases.

1.4 Leasing obligations

Under OR Art. 663b, para. 3 the 'total amount of the lease obligations that are not included in the balance sheet' (as liabilities) has to be disclosed in the notes. It is understood that only finance leases are involved. Operating leases do not need to be disclosed, as they are not reckoned as liabilities in the balance sheet.

The sum of the outstanding future lease fee payments must be disclosed (the number of outstanding payments times the leasing fee).

1.5 Fire insurance value of fixed assets

Under the OR Art. 663b, para. 4 the 'fire insurance value of fixed assets' must be disclosed in the notes. Sometimes two values are shown:

- Fire insurance value of buildings.
- Fire insurance value of machinery, equipment, etc.

Disclosure is not required by international standards and this requirement is therefore unique to Switzerland.

1.6 Commitments to pension funds

Under OR Art. 663b, para. 5 'obligations towards pension funds' must be disclosed in the notes. Pension funds have to be established as separate legal units such as foundations or co-operatives to protect the beneficiaries. Such organizations are subject to considerable regulation and government scrutiny. The pension contributions of employees and employers (normally contributory, defined contribution plans) have to be funded, i.e. cash payments must be transferred to the pension fund. Strict investment policies apply. Share, bond and loan investments in the employer company are restricted or must be secured (normally against property).

Investments by the pension fund in the employer company are normally financial debts which have to earn interest.

The employer has to disclose the total debt to the fund. It may consist of:

- Current accounts.
- Short-term loans.
- Long-term loans (e.g. a loan secured by a mortgage).

1.7 Debentures

Under OR Art. 663b, para. 6 the following information has to be disclosed in the notes: the 'amounts, interest rates and maturities of outstanding debenture bonds'. Debentures are defined as follows:

- Public bonds issued in relatively small amounts, e.g. CHF 1,000 to CHF 5,000.
- Medium- to long-term maturities.
- Mostly in sizeable amounts of CHF 10 million or more.
- Publicly placed and traded on stock exchanges.
- Companies that have issued such bonds are subject to consolidation and publication requirements in the same way as those with quoted shares.

Not considered as debentures are private placements with the following characteristics:

- In high denominations, e.g. units of CHF 50,000 to CHF 100,000.
- Normally first-class issuers.
- Mostly sophisticated (institutional) investors.
- No special prospectus requirements for creditors.
- Limited consolidation and disclosure requirement.

1.8 List of equity investments

Under OR Art. 663b, para. 7 'each equity investment that is material to an understanding of the financial position and results of the company' must be disclosed in the notes. Important investments are shares held in subsidiaries which are controlled and therefore consolidated or those with a significant influence where the company holds more than 20 per cent of the voting rights of the investee. Investments carrying less than 20 per cent of the voting rights should be included where the investment is material to the balance sheet of the investor. An investment is a long-term financial (fixed) asset which is not held on a merely temporary basis. In the group accounts, obviously, only non-consolidated investments are capitalized in the balance sheet as fixed assets. The consolidated investments are eliminated in the consolidation. For each investment the following information needs to be disclosed:

- Legal name as stated in the articles and the commercial register.
- Legal domicile where the company has its registered office.
- Country of incorporation.
- Percentage of capital held.
- If other than the above, additionally the percentage of voting rights held.

1.9 Release of hidden reserves

Under OR Art. 663b, para. 8 'the total amount of hidden reserves released must be disclosed if the amount exceeds the newly created hidden reserves and if the amount is material to the results as shown in the income statement'. Hidden reserves in this legal sense are:

- The excess of acquisition cost over the book value. The acquisition cost, however, may not exceed the fair market value.

- Not, however, the difference between current value and acquisition cost.

Such hidden reserves can exist under almost any balance sheet heading except nominal share capital. They are created by understating assets (e.g. inventories, fixed assets, receivables) or overstating liabilities (provisions, accrued liabilities, etc.). It is difficult to calculate the extent to which hidden reserves have been released if no sophisticated accounting system is available. The release is calculated by deducting the higher hidden reserves at the beginning of the year from the balance of hidden reserves at the end of the year, making due allowance for deferred taxation. In principle, companies have to prepare shadow financial statements to calculate the hidden reserves. Swiss subsidiaries of groups, foreign or domestic, normally use the accounts they have prepared for consolidation purposes to generate these figures. It is accepted that any accounting standards which require a true and fair view will rule out the creation of significant hidden reserves. Such standards would include:

- FERs.
- Fourth and Seventh EU Directives translated in country law such as:
- German Accounting Directives Act (*Bilanzrichtlinien-Gesetz*, BiRiLiG).
- International Accounting Standards of the IASC.
- United States Generally Accepted Accounting Principles (US GAAP).

The statutory (commercial and tax) accounts (referred to in Germany as *Handelsbilanz* I, HB I) are normally adjusted to group accounting policies on financial statements for consolidation purposes (*Handelsbilanz* II, HB II), after allowing for deferred taxation on the restated amounts. It is accepted practice to take the difference in the equity of HB I to HB II as hidden reserves net of tax. Invariably HB I is lower than HB II. If the opposite is the case, then HB I may be overstated and under Swiss law will be reduced to the HB II amount.

If the HB I profit is higher than the HB II profit in the income statement, the difference would be due to the release of hidden reserves, i.e. a reduction in hidden reserves compared with the previous year. If the amount materially influences the stated profit in the commercial accounts (HB I), the amount of released hidden reserves has to be disclosed in the notes on the commercial accounts. The gross amount of hidden reserves or the net amount (net of deferred tax) can be shown.

However, the creation of hidden reserves or their balance at year end need not be disclosed. This Swiss practice of disclosing the release of hidden reserves is unique. It stems from the fact that Switzerland has not had a currency reform for 100 years, has never allowed the revaluation of real estate above acquisition cost, except as mentioned below, and has enjoyed a liberal income tax regime and generous tax advantages.

1.10 Subject and amount of revaluations

Under OR Art. 663b, para. 5 'information about the subject and amount of revaluations' of certain fixed assets has to be presented in the notes. The information in the notes would normally include:

- The type of asset that has been revalued (property or investment in companies).
- The amount of the revaluation (the same as the amount of the revaluation reserve in the balance sheet).
- The date of the revaluation (normally the balance-sheet date).
- The basis of the valuation (prudent current cost).
- A report by a suitably qualified specialist (e.g. an estate agent or a valuation report for companies).

This procedure is permissible only for property and equity investment in other companies (see VIII).

1.11 Own shares

OR Art. 663b, para. 10 requires:

> information on the acquisition or disposal and on number of own shares held by the company, including those held by a company in which it holds a majority of voting rights;

Table IX.2 Disclosure of own shares

Own shares	No.	Amount (SFr)
Balance at 1 January 1997	1,000	500,000
Additions	50	30,500
Value adjustment		−120,000
Disposals	−60	−20,500
Balance at 31 December 1997	990	390,000

additionally the circumstances must be disclosed in which the company has purchased or sold its own shares.

Hence the number and the amounts in Swiss francs of own shares on hand at year end, and movements in the figures during the year, must be disclosed in the notes. The summarized form of disclosure exemplified in Table IX.2 could be used if the company buys and sells many own shares during the year (e.g. a public company with quoted shares).

Under the old law companies were not allowed to hold their own shares. Inevitably all manner of arrangements were devised to circumvent the restriction. For instance, the own shares might be held by a subsidiary company often domiciled in an offshore tax haven. Sometimes the company's bank would hold these own shares under a trust agreement at the nominal value.

Resort to such devices was necessary because public companies often issued convertible or option debentures and needed own shares in case the holders converted bonds or exercised options to acquire shares. It was not possible under the previous law, as it is now, to have authorized but unissued shares.

Own shares have to be capitalized as an asset in the balance sheet. The same procedure is adopted in the Fourth EU Directive. Under Anglo-American accounting principles own shares are deducted from shareholders' equity at cost. Under US GAAP and IAS SIC 16 own shares are referred to as treasury stock. They are normally shares that have been bought back on the stock exchange. Such transactions reduce the cash position of the company and also the equity. Gains and losses on such transactions are directly debited or credited to additional paid-up capital (*agio*) and therefore do not influence net income. (In the United States this principle is summed up in the saying 'You can make money from dealing in own shares but not a profit'.)

The German or Continental view of own shares is that they are an asset. If their value is lower than cost, a value adjustment is made through the income statement. If own shares are sold again the resultant gain is considered as income. Only if own shares are extinguished by way of a capital reduction is the resultant gain or loss considered an adjustment to additional paid-up share capital (*agio*).

1.12 Authorized and contingent capital

Under OR Art. 663b, para. 11 'the amount of the authorized and contingent capital' must be disclosed in the notes. The concept of authorized and contingent capital was introduced into Swiss law only with the revised company law of 1 July 1992. The notes must state the proportion of the authorized and contingent capital as of the balance-sheet date which has not been utilized (OR Arts. 651 and 653 respectively). Such residual capital arises where the general meeting of shareholders has authorized the board of directors to increase the share capital (authorized capital) and/or where options and conversion rights relating to debentures have not been exercised by the balance-sheet date.

This information is important in assessing the total amount of share capital and the shares outstanding and to assess the dilution of the shareholders' stake through such transactions. They must also be noted in the commercial register and in the company's articles.

1.13 Departures from certain generally accepted accounting principles

Under OR Art. 662a, para. 3 'departures from the principles of a going concern, of consistency in presentation and valuation and from the prohibition of netting amounts in the financial statements are permissible if they can be justified and must be disclosed in the notes on the accounts'. Examples would include:

- Netting financial expenditure and income.
- Failure to record depreciation one year.
- Owing to liquidity problems the going concern principle is in question.
- The method of valuing inventories has been changed.

However, the ability to create hidden reserves will always negate the above principles, including that of consistency. As long as hidden reserves can be created, the company does not need to mention departures from consistency. Only when hidden reserves are released does the amount involved (income effect) have to be disclosed and departures from consistency mentioned in the notes as stated above.

1.14 Disclosure of important shareholders

Under OR Art. 663c, para. 1 'companies whose shares are listed on a stock exchange have to disclose the names of important shareholders and their stake (the percentage held) if their identity is known or should have been known'. OR Art. 663, para. 2 defines important shareholders as those shareholders or groups of shareholders legally connected through a shareholders' agreement that hold more than 5 per cent of the votes of the company. If the articles of the company set a lower limit on the votes one shareholder may dispose of, then that percentage applies. For example, there are companies which limit shareholdings to 2 per cent per shareholder. In that case shareholders with 2 per cent or more of the votes must be disclosed. Swiss public companies often have registered shares of low par value (e.g. CHF 100) which are controlled by the owning families. The investing public hold bearer shares of higher nominal value (e.g. CHF 500) or even non-voting participation certificates.

Shareholders are not obliged to reveal major holdings to the company except as required by the new Swiss Stock Exchange Take Over Regulations. Only the holders of registered shares will be known to the company through the share register. However, material holdings (representing more than 20 per cent of the voting rights) must be disclosed by public companies in the list of investments which has to be included in the notes.

As mentioned above, the disclosure of the stakes of corporate and private shareholders in the company is now a requirement of the new Stock Exchange Act which was enacted in 1997.

1.15 Consolidation and valuation principles in the group accounts

There has been criticism of the fact that the valuation principles need be disclosed only in the consolidated financial statements and not in the individual accounts. It is difficult to assess the financial situation of a company and its results when the main accounting policies are not known. FER 8 contains a long list of recommendations for additional disclosure, especially in group accounts.

1.16 Swiss requirements and international standards

The comparison in Table IX.3 shows some peculiarities in Swiss disclosure requirements. The fire insurance value of fixed assets and the release of hidden reserves have to be disclosed in the notes not only on the individual accounts but also on the group accounts. Neither requirement figures in international accounting standards. The FER recommendations come close to the requirements of the Fourth and Seventh EU Directives and the International Accounting Standards of the IASC. They comfortably exceed the requirements of company law.

1.17 Example of disclosure

Voluntary disclosure is exemplified in the annual report of Prodega for the year 1991 in Table IX.4.

Table IX.3 Swiss requirements for the notes compared internationally

Information to be disclosed	Swiss law	FER	4th EU Directive	IASC	US GAAP
Contingent liabilities	•	•	•	•	•
Pledged assets	•	•		•	•
Leasing obligations	•	•	•	•	•
Fire insurance value of fixed assets	•				
Liabilities to pension funds	•				
Debenture loans	•	•	•	•	•
List of investments (subsidiaries)	•	•	•	•	
Release of hidden reserves	•				
Revaluation of fixed assets	•	•	•	•	
Own shares	•	•	•	•	•
Authorized and contingent capital	•	•	•	•	•
Foreign currency translation	•	•	•	•	•
Valuation principles		•	•	•	•
Disclosure of shareholders	•				
Comparative figures	•	•	•	•	•
Changes of valuation principles	•	•	•	•	•
Changes in fixed assets		•	•	•	•
Cash flow statement		•		•	•
Elements of equity		•	•	•	•
Liabilities due in five years or more			•		
Secured liabilities		•	•	•	•
Net sales by segments		•	•	•	•
Average number of personnel			•		
Tax expenses and deferred tax		•	•	•	•
Payments to board and management			•		
Subsequent events		•	•	•	•
Research and development costs		•	•	•	•
Changes in equity accounts		•	•	•	•
Commitments to equipment ordered				•	•
Pension funds		•		•	•
Pending lawsuits, including tax cases				•	•
Financial instruments		•		•	•
Earnings per share					•
Quarterly financial statements					•

2 Asset movement schedule (*Anlagespiegel*)

The Fourth EU Directive requires that changes in fixed assets are shown from the opening of the year to the closing balance in the individual accounts. This applies to all fixed assets, including tangible, intangible and financial fixed assets. The amounts must be shown gross, i.e. acquisition cost, accumulated depreciation, etc. No such legal requirement exists in Switzerland. FER 8 requires such a listing of property, plant and equipment in the group accounts only.

3 Segmental reporting

There are no legal requirements or professional recommendations on the presentation of segmental information in the notes on the individual financial statements. However, group accounts normally reflect the net sales and number of personnel broken down by geographical area and product line. Few groups present segment operating profits as required by the International Accounting Standards of the IASC.

Table IX.4 Disclosure in the notes on the accounts of Prodega AG

Disclosures required by law under OR Art. 663b

1 Contingent liabilities — None

2 Pledged assets or assets with retention of title:

Real estate at fair market value	CHF 155,178,007
Loans secured against real estate	CHF 44,360,000

The investment in Usego-Trimerco Holding AG of a nominal value of CHF 3.1 million is pledged in favour of a bank loan of CHF 5 million. No receivables have been pledged

3 Lease liabilities not shown in the balance sheet — None

4 Fire insurance value of tangible fixed assets:

Machinery, equipment, etc.	CHF 27,933,300
Real estate	CHF 120,067,325

5 Liabilities to pension funds

Prodega owes the BVG-Personalvorsorgestiftung der Prodega AG CHF 1 million through a loan secured against real estate

6 Issued and outstanding debenture bonds

Private placement of CHF 10 million with Berner Kantonalbank:
Interest 6.75%
Starting and maturity dates 5 June 1991 to 5 June 1996

7 Significant participations

Company	Capital (CHF)	Share (%)
Consolidated companies:		
Growa AG, Moosseedorf BE	500,000	100.00
Non-consolidated companies:		
Usego-Trimerco Holding AG, Volketswil	55,000,000	5.68
EG Dritte Kraft AG, Baar	600,000	6.66
Hopf AG, Basle	625,000	5.00
Hopf Trading AG, Rümlang	600,000	33.33

8 Dissolution of hidden reserves — None

9 Revaluations — None

10 Own shares or participation certificates

No own shares or participation certificates were acquired, held or disposed of during the year

11 Authorized or capital with conditions — None

12 Other information to be disclosed by law:

The non-quoted registered shares are held by Curti & Co. AG, Lucerne (34,000 registered shares, nominal value CHF 100, total value CHF 3.4 million, 50.75% of voting rights).

Source: Prodega AG, annual report, 1991 (when such disclosure was purely voluntary). Author's translation.

4 Earnings per share (EPS)

There are no requirements or professional recommendations on the calculation and disclosure of net income/loss per share. However, multinational groups publish the per share information on a consolidated basis only. The calculation is on a primary basis only and does not take share equivalents into consideration (e.g. options or convertible bonds, etc.). Normally the group profit before minorities is divided by the year end number of shares issued. This is not in line with Anglo-American standards. IAS is followed only by big Swiss groups.

5 Funds flow statement

The Code of Obligations (OR) does not make cash flow or funds flow statements a requirement. This is in line with the Fourth and Seventh EU Directives, which impose no obligation to prepare and publish such information. Normally no such statement is prepared for the individual financial statements or parent company statement. However, in group accounts almost all Swiss companies present consolidated cash flow statements.

According to FER 6, which was issued in May 1992, a funds flow statement may be omitted for the individual company if a funds flow statement is presented for the consolidated group (see Switzerland – Group Accounts).

6 Value-added statement

There is no requirement to prepare a value-added statement (*Wertschöpfungsrechnung*). However, the Swiss format for the income statement makes it very easy to calculate this information. Many public companies therefore present it voluntarily, but only in the group accounts.

7 Human resource accounting

There are no requirements or professional recommendations on the presentation of information in this field (*Sozialbilanz*). Some progressive concerns disclose such information in their annual report for public relation purposes. They are usually large employers (e.g. ABB, Novartis).

8 Green accounting

There are no requirements or professional recommendations that such information should be provided. Some multinational companies do so in their annual report for public relation purposes (e.g. Novartis and Roche as large chemical and pharmaceutical companies, Migros as a large retailer).

9 Directors' report

Under OR Art. 662 the board of directors must prepare for each financial year a business report which consists of the audited annual financial statements, the annual report, and the consolidated financial statements if such statements are required by law. The annual financial statements consist of the profit and loss account (income statement), the balance sheet and the notes on the accounts. All these have to be audited. However, in contrast to the requirements of the EU Directives on the directors' report (*Lagebericht*), in Switzerland the directors' report does not need to be audited.

The contents of the directors' report to the shareholders are not defined. Often the reports are very short, merely repeating facts and figures which can easily be derived from the financial statements.

X Auditing

1 Audit requirement and appointment of auditors

Auditing requirements are to be found mainly in Arts. 727–731a of the Code of Obligations (OR), revised as of 1 July 1992. Some other legislation also refers to audit requirements (e.g. banking law). Only stock corporations are under a requirement to be audited. As stock corporations are by far the most important legal form of business, the requirements for the auditing of stock corporations are treated here.

According to OR Art. 727 the general meeting of shareholders of every stock corporation (there are approximately 170,000) must appoint one or more auditors. It may designate substitute persons. At least one auditor must have his domicile, registered office or a registered branch office in Switzerland. Individual auditors or audit firms may be appointed. Public companies with quoted shares or debentures outstanding have usually one of the Big Five international accounting firms as auditor.

The auditors have to confirm their acceptance of the mandate in writing to the general meeting because the fact that they have done so has to be filed with the commercial registrar.

It is therefore possible to find out who the auditors of a stock corporation are, something which would otherwise remain unknown because the accounts are not published (see XI). The Swiss domicile is important in order that correspondence may be addressed to an address in Switzerland. The same requirement applies to the board of directors.

2 Qualifications of auditors

Under OR Art. 727a auditors must be qualified to fulfil their duty to the company to be audited. This means that they should have at least a basic knowledge of accounting, auditing, business law, taxation, etc. However, under OR Art. 727b public companies and those that are of a certain size have to appoint specially qualified auditors. The auditors must have special professional qualifications if:

- The company has bond issues outstanding.
- Its shares are listed on the stock exchange.
- Two of the following thresholds are exceeded in two consecutive years:
  (i) Balance sheet total of CHF 20million.
  (ii) Revenues of CHF 40 million.
  (iii) Average annual number of employees, 200.

The Swiss Federal Council defined the professional qualifications of specially qualified auditors in a decree of 11 June 1992 as follows:

- Swiss Certified Public Accountant (*dipl. Wirtschaftsprüfer,* previously: *dipl. Bücherexperte*) as registered by the Swiss Institute of Certified Public Accountants.
- Swiss Certified Trust Expert (*dipl. Treuhandexperte*) or Swiss Certified Tax Expert (*dipl. Treuhand-/Steuerexperte*) and Swiss Certified Bookkeeper (*dipl. Buchhalter/Controller*) with five years' practical experience of auditing under the supervision of a Swiss Certified Public Accountant.
- Graduates of business and/or law schools (BWL, VWL, *Ius-Lizenziat*/HWV) with 12 years' experience of auditing practice under the supervision of a Swiss Certified Public Accountant.
- Persons who have acquired an equivalent foreign diploma as certified public accountant with the requisite experience and knowledge of Swiss law. Such qualifications would normally embrace Chartered Accountants (UK, Canada, Australia, etc.), Certified Public Accountants (United States), *Wirtschaftsprüfer* (Germany), *Experts Comptables Diplomés* (France), etc.
- Foreign Certified Public Accountants who are qualified under the Eighth EU Directive and have the requisite acquaintance with Swiss law.

3 Independence requirements

Under OR Art. 727c, auditors must be independent of the board of directors and of any shareholder with a majority vote. In particular they must not be employees of the company to be audited, and they must not perform work for it that is incompatible with the auditing mandate. They must be independent of companies belonging to the same group of companies if a shareholder or creditor so requests. Swiss Certified Public Accountants have to observe the independence requirements of the Institute as published in the *Auditing Handbook* as well. Partners and staff of the Big Five also have to meet the independence requirements of the firm concerned, all of which substantially surpass the Swiss standards (because of SEC requirements).

Under Swiss law, an audit firm's independence would not be jeopardized where it provided the following services:

- Tax advice. The decisions themselves would have to be taken by the company, e.g. by signing the tax return and the accounts, which have to be filed.

- Bookkeeping services. Decisions on accounting policy and estimates would have to be taken by the client company, e.g. by signing the accounts. The bookkeeping may not be done by the auditors involved in the audit.
- Installing computer software. Under OR Art. 727d, commercial companies or co-operatives may also be appointed as auditors. They would have to ensure that the personnel in charge of the assignment met the qualification and independence requirements.

4 Terms of mandate, resignation, etc.

Under OR Art. 727e, the maximum term of the mandate is three years. It ends at the general meeting of shareholders to which the last audit report is to be submitted. Re-election is possible. An auditor who resigns must explain the reason to the board directors, who must communicate it to the next general meeting of shareholders. The general meeting of shareholders may, at any time, remove an auditor. In addition, any shareholder or creditor may, by bringing a suit against the company, request the removal of an auditor unqualified for the position. The board of directors must notify the commercial registrar without delay of the termination of the mandate. If such notification is not lodged within 30 days the outgoing auditor may personally request the termination.

Under OR Art. 727f, the commercial registrar, upon learning that a company has failed to appoint auditors, must set a time limit for the situation to be remedied. If the company fails to observe the deadline, a civil court judge must appoint the auditors for one business year on the request of the commercial registrar selecting the auditors at his or her discretion. Should these auditors resign, the commercial registrar will notify the judge accordingly. Provided the reasons are valid, the company may request the removal of the auditors appointed by the court.

5 Duties

Under OR Art. 728 the auditors have to examine the bookkeeping and the annual accounts as well as the proposed appropriation of retained earnings to satisfy themselves that they comply with the law and with the company's articles of incorporation. The board of directors should release to the auditors all requisite documents and provide them with all necessary information, in writing if so requested. This includes a 'representation letter' from the board attesting to the completeness of the liabilities shown, the truth of the valuations and the existence of the assets.

Under OR Art. 729, the auditors must report the findings of their audit in writing to the general meeting of shareholders. They should recommend approval, with or without qualification, or rejection of the annual accounts. The report should name the persons in charge of the audit, and confirm that their qualifications and independence met the requirements.

6 Standard form of audit report on the annual financial statements

In the 1999 edition of its *Auditing Handbook*, the Swiss Institute of Certified Public Accountants includes a standard form of unqualified audit opinion for use with corporations (Table X.1).

7 Report to the directors

OR Art. 729a requires that where the auditors of a company must be specially qualified within the meaning of OR Art. 727b, they should report not only to the shareholders (the auditors' report) but also to the board of directors. The report to the board should comment on the way they have carried out their task and the results of their examination. The types of company affected by OR Art. 727b are public companies and other companies above a certain size (see X.2). Where companies have a sophisticated system of accounting records and internal reports, the Swiss Institute of Certified Public Accountants proposes (Auditing Handbook) a standard form of report (Table X.2).

Table X.1 Standard form of unqualified opinion to shareholders

Report of the auditors to the general meeting of ABC Ltd, Zurich

As auditors of your company we have examined the books of account and the financial statements presented by the board of directors for the year ended 31 December 1998 according to the requirements of the law. Our audit was conducted in conformity with the standards promulgated by the auditing profession. We confirm that our professional qualifications and independence meet the legal requirements.

Based on our examination we have concluded that the books of account, the financial statements and the proposed appropriation of the available profits are in accordance with the law and with the articles of incorporation.

We recommend approval of the financial statements submitted to you.

XYZ Audit Ltd, Zurich, 28 February 1999
A. Mueller *B. Huber*
Swiss Certified Accountants
Auditors in charge

Enclosures

Financial statements consisting of:

- Balance sheet
- Profit and loss account
- Notes to the financial statements

Proposed appropriation of the available profits

Source: Auditing Handbook, 1999.

Table X.2 Recommended standard form of report to the directors

Report of the auditors to the board of directors of ABC Ltd, Zurich

As auditors of your company we have examined the books of account and the financial statements for the year ended 31 December 1998. We recommend without qualification their approval by the general meeting of shareholders.

In accordance with OR Art. 729a we additionally confirm that our examination encompassed such audit procedures as were considered appropriate in the circumstances. As agreed, these audit procedures are not listed here but are documented in our working papers.

In addition, we have reviewed the internal accounting documentation supporting the financial statements and examined such assertions as are material to the trading situation and the results of your company. It is our opinion that no further comment is called for in consequence.

XYZ Audit Ltd, Zurich, 28 February 1999
A. Mueller *B. Huber*
Swiss Certified Accountants
Auditors in charge

Enclosures

Financial statements consisting of:

- Balance sheet
- Profit and loss account
- Notes to the financial statements

Source: Auditing Handbook, 1999.

8 Overindebtedness

Under OR Art. 729b the auditors must report in writing to the board of directors any infringement of the law or of the company's articles which they uncover in the course of their examination. In serious cases they are required to report the fact to the general meeting of shareholders as well. In the event of obvious overindebtedness, the auditors must notify the court should the board of directors fail to do so. Overindebtedness is defined by OR Art. 725 as the total loss of the shareholders' equity where liabilities exceed assets. In severe cases, auditors have to deposit the balance sheet with the bankruptcy court (OR Art. 729b).

9 Education and certification

Accounting is not a major discipline at Swiss universities. Consequently only about 50 per cent of practising auditors hold a university degree. The trend, however, is increasingly to enter the profession from a base of graduate studies in business administration, economics and/or business law. The only university offering a specialist course in auditing is the University of Saint-Gall. Typically, entry to the profession has been via commercial junior colleges (*Höhere Wirtschafts- und Verkehrsschulen*, HWV), although a small and decreasing percentage of professional entrants hold only high school degrees supplemented by commercial apprenticeship diplomas. Another group are certified bookkeepers who have gained their qualification through non-academic education and on-the-job training as clerks and bookkeepers/controllers.

Once an aspirant has decided to enter the auditing profession, he or she typically enters into an employment contract with an audit firm, where various internal professional training courses are available. This would normally be at the age of about 23 to 25, after university and, in the case of men, National Service (army conscription).

After three years or so in practice, the aspirant will combine work with attendance at a professionally run auditing school (*Kammer-Schule*) during the late afternoon, evening and weekend. Over the course of two or three years he or she completes the curriculum and then takes the publicly administered professional exams. They consist of lengthy written case studies and problems to be solved within three days as well as an extensive oral. Successful candidates are on average 28 to 32 years of age. Passing these demanding exams (the failure rate is some 40 per cent) as Certified Public Accountant (*dipl. Wirtschaftsprüfer*) means advancement to supervising senior or assistant manager (*Prokurist*), involving the supervision of professional assistants and authority to co-sign audit reports.

Once certification has been achieved, there are no continuing education requirements. Most people who qualify as Certified Public Accountants work for the big accounting firms or the internal audit departments of large industrial companies and financial institutions. A separate career structure for Certified Tax Experts (*dipl. Steuerexperte*) was established some years ago. As a result, auditors normally do little tax work. Moreover, the Swiss system is complex, with federal taxes on top of 26 cantonal regimes which vary considerably.

The number of Certified Public Accountants has risen significantly in both number and quality. The future is likely to see more and more Certified Accountants leaving public accounting to take up management positions in industry and/or government.

The Swiss Institute of Certified Accountants (*Treuhand-Kammer*), founded in 1925, is active in training and education. It publishes a monthly periodical (*Der Schweizer Treuhänder*) which is the leading source of information on new developments in accounting and auditing. It also publishes monograph and textbook series. Over 150 books have been published in these series, often PhD theses on finance, accounting, auditing, tax, business law, etc., which advance the profession at the academic level.

XI Filing and publication

In general, users enjoy only limited access to the accounts of Swiss companies, since there is no requirement to publish or file them with the commercial registrar. The only exceptions are quoted companies, banks, insurance companies and mutual investment funds. The financial institutions are subject to strict supervision. They must file accounts with the regulatory authorities (the Swiss Banking Commission, the Swiss Insurance Agency, Swiss Stock Exchange, etc.). In addition, banks and insurance companies are required to publish their balance sheets in the Swiss Trade Journal (*Schweizerisches Handelsamtsblatt*, SHAB) and such other newspapers as their articles of association may require. These must appear annually, half-yearly or quarterly, depending on the size of the company.

Thus the publication of financial statements is very restricted. Under OR Art. 697h, the annual financial statements and the consolidated financial statements, once approved by the general meeting of shareholders, must either be published in the official journal or a copy must be sent to any person requesting it, at his or her own expense, within one year of formal approval, if:

- The company has bond issues outstanding.
- Its shares are listed on a stock exchange.

For the above companies there is no size threshold. All public companies, and only public companies, have to publish their audited individual and (where applicable) consolidated accounts. The deadline is the end of June (six months after the balance-sheet date – virtually all businesses take the calendar year as their financial year). Public companies normally provide a copy to anyone requesting it. Big companies normally make their full annual reports available in French and English as well as in German. The accounts are normally available on the home page through the Internet or in conventional printed form.

Other companies must make their annual financial statements, consolidated financial statements and auditors' reports available for inspection by creditors who can furnish evidence of an interest worthy of being protected. The courts decide in the event of dispute. Normally the company can avoid a creditor's inspection of the audited financial statements by paying the amount outstanding.

As only about 500 to 600 companies have bonds issued and/or quoted shares, extremely limited information is available on all other companies, including the 170,000 stock corporations whose accounts are not open to the public. This is at odds with the practice of most European countries, where all corporations must lodge their accounts in a commercial register to which everyone has access.

XII Sanctions

Given that the duty to disclose their financial statements is limited to public companies, there are few sanctions upon a company that fails to prepare proper financial statements. The normal procedure is as follows where a company neglects to keep proper books of account or to produce financial statements:

Under OR Art. 699, the general meeting of shareholders must be convened by the board of directors. The ordinary meeting should take place annually within six months of the close of the financial year. Under OR Art. 700, para. 1, the general meeting of shareholders must be summoned in the form provided for by the articles of incorporation at least 20 calendar days before the date of the meeting. For companies whose financial year is the same as the calendar year, the latest day for the general meeting is therefore the 30 June following the calendar year end. The audit report and financial statements have to reach the shareholders before 10 June. The auditors thus need to be able to finish their work before 10 June if calendar year companies are to comply with the law.

If no financial statements are made available, the auditors must remind the board, preferably in writing. When all measures have been exhausted, if the board has still not submitted the financial statements, the auditors will call a general meeting to inform the shareholders of the illegality. In one-person companies where there is only one shareholder and one board member at any one time, the shareholder may fail to attend the meeting. The auditors will ask a public notary to attend as well, to witness that no shareholder has shown up, and the auditors will resign at this meeting. At the same time they will communicate their resignation to the board, the commercial registrar and possibly the courts where there is a risk of bankruptcy. The court will then decide what to do next. The commercial registrar can close the company down with a court order and put it into liquidation and bankruptcy, the court appointing a receiver and an auditor.

Similarly, the tax authorities will remind the company to submit its tax return. However, the audit report does not need to be filed. If the return fails to materialize, the company is assessed on an estimated basis. If the assessed tax is not paid, the tax authorities can also have the company declared bankrupt and penalize the board members for failure to comply with the tax laws. Eventually the company will be struck from the commercial register: it will cease to exist and will be unable to transact any further business.

The main articles of the Swiss Penal Code (*Strafgesetzbuch* StGB) relating to accounting and bookkeeping are as follows:

- StGB Art. 148 Fraud.
- StGB Art. 152 False information on businesses.
- StGB Art. 163 Fraudulent bankruptcy.
- StGB Art. 166 Negligent bookkeeping.

There have been only a few court judgements in these areas, most of them relating to bankruptcy cases and tax fraud.

As mentioned, few companies have to publish their accounts (only banks, insurance companies and quoted companies). Where companies whose shares or debentures are listed fail to publish accounts, the Swiss Stock Exchange (SWX) has the right to delist them and so debar errant companies from public trading. Similar measures would be taken by the Swiss Banking Commission or the Federal Insurance Agency. Banks and insurance companies that neglect their duty to file audited accounts may find their licence to operate being withdrawn by the regulatory agency.

XIII Financial reporting institutions and national databases

FER – Swiss Accounting and Reporting Recommendations (ARR)
P.O. Box 892, 8025 Zurich
Fax: 41 01 267 75 85
Internet: http://www.fer.ch
Main publication:
Swiss Accounting and Reporting Recommendations, 1999 Edition, about 160 pages, annual publication of all published standards, booklet available in German, French, Italian and English.

Swiss Institute of Swiss Certified Accountants and Tax Consultants (*Treuhand-Kammer*)
Limmatquai 120, 8025 Zurich
Tel: 41 01 267 75 75
Fax: 41 01 267 75 85
Internet: http://www.treuhand-kammer.ch/
Main publication:
Schweizer Handbuch der Wirtschaftsprüfung (*Swiss Auditing Handbook*)
4 Volumes, approximately 2,000 pages, in German and French only
1998 Edition, ISBN 3-908567-63-7
Volume I, about 500 pages on accounting issues including consolidated financial statements; also available on CD-ROM.

Swiss databases for company information on the Internet

Neue Zuercher Zeitung, Zurich
leading business daily newspaper
http://www.nzz.ch/online/05_service/jahresberichte/cgi-bin/SCindexSite.cgi
Contains links to about 250 Swiss listed companies, many of whom have their annual reports on the Internet, an order form for the paper annual reports of Swiss listed companies, free of charge.

Bilanz, Zürich
Monthly business magazine
http://www.aktienfuehrer.ch
Business information on about 250 Swiss listed companies, free of charge.

http://www.schweizerpresse.ch
Internet links to all major Swiss newspapers, magazines, news services, etc.
very helpful, free of charge.

Schweizerische Handelszeitung (SHZ), Zürich
Weekly business newspaper
http://www.handelszeitung.ch/archiv/welcome.html
Archive of articles that appeared in the newspaper, free of charge.

Cash
Weekly business newspaper
http://www.cash.ch
Connection to Reuters.

Credit Suisse, Bank
Homepage of CS Research Department
http://www.credit-suisse.ch/de/index.html
Wide area of research report, mostly in English, links to other helpful databases, most services free of charge.

United Bank of Switzerland
http://www.ubs.com
General information and data directly from Switzerland's largest bank, mostly free of charge.

Banque Pictet
http://www.pictet.ch/
Good site with research reports.

Zuercher Kantonalbank
http://www.zkb.ch
News, economic and economy reports.

http://www.search.ch/
Very helpful search magazine in English, German, French, Italian, free of charge.

Swissinvest
http://www.swissinvest.com
Free Email service for the latest news on Swiss listed companies.

Teledata
http://www.teledata.ch
Legal information on listed and unlisted companies, access to Dun & Bradstreet, credit card charge.

Creditreform
http://www.creditreform.ch
Legal information on listed and unlisted companies, mergers & acquisitions news.

Bibliography

Albisetti, E., Boemle, M., Ehrsam, P., Gsell, M., Nyffeler, P. and Rutschi, E. (1987). *Handbuch des Geld-, Bank- und Börsenwesens der Schweiz* (Schulthess, Thun).

Amt für berufliche Vorsorge des Kantons Zürich (ed.) (1991). *Handbuch der Personalvorsorge-Aufsicht* (Amt für berufliche Vorsorge des Kantons Zürich, Zurich).

Bertschinger, P. (1985). Accounting in Switzerland. *World Accounting Report,* **10**, 14.

Bertschinger, P. (1991a). Switzerland: accounting aspects. *Handbook of Mergers and Acquisitions in Europe*, 3rd edition, 53–64 (Gee, London).

Bertschinger, P. (1991b). Percentage-of-Completion Methode für langfristige Fertigungsaufträge. Grundlage für eine zuverlässige Budgetierung. *Der Schweizer Treuhänder,* **10**, 472–478.

Bertschinger, P. (1992a). Das neue schweizerische Aktienrecht. Auswirkungen auf Rechnungslegung und Prüfung im Einzel- und Konzernabschluss. *Die Wirtschaftsprüfung,* **45**, 65–73.

Bertschinger, P. (1992b). Rechnungslegung. In R. Roth (ed.) *Das aktuelle schweizerische Aktienrecht* (loose-leaf) (WEKA, Zurich).

Bertschinger, P. (1992c). Revisionsstelle. In Roth, R. (ed.) *Das aktuelle schweizerische Aktienrecht* (loose-leaf) (WEKA, Zurich).

Bertschinger, P. (1992d). Rechnungslegung gemäss den International Accounting Standards (IASC). In H. Siegwart (ed.) *Jahrbuch zum Finanz- und Rechnungswesen* (WEKA, Zurich).

Bertschinger, P. and Schibli, F. (1987). Tendenzen im Konzernrechnungswesen. *Neue Zürcher Zeitung,* 30 October, 7.

Bertschinger, P. and Sommerhalder, C. (1991). *EG 92. Einfluss auf die Rechnungslegung*, 2nd edition (Crédit Suisse, Zurich).

Böckli, P. (1992). *Das neue Aktienrecht* (Schulthess, Zurich).

Bodmer, D., Kleiner, B. and Lutz, B. (1991). *Kommentar zum Bundesgesetz über die Banken und Sparkassen* (Schulthess, Zurich).

Boemle, M. (1993). *Der Jahresabschluss* (SKV-Verlag, Zurich).
Choi, F. D. S. (ed.) (1991). *Handbook of International Accounting* (Wiley, New York).
Choi, F. D. S. and Mueller, G. G. (1978). *International Accounting* (Prentice-Hall, Englewood Cliffs, N.J.).
Coopers & Lybrand (ed.) (1991). *International Accounting Summaries: a Guide for Interpretation and Comparison* (Coopers & Lybrand, New York).
Coopers & Lybrand (ed.) (1992). *Switzerland: a Guide for Businessmen and Investors* (Coopers & Lybrand, Zurich).
Dessemontet, F. and Ansay, T. (eds.) (1981). *Introduction to Swiss Law* (Kluwer, Deventer).
Erb, H. (1986). *Grundzüge des Versicherungswesens. Leitfaden für das Versicherungswesen*, 5th edition (SKV-Verlag, Zurich).
Ernst & Young (ed.) (1991). *Doing Business in Switzerland* (Ernst & Young, Zurich and New York).
Fachempfehlungen zur Rechnungslegung (1999). Collected standards (Schweizerische Stiftung, Zurich).
Forstmoser, P., Meier-Hayoz, A. and Nobel, P. (1996). *Schweizerisches Aktienrecht* (Verlag Stämpfli+Cie AG, Bern).
Fox, S. and Rueschoff, N. G. (1986). *Principles of International Accounting* (Texas: Austin Press, Austin, Texas).
Guhl, T. (1991). *Das Schweizerische Obligationenrecht*, 8th edition, A. Koller and J. N. Druey (eds.), (Paul Haupt, Zurich).
Helbling, C. (1988). *Steuerschulden und Steuerrückstellungen* (Swiss Institute of Certified Auditors, Bern and Stuttgart).
Helbling, C. (1989). *Bilanz- und Erfolgsanalyse*, 7th edition (Swiss Institute of Certified Auditors, Bern).
Helbling, C. (1990). *Personalvorsorge und BVG. Gesamtdarstellung der rechtlichen, betriebswirtschaftlichen, organisatorischen und technischen Grundlagen der beruflichen Vorsorge in der Schweiz*, 5th edition (Swiss Institute of Certified Auditors, Bern and Zurich).
Helbling, C. (1991). *Unternehmungsbewertung und Steuern*, 6th edition, Schriftenreihe der Treuhand-Kammer 10 (Treuhand-Kammer, Zurich).
Käfer, K. (1987). *Kontenrahmen für Gewerbe-, Industrie- und Handelsbetriebe*, 10th edition (Paul Haupt, Bern).
Käfer, K. (1991). *Berner Kommentar* 8, Part 2, two volumes, 3rd edition (Stämpfli, Bern).
KPMG Fides Peat (ed.) (1989). *Investment in Switzerland and Liechtenstein*, 2nd edition (KPMG Fides Peat, Zurich).
KPMG Fides Peat (ed.) (1990). *Banking in Switzerland*, 3rd edition (KPMG Fides Peat, Zurich).
Kuhn, M. (1989). *Grundzüge des Schweizerischen Privatversicherungsrechtes.* (Schulthess, Zurich).
Meyer, C. (1993). *Konzernrechnung. Theorie und Praxis des konsolidierten Abschlusses*, Schriftenreihe der Treuhand-Kammer 122 (Treuhand-Kammer, Zurich).
Mueller, G. (1980). Accounting in Switzerland. Unpublished paper (Fides Revision, Seattle, Wash.).
Orsini, L., McAllister, J. and Parikh, R. (1992). *World Accounting* **3**, *Switzerland* (Bender, New York).
Price, Waterhouse (ed.) (1982). *Doing Business in Switzerland* (Price Waterhouse, Zurich).
Schuster, J. B. (1975). *Taschenausgabe des Anlagefondsgesetzes mit Verweisungen, Anmerkungen und Sachregister*, 2nd edition (Orell Füssli, Zurich).
Sterchi, W. (1996). *Kontenrahmen KMU – Schweizer Kontenrahmen für kleine und mittlere Unternehmen in Produktion, Handel und Dienstleistung* (Verlag des Schweizer Gewerbeverbandes und Verlag des Schweizerischen Kaufmännischen Verbandes, Bern).
Swiss American Chamber of Commerce (ed.) (1992). *Doing Business and Living in Zurich. Bibliography*, 9th edition (Swiss American Chamber of Commerce, Zurich).
Swiss American Chamber of Commerce (ed.) (1992). *Swiss Code of Obligations* II, *Company Law Articles 552 to 964* (English translation of the official text) (Swiss American Chamber of Commerce, Zurich).
Swiss Federal Department of Justice (1983). *Botschaft des Bundesrates über die Revision des Aktienrechts* (Department of Justice, Bern).

Swiss Federal Department of Justice (1991). *Obligationenrecht* (*Die Aktiengesellschaft*), official text, changes of 4 October 1991 (Eidgenössische Druck- und Materialzentrale, Bern).
Swiss Stock Guide 98/99 (1998). Verlag Finanz und Wirtschaft, Zurich.
Touche Ross International (ed.), (1978). *Business Study. Switzerland/Liechtenstein* (Touche Ross, New York).
Treuhand-Kammer (1999). *Schweizer Handbuch der Wirtschaftsprüfung* (HWP), Zurich 1999 (Auditing Handbook) (Treuhand-Kammer, Zurich).
Tschopp, F. (1993). *Rechnungslegung von Ertragssteuern im Konzernabschluss in der Schweiz*, Schriftenreihe der Treuhand-Kammer 119 (Treuhand-Kammer, Zurich).
Zünd, A. (1992). Switzerland. In A. David and S. Archer (eds.) *The European Accounting Guide*, 825–84 (Academic Press, London).

SWITZERLAND
GROUP ACCOUNTS

Peter Bertschinger

Acknowledgements

I wish to thank Günther Schulz, retired Senior Partner of KPMG Switzerland and former member of the Executive Committee of the Swiss Accounting Standard Board (FER), for his efforts to improve financial reporting in Switzerland.

CONTENTS

I Introduction and background

1 Introduction

In Switzerland, group accounts play a much more important role for public companies than individual accounts do. Although some 170,000 Swiss stock corporations are required to prepare individual accounts and follow company law provisions, these individual accounts are mostly tax driven. They disclose little and tend to follow conservative accounting policies aimed at avoiding or reducing income tax.

The history of group accounts began with big multinational companies which needed to finance themselves on domestic and international stock exchanges. One of these was Alusuisse (now Alusuisse-Lonza group), a big aluminium and chemical company. Alusuisse sought funds on the London financial markets, in the seventies, in order to finance a large expansion project in Australia. The London bankers requested audited consolidated financial statements for the whole group.

Alusuisse, therefore, has a long history of consolidated statements which have undergone continual improvements since their introduction. The adoption of the Fourth and Seventh EC Directives in 1986 was followed by a switch towards the International Accounting Standards (IAS) of the International Accounting Standards Committee (IASC). In 1992 the Financial Analysts Association of Switzerland rated the group's accounts highest among all Swiss multinationals and awarded it the Merkur prize for best financial reporting.

Many other big Swiss companies followed this trend.

When Switzerland voted against joining the European Economic Area (EEA) in 1992, many public companies started to move their standards away from the EU Directives and towards the Anglo-American accounting standards. Almost all the big Swiss-based industrial multinationals have now adopted the International Accounting Standards (IAS) of the International Accounting Standards Committee (IASC) in London (see Tables I.1 and I.2). Some of the largest companies are also preparing to change to United States Generally Accepted Accounting Principles (US GAAP) as they plan to place shares on US stock exchanges in the near future.

An increasing number of Swiss groups have established holding companies in the last decade. The purpose of the holding company is to separate the operating activities in newly established subsidiary companies. As the Swiss operations have sometimes been limited compared to foreign operations, the main emphasis of a holding company has been to:

- Provide worldwide headquarters with top management and staff.
- Promote investor relations.
- Manage investments in subsidiaries.
- Finance subsidiaries with debenture bonds.

2 Effects on companies

There are between 400 and 600 companies whose shares or debenture bonds are quoted on Swiss stock exchanges. They represent less than one per cent of all Swiss corporations and the majority are holding or parent companies which are required to prepare and publish consolidated financial statements. Publication can involve distributing printed annual reports to existing or potential shareholders or printing the group accounts in the *Swiss Federal Commercial Gazette*. The latter method is, however, rarely used.

It is also estimated that there are about 12,000 privately held holding companies in Switzerland, all of which are too small to be required to prepare or publish group accounts. Approximately 1,500 only are subject to consolidation requirements due to their size. However, only public companies need to publish group accounts.

The requirements for group accounts have some features in common with the Seventh Directive. The law prescribes when a consolidation has to take place but does not strictly regulate the methods of consolidation and valuation principles to be used. These must be disclosed in the annexe to the accounts.

Certainly those groups that utilize US GAAP, IAS, and the Fourth and Seventh Directives or FER recommendations are unlikely to encounter problems in obtaining an unqualified audit report

Table I.1 Swiss listed companies that apply International Accounting Standards (IAS)

	since		*since*
ABB Asea Brown Boveri	1988	Eichhof	1993
Ares Serono	1988	Interdiscount	1993
Inspectorate International	1988	Merkur	1993
Nokia-Maillefer	1988	Siegfried	1993
Société Financière de Genève	1988	Rieter	1993
Spiro International	1988	Cementia	1993
Tecan Holding	1988	Danzas	1993
Brown Boveri	1989	Berner Kantonalbank	1993
Nestlé	1989	Calida	1993
Roche	1990	Esec	1993
Landis & Gyr	1990	IPU	1993
Fotolabo	1990	Phonak	1993
Immuno	1991	Zschokke	1993
Logitech	1991	Zellweger Luwa	1993
Ascom	1991	Elco-Looser	1993
Alusuisse-Lonza	1991	Elektrowatt	1993
Adia	1991	Intershop	1993
Bossard	1992	Keramik Laufen	1993
Dätwyler	1992	Surveillance (SGS)	1993
BSI (*Bank*)	1992	Richemont	1993
LEM	1992	Schweiter	1994
Mikron	1992	Tege	1994
Ciment Portland	1992	Südelektra	1994
Saurer	1992	Allgemeine Finanz	1994
Sika	1992	WMH Walter Meier	1994
Pargesa (*Investments*)	1992	Mövenpick	1994
Swiss Casinos	1992	Hesta	1994
Holderbank	1992	BB Biotech (*Investments*)	1994
Sulzer	1992	BB Industrie (*Investments*)	1994
Swisslog	1992	Big Star	1994
STRATEC	1992	Bossard	1994
Sandoz	1992	Clariant	1994
Forbo	1992	Gurit	1994
Holzstoff (Holvis)	1992	Hesta Tex (Schiesser)	1995
Ems-Chemie	1992	Adecco	1995
Sihl	1992	Bank Julius Baer (*Bank*)	1996
Oerlikon-Bührle	1992	Novartis	1996
Porst	1992	Kraftwerk Laufenburg (KWL)	1996
SRG	1993	SEZ	1996
Attisholz	1993	Interroll	1996
Ciba	1993	Ciba Specialty Chemicals	1996
Mettler-Toledo	1993	Selecta	1997
Jelmoli	1993	Swisscom	1997
Georg Fischer	1993	New Venturetec (*Investments*)	1998
Biber	1993	UBS (*Bank*)	1998
Basler Handels-Gesellschaft	1993		

Table I.2 Swiss companies that use FER (ARR) in their consolidated financial statements

	since		*since*
Prodega	1988	KW Brusio	1993 (and EU)
Walter Rentsch	1988	EG Laufenburg (EGL)	1993 (and EU)
Lindt & Sprüngli	1991 (and EU)	CKW (Kraftwerke)	1993
Sarna	1991 (and EU)	Warteck Invest	1993
Kaba	1991 (and EU)	KPMG Fides	1993
Industrieholding Cham	1991	VPB Bank	1994
Hilti	1992 (and EU)	Starrag	1994
Bossard	1992 (and IAS)	Orell Füssli	1994
ATEL	1992 (and EU)	Maag	1994
Globus	1992	Orior	1994
Pirelli	1992	Rentenanstalt	1994
SMH	1992 (and EU)	von Roll	1994 (and IAS)
Zürcher Ziegeleien	1992 (and EU)	Loeb	1994
Feldschlösschen	1992	Zürich Versicherungen	1994
Sibra	1992	LEM	1994
Hürlimann	1992	Leclanché	1994
Züblin	1992	Stuag	1994
Serum und Impfinstitut	1992	Golay Buchel	1994
Motor Columbus	1993 (and EU)	Clariant	1994 (and IAS)
Hügli	1993	Usego Hofer Curti	1994
Affichage	1993	Banque Cantonal de Genève	1994
Intersport	1993	Scana	1994
APG	1993	Accu Oerlikon	1994
Galactina	1993	AGIE	1994
Infranor	1993	Bâloise	1994
Arbonia-Forster	1993	Haldengut/Calanda	1994
Belimo	1993	Christ	1994
Bell	1993	Asselsa	1994
CKW	1993	Bobst	1994
Valora (ex Merkur)	1993	Bucher	1994
COS	1993	Zehnder	1994 (and EU)
Swissmetal UMS	1993	Agie	1994
Edipresse	1993	Canon (Schweiz)	1995
Schweizerhall	1993	Pargesa	1995
Galenica	1993 (and EU)	Schlatter	1995
Helvetia Patria	1993	Swiss Steel (von Moos)	1995
Huber + Suhner	1993	Straumann	1995
Sopracenerina (Motor Col)	1993 (and EU)	Also	1995
Publicitas	1993	Escor	1995
Vetropack	1993 (and EU)	Schulthess	1995
Perlen	1993	Generali (Schweiz)	1995
NZZ	1993	India Investments	1995
Netstal	1993 (and EU)	Nationalbank (SNB)	1996
National-Versicherung	1993	Schindler	1996
Metallwaren-Holding	1993	Crossair	1996
Winterthur Versicherungen	1993	Grasshopper	1996
LO Holding	1993	Adval Tech	1996
Schweizer Rück (Swiss Re)	1993 (and EU)	Alcopor	1997
Jungfraubahn	1993	Batigroup	1997
Kardex	1993		

from their (specially qualified) auditors, to the effect that their consolidated accounts are in agreement with the law and the disclosed accounting policies.

3 Legislation and standards

Under the provisions of the old Code of Obligations, effective until 30 June 1992, there were no requirements for consolidation (see Zenhäusern and Bertschinger, 1993, p. 28). Certain rules were established after lengthy discussions in parliament; however, there are very few articles of the Swiss Code of Obligations (*Schweizerisches Obligationenrecht,* OR) which relate to group accounts.

3.1 Duty to consolidate (OR Art. 663e, part 1)

(i) If the company has, by way of a majority of votes or other means, one or more companies under common control (a group) it must prepare consolidated accounts (group accounts).

(ii) The company is excluded from the duty to consolidate if, together with its subsidiaries, it does not exceed two of the following criteria in two subsequent years:

- Balance sheet total of CHF 10 million.
- Net sales of CHF 20 million.
- 200 employees per annual average.

(iii) Group accounts still have to be prepared, if:

- The company has debenture bonds outstanding.
- The company's shares are quoted on a stock exchange.
- Shareholders holding at least 10 per cent of the share capital request them.
- It is necessary to receive a picture which is as reliable as possible of the financial position and the results of operations of the company.

3.2 Intermediate holding companies (OR Art. 663f, part 2)

(i) Any company included in the group accounts of a parent company, which is established and audited according to Swiss law or equivalent foreign standards, is not required to prepare separate group accounts if it makes the group accounts of its parent known to its shareholders and creditors in the same way as its own individual accounts.

(ii) Such a company is, however, required to establish separate group accounts if it has to publish its individual accounts (because of debenture bonds outstanding and shares quoted on stock exchanges) or if group accounts are requested by shareholders who hold at least 10 per cent of the share capital.

3.3 Preparation of group accounts (OR Art. 663g, part 3)

(i) The group accounts have to follow accepted (Swiss) accounting principles.

(ii) The company describes the applied consolidation and valuation principles in the annexe to the group accounts. If it deviates from these, it has to disclose this and give additional information necessary for a clear presentation of the financial position and results of operations of the group.

3.4 Publication of group accounts (OR Art. 697h)

(i) After group accounts have been adopted by the general meeting of shareholders they, together with the group audit report, have to be published. Publication can either be in the Commercial Gazette or, if the following conditions apply, by giving a copy (free of charge) to anybody requesting it within a year after adoption if:

- The company has debenture bonds outstanding.
- The company's shares are quoted on a stock exchange.

(ii) The other companies have to show group accounts and the audit report to creditors who have a genuine interest in them. (Where there is a dispute, a judge decides.)

As the legal coverage of this complex topic is limited to the basic requirements, the accounting profession's published interpretations are needed to clarify the position. Such interpretations are included in the Swiss Accounting and Reporting Recommendations (FER) and the *Auditing Handbook*.

Prior to the introduction of the new OR, there was a lack of consolidation standards. As a result many multinational groups experimented with other sets of standards such as US GAAP, IAS, German Commercial Code and the Fourth and Seventh Directives. The majority of these standards are compatible with Swiss law.

II Objectives, concepts and general principles

1 Relevant objectives

Group accounts have long been prepared on a more or less voluntary basis; Swiss company law has only required consolidated accounts since 1 July 1994. Because there was no federal stock exchange law the cantonal stock exchanges in Zurich, Basle and Geneva did not require consolidated financial information. Since 1996 such information, and the adherence to the FER recommendations, has become compulsory under the new regulations of the Swiss Stock Exchange (SWX).

Previously, although not required by company law or stock exchange regulations such information was demanded by market forces. Bankers, accounting professionals, auditors, accountants, financial analysts, business journalists and others have pressured public companies to publish more meaningful information. Once such information was available demand grew for these reports to be audited by internationally experienced auditing firms.

Group accounts are not used for profit distribution: dividends can only be paid out of the current profits and available reserves shown in the holding company's individual accounts. Swiss annual reports always have to disclose the holding company's statutory accounts, as shareholders base their decision on the proposed dividends on these individual accounts. Special allocations have to be made to legal reserves which are not distributable (see Switzerland – Individual Accounts, V).

Financial analysts often calculate the dividends paid out by the parent compared to the consolidated net profit (pay-out ratio). Compared to Anglo-American groups these ratios tend to be low (e.g. 20 to 30 per cent) although they are increasing. Swiss companies only pay out the dividends once per year, normally one day after the general meeting of shareholders, and they are generally paid in cash (stock dividends are rare). The shareholders receive only 65 per cent of the dividend in cash. The 35 per cent withholding tax has to be deducted by the company and paid to the federal tax authorities. The shareholders can claim this tax back if they declare the dividend as income based on their tax return.

The group accounts are not relevant for tax purposes. Switzerland is one of the few industrialized countries where it is not possible to offset the profits of one subsidiary against the losses of another in the same country through a consolidated tax return. The only way to achieve these tax benefits is by merging two Swiss group companies with each other (combination) or with the parent (absorption) in a legal merger (fusion) and therefore establishing head office and branch offices/operations.

After consolidation became compulsory under Swiss law, on 1 July 1994, the tax authorities could access the group accounts of a parent company. Additional information, which would make it more difficult for the parent or the Swiss subsidiaries to hide profits from the tax authorities, could become available.

Historically, the contents of the group accounts in the published annual reports of public companies have been much more comprehensive than the individual parent company accounts. Although these group accounts have been voluntary they have normally included the following:

- Description of the accounting policies applied, such as methods of consolidation, foreign currency translation, valuation, etc. (Parent company statements often omitted valuation methods, or only stated compliance with legal requirements. These accounts were often conservatively stated for tax reasons.)
- Funds flow statements (often shown in group accounts).

- List of consolidated subsidiaries (only shown in group accounts).
- Notes on balance sheet and income statement positions (more detailed in group accounts). Under the old company law (up to 1992) the only additional balance sheet information was given in footnotes and consisted of fire insurance value of tangible fixed assets and contingent liabilities (e.g. from guarantees).
- Comparative financial information for the previous year, or for the last five or 10 years in an annexe. (The old legislation allowed the omission of the previous year figures for the parent.)
- Segmented sales by division and area (only shown in group accounts).

The duty to prepare, audit and publish group accounts under the new company law can be summarized as follows:

There is a duty to consolidate with majority of votes (OR Art. 663e). Exceptions are made for intermediate holdings and small companies. Therefore there are fewer than approximately 1,500 companies which must consolidate and have their group accounts audited by specially qualified auditors.

The rules concerning the preparation of group accounts (OR Art. 663g) are very limited. They have to be drawn up according to commercially accepted accounting principles in Switzerland (*Grundsätze ordnungsmäßiger Rechnungslegung*, GoR). However, those are not specified in detail. FER provides a possible set of acceptable standards. The annexe to the group accounts has to specify the consolidation and valuation principles used.

The disclosure requirements for audited group accounts (OR Art. 697h) are also very limited. The group accounts have to be published only if the company has shares or debenture bonds quoted. Fewer than 600 companies are involved.

Auditing of group accounts (OR Art. 731a) is reserved for qualified auditors (normally Swiss Certified Accountants or those with an equivalent foreign degree or diploma). The auditors have to confirm that the group accounts are in agreement with the law and that the consolidation and valuation principles, as disclosed in the annexe, are actually applied. The abbreviated report is directed to the general meeting of shareholders which has to adopt the group accounts in question.

Specific consolidation principles and methods are not mentioned in the law. However, FER provides minimal requirements.

FER was created in 1984 as a foundation under federal supervision. It consists of representatives from industry, banks, universities, politics etc. It has issued the recommendations shown in Table II.1 in the three Swiss languages – German, French and Italian – and additionally English, after a due consideration process. All these standards apply to group as well as individual accounts (see also Switzerland – Individual Accounts).

The purposes and objectives of the FER standard-setting body are set out in the policy statement as follows:

1.1 FER 0 'Objectives, subject matter, procedures of the FER recommendations'

FER 0 was originally issued in December 1985 and restated as of March 1993. The recommendations are as follows:

- All companies which adopt the FER (English translation: Accounting and Reporting Recommendations in Switzerland, short ARR) in their financial reporting are invited to state this in their financial statements. If this is the case, then the company should have their independent auditors examine and report the degree of compliance.
- All accounting and reporting recommendations should be applied to the individual and consolidated financial statements except where their use is expressly restricted by other FERs.
- The recommendations do not deal with publication of individual or consolidated financial statements. This is dealt with by company law.

1.2 FER 1 'Components of financial statements'

FER 1 describes the components of individual and consolidated financial statements. The recom-

Table II.1 FER recommendations (year of issue) and proposals

0	*Objectives, Subject Matter, Procedures of the Recommendations* (1985, revised 1993)
1	*Components of Individual Company Accounts and Consolidated Financial Statements* (1985, revised 1993)
2	*Consolidated Financial Statements* (1986, revised 1993)
3	*Generally Accepted Accounting Principles* (1990, revised 1993)
4	*The Translation of Financial Statements Expressed in Foreign Currencies for Consolidation Purposes* (1990, revised 1993)
5	*Valuation in the Consolidated Financial Statements* (1990)
6	*Funds Flow Statement* (1992)
7	*Presentation and Format of the Consolidated Balance Sheet and Income Statement* (1992)
8	*Notes to the Consolidated Financial Statements* (1992)
9	*Intangible Assets* (1995)
10	*Off-balance Sheet Transactions* (1997)
11	*Taxes in the Consolidated Financial Statements* (1995)
12	*Presentation of Interim Statements* (1995)
13	*Accounting for Leases by the Lessee* (1997)
14	*Consolidated Financial Statements of Insurance Companies* (1995)
15	*Related Party Transactions* (1997)
16	*Employee Benefit Obligations* (1998)
Under consideration are the following topics:	
17	*Inventories*
18	*Fixed Assets*
19	*Individual Company Accounts*
20	*Impairment of Fixed Assets*
21	*Accounting for Non-Profit Organizations*
22	*Construction Contracts* (PoC-Method)

mendation was issued in December 1985 and restated as of March 1993. Its recommendations are as follows:

- The individual and consolidated financial statements comprise a balance sheet, income statement (profit and loss account), funds flow statement and annexe.
- The funds flow statement shows the funds flow from operations in the individual and consolidated financial statements, investing/divesting activities and financing activities.
- The annexe discloses the accounting policies applied in the individual and consolidated financial statements, further details to other parts of the financial statements and additional information that is not included in the other parts of the financial statements.
- In addition to the current year amounts, the financial statements must show comparative figures for the previous year. If they are not comparable, this fact must be explained.
- The financial statements of a consolidated company can be simplified if consolidated financial statements are prepared and made

available to the users of the company's individual statements. In this case, the company's individual statements can omit the funds flow statement and the additional information contained in the annexe, except for what is legally required.

2 Underlying concepts and general principles

2.1 General requirements

Given that the Swiss Code of Obligations is very basic in respect of group accounts, such group accounts, according to the legal minimum, can omit a lot of the information normally required under international principles. However, group accounts published in Swiss annual reports tend to follow the international true and fair view principle. The management accounts, which top management uses to control the group internally, are often published.

Company law has no specific requirements as to consolidation methods. However, the private sector standard-setting body, FER, has issued standards on how consolidated statements should be prepared. FER 2 'Consolidated financial statements' (*Konzernrechnung*) was originally issued in September 1986 and restated as of March 1993. The general recommendations are that the consolidated financial statements have to present a true and fair view of the group's financial position, its results and its funds flows. The original German wording as used in the Seventh Directive is: '*ein den tatsächlichen Verhältnissen entsprechendes Bild der Vermögens-, Finanz- und Ertragslage*'. Consolidated financial statements are the financial statements of a group of companies as defined by the scope of consolidation (*Konsolidierungskreis*).

2.2 Principles of consolidation

Although the parent and subsidiaries are separate legal entities and Switzerland, unlike Germany, does not have a group company law (*Konzernrecht*), the group is considered a fictitious legal entity. Each consolidated company is initially responsible for its own accounting and finance. However, the group management and board also have a duty to control the subsidiaries. In fact, the parent has to ensure that the subsidiaries have sufficient financing available and that they meet their obligations in Switzerland and abroad. It is extremely rare for Swiss parent companies not to bail out their subsidiaries when they are in financial difficulties. It is also quite unusual for public companies not to be supported by the big Swiss banks and/or industrial groups when they are in financial difficulties.

Other important concepts prevalent in Swiss group accounts are going concern, substance over form, materiality, consistency, completeness, etc. These principles are particularly relevant to group accounts and are therefore expanded here (see also Switzerland – Individual Accounts, III).

Materiality is a very important principle in group accounts. Most Swiss multinational companies round their group account figures to the nearest CHF million. It is considered irrelevant to show thousands or even francs and centimes (*Rappen*) if a group shows a net turnover of CHF five to ten thousand million (five to ten billion) (Zenhäusern and Bertschinger, 1993, pp. 44–5). In addition, Swiss accounting firms use their materiality concepts to decide whether or not cumulative errors should be corrected or might result in qualifications to the group accounts.

The going concern principle is of special importance for a group. If a going concern is no longer maintainable then a consolidation is no longer justified. A holding company where the continuation is no longer assured should immediately cease to prepare consolidated accounts as this would give a misleading impression.

FER 3 'Fundamental accounting principles' (*Grundlagen und Grundsätze ordnungsmässiger Rechnungslegung*) also deals with general principles. It was originally issued in November 1990 and restated as of March 1993. The recommendations are as follows:

- The basic accounting principles for the individual and consolidated financial statements are:
  (i) The continuation of the company as a going concern.
  (ii) The materiality principle.
  (iii) The matching principle.

- In the preparation of individual and consolidated financial statements the following accounting concepts should be observed:
  - (i) Completeness.
  - (ii) Clarity.
  - (iii) Conservatism (prudence).
  - (iv) Consistency in presentation and valuation.
  - (v) Gross principle (prohibition to offset positions, that is, it is forbidden to net out assets, liabilities, expenses and income items. They must be shown on a gross basis).
- The principle of individual valuation applies to assets and liabilities. In the individual financial statements similar assets and liabilities can be valued as a group (portfolio valuation, i.e. netting over and under valuation within the specific group).
- In the individual and consolidated financial statements the previous year figures have to be disclosed next to the actual figures.
- Deviations in the consistent application of presentation and valuation in the individual financial statements have to be disclosed in the annexe. Alternatively, the previous year's figures can be adjusted (restatement).
- The applied accounting policies and valuation principles for the important balance sheet items in the individual and consolidated financial statements have to be disclosed in the annexe.

III Group concept and preparation of group accounts

1 Group accounts

1.1 Duty to consolidate

The duty to consolidate is stated in OR Art. 663e para. 1 as follows. 'If the company, by majority vote or by another method, joins one or more companies under a common control (group of companies), it has to prepare annual group accounts (consolidated financial statements)'. This means, in principle, that only corporations (*Aktiengesellschaft*) have to consolidate as the article applies only to corporations. Also affected are corporations with one or more unlimited partners as shareholders (*Kommanditaktiengesellschaften*). However, these are very rare in Switzerland. Associations (*Genossenschaft*) and small limited companies (*Genossenschaft mit beschränkter Haftung*) do not need to consolidate if they hold subsidiary companies. The latter are rare. However, there are some large retail organizations (e.g. Migros and COOP) which are associations. They, however, consolidate on a voluntary basis (see Zenhäusern and Bertschinger, 1993, pp. 48–9).

An exception to the duty to consolidate, for small groups, is stated in Art. 663e para. 2:

> The company is exempted from the duty to prepare consolidated financial statements if it, during two consecutive business years, together with its subsidiaries, does not exceed two of the following:
>
> 1. A balance sheet total of CHF 10 million.
> 2. Revenue of CHF 20 million.
> 3. An average annual number of employees of 200.

The above rule is reasonable to protect employees of big companies, although the employees might have difficulty in obtaining group accounts from their employers. Even in big private and public groups employees are not represented on the boards of their employers. The figures for the balance sheet total and revenue under OR Art. 663e para. 2 are calculated on an unconsolidated level (gross method).

The consolidated financial statements include the parent company, the fully consolidated subsidiary companies, joint ventures (which are often included according to the proportionate consolidation) and investments in associate companies (which are included under the equity method). The parent's individual accounts have to be established and audited even if there are group accounts.

Small groups have to prepare group accounts if the following exception (of the exception according to OR Art. 663e para. 3) applies:

1. The company has outstanding bond issues.
2. The company's shares are listed on a stock exchange.

3. Shareholders representing at least 10 per cent of the share capital request group accounts.
4. It is necessary for assessing as reliably as possible the company's financial and income situation.

Public companies' duty to consolidate seems feasible as there are many shareholders or creditors who need special protection. It is estimated that there are less than 600 companies which are quoted with their shares or bonds. Although this is a small percentage of Swiss corporations as a whole (some 170,000) these companies are economically important and they are big employers.

The Swiss Stock Exchange, as we have seen above, has requested consolidated accounts since 1995. Under the new company law group accounts have to be established, but only according to the minimal legal requirements. Group accounts which show a true and fair view are only required under the new quotation rules of the Swiss Stock Exchange. Although minority rights are not very well covered in Swiss law, big minorities can now request consolidated financial statements (OR Art. 663e para. 3 point 3).

Point 4 of OR Art. 663e para. 3 is very vague. Company auditors could probably, in critical cases, request their clients to prepare group accounts. If not, they might not be able to form an opinion on the valuation of the investments and loans to subsidiaries in the parent's balance sheet.

Under Swiss law (OR Art. 662) the components of group accounts are as follows:

- Consolidated balance sheet.
- Consolidated income statement.
- Annexe to the consolidated financial statements.

The cash flow statement and the statement of changes of fixed assets are only required for public companies which have to apply FER standards (FER 6 and FER 8; see also Zenhäusern and Bertschinger, 1993, pp. 88–90).

1.2 Preparation of group accounts

Current company law provides little guidance as to how the consolidation should be carried out and presented in the annual report. The only rules available are the general rules given in the commercially accepted accounting principles (GoR). According to OR Art. 662a para. 1, these require that:

> The financial statements (and the consolidated financial statements) have to follow the commercially accepted accounting principles in order to provide as reliable a picture as possible of the financial situation and results of the company. They also contain the previous year's figures.

The 'as reliable a picture as possible' required by para. 1 still allows reserves to remain hidden. This contradicts international accounting standards that require a true and fair view. Group accounts that are only in conformity with Swiss law might therefore not present the financial situation and the results fairly. Thus they might not be equivalent to foreign requirements.

The requested comparative figures are internationally recognized. The regulations are in line with legal principles but are too vague for accounting practice. The materiality principle is mentioned in law for the first time: large groups with large figures in their financial statements can now round their figures to the nearest CHF 100,000 or CHF million, if applicable. Unnecessary detail is therefore eliminated and the statements are easier to read.

According to OR Art. 663h the group accounts may omit data which could damage the interests of Switzerland and/or the company/group. The auditors have to be informed of the reasons. There is insufficient evidence available to interpret the specific meaning of this general article. It could, for example, relate to disclosure of names of subsidiaries and/or non-consolidated participations although these companies would be included in the group accounts.

The law does not distinguish whether the minimal structure of the financial statements and the valuation principles also apply for the group accounts. This was confirmed by practical interpretation.

2 Sub-group accounts

In principle every holding company (ultimate and intermediate parent) which is domiciled in Switzerland as a corporation has to prepare group accounts according to OR Art. 663e para. 1, unless it is exempted from consolidation.

Exception from the duty to consolidate is stated in OR Art. 663f for sub-holding (or intermediate) companies:

> A company included in the consolidated statements of a parent company that are prepared and audited according to the provisions of Swiss or equivalent foreign law need not prepare separate consolidated statements if it communicates the parent's consolidated statements in the same way as its own annual statement to its shareholders and creditors. It is, however, obliged to prepare separate consolidated statements if it is required to publish its annual report or if group accounts are requested by shareholders representing at least 10 per cent of the share capital.

The intermediate holding company's sub-consolidated group accounts have to be included in its parent's consolidation as well as its individual accounts. If the intermediate holding company has minority shares or bonds outstanding on a Swiss stock exchange it has to publish (sub-)consolidated financial statements.

The first question to ask is which foreign standards are equivalent? Possibilities include legal rules of a foreign country or professional standards issued by accountants' organizations. These could include American Generally Accepted Accounting Principles (US GAAP), International Accounting Standards (IASC), German Commercial Code (HGB) and European Directives. However, some additional disclosures required under Swiss law are also necessary (e.g. fire insurance value of property, plant and equipment). A few Swiss groups have been applying US GAAP, such as Carlo Gavazzi Holding, Adecco and Logitech. Those that apply German rules are a few German-dominated groups.

The intermediate holding company auditor will have to examine the foreign consolidated statements. According to OR Art. 663f, if no consolidation is prepared a comment should be made in the annexe to the financial statements of the intermediate holding company referring to these foreign statements and where they are available. The intermediate holding company auditor has to judge not only whether the upper or ultimate consolidation is equivalent to Swiss standards, but also whether the group auditors of such group accounts have an equivalent qualification to the specially qualified Swiss auditors (OR Arts. 727b and 731a). These are normally Swiss Certified Auditors (*dipl. Wirtschaftsprüfer,* previously: *dipl. Bücherexperte*). Obviously all qualified auditors in Europe and in Anglo-American countries would be equivalent (e.g. German *Wirtschaftsprüfer*, British Chartered Accountants (CAs), and United States Certified Public Accountants (US CPAs)). The majority of these certified auditors would be employed by big international accounting firms.

The question of whether this foreign consolidation must be available in one of the official Swiss languages (German, French or Italian) remains to be answered. The currency should normally be Swiss francs.

Foreign regulations (e.g. EU rules) concern the equivalency (mutual recognition) of financial statements. Germany, for example, renounces a duty to sub-consolidate intermediate companies based in Germany if consolidated financial statements, equivalent to those required in Germany according to the German Commercial Code (HGB), are presented.

An English annual report would probably be considered equivalent under Swiss law. However, it is not certain whether this would apply to reports in Japanese or Spanish, even though Spanish is an official EU language.

Minority shareholders of an ultimate parent cannot normally request sub-consolidated group accounts of an intermediate holding company. A special investigation (*Sonderprüfung*) has to be requested and agreed before court (OR Art. 697a) and the minority shareholders have to represent at least 10 per cent of the share capital or a minimum nominal amount of CHF 2 million (OR Art. 697b para. 1). In such cases a full consolidation of the intermediate holding company has to be prepared if this is ordered by the judge deciding the case. An equity consolidation would, however, not be sufficient at the ultimate parent's level. As this special investigation (*Sonderprüfung*) was only introduced in Swiss law as of 1 July 1992 few such cases have yet been decided by the courts.

3 Special problems

In some cases the parent company does not control any wholly- or majority-owned subsidiaries. Here it is not possible to prepare fully consolidated financial statements. If the parent holds investments in associated companies at 20 to 50 per cent or is participating in joint ventures it would normally account for these investments using the equity method. These accounts would still be called consolidated financial statements and be prepared using the true and fair view principle. The equity method is required by FER 2 for such investments, while the legal requirements of the OR do not address the issue. The parent has to prepare individual accounts in addition to these equity accounts.

However, it is rare for independent companies under common control to be grouped together on so-called combined financial statements. Such examples are parallel groups under common control, i.e. with the same shareholders or where a single person owns a number of individual companies managed on a unified basis (*Gleichordnungskonzern*). Swiss law requires only that a corporation consolidates majority- or wholly-owned subsidiaries which it controls directly or indirectly. Sister companies need not be combined. However, banks sometimes require an individual person to prepare combined financial statements to summarize all the interests a person or group of persons controls to assess their financial viability. It is, however, evident that transactions between such companies and persons need not be executed following arm's length principles. These related party transactions might be scrutinized by tax authorities or in special investigations (*Sonderprüfung*) triggered by dissident minority shareholders. Therefore, care should be taken to ensure that such transactions represent bona fide business dealings valued at fair market values, i.e. they should be valued under the same conditions as would apply to third parties. The above-mentioned construction is occasionally used to circumvent the duty of consolidation. This can be achieved if an individual person holds all the companies directly in his or her private portfolio.

In Switzerland only public companies are required to publish audited group accounts, although big private companies still have to prepare group accounts. These are also subject to group audits by specially qualified auditors. Various schemes were discussed in professional circles as to how the obligation to consolidate might be circumvented. One method (mentioned above) would be for a shareholder to hold the subsidiaries directly in his or her portfolio. Some groups have also paid back their debenture bonds or transferred their debenture bonds to subsidiaries to escape publication of the consolidated financial statements. However, there are few known examples of the use of the following practices:

- Delisting of shares and/or debenture bonds: a few companies with public bonds outstanding did not renew them when they matured and refinanced themselves with other resources, e.g. by private placements.
- Merger of subsidiaries into the parent: the merged companies, as a result, have accounts similar to group accounts. For (legal) merger accounting similar principles apply as in consolidations.
- In privately held groups it would be possible to sell all the subsidiaries to the shareholder, therefore avoiding a group structure and creating a flat structure. Every company is directly held by the shareholder and can still be commonly controlled. However, this is often not feasible because of the significant tax consequences which result from restructuring. As the private groups do not have to file their accounts with a public register or otherwise make them available to the public there is little motivation (e.g. tax considerations) for avoiding group accounts.
- Decreasing the voting stake under 50 per cent by selling shares.

IV Consolidated group

1 Definition of a subsidiary

The duty to consolidate is stated in OR Art. 663e para. 1 as follows: 'If the company, by majority vote, or by another method, joins one or more companies under a common control (group of companies), it has to prepare annual group accounts (consolidated financial statements)'.

This wording of OR Art. 663e para. 1 is in line with international standards. As in other countries the majority of voting power is meant to be a *de facto* control; control which is actually exercised, not merely available. Indirect control through other subsidiaries can also lead to this effect. For example, a parent owns two intermediate holding companies A and B at 60 and 50 per cent. The 60 per cent controlled subholding A has an interest of 40 per cent in a subsidiary C. The 50 per cent controlled associate owns 60 per cent of this same subsidiary C. It is assumed that C is controlled by the group and therefore can be fully consolidated with minority interests separated. However, it is very rare in practice for the voting control of loss-making subsidiaries to be transferred to third parties (trustees) to avoid consolidation.

Therefore, consolidation is required if a parent company disposes of more than 50 per cent of the voting shares of a subsidiary company. In Switzerland voting shares that have a smaller par value than the other shares are often used. They combine more voting power with less capital. Therefore, a majority of capital is not required. Normally, however, the voting rights are equal to the percentage of capital held. A company can also be controlled by means other than the majority of votes – for example, by way of contract, statutes, majority of board members and so on.

The provisions of the IAS are normally observed in this respect (IAS 27 para. 10):

> Control is presumed to exist when the parent owns, directly or indirectly through subsidiaries, more than half of the voting power of an enterprise unless, in exceptional circumstances, it can be clearly demonstrated that such ownership does not constitute control. Control also exists when the parent owns half or less of the voting power of an enterprise when there is:
>
> - Power over more than half of the voting rights by virtue of an agreement with other investors.
> - Power to govern the financial and operating policies of the enterprise under a statute or an agreement.
> - Power to appoint or remove the majority of the members of the board of directors or equivalent governing body.
> - Power to cast the majority of votes at meetings of the board of directors or equivalent governing body.
>
> (see International Accounting Standards 1998).

There is no empirical research available on the relationship between parents and subsidiaries in Swiss group accounts. However, in the majority of cases the capital percentage corresponds to voting control in parent/subsidiary relationships. Only public companies (ultimate parents) have complex capital structures (e.g. non-voting shares, different voting rights of shares due to differing nominal values, etc.).

In principle, foreign and Swiss subsidiaries take the form of capital companies with limited liability (AG and GmbH). However, subsidiaries with other legal forms, such as associations and partnerships, are also consolidated.

Pension funds organized as associations and foundations are always excluded from consolidation because they are assets separated from the employer, as prescribed by Swiss law. Excess funds are sometimes capitalized as prepaid pension contributions in the employer's consolidated financial statements according to FER 16, IAS 19 revised and FAS 87.

2 Exclusion of companies

2.1 Overview

Circumstances where the inclusion of subsidiaries is forbidden or optional are not covered in the law. In FER 2 para. 9 the following is given as an optional reason for exclusion: 'Companies which are not material to the consolidated statement can be excluded'. Many Swiss companies also follow the relevant IAS provision in this respect (IAS 27 para. 11). It states that a subsidiary is commonly excluded from consolidation when it is a temporary investment only, or is operating in countries with significant restrictions.

2.2 Temporary investment

Here control is intended to be temporary because the subsidiary is acquired and held exclusively with the view to its subsequent disposal in the near future.

2.3 Restrictions

The subsidiary operates under severe long-term restrictions which significantly impair its ability to transfer funds to the parent.

Other reasons for exclusions are non-homogenous operations and immateriality as follows.

2.4 Non-homogenous operations

Investments in industrial enterprises are normally not included in the group accounts of financial institutions (e.g. banks or insurance companies). This rule is not covered by company law, but is supported by Swiss professional standards: FER 2 para. 9 on group accounts states that 'Companies which might be excluded are those whose activities are so different from those of the consolidated group that their inclusion is misleading'. Although this procedure is in line with European legislation (Seventh Directive) it contradicts the Anglo-American view adopted in IAS 27 para. 12 and FAS 94 para. 13 which requires inclusion of all majority-owned subsidiaries regardless of the type of industry.

Increasing numbers of Swiss companies are following IAS standards and therefore include all subsidiaries even if they have non-homogenous activities. Banks tend to divest themselves from industrial holdings: sometimes in a turnaround banks involuntarily become shareholders of the companies they lent money, but these are not fully consolidated because the holding is of a temporary nature only (see temporary investment under (II.2.2)). The Seventh Directive and its interpretation requires a stringent test as to the exclusion of non-homogenous operations. If the group accounts do not give a true and fair view of the group unless such operations are included, they have to be fully consolidated in the group accounts. Please note that IAS and US GAAP do not allow the exclusion of non-homogenous operations.

2.5 Immaterial operations

Subsidiaries which are not material to the group accounts can be excluded from consolidation (see Zenhäusern and Bertschinger, 1993, pp. 71–2). Materiality is not defined in any Swiss law or standards. Based on foreign professional standards (e.g. Australian) it is assumed that approximately 2 per cent or more of group net sales would be considered material. Therefore, net sales of all immaterial non-consolidated subsidiaries have to be added to determine that combined together they would not exceed 2 per cent of net sales of the group. These would normally be small sales and marketing subsidiaries abroad. The sales they generate would already be included in the inter-company sales of the group's manufacturing or trading subsidiaries, and therefore the effect of non-inclusion is limited.

The test of whether such subsidiaries are immaterial or not has to be determined every year. As soon as large or unusual transactions are booked through such companies the decision of whether they can be excluded from consolidation has to be reconsidered.

V Uniformity of accounts included

1 Overview

It is required by FER 2 para. 3 that the financial statements of the group companies included in consolidation must 'comply with uniform group accounting policies'. This is further explained in FER 2 para. 6: 'If necessary, the individual statements of the consolidated companies have to be adapted to the uniform accounting policies for consolidation purposes'. These adjustments may lead to statements different from those submitted as statutory statements for approval by the shareholders. Swiss public companies use one or all of the following as a set of accounting policies:

- Swiss company law.
- FER.
- Fourth and Seventh EU Directives.
- German Commercial Code (HGB).
- International Accounting Standards of the IASC.
- United States Generally Accepted Accounting Principles (US GAAP).

It is understood that with little adjustment these standards are upwardly compatible. For example, if a group uses IAS standards it will normally comply with FER, FER will cover the minimal legal requirements, etc.

There are normally significant differences between the legal statutory accounts of the parent and subsidiaries and the accounts for consolidation purposes. The legal accounts are mainly tax-driven. They are normally called *Handelsbilanz* I (HB I). The adjustments to the true and fair accounts to be included in the consolidation (*Handelsbilanz* II or HB II) are discussed below (see 4; see also Switzerland – Individual Accounts).

2 Uniformity of formats

It is understood that the structure of balance sheet, income statement and annexe has to be the same for individual accounts as for group accounts (OR Arts. 663, 663a and 663b). The assets in the balance sheet have to start with liquid funds and end with fixed assets and the liabilities with short-term liabilities. The format of the consolidated income statement can be in either the account or report format, but has to show all the cost types as required by OR Art. 663b. The annexe to the group accounts has to disclose the 12 minimal types of disclosure information prescribed in OR Art. 663b.

Those companies which also have to apply FER standards (public companies) have to follow the detailed requirements of FER 7 and 8 which are more in line with international standards. However, they are less detailed than those required by the Fourth and Seventh Directives.

3 Balance-sheet date

In accordance with international standards uniform balance sheet dates are required for all consolidated companies. The *Auditing Handbook* confirms this view. It requires interim financial statements to be prepared for those companies which have other statutory year end dates. It allows inclusion of subsidiary companies with earlier year ends up to three months before the balance sheet date of the group accounts, if these companies combined are of an immaterial nature. Under internationally accepted consolidation standards, however, the parent company has to prepare (audited) interim financial statements as of the balance sheet date of the group accounts. This is sometimes the case when the holding company has a 31 March year end closing date and the operating subsidiaries 31 December year ends. Then it is preferable to prepare the group accounts as of 31 December as it is easier to do an interim closing date for the holding company.

This differing date for the holding company is sometimes chosen to facilitate the dividend flow from the subsidiaries to the parent and its shareholders. The current profits of the operating companies can be distributed as dividends to the parent before its closing date. This means that the dividend income in the parent's individual statements is available for distribution to the shareholders. However, the subsidiaries cannot pay dividends until the books are closed, the individual financial statements prepared and audited and the ordinary shareholders' meetings held in order to declare these dividends. As group accounts take longer to prepare than the parent (holding) company's individual statements, it is still possible to convene the shareholders' meeting early in the year, for example April. Under the new Swiss law, however, the annual report has to be made available at least 20 days before the date of the annual meeting (OR Art. 700) which makes careful planning of the preparation of group accounts crucial.

4 Recognition criteria and valuation rules

4.1 Overview

Swiss company law requires the statutory accounts of the parent and the subsidiaries to be consolidated. However, this adding up of the individual accounts, which are mainly tax-driven, is considered very unprofessional. This method of adding book values (*Buchwertkonsolidierung*) is therefore not in accordance with the true and fair view principle. As we have seen above, the adjustments to the legal accounts are normally necessary. The statutory accounts (HB I) and the adjustments to the true and fair accounts to be included in the consolidation (HB II) are usually as follows:

- The very favourable allowance for doubtful receivables granted by Swiss taxes is decreased.
- The stock relief (allowance on inventories of a third granted by Swiss tax law) is reversed.
- Depreciation expense is reduced from accelerated tax methods to the straight-line method over the useful lives of the assets.
- Tax provisions are reversed if they are not necessary.

These tax-related accounting principles are described in Switzerland – Individual Accounts. A result of this is that the values in the group accounts can substantially deviate from the tax-driven individual accounts.

The items listed above are considered temporary differences and therefore a full deferred tax provision will be deducted (see VIII.1). All the items will materially change the net income and shareholders' equity of a consolidated group (parent and subsidiaries). Due to a lack of guidance in company law, FER has issued standards on the valuation principles to be used in group accounts: FER 5 'Valuation in the consolidated financial statements' was issued in November 1990. The recommendations are as follows:

- Valuation directives applied in consolidated financial statements should ensure valuation uniformity and consistency.
- Valuation bases for consolidated financial statements are:
  (i) Historical cost (acquisition cost and cost to manufacture).
  (ii) Current cost (replacement cost or actual current cost).

  Valuations based on special laws applicable to certain financial statement positions are treated as exceptions.
- The valuation basis is considered uniform if it is used for the valuation of the financial statements of all group companies being consolidated or for the valuation of all financial statement positions in the consolidated financial statements.
- Deviations from the selected valuation basis are possible as long as they are objectively substantiated; these are to be disclosed in the footnotes. For the valuation of related financial statement positions, a uniform valuation basis is always to be applied.
- The valuation principles for each financial statement position should provide for the systematic determination of depreciation and valuation adjustments; these should correspond to the selected valuation basis.
- In the current and prior periods the same valuation principles for the financial statement positions are to be used and disclosed in the footnotes, especially the valuation principles for:
  (i) Inventories.
  (ii) Receivables.
  (iii) Fixed assets.
  (iv) Equity holdings and financial investments.
  (v) Intangible assets.
  (vi) Other financial statement positions which are material to the consolidated financial statements.

Very few Swiss groups use actual values (current cost) in their group accounts. Current cost would usually be restricted to real estate, which would be stated at prudent market values. Some groups (e.g. Georg Fischer) only use this approach for non-operating real estate. These assets are disclosed separately in the group balance sheet. They include land reserves, residential buildings and factory and office buildings which are no longer used for operations. The conservative value is regularly updated after valuations by independent experts. A deferred tax liability on the revalued amounts is shown as deferred tax provisions under liabilities (gross approach). The revaluation is directly credited to a revaluation reserve in consolidated shareholders' equity. It is normally only the accounts for consolidation purposes (HB II) that are revalued, not the individual accounts (see also Zenhäusern and Bertschinger, 1993, pp. 152–6).

In the following sections we will consider the most significant accounting and valuation policies in Swiss group accounts. The group accounts tend to give a true and fair view whereas the individual accounts are tax driven. Significant differences exist between the two (see Bertschinger, 1991c).

4.2 Income and expense recognition

Prudence, encouraged by the tax laws and the interests of the major investors, is the dominant principle. This is intended to ensure the continued prosperity of the company so that income may be smoothed and dividends distributed regularly. The conservative approach is sometimes also taken in the preparation of group accounts but not to excess because these accounts are not used for tax assessments (no consolidated tax returns).

Revenue and profits are not anticipated, but are sometimes recognized when realized in the form of cash or near-cash assets. Provision, however, is made for all known liabilities, whether the amount is known with certainty or is a best estimate in the light of available information.

4.3 Long-term contract work in progress

Although accounting practices permit companies to take credit for ascertainable profit while contracts are in progress, such profits are rarely taken. Losses, however, are recognized as soon as they become apparent (see Bertschinger, 1991b). The percentage of completion (POC) method is normally only applied in the group accounts.

4.4 Taxation

Taxes on profits of any one financial period are generally assessable to tax only in a later year. It is general practice for companies' individual accounts to provide only for the taxes assessable in the accounting period, but for group accounts to provide for full accrual of current taxes.

Full deferred taxation is normally provided for in the group accounts which is required by FER 11. The full provision method (comprehensive liabilitiy method) is most common (full actual tax rate of 20 to 30 per cent depending on the domicile in Switzerland). In the past sometimes the liability was restricted to 50 per cent of the maximum tax rate, i.e. discounted values (half of 30 per cent, i.e. 15 per cent; see also VIII).

4.5 Foreign exchange

In practice there is only one method of translating foreign currency financial statements used in the group accounts: the current rate method. The balance sheet is translated at year end rates and the income statement is translated at the average rate for the year. Resulting differences are taken to reserves (see V.5).

The same method is generally applied in the individual accounts for assets denominated in a foreign currency. Unrealized gains are frequently deferred rather than taken to income, but losses are always expensed.

4.6 Research and development expenditure

In general, research and development costs are directly charged to the profit and loss account as incurred. This principle is applied to both the individual and the group accounts. Occasionally such costs are capitalized and matched against revenues in future periods, but this is rare. Most large groups disclose the amount of their expenditure on research and development as a note to the group accounts.

4.7 Pension liabilities

Most Swiss companies make contributions to a separate retirement benefits fund in the legal form of a foundation, which is often reinsured with an insurance company. These funds are frequently overfunded and may themselves include hidden reserves. Contributions are charged to the profit and loss account of the employer as paid and are tax deductible. Where a company does not have a separate pension fund, it will pay the legally required premium (defined contribution) to an insurance company. Under German law it is possible to make provisions for pension liabilities in the employer's individual and group balance sheet. This would not be allowed under Swiss law, as the pension fund assets have to be separated into foundations under joint control of employer and employee representatives. These funds are also under the strict supervision of governmental authorities (Cantonal supervision).

Swiss groups are increasingly applying IAS 19 revised. This requires additional actuarial calculations to be performed for defined benefit plans. Resulting deficits or surpluses have to be integrated into the group accounts and amortized over the remaining average employment duration of the employees. Some Swiss groups have done such calculations under SFAS 87. A new FER 16 on pension obligations will be applicable from the year 2000 on. However, it is less stringent as the IAS 19 revised and SFAS 87 and SFAS 132.

4.8 Unusual items

There is no exact disclosure requirement for, or consistent treatment of, unusual items, such as gains from the disposal of significant fixed assets. They have to be disclosed under the new company law; however no distinction is made between unusual items and extraordinary items. Non-operating income and expense should, however, be shown separately, i.e. transactions falling outside the ordinary activities of the company. Prior year items are not usually disclosed, particularly if they relate to the normal business of the company. They are expensed in full in the current year.

4.9 Valuation basis

The valuation basis used is a strict version of historical cost accounting in the Swiss individual accounts. As a general rule, companies are required to value their assets at amounts not in excess of the lower of cost or market (current value). The law also gives discretionary powers to the board of directors to value assets at amounts lower than the maximum carrying value prescribed by law, i.e. to understate assets and overstate liabilities to create hidden reserves. This is also allowed in the group accounts, where it is even possible to create additional hidden reserves by way of consolidating journal entries under the Swiss Code of Obligations (OR).

Often, however, no hidden reserve accounting is practised in the group accounts. It is generally accepted that values higher than historical costs can be used. For example, some groups apply current cost accounting to real estate, particularly if it is of a non-operating nature (e.g. rental properties).

4.10 Tangible fixed assets

Tangible fixed assets are valued at historic cost and revaluation of fixed assets is allowed in rare circumstances in the individual accounts. Real estate and investments in other companies can be revalued if a special audit report shows the company to be in financial difficulties. No distributions may be made until the revaluation has been reversed or changed into capital. As we have seen above, FER 5 allows the revaluation of fixed assets in the group accounts. Deferred tax and the resulting timing differences have to be provided for. In most group accounts, however, historical cost accounting is used (benchmark treatment under IAS standards).

The fire insurance value of assets must be shown as a note in the annexe also in the consolidated financial statements. It is sometimes used to calculate the hidden reserves caused by excessive depreciation and/or inflationary trends. It may, however, be misleading, as the disclosure does not take into account the value of land, which does not require insurance coverage, or the age of the buildings.

Depreciation in group accounts is normally taken from acquisition cost using the straight-line method. There is no legal requirement that cost and depreciation should be shown separately in the balance sheet. However, FER 7 and 8 require such information for group accounts. FER 18 on tangible fixed assets will require a comprehensive gross statement of movements in property, plant and equipment from the beginning of the year to the end of the year.

4.11 Intangible assets

Intangible assets, such as trademarks and goodwill, may be shown at a value restricted to their cost less appropriate amortization. Purchased goodwill, if capitalized, must be written off over a reasonable period, usually between five and 20 years. The usual treatment of goodwill is to charge it to reserves in the year of acquisition. This is allowed also under FER 2 but not IAS or US GAAP.

Formation costs, such as legal fees and pre-incorporation costs, together with stamp duty paid on the issue of common stock, may be capitalized, but must be amortized over a period of five years or less. This is very rare in group accounts.

4.12 Leased assets

Leases are not widely used in Swiss companies. Leased assets are occasionally capitalized under FER 16 Leases, together with the corresponding liability, but there is no requirement to do so. Most lessees use the operating method, under which the lease is treated as a rental agreement and a charge, equal to the rental payment, is made to the income statement.

4.13 Stocks

Stocks must be shown at the lower of cost and the generally prevailing market price at the balance sheet date. In applying this principle, the comparison may be made between individual items or based on product groups or stock as a whole. However, FER 3 requires the individual recognition method for the group accounts.

There is no legal requirement to differentiate between raw materials, work in progress and finished goods in the financial statements. However, FER 7 and 8 require such disclosure in the annexe. Purchase or manufacturing cost can be determined by any recognized method, including:

- LIFO ('last in, first out').
- FIFO ('first in, first out').
- Weighted average cost or average method (the most widely used method).
- Selling price less estimated profit margin.

Substantial provisions are often made against stock in the individual accounts, the only limitation being established by tax laws which state that, where sufficiently detailed stock records are maintained to identify individual prices and quantities, the minimum acceptable value is two-thirds of the lower of cost and market value. However, in the group accounts these provisions are reversed to show a true and fair picture (FER 2). A deferred tax provision is taken on the revaluation.

4.14 Debtors

Irrecoverable debts are usually deducted from debtors, although an additional general allowance is taken from the net value. If the allowance for tax and statutory purposes is excessive it is partly reversed to give a true and fair view in the group accounts. A deferred tax provision is taken on the revaluation.

4.15 Investments

Investments are securities (quoted shares and bonds held for investment purposes). In Switzerland quoted securities can be valued at year end quoted prices (not allowed in Germany). Swiss tax authorities define year end prices as the average price of the month of December. However, companies can also state their securities at acquisition cost if it does not exceed quoted share prices.

4.16 Reserve accounting

By law, a legal reserve must be created, with at least 5 per cent of profit being allocated to it annually until it reaches 20 per cent of paid-up share capital. In addition, an allocation of 10 per cent of the value of any dividend must be made to the extent that the dividend exceeds 5 per cent of the paid-up share capital. This legal reserve is not distributable and may be used only to cover losses. The use of legal and revenue reserves is sometimes also governed by the company's articles of association. In group accounts the legal reserves are normally included in retained earnings. Therefore, it should be borne in mind that those are not fully distributable.

While profits and losses are recognized primarily through the profit and loss account, reserve accounting is occasionally used, and is disclosed in a note. Such transactions include, for example, direct write-off of investments and losses to the capital reserves (*agio*). Such treatment is not, however, in line with international accounting standards where such items have to be charged to the income statement.

4.17 Provisions

Excessive provisions are a common feature of Swiss individual accounts. Provision is made for all known liabilities and all diminutions in the value of assets. In a few cases these are also included in the group accounts. They may relate to product liability, tax, litigation, pollution, restructuring and other risks to the domestic and foreign subsidiaries.

4.18 Contingent liabilities

OR Art. 663b point 1 requires that details of contingent liabilities, guarantees and charges, for which no provision has been made, must be disclosed as a footnote to the balance sheet, although no details need be given. This is also required on a consolidated basis.

4.19 Discounted bonds and other innovative financial instruments

Under OR, the full repayment value of the bond must be treated as a liability. The difference between the issue price and the amount repayable may be treated as an asset, but must be amortized over the period to redemption. However, discounted bonds are rare, for tax reasons.

5 Translation of foreign currency

Swiss company law does not specify any specific method of translation of foreign currency financial statements into the Swiss franc group accounts. OR Art. 663g para. 2 only requires that the accounting policies must be disclosed in the annexe. Therefore professional guidance on this subject was needed. Some guidelines were issued by FER as a private standard setter: FER 4 'Foreign currency translation' was issued in February 1990 and restated as of March 1993.

The restatement related to the current and non-current rate method which has now been disallowed. The recommendations are as follows:

- For the purpose of consolidation, the individual local currency accounts must be translated into the currency in which the group accounts are expressed (the reporting currency). This translation can be achieved by means of one of the following methods:
  - (i) Closing/current rate method: translation of the balance sheet of foreign subsidiaries at the year end rate and the income statement at the annual average rate, translation differences are taken to equity.
  - (ii) Monetary/non-monetary method: tangible fixed assets and inventories are translated at historical rates, other balance sheet items at current rates, depreciation and material expenses are also translated at historical rates. The other income statement items are translated at the annual average rate and the translation differences are taken to the income statement.
  - (iii) Temporal method: nominal values are translated at current rates, all other values at historical rates. Depreciation and material expenses are also translated at historical rates. The other income statement items are translated at the annual average rate and the translation differences are taken to the income statement.

  A combination of these methods in the group accounts should be avoided. The only exception permitted is where the net investment approach is used.
- The chosen method of translation is to be disclosed in the notes to the accounts.
- The chosen method of translation should be in agreement with the principles of valuation applied in the group accounts. For example, if the valuation is done with actual values (inflation accounting) currency translation can only be done with the current rate method.
- The translation of local currency accounts into the reporting currency of the group will result in translation differences. The treatment of such translation differences has to be disclosed in the notes of the accounts.

The choice of methods above is theoretical in nature. More than 90 per cent of Swiss companies with foreign operations have adopted the Anglo-American current rate method of currency translation (see Meyer, 1993, p. 44). This is in line with IAS 21 'The effects of changes in foreign exchange rates', which is based on the US standard SFAS 52 'Foreign currency translation'. Those standards require that:

- Balance sheets are translated at current, i.e. at year end rates (*Stichtagsmethode*).
- Income statements are translated at the average exchange rate for the year.

- Translation differences are taken into a separate section of shareholders' equity called translation adjustment (*Umrechnungsdifferenz*) but do not influence profit as long as the subsidiary company belongs to the group.

The classification under the net investment approach is rarely applied as almost all subsidiaries are independent companies where the functional currency does not differ from the reporting currency. Few Swiss groups have subsidiaries in high inflation countries where a historic translation of fixed assets is normally applied.

Swiss groups publish a list of the year end and average rates that are used compared with the prior year in the annexe to the group accounts. Year end rates are normally those currency rates published by the major international banks (e.g. 31 December rates as published by the *Wall Street Journal* or *Financial Times* for the last trading day of the year) or those available on financial database services. The rates published by the Swiss tax authorities in the price/rate list (*Kursliste*) are not widely used for various reasons:

- The rates for the main currencies are published late in the second week of January.
- The rates for the more exotic currencies are not available until the end of January.
- The rate is the average exchange rate for the month of December (according to OR Art. 667) which avoids unrepresentative year end rates (ultimo rates). This average rate is not internationally accepted.

Average annual rates are derived by a number of methods. Database services provide annual rates on hourly, daily or monthly averages. Some more sophisticated companies use weighted averages based on monthly sales figures. With this method monthly sales are translated into Swiss francs at the average rate of the respective month. Then the Swiss franc amounts of all 12 months are added together. This new sales-weighted average rate is then applied to the whole income statement including net income (or loss). Of course, this necessitates the monthly availability of net sales figures.

According to the FER 4 requirements, the method of foreign currency translation under accounting policies and the balance and change of translation adjustment directly booked into equity both have to be disclosed in the annexe to the consolidated financial statements. These amounts are normally disclosed as part of the statement of shareholders' equity which is included in the annexe. It is usually presented in a matrix form with the elements of shareholders' equity shown horizontally and the changes to the equity positions shown vertically.

VI Full consolidation

1 Consolidation of capital

1.1 Overview

Swiss company law does not prescribe specific consolidation methods. However, the method applied has to be disclosed in the annexe to the group accounts (OR Art. 663g para. 2). It is understood that the group accounts are prepared under the full consolidation method using the purchase method. This is confirmed in FER 2 'Consolidated financial statements'. Para. 4 of FER 2 states: 'Consolidated financial statements must be prepared according to the method of full consolidation. Any exceptions must be explained and justified in the annexe'. When applying this method all assets and liabilities, as well as all income and expenses, are fully included (at 100 per cent) in the group accounts even if third parties hold a minority stake in the subsidiary. These minority interests have to be shown separately.

1.2 Purchase method, without minority interests

1.2.1 The German method

The so-called German method of capital consolidation has long been criticized in professional circles. This method was forbidden in Germany after the adoption of the Seventh EU Directive into German law in 1986. Unfortunately it is still used in some Swiss consolidations. Although its use is decreasing it is necessary to explain how it works.

The elimination of capital is done by calculating, at every balance sheet date, the book value of investments in consolidated companies of the parent, less the combined shareholders' equity of the subsidiary companies, resulting in a consolidation difference. This consolidation difference can be negative or positive. If the value is positive it would be a debit and could be qualified as an asset (goodwill). In Switzerland it was sometimes charged directly to reserves (netted with equity). If it is negative it can be a reserve or a provision (negative goodwill, also known as badwill). These items were normally not further analyzed and/or amortized in Switzerland. These positions were called positive consolidation difference (*aktivische Konsolidierungsdifferenz* or *Kapitalaufrechnungsdifferenz* or AKAD) or negative consolidation difference (*negative Konsolidierungsdifferenz* or *Konsolidierungsreserve*). The change in this reserve from one year to another could not normally be satisfactorily explained.

Very few Swiss groups use this German method today, although many did when they first published group accounts. It was often impossible to restate the acquisitions of the last 20 to 50 or even more years according to the purchase method, as the information was no longer available. This transitory method was also allowed and applied when the European countries had to adopt the Seventh Directive.

1.2.2 Purchase method according to FER

Swiss law does not require a specific full consolidation method. However, as public companies have to follow FER according to stock exchange regulations, the more professional purchase method is the only one allowed. It is required by FER 2 para. 5 that: 'The interests in the equity of consolidated subsidiaries must be consolidated using the Anglo-Saxon method.' Paras 15 to 18 explain the method in more detail:

The equity of consolidated subsidiaries has to be established as of the acquisition date using uniform group accounting principles. Therefore hidden reserves are normally uncovered. Assets and liabilities of the acquired company are stated at their fair value to the acquiring group (calculated separately for each major asset), which may include the release of the more significant hidden reserves, especially in real estate and inventories. A deferred tax provision is normally calculated on the revalued amounts using the full actual tax rate (usually about 30 per cent). Any difference between the aggregate fair values of the assets (and liabilities) and the consideration paid, represents positive, or negative, goodwill.

In practice, the revaluation is done by applying the specific uniform accounting methods of the acquiring group. Intangible assets are not normally revalued. If there are significant hidden reserves in real estate (especially non-operating) it might be revalued above the historic acquisition cost of the subsidiary. If the target company has to be restructured, often provisions are established and then used in subsequent periods. However, the revalued equity after deferred taxes cannot exceed the purchase price of the acquired company, as the historical acquisition cost principle is usually followed by Swiss groups.

This equity calculated at fair market value has to be eliminated against the purchase price at acquisition date. Therefore only the profit since the acquisition date can be included in the group profit (post-acquisition earnings). After the first consolidation the retained earnings of the subsidiary are included in group retained earnings. If the purchase price exceeds the equity value of the subsidiary restated at fair market value the difference is called goodwill.

The principal method of accounting for takeovers (legal mergers) and acquisitions is the purchase method. The pooling-of-interest method is rarely used, mainly because the complexity of the restatement and because of the unfavourable tax implications of some legal mergers (fusion) (see Bertschinger, 1991c). A new merger law which is discussed in parliament would alleviate the tax consequences of corporate restructurings in Switzerland (*Fusionsgesetz*, abbreviated to *FusG*).

Under the interpretation of the purchase method by a few public companies, the results of the acquired company are brought into group accounts from the beginning of the year in which the acquisition is made, and disposals eliminated from the beginning of the year in which the disposal is made. This may lead to a material distortion of trading performance (see also 1.2.9. and 1.2.10).

1.2.3 Treatment of goodwill

Accounting standards permit two possible treatments (see Bertschinger, 1991c):

- Immediate write-off of the goodwill balance against retained earnings (instead of a charge to the income statement).
- Carrying the goodwill balance as an asset in the balance sheet, which is written off over its useful life as an annual amortization expense through the profit and loss account.

The former treatment, although not permitted any longer under IAS 22 revised and US GAAP, is currently regarded in Switzerland as the preferred approach. Goodwill is occasionally eliminated against the share premium account (*agio*). Few companies adopt the latter treatment. The useful life chosen can vary from five to 20 years as practised by some groups (see Meyer, 1993, p. 43).

The accounting impact of the two treatments may be characterized as follows. Immediate write-off can cause a reduction in a group's net assets and may create doubts over the adequacy of its resources. Capitalization and amortization will maintain net assets, but will have the effect of reducing future reported earnings. A group is not precluded from using both methods in relation to different acquisitions, provided that the notes to the accounts adequately describe the situation.

It is important to bear in mind that the effect of the varying treatments of the goodwill in the consolidated accounts is to some extent more apparent than actual. The group's ability to distribute profits to its shareholders is dependent on the distributable profits of the holding company (parent). Since goodwill is a balance which arises on consolidation, its write-off in the consolidated accounts has no impact on the holding company's own reserves. From a tax point of view it is difficult to amortize goodwill in the investment account of the parent unless the subsidiary continues to show big losses.

Capitalized goodwill is rarely allocated to intangible assets (brands, licences, copyrights etc.) in Swiss group accounts (see Zenhäusern and Bertschinger, 1993, pp. 285–9). Some large Swiss groups have charged big parts of goodwill immediately to income (in the form of 'in process research and development costs'). However, the accounting profession and regulators are increasingly frowning at such practice.

Negative goodwill in acquisitions is relatively rare. If a loss-making operation is acquired at a low price or at a nominal amount (e.g. CHF 1) any resulting negative goodwill is used to either write off (overstated) assets or set up as a provision. Any subsequent restructuring costs are then charged to these provisions. In a very few cases the negative goodwill represents reserves, but this only occurs if the acquisition turns out be a lucky buy. These reserves are then sometimes classified as capital reserves.

If shares are issued by the parent company, the premium on issue or *agio* (the difference between the nominal value and the market value of the shares issued) must be credited to legal reserves in the individual accounts. These are non-distributable reserves against which costs (including goodwill) may be eliminated (OR Art. 671 para. 2.1). If this treatment is applied in the individual accounts of the parent company (holding company), it is normally also taken in the group accounts. The *agio* included in the legal reserves is included in an account called capital reserve (*Kapitalreserven*) in the group accounts.

1.2.4 Write-down of the participation in the individual accounts

Investments in subsidiary companies are sometimes written down in the individual accounts if the value is permanently impaired. If this is the case the write-down is normally reversed in the group accounts. As goodwill was normally eliminated directly against equity in the year of acquisition this write-down in the individual accounts does not affect the group income statement. If, however, goodwill has been capitalized and amortized over its estimated useful life an extraordinary write-off of the still-capitalized goodwill might be necessary in the group accounts. This unamortized balance of goodwill is likely to be different than that in the individual accounts (see also IAS 22 revised para. 47 which states: 'The unamortized balance of goodwill should be reviewed at each balance sheet date and, to the extent that it is no longer likely to be recovered from the expected future economic benefits, it should be recognized immediately as an expense').

1.2.5 Minority interests

The term subsidiary undertaking covers those companies in which the investing company has a majority shareholding, a majority of the voting power, or in which it controls the composition of the board of directors. There can also be third-party shareholders who have a stake in the subsidiary company. These are called minority or outside shareholders' interests (*Minderheitsanteile*).

Where subsidiary undertakings in which the holding company has less than a 100 per cent interest are consolidated into group accounts, it is normal to include 100 per cent of the subsidiary undertakings' assets and liabilities and post-acquisition profits, and to present the extent to which these are attributable to outside shareholders as a separate balance described as minority interests. The balance is almost always measured by reference to the fair value of the subsidiary undertakings' equity and then included in the group accounts as minority interests. This treatment is not prescribed by professional standards, US GAAP requires that minority interest must not be revalued.

The presentation of such minority interests in the consolidated balance sheet and income statement is not uniform in Switzerland. There are two main schools of thought in this respect:

- A German point of view claims that the board and management of the parent company control minority interests too by controlling the subsidiary. Therefore this results in the minority interests in the equity of subsidiaries being considered group equity and minority interests in the net income of subsidiaries being considered part of consolidated net income/loss (*Einheitstheorie*).
- An Anglo-American point of view claims that the shareholders of the parent have no claims on the minority interests in the subsidiary. Therefore this results in the opposite: minority interests in the equity of subsidiaries are not considered group equity and are excluded from equity, and minority interests in the net income of subsidiaries are not considered part of consolidated net income/loss but as an expense/income reducing group results (*Interessentheorie*). Minority losses are sometimes attributed to the group especially if the subsidiary needs restructuring and minorities are not willing to participate.

The Anglo-American way of calculating net income after deducting (increasing) minority interests as an expense (income) is related to the way earnings per share (EPS) figures are calculated. It is argued that the minority profits can never be distributed as dividends to the parent company's shareholders. They are paid as dividends to the minority shareholders. The German argument is that the parent can control the dividend payments to minorities, i.e. it can prevent these payments being made. Increasing numbers of companies are adopting the Anglo-American method because they are adopting IAS standards, although management normally prefer the first method because it creates normally more profit and equity.

Some companies do not take a position and show the minority in between equity and liabilities (in 'no man's land'). They show two consolidated profit figures before and after minorities. This confuses the users of financial statements. For investors it is evident that the Anglo-American method shows more relevant information. Shareholders of the parent have no access to the profit attributable to minorities. Minority dividends and liquidation proceeds will have to be paid to minority shareholders of those subsidiaries.

The minority treatment is addressed in FER 2 para. 4, where it is stated that consolidated financial statements have to be prepared according to the method of full consolidation. Paras. 13 and 14 expand: 'Under full consolidation, all assets and liabilities and all expenses and income of the subsidiaries, in which there is a minority interest, have to be included in the consolidated financial statements. Minority interests in the equity and income or loss of consolidated companies are to be disclosed separately'. The presentation and format of the consolidated balance sheet and income statement is described in FER 7, which requires that the minority interests are shown as a separate item between long-term liabilities and shareholders' equity in the balance sheet. However, it does not specifically state where the minority interests have to be shown in the income statement (for additional discussions on minorities refer to Zenhäusern and Bertschinger, 1993, p. 236).

1.2.6 Multi-level groups of companies

If minority stakes exist on various levels of groups, the question of how to calculate these interests in consolidation arises. Since there is an increasing trend to treat minority stakes as liabilities and expenses the problem can be automatically resolved by applying the Anglo-American method of stating minority interests. As most Swiss groups eliminate goodwill against equity in the year of acquisition there are no problems of goodwill allocation to minority shareholders of intermediate holding companies.

The subholding company usually prepares its own group accounts. The ultimate holding company takes these in at the date of acquisition. Therefore only the sub-consolidated profits from the date of acquisition are included in the group profit statements.

1.2.7 Own shares

1.2.7.1 *Own shares of the parent*

Under the new Swiss company law the company is now allowed to acquire its own shares if freely disposable equity (i.e. reserves) of the amount necessary for this purpose is available and if the total par value of these shares does not exceed 10 per cent of the share capital (OR Art. 659).

According to OR Art. 659a the company must, at the same time, create a separate reserve in the parent company's individual accounts for an amount corresponding to the acquisition value of its own shares held. This should ensure that no dividends can be paid out in the respective amounts. This also applies to the parent if subsidiaries acquire the shares of its parent.

In cases of disposal or cancellation of own shares, the reserve for own shares may be dissolved into non-restricted reserves to a maximum of their acquisition value (OR Art. 671a).

This has the following implications in the group accounts. The reserve for own shares is the same for the parent and the group. This reserve is directly shown in the consolidated equity (sometimes only in the equity of the individual accounts). It is a reserve in the amount of acquisition costs for all shares of the parent, held by both the parent and its majority-owned subsidiaries, so that the asset position and the reserve do not correspond. However, the asset side of the balance sheet includes in the parent's individual accounts, only those own shares held by the parent. In the group accounts it includes the own shares held by both the parent and subsidiaries. Normally the own shares are included under 'other financial assets', or if it is a public company under 'marketable securities'. If the share price falls below acquisition cost, the value has to be written down to market value. However the company is not allowed to revalue its own shares to market value if this is higher than acquisition cost (see Bertschinger, 1993c).

This treatment of capitalizing own shares was adapted from the EU Directives. It is not consistent with the Anglo-American treatment (US GAAP and IAS SIC-16), which requires that treasury stock (own shares) is openly deducted from shareholders' equity. Therefore, own shares reduce shareholders' equity and are not considered an asset. Consequently, dealings in own shares do not affect the income statement but only additional paid-in capital. In other words: 'You can make cash with own shares but not profit'. Under Swiss company law the sale of own shares will generate a (taxable) profit. Writing down own shares will reduce profits.

Shares of the parent held by non-consolidated companies (e.g. investments between 20 and 50 per cent) do not need to be disclosed. A reserve for own shares must only be made and disclosed for shares in subsidiaries which are at least owned with more than 50 per cent of the voting rights.

1.2.7.2 *Own shares of a subsidiary*

It is possible for a consolidated company other than the holding company to hold shares of another subsidiary. For example, a subsidiary B of subsidiary A holds 10 per cent of shares in its intermediate holding A. This is also called a cross participation. In theory the capital elimination is done by applying mathematical methods (solved problems in Zenhäusern and Bertschinger, 1993, pp. 267–73). In practice, however, these cases are very rare and treated differently. First of all, under the new Swiss company law such participations are discouraged. Cross participations are limited to 10 per cent of the share capital (OR Art. 659). Secondly, reserves for own shares have to be set up in the same amounts. Thirdly, a serious group management tries to untangle complex webs of holdings because they are difficult to control.

The most practical approach used in Swiss group accounts is to debit the amounts of investments of cross holdings directly to the group's equity. This method is also the most conservative, because goodwill is, in the same way, directly deducted from the group's equity. It also reflects the US GAAP and IAS SIC-16 method where own shares (treasury stock) are not capitalized as an asset but are deducted from the equity of the group. Similar eliminations take place in horizontal holdings where a sister company holds shares in another sister company.

1.2.8 Capital not fully paid in

In the individual accounts capital not paid in is treated as an asset. If such a company's financial statements are consolidated as a subsidiary, this amount is normally deducted from the capital and then the purchase method applied. If it is intended to call the money a receivable is set up. Accordingly, the parent has to set up a liability which is then eliminated in debt consolidation. If minorities are involved, the receivable can be taken into the group accounts and need not be eliminated if the minority shareholders are able to pay these amounts.

1.2.9 Acquisition of a subsidiary

The purchase method under international standards (e.g. IAS 22 revised 'Business combinations') requires that profits are only included in the group accounts as of date of acquisition. This is defined in IAS 22 revised para. 21 as the date on which control of the net assets and operations of the acquiree is effectively transferred to the acquirer. This is not always easy to determine as acquisition negotiations can often last a long time and the purchase date cannot always be determined exactly because of many contingencies. Such problems can include, but are not limited to, the following:

- A letter of intent is signed.
- Purchase contracts provide for retroactive assumption of risks and rewards mostly as of the beginning of the year.
- The acquisition and price are contingent on a due diligence review.
- The acquisition is contingent on a successful public tender offer.
- The target (company) is not able to provide an interim balance sheet as of purchase date which is compatible with the accounting policies of the acquirer. It might take more than a year to adjust the accounting system of the target so it can be consolidated.
- The purchase price is paid in instalments contingent on future events and/or earnings (earn-out agreement).

Swiss companies have on occasions included the whole net income of the year even if a target was acquired in the second half of the year. Such practices have been frowned on (see Bertschinger, 1994, p. 334 and Zenhäusern and Bertschinger, 1993, p. 81).

Although Swiss law does not address the above problems, FER 2 para. 5 requires the application of the Anglo-American purchase method. It is explicitly required by FER 2 para. 15 that this method should be applied at the date of acquisition or incorporation of a subsidiary. Therefore, the group is not allowed to include the results of the subsidiary retroactively as of 1 January of the year of acquisition or proactively as of 31 December of the year of acquisition. A reason for later inclusion (year end or following year) might be that it is not possible to generate the necessary figures for consolidation purposes (HB II) because the existing accounting systems do not allow for the quality or timeliness.

As many of the Swiss groups apply IAS Standards they also have to follow IAS 27 'Consolidated financial statements and accounting for investments in subsidiaries'. Para. 18 states explicitly in this respect:

> The results of operations of a subsidiary are included in the consolidated financial statements as from the date of acquisition, which is the date on which control of the acquired subsidiary is effectively transferred to the buyer in accordance with IAS 22.

The successive purchase of subsidiaries is not treated in FER. Although this is very rare in practice, Swiss groups tend to follow IAS 22 'Business combinations'. IAS 22 revised para. 35 describes the applicable method as follows:

An acquisition may involve more than one exchange transaction, as for example when it is achieved in stages by successive purchases on a stock exchange. When this occurs, each significant transaction is treated separately for the purpose of determining the fair values of identifiable assets and liabilities acquired and for determining the amount of any goodwill or negative goodwill on that transaction. This results in a step-by-step comparison of the cost of the individual investments with the acquirer's percentage interest in the fair values of the identifiable assets and liabilities acquired at each significant step.

(See also examples in Zenhäusern and Bertschinger, 1993, pp. 256–8.)

1.2.10 Sale of a subsidiary

Deconsolidations of loss-making subsidiaries, on the other hand, have been made by not including the subsidiary in the consolidation in the year of disposal. The respective losses are charged to equity. Sometimes it was argued that the acquirer had taken over the company retroactively as of the beginning of the year and that the loss was reflected in the modest sale price. Such practices are also not fully compatible with international standards (see Bertschinger, 1994).

As many of the Swiss groups apply IAS standards they also have to follow IAS 27 on 'Consolidated financial statements and accounting for investments in subsidiaries'. Para. 18 states explicitly in this respect:

> The results of operations of a subsidiary disposed of are included in the consolidated income statement until the date of disposal which is the date on which the parent ceases to have control of the subsidiary. The difference between the proceeds from the disposal of the subsidiary, and the carrying amount of its assets less liabilities as of the date of disposal, is recognized in the consolidated income statement as the profit or loss on the disposal of the subsidiary.

The statement above means that the profit or loss on sale of a subsidiary in the parent company should be reversed and calculated on a group level as follows:

Sale price for shares of the subsidiary
– net assets (equity) of the sold subsidiary at the date of disposal as included in consolidation (HB II)
– any unamortized capitalized goodwill relating to this subsidiary, if the group has capitalized goodwill in its group accounts
– any cumulative translation adjustment relating to a foreign subsidiary sold
= gain or loss on sale of subsidiary in the group income statement

(See also examples in Zenhäusern and Bertschinger, 1993, pp. 84–7 and 249–56).

1.3 Pooling-of-interests method

No specific legal standards, FER standards or interpretations in the *Auditing Handbook* exist for merger accounting (see Zenhäusern and Bertschinger, 1993, pp. 108, 192, 221–34).

The pooling-of-interests method (merger accounting, uniting of interests) is not often used. First of all, there are few acquisitions where shares are issued or own shares are used by the acquirer to pay the shareholders of the target. IAS SIC-9 makes pooling most difficult. Secondly, within Switzerland, these deals would be the first step to a legal merger. In fact legal merger accounting in Switzerland follows the rules which are prescribed by the pooling-of-interests method (refer for example to IAS 22 revised paras 61–7 for the techniques). In Switzerland there are two legal forms of mergers described as fusion: absorption and combination.

1.3.1 Absorption

OR Art. 748 describes the merger where one company (normally parent) absorbs the other (normally subsidiary). This is called the absorption. The assets and liabilities are taken over and assumed retroactively (to the start of the year) by the parent. This is a similar problem as in a consolidation. If the equity of the subsidiary exceeds the book investment then a merger gain results which is normally credited to reserves. If the investment exceeds the equity, goodwill results (merger loss) which is treated as goodwill in consolidation and then capitalized and amortized or charged to reserves.

The difference to consolidation under the purchase method for absorption is as follows:

- The merger is booked into the general ledger of the parent as of the merger date and therefore has tax implications.
- The amortization of capitalized goodwill is normally tax-deductible over five years.
- The income statement of the subsidiary is included as of the merger date. Mergers are often carried out retroactively as of 1 January of the merger year. Then the whole year is combined with the parent.
- There is normally no revaluation of assets and liabilities of the subsidiary. Sometimes real estate is revalued to avoid goodwill.

1.3.2 Combination

OR Art. 749 deals with the combination of two corporations into a new company. This new company takes over all the assets and assumes all the liabilities of the two old companies which are dissolved. As in the pooling method, no goodwill is created. The shareholders of the old companies hold the shares of the new company.

The difference between the legal combination and consolidation under the purchase method for the acquisitions is:

- The merger is booked into the general ledger of the new company as of the merger date and therefore has tax implications.
- The income statement of the companies is included as of the merger date. Mergers are often carried out retroactively as of 1 January of the merger year. Then the whole year is combined with the parent. There is normally no revaluation of assets and liabilities of the subsidiary. Sometimes real estate is revalued, which has tax implications. It would also be possible to revalue all the assets and liabilities under new accounting standards (new entity approach).
- The equity is normally the combined equity of the two companies at book value.

2 Consolidation of debt

It is required by FER 2 para. 3 that inter-company assets and liabilities in the individual financial statements be eliminated on consolidation. Accounts receivable and payable must be eliminated in the balance sheet. Although this sounds logical and simple, in practice it can create many problems. Therefore it is recommended that differences in inter-company accounts are reconciled long before the balance sheet date of the group. In international groups it is necessary to define procedures for reconciliation and solve differences in an accounting manual.

Practical problems which are encountered include the following:

- The receivable is in another currency: sometimes both receivable and payable currency are different from the group accounts' reporting currency (normally Swiss francs). The normal procedure is for the headquarters to send a list of currency rates to the subsidiaries to be applied to value inter-company items. These rates are usually the year-end rates which are used to translate foreign currency balance sheets. If the same rates are applied to value inter-company receivables and payables by the consolidated company, these items will match in the consolidation. Sometimes these rates are, for various reasons, not permitted under local tax or statutory law or professional standards, for example because unrealized currency gains have to be booked in the income statement. In such cases it might become necessary to have different treatments in the local statutory accounts (HB I) than in the accounts for consolidation purposes (HB II) which might also lead to deferred taxes on these temporary differences.
- For a number of reasons the inter-company items do not match: for example one company has shipped merchandise and therefore set up an account receivable, the other company has to set up accounts for inventories in transit and accounts payable in order that these inter-company accounts do match. The same applies to cash in transit if a payment has left one company but not reached the other. If the receiving company does not enter it into a transit account, a consolidation entry has to be entered at consolidation level to enable the matching of receivables and payables.
- Matching is impossible: the classification may

not be the same for accounts receivable and payable (trade) from deliveries of goods, short-term financial receivables and payables, long-term financial receivables and payables, prepaid expenses and accrued liabilities, etc. In these circumstances one of the companies has to adjust its reporting to enable a match to be made on a group level. This adjustment can also be made through consolidation entries.

- In restructuring situations the parent has to write off a loan to a subsidiary: the subsidiary has to include the debt at nominal value. In consolidation the written-off amount has to be reversed. Normally a deferred tax provision (see VIII) is taken if the write-off was tax deductible in the local accounts.
- Inter-company notes have been discounted with a bank: the debtor accounts for it as a note payable. The other party has to reclassify the respective amount from the bank account to notes payable. Back to back loans to and from banks are sometimes eliminated if the right of offset is possible under the financing arrangement.

(See also Zenhäusern and Bertschinger, 1993, pp. 289–95).

3 Elimination of intragroup profits

Swiss company law does not prescribe any specific consolidation method such as elimination of inter-company profits. However, it requires that the consolidation follows generally accepted accounting principles. These have to be disclosed in the annexe to the group accounts (OR Art. 663g paras. 1 and 2).

It is required by FER 2 para. 7 that interim profits from inter-company transactions must be eliminated and FER 2 paras. 21 and 22 further comment:

> Based on transactions between consolidated companies unrealized profits can be included in inventories or fixed assets from inter-company deliveries (interim profits). Those interim profits need not be eliminated if the group deliveries were made at market prices, or the elimination would result in significant costs.

In practice inter-company profits are usually eliminated from inter-company inventories still on hand within the group in the balance sheet. The consolidated companies report the inter-company inventories that they have in their books to headquarters, together with quantities and stated values (normally acquisition cost). Headquarters reduces these amounts by a margin that was supplied by the consolidated selling company. These margins are deducted from inventories. Sometimes a deferred tax asset is set up against this deduction (see VIII). The change of the net deduction compared to the previous year is charged (credited) to material expense in the income statement.

In Swiss practice very few inter-company profits on fixed assets deliveries are eliminated, because they are not normally material and these eliminations have long-term implications on depreciation expense, which has to be adjusted over the useful life of the asset. Most such transactions are executed at fair market value where an elimination is not required according to FER 2 para. 22. Additionally, such transactions are often not material to the group accounts.

If the elimination were required it could be easily circumvented. For example, real estate or inventories (e.g. commodities) are sold to a third party and then bought back at a higher price later, e.g. after the year end. These transactions are frowned on and should be disclosed especially if they are transacted with non-consolidated related parties (e.g. pension funds of the company). They also include sale and lease-back transactions, for example for administrative buildings. However, such transactions are often found in groups in financial difficulty and can be difficult to detect from the annual report.

4 Consolidation of intragroup revenues and expenses

It is required by FER 2 para. 3 that inter-company expenses and income in the individual financial statements must be eliminated in the consolidation. Internal deliveries of goods and services have to be eliminated, together with internal financial interests, royalties, fees and dividends.

Internal deliveries of goods are normally eliminated by charging inter-company sales to material expense. It is argued that if inter-company accounts receivable and payable match, then the income statement elimination will be correct. It is often extremely difficult to calculate inter-company material expense. If there are internal deliveries for half-finished products they have to be eliminated against changes of half-finished and finished products on a group level (see Zenhäusern and Bertschinger, 1993, pp. 162–4). For other inter-company expenses and income, such as interest and fees, a matching at individual company level is performed.

The elimination of inter-company dividends is normally carried out by charging the dividend income of the parent company directly to group reserves. An individual treatment of each dividend payment is often necessary. Care should be taken, for example, in dealing with withholding tax, non-recoverable withholding tax and other fees related to dividend payments, although this is not covered by any professional standards. Best practice in this respect is to enter the gross dividend in the individual accounts of the parent as soon as the dividends are declared by the subsidiary. The non-recoverable taxes and fees are directly entered into expenses. The recoverable part of the dividend is capitalized as receivable from tax authorities. In using this treatment the same dividend is entered as income as the amount that left the subsidiary's reserves when the dividend distribution was decided. If the non-recoverable part is not expensed, the statement of changes in equity shows a debit which represents an expense to the group but not to the subsidiary. Although this treatment is not required by FER or any professional recommendations it is widely used in practice (see Zenhäusern and Bertschinger, 1993, pp. 167–70).

VII Proportional consolidation and equity method

1 Overview

Accounting for related parties such as associates has not been consistent in the past. Until recently equity accounting was not used in Switzerland for accounting for investments in associated companies in group accounts. Such investments were shown at acquisition cost or written-down book values. However, now the results of companies, where the group has an interest of over 20 per cent of the share capital but does not exercise management control, are usually incorporated into the group accounts under the equity method of accounting (see Meyer, 1993, p. 40). This method presents the group's share of the related company's profit for the period in the profit and loss account. The equity method is not allowed in the individual accounts as these investments may not be stated at a value higher than acquisition cost or lower intrinsic value. The equity method, however, can show values higher than acquisition cost.

In the case of joint ventures, partial consolidation is often used. There is no exact definition of joint venture and methods applied are not used consistently.

2 Proportional consolidation

Swiss company law does not refer to proportionate or proportional consolidation. A controlled company has to be fully consolidated. However, certain business activities are performed together with partners, where one partner alone cannot dominate the venture and the partners can only act jointly. In that case FER 2 states in para. 25 that joint ventures (*Gemeinschaftsunternehmen*) can be accounted for by using quota consolidation. Accordingly, all assets, liabilities, equity, income and expense positions are included in the group accounts of each joint venture partner (investor) at the respective percentage of holding, e.g. 50 per cent. Loans to the joint venture would be eliminated only at 50 per cent, assuming that the other 50 per cent are loaned to the other half of the joint venture (which is considered a third party).

In Switzerland the legal form of partnership is very rarely used because of the unlimited liability of the partners. In certain industry sectors joint venture activities are performed through corporations, particularly in the energy sector. The majority of power plants are owned by more than one power company. There are normally two or more partner shareholders which form a corporation to build a power plant. They not only provide the initial share capital but also partner loans for construction. Those power plants are cost plus operations. The costs are charged to partners at year end with a mark-up. This mark-up ensures a modest profit to pay a dividend to the partners.

Almost all major power plants in Switzerland (hydro and nuclear) are accounted for by the power companies by the proportionate consolidation method. The method is also called quota consolidation (*Quotenkonsolidierung*). Other utility companies such as gas distribution have followed this practice.

Other than this the method is not very widely used (see Meyer, 1993, p. 39). Some chemical and pharmaceutical groups have research and development (R&D) affiliates which are included by using this method. Typical examples are jointly-owned R&D companies held 50:50.

This option of FER 2 para. 25 is in line with the provisions of the Seventh Directive and IAS 31 which also allows this method.

3 Equity method

Swiss company law does not refer to the equity method. In the individual financial statements the application is not permitted as long-term investments can only be valued at cost or lower intrinsic value. There were also professionals who argued that the equity method should not be used in group accounts under the new Swiss company law, as investments and earnings could be overstated. There have been some spectacular crashes in the last few years of groups built up by financial speculators which made extensive use of the equity method. Earnings were inflated but did not accrue to the parent in cash because the dividend payments in Switzerland are only a fraction of consolidated profits. The balance sheet also showed significant goodwill positions from equity investments which were only amortized over a long period.

However, most Swiss public companies do use the equity method for minority investments which are not generally material to group accounts as a whole. These groups make considerable use of international accounting standards. The equity method can also be used for joint venture investments as an alternative to the proportionate consolidation method (required by US GAAP). Temporary investments intended for resale or where severe restrictions limit the influence of the investor are accounted for at cost or lower intrinsic value.

It is required by FER 2 as restated in March 1993 (para. 6) that 'The equity and net income of significant non-consolidated investments must be accounted for using the equity method'. Paras. 19 and 20 explain their application as follows. Participations are equity investments in other companies with voting rights of at least 20 per cent. In addition there must be the possibility of exerting some influence on the company by way of a supervisory or executive board. Such a company does not need to be fully consolidated. However, in the group accounts it should be included at the respective share of the investee's equity and net income (loss) even if the equity value exceeds the acquisition cost of the participation.

For other equity investments which are not participations as defined above and in the law (OR Art. 665a also refers to 20 per cent of voting rights) the acquisition is included in the group accounts at cost. If there is a long-term value impairment the investment has to be written down both for the group accounts and the individual statements (OR Art. 665). A full write-down becomes necessary if the investments become worthless (e.g. in case of bankruptcy of the investee).

The Swiss practice in using the equity method is normally the following: goodwill on acquisition of an equity investment is directly charged to reserves in the year of acquisition. No debt elimination or intragroup profits are eliminated. If the shareholders' equity of the investment becomes negative due to accumulated losses, the investment is normally written off to zero. If a restructuring is planned by the investors, provisions might be necessary in the group accounts of the investor. Loans receivable from the investee might also need partial write-down or full write-off if the loan value is impaired, i.e. recoverability is uncertain or impossible.

Swiss groups frequently account for the equity investment only by establishing the value at year end without following the movements (net income or loss and dividends). The value of the investment is established based on the latest (audited) financial statements of the investee. The percentage of the stake held is to be disclosed in the annexe and applied to the stated equity at year end.

If the investment is a foreign participation its equity is translated at year end rates (current rate method). This amount is taken into the group balance sheet. The difference compared to the previous year is taken to the income statement as 'Net earnings (losses) from equity method'. Dividend income from the investment is included in 'Other investment income'. The effect of the translation is taken by this method to the income statement instead of to equity. The movements of shareholders' equity of the investment are not analyzed in detail. This simple method is acceptable if the investments in equity companies are not material. However, if an investment is material in relation to the group accounts of the investor, the changes should be analyzed in detail (see Meyer, 1993, p. 71 and Zenhäusern and Bertschinger, 1993, p. 331).

The equity method can be demonstrated by the following example – the acquisition of an investment of 30 per cent which includes a goodwill payment:

  Acquisition cost of investment for 30 per cent of investment
− goodwill paid (if directly charged to reserves)
= Equity at acquisition date 30 per cent
+ share of net income of investee since acquisition
− dividends received from investee
+ capital increase paid
± translation adjustment directly taken to reserves of the investor
= Equity value of the investment at end of the year

The equity method can, in some circumstances, be more conservative than stating the investment at acquisition cost. First, if the goodwill is separated and amortized the value of the investment will decrease. It will only increase through the share of retained earnings which are capitalized proportionally. Many Swiss groups eliminate goodwill directly against group equity in the year of acquisition of the investment. If the value of the investment is written down in the parent company this is normally reversed in the group accounts. If the value is impaired in the group accounts and the capitalized goodwill has to be written down, this would be charged to expense from the equity method.

The equity method will normally also be applied for subsidiaries which are controlled through more than 50 per cent of voting shares but which are not fully consolidated. This could relate to temporary investments where a sale is intended or for non-homogenous operations. It does not usually apply for immaterial subsidiaries. They are not included in the group accounts at the equity method because they are also immaterial for the consolidated balance sheet and income statement.

When an equity company is acquired there is normally no restatement of the fair value of the investment's assets and liabilities. The amount of hidden reserves is not usually known to the acquirer. Special problems arise if a minority investment in a group is taken (minority stake in a holding company). In this case the group accounts are used to calculate the equity stake.

VIII Deferred taxation

1 Recommendation for deferred taxation

Deferred taxation is not normally used in the individual accounts which are usually the same as the tax return. Therefore there are no temporary differences which would require deferred tax accounting.

Thus, deferred taxation normally relates only to group accounts. FER 5 'Valuation directives for consolidated financial statements' referred to the tax effect of accounting, although it was very vague in para. 9 where it says: 'Tax implications from revaluations should be taken into consideration'. Due to the lack of guidance Swiss companies followed IAS and US GAAP standards in this respect.

FER 11 on income taxes in the group accounts became effective in 1997. It describes the accrual of current and deferred income taxes. The accrual of current income taxes is derived from the requirements of calculating taxable net income. This includes the liability from assessed and non-assessed taxes, possible tax penalties and possible tax prepayments and tax credits.

Capital taxes (based on equity) and other fees have to be separated from income taxes. They are not related to the pre-tax profit, i.e. because they are not income taxes they are normally included in other operating expenses. The above-mentioned tax accruals are mostly carried out in the individual accounts. Sometimes, however, they have to be adjusted for the balance sheet for consolidation purposes (HB II).

2 Temporary and permanent differences

If the tax basis is different from the financial accounting valuation method, deferred taxes are accrued. The consolidated financial statements should give a true and fair view of the economic situation of the group but do not reflect the tax situation. This can lead to differences between the tax values and values used in the group accounts.

Permanent differences can occur if certain transactions are included only in either the tax return or the group accounts. Therefore permanent differences do not reverse over the course of time. Typical examples are items that are not allowed for the tax return but are included in the group accounts. In Switzerland these are very rare. Temporary differences are those that are included at the same amount in the tax return and group accounts but not in the same time period. Typical examples are differences arising from the application of different depreciation methods and useful lives. By accounting for deferred tax effects on temporary differences, a relationship between profit before income taxes and the related tax expense is established.

3 Liability method

The annual accrual of deferred taxes provides for a comprehensive allocation of all income taxes from a balance sheet point of view. This liability approach refers to the individual balance sheet positions. These have the character of future tax-deductible assets (debit temporary differences) or future taxable liabilities (credit temporary differences). Accordingly they are classified as assets or liabilities.

Deferred taxes are accounted for by using the full tax rate (future tax rate if known or else the actual rate). This is normally around 20 to 30 per cent in Switzerland. Changes in the tax rate or system are to be accounted for as deferred tax expenses or income of the year when they become known. All future income tax effects are included in the annual calculation of deferred taxes (comprehensive allocation method) independent from the time of reversal.

Accounting for deferred taxes should be governed by the principles of consistency and materiality. This means particularly that the method of calculation and presentation should be applied consistently. It should be explained in the annexe to the consolidated financial statements.

4 Tax jurisdiction

Income taxes should be accounted for separately in each accounting period for each tax jurisdiction. Debit and credit deferred tax items can only be netted in the same tax jurisdiction. Various levels can be distinguished when deferred taxes are calculated:

- Temporary differences in the individual financial statements for consolidation purposes arise because of differences between tax and group valuation. If a subconsolidation is prepared and a consolidated tax return prepared and filed, temporary differences are calculated at this subconsolidation level.
- Temporary differences arise from profit and loss consolidation entries (e.g. elimination of inter-company profits in inventories or matching of inter-company accounts receivable and payable).
- If profits are retained in the consolidated subsidiary companies or associated companies are accounted for under the equity method unless the retained earnings are not to be paid out in the foreseeable future.

If the temporary differences are calculated on an individual company or subconsolidation level existing accumulated tax losses and deferred tax assets can be applied to deferred tax liabilities. Deferred tax assets are only capitalizable if sufficient future profits can be generated. Therefore care should be taken when such assets are set up.

Deferred tax assets from the elimination of intercompany profits on intragroup inventory transfers are always realized when these profits are realized. Swiss groups normally do not extensively capitalize deferred tax assets as their value is often questionable.

Under the comprehensive allocation method all temporary differences must be included in the calculation of deferred taxes independent of the time of reversal. Exceptions are temporary differences arising from goodwill and as unremitted retained earnings which are not paid out.

5 Tax rate

The annual calculation of deferred taxes is based on actual tax rates. For the individual company or subconsolidated group the actual tax rate of the relevant tax jurisdiction applies. For the calculation of deferred tax on consolidation entries affecting profit and loss the respective average tax rate of the group applies. It is also possible to apply a group average tax rate which has to be disclosed in the annexe. For unremitted retained earnings of consolidated subsidiaries and associated companies under the equity method the non-recoverable part of the withholding tax (so-called *Sockelsteuern*) has to be accounted for. Discounting deferred income taxes is not allowed. Under a method used in the past in Switzerland half of the maximum tax rate of approximately 30 per cent, i.e. 15 per cent, was used.

If income tax rates have changed compared to the prior year the tax assets and liabilities have to be adjusted accordingly.

6 Disclosure

Deferred tax liabilities are shown separately under provisions. Deferred tax assets are shown separately under other assets. These items can be shown separately in the balance sheet or summarized with other positions (in which case the amount is disclosed in the annexe). The capitalization of deferred tax assets is allowed, if it is very likely that the future tax benefit can be used. Deferred taxes arising from revaluations of assets to current cost (inflation accounting) also have to be calculated and provided for.

The deferred tax expense (income) results from the change of the deferred tax assets and liabilities in the balance sheet. The deferred tax expense (income) is shown separately in the income statement or is disclosed in the annexe. A netting with current tax expense has to be disclosed if the amount is separately disclosed in the annexe.

The annexe to the consolidated financial statements has to disclose unused accumulated tax losses and the method for accounting for any deferred tax assets. The unused accumulated tax losses give an indication of possible tax savings in the future. If deferred tax assets are capitalized the calculation of this item has to be disclosed and commented on. Increasingly, Swiss companies disclose these accumulated tax losses or the tax effect thereon.

Swiss groups using IAS standards would normally comply with the above-mentioned provisions of FER 11. This standard is broadly in line with the provisions of IAS 12 which itself is broadly compatible with FAS 109 of the United States.

However, as we have seen above, Swiss groups do not normally capitalize deferred tax assets. Such capitalization is not considered prudent accounting by Swiss financial analysts and accountants. It is normally very difficult to predict whether a subsidiary will generate sufficient taxable profits in the near future to enable the group to amortize the deferred tax asset. This asset is considered to be a contingent asset whose value is uncertain. There is only one company which has capitalized deferred tax assets in the group accounts – Carlo Gavazzi – this being the only Swiss-based group which follows US GAAP. Some groups such as Nestlé have netted deferred tax assets against deferred tax liabilities. However, in the notes (annexe) the gross amounts are disclosed.

IX Formats

1 Overview

In general the group accounts have to adopt the same format as the individual accounts (see Switzerland – Individual Accounts). The relevant articles are OR Art. 663, 'Minimal contents of the income statement', OR Art. 663a, 'Minimal contents of the balance sheet' and OR Art. 663b, 'Minimal contents of the annexe'.

OR Art. 663g requires only that the group accounts have to be established in accordance with generally accepted accounting principles (*Grundsätze ordnungsmäßiger Rechnungslegung*, GoR). These are not further specified in the law. Additional requirements have therefore been defined in FER 7 and 8. Additionally, a format for the compulsory funds flow statement has been defined in FER 6, which is only required in the group accounts. Standards for the presentation of typical group positions such as goodwill, minority interests, non-consolidated investments, translation adjustments, etc. are included in FER 7 and 8.

2 Format according to Swiss company law

Every Swiss company preparing consolidated financial statements must use the minimal legal formats given by OR Art. 663, 663a and 663b. These also apply to group accounts. OR Art. 663a describes the minimal balance sheet contents and OR Art. 663 the minimal income statement contents for individual and group accounts (see Switzerland – Individual Accounts).

3 Formats for public companies

3.1 Introduction

Public companies, regardless of size, have to prepare and publish group accounts. Public companies are defined as those with quoted shares or debenture bonds outstanding. They are listed in the annual price list (*Kursliste*) published by the Swiss tax authorities. This list gives details of outstanding titles, security numbers (*Valoren-Nummer*), nominal values, interest rates for bonds and the year end valuation (average price for the month of December).

Public companies have to follow the standards issued by FER. Formats for the consolidated balance sheet and income statement (two formats allowed) are described in FER 7, the format of the required consolidated funds flow statement in FER 6 and the minimal contents of the annexe in FER 8. 'Presentation and format of the consolidated balance sheet and income statement' (FER 7) was issued in May 1992 and applies to the financial years 1996 onwards. The financial statements may be presented in the minimum format presented in the standard or another appropriate form. Most Swiss multinational companies comply with FER 7. This practice is more or less in line with the requirements of the Seventh Directive and the IAS.

3.2 Consolidated balance sheet

The minimal contents of the consolidated balance sheet are described in FER 7. Table IX.1 shows the format and descriptions, details and contents of the individual balance sheet items.

Table IX.1 Consolidated balance sheet (minimal contents)

Assets	
A	Current assets
	Cash and marketable securities
	Receivables
	Inventories
	Prepayments and accrued income
B	Non-current assets
	Fixed assets
	Investments
	Intangible assets
Liabilities and shareholders' equity	
C	Current liabilities
	Loans payable
	Trade and other payables
	Accrued liabilities and deferred income
D	Non-current (long-term) liabilities
	Loans payable
	Other liabilities
	Provisions
E	Minority interests
F	Shareholders' equity
	Share capital
	Capital reserves
	Reserve for own (treasury) shares
	Revaluation reserves
	Retained earnings

According to FER 7 the following items in brackets must be separately disclosed in the balance sheet or in the notes. Other material items are to be separately disclosed. General provisions against current assets, accumulated depreciation on fixed assets and amortization of intangible assets are to be separately disclosed either under the corresponding asset or in the notes.

Cash and marketable securities
 (Amount of marketable securities)
Receivables and liabilities
 (Trade-related accounts receivable and payable)
 (Amounts due from/to non-consolidated subsidiaries and other related parties)
Fixed assets
 (Land and buildings)
 (Machinery and equipment)
 (Other fixed assets)
Investments
 (Securities)
 (Non-consolidated investee companies)
 (Loans receivable from non-consolidated subsidiaries and other related parties)
Intangible assets
 (Goodwill, if capitalized)
 (Capitalized research and development costs, if capitalized)
Provisions
 (For taxes)
 (For pension costs)
Marketable securities/investments
 (Own shares (treasury stock) held)
Shareholders' equity
 (Amount of shares in each share category)
 (Net profit/loss)

3.3 Consolidated income statement

Minimal contents of public companies' consolidated income statements are described in FER 7. They may be prepared according to either of the following:

- The Continental-European period-based costing method (production-based income statement, in German *Gesamtkosten-* or *Produktionskostenverfahren*) presented in Table IX.2.
- The Anglo-American sales type (activity-based income statement, in German *Umsatzkostenverfahren*) presented in Table IX.3.

Table IX.2 Consolidated income statement (period-based costing method)

Net sales of goods and services
Changes in inventory of finished and unfinished goods as well as unbilled goods and services
Capitalized internal services
Other operating income
= Subtotal
Raw material expense
Personnel expense
Depreciation of fixed assets and amortization of intangible assets
Other operating expenses
= Subtotal
Net financing income/expense
Other income/expense
= Profit/loss before taxes
Taxes
= Profit/loss

The latter is permitted if the additional (legally required) information is disclosed in the annexe (see Table IX.4). Such additional information, according to OR Art. 663, includes material expense, personnel expense and depreciation expense (also disclosed in the funds statement under the indirect method of calculating funds from operations).

The following additional disclosures are necessary for the consolidated income statement under FER 7:

- Minority interests in the profit/loss are to be separately disclosed under both of the above methods.
- The following items must also be disclosed separately in the income statement or in the notes:
  (i) Financial income and expenses gross.
  (ii) Extraordinary income and expenses gross.
  (iii) Net result from investments in non-consolidated companies.
  (iv) Interest income and expenses from receivables and liabilities to non-consolidated subsidiaries and other related parties.
- The following items must be disclosed in the notes if the Anglo-American activity-based income statement presentation method is selected under FER 7:
  (i) Personnel expenses.
  (ii) Depreciation of fixed assets and amortization of intangible assets.

Table IX.3 Consolidated income statement (activity-based method)

Net sales of goods and services
Cost of goods sold or services performed
Administrative expenses
Selling expenses
Other operating income
Other operating expenses
= Subtotal
Net financing income/expense
Other income/expense
= Profit/loss before taxes
Taxes
= Profit/loss
The minority interests have to be shown separately.

Table IX.4 Consolidated income statement (minimal reporting format)

Unofficial English translation:	
Net sales	
−	Material usage
−	Personnel expense
−	Depreciation
+	Non-operating income
−	Non-operating expenses
+	Financial income
−	Financial expenses
+	Gain from sale of fixed assets
+	Extraordinary income
−	Extraordinary expenses
=	Consolidated annnual net profit (annual net loss)
The minority interests have to be shown separately.	

- Banks and insurance companies may present equivalent financial statements taking specific industry norms into account.

3.4 Consolidated funds flow statement

Swiss groups do not consider the funds flow statement as a note to the annexe, but as a separate component of the group financial statements. It is normally presented after the income statement and before the notes. Minimal contents of the compulsory funds flow statement are described in FER 6, 'Funds flow statements' (*Mittelflußrechnung*), issued in May 1992. The recommendations are as follows.

The funds flow statement presents the flow of funds from operations, investing activities and financing activities. The flow of funds generated from operations is to be presented separately from the other funds flows. Either the direct or the indirect method of presentation may be used.

The funds flow statement is to be prepared in accordance with GAAP. The composition of the fund selected is to be disclosed. Translation differences are either to be shown separately in the funds flow statement or disclosed in the annexe. A funds flow statement may be omitted from the individual company accounts if one is presented for the consolidated group.

A funds flow statement with liquid assets (cash flow statement) as funds under the indirect method could have the following format under FER 6:

Consolidated net income (loss)
Add (deduct) items not requiring cash:
Depreciation of tangible fixed assets
Amortization of intangible assets
Increase (decrease) in deferred tax provisions
Losses (income) from equity method
Loss (gain) on disposal of fixed assets

Add (deduct) changes in net current assets:
Decrease (increase) of inventories
Decrease (increase) of accounts receivable
Decrease (increase) in other current assets
Increase (decrease) in accounts payable
Increase (decrease) in other current liabilities
Cash from operations

Disposal (purchase) of fixed assets
Disposal (acquisition) of consolidated companies (net of cash acquired or disposed of)
Sale (purchase) of marketable securities
Cash from (for) investing activities

Capital increase
Increase (decrease) of short-term debt
Increase (decrease) of long-term debt
Dividends paid
Cash from (for) financing activities

Net cash inflow (outflow)
Effect of currency translation on cash
Opening cash balance
Closing cash balance

4 Statement of movements in fixed assets

Many Swiss groups voluntarily show the changes in all fixed assets on a gross basis as a separate component of the financial statements. Under FER 8 the only changes which have to be shown are those in property, plant and equipment. Under the Directives the changes in financial and intangible fixed assets also need to be disclosed, although not on a group basis. The components of fixed assets are normally shown vertically and the movements horizontally in a matrix form. The movements, after consolidation of inter-company items, usually include:

(i) Cost:
- Opening balance 1 January.
- Currency translation difference.
- Change in the scope of consolidation (net effect of companies acquired and disposed of during the year).
- Additions (investments in fixed assets).
- Disposals (retirement of cost of sold and scrapped fixed assets).
- Transfers (from construction in progress to buildings, machinery, etc.).
- Closing balance 31 December.

(ii) Accumulated depreciation:
- Opening balance 1 January.
- Currency translation difference.
- Change in the scope of consolidation (net effect of companies acquired and disposed of during the year).
- Additions (depreciation for the year).
- Disposals (retirement of accumulated depreciation of sold and scrapped fixed assets).
- Transfers (between asset categories).
- Closing balance 31 December.

(Bertschinger, 1993a.)

5 Statement of changes in equity

Shareholders' equity is required by FER 2 to be reconciled from the beginning to the end of the year. Para. 9 specifies that the elements of changes of shareholders' equity must be disclosed. In particular, the following have to be presented:

- Consolidated net income (loss).
- Distribution of dividends by parent.
- Changes of share capital of parent.
- Changes in scope of consolidation (companies included in consolidation), i.e. goodwill directly charged to equity.
- Influence of currency translation.

X Notes and additional statements

1 Annexe to the consolidated financial statements

OR Art. 663b describes the minimal contents of the annexe (see Switzerland – Individual Accounts).

2 Annexe to the consolidated financial statements for public companies

FER 2 para. 8 requires the following disclosure of accounting policies in the annexe: the consolidation policies must be disclosed in the annexe to the consolidated financial statements. In particular, the following must be disclosed:

- Criteria for inclusion in consolidation (scope of consolidation).
- Capital consolidation method (purchase method).
- Accounting for non-consolidated investments.
- Accounting for joint ventures.
- Foreign currency translation.
- Treatment of inter-company profits.
- Changes of shareholders' equity.
- Valuation principles.

'Notes to the consolidated financial statements', FER 8, was issued in May 1992. The recommendations are more detailed than FER 2 and state that the notes are an integral part of the consolidated financial statements. They provide additional comments on the balance sheet, income statement and funds flow statement, and have to disclose the following:

2.1 Accounting principles used in consolidation

These comprise in detail:

(i) Accounting principles used in consolidation.
- Valuation principles.
- General basis of valuation (e.g. historic cost or current cost).
- Valuation principles for balance sheet items (e.g. for inventories).
- Disclosure and effect of changes of valuation principles.

(ii) Consolidation principles.

(iii) Consolidation method, especially concerning the method of consolidating the shareholders' equity of subsidiaries.
- Method of foreign currency translation and treatment of foreign currency differences.
- Treatment of non-consolidated investments (e.g. equity method) as well as other joint ventures (e.g. proportionate consolidation).

(iv) Treatment of intragroup profits (intercompany profits).

(v) Information on the scope of consolidation (what is included and what is not).
- Treatment of companies included in the consolidated financial statements (method applied).
- Name and domicile of the consolidated companies (including investments in non-consolidated companies accounted for using the equity or other method).
- Percentage ownership of the share capital of these companies; if the percentage share of voting rights differs from the share capital, then both percentages must be disclosed.
- Changes in the scope of consolidation compared with the prior year as well as the date of applicability of the changes.

2.2 Other components of the consolidated financial statements

This comprises in detail:

(i) Explanations about other components of the consolidated balance sheet.
- Encumbered assets and the type of encumbrance.
- The change in the gross fixed asset category values and accumulated depreciation, preferably in chart form.
- Treatment of goodwill and other intangible assets.
- Material balance sheet information for non-consolidated investments in subsidiaries accounted for under the equity method, if the value of this subsidiary exceeds 20 per cent of consolidated shareholders' equity.
- Information on long-term liabilities including the type and form of security given.
- Changes in shareholders' equity.

(ii) Disclosure on the consolidated funds flow statement.
- Composition of fund selected.
- Effect of currency translation differences.

2.3 Additional information

Information not included in other parts of the consolidated financial statements comprises in detail:

(i) Further information which is not included in other parts of the consolidated financial statements.
- Evolution of gross fixed asset category values and accumulated depreciation.
- Net sales of goods and services by geographical market and business sector.
- Details of net finance income/expenses and other income/expenses.

(ii) Unusual open business transactions and related risks.
- Contract purchase and delivery obligations.
- Foreign exchange forward contracts.
- Open transactions from financial instruments.

(iii) Further disclosures include information about research and development as well as events subsequent to the balance sheet date. Information about research and development and on events subsequent to the balance sheet date can be omitted if it is already included in the annual report.

(iv) Further items which are mentioned in other FER recommendations have to be disclosed.

3 Interim financial statements

The supervisory body of the Swiss stock exchanges (*Zulassungsstelle*) has issued guidelines for detailing the requirements of interim financial statements in accordance with FER 12. Such interim information for public companies includes:

- Balance sheets and income statements for individual and group accounts for the six months after the balance sheet date.
- Comparative figures for the previous year.

The interim balance sheets must show at least the following summarized information:

- Current assets.
- Fixed assets.
- Liabilities.
- Shareholders' equity.

The interim income statements must show at least the following summarized information:

- Net sales.
- Extraordinary expense and income.
- Net income (loss).

4 Value-added statement

Many Swiss groups publish income statements in the format of a value-added statement (*Wertschöpfungsrechnung*) voluntarily, as supplementary information to the group financial statements. Normally the following format is used:

	Net sales	Paid to:
–	Material expense and other services provided	Suppliers of materials and services
=	Value added gross	
–	Depreciation	Suppliers of fixed assets
=	Value added net	

Distributed to the following parties:

Personnel expenses	to employees
Financial (interest) expenses	to creditors (banks, bond holders)
Taxes	to the government
Dividends	to the shareholders
Retained earnings	to the company

The distribution of the value-added net is often also calculated and presented by employee. This would demonstrate the average compensation per employee.

XI Auditing

1 Election

Auditing requirements are to be found mainly in the Swiss Code of Obligations (*Obligationenrecht*, OR) Arts. 727–731a, revised as of 1 July 1992. Some other legislation (e.g. banking law) also refers to audit requirements. Stock corporations alone are under a requirement to be audited. As they are by far the most important legal form of business, only their auditing requirements are considered here. There are no exemptions for small corporations: all corporations need statutory auditors.

According to OR Art. 727 the general meeting of shareholders must elect one or more auditors. It may designate substitute persons. At least one of the auditors must have his or her domicile, registered office, or a registered branch office in Switzerland. Either individual or audit firms may be appointed. Public companies with quoted shares or debentures outstanding usually have one of the Big Five international accounting firms as auditor.

The auditors have to confirm their acceptance of the mandate in writing to the general meeting because the fact that they have done so has to be filed with the commercial registrar. This makes it possible to find out who the auditors of a Swiss corporation are; a fact which would otherwise remain unknown because the accounts are not published. Swiss domicile is important in order that correspondence may be sent to an address within Switzerland. The same requirement applies to the board of directors.

2 Qualifications

Under OR Art. 727a group auditors must be qualified to fulfil their duty to the company to be audited. This means that they should have knowledge of accounting, auditing, business law, taxation, etc. However, under OR Art. 727b public companies and those which are of a certain size have to elect specially qualified auditors. The auditors must have special professional qualifications if:

- The company has bond issues outstanding.
- Its shares are listed on the stock exchange.
- Two of the following thresholds are exceeded in two consecutive years:
  (i) A balance sheet total of CHF 20 million.
  (ii) Revenues of CHF 40 million.
  (iii) An average annual number of employees, 200.

The Swiss Federal Council has defined the professional qualifications of the specially qualified auditors in a decree of 11 June 1992 as follows:

- Swiss Certified Public Accountant (*dipl. Wirtschaftsprüfer,* previously named: *dipl. Bücherexperte*) as registered by the Swiss Institute of Certified Public Accountants.
- Swiss Certified Trust Expert (*dipl. Treuhandexperte*) or Swiss Certified Tax Expert (*dipl. Treuhand-/Steuerexperte*) and Swiss Certified Bookkeeper (*dipl. Buchhalter/Controller*) with five years' practical experience in auditing under the supervision of a Swiss Certified Public Accountant.
- Graduates of business and/or law schools (BWL, VWL, *Ius-Lizenziat*/HWV) with 12 years' practical experience of auditing practice under the supervision of a Swiss Certified Public Accountant.
- Persons who have acquired an equivalent foreign diploma as certified public accountants with the requisite experience and knowledge of Swiss law. Such qualifications would normally embrace Chartered Accountants (UK, Canada, Australia, etc.), Certified Public Accountants (CPA of the United States), *Wirtschaftsprüfer* (Germany), *Expert Comptables Diplomés* (France), etc.
- Foreign certified public accountants who are qualified under the Eighth Directive and have the requisite acquaintance with Swiss law.

3 Independence requirements

Under OR Art. 727c group auditors must be independent of the board of directors and of any shareholder with a majority vote. In particular they must not be employees of the company to be audited, and they must not perform work for it that is incompatible with the auditing mandate. They must be independent of companies belonging to the same group of companies if a shareholder or creditor so requests. Swiss certified public accountants also have to observe the independence requirements of the institute as published in the *Auditing Handbook*. Partners and staff of the Big Five firms also have to meet the independence requirements of the firm concerned, all of which substantially surpass the Swiss standards (because of SEC and EU requirements).

Under OR Art. 727d commercial companies or co-operatives may also be appointed as group auditors. They would have to ensure that the personnel in charge of the assignments met the qualification and independence requirements. To date there have been few law suits to define the position under the new independence requirements. The only noticeable effect is that public companies are choosing to elect big international accounting firms rather than smaller firms or individual auditors.

4 Terms of mandate, resignation etc.

Under OR Art. 727e the maximum term of the mandate is three years. It ends at the general meeting of shareholders to which the last audit report is to be submitted. Re-election is possible. A group auditor who resigns must give the reason to the board of directors, which must communicate it to the next general meeting of shareholders. The general meeting of shareholders may, at any time, remove a group auditor. In addition, any shareholder or creditor may, by bringing a suit against the company, request the removal of a group auditor unqualified for the position, although this is very rare. The board of directors must notify the commercial register without delay of the termination of the mandate. If such notification is not lodged within 30 days, the outgoing group auditor may request the cancellation.

Under OR Art. 727f, should the commercial registrar learn that a company has failed to appoint group auditors, he or she must set a time limit for the situation to be remedied. If the company fails to observe the deadline a civil court judge must appoint the group auditors for one business year on the request of the commercial registrar. The judge selects the auditors at his or her discretion. Should these auditors resign, the commercial registrar will notify the judge accordingly. Provided there are valid reasons, the company may request the removal of group auditors appointed by the court.

5 Duties

Under OR Art. 731 the group auditors are to examine the consolidated financial statements to ensure they comply with the law and principles of consolidation. The following requirements are also applicable to group auditors.

The board of directors has to deliver all required documents to the group auditors and provide them with the necessary information on request, in writing. This includes a representation letter (OR Art. 728 para. 2). According to OR Art. 729b para. 1 the group auditors have to report non-compliance with the law or statutes in writing to the board of directors. In serious cases the group auditors have to report this to the general meeting of shareholders.

According to OR Art. 729c para. 1 the general meeting of shareholders can only adopt the consolidated financial statements if the group audit report has been submitted and the group auditor is present. If no group audit report is available the respective decisions of the general meeting of shareholders are not valid. If the group auditor is absent the decisions can be challenged by a shareholder. However, the group auditors can be absent, provided the shareholders make a unanimous decision. According to OR Art. 730 the group auditors have to keep the group's secrets and are not allowed to report confidential facts to individual shareholders or third parties.

According to OR Art. 729 the group auditors report the results of their audit in writing to the general meeting of shareholders. They recommend approval of the consolidated financial statements with or without qualifications, or rejection of the annual group accounts. The group audit report names the persons who managed the audit and confirms that the requirements concerning their qualifications and independence are fulfilled.

The Swiss Institute of Certified Public Accountants has published standard texts for unqualified group audit opinion for corporations in the following languages: German, French, Italian and English.

6 Standard form of audit report on the annual consolidated financial statements

The Swiss Institute has issued two versions of audit reports on consolidated financial statements.

In their group audit opinion group auditors normally refer to the set of standards applied. One or more of the following true and fair standards are referred to in addition to compliance with the Swiss law:

- FER recommendations.
- German Commercial Code (HGB).
- International Accounting Standards of the International Accounting Standards Committee (IASC).
- United States Generally Accepted Accounting Principles (US GAAP).

The first type of report is shown in Figure XI.1. The second type of report, shown in Figure XI.2, will be required for public companies under stock exchange requirements. Big groups have already anticipated this development and give true and fair view opinions generally in accordance with the IASC standards.

7 Report to the directors

OR Art. 729a requires that where the auditors of a group must be specially qualified within the meaning of OR Art. 731a they should report not only to the shareholders (the auditors' report), but also to the board of directors. The report to the board should comment on the way they have carried out their task and the examination. Where companies have a sophisticated system of accounting records and internal reports, the Swiss Institute of Certified Public Accountants

Figure XI.1 Standard form of unqualified audit opinion (in compliance with Swiss law)

Report of the group auditors to the general meeting of ABC Ltd., Zurich

As group auditors, we have audited the consolidated financial statements of ABC Ltd, Zurich for the year ended 31 December 1998.

These consolidated financial statements are the responsibility of the company's board of directors. Our responsibility is to express an opinion on these consolidated financial statements based on our audit. We confirm that we meet the Swiss legal requirements concerning professional qualifications and independence.

Our audit was conducted in accordance with auditing standards promulgated in Switzerland by the profession. Those standards require that we plan and perform the audit to obtain reasonable assurance about whether the financial statements are free of material misstatement. An audit includes examining on a test basis, evidence supporting the amounts and disclosures in the financial statements. An audit also includes assessing the accounting principles used and significant estimates made by management, as well as evaluating the overall financial presentation. We believe that our audit provides a reasonable basis for our opinion.

In our opinion, the consolidated financial statements are in accordance with the law and the principles of consolidation and valuation described in the notes to the consolidated financial statements.

We recommend that the consolidated financial statements submitted to you be approved.

Zurich, 27 February 1999 XYZ Audit Ltd.

A. Mueller *B. Huber*

Swiss Certified Accountants
Auditors in charge

Enclosures

Consolidated financial statements consisting of:

- Consolidated balance sheet.
- Consolidated profit and loss account.
- Notes to the consolidated financial statements.

Source: Auditing Handbook (1999).

proposes (Auditing Handbook, 1999) a standard form of report (Figure XI.3).

Group management and boards often request the above-mentioned report in a management letter format in which important issues as identified by local and group auditors are highlighted and discussed. Comments of financial management are normally included. These group management letters are discussed by the board's audit committees. Although Swiss companies are not required to have such committees, there is a strong trend towards their establishment, following Anglo-American examples.

8 Interpretation of legal requirements

Auditing of consolidated financial statements is extensively described in the *Auditing Handbook*. It has been reissued to take into consideration the new company law which became effective on 1 July 1992. An updated and expanded version of the handbook was published at the beginning of 1999.

XII Filing and publication

If the company has its shares quoted on a stock exchange or debenture bonds outstanding, OR Art. 697h states that it must publish consolidated financial statements after their approval by the general

Figure XI.2 Standard form of unqualified audit opinion (in compliance with Swiss law and true and fair view)

Report of the group auditors to the general meeting of ABC Ltd., Zurich

As group auditors, we have audited the consolidated financial statements of ABC Ltd, Zurich for the year ended 31 December 1998.

These consolidated financial statements are the responsibility of the company's board of directors. Our responsibility is to express an opinion on these consolidated financial statements based on our audit. We confirm that we meet the Swiss legal requirements concerning professional qualifications and independence.

Our audit was conducted in accordance with auditing standards promulgated in Switzerland by the profession and with the International Standards on Auditing issued by the International Federation of Accountants (IFAC). Those standards require that we plan and perform the audit to obtain reasonable assurance about whether the financial statements are free of material misstatement. An audit includes examining on a test basis, evidence supporting the amounts and disclosures in the financial statements. An audit also includes assessing the accounting principles used and significant estimates made by management, as well as evaluating the overall financial presentation. We believe that our audit provides a reasonable basis for our opinion.

In our opinion, the consolidated financial statements give a true and fair view of the consolidated financial position of ABC Ltd., Zurich, and subsidiaries as of 31 December 1998, and the consolidated results of its operations and their cash flows for the year then ended in accordance with International Accounting Standards of the International Accounting Standards Committee (IASC) and are in accordance with the provisions of the Swiss law.

We recommend that the consolidated financial statements submitted to you be approved.

27 February 1999 XYZ Audit Ltd.

A. Mueller *B. Huber*

Swiss Certified Accountants

Auditors in charge

Enclosures

Consolidated financial statements consisting of:

- Consolidated balance sheet.
- Consolidated profit and loss account.
- Consolidated statement of cash flows.
- Notes to the consolidated financial statements.

Source: Auditing Handbook (1999).

meeting of shareholders. It can either publish them in the Commercial Gazette or send copies to everybody that requests them. The report must be published 20 days before the general meeting of shareholders, which must take place less than six months after the balance sheet date. Virtually all businesses take the calendar year as their financial year; this means that annual reports have to be published before 10 June and the general meeting held before 30 June of the subsequent year.

All creditors (including employees) who have a valid claim against the company can claim access to the audited (consolidated) financial statements. To avoid disclosure, however, the company can settle the claim by paying the creditor.

As only about 600 companies have bonds issued and/or quoted shares, extremely limited information is available on other companies including 170,000 stock corporations whose accounts are not publicly available. This is at odds with the practice of most European countries, where all corporations must lodge their individual and group accounts in a commercial register to which everyone has access.

Figure XI.3 Recommended standard form of report to the directors (in compliance with OR Art. 729a)

Report of the group auditors to the board of directors of ABC Ltd., Zurich

As group auditors of your company we have examined the consolidated financial statements for the year ended 31 December 1998. We recommended without qualification their approval by the general meeting of shareholders.

In accordance with OR Art. 729a we additionally confirm that our examination encompassed such audit procedures as were considered appropriate in the circumstances. As agreed these audit procedures are not listed here but are documented in our working papers.

In addition, we have reviewed the internal accounting documentation supporting the consolidated financial statements and examined such assertions as are material to the trading situation and the results of your group. It is our opinion that no further comment is called for in consequence.

27 February 1999 XYZ Audit Ltd.

A. Mueller *B. Huber*

Swiss Certified Accountants
Auditors in charge

Source: Auditing Handbook (1999).

XIII Sanctions

Under OR Art. 699 the general meeting of shareholders must be convened by the board of directors. The ordinary meeting should take place annually within six months of the close of the financial year. Under OR Art. 700 para. 1 the general meeting of shareholders must be summoned in the form provided for by the articles of incorporation at least 20 calendar days before the date of the meeting.

As mentioned above, for companies whose financial year is the same as the calendar year the latest day for the general meeting is therefore 30 June following the calendar year end. The audit report and financial statements, have to reach the shareholders before 10 June. The auditors thus need to finish their work before 10 June if calendar year companies are to comply with the law.

The sanctions for not publishing group accounts are virtually the same as those which apply to individual accounts (see Switzerland – Individual Accounts). The duty to publish financial statements is confined to public companies and banks/insurance companies. In April 2000 a new rule of the Swiss Stock Exchange (SWX) became effective. The SWX will sanction listed companies and their auditors if the accounting and disclosure standards (normally FER or IAS) are not properly followed. Except for publicly listed companies as mentioned above, there are few sanctions upon a (privately held) company that fails to prepare proper financial statements. The normal procedure is as follows where a company neglects to keep proper books of account to produce consolidated financial statements.

If no financial statements are made available the auditors must remind the board, preferably in writing. When all measures have been exhausted, if the board has still not submitted the consolidated financial statements, the auditors will call a general meeting to inform the shareholders of the illegality. In one-person companies where there is only one shareholder and one board member at any one time, the shareholder may fail to attend the meeting. The group auditors will ask a public notary to attend as well, to witness that no shareholder has shown up, and the auditors will resign at this meeting. At the same time they will communicate their resignation to the board, the commercial registrar and possibly the courts where there is a risk of bankruptcy. The court will then decide what to do next. The commercial registrar can close the company down with a court order and put it into liquidation and bankruptcy, the court appointing a receiver and an auditor.

Similarly, the tax authorities will remind the company to submit its tax return. However, the audit report does not need to be filed. If the return fails to materialize the company is assessed on an estimated basis. If the assessed tax is not paid the tax authorities can have the company declared bankrupt and penalize the board members for failure to comply with the tax laws. Eventually the company will be struck off the commercial register, it will cease to exist and will be unable to transact any further business.

The main articles of the Swiss Penal Code *Strafgesetzbuch* (StGB) relating to accounting and bookkeeping are as follows:

- StGB Art. 148: Fraud.
- StGB Art. 152: False information on businesses.
- StGB Art. 163: Fraudulent bankruptcy.
- StGB Art. 166: Negligent bookkeeping.

There have been few court judgements in these areas, most of them relating to bankruptcy cases and tax fraud.

Bibliography

Bavishi, V. B. (ed.) (1993). *International Accounting and Auditing Trends*, two volumes, 3rd edition (Center for International Financial Analysis & Research (CIFAR), Princeton, NJ).

Behr, G. (1984). Accounting and Reporting in Switzerland. *Der Schweizer Treuhänder*, March, 10–12.

Bertschinger, P. (1985). Accounting in Switzerland. *World Accounting Report*, **10,** 14.

Bertschinger, P. (1988). Organisation des Konzernrechnungswesens. *Der Schweizer Treuhänder*, May, 162–5.

Bertschinger, P. (1989). Goodwill-Behandlung in der Konzernrechnung. *Der Schweizer Treuhänder*, September, 403–5.

Bertschinger, P. (1990a). Fast alles erlaubt – Wohin mit dem Goodwill in der Konzernrechnung. *FOCUS*, September, 408–10.

Bertschinger, P. (1990b). Praxis der schweizerischen Konzernrechnungslegung. *Die Betriebswirtschaft*, 199–203.

Bertschinger, P. (1991a). Konzernrechnung und Konzernprüfung nach neuem Aktienrecht. *Der Schweizer Treuhänder*, November, 564–72.

Bertschinger, P. (1991b). Percentage-of-Completion Methode für langfristige Fertigungsaufträge: Grundlage für eine zuverlässige Budgetierung. *Der Schweizer Treuhänder*, October, 472–8.

Bertschinger, P. (1991c). Switzerland – accounting aspects. *Handbook of Mergers and Acquisitions in Europe*, 3rd edition, 53–64 (Gee, London).

Bertschinger, P. (1992a). Bewertung von ganzen Konzernen. In Zünd A., Schultz G. and Glaus U. B. (eds.) *Bewertung, Prüfung und Beratung*, Festschrift für Carl Helbling, Schriftenreihe der Treuhand-Kammer, **107**, 77–90 (Treuhand-Kammer, Zurich).

Bertschinger, P. (1992b). Das neue schweizerische Aktienrecht. Auswirkungen auf Rechnungslegung und Prüfung im Einzel-und Konzernabschluss. *Die Wirtschaftsprüfung*, **45**, 65–73.

Bertschinger, P. (1992c). Rechnungslegung. In Roth, R. (ed.) *Das aktuelle schweizerische Aktienrecht*, Chapter 7 (loose-leaf) (WEKA, Zurich).

Bertschinger, P. (1992d). Rechnungslegung gemäß den International Accounting Standards (IASC). In Siegwart, H. (ed.), *Jahrbuch zum Finanz-und Rechnungswesen*, 11ff (WEKA, Zurich).

Bertschinger, P. (1992e). Revisionsstelle. In Roth, R. (ed.) *Das aktuelle schweizerische Aktienrecht*, Chapter 12 (loose-leaf) (WEKA, Zurich).

Bertschinger, P. (1993a). Anlagespiegel im Konzern – Praktische Probleme beim Ausweis des Anlagevermögens in der konsolidierten Rechnung. *Der Schweizer Treuhänder*, **12**, 757.

Bertschinger, P. (1993b). Der Anhang im Konzernabschluss nach neuem Aktienrecht. In Siegwart, H. (ed.), *Jahrbuch des Finanz-und Rechnungswesens*, 193 (WEKA, Zurich).

Bertschinger, P. (1993c). Eigene Aktien – Rechnungslegungsaspekte. In Roth, R. (ed.) *Das aktuelle schweizerische Aktienrecht*, Vol. 2, ch. 15/4 (loose-leaf) (WEKA, Zurich).

Bertschinger, P. (1994). Aussagefähigkeit der Konzernrechnungslegung. *Der Schweizer Treuhänder*, May, 10–12.

Bertschinger, P. and Schibli, F. (1987). Tendenzen im Konzernrechnungswesen. *Neue Zürcher Zeitung*, **30**, 7.

Bertschinger, P. and Sommerhalder, C. (1991). *EG 92: Einfluss auf die Rechnungslegung*, 2nd edition (Credit Suisse, Zurich).

Böckli, P. (1992). *Das neue Aktienrecht*, (Schulthess Polygraphischer Verlag, Zurich).

Boemle, M. (1993). *Der Jahresabschluss* (SKV-Verlag, Zurich).

Botschaft des Bundesrates Über die Revision des Aktienrechts (1983). Berne, 23 February.

Busse von Colbe, W. and Ordelheide, D. (1993). *Konzernabschlüsse*, 6th edition, (Betriebswirtschaftlicher Verlag Dr. Th. Gabler GmbH, Wiesbaden).

Canepa Ancillo (1994). Neueste Entwicklungen in der schweizerischen Rechnungslegung. *Praxis* **4/93**, 5–7 (ATAG Ernst & Young, Basle).

Choi, F. D. S. (ed.) (1991). *Handbook of International Accounting* (John Wiley & Sons, New York).

Choi, F. D. S. and Mueller, G. G. (1978). *International Accounting* (Prentice Hall, Englewood Cliffs, New Jersey).

Coopers & Lybrand (ed.) (1991). *International Accounting Summaries – A Guide for Interpretation and Comparison* (Coopers & Lybrand, New York).

Coopers & Lybrand (ed.) (1992). *Switzerland – A Guide for Businessmen and Investors* (Coopers & Lybrand, Zurich).

Dessemontet, F. and Ansay, T. (eds.) (1981). *Introduction to Swiss Law* (Kluwer, Deventer).

Ernst & Young (ed.) (1991). *Doing Business in Switzerland* (Ernst & Young, Zurich and New York).

Fachempfehlungen zur Rechnungslegung (FER), available from FER, P.O. Box 892, CH-8025 Zurich.

Fox, S. and Rueschoff, N. G. (1986). *Principles of International Accounting* (Austin Press, Austin, Texas).

Helbling, C. (1989). *Bilanz-und Erfolgsanalyse*, 7th edition (Swiss Institute of Certified Auditors, Berne).

International Accounting Standards (1998) (International Accounting Standards Committee, London).

Käfer, K. (1987). *Kontenrahmen für Gewerbe-, und Industrie-und Handelsbetriebe*, 10th edition (Paul Haupt, Berne).

Käfer, K. (1991). *Berner Kommentar*, **8**, part 2, two volumes, 3rd edition (Stämpfli, Berne).

KPMG Fides Peat (ed.) (1989). *Investment in Switzerland and Liechtenstein*, 2nd edition (KMPG Fides Peat, Zurich).

KPMG Fides Peat (ed.) (1990). *Banking in Switzerland*, 3rd edition (KMPG Fides Peat, Zurich).

Meyer, C. (1993). Konzernrechnung – Theorie und Praxis des konsolidierten Abschlusses. *Schriftenreihe der Treuhand-Kammer*, **122** (Treuhand-Kammer, Zurich).

Orsini, L., McAllister, J. and Parikh, R. (1992). *World Accounting*, **3**, *Switzerland* (Bender, New York).

Price Waterhouse (ed.) (1982). *Doing Business in Switzerland* (Price Waterhouse, Zurich).

Schultz, G. (1988). Die Konzernrechnung gemäss FER – Einflüsse des neuen Aktienrechts und der Siebten EG-Richtlinie. *Der Schweizer Treuhänder*, May, 166.

Schweizerische Zulassungsstelle (1993). *Jahresbericht* 1992, Zurich.

Swiss American Chamber of Commerce (ed.) (1988). *Doing Business and Living in Zurich*, Bibliography, 9th edition (Swiss American Chamber of Commerce, Zurich).

Swiss American Chamber of Commerce (ed.) (1992). *Swiss Code of Obligations*, II, *Company Law Articles 552 to 964* (English Translation of the Official Text) (Swiss American Chamber of Commerce, Zurich).

Touche Ross International (ed.) (1978). *Business Study. Switzerland/Liechtenstein* (Touche Ross, New York).

Treuhand-Kammer (1999). *Schweizer Handbuch der Wirtschaftsprüfung* (HWP), Zurich 1999 (Auditing Handbook), (Treuhand-Kammer, Zurich).

Zenhäusern, M. and Bertschinger, P. (1993). *Konzernrechnungslegung*, (SKV-Verlag, Zurich).

Zünd, A. (1992). Switzerland. In David, A. and Archer, S. (eds.) *The European Accounting Guide*, 825–84 (Academic Press, London).

UNITED KINGDOM
INDIVIDUAL ACCOUNTS

Terry E. Cooke
M. Choudhury
R. S. Olusequn Wallace

CONTENTS

I Introduction and background

1 Financial reporting in the United Kingdom

This contribution looks at the state of corporate financial reporting in the United Kingdom and the deliberate efforts that have been made and are being made to improve the level of reporting. Corporate financial reporting is an abstract concept which means a variety of things for a variety of users. In a competitive market, such as the business environment of the United Kingdom, customers, suppliers, employees, managers, investors, lenders and government (including tax authorities) need constant streams of information to assist them in their respective roles. In a regulated market, the regulator that serves as a surrogate for competition requires similar flows of information so as to regulate the market and stimulate customers, suppliers, employees, managers, investors and lenders to respond in that market in a way that is not dissimilar from a competitive market. Financial reporting is one constant stream of information available to these different constituencies.

From the accounting standard setters' point of view financial reporting means the provision of transparent and useful information to users of annual reports and accounts about the performance, financial position and efficiency of the reporting company. This information is then used by the accounting standard setting body and the monitoring and compliance body in checking the effectiveness of extant rules on financial reporting and the extent and nature of departure from such rules; by investors in determining their exposure to risks, and the amount and funding of future rewards; by suppliers and customers in checking whether the reporting company is a going concern; by employees in making judgements about the future of their employment by the company and the relationship between their share of the company's added value and those that are due to the shareholders; and by competitors in setting benchmarks for determining their relative strengths and weaknesses in the industry.

Financial reporting is a process that serves every nook and cranny of the business and industrial sectors, the individual company, public sector and regulatory bodies and the trade unions. The information that is disclosed in the annual report and accounts is expected to represent a true and fair view of the company's state of affairs. The effectiveness of the process lies in the integrity of the practical application of the true and fair view and the ability of different parties to understand and use that information. The concept of 'true and fair view' is explained below (see III.5).

2 Development of accounting law

2.1 General remarks

The history of United Kingdom financial reporting that is described is a distillation of more comprehensive analyses that can be found in books devoted to the subject by Brown (1905), Lee and Parker (1979) and Edwards (1990). The development of financial reporting in the United Kingdom started when economic power began to shift from Italy to England in the early fifteenth century. For the next two centuries a mercantilist school of thought prevailed in England in which central government tried to influence and exercise some control over business activity. In the seventeenth century official attitudes moved in favour of allowing business to develop itself rather than requiring it to respond to the direction of the state. This laissez-faire philosophy promoted the growth and expansion of business.

As business expanded, the need to raise large sums of money for ventures led to the formation of unlimited liability companies such as the East India Company (1600) and the Hudson's Bay Company (1670), as well as the formation of the Bank of England in 1694. However, the formation of companies was associated with a number of frauds involving speculative ventures and company flotations so the government hastily introduced the Bubble Act in 1720. The Act prohibited the use of corporations unless specifically formed either by Royal Charter or by Act of Parliament. In practice, the Act failed to limit the formation of companies even though creative methods had to be adopted. Some associations, structured as large partnerships, were in substance a form of company with transferable shares.

By 1825 when the Act was repealed, what had developed were companies that had been formed by deeds of settlement as well as companies formed by Royal Charters and Acts of Parliament. The latter two types of company remain today, but the deed of settlement company was found to be somewhat cumbersome. This type of company was effectively an unincorporated entity formed with a number of shareholders and a trustee(s) in which the shareholders agreed to comply with the terms of the deed. However, rather than permit the further development of these enterprises, which had ambiguous legal status, the government experimented with the chartered company whose arrangements were authorized by the Crown through the Chartered Companies Act 1837. This form of company did not become popular and in 1844 the government passed the Joint Stock Companies Act.

2.2 The Joint Stock Companies Acts 1844 and 1856

The 1844 Act allowed companies to be formed by registration and thereby hold property and enter into legal proceedings and from 1855 limited liability of a company's members was permitted. The 1856 Joint Stock Companies Act superseded the previous year's Act and adopted a liberal approach. Developments in England were mirrored in Scotland through its own laws.

2.3 The Companies Acts 1862 up to 1948

The Companies Act 1862 consolidated a number of Acts including the Joint Stock Companies Act 1856. During the later part of the nineteenth century a number of scandals occurred involving misleading documents including prospectuses and, as a result, amendments were made culminating in a new consolidated Companies Act in 1908. The Act introduced the concept of the private company and improved on requirements concerning the publication of financial statements, the registration of charges and mortgages and prospectuses. Further amendments were made in 1929 and of particular importance was the requirement for holding companies to show how the profits and losses of subsidiaries were dealt with in the accounts of the company.

The majority of the recommendations incorporated into the Cohen Committee Report of 1945 were introduced in the 1948 Consolidation Act. Accountability to the public was the foundation of the Act with new measures on principles of accountancy, the profit and loss account, balance sheet and group accounts. In addition, auditor independence from directors was strengthened and auditors with professional status were required as well as protection for minorities, and the provision to remove directors before the end of their period of office. The 1947 Companies Act (which was re-enacted in 1948) introduced the important concept that the financial statements should show 'a true and fair view' (see III.5) rather than a 'true and correct' view as was previously required.

2.4 Amendments to the Companies Act 1948

A review of the working of the 1948 Act was completed by the Jenkins Committee in 1962 (see the Jenkins Committee Report 1962) which acknowledged the effectiveness of the legislation. As a consequence only minor changes were made. (The historical development of accounting law up to 1967 is presented in Table I.1.)

The United Kingdom joined the European Union on 1 January 1973 and became subject to the terms of the European Communities Act 1972. This Act requires that directives, drafted by the Commission and ratified by the Council of the EU, should be incorporated by member states into their own legal systems within 18 months and brought into effect within 30 months. This has not always been achieved by all member states but the United Kingdom has a relatively good record compared with other members of the Union. The effect of European harmonization will be dealt with in the respective legislation covered below.

Further amendments to the 1948 Act were

Table I.1 The development of financial reporting requirements by company law

1600–1718	Formation of companies by Royal Charters. Popular companies formed during this period include (a) the East India Company; (b) the Levant Company; (c) The Hudson's Bay Company; (d) the Royal African Company; (e) the South Sea Company (of the 'bubble' fame); and (f) the Bank of England. The main objections to formation by charter were that it was very costly, the members were not liable for the company's debt and no enterprise, however promising, could be sure of obtaining a charter. As a result, very large partnerships were formed, which could be described as 'common law companies'. The associations of this kind multiplied during this period and a mania of speculation set in.
1719–1825	To offset the mania of speculation arising from the freedom to float partnerships, the Bubble Companies Act 1720, which limited any form of association to six members, was passed. The Act was not very effective and it restricted the growth of companies in Great Britain, so it was repealed by the Bubble Companies Act 1825 (see Lee, 1979, p. 16) which restored the common law position. The Bubble Companies Act 1825 also introduced the principle of limited liability in the case of chartered companies.
1837	The Chartered Companies Act 1837 empowered the Crown to grant letters of patent to a body of persons associated together for trading purposes. These bodies had to be formed by a deed of partnership of association, or agreement in writing of that nature, by which the company was divided into a specified number of shares. The deed of agreement had to set forth the date of its commencement, its objects, its principal place of carrying on its business and names of two officers of the company to sue or be sued on its behalf. The company was not, however, a body corporate.
1825–44	Various Acts of Parliament were passed relating to the formation of several railway companies. Early Acts required each railway company to maintain accounting records but no provisions were made for shareholders to inspect such accounts nor to be given copies of the financial statements. However, the Great Western Railway Act 1835 required half-yearly financial statements to be laid before the shareholders in general meeting.
1844–54	The Joint Stock Companies Act 1844 introduced the concept of incorporation of companies by registration (as distinct from the formation by an Act of Parliament). The Act did not provide for incorporation by limited liability. Each company was required to maintain adequate accounting records and to present a balance sheet (no indication was given of the content of the balance sheet), which should be sent to each shareholder, to each ordinary meeting. The Act made no provision for the publication of a profit statement but required the audit of the accounts and the balance sheet.
1855	The Joint Stock Companies Act 1855 introduced the concept of limited liability. This Act was necessary to curtail the tendency to over-extend lending and credit facilities to companies.
1856–61	The Joint Stock Companies Act 1856 replaced the 1844 and 1855 Acts and removed the requirements that accounts be maintained, a balance sheet be published and that both be audited. This Act included model articles of association recommended for adoption by persons forming themselves into a company. The model articles of association included provision that proper accounting records would be kept, that a profit and loss statement and balance sheet would be presented to each annual general meeting and an audit of these statements would be carried out. Incorporation and limited liability of banks made available in 1857/58.
1862–1907	The consolidation of Companies Acts 1844, 1845, 1855, 1856, 1857/58. Compulsory annual audit for banks introduced in 1879. Auditors required to report whether audited balance sheet is 'full and fair'. Annual audit made obligatory for all registered companies in 1900. As the facility for public incorporation was available to any business association, there developed a type of limited company described as the 'one-man company' because the majority of the shares were held by one person, the remaining six shares being held singly by other members (the minimum number of members required under the Act was seven and a member was to hold no less than one share).

Table I.1 Continued

1907–28	Companies Act 1907 was promulgated. The Act distinguished, for the first time, between public and private companies, requiring the former to file an annual balance sheet giving a summary of the company's capital, liabilities and assets and giving such particulars as would disclose the general nature of such liabilities and assets and how the values of the fixed assets had been arrived at. The Act had exempted private companies from including balance in the annual return filed with the Registrar, imposed certain conditions as to membership and restriction on transfer and forbade public subscription to its capital. One of the consequences which followed legal recognition of private companies was the development of trading by means of subsidiary companies. Under this system, if the subsidiary were a private company, the law imposed no obligation on the parent company to give its shareholders information as to the financial position or trading results of the subsidiary (private) company, while if the subsidiary were a public company, the published accounts did not have to be combined with those of the parent company. The Companies (Consolidation) Act 1908 was also passed during this period. This Act was amended by the Companies Act 1913, the Companies (Particulars of Directors) Act 1917 and the Companies Act 1928.
1929–47	Companies Act 1929 consolidating all the previous Acts was passed. The Act included a lot of innovation and required, for the first time, the publication by a reporting company of a profit and loss account in addition to a balance sheet. The profit and loss account did not have to state the turnover. The Act also gave the Board of Trade power to investigate the affairs of a company and the shareholders in general meeting the power to appoint inspectors to investigate the company's affairs.
1943	A Committee was appointed, in June 1943, by the President of the Board of Trade to examine the defects of the 1929 Companies Act under the chairmanship of Mr Justice Cohen (as he then was) and to report on what major amendments to the Companies Act 1929 were desirable to safeguard investors and protect the public interest. The Committee reported in June 1945 (Cmd. 6659). This led to the amendment of the Companies Act 1929 by the Companies Act 1947.
1948	Companies Act 1948, which consolidated all the existing Acts to that time, was promulgated and it made the publication of group accounts compulsory. It also distinguished between 'reserves' and 'provisions'. It required auditors to report whether the accounts were 'true and fair' rather than 'true and correct'.
1959	On 10 December 1959 the President of the Board of Trade appointed a Committee under the chairmanship of Lord Jenkins to review and report upon the provisions and workings of the Companies Act 1948, the Prevention of Fraud (Investments) Act 1958 '. . . and to consider in the light of modern conditions and practices, including the practice of take-over bids, what should be the duties of directors and the rights of shareholders, and generally to recommend what changes in the law are desirable.' The Committee Report was submitted on 30 May 1962 and published in June 1962 (Cmd. 1749).
1967	The Companies Act 1967 revised the 1948 Act. It requested that 'turnover' be included as additional information in the profit and loss account. It also requested that Directors' Reports and Notes should include information on the number of employees, charitable and political donations, directors' interests and market values of investments and properties.

Note: A number of sources have been used to compile this chronology. We are indeed indebted to the articles in Lee & Parker (1979).

made in 1976, 1980, 1981 and 1983 and arose for two main reasons. First, increased public accountability of companies was thought to be necessary to respond to changes in the economy. Second, provision had to be made for the harmonization of company law in response to European directives.

The 1980 Companies Act implemented the Second EC Directive on company law harmonization, as well as other reforms including legislation on insider dealing. The Act amended the distinction between public and private companies. In addition, for the first time British company law defined distributable profits as the aggregate of its undistributed realized profits less its accumulated realized losses. Thus, only realized profits are distributable, and only after deducting realized losses. In addition, public companies had a further restriction that any excess accumulated unrealized losses over unrealized profits be set against distributable profit before a distribution is made.

Another important measure was the introduction of regulations on insider dealing which made it a criminal offence to use inside information to purchase shares for personal gain. The Act defined inside information as unpublished price-sensitive information that an individual obtains by virtue of being connected with the company and that the person would not disclose normally in the course of his or her work.

Up until the Companies Act 1981, most of the requirements of law were directed mainly at disclosure rather than measurement. The 1981 Act implemented the Fourth EC Directive on company law and introduced detailed rules on the measurement of items in the financial statements as well as specifying formats for accounts. In addition, the government took the opportunity to close certain loopholes, such as those relating to interests in shares and investigations. The Act prescribed four alternative formats for the profit and loss account and two alternative formats for the balance sheet. A fixed order for items with headings and subheadings was introduced.

As well as prescribing formats for the financial statements, the Companies Act 1981 introduced requirements with respect to the measurement of balance sheet items, the content of the directors' report, special rules for small and medium-sized companies, and additional requirements for auditing and publishing the financial statements.

The Act also introduced provisions relating to:

- Company names and business names.
- The setting up of a share premium account as a result of a merger or group reconstruction.
- Companies dealing in their own shares.
- The disclosure of interests in shares to overcome problems of 'concert parties', i.e. a number of individuals acting together.

2.5 The Companies Act 1985

The aim of this Act was to consolidate the Companies Acts passed between 1948 and 1983 to produce one Act. The main Act was underpinned by the Companies Securities (Insider Dealing) Act, the Business Names Act and the Companies Consolidation (Consequential Provision) Act. The insider trading regulations introduced in 1980 were included in a separate Act both because the provisions of the law extend to company securities rather than companies themselves, and because they affect all companies whether or not they are registered under the Companies Act 1985.

The Business Names Act 1985 was issued separately for the same reasons as the insider trading legislation. The Act restricts the use of names to those permitted by the Secretary of State or appropriate statutory body. The Act applies to those registered under the 1948 Companies Act and to other forms of business organization.

2.6 The Companies Act 1989

This Act introduced the requirements of the Seventh EC Directive on group accounts and the Eighth Directive on the regulation of auditors. Whilst the Companies Act 1985 remains the main Act, the Companies Act 1989 incorporates certain accounting provisions that have been rewritten.

3 The development of accounting standards

Accounting standards in Britain were developed within the accounting profession. However, it was not until the 1920s and 1930s that weaknesses in financial reporting were becoming apparent, particularly in relation to the diversity in accounting practice. Starting in 1944, the Institute of Chartered Accountants in England and Wales (ICAEW) responded to these weaknesses by issuing a series of Recommendations on Accounting Principles. The aim of the series was to identify key areas that were causing problems and to recommend an appropriate method of accounting. As the name suggests, the solutions offered were not mandatory but served to assist practice. The problem with the recommendations was that they were slow in being issued and, since they were not mandatory, were of limited use to an auditor who was trying to persuade a client to adopt a particular accounting approach. In addition, the recommendations sometimes provided alternative accounting approaches to the same problem, all of which might lead to a true and fair view. By 1969, 29 recommendations had been issued.

The takeovers of Associated Electrical Industries (AEI) by General Electric Corporation (GEC) and of Pergamon by Leasco highlighted that the position at that time was not adequate (Cooke, 1986). It was apparent in both takeovers that using different accounting policies could lead to substantial differences in accounting profits. For example, AEI had produced a forecast of profits in the tenth month of their financial year of GBP 10 million, but following the takeover GEC decided that these 'profits' were a loss of GBP 4.25 million. Some of the differences related to fact, but GBP 9.5 million related to differences in judgement. As a result, the ICAEW published its 'Statement of Intent on Accounting Standards' in the 1970s.

The aim of this statement was to reduce the number of alternative accounting practices without attempting rigid uniformity. This was to be achieved by publishing authoritative accounting statements which were to be referred to as accounting standards. Where the outcome of an event was uncertain and its recorded value was therefore subjective, the accounting base should be disclosed. It was also anticipated that non-compliance with accounting standards, which should be disclosed in the accounts, was not acceptable normally and that procedures would be developed to tackle this problem.

The outcome of the Statement of Intent was the formation of the Accounting Standards Steering Committee (ASSC) by the ICAEW in 1970. In that year, the Scottish and Irish Institutes joined the ASSC, followed in 1971 by the Certified Accountants and the Institute of Cost and Management Accountants, and in 1976 by the Chartered Institute of Public Finance and Accountancy. Membership of the ASSC was drawn from the various institutes but individuals did not act as institute representatives (Leach and Stamp, 1981, p. 5). A five-year plan was agreed which had the support of the accountancy bodies, the Stock Exchange and, in general, the media.

The ASSC was reconstituted from 1 February 1976 as a joint committee of the six bodies listed above (the Consultative Committee of Accounting Bodies (CCAB)) and renamed the Accounting Standards Committee (ASC). The six bodies nominated members to the committee but the ICAEW was able to appoint 12 out of the 23 members. Both the chairman and the vice-chairman were appointed by the chairman of the CCAB on the basis of a recommendation from the ASSC.

The constitution of the ASC was:

> to keep under review standards of financial accounting and reporting; to publish consultative documents with the object of maintaining and advancing accounting standards; to propose to the Council of each CCAB member statement of standard accounting practice (SSAP) and interpretations of such statements; and to consult with representatives of finance, commerce, industry and government and others concerned with financial reporting.

Accounting standards describe methods of accounting approved by the member bodies which should be used by companies to show a true and fair view of the financial statements. The ASC did not intend to issue a rigid code but rather to offer accepted methods of accounting to achieve a true and fair view. However, 'true and fair' was the overriding consideration so that it was acceptable to diverge from an accounting standard where necessary to achieve this. Significant departures from accounting standards should be explained and where possible the financial effects of departure should be disclosed. If estimation was not practicable the reasons for this should be stated.

In general, the ASC identified a subject area that required resolution by an accounting standard. Research into the area was undertaken, a preliminary draft statement was prepared for discussion among the CCAB members and discussion with the technical departments of large firms of accountants, and meetings were sometimes held with interested parties. On the basis of these discussions an exposure draft was issued and a deadline set for response. The ASC then prepared a SSAP and submitted it to the CCAB. The council of each member body then considered the SSAP and, if each approved, it would issue the SSAP, with a technical release outlining any changes that had been made from the original exposure draft.

In 1978 a review group, under the chairmanship of T. R. Watts, was set up to consider the setting of accounting standards. In September of that year a consultative document entitled 'Setting Accounting Standards' was published and comments were sought. As a result the ASC published a report in 1981 which raised a number of issues that were not really addressed until the Dearing Committee was formed in 1987.

A further review of the accounting standard setting procedures was undertaken in 1983, but the issues relating to the funding and structure of standard setting raised by the Watts Report were not addressed. It was not until the Dearing Committee was set up in November 1987, in the face of increasing criticism directed at the ASSC, that the issues were addressed. (The development of financial reporting standards is presented in Table I.2.)

4 The current regulatory framework

4.1 Overview

The Dearing Report, which was issued in November 1988, recommended a new regulatory framework. As a result, in 1990 a new standard setting body (Accounting Standards Board, ASB) was set up under the auspices of the Financial Reporting Council (FRC). The ASC remained until the ASB came into effect. By that time the ASC had published 25 SSAPs, 22 of which were extant when the ASB came into being, and 55 exposure drafts. The SSAPs were adopted by the ASB until superseded by the ASB's own accounting standards, called Financial Reporting Standards (FRS). At the end of 1999 there were 17 SSAPs still effective (see Table I.3) with the others having been replaced, largely by FRSs.

The first chairman of the Financial Reporting Council (FRC) was Sir Ronald Dearing (with the chairman of the Accounting Standards Board being Professor David Tweedie). An early achievement of the FRC was that the Companies Act 1989 incorporated the recommendation of the Dearing Committee that the accounts of large companies should be prepared using applicable accounting standards and that any material departures from them should be explained. However, small and medium-sized companies are exempt from this provision. This was the first time that disclosure of non-compliance with accounting standards was specifically required by law. Another recommendation of the Dearing Committee introduced by law was that the Secretary of State or authorized persons should be able to insist that revisions be made to accounts if they are considered to be defective. Voluntary changes may be made but if an offending company refuses to make changes voluntarily, court action can be taken. The current regulatory framework is shown in Figure I.1 and the functions of each body are discussed below.

4.2 The Financial Reporting Council (FRC)

The Financial Reporting Council is the overarching and facilitating body. Its initial role was to bring into being, guide, and secure financing for, the various executive organs proposed in the 1988 report (The Making of Accounting Standards) of the Dearing Committee set up to review accounting standard setting in the United Kingdom. The Council's ongoing role, which is to produce guidance on broad policy issues (but not to approve or monitor compliance with individual accounting standards), includes:

- Promoting good financial reporting and in that context, from time to time, making public its views on reporting standards. In that role it will, when appropriate, make representations to government on the current working of legislation and on any desirable development of it.
- Providing guidance to the Accounting Standards Board on work programmes and on broad policy issues.
- Verifying that the new arrangements are conducted with efficiency and economy and that they are adequately funded.

Table I.2 The development of financial reporting standards

1942	The Institute of Chartered Accountants in England and Wales established a Taxation and Financial Relations Committee (name changed to Taxation and Research Committee, 1949) with a remit to issue recommendations on accounting principles for solving practical accounting problems.
1943–69	ICAEW (Taxation and Research Committee) issued 29 recommendations on diverse accounting problems including tax reserve certificates, war damage contributions, rising price levels, form and content of balance sheet and profit and loss account, the form and contents of accounts of estates of deceased persons and similar trusts, hire purchase, credit sale and rental transactions, accountants' liability to third parties, accounting treatment of investment grants and trust accounts.
1969	As a result of the doubts created concerning the representational faithfulness of profits reported in annual reports and accounts following the takeover of Associated Electrical Industries by General Electric Corporation and of Pergamon Press Ltd by Leasco Data Processing Equipment Corporation, commentaries on the inadequacies of accounting and auditing practices appeared frequently in the Press. One of these commentators was Professor Edward Stamp who argued for a reform of accounting and auditing practices and for more accounting research. Although the then President of the Institute (ICAEW) defended the policies and record of ICAEW, the general opinion was that it would be more appropriate for the accounting profession to define what was correct practice and for this to be followed by all accountants. In response, the ICAEW appointed its first Technical Director (Michael Renshall) and published a Statement of Intent on Accounting Standards in the 1970s. The statement announced an intention to narrow the areas of difference and variety in accounting practice.
1970	ICAEW formed (in January) an Accounting Standards Steering Committee (ASSC) with 11 members including Ronald Leach, its first chairman. In April, three representatives of the Institute of Chartered Accountants of Scotland (ICAS) and one representative of the Institute of Chartered Accountants in Ireland (ICAI) joined the ASSC.
1971	In November, the Association of Certified and Corporate Accountants (ACCA), now known as the Chartered Association of Certified Accountants, and the Institute of Costs and Works Accountants (ICWA), now known as the Chartered Institute of Management Accountants (CIMA), joined as associate members of the ASSC. This increased the membership of the ASSC to 19.
1973	Companies Bill (Bill 52) which was not passed by Parliament, proposed in clause 69 to empower the Secretary of State '. . . to prescribe by regulations the matters to be disclosed in accounts, directors' and auditors' reports and annual returns and to make different provisions for different classes of companies' (Companies Bill 52, p. viii). If this bill had been passed, accounting standard setting might have been in the hands of government in the United Kingdom.
1976	ASSC changed its name to Accounting Standards Committee (ASC).
1978	A review of the standard setting process in the United Kingdom chaired by Mr T. R. Watts was carried out. The review report led to a number of recommendations by the ASC to the CCAB for fundamental changes to the standard setting process. These included the need for a conceptual framework, the establishment of a monitoring committee, the limitation of the scope of application of standards to certain large companies and the need for more financial and staff resources. The McKinnon Report recommended the introduction of Statements of Recommended Practices (SORPs).
1987	The CCAB appointed another panel under the chairmanship of a non-accountant but a well-respected entrepreneur, Sir Ronald Dearing, to review the standard-setting process. The committee reported in November 1988. The acceptance of the committee's recommendations led to the creation of the Financial Reporting Council, the Accounting Standards Board and the Financial Reports Review Panel, and the Urgent Issues Task Force.
1971–90	ASC issued 25 Statements of Standard Accounting Practice (SSAPs). See Table I.3 for details.
1990	The Accounting Standards Board took over from the ASSC.

Table I.3 Statements of Standard Accounting Practice (SSAPs)

Number	Title
SSAP 2	Disclosure of Accounting Policies
SSAP 3	Earnings per Share
SSAP 4	Accounting for Government Grants
SSAP 5	Accounting for Value Added Tax
SSAP 8	The Treatment of Taxation under the Imputation System in the Accounts of Companies
SSAP 9	Stocks and Long-term Contracts
SSAP 12	Accounting for Depreciation
SSAP 13	Accounting for Research and Development
SSAP 15	Accounting for Deferred Tax
SSAP 17	Accounting for Post-Balance Sheet Events
SSAP 18	Accounting for Contingencies
SSAP 19	Accounting for Investment Properties
SSAP 20	Foreign Currency Translation
SSAP 21	Accounting for Leases and Hire Purchase Contracts
SSAP 23	Accounting for Acquisitions and Mergers
SSAP 24	Accounting for Pension Costs
SSAP 25	Segmental Reporting

The Council comprises:

- A chairman and three deputy chairmen drawn from accountancy, industry and commerce, and the City who are appointed by the Secretary of State for Trade and Industry and the Governor of the Bank of England.
- The Chairman of the Accounting Standards Board and the Chairman of the Financial Reporting Review Panel (FRRP) as ex officio.
- Twenty-four other members and observers drawn from the most senior levels in accountancy, industry and commerce, the City and others having an interest in good financial reporting.

The Council has, through moral suasion, invited institutional investors to act strongly in support of high standards of financial reporting. It has also written to the chairmen of all listed companies seeking support for high standards, drawing attention to the legal obligations and liabilities of directors, and providing a short guide to the new framework for accounting standards and the arrangements for securing compliance with them.

4.3 The Accounting Standards Board (ASB)

The ASB is able to introduce accounting standards under its own authority which contrasts with the previous position under which the ASC had to leave the publication and approval to each member of the CCAB. However, the ASB consults widely on its proposals and its due process attests to this statement of intent. The ASB has a full-time chairman, a technical director and a membership that does not exceed 10 (including its chairman).

The procedure for the formulation of accounting standards is designed to ensure the participation of various groups of persons and organizations interested in financial reporting. After deciding on a topic on which accounting standards should be issued, the ASB designates a member of its staff as a project director. The project director is responsible for studying existing practices on the subject, liaising with interested persons and organizations and preparing the preliminary draft of the standard. In addition, a group of consultants with a special interest and expertise in the subject is formed to advise the ASB and the project director. The preliminary draft is considered by the ASB and, after necessary modifications, the revised draft is approved for publication as a Financial Reporting Exposure Draft (FRED) which invites comments from interested constituencies and the public. Comments received on the FRED are considered by the staff and members of the ASB. The draft is then revised and finalized, and the final draft is considered and approved by the ASB, provided that a majority of seven agree. The approved standard is then issued with an indication of the date on

Figure I.1 Regulatory Framework

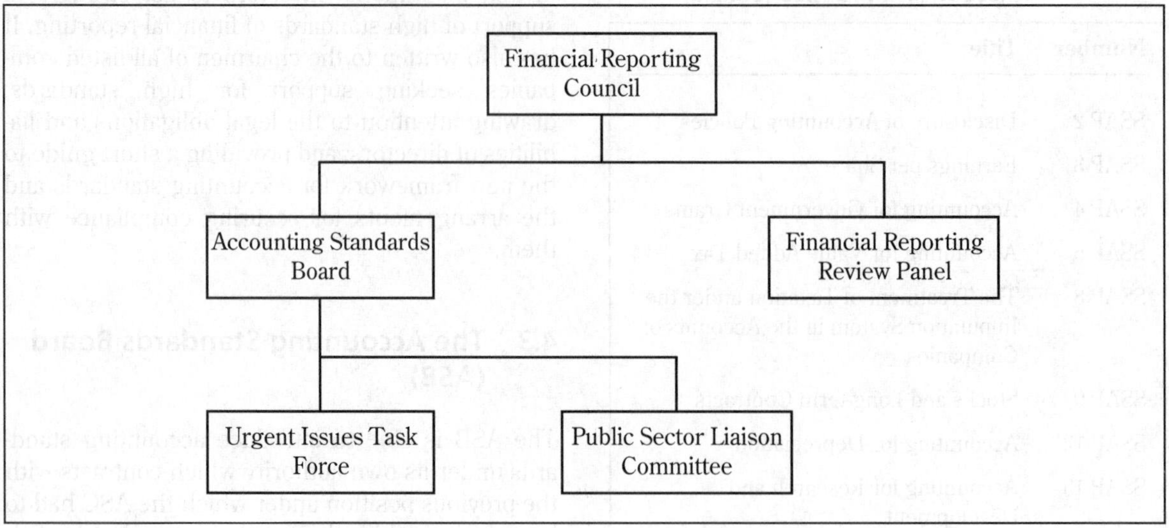

which its provisions are to become effective. In certain circumstances, a transitional period is specified if it is considered necessary to phase in its implementation.

Unlike its predecessor organization, the ASC, the ASB is well staffed with high-calibre project directors, each of whom is responsible for the production of at least one of the Board's major pronouncements. The ASB has issued its 'Statement of aims' which outlines the approach to standard setting and includes developing principles. The first stage of developing a framework was the Statement of principles in the form of seven draft chapters. The draft was revised in 1995. At the initial stage the ASB focused on eliminating abuses but the Statement of principles project moves the emphasis to underlying concepts.

The ASB still retains 17 of the ASC's SSAPs and has added 14 FRSs. In addition, the Statement on Operating and Financial Review (see IX.8), published in July 1993, calls upon the management of reporting companies voluntarily to provide commentaries in their annual reports that interpret the accounts and highlight the main factors underlying their companies' financial performance and position; the ASB has issued the pronouncements (Financial Reporting Standards, FRSs) and unresolved exposure drafts (Financial Reporting Exposure Drafts, FREDs) listed in Table I.4.

4.4 The Urgent Issues Task Force (UITF)

An Urgent Issues Task Force (UITF) was created in March 1991 to assist the ASB. The assistance relates to the issue of authoritative guidance on the application of the provisions of extant companies' legislation and accounting standards to important and significant accounting issues where conflicting or unsatisfactory interpretations have developed or seem likely to develop, so as to safeguard against opportunism or lack of wisdom.

The UITF's main role is to assist the ASB in areas where an accounting standard or Companies Act provision exists, but where unsatisfactory or conflicting interpretations have developed or seem likely to develop. Now removed from the UITF's remit was the indication by the ASB that it may from time to time seek the UITF's view on significant developments in accounting and financial reporting in areas where there is no legal provision or accounting standard. In reaching its decision, the UITF is expected to adopt a consensus approach like its counterparts in the United States and Canada. Such a consensus is arrived at when not more than two members of the UITF present at the meeting (which shall have a quorum of 11 members) have voted against the accounting treatment in question.

Table I.4 Financial Reporting Standards (FRS) and Exposure Drafts (FRED)

Financial Reporting Standards:	
FRSSE	Financial reporting standard for smaller entities
FRS 1 (revised)	Cash flow statements
FRS 2	Accounting for subsidiary undertakings
FRS 3	Reporting financial performance
FRS 4	Capital instruments
FRS 5	Reporting the substance of transactions
Amendment to FRS 5	Reporting the substance of transactions – the private finance initiative
FRS 6	Acquisitions and mergers
FRS 7	Fair values in acquisition accounting
FRS 8	Related party disclosures
FRS 9	Associates and joint ventures
FRS 10	Goodwill and intangible assets
FRS 11	Impairment of fixed assets and goodwill
FRS 12	Provisions, contingent liabilities and contingent assets
FRS 13	Derivatives and other financial instruments
FRS 14	Earnings per share
FRS 15	Tangible fixed assets
FRS 16	Current tax
Financial Reporting Standards Exposure Drafts:	
FRED 18	Current tax
FRED 19	Deferred tax
FRED 20	Retirement benefits
FRED 21	Accounting policies

If a resolution by consensus is reached, the UITF issues an Abstract discussing the matter, the accounting issues identified and reference sources, as well as a summary of the UITF's deliberations and a clear indication of the conclusion it has reached on the issue. Abstracts are also published where a consensus has not been reached or where a consensus has been overturned by the ASB. A consensus-bearing UITF abstract that does not conflict with the law, accounting standards, or the ASB's policy or plans, is regarded as accepted accounting practice on the topic and is normally accepted by the ASB. Whilst the consensus-bearing UITF abstract does not have the status of an accounting standard, it is considered as a part of the body of practices forming the basis for what determines a true and fair view. As a result, it is expected that reporting companies will conform to it and if necessary change their previously adopted accounting policies. Furthermore, the Financial Reporting Review Panel (see I.4.6) may take it into account when considering whether financial statements should be revised.

Items to be included in the agenda of the UITF are determined by a committee consisting of the Chairman of the UITF together with two ASB members and two members of the UITF selected by the Chairman. The UITF may not consider any issue which the ASB considers falls within its own agenda unless specifically requested to do so by the ASB.

The consensus pronouncements listed in Table I.5 have been issued by the UITF.

4.5 Public Sector Liaison Committee and special industry committees for approving Statements of Recommended Practice (SORPs)

The recognition of public sector accounting guidelines by the ASB is based on the advice of the Public Sector Liaison Committee that is a subcommittee of the ASB. The formation of this subcommittee was recommended in the Dearing Report. Representatives on the Public Sector Liaison Committee are appointed from academe, from government organizations such as the Post

Table I.5 Consensus pronouncements of the UITF (UITF abstracts)

No.	Subject/Title & Short Comment
4	**Presentation of long-term debtors in current assets** This abstract requires disclosure of long-term debtors on the face of the balance sheet in those instances where the figure of debtors due after more than one year is so material in the context of the total net current assets that in the absence of such disclosure readers may misinterpret the accounts.
5	**Transfers from current assets to fixed assets** This requires transfers from current to fixed assets to be made at the lower of cost or net realizable value. Where the transfer is made at net realizable value at the date of transfer (because this is below cost) the asset should then be accounted for as a fixed asset at a valuation under the alternative rules in the Companies Act.
6	**Accounting for post-retirement benefits other than pensions** The abstract provides that post-retirement health care and other benefits accruing to former employees should be recognized in financial statements as liabilities according to the principles enunciated in SSAP 24, Accounting for Pension Costs. This abstract will be of particular relevance to UK groups with operations in countries where the provision for such benefits is particularly significant, such as the United States. Although approved in November 1992, this consensus-bearing UITF Abstract was allowed a long transitional period before coming into operation for accounting periods ending on or after 24 December 1994. There are significant deferred tax implications that would arise from this pronouncement. As a result, the UITF recommended a limited amendment to SSAP 15, Accounting for Deferred Tax. The ASB accepted this recommendation and has accordingly amended SSAP 15.
7	**The true and fair view override disclosures** The abstract seeks to clarify the requirements where the true and fair view override is involved. In such a case, the report should state clearly and unambiguously the particulars of any departure from the Companies Act, the reasons for such departure, and the effect (including quantification in monetary terms if feasible) that is, how the position shown in the accounts is different as a result of the departure. Where a departure continues in subsequent years, the disclosures should be made in all such subsequent statements and should include corresponding amounts for the previous year(s).
9	**Adjustments for the distortion effects of inflation from hyper-inflationary economies** This abstract requires adjustment where the distortions caused by hyper-inflation are such that they affect the true and fair view given by the group financial statements. In any event, adjustments are required where the cumulative inflation rate over three years is approaching or exceeds 100 per cent and the operations in the hyper-inflationary economy are material. The distortions can be eliminated in one of two ways: (i) restate and then translate adjust the local currency financial statements to reflect current price levels before the translation process is undertaken as suggested in SSAP 20. Gains and losses on net monetary position can be taken to the profit and loss account. (ii) Use a relatively stable currency (not necessarily sterling) as the functional currency (i.e. the currency of measurement) for the relevant foreign operations. If local currency transactions are not recorded initially in the stable currency, they must be remeasured into that currency by applying the temporal method. If neither of the above methods is appropriate, the reasons should be stated and alternative methods to eliminate the distortions should be adopted.
10	**Disclosure of directors' share options** Information concerning the option prices applicable to individual directors, together with market price information at year end and at the date of exercise, should be disclosed. This is in accord with the request by the Cadbury Committee that shareholders are entitled to a full and clear statement of directors' present and future benefits and how they have been determined.
11	**Capital instruments: issuer call options** Where a capital instrument includes a call option that can be exercised only by the issuer, the payment required on exercise of that option does not normally form part of the finance costs of the instrument in

Table I.5 Continued

No.	Subject/Title & Short Comment
	accordance with the requirements of FRS 4. As a result, any gain or loss arising on repurchase or early settlement will reflect the amount payable on exercise.
12	**Lessee accounting for reverse premiums and similar incentives** The abstract requires benefits received and receivable by a lessee as an incentive to sign a lease to be spread on a straight-line basis over the lease term, or if shorter than the full lease term, over the period to the review date on which the rent is first expected to be adjusted to the prevailing market rate. The rationale for this requirement is the presumption that an incentive (however structured) is in substance part of the market return which the lessor is for the time being prepared to accept in order to let a particular property.
13	**Accounting for ESOP trusts** The sponsoring company of an ESOP trust should recognize certain assets and liabilities of the trust as its own where de facto control is a reality e.g. where shares are held for an employee remuneration scheme.
14	**Disclosure of changes in accounting policy** Disclose the effect on current year's results of a change in accounting policy as well as complying with FRS 3. If the effect on the current year's results is either immaterial or similar in a quantified effect on the prior year a statement to that effect would be sufficient.
15	**Disclosure of substantial acquisitions** In effect, any transaction involving a business combination in which any of the ratios set out in the London Stock Exchange Listing Rules exceeds 15 per cent, requires disclosure.
16	**Income and expenses subject to non-standard rates of tax** Income and expenses subject to non-standard rates of tax should be included in pre-tax results as income/expenses actually receivable/payable, without any notional tax paid/relieved in respect of the transaction if it had been dealt with on a different basis.
17	**Employee share schemes** The cost of awards to employees should be recognized over the period of employees' performance. Annual bonuses should be dealt with in the year to which they relate. Long-term incentive schemes should be dealt with over the period to which the performance criteria relate. Recognition should be made on the basis of the fair value of the shares at the date of the award.
18	**Pension costs following the 1997 tax changes in respect of dividend income** Changes to tax legislation so that the pension schemes cannot reclaim a tax credit on dividend income does not fall outside the normal scope of the actuarial assumptions as set out in SSAP 24. It should be regarded as a change in the expected return on assets in a similar manner to those arising from changes in the tax rates. The loss should be spread over the remaining service lives of current employees in the scheme, regardless of the financial position of the scheme and regardless of any additional contributions that are made.
19	**Tax on gains and losses on foreign currency borrowings that hedge an investment in a foreign enterprise** Exchange differences on foreign currency borrowings used as a hedge an investment in a foreign enterprise should be taken to reserves and reported in the statement of total recognized gains and losses. Tax charges or credits attributable to such transactions should also be taken to reserves.
20	**Year 2000 issues: accounting and disclosures** Costs involved in ensuring that computer software is 2000 compliant should be written off to profit and loss account except in cases where an entity already has an accounting policy for capitalizing software costs and to the extent that the expenditure represents an improvement in an asset beyond that originally assessed.

Table I.5 Continued

21	**Accounting issues arising from the proposed introduction of the euro** The costs of making the necessary changes to assets to deal with the euro should be written off to profit and loss account except where an entity has an accounting policy to capitalize such expenditure or where the asset is enhanced. Any other costs should be written off to profit and loss account.
22	**The acquisition of a Lloyd's business** On the acquisition of a Lloyd's managing agent, the business's assets and liabilities should be recognized and include all profit commissions receivable in respect of periods before the acquisition. The profit commissions should be recognized at their fair value on the basis of the best estimate of the likely outcome based on profits earned before the date of the acquisition.

Office, from public accountancy practice and from central government. The role of the committee is to appraise the ASB of developments in the public sector so that differences between public and private sector practices are minimized. In addition, the committee advises the ASB on proposed public sector Statements of Recommended Practice (SORPs), provides comments on material issued by the ASB that may have an impact on accounting in the public sector, and seeks to foster a common philosophy for development of public and private sector reporting.

The Foreword to Accounting Standards states that the prescription of accounting requirements for the public sector is the responsibility of the government. Where public bodies prepare accounts on commercial lines, the government may require the use of accounting standards to show a true and fair view. Where accounts are prepared on commercial lines it will be normal for the ASB's pronouncements to be adhered to unless the government deems otherwise.

The ASB has also recognized several bodies as being capable of issuing Statements of Recommended Practice (SORPs). The ASB has stated that it does not believe it necessary to adopt any existing SORPs. If the ASB believes that its own authority is required to standardize practice within a specialized industry, it would issue an industry standard. However, the ASB will recognize bodies developed within an industry or public sector to provide guidance on the application of accounting standards of broad scope that it issues to specific industries. Recognition will depend on the size of the industry or sector in question, the representative nature of the body proposing to produce the SORP and the body agreeing to follow the ASB's code of practice for the SORP development process. Instead of giving a nod to (or 'franking') an SORP, the ASB would require a 'negative assurance statement' to be appended to the SORP. In essence the negative assurance statement will make it clear that the ASB is not approving the SORP but is rather confirming that it contains no fundamental points of principle which the Board considers unacceptable in the context of current accounting practice, that the SORP is not in conflict with any existing or currently contemplated accounting standard, and that the SORP has been prepared in accordance with the ASB's code of practice.

The ASB has so far approved the application by the following 10 specialized industrial bodies:

- Association of British Insurers.
- Association of Investment Trust Companies.
- British Bankers' Association (BBA) and the Irish Bankers' Federation (IBF).
- Charity Commission.
- The Chartered Institute of Public Finance and Accountancy (CIPFA) – for local authorities.
- Committee of Vice-Chancellors and Principals of the Universities of the United Kingdom (together with the Standing Conference of Principals and the Conference of Scottish Centrally Funded Colleges) – in respect of accounting for UK Universities.
- Finance & Leasing Association.
- Investment Management Regulatory Organization (IMRO).
- National Federation of Housing Associations.
- Oil Industry Accounting Committee (OIAC).

The SORPs which have been issued at the time of writing are included in Table I.6.

Table I.6 Statements of Recommended Practice

BBA/IBF
- Advances
- Contingent liabilities and commitments
- Segmental reporting
- Derivatives

IMRO
- Financial statements of authorized unit trust schemes

Local Authorities
(developed jointly by the Chartered Institute of Public Finance and Accountancy (CIPFA) and the Local Authority (Scotland) Accounts Advisory Committee and 'franked' by the ASC, the predecessor body of the ASB, at its final meeting in July 1990)
- The application of accounting standards (SSAPs) to local authorities in Great Britain
- CIPFA – Local authorities accounting (the Code of Practice)

UK universities
- Accounting in UK universities

NHF/SFHA/WFHA
- Accounting by registered social landlords

Charity Commission
- Accounting by charities

PRAG
- The financial statements of pension schemes

AITC
- Financial statements of investment trust companies

ABI
- Accounting for insurance business

OIAC
- Accounting for various financing, income and other transactions of oil and gas exploration and production companies
- Accounting for the value discovered reserves of oil and gas

4.6 Financial Reporting Review Panel (FRRP)

4.6.1 Establishment and scope of duty of the FRRP

The Financial Reporting Review Panel (FRRP) was established and authorized in February 1991 by the Secretary of State for Trade and Industry (through the Statutory Instrument 1991 No. 13 for the purposes of Companies Act 1985, section 245[B]) to examine apparent material departures by public and large private companies from the accounting requirements of the Companies Act 1985, whether or not they also involve departures from accounting standards, and if necessary to seek a court order to remedy them. The Department of Trade and Industry (DTI) is responsible for monitoring and dealing with the departures by small and medium-sized private companies. An objective of the introduction of an institution for monitoring and enforcing compliance with applicable Companies Act and accounting standards 'is to remove much of the onus from auditors and place it instead on a well-equipped investigating agency in the private sector (or public sector, for small and medium-sized companies), but with the authority granted by the public sector' (Turley, 1992; Zeff, 1993). A further objective is to provide legitimacy for the accounting standards issued by the ASB and to let preparers and users know that corporate annual reports are expected to be based on established rules.

Although the FRRP started operations on 10 January 1991, its remit included the examination of accounts for reporting periods beginning on or after 23 December 1989. By the end of December 1999, the FRRP's attention had been drawn to the accounts of 368 companies that were perceived to have departed from the requirements of the Companies Act 1985 and/or the provisions of extant accounting standards. The FRRP does not monitor or actively initiate scrutinies of companies for possible defects. However, it examines matters drawn to its attention, either directly through the complaints it receives or through audit reports published in the financial reports of companies within its remit, or indirectly through press comment.

4.6.2 Cases of apparent substance

The FRRP usually concentrates on matters drawn to its attention. While its enquiries may extend beyond the initial issue it does not necessarily involve an examination of all aspects of the company accounts in question. The FRRP's detailed approach involves making a preliminary decision, after consultation with the accountants on the staff of the Financial Reporting Council Limited (FRC) (the parent organization whose staff the FRRP shares), about whether the issue is one of substance. If the decision is that the matter is not substantial but that there are minor matters which should be explored further, the FRRP will invite the comments of the company concerned. On receipt of the reply from the company, the Chairman of the FRRP will decide whether to drop the matter, to pursue it as a case of apparent substance or to treat it as a minor matter. If the minor matter requires remedy, the FRRP will hold informal discussions with the company to agree on the desirable remedy.

When a case is classified as apparently substantive, the FRRP will invite the directors of the erring company to comment, within one month, on the departure from the Companies Act and/or the affected accounting standard after having been cautioned about its obligation under Companies Act 1985, section 245A (1).

The Secretary of the FRRP is obliged to meet with the company, on its request, and with its auditors for informal discussion and further clarification of the content of the FRRP's letter. After reviewing the company's response, the FRRP may decide to continue with the investigation, if it is not satisfied with the company's explanation, by forming a group of five or more persons drawn from its total membership to enquire into the matter. The group will represent the FRRP for all purposes for that particular case and the full FRRP membership will not operate as a collegiate body. The membership of the group shall be communicated to the affected company in order that the company may advise the FRRP of any conflict of interest that might arise were a particular member of the advisory group to be engaged. The hearing and the enquiry of the group shall be in private and all those present are identified to the hearing by name, function and connection with the parties concerned. The group may also on occasion give a separate hearing to third parties who appear to have useful and relevant information to contribute. This provision enables the group to hear the auditors of the affected company where they do not appear with the company.

The decision of the group depends on whether it has received satisfactory explanations from the company or whether it is satisfied with the corrective action volunteered by the company. If the group is not satisfied by either or both of these matters, it will inform the company and advise it that it intends to proceed to obtain a court order in accordance with Companies Act 1985, section 245(B) (1)(b). If the group is still not satisfied with the company's response, it will then proceed to obtain the order.

Companies Act 1985, section 245(C) empowers the courts, where they are satisfied that the annual accounts of the affected company do not comply with the provisions of the Act, notably where a material departure from an accounting standard results in the accounts not giving a true and fair view, to require a company to prepare and re-issue revised accounts, and to require directors of the company personally to bear the costs of the application and the cost of revising and re-issuing the accounts.

On application to the court, the FRRP may, after taking account of market sensitivities, make a public announcement of the details of the case. However, the company is free to make an announcement about the contact with the FRRP and the outcome of that matter at any stage.

Since its inception, the FRRP has not sought the order of the court for a declaration that the annual accounts of a company do not comply with the requirements of the Companies Act 1985 and for an order requiring an erring company to prepare revised accounts. When a company's accounts are defective, the FRRP has endeavoured to secure their revision by voluntary means.

It is not the practice of the FRRP to provide advice to a company or its auditors as to whether a particular accounting treatment would or would not meet the requirements of the Act. However, in order to assist those responsible for the regulation of auditors and professional bodies responsible for the supervision of the professional conduct of individual members, the FRRP has regularly drawn the attention of the public to those cases where, at its insistence, a company has voluntarily accepted (and will, where the court has declared) that the company's accounts were defective and where the audit report had not been qualified in that respect. The results of the deliberations on the 368 cases processed by the FRRP are summarized below:

• Not pursued beyond an initial examination either because they did not fall within the ambit of the legislation or because no point of substance arose	134
• Suspended after initial and further investigation	22
• Concluded after correspondence/discussion with the companies concerned	202
• Currently under consideration	10
Total	368

Table I.7 summarizes the concluded cases published by the FRRP in several press releases. It should be stressed, however, that a case normally deals with a variety of complaints about a set of departures from the Companies Act and accounting standards by a reporting company, and a voluntary revision of the annual accounts or re-issue of the accounts does not imply that the FRRP has laid down specific rules on the matter. It all depends on the circumstances of each case and the outcome of a case is not likely to be known to the complainants, the management of the reporting company when it prepares the accounts, or the auditors of the reporting company when they issue an audit opinion.

A problem with the FRRP's confidential procedure of requiring erring companies to revise their annual accounts is that it denies a court of law the opportunity to determine whether or not a company's account has been presented in a manner which would portray a true and fair view of the company's state of affairs. Some of the complaints that have been examined by the FRRP have related to departures from the letter of the law and accounting standards, and have not been directly concerned with a true and fair presentation of the accounts nor with the issue of economic substance. Such a procedure that emphasizes adherence to legal form may gradually make it difficult to reflect economic substance in financial statements. According to Zeff (1993, p. 406):

> since the law does not require companies to adhere to applicable accounting standards, but to report whether their financial reports do so, it has not yet been determined whether, and to what extent, the failure of such adherence detracts from 'a true and fair view'.

The monitoring system of the FRRP is atrophied in the desire to uphold legislation and accounting standards and to question non-reporting of the extent of compliance of corporate annual reports with the Companies Acts and accounting standards rather than to query whether, in actual fact, the disclosure in annual reports presents 'a true and fair view'.

Using the spectre of court action to persuade 'erring' companies to revise their apparently non-complying annual reports both prevents the judicial determination of the provisions of the relevant law and the applicable accounting standards, and has the tendency of evolving a reporting practice that may be contrary to acceptable accounting practice and may not be in the public interest.

Table I.7 Summary of concluded cases published by the FRRP

Issue number, date	Company, accounts date	Nature of departure
PN04, 28-1-92	Ultramar plc, 31-12-90	Irrecoverable ACT included as part of the cost of dividends
PN05, 28-1-92	William Holdings plc, 31-12-90	Exceptional items stated in the profit and loss account net of tax. EPS stated net of exceptional items. Failure to disclose the names of acquired companies.
PN06, 31-1-92	The Shield Group plc, 31-3-91	Reduced carrying values of certain assets shown as prior year adjustment.
PN07, 4-2-92	Forte plc, 31-1-91	Inadequate information on accounting policies on capitalization of interest, certain expenses, and absence of depreciation on properties.
PN10, 10-8-92	Williamson Tea Holdings plc, 31-3-91	Inadquate disclosure and explanation of accounting policies, relating to certain assets such as non-depreciation of certain properties, depreciation on revalued assets.
PN11, 10-8-92	Associated Nursing Services plc, 30-3-91	Inadequate explanation of the accounting policy in respect of start-up costs
PN12, 7-10-92	GPG plc, 30-9-91	Adoption of FRED 1 resulting in non-compliance with SSAP3 and SSAP6.
PN13, 15-10-92	Trafalgar House plc, 30-9-91	Treatment of current assets and ACT carried forward. Compliance with profit and loss accounts format.
PN014, 26-10-92	British Gas plc, 31-12-91	Presentation of the profit and loss account after change in year end.
PN15, 26-10-92	S.E.P. Industrial, 30-9-91	Failure to depreciate properties.
PN16, 22-2-93	Eurotherm plc, 31-10-91	Treatment of exceptional/extraordinary items.
PN17, 17-3-93	Foreign and Colonial Investment Trust plc, 31-12-91	Inadequate disclosure of directors' remuneration. Non-consolidation of subsidiaries.
PN18, 1-4-93	Warnford Investments plc, 25-12-91	Failure to comply with SSAP19.
PN19, 5-4-93	Penrith Farmers' and Kidds' plc, 31-3-92	Failure to provide reasons for restating the 1991 comparative figures.
PN20, 27-7-93	Breverleigh Investments, 30-6-92	Failure to produce a cash flow statement. Non-consolidation of subsidiary.
PN21, 11-8-93	Royal Bank of Scotland Group, plc, 30-9-92	Change in accounting policy not treated as a prior year adjustment.
PN22, 24-9-93	Control Techniques, plc, 30-9-92	A classification error in the group cash flow statement.

Table I.7 Continued

Issue number, date	Company, accounts date	Nature of departure
PN24, 19-10-93	BM Group plc, 30-6-92	Incorrect description of profit on sale of shares in subsidiary. Error in cash equivalents.
PN25, 25-10-93	Ptarmigan Holdings plc, 30-6-92	Failure to disclose the reasons for a change in accounting policy with respect to goodwill.
PN26, 29-11-93	Chrysalis Group plc, 31-8-92	Treatment of associate which was carried at valuation in the group accounts.
PN27, 28-1-94	The Intercare Group plc 31-10-92	Mistake in the cash flow statement as shares issued as consideration were shown as cash flows.
PN28, 11-2-94	Pentos plc, 31-12-92	Disclosure of treatment of reverse lease premiums.
PN29, 24-4-94	BET plc, 27-3-93	Incorrect presentation of operating exceptional items under statutory categories.
PN30, 2-11-94	Butte Mining plc, 30-6-93	Incorrect classification of a bank overdraft, incorrect treatment of a cash equivalent and unacceptable use of the true and fair override.
PN31, 23-11-94	Clyde Blowers plc, 31-8-93	Presentation of provision for loss on disposal and no disclosure of the cash flow effect of the partial disposal.
PN32, 20-3-95	Alliance Trust plc, 31-1-94	Subsidiary not consolidated on grounds of dissimilar activities and materiality, not accepted.
PN33, 21-6-95	Courts plc, 31-3-94	Accounting treatment of long-term credit sales.
PN34, 8-11-95	Caradon plc, 31-12-94	Failure to include reduced share premium under non-equity.
PN35, 13-12-95	Ferguson International Holdings plc, 28-2-95	Incorrect disclosure of goodwill.
PN36, 15-2-96	Securicor Group plc, 30-9-94	Participating preference shares calculated on the basis of rights to profits.
PN37, 8-3-96	Newartill plc, 31-10-94	Failure to accrue for minority interest finance relating to premium comparative figures.
PN38, 28-3-96	Brammer plc, 31-12-94	Incorrect classification of assets held for rental. Correct treatment as fixed assets not stock.
PN39, 9-4-96	Foreign and Colonial Investment Trust plc, 31-12-94	Inadequate explanation for departure from SSAP1.
PN40, 1-3-96	Alexon Group plc, 28-1-95	Incorrect sub-classification between equity and non-equity interests.

Table I.7 Continued

Issue number, date	Company, accounts date	Nature of departure
PN41, 30-5-96	Ransomes plc, 30-9-95	Incorrect sub-classification between equity and non-equity interests.
PN42, 24-7-96	Sutton Harbour Holdings plc, 30-3-95	Incorrect use of the true and fair override regarding the treatment of government grants and non-depreciation of investment properties.
PN43, 2-10-96	Butte Mining plc, 30-6-95	Incorrect treatment of shares received in consideration of services and treated as realized profit.
PN44, 17-2-97	Associated Nursing Services plc, 31-3-95 and 31-3-96	Incorrect treatment of certain joint ventures. Treated as quasi-subsididaries instead of associates. Also incorrect treatment of sale and leaseback arrangements.
PN45, 15-4-97	Reckitt and Coleman plc, 30-12-95	Insufficient disclosure of adjustments to fair values in the year after acquisition.
PN46, 29-8-97	M and W Mack Ltd, 26-4-96	Failure to comply with UITF 13 regarding an employee share scheme.
PN47, 2-10-97	Burn Stewart Distillers plc, 30-6-96	Failure to provide sufficient information on the effect of a transaction excluded from the accounts under FRS 5.
PN48, 10-11-97	Strategem Group plc, 31-8-96	Lack of appropriate analysis of FRS 7 fair value.
PN49, 25-2-98	Guardian Royal Exchange plc, 31-12-96	Incorrect treatment of insurance equalization reserves in the consolidated accounts.
PN50, 27-4-98	Strategem Group plc, 31-8-97	Incorrect presentation of non-operating exceptional items
PN51, 12-5-98	RMC Group plc, 31-12-95	Incorrect disclosure of fines.
PN52, 27-7-98	Reuters Holdings plc, 31-12-97	Failure to amortize goodwill to arrive at operating profit.
PN53, 7-8-98	H&C Furnishings plc, 26-4-97	Insufficient disclosure. Fair value of shares used for acquisition.
PN54, 2-9-98	Photo-Me International plc, 30-4-97	Included intragroup sales in sales figure in the consolidated accounts.
PN55, 20-10-98	Concentric plc, 30-9-97	Lack of disclosure with regard to an acquisition contrary to FRS 6 and FRS 1.

5 Relationship between legal rules and standards

Prior to the Companies Act 1989, accounting standards had no force of law but were simply persuasive documents and reporting companies were not obliged to comply with the provisions of accounting standards. The Companies Act 1989 requires reporting companies, other than small and medium-sized companies as defined by the law, to state whether their accounts have been prepared in accordance with applicable accounting standards, and to provide details and reasons for any material departure from those standards. The accounting standards that the Act refers to are the statements of standard accounting practice issued by such body or bodies as may be prescribed by regulations. The body currently recognized for that purpose is the ASB. Accounting standards can be said to represent important statements of accounting practice which standard-setters in the UK expect reporting companies to follow so as to present accounts which reflect a true and fair view.

The accounting requirements in the Companies Acts contain some detailed specifications of the form and content of financial statements and, in some cases, the accounting rules to be followed as well as emphasizing the necessity for the accounts to present a true and fair view of the state of the company's affairs. What constitutes a true and fair view is not defined by law. This issue of the true and fair view and its operationalization is discussed below (see III.5). In a similar manner, the law is not explicit with regard to the mechanics of accounting and of measuring many of the expenses, revenue, gains, losses, assets and liabilities that it requires reporting companies to disclose in their annual reports and accounts. In this circumstance, the provisions of financial reporting standards seek to supplement the legal requirements in the Companies Acts but they do not go outside the bounds of the law. In paragraph 26 of its Foreword to Accounting Standards, the ASB stated that financial reporting standards (FRSs) are drafted in the context of current law and EC Directives with the aim of ensuring consistency between accounting standards and the law.

The legitimacy which the law now gives to financial reporting standards, as well as the prospect that reporting companies would no longer have the freedom to override those standards without publishing their reasons for such departure, and the constant surveillance of the FRRP of the basis for such departure, suggest that the ASB would have to exercise due diligence in the development and publication of its financial reporting standards. It is not entirely clear whether the ASB can disregard the law in order to develop a standard that would go further in promoting the attainment of a true and fair view by financial statements presented on the basis of such a standard than those presented on any other basis. However, there is the prospect that the ASB may be challenged in the court of law if it pursues a strategy which is not in accordance with the law. Such a situation has not arisen as yet and the ASB has always made reference to the implication that compliance with any of its standards ensures compliance with the provisions of the law by referring in each standard to the relevant sections of the law which the standard deals with.

The provisions relating to accounts in the Companies Acts can be amended by the Department of Trade and Industry through the power of statutory instruments. Statutory instruments are secondary legislative instruments which enable a Minister to alter existing rules with or without reference to Parliament. The new requirements introduced by a statutory instrument may be more or less onerous than those in operation under the Companies Act. However, a different procedure applies to either case. If the new requirements are more onerous, a draft of the statutory instrument containing the new regulations has to be laid before Parliament and has to be approved by resolution of each House of Parliament. However, if the proposed new regulations are less onerous than those in operation under the Acts, the approval by the Houses of Parliament of the statutory instrument introducing the new regulations is not a condition precedent but a condition subsequent, and the statutory instrument is valid subject to annulment by either House of Parliament.

In determining whether the proposed new regulations are more or less onerous, they have to be compared with those in force when the regulations are made, and not with those originally set out in the Act. It is believed that standard setters in the United Kingdom can avail and have often availed themselves of the facility provided by statutory instruments to bring the Companies Act into line with accounting standards, especially when the standard is dealing with matters not covered by the Act or matters dealt with in the Act but which have been overtaken by changing business conditions.

II Forms of business organization

By far the most important form of business in the United Kingdom is that of registered company. However, it is not the only form of business which the law places at the disposal of the business person. According to their legal characteristics, the forms of business organization in use are those summarized in Figure II.1. The main forms of business organization in the United Kingdom are corporations, including companies registered under the Companies Act, partnerships, branches of foreign companies and sole proprietorships.

1 Company

The principal Act affecting companies in Great Britain is the Companies Act 1985, although some amendments were made in the Companies Act 1989. In Northern Ireland the main Act is the Companies (Northern Ireland) Order 1986 which makes Companies Acts 1985 and 1989 applicable to Northern Ireland. Companies incorporated under the Companies Act form the greater proportion of those companies describable as corporations aggregate, which are corporations consisting of more than one member (different from corporations sole that are constituted by a single person who has corporate status, such as the Sovereign, an Archbishop or a Minister or Officer of the Crown). As Figure II.1 indicates, corporations may be created by royal charter and by the authority of Parliament. Companies are a form of organization created by the authority of Parliament since the Acts which govern their creation and continued existence are created by Parliament.

While the Companies Acts affect those companies described in Figure II.1 as 'other types', these companies also operate under the provisions of special Acts of Parliament. Most companies have limited liability for their members but there are a few unlimited liability companies as well as some whose liability is limited by guarantee rather than by shares. The limitation of liability refers to the debts arising from the assets of the company. The concept of limited liability provides protection for a member of a company against debts not covered by the assets of the company or the activities of the company that is outside the powers of the directors. A company limited by guarantee is defined as one having the liability of its members limited by the memorandum to such amount as the members may respectively thereby undertake to contribute to the assets of the company in the event of its being wound up.

The main difference between a company limited by guarantee and one limited by shares is that in the latter case, each member's liability is limited to the amount of nominal shares held and the liability may be called upon at any time during the life of the company. In the case of the guarantee company on the other hand, the liability of a member is limited to the amount of the guarantee undertaken by the member and this liability can only be called upon after the commencement of the winding up. As a result, companies limited by guarantee do not obtain their initial funding from their members but from other sources. This type of company is therefore only suitable if no initial funds are required or those funds are obtainable from other sources such as endowments, fees, charges, donations or subscriptions. Examples include professional associations, trade associations and research associations. However, two types of companies limited by guarantee are permitted: those with no share capital (guarantee company in the pure form) and those with share capital (guarantee company of the hybrid type). The latter type of guarantee company is suitable for an enterprise which is sustainable by funds from other sources but requires initial contributions from members to meet the purchase of initial assets but not normal working capital.

Figure II.1 The organization of business in the United Kingdom

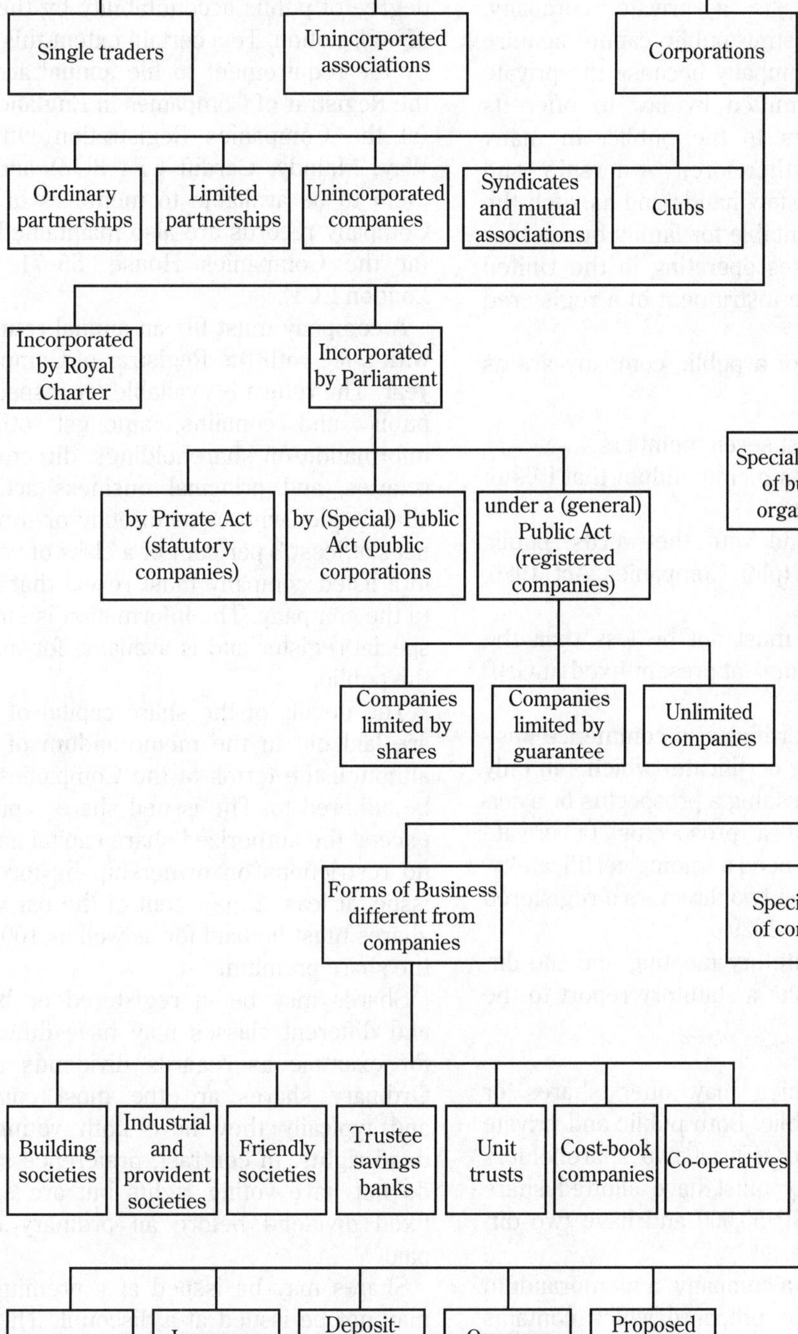

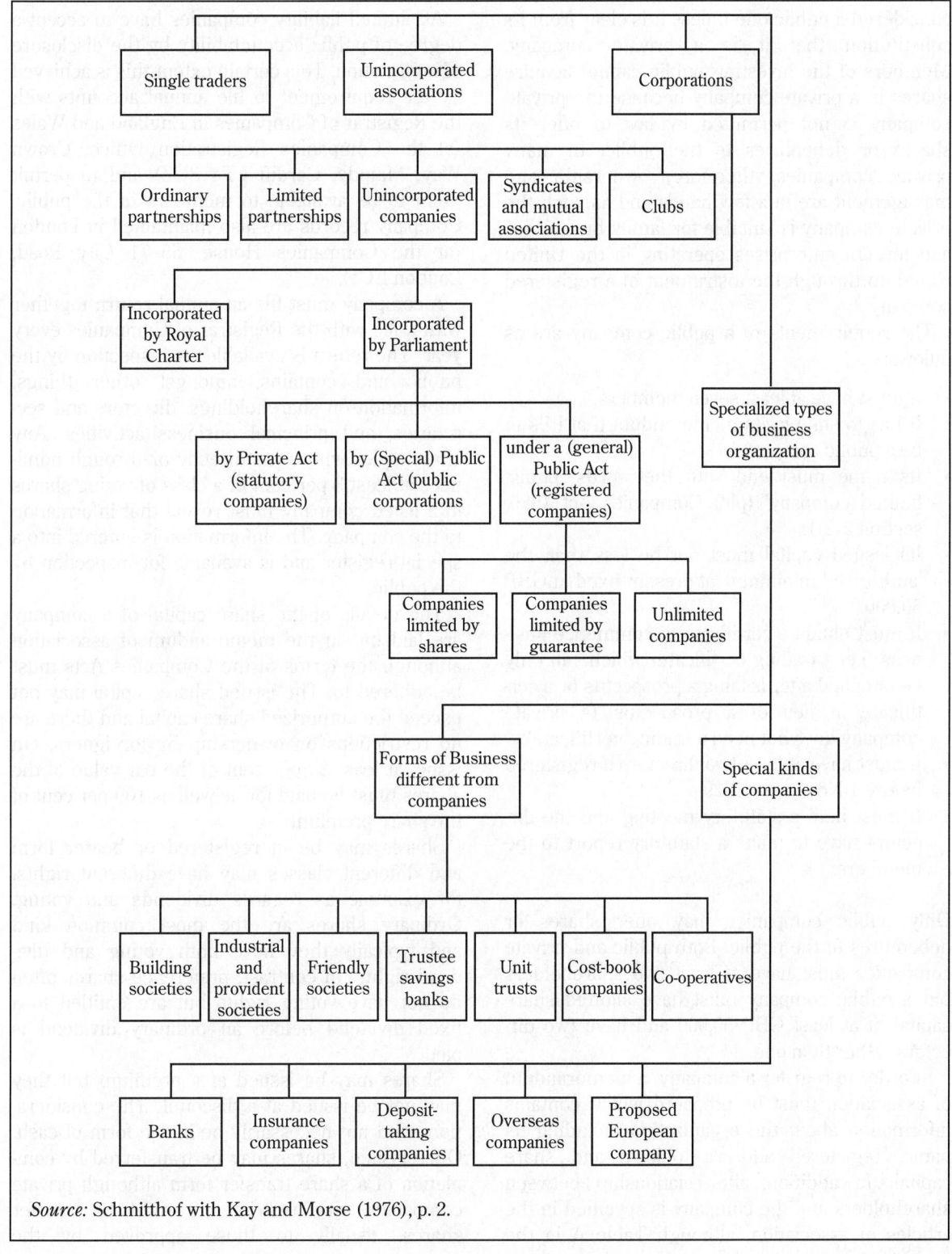

Source: Schmitthof with Kay and Morse (1976), p. 2.

A company limited by shares may be formed as a private or as a public company. A company is considered a public one unless it is clear from its constitution that it is a private company. Members of the investing public cannot acquire shares in a private company because the private company is not permitted by law to offer its shares or debentures to the public. In many private companies, therefore, ownership and management are in a few hands and as such the private company is suitable for family businesses and foreign enterprises operating in the United Kingdom through the instrument of a registered company.

The requirements of a public company are as follows:

- It must have at least seven members.
- It has to state in its memorandum that it is to be a public company.
- Its name must end with the words 'public limited company' (plc) Companies Act 1980, section 2(2)).
- Its issued capital must not be less than the 'authorized minimum', at present fixed at GBP 50,000.
- It must obtain a certificate to commence business (i.e. a trading certificate) which can only be obtained after issuing a prospectus or a certificate in lieu of a prospectus (a private company does not need a trading certificate).
- It must have at least two directors if registered before 1 November 1929.
- It must hold a statutory meeting and the directors have to make a statutory report to the members.

Only public companies may offer shares or debentures to the public. Both public and private companies must have at least two shareholders but a public company must have allotted share capital of at least GBP 50,000 and have two directors rather than one.

In order to register a company, a memorandum of association must be prepared which contains information about the organization including its name, registered address, objects and share capital. In addition, the relationship between shareholders and the company is specified in the articles of association, although Table A in the Companies Regulations 1985 may be used for this purpose.

All limited liability companies have to accept a degree of public accountability by the disclosure of information. To a certain extent this is achieved by the requirement to file annual accounts with the Registrar of Companies in England and Wales (at the Companies Registration Office, Crown Way, Maindy, Cardiff CF4 3UZ) and to permit them to be available to members of the public. Company records are also maintained in London (at the Companies House, 55–71 City Road, London EC1).

A company must file an annual return together with a fee with the Registrar of Companies every year. The return is available for inspection by the public and contains, amongst other things, information on shareholdings, directors and secretaries, and principal business activities. Any shareholder who owns directly or through nominees at least 3 per cent of a class of voting shares in a listed company must reveal that information to the company. The information is entered into a special register and is available for inspection by the public.

The details of the share capital of a company are laid out in the memorandum of association although the terms of the Companies Acts must be adhered to. The issued share capital may not exceed the authorized share capital and there are no restrictions on ownership by foreigners. On issue, at least 25 per cent of the par value of the shares must be paid for as well as 100 per cent of the share premium.

Shares may be in registered or bearer form and different classes may have different rights, for example as regards dividends and voting. Ordinary shares are the most common kind and typically they have both voting and dividend rights. In contrast, preference shares often do not have voting rights but are entitled to a fixed dividend before an ordinary dividend is paid.

Shares may be issued at a premium but they may not be issued at a discount. The consideration need not necessarily be in the form of cash. Once issued, shares may be transferred by completion of a share transfer form although private companies may restrict the right to transfer shares, usually to those approved by the directors. The liability of shareholders is limited to the extent of uncalled-up capital.

Debentures are a kind of loan which can take a number of forms sometimes with security being based on a charge on specified assets. If a charge is attached it must be registered with the Registrar of Companies so that the public have access to the information. Debentures can be irredeemable although most are in fact redeemable at a date and on terms specified at the issuance of the security. Debentures, like shares, can be traded on the Stock Exchange, provided the company issuing them is a registrant with the Stock Exchange.

A director need not be a shareholder in law although sometimes this is required by the articles of association. The shareholders appoint the directors for a fixed period of time and in turn the directors appoint the chairman and managing director from among the body of directors. Directors' interests in the company or an associated company must be recorded in a register and this must be available for public inspection.

The responsibilities of directors are specified in the Companies Act and the company's articles of association and the directors must consider the interests of employees as well as those of the shareholders. The directors also have a responsibility to the public and investing community through the provision of information about the company. Each company must appoint a secretary who acts as the officer responsible for ensuring that legal requirements are met. It is the responsibility of the secretary to maintain the registers although these duties may be delegated to a registrar. The secretary may also be a director except when there is only one director.

Tables II.1, II.2 and II.3 provide statistics on the distribution of companies incorporated in the United Kingdom (Table II.1 historically, Table II.2 distinguishing between public and private, and Table II.3 according to amount of issued share capital).

Over a period of 136 years of company incorporation, 3,712,200 companies have been incorporated in the United Kingdom. A point of note is the high level and increasing number of registrations in the 1990s. Of equal significance is the small number of companies which were incorporated between 1862 and 1869 which survive today. Out of the 5,000 incorporations, less than 100 companies are effectively still on the register.

Public limited liability companies seeking capital often access the capital market by selling their shares to the public or placing them privately through stockbrokers. In the United Kingdom, the sale of shares of listed companies to the public is usually undertaken through the International Stock Exchange (ISE), London. A company does not need to reside in the United Kingdom for its shares to be listed on the ISE. Access to UK capital is therefore open to many foreign registered companies. However, all companies whose shares and loan stocks are listed on the ISE must be registered with the ISE.

Where a company's shares or debentures are listed on the Stock Exchange, the directors of the company and of any associated companies are prohibited from dealing in options to buy or sell any quoted shares or debentures of the company, and the company must maintain a register of substantial interests in one-tenth or more of any class of share capital carrying unrestricted voting rights. Listed companies must also disclose information relating to significant contracts (that is those over 1 per cent of aggregate capital and reserves for a capital transaction or the granting of credit or 1 per cent of total purchases, sales, payments or receipts in respect of other contracts). If no such contracts exist, the company must make such disclosure. The listed company must also disclose details of any arrangement relating to a director's waiver or agreement by a director to waive any emoluments.

A general meeting must be held on an annual basis and not more than 15 months may elapse between one meeting and another. The company must give 21 days notice in writing to shareholders, although a shorter period is acceptable provided the shareholders agree. In general, the annual general meeting is administered in accordance with the articles of association, but there is a statutory requirement to consider the annual report including the accounts, the appointment of auditors and declaration of dividends. Meetings in addition to the annual general meeting may be held to consider specific issues.

The maximum amount that can be paid out as dividends is determined by law and amounts to the sum of the company's accumulated realized

Table II.1 Analysis of companies on the register by period of incorporation

Thousands of companies
% of incorporations in period

Period of incorporation (calendar years)	Incorporations in period	On register at 31 March 1997	On register at 31 March 1998: All companies		On register at 31 March 1998: In liquidation/course of removal		On register at 31 March 1998: Effective register	
	000s	000s	000s	%	000s	%	000s	%
1862–69	5.0	0.1	0.1	2.0	–	0.0	0.1	2.0
1870–79	9.9	0.2	0.2	2.0	–	0.0	0.2	2.0
1880–89	18.0	0.6	0.6	3.3	–	0.0	0.6	3.3
1890–99	36.6	1.8	1.8	4.9	0.1	0.3	1.7	4.6
1900–09	45.0	2.8	2.7	6.0	0.3	0.7	2.4	5.3
1910–19	58.9	4.0	4.0	6.8	0.4	0.7	3.6	6.1
1920–29	87.3	7.8	7.6	8.7	0.7	0.8	6.9	7.9
1930–39	119.1	12.6	11.5	9.7	1.1	0.9	10.4	8.7
1940–44	35.4	4.8	4.8	13.6	0.6	1.7	4.2	11.9
1945–49	88.7	12.4	11.7	13.2	1.2	1.4	10.5	11.8
1950–54	68.9	11.9	11.7	17.0	1.2	1.7	10.5	15.2
1955–59	107.3	20.1	19.7	18.4	1.9	1.8	17.8	16.6
1960–64	193.3	35.6	33.6	17.4	3.3	1.7	30.3	15.7
1965–69	141.7	27.3	27.0	19.1	3.0	2.1	24.0	16.9
1970–74	234.0	49.8	48.3	20.6	6.0	2.6	42.3	18.1
1975–79	287.0	69.4	62.8	21.9	8.1	2.8	54.7	19.1
1980	69.4	16.7	15.8	22.8	2.0	2.9	13.8	19.9
1981	72.4	18.3	17.5	24.2	2.2	3.0	15.3	21.1
1982	87.2	22.9	21.7	24.9	2.8	3.2	18.9	21.7
1983	96.2	25.3	23.8	24.7	3.1	3.2	20.7	21.5
1984	97.9	27.4	25.9	26.5	3.2	3.3	22.7	23.2
1985	104.6	30.7	29.0	27.7	3.6	3.4	25.4	24.3
1986	117.3	37.1	34.6	29.5	4.1	3.5	30.5	26.0
1987	128.0	41.9	39.3	30.7	4.3	3.4	35.0	27.3
1988	128.9	44.5	41.1	31.9	4.6	3.6	36.5	28.3
1989	130.3	46.8	43.3	33.2	5.1	3.9	38.2	29.3
1990	120.9	47.0	43.1	35.6	4.9	4.1	38.2	31.6
1991	109.8	50.4	45.8	41.7	5.5	5.0	40.3	36.7
1992	108.5	57.8	52.2	48.1	6.5	6.0	45.7	42.1
1993	113.6	71.5	63.7	56.1	7.7	6.8	56.0	49.3
1994	127.5	95.7	81.9	64.2	10.0	7.8	71.9	56.4
1995	143.9	138.8	106.9	74.3	12.9	9.0	94.0	65.3
1996	170.2	163.4	148.7	87.4	24.0	14.1	124.7	73.3
1997	199.0	46.6	190.2	95.6	3.8	1.9	186.4	93.7
1998*	50.5	0.0	50.5	100.0	–	0.0	50.5	100.0
All companies	3,712.2	1,244.0	1,323.1	35.6	138.2	3.7	1,184.9	31.9

* To 31 March 1998
– Fewer than 50 companies
Source: Department of Trade & Industry (1998), p. 28.

Table II.2 Public and private companies incorporated and on the GB register 1993–94 to 1997–98

	Thousands of companies				
	1993–94	1994–95	1995–96	1996–97	1997–98
Public Companies					
New incorporations	1.3	0.9	0.9	1.1	1.1
Conversions from private	2.6	2.9	3.1	3.4	3.6
Dissolved	0.8	1.1	0.9	1.3	0.7
In liquidation/course of removal	1.8	1.8	2.3	1.7	1.7
Effective number on register at end of period	12.0	11.9	11.5	11.7	11.9
Public companies as percentage of effective register	1.2%	1.2%	1.1%	1.1%	1.0%
Private Companies					
New incorporations	114.0	130.9	145.7	169.1	204.2
Conversions from public	1.2	1.4	1.6	2.0	2.3
Dissolved	132.9	127.4	105.6	91.5	126.7
In liquidation/course of removal	161.1	140.6	124.4	150.4	136.5
Effective number on register at end of period	944.7	969.9	1,027.3	1,080.2	1,173.0
Of which: Unlimited	3.7	3.8	3.8	3.9	3.8
GB total of effective numbers of public and private companies	956.7	981.8	1,038.8	1,091.9	1,184.9

Source: Department of Trade & Industry (1998), p. 27.

profits (see III) less its accumulated losses and is often referred to as a 'revenue reserve'. In contrast, non-distributable profits are sometimes referred to as a 'capital reserve' although neither of these 'reserve' terms are of legal importance.

2 Partnership

Whereas companies usually limit the liability of their members, in a partnership the partners are jointly liable for the debts of the organization. Arrangements for the sharing of profits, interest and capital are usually specified in a partnership agreement, but if there is no such arrangement the terms of the Partnership Act 1890 prevail. It is possible for some of the partners to have limited liabilities provided that others accept unlimited liability. Partnerships are normally limited to not more than 20 partners but there are exceptions to this, notably in accounting and the legal profession.

3 Branch of a foreign company

A foreign company which has an established place of business in the United Kingdom is described as an overseas company. The UK operations of such a company may be constituted

Table II.3 Analysis of companies on the register at 31 March 1998 by issued share capital

	England & Wales		Scotland		Great Britain	
Issued share capital	No. of companies 000s	Issued capital GBP m	No. of companies 000s	Issued capital GBP m	No. of companies 000s	Issued capital GBP m
No issued share capital	42.2	0.0	2.7	0.0	44.9	0.0
Up to GBP 100	841.1	29.8	43.6	1.4	884.7	31.2
Over GBP 100 and under GBP 1,000	54.2	19.1	2.1	0.8	56.3	19.9
GBP 1,000 and under GBP 5,000	108.0	165.3	5.6	9.5	113.6	174.8
GBP 5,000 and under GBP 10,000	28.1	173.9	2.4	15.2	30.5	189.1
GBP 10,000 and under GBP 20,000	39.4	463.3	3.8	46.0	43.2	509.3
GBP 20,000 and under GBP 50,000	35.2	1,028.3	3.9	114.4	39.1	1,142.7
GBP 50,000 and under GBP 100,000	28.1	1,732.3	2.7	170.2	30.8	1,902.5
GBP 100,000 and under GBP 200,000	22.1	2,746.1	2.2	279.4	24.3	3,025.5
GBP 200,000 and under GBP 500,000	17.8	5,295.1	1.6	467.3	19.4	5,762.4
GBP 500,000 and under GBP 1m	9.5	6,216.5	0.8	516.9	10.3	6,733.4
GBP 1m and over	24.6	650,479.5	1.4	24,074.7	26.0	674,554.2
Total	1,250.3	668,349.2	72.8	25,695.8	1,323.1	694,045.0

Source: Department of Trade & Industry (1998), p. 30.

in the form of a branch or a subsidiary, but its residence must be intended to have more than a fleeting character. The provisions in the Companies Acts relating to the registration of a foreign company include:

- The approval of the name of the business by the Secretary of State.
- The filing of a number of documents with the Registrar of Companies.

The documents must include a copy of the memorandum of association and articles of association, a list of the directors and secretary of the company, a declaration by a director or secretary about the date of establishment of the place of business and the names of any other persons who are authorized to accept notices served on the company. All foreign companies must display the name of the company and must provide details of incorporation, country of origin and name of directors.

With respect to accounting information, the company must file audited accounts with the Registrar although there is no requirement to file a directors' report and an auditors' report.

4 Sole proprietorship

There are no restrictions with respect to who can operate as a sole practitioner and no regulations concerning the production of financial statements. However, the Inland Revenue and the Customs and Excise require adequate information to be supplied for assessment purposes. Sole proprietors have unlimited liability which differs from trading as a company.

III Objectives, concepts and general principles

1 Leading the way to a conceptual framework

1.1 Setting accounting standards

The aim of a conceptual framework is to develop a coherent set of generally accepted accounting principles which may guide the solution of accounting problems in a consistent manner and against which the chosen solutions may be tested. This process requires, inter alia, the identification of users and the type and form of information that user groups need. Compared with the United States, the United Kingdom was relatively slow in considering the desirability of a conceptual framework.

It was not until 1978 that the ASC published a consultative document entitled 'Setting Accounting Standards' (ASC, 1978). Basically the document recognized that the ASC had been slow to contemplate the issue but considered that it would not be possible to develop an accepted framework because of the demands of different user groups. However, the consultative document did not dismiss the issue completely as it raised two questions on which views of interested parties were sought:

- Is it accepted that there is at present no single 'model' or 'agreed conceptual framework' which can be used as the touchstone for accounting standards?
- Should the ASC encourage research into the possibility of finding an acceptable 'model'?

The answer to both questions was 'yes'. As a result of the answer to the second question, the ASC commissioned Professor Richard Macve to review the international literature, to express an opinion as to the feasibility of developing an agreed conceptual framework, and to advise on the desirability of further research work. Before this report (the Macve Report) is considered (see III.1.5) it is desirable to look at earlier work that had been undertaken in the United Kingdom that is of relevance to users and user groups.

1.2 SSAP 2, Disclosure of Accounting Policies

SSAP 2, Disclosure of Accounting Policies, was issued in November 1971 and dealt with fundamental issues rather more than any of the other accounting standards. SSAP 2 attempted to improve the quality of information disclosed in order to assist understanding and interpretation of financial statements. It attempted to do this by describing four fundamental accounting concepts, accounting bases and the selection of accounting policies.

The four concepts identified as being fundamental were the going concern concept, the accruals or matching concept, the consistency concept and the prudence concept. The going concern concept assumes that a company will exist for the foreseeable future and that there is no necessity or intention to liquidate or substantially reduce the extent of the business. The accruals concept relates to the matching of revenues with the appropriate costs. The consistency concept expects like items to be treated in the same manner within an accounting period and from one accounting period to another. The prudence concept is one of conservatism in which revenues and profits are not anticipated but are recognized on realization. In contrast all known liabilities, expenses and losses must be provided for regardless of whether the liability is known with certainty. If there is a conflict between the matching concept and the prudence concept then the latter must prevail.

The predominant behavioural disposition of corporate reporting in the United Kingdom before the advent of the revolutionary changes being introduced by the ASB was conservatism. This is an attribute of financial reporting that refers to a preference for accounting and disclosure practices that tend to minimize reported profits (or maximize reported losses). SSAP 2 and the ASB associate prudence with conservatism though the ASB distances itself from conservatism's idea of deliberate, consistent understatement of net assets and profits. The ASB defines prudence as the inclusion of a degree of caution in the exercise of the judgements needed in making the estimates required under conditions of uncertainty, such that assets or income are not overstated and liabilities or expenses are not understated. However, as the ASB states in its proposed Statement of Principles (Chapters 1 and 2, para. 32):

> the exercise of prudence does not allow, for example, the creation of hidden reserves or excessive provisions, the deliberate understatement of assets or income, or the deliberate overstatement of liabilities or expenses, because the financial statements would not be neutral and, therefore, not have the quality of reliability.

Examples of prudent behaviour include the choice of lower (rather than higher) revenues, higher (rather than lower) expenses, the lower of (historical) cost or market values (for inventories and for marketable securities), and the recognition of contingent losses and unrealized losses but not contingent gains and unrealized gains. In essence, the desire is to recognize expenses and losses as quickly as possible and to delay the recognition of revenues and gains as long as possible.

This operational behaviour is at variance with the expectation of agency theory, according to which management are viewed as being inclined to present the most optimistic picture of their organization (Watts and Zimmerman, 1986). Management, it is assumed, tend to report good rather than bad news. Given this managerial disposition, why has conservatism held its sway in accounting?

Several reasons have been suggested for this prudent disposition in accounting. One is accounting's desire to respond to the potential asymmetric treatment of gains and losses by risk averse investors. Another is the fact that users can more easily verify the unduly conservative claim than the unduly optimistic claims of management. Yet another is the desire of accountants to offset management's natural tendency to focus on the positive and to postpone contemplation, let alone revelation, of unpleasant facts as long as possible (Devine, 1963). There is also the desire to facilitate contracts and to use conservatism as a means of maintaining capital.

Another source of the conservative behaviour is the national temperaments of accounting regulators and their desire to use the concept as a means of regulating corporate behaviour. For example, the high degree of conservatism in Germany arises from the legislative imperatives of the German Companies Act of 1844. Following speculative episodes in the 1870s, the German Companies Act of 1884 (unlike the United Kingdom Companies Act) required companies not to take inflation into account in determining their profits and to value assets recorded in balance sheets at the lower of cost or market values. Another reason for German conservatism and high hidden reserves is that there is little distinction between financial and tax accounting in Germany.

In contrast, inflationary experiences in the United Kingdom have led to the preference for periodic revaluation of balance sheet assets over historical costs. SSAP 2 defines accounting bases as those methods developed for applying fundamental concepts to the company's financial accounts. The bases selected will determine the accounting periods in which revenue and costs are recognized in the profit and loss account and the amounts at which items in the balance sheet should be stated. There is not a unique accounting basis.

Accounting policies are those accounting bases adopted by the company which the directors consider appropriate to show a true and fair view of the state of affairs of their company. The accounting policies should be disclosed as they are fundamental to an understanding of the financial statements.

1.3 The Corporate Report

In 1975 the then ASSC published a discussion paper entitled 'The Corporate Report' (ASSC, 1975) which was intended to generate debate about the fundamental aims of published financial reports and how they are to be achieved. The discussion paper defined the corporate report as 'the comprehensive package of information of all kinds which most completely describes an organization's economic activity'. The document suggested that the corporate report should be considered as embodying the notes and explanations to the accounts as well as the basic financial statements of the profit and loss account, the balance sheet and the funds statement.

The document also identified the types of

organization that should publish regular financial statements, the main users of such information and their needs and the form of the report which would best fulfil those needs. Users were defined as those having a right to information on the entity. Seven user groups were identified: the equity investor group, the loan creditor group, the employee group, the analyst-adviser group, the business contact group, the government and the public. Thus, a pluralistic approach was taken to the identification of user groups rather than the US approach of identifying investors as the main user group.

The committee stated that the 'fundamental objective of corporate reports is to communicate economic measurements of and information about the resources and performance of the reporting entity useful to those having reasonable rights to such information.' To fulfil these objectives reports should be relevant, understandable, reliable, complete, objective, timely and comparable.

Based on a survey of the chairmen of 300 of the largest UK listed companies, the report also stated that distributable profit was not the only measure of performance and that other measures may be appropriate.

In order to achieve the aims of the corporate report there was consideration of additional financial statements such as a value added statement, an employment report, a statement of money exchanges with government, a statement of foreign currency transactions, a statement of future prospects and a statement of corporate objectives. The report also recommended that further work should be undertaken on disaggregation of financial information and social accounting.

The committee also gave consideration to concepts underlying profit measurement and income distributability. There was recognition that there were considerable deficiencies in historical cost accounting and that this limited the extent to which the objectives of financial reporting could be achieved. The concept of maintaining capital in an inflationary environment was considered and a number of bases, including current purchasing power and current value accounting, both replacement cost and net realizable value, were discussed. It was accepted that no one system of accounting would meet the needs of all the user groups identified but that multi-column reporting and current value accounting might be developed in the future.

1.4 The Sandilands Committee

To a large extent the proposals put forward in The Corporate Report (see III.1.3) were obscured by the government's Inflation Accounting Committee, the Sandilands Committee, that was set up in 1974. The remit of this committee was to consider whether and how companies should account for relative changes in costs and prices. Like The Corporate Report, the Sandilands Committee focused on the information needs of user groups. The essence of the inflation accounting committee's findings was that money is the unit of measurement rather than 'purchasing power units', that balance sheet amounts should be stated at their 'value to the business', and that operating profit should be stated after charging the 'value to the business' of assets consumed during the year. As a result operating profit would be current cost profit and would exclude holding gains.

1.5 The Macve Report

The overall result of the prevailing enquiry about the purpose of accounting and the bases for measuring accounting profit was to set up a study into the conceptual nature of accounting. The study, which was commissioned by the ASC, was to review the conceptual framework projects in the United States and Canada and to consider whether the United Kingdom should embark upon such a project. The Macve Report, which was published in August 1981, defined a conceptual framework as an attempt to clarify the objectives of financial reporting and how alternative practices are likely to help achieve those objectives. Macve concluded that various attempts in Canada and the United States to establish an agreed conceptual framework have not been successful. He concluded that user needs are not homogeneous and, as a consequence, seeking agreement on the form and content of financial statements is just as much a political process as the pursuit of the technically best method. According to Macve (1981, p. 14):

> ... it seems unlikely that searching for an agreed conceptual framework of theory in abstraction from individual problems of disclosure and method will be successful. What is important is that in identifying and considering particular problems there should be questioning of purpose, of likely consequences, and of how user needs and different interests will be served ... A conceptual framework for accounting should be regarded rather as a common basis for identifying issues, for asking questions and for carrying out research than as a package of solutions.

The report called for further research into the conceptual nature of financial reporting and the politics of setting accounting standards, but this advice was not heeded. However, several incursions into these problems were made by recent research studies: one commissioned by the Institute of Chartered Accountants of Scotland and two by the Institute of Chartered Accountants in England and Wales. The results of these studies point to the direction in which British accounting may be expected to move in the future. Some of the ideas contained in these documents seem, in the current state of corporate reporting, more like accounting in Utopia, whilst others may be described almost as revolutionary.

1.6 Making Corporate Reports Valuable

In 1988 the Institute of Chartered Accountants of Scotland (ICAS) issued a discussion document entitled 'Making Corporate Reports Valuable' (McMonnies, 1988) with the object of stimulating debate. It focuses more on the future of financial reporting and does not give any idea of the current state of financial reporting in the United Kingdom. The committee identified many weaknesses in current financial reporting including a perceived deficiency that financial statements are too influenced by the law rather than by economic reality, that corporate reports are not made public sufficiently quickly, and the assumption of reporting companies that the information needs of investors are similar to the needs of management in running their businesses. The report considered the identification of user groups but whilst a pluralistic approach was adopted, four main groups were identified rather than the seven identified in The Corporate Report. The four groups were the equity investor group, the loan creditor group, the employee group and the business contact group including creditors.

Having identified the user groups, their information needs were specified. They were in fact very similar to the needs of equity investors identified in The Corporate Report but included some additional information that might be disclosed to help users compare performance with budget, on management's explanations of divergences and on future plans together with any assumptions that are made.

The committee also considered valuation bases for assets and liabilities and decided that both historical cost and economic value had considerable problems, that current replacement cost and net realizable value had advantages and that the latter better reflected value rather than cost. As a consequence, the committee advocated four main financial statements:

- An assets and liabilities statement based on net realizable value.
- An operations statement which is a statement of additions to financial wealth during the year.
- A statement of changes in financial wealth from one period to another.
- A distributions statement reflecting the change in distributable wealth for the current and previous period.

As well as the four main financial statements, additional information was advocated in a number of areas including cash flows and segmental information.

1.7 The Solomons Report

In 1989 David Solomons introduced a report (Solomon, 1989) that had been commissioned in May 1987 by the Research Board of the ICAEW. The report considered the purposes of financial reporting, the users and their needs and how these needs are being fulfilled. He followed an asset and liability approach to income determination and then went on to define the elements of financial statements. Once defined, an item should be recognized when it conforms with his definitions of assets and liabilities, can be measured and verified with reasonable certainty and the amounts are material.

Solomons then went on to consider current financial reporting in the United Kingdom and was critical of the historical cost/modified historical cost approaches. He advocated the concepts of value to the business and the maintenance of real financial capital but dismissed an approach based on net realizable values as advocated by the ICAS (see III.1.6). His method links the approaches of current purchasing power (CPP) and current value accounting without introducing the gearing adjustment and monetary working capital adjustments (discussed below, see VIII) that were not very popular in practice.

1.8 The Future Shape of Financial Reports

In 1991 the Research Board of the ICAEW and the Research Committee of the ICAS issued a discussion paper entitled 'The Future Shape of Financial Reports' (Arnold *et al.*, 1991). The report identifies the purpose of financial reporting as the provision of information for shareholders, lenders and others so that they can evaluate the performance of the entity in order to make decisions about future expectations of performance and also to enforce contracts. The paper advocated a statement of objectives and related strategic plan, a statement of assets and liabilities, an income statement, a gains statement, a cash flow statement and information on future prospects.

The paper adopted a compromise position to accommodate previous attitudes because the Scottish report had favoured realizable values whereas the Solomons report advocated current cost. The outcome was a mishmash of the two systems that, in our opinion, has proved to be less than satisfactory.

2 Development of a conceptual framework by the ASB

2.1 Statement of Aims

The genesis of the conceptual framework 'Statement of Principles' project lies in the recommendations of the committee that also recommended the setting up of the ASB. Under the Chairmanship of Sir Ron Dearing, a report entitled 'The Making of Accounting Standards' stated that:

> a lack of a conceptual framework is a handicap to those involved in setting standards as well as to those applying them. Work on its development should, therefore, be pursued at a higher rate than hitherto but consistent with the perceived scope for progress .... We believe that work in this area will assist standard setters in formulating their thinking on particular accounting issues, facilitate judgements on the sufficiency of the disclosures required to give a true and fair view, and assist preparers and auditors in interpreting accounting standards and in resolving accounting issues not dealt with by specific standards.

In 1990, upon taking office, the ASB adopted the 22 extant standards (see Table 1.3) which thereby were given the status of 'accounting standards' within the terms of Part VII of the Companies Act 1985. In 1991, the ASB issued a Statement of Aims which specified that the Board will attempt '... to establish and improve standards of financial accounting and reporting, for the benefit of users, preparers, and auditors of financial information'. These aims will be achieved by developing principles which will guide the establishment of accounting standards and provide a framework which will assist in resolving accounting issues. This will be operationalized by issuing or amending accounting standards in the light of changes and by addressing urgent issues promptly.

2.2 Statement of Principles – overview

2.2.1 Positioning the Statement of Principles

The ASB published the first draft of the Statement of Principles in the early 1990s as a series of discussion drafts with a revision being published in November 1995 as an Exposure Draft. The exposure draft attracted more letters of comment than any other document published by the Board and several of them were critical (See Ernst and Young's paper – The ASB's Framework: Time to

Table III.1 Summary of responses to the Statement of Principles

Response type	All responses	Number of respondents expressing this view	% of 175 respondents
1	A statement is in principle a good idea	68	39%
2	This SOP is acceptable	12	7%
3	This SOP is not acceptable	134	77%
4	Current values/discounting not acceptable	125	71%
5	Current values/discounting acceptable	9	5%
6	Cash aspects not adequate	12	7%
7	Presentation of SOP inadequate	46	26%
8	Information overload for users	10	6%
9	New concepts not acceptable	12	7%
10	Do not abandon established concepts	82	47%
11	STRGL not viable	51	29%
12	Goodwill treatment inadequate	16	9%
13	Overwhelming practical problems	60	34%
14	Downgrading of P&L not acceptable	71	41%
15	New definitions inadequate	39	22%
16	Recognition/realization criteria inadequate	39	22%
Total no. of summarized points recorded		786	

Source: Wilkinson-Riddle G. J., Holland L., JAAR, July 97.

Decide, published in February 1996, and the analysis to responses to the Statement of Principles carried out by two UK academics, published in July 1997, Table III.1).

Some comments highlighted fundamental differences of opinion over key issues dealt with in the draft statement. The content of other comments, according to the ASB, suggested that the 'debate over the exposure draft had become obscured by misunderstanding and confusion'. In July 1996 the ASB effectively withdrew the draft stating that the Board has concluded that it would not be appropriate to proceed directly to the development of a final document. A revised exposure draft, although promised for 1997, was eventually published in March 1999 together with an introductory booklet and a Technical Supplement.

The introductory booklet aims to address the common misconceptions that arose in respect of the 1995 draft. The Technical Supplement explains the rationale underlying key aspects of the draft statement and, in so doing, seeks to address some of the most common criticisms made of the approach adopted.

2.2.2 The purpose of the Statement of Principles

The ASB still believes that the main role of the Statement of Principles for Financial Reporting is to provide a framework for the consistent and logical formulation of individual accounting standards and the statement will provide valuable conceptual input into its (the ASB's) work on the development and review of accounting standards. The principal use of the Statement will be by standard setters in conceptually underpinning their work. The revised Statement stresses that it is not to be considered as either an accounting standard nor any other form of mandatory document and that the main impact on accounting practice will be through its influence on the standard setting process. The Board further clarifies the point that the Statement will not be the sole factor to be taken into account in developing and reviewing standards. Other factors such as legal requirements, implementation issues, industry-specific issues, cost–benefit consideration and the desirability of evolutionary change will also be important. The ASB concedes that these other factors may result in an accounting standard adopting an approach that is different from the approach suggested by the Statement.

2.2.3 The Statement of Principles and current accounting practice

To emphasize the not-so-revolutionary nature of the Statement, the ASB cited the following examples:

- The draft Statement is based on the IASC's 'Framework for the Preparation and Presentation of Financial Statements', which was issued almost 10 years ago. That framework document was itself based on the Statements of Financial Accounting Concepts issued in the USA between 1978 and 1985 by the Financial Accounting Standards Board.
- The principles in the draft Statement are already being used by the leading financial reporting standard setters, including the IASC and those in Australia, Canada, New Zealand and the United States.
- Many of the key principles already play a prominent role in company reporting. For example:
  FRS 2 'Accounting for Subsidiary Undertakings' uses the reporting entity concept described in Chapter 2,
  FRS 3 'Reporting Financial Performance' draws on the principles of good presentation described in Chapter 7,
  FRS 4 'Capital Instruments' uses the definitions of assets and liabilities set out in Chapter 4,
  FRS 5 'Reporting the Substance of Transactions' uses and operationalizes the definitions of assets as liabilities set out in Chapter 4,
  FRS 11 'Impairment of Fixed Assets and Goodwill' uses the recoverable amount notion described in Chapter 6.
- The draft statement recognizes that the concept of a true and fair view (discussed below) is fundamental to the whole system of financial reporting and represents the ultimate test of financial statements.
- SSAP 2 'Disclosure of accounting policies' and its fundamental accounting concepts remains of importance and has been integrated within the framework it describes. To add greater weight to these citations Sir David Tweedie, Chairman of the ASB said:

> We have, over the last nine years, published 15 accounting standards that are all, to a greater or lesser extent, based on principles in the draft Statement. Those standards have generally been well received, which shows that the Statement of Principles is not the harbinger of revolutionary change that it is sometimes characterized as. On the contrary, they have proved an effective means of restoring sound accounting practice by stamping out the appalling abuses of the 1980s. The dog-eared accounting concepts used in the 1980s were simply not up to the task of dealing with the transactions of the late 20th century and had to be replaced – their replacement is the Statement of Principles (ASB's PN134, March 1999).

2.3 Summary of the Statement of Principles for Financial Reporting

The new draft Statement of Principles for Financial Reporting consists of a compendium of eight chapters. They can be summarized as follows:

2.3.1 Chapter 1 – The objective of financial statements

The objective of financial statements is to provide information about the reporting entity's financial performance and financial position that is useful to a wide range of users for assessing the stewardship of management and for making economic decisions. The chapter goes on to explain that this objective can usually be met by focusing exclusively on the needs of present and potential investors. Such investors need information about financial performance and financial position that is useful to them in evaluating the reporting entity's ability to generate cash (including the timing and certainty of its generation) and in assessing the entity's financial adaptability. The Chapter emphasizes that financial statements will not provide all the information needed by users; they will, however, provide a frame of reference against which users can evaluate the more specific information they obtain from other sources.

2.3.2 Chapter 2 – The reporting entity

Formerly Chapter 7, the reporting entity is now encapsulated within Chapters 2 and 8 of the new draft statement. Chapter 2 differentiates between an entity's reporting activities and resources under its direct control in the form of single entity financial statements and activities and resources under the entity's direct and indirect control; in other words, on the activities it carries out itself and its own assets and liabilities and also the activities, and liabilities of its subsidiaries, in the form of consolidated financial statements.

2.3.3 Chapter 3 – The qualitative characteristics of financial information

In deciding what information should be included, when it should be included and how it should be presented, the aim is to ensure that financial statements yield useful information – therefore, all financial information must contain certain qualitative characteristics. Chapter 3 identifies the main qualitative characteristics for information to be useful viz: relevance, reliability, comparability and understandability (see Figure III.1). Looking at each of these in turn:

- Relevance – the quality characteristic of relevance requires information to have the ability to influence the economic decision of users.
- Reliability – information presented can be depended upon to represent faithfully what it either purports to represent or could reasonably be expected to represent, is complete and is free from deliberate or systematic bias and material error, and under conditions of uncertainty – involves a degree of caution.
- Comparability – for information to be comparable it must enable the users to discern and evaluate similarities in, and differences between, the nature and effects of transactions and other events over time and across different reporting entities.
- Understandability – a user with reasonable knowledge of business, economic activities and accounting displaying a willingness to study with reasonable diligence the information provided should be able to appreciate the significance of reported information.

The chapter acknowledges that conflicts may arise in applying the characteristics. It suggests that a trade-off needs to be found that still enables the objective of financial statements to be met. It cites as an example that, if the information that is the most relevant is not the most reliable and vice versa, the item of information to use is that which is the most relevant of those that are reliable. Having used the qualitative characteristics to maximize the usefulness of financial information the materiality test should then be used to determine whether the information's usefulness is of such significance as to require it to be given in the financial statements. An item of information is material to the financial statements if its misstatement or omission might reasonably be expected to influence the economic decisions of users of those financial statements.

2.3.4 Chapter 4 – The elements of financial statements

Taking a largely balance sheet approach in this chapter the ASB has retained the emphasis it places on assets and liabilities, even defining the items that are to be included in the profit and loss account (income statement) in terms of assets and liabilities. It cites this approach as being the one adopted in the framework documents of all the major financial reporting standard-setters around the world. As the effects of transactions and other events on the reporting entity's financial performance and financial position have to be reflected in the financial statements in a highly aggregated form – order is imposed on this process by specifying and defining the classes of items (the elements) that encapsulate the key

Figure III.1 The qualitative characteristics of accounting information

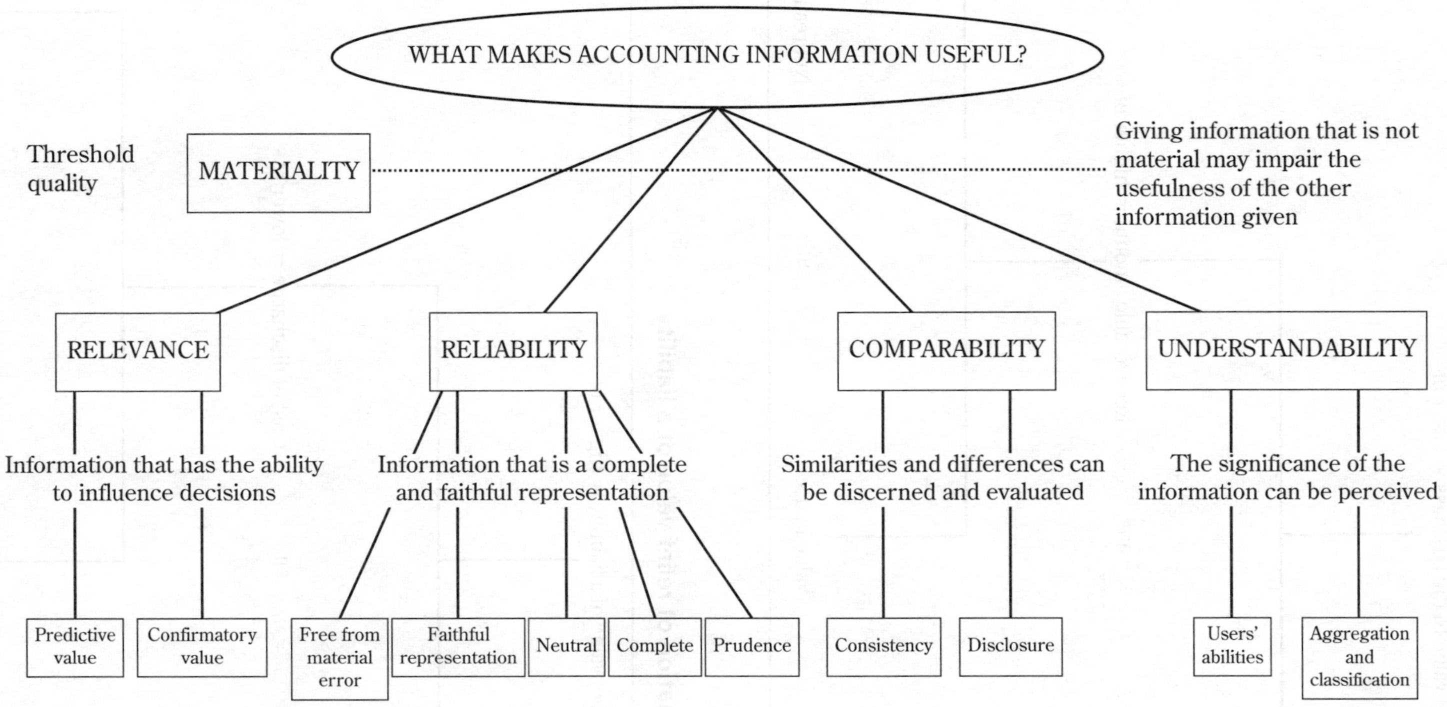

Source: ASB, Exposure Draft, Statement of Principles, The Objective of Financial Statements and The Qualitative Characteristics of Financial Information.

Figure III.2 Determination of value to the business

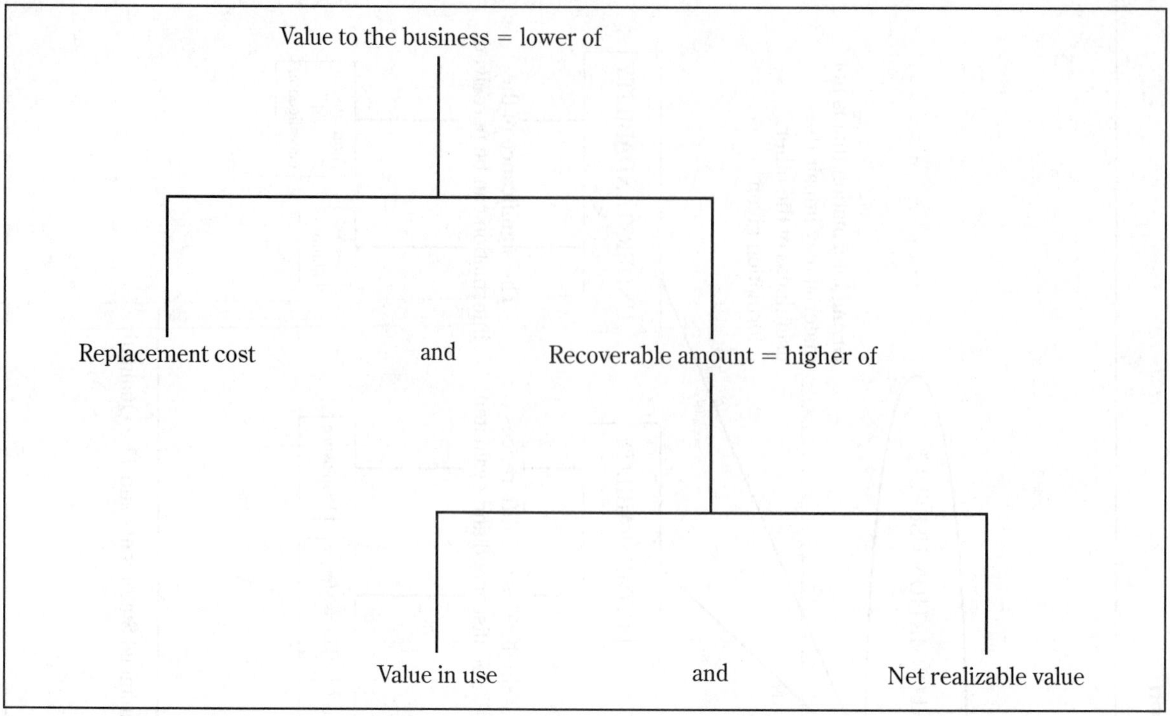

Figure III.3 Determination of relief value of a liability

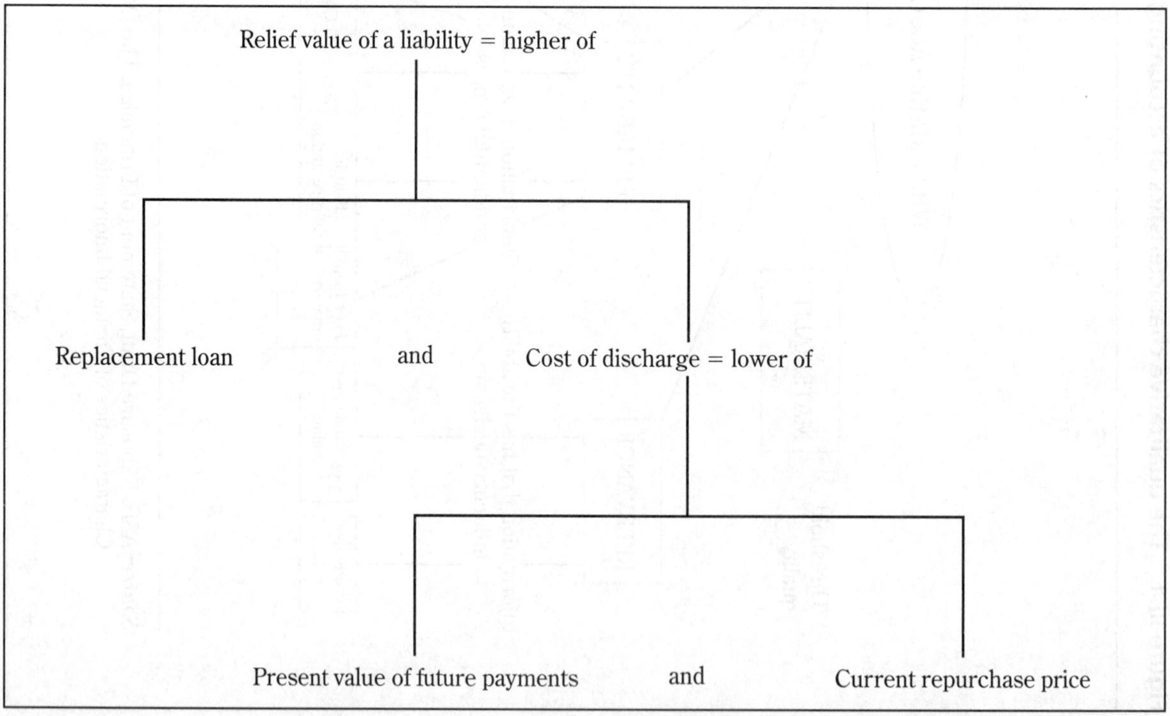

aspects of the effects of those transactions and other events. The chapter lists the main elements and their definitions as follows:

- Assets – rights or other access to future economic benefits controlled by an entity as a result of past transactions or events.
- Liabilities – obligations to transfer economic benefits as a result of past transactions or events.
- Ownership interest – residual amount found by deducting all of the entity's liabilities from all of the entity's assets.

The chapter requires the above elements to be included in the balance sheet (or statement of position). The elements to be included in the profit and loss account and any other statement of financial performance are as follows:

- Gains – increases in ownership interest not resulting from transfers from owners in their capacity as owners (for example, new capital investment by owners).
- Losses – decreases in ownership interest not resulting from transfers to owners in their capacity as owners (for example, transfers in the form of dividend payments).

2.3.5 Chapter 5 – Recognition in financial statements

The effects that transactions and other events have on the elements of financial statements can largely be achieved through depicting the same, in words and by a monetary amount, in the primary financial statements. This process is termed recognition in Chapter 5. For example, if the effect of a transaction is to create a new asset or liability or to add to an existing asset or liability, that new asset or liability or addition will be recognized in the balance sheet if there is sufficient evidence that it exists and it can be measured reliably enough at a monetary amount. A gain or loss will be recognized at the same time, unless there has been no change in the total net assets or the whole of the change is the result of capital contributions or distributions. Sometimes it is easier to identify the effect that a transaction or other event has had on the elements of financial statements by considering whether the critical event in the operating cycle has occurred, because it is only then that there will be sufficient evidence that a gain exists and that a reliable measure of the gain can be determined. Even though matching is not used by the draft statement to drive the recognition process, it still plays an important role in the approach described in the draft in allocating the cost of assets across reporting periods and in telling preparers where they may find assets and liabilities.

2.3.6 Chapter 6 – Measurement in financial statements

From normal activities most assets and liabilities arise from arm's length transactions. For such items, regardless of the measurement basis used, the carrying amount assigned on initial recognition will be the transaction costs. Chapter 6 explains that historical cost and current value are alternative measurement bases. After initial recognition, however, the carrying amounts derived from the two bases will usually diverge. To deal with this divergence the chapter adopts the approach followed by the majority of large UK listed companies involving carrying some categories of balance sheet items at historical cost and others at current value (see Figure III.2). This is referred to in the chapter as the mixed measurement system, although it is often referred to elsewhere as the modified historical cost basis.

The draft chapter envisages that the measurement basis used for a category of assets or liabilities will be determined by reference to factors such as the objective of financial statements, the nature of the assets or liabilities concerned and the particular circumstances involved. Whatever the measurement basis chosen, the carrying amount may need to be changed from time to time. This process is termed 'remeasurement'. For example, when a historical cost measure is used, remeasurement is necessary to ensure that all assets are stated at the lower of cost and recoverable amount and all monetary items denominated in a foreign currency are stated at an amount that is based on up-to-date exchange rates. Remeasurements will be recognized only if there is sufficient evidence that the monetary amount has changed and the new amount can be measured with sufficient reliability.

The draft chapter then goes on to describe a framework that would guide the choice of an appropriate measurement basis for each balance sheet category (see Figure III.3).

2.3.7 Chapter 7 – Presentation of financial information

The chapter explains that as the primary objective of financial statements is to communicate clearly and effectively the financial statements should be as simple, straightforward and brief as is possible, while retaining their relevance and reliability. Good presentation, it explains, avoids adding a mass of material and significantly lengthening the documents.

As the presentation of information in financial statements will always involve a fair amount of interpretation, simplification, abstraction and aggregation, the process requires integrity and clarity in order for the users of financial statements to benefit and for greater knowledge and understanding to be created at all levels. To this end, the chapter argues, the presentation of financial performance information should focus on the components of financial performance and their characteristics including, for example, their nature, cause, function, relative continuity or recurrence, stability, risk, predictability and reliability. All components are relevant to an assessment of financial performance, although their individual characteristics mean that some will carry more weight in certain assessments – and in the assessment of some entities – than others. The presentation of information and financial position should focus on the types and functions of assets and liabilities held and the relationships between them. For example, the cash flow information should be presented in a way that distinguishes between those cash flows that result from operations and those that result from other activities. It should also show the extent to which the entity's various activities generate and use cash.

2.3.8 Chapter 8 – Accounting for interests in other entities

The chapter explains that interests in other entities can have an important effect on the reporting entity's own financial performance and financial position. They therefore need to be fully reflected in the financial statements. How this is done will depend on whether single entity or consolidated financial statements are being prepared (see Chapter 2 above) and the degree of influence, if any, exerted over the activities and resources of the investee and the nature of that interest.

In a single entity financial statement, the chapter explains, an interest in another entity is treated like any other asset. In consolidated financial statements, however, an interest that involves control of that other entity's operating and financial policies is dealt with by making the controlled entity part of the reporting entity itself.

Further, an interest that involves joint control of, or significant influence over, that other entity's operating and financial policies is dealt with by recognizing the reporting entity's share of that other entity's results and resources in the financial statements in a way that does not imply that they are controlled by the reporting entity. Finally, where the interest involves lesser or no influence over that other entity's operating and financial policies the appropriate treatment is the same as any other asset.

The chapter briefly looks at business combinations and states that an amalgamation of two or more reporting entities will be accounted for in accordance with its character. All amalgamations can be characterized as either a purchase (or acquisition) or a uniting of interests (or merger).

3 Realizable or realized profit

United Kingdom accounting is primarily based on transactions. Thus, one of the main areas for decision making is the allocation of part-completed transactions to accounting periods. However, SSAP 2 identifies two of the four fundamental accounting concepts as accruals and prudence. The accruals or matching concept attempts to match costs with revenues it has assisted in generating. Revenue and costs are accrued rather than being recognized when cash is received. In contrast, prudence dictates that revenue and profits should not be anticipated but recognized only when realized in the form of cash or other assets.

In terms of company profits the concept of 'realized profits' was introduced into the Companies Act 1980 (see I.2.4). The Act was further clarified by the 1981 and 1985 Companies Acts and subsequently by the 1989 Act. The Act states that only realized profits at the balance sheet date should be included in the profit and loss account. Profits or losses determined in accordance with accounting standards will be realized profits or losses. A provision should be treated as a realized loss. Development costs must be treated as a realized loss even if carried forward as an asset. However, it is possible, in certain circumstances, for the directors to decide that the amount should not be treated as a realized loss, in which case this should be stated in the notes to the accounts and the circumstances should be explained to justify the directors' decision.

One of the problems with the requirements of the Companies Acts is that it is not clear exactly what the generally accepted accounting principles are for the determination of realized profits and losses. However, the ASB through Chapter 5 of its revised exposure draft – Statement of Principles – has attempted to conceptualize this problem. The general recognition criteria in Chapter 5 are similar to those given by the IASC with the exception that more evidence is required for recognition of gains than for recognition of losses. The three types of event that may trigger recognition are specified as transactions, contracts for future performance and other events. A transaction is defined as a transfer of assets or liabilities to or from an external party. Such a definition includes both fully performed contracts and partly formed contracts since they involve a transfer of assets or liabilities. The chapter cites, as examples, cash and credit sales as transactions but not the placing of an order with no payment being made.

A contract is defined as an enforceable, but as yet unperformed, promise given to or by an external party to transfer assets and/or liabilities in the future. As a result, contracts such as leases, in which the asset and first rental have not taken place, or those unperformed parts of contracts are included as contracts.

Other events are defined as changes in assets and liabilities other than transactions and contracts such as changes in market prices, the elapse of time and the manufacture of goods.

4 Relevance of tax accounting to financial reporting

In the United Kingdom there is a clear distinction between the tax accounts and the financial statements. While the underlying concepts and principles in both sets of accounts are identical, it often occurs that the amount of tax payable on the profits of a period may not appear to be related to the income and expenditure for that period. This is because the basis for arriving at profits in the financial accounts is not the same as the basis used for tax purposes. There are two main reasons for this. First, some income is not taxable and some expenditure is not allowable for tax. As a result, permanent differences may arise where there are income or expenses in the financial accounts that are not present in the tax accounts. Second, there are items in the financial statements which are not dealt with in the same period as in the tax accounts. This leads to timing differences and a potential deferred tax liability or asset.

Timing differences are important because performance is often measured on an after-tax basis. As a result, the extent of recognition of deferred tax liabilities is important. Recognition of deferred taxes will affect not only the profit and loss account but also the balance sheet to the extent that a liability or asset will crystalline. Deferred taxation is considered below (see VII.2).

Finally, the increasing influence of global trade, particularly through the Internet, will compound the differences between tax accounting and financial reporting. This is in addition to the specific problems faced within both tax accounting and financial reporting in relation to revenue recognition and revenue capture.

5 True and fair view

5.1 General

The term 'true and fair view' is a means by which users of annual reports are informed about the overall quality of disclosure in annual reports (Lee, 1994, p. 30). All the Companies Acts since 1948 expect auditors to inform users that the accounts present a 'true and fair view'. Companies Act 1985 also expects reporting companies to override the accounting provisions contained in the law if that departure would enable the accounts to present a 'true and fair view'. The term 'true and fair view' was introduced into company legislation by the Companies Act 1948, but different variations of the term can be found in earlier Companies Acts. In the Companies Act 1844, auditors were expected to examine the balance sheet to be presented to shareholders at general meetings and were required to state whether the balance sheet was 'full and fair' and whether it exhibited a 'true and correct' view of the company's state of affairs. Although the Companies Act 1856 did not mandate an audit, it required the balance sheet to be presented to shareholders at general meetings to convey a 'true and correct view' of the company's state of affairs. It was suggested in the model articles of association in the Companies Act 1856 that the auditor was to ascertain the 'truth and correctness' of the state of affairs described in the balance sheet, although an audit was not mandated by that Act. In the Railways Companies Act 1867, section 30 provided for an audit in which the auditor had to ascertain that the balance sheet of the company represented a 'full and true' statement of the financial condition of the company. The Companies Act which mandated the audit of a company's account for the first time since 1856 required the auditor to state whether, in his opinion, the balance sheet conveyed a 'true and correct view' of the state of the company's affairs.

The insistence that published accounts be 'full and fair', 'true and correct' or 'full and true' was to prevent fraudulent financial reporting. The truthfulness and correctness were meant to refer to the relationship between the books of account and supporting records kept by the company and the balance sheet that was presented to the shareholders. It is an indication that the data was free from any deliberate act of dishonesty or from inadvertent error (Lee, 1979). Although the term 'true and fair view' was introduced into the law by Companies Act 1948, the term has no legal definition or a uniquely singular judicial interpretation in the United Kingdom (Lee, 1994, p. 34). It is neither defined by statute nor by case law in the United Kingdom, nor defined in any member country of the European Union. Also, it is not defined in practice and as such has no absolute meaning. The law does not require that 'the' true and fair view be shown but that 'a' true and fair view be shown. Therefore the truth and fairness of a view given at a point in time may change over time.

5.2 The meaning of 'true and fair view'

Literally, the phrase can be interpreted as follows:

- True – information is not false but factual and conforming with reality. In addition, information conforms with required standards and law. In practical terms the accounts have been correctly extracted from the books and records.
- Fair – information is free from discrimination and bias and in compliance with expected standards and rules. Practically the accounts should reflect the commercial substance of the company's underlying transactions.

On a more theoretical basis many attempts have been made to make sense of the term 'true and fair view'. Walton (1993) considers that the term can be construed as representing one or more of three basic ideas:

- A legal residual clause.
- All independent concept.
- Generally accepted accounting principles (GAAP).

The legal residual clause is the usual *ejusdem generis* ('catch all') clause which seeks to embrace those circumstances not covered and foreseen by the situations identified in the legal document. The true and fair view as an independent concept refers to the notion of a higher objective (independent of accounting rules) which accountants and auditors are to seek from a set of accounts. The third interpretation is that the construct can be operationalized by the observance of generally accepted accounting principles.

The use and understanding of the term have been changing over time as a result of the pragmatic approach to accounting in the United Kingdom. The overall intent of the 'true and fair view' construct is for the information disclosed to represent reality. This is the desire to let accounting and financial reports reflect the current and real issues rather than the historical figures they are prone to reflect. Central to this concern is the notion of income and value as concepts rather than the outcome of the system of accounting. Accounting rules contained in the Companies Act and those in accounting and financial reporting standards are concerned with how reporting companies should communicate information contained in their books of accounts. As a result, the application and interpretation of the 'true and fair view' construct presume that corporate annual reports would communicate details of the books of accounts and represent reality at the same time.

There are two aspects to the meaning of true and fair view:

- The aspect of communication connecting reporting companies and the users of annual reports and accounts.
- The aspect of representation connecting the language of disclosure and information.

The first aspect (communication of details of books of accounts) is associated with factual disclosure of what is reported in the books of accounts. It is in this respect that many regulators believe that the observance of accounting rules will ensure the presentation of a true and fair view.

The second aspect (representation of reality) is associated with fairness – relating the reported numbers to reality in the best and most neutral manner possible. In this respect, discussion in the literature has centred around the correspondence between the signifier ('true and fair view') and the signified (the underlying reality) (Walton, 1993; Lee, 1994). The major problem is ensuring that corporate annual reports achieve a balance between four 'worlds': the world as a reservoir of figures and words; the world as representable by figures and words; the world as a source of motivation; and the world as a field of action (Wallace, 1987, p. 137). As an instrument of communication, a company's annual report and accounts represents the activities of the reporting company. Apart from the fact that the figures and words tend to (but do not of their own) mirror commercial reality and often create or nurture, in a subliminal manner, a world view different from the world the annual report and accounts purport to represent, the images (such as 'true and fair view') of the world the annual report and accounts purport to represent are far from being stagnant. As a result, a term such as 'true and fair view' could end up meaning many more things than the various things it represents to different groups of the same or different societies. On the other hand, the scope and meaning of and respect accorded GAAP, including accounting and financial reporting standards, have changed with time and space.

A distinction should be drawn between disclosure in the annual report of a company (the signifier) and the world view about the company (the signified). Different types of world views can be constructed about a company (rather than the one represented in a particular annual report and accounts). This may be because accounting regulators and standard setters, reporting companies and their auditors and the users of accounts that live in the world (serving, for all, as a source of motivation and a field of action) are different groups with different aspirations for corporate reporting. In addition, these groups or parties interested in annual reports and accounts and the world in which they live are not often exactly the same before and after the annual report and accounts have been issued. However, the duty to ensure that accounts portray a 'true and fair view' is left with directors of reporting companies and their auditors. As a result, 'true and fair view', like beauty, is packaged by the preparer (reporting company) but, unlike beauty, it is in the eye of the preparer and its auditor(s), not in the eye of the beholder or user. To make the 'true and fair view' construct understandable to all, Alexander (1993, p. 74) suggests two alternative solutions:

- A situation where preparers and users of accounting information have a full understanding of the differences between each other's general and accounting culture.
- A situation where both have identical general and accounting cultures (i.e. there are no differences needing to be understood).

According to Alexander (1993, p. 74), neither of these is attainable.

5.3 The 'true and fair view' outside the United Kingdom

The concept of 'true and fair view' is no longer restricted only to the United Kingdom. The term has been introduced into the American scene (though identified by a different label 'present fairly'; Lee, 1994, p. 35), into Australia (where its operationalization has resulted in controversy; Parker, 1994) and into European legislation. In the United States, although there is an SEC Rule 203 that permits reporting companies registered with the SEC to depart from GAAP, the operational understanding of the concept of 'present fairly' is in terms of compliance with GAAP (Lee 1994, p. 36). In Australia, practice is not settled on whether 'true and fair view' should mean compliance with accounting and disclosure standards in the country in line with the practice in the United States, compliance with both accounting standards and conceptual statements issued by the standard setting board, or whether it should be a basis on which reporting companies could escape from legal control without actually violating the law (Parker, 1994).

The need to observe the principles of the true and fair view has been incorporated in the Fourth Directive Art. 2 (3)–(5). The Fourth Directive permits departure from the accounting rules incorporated in company law and financial reporting standards if following such rules and standards would prevent the presentation of a true and fair view. Member states of the European Union which have incorporated the true and fair view provision as specified by the Fourth Directive in their laws have drawn up rules relating to when a case is sufficiently exceptional that the need to give a true and fair view should override the requirement to comply with the detailed treatment (measurement, and sometimes disclosure) rules of the Fourth Directive.

The implementation and interpretation of the true and fair view provision in each member country of the European Union have varied from a requirement to comply with accounting rules drawn up by each country in observance of the provision of the Fourth Directive to the requirement in some countries (e.g. the United Kingdom) that reporting companies must override accounting rules enshrined in the law if, by doing so, they would be presenting a true and fair view of the state of their companies' affairs (Alexander, 1993 and 1994). Some have even suggested that a European operational meaning of 'true and fair view' different from the British operational meaning may develop (Ordelheide, 1993). In the United Kingdom, Chopping and Skerratt (1994, p. 30) have suggested that the principle of 'true and fair view' can be implemented through:

- Current accounting rules, i.e. rules in the Companies Act, including financial reporting standards.
- Current accepted practice, i.e. rules in the Companies Act including evolving practice.
- Justified deviations from accepted practice.

The role of rules (i.e. those in the Companies Acts and in financial reporting standards) requiring corporate reports to portray a 'true and fair view' is unclear in the EU. According to Parker (1994, p. 52) 'rules might serve (1) as a necessary and sufficient condition for truth and fairness; or (2) as a necessary but not sufficient condition for truth and fairness or (3) as sufficient in themselves without any reference needed to truth and fairness'. All three interpretations can be found to subsist in the EU. It is also not clear whether reported income or loss and reported values of assets and liabilities should be based on observance of rules or on the overriding desire to present a 'true and fair view'. Nor is it clear whether information which would enable users of accounts to be presented with a true and fair view, that is different from information based on figures derived from the observance of rules, should be reported in the notes to the accounts.

5.4 The operationalization of the 'true and fair view' in the United Kingdom

The theoretical presumption, in the United Kingdom and Ireland, is that financial statements must not depart from financial reporting standards and that those that do depart do not give a true and fair view. This was the opinion of the ICAEW-appointed legal counsel on the status of accounting standards, though the legal counsels (see Hoffman and Arden, 1983, paras. 4 and 14) concluded that the courts ultimately have to decide whether accounts prepared on the basis of accounting standards issued by the ASC would be sufficient to portray a true and fair view. This is because SSAPs have no direct legal effect; they are simply rules of professional conduct for accountants (Hoffman and Arden, 1983, para. 11). However, a recent council opinion obtained by the ASB has suggested that accounts prepared on the basis of financial reporting standards issued by the ASB are more likely to be construed by the courts as meeting the true and fair view requirement. This is because the ASB, unlike its predecessor the ASC, no longer reflects the views of the accounting profession given its broad membership, its partial funding by government, and its statutory recognition (Arden, 1993).

In practice, there is the understanding that to conform to the requirements of company law relating to the form and content of financial statements, the notes thereto and the financial reporting standards do not necessarily result in the presentation of a true and fair view. While extant financial reporting standards are authoritative descriptions of the state of accounting thought, such standards may not reflect changing situations. Whether the information provided in the annual report and accounts would be enough to give a true and fair view is, to some extent, a matter of opinion of the directors based on the circumstances of a particular company. The opportunity to use the desire to attain a 'true and fair view' as a basis for overriding the requirements of financial reporting standards is a reflection of the UK and Irish preference for a dynamic interpretation of the accounting provisions of the law and of extant financial reporting standards. In effect, the 'true and fair view' requirement provides reporting companies, auditors and lawyers with room to manoeuvre in different financial reporting situations (Lee, 1994, p. 36).

The greater emphasis on judgement, which distinguishes the UK and Irish approach from that of much of continental Europe, requires the particular circumstance of each individual company to be balanced against the standard practice, whether prescribed by company law, financial reporting standard or otherwise (ASC, 1978, p. 17). As a result, the law recognizes that a departure from the express requirements of the law and accounting standards, is not only desirable but essential in those special circumstances where it is necessary in order to give a true and fair view. In short, what is a 'true and fair view' can change with time and it is possible for more than one set of financial accounts relating to a company's financial transactions and activities in one year to equally show a true and fair view of the company's state of affairs and operations. This is why the UITF Abstract no. 7 (see Table I.5) requires a company to justify its departure from the provisions of the law and financial reporting standards in an explanatory note to the accounts. Reporting companies must disclose the particulars of such a departure and the reasons for it, and its effect on the published results must be given in the notes.

IV Bookkeeping and preparation of financial statements

There is no uniform set of accounts which companies are required to follow as there are in France (the *plan comptable*) or in Germany (the branch specific *Kontenrahmen*). There are no financial reporting standards and accounting standards dealing with bookkeeping and the taking up of inventory. The accounting rules in the United Kingdom operate under the principle that the accounting systems which a company adopts are entirely a matter for the company. However, there are requirements that a company must maintain proper books of accounts. What are proper books of accounts is not defined, but they must enable a company to prepare its financial statements within the period specified by law and must be in a form that is verifiable by auditors. The Companies Act 1985 (s.221) requires a company to keep accounting records that would enable it to show and explain the company's transactions and to:

- Disclose with reasonable accuracy, at any time, the financial position of the company at that time.
- Enable the directors to ensure that any balance sheet and profit and loss account prepared by them under Part VII of Companies Act 1985 complies with the requirements of the Act.

Under s.237 of Companies Act 1985, unless expressly mentioned in the auditors' report, there will be an implication that:

- Proper accounting records have been kept by the company and proper returns adequate for the audit have been received from branches not visited.
- The company's balance sheet and profit and loss account are in agreement with these.
- The auditor has obtained all the information and explanations which are necessary in order for him to form his opinion.
- The disclosure requirements under Schedule 6 of the Companies Act 1985 in relation to transactions with directors have been met.

The accounting records must in particular contain the following:

- Entries from day to day of all sums of money received and expended and the matters in respect of which the receipt and expenditure takes place.
- A record of assets and liabilities.
- If the company's business involves dealing in goods, statements of stock held by the company at the end of each financial year. The company must retain statements of stocktakings from which the statements of stock are prepared and, except in the case of goods sold by way of ordinary retail trade, records of goods sold and purchased identifying the goods and the buyers and sellers.

The presumption in Companies Act 1985, Schedule 3 (1), para. 4A and Companies Act 1989 Schedule 2 is that the accounting policies for assets and liabilities of individual companies that are part of a group of companies should be in accordance with the accounting policies for assets and liabilities of the group. If there are different reasons for departing from the requirements for uniform accounting policies, for example where in exceptional cases it is impracticable to make adjustment, there shall be disclosed (a) the particulars of the different accounting policies used; (b) the reasons for the different treatment; and (c) the effect of the departure, indicating the amount of the assets and liabilities involved and, where practicable, the effect on the assets and net assets.

The Fourth Directive permitted small and medium-sized companies to file modified financial statements and this has been introduced by providing some exemptions from the Companies Acts 1948 to 1981. A curious aspect is that a full set of annual accounts (although not filed) must still be sent to all members of a company so that companies taking advantage of the concessions have to undertake more rather than less work. Small and medium-sized companies are defined by Companies Act 1985, section 247(3) as those that satisfy two of the conditions specified in Table IV.1 for both their current and preceding financial year. The thresholds may be revised from time to time by Statutory Instruments (SI)

Table IV.1 Size criteria for classification as a small or medium-sized company for individual (not group) companies

	Small	Medium-sized
Turnover (proportionately adjustable for accounting period less than one year)	≤GBP 2.8 million	≤GBP 11.2 million
Balance sheet totals (of items A to D in Format 1 (see Table V.1) or total of assets before deducting liabilities in Format 2 (see Table V.2))	≤GBP 1.4 million	≤GBP 5.6 million
Average number of persons employed, determined on a weekly basis	≤50	≤250

issued by the Department of Trade and Industry. They have been so amended twice – by Companies Act 1989, section 13 and by Statutory Instrument 1992 No. 2452.

It is important to note that the Act permits the filing of modified financial statements, although the requirement to prepare full accounts still applies. The ASB issued a Financial Reporting Standard for Smaller Entities (FRSSE) in November 1997 that was revised in December 1998. The FRSSE focuses on a reporting standard for small companies although it is voluntary. In general the same measurement rules apply to smaller entities as for larger companies although disclosure for smaller companies is not so onerous. Revisions to the FRSSE will be made in the light of changes in financial reporting.

In practice, a small company can prepare and file full accounts; can prepare accounts in accordance with the FRSSE; or prepare accounts on either of the previous bases and file abbreviated accounts in accordance with Companies Act 1985.

V Balance sheet and profit and loss account formats

1 Form and content according to the law

The Companies Act 1985, in line with the expectations of the Fourth EC Directive, specified that the balance sheets to be published by reporting companies can take one of two indicated formats (see Tables V.1, V.2 and V.3) and the profit and loss accounts can take one of four formats (see Tables V.4, V.5, V.6, V.7 and V.8). The choice of format is left to the directors but once a choice is made that format should not be changed in normal circumstances.

2 Reporting financial performance

2.1 Review of FRS 3 'Reporting financial performance'

In one of its efforts to reshape financial reporting in the United Kingdom, the ASB issued FRS 3, Reporting Financial Performance. This accounting standard introduces a change to the current form and content of the profit and loss account suggested by the Companies Act and the Fourth Directive. The profit and loss account before the issue of FRS 3 emphasized a bottom-line figure, the profit after tax. According to the ASB, a company's results cannot realistically be encapsulated in a single number such as earnings per share or profit before tax. For a proper understanding of the position a much wider perspective is needed, and that is what FRS 3 attempted to introduce. Although FRS 3 has been the culmination of several years of research on how to make corporate annual reports more meaningful and useful to the average user it is currently under review (see below).

The evolution of FRS 3 started with many of the research publications described earlier, such as the Corporate Report (ASSC, 1975), Macve Report (1981), Making Corporate Reports Valuable (ICAS, 1988), Solomons Report (1989), and the Future Shape of Financial Reports (1991) and the Statement of Principles published by the ASB at different times between 1991 and 1994. FRS 3 supersedes SSAP 6, Extraordinary Items and Prior Year Adjustments, issued in 1974 and revised in 1986. SSAP 6, was based on an all-inclusive concept of profit, and permitted different treatments of apparently similar events and transactions as either ordinary or extraordinary items in the profit and loss account. The standard also permitted the reporting companies the opportunity to pass some similar items through profit and loss accounts (above the bottom-line figure) or through reserves (below the bottom-line figure). The ASB's intention to alter this and to narrow the differences and variety of accounting practice was announced in a discussion draft issued in April 1991 which was developed into an exposure draft, FRED 1, The Structure of Financial Statements – Reporting of Financial Performance, published in December 1991.

The key features of FRS 3 are as follows:

- The profit and loss account should distinguish between the results of continued (including the results of acquisitions) and discontinued operations to the level of operating profit.
- Exceptional items should include any material

Table V.1 Information to be disclosed in the balance sheet (Format 1)

A	**Called up share capital not paid[1]**		
B	**Fixed assets**		
	I	Intangible assets	
		1	Development costs
		2	Concessions, patents, licences, trademarks and similar rights and assets[2]
		3	Goodwill[3]
		4	Payments on account
	II	Tangible assets	
		1	Land and buildings
		2	Plant and machinery
		3	Fixtures, fittings, tools and equipment
		4	Payments on account and assets in course of construction
	III	Investments	
		1	Shares in group companies
		2	Loans to group companies
		3	Shares in related companies
		4	Loans to related companies
		5	Other investments other than loans
		6	Other loans
		7	Own shares[4]
C	**Current assets**		
	I	Stocks	
		1	Raw materials and consumables
		2	Work in progress
		3	Finished goods and goods for resale
		4	Payments on account
	II	Debtors[5]	
		1	Trade debtors
		2	Amounts owed by group companies
		3	Amounts owed by related companies
		4	Other debtors
		5	Called up share capital not paid[1]
		6	Prepayments and accrued income[6]
	III	Investments	
		1	Shares in group companies
		2	Own shares[4]
		3	Other investments
	IV	Cash at bank and in hand	
D	**Prepayments and accrued income[6]**		
E	**Creditors: amounts falling due within one year**		
	1	Debenture loans[7]	
	2	Bank loans and overdrafts	
	3	Payments received on account[8]	
	4	Trade creditors	

5 Bills of exchange payable
6 Amounts owed to group companies
7 Amounts owed to related companies
8 Other creditors including taxation and social security[9]
9 Accruals and deferred income[10]

F Net current assets (liabilities)[11]

G Total assets less current liabilities

H Creditors: amounts falling due after more than one year

1 Debenture loans[7]
2 Bank loans and overdrafts
3 Payments received on account[8]
4 Trade creditors
5 Bills of exchange payable
6 Amounts owed to group companies
7 Amounts owed to related companies
8 Other creditors including taxation and social security[9]
9 Accruals and deferred income[10]

I Provisions for liabilities and charges

1 Pensions and similar obligations
2 Taxation, including deferred taxation
3 Other provisions

J Accruals and deferred income[10]

K Capital and reserves

I Called up share capital[12]

II Share premium account

III Revaluation reserve

IV Other reserves
- 1 Capital redemption reserve
- 2 Reserve for own shares
- 3 Reserves provided for by the articles of association
- 4 Other reserves

V Profit and loss account

Source: Companies Act 1985, Schedule 4.
(The footnotes are described in Table V.3 and are from the Companies Act 1985, Schedule 4.)

Table V.2 Information to be disclosed in the balance sheet (Format 2)

ASSETS

A Called up share capital not paid[1]

B Fixed assets

I Intangible assets
1 Development costs
2 Concessions, patents, licences, trademarks and similar rights and assets[2]
3 Goodwill[3]
4 Payments on account

II Tangible assets
1 Land and buildings
2 Plant and machinery
3 Fixtures, fittings, tools and equipment
4 Payments on account and assets in course of construction

III Investments
1 Shares in group companies
2 Loans to group companies
3 Shares in related companies
4 Loans to related companies
5 Other investments other than loans
6 Other loans
7 Own shares[4]

C Current assets

I Stocks
1 Raw materials and consumables
2 Work in progress
3 Finished goods and goods for resale
4 Payments on account

II Debtors[5]
1 Trade debtors
2 Amounts owed by group companies
3 Amounts owed by related companies
4 Other debtors
5 Called up share capital not paid[1]
6 Prepayments and accrued income[6]

III Investments
1 Shares in group companies
2 Own shares[4]
3 Other investments

IV Cash at bank and in hand

D Prepayments and accrued income[6]

LIABILITIES

A Capital and reserves

I Called up share capital[12]

II Share premium account

III Revaluation reserve

IV Other reserves
1 Capital redemption reserve
2 Reserve for own shares
3 Reserves provided for by the articles of association
4 Other reserves

V Profit and loss account

B Provisions for liabilities and charges

1 Pensions and similar obligations
2 Taxation including deferred taxation
3 Other provisions

C Creditors[13]

1 Debenture loans[7]
2 Bank loans and overdrafts
3 Payments received on account[8]
4 Trade creditors
5 Bills of exchange payable
6 Amounts owed to group companies
7 Amounts owed to related companies
8 Other creditors including taxation and social security [9]
9 Accruals and deferred income[10]

D Accruals and deferred income[10]

Source: Companies Act 1985, Schedule 4.
(The footnotes are described in Table V.3 and are from the Companies Act 1985, Schedule 4.)

profits or losses on sale or termination of operations, costs of fundamental reorganization or restructuring and profits and losses on disposal of fixed assets.

- Extraordinary items, which would now be extremely rare given the very broad definition of ordinary activities adopted in FRS 3, should be disclosed.
- Earnings per share should be disclosed and calculated on the profit attributable to equity shareholders of the reporting entity, after accounting for minority interests, extraordinary items, preference dividends and other appropriations in respect of preference shares. A reporting entity can disclose an additional earnings per share based on another level of earnings, but that additional indicator should be given at least as prominent a disclosure as the one required by FRS 3, be reported consistently from year to year and should be reconciled to the FRS 3 required indicator (see IX.6).
- A memorandum note of historical cost profits and losses should be presented by reporting entities that have revalued assets if there is material difference between the results reported in the profit and loss account and the result on an unmodified historical cost basis. Such a memorandum should be presented immediately following the profit and loss account or the statement of recognized gains and losses.
- A statement of recognized gains and losses should be presented. The statement should include profit or loss for the period together with all other movements on reserves reflecting gains and losses attributable to shareholders.
- There should be a reconciliation of movements in shareholders' funds which brings together the performance of the period, as shown in the statement of total recognized gains and losses, with all the other changes in

Table V.3 Notes on information to be disclosed in the balance sheet (Formats 1 and 2)

(1) Called up share capital not paid
(Formats 1 and 2, items A and C.II.5.)
This item may be shown in either of the two positions given in Formats 1 and 2.

(2) Concessions, patents, licences, trademarks and similar rights and assets (Formats 1 and 2, item B.I.2.)
Amounts in respect of assets shall only be included in a company's balance sheet under this item if either –
(a) the assets were acquired for valuable consideration and are not required to be shown under goodwill; or
(b) the assets in question were created by the company itself.

(3) Goodwill (Formats 1 and 2, item B.I.3.)
Amounts representing goodwill shall only be included to the extent that the goodwill was acquired for valuable consideration.

(4) Own shares (Formats 1 and 2, items B.III.7 and C.III.2.)
The nominal value of the shares held shall be shown separately.

(5) Debtors (Formats 1 and 2, items C.II.1 to 6.)
The amount falling due after more than one year shall be shown separately for each item included under debtors.

(6) Prepayments and accrued income (Formats 1 and 2, items C.II.6 and D.)
This item may be shown in either of the two positions given in Formats 1 and 2.

(7) Debenture loans (Format 1, items E.1 and H.1 and Format 2, item C.1.)
The amount of any convertible loans shall be shown separately.

(8) Payments received on account (Format 1, items E.3 and H.3 and Format 2, item C.3.)
Payments received on account of orders shall be shown for each of these items in so far as they are not shown as deductions from stocks.

(9) Other creditors including taxation and social security (Format 1, items E.8 and H.8 and Format 2, item C.8.)
The amount for creditors in respect of taxation and social security shall be shown separately from the amount for other creditors.

(10) Accruals and deferred income (Format 1, items E.9, H.9 and J and Format 2, items C.9 and D.)
The two positions given for this item in Format 1 at E.9 and H.9 are an alternative to the position at J, but if the item is not shown in a position corresponding to that at J it may be shown in either or both of the other two positions (as the case may require).

(11) Net current assets (liabilities) (Format 1, item F.)
In determining the amount to be shown for this item any amounts shown under 'prepayments and accrued income' shall be taken into account wherever shown.

(12) Called up share capital (Format 1, item K.I and Format 2, item A.I.)
The amount of allotted share capital and the amount of called up share capital which has been paid up shall be shown separately.

(13) Creditors (Format 2, items C.1 to 9.)
Amounts falling due within one year and after one year shall be shown separately for each of these items and their aggregate shall be shown separately for all of these items.

Source: The Companies Act 1985, Schedule 4.

Table V.4 Information to be disclosed in the profit and loss account (Format 1 vertical presentation using the cost of sales method)

(See note (17) in Table V.8)

1	Turnover
2	Cost of sales[14]
3	Gross profit or loss
4	Distribution costs[14]
5	Administrative expenses
6	Other operating income
7	Income from shares in group companies
8	Income from shares in related companies
9	Income from other fixed asset investments[15]
10	Other interest receivable and similar income[15]
11	Amounts written off investments
12	Interest payable and similar charges[16]
13	Tax on profit or loss on ordinary activities
14	Profit or loss on ordinary activities after taxation
15	Extraordinary income
16	Extraordinary charges
17	Extraordinary profit or loss
18	Tax on extraordinary profit or loss
19	Other taxes not shown under the above items
20	Profit or loss for the financial year

Source: Companies Act 1985, Schedule 4.
(The footnotes are described in Table V.8 and are from the Companies Act 1985, Schedule 4.)

Table V.5 Information to be disclosed in the profit and loss account (Format 2 vertical presentation using total cost of production method)

1	Turnover
2	Change in stocks of finished goods and in work in progress
3	Own work capitalized
4	Other operating income
5	(a) Raw materials and consumables
	(b) Other external charges
6	Staff costs:
	(a) Wages and salaries
	(b) Social security costs
	(c) Other pension costs
7	(a) Depreciation and other amounts written off tangible and intangible fixed assets
	(b) Exceptional amounts written off current assets
8	Other operating charges
9	Income from shares in group companies
10	Income from shares in related companies
11	Income from other fixed asset investments[15]
12	Other interest receivable and similar income[15]
13	Amounts written off investments
14	Interest payable and similar charges[16]
15	Tax on profit or loss on ordinary activities
16	Profit or loss on ordinary activities after taxation
17	Extraordinary income
18	Extraordinary charges
19	Extraordinary profit or loss
20	Tax on extraordinary profit or loss
21	Other taxes not shown under the above items
22	Profit or loss for the financial year

Source: Companies Act 1985, Schedule 4.
(The footnotes are described in Table V.8 and are from the Companies Act 1985, Schedule 4.)

Table V.6 Information to be disclosed in the profit and loss account (Format 3 horizontal presentation using the cost of sales method)

(See note (17) in Table V.8)

A	**Charges**	
	1	Cost of sales[14]
	2	Distribution costs[14]
	3	Administrative expenses[14]
	4	Amounts written off investments
	5	Interest payable and similar charges[16]
	6	Tax on profit or loss on ordinary activities
	7	Profit or loss on ordinary activities after taxation
	8	Extraordinary charges
	9	Tax on extraordinary profit or loss
	10	Other taxes not shown under the above items
	11	Profit or loss for the financial year
B	**Income**	
	1	Turnover
	2	Other operating income
	3	Income from shares in group companies
	4	Income from shares in related companies
	5	Income from other fixed asset investments [15]
	6	Other interest receivable and similar income [15]
	7	Profit or loss on ordinary activities after taxation
	8	Extraordinary income
	9	Profit or loss for the financial year

Source: Companies Act 1985, Schedule 4.
(The footnotes are described in Table V.8 and are from the Companies Act 1985, Schedule 4.)

shareholders' funds in the period, including capital contributed by or repaid to shareholders.

- Prior period adjustments should be accounted for by restating the comparative figures for the preceding period in the primary statements and notes and adjusting the opening balance of reserves for the cumulative effect.

In addition to the structural improvement in the profit and loss account, two additional improvements that would enable the reader to understand the financial performance of a reporting entity from reading its annual report are the statement on Operating and Financial Review issued by the ASB (see IX.8) and the adoption by the International Stock Exchange in London of the recommendations of the Cadbury Committee on the improvement of corporate governance (see IX.7). According to these recommendations, reporting companies should indicate the extent to which they have adopted the 'code of best practice' for corporate governance suggested by the committee. Table V. 9 is an illustration of a profit and loss account that blends the requirements of the Companies Act and FRS 3. The new format provides more information and seeks to disaggregate the profit and loss number in such a way that the reader can determine the extent of continuity of the earnings number and the potential for its improvement.

2.2 A review of reporting financial performance

The G4+1 (consisting of the United Kingdom, United States, Canada, Australia and the IASC, but also now including New Zealand) published a

Table V.7 Information to be disclosed in the profit and loss account (Format 4 horizontal presentation using total cost of production method)

A Charges

1 Reduction in stocks of finished goods and in work in progress
2 (a) Raw materials and consumables
(b) Other external charges
3 Staff costs:
(a) Wages and salaries
(b) Social security costs
(c) Other pension costs
4 (a) Depreciation and other amounts written off tangible and intangible fixed assets
(b) Exceptional amounts written off current assets
5 Other operating charges
6 Amounts written off investments
7 Interest payable and similar charges[16]
8 Tax on profit or loss on ordinary activities
9 Profit or loss on ordinary activities after taxation
10 Extraordinary charges
11 Tax on extraordinary profit or loss
12 Other taxes not shown under the above items
13 Profit or loss for the financial year

B Income

1 Turnover
2 Increase in stocks of finished goods and in work-in-progress
3 Own work capitalized
4 Other operating income
5 Income from shares in group companies
6 Income from shares in related companies
7 Income from other fixed asset investments[15]
8 Other interest receivable and similar income[15]
9 Profit or loss on ordinary activities after taxation
10 Extraordinary income
11 Profit or loss for the financial year

Source: Companies Act 1985, Schedule 4.
(The footnotes are described in Table V.8 and are from the Companies Act 1985, Schedule 4.)

Table V.8 Information to be disclosed in the profit and loss account (Format 1 to 4)

(14) Cost of sales; distribution costs; administrative expenses (Format 1, items 2, 4 and 5 and Format 3, items A.1, 2 and 3.)
These items shall be stated after taking into account any necessary provisions for depreciation or diminution in value of assets.

(15) Income from other fixed asset investments; other interest receivable and similar income (Format 1, items 9 and 10; Format 2, items 11 and 12; Format 2, items B.5 and 6; Format 4, items B.7 and 8.)
Income and interest derived from group companies shall be shown separately from income and interest derived from other sources.

(16) Interest payable and similar charges (Format 1, item 12; Format 2, item 14; Format 3, item A.5; Format 4, item A.7.)
The amount payable to group companies shall be shown separately.

(17) Formats 1 and 3
The amount of any provisions for depreciation and diminution in value of tangible and intangible fixed assets falling to be shown under items 7(a) and A.4(a) respectively in Formats 2 and 4 shall be disclosed in a note to the accounts in any case where the profit and loss account is prepared by reference to Format 1 or Format 3.

Source: Companies Act 1985, Schedule 4.

Table V.9 Examples of a profit and loss account and statement of recognized gains and losses according to FRS 3

Consolidated profit and loss account
Year ending 31 March

	Note	1999 Continuing operations GBP m	1999 Discontinued operations GBP m	1999 Total GBP m	1998 Continuing operations GBP m	1998 Discontinued operations GBP m	1998 Total GBP m
Turnover							
– Group and share of joint ventures		**1,402.8**	–	**1,402.8**	1,410.3	5.9	1,416.2
– Share of joint ventures		**(77.3)**	–	**(77.3)**	(27.2)	–	(27.2)
– Group	3	**1,325.5**	–	**1,325.5**	1,383.1	5.9	1,416.2
Operating costs	4	**(825.3)**	**(0.3)**	**(825.6)**	(916.5)	(14.8)	(931.3)
Other operating income	7	**12.7**	–	**12.7**	10.8	3.3	14.1
Utilization of discontinued business provision	2	–	**1.4**	**1.4**	–	5.5	5.5
Group operating profit		**512.9**	**1.1**	**514.0**	477.4	(0.1)	477.3
Share of operating profit in:							
– joint ventures		**8.1**	–	**8.1**	3.3	–	3.3
– associates		**(0.8)**	–	**(0.8)**	–		–
Operating profit		**520.2**	**1.1**	**521.3**	480.7	(0.1)	480.6
Non-operating exception items:							
– Business centre redevelopment		–	–	–	(8.3)	–	(8.3)
– Loss on closure of discontinued operations	2	–	–	–	–	(0.7)	(0.7)
– Release/ultilization of discontinued business provision	2	–	**7.0**	**7.0**	–	20.7	20.7
Profit on ordinary activities before interest and taxation	3	**520.2**	**8.1**	**528.3**	472.4	19.9	492.3
Net interest expense							
Group	8	**(106.0)**			(73.7)		
Share of joint ventures		**(2.9)**			–		
Share of associates		**(0.8)**		**(109.7)**	–		(73.7)
Profit on ordinary activities before taxation	3			**418.6**			418.6
Taxation on profit on ordinary activities	9			**(34.2)**			(62.2)
Windfall tax	10			–			(230.7)
				384.4			125.7
Non-equity dividend	13			**(0.9)**			–
Profit for the financial year	26			**383.5**			125.7
Ordinary dividends	13/26			**(148.8)**			(146.2)
Retained profit/(deficit) transferred to/(from) reserves				**234.7**			(20.5)
Earnings per ordinary share – basic	14			**106.3p**			33.1p
Non-operating exception items				**(1.9p)**			(3.1p)
Windfall tax				–			60.7p
Adjusted earnings per ordinary share	14			**104.4p**			90.7p
Earnings per ordinary share – diluted	14			**105.5p**			32.8p

There is no difference between the profit on ordinary activities before taxation and the retained profit/(deficit) transferred to/(from) reserves for the year stated above and their historical cost equivalents.

Statement of recognized gains and losses
Year ending 31 March

	1999 GBP m	1998 GBP m
Profit for the financial year	**383.5**	125.7
Share capital related costs	**(7.1)**	–
Foreign exchange adjustments	**(1.3)**	(2.2)
Advance corporation tax on the repurchase of shares	**(0.7)**	–
Profit for the financial year	**383.5**	125.7
Total recognized gains for the financial year	**374.4**	123.5

special report entitled 'Reporting Financial Performance: Current Developments and Future Directions' in January 1998. This was soon followed by a UK discussion paper, published in June 1999, in conjunction with the other members of the G4+1. The main proposals of the Paper is that a single performance statement should replace the profit and loss account and the statement of total recognized gains and losses, effectively combining both in one statement. All recognized gains and losses attributable to shareholders should be reported in the single statement in one of three major components:

- The results of operating (or trading) activities.
- The results of financing and other treasury activities.
- Other gains and losses.

The basis on which gains and losses (taken here to include all revenues and expenses) should be reported in the different components of financial performance are laid out in the Paper.

The Paper takes the view that gains and losses should be reported only once and in the period in which they arise, and therefore they should not be reported again in another component at a later date (a practice sometimes called 'recycling'). However, the paper also recognizes that recycling may be permitted or required by standard-setters in certain situations, in the short term, where this represents an improvement on existing practice.

Although many other aspects of FRS 3 are retained the following are some of the main proposed changes:

- The Board believes that similar gains and losses should be disclosed in the same section of the performance statement. The Paper's proposals would accordingly change the treatment of those items that, as required by paragraph 20 of FRS 3, are shown separately after operating profit. Most gains on asset disposals would be shown in the 'other gains and losses' section.
- The Paper calls for additional disclosures to be put in context presentations of 'pre-exceptional results', which are sometimes made on a voluntary basis.
- The proposals in the Paper would express somewhat differently from UK standards the treatment of the correction of errors that are discovered after the financial statements have been issued. The view of the G4+1 is that it would not be practicable to distinguish fundamental from material errors.
- Dividends would not be treated as items of financial performance, but rather as appropriations of profits and so would not be shown on the face of the performance statement.

3 Corresponding amounts

The Companies Act 1985, Schedule 4 requires corresponding amounts to be given for each item in the balance sheet and profit and loss account and notes to the accounts (adjusted if necessary to make them comparable with particulars and reasons for such adjustment to be given).

VI Recognition criteria and valuation

1 Fixed assets

The Companies Act 1985 defines fixed assets as those that 'are intended for use on a continuing basis in the company's activities'.

Schedule 4 of the Act requires fixed assets to be classified (and sub-classified as required) under the following headings:

- Intangible assets.
- Tangible assets.
- Investments.

1.1 Intangible fixed assets

1.1.1 General remarks

Accounting for intangible fixed assets, and in particular brands and goodwill, has been one of the major issues facing the accounting profession for many years. Many commentators believe brands to be subsumed within goodwill whilst others believe it is distinct from goodwill and its value should be independently valued. In many acquisitions, the consideration paid far exceeds the fair value of the acquired company's recognized net assets thereby giving rise to large amounts of goodwill. This goodwill cannot, however, be recognized by the acquired company in its own balance sheet as it is internally generated.

Schedule 4 of Companies Act 1985 specifically permits the inclusion of the following intangible assets:

- Concessions, patents, licences, trademarks, and similar rights and assets.
- Development costs.
- Goodwill.
- Payments on account.

The Act specifies that concessions, patents, licences, trademarks and similar rights and assets may only be included if they were either acquired for valuable considerations in circumstances that do not qualify them to be shown as goodwill or they were created by the company itself. The Act also permits intangible assets, other than goodwill, to be included in the financial statements at their current value (although FRS 10 'Goodwill and intangible assets' is much more restrictive on this issue).

FRS 10, the long-awaited replacement of SSAP 22, on goodwill and intangible assets, has attempted to provide proper guidance and standard on accounting for intangibles. The standard deals with the following issues in respect of intangible assets:

- Recognition rules for intangible assets
  - developed internally;
  - purchased separately;
  - acquired as part of the acquisition of a business.
- Valuation rules.
- Amortization and impairment rules.
- Financial statement disclosures.

The standard defines intangible assets as:

> non-financial fixed assets that do not have physical substance but are identifiable and are controlled by the entity through custody or legal rights (FRS 10 para 2).

The application of this definition is important not only to distinguish intangible from tangible assets, but also to distinguish an intangible asset from goodwill. These distinctions are important as different accounting rules apply to each category. In particular, the rules for capitalizing and revaluing intangible assets are much more prudent than those for tangible assets.

1.1.2 Intangible assets other than goodwill

FRS 10 categorizes intangible assets between those that are 'unique to the business' and those that have a 'readily ascertainable market value'. The latter is accounted for in the same way as tangible fixed assets (see below), whereas more restrictions are placed on recognition and valuation of the former. The ASB's rationale for the distinction is twofold: first, it considers that many unique intangible assets are similar to goodwill and should be treated in the same way; and second, it considers that the valuation of unique intangible assets is too subjective to allow them to be recognized (other than where they have actually been purchased) or revalued.

An intangible asset purchased separately from a business should be capitalized at cost. Internally generated intangible assets, other than goodwill, may only be capitalized if it has a readily ascertainable market value, otherwise the expense should be written off as incurred. Intangible assets capitalized should be systematically amortized through the profit and loss account (usually over 20 years or less). An important aspect is that impairment reviews must be undertaken, particularly if the intangible asset is regarded as having an infinite life and is therefore not amortized.

1.1.3 Goodwill

Intangible assets, unless they can be separately identified from goodwill is required to be treated as goodwill. FRS 10 states:

> An identifiable asset is defined by companies legislation as one that can be disposed of separately without disposing of a business of the entity. If an asset can be disposed of only as part of the revenue-earning activity to which it contributes, it is regarded as indistinguishable from the goodwill relating to that activity and is accounted for as such. (FRS 10 para 2).

This test of separability resulted in many intangibles, including brands, that could not be separately identified to be incorporated within reported goodwill.

FRS 10 takes the view that goodwill arising on an acquisition (i.e. the cost of acquisition less the aggregate of the fair value of the purchased entity's separable assets and liabilities) is neither an asset like other assets nor an immediate loss in value. Rather, it forms a bridge between the cost of an investment shown as an asset in the acquirer's own financial statements and the values attributed to the acquired assets and liabilities in the consolidated financial statements. Even though purchased goodwill is not in itself an asset, its inclusion amongst assets of the reporting entity, rather than as a deduction from shareholders' equity, recognizes that goodwill is part of a larger asset, the investment, for which management remains accountable. The standard, therefore, requires purchased goodwill to be capitalized and subsequently amortized systematically through the profit and loss account (usually over 20 years or less). Impairment reviews must be undertaken, particularly if the goodwill is regarded as having an infinite life and is therefore not amortized.

Internally generated goodwill may not be capitalized and must, therefore, be written off as incurred.

Excerpt from Annual Report of SCAPA Plc, 1999

> **Basis of consolidation**
>
> The Group accounts consolidate the accounts of the company and its subsidiaries made up to 31 March in each year. The results of companies acquired or disposed of during the year are included from the date of acquisition or up to the date of disposal.
>
> The Group has adopted the requirements of FRS 10 *Goodwill and intangible assets*. As a result, goodwill arising on acquisitions remains written off against reserves.
>
> The effect of this change in accounting policy has had no impact on the financial results.

1.1.4 Research and development

As mentioned above, Companies Act 1985 permits development expenditure to be capitalized in certain circumstances but fails to define the term development. SSAP 13 'Accounting for research and development' provides a definition and categorizes research and development into one of the following three broad categories:

- Pure (or basic) research concerned with experimental or theoretical work designed to acquire new scientific or technical knowledge for its own sake rather than directed towards any specific aim or application (SSAP 13, para. 21(a)).
- Applied research meaning original or critical investigation undertaken in order to gain new scientific or technical knowledge and directed toward a specific practical aim or objective (SSAP 13, para. 21(b))
- Development concerned with the use of scientific or technical knowledge in order to produce new or substantially improved materials, devices, products or services, to install new processes or systems prior to the commencement of commercial production or commercial applications, or to improving substantially those already produced or installed (SSAP 13, para. 21(c)).

If any of the above types of activities relates to the location or exploitation of oil, gas or mineral deposits or is reimbursable by third parties either directly or under the terms of a firm contract to develop and manufacture at an agreed price calculated to reimburse both elements of expenditure, it cannot be treated as a research and development expenditure for the purpose of financial reporting (SSAP 13, para. 21).

Expenditure on pure and applied research (other than that relating to assets acquired or constructed in order to provide facilities for research and development) should be written off in the year of expenditure through the profit and loss account (SSAP 13, para. 24). Development expenditure should be written off in the year of expenditure except in the following (mutually inclusive) circumstances when it may be deferred to future periods (SSAP 13, para. 25):

- There is a clearly defined project.
- The related expenditure is separately identifiable.

- The outcome of such a project has been assessed with reasonable certainty as to (i) its technical feasibility, and (ii) its ultimate commercial viability considered in the light of factors such as likely market conditions (including competing products, public opinion, consumer and environmental legislation).
- The aggregate of the deferred development costs, any further development costs and related production, selling and administration costs is reasonably expected to be exceeded by related future sales or other revenues.
- Adequate resources exist, or are reasonably expected to be available, to enable the project to be completed and to provide any consequential increases in working capital.

Development expenditure may be deferred only to the extent that its recovery can reasonably be regarded as assured (SSAP 13, para. 26).

1.2 Tangible fixed assets

1.2.1 General remarks

Despite its importance within the context of a balance sheet, a standard on tangible fixed assets, FRS 15, was only issued in February 1999. The definition of a fixed asset was originally provided in s.262 of the Companies Act as one which is 'intended for use on a continuing basis in the company's activities'. All assets, therefore, not intended for continuing use are current assets. FRS 15 (para. 2) defines a tangible fixed asset as:

> assets that have physical substance and are held for use in the production or supply of goods or services, for rental to others, or for administrative purposes on a continuing basis in the reporting entity's activities.

FRS 15, although devoted to assets and issued within the wider debate of the ASB's Statement of Principles, fails to define what an asset constitutes. One has to refer back to FRS 5, Reporting the substance of transactions, to obtain a definition. Assets are defined as:

> rights or other access to future economic benefits controlled by an entity as a result of past transactions or events.

1.2.2 Statutory Headings

Under the generic statutory heading of 'Tangible assets' are subheadings that may either be shown on the face of the balance sheet or in the notes. These are:

- Land and buildings.
- Plant and machinery.
- Fixtures, fittings, tools and equipment.
- Investments.
- Payments on account and assets in course of construction.

In accordance with Schedule 4 of Companies Act 1985 for each of the subheadings movements in respect of gross amounts and depreciation or diminution thereof must be shown. Land and building require further analysis between freeholds, long leaseholds (not less than 50 years unexpired) and short leaseholds.

1.2.3 Initial measurement

Acquired or constructed tangible fixed assets should initially be measured at such cost. These costs are those that are directly attributable to bringing the asset into working condition for its intended use. Examples of these costs, though not exhaustive, include acquisition costs, cost of site preparation and clearance, installation costs, professional fees, labour costs of own employees and incremental costs to the entity that would have been avoided if only the asset had not been constructed. Abnormal costs, such as those relating to industrial disputes or wasted materials are not directly attributable and therefore excluded from initial measurement. Paragraph 12 of FRS 15 states that 'capitalization of directly attributable costs should cease when substantially all the activities that are necessary to get the tangible fixed asset ready for the user are complete, even if the asset has not yet been brought into use'.

1.2.4 Finance cost

Although an entity need not capitalize finance costs, where it adopts such a policy it should capitalize only those that are directly attributable to the asset's construction. The total amount of finance costs capitalized during a period should not exceed the total amount of finance costs incurred during that period.

1.2.5 Subsequent expenditure

Often referred to as 'repairs and maintenance', subsequent expenditure to ensure that the asset maintains its intended level of performance should be charged to the profit and loss account as incurred. An exception to this rule can only be applied in the following circumstance: where

(a) The subsequent expenditure provides an enhancement of the economic benefits of the tangible fixed assets in excess of the previously assessed standard performance.
(b) A component of the tangible fixed asset that has been treated separately for depreciation purposes and depreciated over its individual useful economic life, is replaced or restored.
(c) The subsequent expenditure relates to a major inspection or overhaul of a tangible fixed asset that restores the economic benefits of the asset that have been consumed by the entity and have already been reflected in depreciation.

The accounting treatment under such circumstances should be the same as that adopted on the initial recognition of the asset and the depreciation profile adopted (para. 40, FRS 15).

1.2.6 Depreciation

Depreciation (measure of the amount of economic benefits consumed) of a tangible fixed asset should be allocated on a systematic basis over its useful economic life (UEL) and charged to the profit and loss account. A simple calculation of depreciation would be to take the cost of the asset and deduct any residual value that may exist at the end of the asset's UEL, and then allocate the difference through a variety of allocation methods. More commonly – methods of allocating the depreciable amount include:

- Straight-line – where it is assumed that equal amounts of the asset's economic benefits are consumed in each year of the asset's estimated useful economic life.
- Reducing balance – which more closely reflects the pattern of consumption of the economic benefits of assets that clearly provide greater benefits when new than as they become older.

Whichever method of allocation is adopted it should be applied consistently. A change from one method to another is permissible only on the grounds that the new method will give a fairer presentation of the results and of the financial position of an entity.

Excerpt from Annual Report of National Power Plc, 1999

Tangible fixed assets

Tangible fixed assets are stated at original cost less accumulated depreciation. In the case of assets constructed by the Group, related works and administrative overheads, commissioning costs and borrowing costs as per FRS 15 are included in cost. Assets in the course of construction are included in tangible fixed assets on the basis of expenditure incurred at the balance sheet date.

Depreciation is calculated so as to write down the cost of tangible fixed assets to their residual value evenly over their estimated useful lives. Estimated useful lives are reviewed periodically, taking into account commercial and technological obsolescence as well as normal wear and tear, provision being made for any permanent diminution in value.

The depreciation charge is based on the following estimates of useful lives:

	Years
Power stations under operating leases (primary lease term)	7
Combined cycle gas turbine power stations	20
Other power stations	20–40
Non-operational buildings	40
Fixtures, fittings, tools and equipment	4–5
Computer equipment and software	3–5
Hot gas path CCGT turbine blades	4

1.2.7 Valuation

Tangible fixed assets need not be revalued, but where such a policy is adopted by an entity it should be applied to individual classes of assets. The policy need not be applied to all classes of tangible fixed assets held by the entity. The revaluation policy should reflect current values at the balance sheet date although FRS 15 does not insist on annual revaluations.

Properties should be professionally valued every five years with an interim, internal or external, valuation in year three although more frequent valuations may be advisable under the circumstances. FRS 15 (para. 53) details the valuation bases for revalued properties that are not impaired as follows:

(a) Non-specialized properties should be valued on the basis of existing use value (EUV), with the addition of notional directly attributable acquisition costs where material. Where the open market value (OMV) is materially different from EUV, the OMV and the reasons for the difference should be disclosed in the notes to the accounts.
(b) Specialized properties should be valued on the basis of depreciated replacement cost.
(c) Properties surplus to an entity's requirements should be valued on the basis of OMV, with expected directly attributable selling costs deducted where material.

Tangible fixed assets other than properties should be valued using market value, where possible. Where market value is not obtainable, assets should be valued on the basis of depreciated replacement cost. For some types of tangible fixed assets such as company cars, an active second-hand market may exist and in such circumstances directors' valuation, using appropriate indices, is acceptable.

1.2.8 Reporting gains and losses on revaluation

Revaluation gains should be recognized in the Statement of Total Recognized Gains and Losses (STRGL) unless they reverse a previous loss on the same asset, charged through the profit and loss account, when it may be recognized in the profit and loss account. Gains recognized in this manner must be adjusted for any depreciation that would have been charged had the loss previously taken to the profit and loss account not been recognized in the first place.

All revaluation losses that are caused by a clear consumption of economic benefits should be charged to the profit and loss account. Other revaluation losses should be recognized in the STRGL until the carrying amount reaches its depreciated historical cost; and thereafter in the profit and loss account.

The example given in FRS 15 clearly illustrates the reporting of revaluation gains and losses.

2 Investment properties

Whilst the Companies Act 1985 does not use the term 'investment property', SSAP 19 'Accounting for investment properties' defines it:

> as an interest in land and/or buildings, where: the construction work and development have been completed; and the interest is held for its investment potential, with any rental income being negotiated at arm's length (SSAP 19 para 7).

The reason there is a separate accounting standard on investment properties (SSAP 19, Accounting for Investment Properties) is because the property industry lobbied extensively against the requirement in SSAP 12 to depreciate buildings. The industry has argued that the current value of properties is usually greater than its historical cost. As a result of its lobbying, the industry was provided with an exemption to SSAP 12, issued in December 1977, which allowed investment properties not to be depreciated. This exemption continued until SSAP 19 was introduced in November 1981. This standard exempted companies holding investment properties from the need for depreciation as specified in SSAP 12, except for those held on lease with an unexpired term of 20 years or less. Another aspect of the standard is that investment properties should be stated in the balance sheet at open market value, although the valuation need not be undertaken by qualified or independent valuers. However, the names or qualifications of the valuers, the bases used by them and whether the person making the valua-

Example – Reporting revaluation gains and losses

Assumptions

A non-specialized property costs GBP 1 million and has a useful life of 10 years and no residual value. It is depreciated on a straight-line basis and revalued annually. The entity has a policy of calculating depreciation based on the opening book amount. At the end of years 1 and 2 the asset has an EUV of GBP 1,080,000 and GBP 700,000 respectively. At the end of year 2, the recoverable amount of the asset is GBP 760,000 and its depreciated historical cost is GBP 800,000. There is no obvious consumption of economic benefits in year 2, other than that accounted for through the depreciation charge.

Accounting treatment under modified historical cost

	Year 1 GBP 000	Year 2 GBP 000	
Opening book amount	1,000	1,080	
Depreciation	(100)	(120)	*
Adjusted book amount	900	960	
Revaluation gain (loss)			
• recognized in the STRGL	180	(220)	
• recognized in the profit and loss account	—	(40)	
Closing book amount	1,080	700	

* *As the remaining useful economic life of the asset is nine years, the depreciation charge in year 2 is 1/9th of the opening book amount (GBP 1,080,000/9 = GBP 120,000).*

In year 1, after depreciation of GBP 100,000, a revaluation gain of GBP 180,000 is recognized in the statement of total recognized gains and losses, in accordance with paragraph 63.

In year 2, after a depreciation charge of GBP 120,000, the revaluation loss on the property is GBP 260,000. According to paragraph 65, where there is not a clear consumption of economic benefits, revaluation losses should be recognized in the statement of total recognized gains and losses until the carrying amount reaches its depreciated historical cost. Therefore, the fall in value from the adjusted book amount (GBP 960,000) to depreciated historical cost (GBP 800,000) of GBP 160,000 is recognized in the statement of total recognized gains and losses.

The rest of the revaluation loss, GBP 100,000 (ie the fall in value from depreciated historical cost (GBP 800,000) to the revalued amount (GBP 700,000)), should be recognized in the profit and loss account, unless it can be demonstrated that recoverable amount is greater than the revalued amount. In this case, recoverable amount of GBP 760,000 is greater than the revalued amount of GBP 700,000 by GBP 60,000. Therefore GBP 60,000 of the revaluation loss is recognized in the statement of total recognized gains and losses, rather than the profit and loss account – giving rise to a total revaluation loss of GBP 220,000 (GBP 60,000 + GBP 160,000) that is recognized in the statement of total recognized gains and losses. The remaining loss (representing the fall in value from depreciated historical cost of GBP 800,000 to a recoverable amount of GBP 760,000) of GBP 40,000 is recognized in the profit and loss account.

Source: FRS 15, ASB Feb. 1999, pp. 38–39.

tion is an employee or officer of the company should all be disclosed. If the investment properties form a substantial part of total assets, the valuation should be undertaken at least every five years and by a recognized valuer.

3 Investments

Companies Act 1985, Schedule 4 states that investments may be classified as either fixed asset investments or current asset investments. Fixed asset investments include shares in group undertakings; loans to group undertakings; participating interests; loans to undertakings in which the group has a participating interest; other investments other than loans; other loans; and own shares. Investments that come under the current assets heading include shares in group undertakings; own shares; and other investments. Investments made in the nature of trade investments or other investments that are held on a long-term basis should be categorized as fixed in normal circumstances, whereas those of a temporary nature should be treated as current assets.

Listed investments are those with a quotation on a recognized investment exchange (Companies Act 1985, Schedule 4, para. 84) other than an overseas investment exchange within the meaning of the Financial Services Act 1986. As far as the United Kingdom is concerned, this means securities listed on the International Stock Exchange and excludes those on the Unlisted Securities Market. The Companies Act 1985, Schedule 4, para. 45 states that the amount attributable to listed investments and an analysis in terms of those on a recognized exchange and those overseas should be disclosed. The specific information to be disclosed includes the aggregate market value of investments where it differs from the amount included in the balance sheet, and the market value and stock exchange value where the former is higher. Where investments such as listed investments are held for trading purposes, there is a recent trend of treating their market values as carrying amounts instead of their historical costs. The 'marked to market' practice requires gains and losses to be recognized in the profit and loss account of the current period.

4 Stocks

Stocks are the outstanding portion, at the end of the year, of goods or other assets purchased for resale, finished goods produced for sale, raw materials and components purchased for incorporation into products for sale, products and services in intermediate stages of completion (work in progress) including balances on long-term contracts. Essentially, uncompleted long-term contracts (i.e. contracts stretching over one financial year) are a kind of stocks. They are different, however, in several respects because of the many contractual conditions surrounding them, such as the conditions for payment in line with work performed to date and the treatment of earned profit while the contract is in progress. As a result, the accounting treatment of long-term contracts is considered separately from that of other stocks.

Total stocks should be stated in the balance sheet in accordance with the valuation rules at the lower of cost or net realizable value, and subclassified so as to indicate the amounts held in each of the main categories in the balance sheet (Companies Act 1985, Schedule 4, para. 23 (1) and SSAP 9, paras 26 and 27). The disclosure should include the methods for valuing different types of stocks and work-in-progress and the amount included in the accounts under each method.

Companies Act 1985, Schedule 4, para. 26 permits the purchase cost or the production cost of stocks to be determined by whichever of the following methods appears appropriate to the directors:

- First in, first out (FIFO).
- Last in, first out (LIFO).
- Weighted average method.
- Other methods similar to any of the above.

Although the LIFO method is permitted by law (Companies Act 1985, Schedule 4, para. 27(2)), it is not permitted by SSAP 9 or used in practice because it does not provide a fairest approximation to actual cost and because the Inland Revenue has never permitted its use in the United Kingdom. In the United States, however, the Internal Revenue Codes of 1938 and 1939 permitted its use for the computation of tax. This is an interesting, though not often admitted, influence of taxation on accounting in theUnited Kingdom. If the reporting companies insist on using the LIFO method contrary to the expectation of the Inland Revenue, this means that the accounting profit chargeable to tax would have to be recomputed using another inventory valuation method acceptable to the Revenue. This would be an onerous proposition and so reporting companies have chosen to follow the Inland Revenue approach.

Although not permitted by SSAP 9 (para. 25), the use of a fixed quantity and value (base stock) to represent the value of items included in raw materials and consumables which are constantly being replaced is permitted by the Companies Act 1985 (Schedule 4, para. 25). Base stock is permitted provided the value of the items of stock to which they relate is not material and their quantity, value and composition are not subject to material variation. The method is not permitted for tax purposes in the United Kingdom or United States and is rarely seen in practice (Davies, Paterson and Wilson, 1994, p. 696).

As mentioned earlier, stocks should be recorded at the lower of cost or net realizable value. Net realizable value is a surrogate for market price where no competitive market such as the London Metal Exchange exists. Net realizable value is based on the estimated selling price less any further cost expected to be incurred in selling the stock items or bringing them to a saleable point. If there is a material difference between the value of stocks included in the accounts and the value of stocks at replacement cost or most recent purchase or production cost (whichever the directors consider the more appropriate comparison), the Companies Act requires the amount of the difference to be stated in a note to the accounts.

Purchase cost includes purchase price, import duties, transport and handling costs and any other attributable costs less trade discounts, rebates and subsidies (SSAP 9, para. 18).

Production cost is defined by SSAP 9 (para. 19) as comprising:

- Costs which are specifically attributable to units of production, e.g. direct labour, direct expenses and sub-contracted work.
- Production overheads.
- Other overheads, if any, attributable in the particular circumstances of the business to bringing the product or service to its present location and condition.

5 Capital instruments

5.1 Background

In November 1987 the ICAEW issued a technical release (TR 677) on complex capital issues. This was followed by a discussion paper issued by the ASB in 1991, FRED 3, Accounting for Capital Instruments, in December 1992 and FRS 4, Financial Reporting Standard on Capital Instruments, issued in December 1993.

5.2 FRS 4 'Capital instruments'

The objective of FRS 4 is to ensure that financial statements provide a clear, coherent and consistent treatment of capital instruments, in particular as regards the classification of instruments as either debt or equity. The presentation of the instruments in the financial statements should be in a way that reflects the obligation of the issuer. The standard prescribes the methods to be used to determine the amounts to be ascribed to capital instruments and their associated costs, commitments and potential commitments.

The amount of shareholders' funds attributable to equity interests, non-equity interests and (for consolidated financial statements) minority interests is to be disclosed. The key distinctions are summarized in Table VI.1.

To aid application, FRS 4 contains a number of application notes that show how its requirements apply to transactions with certain features.

Other aspects of the standard include the need to account for costs associated with capital instruments in a manner consistent with their classification, and, for redeemable instruments, allocated to accounting periods on a fair basis over the period the instrument is in issue. The standard also requires financial statements to provide relevant information concerning the nature and amount of the entity's source of finance and the associated costs, commitments and potential commitments.

Table VI.1 Division of capital instruments between equity and non-equity

Item	Analyzed between	
Shareholders' funds	Equity interests	Non-equity interests
Minorities interest (MI) in subsidiaries	Equity interests in subsidiaries	Non-equity interests in subsidiaries
Liabilities	Convertible liabilities	Non-convertible liabilities

6 Liabilities

6.1 The legal requirements

Companies Act 1985 requires creditors to be analyzed between amounts that will fall due within one year of the balance sheet date and amounts that will fall due after more than one year. The Act further requires that creditors should be analyzed, amongst others, into the following categories:

- Debenture loans.
- Bank loans and overdraft.
- Payment and overdraft.
- Payments received on account.
- Trade creditors.
- Bills of exchange payable.
- Amounts owed to group undertakings.
- Other creditors including taxation and social security.
- Accruals and deferred income.

6.2 Accounting practice

Relevant accounting standards relating to liabilities include:

- FRS 4 'Capital instruments' which differentiates debt (liability) from equity (ownership interest).
- FRS 5 'Reporting the substance of transactions' defines a liability as 'obligations to transfer economic benefits as a result of past transactions or events'.
- FRS 12 'Provisions, contingent liabilities and contingent assets' which strictly regulates the creation of provisions and in recognizing or accruing for contingent liabilities.
- FRS 13 'Derivatives and other Financial Instruments: Disclosures' which requires both narrative and numerical disclosure of the entity's exposure to derivatives and other financial instruments.

The prudent nature of accountancy has always placed greater importance on recognizing and providing for all liabilities, be they at management discretion or true obligations of a business entity. This has allowed a choice of treatments which has affected reported position and performance. For example, classifying debt as equity has decreased the gearing ratio, long-term financial standing of an entity and therefore the perceived financial risk. FRS 4, assisted by FRS 5 definition of a liability, now has strict regulations in place in this area (see above).

6.3 Provisions, contingent liabilities and contingent assets

The accounting treatment of provisions has been a matter of some debate, particularly the timing of such provisions. FRS 12's objective is to ensure that a provision (a liability that is of uncertain timing or amount) is recognized only when it actually exists at the balance sheet date. A provision should be recognized therefore only when:

- An entity has a present obligation (legal or constructive) as a result of a past event.
- It is probable that a transfer of economic benefits will be required to settle the obligation.

- A reliable estimate can be made of the amount of the obligation.

The amount recognized as a provision should be the best estimate of the expenditure required to settle the present obligation at the balance sheet date. Contingent liabilities and contingent assets are not recognized as liabilities or assets. However, a contingent liability should be disclosed if the possibility of an outflow of economic benefit to settle the obligation is more than remote. FRS 12 provides a number of examples relating to its detailed provision. These are sometimes more helpful and accurate than the standard which can give a false understanding.

7 Derivatives and other financial instruments

Some financial instruments, such as cash, debtors and creditors, generally arise as part of an entity's operating and financing activities and tend to be highly visible in the financial statements. Others (such as swaps, forwards, caps and collars, and other derivatives) are entered into in order to manage the risks arising from the operating and financing activities of the entity and are generally less visible. In February 1996, the British Bankers' Association and the Irish Bankers' Federation jointly issued an SORP setting out recommendations for the accounting treatment and disclosure of derivatives in the financial statements of banks. FRS 13 'Derivatives and other financial instruments: disclosures', coming into effect almost three years later, seeks to improve the disclosures provided in respect of all financial instruments and it does this by focusing on the way in which they are used by the reporting entity.

The objective of its disclosures is to provide information about:

- The impact of the instruments upon the entity's risk profile.
- How the risks arising from financial instruments might affect the entity's performance and financial condition.
- How these risks are managed.

The ASB has defined a financial instrument as:

> . . . any contract that gives rise to both a financial asset of one entity and a financial liability or equity instrument of another entity (FRS 13 para 2).

This definition is almost identical to that in IAS 32.

FRS 13 applies to all entities, other than insurance companies and groups that have one or more of their capital instruments listed or publicly traded on a stock exchange or market and all banks and similar institutions. The standard requires both narrative and numerical disclosures.

The standard requires narrative disclosures to include an explanation of the role that financial instruments play in creating or changing the risks that the entity faces in its activities. The directors' approach to managing each of those risks also needs be explained, including a description of the objectives, policies and strategies for holding and issuing financial instruments.

The numerical disclosures are intended primarily to show how these objectives and policies were implemented in the period, focusing on:

- Interest rate risk.
- Currency risk.
- Liquidity risk fair values.
- Hedging activities.

Although all entities falling within the scope of FRS 13 are required to provide the same type of narrative disclosures, the standard requires different numerical disclosures for each of:

- Entities that are not financial institutions.
- Banks and similar institutions.
- Other types of financial institution.

Current developments:
The ASB, with eight other national standard setters and the IASC, is participating in the Financial Instruments Joint Working Group, in developing proposals for an accounting standard that comprehensively addresses recognition and measurement of financial instruments. The aim is to develop an internationally harmonized accounting approach to financial instruments. Hedge accounting will also be considered, not least to simplify IAS 39 and FAS 133 on the issue.

VII Special accounting areas

1 Accounting for the costs of pensions and other retirement benefits

1.1 General remarks

Accounting for the costs of pensions and other retirement benefit schemes can be a complicated exercise. It involves recognition in the annual report and accounts of the costs of future obligations to employees. Some of these obligations are specified and predetermined amounts such as cash pensions; others are indeterminable future benefits such as health and welfare schemes. While only a few UK companies provide them, post-retirement benefits other than pensions can amount to huge liabilities despite the existence of national social security covering a large part of medical, health and welfare expenses. Pensions and other retirement benefits may be payable directly by the employer or indirectly by the employer out of a specially created fund. In addition, there is the problem of how retirement benefits are funded and managed and whether or not the retirement benefit fund management should be regarded as a related party of the employer organization. Funding is the setting aside of assets by transferring them to a distinct entity to which the beneficiaries have enforceable rights ('the fund'), often one separate from the employer, to meet future obligations for the payment of retirement benefits.

The current standard (SSAP 24), Accounting for Pension Costs, issued in May 1988 derives its regulatory perspectives from the report of a study on the subject commissioned by the Institute of Chartered Accountants in England and Wales (Napier, 1983).

1.2 Pension schemes

SSAP 24 identifies two types of pension scheme: defined contribution schemes and defined benefit schemes. In practice, some pension schemes combine the features of the two defined schemes. Some UK companies maintain foreign pension schemes or are part of a group pension scheme. Each of these schemes is described below.

1.2.1 Defined contribution schemes

Under a defined contribution scheme, the employer's obligation is the amount that the employer agrees to contribute to cover payment of retirement benefits to the employees. The benefits payable under the scheme will depend upon the funds available from these contributions and investment earnings thereon. When the fund arising from a defined contribution scheme is managed by a third party, the third party (e.g. a trustee or an insurance company) would undertake to meet the scheme's obligations without recourse to the employer. The cost to the employer can, therefore, be measured (or ascertained) with reasonable certainty. Although the SSAP suggests that there are a number of types of pension scheme (the other principal types are described below), many pension schemes in the United Kingdom are defined contribution schemes and the proportion of such schemes is increasing.

1.2.2 Defined benefit schemes

In contrast to the defined contribution scheme that sets the employer's obligation at a well-defined annual contribution over the life of the pension scheme, the defined benefit scheme focuses on the benefits payable to the employees. The benefits payable to employees are usually based on multiples of employees' average pay during the entire employment or their terminal pay. As a result, it is impossible to be certain in advance that the contributions to the pension scheme, together with the investment return thereon, will equal the benefits to be paid. The employer sponsoring the defined benefit scheme is usually legally obliged to contribute any unforeseen shortfall in the funds. If not legally obliged to do so, the employer may find it necessary to meet any shortfall in the interest of maintaining good employee relations. Conversely, if a surplus arises, the employer may be entitled to a refund of, or reduction in, contributions paid into the scheme.

1.2.3 Hybrid schemes

In practice, a few pension schemes combine the features of defined contribution and defined benefit scheme. In such instances, SSAP 24 (para. 39) recommends that the rules or trust deed should be carefully studied and the operation of the hybrid scheme, or its proposed method of operation, should be taken into account when determining the appropriate accounting treatment. The accounting treatment should be in accordance with the underlying substance of the scheme.

1.2.4 Group schemes

These refer to the operation by a number of companies within the same group of a single pension scheme. SSAP 24, in recognition of the practice of using a common group rate for contributions payable by sponsoring employers, even though on an individual basis the rate payable might differ, permitted the use of a common group rate. As a result, SSAP 24 does not require the frill disclosure of the details relating to the group pension scheme in the financial statements of individual sponsor companies, but it does require that the frill details be given in the financial statements of the holding company.

1.2.5 Foreign schemes

Some UK companies are obliged to maintain foreign pension schemes for employees working in foreign countries. In such situations, the commitment of the employer regarding the provision of pensions may be different from that which is customary in the United Kingdom so that it would be inappropriate to adjust the pension cost charged. As a result, SSAP 24 (para. 91) permits the use of the pension cost as determined for local purposes as the basis of the charge in respect of foreign schemes. If this charge is not in accordance with the provisions of SSAP 24, this fact should be stated, the circumstance explained and amount charged in the profit and loss account. The basis of the charge should, as a minimum, be disclosed.

1.3 Actuarial valuation

The essence of a pension scheme is the provision of sufficient funds to finance the costs of liabilities that may arise from catering for the retirement needs of retirees until death or the end of the benefits due to them. In the case of a defined benefits scheme, this involves an estimation of the duration of benefits payable to retirees and the costs of those benefits (taking account of inflation). The calculation of the average duration of benefits payable to retirees may be derived from the average age of the workforce and their average life expectancies. The actuarial valuation entails the calculation of the costs of future obligations under the pension scheme discounted to present values and the comparison of that present value of obligation with the current value of the pension fund. The aim is to find out whether the current level of funding under the pension fund would be sufficient to pay for all the costs of pension obligations if they were to be due at the end of the financial year or date of the actuarial valuation. The intention is to make funds available to meet the obligations under a pension scheme as and when they become due for payment. The problem is one of cash flow management rather than profit measurement. The actuarial valuation may also contribute to the timing of funding in an orderly manner to meet the future costs of a given set of benefits under a pension scheme.

There are two methods of actuarial valuation recommended by SSAP 24: the accrued benefits method and the prospective benefits method. The accrued benefits method is a valuation method that relates actuarial value of liabilities at a given date to:

- The benefits, including future increases promised by the rules, for current and deferred pensioners and their dependants.
- The benefits which the members assumed to be in service on the given date will receive for service up to that date only.

According to SSAP 24 (para. 57), allowance may be made for expected increases in earnings after the given date, and/or for additional pension increases not promised by the rules. The given date may be a current or future date. The further into the future the adopted date lies, the closer the results will be to those of a prospective benefits valuation method.

The prospective benefits method is a valuation method that relates the actuarial value of liabilities to:

- The benefits for current and deferred pensioners and their dependants, allowing where appropriate for future pension increases.
- The benefits which active members will receive in respect of both past and future service, allowing for future increases in earnings up to their assumed exit dates, and where appropriate for pension increases thereafter.

1.4 Accounting rules

SSAP 24 (para. 77) states that the accounting objective is that the employer should recognize the expected cost of providing pensions on a systematic and rational basis over the period during which the company derives benefit from the employees' services.

1.4.1 Defined contribution schemes

The charge against profit should be the amount of contributions payable to the pension scheme in respect of the accounting period. If the amount actually paid is more or less than the amount payable, a prepayment or accrual will appear in the balance sheet in accordance with normal accounting practice. The following disclosures should be made in respect of defined contribution schemes:

- The nature of the scheme (i.e. defined contribution).
- The accounting policy.
- The pension cost charged for the period.
- Any outstanding or prepaid contributions at the balance sheet date.

1.4.2 Defined benefit schemes

The pension cost should be calculated using actuarial valuation methods designed to provide the actuary's best estimate of the cost of providing the pension benefits promised. The method should be such that regular pension cost is a fairly level percentage of current and expected future pensionable payroll in the light of the current actuarial assumptions. The regular pension cost is defined by SSAP 24 (para. 72) as the consistent ongoing cost recognized under the actuarial method used. Variations from the regular cost should be allocated over the expected remaining service lives of current employees in the scheme except as stated in the following:

- Where a significant reduction in the number of employees is related to the sale or termination of an operation, the associated pension cost or credit should be recognized immediately to the extent necessary to comply with the requirements of FRS 3. In all other cases where there is a reduction in contributions arising from a significant reduction in employees, the reduction of contributions should be recognized as it occurs.
- Where a company takes a cash refund as part of a scheme to reduce a surplus in accordance with the provisions of the Finance Act 1986, or equivalent legislation, the refund may be credited to the profit and loss account in the year of receipt, rather than spreading forward the effects of the variation which has given rise to the surplus.

Where ex gratia pensions or ex gratia pension increases not allowed for in the actuarial assumptions are granted, the capital cost, to the extent not covered by a surplus, should be recognized in the accounting period in which they are granted. As earlier discussion indicates, UITF Abstract No. 6 requires post-retirement benefits other than pensions to be treated as liabilities and to be governed by the requirements of SSAP 24.

1.4.3 Illustrations

Below are extracts from the annual reports and accounts of three major UK companies on their accounting policies on pensions as well as particulars about the pensions schemes.

Excerpt from the Annual Report of Anglia Water Plc, 1999

Pension costs

Contributions to the group's defined benefit pension schemes are charged to the profit and loss account so as to spread the regular cost of pensions over the average service lives of employees, in accordance with the advice of an independent qualified actuary. Actuarial surpluses and deficits are amortized, where appropriate, over the average remaining service lives of employees in proportion to their expected payroll costs. The cost of defined contribution schemes is charged to the profit and loss account in the year in respect of which the contributions become payable.

Excerpt from the Annual Report of Fuller Smith and Turner Plc, 1999

Pension benefits

The Group operates pension plans for eligible employees administered by Trustees. Contributions to these schemes are charged to the profit and loss account so as to spread the cost of pensions over the employees' working lives within the Group. The schemes are subject to periodic valuation by a qualified actuary. Variations identified as a result of the valuations are amortized over the average expected remaining working lives of the employees in proportion to their expected payroll costs or within the period permitted by Minimum Funding legislation.

Excerpt from the Annual Report of Tesco Plc, 1999

Pension commitments

The Group operates a funded defined benefit pension scheme for full-time employees, the assets of which are held as a segregated fund, administered by trustees.

The pension cost relating to the scheme is assessed in accordance with the advice of an independent qualified actuary using the projected unit method. The latest assessment of this scheme was at 5 April 1996. The assumptions which have the most significant effects on the results of the valuation are those relating to the rate of return on investments and the rate of increase in salaries and pensions. It was assumed that the investment return would be 8½% per annum with dividend growth of 4% per annum, that salary increases would average 5½% per annum and that pensions would increase at the rate of 3½% per annum.

At the date of the latest actuarial valuation, the marker value of the scheme's assets was GBP 792m and the actuarial value of these assets represented 108% of the benefits that had accrued to members, after allowing for expected future increases in earnings.

Benefit improvements to members have been agreed with the trustees which have resulted in an increased company cost. This increasing ongoing cost has been offset by the amortization of the surplus as a level percentage of pay over nine years. The pension cost of this scheme to the Group was GBP 55m (1998 – GBP 44m).

The Group also operates a defined contribution pension scheme for part-time employees which was introduced on 6 April 1988. The assets of the scheme are held separately from those of the Group, being invested with an insurance company. The pension cost represents contributions payable by the Group to the insurance company and amounted to GBP 17m (1998 – GBP 15m). There were no material amounts outstanding to the insurance company at the year end.

The Group also operates defined contribution schemes in the Republic of Ireland and Hungary. The contributions payable under these schemes of GBP 1m (1998 – GBP 3m) have been fully expensed against profits in the current year.

1.4.4 A review of SSAP 24

Following recent failures and enacted legislation, SSAP 24 has come under increasing criticism for allowing too many options and inadequate disclosure. Following the ASB's publication of a Discussion Paper in April 1995, a revised paper was issued in July 1998 entitled 'Aspects of Accounting for Pension Costs', which set out its reaction to IAS 19 (revised) and sought views on four specific issues:

- The use of market values.
- The discount rate.
- The treatment of actuarial gains and losses.
- The treatment of past service costs.

FRED 20 on retirement benefits was issued in November 1999 and covers pensions and other forms of retirement benefits. The exposure draft proposes substantial changes to SSAP 24 based on the previous discussion documents and in light of the revised IAS 19 on employee benefits issued in 1998. The proposed changes are:

- When measuring the assets of pension schemes, market values should be used rather than actuarial valuations.
- The discount rate for scheme liabilities should be based on a rate that reflects the characteristics of the liabilities rather than the expected rate of return.
- In recognizing actuarial gains and losses there should be a move from gradual recognition in the profit and loss account to immediate recognition in the statement of total recognized gains and losses.
- As a consequence of the above the balance sheet would show a pension asset or liability equal to the recoverable surplus or deficit in the scheme.

Comments on FRED 20 had to be submitted to the ASB by 5 February 2000.

2 Accounting for deferred taxation

2.1 Different rules for the calculation of profit

Accounting for deferred taxation attempts to deal with the differences between the principles of taxation and the principles of accounting. The rules for calculating profit under the two sets of principles differ.

Expenses which are deductible from income under generally accepted accounting principles may not be allowable under tax rules; and some deductions allowable under tax rules may not be allowable under generally accepted accounting principles, such as capital allowances for specific assets in the calculation of taxable profits. In short, the differences between accounting profit and taxable profit would arise from differences due to:

- Whether an item of expense is deductible (or an item of income is chargeable) in arriving at profit or loss at all.
- If allowable (or chargeable) by both accounting and taxation rules, whether they both allow the deduction (charge the income) within the same accounting period.

The first type of difference describes what is known as 'permanent differences' between the two systems; the second relates to 'timing differences'. The difference between permanent difference and timing difference is in the expectation that the difference may correct itself over time. In the case of a permanent difference, the difference between accounting profit and taxable profit in one year cannot self-adjust over time, whereas the timing difference will self-adjust itself over time. Taxation that is saved in one period, may become payable in a future period. Examples of permanent differences include:

- Incomes incorporated in the accounting profit (such as government subsidy) that are not taxable.
- Expenses deducted before arriving at the accounting profit (such as entertainment expenses) that are not allowable under the tax rules.

These permanent differences do not have tax implications. This is because the accounting treatment need not be reconciled with the tax treatment and is not dependent on tax considerations. Whatever tax is payable cannot be corrected in the future since the permanent difference is itself not self-adjusting. Therefore, there is no need to adjust the financial statements for permanent differences.

Timing differences are those requiring adjustment for deferred taxation. A typical example of a timing difference is that arising as a result of the difference between the charge for depreciation in the profit and loss account and the capital allowance that is granted by the Inland Revenue in the computation of taxable profits. Although the annual depreciation and the annual capital allowance may differ, the overall total of depreciation relating to an asset and the total capital allowance granted on the asset should be equal over the life of the asset.

There are two methods of accounting for the effect of timing differences: the deferral and the liability method. There are two variants of the liability method: the full and the partial.

2.2 The deferral method

Under this method, the actual tax liability for an accounting period is compared with the tax liability that would have arisen had there not been any timing differences in the accounting year (i.e. the true tax liability). If this true tax liability is higher than the actual tax liability, the difference is treated as a tax deferral. The profit and loss account for that period is debited with the deferred tax, so that the tax liability that is charged to the profit and loss account is the true tax liability. The corresponding debit or credit is taken to the balance sheet as a deferred debit or credit. In the year when the true tax liability is less than the actual tax liability, the difference is sometimes referred to as tax acceleration. The tax acceleration is deducted from the actual tax liability and the corresponding entry in the balance sheet is a deferred debit.

In future years, when the timing differences reverse, the deferred tax debit or credit balance would be released. The amount that can be taken out of the deferred tax balance would be the one that was originally credited/debited when the timing difference was provided for in the first instance. In essence, the reversal entry would be at the old rate of tax on which the original deferred tax was calculated. The problem with this approach is that when the tax liability that was deferred falls due, it is going to be based on the rate of tax prevailing at that time, not the rate of tax at the time of deferral.

2.3 The liability method

Whereas the deferral method focuses on the profit and loss account as the pivot for the adjustment for timing difference, the liability method focuses on the balance sheet. The perspective of the deferral method is the past, whilst that of the liability method is the future. This means that the reversal entries would be based on the rates of tax prevailing at the time of the reversal, not at the time the deferred tax was created. In this process, the deferred tax balances are always updated to reflect the most recent tax rate.

Accounting for deferred tax under both the deferred and the liability methods can either take full cognizance of all timing differences or take partial cognizance of the cumulative timing differences. Partial provision for timing differences is the approach adopted in the United Kingdom. The full provision is adopted in practice in the US, Australia and the Netherlands.

SSAP 15, Accounting for Deferred Taxation, issued in October 1978, revised May 1985, adopts the liability method and requires disclosure of the following:

- The amount of deferred tax charged or credited in the profit and loss account for the accounting period, separated between that relating to ordinary activities and that relating to any extraordinary items.
- The amount of any unprovided deferred tax in respect of the period, analyzed into its major components.
- Any adjustments to deferred tax passing through the profit and loss account which relate to changes in tax rates or in tax allowances. If the change in the tax system is regarded as sufficiently fundamental, this fact may be separately disclosed (FRS 3).
- The deferred tax balance, and its major components, and the total amount of any unprovided deferred tax analyzed into its major components.
- Transfers to and from the deferred tax balance.
- Movements in reserves which relate to deferred tax.
- Where the value of an asset is shown in a note (or in a directors' report) because it differs

materially from its book amount, the note should also show, if material, the tax effects, if any, that would arise if the asset were realized at the balance sheet date at the noted value.

- Any assumptions regarding the availability of group relief and the payment therefore which are relevant to an understanding of the company's deferred tax provision.
- The fact (if applicable) that provision has not been made for tax which would become payable if retained earnings of foreign subsidiaries were remitted to the United Kingdom.
- Where a company changes its accounting policy to account for the deferred effects of pensions and other post-retirement benefits on a full provision basis, the effects of this change in the basis of taxation should be included in the tax charge or credit for the period and separately disclosed on the face of the profit and loss account in accordance with FRS 3.

The following illustration provides information on disclosure practice relating to deferred taxation:

Excerpt from the Annual Report of British Telecom Plc, 1999

> **Taxation**
>
> The charge for taxation is based on the profit for the year and takes into account deferred taxation. Provision is made for deferred taxation only to the extent that timing differences are expected to reverse in the foreseeable future, with the exception of timing differences arising on pension costs where full provision is made irrespective of whether they are expected to reverse in the foreseeable future.

2.4 A review of SSAP 15

Substantial changes both in the tax system and in general economic conditions have increased criticism of SSAP 15. Commentators highlighted its inconsistency with international practice and other UK standards, the fact that it takes account of future transactions and its subjectivity. Some of these criticisms have been partly addressed through an amendment to SSAP 15 but having issued a Discussion Paper 'Accounting for Tax' in March 1995 the ASB issued an exposure draft, FRED 19 'Deferred Tax'.

The proposed accounting standard would replace SSAP 15 and would principally require entities applying accounting standards to account for deferred tax on a 'full provision' basis as opposed to the current practice of using the 'partial provision' method. This change would bring UK practice into line with the requirements of other standard setters including the United States and IASC's IAS 12.

However, some differences would remain. In particular:

- IAS 12, in contrast, requires deferred tax to be provided for on revaluation gains, even if there is no prospect that the asset will be sold whereas FRED 19 does not.
- For the same reason, deferred tax would not be provided for in respect of the tax that would be payable if overseas profits were remitted to the United Kingdom. IAS 12 requires deferred tax to be provided for if the reporting entity cannot control the overseas business's dividend policy.
- The FRED proposes that balances should be discounted where the effect is material. IAS 12 prohibits discounting of long-term deferred tax balances on the grounds that it could be too difficult to do in practice.

3 Foreign currencies

3.1 General remarks

A company may engage in two types of foreign activity. Both types are recognized in SSAP 20. The two types are now defined and discussed sequentially:

- Direct transactions by an individual company involving one or more foreign currencies such as importing and exporting goods, lending and borrowing money. The rates at which foreign currencies are exchanged are not generally constant; it is therefore likely that transactions may be originated and settled at different exchange rates. The extent of variation between the settlement and originating rates of exchange is called 'transaction exposure'. Foreign currency transaction gains (losses) reflect the effects on the company of exposure to changes in exchange rates.

- Indirect transactions, where foreign operations are carried out through an intermediary such as a subsidiary, an associated company, a branch or an agent. To prepare financial statements a reporting company must translate the foreign currency financial statements of its foreign operations. Such a translation would result in gains (losses). No economic exchange per se underlies the translation process: gains (losses) arise due to changes in exchange rates which have occurred since the previous financial statement (translation) date. Because different translation methods affect figures in the financial statements differently, 'translation exposure' refers to the amounts which are at risk of producing translation gains or losses.

3.2 Accounting for transactions in foreign currencies

SSAP 20, Accounting for Foreign Currency Translations, specifies the following general requirements:

- All transactions in a foreign currency should be translated and recorded at the exchange rate ruling on the date on which the transaction occurred (i.e. actual rate).
- Underlying items of each transaction are distinguishable for accounting purposes; at the balance sheet date, there will be no retranslation of non-monetary items (i.e. fixed assets, equity investments and stocks). This means the non-monetary items acquired in foreign currency will always be translated at the actual rate ruling on the date of the transaction and no exchange gain (loss) will arise.
- At the balance sheet date, all monetary items (e.g. cash and bank balances, loans and amounts receivable and payable) should be translated at the rate ruling on the balance sheet date (i.e. closing date, described as 'current rate' in the United States) or, where applicable, the rates of exchange fixed under the terms of the relevant transactions. Where there are related or matching forward contracts in respect of trading transactions, the rates of exchange specified in those contracts may be used.

Exchange differences arising are usually included in the profit or loss from ordinary activities unless they arise from events which themselves would fall to be treated as extraordinary items. In short, an exchange gain or loss would arise:

- When the rate ruling on the settlement date is different from the actual rate.
- At the balance sheet date, where monetary items originally translated at the rate ruling on the date of the transaction are now being retranslated at the closing rate.

3.3 Forward contracts

Forward contracts are agreements to exchange different currencies at a specified rate in the future. SSAP 20 distinguishes between forward exchange contracts that are speculative and those that hedge or cover existing commitments. If the hedged commitment is an identifiable foreign currency transaction to take place in the future (for example an agreement to purchase an inventory item), it is most likely that the gain or loss on the forward contract will be deferred and included in the measurement of the related foreign currency transaction. But if the hedged commitment is an identifiable executory foreign currency transaction (i.e. an underlying part of a transaction has already taken place; for example, stock items purchased with payment to be made at a later date), any gain or loss on the forward contract should be recognized in the profit and loss account of the current period.

A UK importer may choose one of the following options to pay for any imported item:

(i) Wait until the settlement date to buy and remit foreign currency to the exporter.
(ii) Buy foreign currency spot and invest the funds in the money market until payment is due.
(iii) Enter into a forward contract for foreign currency to coincide with the delivery date of the goods.

The importer would most probably choose the option that appears cheaper. Given interest rate differentials between the two countries, the cost of option (i) is the value of imports in foreign currency multiplied by the expected exchange rate at the settlement date less interest earned on this amount from the delivery date to the settlement date. The importer may not choose this option even if it is the cheapest as the importer does not want the foreign exchange risk. The cost of option (ii) is the foreign currency value of the import multiplied by the spot exchange rate at the date of the transaction less interest earned on this amount in foreign currency or in Euromoney accounts and the cost of the option plus interest foregone on the sterling funds used to buy the foreign exchange. Option (iii) is the foreign currency value of the imports multiplied by the forward exchange rate less interest earned on this amount in the United Kingdom.

Both options (ii) and (iii) involve hedging against changes in exchange rates that could otherwise alter the magnitude of the liability created as a result of agreeing to pay for imports in foreign exchange. Therefore under both options exchange rate fluctuations would henceforth be inconsequential and could be expunged from the accounts. If the exchange rate changes are to be shown under either of these two options, the gain (loss) on foreign exchange purchased would be offset by the loss (gain) on the liability to the exporter. SSAP 20 acknowledges that a rate specified in a forward hedging contract may be used for the translation of trading transactions denominated in a foreign currency (and for any monetary assets or liabilities arising from such transactions) where these have been covered by a related or matching forward contract. As a result, SSAP 20 recognizes that such contracts are used to hedge any exchange risks involved in foreign currency operations, and that no economic gain or loss will arise.

4 Long-term contracts

A long-term contract is defined by SSAP 9 (para. 22) as one where the contract activity falls into different accounting periods, usually, but not necessarily, extending for a period exceeding one year. Accounting for long-term contracts should be on a contract by contract basis and reflected in the profit and loss account by recording turnover and related costs as contract activity progresses. Turnover is ascertained in accordance with the stage of completion of the contract, the business and the industry in which a reporting company operates.

Where the outcome of a long-term contract can be assessed with reasonable certainty before its conclusion, the prudently calculated attributable profit is required by SSAP 9 (para. 29) to be recognized (that is deemed realizable) in the profit and loss account as the difference between the reported turnover and related costs for that contract. If the profitable outcome cannot be ascertained with reasonable certainty, there can be no attributable profit. Attributable profit is defined by SSAP 9 (para. 23) as that part of the total profit currently estimated to arise over the duration of the contract, after allowing for estimated remedial and maintenance costs and increases in costs so far as not recoverable under the terms of the contract, that fairly reflects the profit attributable to that part of the work performed at the accounting date.

SSAP 9 (para. 30) stipulates the rules relating to the presentation in the financial statements in respect of long-term contracts as follows:

- The amount by which recorded turnover is in excess of payments on account should be classified as 'amounts recoverable on contracts' and separately disclosed within debtors.
- The cost of long-term contracts, net of amounts transferred to cost of sales, should be classified as 'long-term contract balances' separately disclosed within the balance sheet heading 'stocks', after deducting foreseeable losses and payments on account not matched with turnover. The balance sheet note should disclose separately the balances of:
  (i) net cost less foreseeable losses.
  (ii) applicable payments on account.
- If provisions or accruals for foreseeable losses exceed costs incurred (after transfers to cost of sales), that excess should be included within either provisions for liabilities and charges or creditors as appropriate.
- The balance of payments on account (in

excess of amounts matched with turnover and offset against long-term contract balances) should be classified as payments on account and separately disclosed within creditors.

In line with the consensus reached by the UITF and reported in the UITF Abstract 4, significant long-term contract debtors balances may be shown separately on the face of the balance sheet rather than as part of current assets.

Excerpt from Annual Report of Morrison Construction Group Plc, 1999

Accounting for profits

(i) Building and Civil Engineering Operations

Profits on short-term contracts are included in the financial statements upon substantial completion of those contracts. Profits on long-term contracts are included in the financial statements when the outcome of a contract can be assessed with reasonable certainty and are determined by reference to an internal valuation of measured work carried out less related costs of production. Provision is made in full for foreseeable losses.

(ii) Property Developments

Profit is included in the financial statements in connection with property developments when a legally binding contract for the sale of the development has been entered into and legal completion has taken place before or shortly after the year end. Where legally binding contracts exist, profits on the construction and refurbishment elements of the development are determined on the same basis as for building and civil engineering operations. Other profits arising from developments are included in the financial statements only when legal completion of the sale of the development has been effected.

(iii) Claims

In establishing turnover and profit, credit is taken for claims only when agreed in writing by the client. Having taken such claims to income, provision is made whenever ultimate payment becomes doubtful.

5 Lease and hire purchase contracts

A company can acquire the use of an asset by the following means:

- Buying it outright, and paying cash for it immediately or within a reasonable time afterwards, i.e. on open account terms.
- Buying it on hire purchase agreement whereby the purchase price is paid for by instalment together with interest due on the amount outstanding and legal title to the asset is assured when the last instalment is paid.
- Hiring it on a lease contract where legal title to the asset is never assumed.

The assets under the leasehold and hire purchase heading exclude assets which are bought outright and those which relate to royalty agreements, such as those arising from the use of natural resources to produce oil, gas, timber, metals and other mineral rights, and to licensing agreements for such items as motion picture films, video recordings, plays, manuscripts, patents and copyrights.

The difference between a hire purchase and a lease agreement is that under a hire purchase agreement, the hirer has a potential right to own the underlying asset on the payment of the final instalment. Under a lease contract, the legal title does not pass to the lessee. Because the lessor retains ownership of leased assets, while the lessee pays a rental, the accounting treatment in the lessee's books has traditionally been merely to show the periodic payments as expenses, without recording the assets in the balance sheet. On this basis, one would expect the lessor to show the leased assets as such in the balance sheet and all the rental as income in the profit and loss account.

Leasing has become much more common in recent years and the types of lease much more diverse. Many leasing arrangements are in reality more akin to hire purchase buying than to a renting agreement over a limited period. Many leases give the sole right of use to the lessee over the entire life of the asset, with the option to purchase at the end of the leasing period at a purely nominal amount. If the lessee also has full obligation for maintenance, this type of lease is, in reality, similar to hire purchase or buying on credit over a period. Such a lease, is therefore, tantamount to the provision of a loan to the lessee by the lessor, enabling the lessee to use the asset for all of its life and requiring the lessee to pay for the full value of the asset with interest over a period. This type of lease should be distinguished from a hire (not hire purchase) contract. In this case the lessee acquires possession and use of the asset for a short period, and the asset is then returned to the lessor who may lease the same asset to another user in future.

In general, a lease is a legal instrument that transfers property rights from lessor to lessee and creates obligations between them. A finance lease is one whereby the lessee obtains exclusive rights to the leased properly in return for a rent. The rent may consist of two elements: the primary period, which is that period which enables the asset to be fully amortized (i.e. paid for), and the secondary period, which is that time over which the asset is still in use for a peppercorn rent (minimal rental). The legal nature (i.e. ownership of the asset by the lessor) of the agreement is maintained which enables the lessor to obtain capital allowances and investment grants (where applicable). In the United Kingdom, for a lessor to obtain the capital allowances there must not be an option for the lessee to buy the asset.

The tax position has been important in influencing the growth of leasing. In the early 1980s much of the manufacturing sector in the United Kingdom was recording tax losses even though many had accounting profits. This arose because of differences between profits for financial accounts purposes and for tax purposes. The reason why there were huge tax losses at that time was because manufacturing companies could obtain 100 per cent first year allowances on their capital expenditure and because companies could claim stock relief to avoid having to pay tax on the unrealized gains on stocks during periods of rising prices. In these circumstances, if a manufacturing company with sizeable tax losses made further capital purchases, no immediate tax relief would have been obtained. Instead the relief was added to the carried forward tax losses. However, if a company with taxable profits purchased the asset and leased it to the manufacturing company, immediate tax relief could be obtained which could be shared between the parties. The lessor with taxable profits could get relief for the capital purchase by obtaining the capital allowance and the lessee, the manufacturing company, could obtain some benefit in the form of reduced rentals. These taxation aspects provided considerable incentives for the development and growth of leasing in the United Kingdom.

In 1984 the taxation incentives for leasing were reduced considerably. No stock relief was to be made available for periods of account beginning after 13 March 1984 and the 100 per cent first year allowance was reduced over the next two years and eliminated altogether from 1 April 1986. The current tax position is that the lessor may obtain 25 per cent writing down allowance each year (a first year allowance of 40 per cent is applicable to expenditure incurred after 31 October 1992 and before 1 November 1993) on a reducing balance basis and pays tax on the total rentals receivable. The lessee obtains tax relief on the rentals payable, although there are some maximum restrictions in the case of leased cars and leases where all the rentals are front end loaded.

In contrast to the leasing position, a company that acquires the use of an asset through hire purchase is treated for tax purposes in the same way as if the asset had been purchased for cash. Capital allowances may be obtained even though legal ownership does not arise until the final payment is made.

In August 1984 the ASC issued SSAP 21, Accounting for Leases and Hire Purchase Contracts, which took effect for periods commencing on or after 1 July 1984. The introduction of the standard was phased but since 1987 it has been fully operational. The main reason for the introduction of SSAP 21 was to distinguish between the economic substance and legal form of a lease contract. Prior to SSAP 21, most companies treated all leases in the same way – according to the legal formality whereby the lessor shared the underlying asset in its balance sheet and recorded all rentals as income, while the lessee treated only the periodic rental payments as expenses without recording the asset in the balance sheet. The effect of showing neither the asset nor the leasing obligation in the balance sheet of the lessee is that the return on capital employed will be overstated and the firm is provided with a kind of hidden borrowing which decreases its overall debt raising capability.

SSAP 21 made a distinction between a finance lease and an operating lease. A finance lease is one that transfers 'substantially all the risks and rewards of ownership' of an asset to a lessee. Thus, economic substance takes precedence over legal form. Generally, if the minimum lease payments amount to substantially all of the fair value of the leased asset, the lease constitutes a finance lease. The standard offers guidelines for deciding whether substantially all the risks and rewards have passed to the lessee. The guidelines state that, at the start of the lease, if the minimum lease payments including any initial payment amount to substantially all (90 per cent or more) of the fair value of the leased asset, it should be presumed that the lease is a finance lease. In undertaking this present value calculation the interest rate implicit in the lease should be used.

Fair value is defined as the price at which an asset could be exchanged in an arm's length transaction less, where applicable, any grants receivable towards the purchase or use of the asset. If this is not known, an estimate should be used.

The standard also defines the implicit interest rate as the rate that, when applied to the amounts which the lessor expects to receive and retain, at the inception of a lease, produces an amount (present value) equal to the fair value of the leased asset.

The amounts the lessor expects to receive and retain comprise the minimum lease payments, plus any unguaranteed residual values less any amount for which the lessor will be accountable to the lessee. The minimum lease payments consist of the minimum payments over the remaining part of the lease term, any residual amount guaranteed by the lessee or a party related to it and any residual amounts guaranteed by any other party. An example in Table VII.1 will illustrate how the calculations work.

The ASC argues that a fairer view of leased assets is obtained if those who have rights to the asset together with the obligations to pay rentals capitalize those assets. In implementing this there should be recognition of the interest elements in lease payments by recognizing separately the finance charge in the profit and loss account of the lessee. The rules for dealing with operating and finance leases are as follows:

- Operating leases:
  - (i) Lessee – payments should be charged to the profit and loss account on a straight line basis over the lease term.
  - (ii) Lessor – capitalize and depreciate the asset held, recognize income on a straight line basis over the period of the lease.
- Finance leases:
  - (i) Lessee – capitalize leased asset as a balance sheet asset along with the associated liability; in the profit and loss account the asset should be depreciated and the interest charge on the outstanding debt should be expressed at a rate that produces a constant periodic rate of return on the remaining balance of the obligation over the lease term.
  - (ii) Lessors – show the leased asset in the balance sheet as a debtor (for lease rentals) equal to the net cash investment in the lease in each period; recognize lease income in the profit and loss account in a way to produce a constant rate of return on the balance outstanding

Table VII.1 Determining the underlying rate in a lease – an example

Fair value of the asset = GBP 20,000

Five annual rentals payable in advance = GBP 4,000 per annum

Total estimated residual value at the end of five years = GBP 6,000 of which GBP 4,000 is guaranteed by the lessee

Calculation of the rate implied in the lease

	t_0	t_1	t_2	t_3	t_4	t_5
Rentals	4,000	4,000	4,000	4,000	4,000	
Residual value						6,000
Fair value*	(20,000)					
	(16,000)	4,000	4,000	4,000	4,000	6,000
Discount factor at 10%	1	0.9091	0.8264	0.7513	0.6830	0.6209
Discounted amount	(16,000)	3,636	3,306	3,005	2,732	3,725

NPV = +GBP 404

Discount factor at 12%	1	0.8929	0.7972	0.7118	0.6355	0.5674
Discounted amount	(16,000)	3,572	3,189	2,847	2,542	3,404

NPV = −GBP 446

IRR (rate implied in the lease) = 10% + (404)/(404 + 446) × 2% + 10.95%

*Net amount the lessor expects to receive and retain.

Calculation of the present value of the minimum lease payments

	t_0	t_1	t_2	t_3	t_4	t_5
Rentals	4,000	4,000	4,000	4,000	4,000	
Guarantee						4,000
Minimum lease payments	4,000	4,000	4,000	4,000	4,000	4,000
10.95%	1	0.9013	0.8124	0.7322	0.6599	0.5948
Discounted amount (Present value of minimum lease payments)	4,000	3,605	3,205	2,929	2,640	2,379

Present value of the minimum lease payments = GBP 18,803

Percentage of the asset's value = (18,803/20,000) × 100 = 94.015%

5.1 A review of SSAP 21

The ASB has been developing, in consultation with other standard setters within the G4+1, a discussion paper based on the recommendations made in a special report 'Accounting for Leases: A New Approach Recognition of Lessees of Assets and Liabilities Arising under Lease Contracts' published in 1996.

Criticism of SSAP 21 includes:

- The omission of material assets and liabilities arising from operating lease contracts.
- There being marginal differences from one lease being treated as a finance lease and one as an operating lease, thereby failing to secure the same accounting treatment for similar transactions.
- The 'all or nothing' approach of SSAP 21 to capitalization of leased assets does not adequately reflect modern complex transactions.

Other issues, such as residual value arrangements, contingent rentals and options to cancel or renew leases, the distinction between leases and executory contacts, the treatment of sale and leaseback transactions and asset and income recognition principles for lessors are also under review.

6 Accounting for government grants

6.1 General remarks

Government grants to companies take many forms. They come in the form of capital grants to assist in the location of companies to designated development areas and in the attraction of foreign companies to the country (e.g. the investment grants granted to Japanese automobile companies), or in the form of revenue grants to assist companies to offer their services to the public at a discount. These grants may be made contingent upon the occurrence of a future event or be free from any encumbrance. Grants are not usually paid for singular purposes. They may come in the form of a multi-purpose offer relating to both capital and revenue subsidies at the same time. In such situations, reporting companies may need to make a judgement on which part of the grant belongs to capital and which to revenue.

The current accounting standard on this topic is SSAP 4, The Accounting Treatment of Government Grants, issued in 1974 to deal with the predominantly capital grants made available by the government in the early 1970s. This standard was revised in July 1990 to cater for the many forms of government grant that became available between 1980 and 1990.

6.2 Capital grants

In the original SSAP 4, grants made available to motivate investment in capital expenditure were to be spread over the expected life of the assets they helped to finance in one of two ways:

- Netting off the grant from the cost of the asset in the balance sheet and charging depreciation on the net figure.
- Treating the grant as a deferred credit in the balance sheet and releasing it to the credit of the profit and loss account over the life of the asset to offset the depreciation charge.

The revised SSAP 4 (1990) proscribed the first option because the ASC, in the opinion of counsel, considered that the option was in violation of the Companies Act 1985, Schedule 4 which required fixed assets to be carried at their purchase price or production cost. This prohibition only applies to companies governed by Companies Act 1985, Schedule 4.

6.3 Revenue grants

Government grants may be given to assist a company to compete favourably in international markets. They may come in the form of subsidies to finance the general activities of an enterprise over a specific period or to compensate for the loss of the current or future periods. Some of these grants are released after the revenue cost towards which the grants are contributing have been incurred. In this case grants are credited to the profit and loss account in the period the cost is incurred. Some grants may be released before the revenue costs related to it are incurred, such as immediate financial assistance or support. In this case, the grant should be credited to a deferred credit account and released in the period the attributable revenue costs are incurred.

6.4 Conditional grants

When grants are repayable if certain conditions are not met within a qualifying period, the recipient company should make provision for such repayment. For example, if 40 per cent of a grant of GBP 400,000 to purchase an asset with a life of ten years is repayable if a certain violation occurs within five years, it would be prudent for the recipient company to delay accruing to income any portion of the 40 per cent (i.e. GBP 160,000) within the first five years. In this case only the unencumbered 60 per cent would be available for releasing to the profit and loss account at GBP 24,000 per annum.

When the grant becomes payable, SSAP 4 (revised) requires that the repayment should be deducted from any unamortized deferred credit relating to the grant with any excess being charged to the profit and loss account.

6.5 Disclosure

The disclosure requirement is that the effects of government grants on the financial position of a reporting company should be shown where the results of future periods are expected to be affected materially by the recognition in the profit and loss account of the grants already received. The accounting policies adopted for government grants and the specified circumstances where there are potential liabilities to repay grants already received should also be disclosed.

7 Reporting the substance of transactions

7.1 Background

Having issued SSAP 21 'Accounting for leases and hire purchase contracts' in August 1984, the next venture into the area of off balance sheet finance (OFSB) was in December 1985 when the Technical Committee of the ICAEW issued Technical Release 603 (TR 603). For the first time a professional statement recommended that in order to give a 'true and fair view' the economic substance of transactions should be considered rather than mere legal form when determining their nature and, in turn, their accounting treatment. The English Law Society argued that applying substance over form would introduce 'unacceptable subjectivity' making legal analysis of financial statements impossible. With increasing use of OBSF the ASC issued an exposure draft, ED 42 titled 'Accounting for special purpose transactions' in March 1988. The ASC avoided using detailed accounting rules to regulate what might be termed 'off balance sheet' or 'window dressing', opting instead to set out general concepts that could be applied to a variety of situations. It adopted wide descriptions of assets and liabilities. Assets were described in terms of control of future economic benefits. Liabilities were described in terms of present obligations entailing a probable future sacrifice of benefits involving the transfer of assets or provision of services to another party.

The Companies Act 1989 introduced the first legal rule in redefining subsidiaries for consolidation purposes. Spurred on by this and taking on board comments on ED 42 the ASC issued a revised exposure draft, ED 49, entitled 'Reflecting the substance of transactions in assets and liabilities' in May 1990. The general concepts in ED 42 were reinforced by a series of detailed application notes addressing a number of arrangements that had caused difficulty in practice. These were:

- Consignment stock.
- Sales and repurchase agreement.
- Factoring of debts.
- Securitized mortgages.
- Loan transfers.

The ASC believed that ED 49 achieved the right mixture of general concepts and detailed guidance on application that satisfied those who argued that an over-conceptual approach was too subjective and those who believed too many rules led to literal interpretation and to scope for avoidance. Following heavy debate on the issue of securitization the ASB, successor body to the ASC, published another exposure draft, FRED 4, titled 'Reporting the substance of transactions', which largely forms the basis for the current accounting standard on the subject, FRS 5. The two main differences between FRED 4 and FRS 5 were:

- Derecognition – FRED 4 took the position that an item either remained on, or was taken off the balance sheet in its entirety according to whether an asset or liability was, or was not, present and the recognition criteria met. FRS 5 adopted a new category which it referred to as 'partial derecognition'.
- Offset rule – Under FRS 5 it is still possible to offset assets and liabilities denominated in different currencies where there is a legal right of set-off and the conditions in the FRS are met.

7.2 FRS 5 'Reporting the substance of transactions'

As introduced above, FRS 5 addresses the problem of what is commonly referred to as 'off balance sheet financing'. One of the main aims of such arrangements is to finance a company's assets and operations in such a way that the finance is not shown as a liability in the company's balance sheet. A further effect is that the assets being financed are excluded from the accounts, with the result that both the resources of the entity and its financing are understated.

FRS 5 requires that the substance of an entity's transactions is reported in its financial statements. This requires that the commercial effect of a transaction and any resulting assets, liabilities, gains and losses are shown and that the accounts do not merely report the legal form of a transaction.

For example, a company may sell (i.e. transfer the legal title to) an asset and enter into a concurrent agreement to repurchase the asset at the sales price plus interest. The asset may remain on the premises of the 'seller' and continue to be used in its business. In such a case, the company continues to enjoy the economic benefit of the asset and to be exposed to the principal risks inherent in those benefits. FRS 5 requires that the asset continues to be reported as an asset of the seller, notwithstanding the transfer of legal title, and that a liability is recognized for the seller's obligation to repay the sales price plus interest.

FRS 5 incorporated the five application notes contained in ED 49 and in September 1998 included an application note on 'Private finance initiative and similar contracts'.

8 Post-balance sheet events and contingent liabilities

8.1 Background

The principal accounting and disclosure requirements in relation to post-balance sheet events and contingencies are contained in two SSAPs, namely SSAP 17 'Accounting for post-balance sheet events' and SSAP 18 'Contingencies'. In addition, Schedule 4 and 7 of the Companies Act 1985 dealt with commitments and contingent liabilities and disclosures required in directors' reports in relation to post-balance sheet events.

As mentioned above, FRS 12 'Provisions, contingent liabilities and contingent assets' – issued in September 1998, effective for accounting periods ending 23 March 1999 – supersedes SSAP 18. This section therefore deals principally with SSAP 17.

8.2 Post-balance sheet events

Many numbers reported in financial statements are the result of assumptions and estimates about future events and outcomes such as future sales of inventory items, collectibility of accounts receivable, outcomes of pending litigations, useful lives of fixed assets and their salvage values. Because many of these uncertain items of financial information are presented in numerical form, their characteristics and potential consequences are masked so that people (users and preparers alike) fail to appreciate the uncertainty underlying them and as a result, tend to treat the numbers as precise. This illusion may create user confidence that may later be destroyed by future events. In addition, the operational toolkit that has grown around the desire to present information that is true and fair representation of the state of affairs reported tends to heighten the illusion of reality. For example, an event occurring immediately after a balance sheet date but before the accounts have been signed, could cause the reported equity, income and earnings per share to be materially misstated.

The real problem is how to communicate information relating to uncertain events and outcomes. Such a message can be given numerically (e.g. GBP 2 million) or non-numerically by an adjectival-qualitative description (extremely large). Because the use of numbers implies measurement, there is the presumption that a numerical message may be more precise and authoritative than a non-numerical message. However, this does not mean that all measurements have to be expressed numerically or that all numerical presentations do reflect precise measurements. It all depends on how one perceives the message rather than the absolute precision of numbers of words. This reduces to the problem of whether the effect of probable events should be incorporated into the profit and loss account or should be reported as additional verbal information in the operating and financial review, directors' report or footnote to the financial statements. If the outcome of a probable event is likely to affect the value of an asset negatively, the estimated monetary value of the loss is reflected in the accounts immediately. If there is an uncertainty about the probable outcome (like the result of a pending 'path-breaking' litigation), the practice is not to report the estimated monetary value of the loss from the probable event but to give a non-numerical explanation of the nature of the uncertainty. These future situations with potentially significant consequences for accounting numbers are referred to, in accounting, as contingencies and post-balance sheet events.

Contingencies refer to events that have taken place whose outcomes are not known at the balance sheet dates. Post-balance sheet events are those that occur after the balance sheet date and are likely to undermine the figures in the financial statements for the year just ended.

The issue of how to treat future events in the accounts has recently engaged the attention of four national standard setting bodies (from Australia, Canada, the United States and the United Kingdom) and the IASC who are collaborating on the study of the subject. These standard-setting bodies have jointly issued a discussion document on Future Events: A Conceptual Study of their Significance for Recognition and Measurement. The discussion document examines the question of whether and when accounts should reflect future events through a series of case studies, some of which are controversial (ASB, October 1994, p. 1). For example, should actions that management intends to take in the future be reflected in today's accounts? Should tax losses be treated as assets, if they will have value only if and when a profit is earned in the future? Should the cost for pensions included in accounts reflect the increase in their cost that will be caused by future salary increases over the working lifetime of the employees? (ASB, October 1994).

SSAP 17 specifies the basis for dealing with post-balance sheet events. If events arising after the balance sheet date are likely to provide additional evidence of conditions that existed at the balance sheet date and would materially affect the amounts to be included, they are usually reflected in the financial statements. A post-balance sheet event is one that occurs between the balance sheet date and the date on which the financial statements are approved by the board of directors. This excludes matters which are currently being discussed but may lead to a decision by the board of directors in the future. Events which occur after the balance sheet date may be classified as adjusting events (i.e. events which are likely to provide additional evidence relating to conditions existing at the balance sheet date and require changes in the amounts to be included in the financial statements); and non-adjusting events (i.e. events which arise after the balance sheet date and concern conditions which did not exist at that time and so would not result in changes in amounts in financial statements). Adjusting events do not require special disclosure but non-adjusting events do by way of a note to the accounts if material.

VIII Revaluation accounting

1 The effect of inflation on accounting numbers

Accounting figures in the annual accounts of reporting companies are often based on historical costs modified by valuations to reflect changes in the market values of assets. A change in the value of the pound, that is, a change in price levels, may therefore seriously lessen the validity of the values reported in the financial statement. This is most easily understood in relation to fixed assets and depreciation. If fixed assets are valued at historical cost, depreciation will usually be based on historical cost as well. A fixed asset possesses many attributes such as production capacity, future cash flows, realizable values and so on.

Suppose each of two companies owns a fixed asset with identical attributes, it is most unlikely that their values on the balance sheets of the two companies will be equal. If the fixed asset of one company was purchased three years ago when prices were low, but the other company purchased its fixed asset (of identical age and productive capacity) from the second-hand market only recently when the price of new fixed assets was higher than three years ago, there is likely to be a difference between the values recorded by the two companies for the seemingly identical fixed asset. This is because the current replacement cost of the fixed asset will be higher in times of rising prices than the book value. Even when prices are not rising, it is most unlikely that the book value of the asset in use will be equal to the replacement value of an identical asset. This is because the figure chargeable in each year's accounts for the use of the fixed asset may not be identical to the fall in the economic value of the fixed asset. It can, in fact, be strongly argued that the use of historical cost during a period of inflation can lead to the publication of profit figures which are, in part, fictitious.

The effect of inflation on a company's declared results can be better illustrated with the sale of the company's product. The accounting profit on the sale of a product is the difference between the sale proceeds and the cost of the product. Assume a company maintains a policy of holding ten items of a finished product which costs GBP 10 (in material, labour and expenses) per unit to produce in a previous period and were sold in the current period at GBP 15 per unit. The profit per item will be recorded as GBP 5, thus reporting a total profit of GBP 50. But if it now costs GBP 12 to replace each item, it means that the cost of maintaining the stock intact (i.e. ten units) would now be GBP 120 (i.e. GBP 12×10) which is higher than the original cost of GBP 100. To this extent, the GBP 50 profit declared in the current period would have included the element of inflation (i.e. the amount needed to maintain the stock intact). The declaration and distribution of the historical cost profit figure of GBP 50 would have meant a running down of the capital of the company. The lack of consensus on the operational definition of the capital not to be impaired has led to the development of various methods of accounting for inflation. Five competing definitions of the capital to be maintained are as follows:

- Money capital (as in traditional historical cost accounting).
- Physical capital (as in current-cost accounting).
- Real capital (as in constant purchasing power accounting).
- Geared capital (a version of cost accounting peculiar to the United Kingdom).
- Maintenance of future consumption or dividends.

2 The inflation accounting debate

Interest in accounting for the effect of inflation has varied in the United Kingdom in line with the growth in the rates of inflation in the country. When double digits inflation was experienced in the late 1970s and early 1980s, interest in the methods for accounting for the effect of inflation was intense. When the rate of inflation fell to a single digit level, the noticeable interest in accounting for inflation fell substantially to the point where the standard on accounting for inflation was withdrawn. Since then, attempts to incorporate the effects of inflation into accounts have either been through the revaluation of fixed assets or through the provision for the potential losses that may arise from the excessive (or fictitious) profits which may accrue from the use of historical cost accounting.

There are no extant regulatory provisions on accounting for the effect of changing price levels in the United Kingdom at the present time. What is discussed below are the inflation accounting methods that have been tried or suggested for adoption in the United Kingdom. These are presented not only for their historical interest, but also for their potential value in the future.

Two principal methods of accounting for inflation which have been suggested are:

- Adjusting for changes in the general price level only, i.e. current purchasing power (CPP) accounting.
- Adjusting for changes in specific prices, i.e. current cost accounting (CCA).

No attempt is made to compare these two methods. However, it is of interest to note that both methods have been tried in the United Kingdom. Prior to 1980, CPP was the basis of accounting for inflation as required by SSAP 7, issued in 1974. With increasing inflation in the 1970s, the inadequacy of SSAP 7 became apparent and several committees were set up by the government (Sandilands Committee) and the profession (e.g. Hyde Committee) to consider alternative methods of accounting for inflation. The work of these committees culminated in the issue of SSAP 16 in March 1980 which adopted the CCA method of inflation accounting. The scope of SSAP 16 was limited to all listed companies and those companies which satisfy at least two of the following criteria:

- A turnover of GBP 5 million per annum or above.
- A balance sheet total in the historical cost accounts of GBP 2.5 million or more.
- An average number of employees in the United Kingdom of 250 or more.

SSAP 16 was not prescriptive as it allowed reporting companies the choice of one out of three methods of accounting for inflation:

- The main accounts could be based on historical cost accounts and supplementary current cost accounts can be presented in addition.
- The main accounts could be based on the current cost accounts while historical cost accounts could be offered as a supplement.
- Current cost accounts could be published as the only accounts supported by adequate historical cost information.

The discretion allowed management in the choice of one of the above methods and the indication given by the ASC that it would review the case of accounting for inflation only after three years from issue probably contributed to the low level of adoption of accounting for inflation. In addition, the double digit inflation of the 1970s declined to single digit in the 1980s. SSAP 16 was later withdrawn. Apart from the above problems, there other substantive and practical difficulties with inflation accounting. One problem refers to the underlying perspectives of inflation accounting. Three differing perspectives can be identified:

- The need to correct for changes arising from specific price levels.
- The need to correct for changes arising from general price levels.
- The need to correct for changes in relative price levels.

Specific price changes relate to sectoral inflation, i.e. changes in the price levels of inputs peculiar to particular businesses. General price level changes relate to changes in the overall trend of all prices people face in the marketplace. This sort of price change can be measured only by constructing indices of the prices of a set of goods which represents what an average individual or family will purchase at different dates. There are different methods of compiling such indices and the commodities in the basket which forms the basis of the indices may be of doubtful relevance to a company's situation, even if they substantially measure the real feature of the life of individuals. However, the rise in general price level is usually referred to as 'inflation'. If prices of specific items change disproportionately, their relative prices will alter leading to a change in the relative price level.

Another problem which may explain the diversity in inflation accounting and the difficulty in regulatory choice arises from the fact that money serves as a store of value, a medium of exchange and a basis for measuring wealth, and traditionally accountants have used the unit of currency to keep track of all these functions. While money may serve effectively as a medium of exchange because of its acceptability as a legal tender, the inconsistencies in its value (especially as concerns its purchasing power) make its roles as a store of value and a basis of measurement of doubtful validity. Since the unit of currency does not usually alter, money as a store of value may be defended as long as one does not consider what such money may buy. The most critical function of money which has come under severe aback is its function as a basis for measuring wealth. If the item to be measured is monetary and short term (i.e. near money time deposits), then the expected unit of currency from such items is unlikely to be different from those indicated by an earlier measurement. But if the item to be measured is non-monetary (i.e. a fixed asset) and long-term monetary items, the unit of currency reflected by such measurement may not be identical with the expected unit of currency from such a sale.

There is also the issue of whether balance sheet items (especially assets) refer to legal rights, bundle of costs, physical things, service or cash flow potential. The meaning different people get from the items in the balance sheet depends on their perception of each of the items. In the context of the effect of inflation on the present values of the items in the balance sheet, a major distinction is usually drawn between monetary and non-monetary items.

The major significance of the monetary and non-monetary distinction is its usefulness for classifying financial items on the basis of the degree to which they make the estimation of cash flows easier and more certain. Monetary items are distinguished from their non-monetary counterparts because their amounts are fixed in terms of units of currency whereas the lager are not. Information about the underlying variables that affect the value of monetary items (i.e. the periodic interest accruals and maturity values) is readily available. In contrast, the amounts at which non-monetary financial assets and liabilities will be sealed are not fixed in terms of units of currency. They can be realized and liquidated respectively by the future receipt or disbursement of a currently undetermined amount of cash or by the future receipt or transfer of other forms of consideration, such as economic goods and services, whose prices may fluctuate over time. Thus, their present values (i.e. the amounts reflected in the balance sheet) are heavily dependent on many variables over which an enterprise has no control. The amounts at which they will ultimately be sealed (or sold) can be affected by external events such as:

- Changes in interest and tax rates.
- Changes in other specific price levels due to changes in supply/demand relationships, technology, and other variables.
- Changes in general price levels.

Therefore, values assigned to non-monetary items are probabilistic, compared with the deterministic character of cash flows, interest rates and values associated with monetary assets and liabilities.

3 Current cost accounting (CCA)

Although there are many inflation accounting (or value accounting) methods, only the current cost accounting method is discussed here since it was the one recommended by SSAP 16 and formed the basis of practice for over seven years in the United Kingdom. Other inflation accounting methods are merely theoretical and have not been put into practice in any country in a systematic manner apart from a variant of replacement cost accounting based on current entry values that is common in the Netherlands. Two other theoretical constructs for inflation accounting are present value accounting and continuously contemporary accounting.

Since there are several variants of CCA, the one introduced by SSAP 16 is now described in detail. It should be remembered that the use of the CCA described below was effectively optional.

In a current cost balance sheet, assets are not valued at historical cost but the 'value to the business'. In practice, this is likely to mean that fixed assets are included at the lower of net current replacement cost and net realizable value (rather than the lower of historical cost and net realizable value). Monetary assets such as bank balances, debtors and creditors are valued as in the historical cost accounts. Liabilities are valued also as in historical cost accounts and no change is made to the share capital. Normally, the only balance sheet items in need of adjustment are fixed assets and inventories. A new reserve is introduced to effect a balance to the new adjusted historical cost balance sheet. This balancing figure was called 'current cost reserve' by SSAP 16 or 'capital maintenance reserve' by ED 24. This reserve was part of the equity section of the balance sheet and represented the amount required by a company to maintain the operating capacity of its business, to the extent that this had not already been allowed for in the historical cost accounts. This current cost reserve can also be explained as the net result of all the adjustments necessary to convert historical cost accounts to current cost accounts, that is, the surplus on revaluations of fixed assets and stocks plus all the adjustments for gearing, cost of sales and monetary working capital, but without incorporating the depreciation adjustment which is an offsetting entry in the profit and loss account for the increase in accumulated depreciation on the current cost balance sheet. In the profit and loss account, four adjustments were usually made:

- A cost of sales adjustment (COSA).
- A monetary working capital adjustment (MWCA).
- A depreciation adjustment.
- A gearing adjustment.

These adjustments were applied in two stages:

- To arrive at a current cost operating (or trading) profit after adjusting for the first three items: COSA, MWCA and depreciation.
- To arrive at current cost profit after taxation after adjusting for gearing.

The cost of sales adjustment is made in order to allow cost of sales to reflect the cost current at the time of using materials instead of the time of purchase. This can be done in several ways including on an individual sales basis and by adjusting the monetary values for inventories, using what is known as the 'averaging method'. This method adjusts the amounts of opening and closing inventories (at historical costs) to the average prices for the year, so as to derive the real change in inventories. The difference between the absolute change and the real change in inventory values is referred to as COSA. If the cost of sales was accounted for on an individual basis, it is quite unlikely that increased prices would be tied up in inventory. But if it was not practical to do so, and cost of goods sold were accounted for by prices other than current costs, the inventory values would capture some (if not all) of the increased prices not reflected in the cost of goods sold. That was why it was thought necessary to distinguish real changes in inventory levels from money changes due to changes in price levels. An illustration of the determination of the current cost of sales is provided in Table VIII.1.

Increased prices included in current costs are not only traceable to unsold stocks but also to monetary working capital (in essence, bank balance + debtors − creditors). The monetary working capital adjustment (MWCA) can be calculated by the averaging method in a way similar to that used for the cost of sales adjustment. Just as the change to the historical cost figures of inventory at the two balance sheet dates is compared to the change in values of inventory at the two balance sheet dates based on average price index, so the change in MWC on a historical cost basis is compared with the change in MWC when the opening and closing figures have been adjusted to average prices during the year.

The depreciation adjustment is made in order to base depreciation on current replacement cost instead of historical cost. It is the excess of current cost depreciation over historical cost depreciation.

With respect to gearing adjustment, the three current cost adjustment – cost of sales, monetary working capital and depreciation – relate to assets which may be financed partly by loan capital. If this is the case, then the equity shareholders would gain at the expense of lenders when prices rise. This is so because as the value of assets is rising, the contractual amounts payable (interest and principal) to lenders do not rise and the gain resulting therefrom becomes attributable to the equity shareholders. However, the treatment of this purchasing power gain (or loss) on debt in current cost accounts is a controversial maker. The controversy relates to the basis of calculation and also whether such a gain should form part of

Table VIII.1 Determination of the current cost of sales

As an illustration, suppose the cost of goods sold based on historical cost was derived as follows:

	GBP
Stock 1 January 1998	8,000
Purchases	30,000
	38,000
Less stock 31 December 1998	12,000
Cost of goods sold for the year	26,000

If the price index of goods of this class was 192 on 1 January 1998, 240 on 31 December 1998 and 196 on average during 1998, it is possible to calculate the real (i.e. physical) change in stock and so the adjustment needed to convert the historical cost of sales to current cost of sales:

	GBP	GBP
Increase in stock as per historical cost accounts (GBP 12,000 − GBP 8,000)		4,000
Closing stock at average prices GBP 12,000 × (196/240)	9,800	
Opening stock at average prices GBP 8,000 × (196/192)	8,167	
Real increase in stock		1,633
Cost of sales adjustment		2,367

The adjusted cost of sales will become GBP 26,000 + GBP 2,367 = GBP 28,367 which equals purchases GBP 30,000 less real increase in stock GBP 1,633.

the income attributable to the equity shareholders.

There are two ways of calculating the purchasing power gain (or loss). One is by reference to a general price index (FASB, 1979, SFAS 33). Another is to derive the proportion of the current cost adjustment due to net borrowing (UK SSAP 16), that is:

(Net borrowing net operating assets) × total current cost adjustment

Net borrowing refers to all liabilities fixed in money terms other than those included in monetary working capital less any current assets which have not been taken into account in the cost of sales or monetary working capital calculation. Net operating assets comprise fixed assets, stock and monetary working capital. Usually, the averages of the relevant figures on the current cost balance sheets at the start and end of the relevant period form the basis of this calculation.

As far as the impact on net current cost income is concerned, the proper treatment of the purchasing power gain (or loss) on debt in the computation of current cost income has been the subject of considerable debate in the literature and among standard-setters across the world. While some (e.g. SSAP 16) would prefer to include such a gain as part of the net current cost income, others would specifically prefer to report the gain or loss in the income statement but as a separate item not added to (or subtracted from) net current cost income (FASB, SFAS 33). Still others would wish to see such a profit or loss disclosed as a note to the accounts rather than as an item in the income statement (Australia's Statement of Accounting Principles No. 1).

IX Cash flow statements, notes and additional statements

1 Cash flow statements

1.1 Introduction and history

Despite the importance of cash both in the success, growth and survival of every reporting entity and the understandability of the cash concept, the provision of cash flow information by UK companies, as part of their external reporting function, is a relatively new phenomenon.

The preparation of funds statements can be traced back to the Assam Company in England in 1862. During the 1970s and 1980s the funds statement (or statement of changes in financial position or source and application of funds statement) became popular throughout the world. The impetus for this development came from professional accounting associations in individual countries and also by the issuance of IAS 7 in 1977. During this period of time the underlying fundamental concept of funds statements changed from working capital and all financial resources to cash.

In July 1975 the ASC issued SSAP 10, Statements of Source and Application of Funds, which required the preparation of a funds statement that was subject to audit. The principal shortcoming of SSAP 10 was that it simply provided an analysis of the sources and application of funds (however defined) in terms of movements in assets, liabilities and capital that had taken place during the year, rather than in terms of how the various activities of the business had either generated or absorbed funds. The result was that the funds statement merely listed the changes in balance sheet totals, thereby giving little, if any, additional information and so explained little about a company's ability to meet obligations or to pay dividends or about its need for external financing. Another inadequacy of the standard was the lack of a precise definition of 'funds'. As a result numerous interpretations of the word were used in practice, not exactly enhancing comparability.

In order to address these shortcomings and also to keep pace with the significant international developments that had already taken place in cash flow reporting, particularly in the USA and Canada, the ASC, prior to its demise, published exposure draft ED 54 'Cash flow statements', in July 1990.

In September 1991 the new regime of the ASB issued its first accounting standard, Financial Reporting Standard (FRS) 1, Cash Flow Statements. The standard was issued against a background of recession and company failure which made the disclosure of cash flow all the more relevant. FRS 1 remained in operation for about five years and was generally well received and many listed companies saw advantages in FRS 1 and adopted the statement early. The standard was also, largely, in line with the pronouncement of other standard-setters on the same issue. However, as preparers and users became accustomed to the standard, a number of practical matters, some more fundamental than others, arose.

The principal concern related to the concept of cash equivalents. The standard classed deposits with more than three months to maturity when acquired as investments, not cash equivalents. Critics argued that their exclusion from cash equivalents failed to capture their substance and that a narrow definition of cash equivalents was not consistent with the objective of FRS 1, which asserted that the purpose of the cash flow statement was to assist users of the financial statements in their assessment of the reporting entity's liquidity, viability and financial adaptability.

In light of the growing concerns and having invited comments in March 1994 the ASB issued FRED 10 'Revision of FRS 1 Cash flow statements' which proposed significant changes to FRS 1.

1.2 Current accounting practice

Moving away from international practice the ASB reissued FRS 1 in October 1996 to take effect for accounting periods ending 23 March 1997.

The original standard had categorized cash flows under five headings, compared to IAS 7 and SFAS 95's three headings, but the revised standard increased the headings to nine (increased from the revised eight headings to nine in November 1997 following publication of FRS 9 'Associates and joint ventures') and moved to a 'pure' cash flow statement, abandoning cash equivalents.

The principal objective of FRS 1 (revised), FRS 1 hereinafter, is to require reporting entities falling within its scope to:

- Report their cash generation and cash absorption for a period by highlighting the significant components of cash flow in a way that facilitates comparison of the cash flow performances of different businesses.
- Provide information that assists in the assessment of their liquidity, solvency and financial adaptability.

These objectives were consistent with the draft 'Statement of Principles for financial reporting' being developed at the time by the ASB.

1.3 The detailed requirements of FRS 1

1.3.1 Definitions

As the cash flow statement only reflects movements in cash, the definition of cash is central to its proper preparation. The standard defines cash as being:

> Cash in hand and deposits repayable on demand with any qualifying financial institution, less overdrafts from any qualifying financial institution repayable on demand. ... Cash includes cash in hand and deposits denominated in foreign currencies (FRS 1 para. 2).

In order to qualify as cash, deposits must be repayable on demand, which they are if they meet one of the following criteria:

> They can be withdrawn at any time or demanded without notice and without penalty, and ... a period of notice no more than 24 hours or one working day has been agreed (FRS 1 para. 2).

1.3.2 Formats

The standard requires cash flows to be classified under the following headings, in the order listed:

- Operating activities.
- Dividends from joint ventures and associates.
- Returns on investments and servicing of finance.
- Taxation.
- Capital expenditure and financial investment.
- Acquisitions and disposals.
- Equity dividends paid.
- Management of liquid resources.
- Financing

The last two headings may be combined under one heading provided the cash flows relating to each are shown separately and a separate subtotals for each given. Striking a subtotal after any of the other headings is neither required not prohibited.

See example 'Cash flow statement of British Telecom Plc, 1999'.

1.3.3 Reconciliations to profit and loss account and the balance sheet

The standard requires the following reconciliations:

- Reconciliation of operating profit to net cash flow from operating activities.
- Reconciliation to net debt.
- Analysis of changes in net debt.

See example 'Reconciliation in cash flow statement of Anglia Water Plc, 1999'.

1.3.4 Consolidated cash flow statements

The form and content of cash flow statements for a single entity, discussed above, also apply equally to any group of enterprises where consolidated financial statements are prepared. But some additional issues such as cash flows from/to:

- minority interests.
- investments accounted for using the equity method.
- investments in joint arrangements;

need to be taken into account. Also, where there has been an acquisition and/or disposal of a subsidiary additional notes need to be provided to show the effects of such transactions.

See example 'Analysis of acquisitions/dispositons of subsidiaries in cash flow statement of Anglia Water Plc, 1999'.

1.3.5 Other issues

Other issues such as foreign currency for both individual companies with borrowings used for hedging equity investments and groups using either the temporal or net investment methods; and hedging transactions have been considered by the standard.

There has also been some continuing debate on issues such as:

- reporting of operating cash flows on a direct (as mandated in Australia and New Zealand) as opposed to an indirect basis (required by FRS 1);
- non-articulation of the cash flow figures to the balance sheet (i.e. movements in stocks, debtors, trade creditors not articulating with the reconciliation of operating profit to operating cash flows);

EXAMPLE: Group cash flow statement of British Telecom Plc, 1999

		1999 GBP m	1998 GBP m	1997 GBP m
Net cash inflow from operating activities	a	**6.035**	**6.071**	**6.185**
Dividends from joint ventures and associates	b	**2**	**5**	**7**
Returns on investments and servicing of finance				
Interest received		111	168	196
Interest paid, including finance costs		(439)	(328)	(342)
Premium paid on repurchase of bonds		0	0	(60)
Dividends paid to minorities		0	0	(14)
Net cash outflow for returns on investments and servicing of finance	c	**(328)**	**(160)**	**(220)**
Taxation				
UK corporation tax paid		(359)	(1.625)	(1.032)
Windfall tax paid		(255)	(255)	0
Overseas tax paid		(16)	(6)	(13)
Tax paid	d	**(630)**	**(1.886)**	**(1.045)**
Capital expenditure and financial investment				
Purchase of tangible fixed assets		(3.220)	(3.020)	(2.823)
Sale of tangible fixed assets		143	127	124
Purchase of fixed asset investments		(103)	(265)	(172)
Disposal of fixed asset investments		4.226	50	51
Net cash inflow (outflow) for capital expenditure and financial investment	e	**1.046**	**(3.108)**	**(2.820)**
Acquisitions and disposals				
Purchase of subsidiary undertakings		(672)	(121)	(126)
Investments in joint ventures		(1.038)	(323)	(131)
Investments in associates		(288)	(1.057)	(17)
Sale of subsidiary undertakings		14	0	11
Sale of investments in joint ventures and associates		17	0	11
Net cash outflow for acquisitions and disposals	f	**(1.967)**	**(1.501)**	**(252)**
Equity dividends paid	g	**(1.186)**	**(3.473)**	**(1.217)**
Cash inflow (outflow) before management of liquid Resources and financing (h = a + b + c + d + e + f + g)	h	**2.972**	**(4.052)**	**638**
Management of liquid resources	f	**(2.447)**	**2.247**	**(504)**
Financing				
Issue of ordinary share capital		161	144	160
Minority shares issued		13	48	51
New loans		10	1.637	35
Loan repayments		(457)	(338)	(670)
Net increase (decrease) in short-term borrowings		(185)	303	200
Net cash inflow (outflow) from financing	j	**(458)**	**1.794**	**(224)**
Increase (decrease) in cash in the year (k = h + i + j)	k	**67**	**(11)**	**(90)**
Decrease (increase) in net debt in the year		**3.146**	**(3.860)**	**849**

Example: Reconciliation of operating profit to net cash inflow from operating activities of Anglia Water Plc, 1999

	1999	1998
	GBP m	GBP m
Operating profit	**322.3**	352.9
Dividends received from trade investments	**(0.5)**	(0.4)
Profit on disposal of tangible fixed assets	**(0.8)**	(2.0)
Depreciation (net of amorti`ation of deferred grants and contributions)	**126.4**	119.8
Net movement on pensions balances	**(2.4)**	(4.5)
Net movement on other provisions	**5.0**	(14.4)
	450.0	451.4
(Increase)/decrease in working capital:		
Stocks	**(3.5)**	0.9
Debtors	**(23.2)**	(19.3)
Creditors	**14.4**	7.0
	(12.3)	(11.4)
Net cash inflow from operating activities	**437.7**	440.0

Net cash inflow from operating activities for the year ended 31 March 1999, is arrived at after deducting cash outflows to meet pension enhancements incurred and provided for as exceptional operating charges for restructuring in prior years of GBP 8.7 million (1998 – GBP 17.6 million). The movement on the provision is shown in note 24. Cash outflows of GBP 2.1 million relate to 1999 exceptional charges.

Analysis of net debt	1 April 1998 GBP m	Cash flows GBP m	Acquisitions GBP m	Disposals GBP m	Non-cash movements GBP m	Exchange movement GBP m	31 March 1999 GBP m
Cash	14.0	(12.6)	0.9	(0.8)	–	–	1.5
Bank overdrafts	(42.5)	36.1	–	–	–	–	(6.4)
	(28.5)	23.5	0.9	(0.8)	–	–	(4.9)
Deposits and investments	40.4	109.5	–	–	–	–	149.9
Debt due within 1 year	(90.1)	83.8	–	6.7	(59.8)	–	(59.4)
Debt due after 1 year	(1.070.4)	(403.1)	–	–	38.0	(0.2)	(1.435.7)
	(1.148.6)	(186.3)	0.9	5.9	(21.8)	(0.2)	(1.350.1)

Example: Reconciliation of operating profit to net cash inflow from operating activities of Anglia Water Plc, 1999 – Continued

Non cash movements comprise amortization of discounts and expenses relating to debt issues, indexation of loan stock, transfers between categories of debt and inception of finance leases.

Management of liquid resources shown in the cash flow statement comprises movements in short-term deposits, which have maturity dates of up to one year.

movement in group net debt	**1999 GBP m**	1998 GBP m
At beginning of year	**(1,148.6)**	(938.6)
Increase/(decrease) in cash	**23.6**	(26.6)
(Increase)/decrease in short-term deposits and investments	**109.5**	(20.1)
Increase in loans and finance lease arrangements	**(403.1)**	(177.1)
Repayment of amounts borrowed	**76.6**	31.0
Disposals/(acquisitions)	**6.7**	(4.9)
Non cash finance lease inceptions	**(16.6)**	(13.1)
Indexation of loan stock	**(4.6)**	(4.3)
Amortization of discount and expenses relating to debt issue	**(0.6)**	(0.6)
Exchange translation	**(0.2)**	0.4
Capital element of finance lease rental payments	**7.2**	5.3
At end of year	**(1,350.1)**	(1,148.6)

- free cash flows;
- cash flow per share;
- segmental cash flow reporting, and classification into continued and discontinued cash flows;

combined with the disharmony of cash flow reporting in the UK *vis-à-vis* other countries and the adoption of IAS 7 by IOSCO to fulfil listing requirements – will no doubt mean cash flow statements will need to be revisited again.

2 Notes to the annual accounts

Corporate annual reports convey more than financial information. In addition to the notes to the financial statements, there is other non-financial information in verbal and graphical forms. The notes to the financial statements come in several forms; some are general and refer to all the figures which appear in all the statements contained in the corporate annual reports; some are specific and restricted to particular financial statements such as the cash flow statement. Some notes convey information about the accounting principles (e.g. measurement and valuation bases chosen by the reporting company) that form the basis of the reporting numbers; others provide more disaggregated information on specific accounting numbers.

According to the Companies Act 1985, section 228(1), a company's accounts shall comply with the requirements of Schedule 4 (so far as applicable) with respect to the form and content of the balance sheet and profit and loss account and any additional information to be provided by way of notes to the accounts. The notes which are expected to accompany the accounts are detailed in Companies Act 1985, Schedule 4, Part III and in several other sections of Schedule 4 and are summarized in Table IX.1. The notes constitute a part of the accounts by virtue of the fact that they elucidate them. The notes also form part of the details that are required to be given in order for

Example: Analysis of acquisitions/dispositions of subsidiaries in cash flow statement of Anglia Water Plc, 1999

Acquisition of subsidiary undertakings	**1999 GBP m**	1998 GBP m
Net assets/(liabilities) acquired:		
Fixed assets	**6.3**	7.9
Stocks	**0.5**	0.3
Debtors	**4.7**	2.9
Cash at bank	**0.9**	3.1
Creditors	**(4.1)**	(3.6)
Loans	**–**	(1.8)
	8.3	8.8
Less minority interest share of net assets acquired	**(0.9)**	–
	7.4	8.8
Goodwill	**0.7**	11.2
	8.1	20.0
Satisfied by:		
Cash	**7.8**	16.9
Deferred consideration	**0.3**	–
Loan notes issued as part consideration	**–**	3.1
	8.1	20.0
Analysis of the net outflow of cash in respect of the acquisition of subsidiary undertakings:		
Total cash paid	**7.8**	16.9
Less cash paid in previous accounting period	**(3.5)**	–
Cash at bank of acquired subsidiaries	**(0.9)**	(3.1)
Net outflow of cash in respect of the acquisition of subsidiaries	**3.4**	13.8

Disposal of subsidiary undertakings	**1999 GBP m**	1998 GBP m
Net assets/(liabilities) disposed of:		
Fixed assets	**4.1**	0.9
Stocks	**5.3**	1.4
Debtors	**4.2**	3.2
Cash at bank	**0.8**	1.8
Bank overdrafts	**–**	(0.1)
Creditors	**(5.0)**	(4.2)
Loans	**(6.7)**	–
	2.7	3.0
Goodwill (previously eliminated against reserves)	**16.2**	0.2
Costs associated with disposal	**1.8**	0.9
Profit on disposal	**2.2**	0.4
Satisfied by cash	**22.9**	4.5
Analysis of the net inflow of cash in respect of the disposal of subsidiary undertakings:		
Cash received upon completion	**22.9**	4.5
Cash at bank of disposed subsidiary undertakings	**(0.8)**	(1.8)
Bank overdrafts of disposed subsidiary undertakings	**0**	0.1
Costs of disposal during the year	–0.7	0
Cash received in respect of prior year disposal	0	1.2
Net inflow of cash in respect of disposal of subsidiaries	21.4	4.0

Table IX.1 Items to be disclosed in the notes

No.	Information to be disclosed in notes	Source in Companies Act 1985
I	**General information**	**Companies Act 1985**
1	General rule on notes	S. 228 (1)
2	Information necessary to enable true and fair view to be attained	S. 228 (4)
3	Details of departure from the requirements of Schedule 4 and matters to be included in accounts and notes to those accounts in order to give a true and fair view	S. 228(6)
4	Basis of translating foreign currencies brought into account	Schedule 4, Part III, para. 58(1)
5	Comparative figures relating to previous financial year	Schedule 4, Part III, para. 58(2), clause 1
6	Adjustment of previous year's figures	Schedule 4, Part III, para. 58(2), clause 2
7	Particulars of shares held in companies other than subsidiaries	Schedule 5, Part II
8	Details of the emoluments of chairman and other directors including pensions and compensation for loss of office	Schedule 5, Part V
9	Particulars relating to number of employees remunerated at higher rates	Schedule 5, Part VI
10	Particulars of loans and other transactions favouring directors and officers	Schedule 6
II	**Accounting principles and rules**	
11	Departure from accounting principles of going concern, consistency, prudence, accruals and individuality of each asset and liability	Schedule 4, Part II, para. 15
12	Period for writing off development costs and reasons for capitalizing development costs	Schedule 4, Part II, para. 20 (2)
13	In case of departure from historical cost accounting rules, the items affected and the basis of valuation, and comparable amount determined according to historical cost accounting rules or the difference between those amounts and the corresponding amounts shown in the balance sheet	Schedule 4, Part II (C) paras 33 (2) and 33 (3)
14	The treatment for taxation purposes of amount credited or debited to the revaluation reserve	Schedule 4, Part II (C), para. 34(4)
15	Disclosure of accounting principles	Schedule 4, Part III, para. 36
III	**Information supplementing the balance sheet**	
16	The authorized share capital and the number and aggregate nominal value of shares of each class allotted	Schedule 4, Part III, para. 38(1)
17	Details of redeemable preference shares	Schedule 4, Part III, para. 38(2)
18	Details of any shares allotted	Schedule 4, Part III, para. 39
19	Details of contingent right to the allotment of shares	Schedule 4, Part III, para. 40(1)
20	Details of any debentures issued during the financial year	Schedule 4, Part III, para. 41(1)
21	Particulars of any redeemed debentures in which the company has power to reissue	Schedule 4, Part III, para. 41(2)

Table IX.1 Continued

No.	Information to be disclosed in notes	Source in Companies Act 1985
22	Details of debentures held by nominee	Schedule 4, Part III, para. 41(3)
23	Schedule of fixed assets	Schedule 4, Part III, para. 42
24	Details of any valuation of fixed assets other than by historical cost accounting rules	Schedule 4, Part III, para. 43
25	Separation of amount ascribable to land into freehold and leasehold and of the leasehold portion into those ascribable to long and short leases	Schedule 4, Part III, para. 44
26	Details of investments owned distinguishing listed investments on recognized stock exchanges from others and their market values if different from the amount so stated	Schedule 4, Part III, para. 45
27	Details of any movements on reserves and provisions	Schedule 4, Part III, para. 46
28	Provision for taxation other than deferred taxation	Schedule 4, Part III, para. 47
29	Details of indebtedness distinguishing between those falling due within 12 months and those falling due after five years	Schedule 4, Part III, para. 48
30	Arrears of cumulative dividends and period and class for which the arrears are due	Schedule 4, Part III, para. 49
31	Particulars relating to guarantees, contracts for capital expenditure, pensions and any other financial commitments	Schedule 4, Part III, para. 50
32	Information on amount recommended for distribution by way of dividend	Sch 4, Part III. para. 51(3)
33	Details of outstanding loans granted to assist employees and directors in the purchase of the company's shares	Schedule 4, Part III, para. 51(2)
IV	**Information supplementing the profit and loss account**	
34	Interest on bank loans and overdrafts distinguishing between interest on loans falling due within one year and before the end of five years; and interest on loans from any other source	Schedule 4, Part III, para. 53(2)
35	Amounts set aside respectively for redemption of share capital and for redemption of loans	Schedule 4, Part III, para. 53(3)
36	The amount of income from listed investments	Schedule 4, Part III, para. 53(4)
37	The amount of rents from land (after deduction of ground rents, rates and other outgoings) if substantial	Schedule 4, Part III, para. 53(5)
38	Amount in respect of hire of plant and machinery	Schedule 4, Part III, para. 53(6)
39	Amount of auditors' remuneration and expenses	Schedule 4, Part III, para. 53(7)
40	Particulars of tax, including basis on which tax is computed	Schedule 4, Part III, para. 54
41	Information on turnover, profit or loss before taxation attributable to two or more classes or markets of business that differ substantially from each other	Schedule 4, Part III, para. 55

Table IX.1 Continued

No.	Information to be disclosed in notes	Source in Companies Act 1985
42	Information about non-disclosure of segmental information on basis of geography or line of business	Schedule 4, Part III, para. 55(5)
43	Particulars of staff	Schedule 4, Part III, para. 56
44	Particulars of the effect of any amount included in the profit and loss account relating to the preceding financial year	Schedule 4, Part III, para. 57(1)
45	Details of any extraordinary income or charges arising in a financial year	Schedule 4, Part III, para. 57(2)
46	The effect of any exceptional transactions	Schedule 4, Part III, para. 57(3)

accounts to present a true and fair view (see V.1).

Small companies presenting modified individual accounts need not give the information contained in Table IX.1 apart from that relating to accounting policies, details of share capital, particulars of allotments, particulars of debt and the basis of translation of foreign currency amounts into sterling and corresponding amounts for the preceding financial year.

3 Disclosure rules for pension schemes

Somewhat surprisingly, Companies Act 1985 does not contain detailed disclosure requirements about pensions. The Act requires disclosure of:

- The pension costs charged (4 Sch. 56[4]).
- Any pension commitments included under any provision shown in the company's balance sheet.
- Any such commitments for which no provision has been made (4 Sch. 50[4]).

SSAP 24, Accounting for pension costs, specifies the following disclosures in respect of a defined benefit scheme:

- The nature of the scheme (i.e. defined benefit).
- Whether it is funded or unfunded.
- The accounting policy and, if different, the funding policy.
- Whether the pension cost and provision (or asset) are assessed in accordance with the advice of a professionally qualified actuary and, if so, the date of the most recent formal actuarial valuation or later formal review used for this purpose. If the actuary is an employee or officer of the reporting company, or the group of which it is a member, this fact should be disclosed.
- The pension cost charge for the period together with explanations of significant changes in the charge compared to that in the previous accounting period.
- Any provisions or prepayments in the balance sheet resulting from a difference between the amounts recognized as cost and the amounts funded or paid directly.
- The amount of any deficiency on a current funding level basis, indicating the action, if any, being taken to deal with it in the current and future accounting periods.
- An outline of the results of the most recent formal actuarial valuation or later formal review of the scheme on an ongoing basis. This should include disclosure of:
  - (i) The actuarial method used and a brief description of the main actuarial assumptions.
  - (ii) The market value of scheme assets at the date of the valuation or review.
  - (iii) The level of funding expressed in percentage terms.
  - (iv) Comments on any material actuarial surplus or deficiency indicated by (iii) above.
- Any commitment to make additional payments over a limited number of years.

- The accounting treatment adopted in respect of a refund made under deduction of tax, where a credit appears in the financial statement in relation to it.
- Details of the expected effects on future costs of any material changes in the company's pension arrangements.

SSAP 24 was reviewed in June 1995 and again in July 1998 and suggests a move towards a market value approach as advocated in IAS 26, Accounting and reporting by retirement benefit plans.

4 Disclosures relating to capitalization of borrowing costs

FRS 15 (para. 31) stipulates that where a policy of capitalization of finance costs is adopted, the financial statements should disclose:

- The accounting policy adopted.
- The aggregate amount of finance costs included in the cost of tangible fixed assets (4 Sch. 26[3]).
- The amount of finance costs capitalized during the period.
- The amount of finance costs recognized in the profit and loss account during the period.
- The capitalization rate used to determine the amount of finance costs capitalized during the period.

5 Segmental reporting

The first regulation on segmental disclosure to be provided for UK companies came in 1965 in the form of a stock exchange requirement. This regulation, which applied to listed companies only, stipulated that turnover and profits by line of business should be disclosed as well as turnover and profit by geographical segments.

The stock exchange requirement was followed in 1967 by a Companies Act (section 17) requirement for segmental turnover and profit before tax to be disclosed where a company carries on a business of two or more classes which 'in the opinion of the directors, differ substantially from each other.' In effect, disclosure was at the discretion of directors (Cooke and Whittaker, 1983, p. 2). An additional requirement, for the value of exports to be disclosed, was introduced for companies with a turnover in excess of GBP 50,000. Where no goods were exported the company had to disclose that fact. These requirements could be waived on application to the then Board of Trade (Companies Act 1967, section 20).

The Companies Act 1981 (para. 55) extended the legal requirements by requiring disclosure of turnover by export market. This requirement, together with turnover and profit before tax, had to be split by class of business and included in the notes to the accounts and not in the directors' report as previously required. However, the information need not be disclosed if the directors consider it 'would be seriously prejudicial to the interests of the company', provided that the claim for exemption is clearly stated (para. 55). The above legal requirements were consolidated by the Companies Act 1985 and were not affected by the Companies Act 1989.

Small companies are exempt from the disclosure requirements of the Act. However, if a company or group supplies goods outside the United Kingdom, the notes must disclose the percentage of turnover going overseas.

There is no definitive guideline on how to identify a segment; this task is essentially left to the discretion of the directors by the Companies Act. SSAP 25 provides some minimum guidance for determining the presence of a separate and, therefore, reportable segment. A business line or geographical area of business is separable from the rest and is reportable if it is significant and it fulfils any of the following:

- It earns returns on investment that are different from the rest of the company.
- It is subject to different degrees of risk.
- It experiences different rates of growth.
- It reveals potential for growth different from the rest of the company.

To arrive at such a decision, directors need to consider many factors including the nature of the products or services, their production processes, markets, distribution channels, the legal framework affecting the products and services and the organization structure of the company (SSAP 25, para. 8). In determining the presence of a distinguishable and separable geographical segment, directors need to consider the economic and political environments in which the businesses are situated, exchange control regulations and exchange rate fluctuations (SSAP 25, para. 15). A separable segment is significant for reporting purposes if its share of the total revenue of the company or group from third parties is 10 per cent or more; or its share of the total corporate or group's profit/loss is 10 per cent or more; or its share of the total net assets is 10 per cent or more (SSAP 25, para. 9). In practice, many reporting companies have reported segments whose revenue, profits or total assets are below the 10 per cent threshold.

In essence, the requirements of SSAP 25 are the same as the US standard, SFAS 14 (Financial Reporting for Segments of a Business Enterprise, issued December 1986), with respect to disclosure of turnover, profits and assets identified by industry and geographical segments. In terms of scope of the relative requirements, SSAP 25 applies to all public companies whereas SFAS 14 applies to those with an SEC filing or those with publicly traded securities (para. 41). However, whilst SSAP 25 permits disclosure to be at the discretion of directors, the requirements of SFAS 14 are mandatory.

Another difference between SSAP 25 and SFAS 14 is that there is no requirement in the United Kingdom for the disclosure of inter-segment pricing policies. In addition, sales to other segments should be accounted for on a basis consistent with other transfers and inter-segment sales in the United States, whereas there is no similar requirement in the United Kingdom.

Disclosure of operating profits is similar in the United States and the United Kingdom and therefore the operating profit or loss of each reportable segment will be similar. However, whilst SFAS 14 requires the results of associated companies to be excluded from operating profit, SSAP 25 stipulates that the enterprise's share of the profits and losses of associated companies as well as its share of the net assets should be disclosed if they are material.

With respect to the assets test, SFAS 14 requires the aggregate carrying amount of identifiable assets to be disclosed whereas SSAP 25 refers to segmental analysis of net assets, i.e. non-interest bearing operating assets less non-interest bearing operating liabilities.

There are no major differences between the US and the United Kingdom standards in identifying reportable industry segments and geographical segments. There is no requirement in SFAS 14 for a geographical analysis of turnover to be disclosed except for sales to unaffiliated customers by an enterprise's foreign operations. In contrast, SSAP 25 includes a more general requirement that the reporting entity should disclose the geographical segmentation of turnover by origin.

The Listing Rules of the Stock Exchange were revised in 1994 and do not require any segmental disclosures. However, listed UK companies are required to comply with UK GAAP.

6 Earnings per share

SSAP 3, Earnings per Share (EPS), was issued in February 1972, revised in August 1974 as a result of the introduction of the imputation system for taxing dividends in April 1973, and revised again in October 1992 on the introduction of FRS 3, Reporting financial performance. In the case of listed companies considerable emphasis has been placed on the EPS figure as a measure of assessment of performance. The price earnings ratio has become important when comparing one country with another. The aim of the standard is to ensure that listed companies compile their EPS figures on a consistent basis.

The aim of the ASB, in passing FRS 3, has been to attempt to reduce the importance attached to the EPS so that users will look at the accounts as a whole. It mandates that a range of items be displayed on the face of the statement, but that the EPS be calculated after taking account of all these items, including capital profits and losses and extraordinary items. However, the Institute of Investment Management and Research (IIMR) has argued that it is clearly desirable to define an earnings figure which can be used as a reference point to assess 'trading performance'. The IIMR use the trading performance which excludes capital items, is robust and is a matter of fact rather than theoretical adjustments. This is not particularly accurate since the figures in the profit and loss account are often subject to judgement and even in calculating EPS judgement has to be made in dealing with capital changes. However, the IIMR argues that the measure is justified by its practical usefulness even if the one number does not encapsulate the company's performance.

The IIMR argues that it has a role to play here similar to that of investment analysts in Germany. The organization defines what it refers to as 'headline earnings'. Headline earnings include all trading profits and losses of the company for the year (including interest). Items that are abnormal in size or nature are also included but should be prominently displayed if significant. Profits and losses on the sale of fixed assets or of businesses should be excluded, although this does not apply to assets acquired for resale. Profits and losses arising from activities that are discontinued during the year or from activities acquired during the year should remain in the earnings figure.

The IIMR adds that prior period items and the effects of changes in accounting policies and of past fundamental accounting errors should not affect the current year's calculation of earnings. Headline earnings should include tax adjustments to reflect the fact that certain items are excluded from the headline figure. An example of the IIMR headline earnings adjustment (based on the IIMR publication) is shown in Table IX.2.

Table IX.2 IMR headline earnings adjustments – 1994

	Notes	Profits on ordinary activities GBP m	Tax GBP m	Minority GBP m	Profit interests GBP m
Per accounts adjustments	(i)	45	(14)	(2)	29
Less 1993 provision	(ii)	(10)	3	–	(7)
Exceptional items:					
Loss on disposal of discontinued operations	(iii)	17	(5)	–	12
Less 1993 provision	(iv)	(20)	6	–	(14)
Provision for loss on operations to be discontinued	(v)	1	–	–	1
Profit on sale of properties	(vi)	(12)	3	1	(8)
Goodwill amortization	(vii)	2	–	–	2
		23	(7)	(1)	
IIMR headline earnings					15
IIMR headline earnings per share					20p

Notes

(i) Figures as shown in the published profit and loss account as 'profit on ordinary activities'.
(ii) This adjustment relates to the provision made in the 1993 accounts in respect of loss on operations to be discontinued and relates only to the operating losses on operations in the 1994 financial year up to the date of termination/sale (and not the direct costs of termination/sale).
(iii) As shown in the published profit and loss account.
(iv) This adjustment is similar to that described in Note (ii) above: in this case it relates to a provision made in the 1993 accounts in respect of the loss on disposal of operations discontinued in 1994.
(v) The adjustment relates to the provision made in the 1994 financial year in respect of loss or an operation to be discontinued in the next financial year – 1995. This provision, when released in 1995, will be adjusted for in arriving at 1995 IIMR headline earnings.
(vi) As shown in the published profit and loss account.
(vii) It is assumed that this charge to the profit and loss account is disclosed in the notes to the accounts.
(viii) It is assumed that the associated tax credit or charge and also the minority charge in respect of each adjustment are disclosed in the notes to the accounts as required by FRS 3.
(ix) It is assumed that a review of the remaining exceptional items (i.e. those other than those disclosed on the face of the profit and loss account) does not necessitate any further adjustment to the headline earnings figure.

Source: Statement of Investment Practice No. 1, The Institute of Investment Management and Research.

While SSAP 3 was proving reasonably adequate in the United Kingdom, the IASC issued E 52, Earnings per share, to try to improve international harmonization in this area. The ASB encouraged comments on E 52 and in February 1997 IAS 33 was issued. The US also provided detailed guidance in FAS 128. In response to these developments the ASB issued FRED 16 in June 1997 which adopted nearly all the requirements of IAS 33. However, some amendments were made to the exposure draft before FRS 14, Earnings per share, was issued on 1 October 1998 which replaced SSAP 3.

The objective of FRS 14 is to improve comparison of the performance of different entities in the same period and of the same entity in different accounting periods by prescribing methods to determine the number of shares to be included in the calculation of EPS (para. 1). The standard applies to companies that are publicly traded. Paragraph 9 states that basic EPS should be calculated by dividing the net profit or loss for the period attributable to ordinary shares by the weighted average number of ordinary shares outstanding during the period. The weighted average number of ordinary shares outstanding should be adjusted for events that change that number (para. 21).

In terms of disclosure, an entity should present both basic and diluted EPS on the face of the profit and loss account for each class of ordinary share with different rights to profit shares (para. 69). Both should be disclosed with equal prominence (para. 69). If a loss is incurred both basic and diluted EPS should still be disclosed (para. 70). In addition, the amounts used as the numerators in calculating basic and diluted EPS and a reconciliation of those amounts to the net profit or loss for the period should be disclosed (para. 71a). Both basic and diluted EPS will require a calculation of the weighted average number of shares used in the denominator and this should be disclosed for each class of ordinary share as well as a reconciliation of these denominators to each other (para. 71b).

The standard is effective for all accounting periods ending on or after 23 December 1998.

7 Corporate governance

A committee was set up in May 1991 by the Financial Reporting Council, the London Stock Exchange and the accounting profession to review corporate governance in the United Kingdom following the growing dissatisfaction with the level of corporate governance in large UK companies brought about by the spate of corporate failures. The committee was known as the Cadbury Committee after its chairman, Sir Adrian Cadbury. The terms of reference of the committee included related aspects of financial reporting and corporate accounting because of the perceived gap between what the users of published financial reports expected and what they were getting and the perceived inability of the auditing profession to give users the assurance that financial reporting would meet their needs for transparency and probity. The Committee issued an interim report in May 1992 and invited comments on the report. The Committee's final report was issued in December 1992.

The recommendations in the Cadbury Report include a Code of Best Practice ('the Code') which boards of directors of all listed companies are expected to observe in order to achieve the necessary high standards of corporate behaviour. The Cadbury Report requires reporting companies to disclose in their annual accounts and report in respect of years ending after 30 June 1993 whether they comply with the Code of Best Practice and identify and give reasons for any areas of non-compliance. In addition, the Cadbury Report expects that a reporting company's statement of compliance with the Code should be subject to review by its auditors before publication and that this review should cover only those parts of the compliance statement which relate to provisions of the Code where compliance can be objectively verified.

The London Stock Exchange has adopted this recommended Code of Best Practice. On 23 April 1993, it included as part of its listing regulations the following information relating to the continuing obligations for listed companies:

- A company incorporated in the United Kingdom must state in its annual report and accounts for accounting periods ending after 30 June 1993 whether or not it has complied throughout the accounting period with the Code of Best Practice published in December 1992 by the Committee on the Financial Aspects of Corporate Governance.

- A company that has complied with only part of the Code, or has complied (in the case of requirements of a continuing nature) during only part of an accounting period, must specify the paragraphs with which it has not complied, and (where relevant) for what part of the period it has not complied, and give reasons for any non-compliance.
- A company's statement of compliance must be reviewed by the auditors before publication insofar as it relates to paragraphs 1.4, 1.5, 2.3, 2.4, 3.1 to 3.3 and 4.3 to 4.6 of the Code (see below).

Companies reporting on their first financial year ending after 30 June 1993 may, according to the London Stock Exchange, limit their statement of compliance to that part of the financial year which begins after that date. However, they are encouraged to give the statement of compliance for the financial year as a whole.

The provisions of the Code with which the Cadbury Report suggests compliance can be objectively verified, and which therefore are within the scope of the auditors' review, are specified as the following:

1.4 The board should have a formal schedule of matters specifically reserved to it for decision to ensure that the direction and control of the company is firmly in its hands.

1.5 There should be an agreed procedure for directors in the furtherance of their duties to take independent professional advice if necessary, at the company's expense.

2.3 Non-executive directors should be appointed for specified terms and reappointment should not be automatic.

2.4 Non-executive directors should be selected through formal process and both this process and their appointment should be a matter for the board as a whole.

3.1 Directors' service contracts should not exceed three years without shareholders' approval.

3.2 There should be full and clear disclosure of directors' total emoluments and those of the chairman and highest-paid UK director, including pension contributions and stock options. Separate figures should be given for salary and performance related elements and the basis on which performance is measured should be explained.

3.3 Executive directors' pay should be subject to the recommendations of a remuneration committee made up wholly or mainly of non-executive directors.

4.3 The board should establish an audit committee of at least three non-executive directors with written terms of reference which deal clearly with its authority and duties.

4.4 The directors should explain their responsibility for preparing the accounts next to a statement by the auditors about their reporting responsibilities.

4.5 The directors should report on the effectiveness of the company's system of internal control.

4.6 The directors should report that the business is a going concern, with supporting assumptions or qualifications as necessary.

The board of directors may, at their discretion, provide a description of other features of the company's corporate governance arrangements.

Practice relating to the disclosure of corporate governance arrangements is evolving. These matters may be dealt with in a statement on corporate governance separate from the financial statements and Operating and Financial Review (OFR); or they may be included in the OFR; or they may be given as part of notes to financial statements.

7.1 The Greenbury Report

Sir Richard Greenbury chaired a committee to study directors' remuneration and reported in July 1995 (the Greenbury Report). The committee recommended increased disclosure of directors' remuneration and this was adopted by the Stock Exchange relating to listed companies. In addition, an amendment was made to Companies Act 1985 relating to disclosure of directors' remuneration and pensions, with effect from 31 March 1997. In 1998 a Committee on corporate Governance was formed under the chairmanship of Sir Ronald Hampel.

7.2 The Hampel Report

The Hampel Report, issued in January 1998, considered both the Cadbury and Greenbury Reports and generally endorsed the majority of their recommendations. However, greater emphasis was given to the relationship between corporate governance and prosperity rather than on accountability aspects of previous reports. In addition, smaller listed companies found that some of the disclosure information required was burdensome and Hampel responded by recommending greater flexibility. The Report recommended that companies should issue a narrative statement in their annual reports on the application of principles of corporate governance to their businesses. Companies should also report if they deviate from best practice.

7.3 The new Combined Code

Listed companies are now required to indicate compliance with the new Combined Code which is based on the recommendations of Cadbury and Greenbury. The Code has both principles and detailed provisions and consists of a Part 1 on principles of good governance and Part 2 on best practice. Part 1 consists of 17 principles based on transparency and openness in disclosure practice. However, Hampel considered that such concepts have limited effectiveness if the process is one-way. As such, the Combined Code incorporates principles relevant to institutional shareholders and which focus on voting, dialogue with companies and evaluation of governance disclosures. Each Part of the Code is split into two sections:

Section 1	Companies
A	Directors
B	Directors' remuneration
C	Relations with shareholders
D	Accountability and audit
Section 2	Institutional shareholders
E	Institutional investors

The new code applies to the annual reports of listed companies ending on or after 31 December 1998. In effect, the following provisions should be adhered to:

1 All listed companies should have an effective board, meet regularly and have a formal list of matters for decision-making. Directors should be properly trained.

2 The roles of the chairman and chief executive should be separate. The board should be run by the chairman and the running of the business should be the responsibility of the chief executive officer. Combining these roles is not recommended but if a decision is made otherwise it is essential to provide public justification.

3 The board should have a balance of executive and non-executive directors to avoid domination of the board by an individual or group of individuals. Normally the non-executive directors should form at least one-third of the board. Shareholders' concerns should be dealt with by a senior non-executive director.

4 The board should have appropriate and timely information to discharge their duties adequately. Where a director considers the information not to be adequate it is essential that further enquiries are made. Directors must be fully briefed for board meetings.

5 Appointments to the board should be transparent. Directors should be available for re-election at least once every three years. Non-executive directors should also have specific periods of appointment and can be re-elected but this is not to be automatic.

6 The remuneration of directors should be sufficient to attract and retain individuals to successfully run the company with some element of remuneration being linked to performance. The Remuneration Committee should consist of non-executive directors. The annual report should contain a policy statement about directors' remuneration.

7 The board has the responsibility to report financial information to shareholders and this should be presented in a balanced and understandable manner. The board also has the responsibility to ensure that the system of internal control is sound. The Audit Committee should monitor the principles of internal control.

An illustration of a current disclosure by Associated British Foods plc on corporate governance and board committees that is in the spirit of what is required by the Combined Code is reproduced.

Disclosure about corporate governance by Associated British Foods Plc, 1999

Corporate governance

Corporate governance has been and remains the responsibility of the whole Board. The Combined Code – Principles of Good Governance and Code of Best Practice ('the Combined Code') was published by the London Stock Exchange in June 1998. This statement describes how the company applies the principles and complies with the provisions of the Combined Code.

Compliance

Following publication, the Board took steps to achieve compliance with the Code provisions set out in Section 1 of the Combined Code. In particular:

- Peter J Jackson was appointed chief executive on 1 June 1999.
- Professor Sir Roland Smith was appointed senior independent director on 30 June 1999.
- Martin G Adamson was appointed as an independent non-executive director on 11 October 1999.
- A Nomination committee was established on 30 June 1999 comprising any two of the three independent non-executive directors and is chaired by Garry H Weston.
- A resolution proposing an amendment to the company's Articles of Association, requiring all directors to be subject to re-election every three years, will be put to the forthcoming annual general meeting. The results of proxy votes on each resolution will be announced at the forthcoming annual general meeting.

After implementation of these changes on 11 October 1999, the Board considers that it was and continues to be in full compliance with the Code provisions set out in Section 1 of the Combined Code with the following exceptions:

- The Combined Code recommends that the Audit and Remuneration committees should only comprise non-executive directors. The Board does not accept this recommendation as it considers that Garry H Weston, executive chairman, should serve on both committees in view of his unique knowledge of the business and its people.
- The Combined Code recommends that the performance related elements of remuneration should form a significant proportion of the total remuneration of executive directors. The Board does not accept this recommendation as it considers its existing policies in this regard to be in the best interests of the company and its shareholders.

The statement of directors' responsibilities for preparing the financial statements is set out on page 32.

The Board

The Board of directors generally meets on a quarterly basis, with additional meetings to consider specific issues when required, and concentrates mainly on strategy, direction and financial performance. The Board is chaired by Garry H Weston and Peter J Jackson is chief executive. Details of the full Board are set out on page 18. Professor Sir Roland Smith is the recognized senior independent director. For the purposes of the Combined Code, WG Galen Weston is not regarded as independent. The Board has a formal schedule of matters reserved for its decision, but also delegates specific responsibilities to Board committees, notably the Audit, Remuneration and Nomination committees. Directors receive Board and committee papers in advance of Board and committee meetings and also have access to the advice and services of the company secretary. The Board has adopted a procedure whereby directors may, in the furtherance of their duties, take independent professional advice on any matter at the company's expense.

The group's organizational structure is decentralized, based on short lines of communication. Management responsibilities for major operating divisions are devolved to divisional chief executives reporting directly to the chief executive. George Weston Foods Limited is a quoted Australian public company having its own board chaired by Garry H Weston.

Board committees

Remuneration committee

The Remuneration committee sets the remuneration and other terms of employment of executive directors and the company's policy on remuneration of the senior executives within terms of reference agreed by the Board.

Nomination committee

The Nomination committee reviews the composition of the Board and recommends to the Board appointments of new executive and non-executive directors. Garry H Weston chairs the committee, which also comprises any two of the independent non-executive directors of the company.

Group audit committee

The group Audit committee has terms of reference modelled closely upon those recommended in the Combined Code. It comprises the independent non-executive directors and Gary H Weston and is chaired by Professor Sir Roland Smith. It meets regularly to receive and review reports from the external auditors and from management. As part of its duties, the committee receives and considers reports on the system of internal financial control.

Disclosure about corporate governance by Associated British Foods Plc, 1999 – Continued

Communication with shareholders

Apart from the annual general meeting, the company communicates with its shareholders by way of the annual report and accounts, the half-yearly interim report and the company's web site.

Internal financial controls

The Combined Code introduced a new requirement that the directors review and report on the effectiveness of the group's internal controls, including operational controls. This affects the existing requirements to report on internal financial controls. However, as permitted by the London Stock Exchange, the directors have restricted their review to internal financial controls in accordance with existing guidelines.

The directors are responsible for the group's systems of internal financial control, which are directed to safeguarding the assets of the group, ensuring proper accounting records are maintained and that financial information used within the business or for publication is reliable. Any system of internal financial control can, however, only provide reasonable and not absolute assurance against mistakes or loss.

The group's system of internal financial controls includes:

- Standards
  There are group-wide guidelines on the minimum level of internal control that each of the divisions should exercise over specified processes. Each business has developed and documented policies and procedures in order to comply with the minimum control standards established, including procedures for monitoring compliance and taking corrective action. The board of each business is required to confirm annually that the policies and procedures it has established have been complied with.
- High-level controls
  All businesses prepare annual plans and budgets for operational and cash performance, which are updated regularly. Performance against budget is monitored at operational level and centrally, with variances being reported promptly. The cash position at group and operational level is monitored constantly and variances from expected levels thoroughly investigated.
  A significant part of the group's cash reserves is managed by independent fund managers operating within detailed guidelines specified by the group relating to, inter alia, permitted investments and counter parties, currency exposures and approved instruments. The balance of the group's cash reserves is managed by its treasury function in accordance with guidelines referred to in the financial review.
  There are clearly defined guidelines for capital expenditure and investment decisions encompassing budgets, appraisal and review procedures and levels of authority.
- Review
  The detailed policies and internal financial control procedures established at operational level are reviewed by group personnel. The Audit committee receives reports on internal financial control issues from management and from the external auditors. The directors confirm that they have reviewed the effectiveness of the system of internal financial control utilising the review process set out above.

Going concern

After making due enquiries, the directors have a reasonable expectation that the group has adequate resources to continue in operational existence for the foreseeable future. For this reason they continue to adopt the going concern basis for preparing the financial statements on pages 34 to 56.

By order of the Board
THM Shaw, Secretary

8 November 1999

Company law requires the directors to prepare financial statements for each financial year which give a true and fair view of the state of affairs of the company and of the group and of the profit or loss for that period. In preparing those financial statements, the directors are required to:

- select suitable accounting policies and then apply them consistently;
- make judgements and estimates that are reasonable and prudent;
- state whether applicable accounting standards have been followed, subject to any material departures disclosed and explained in the financial statements; and
- prepare the financial statements on the going concern basis unless it is inappropriate to presume that the group will continue in business.

The directors are responsible for keeping proper accounting records which disclose with reasonable accuracy at any time the financial position of the company and to enable them to ensure that the financial statements comply with the Companies Act 1985. They have general responsibility for taking such steps as are reasonably open to them to safeguard the assets of the group and to prevent and detect fraud and other irregularities.

8 Operating and financial review (OFR)

8.1 General remarks

As part of the significant changes to financial reporting in the United Kingdom, the ASB issued a statement in July 1993, which was intended to have a persuasive rather than a mandatory force. This statement requires reporting companies to include in their annual reports and accounts an objective discussion that analyzes and explains the main features underlying their reported results and financial position. The operating and financial review should, essentially, offer a discussion and interpretation of the results of the reporting business. It should also provide a discussion of the main factors, features as well as uncertainties that underlie it and the structure of its financing. The ASB also encourages reporting companies to draw the attention of readers to those aspects of the year under review that are relevant to an assessment of future prospects, though the OFR is not expected to be a forecast of future results. The OFR is, therefore, principally a review of past events. The intention is for the OFR to give users of the annual report and accounts a more consistent foundation on which to make investment decisions regarding the company. It is not intended that directors should provide information of a confidential and sensitive nature but that the OFR should be reasonably comprehensive and informative.

It is intended that the OFR statement should apply to all listed companies and those corporations for which there is legitimate public interest in their financial statements. The OFR is similar to the North American Management Discussion and Analysis (MD&A) (Collins, Davie and Weetman, 1993). In its evolution, the OFR would locate itself strategically within the annual report and accounts over time. In the interim, the OFR can be a separate stand-alone statement or a part of an already existing statement such as the chief executive's report.

The essential features of an OFR are as follows:

- A clear and succinct style of writing that can be understood by the general reader of annual accounts.
- An objective and balanced analysis that deals with both good and bad aspects of the company.
- Reference to matters from previous statements not borne out by events in the current year.
- Analytical discussion rather than mere numerical analysis.
- Discussion of individual aspects of business should be in the context of the business as a whole.
- Changes in accounting policies should be identified and explained.
- Relationships of any reported ratios or other numerical information to financial statements should be made clear.
- Discussion of the significance to the business of trends and factors affecting results but are not expected to continue in the future, and of known events, trends and uncertainties that are expected to have an impact on the business in the future.

The ASB statement divides the OFR into two parts: the operating review and the financial review. Each of these is discussed below.

8.2 Operating review

The operating review is intended to provide information on the main influences on the overall results and how these interrelate, as well as indicating those factors that have varied in the past or are expected to change in the future. The factors affecting the current reported results for the business as a whole or for a segment of the business should be reported. These factors may relate to the industry or the environment in which the reporting company operates and to developments within the business and their effect on the results. Examples of these factors include changes in market conditions, product and service innovations, changes in market share or positions, changes in turnover and margins, changes in exchange rates and inflation rates, new and discontinued business activities and other acquisitions and disposals.

The operating review should discuss the main factors and influences that may have a major effect on future results, whether or nor they were significant in the current period. These factors relate to business risks such as scarcity of raw materials, skill shortages and expertise of uncertain supply, patents, licences and franchises, dependence on major suppliers or customers, product liability, health and safety, environmental protection costs and potential environmental liabilities, self insurance, exchange rate fluctuations and rates of inflation differing between costs and revenue or between different markets.

In addition, the operating review should include information on the extent to which directors have sought to maintain and enhance future income or profits by capital expenditure and investment in the future of the business. This information should include activities such as marketing and advertising campaigns, training programmes, refurbishment and maintenance programmes, development of new products and services, and technical support to customers.

The operating review should also include a discussion of the overall return attributable to shareholders, in terms of dividends and increases in shareholders' funds, and a comparison between profit and dividends for the current year, both in aggregate and per share terms. There should also be a discussion of other measures of performance in the annual report and accounts, such as earnings per share.

8.3 Financial review

This section of the OFR is concerned with the explanation of the capital structure of the reporting company, its treasury policy and the dynamics of its financial position – its sources of liquidity and their application, including the implications of the financing requirements arising from its capital expenditure plans. The following matters should be included in the discussion if they are considered important to an understanding of the business by the user of the annual report.

8.3.1 Capital structure and treasury policy

Details of the maturity profile of debt, type of capital instruments used, currency and interest rate structure should be provided including comments on relevant ratios such as interest cover and debt/equity ratios.

Information on the funding of capital and treasury policies and objectives is encouraged. Such information should include the management of interest rate risk, the maturity profile of borrowings and the management of exchange rate risk. Details of the implementation of treasury policies and strategies are expected to be provided in this section in terms of the following:

- The manner in which treasury activities are controlled.
- The currencies in which borrowings are made and in which cash and cash equivalents are held.
- The extent to which borrowings are at fixed interest rates.
- The use of financial instruments for hedging purposes.
- The extent to which foreign currency net investments are hedged by currency borrowings and other hedging instruments.

8.3.2 Taxation

The reconciliation between actual and 'standard' tax charges should be discussed.

8.3.3 Cash flows

Information relating to cash generated from operations and other cash inflows during the period under review should be discussed, with comments on any special factors that influenced these. The significant variation between segmental profits and cash flows should also be reported upon.

8.3.4 Current liquidity

Discussion on the business's liquidity at the end of the period should include comment on the level of borrowings, the seasonality of borrowing requirements, indicated by the peak level of borrowings during the period, and the maturity profile of both borrowings and committed borrowing facilities. Discussion should also include reference to any restrictions on the ability to transfer

Example: Disclosure about financial instruments in the operating and financial review of Hogg Robinson Plc, 1999

Financial Instruments

The Company's treasury policy and foreign exchange risk management is described in the Financial Review on page 26.

a Short-term debtors and creditors

Short-term debtors and creditors have been excluded from all the following disclosures, other than the currency risk disclosures.

b Interest rate risk profile and financial liabilities

The interest rate risk profile of the group's financial liabilities at 31 March 1999, after taking into account the interest rate and currency swaps used to manage the interest and currency profile, was:

Currency	**Total GBP '000**	**Floating rate financial liabilities GBP '000**	**Fixed rate financial liabilities GBP '000**
Sterling	**40,582**	**16,277**	**24,305**
French Francs	**2,039**	**2,039**	–
Italian Lira	**2,072**	–	**2,072**
Swedish Krona	**4,386**	**3,630**	**756**
Norwegian Krone	**3,983**	–	**3,983**
At 31 March 1999	**53,062**	**21,946**	**31,116**
Sterling	6,426	2,100	4,326
French Francs	964	964	–
Italian Lira			
Swedish Krona	5,293	–	5,293
Norwegian Krone	**3,893**	–	**3,893**
At 31 March 1998	**16,576**	**3,064**	**13,512**

All the group's creditors falling due within one year (other than bank overdrafts and borrowings) are excluded from the above tables either due to the exclusion of short-term items or because they do not meet the definition of a financial liability, such as tax balances.

During the year, the Company negotiated a GBP 20 million, ten year, floating rate facility (LIBOR + 0.45%) which was concurrently swapped, via a separate agreement, into a 6.5% fixed rate contract.

The original floating rate facility has a ten year, staged repayment structure commencing October 2000. The interest swap agreement has matching quarterly roll-over dates. At 31 March 1999 the whole of this facility had been utilised (included within the GBP 24,305.00 fixed rate liability above).

Fixed rate financial liabilities as at 31 March 1999

Currency	Weighted average interest rate %	Weighted average period for which rate is fixed Years
Sterling – Bank Loans	6.5	10.0
Sterling – HP & Finance Leases	8.6	1.6
Norwegian Krone	6.4	1.0
Italian Lira	4.0	0.3

Floating rate financial liabilities bear interest rates, based on relevant national LIBOR equivalents, which are fixed in advance for periods of between 3 days and 10 years. Comparative rates at 31 March 1998 are not available.

Example: Disclosure about financial instruments in the operating and financial review of Hogg Robinson Plc, 1999 – Continued

c Currency denominations and interest rate risk of financial assets

	1999 Cash at Bank and in hand GBP '000	1999 Short-term deposits GBP '000	1998 Cash at Bank and in hand GBP '000	1998 Short-term deposits GBP '000
Currency				
Sterling	**11,477**	**192**	18,697	1,023
Italian Lira	**294**	–	223	–
French Francs	**301**	–	221	–
Belgian Francs	**387**	–	423	–
Danish Krone	**186**	–	713	339
Finish Mark	**1,387**	–	698	–
Norwegian Krone	**4,110**	–	5,692	–
Swedish Krona	**11,288**	–	7,028	–
Russian Roubles	**35**	–	–	–
Canadian Dollars	**2,750**	–	–	–
US Dollars	**175**	–	–	–
Dutch Guilders	**9**	–	13	–
At 31 March	**32,399**	**192**	33,708	1,362
Floating rate	**32,399**	–	33,708	–
Fixed rate	–	**192**	–	1,362
At 31 March	**32,399**	**192**	33,708	1,362

Any surplus cash is invested for periods up to one month. The average interest earnings achieved on Sterling deposits over the year ended 31 March 1999 was 6.9%.

d Maturity of financial liabilities

The maturity profile of the carrying amount of the Group's financial liabilities, other than short-term creditors such as trade creditors and accruals, at 31 March was as follows:

Group	**Debt GBP '000**	**Finance Leases GBP '000**	**1999 Total GBP '000**	Debt GBP '000	Finance Leases GBP '000	1998 Total GBP '000
Within 1 year or on demand	**28,001**	**2,700**	**30,701**	3,064	2,387	5,451
Between 1 and 2 years	–	**1,296**	**1,296**	–	1,367	1,367
Between 2 and 5 years	–	**1,065**	**1,065**	8,207	1,551	9,758
Over 5 years	**20,000**	–	**20,000**	–	–	–
Total	**48,001**	**5,061**	**53,062**	11,271	5,305	16,576
Company						
Within 1 year or on demand	**19,560**	**182**	**19,742**	–	147	147
Between 1 and 2 years	–	**140**	**140**	–	80	80
Between 2 and 5 years	–	**49**	**49**	–	22	22
Over five years	**20,000**	–	**20,000**	–	–	–
Total	**39,560**	**371**	**39,931**	–	249	249

e Borrowing facilities

The Group has various committed, undrawn, floating rate overdraft facilities at 31 March 1999 amounting to GBP 21,578,000. These facilities expire within one year and are the subject of annual review at various dates during 1999/2000.

f Fair values of financial assets and financial liabilities

Set out below is a comparison by category of book values and fair values of all the Group's financial assets and financial liabilities as at 31 March 1999.

Example: Disclosure about financial instruments in the operating and financial review of Hogg Robinson Plc, 1999 – Continued

	1999 Book value GBP '000	1999 Fair value GBP '000
Primary financial instruments held or issued to finance the Group's operations:		
Financial liabilities:		
Overdrafts and short-term borrowings	30,701	30,701
Long-term borrowings	22,361	14,818
Total	53,062	45,519
Financial assets:		
Cash deposits	192	192

The fair value of short-term deposits, loans and overdrafts approximates to the carrying amount because of the short maturity of these instruments.

The difference between book values and fair values of long-term borrowings takes into account the present value of the staged repayments of the ten year loan facility described in note 21b. Fair value comparatives are not available.

g Currency exposures

To mitigate the effect of the currency exposures arising from its net investments overseas the group either borrows in the local currency of its main operating units or swaps other borrowings, using currency swaps, into such local currencies. Gains and losses arising on net investments overseas and the financial instruments used to hedge the currency exposures are recognized in the statement of recognized gains and losses.

The table below shows the extent to which group companies have monetary assets and liabilities, excluding long-term equity loans, in currencies other than their local currency. Foreign exchange differences on retranslation of these assets and liabilities are taken to the profit and loss account of the group companies and the group. As at 31 March 1999 these exposures were as follows:

	Net foreign currency monetary assets/(liabilities)					
1999	Sterling GBP '000	US & Can. Dollars GBP '000	Scandinavian currencies GBP '000	Other European currencies GBP '000	Australian Dollars GBP '000	Total GBP '000
Functional currency of group operations:						
Sterling	–	267	–	816	161	1,244
Canadian Dollars	(315)	–	–	–	–	(315)
Scandinavian currencies	62	467	(18)	–	–	511
Other European currencies	(1,641)	(640)	–	47	–	(2,234)
Other currencies	1,015	–	–	–	–	1,015
Total	(879)	94	(18)	863	161	221

h Hedges

It is policy to use, where appropriate, currency forward contracts to mitigate exchange fluctuations on funds remitted between currencies. Consequently, three short-term contracts, in respect of remitted Scandinavian dividends and interest, were taken-out during the year. These forward contracts matured during the year ended 31 March 1999 and GBP 30,000 gain was recognised. There were no outstanding forward contracts at either 31 March 1999 or 31 March 1998.

i Financial instruments held for trading purposes

The group does not trade in financial instruments.

funds from one part of the group to meet the obligations of another part of the group, where these represent, or might foreseeably come to represent, a significant restraint on the group. Such restraint would include exchange controls and taxation consequences of transfers. Discussion relating to the restricting effects of debt covenants on current and future credit facilities are expected to be included in the financial review.

8.3.5 Going concern

Directors are expected by the Combined Code of Best Practice to indicate whether or not they consider the reporting company to be a going concern. The form and procedure for ascertaining the 'going concern' status of a company are yet to be developed.

8.3.6 Balance sheet value

The financial review should also provide commentary on the strengths and resources of the business if their value is not reflected in the balance sheet (or only partially shown in the balance sheet). Examples include brands and similar intangible items. Increases and decreases in such values may be discussed. However, it is not the intention of the ASB that the overall valuation of a reporting company should be provided or that the asset value be reconciled with the market capitalization.

8.4 Presentation of the OFR

Practice in the publication of the OFR is evolving and there is diversity in the style and pattern of reporting. A comprehensive and informative OFR was published by Tesco plc and is reproduced here.

9 Interim financial statements

Part of the continuing obligations accepted by companies listed on the International Stock Exchange (ISE) and the Unlisted Securities Market (USM) is the publication of half-yearly reports and preliminary statements popularly known as interim financial statements.

Interim statements should be circulated by listed ISE companies to holders of listed securities and inserted as a paid advertisement in two leading daily newspapers no later than four months after the end of the period to which it relates. For USM listed companies, interim financial statements should be circulated to shareholders or published in one daily newspaper.

Interim statements must consist of information in table form relating to:

- Net turnover;
- Profit or loss before taxation and extraordinary items.
- Taxation on profits (UK taxation and, if material, overseas and share of associated undertakings' taxation to be shown separately).
- Minority interests.
- Profit or loss attributable to shareholders, before extraordinary items.
- Extraordinary items (net of taxation).
- Profit or loss attributable to shareholders.
- Rates of dividend(s) paid and proposed and amount absorbed thereby.
- Earnings per share expressed as pence per share (computed on the figures shown for profits after taxation as defined in FRS 14).
- Comparative figures in respect of the above points inclusive for the corresponding previous period.

Interim financial statements (reports) are expected to incorporate explanatory statements on the company's activities for the reporting period:

- To enable investors to make an informed assessment of the trend of the company's activities and profit or loss together with an indication of any special factor which has influenced those activities and the profit or loss for the year in question.
- To enable a comparison to be made with the corresponding period of the preceding financial year.
- As far as possible, to refer to the company's prospects in the current financial year.

Information relating to the audit status of the figures included in the interim report should be disclosed. In short, audit reports, including any qualifications, should be published if the interim accounts are audited. Where the interim accounts are not audited, this fact should be stated.

OFR of Tesco Plc, 1999

Operating and financial review

This operating and financial review analyses the performance of Tesco in the financial period ended 27 February 1999. It also explains certain other aspects of the Group's results and operations including taxation and treasury management.

Group summary

	1998 GBP m	1998* GBP m	Charge %
Group sales (including value added tax)	**18,546**	17,447	6.3
Group operating profit (prior to integration costs and goodwill amortization)	**965**	895	7.8
Profit on ordinary activities before tax†	**881**	817	7.8
Adjusted diluted earnings per share†	**9.37p**	8.70p	7.7
Dividend per share	**4.12p**	3.87p	6.5

* 52 weeks proforma

† Excluding net loss on disposal of fixed assets and discontinued operations, integration costs and goodwill amortisation

The financial period to 28 February 1998 was a 53-week trading year compared to a 52-week trading period this financial year. All comparisons in this operating and financial review are based on a 52-week proforma profit and loss account for 1998.

Group performance

Group sales including VAT increased by 6.3% to GBP 18,546m (1998 – GBP 17,447m). Group sales from continuing businesses increased by 10.1% (1998 – GBP 16,847m).

Group operating profit (prior to integration costs and goodwill amortization) rose by 7.8% to GBP 965m (1998 – GBP 895m).

Group profit before tax rose by 7.8% to GBP 881m (1998 – GBP 817m). This excludes the net loss on disposal of fixed assets and discontinued operations of GBP 8m (1998 – GBP 9m), integration costs of GBP 26m (1998 – GBP 63m) and goodwill amortization of GBP 5m (1998 – nil).

Group capital expenditure was GBP 1,067m (1998 – GBP 841m) with GBP 848m in the UK, GBP 27m in Thailand, and GBP 192m in Europe. In the current year, supported by our strong cash flows, Group capital expenditure will rise to around GBP 1.3bn. The increase in 1999 relates mainly to our development plans in the rest of Europe and Asia.

Change in net debt

Total net debt at the year end amounted to GBP 1,720m (1998 – GBP 1,191m). This reflects the cash outflow from our net capital expenditure and acquisitions of GBP 1,260m (1998 – GBP 1,082m), partly offset by strong cash generation from the main business of GBP 1,321m (1998 – GBP 1,156m). As a result, gearing at the year end has increased to 39% (1998 – 31%).

OFR of Tesco Plc, 1999 – Continued

Interest and taxation

Net interest payable was GBP 90m (1998 – GBP 72m) with the increase on last year primarily due to the financing costs of our acquisition in Thailand.

Corporation tax has been charged at an effective rate of 28.1% (1998 – 30.0%). Prior to accounting for the net loss on disposal of fixed assets, integration costs and goodwill amortization, our underlying tax rate was 27.8% (1998 – 28.7%).

Profit-sharing increased by 8.6% to GBP 38m (1998 – GBP 35m). In addition staff have continued to benefit from profit-related pay.

Shareholder returns and dividends

Adjusted diluted earnings per share (excluding the net loss on disposal of fixed assets and discontinued operations, integration costs and goodwill amortisation) increased by 7.7% to 9.37p (1998 – 8.70p).

The Board has proposed a final net dividend of 2.87p giving a total dividend for the year of 4.12p (1998 – 3.87p). The dividend is covered 2.27 times by earnings.

Shareholders' funds, before minority interests, increased by GBP 479m. This was due to retained profits of GBP 329m and issue of new shares less expenses of GBP 169m, offset by losses on foreign currency translation of GBP 19m. As a result, return on shareholders' funds was 21.3%.

The share price rose from 172p at the start of the financial year to 177p on 27 February 1999, giving a market capitalisation of approximately GBP 11.8bn (1998 – GBP 11.4bn). The share price reached a high of 202p on 2 July 1998.

Total shareholder return, which is measured as the percentage change in the share price plus the dividend, has been 22.7% over the last five years, compared to the market average of 15.1% and has been 30.7% over the last three years, compared to the market average of 19.2%. In the last year, total shareholder return in Tesco has been 5.5% compared to the market average of 8.0%. This reflects our efforts to grow the business while ensuring returns to shareholders are improved.

UK performance	**1999 GBP m**	1998* GBP m	Change %
Food retail sales			
(including value added tax)	**17,070**	15,799	8.0
Operating profit	**919**	859	7.0

* 52 weeks proforma

Sales growth for the industry has again been slow, as expected, reflecting lower inflation and a modest slowdown in volume growth from the very high levels of recent years. Our market share, based on estimates of IGD data, increased again to 15.8% in the year to December 1998, from 15.2% last year. Overall, it has been a challenging year for the industry, but our business has remained focused on our customers and has achieved good results. Over the last six years, sales volumes have grown by 22%.

OFR of Tesco Plc, 1999 – Continued

UK retail sales have grown by 8.0% to GBP 17,070m (1998 – GBP 15,799m), of which 4.0% came from existing stores including volume growth of 2.5%. New stores contributed a further 4.3% to total sales growth before closures of 0.3%.

UK operating profit was 7.0% higher at GBP 919m (1998 – GBP 859m) and the operating margin fell 0.1% to 5.8%. This reflects our strong trading performance in a very competitive and challenging environment. We continue to invest to cut prices and increase customer service, particularly through longer opening hours and more service counters.

Store development and capital expenditure

In the United Kingdom we spent GBP 350m on opening 25 new stores with a total sales area of 635,000 sq ft. This comprised 20 superstores and compact stores, one Metro, two Expresses and two Extras. As part of our programme to improve our stores and introduce non-food products, we have added over 200,000 sq ft through extensions. We also completed major refits at an additional 18 stores.

In the year, we continued to build on the success of our first Extra in Pitsea, Essex. We now have five Extras, including the latest store at Peterborough, providing one-stop shopping convenience for customers.

Express is also proving to be a promising format. We have 17 stores and during the year we announced our joint venture with Esso to develop more petrol sites, the first of which opened recently on the Fulham Road, Chelsea.

In 1999/2000, we plan to open around 26 stores with over 700,000 sq ft of new space, including 23 superstores and compact stores, one Metro store and two Express stores.

European performance	**1999 GBP m**	1998* GBP m	Change %
Retail sales			
(including value added tax)	**1,285**	1,029	24.9
Operating profit	**48**	38	26.3

* 52 weeks proforma

Europe

In the rest of Europe (Central Europe and Republic of Ireland), total sales rose by 24.9% to GBP 1,285m (1998 – GBP 1,029m).

In Central Europe, a fast growing market, we successfully opened six more hypermarkets during the year, creating almost 600,000 sq ft of new modern retailing space. This included three hypermarkets in Hungary, two in the Czech Republic and one in Poland. In 1999/2000 we plan to open 10 more hypermarkets, adding a further one million sq ft.

In the Republic of Ireland our acquisition was nearly two years ago and we have achieved the milestones we set ourselves whilst meeting the undertakings made to the Government. We have rebranded 30 stores and our customers have responded well. Next year we expect to rebrand a further 20 stores and start to build two new stores.

OFR of Tesco Plc, 1999 – Continued

The business is performing strongly and remains full of potential for the future.

Asian performance	**1999 GBP m**	1998 GBP m	Change %
Retail sales			
(including value added tax)	**170**	–	–
Operating loss	**2**	–	–

Asia

In May we acquired a controlling interest in Lotus, a chain of 13 hypermarkets in Thailand with 1.6m sq ft of selling space. In the 32 weeks to 31 December 1998, Lotus contributed GBP 170m to Group sales and reported a small operating loss of GBP 2m. In December 1998 we added a 14th store and over the next three years we will develop the business further by doubling the number of stores and providing a strong base for profitable long-term growth.

The purchase price of our 75% stake was GBP 206m, including GBP 89m debt. Goodwill amounted to GBP 117m after fair value adjustments of GBP 38m on net assets acquired of GBP 127m. Goodwill will be amortised over 20 years in accordance with FRS10, resulting in a charge to the Group profit and loss account of GBP 5m this year.

After the financial year end, we announced on 23 March 1999 that we were to form a partnership company with Samsung Corporation to develop hypermarkets in South Korea. This company will initially have net assets of GBP 160m, comprising two existing Homeplus hypermarkets, three development sites and GBP 80m cash. Tesco will invest a total of GBP 130m, prior to costs, for an 81% controlling interest.

South Korea is a large developed market of 46 million people with GDP per capita already 70% of that in the UK but, in contrast, modern retailing is underdeveloped with only around 25 hypermarkets. The potential for us to build on the two successful Homeplus stores we now own is significant, and we plan to open a further 12 stores by the end of 2002.

We are continuing with our research in Taiwan and Malaysia.

Joint ventures

The Group operates several businesses as joint ventures with external partners including property joint ventures and Tesco Personal Finance, our financial services joint venture. The total share of profits of our joint ventures was GBP 6m (1998 – loss of GBP 6m).

Tesco Personal Finance has been an important part of our business strategy now for two years and is in good shape. Products launched to date, including savings, loans, visa, insurance and pensions have all been popular and over one million customers now use our financial services. This year our share of the Tesco Personal Finance operating loss was GBP 12m. This was slightly better than expected reflecting significant improvements in efficiency and we aim to break even towards the end of the 1999/2000 financial year.

Property joint ventures contributed an operating profit of GBP 18m principally comprising rental income on properties owned by our joint ventures with British Land and Slough Estates.

OFR of Tesco Plc, 1999 – Continued

Treasury management and financial instruments

The Group's treasury operations are managed by Group Treasury within parameters defined formally and regularly reviewed by the Board. Group Treasury's activity is routinely reported to members of the Board and is subject to review by the internal and external auditors.

Consistent with Group policy, Group Treasury does not engage in speculative activity. Financial instruments, including derivatives, are used to raise finance and to manage financial risk arising from the Group's operations.

The main financial risks faced by the Group relate to credit, interest and foreign exchange. The Board reviews and agrees policies for managing these risks as summarised below.

The Board establishes annually the policy which Group Treasury follows in managing credit risks. Limited exposures are permitted only with banks or other institutions meeting required standards as assessed normally by reference to the major credit rating agencies. Deals are authorized only with banks with which dealing mandates have been agreed.

Finance and interest rate risk

The Group's policy is to finance its operations by a combination of retained profits, bank borrowings, commercial paper, medium term notes, long-term debt market issues and leases.

Derivatives, predominantly forward rate agreements and interest rate swaps and caps, are used to manage the mix of fixed and floating rate debt. The policy is to fix or cap between 30% and 70% of the interest cost on outstanding debt, although a higher percentage may be fixed within a 12 month horizon. At the year end, after taking account of interest rate swaps, GBP 649m (1998 – GBP 614m) or 38% of our net debt was fixed at an average rate of 8.2% for a period of five years. A further GBP 100m (1998 – GBP 170m) or 6%, was covered by interest caps at an average rate of 8.3% for a period of three years.

The average rate of interest paid during the year was 7.1% (1998 – 8.1%). Excluding capitalized interest, interest is covered 7.8 times by profit before interest (1998 – 8.5 times). A 1% rise in market interest rates would reduce profit before tax by less than 2%.

The Group ensures continuity of funding by arranging for short-term bookings and commercial paper issuance to be fully backed by committed bank facilities, by limiting the amount of debt repayable in any one year, and by managing the average debt maturity in line with gearing levels. At the year end undrawn committed facilities amounted to GBP 510m (1998 – GBP 645m) and the average debt maturity of net debt, including these facilities, was over five years.

Foreign currency risk

The Group's policy is to use foreign currency borrowings, forward foreign currency transactions and swaps to offset part of the impact on the Group's balance sheet of exchange rate movements on the 12% of its net assets before financing which are held overseas.

The Group does not hedge exposure to currency movements on the translation of the 4.8% of profits made overseas except to the extent that those profits are matched by foreign currency interest costs.

Significant transactional currency exposures resulting predominantly from purchases in currencies other than the subsidiaries' reporting currencies are hedged by forward foreign currency transactions, currency options and by holding foreign currency cash balances.

OFR of Tesco Plc, 1999 – Continued

Year 2000

Tesco has been working on the Year 2000 issue for over three years with the specific objective of ensuring business continuity under the banner of 'Shopping as Normal for our Customers'.

A central dedicated team has been co-ordinating the project across all countries and reports to the Board every month. Accountability has been firmly placed with line directors for defining and actioning their work programme, co-ordinated through this central team.

The work has concentrated on taking corrective action across systems, embedded chips and working with our many suppliers to ensure product supply and continuity of services.

The activities involved have been to:

- identify the problem
- take corrective action
- re-test new systems and processes
- validate external suppliers of goods and services
- categorize risk and develop contingency plans
- ensure good communication within the business and to external suppliers and stakeholders

The main areas of work lie within:

- computer systems
- suppliers of products for re-sale
- suppliers of equipment and services
- the supply chain and our distribution network

There is a risk to the Group, as with all companies, that suppliers may experience Year 2000 failures. We have held five major supplier conferences involving over 2,000 firms and are currently fine-tuning contingency plans to ensure the transition is smooth and trade is unaffected.

We have made the necessary changes and re-tested all our business critical computer systems.

The Board have agreed store trading times over the millennium period and all areas of the business will have created staffing and contingency plans by the end of August 1999.

The approach developed within the United Kingdom has been used in Europe and Asia where compliance is also well advanced.

The programme was estimated to cost GBP 30m over three years and we expect to spend within that budget.

We have worked closely with Government and Action 2000 as well as the Retail Industry bodies, sharing information for the good of the consumer in order to achieve 'Shopping as Normal'

Economic Monetary Union

Our aim is for all the relevant parts of the Group to be able to handle business in euros when required. Tesco has project groups addressing the issues arising from EMU and is working with external consultants.

Going concern

The directors consider that the Group and the company have adequate resources to remain in operation for the foreseeable future and have therefore continued to adopt the going concern basis in preparing the financial statements. As with all business forecasts the directors' statement cannot guarantee that the going concern basis will remain appropriate given the inherent uncertainty about future events.

10 Summary financial statements

For several reasons, not least of which are costs to the users and preparers in terms of both production and printing and information overload, publicly listed companies are permitted to send to those entitled to receive the full set of annual accounts and reports a summary financial statement instead. However, the full set of annual accounts and reports should be sent to any member of the company who wishes to receive them.

The summary financial statement shall be derived from the company's full set of annual accounts and reports and shall:

- State that it is only a summary of information in the company's annual accounts and the directors' report.
- Contain a statement by the company's auditors of their opinion as to whether the summary financial statement is consistent with those of the full set of annual accounts and report and complies with the requirements governing summary financial statements.
- State whether the auditors' report on the full set of annual accounts was unqualified or qualified, and if it was qualified, set out the report in full together with any further material needed to understand the qualification.
- State whether the auditors' report on the full set of annual accounts contained a statement about the inadequacy of accounting records, the lack of agreement between the full set of annual accounts and the accounting records and about the auditors' failure to receive necessary information and explanation.

11 Related party transactions

11.1 General remarks

The discussion to this point has assumed that the transactions that are reported in the annual report and accounts are the result of negotiations between the reporting company and outside independent and unrelated parties. However, the company may enter into transactions with related parties. When two or more parties are related, they may enter into transactions that unrelated parties would not undertake or would undertake only on different terms. Such transactions and their values may be material and may affect the results reported in the financial statements of the related parties.

Companies Act 1985 requires some disclosures concerning related party transactions such as emoluments and other benefits of directors and guarantees and other financial commitments and the Stock Exchange also have requirements with regard to transactions with substantial shareholders but together they are not considered to be comprehensive. In the light of this weakness the ASB issued FRS 8, Related party transactions, in October 1995.

11.2 Definition of a related party

Two or more parties are related when at any time during the financial period (FRS 8, para. 2.5):

(i) One party has either direct or indirect control of the other party; or
(ii) The parties are subject to common control from the same source; or
(iii) One party has influence over the financial and operating policies of the other party to an extent that the other party might be inhibited from pursuing at all tomes its own separate interests; or
(iv) The parties, in entering a transaction, are subject to influence from the same source to such an extent that one of the parties to the transaction has subordinated its own separate interests.

To avoid doubt, the following are related parties:

(i) The ultimate and intermediate parent undertakings, subsidiary undertakings, and fellow subsidiary undertakings.
(ii) Associates and joint ventures.
(iii) The investor or venturer in respect of which the reporting entity is an associate or a joint venture.
(iv) Directors (including shadow directors) of the reporting entity and the directors of its ultimate and intermediate parent undertakings.
(v) Pension funds for the benefit of employees of the reporting entity or of any entity that is a related party of the reporting entity.

Parties presumed related to a reporting entity unless proved otherwise include:

(i) A person owning or able to exercise control over 20 per cent or more of the voting rights of the reporting entity, whether directly through nominees.
(ii) The key management of the reporting entity and the key management of its parent undertaking or undertakings.
(iii) Each person acting in concert in such a way as to be able to exercise control or influence over the reporting entity.
(iv) An entity managing or managed by the reporting entity under a management contract.

11.3 Disclosure requirements

Related party disclosures are required for all material related party transactions whether or not a price is charged. The disclosure should include (FRS 8, para. 6):

(i) The names of the related parties.
(ii) A description of the relationship between the parties.
(iii) A description of the transactions.
(iv) The amounts involved.
(v) The amounts due to or from related parties at the balance sheet date and provisions for doubtful debts due from such parties at that date.
(vi) Any other elements of the transactions necessary for an understanding of the financial statements.
(vii) The amounts written off in the period in respect of debts due to or from related parties.

12 Directors' report

12.1 Political and charitable contributions

Companies Act 1985 requires that companies should disclose in the directors' report attached to their annual accounts information about contributions for political and charitable purposes, if together they exceed GBP 200. The name of the recipient must be given in the case of individual political contributions exceeding GBP 200.

The term 'political purposes' is defined broadly to include contributions to political parties and to quasi-political organizations which may be expected to affect, directly or indirectly, the finance or public support of a political party in the United Kingdom. Payments for charitable purposes do not include sponsorships involving a fairly direct commercial interest and payments to charitable organizations for services rendered.

The Companies Act does not require that amounts given to a political party should be specified alongside the name of the political party receiving a donation.

12.2 Disabled persons

Companies employing more than 250 persons must state in their directors' report their policies for the employment of disabled persons, the continued employment and training of persons who become disabled whilst employed by the company, and the training, career development and promotion of disabled persons.

12.3 Employee involvement in the company

Large companies employing more than 250 persons are also required to include in their directors' report information on the action they have taken during the year covered by the report to introduce, maintain and develop arrangements aimed at corporate reporting for employees, to consult with employees on the future of the company likely to affect them, to encourage the involvement of employees through the employees' share scheme or other incentives, and to make employees aware of the financial and economic factors affecting the performance of the company.

12.4 Directors

The names of the directors of the company during the financial year should be disclosed. Companies listed on the International Stock Exchange (ISE) should include in the directors' report the unexpired period of any service contract relating to any director proposed for re-election at the forthcoming annual general meeting. If a director is available for re-election and does not have a service contract then this must be stated. It is now common practice, following the recommendation of the Cadbury Committee, to obtain shareholders' approval for directors' service contracts of three years and over.

Information relating to directors' interests in shares and debentures of the company, including the number and amount of the shares and debentures, should be included in the directors' report. A statement should also be given in the accounts of a company listed on the International Stock Exchange about persons (other than directors) holding substantial interest (3 per cent or more) in the equities of the company.

12.5 Substantial shareholders

Companies listed on the ISE with substantial shareholders' interests (normally 3 per cent or more but sometimes 10 per cent or more) are required to give particulars in their annual report and accounts of:

- Any contract of significance (i.e. contracts of 1 per cent or more of net assets (for capital transactions) or of total purchases, sales, payments or receipts) between the company and a controlling shareholder (i.e. 30 per cent or more).
- Any contract for the provision of services to the company or any subsidiary by a corporate substantial shareholder (except if the services provided are the shareholder's principal business and it is not a contract of significance).

12.6 Directors' responsibility for financial statements

Before the Auditing Practice Board's (APB's) statement of principles project was commenced, many users of financial statements had operated under the illusion that the audited annual reports were prepared by auditors rather than managers. The Cadbury Committee's Code of Best Practice suggests that the directors should make a brief statement explaining their responsibilities for the preparation of the financial statements. Indeed, SAS 600 stipulates that the auditors' report should distinguish between the responsibilities of the directors and the auditors.

X Auditing

1 Auditing in historical perspective

The requirement that joint stock companies should provide shareholders with audited accounts was introduced by the Joint Stock Companies Act 1844. The Act required that audited balance sheets be provided to shareholders. The requirement had a brief life since it was not included in the 1856 Joint Stock Companies Act. It was, however, reintroduced by the Companies Act 1900. While the Companies Act 1929 required that the profit and loss account should be sent to shareholders in addition to the audited balance sheet, it did not require that the profit and loss account be audited. The profit and loss account was required to be audited for the first time by the Companies Act 1948 (see Table X.1 for full details of the historical development of auditing).

Until 1994, the United Kingdom, unlike other advanced industrialized countries, required a compulsory annual audit for all registered companies irrespective of size. Many have suggested that the requirement that all companies be audited imposes an unnecessary burden on small businesses because there are other less onerous ways of providing assurance and because the resources of the auditing profession can be better concentrated on those enterprises where there is a real public interest, i.e. public companies and companies in regulated sectors such as financial services. The small company may be owner-managed and so may not possess the critical need for external auditing that arises from the separation of ownership from management.

The situation has now changed. With the issue by the Board of Trade of Statutory Instrument No. 1994/1935, The Companies Act 1985 (Audit Exemption), Regulations 1994, which came into force on 11 August 1994, no audit is now required for companies with turnover under GBP 90,000 and balance sheet total of less than GBP 1.4 million. Companies with turnover between GBP 90,000 and GBP 350,000 (GBP 250,000 gross income for companies that are charities) will be exempt from audit if the directors ensure that an exemption report is prepared, though such com-

Table X.1 Historical development of auditing

1844	The Companies Act 1844 required balance sheets to be audited and filed with the Registrar of Companies. The Joint Stock Banking Act also required annual balance sheets of banks to be audited.
1856	The Joint Stock Companies Act replaced the 1844 Acts and abandoned the need to audit balance sheets of registered companies. The law made the appointment of auditors voluntary and suggested that companies may make provision for the appointment of an auditor who was to ascertain that the balance sheet represented the 'true and correct view' of the state of affairs.
1867	The Railways Companies Act mandated the audit of the balance sheets of railways companies and required the auditors to ascertain whether the balance sheet of the company contained a 'full and true' statement of the financial condition of the company.
1879	The requirement to audit banking companies registered as limited liability was reinstated by the Companies Act 1879. Auditors were to report whether the balance sheet was 'true and fair'.
1900	Annual audits were made obligatory for all companies by the Companies Act 1900.
1948	The Companies Act 1948 mandated the audit of profit and loss accounts in addition to the existing audit requirement for balance sheets and required auditors to state whether accounts were 'true and fair' rather than 'true and correct'. The Act also requires all auditors to be professionally qualified.
1961–76	The ICAEW approved and issued twenty-three auditing statements (known as the 'U' series) on *ad hoc* matters including Statement of Auditing No. 17, Effect of Accounting Standards on Auditors' Reports, which required auditors to refer in their auditors' reports to 'all significant departures from accounting standards made by the directors in preparing the accounts', whether or not they are disclosed in notes to the accounts.
1976	The Auditing Practices Committee was formed to produce comprehensive auditing standards covering both personal standards, operating standards and reporting standards. The Companies Act 1976 recognized the Institutes of Chartered Accountants in England and Wales, of Scotland and in Ireland and the Association of Certified and Corporate Accountants as those that can issue certificates of competence to persons (their members) who can serve as auditors of companies registered under the Companies Act.
1983	The ICAEW published a legal opinion on 'true and fair view' with particular reference to the role of accounting standards, where it was stated that accounts which are prepared on the basis of accounting standards would be presumed to portray a 'true and fair view' although 'true and fair view' is a legal concept that can only be decided by a court of law.
1989	The Auditing Practices Committee recommended that, when reporting that a 'true and fair view' existed, the auditor should verify that the accounting policies used were appropriate, consistent and adequately disclosed.

panies will require a compilation report prepared by an accountant. However, holders of 10 per cent or more of a company's issued share capital may require an audit but they must notify the company at least one month before the year end. Where a company decides to take advantage of the audit exemption, the financial statement must include a statement by the directors that:

- The company is eligible to take advantage of the audit exemption.
- The requisite number of members have not required an audit.
- They are aware of their obligation to keep proper records and to prepare accounts that give a true and fair view of the company's position and its profit or loss.

The exemption does not apply if a small and medium-sized company is a public company, a banking or insurance company, a registered insurance broker, an operator under the Financial

Services Act or a registered trade union body.

In a press notice (No. P/94/447) issued by the Department for Enterprise of the Department of Trade and Industry, the Corporate Affairs Minister was quoted as saying:

> These changes represent a major step in the Government's drive to free business from unnecessary burdens. I hope that small businesses and the accountancy profession will warmly welcome them. We have created a new environment for small companies and I want them to gain maximum benefit from these new measures.

Some disquiet has been expressed by the banking community about the blanket nature of this rule. This, according to the British Bankers' Association, is because there may be some instances where a bank may wish to consider retaining the audit, for example, where small or medium-sized companies are not in a position to provide regular and reliable management information or where the bank has no account history against which to measure a company's ability to repay facilities (Chisnall, 1994, p. 76).

The primary role of the auditor is to provide the users of financial information with assurance as to the integrity and reliability of the financial statements on which vital decisions affecting the allocation of scarce resources are based. As a result, there is increasing concern about the independence of auditors and the quality of the audit. The issues of audit quality and auditors' independence are taken up later (see X.4).

As noted above, the regulatory requirement that published financial statements be audited dates back to 1900 when published balance sheets were expected to be audited. The rights and duties of auditors were gradually extended. By the Companies Act 1948, the scope of auditors' work was extended to include reporting on the profit and loss account and on the group accounts as well as on the balance sheet of the accounts examined by them. The auditors' report must be signed and must state the auditors' names and whether in the auditors' opinion the annual accounts have been prepared in accordance with the Companies Act 1985 (revised by Companies Act 1989, section 9), and in particular whether a true and fair view is given:

- In the case of an individual balance sheet, of the state of affairs of the company as at the end of the financial year.
- In the case of an individual profit and loss account, of the profit or loss of the company for the financial year.
- In the case of group accounts, of the state of affairs as at the end of the financial year, and profit or loss for the financial year, of the undertakings included in the consolidation as a whole, so far as concerns members of the company.

The auditors shall also consider whether the information given in the directors' report for the financial year for which the annual accounts are prepared is consistent with those accounts. If they are of the opinion that it is not, they shall state that fact in their report.

A company's auditors shall, in preparing their report, carry out such investigation as will enable them to form an opinion as to:

- Whether proper accounting records have been kept by the company and proper returns adequate for their audit have been received from branches not visited by them.
- Whether the company's individual accounts are in agreement with the accounting records and returns.

The auditors' report shall state the facts relating to the above matters if the findings were negative.

Auditors usually hold office until the conclusion of the next annual general meeting and, unless the auditors cease to be qualified, retiring auditors are usually automatically reappointed unless they have been superseded by an affirmative resolution or have given written notice of their unwillingness to be reappointed. The remuneration of the auditors is usually fixed by the company in general meeting, or in such manner as the company in general meeting shall determine, or by the Secretary of State for Trade and Industry or by the directors.

Information must be disclosed in the company's accounts in respect of fees, including expenses and estimated money value of any benefits in kind, payable to the auditor for the discharge of audit functions (Companies Act 1985, section 390A and Companies Act 1989, section 121). In addition, as a result of Statutory Instrument No. 2128, issued by the Board of Trade on 1 October 1991, disclosure is also required in the notes to the annual accounts of large (i.e. not small and medium-sized) companies of the amount of remuneration (including expenses) payable to the company's auditors or their associates for non-audit services provided to the company and any subsidiary undertakings audited by the same auditors (or by the auditor's associates) as the company. Non-audit fees include, for example, fees for accountancy or taxation services, investigation or other company finance work or management consultancy.

Officers or servants of a company, their partners or persons in their employment, and body corporates cannot be appointed as auditors of the company. The requirement that auditors be professionally qualified was introduced for the first time in the United Kingdom by the Companies Act 1948. Only persons who are members of a body of accountants established in the United Kingdom, and for the time recognized for the purposes by the Secretary of State for Trade and Industry, shall be qualified for appointment as auditors of a company. In addition, persons authorized by the Secretary of State for Trade and Industry as having similar qualifications obtained outside the United Kingdom, or persons granted authorization formerly granted by the Board of Trade or the Secretary of State under the Companies Act 1948, section 161 (1) (b), (on the basis of adequate knowledge and experience, or pre-1947 practice) may also be permitted to be appointed as auditors of the company. The recognized supervisory bodies (RSB) authorized by the Secretary of State for Trade and Industry whose members may be appointed as auditors of companies are:

- The Institute of Chartered Accountants in England and Wales.
- The Institute of Chartered Accountants in Scotland.
- The Chartered Association of Certified Accountants.
- The Institute of Chartered Accountants in Ireland.

2 Standards of auditing practice

The development of formal auditing standards started about the same time (in 1971) as the development of formal accounting standards. While accounting standards were being developed by the Accounting Standards Committee of the CCAB, auditing standards and guidelines were being developed by the Auditing Practices Committee (APC) of the CCAB. Following the public perception of audit failures in respect of recent corporate scandals (such as those relating to the Bank for Credit and Commerce International (BCCI), the Maxwell Group of Companies and the London United Investment) and the criticisms of the inspectors from the Department of Trade and Industry on the auditing standards that some of the Big Six firms have brought to bear on their works, there was a clamour for change in the institutional arrangements for audit regulation. This was accomplished when the APC was superseded by the Auditing Practices Board (APB). Before 1991, when the APC was disbanded and replaced by the APB with membership including non-accountants, two auditing standards (the auditor's operational standard and the audit report standard) and 38 audit guidelines were issued (see Table X.2).

The objectives of the APB are to establish high standards of auditing, to meet the needs of users of financial information and to ensure that the public has confidence in the auditing process. The APB tries to achieve its objectives by issuing Statements of Auditing Standards (SASs), Practice Notes and Bulletins. SASs contain basic principles and procedures that an auditor should comply with. The accountancy bodies in the UK have undertaken to adopt all auditing standards and failure could lead to internal disciplinary measures. As part of its endeavours, the APB supports the International Auditing Committee of the International Federation of Accountants (IFAC) with its aim of improving harmonization.

Practice notes try to offer guidance to auditor in the application of auditing standards and bulletins provide guidance on emerging issues. Both are considered to be part of good practice although not mandatory.

Table X.2 Pronouncements by the APC and APB

Audit standards	
100	Objective and general principles governing an audit of financial statements
110	Fraud and error
120	Consideration of law and regulations
130	The going concern basis in financial statements
140	Engagement letters
150	Subsequent events
160	Other information in documents containing audited financial statements
200	Planning
210	Knowledge of the business
220	Materiality and the audit
230	Working papers
240	Quality control for audit work
300	Accounting and internal control systems and audit risk assessments
400	Audit evidence
410	Analytical procedures
420	Audit of accounting estimates
430	Audit sampling
440	Management representations
450	Opening balances and comparatives
460	Related parties
470	Overall review of financial statements
480	Service organizations
500	Considering the work of internal audit
510	The relationship between principal auditors and other auditors
520	Using the work of an expert
600	Auditors' reports on financial statements
601	Imposed limitation of audit scope
610	Reports to directors or management
620	The auditors' right and duty to report to regulators in the financial sector
APB practice notes	
1	Investment Businesses
2	The Lloyd's Market
3	The auditors' right and duty to report to the Bank of England
4	The auditors' right and duty to report to the Building Societies Commission
5	The auditors' right and duty to report to SIB and other regulators of investment business
6	The auditors' right and duty to report to the DTI in relation to insurers authorized under the Insurance Companies Act 1982
7	The auditors' right and duty to report to the Friendly Societies Commission
8	Reports by auditors under company legislation in the UK
9	Reports by auditors under company legislation in the Republic of Ireland
10	Audit of central government financial statements in the UK
11	The audit of charities
12	Money laundering
13	The audit of small businesses
14	The audit of registered social landlords in the UK
15	The audit of occupational pension schemes in the UK
16	Bank reports for audit purposes
17	The audit of regularity on the central government sector
18	The audit of building societies in the UK
19	Banks in the UK
20	The audit of insurers in the UK

Table X.2 Continued

APB practice bulletins	
1993/1	Review of interim financial information
1995/1	Disclosures relating to corporate governance (revised)
1996/2	The auditors' right and duty to report to regulators: implementation of the Post-BCCI Directive in the UK
1996/3	Disclosures relating to corporate governance (supplement)
1996/4	Equalization reserves
1997/1	The special auditors' report on abbreviated accounts in GB
1997/2	Disclosure of directors' remuneration
1997/3	The FRSSE: guidance for auditors
1998/1	The year 2000 issue: preliminary guidance for auditors
1998/2	Using the work of an actuary with regard to insurance technical provisions
1998/3	Auditors' reports on regulatory returns made under the Insurance Companies Act 1982
1998/4	The duty of recognized accountants to report to Lloyds
1998/5	The year 2000 issue: supplementary guidance for auditors
1998/6	Review of interim financial information – supplementary guidance for auditors
1998/7	The auditors' association with preliminary announcements
1998/8	Reporting on pro forma financial information pursuat to the Listing Rules
1998/9	The introduction of the euro: guidance for auditors
1998/10	Corporate governance reporting and auditors' responsibilities statements
1999/1	Reports by auditors under sections 156(4) and 173(5) Companies Act 1985 and the year 2000 Issue
1999/2	Corporate governance reporting and auditors' responsibilities statements in the Republic of Ireland
1999/3	Reports by reporting accountants on working capital and the year 2000 Issue
1999/4	Review of interim financial information
1999/5	The Combined Code: requirements of auditors under the listing rules of the stock exchange

3 Audit qualifications

The one way in which auditors promote good corporate reporting is in their preparedness not to issue a clean audit report if they are convinced that there is sufficient reason to believe that the accounts do not present a true and fair view of the company's state of affairs. Instead they may issue an audit qualification on the status of the client company as a going concern, or on the validity of the underlying accounting policies and their conformity with Companies Acts and accounting standards.

The Auditing Practices Board (APB) has issued SAS 130 on how auditors should assess whether it is appropriate for directors of client companies to prepare financial statements on the going concern basis and on how auditors should prepare reports that take into account that assessment. To reach a conclusion on whether the accounts have been prepared correctly on the going concern basis, auditors are required to do the following:

- Consider a written assertion from the directors as to their assessment of the appropriateness of the going concern basis of preparing the financial statements.
- Consider the adequacy of disclosures in the financial statements relevant to the appropriateness of the going concern basis.
- Take account of their findings in preparing their report.

The going concern concept referred to in SAS 130 is the one intended by company legislation and applies to the values at which assets, liabilities, profits, gains and losses are stated in the financial statements and other disclosures in the financial statements which the audit report is intended to cover. The underlying assumption of the going concern is that the audited company will continue in operational existence for the foreseeable future and will be able to recover in the normal course of business (through use or realization) the amounts at which assets are recorded and discharged in the normal course of business the recognized and recorded liabilities. This means that the profit and loss account and the balance sheet assume no intention or necessity to liquidate or curtail significantly the scale of operations. This is implied to mean that the audited company has neither the intention nor the necessity of:

- Entering into a scheme of arrangement, under the Companies Act, with its creditors.
- Making an application for an administrative order.
- Placing itself in administrative receivership.
- Liquidation.

Where the auditors disagree with the presumption that the company is a going concern, SAS 130 stipulates that an adverse audit opinion should be given. An adverse opinion should be given even when disclosure in the financial statements of the matters giving rise to the conclusion is not sufficient for them to give a true and fair view (SAS 130, para. 8). The same is true when the effect on financial statements prepared on that basis is so material or pervasive that the financial statements are seriously misleading (SAS 130, para. 8).

Before 1993, guidance on qualifications in audit reports can be summarized into two categories:

- Uncertainty – where there is an uncertainty which prevents the auditor from forming an opinion on a matter.
- Disagreement – where the auditor is able to form an opinion on a matter but this opinion conflicts with the view given by the financial statements.

The wording of the opinion paragraph will depend upon whether the subject matter of the uncertainty or disagreement is or is not considered fundamental to the view shown by the financial statements. The types of opinions which are appropriate in different circumstances and the operational procedure for reaching the decisions are encapsulated respectively in Table X.3 and Figure X.1.

Adverse opinion and the disclaimer of opinion are the extreme forms of the categories of audit qualification that can arise respectively from uncertainties and disagreements. Where disagreements are such that a reader of financial statements will not be able to obtain readily a proper appreciation of the results of business operations, financial position and cash flow statements by reading the statements in conjunction with the information contained in the audit report, the auditor would express an adverse opinion.

Empirical evidence suggests that there is a low rate of qualification on a going concern basis of the accounts of companies prior to bankruptcy (Citron and Taffler, 1992, p. 337). The reasons for this rather low incidence of qualifications in the face of apparent dissatisfaction with corporate governance (e.g. the setting up of the Cadbury Committee and the critical reports from the DTI inspectors on poor standards of auditing) and the phenomenon of the failure of companies that have recently been given clean audit reports include some economic factors. The economic factors that may reduce auditors' independence include the value of the auditor's economic interest in the client, the likelihood that this economic interest will be lost due either to the client switching auditors or to failure brought about by the auditor giving a going concern qualification (self-fulfilling prophecy), or the likelihood that the client will sue if the report is qualified and the client does not fail (Citron and Taffler, 1992, p. 337).

The economic factors likely to help auditors' independence include the loss of future revenues due to loss of reputation should the auditor not qualify and the client fail, and the likelihood of lawsuits by third parties if there is no going concern qualification and the client fails. With the issue of SAS 600, no 'subject to' opinion now exists. Auditors are required to either qualify or not qualify.

4 The independence and integrity of auditors

The concern about the adequacy of safeguards of independence and integrity led to discussions about possible restrictions on the way in which auditors provide their services to clients. The Department of Trade and Industry, in a consultative document on the regulation of auditors and the implementation of the Eighth Company Law Directive, suggested, among other things, that client companies should rotate their auditors every five years. The accounting profession

Table X.3 Types of audit opinions

Nature of circumstances	Material but fundamental	Fundamental
Uncertainty	'Subject to' opinion	Disclaimer of opinion
Disagreement	'Except for' opinion	Adverse opinion

Figure X.1 Going concern and reporting on the financial statements

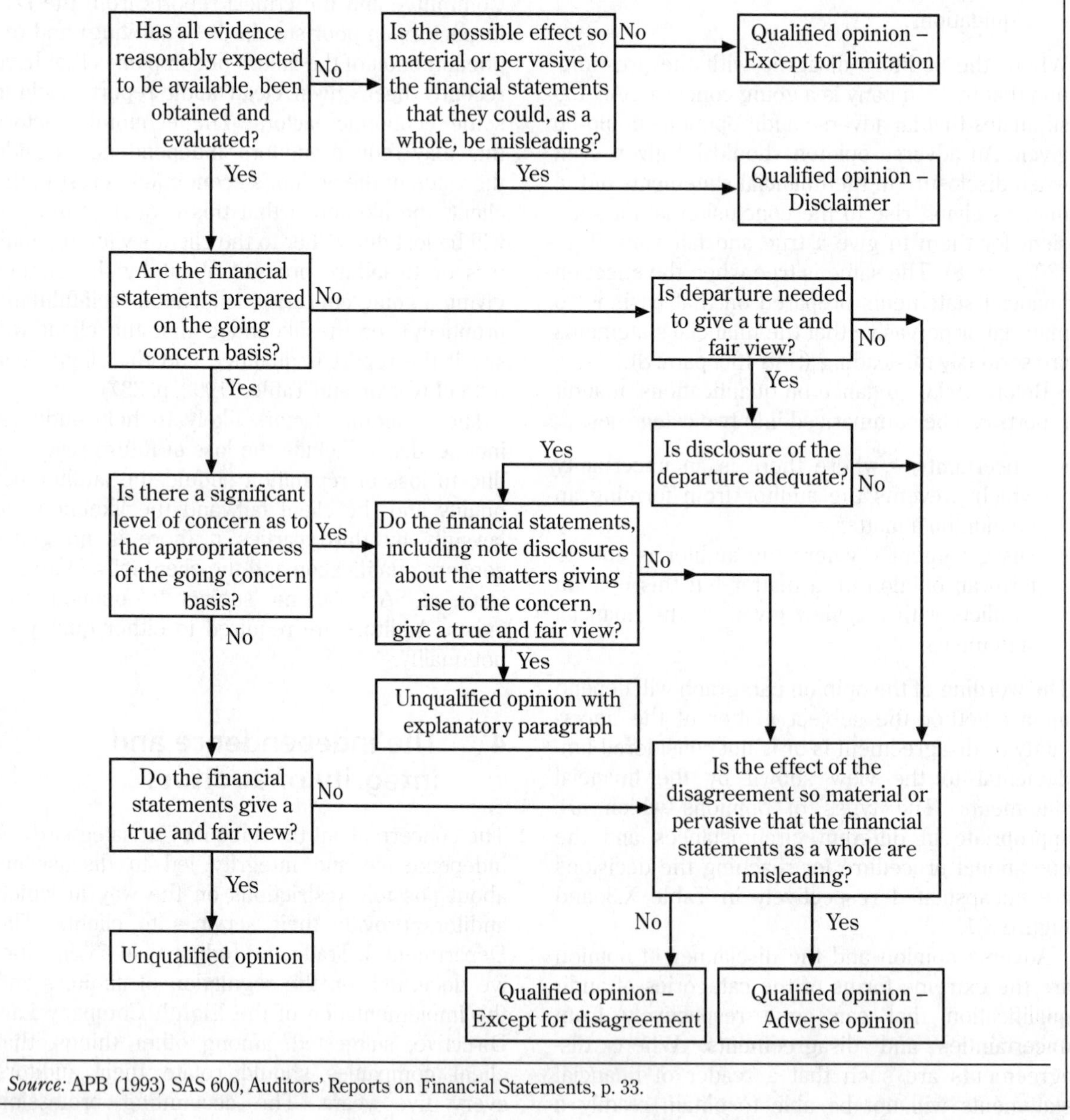

Source: APB (1993) SAS 600, Auditors' Reports on Financial Statements, p. 33.

lobbied against this suggestion on the grounds that audit failures are more likely in the first few years of an auditor's appointment. An auditor is therefore more likely to incur higher costs in the earlier period of appointment and if these costs are to be recovered over five years, then the cost of auditing will inevitably increase significantly. The mandatory rotation of auditors in every country in the European Union would be impracticable if it applied to multinational European companies.

Another recommendation of the Department of Trade and Industry was to limit the extent of non-audit work which auditors perform for their clients. The argument of the accounting profession against this potentially constraining recommendation is that there has not been any evidence to suggest that the provision of non-audit work by auditors for their clients has caused any problems of independence. Furthermore, an auditor's knowledge of a client's business would enable the auditor to provide additional services at lower cost than would otherwise be possible. In addition, it has been argued that the provision of non-audit services by the auditor to the client would mean the audit task could be performed more effectively because of the auditor's additional knowledge of relevant aspects of the business's affairs. If the objective of auditor independence is to support the quality of audit services provided, a ban on the provision of other services, some may argue, may be counter-productive.

An auditor must be registered before accepting an appointment as a company auditor in accordance with Companies Act 1989. Either an audit firm or individual must be registered with a RSB. Such organizations must be able to demonstrate that there are adequate rules for proper regulation, that members are monitored, and that there should be an investigation into complaints and discipline. Proper regulation includes professional integrity and independence; qualifications of auditors; and individuals should be fit and proper persons. In addition there should be control of audit firms by qualified individuals; appropriate technical standards; procedures to maintain competence; and be able to meet claims.

The Institutes of Chartered Accountants in consultation with the Association of Chartered Certified Accountants formed the Chartered Accountants' Joint Ethical Committee to consider issues of integrity, objectivity and independence. The guidance (effective from 1 September 1997) identifies threats to objectivity which are categorized as the self-interest threat; the self-review threat; the advocacy threat; the familiarity or trust threat; and the intimidation threat. Guidance on specific areas that may affect risks to auditors' objectivity and independence and include: fees; loans to or from a client as well as guarantees and overdue fees; hospitality or other benefits; actual or threatened litigation; participation in the affairs of a client; principal or senior employee joining client; mutual business interest; beneficial interests in shares and other investments; beneficial interests in trusts; trusteeships; nominee shareholdings; voting on audit appointments; connections and associated parties; provision of other services.

The ethical statement requires that no audit engagement partner should be in charge of an audit of a listed company for more than seven consecutive years and should not return to the audit for at least five years (para 4.80). Specific advice is also given where an auditor is asked to provide certain services, such as the valuation of assets.

5 The expectations gap

In addition to the issues relating to auditor independence and integrity, there is the societal concern about an apparent gap between what society expects from auditors and what auditors think that they can accomplish (see, for example, Power, 1997). Fanning the embers of societal concern for the work of auditors are the use of creative accounting and several business failures that have resulted in an enormous loss of shareholders' funds. This gap is referred to as the expectations gap.

There is evidence of an expectations gap between what users and auditors believe or perceive to be auditors' responsibilities. Although the nature of such a gap has not been articulated in the United Kingdom, the description of the components of audit expectation by the Canadian Institute of Chartered Accountants (CICA) in 1988 provides a basis for understanding the make-up of the expectations gap in the United Kingdom. Figure X.2 (adapted from the CICA) describes the nature of the gap.

It reveals its two components: a performance gap and a standards gap. The performance gap can be analyzed into the gap between what the auditing profession expects auditors to deliver and what auditors are really delivering (audit performance shortfall), and the gap between what auditors are delivering and what the public thinks they are actually delivering (audit performance deficiency). Audit performance deficiency may be real or fictional. In most cases, the public may not be a better judge than the auditor of the value of auditor's work.

The standards gap describes the difference between what the public expects the work of auditors should be and what the auditing profession thinks it should be. What the public expects as the extent of audit services may include services for which there is no effective demand (i.e. there is a demand for services for which the public is not prepared to pay). This gap is described in Figure X.2 as the unmet and unreasonable expectations of the public. The gap between audit services the public can effectively demand and what the auditing profession thinks it should provide is described as an unmet but reasonable gap.

Some of the components of this gap may be removed through the improvement of the professional services provided by auditors (unmet but reasonable public expectations, audit performance shortfall, and 'real' audit performance deficiency) and some may be reduced by the improvement of the communication between the audit profession and the public (unmet and unreasonable public expectations and fictional audit performance deficiency).

In response to the clamour for changes brought about by pressures from the expectations gap, the APC has issued SAS 600 that seeks to expand the audit report. In addition, one of the recommendations of the Cadbury Committee on corporate governance is that directors of reporting companies should indicate the extent of their responsibilities for the finan-

Figure X.2. Components of the expectations gap

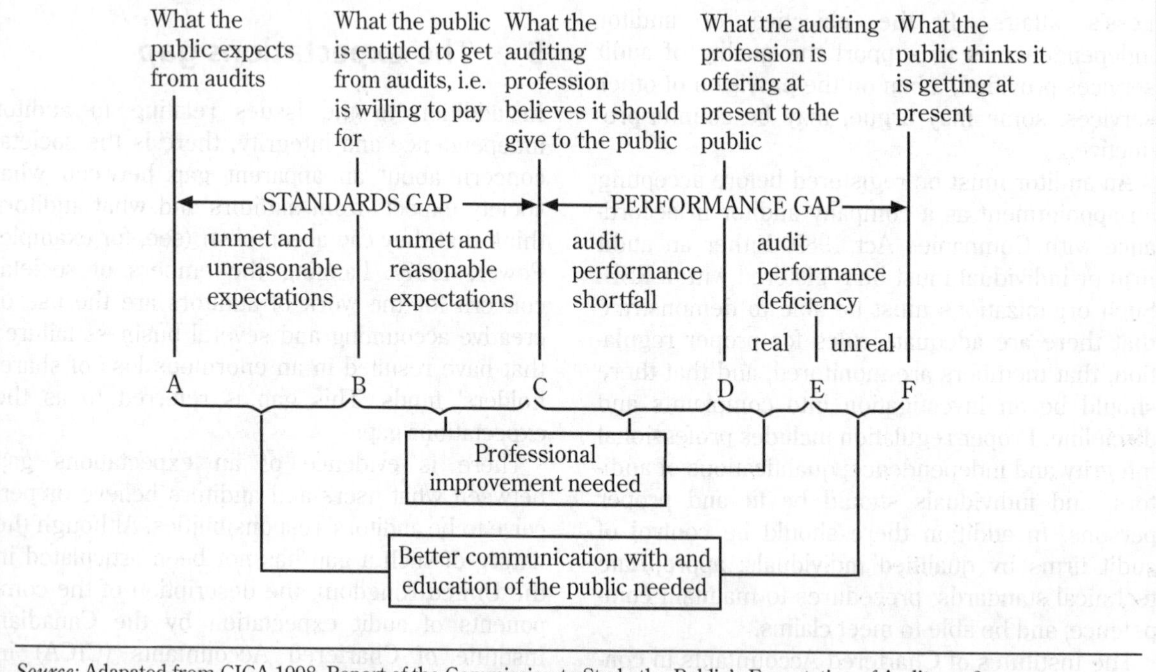

Source: Adapated from CICA 1998, Report of the Commission to Study the Public's Expectation of Audits, p. 6.

Table X.4 Illustration of statements of responsibilities and auditors' report

Report of the Auditors
Members of British Energy Plc

We have audited the financial statements on pages 33 to 56. We have also examined the information specified for our review by The London Stock Exchange which is included in the Remuneration Report on pages 26 to 29.

Respective Reponsibilities of Directors and Auditors
The Directors are responsible for preparing the Annual Report including, as described on page 31, the financial statements. Our responsibilities, as independent auditors, are established by statute, the Auditing Practices Board, the Listing Rules of the London Stock Exchange and our profession's ethical guidance.

We report to you our opinion as to whether the financial statements give a true and fair view and are properly prepared in accordance with the Companies Act. We also report to you if, in our opinion, the Directors' report is not consistent with the financial statements, if the Company has not kept proper accounting records, if we have not received all the information and explanations we require for our audit, or if information specified by law or the Listing Rules regarding Directors' remuneration and transactions is not disclosed.

We read the other information contained in the Annual Report and consider the implications for our report if we become aware of any apparent misstatements or material inconsistencies with the financial statements.

We review whether the statement on pages 24 and 25 reflects the Company's compliance with those provisions of the Combined Code specified for our review by the London Stock Exchange, and we report if it does not. We are not required to form an opinion on the effectiveness of the Company's or Group's corporate governance procedures or its internal controls.

Basis of Audit Opinion
We conducted our audit in accordance with Auditing Standards issued by the Auditing Practices Board. An audit includes examination, on a test basis, of evidence relevant to the amounts and disclosures in the financial statements. It also includes an assessment of the significant estimates and judgements made by the Directors in the preparation of the financial statements, and of whether the accounting policies are appropriate to the Company's circumstances, consistently applied and adequately disclosed.

We planned and performed our audit so as to obtain all the information and explanations which we considered necessary in order to provide us with sufficient evidence to give reasonable assurance that the financial statements are free from material misstatement, whether caused by fraud or other irregularity or error. In forming our opinion we also evaluated the overall adequacy of the presentation of information in the financial statements.

Opinion
In our opinion the financial statements give a true and fair view of the state of affairs of the Company and the Group at 31 March 1999 and of the profit and cash flows of the Group for the year then ended and have been properly prepared in accordance with the Companies Act 1985.

PricewaterhouseCoopers
Chartered Accountants
and Registered Auditors
Edinburgh
11 May 1999

cial statements and their underlying accounting records. The Combined Code modified this so that 'directors should explain their responsibility for preparing the accounts and there should be a statement by the auditors about their reporting responsibilities' (Section 1D 1.1 of the Combined Code).

SAS 600 stipulates that the directors' responsibilities should be made clear in the audited financial statements and is similar to the recommendation in the Cadbury Report (see Table X.4).

XI Filing and publication

Provisions about the publication of accounts are contained in the Companies Act 1985. By publication, the Act (Companies Act 1985, section 254(1)) refers to the following:

- Laying of accounts before the company in general meeting.
- Distribution to every member of the company (whether or not entitled to receive notice of general meetings), every holder of the company's debentures (whether or not so entitled), and all persons other than members and debenture holders so entitled (Companies Act 1985, section 241).
- Filing of such accounts with the Registrar of Companies.

The published accounts shall include the profit and loss account and balance sheet, the directors' report and the auditors' report (Companies Act 1985, section 239).

Companies that are permitted to publish abridged (summarized) accounts must publish, in addition, a statement indicating:

- That the accounts are not full accounts.
- Whether the full accounts have been delivered to the registrar of companies.
- Whether the company's auditors have reported.
- If the auditors have reported whether the report was unqualified (meaning that the company's accounts had been properly prepared). Published abridged accounts need not include unqualified audit reports. Unlimited companies which are exempt from publishing accounts, must publish the fact of that exemption.

Small and medium-sized companies taking advantage of the provisions of the Companies Act that they may file modified accounts must also include with their filing with the Registrar of Companies a statement by the directors that the company is entitled to deliver modified accounts. The filing must be accompanied by a special report of the auditors stating that in their opinion the directors are entitled to deliver modified accounts and indicating whether the modified accounts have been properly prepared.

XII Sanctions

1 The principal regulatory obligations

Reporting companies have three principal regulatory obligations relating to the issue of annual accounts and reports:

- A copy of the annual accounts, together with a copy of the directors' report for the financial year and of the auditors' report on those accounts, shall be sent to (a) every member of the company; (b) every holder of the company's debentures; and (c) every person who is entitled to receive notice of general meetings of the company not less than 21 days before the date of the meeting at which copies of those documents are to be considered.
- A copy of the company's annual accounts with the directors' report for that year and the auditors' report on those accounts shall be delivered to the Registrar of Companies. Such documents shall be in English or accompanied by a certified translation of the documents into English.
- The annual accounts are expected to comply with the requirements of applicable Companies Acts and accounting standards.

The annual accounts shall be considered at the relevant annual general meeting and delivered to the Registrar of Companies within the applicable period. The period allowed for laying (at the annual general meeting) and delivering (to the Registrar) annual accounts and reports is seven months and ten months after the end of the relevant accounting reference period (i.e. financial year) respectively for a public company and a private company.

2 Failure to lay and deliver annual accounts on time

Failure to lay and deliver the annual accounts within the allowable period would result in every person who immediately before the end of that period was a director of the company becoming guilty of an offence and liable to a fine and, for continued contravention, to a daily default fine. The court may, on application of an aggrieved party, order the directors (or any of them) to make good the default within such a time as may be specified in the order.

In addition to any liability of the directors to fines, the erring company is liable to a civil penalty calibrated according to the length of the delay (i.e. the time between the end of the period allowed for laying and delivering the accounts and reports and the day on which the requirements are met) and according to whether the company is public or private. The relevant penalties are detailed in Table XII.1.

3 Failure to comply with accounting requirements

Three avenues are available for ensuring that a failure by a reporting company to comply with accounting requirements is rectified:

- The reporting company may, without prompting by an external agency responsible for ensuring compliance with accounting requirements, volunteer to revise the offending annual accounts or directors' report. The revisions may be confined to the correction of those aspects of the accounts that did not comply with the requirements of the Companies Acts and/or the applicable accounting standards and the making of any necessary consequential alterations. No rules of procedure for making voluntary corrections have been laid down. As a result, it is not clear whether the previous (incorrect) accounts are to be replaced or whether supplementary documents indicating the corrections made to the books of accounts are to be issued. The roles of auditors and directors in the correction process are not clear. In addition, there is no definite information on how notice on the voluntary corrections should be circulated and who should receive it.
- The reporting company may be informed by the Secretary of State for Trade and Industry (or the delegated representative) of any departure from the requirements of the Companies Acts and applicable accounting standards by its annual accounts and reports and may be requested to revise the annual accounts and reports so as to comply with the requirements if it offers no satisfactory explanation.
- An application may be made to the Court, by the Secretary of State for Trade and Industry or a person authorized by the Secretary of State after the preceding procedure has been complied with, for a declaration (or declarator) that the annual accounts of the company do not comply with the requirements and for an order requiring the directors of the company to prepare revised accounts. As earlier discussion suggests (see I.4.6.1), the Secretary of

Table XII.1 Penalties according to the length of contravention

Length of contravention	Public company	Private company
3 months	GBP 500	GBP 100
>3 months but <6 months	GBP 1,000	GBP 250
>6 months but <12 months	GBP 2,000	GBP 500
>12 months	GBP 5,000	GBP 1,000

State for Trade and Industry has delegated the monitoring and compliance duties under the Companies Acts to the Financial Reports Review Panel (FRRP) for public and large private companies and to the Board of Trade for small and medium-sized private companies.

Where the court finds that the accounts have departed from the requirements, it may order that all or part of (a) the costs of and incidental to the application and any reasonable expenses incurred by the company in connection with or in consequence of the preparation of revised accounts and (b) the costs of litigation both of the plaintiff (either the Financial Reports Review Panel or the Board of Trade) and of the company be borne by such of the directors as were party to the approval of the defective accounts. The affected directors of the company may also on conviction be required to serve a sentence of imprisonment. For this purpose, every director of the company at the time the accounts were prepared shall be taken to have been a party to their approval unless the director shows that she or he took all reasonable steps to prevent the accounts being approved.

XIII Financial reporting institutions and databases

Body	Category	Address	Telephone	FAX	e-mail	Internet
Accounting Standards Board (ASB) Holborn Hall,	R	7th Floor, 100 Gray's Inn Road, London WC1X 8AL	44 20 7404 8818	44 20 7404 4497	mailbox@frc-asb.demon.co.uk	http://www.asb.org.uk
Association of International Accountants (AIA)	P	South Bank Building Kingsway, Team Valley Newcastle upon Tyne NE11 0JS	44 191 482 4409	44 191 482 5578	aia@a-i-a.org.uk	http://www.a-i-a.org.uk
Auditing Practices Board (APB)	R	PO Box 433 Moorgate Place London EC2P 2BJ	44 20 7920 8408	44 20 7638 6009	No e-mail address as yet	http://www.apb.org.uk
Association of Chartered Certified Accountants (ACCA)	P	29 Lincoln's Inn Fields London WC2A 3EE	44 20 7242 6855	44 20 7831 8054	technic@acca.org.uk	http://www.acca.co.uk
Companies House	D	Crown Way, Cardiff CF4 3UZ	44 29 2038 8588	44 29 2038 0900	NBARNES@companies-house.gov.uk	http://www.companies-house.gov.uk
Company Reporting	S	68 Dundas Street Edinburgh EH3 6QZ	44 131 558 1400	44 131 556 0639	Info@comrep.co.uk	No web address
Financial Reporting Council (FRC)	R	As for ASB	As for ASB	As for ASB	As for ASB	http://www.frc.org.uk
Financial Reporting Review Panel		As for ASB	As for ASB	As for ASB	As for ASB	http://www.frrp.org.uk
Institute of Chartered Accountants in England and Wales (ICAEW)	P	Chartered Accountants' Hall PO Box 433 Moorgate Place London EC2P 2BJ	44 20 7920 8100	44 20 7920 0547	comms@icaew.co.uk	http://www.icaew.co.uk
Institute of Chartered Accountants in Ireland (ICAI)	P	CA House, 87–89 Pembroke Road, Dublin 4.	353 1668 0400	353 1668 5685	ca@icai.ie	http://www.icai.ie
Institute of Chartered Accountants in Scotland (ICAS)	P	27 Queen Street, Edinburgh EH2 1LA	44 131 247 4869	44 131 225 381	icas@icas.org.uk	http://www.icas@icas.org.uk
London Stock Exchange (LSE)	R	Old Broad Street, London EC2N 1HP	44 20 7797 1000	44 20 7410 6861	Corporateaffairs@londonstockex.co.uk	http://www.londonstockex.co.uk

Key: R = regulatory body, P = professional body, D = source of company data, S = survey of company practices.

Bibliography

Accounting Standards Board (ASB) (September 1991). Financial Reporting Standard No. 1 – Cash Flow Statements, ASB, London.

ASB (July 1992), Financial Reporting Standard No. 2, Accounting for Subsidiary Undertakings (ASB, London).

ASB (October 1992), Financial Reporting Standard No. 3, Reporting Financial Performance (ASB, London).

ASB (July 1993), Statement: Operating and Financial Review (ASB, London).

ASB (December 1993a), Financial Reporting Standard No. 4, Capital Instruments (ASB, London).

ASB (December 1993b), Discussion Paper: Goodwill and Intangible Assets (ASB, London).

ASB (April 1994), Financial Reporting Standard No. 5, Reporting the Substance of Transactions (ASB, London).

ASB (July 1994a), Amendment to SSAP 19, Accounting for Investment Properties (ASB, London).

ASB (July 1994b), Discussion Paper: Associates and Joint Ventures (ASB, London).

ASB (September 1994a), Financial Reporting Standard No. 6, Acquisitions and Mergers (ASB, London).

ASB (September 1994b), Financial Reporting Standard No. 7, Fair Values in Acquisition Accounting (ASB, London).

ASB (December 1994), Amendment to FRS 5, Reporting the Substance of Transactions: Insurance Broking Transactions and Financial Reinsurance (ASB, London).

ASB (October 1994), 'Future events' – publication of collaborative research study, Press Notice ASB PN 53 (ASB, London).

Accounting Standards Committee (1978). Setting Accounting Standards Report and Recommendations by the Accounting Standards Committee, ASC, London.

Accounting Standards Steering Committee (1975). Corporate Report: A Discussion Paper (ASSC, London).

Alexander, D. (1993). A European true and fair view?. *European Accounting Review*, 2, No. 1, pp. 59–80.

Alexander, D. (1994). The true and fair view concept – towards an international perspective. Working Paper Series No. HUSM/DA/21 (The University of Hul)l.

Aliber, R. L. and Stickney, C. P. (1975). Accounting measures of foreign exchange exposure: the long and short of it. *The Accounting Review*, 50, no. 1, January, pp. 44–57.

Arden, M. H. (1993). Legal opinion on financial reporting standards and true and fair view. *Accountancy*, July, pp. 12–23.

Arnold, J., Boyle, P., Carey, A., Cooper, M. and Wild, K. (1991). The Future Shape of Financial Reports (ICAEW, London).

Barwise, P., Higson, C., Likierman, A. and Marsh, P. (1989). *Accounting for Brands* (ICAEW and London Business School, London).

Brown, R. (1905). *A History of Accounting and Accountants* (Edinburgh).

Canadian Institute of Chartered Accountants (1988). Report of the Commission to Study the Public's Expectation of Audits (CICA, Toronto).

Chisnall, P. (1994). Banks and the small company audit reforms. *Accountancy*, 114, (1216), December, pp. 76–77.

Chopping, D. and Skerratt, L. (1994). Applying GAAP 1994/95: A Practical Guide to Financial Reporting, third edition (The Institute of Chartered Accountants in England and Wales, London).

Citron, D. B. and Taffler, R. J. (1992). The audit report under going concern uncertainties: an empirical analysis. *Accounting and Business Research*, 22, no. 88, Autumn, pp. 337–345.

Collins, W., Davie, E. S. and Weetman, P. (1993). Management discussion and analysis: an evaluation of practices in UK and US companies. *Accounting and Business Research*, 23, no. 90, Spring, pp. 123–137.

Cooke, T. E. and Whittaker, J. (1983). Segment reporting – directors' discretion or SSAP? *Accountancy*, 94, no. 1076, April, pp. 77–81.
Cooke, T. E. (1986). *Mergers and Acquisitions* (Blackwell Publishing, Oxford).
Davies, M., Paterson, R. and Wilson, A, (1994). UK GAAP: Generally Accepted Accounting Practices in the United Kingdom (Macmillan, Basingstoke).
Dearing, R. (Sir) (1988). The Making of Accounting Standards: Report of the Review Committee under the Chairmanship of Sir Ron Dearing (The Institute of Chartered Accountants in England and Wales, London).
Department of Trade and Industry (July 1994). Small Company Audit: Implementing Regulations Come into Force in August (Department of Enterprise of the DTI, London).
Department of Trade and Industry (September 1994). Companies in 1993–94: Report for the year ended 31 March 1994, presented pursuant to the Companies Act 1985 Section 729 (HMSO, London).
Devine, C. (1963). The rule of conservatism reexamined. *Journal of Accounting Research*, 1, pp. 127–138.
Edwards, J. R. (ed.) (1990). British Company Legislation and Company Accounts, 1844–1976, Vols. 1 and 2 (Arno Press, New York).
Financial Reporting Review Panel (1992). Press Release No. 9: Progress Report to 30 June 1992 (FRRP, London).
Fisher, L. (1994). Big GAAP/Little GAAP. *Accountancy*, 114, no. 1216, p. 23.
Hoffman, L. (QC) and Arden, M. H. (1983). Joint Legal Opinion on True and Fair View. Addressed to the Accounting Standards Committee, reproduced in *Accountancy*, November, pp. 154–156.
Johnson, L. T. (1994). *Future Events: A Conceptual Study of their Significance for Recognition and Measurement*, FASB (Norwalk, CT).
Leach, R. and Stamp, E. (1981). *British Accounting Standards: The First 10 Years* (Woodhead-Faulkner), London.
Lee, T. (1979). A brief history of company audits: 1840–1940. In T. A. Lee and R. H. Parker (eds.) *The Evolution of Corporate Financial Reporting* (Thomas Nelson and Sons Ltd, Middlesex), pp. 153–163.
Lee, T. (1994). Financial reporting quality labels: The social construction of the audit profession and the expectations gap. Accounting, Auditing & Accountability Journal, 7, no. 2, pp. 30–49.
Lee, T. A. and Parker, R. H. (1979). *The Evolution of Corporate Financial Reporting* (Thomas Nelson and Sons Ltd, Middlesex).
Lorensen, L. (1972). Reporting Foreign Operations of US Companies in US Dollars, Accounting Research Study 12 (AICPA, New York).
Macve, R. (1981). A Conceptual Framework for Financial Accounting and Reporting: The Possibilities for an Agreed Structure, a Report prepared at the request of the Accounting Standards Committee (ICAEW, London).
McMonnies, P. N. (ed.) (1988). *Making Corporate Reports Valuable* (ICAS and Kogan Page, London).
Napier, C. J. (1983). *Accounting for the Cost of Pensions*, ICAEW, London.
Nobes, C. W. (1980). A review of the translation debate. *Accounting and Business Research*, 10, no. 40, Autumn, pp. 411–431.
Ordelheide, D. (1993). True and fair view: A European and a German perspective. *European Accounting Review*, 2, no. 1, pp. 81–90.
Parker, R. H. (1994). Debating true and fair in Australia: An exercise in deharmonization?' *Journal of International Accounting Auditing & Taxation*, 3, no. 1, pp. 41–69.
Power, M. K. (1992). The politics of brand accounting in the United Kingdom. *The European Accounting Review*, 1, no. 1, May, pp. 39–68.
Power, M. K. (1997). *The Audit Society: Rituals of Verification* (Oxford University Press, Oxford).
Schmitthoff, C. M. with Kay, M. and Morse, G. K. (1976). Palmer's Company Law, Volume I, twenty-second edition (Stevens & Sons, London).
Solomons, D. (1989). Guidelines for financial reporting standards, a paper prepared for The Research

Board of the Institute of Chartered Accountants in England and Wales and addressed to the Accounting Standards Committee (ICAEW, London).

Turley, S. (1992). Developments in the structure of financial reporting regulation in the United Kingdom. *The European Accounting Review*, 1, May, pp. 105–122.

Wallace, R. S. O. (1987). Disclosure of Accounting Information in Developing Countries: A Case Study of Nigeria, Ph.D. thesis (University of Exeter, Devon).

Wallace, R. S. O. and Cooke, T. E. (1990). The diagnosis and resolution of emerging issues in corporate disclosure practices. *Accounting and Business Research*, 20, No. 78, Spring, pp. 143–151.

Walton, P. (1993). Introduction: The true and fair view in British accounting. *European Accounting Review*, 2, no. 1, pp. 49–58.

Watts, R. L. and Zimmerman, J. J. (1986). Positive Accounting Theory (Prentice-Hall Inc., Englewood Cliffs, NJ).

Zeff, S. A. (1993). International accounting principles and auditing standards. *The European Accounting Review*, 2, no. 2, September, pp. 403–410.

UNITED KINGDOM
GROUP ACCOUNTS

Paul A. Taylor

Acknowledgements

I would like to thank Pelham Gore, Michael Mumford, John O'Hanlon, Ken Peasnell and Stephen Young from Lancaster University, and Cliff Taylor for help on the way to preparing this contribution; also various members of professional bodies, standard setting bodies and accounting firms who did their best to respond to sometimes obscure technical and interpretive queries. Thanks also to my wife, Trish, for helping me keep a sense of perspective. Any errors remaining are, of course, my own responsibility.

CONTENTS

I Introduction and background

The use of investment holding companies had existed in the United Kingdom since the mid-nineteenth century. However, Edwards and Webb (1984, p. 37) found the earliest use of consolidation was delayed until 1910, and as late as 1933 a full set of consolidated financial statements by the Dunlop Rubber Co. was still newsworthy. Bircher's survey results on the rate of adoption of consolidation in the largest listed UK companies from the late 1930s to the late 1940s are shown in Table I.1. He found the main driving force was the imminence of the first relevant UK legal pronouncement, the Companies Act 1947, with lenders' information demands and the relaxation of wartime taxation restrictions also having some impact (Bircher, 1988, p. 12).

By the 1960s the use of associates was common and was addressed by the first UK accounting standard in 1971. Many major group accounting areas waited until the late 1980s and 1990s to be addressed by accounting standards. Merger [pooling-of-interests] accounting was used in the early 1970s, but then fell into disuse because of doubts over its legality. The enactment of the Fourth and Seventh EU Directives enabled a review of UK company law relating to group accounting generally. The Companies Act 1981 clarified the situation and led, after much debate, to accounting standards on accounting for acquisitions and mergers and on goodwill, in 1985. The former spawned many artificial schemes to enable merger accounting to be used. The latter allowed two radically opposed goodwill accounting approaches to co-exist. The absence of any standard on fair values at acquisition, gave corporate management considerable scope to manipulate reported profits and gearing. Table I.2 summarizes developments.

Table I.1 Percentages of large listed UK companies adopting consolidated reporting

Year ends	Consolidated balance sheet	Consolidated balance sheet and consolidated profit and loss
1938/9	22.5	12.5
1944/5	32.5	17.5
1947/8	74.0	64.0

Source: Birchner (1988), p. 3

Table I.2 Development of the regulation of group accounting

Date	Pronouncement	Comments
1947	Companies Act	Consolidated balance sheet and profit and loss account required.
1967	Companies Act	Introduced minimal segmental disclosures.
1971	SSAP 1	Equity accounting required for 20–50 per cent affiliates and for joint ventures.
1981	Companies Act	Merger relief provisions allow merger accounting to be reconsidered. Fourth EU Directive enacted.
1985	SSAP 22	Gradual amortization of goodwill to profit and loss or immediate write-off to reserves allowed.
1985	SSAP 23	Merger accounting allowed as an option when certain qualifying conditions were met.
1989	Companies Act	Seventh EU Directive enacted into UK law. Criteria for consolidation extensively changed.
1990	Accounting Standards Committee	Series of exposure drafts on group accounting issued prior to disbanding of the ASC, standard on segmental reporting.
1990	Accounting Standards Board	More conceptually-based discussion papers, exposure drafts and standards on business combinations, fair values in acquisition accounting, quasi-subsidiaries, related parties, subsidiaries, associates and joint ventures, goodwill and intangibles.

The use of quasi-subsidiaries was also common. When enacted into UK law in 1989, the Seventh EU Directive's options on defining subsidiaries were used to outlaw many such schemes. The first UK accounting standard on segmental reporting was not issued until 1990. By the mid- to late 1990s a series of standards by a more powerful standard setting body, the Accounting Standards Board, had addressed many of the perceived abuses discussed above. Table I.3 summarizes UK accounting standards relating to group accounting matters at December 1998; Table I.4 summarizes the main statutory requirements.

Recently, considerable debate has taking place concerning the degree to which UK standards should converge with IASC standards. The London Stock Exchange will accept annual reports and accounts prepared in accordance with the issuer's national law and, in all material respects, with UK GAAP, US GAAP or International Accounting Standards. They must be independently audited to equivalent auditing standards – UK, International or US. The Exchange has the discretion to recognize additional categories of standards and may require further information to ensure that the report and accounts give a true and fair view. Company Reporting (December 1998, p. 7) reports a steady increase in the number of UK companies reconciling their financial statements to US GAAP. Table I.5 illustrates major differences between UK and IASC group accounting standards.

Table I.3 UK accounting standards relating to group accounting

Reference	Type of pronouncement
	Accounting Standards Committee (ASC)
SSAP 20	Foreign Currency Translation (1983)
SSAP 25	Segmental Reporting (1990)
	Accounting Standards Board (ASB)
FRS 1	Cash Flow Statements (1991, revised 1996)
FRS 2	Accounting for Subsidiary Undertakings (1992)
FRS 3	Reporting Financial Performance (1993)
FRS 4	Capital Instruments (1993)
FRS 5	Reporting the Substance of Transactions (1994)
FRS 6	Acquisitions and Mergers (1994)
FRS 7	Fair Values in Acquisition Accounting (1994)
FRS 8	Related Party Disclosures (1995)
FRS 9	Associates and Joint Ventures (1997)
FRS 10	Goodwill and Intangible Assets (1997)
FRS 11	Impairment of Fixed Assets and Goodwill (1998)
FRS 12	Provisions, Contingent Liabilities and Contingent Assets (1998)
FRS 13	Derivatives and Other Financial Instrument Disclosures (1998)
FRS 14	Earnings Per Share (1998)
FRSSE	Financial Reporting Standard for Smaller Entities (1998)
UITF Abstract 9	Accounting for Operations in Hyper-Inflationary Economies (1993)
UITF Abstract 15	Disclosure of Substantial Acquisitions (1995)
UITF Abstract 19	Tax on gains and losses on Foreign Currency Borrowings that Hedge an Investment in a Foreign Enterprise (1998)
UITF Abstract 21	Accounting Issues arising from the proposed introduction of the euro (1998)

Table I.4 Main Companies Act 1985 provisions relating to group accounting

Source	Content
S 131	Merger relief provisions.
S 223	Financial year definition – subsidiary year end to coincide with parent in the absence of good reasons.
S 227	Main provisions relating to duty to prepare group accounts.
S 228	Exemption for parent included in larger group from preparing group accounts.
S 229	Criteria for exclusion of subsidiaries from consolidation.
S 230	Exemption from preparing full parent company profit and loss account where group accounts are prepared.
S 231	Disclosures required in group accounts and exemptions granted.
S 243	Appending of subsidiary's accounts where excluded on too dissimilar activities criterion (see S 229).
S 244	Period allowed for laying and delivering reports and accounts.
S 248–249	Exemptions from preparing group accounts for small and medium-sized groups.
S 258–260	Definition of parent and subsidiary undertakings and interpretation provisions.
S 430	Rights of minority interests to be bought out in takeovers.
S 691–698	Items to be delivered to registrar when companies incorporated outside UK establish a place of business in the UK ('oversea companies') and details to be included in such companies' documents.
S 700–703	Delivery of accounts and reports by oversea companies and registration of charges.
S 736	Definition of 'subsidiary' as opposed to 'subsidiary undertaking' for general legal purposes.
S 736A	Detailed interpretation of the meaning of 'voting rights' for determining subsidiaries and subsidiary undertakings.
Schedule 4A	Main provisions relating to the form and content of group accounts including accounting principles: elimination of group transactions, acquisition and merger accounting, minority interests, joint ventures and associated undertakings.
Schedule 5	Disclosure of information about related undertakings including subsidiary undertakings, merger relief, joint ventures, associated undertakings. Separate section for exempt groups.
Schedule 10A	Further interpretation provisions relating to the definition of parent and subsidiary undertakings (S 258) including the meaning of 'rights' and when they are deemed exercisable.
Schedule 24	Penalties and punishments under the Act.

II Objectives, concepts and general principles

1 Status of consolidated statements

The Accounting Standards Board's (ASB) Statement of Principles Exposure Draft in its Chapter 7 entitled 'The Reporting Entity', issued in November 1995, comments on the status of the consolidated financial statements as follows:

> Consolidated financial statements recognize the parent's control of its subsidiaries. Consolidation is a process that aggregates the total assets, liabilities and results of the parent and its subsidiaries (the group) so that the consolidated financial statements present financial information about the group as a single reporting entity. (para. 7.21)

In the UK consolidated financial statements are produced in addition to the parent's and the subsidiaries' individual company financial statements, and the Companies Act 1985 (S 227(3)) states that '(they) shall give a true and fair view . . . so far as concerns members of the (parent) company'. However, the parent's individual company profit and loss account is not required to be published

Table I.5 Major differences between UK and International Accounting Standards

Area	UK / IAS pronouncement	UK accounting treatments which differ from benchmark treatment in International Accounting Standards
Statement of principles	ED	UK exposure draft includes section on 'The Reporting Entity', which addresses group accounting issues.
Business combinations	FRS 7/IAS 22	Fair values at acquisition not based on acquirer's intentions. Provisions/accruals for reorganization costs as a result of acquisition, relating to acquired entity or acquirer, are post-acquisition items.
Goodwill and intangibles	FRS 10/IAS 22	Useful economic life of goodwill/intangibles is period over which goodwill continues to exist, not just originally purchased goodwill/intangibles. This can be indefinite and if so, gradual amortization is not required, just annual impairment reviews.
Joint ventures	FRS 9/IAS 31	Gross equity method used for 'joint ventures'. 'Joint ventures' must carry on a trade or business of their own, rather than just being a part of the participants' trades or businesses. 'Joint arrangements which are not entities' accounted for in participants' own financial statements.
Associates	FRS 9/IAS 28	'Significant influence' requires actual exercise of significant influence rather than just ability to exercise such influence.
Minority interests	FRS 7/IAS 22	Measured at fair value at acquisition – IAS 22 alternative treatment
Segmental reporting	SSAP 25/IAS 14	Business and geographical segments have equivalent disclosure. No primary/secondary structure.
Deferred taxation	SSAP 15	Partial provision approach to deferred tax liabilities.
Cash flow statements	FRS1/IAS 7	Based on cash not cash and cash equivalents. Greater analysis.
Foreign currency translation	SSAP 20 UITF Abstract 9/IAS 21	Cumulative translation adjustment need not be separately disclosed. Such adjustments not be included on disposals. In hyper-inflationary economies the use of a relatively stable currency as the functional currency allowed to proxy for inflation.

provided the company's profit or loss determined in accordance with the Act is disclosed in the notes to the group accounts (S 230). Group accounts also include information about the economic segments of a group.

For most UK legal purposes, the individual company is the relevant entity, for example:

- Distributability of profits: the treatment of consolidated goodwill does not affect realized reserves and hence profits available for distribution.
- Legal liability: creditors' claims are on companies within the group, though in practice the existence of loan cross-guarantees blurs such boundaries.
- Taxation: the basis of computation starts with the individual company profits. Different 'group' reliefs are available; for example, tax losses can be surrendered between group companies, but the term 'group' is defined differently for taxation purposes (see VIII).

2 Group concepts

Group accounts in the United Kingdom include all affiliates within the sphere of influence of the parent undertaking, and the accounting approach adopted broadly depends on long-term participation in benefits and the degree of influence exercisable by the parent undertaking. The key distinctions in current practice are between:

- Unilateral control (subsidiary undertakings) where full consolidation is used.
- Joint control of an entity which carries on its own trade or business (joint ventures) where the gross entity method is used. Other joint arrangements (joint arrangements that are not entities) are accounted for in the accounts of each party to the arrangement for their own share of the assets, liabilities and cash flows.
- Significant influence (associates) where equity accounting must be used.

Subsidiaries: A key conceptual issue is the accounting characterization of the relationship between controlling and non-controlling (minority) interests. Three main consolidation concepts have been proposed which take different stances on this relationship:

- The entity/economic unit concept. The group is viewed as a reporting entity in its own right with different claims on its assets, liability and results. The ASB's 1995 Statement of Principles Exposure Draft (SOPED) considers that under this approach 'the parent's ability to control its subsidiaries is all-important, regardless of the size of the entity that it directs' (para. 7.18). The residual equity of the entity comprises both controlling interests, i.e. parent shareholdings, and non-controlling ones, usually termed minority interests. Non-controlling interests are considered to be 'inside' the group for consolidation valuation adjustment purposes. Therefore, both controlling and non-controlling portions of assets and liabilities are restated to fair values at acquisition and where necessary adjusted to group cost, for example in unrealized intragroup stock profit elimination. Minority interests are therefore based on consolidated carrying values. Some argue that such a perspective strictly followed means that no gains or losses should be recognized on partial disposals of subsidiaries where control is retained since these would be merely a reorganization of ownership claims between controlling and non-controlling interests. Others argue that such a perspective requires goodwill to be imputed to minority interests so they are treated consistently with controlling interests.
- The proprietary/proportional consolidation concept. Only controlling-interest proportions of assets and liabilities of affiliates are consolidated. There is no minority interest. Only the controlling interest's share of intragroup stock profits would therefore be eliminated. The 1995 SOPED argues this view considers that ownership is basis for consolidated financial statements that, 'on a strict proprietary view, the investor's influence ... is irrelevant ..., and all its interests in other entities should be proportionately consolidated to give a picture of the parent's access to benefit from all of its investments'. It notes that such a strict proprietary approach has never been applied. The equity approach used for associates is closely related to this approach.
- The parent concept. Some commentators (e.g. the 1995 SOPED (pp. 119–120), and the Financial Accounting Standards Board 1991 Discussion Memorandum, 'Consolidation Policy and Procedures', pp. 23–24 and 32–33) use this term to describe consolidation approaches combining features of the above two approaches. Consolidation is seen as an amplification of information in the parent's own accounts, with group equity being that of the parent. All the subsidiary's assets and liabilities are included counterbalanced by minority interests, which sits uneasily astride liabilities and equity. Some argue that minority interests should be treated as 'outsiders' and only the controlling interests in acquired assets and liabilities should be restated to fair values at acquisition and only their share of unrealized intragroup stock profits eliminated. Minority interests would therefore be based on original individual company carrying values. Others argue that the concept requires both controlling and non-controlling portions of assets and liabilities to be restated to fair values and adjusted to group cost, thereby measuring minority interests at consolidated carrying values. The 1995 SOPED does use the term 'parent approach', but observes that its preferred approach uses both entity and proprietary perspectives. It favours the full consolidation of assets and liabilities of entities controlled by the parent and keeps separate account of outside equity interests. (paras. 7.19–7.20)

UK practice is based on one variety of the 'parent' concept. Unlike the United States, fair values of identifiable assets at acquisition including the minority proportion are adjusted. Unrealized profits on intragroup transactions are fully eliminated and apportioned according to the ownership proportions of the originating company. Profits or losses on partial disposals of controlling shareholdings in subsidiaries where control is retained are recognized. On the other hand, minority goodwill at acquisition is not. The consolidation concepts above are simplistic and other fundamental concepts such as reliability of asset values and contracting cost effects may explain better why such a seemingly mixed approach is adopted (see Taylor, 1996, Ch. 6).

Associates and Joint Ventures: The ASB's Statement of Principles Exposure Draft (SOPED), Chapter 7, The Reporting Entity (1995), argues that full consolidation is more appropriate than proportional consolidation for subsidiaries, since the parent can deploy all the subsidiary's resources even though its benefits may be limited by the existence of outside equity interests (para. 7.22). It also argues against proportional consolidation and in favour of equity accounting for associates because it considers that the investor unilaterally controls an interest in the investment in its associate rather than a proportional share of the associate's underlying assets and liabilities. A similar argument is mounted for the use of equity accounting approach for investments in joint ventures where venturers share in common the risks and rewards of their joint venture as a separate business (para. 7.40). Proportional consolidation would be appropriate for joint ventures where each venturer has 'its own separate interests in the risks or rewards that derive from its particular share of the fixed assets of the venture, or by its having a distinct share of [its] output..., service, and, in many cases, of its financing'. (para. 7.41)

FRS 9, issued subsequently in 1997, narrows the use of the term 'joint venture' to include only jointly-controlled entities which carry on a trade or business of their *own*, rather than just being a part of the participants' trades or businesses. For these the gross equity approach is required. It uses the term 'joint arrangement that is not an entity' for other joint arrangements, which are to be accounted for in participants' *own* financial statements by including their own assets, liabilities and cash flows. This is not a consolidation procedure but the consolidated statement effect is usually similar to proportional consolidation. A revised ASB statement, *Statement of Principles for Financial Reporting*, issued in December 1999, does not discuss consolidation concepts per se. Consistent with FRS 9, it supports using equity accounting variants for associates and joint ventures. For further discussion see VII.

3 Principles of group accounting

The Companies Act 1985 states that:

> Group accounts shall comply so far as is practicable with the ... (provisions of the Act) ... as if the undertakings included in the consolidation ('the group') were a single company. (Sch 4A, para. 1(1)).

FRS 2 elaborates:

> Consolidated financial statements ... are intended to present financial information about a parent undertaking and its subsidiary undertakings as a single economic entity to show the economic resources controlled by the group, the obligations of the group and the results the group achieves with its resources (para. 1) (and) are required in order to reflect the extended business unit that conducts activities under the control of the parent undertaking (para. 59).

Deduced from this is the need for adjustments to reflect the change in scope of the financial statements from a company to a group perspective, with requirements relating to the elimination of intragroup balances, transactions and unrealized intragroup profits, the adoption of uniform group accounting policies, and usually coterminous year ends (Companies Act 1985, Sch 4A, paras. 3(1) and 6, FRS 2, paras. 39, 40 and 42). Materiality is to be determined from a group perspective, e.g. in deciding whether to adjust assets and liabilities to a group basis (Companies Act, Sch 4A, para. 3 (5)), in deciding the detail of excluded subsidiary disclosures (FRS 2, paras. 31 and 32)), and for group auditors in deciding whether to report in the group accounts, qualifications in subsidiaries' financial statements. The 'true and fair view' is based on 'the undertakings included in the consolidation as a whole, so far as concerns the members of the (parent) company' (Companies Act 1985, S 227(3)).

III Group concept and preparation of group accounts

1 Prior to adoption of the Seventh EU Directive

The Seventh EU Directive changed radically the UK group definition and its consolidation criteria. This was effected through the Companies Act 1989 that amended the Companies Act 1985. The amended Act is still called the Companies Act 1985. Where there is potential for ambiguity, the unamended Act is denoted 'Companies Act 1985u' and the amended one 'Companies Act 1985a'. Otherwise the term 'Companies Act 1985' refers to the amended Act. Prior to revision, the UK group definition was legalistic, based on three criteria for determining the existence of a holding company – subsidiary relationship (Companies Act 1985u, S 736, deriving from the Companies Act 1947):

(i) membership with the ability to control the composition of the board of directors; or
(ii) holding more than half the nominal value of the subsidiary's equity (not votes); or
(iii) the company is a subsidiary of a subsidiary.

'Subsidiaries' only included companies. Criterion (i) recognized one means of control, but (ii) dealt with participation only and if shares had different voting rights might not give voting control. Differential directors' voting rights allowed groups to achieve control with a minority of directors. Such loopholes spawned an off-balance sheet financing 'industry', setting up controlled affiliates which did not meet the definition and so did not need to be consolidated (see Peasnell and Yaansah, 1988, pp. 9–15). Consolidation, though the most common approach, was only one of the options for group accounts allowed by the Companies Acts from 1947 to 1985u. Other options had included, at the directors' discretion, separate accounts for each subsidiary or subconsolidations for parts of the group. SSAP 14, Group Accounts, issued in 1978, banned these and required consolidation.

2 Regulation after adoption of the Seventh EU Directive

The enactment of the Seventh EU Directive into the Companies Act 1985a outlawed all but a handful of off-balance sheet financing schemes. Consolidated financial statements are now the only permitted form of group accounts. The Directive's impact was sharpened by subsequent standards and exposure drafts, especially FRS 2, Accounting for Subsidiary Undertakings (1992) and FRS 5, Reporting the Substance of Transactions (1994).

The UK moved from its previous, legalistic criteria for consolidation to more control-based ones. FRS 2 considers control as the accounting concept that underlies consolidated financial statements. Elements of an economic group concept are contained within an overall control-based framework in the United Kingdom. Unified management is also one criterion, but only if the parent also holds a participating interest ('held long-term for the purpose of securing a contribution to its activities through control or influence').

The concept of control is developed in the 1995 SOPED, Chapter 7, The Reporting Entity (1995). A 'reporting entity' is to be identified by the existence of demand for its statements from potential users with legitimate interests and by the ability of the entity to supply useful financial statements. Ability to supply is predicated on the entity being a 'cohesive economic unit', which usually necessitates a 'unified control structure'. Its boundaries are based on its ability to control (para. 7.2). Control is the highest degree of influence an investor can have over its investee and is the power to direct. Control of assets or entities has two aspects, both of which must be present:

- 'the ability to deploy economic resources, or direct the entities; and
- the ability to ensure that any resulting benefits accrue to itself (with corresponding exposure to losses) and to restrict the access of others to those benefits' (para. 7.10).

Control is dependent on the relationship between entities in practice and is presumed where an entity has by right the ability to control another. It can only be rebutted where it can be shown that another entity is actually exercising control and is deploying the controlled entity's resources on its own behalf and benefiting from them. Apparent independence is not enough for rebuttal because of the possibility of latent or implicit control. The determination of control depends on three interrelated factors: the respective rights held, the inflows and outflows of benefits and exposure to risk (characterized as exposure to variability in outcomes). Whilst powers of veto may form part of rights underlying control, they are unlikely to be sufficient in themselves. Management of an entity is not sufficient to constitute control, benefit is also necessary. Control also exists where an entity uses the above two abilities to set up a continuing arrangement predetermining the operating and financial policies of an entity and where the investor will gain any benefits arising (paras. 7.12–7.15).

In the United Kingdom the means of moving from the group, as legally defined, to the undertakings to be consolidated, is effected as follows:

- Start with the parent and the complete set of all subsidiary undertakings (the group), then
- Exclude non-controlled subsidiary undertakings, and
- Include quasi-subsidiaries (defined by FRS 5, Reporting the Substance of Transactions, issued in April 1994).

Multiple parents. FRS 2 states that where more than one undertaking, is identified as a parent of one subsidiary undertaking, not more than one of those parents can have control (para. 62). Either the others are deemed to have severe long-term restrictions over their rights, a ground for exclusion from consolidation (see IV.3.2), or the control is joint, in which case both 'parents' should treat the subsidiary as a joint venture.

3 Obligation to prepare and file consolidated financial statements

The Companies Act 1985 (S 227 (1) and (2)) requires that, 'if at the end of a financial year a company is a parent company the directors shall, as well as preparing individual accounts for the year, prepare group accounts (taking the form of consolidated accounts)' unless the group is exempt from preparing consolidated accounts at all. In practice such exemption usually occurs because the group is unlisted and is either small/medium-sized, or has an immediate EU parent and its minority shareholders acquiesce. Group accounts must be prepared in consolidated form comprising a consolidated balance sheet, profit and loss account (S 227), statement of total recognized gains and losses (FRS 3, para. 27), and cash flow statement (FRS 1). All subsidiary undertakings must be consolidated unless they individually satisfy exclusion criteria, which are framed mainly in terms of lack of control. The term 'subsidiary undertaking', defined below (see IV.1), includes non-corporate entities such as partnerships and unincorporated associations whether trading or in business for profit or not for profit (S 258 and 259).

The Act refers to parent *companies*, but FRS 2 requires that all 'parent undertakings that prepare consolidated financial statements intended to give a true and fair view of the financial position and profit and loss (or income and expenditure) of their group should prepare such statements in accordance with its requirements' (para. 18). If the parent undertaking uses one of FRS 2's exemptions, described below (see III.4), but prepares individual accounts which show a true and fair view, it should state that its financial statements are of an individual undertaking and not a group, giving the grounds on which it claims exemption (para. 18). Compliance with FRS 2 is not required to the extent that such compliance is not permitted by any statutory framework under which such undertakings report (para. 19).

FRS 2 does not therefore require all parent undertakings of any description to prepare consolidated accounts. Only those which already prepare consolidated statements intended to give a true and fair view must follow its requirements. Thus it does not apply partnerships since the Partnership Act 1890 has no requirement for consolidated accounts which show a true and fair view. Unlike Germany there is no requirement for large groups headed by non-corporate parents to prepare group accounts. Voluntary publication is virtually non-existent. There is little debate in the United Kingdom about the German concept of 'equal ranking' ('horizontal') groups controlled by private persons or a family. These are not 'parent undertakings' in terms of the Companies Act 1985 or FRS 2, and so would not be required to prepare group accounts unless they fall within specific Companies Act or other legal frameworks. FRS 8, Related Party Transactions, issued in October 1995, does however require, for entities which prepare financial statements intended to give a true and fair view of its financial position and profit or loss (or income and expenditure) for a period, details of its controlling parties and of material transactions with its related parties, which include family members of its significant shareholders, key management and directors (see X.3).

One set of companies/groups affected by FRS 2 are so-called 'oversea companies', defined under the Companies Act 1985 as 'a company incorporated elsewhere than in Great Britain which ... establishes a place of business in Great Britain' (S 744). Under the Act (S 700(1)) 'every oversea company shall in respect of each financial year ... prepare the like accounts ... as would be required if the company were formed and registered under the Act'. Such accounts must be delivered to the Registrar of Companies, with a certified English translation if any part is not in English; the period for delivering the accounts is 13 months after the end of the relevant accounting reference period (S 702). Under Statutory Instrument (SI 1990/440), The Oversea Companies (Accounts) (Modifications and Exemptions) Order 1990, the Act's provisions are modified so that an auditor's report and directors' report need not be given. Also not necessary are, for example, details of the basis of the UK tax charge, the amount of the UK and non-UK tax charge, the amount and method of computing turnover, details of subsidiaries and shareholding in the company of more than 10 per cent, and the identity of the ultimate holding company.

Examples of other types of bodies which are subject to particular legislation and are required or recommended to produce consolidated financial statements are given in Table III.1. The Companies Act 1985 filing requirements and time limits for the preparation of group accounts are broadly the same as for individual companies; references to 'accounts' therein include consolidated accounts (see XI).

Charities. SORP 2, Accounting by Charities, issued in October 1995, recommends that a charity with non-charitable subsidiary undertakings should prepare consolidated accounts for itself and its subsidiaries, in addition to separate accounts for the charity. Assets and liabilities of all subsidiaries should be consolidated using the normal line-by-line basis (para. 66). In presenting

Table III.1 Examples of UK bodies publishing consolidated accounts

Name of undertaking	Governing statute/regulation
Companies	Companies Act 1985 – includes special provisions for banking groups (Sch 9) and insurance groups (Sch 9A).
Charities	SORP 2, *Accounting for Charities* issued by The Charities Commission in October 1995. Also Part VI of Charities Act 1993 and The Charities (Accounts and Reports) Regulations 1995.
Pension schemes	SORP 1, *Pension Scheme Accounts*, recommended practice.
Industrial and Provident Societies	Friendly and Industrial and Provident Societies Act 1968 S 13(1).
City and County Councils and local authorities	Code of Practice on Local Authorities in Great Britain – Chartered Institute of Public Finance and Accountancy.
Certain public sector bodies (e.g. British Rail, British Coal, The Post Office, British Nuclear Fuels)	Statutory Instrument by the Secretary of State for Trade and Industry based on the enabling Act setting up the body, e.g. the Coal Industry Act 1971 S 8 and the British Telecommunications Act 1981 S 75(3).
Building Societies	Building Societies Act 1986 S 73(5) and Building Society Accounts regulations 1991.

the results of trading activities where subsidiaries carry out non-charitable trading activities, their results are best presented segregated and summarized on the face of the consolidated Statement of Financial Activities, and Summary Income and Expenditure Account, backed up by full disclosure in the notes. Where subsidiaries carry out charitable trading activities their results are to be consolidated on a line-by-line basis in these statements. Where a charity has a participating interest in an entity and exercises significant influence over its operating and financial policy, then the equity method of accounting should be used. The SORP contains a rebuttable presumption that more than 20 per cent of the voting rights signifies significant influence. Reported information must be sufficiently detailed to distinguish the charity and its subsidiaries. The SORP also recommends further note disclosures not discussed here, which include details of subsidiaries and associates, and notes to be provided relating to subsidiary undertakings if consolidated accounts are not prepared.

Pension schemes. The industry SORP, Financial Reports of Pension Schemes, issued in 1996, deals with the case where pension schemes have investments in subsidiaries, associates and joint ventures. It recommends that investments in subsidiaries should be accounted for according to FRS 2, Accounting for Subsidiary Undertakings, discussed in subsequent sections here. In the context of the SORP the term 'subsidiary' includes both 'subsidiary undertakings' (IV.1) and 'quasi-subsidiaries' (IV.2). The format of such consolidated statements should follow those of the SORP and additional disclosures should be provided by way of segmental information where activities of subsidiaries are both significantly different from those of the scheme and sufficiently significant to require such amplification. Minority interests should be disclosed as a separate item after borrowings in the net assets statement, analyzed between equity and non-equity interests. If activities are conducted through associated undertakings or joint ventures, the SORP states that SSAP 1, Accounting for Associated Companies, should be followed [though it presumably would be better practice now to follow FRS 9, Associates and Joint Ventures (VII.2 and VII.3)]

Other bodies. The group accounting requirements of other bodies shown in Table III.1 are not discussed further here (see Johnson and Patient, 1990). Banking companies and insurance companies are covered by the Companies Act 1985 and its Schedule 9 contains special provisions relating to them.

4 Exemption criteria

These criteria exempt a parent of a group from preparing consolidated financial statements at all, unlike exclusion criteria which are concerned with inclusion or exclusion of individual group companies from consolidated statements (see IV.3). Two main criteria exist for exemption:

- The parent's immediate parent is established in a member state of the European Union, publishes group accounts, and its own minority shareholders acquiesce.
- The group falls within the Act's size exemptions.

The United Kingdom did not incorporate a Seventh EU Directive option to exempt certain financial holding companies. Disclosure requirements are dealt with below (see X). A parent undertaking of a group is also exempt from preparing consolidated financial statements (S 229(5)) if all of its subsidiary undertakings are permitted or required to be excluded from the consolidation under S 229 of the Companies Act 1985 (see IV.3).

4.1 Parent embedded in a larger group

A UK company which is a parent undertaking of a subgroup is exempted from preparing group accounts only if:

- None of its securities are listed on any EU member state stock exchange.
- It has an immediate parent undertaking established under the law of an EU member state, which must own more than 50 per cent of its shares.
- A veto is not exercised by minority holdings of more than half of the remaining shares or 5 per cent of the total shares, by serving notice requesting group accounts for the subgroup to be prepared (S 228(1a)).

The exemption also applies for wholly-owned parents, but this is subsumed in the more general rule described above.

There are further conditions. The embedded parent company must be included in consolidated financial statements for a larger group drawn up to the same date or earlier in the same financial year, by a parent undertaking, not necessarily the immediate parent, established under the law of an EU member state, which are audited and comply with the Seventh EU Directive. These audited consolidated financial statements, which in practice will often be in a different reporting currency from sterling, must be delivered by the UK parent of the subgroup (the 'embedded' parent) to the British Registrar of Companies together with an appropriate certified translation of the text into English if necessary. In its individual accounts, the embedded parent company must disclose the fact of the exemption and, concerning the parent of the larger group drawing up group accounts, its name and location, or if incorporated, its principal place of business (S 228).

In the second criterion above, ownership of 'shares' by the immediate parent undertaking is specified, so the criterion may not apply if the immediate parent exercises control by other means (see IV.1). If the embedded parent has, for example, a Swiss parent which itself has a French parent producing consolidated accounts, this will not qualify for the exemption because the immediate parent is not established in an EU member state. The fact that the embedded parent company must be included in the consolidated financial statements of a larger group normally means that the whole subgroup would have to be included in the higher level consolidation since subsidiary undertakings of a subsidiary of any parent are themselves regarded as subsidiaries of that parent (see IV.1). It seems that the immediate parent could be non-corporate, though unless such an immediate parent were itself part of a larger group drawing up group accounts, it would need to make its accounts publicly available by allowing the embedded parent to file them with the Registrar of Companies. The consolidated financial statements for the 'larger' group for exemption purposes will often not be those of the ultimate group.

The UK position described above, derived from the Seventh EU Directive, is more relaxed about percentage ownership (more than 50 per cent) than is IAS 27 and requires the minority to 'opt out' by use of its power of veto rather than to 'opt in'. According to IAS 27 (para. 27):

> A parent that is a wholly-owned subsidiary, or is virtually wholly-owned (often taken to mean that the parent owns 90 per cent or more of the voting power) need not present consolidated financial statements provided, in the case of one that is virtually wholly-owned, the parent obtains the approval of the owners of the minority interest. Such a parent should disclose the reasons why consolidated financial statements have not been presented together with the bases on which the subsidiaries have been accounted for in its separate financial statements. The name and registered office of its parent that publishes consolidated financial statements should also be disclosed.

4.2 Small and medium-sized groups

Groups have the option not to prepare group accounts provided they do not exceed for two consecutive years, or in the parent's first financial year, the criteria for small or medium-sized groups under the Companies Act 1985 unless they include:

- A public company or other body corporate which has the power under its constitution to offer its shares or debentures to the public.
- An authorized institution under the Banking Act 1987.
- An insurance company to which Part II of the Insurance Companies Act 1982 applies.
- An authorized person under the Financial Services Act 1986 (S 248).

The aggregate criteria for medium-sized groups provide the ceiling for this exemption and are shown in Table III.2. Each set can be met either in net form or gross form. The gross form is before consolidation adjustments, allowing exemption to be claimed without preparing consolidated figures for turnover and balance sheet totals, whereas the net form is after such adjustments.

Table III.2 Criteria for size exemption from producing group accounts (S 249(3))

Aggregate criteria for group	Net amounts	Gross amounts
Turnover	= GBP11.2 million	= GBP13.44 million
Balance sheet total	= GBP5.6 million	= GBP6.72 million
Number of employees	= 250	= 250

The term 'balance sheet total' is defined as, under Format 1 the total of items A to D, and, under Format 2 the aggregate of the amounts shown under the general heading 'Assets' (S 247). The accounts to be used are for the financial year ending on the same date as the parent's financial year or the last financial year ending before this date (S 249). The directors of the parent must ask the auditors to include in their individual company audit report a statement that in the auditors' opinion the company is entitled to the exemption. When a group ceases to meet the requirements, it is allowed a 'grace' year before having to prepare group accounts. IAS 27 has no specific size exemptions. There is no statutory requirement that would enable small or medium-sized parent companies to file abbreviated consolidated financial accounts in the same style as abbreviated company accounts. The group accounts for such companies would have to be disclosed in full unless the group meets the small or medium-sized group exemptions, in which case no group accounts would be required, and any supplementary information disclosed would be at the directors' discretion provided it was disclosed as such.

Small groups that voluntarily choose to prepare consolidated accounts are allowed to follow the Financial Reporting Standard for Smaller Entities (known as the 'FRSSE'). In respect of the consolidated financial statements only, such small groups must follow the relevant full standards, FRS 2, 6 and 7 and, as they apply to consolidated financial statements, FRS 5, 9, 10 and, as directed by FRS 10, FRS 11, which need only be applied in respect of purchased goodwill arising on consolidation. The FRSSE contains simplified requirements where goodwill is purchased as part of an unincorporated entity, and for intangible assets and impairment tests in a non-group context, which are not discussed further here. Where the reporting entity is itself part of a group that prepares publicly available consolidated financial statements, it is entitled to the exemptions in FRS 8 relating to groups and consolidated financial statements, discussed in X.3.2, later.

IV Consolidated group

1 Definition of a subsidiary undertaking

1.1 Overview

The Companies Act 1985 distinguishes between the terms 'subsidiary undertaking', used for accounting purposes only, and 'subsidiary', used for other legal purposes. The distinction was made to attack off-balance sheet accounting schemes whilst disturbing other company law as little as possible.

A subsidiary undertaking is one where the parent undertaking (S 258):

- Holds a majority of voting rights; or
- Is its member and has the right to appoint a majority of its board of directors; or
- Has the right to exercise a dominant interest via its memorandum or articles or through a control contract; or
- Is its member and controls alone a majority of its voting rights by means of an agreement with other shareholders or members; or
- Has a participating interest and actually exercises a dominant interest, or it and the subsidiary undertaking are managed on a unified basis; or
- Is the parent undertaking of undertakings of which any of its subsidiary undertakings are, or are to be treated as, parent undertakings.

The first four, which define the term 'subsidiary', were criteria compulsory to member states in the

Seventh EU Directive. They changed the focus of the previous UK definition from equity held to voting rights held, and also introduced clauses on the use of control contracts and concert party agreements. The addition of the fifth, more economic-based optional criterion in the Seventh EU Directive, when combined with the first four, defines the term 'subsidiary undertaking'. Thus, all subsidiary undertakings are subsidiaries, but not vice versa. Subsidiaries of subsidiaries are subsidiaries, and subsidiary undertakings of subsidiary undertakings are subsidiary undertakings. Further Seventh EU Directive options available to member states such as requiring consolidation where a parent has appointed the majority of the board solely as a result of its voting rights, and where the same personnel are a majority on the boards of both companies throughout the year, were not adopted. Many former off-balance sheet schemes engendering control, for example, through issuing shares with different voting rights, temporary voting rights and distribution rights, or through granting options exercisable at advantageous prices to acquire instant control, are now precluded. IAS 22 bases its consolidation criteria on control and does not include any unified management element. Each criterion is now discussed in turn.

1.2 Majority of voting rights

The definition of voting rights is given in Schedule 10A: 'rights conferred on shareholders in respect of their shares . . . to vote at general meetings . . . on all, or substantially all, matters' and is elaborated:

- To cover undertakings with different legal structures, for example entities without a share capital, or an entity which does not decide matters by voting at general meetings.
- To include rights held by subsidiaries, but not those held by an undertaking in itself.
- To include conditional rights where the conditions currently render the rights operational, or are under the control of the holder, or are only temporarily non-operational.
- To exclude rights held in a fiduciary capacity or merely as security where they are exercised only to preserve the investment's value or in the normal course of a loan-granting business. Consistent with this general approach, rights held by a nominee for the undertaking itself are included.

Unlike in UK regulations, majority voting rights in IAS 27 is a rebuttable presumption.

1.3 Rights to control the board of directors

This is defined as 'the right to appoint or remove . . . directors holding a majority of the voting rights at meetings of the board on all, or substantially all matters' (Sch 10A, para. 3). Davies, Paterson and Wilson (1997, p. 228) point out potential anomalies in 50/50 joint ventures where the chair is rotated, but these are not discussed further here.

1.4 Control by contract or constitution

Though this criterion is common in certain other member states (e.g. Germany), FRS 2 notes that:

> in the UK directors are bound by a common law duty to act in the best interests of their company. For this reason there may, in some cases, be a risk that accepting . . . (such a contractual right of 'dominant influence' to be exercised over them) would be in breach of the above duty. (para. 70)

For the purposes of this criterion 'dominant influence' is the right to give directions with respect to the operating and financial policies of that undertaking which its directors are obliged to comply with whether or not they are for the benefit of that other undertaking (Sch 10A, para. 4(1)). Such a contract must be in writing, authorized by the 'controlled' undertaking's memorandum or articles, and permitted by the law of the country in which it is established. Thus it may be possible for German subsidiaries to be controlled in this way.

1.5 Control by agreement with other members

An actual agreement to control a majority of voting rights is required. The Seventh EU Directive allows member states to introduce more detailed provisions as to the form and content of such agreements, but this has not yet happened in the United Kingdom. Another Seventh EU Directive option where the majority of the board has in fact been appointed by a 'parent' through the exercise of its voting rights and where a large minority might exercise *de facto* control because the rest of the shareholdings are scattered, was not enacted.

1.6 Participating interest and dominant influence or unified management

This criterion caused a radical change to the previous UK definition of a subsidiary and was seen as a broad scope anti-avoidance measure to prohibit off-balance sheet schemes using quasi-subsidiaries. The 'unified management' element introduces aspects of an economic group concept, but requires concomitantly the holding of a participating interest. The terms are explored below.

A participating interest is defined as 'an interest ... held on a long-term basis for the purpose of securing a contribution to (the holding entity's) activities by the exercise of control or influence arising from or related to ... that interest' (S 260). A holding of 20 per cent of shares is presumed to be a participating interest unless rebutted, but smaller holdings can also be participating interests. Convertibles and options are to be included in voting rights even if not yet exercised. Unlike in the United States, there is no exclusion for 'worthless' options or for convertibles that are unlikely to be exercised because of adverse prevailing market circumstances.

Dominant influence must be actually exercised, but the expression 'actually exercises a dominant influence' is left undefined by the Act. Although the term 'dominant interest' is defined in the context of control contracts, the latter 'shall not be read as affecting the construction of the (former) expression ...' (Sch 10A 3). It is FRS 2 therefore that defines the 'actual exercise of dominant influence' as '... the exercise of an influence that achieves the result that the operating and financial policies of the undertaking influenced are set in accordance with the wishes of the holder of the influence and for the holder's benefit whether or not those wishes are explicit. The actual exercise of dominant influence is identified by its effect in practice rather than by the way it is exercised' (para. 7b). It states further that:

- A power of veto will usually need to be held in conjunction with other rights or powers, or be related to day-to-day activities with no similar power being held by unconnected parties.
- Normal commercial relationships *per se* are not sufficient.
- A rare intervention on a critical matter can be sufficient and will be assumed to continue until there is evidence to the contrary.
- Each year the status of the subsidiary should be reassessed (paras. 72 and 73).

Unified management is interpreted by FRS 2 to mean:

> (that) the whole of the operations of the undertakings are integrated and they are managed as a single unit. Unified management does not arise solely because one undertaking manages another ... The operations ... (must be) integrated. (para. 74)

Such a definition would justifiably exclude most pure agency relationships.

2 Quasi-subsidiaries

2.1 Characteristics

The accounting profession has attempted since 1985 to prevent off-balance sheet financing using what has been progressively termed 'non-subsidiary subsidiaries', 'controlled non-subsidiaries' and 'quasi-subsidiaries'. Technical Release 603 (1985) and ED 42, *Accounting for Special Purpose Transactions (1988)*, both proposed that the substance of transactions should be accounted for rather than their legal form. Some claimed that the Companies Act 1985 allowed the legal form of transactions to be recorded with their economic substance disclosed in the notes. The profession felt that economic substance had to be reflected in the accounting treatment. The Companies Act 1989 revision removed any ambiguity, following the profession's interpretation. *Financial Report-*

ing 1990–91 provides examples of actual quasi-subsidiary schemes prior to the changes, including differential voting and directors' voting rights, diamond structures and options (Skerratt and Tonkin, 1991, pp. 146–161). ED 49, *Reflecting the Substance of Transactions in Assets and Liabilities (1990)*, updated ED 42 to incorporate the Companies Act changes which themselves prohibited most extant off-balance sheet schemes. The ASB issued FRS 5, *Reporting the Substance of Transactions (April 1994)*, with very similar requirements.

FRS 5 defines a quasi-subsidiary as

> . . . a company, trust, partnership or other vehicle which, though not fulfilling the definition of a subsidiary, is directly or indirectly controlled by the reporting entity and gives rise to benefits for that entity that are in substance no different from those that would arise were the vehicle a subsidiary. (para. 7)

'Control of another entity' is correspondingly defined as '. . . the ability to direct the financial and operating policies of that entity with a view to gaining economic benefit from those activities' (para. 8). The parent usually only accesses the benefit flows from the vehicle's *net* assets because of, for example, limited liability and the possible existence of liabilities with prior claims over gross assets (para. 96).

Further criteria are proposed to aid identification (paras. 32–4):

(i) The disposition of benefit flows arising from the net assets of the vehicle including the risks inherent in those flows.
(ii) Whether the reporting entity in practice directs the financial and operating policies of the vehicle through
- Ownership or other rights; or
- The ability to prevent others' direction of those policies or from enjoying the benefits arising from the vehicle's net assets (the power of veto is not sufficient *per se* unless its effect is that major policy decisions are taken in accordance with the wishes of the veto holder); or
- Evidence of the actual exercise of such abilities; or
- Predetermining, by contract or otherwise, that the reporting entity gains the benefits from the vehicle's net assets, evidence of which is given by which party is exposed to the risks inherent in them.

Control need not be interventionist and will not be present where a third party has the ability to determine all major policy issues. The example given in FRED 4, precursor of FRS 5, is of an entity's pension fund not being a quasi-subsidiary because fund policy is decided by independent trustees (para. 75).

2.2 Accounting

Quasi-subsidiaries should be accounted for as if subsidiaries. If the parent is not otherwise required to produce group accounts, it should produce pro forma supplementary consolidated accounts of it and its quasi-subsidiary, which should be given equal prominence with its individual accounts (para. 35). If a single item or a single portfolio of similar items, and its finance, are held in a quasi-subsidiary as a means of 'ring fencing' them on a non-recourse basis to the wider group, and meeting FRS 5's 'linked presentation' criteria from the group perspective, linked presentation can be used for the quasi-subsidiary (para. 37). However, similar ring fencing within a subsidiary undertaking itself as legally defined cannot lead to 'linked presentation' as the asset and liability would be deemed assets and liabilities of the group, rather than merely financing arrangements (para. 102).

2.3 Disclosure

FRS 2 effectively penalizes companies using quasi-subsidiaries by requiring the additional disclosure of summarized financial statements of each quasi-subsidiary showing each major balance sheet, profit and loss, statement of total recognized gains and losses and cash flow statement heading for which there is a material item, plus comparatives. Information about similar nature quasi-subsidiaries can be combined (para. 38).

3 Exclusion criteria

3.1 Overview

Under the Companies Act 1985, individual subsidiary undertakings may be excluded from the consolidation and alternative information given if (S 229):

- Severe long-term restrictions exist over the parent's rights.
- The parent's interest is held exclusively with a view to subsequent resale.
- The relevant information cannot be obtained without disproportionate expense or undue delay.
- Subsidiaries are excluded on materiality grounds and the aggregate of such exclusions is itself immaterial.
- The activities of one or more subsidiaries are too different from the rest of the group.

Under the Act the first four are optional and the last is mandatory. FRS 2, *Accounting for Subsidiary* Undertakings, pre-empts these options, prohibiting exclusion because of disproportionate expense or undue delay. 'Severe long-term restrictions' and 'held exclusively for resale' are made compulsory and their scope tightened. It defines 'too different activities' so stringently that it is unlikely to be usable in practice. The combined effect of the Act and FRS 2's pre-emptions makes the UK position in practice close to that in IAS 27. It is no longer possible to exclude subsidiaries on the grounds that the result would be misleading or harmful to the business of the company or any of its subsidiaries. Disclosure requirements are considered later (see X). FRS 5 comments that the only exclusion criterion generally appropriate for quasi-subsidiaries is where they are 'held exclusively for resale'. 'Severe long-term restrictions' will result in a lack of control which will prevent the definition of quasi-subsidiary being met, and as discussed below the 'too dissimilar activities' criterion is defined by FRS 2 to be effectively empty (FRS 5, para. 101).

3.2 Too dissimilar activities

An analogous exemption in earlier Companies Acts was used in the 1980s to exclude finance and insurance subsidiaries from consolidation. Over the gestation period of the Seventh EU Directive, policy makers' attitudes to the consolidation of dissimilar subsidiaries changed considerably, influenced by changes in US practice. The mandatory exclusion in the Directive, a residue from earlier perspectives, became an embarrassment and a potential obstacle to international comparability.

The United Kingdom issued SSAP 25, *Segmental Reporting*, in June 1990, making it less necessary to exclude dissimilar subsidiaries from consolidation. In 1987, SFAS 94, *The Consolidation of All Majority Owned Subsidiaries*, had been issued in the United States, predicated on the pre-existence of a standard on segmental reporting (SFAS 14), and requiring the consolidation of all 'controlled' subsidiaries. It prohibited exclusion because of dissimilarity, just at a time when the Seventh EU Directive was requiring such exclusion!

FRS 2 tackles this dissonance by defining 'dissimilarity' in such an extreme manner as to make the mandatory exclusion effectively unusable, only applying where the activities, 'are so different ... that (the subsidiary's) inclusion would be incompatible with the obligation to give a true and fair view. It is exceptional for such circumstances to arise and it is not possible to identify any particular contrast of activities where the necessary incompatibility with the true and fair view generally occurs' (para. 25). Its predecessor, ED 50, *Consolidated Accounts*, issued in 1990, had proposed that 'special category companies', covered by Schedule 9 of the Companies Act 1985 which required special accounts formats, such as banking and insurance companies, would be dissimilar enough. However, FRS 2 now comments that such companies do not exhibit sufficient difference to warrant exclusion.

Prior to FRS 2's implementation, *Company Reporting* in its December 1990 issue found that of 425 companies examined, nine excluded subsidiaries on the grounds of dissimilar activities, of which eight were financial services subsidiaries (Company Reporting Limited, 1990, p. 6). Later, Skerratt and Tonkin (1994, p. 166) found in a stratified sample of 300 companies that none excluded subsidiary undertakings on these grounds. The economic effects of such definitional changes have not been examined in any depth in the United Kingdom.

Accounting treatment. If a 'too dissimilar' subsidiary were to be excluded, the equity approach should be used in the consolidated statements (Sch 4A, para. 18). FRS 2 requires, in addition, that the separate financial statements of such subsidiaries be appended to the consolidated financial statements. These can be provided in summarized form, provided such undertakings, individually or combined with similar operations, do not comprise more than 20 per cent of any of operating profits, turnover or net assets of the group (para. 31d). FRS 2 requires unrealized intragroup profits to be eliminated on the same basis as if the subsidiary were included in the consolidation (para. 39), i.e. complete elimination. This reflects the fact that the reason for exclusion is not loss of control (see X).

3.3 Severe long-term restrictions

This is a compulsory exclusion under FRS 2 (para. 25). Under the Act (S 229) such restrictions must 'substantially hinder the exercise of the rights of the parent company over the assets or management of (the subsidiary) undertaking ... in the absence of which (the holder of those rights) would not be the parent company'. FRS 2 requires such restrictions to be identified by their effects in practice and should lead to an actual not just threatened loss of control (para. 78a). Disclosure of the restrictions and the consolidation of affected subsidiaries is the preferred option unless such consolidation would be misleading. A common example in practice is that of political restrictions over remittances from overseas countries or over management therein. FRS 2 provides examples:

> Where a subsidiary undertaking is subject to an insolvency procedure in the UK, control of that undertaking may have passed to a designated official (e.g. an administrator, administrative receiver or liquidator) with the effect that severe long-term restrictions are in force. A company voluntary arrangement does not necessarily lead to a loss of control. In some overseas jurisdictions even formal insolvency procedures may not amount to a loss of control. (para. 78)

Skerratt and Tonkin (1994, p. 166) found in a sample of 300 groups that only three excluded subsidiaries on the grounds of lack of control and one because of severe restrictions. IAS 22 defines severe long-term restrictions in terms of significant impairment in ability to transfer funds to the parent, rather than in terms of loss of control. It is difficult to assess the effects of this difference, but there may well be overlap in practice.

Accounting treatment. This depends on the extent of the restrictions. FRS 2 requires that:

- If significant influence can still be exercised, the equity approach is to be used.
- If even significant influence is not exercisable, the investment should be frozen at its carrying amount under 'the equity approach at the date the restrictions came into force or at cost if this is at the acquisition date'. It should be treated as a fixed asset investment, and its carrying amount should be written down for any permanent diminution in value.

When such restrictions are lifted, the amount of unrecognized profits or losses accrued and any write-backs of previous write-downs for permanent diminution in value should be separately disclosed in the consolidated profit and loss account (para. 28). FRS 2 requires that if significant influence is exercisable 'it is important to consider whether it is prudent to record any profits arising from transactions with subsidiaries excluded on these grounds' (para. 83). If even significant influence is not exercisable, unrealized intragroup profits or losses need not be eliminated (para. 83). Where significant influence is exercisable, consistent with the excluded subsidiary being treated as an associate, proportional elimination of unrealized intragroup profits should probably be used.

3.4 Interest held exclusively with a view to resale

This is a compulsory exclusion under FRS 2 (para. 25). It only applies to subsidiary undertakings not previously included in the parent's consolidated financial statements (S 229). FRS 2 further narrows the interpretation of the term (para. 11) to:

(i) ... (where) a purchaser has been identified or is being sought, and (such an interest) is reasonably expected to be disposed of within approximately one year of its date of acquisition; or

(ii) an interest ... acquired as a result of the

enforcement of a security unless the interest has become part of the continuing activities of the group or the holder acts as if it intends the interest to become so.

Criterion (i) is deemed to be satisfied, '... if the sale is not completed within a year of acquisition ... but if, at the date the accounts are signed, the terms of the sale have been agreed and the process of disposing of the interest is substantially complete...' (para. 78a). Skerratt and Tonkin (1994, p. 166) found only three groups disclosing under this criterion. IAS 27 merely requires exclusion where control is intended to be temporary with a view to subsequent disposal in the near future, but gives no timescale (para. (13(a)).

Accounting treatment. The investment must be treated as a current asset at the lower of cost and net realizable value (FRS 2, para. 29). FRS 2 comments that intragroup profits or losses need not be eliminated (para. 83). FRS 3, Reporting Financial Performance (1993), requires the results of *previously* consolidated subsidiaries now held for resale to be included in the continuing operations category in the consolidated profit and loss account until they meet the qualifying criteria for discontinued operations, when their results should be reported in the discontinued operations category (see VI.1.5.1).

4 Remaining group off-balance sheet avenues

The effects of the Companies Act 1984a, FRS 2 and FRS 5 were to change balance sheet ratios because subsidiaries formerly excluded and equity accounted on the grounds of dissimilarity of activities (such as high-geared finance subsidiaries), and quasi-subsidiaries, had to be consolidated. Little systematic empirical work is available in the United Kingdom on the effects of this change beyond a few widely quoted examples (see for example, Smith, 1992, pp. 76–91). Pimm (1990, p. 88) concludes since the change in regime that only if control is temporarily relinquished over the period in which the right to a majority stake in benefits is retained can undertakings now be kept off balance sheet. This could only work where there are congruent commercial interests. If there are any other devices to enable 'arm-twisting', it is likely that the dominant interest provisions of S 258(4) of the Companies Act 1985 would crystallize (see IV.1.6). Pimm cites the example of the so-called diamond structure, where a majority share of benefits is obtained, but the potential quasi-subsidiary is jointly controlled by an independent subsidiary and a third party (see Figure IV.1). He assumes that no extraneous considerations trigger the dominant interest provisions. He gives another example where a controlling stake is potentially held through options that are exercisable only at some future date and considers that consolidation is unnecessary until the earliest exercise date. However, if prior to this the parent also used its muscle as a dominant trading partner to exercise 'control', it could be argued that this could trigger dominant influence criteria.

Figure IV.1 Diamond structure

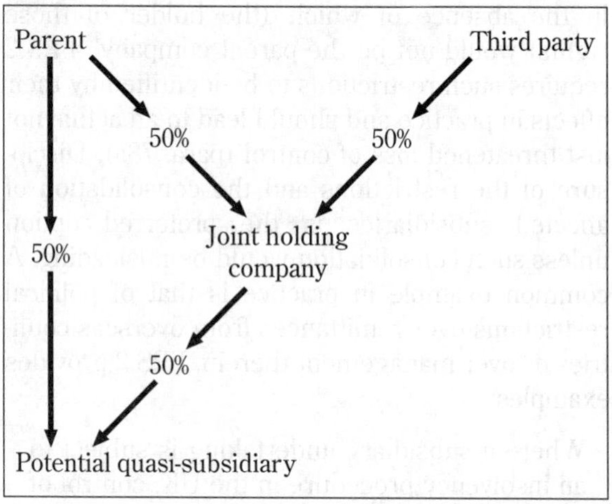

V Uniformity of accounts included

1 Uniformity of formats

Many of the group accounting and consolidation requirements are found in Schedule 4A of the Companies Act 1985, and based on the Seventh EU Directive. Under it group accounts must comply with the individual company provisions in Schedule 4 concerning formats, disclosures and accounting rules, as far as is practicable, as if the consolidated undertakings were a single company (see also IX).

2 Balance sheet date of companies included

If the subsidiary undertakings' balance sheet dates do not end with that of the parent, the Act allows either the subsidiary's accounts to be used for consolidation, if its year end is not more than three months prior to the parent's, or interim accounts made up to the parent's year end (Sch 4A 2(2)). However, FRS 2 requires:

- Wherever practicable, all subsidiary undertakings' financial statements should be based on the parent's financial year end and accounting period (para. 42); or
- If such uniformity is not practicable, interim financial statements to the parent's year end should be used for consolidation purposes; or
- If interim statements are themselves not practicable, a subsidiary's financial statements prepared to a date within three months of the parent's year end can be used, provided any material changes over the intervening period are included by adjustment in the consolidated financial statements (para. 43). Disclosure must be made for such undertakings of their name, accounting date and period, and the reasons for using such a different date or period (para. 44).

Normally such interim accounts are not required to be audited, but if the subsidiary undertaking is material to the consolidated financial statements, the main group auditors may well decide that they cannot form an opinion on the consolidated statements unless such interim statements or consolidation adjustments over the intervening period are audited.

FRS 9 has similar requirements in the case of associates and joint ventures – where there are non-coterminous period accounts such entities should be included on the basis of accounts prepared to the investor's period end or, if not practicable, for a period ending not more than three months prior to this date. However, FRS 9 recognizes that using such financial statements could release restricted, price-sensitive information. Where this is the case, financial statements prepared to a date not more than six months prior to the investor's period end may be used. Material changes between the financial statement date and the investor's period end date should be taken into account by adjustment (para. 31 (d)).

3 Recognition criteria and valuation rules

Uniform accounting policies must be used across all group undertakings for determining consolidated amounts, if necessary by consolidation adjustment. The directors may depart from this if there are special reasons in exceptional cases, when the departure, reasons and effects must be disclosed including the accounting policies used (Companies Act 1985, Sch 4A, para. 3 and FRS 2, para. 41). The Act does not require such uniform policies or account formats to be the same as in the parent's individual accounts. Differences between the accounting rules adopted in the parent's individual accounts and group accounts must be disclosed with reasons (Companies Act 1985, Sch 4A, para. 4).

4 Translation of foreign currency financial statements

The consolidation of foreign currency financial statements can be viewed as a three-stage process:

(i) Translation of foreign currency transactions within each foreign subsidiary's financial statements.
(ii) Translation of these financial statements into the reporting currency.
(iii) Consolidation of the translated and reporting currency financial statements.

Translation of transactions is dealt with elsewhere in United Kingdom – Individual Accounts VII.3, including accounting for Derivatives and Other Financial Instruments and other hedging matters. Here only steps (ii) and (iii) are considered.

4.1 Background

The main UK approach for translating foreign currency financial statements has traditionally been the closing rate approach, with less common use of the current non-current approach disappearing by the early 1970s (Nobes, 1980, pp. 424–425). The temporal approach was used by a small minority of UK multinationals until SFAS 52 was issued in the United States, but not subsequently. SSAP 20, Foreign Currency Translation, issued in 1983, broadly harmonized UK with US practice in SFAS 52 and international standards in IAS 21, requiring the closing rate approach unless the trade of the foreign enterprise is more dependent on the economic environment of the investing company's currency than its own reporting currency, in which case the temporal approach should be used (paras. 52 and 54). Further discussion of the debate over approach is contained for example in Nobes (1980, pp. 424–429) or Taylor (1996, pp. 324–328). The two approaches are now examined before SSAP 20's criteria for deciding the circumstances under which each is appropriate (see V.4.4).

4.2 Closing rate approach

4.2.1 Conceptual underpinnings

Under the closing rate approach all balance sheet asset and liability balances are translated at the exchange rate ruling at the balance sheet date. One justification for the approach is that it preserves the relationships in the original foreign currency financial statements on translation into sterling. Another is the net investment rationale, under which the group is viewed as a series of investments in autonomous subsidiaries transacting in local currencies, rather than directly in foreign assets and liabilities. IAS 21 terms such arrangements 'foreign entities'. The components underlying the net investment, the individual assets and liabilities, are deemed to hedge each other and so are all translated at the same rate.

4.2.2 Translation of foreign currency financial statements

Balance sheet. The exchange rate ruling at the balance sheet date is used to translate asset and liability balances (para. 16) (see Table V.1).

Profit and loss account. SSAP 20 allows all profit and loss items to be translated either at the closing rate (more consistent with the preservation of relationships justification) or the average rate for the period (more consistent with the net investment rationale), the latter being calculated by the most appropriate method for the circumstances of the foreign enterprise (para. 54; see

Table V.1 Rates used for translating balance sheet items (SSAP 20)

Balance sheet item	Closing rate	Temporal
Cash, current debtors and current creditors	C	C
Stocks and short-term investments		
• at cost	C	H
• at net realizable value	C	C[1]
Fixed assets and fixed asset investments		
• at cost	C	H
• revalued or written down for permanent diminution in value	C	C[1]
Long-term debtors, creditors and loans	C	C
Equity	Residual[2]	Residual[2]

Notes: 1 The rate used under the temporal approach would be the exchange rate ruling at the date of valuation.
2 See V.4.6.1 on how this total translated equity is to be split into pre- and post-acquisition amounts for the determination of goodwill and consolidated reserves.

Table V.2). No definitive method is prescribed, but 'the use of a weighting procedure will in most cases be desirable' based on a consideration of 'the company's internal accounting procedures and the extent of seasonal trade variations' (para. 18). IAS 21 requires that transaction rates are used, but in practice this is very similar to the average rate discussed above. *Company Reporting* (April 1998, p. 4), based on a sample of over 500 UK groups, reports a steady increase in those using the average rate from 57 per cent in 1991 to 75 per cent in 1998. Over the same period the use of the closing rate declines from 32 per cent to 18 per cent.

Cash flow statement. FRS 1, *Cash Flow Statements (1996)*, requires the same basis to be used in translating foreign subsidiaries' cash flow statements as for translating the subsidiaries' results in the profit and loss account of the reporting entity (para. 41). This is not the same as IAS 7, which requires translation using exchange rates at the dates of the cash flows or a weighted average approximating to those. FRS 1, however, does make an exception for the translation of separately identifiable intragroup cash flows, where the actual rate or an approximation may be used to ensure cancellation on consolidation. For such intragroup cash flows, FRS 1 states that if the rate used is not an actual rate, exchange differences on cancellation should be included as the effect of exchange rate movements as part of FRS 1's reconciliation to net debt (para. 41).

Georgiou concludes that cash flow translation under FRS 1's requirements is technically much simpler when the closing rate is used for the consolidated profit and loss account, and thus for the cash flow statement. Thereby, for example, the translated cash payment for a fixed asset will be the same as its translated amount capitalized in the consolidated balance sheet. The translated cash flow statement can be derived from consolidated balance sheet changes. Use of the average rate for foreign subsidiaries' profit and loss translation would not allow such derivation of translated cash flow statements from balance sheet changes and would necessitate subsidiaries providing local currency cash flow statements to be translated and then consolidated directly (see Georgiou, 1993, pp. 235–236). Georgiou feels that using the closing rate to translate cash flows 'will indiscriminately distort the actual cash flows of a subsidiary' (p. 230). However, those who support the closing rate approach on the grounds that relationships between local currency amounts are preserved on translation would see only an advantage.

4.2.3 Treatment of translation differences

Exchange differences arise from the retranslation of the opening balance sheet amounts to the exchange rate at the closing balance sheet date, and in addition, if the average rate is used for profit and loss account translation, a further source of such differences arises. Under the closing rate approach both should be recorded as a movement on reserves (paras. 53 and 54), analogous to 'revaluations' of fixed asset investments.

Table V.2 Translating profit and loss accounts (SSAP 20)

Translation approach	Rates used	Disposition of financial statement exchange differences
Closing rate	All items at closing rate, or All items at average rate	Direct to reserves
Temporal	At transaction rate, items corresponding to:	
	• Current dated balances actual or average rate	Profit and loss account
	• Historical dated balances original transaction (historical) rate	

Unlike in IAS 21, such differences do not have to be taken to a separate disclosed cumulative translation adjustment reserve. FRS 3, *Reporting Financial Performance*, implies in its examples that such gains and losses must be reported in the Statement of Total Recognized Gains and Losses (see VI.4). . . .UITF Abstract 21, *Accounting issues arising from the proposed introduction of the Euro*, issued in 1998, addresses the issue of what happens to cumulative exchange differences previously taken direct to reserves under the closing rate approach where the currencies of the both parent and foreign subsidiaries become tied to the euro. After this point no future exchange gains or losses can arise. Consistent with FRS 3, the Abstract requires such differences to remain in reserves rather than being transferred to the profit and loss account.

Disposals. SSAP 20 does not cover the treatment of cumulative financial statement translation gains and losses on the disposal of a foreign enterprise. Dominant UK practice has always been that accumulated translation differences under the closing rate approach have not been transferred from reserves and recognized in the profit and loss account on disposal and so are not used to determine the gain or loss (see Pereira, Paterson and Wilson, 1992, p. 122). This has been confirmed by FRS 3's stance that gains and losses, once reported in the Statement of Total Recognized Gains and Losses, should not subsequently be reported in the consolidated profit and loss account. Thus in the United Kingdom such closing rate translation differences never go through the consolidated profit and loss account. In this, the United Kingdom differs from IAS 21 and SFAS 52. In its 1995 Statement of Principles Exposure Draft, the ASB rejects the view that the profit and loss account should show realized gains and losses and the statement of total recognized gains and losses unrealized ones (para. 6.24). A general discussion of disposals of subsidiaries is contained in VI.1.5.3 and VI.4.

4.3 Temporal approach

4.3.1 Conceptual underpinnings

Under the temporal approach, the foreign enterprise is viewed as a direct extension of the parent – both are characterized as being part of a single unified entity. The temporal approach achieves the same result as if transactions of foreign enterprises are accounted for as though they were direct transactions of the parent – SFAS 52 calls this the use of 'a single functional currency'.

4.3.2 Translation of foreign currency financial statements

Balance sheets. Translation rates are shown in Table V.1. Items are translated according to the measurement basis in the local currency financial statements. Accordingly, items stated at historical costs are translated at their original transaction rates, impaired items at the rate ruling at the date of the impairment, and items stated at current values at the rate ruling at the current balance sheet date (paras. 4–6).

Profit and loss account. Each item should be translated at the rate ruling on the date when the transaction occurred – depreciation will be translated at the rate used for translating the corresponding fixed asset in the balance sheet, and cost of goods sold will be based on a historical transaction rate, a composite of the transaction rates for purchases and balance sheet rates for opening and closing stocks (see Table V.2). Other items are translated at rates corresponding to when the transactions took place during the year or approximating averages. IAS 21 requires that transaction rates be used.

Cash flow statement. FRS 1, *Cash Flow Statements (1996)* requires the same rate to be used as for translating the results of such foreign subsidiaries in the reporting entity's profit and loss account (para. 41). It is not really clear what this means where the temporal approach is used. SSAP 20 comments 'the mechanics of the [temporal] method are identical with those used in preparing the accounts of an individual company' (para. 22); the correct approach would therefore seem to be to use the exchange rates at the date of the cash flows. However, FRS 1's requirement seems to imply transaction rates at the dates of purchases, sales and expenses, rather than payments and receipts.

4.3.3 Treatment of translation differences

Because the translation differences are treated as if they were transaction gains and losses of the parent, they are recorded in the consolidated profit and loss account with other transaction gains and losses. Therefore temporal translation differences will not be recorded separately in the statement of total recognized gains and losses, but merely as part of the profit or loss for the year (see VI.4). Under this approach no adjustments are necessary on disposals of foreign enterprises.

4.4 Choice of translation method

The choice of translation approach is regulated by SSAP 20 and its criteria apply equally to foreign branches (para. 25). It cites foreign currency financial statement translation objectives as to

> produce results ... generally compatible with the effects of rate changes on a company's cash flows and equity ... Consolidated statements should reflect the financial results and relationships as measured in the foreign currency financial statements prior to translation. (para. 2)

Such objective is more consistent with the closing rate approach, so it is not surprising that SSAP 20 requires that 'in most circumstances the closing rate/net investment method should be used' (para. 14) unless

> the affairs of a foreign enterprise are so closely interlinked with those of the investing company that its results may be regarded as being more dependent on the economic environment of the investing company's currency than on that of its own reporting currency (para. 22),

in which case the temporal approach should be used. The following factors, no single one by itself being sufficient, must be taken into account in determining the choice of approach:

- The extent to which the cash flows of the enterprise have a direct impact on the investing company.
- The extent to which the functioning of the enterprise is dependent directly upon the investing company.
- The currency in which the majority of trading transactions are denominated.
- The major currency to which the operations are exposed in its financing structure (para. 24).

SSAP 20 gives illustrative examples where the temporal method could be appropriate:

- Selling agencies where stock is sent out from the investing company and sale proceeds are remitted.
- Raw material producers or intermediate component manufacturers whose products are shipped to and incorporated by the parent in its own products.
- Haven-type companies located abroad for tax, exchange control or other financial reasons.

The accounting treatments under each approach are summarized above (see Tables V.1 and V.2). There is little evidence in surveys of the use of the temporal approach in practice. *Company Reporting* (April 1998, p. 6) reports the use of the temporal approach is rare – rare enough to draw attention to a single company actually using it! SSAP 20's requirements on choice of approach are very similar to those contained in IAS 21.

4.5 Other matters

4.5.1 The cover concept: net investments hedged by foreign borrowings

Under the closing rate approach, SSAP 20 allows transaction gains and losses on currency borrowings of the parent used to finance group net investments in foreign enterprises, or to hedge exchange risks with similar existing investments (para. 30) to be offset against gains and losses on the translation of such net investments, subject to certain qualifying conditions (para. 31). If this were not the case, transaction gains or losses on translating the parent's loan in its individual company statements would have been shown in the profit and loss account, but financial statement translation losses or gains on [the balances underlying] the net investment would have been taken direct to consolidated reserves and shown in the statement of total recognized gains and losses. The 'gain' on one aspect of this linked transaction would have been accounted for in a different place to the 'loss' on the other, despite the intention to hedge. UITF Abstract 19, issued in February 1998, requires that where exchange differences on foreign currency borrowings used to finance or hedge equity investments in foreign enterprises are taken to reserves and reported in the statement of total recognized gains and losses, tax charges or credits directly and solely attributable to such exchange differences should also be taken to reserves and reported in that statement (para. 7).

The qualifying conditions are (para. 31 and UITF Abstract 19 para. 8):

- The use of the closing rate to translate the foreign enterprise's statements is justified.
- In any accounting period, the exchange differences on translating net investments, after taking into account any tax charge or credit directly and solely attributable to the borrowings, provide a ceiling for such offsetting.
- The currency borrowings used in the offsetting process should not exceed in aggregate the total amount of cash in after-tax terms that the net investments are expected to generate, whether from profits or otherwise.
- The treatment adopted must be applied consistently through time.

The meaning of 'the total amount of cash that the net investments are expected to generate' is unclear. Davies, Paterson and Wilson (1998, p. 554) consider it equal to the immediate sale value of the investment – but on prudence grounds suggest its carrying value of the investment is more appropriate. They prefer such an approach to attempting to forecast future dividend streams and ultimate disposal proceeds for the investment, on pragmatic grounds.

The Polly Peck affair revealed a loophole in SSAP 20. The parent borrowed in 'hard' low-interest rate currencies and used foreign subsidiaries to lend in depreciating high-interest rate currencies. In economic terms the 'true' interest charge on lending is the actual interest charge net of the decline in the translated debt value. However, under SSAP 20 the decline on the subsidiary's long-term receivable [part of the financial statement translation loss] is taken direct to reserves. Therefore, in the profit and loss account, the high translated interest received [on the subsidiaries' lending] for Polly Peck exceeded the lower translated interest paid [on the parent's loan transaction]. Because the huge translation loss was diverted to reserves, Polly Peck showed net interest receivable at a time when its net monetary liabilities were increasing. See Smith (1992, pp. 169–180), or Gwilliam and Russell (1991, pp. 25–26).

4.5.2 Subsidiaries in hyper-inflationary economies

The closing rate approach also suffers acute problems when the inflation rate in overseas economies is very high, leading to the 'disappearing assets syndrome' – rapidly declining exchange rates which reflect the inflationary spiral are applied to historical costs which do not. Internationally, two solutions have been proposed to counteract this problem which arises under historical cost, but not current value or current purchasing power accounting:

- SSAP 20 requires that 'the local currency financial statements should be adjusted where possible to reflect current price levels before the translation process is undertaken' (para. 26). No guidance is given whether this should be for general or specific price levels.
- In the United States, SFAS 52 requires the use of the temporal approach in such circumstances. However, this is inconsistent with the net investment rationale.

UITF Abstract 9, *Accounting for Operations in Hyper inflationary Economies*, recognizes the difficulties inherent in defining a hyper-inflationary economy (para. 3), but requires adjustment 'where the cumulative exchange rate over three years is approaching, or exceeds, 100 per cent and the operations in the hyper-inflationary economy are material' (para. 5), or if the inflationary distortions affect a true and fair view. This is consistent with a maintained inflation rate of about 26 per cent per annum.

UITF Abstract 9 tries to reconcile the two above approaches by interpreting SSAP 20 as allowing:

- Adjustment to current price levels before translation: the abstract interprets this in a general price level sense, requiring the gain or loss on net monetary items to be taken to the profit and loss account; or

- The re-measurement of the subsidiary's balances and flows into a relatively stable currency, which can be different from the parent's, before translation into the reporting currency. This pseudo-temporal approach is similar to the US solution.

It links the two approaches, rationalizing 'the movement between the original currency of record and the stable currency ... as a proxy for an inflation index' (para. 6). The method used must be disclosed. Other approaches can be used but the reasons for using them must be stated. IAS 29 supports the first approach, specifying that a general price level adjustment should be made using a local inflation index, the resulting gain or loss on net monetary items being taken to profit and loss. The Abstract's conceptual linking of the allowed alternatives seems rather forced.

4.5.3 Calculation of exchange rates

Where multiple exchange rates exist, SSAP 20 says little about which exchange rate to use to translate foreign currency statements except that it defines the closing rate as a spot rate, the mean of buying and selling rates at the close of business on the relevant date (para. 41). In the United States, SFAS 52 requires, where there is more than one type of exchange rate, that the dividend remittance rate should be used except in unusual circumstances, as it will be the rate at which the net investment will be realized as cash flows over time (para. 138). SSAP 20 does not address this issue nor does IAS 21.

Neither does SSAP 20 discuss the basis for calculating the average rate in the translation of profit and loss items, although it does say 'the use of a weighting procedure will in most cases be desirable' based on a consideration of 'the company's internal accounting procedures and the extent of seasonal trade variations' (para. 18). Under the temporal approach and the net investment rationale for the closing rate, the theoretically correct approach would be to translate transactions at the actual rates when they occurred, the use of averages being seen as an approximation. SFAS 52 allows a weighted-average, where the level of detail is such that its effects are not likely to be materially different from actual rates. From published accounts it is difficult to discern the averaging approach used by most UK groups in practice.

4.6 Consolidation of translated statements

4.6.1 Goodwill

A UK group's share in the foreign subsidiary's equity at acquisition is determined after completely restating the foreign subsidiary's identifiable assets and liabilities to fair values at acquisition by consolidation adjustment (see VI.1.3). The parent's share of such fair value adjusted equity is translated into sterling at the rate ruling at acquisition, and is then cancelled against the parent's investment to determine consolidated goodwill at the acquisition date.

4.6.1.1 Closing rate approach

Under the closing rate approach, if goodwill is capitalized and gradually amortized issues arise then relating to its translation in subsequent financial statements. Three alternatives are possible. SSAP 20 does not clearly specify which is to be used and IAS 21 allows only the first or second:

(i) Historical rate translation: pre-acquisition equity is translated in subsequent statements at the date of acquisition exchange rate and cancelled against the sterling cost investment at that date. Thus the cost of goodwill is fixed at the historical rate for future periods. IAS 21 describes this as treating goodwill as a reporting currency balance.

(ii) Closing rate translation: the sterling investment is treated as if a foreign currency amount and it and the pre-acquisition equity are retranslated at the current closing rate each period. Thus goodwill cost is translated at the closing rate. IAS 21 describes this as treating goodwill as a reporting currency balance. Exchange differences on goodwill translation would be taken direct to reserves. This is also the US position in SFAS 52.

(iii) Mixed translation: pre-acquisition equity is retranslated in subsequent financial statements at the current closing rate each year, but cancelled against the sterling investment cost at acquisition.

EXAMPLE

Goodbuy Plc purchases a 60 per cent stake in So-Long Inc. on 1 January 1995 for GBP105,000 million. Goodbuy Plc amortizes consolidated goodwill over a ten-year period. It uses the closing rate approach with the average rate for translating the profit and loss account. The current year end is 1 January 1999. Capital and Reserves of So-long Inc. (in $ million) at the date of acquisition were as follows.

	So-long Inc ($ million)
Equity/capital	100,000
Profit & loss	120,000
	220,000

Exchange rates over the period included the following (in the United Kingdom these are indirect quotes, i.e. US dollars to the GBP sterling):

1 January 1995	1.5526 [Implicit direct quote = 1/1.5526 = 0.6441 GBP sterling to the dollar]
Average 1995–96	1.5380
1 January 1998	1.6454
Average 1998–99	1.6593
1 January 1999	1.6638

Balance sheet and profit and loss translation of goodwill in GBP million:

1. As a reporting currency balance / historical rate:

Goodwill cost at acquisition $= 105{,}000 - 60\% \times \frac{220{,}000}{1.5526}$ = GBP19,981 million

Amortization 1995–96 $= 19{,}981 / 10$ = GBP1,998 million

Amortization 1998–99 $= 19{,}981 / 10$ = GBP1,998 million

Net book amount at 1 January 1999 $= \frac{6}{10} \times 19{,}981$ = GBP11,989 million

2. As a foreign currency balance / closing rate:

Goodwill cost at acquisition $= 105{,}000 \times 1.5526 - 60\% \times 220{,}000 = \$31{,}023$ million

$= \frac{31{,}023}{1.5526}$ = GBP19,981 million

Amortization 1995–96 $= \frac{31{,}023}{1.5380} \times \frac{1}{10}$ = GBP2,017 million

Amortization 1998–99 $= \frac{31{,}023}{1.6593} \times \frac{1}{10}$ = GBP1,870 million

Net book amount at 1 January 1999 $= \frac{31{,}023}{1.6638} \times \frac{6}{10}$ = GBP11,188 million

3. As a mixed balance

Goodwill cost at acquisition $= 105{,}000 - 60\% \times \frac{220{,}000}{1.5526}$ = GBP19,981 million

Amortization 1995–96 $= [105{,}000 - 60\% \times \frac{220{,}000}{1.5380}] \times \frac{1}{10}$ = GBP1,917 million

Amortization 1998–99 $= [105{,}000 - 60\% \times \frac{220{,}000}{1.6593}] \times \frac{1}{10}$ = GBP2,545 million

Net book amount at 1 January 1999 $= [105{,}000 - 60\% \times \frac{220{,}000}{1.6638}] \times \frac{6}{10}$ = GBP15,398 million

Davies, Paterson and Wilson (1997, p. 534) interpret SSAP 20 as possibly requiring historical rate translation based on a painstaking analysis of its wording. Their interpretation is persuasive, but not conclusive. SFAS 52 requires the closing rate translation alternative (para. 101) and Davies, Paterson and Wilson prefer this because 'the value of the foreign company as a whole is likely to be based on the expected future earnings stream expressed in the foreign currency and the goodwill relates to a business which operates in the economic environment of that currency' (p. 534). This author prefers the closing rate translation approach on the grounds of consistency with the net investment rationale for the closing rate approach. This is because in terms of mathematical relationship (see Taylor 1996, pp. 85–87) it can be shown that a net investment necessarily must equal in component terms the subsidiary's attributable net assets *and* its unamortized goodwill. Further it can also be demonstrated that only where goodwill is translated at the closing rate is total foreign currency exposure in component terms equal to exposure of the subsidiary's attributable net assets *and* its goodwill. When goodwill is translated at the historical rate, exposure is limited only to its attributable net assets excluding goodwill.

Chopping and Skerratt argue for the historical rate translation, so that goodwill 'reflects the value of the business over that of its separable assets as assessed at a particular moment in time and recorded in the reporting currency ... [and that retranslation] ... seems to take account of subsequent changes in a way that is inconsistent with the logic of SSAP 22, Accounting for Goodwill ...' (Chopping and Skerratt, 1998, p. 241). The mixed rate approach reflects that the investment is a sterling balance and foreign equity a foreign currency balance, but results in goodwill cost and amortization increasing when the overseas currency is weakening, which seems counter-intuitive. It also results in more volatile charges. Such disagreements reveal why there are critics of the closing rate approach (see Taylor, 1996, pp. 324–328).

4.6.1.2 Temporal approach

If acquisition [purchase] accounting is used and a dependent-type subsidiary is acquired, acquisition date exchange rates should be used to translate all the subsidiary's fair valued balances at acquisition as their group costs are established at that date. Translated equity at acquisition is the residual balance at that date and goodwill cost is determined by deducting this balance from the sterling investment cost. Goodwill original cost is fixed in sterling and not retranslated in subsequent financial statements.

4.6.2 Intragroup balances, transactions and profits

These are dealt with in VI.2.1.2, VI.2.1.3 and VI.2.2.2.

4.7 Disclosures

SSAP 20's disclosures (see X) focus on foreign currency borrowings, net of deposits, showing separately the amounts offset to reserves under the various cover provisions, and the net charge or credit to profit and loss. Also required is the method and the net movement on reserves arising from foreign exchange differences (para. 60). A notable omission is the profit and loss disclosure of exchange differences (a) on transactions, and (b) on using the temporal approach for translating financial statements.

IAS 21, The Effects of Changes in Foreign Exchange Rates (revised in 1993) covers a number of SSAP 20's omissions and requires disclosure of:

- The amount of exchange differences dealt with in the net profit or loss.
- The (cumulative) exchange differences classified as equity, disclosed separately.
- A reconciliation of the beginning and ending balances in the exchange differences.
- Specified details relating to any changes in the classification of any significant foreign operations, including the impact of the restatement of shareholders' equity and prior periods profit and loss figures. (paras. 42–4)

IAS 21 also encourages 'disclosure ... of an enterprise's foreign currency risk management policy' (para. 47). The ASB Statement (of best practice), Operating and Financial Review (1993), recommends narrative commentary on the management of exchange rate risk (para. 26) and on restrictions on the ability to transfer funds from one part of a group to meet another's obligations, where these form or forseeably could form a significant restraint, e.g. exchange controls (para. 34).

FRS 13, Derivatives and Other Financial Instruments: Disclosures, issued in September 1998, requires an analysis of the net amount of monetary assets and liabilities at the balance sheet date showing the amount denominated in each currency analyzed by reference to the functional currencies of the operations involved. The term 'functional currency' is defined as 'the currency of the primary economic environment in which an entity operates and generates net cash flows' (para. 2). Assets and liabilities not denominated in the functional currency of the entity that uses them (for example its foreign currency transactions or its foreign operations integral to its operations) will need to be remeasured into that functional currency. They are recognized by the fact that such gains and losses will be shown in its profit and loss account. FRS 13 notes that the disclosures do not focus on gains and losses arising on the financial statement translation of net investments/foreign entities in which exchange gains or losses are taken direct to reserves. Borrowings used to finance or provide a hedge against such net investments/foreign entities (whose exchange gains and losses are included in the statement of total recognized gains and losses) should also not be included in the analysis. However, other derivatives including currency swaps, forward contracts and other instruments that contribute to the matching of foreign currency exposures should be taken into account, or a summary of the main effect of any instruments not so taken into account should be given (para. 34). Special disclosures relating to Banking and Similar Institutions, and Financial Institutions are not discussed here. Other aspects of accounting for derivatives and other financial instruments and disclosures is considered in United Kingdom – Individual Accounts VI.6.4.

VI Full consolidation

1 Consolidation of capital

1.1 Merger accounting

1.1.1 Development of regulation

Merger accounting, also known as 'pooling-of-interests', is not widely used in the United Kingdom. The ASB's Draft Statement of Principles, Chapter 7, The Reporting Entity, issued in November 1995, distinguishes between the situation of a group 'when one reporting entity acquires another by obtaining control over its operating and financial policies or disposes of one of its entities by ceding control' (an acquisition), and a group where 'a new economic unit is formed by entities merging together to form a new group ... where no party to the combination in substance obtains control over any other or is otherwise seen to be dominant' (a merger). In the latter case continuing interests are in the whole entity and not just part of it (paras 7.28 and 7.29). IAS 22 uses the terms 'acquisition' and 'uniting of interests' for these two situations. There is no term in British law equivalent to the term 'legal merger' in a continental law sense. The situation where 100 per cent of the shares in a company are acquired and that company is wound up and its assets transferred to the parent is not legally termed a 'merger' in the United Kingdom, and would merely be treated in accounting terms in the consolidated statements as a purchase of its individual assets and liabilities by the acquirer.

Towards the end of the 1960s a few major UK companies used merger accounting and the Accounting Standards Committee issued ED 3, *Accounting for Acquisitions and Mergers* in 1971. This never became a standard because of doubts over the legality of the merger accounting approach under then extant company law, and this was confirmed by a tax case in 1981. Changes to UK company law that year during the implementation of the Fourth EU Directive enabled it to be contemplated again and SSAP 23, *Accounting for Acquisitions and Mergers*, issued in 1985, allowed it as an option if certain criteria were satisfied, otherwise acquisition [purchase] accounting was to be used. These emphasized the form not the substance of the combination transaction and schemes were invented to allow acquisition-type combinations to meet SSAP 23's merger accounting criteria. However, the wide scope then allowed in acquisition accounting, e.g. over goodwill and the fair values at acquisition (see VI.1.3 and VI.1.4) made that approach more attractive than merger accounting. Therefore few qualifying groups opted for it.

In 1990 the ASC issued exposure drafts tackling the connected issues of: redefining merger accounting (ED 48), removing the options in accounting for goodwill (ED 47) and other intangibles (ED 52), and standardizing the treatment of fair values at acquisition (ED 53). Subsequently the ASB has developed standards in these areas. ED 48, Accounting for Acquisitions and Mergers, defined when merger accounting must be used, using a more substance-based definition. FRS 6, Acquisitions and Mergers, issued by the ASB in 1994 followed it quite closely. The ASB had contemplated the consequences of abolishing merger accounting (in its discussions in FRED 6) but FRS 6 decided against this. FRS 6 requires the choice and presentation of accounting approach to depend on the substance rather than the form of the transaction through which the combination is effected. To simplify discussion, unless otherwise stated in what follows it is assumed that one company, the parent, obtains a controlling interest in the shares of another, the subsidiary undertaking, though the principles discussed apply equally to other structures.

The ASB's 1998 Discussion Paper, Business Combinations, reprints the text of a Position Paper prepared for the G4+1 group of representatives of standard setting bodies from Australia, Canada, New Zealand, the United Kingdom, the United States and the IASC. It concludes that a single method, the purchase/acquisition method, is appropriate for all business combinations other than group reconstructions. It rejects another alternative, the 'fresh-start' method, in which a new entity would be regarded as coming into existence at the date of the combination, with fair values being assigned to the assets and liabilities of all combining parties and its history commencing at that date. The Discussion Paper states that 'the [UK] Board is not aware of a general demand for a revision to FRS 6; however, concerns about the application of FRS 6 have been expressed overseas'. In its accompanying Bulletin, the ASB expresses concerns that '... if other countries were to ban merger accounting, there would be significant implications for the financial community in the UK and the Republic of Ireland'.

1.1.2 Defining a merger

1.1.2.1 Individual accounts: merger relief

Prior to the Companies Act 1981, in the individual accounts of parent undertakings, the purchase consideration for any investment in another undertaking had to be recorded at the amount of cash consideration and the fair value of any other consideration. This meant that the excess of the fair value of any share-based consideration over its nominal (par) value had to be accounted for as share premium. The Companies Act 1981 removed the requirement for a share premium to be recorded on a share issue where the issuing company has secured at least a 90 per cent equity holding in another company. This enabled the share element of the purchase consideration to be recorded at nominal (par) value when merger accounting was used. Analogous criteria are now embodied in the Companies Act 1985 S 131, which should be consulted for detailed criteria.

1.1.2.2 Relationship between merger relief and group accounting treatment

Whilst it is true that 'merger relief' provisions facilitated the path for merger accounting by allowing shares used to purchase certain investments to be recorded at nominal value, merger relief is an *individual* accounts matter and is different from merger accounting. Many companies satisfying S 131 and qualifying for merger relief will not meet FRS 6's criteria for merger accounting. Under these circumstances acquisition accounting must be used.

Prior to the revision of the Companies Act 1985, some groups, whilst adopting acquisition accounting, took advantage of merger relief provisions to record the investment in the subsidiary at nominal amount, and using this artificially low value, understated goodwill at acquisition. However, Sch 4A, para. 9(4) now makes it clear that under acquisition accounting, the acquisition cost 'means the amount of cash consideration and the fair value of any other consideration...'. FRS

4, Capital Instruments, also requires the investment in the parent's individual accounts to be consistent with the consolidation approach adopted. If merger accounting is used, the investment can be recorded at nominal amount (FRS 4, para. 21 (c)). In all other circumstances, where a company issues shares to acquire a subsidiary, the net proceeds (i.e. the fair value of the consideration received) must be credited to shareholder funds (FRS 4, para. 10). If merger relief criteria are met and acquisition accounting is used, the excess over nominal amount is credited to a separate reserve, often termed a merger reserve, not share premium (FRS 6, para. 43). In consolidated accounts such a reserve was used for the immediate write-off of goodwill since, unlike the share premium account, it did not require the court's permission. However, FRS 10 no longer allows such immediate write-off and Company Reporting (December 1998, p. 6) comments that the Act gives no indication of what else to do with this reserve and finds that a small number of companies are now releasing it to the profit and loss account reserve. Davies, Paterson and Wilson comment that there are legal differences of opinion over whether a share premium account can be set up when S 131's merger relief criteria are met, or whether a merger reserve or a nominal amount investment is compulsory (see Davies, Paterson and Wilson, 1998, p. 199).

1.1.2.3 Criteria prior to FRS 6

SSAP 23, Accounting for Acquisitions and Mergers, issued in 1985, had defined conditions under which merger accounting was available as an optional treatment:

- The offer has to be made to all holders of equity and voting shares not already held; and
- The minimum ending stake must be at least 90 per cent of all equity shares (each class taken separately) and 90 per cent of the votes; and
- The maximum permissible starting stake must be 20 per cent of all equity shares (taking each class separately) and 20 per cent of the votes.
- A minimum final offer proportion of 90 per cent of the fair value given for equity capital to be in equity, and a minimum final offer proportion of 90 per cent of the fair value given for voting non-equity in equity or voting non-equity.

In the 1989 revision of the Companies Act 1985, statutory criteria for allowing the option of merger accounting where they are all met were included in the Act, (Sch 4A, para. 10). They are still in force and must be read in conjunction with FRS 6. They are as follows, that:

- The final stake held by the parent company and its subsidiaries is at least 90 per cent in nominal value of shares with unlimited participation rights in both distributions and assets on liquidation.
- This limit was passed by means of an equity share issue by the parent or subsidiaries.
- In the offer mix, the fair value of the consideration given by parent or subsidiaries other than in equity shares should be less than 10 per cent in nominal amount of the equity shares issued.
- The adoption of merger accounting accords with generally accepted accounting principles or practice.

The limit for the non-equity element in the offer was tightened from 10 per cent of fair value to 10 per cent of nominal value of equity issued. The Act's restriction can be made less binding by the parent capitalizing its reserves by a bonus issue of its shares (Davies, Paterson and Wilson, 1998, p. 204). This increases the nominal amount of shares in issue, but not necessarily market capitalization, thus increasing the ratio of nominal amount to market value for each share issued.

Critics pointed out that SSAP 23's criteria for merger accounting did not correspond to intuitive notions of a merger. Instead they focused on the form of the transaction – share for share exchanges without significant resources leaving the combining companies. This led to schemes enabling what were not 'true' mergers to use merger accounting, for example:

- Vendor rights: although the parent's shares were technically offered to the subsidiary's shareholders, an intermediary agreed to purchase them for cash and immediately offer them back to the parent's shareholders as a rights issue. However, if the parent had got to the same end result by itself making a rights issue and then offering cash, the combination would have had to be accounted for as an acquisition.

- Vendor placings: as above except the intermediary placed the shares with outsiders.
- Temporary placement of starting holdings: Selling a part of a 20+ per cent holding to a friendly outsider just prior to the offer and buying it back just afterwards.

Using intermediaries could make what might have been in substance 100 per cent cash offers use merger accounting. Surprisingly, however, merger accounting never became popular. Higson reported that in the years 1982–87, out of 227 combinations 13 per cent used merger accounting, and that between 1976 and 1987 it tended to be used where the 'target' company historically had a greater profit than the bidder and where the 'target' was relatively large (see Higson, 1990, pp. 44–46). But *Company Reporting* reported its declining use over a later five-year period to 1993, when only 1.4 per cent of business combinations including equity consideration used it. It questioned whether the ASB should have devoted resources to resolving a non-problem (*Company Reporting,* July 1993, pp. 6–8). However, now that the opportunities for manipulating acquisition accounting have been removed by the ASB through FRS 7, 10 and 11, it may become popular again and so, short of abolition, a tight definition such as FRS 6's may be necessary.

1.1.2.4 Current criteria for merger accounting – FRS 6

FRS 6, Acquisitions and Mergers, issued in September 1994, has criteria similar to those used in IAS 22 [to define what that standard terms a 'uniting of interests'], but they are more explicit and detailed. It defines a merger as:

> A business combination which results in the creation of a new reporting entity formed from the combining parties, in which the shareholders of the combining entities come together in a partnership for the mutual sharing of the risks and benefits of the combined entity, and in which no party to the combination in substance obtains control over any other, or is otherwise seen to be dominant, whether by virtue of the proportion of its shareholders' rights in the combined entity, the influence of its directors or otherwise (para. 2).

All other business combinations except certain group reconstructions are acquisitions (para. 5). FRS 6 then sets out detailed merger accounting criteria to make this definition operational to prevent what are not mergers in substance being accounted for as mergers. If the criteria are met merger accounting is mandatory. Taylor (1996, pp. 54–62) gives a conceptual appraisal of the criteria and international differences. Merger accounting must be used (para. 5) if its use is not prohibited by the Companies Act and if, but only if, all the following criteria are met (paras. 6–11). Following each criterion is a short summary of the standard's commentary (paras. 60–77):

- Criterion 1: Portrayal of parties. No party to the combination is portrayed as either acquirer or acquired, either by its own board or management or by that of another party to the combination.There is a rebuttable presumption that the combination is an acquisition if a premium is paid over the market value of the shares acquired. Other factors suggestive of the nature of the combination, whilst not individually conclusive (paras. 60–62) are its form, plans for the combined entity's future operations (including whether closures or disposals were unequally distributed between parties), content of communications of a publicly quoted party with its shareholders, and the proposed corporate image (name, logo, location of headquarters and principal operations).
- Criterion 2: Consensus management structure. All parties to the combination, as represented by the boards of directors or their appointees, participate in establishing the management structure for the combined entity and in selecting the management personnel, and such decisions are made on the basis of a consensus between the parties to the combination rather than purely by the exercise of voting rights. FRS 6 recognizes even in a genuine merger that the parties should be free to choose their management, and equal participation on the combined board is not necessary. Such management could possibly come from a single party, but genuine participation must be demonstrated. However, consensus decision-making rather than voting power

against the wishes of one of the parties to the merger is necessary. Informal and formal management structures must be considered. Only management structure decisions made in 'the period of initial integration and restructuring at the time of the combination' need be considered, taking into account their short- and long-term effects (paras. 63–66).

- Criterion 3: Relative sizes. The relative sizes of the combining entities are not so disparate that one party dominates the combined entity merely by virtue of its relative size. Such domination would be presumed if one party is substantially larger – this would be inconsistent with the concept of a merger as a substantially equal partnership between combining parties. A rebuttable presumption of dominance is made if any party is more than 50 per cent larger than each of the others, when considering the proportions of the combined equity attributable to the shareholders of the combining parties. Factors, for example voting or share agreements, blocking powers or other arrangements, can be deemed to reduce or increase this relative size influence. The reasons must be disclosed if rebutted (paras. 67–68).
- Criterion 4: Offer mix. Under the terms of the combination or related arrangements, the consideration received by equity shareholders of each party to the combination, in relation to their shareholding, comprises primarily equity shares in the combined entity; and any non-equity consideration, or equity shares carrying substantially reduced voting or distribution rights, represents an immaterial proportion of the fair value of the consideration received by the equity holders of that party. Where one of the combining entities has, within the period of two years before the combination, acquired equity shares in another of the combining entities, the consideration for this acquisition should be taken into account in determining whether this criterion has been met. FRS 6 requires that all but an immaterial portion of the fair value of the consideration should be in the form of equity shares – to prevent shares with unusual rights getting round the Companies Act restriction that non-equity consideration should not exceed 10 per cent of the nominal amount of equity shares issued. It defines equity shares more rigorously than the Act (following FRS 4, Capital Instruments), excluding shares with limited rights to receive payments not calculated on underlying profits, assets or equity dividends, or with effective limitations on participation rights in any winding up surplus, or redeemable contractually or at the option of parties other than the issuer (para. 2). Cash, other assets, loan stock and preference shares are cited as examples of non-equity consideration (para. 69). All arrangements made in conjunction with the combination (including e.g. vendor placings) must be taken into account unless made independently by shareholders. Substantially reduced voting or distribution rights indicate an acquisition, though some reduction may be compatible with a normal merger negotiating process. Where a peripheral part of one of the businesses of one of the combining parties (i.e. one disposable without material effect on the nature and focus of its operations) is excluded from the combined entity, shares or proceeds of sale distributed to its shareholders are not counted as consideration in determining offer mix (paras. 69–74).
- Criterion 5: No protected holdings. No equity shareholders of any of the combining entities retain any material interest in the future performance of only part of the combined entity. Mutuality in sharing risks and rewards in the combined entity is deemed absent where one party's equity share depends on post-combination performance of the entity previously controlled by it; where earnouts or similar performance-related schemes are included in the merger arrangements; or where the statutory ending stake (90 per cent) is not achieved. It is, however, permissible to allocate holdings based on the subsequently determined value of a specific asset or liability (paras. 75–77).
- Other criteria. The combination transaction must be considered as a whole, including any related arrangements in contemplation of the combination, or as part of the process to effect it (para. 56). FRS 6 defines the parties to the combination to include the management of

each entity and the body of its shareholders as well as its business (para. 57). Financial arrangements in conjunction with the transaction are to be included, which precludes vendor rights and vendor placings. If convertible shares or loan stock is converted into equity as part of the combination, they are to be treated as equity (para. 12). The acquisition cost and consideration includes shares issued and owned by subsidiary undertakings (Companies Act 1985, Sch 4A, para. 9(4)). FRS 6 applies analogously to entities without share capital (para. 59). Divested elements of larger entities are not eligible for merger accounting since they are not independent enough to be considered separate from their former owners until they have a track record of their own. An exception is if the divestment can be shown to be peripheral (see Criterion 4 above). Shareholdings acquired within the two years before the combination for non-equity consideration or for consideration with reduced equity rights must be included in assessing the criteria (para. 73).

1.1.2.5 Use of a new parent company

FRS 6 comments that

> the legal form of a business combination will normally be for one company to acquire shares in one or more others. This fact does not make that company an acquirer (in an accounting sense).... Similarly the question of whether the combined entity should be regarded as a new reporting entity (i.e. a merger) is not affected by whether or not a new legal entity has been formed to acquire shares in others. (para. 46)

Where a newly formed parent company is set up to acquire the shares of the combining parties, 'the accounting treatment depends on the substance of the combination effected', as discussed below.

1.1.2.6 Group restructurings

Merger accounting is an option for various types of restructurings, the transfer of ownership of a subsidiary between group companies, the addition of a new parent to a group, transfer of shares in subsidiaries to a new non-group company with the same shareholders as the group's parent, and the combination into a group of two companies previously under common ownership (para. 2). Merger accounting must not be prohibited by companies' legislation, the ultimate shareholders must remain the same and rights relative to each other be unchanged, and no minority interest must be altered by the transfer (para. 13).

1.1.3 Merger accounting consolidation procedures

Under merger accounting the combination transaction is seen as a change in the scope of the accounts, showing the combined results of two pre-existent entities now viewed together. FRS 6 terms this a 'new reporting entity' (para. 2). Comparatives and notes are adjusted to reflect the new group structure. The carrying values of the subsidiary's assets and liabilities are not adjusted except to achieve uniform group accounting policies (para. 16). The parent's investment in its individual company accounts is recorded at nominal amount. Consolidated retained profits include both the subsidiary's pre- and post-acquisition profits and in the year of combination the whole of the results of all the combining entities are included. Goodwill *per se* does not arise, and the difference between the consideration given, i.e. the nominal value equity plus the value of other consideration, and the nominal value of the shares received in exchange, should be shown as a movement on other reserves in the consolidated financial statements (para. 18). FRS 6 also requires that any existing balances on the new subsidiary's share premium account or capital redemption reserve should also be included as a movement on reserves and that both sets of movements above should be shown in FRS 3's reconciliation of movements in shareholders' funds (para. 18). Thus reported consolidated share premium or capital redemption reserves under merger accounting will only be that which relates to the share capital of the reporting entity (para. 41). Merger expenses must be charged to the profit and loss account of the combined entity at the effective merger date as reorganization or restructuring expenses in accordance with para. 20 of FRS 3 (para. 19).

Where a newly formed parent company is set up to acquire the shares of the combining parties, if 'a combination of the companies other than the new parent would have been an acquisition' the party identified as in substance the acquirer should be merger accounted with the newly formed parent company and all the acquired companies must be acquisition accounted with the new parent. If the 'combination' would have been a merger, all parties are merger accounted with the newly formed parent company (para. 14). FRS 6's provisions apply equally to any other arrangements achieving similar results (para. 15).

Disclosures contained in FRS 6 are summarized below (see X). The profit and loss note disclosures are more extensive than for acquisition accounting. For each party to the merger, an analysis of the principal components of the profit and loss account (including at a minimum turnover, profit and loss and exceptional items, split between continuing operations, discontinued operations and acquisitions; profit before taxation; taxation and minority interests; and extraordinary items) and statement of total recognized gains and losses, is required for the period of the combination up to the merger date, and separately for the previous financial year. A similar analysis is required for the merged entity as a whole for the current period after the merger date (para. 22). Analysts cannot quite determine how the combination would have looked under acquisition accounting – whilst the fair value of the consideration is to be disclosed, the fair value of identifiable net assets is not.

1.2 Acquisition accounting – introduction

Acquisition accounting must be used where the criteria for merger accounting are not met (see VI.1.1.2.4). The logic of acquisition accounting treats the subsidiary as if its individual assets, liabilities and goodwill had been purchased at the combination date.

The Companies Act 1985a, Sch 4A, para. 9 requires in the consolidated accounts:

- (The entire) identifiable assets and liabilities (i.e. those capable of being disposed of without disposal of the business as a whole) should be included at fair value at the acquisition date.
- Income and expenditure should be brought in only from the acquisition date.
- Attributable capital and reserves at acquisition (including fair value adjustments) are to be deducted from the total purchase consideration (cash, the fair value of non-cash consideration, plus, at the directors' discretion, fees and other acquisition expenses) of parent and subsidiaries; if positive to be treated as goodwill, if negative as a negative consolidation difference.

FRS 6's definition of acquisition accounting reinforces this, adding that under acquisition accounting, comparatives are not adjusted and that goodwill (positive or negative) is the difference between the fair value of the consideration given and the fair value of the net identifiable assets acquired (para. 20). The fair valuation of identifiable assets and liabilities at acquisition restates them to group costs. Both controlling and non-controlling interests' proportions of such assets and liabilities are restated even if such restated values in aggregate exceed acquisition costs. This is the 'alternative' treatment allowed by IAS 22, *Business Combinations*, revised in 1998 (para. 34). The equity of the subsidiary not owned by the parent is reported as 'minority interests' and stated at acquisition as the minority stake in the fair values of the subsidiary's assets and liabilities. The UK fair value adjustment to minority interests at acquisition angles the United Kingdom towards the entity consolidation concept in this respect (see II.2) whereas US and German approaches are closer to the 'pure' parent position – only the controlling interests' portion of identifiable assets and liabilities at acquisition are restated to fair value (see VI.1.3 and Crichton, 1990b, pp. 30–31). FRS 7 requires only direct incremental acquisition costs to be included in the total purchase consideration, tightening the Act's requirements above.

FRS 3, *Reporting Financial Performance (1993)* requires acquisitions during the year to be distinguished in the profit and loss account within continuing operations. For such acquisitions it requires a separate analysis of profit and loss disclosures since acquisition – on the face of the profit and loss account itself turnover and operating profits must be analyzed as a minimum, with the rest disclosed in notes (para. 14) (see, for example, Figures IX.1 and IX.2). In the comparatives for continuing operations there is no need to distinguish last year's acquisitions separately (para. 64). They are to be restated to include only the results of those operations included in the current period's continuing operations (para. 30), If it is not practicable to determine post-acquisition results for the current period, an indication should be given of the contribution of the acquisition to the turnover and operating profit of the continuing operations of the current period, together with standard disclosures required by the Companies Act 1985, Sch 4A, para. 13, relating to acquisitions (see Table X.1). If such an indication cannot be given, this fact and the reason must be explained (para. 16).

Prior to FRS 7 (1994) neither the Act nor accounting standards specified how fair values at acquisition were to be determined. SSAP 22, *Accounting for Goodwill*, had allowed free choice between its preferred option of immediate write-off of purchased goodwill direct to reserves and its allowed alternative of gradual amortization through the profit and loss account. Such a loose framework encouraged understatement of assets and overstatement of liabilities and provisions at acquisition, combined with the immediate goodwill write-off direct to reserves. This was because the resulting overstatement of goodwill bypassed the profit and loss account, but biased asset and liability valuations increased post-acquisition profits through, for example, lower depreciation charges and the writing-back of excess 'provisions' (see Smith 1996, pp. 27–33). FRS 7 and FRS 10 together address such creative accounting (see VI.1.3.1). The determination of fair values at acquisition and the treatment of goodwill, brands and other intangibles are examined below (see VI.1.3 and VI.1.4).

The date of acquisition for accounting purposes determines when the subsidiary's results can first be included in the consolidation under acquisition accounting. FRS 2, Accounting for Subsidiary Undertakings, defines it, and analogously the date of merger, as 'the date on which control of (a subsidiary) undertaking passes to its new parent undertaking' (para. 45), and amplifies this (para. 85) by specifying the date as one of the following:

- When the offer becomes unconditional, for public offers, usually through sufficient acceptances.
- Generally the date when an unconditional offer is accepted, for private treaties.
- That of share issue, where this means is used.

The date control passes to the new parent may be indicated by the date when the acquiring party commences its direction of operating and financial policies, or the flow in economic benefits changes, or when the consideration for transfer of control is paid (para. 85).

1.3 Fair values at acquisition

1.3.1 Overview

The ASC issued a discussion paper in June 1988 followed by an exposure draft ED 53, *Fair Value in the Context of Acquisition Accounting*, in 1990. Its successor body, the ASB, moved cautiously, also issuing a discussion paper, *Fair Values in Acquisition Accounting*, in July 1993, followed by an exposure draft then FRS 7, *Fair Values in Acquisition Accounting*, in September 1994. FRS 7 also applies to acquisitions of businesses other than subsidiary undertakings (para. 4). Fair value is defined by FRS 7 as 'the amount at which an asset or liability could be exchanged in an arm's length transaction between informed and willing parties, other than in forced or liquidation sale' (para. 2). FRS 7 considers the valuation of:

- Purchase consideration at acquisition, including dealing with abnormal market fluctuations, contingent consideration (based on post-acquisition performance) and deferred consideration (paid subsequent to the acquisition date).
- Identifiable assets and liabilities at acquisition by:
  - (i) Specifying valuation bases, consistent with the value to the business basis in the ASB's Draft Statement of Principles;
  - (ii) Providing 'cut-off' guidance in identifying which assets and liabilities/provisions should be included or excluded;
  - (iii) Deciding the level of detail for valuations and the period over which they and goodwill can be adjusted to a final estimated amount – the investigation period;

(iv) Requiring discounting for certain long-term assets and liabilities.

FRS 7's objectives seem at first sight unexceptional:

> when a business is acquired by another, all the assets and liabilities that existed in the acquired entity at the date of acquisition are recorded at fair values reflecting their condition at that date; and (...) all changes to the acquired assets and liabilities, and the resulting gains and losses, that arise after control ... has passed to the acquirer, are reported as part of the post-acquisition financial performance of the acquiring group. (para. 1)

However, certain of its proposals based on those objectives are controversial, particularly its two-stage approach to valuing identifiable assets and liabilities at acquisition:

- The identifiable assets and liabilities, i.e. those capable of being disposed of or settled separately from disposal of a business of the entity (para. 2), to be recognized are those existing at the date of acquisition, to be measured using fair values that reflect the conditions at the date of acquisition (paras. 5 and 6).
- They should not reflect the acquirer's intentions or future actions; impairments or other changes resulting from events subsequent to the acquisition; or provisions for future operating losses or provisions for reorganization and integration costs expected to be incurred as a result of the acquisition, whether relating to the acquired entity or to the acquirer (para. 7). These are to be accounted for as post-acquisition events. This has been termed a 'neutral' rather than an acquirer's perspective.

This considerably changed practice and its implications are explored later (see VI.1.3.3.4).

1.3.2 Acquisition cost of the investment in the subsidiary undertaking

The cost of acquisition is 'the amount of cash paid and the fair value of other purchase consideration given by the acquirer, together with the expenses of the acquisition' (FRS 7, para. 26). In the case of multi-stage acquisitions, it is the aggregate of the costs determined at each date (FRS 2, para. 30). Table VI.1 shows FRS 7's provisions for determining the fair value of purchase consideration for the investment where such consideration is not in the form of immediate cash. Acquisition date market prices are to be used as far as possible, particularly where shares and capital instruments are quoted on a ready market. Discounting should be used where payment is later than the acquisition date. Reasonable estimates of the fair value of amounts expected to be payable, or reasonably expected to be payable, are to be used where consideration is contingent on uncertain future events, e.g. 'earnouts' where part of the consideration is based on multiples of future profits, and revised as information unfolds. Direct incremental acquisition costs such as merchant bank costs, but not allocated internal costs, such as the costs of an internal acquisitions department, are to be included (para. 85). FRS 7 comments that where it is not possible to value the consideration by using the above methods, the best estimate may be given by valuing the entity acquired (para. 79). FRS 7 states that an averaging period would need to be considered where there are unusual fluctuations.

1.3.3 Identifiable assets and liabilities acquired

1.3.3.1 *Overall valuation basis*

The fair values of the individual identifiable assets and liabilities acquired are included in the consolidated financial statements usually by consolidation adjustment. This, as noted above (see VI.1.2) includes the minority interests' share of such fair value adjustments regardless of whether the acquisition cost of the investment in the subsidiary exceeds the parent's share of the subsidiary's restated pre-acquisition equity or not. In the latter case, such a difference would be treated as negative goodwill (see VI.1.4.2.3). The term 'fair value' is defined as 'the amount at which an asset or liability could be exchanged in a transaction between informed and willing parties, other than in forced or liquidation sale' (para. 2).

FRS 7's requirements are broadly consistent with the value to the business principle in Chapter 5 of the ASB's Statement of Principles Exposure Draft (1995). Conceptually this is the lower of replacement cost and recoverable amount, i.e.

Table VI.1 Determining the fair value of the investment/purchase consideration (FRS 7)

Type of consideration	Valuation basis
Shares and other capital instruments quoted on a ready market	• Market price on acquisition date (for public offers the date when offer or successful bid becomes unconditional) • If unreliable because of unusual fluctuations, market prices for a reasonable period prior to acquisition during which acceptances could be made need to be considered
Unquoted instruments, or quoted with an inactive market in the quantities involved	Estimate using e.g. value of similar quoted securities, present value of future cash flows of the issued instrument, any cash alternative, or the value of any underlying security into which there is an option to convert.
Cash and other monetary items	• Amount paid or expected to be payable discounted to present value if deferred, using the acquirer's rate for similar borrowing allowing for credit standing and any security given.
Non-monetary assets transferred	Market prices, estimated realizable values, independent valuations or based on other evidence
Contingent consideration	• Reasonable estimate of fair value of amount *expected* to be payable in the future, or of at least those amounts reasonably expected to be payable (e.g. minimum amounts if any), if total expected amounts too uncertain. • Initial estimates to be revised as information unfolds. • If to be in the form of shares, credit separate heading in shareholders' funds, disclosing equity and non-equity interests per FRS 4, and then transfer to share capital and premium on issue. Similar initial treatment if *acquirer* has the option over whether consideration is in shares or cash, until irrevocable decision taken. • If *vendor* has the option over form of consideration, credit should be to liabilities until consideration is issued or paid. • Care is needed over whether acquisition agreement contingent payments are in substance payments for the acquired business, or other expenses (e.g. compensation for services or profit sharing) to be accounted for in the period to which they relate.
Acquisition expenses	Only *incremental* costs (e.g. professional fees) and *not* the allocation of non-incremental costs (e.g. acquisitions department costs or management remuneration) should be added to the cost of the consideration (however, qualifying issue costs for shares and capital instruments should be dealt with according to FRS 4 and are not added to the cost of acquisition).

- Replacement cost.
- If the asset is impaired, its recoverable amount. This is the greater of net realizable value or value in use.

Fair values will be the market price where similar assets are bought and sold on a readily accessible market, but care must be taken that the price is 'appropriate to the circumstances of the acquired business'. Counter-examples are given where the second-hand market deals in low volumes or excludes normal maintenance or technical support, when depreciated replacement cost of an equivalent new asset may be more appropriate (para. 44).

Net realizable value in a ready market is defined as market price less realization costs and dealer's margin (para. 45). Where quoted prices are not available, subsequent sales may provide the most reliable evidence of fair values at acquisition where similar assets are bought and sold on a readily accessible market (para. 43), but these should not reflect any impairments from post-acquisition events. Subsequent reduced price disposals from, for example, group reorganizations, are post-acquisition.

Value in use is 'the present value of the future cash flows obtainable as a result of the asset's continued use, including those resulting from the ultimate disposal of the asset' (para. 2). Aggregation at an appropriate level for jointly used assets may facilitate the determination of attributable cash flows and recoverable amounts (paras. 47–49). The calculation of value in use is more often used in practice in deciding between replacement cost or net realizable value as the appropriate basis, rather than as a valuation basis in its own right. FRS 7 considers value in use is applicable to fixed assets but not to stocks (paras. 11, 12 and 45). Net realizable value and value in use are unaffected by the acquirer's intentions; the former is measurable whether or not a sale is intended, and the latter based on most profitable possible use rather than intended use (para. 46).

1.3.3.2 Application to individual assets and liabilities

Table VI.2 summarizes FRS 7's application to specific assets and liabilities. Quoted market prices are preferred where available, and subsequent sale prices may provide the best evidence of fair values. Otherwise independent valuations and techniques such as present value discounting methods are suggested, together with other approximation methods where there are no active markets, e.g. for specialized or semi-completed assets or assets still under development. Such approximations are:

- Indices, for example for plant and machinery.
- The current cost of reproducing an asset of a similar type, for example for development land or work-in-progress.
- Adjusted historical cost for interest in the case of maturing stocks with no intermediate market, to reflect holding costs, or for prudently estimated attributable profits in the case of long-term contracts.
- Discounted future cash flows at appropriate current market rates, for unlisted long-term receivables and liabilities, subsequently adjusting the post-acquisition interest charge to a constant rate based on the fair value carrying amounts.

Unlike ED 53 which had proposed acquired stocks should include profits earned to the date of acquisition, FRS 7 follows the 'value to the business' basis closely – such stocks should be stated at the lower of replacement cost or net realizable value, plus current costs of manufacturing for work-in-progress. This is consistent with the purchase of an independent entity. Contrast this with FRS 7's proposed basis for determining the fair value of long-term contracts, where attributable profit is to be included. FRS 7 appears to consider that the normal prudent estimation procedure in this case, see for example SSAP 9, Stocks and Long-term Contracts (para. 29), produces a value closer to replacement cost than a 'selling' price.

1.3.3.3 Potential conflicts with other accounting standards

The recognition of the fair values of certain assets and liabilities at acquisition potentially conflicts with the bases required by other accounting standards for ongoing businesses, for example:

- Pension surpluses or deficits. Under SSAP 24, Accounting for Pension Costs, they are smoothed to income over an extended period. Under FRS 7's approach, the recognizing of fair values of all the acquired entity's assets and liabilities at acquisition is given primacy over SSAP 24's 'smoothing' of pension cost allocations to profit and loss (para. 36). However, some caution is necessary in estimating any pension assets the surplus of which is reasonably expected to be realized in the future in cash terms (paras. 41–43).
- Contingent assets and liabilities. FRS 7 requires that these are to be included based on reasonable estimates of contingencies. Its previous exposure draft had justified such a treatment on the grounds that recognizing such assets at acquisition does not anticipate future gains, but reflects the expectation that

Table VI.2 Determining the fair value of identifiable assets and liabilities acquired (FRS 7)

Type of asset or liability	Special characteristics in determining fair value
Tangible fixed assets	• Market value if assets of a similar type are bought and sold on an open market, where obtainable with reliability e.g. certain property and quoted investments. • Depreciated replacement cost, reflecting the business's normal buying process and the sources of supply and prices available to it, unless this exceeds the recoverable amount. This would be used e.g. for most plant and machinery, and specialized properties specific to the business. Price indices can be used where it is difficult to measure directly and it is not possible to estimate the value of future services because of inherent subjectivity. Historical carrying value can be used if prices have not changed materially. • Recoverable amounts, depreciation rates, asset lives and residual amounts should be consistent with the acquirer's policies for similar assets, without any change in the asset's use or intended use.
Intangible fixed assets	• Based on replacement cost, normally its estimated market value.
Stocks and work-in-progress	• Stocks in markets in which acquired entity trades as *both* buyer and seller (e.g. commodities) at current market price. • Other stocks at the lower of the acquired entity's replacement cost, reflecting its normal buying process and sources of supply, and net realizable value. Replacement cost excludes unrealized profit. • Market values to be used to estimate replacement costs where a ready market exists (e.g. commodities, dealing stocks, certain land and buildings held as trading stock, and certain maturing stocks readily tradable at similar completion states). • If no ready market, replacement cost is the acquired company's current cost of reproduction, e.g. using current standard costs for manufactured stocks and work-in-progress. • For maturing stocks with no intermediate or a thin market and where replacement costs are difficult to find because of impossibility of short-term replacement, historical cost plus an interest holding cost may be used. • Long-term contracts require no further adjustment to reflect fair value other than to reflect reassessments of the contract outcome or change to acquirer's accounting policies. • Write-downs to NRV may reflect acquirer's judgements but based on the *acquired* entity's circumstances at acquisition. Fair values at acquisition must be re-examined and adjusted (including goodwill) if exceptional post-acquisition profits do not result from post-acquisition events. Even then, exceptional item disclosure may be necessary.
Quoted investments	• Market price adjusted for unusual price fluctuations or for size of holding.
Short-term monetary items	• Usually at settlement or redemption amount.
Long-term receivables and liabilities	• If quoted, market price or market price of similar items. A lower market value reflecting market concerns about risks of non-fulfilment of repayment obligations, no longer applicable because of the acquisition, should *not* be recognized.

Table VI.2 (contd.)

Type of asset or liability	Special characteristics in determining fair value
	• If not quoted, fair values determined by considering similar monetary assets or liabilities or by *discounting* total amounts to be received or paid where effect is material. • Discounting rates: i. borrowing – must take into account equivalent term current borrowing rates, the issuer's credit standing and nature of security given. ii. lending – after making necessary provisions, current lending rates. • Differences between fair values and total undiscounted amounts receivable or payable, i.e. representing fair value discounts or premiums on acquisition, should be allocated to adjust post-acquisition interest charges to a constant rate based on new carrying amounts.
Pension schemes & other post-retirement benefits	• The fair value of a deficiency or, to the extent that it is reasonably expected to be realized (e.g. considering reductions in future contributions and the timescale of such potential realizations), a surplus in funded pension or other post-retirement benefit schemes, or accrued obligations in unfunded schemes, should be recognized as a liability or an asset of the acquiring group. These substitute for amounts calculated under SSAP 24. They should be calculated using the acquirer's actuarial methods. • Changes in arrangements following acquisition are post-acquisition, including harmonization of benefits throughout the expanded group. SSAP 24's requirements for variations in costs must be followed.
Reorganization/integration provisions	• For all practical purposes prohibited, unless the acquired entity was already demonstrably committed, and unable realistically to withdraw. If there was evidence of acquirer's influence, control may be deemed to have been yielded at an earlier date, making such commitments post-acquisition.
Business held exclusively with a view to resale	• Includes any business operation (including subsidiaries or divisions) where i. assets, liabilities and results from operations are distinguishable, physically, operationally and for financial reporting purposes; and ii. a purchaser has been identified or is being sought; and iii. disposal is reasonably expected within approximately one year of acquisition date. • If sold or is expected to be sold as a single unit within one year of acquisition, it should be treated as a single asset, provided it had not been previously consolidated or formed part of the acquiring group's continuing activities. • To be shown as a current asset within the acquirer's consolidated accounts if a subsidiary undertaking. The business's results are excluded from the acquiring group's profit and loss over the holding period. The same principles are to be applied to disposals of other business operations that are not subsidiary undertakings.

Table VI.2 (contd.)

Type of asset or liability	Special characteristics in determining fair value
	• Normally actual net realized value is the most reliable evidence of fair value at acquisition, discounted to the acquisition date if effect material and taking into account any distribution of profits from the business. If sale not completed by first post-acquisition financial statements, estimated sale proceeds to be used, adjusted to actual within FRS 7's investigation period. No further adjustment for goodwill on disposal is necessary to comply with UITF Abstract 3. If not in fact sold within approximately one year of acquisition it should be consolidated normally. • Net realized value cannot be used where: i. acquirer changes materially the acquired business prior to disposal; or ii. specific post-acquisition events materially change the fair value from the estimated acquisition value; or iii. a reduced price is obtained for a quick sale.
Deferred taxation	• Deferred tax assets and liabilities recognized in the fair value exercise should be determined by considering the enlarged group as a whole – see VIII. 3.2 • Unrelieved tax losses of the acquired entity to be treated consistently with the standard on deferred tax – see VIII. 3.2
Contingencies	• Contingent assets and liabilities to be measured at fair values where these can be determined. Reasonable estimates of the expected outcome (i.e. best estimates of the likely outcome) are to be used. In rare cases where the commitment or contingent asset is of the type assumed or acquired in an arm's length transaction, fair value would reflect market price.

amounts expended will be recovered (para. 44). Under SSAP 18, *Accounting for Contingencies*, contingent gains are generally not accrued in financial statements.

1.3.3.4 Cut-off: the two-stage perspective

Whereas ED 53 allowed the acquirer's intentions for using assets or incurring future costs to be incorporated in fair values at acquisition, as does the revised IAS 22 and to a significant extent US generally accepted accounting practice, FRS 7 does not, and distinguishes between:

- The fair values of assets, obligations, commitments and contingencies reflecting the conditions at acquisition; and
- Changes arising from subsequent events, e.g. from subsequent impairments in value or arising from the acquirer's intentions for future actions.

The crucial decision taken by the Accounting Standards Board and outlined in FRS 7 is that:

> . . . management intent is not a sufficient basis for recognizing changes to an entity's assets and liabilities. It is events, not intentions for future actions that increase or decrease an entity's assets or liabilities. When intentions are translated into actions that commit the entity to particular courses of action, the accounting should then reflect any obligations or changes in assets that arise from those actions . . . (Thus) events of a post-acquisition period that result in the recognition of additional liabilities or the impairment of existing assets of an acquired entity should be reported as events of that (post-acquisition) period. (Appendix III, para. 14)

The ASB considers fair value to be a neutral concept, the result of a bargaining process and independent of the acquirer or acquiree's circumstances (Appendix III, para. 36). In FRS 7's view, reorganizations intended by the acquirer are ultimately discretionary and so do not constitute 'liabilities' at acquisition, as obligation is not present. So, where assets are disposed of as a result of reorganizations, unless the values had already been impaired at the acquisition date, any losses would be attributable to the post-acquisition reorganization (para. 48). Such a view is consistent with the stance taken on provisions generally in FRS 12, *Provisions, Contingent Liabilities and Contingent Assets*, issued in September 1998.

Provisions for future losses. Consistent with the revised IAS 22, FRS 7 (and FRS 12) does not allow the setting up of provisions for future losses. These may be probable, but again, they are not obligations, at acquisition or any other date.

Reorganization provisions: FRS 7 requires that provisions or accruals for reorganization costs expected to be incurred as a result of the acquisition, whether they relate to the acquired entity or to the acquirer, should be treated as post-acquisition items. Prior to FRS 7, ED 53 had proposed allowing reorganization provisions at acquisition, but only where there was evidence that the offer took into account such plans and they were costed in reasonable detail. However, FRS 7 felt that such a definition could not be enforced (Appendix III, para. 12c). If such provisions are overstated, goodwill will necessarily be overstated. What are really post-acquisition costs would be written off against such a 'fattened' provision, which would eventually be written back. Though 'fattened' goodwill would be gradually amortized, this would usually be over a longer period than the diverted expenses. The combined effect would increase immediate profits whilst reducing later ones. The availability of immediate write-off of goodwill to reserves prior to FRS 10 had meant that overstatements of goodwill had no profit and loss impact and the combined effect increased post-acquisition profits without later charges, compared to correctly measured provisions. This had made acquisition accounting more popular than merger accounting as a tool for creative accounting! Claiming indefinite life for goodwill under FRS 10 may provide similar temptations.

Company Reporting (September 1998, p. 5) finds that in 1994 29 per cent of companies with evidence of business combinations included reorganization provisions at acquisition, but subsequent to FRS 7, in 1997, this had dropped to 2 per cent. The ASB considers that only by prohibiting such provisions could it create a level playing field between companies that develop through acquisitions and companies that develop organically. It argues that allowing the creation of reorganization provisions at acquisition increases purchased goodwill, but that in companies that develop organically such reorganization expenditures are expensed. The increased goodwill resulting from the use of reorganization provisions at acquisition is created by the acquirer and is, as such, internally generated goodwill not recognized under generally accepted accounting practice. While such expenditures are expected to produce future benefits, the ASB does not consider this gives sufficient justification for their effective capitalization as goodwill (FRED 7, Appendix III, paras. 9–12 and 25).

Opponents argue that acquirers assess takeover possibilities as a single integrated decision and FRS 7's requirements do not reflect this (para. 21), that the requirements will make UK companies less willing to engage in necessary restructuring of inefficient businesses (para. 29), and that they make financial statements more difficult to understand (para. 30). FRS 7 points to considerable user enthusiasm for its proposed changes! The only dissenting ASB member agrees with the concerns of the preparer. He suggests that reaction to past abuses leads to user enthusiasm for change, but only to seeming support for what he considers a considerable over-reaction by the Board. In his view a better solution would be more strictly defined limits on such provisions at acquisition. Some argue that FRS 7's requirements disadvantage the competitive position of UK companies. This is interesting in the light of earlier arguments that previous laxity in UK standards had facilitated competitive advantages in UK bids for overseas companies. The ASB responds that accounting standards should have neutral effect and believes that greater transparency in financial reporting will allow better economic decisions.

The IASC moved from being consistent with FRS 7, in its exposure draft E61, to a position more consistent with ED 53, in its revision of IAS 22, Business Combinations, in 1998. While this standard prohibits liabilities being recognized at acquisition if they result from the acquirer's intentions or actions, it makes a specific exception for provisions for costs resulting from the acquirer's plans relating to an acquisition. Such provisions are allowed only if the main features of such a plan were developed and announced at or before acquisition, thereby raising valid expectations by those affected that it will be implemented; and also a detailed formal plan with certain characteristics specified in the standard must be developed by the earlier of three months after the date of acquisition or the date the annual financial statements are approved (para. 31).

FRS 6 suggests management might wish to provide optional note disclosure of the total investment in acquiring a business (see VI.1.3.5). FRS 7 also argues that the additional exceptional item disclosures under FRS 3, *Reporting Financial Performance*, would enable analysts to properly assess the effects of reorganizations even if they were treated as post-acquisition (see VI.3). Other restructuring provisions are examined in VI.1.5.1.

Pension assets or liabilities. Changes in estimation methods to harmonize with the acquirer's accounting policies are allowed in determining the fair values of pension assets and liabilities at acquisition, for example to the acquirer's actuarial assumptions. However, changes caused by the acquirer's intentions or to adjust to group management policies, for example to standardize pension rights across the group, are not.

Other items. The net realizable value of stocks can be based on the acquirer's judgements, but on the circumstances of the acquired entity before the acquisition. Fixed asset lives must be estimated not taking into account the acquirer's plans, but can reflect the acquirer's accounting policies on useful lives. Identifiable liabilities do include provisions for onerous contracts or other liabilities that existed at acquisition whether or not they were recognized in the acquired entity's financial statements. Commentators disagree whether such provisions should be recognized only if similar contracts would be recognized under the acquirer's accounting policies and extant standards (Lennard and Peerless (1995), p. 129), or under wider circumstances (Bircher, 1988, p. 82). Davies *et al.* (1997, p. 322) support the wider view, giving the example of operating leases taken out on now unfavourable terms, since FRS 7 allows recognition in other cases of items at acquisition not otherwise allowed by relevant accounting standards. The treatment of deferred taxation is dealt with below (see VIII).

Policing cut-off in uncontested takeovers. FRS 7 allows provisions which are obligations entered into by an acquired company prior to acquisition to be included, but requires particular attention to be paid to the circumstances in which such provisions were made. This is to ensure that the acquired entity was demonstrably committed to the expenditure such that it would have had liability whether or not the acquisition had been completed, and to assess if there had been undue influence, which might mean 'control of the entity had been transferred at an earlier date' (para. 40). FRS 6 requires in the fair value table described below (see VI.1.3.5) that such provisions and related asset write-downs made in the 12 months up to the acquisition date by the acquired company should be separately identified (para. 26).

1.3.4 Investigation period and goodwill adjustments

FRS 7 wishes the determination of fair values at acquisition to be completed, if possible, by the date on which the first post-acquisition financial statements of the acquiring group are approved by its directors. If this is not possible, provisional estimates can be used. These can be adjusted up to the date on which the financial statements for the acquiring group's first full financial year following the acquisition are approved, with consequent amendments to goodwill at acquisition. After this time adjustments should be treated as normal accounting corrections in the period they occur (FRS 3), and only as prior year adjustments if they are as a result of fundamental errors (paras. 23–25). The fact that provisional values have been used should be disclosed, as should subsequent adjustments (FRS 6, para. 27). Exceptional profits and losses arising in subsequent financial statements from the determination of fair values at acquisition should be disclosed in accordance with FRS 3 (i.e. as exceptional and attributed to the statutory profit and loss headings to which they relate) and identified as relating to the acquisition (FRS 6, para. 30).

1.3.5 Disclosures

Required disclosures relating to the determination of fair values at acquisition and generally to combinations accounted for as acquisitions during the year are shown in Table X.1. They are now collected together in FRS 6. Such disclosures include details about the composition and fair value of the purchase consideration for each material acquisition, and a fair value table showing original book values of each major category of assets and liabilities acquired, their fair values and a categorized analysis of adjustments. Disclosure of provisional fair values and of subsequent material adjustments to them and goodwill, information about deferred and contingent consideration, exceptional post-acquisition profits and losses arising from fair values at acquisition, and movements on provisions for liabilities relating to the acquisition (paras. 34–38) must be given. Reorganization and restructuring costs included in the acquired entity's liabilities at acquisition and related asset write-downs made in the 12 months to the date of acquisition must be separately identified in the fair value table (FRS 6, para. 26). Also, details must be given in subsequent periods of reorganization, restructuring and integration costs incurred in those periods relating to acquisitions. Conditions are identified which ensure that only costs directly relating to the acquisition, set up at that time or in an immediate post-acquisition review, are disclosed (para. 31; see Table X.1).

Profit and loss note disclosures relating to acquisitions are less extensive than mergers (see VI.1.1.3). For material acquisitions the required disclosure is based around only the profit after tax and minority interests of the acquired entity from the start of its financial year to the date of acquisition, and for its previous financial year. Even in the case of substantial acquisitions (i.e. where the transaction is one in which any of the ratios set out in the London Stock Exchange Listing Rules defining Super Class I transactions exceeds 15 per cent, for listed companies, or for other entities, where net assets or operating profits of the acquired entity are more than 15 per cent of the acquirer's, or where the consideration is more than 15 per cent of the acquirer's net assets), a summarized profit and loss account for the acquired entity only is required from the beginning of its financial year to the date of acquisition, and only its profit after tax and minority interests for its previous financial year. Both should be based on the acquired entity's accounting policies. This reflects the fact that the acquired company is joining an ongoing group, unlike under merger accounting where the scope of the accounts is changed to reflect all merger parties, and comparatives are restated (see VI.1.1.3).

FRS 6 suggests (para. 87) optional disclosures relating to reorganization, restructuring and integration costs (costs, which FRS 7 does not allow to be adjusted in determining the identifiable net assets at acquisition). It suggests that management could provide note disclosure of such costs expected to be incurred in relation to an acquisition (including asset write-downs) indicating the extent to which they have been charged to profit and loss. An illustrative example (FRS 6, Appendix IV) shows a statement of 'Costs of reorganizing and integrating acquisitions' under the following headings:

- Announced but not charged at the previous year end.
- Announced in relation to acquisitions during the year.
- Adjustments to previous years' estimates.
- Charged in the year to profit and loss, and elsewhere.
- Announced but still to be charged at the current year end.

A further illustrative optional disclosure shows the 'total investment', made up of the acquisition cost as reported in the financial statements, plus reorganization and integration expenditure announced. This gives prominence to what opponents of FRS 7 see as the consequences of the acquisition decision, but only as a note disclosure and not in the statements themselves. See Table X.1 for detailed disclosure requirements.

1.3.6 Push-down accounting

The term 'push-down accounting' refers to the practice of incorporating fair value adjustments into the company records and published individual company financial statements of the acquired company. It is unusual in the United Kingdom. Whilst pushing down certain fair value adjustments could be justified under the Act's alternative accounting rules, others could not. For example in the case of pension surpluses, the company would be contravening SSAP 24; in others, adjustments would reflect the acquirer's accounting policies, not the subsidiary's. Any such adjustments would be subject to the requirements of the Companies Act 1985 relating to revaluations.

1.4 Goodwill and other intangibles

Accounting for goodwill and intangibles in the United Kingdom has been a battleground, probably the most controversial area in UK financial reporting. For many years the United Kingdom had chosen to sidestep mainstream international practice by predominantly using the immediate write-off of goodwill direct to reserves (although capitalization and gradual amortization had also been tolerated). Using an empirical 'index of harmonization', Weetman *et al.* (1998, p. 200) found that differences in goodwill accounting treatments caused the most striking incompatibility between UK GAAP, US GAAP and international standards between 1988 and 1994, with the later year showing even greater incompatibility. A variety of practices had developed for accounting for intangibles in the absence of a standard including, for a small number of groups, valuation of internally created brands. In December 1997, FRS 10, Goodwill and Intangible Assets, was issued, moving the United Kingdom closer to international practice, though significant and interesting differences remain. The ASB has broken new ground in conceptualizing how goodwill should be treated. Since FRS 10 contains extensive transitional arrangements which allow, but do not require, the restatement of the goodwill accounting treatment for acquisitions prior to the date it comes into force, i.e. for years ending after 23 December 1998, it is also necessary to understand SSAP 22, the preceding standard. Such transitional effects will affect UK financial statements for a considerable period.

1.4.1 Context and recent proposals

1.4.1.1 UK perspectives on goodwill

Controversy has characterized the goodwill debate in the United Kingdom. Purchased goodwill has been considered, for example, as:

- An asset, analogous to other fixed assets, to be recognized and amortized. Grinyer, Russell and Walker (1990, p. 228), for example, argue that managers should be accountable for any costs they incur, and that all such costs including purchased goodwill should be expensed over the life of the investment. This asset-based view is required by APB Opinion No. 17 (1970) in the United States and by the IASC's revised IAS 22 (1998).
- Primarily an aggregation level difference, since the entity as a whole has been acquired, but the individual assets and liabilities are consolidated. Supporters deem goodwill different in nature from other assets (see, for example, Ma and Hopkins, 1988, p. 84) since it cannot be realized separately from the business, and conclude that it should be written off immediately to reserves. Some claim this treatment ensures comparability with internally created goodwill.
- A link between the cost of the investment in the acquirer's individual financial statements and the identifiable net assets in the consolidated statements (FRS 10, p. 49). Though not strictly an asset under Statement of Principles definitions in the United Kingdom, reliability of measurement is deduced from the fact that the investment in the subsidiary in the acquirer's own accounts is reliable enough to report (Arnold *et al.*, 1992). Subsequent treatment of goodwill is analogous to impairment of the investment. Therefore goodwill is capitalized and is subject to regular impairment checks, but not necessarily gradually amortized.

While international standards are based more on the first rationale, UK practice has moved from the second towards the third rationale. The ASB admits this is as much for pragmatic as conceptual reasons, and acknowledges that goodwill is an 'accounting anomaly' (para. 11). However, given the way that costs and benefits of impairment tests have been set up in FRS 10 and 11, it is likely that many UK preparers will choose to capitalize and gradually amortize goodwill.

1.4.1.2 Development of UK accounting for goodwill and intangibles

Prior to the first UK goodwill standard, preparers were divided over which approach to use. For example, the *Survey of Published Accounts 1979* (Skerratt, 1979, p. 158) showed that in 1976–77, 38 per cent of companies disclosing a goodwill accounting policy carried goodwill at original cost, 42 per cent used immediate write-off to reserves and 20 per cent capitalized it and amortized it. The Fourth EU Directive in the late 1970s in initial draft versions had allowed a maximum life for goodwill in company accounts of five years. Its final version, enacted in the UK Companies Act 1981, gave member states the option, where goodwill was capitalized, to extend this period up to its economic life. Permanent retention and dangling debit were prohibited. In 1980 the ASC issued its first discussion paper which opted for gradual amortization. However, by this time, the 1980–81 *Survey of Published Accounts*, found that 77 per cent of 254 companies used immediate write-off.

In 1984, SSAP 22, the first goodwill standard, stated that immediate write-off to reserves should be the normal treatment, but allowed capitalization and gradual amortization, with no maximum life being specified. Groups were allowed to change treatments from acquisition to acquisition, because the ASC was concerned that groups with 'weaker' balance sheets might not use its preferred treatment if they had to stick with immediate write-off for all future acquisitions. It prohibited permanent retention of capitalized goodwill at cost. Immediate write-off to reserves, which became the dominant approach in the United Kingdom from 1984 to 1998, improves reported profit relative to capitalization and amortization, because there is no amortization charge. For example, Skerratt and Tonkin (1994, p. 185) showed that out of 300 UK listed and unlisted companies, only 5 per cent capitalized goodwill. It was argued that this gave an unfair competitive advantage to UK companies in making international bids (Hodgkinson, 1989, p. 20). Collinson, Grinyer and Russell (1993, p. 21) found in interviews of top managers from 246 of The Times Top 1000 UK companies in May 1991 that 55.8 per cent believed that top management decisions on whether or not to take over another company are influenced by the accounting treatment of goodwill whereas 34.8 per cent did not. Choi and Lee (1991, pp. 233–236) found that between 1985 and 1989, after controlling for other possible influences, amounts paid by UK groups for acquiring US firms were consistently higher than those offered by similar US groups. Such premiums were positively and significantly associated with the income statement effect of immediate write-off for UK groups. They suggest that it may have been possible that UK groups overpaid for US targets. However, the downside of immediate write-off is its explosive gearing effect as consolidated equity is hit by large immediate write-offs. According to Barwise *et al.* (1989, p. 21), average goodwill to bidders' net worth rose from 1 per cent in 1976 to 44 per cent in 1987 in the then exuberant market conditions, and from 26 per cent to 70 per cent over the same period in terms of the target's net worth.

This started the 'brands' debate. Popular sentiment had attributed the 1988 Nestlé takeover of Rowntree-Mackintosh to important intangibles being omitted from Rowntree's balance sheet (see Mather and Peasnell, 1991, p. 152). Some service sector groups, for example the advertising agency Saatchi & Saatchi, actually reported negative consolidated equity as a result of immediate goodwill write-offs. This would not normally have affected distributable profits, which in the United Kingdom are based on individual company not consolidated accounts. Neither UK standards nor company legislation prescribed an accounting treatment for intangibles. This opened the way for 'accounting arbitrage' between the treatment of goodwill and intangibles to gain the most favourable overall financial statement effects. In the same year as the Nestlé takeover, Grand Metropolitan Hotels capitalized certain acquired brands and Rank Hovis MacDougall capitalized in addition internally created brands. Smith (1996, pp. 91–100) gives examples and discusses the methodologies used to value brands. Power (1992) provides a good review of the debate and professional responses.

1.4.1.3 Professional responses

An ASC-commissioned report by Barwise *et al.* (1989) recommended against brand valuations in financial statements because of the lack of separability of many brands from the underlying entity and also measurement reliability problems. In 1990, the ASC tried to harmonize UK practice with international standards. ED 47, *Accounting for Goodwill*, and ED 52, *Accounting for Intangible Fixed Assets*, proposed that only purchased goodwill and separable intangibles could be capitalized and amortized over their useful life. A maximum life of 20 years was specified, or 40 years if justified. 73 per cent of all respondents and 93 per cent of preparers opposed the draft! Nobes (1992, p. 146), provides an excellent review of the UK goodwill debate.

The newly established Accounting Standards Board commissioned a report (Arnold *et al.*, 1992, pp. 74–75) which introduced the concept of impairment testing, found later in FRS 10 (see also Egginton, 1990, pp. 201–204) and also consolidated goodwill accounting approaches based on the link with the investment in the parent's accounts. In December 1993, the ASB, admitting disagreement amongst its board members, issued a Discussion Paper (DP), Goodwill and Other Intangibles, outlining six alternatives for accounting for goodwill including capitalization with predetermined life amortization, capitalization subject only to impairment tests, immediate write-off to designated or undesignated reserves and various combination approaches including immediate write-off to reserves with impairment tests (see Arnold *et al.*, 1992, or Taylor, 1996, pp. 127–141) for a further discussion of the UK debate at this time). The DP indicated that the ASB would adopt a single approach and introduced the idea of impairment test-based approaches into its pronouncements for the first time – the ASB had included in its thinking the third view of goodwill discussed in 1.4.1.1. Controversially it recommended that intangibles should not be separately measured from goodwill.

UK research considered whether the ASB's choice of goodwill and intangibles accounting approach mattered. The 'efficient markets hypothesis' (EMH) suggests that the stock market can see through and will not react to 'cosmetic' accounting changes with no real cash flow effects. However, financial statement effects may also have indirect cash flow consequences, for example, causing renegotiation costs of violating debt covenant restrictions based on, for example, gearing ratios or interest cover ratios where these are based on current GAAP, or in triggering costly stock market requirements (in the early 1990s where the target's book equity was more than 15 per cent of the acquirer's, the acquirer was required to issue a circular to its shareholders, costly in management time and effort. The acquirer's equity figure could be augmented by brands and intangibles only if they were included in consolidated balance sheets). Accounting-based management compensation might also be affected by such accounting changes. Mather and Peasnell (1991) found between 1986 and 1989 evidence to support debt covenant and stock market circular cost reasons for capitalizing brands, but inconclusive evidence on whether brand valuations affected share prices. Grinyer *et al.* (1991, pp. 51–52), found between 1982 and 1986 that management had less tendency to overstate goodwill (reducing the effect of immediate write-off on gearing) the higher the post-acquisition gearing of the group (i.e. debt covenant reasons). Gore, Taib and Taylor (1998) examined which factors influenced corporate finance directors' preferences for which approach the ASB should adopt in a new standard. Using a survey together with publicly available financial data, they found that companies which had binding gearing-based debt covenant restrictions preferred capitalization-based approaches, which would ease such restrictions compared to immediate write-off-based approaches. Companies with affected profit-based management compensation plans preferred immediate write-off based approaches, which would increase compensation relative to amortization approaches. However, corporate preferences seemed to be even more strongly associated with managerial beliefs about how financial analysts reassess company characteristics as a result of the ASB's choice.

The ASB issued a Working Paper (WP), Goodwill and Intangible Assets in 1995 and after public hearings an exposure draft in 1996 and FRS 10 in December 1997. Each contained similar proposals. Unlike the earlier discussion paper they allowed the recognition of intangibles other than goodwill. FRS 10 requires capitalization of both goodwill and intangibles, with similar accounting treatment for both to prevent 'accounting arbitrage' between them. For both there is a rebuttable presumption that their useful economic life will not exceed 20 years. Both should be amortized over their useful economic lives with impairment testing when economic conditions suggest their carrying amount is impaired. However, unlike IAS 22, FRS 10 allows capitalization with no gradual amortization where goodwill or intangibles are demonstrated to have an indefinite life and are capable of continued measurement. In such a case and also where the presumption of lives not exceeding 20 years is rebutted, annual impairment tests are required to assess whether carrying amounts need to be written down to recoverable amount. Useful economic life in FRS 10 is based on the concept of the 'link between the carrying value of the goodwill and the continuing value of the goodwill in the acquired investment' (Appendix 3 para. 30), rather than IAS 22's definition where, analogous to other fixed assets, it is the period expected to provide economic benefits from the originally purchased goodwill. FRS 10 also contains extensive transitional provisions which encourage, without requiring, the restatement of prior goodwill accounting treatments. This means that international users need to understand prior UK goodwill accounting treatments.

1.4.2 Current requirements for goodwill

1.4.2.1 Companies Act 1985

Goodwill can only be included 'to the extent that (it) . . . was acquired for valuable consideration' (Sch 4, Formats Note 3). It arises mainly in consolidation under acquisition accounting, the difference between the fair value acquisition cost and the attributable capital and reserves of the subsidiary after fair value adjustments at acquisition to identifiable assets and liabilities (Companies Act 1985, Sch 4A, para. 9). If goodwill is treated as an asset, it should be written off to the profit and loss account over a period chosen by directors, not exceeding its useful economic life (Sch 4, para. 21). General rules applying to fixed assets require impairment write-downs of goodwill or intangibles if there has been permanent diminution in value (Sch 4, para. 19). Goodwill may not be revalued or included at current cost under the alternative accounting rules, though other intangibles may (Sch 4, para. 31(1)). Schedule 4's balance sheet formats require that purchased goodwill, to the extent that it has not been written off, should be shown separately from other intangible assets, under the intangible assets heading. The immediate write-off of goodwill direct to reserves is possible under the Act; Schedule 4, para. 21 above then does not apply since goodwill is never treated as an asset.

1.4.2.2 Accounting for positive goodwill (FRS 10)

FRS 10, *Goodwill and Intangible Assets*, issued in December 1997, applies to all financial statements that are intended to give a true and fair view of financial position and profit and loss or income and expenditure. It also applies to where equity accounting is used. [Small] reporting entities subject to the FRSSE are exempt unless they prepare consolidated statements, in which case they should refer to that standard. FRS 10 does not allow capitalization of internally generated goodwill (para. 8). Purchased goodwill is 'the difference between the cost of an acquired entity and the aggregate of the fair value of that entity's identifiable assets and liabilities'. When the former exceeds the latter positive goodwill arises – in the converse case, negative goodwill.

Accounting treatment: positive purchased goodwill must be capitalized and classified as an asset. Where it is regarded as having a limited useful economic life, it should be amortized on a systematic basis over that life. Where it is regarded as having an indefinite useful economic life, it should not be amortized. There is a rebuttable presumption that useful economic life is 20 years or less, which can be rebutted only if:

- The durability of the acquired business can be demonstrated and a greater useful economic life can be justified; and
- Goodwill is capable of continued measurement [so that annual impairment reviews are feasible].

FRS 10 enumerates circumstances where there may be grounds for regarding the premium on acquisition as more durable and assigning it a longer or even indefinite economic life. Factors include:

- The nature of the business.
- The stability of the industry in which the acquired business operates.
- Typical lifespan of the products to which the goodwill attaches.
- The extent to which the acquisition overcomes

market entry barriers that will continue to exist.

- The expected future impact of competition on the business.

Uncertainty of economic life does not provide grounds for adopting a 20-year or indefinite life by default, nor for choosing a life that is unrealistically short. Continued measurement is not deemed possible if measurement costs are unjustifiably high, for example, because purchased goodwill cannot continue to be tracked after merging acquired and existing businesses, or management information systems cannot identify and allocate cash flows at a detailed enough income generating level, or amounts are too immaterial to justify the costs of such reviews.

Useful economic life: this is defined as 'the period over which the value of the underlying business is expected to exceed the values of its identifiable net assets' (para. 2). As alluded to above, this is the period over which goodwill continues to exist, not only originally purchased goodwill. FRS 10 acknowledges this differs from IAS 22, which comments that '[though] the value of goodwill may appear not to decrease over time [, this is] because the ... goodwill that is purchased is being replaced by internally generated goodwill' and underlines the fact that IAS 38 on *Intangible Assets*, prohibits the recognition of internally generated goodwill as an asset (para. 47). Thus 'useful economic lives' under FRS 10 are likely to be considerably longer than under IAS 22. The fact that IAS 22 (1998) also has a rebuttable presumption of 20 years for useful life may therefore be a deceptive similarity. IAS 22 does not recognize indefinite life goodwill (para. 51). FRS 10 requires useful economic life to be reviewed at the end of each period and, if revised, the carrying value must be amortized over the revised remaining economic life. If this becomes more than 20 years from the acquisition date the second or third amortization and impairment requirements below apply.

Amortization and impairment: Figure VI.1 shows the relationship between useful economic life, amortization and impairment testing required by the standard:

- Goodwill with a useful economic life of not more than 20 years is to be amortized choosing a method that reflects the estimated depletion pattern. Straight-line depreciation is the benchmark unless another method can be shown to be more appropriate. An abbreviated impairment review is to be carried out at the end of the first full year following the acquisition, and in other periods if events or circumstances indicate that the carrying value may not be recoverable.
- Goodwill with a definite useful economic life of more than 20 years is to be similarly amortized, but a full impairment review must be carried out at the end of the first full year following the acquisition and at the end of each reporting period.
- Goodwill with an indefinite useful economic life is not to be amortized, but a full impairment review must be carried out at the end of the first full year following the acquisition and at the end of each reporting period.

If the conceptual perspective were taken that the primary justification for accounting for goodwill is to be based on the link between consolidated goodwill and the continuing goodwill element in the investment in the parent's individual financial statements, the use of impairment tests without amortization would seem to be the primary approach to preserving such a link. The rebuttable presumption of gradual amortization over a period of 20 years or less would then be seen only as a way of approximating to the actual impairment pattern, a 'cheaper' and more conservative alternative for groups which do not wish the expense of annual impairment tests. Though companies adopting this 'default' amortization approach would seem to meet IAS 22's requirements, one should note that FRS 10's useful economic life is differently defined to IAS 22.

First year impairment review: where the 20-year life presumption is not rebutted, only an abbreviated first year impairment review is required. This merely confirms that post-acquisition performance in the first year is not worse than pre-acquisition forecasts of the same period used to support the purchase price, and that any other previously unforeseen events or changes in circumstances do not indicate that the carrying values may not be recoverable. If this is not con-

Figure VI.1 Subsequent treatment of goodwill and intangible assets (FRS 10)

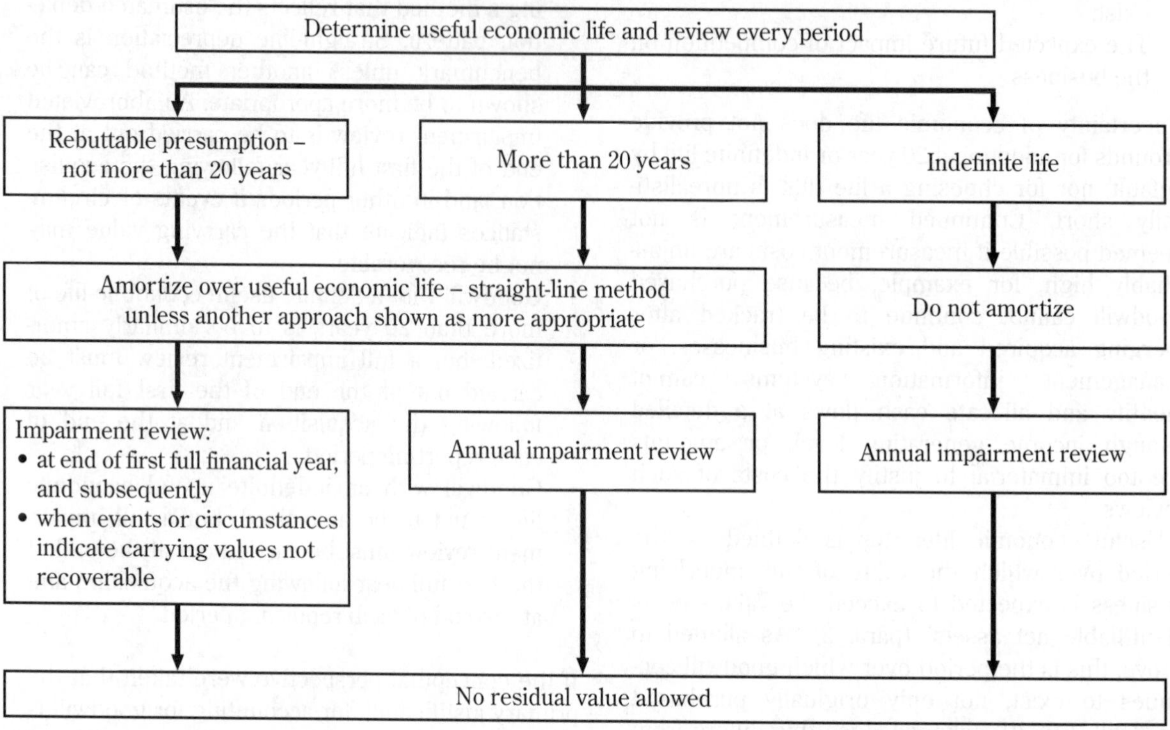

firmed by the abbreviated review and further, in all cases where the presumption is rebutted, the first year review must satisfy the full requirements of FRS 11, *Impairment of Fixed Assets and Goodwill.* According to FRS 10, impairments at this first year review reflect overpayment, events between acquisition and the review, or actual depletion exceeding the amortization charge. The standard comments that a belief that the value of goodwill will not be capable of continued measurement does not justify writing off the whole balance at the first year review and that under such circumstances it should be possible to carry out the review by updating investment appraisal calculations.

Subsequent impairment reviews: where the 20-year life presumption is not rebutted, subsequent impairment reviews are required only if events or changed circumstances indicate that carrying values may not be recoverable. Where it is rebutted they must be conducted annually. After the first year review subsequent reviews only need to be updated, and need not be onerous if expectations of future cash flows and discount rates have not changed significantly. Indeed FRS 10 comments that if there have been no adverse changes in key assumptions or substantial leeway between carrying value and estimated value in use, 'it may be possible to ascertain immediately that an income generating unit is not impaired' (para. 38).

True and fair departures for indefinite life goodwill: the use of impairment tests only for indefinite life goodwill under FRS 10 does not meet Companies Act requirements that goodwill treated as an asset must be amortized systematically over a finite period. However, the Act does allow departures from its requirements in order that a true and fair view may be given, provided certain disclosures are made. Therefore for all indefinite life goodwill, FRS 10 requires the necessary disclosures required by the Act and UITF 7, *True and Fair Override Disclosures*, be given. These include particulars of the departure, reasons for it, and its effect, in sufficient detail to convey to readers the circumstances justifying the use of the true and fair override. The reasons should include the explanation of special factors contributing to the durability of the acquired business or intangible asset (para. 58).

1.4.2.3 Accounting for negative goodwill

FRS 10 allows the recognition of negative goodwill when it exists but requires that, when it appears to arise, the fair values of assets at acquisition should be checked for impairment and liabilities at that date should be checked for omission and understatement. Negative goodwill should be disclosed on the face of the balance sheet immediately below positive goodwill, subtotaling to the net amount of positive and negative goodwill. Goodwill on a single transaction must not be divided into positive and negative components. The fair value of intangible assets at acquisition which do not have a readily ascertainable market value must be limited so as not to create or increase negative goodwill.

Negative goodwill is to be subsequently recognized in the profit and loss account on a two-tier basis:

- Up to the amount of the fair values of the non-monetary assets acquired, in the periods in which such assets are recovered through depreciation or sale.
- In excess of this amount, in the periods expected to be benefited. However, negative goodwill in excess of the fair values of non-monetary assets is expected to occur extremely rarely and in unusual circumstances which are unspecified.

FRS 10 considers that negative goodwill arises either because there has been a genuine bargain purchase, or because the purchase price has been reduced to take account of future costs or losses that do not yet represent identifiable liabilities at the acquisition date. Whereas IAS 22 considers that these two causes of negative goodwill can be separately identified and accounted for, FRS 10 adopts a common accounting treatment for both situations to prevent creative accounting. FRS 10 justifies its treatment on the following grounds. In genuine bargain purchases negative goodwill is an unrealized gain. The exposure draft, FRED 12, had proposed its immediate recognition in the statement of total recognized gains and losses. FRS 10 takes the more conservative view that the 'gain' should be recognized by releasing negative goodwill to the profit and loss account over the periods when the 'bargain' non-monetary assets are expensed. Respondents had pointed out that recognition of unrealized gains is not required under other UK standards. Where negative goodwill arises from the expectation of future reorganization costs, FRS 10 comments that matching negative goodwill to such reorganization costs would be inconsistent with the treatment of similar costs included in positive goodwill and that two distinct treatments could encourage creative accounting. It rationalizes releasing negative goodwill over the lives of the non-monetary assets in this case also, by regarding such assets as being encumbered or impaired. Further, it considers that most negative goodwill arising from this cause will disappear as a result of impairment reviews.

IAS 22, on the other hand, requires that where negative goodwill relates to expectations of future losses identified in the acquirer's plan for the acquisition but does not represent identifiable liabilities at acquisition, it should be recognized in income when the future losses and expenses are recognized. Otherwise negative goodwill not exceeding the fair values of acquired identifiable non-monetary assets should be recognized as income over the remaining weighted average useful life of the identifiable acquired depreciable or amortizable assets. Any excess should be recognized as income immediately.

1.4.2.4 Transitional arrangements

FRS 10 encourages but does not require that the accounting treatment of goodwill prior to its issue, predominantly immediate write-off direct to reserves, be adjusted. Under SSAP 22 (1984) permanent retention of goodwill had been expressly prohibited, and two accounting treatments allowed:

- Immediate write-off directly against reserves (its normal accounting treatment).
- Gradual amortization through the profit and loss account over its useful economic life (the allowed alternative).

Different treatments could be chosen for different acquisitions. SSAP 22 did not specify which reserves should be used for immediate write-off and a number of controversial practices had developed, including the use of a separate (debit) reserve for writing off goodwill. Most companies meeting Section 131 merger relief criteria, discussed earlier, wrote off goodwill against a merger reserve. Despite statutory restrictions on the use of the share premium account (Companies Act 1985, S 130), a number of companies had obtained court permission to write off goodwill against this reserve. The use of consolidated revaluation reserves for immediate goodwill write-off was specifically outlawed by the Companies Act 1985 revision (Sch 4, para. 34, extended to group accounts by Sch 4A, para. 1). Many groups also used the consolidated profit and loss reserve. FRS 10 does not require restatement, except that it prohibits goodwill being shown as a debit balance on a separate reserve – it requires any remaining former debit reserves to be offset against the profit and loss account (presumably balance sheet reserve) or another appropriate reserve. The amount offset should not be shown on the face of the balance sheet because respondents had criticized goodwill being shown in two different locations on the balance sheet with different amortization requirements for each. The cumulative amounts of positive goodwill and negative goodwill net of disposals, adjusted against reserves, must be disclosed as a note to the accounts.

FRS 10 encourages restatement by enabling restatement of any of the following categories of goodwill previously eliminated:

- All goodwill; or
- Only goodwill relating to acquisitions after 23 December 1989 where the necessary information is available and can be obtained without unreasonable expense or delay; or
- Only goodwill after the implementation of FRS 7.

The standard, however, expresses the categories above in terms of goodwill allowed to remain eliminated against reserves. If there is no restatement, all previously eliminated goodwill can remain eliminated against reserves. The 1998 revision of IAS 22 contains similar transitional requirements but relating to periods beginning before 1 January 1995, the date its previous incarnation, which required capitalization and amortization, came into force. It requires capitalization and amortization for all goodwill after that date. To the extent that a UK company does not choose to reinstate goodwill for periods beginning after 1 January 1995, its accounting treatment would still not accord with IAS 22.

FRS 10 requires that, for any goodwill reinstated:

- Any impairment loss attributed to prior periods must be determined according to the procedures in FRS 11, discussed below; and
- The notes to the statements must disclose original goodwill cost, amounts attributed to prior period amortization, and separately to prior period impairment; and
- It is not necessary to separately identify any intangibles that were previously subsumed in goodwill.

However, impairment losses that are recognized on implementing FRS 10 relating to previously capitalized goodwill or intangibles must be charged as an expense in the period. Unlike the treatment above of reinstated goodwill, impairment of which can be attributed to prior periods, the carrying value of previously capitalized items would just be deemed to be misstated.

For any goodwill not reinstated, the notes must state:

- The accounting policy followed in respect of that goodwill.
- The cumulative amounts of positive and negative goodwill adjusted against reserves, net of goodwill attributable to disposals before the balance sheet date.
- The fact that this goodwill had been eliminated as a matter of accounting policy and would be charged or credited in the profit and loss account on the subsequent disposal of the business to which it related (see 1.5.2.1).

Where for businesses acquired before 1 January 1989 it is not possible to determine attributable goodwill, this must be disclosed and reasons given.

1.4.2.5 *Dividends from pre-acquisition profits*

Either at the acquisition of a subsidiary undertaking or subsequently, it is possible for a subsidiary undertaking to distribute dividends in excess of its post-acquisition retained profits, or circumstances to exist such that its dividends are connected to its pre-acquisition retained profits. From a consolidated accounts perspective, such dividends out of pre-acquisition profits relate to the period when the undertaking was not a part of the group. Therefore they should be treated as a repayment of the acquisition cost. However, the Companies Act 1985 requirements relating to the recognition of investment income in individual company accounts ignore the group perspective. This may necessitate a consolidation adjustment to adjust what may be legitimately recognized as income by the parent in its own accounts to reflect the consolidated accounts perspective.

Legal requirements for determining investment income in the parent's individual company accounts do not require it to ascertain whether a dividend receivable is declared out of pre- or post-acquisition profits. However, they require an assessment whether, as a consequence of the dividend, it is necessary in the parent's individual company accounts to assess whether there is any impairment in value of the (fixed asset) investment in the subsidiary – per FRS 11's impairment requirements discussed in VI.1.4.4.2. To the extent that there is no impairment, the dividend can be legitimately regarded as the parent company's income. Indeed, Davies, Paterson and Wilson (1997, p. 290) consider that such legal requirements have been interpreted so as to allow in principle an acquiring company 'to distribute immediately all the pre-acquisition profits shown in the subsidiary's balance sheet provided that it could foresee that the subsidiary would earn an equivalent amount of profits in the future', though they question that such an interpretation could be regarded as good accounting.

To the extent that in the parent's individual accounts the parent deems it necessary to write down its impaired investment in the subsidiary undertaking by part or all of the amount of the pre-acquisition dividend, no further consolidation adjustment is necessary. However, to the extent that the parent treats any part of such dividends from pre-acquisition profits as its individual company investment income, a consolidation adjustment is necessary.

Because the dividend is merely a cash transfer between group companies, it should have no effect on the consolidated accounts. The declaration of the dividend by the subsidiary reduces its pre-acquisition retained profits. To the extent that the investment is not written down in the parent's individual accounts as a result of the dividend, consolidated goodwill will increase as a result of the declaration, since pre-acquisition equity has been reduced but the acquisition cost of the investment has remained unchanged. Further, the dividend, if treated as part of the parent's income, will increase consolidated retained profits. Therefore the consolidation adjustment necessary to remove this effect is to reduce (credit) consolidated goodwill and also to reduce (debit) consolidated retained profits by the amount of any dividend from pre-acquisition profits treated as investment income by the parent.

Prior to the Companies Act 1981 there was a requirement for the accounting treatment to be treated consistently in the individual company and group accounts. Wording changes led Martindale to speculate whether the change could have been caused by unforeseen drafting errors, or may have been a result of a policy decision that policing of the consistency between individual company and consolidated accounts was an accounting profession matter, or even could have been linked to the fact that merger accounting was first enabled by that Act. The last reason would have necessitated allowing groups to distribute pre-acquisition profits if merger accounting was used (Martindale, 1982, p. 130). It seems that as the 'error' was carried forward in subsequent Companies Acts, the last two explanations are the most likely, which necessitates no consolidation adjustment under merger accounting, but under acquisition accounting, the consolidation adjustments described above.

1.4.3 Accounting for intangibles

1.4.3.1 Companies Act requirements

For intangible assets other than goodwill, Companies Act requirements allow considerably more flexibility in accounting treatment. For example, other intangibles can be carried at current cost under the alternative accounting rules of the

Companies Act 1985, whereas goodwill cannot. The balance sheet formats in Schedule 4 of the Companies Act (see United Kingdom – Individual Accounts V) distinguish under 'Intangible assets':

- Development costs.
- Concessions, patents, licences, trademarks and similar rights and assets.
- Goodwill.
- Payments on account.

In the Act's 'Notes on balance sheet formats', the note to heading 2 comments that 'amounts in respect of assets shall only be included in a company's balance sheet if either

- The assets were acquired for valuable consideration and are not required to be shown under goodwill; or
- The assets in question were created by the company itself.

These requirements make clear that, under the Act, other intangibles are a separate category from goodwill. The general accounting rules relating to fixed assets also apply. Immediate write-off is justifiable under the Act for reasons similar to those used for goodwill (see VI.1.4.2.1).

1.4.3.2 FRS 10's requirements

FRS 10 on intangibles also does not apply to [small] reporting entities subject to the FRSSE, nor to oil and gas exploration and development costs, research and development costs, and any other intangible asset specifically addressed by another [i.e. including a potential future] accounting standard.

Intangible assets are defined as 'non-financial assets that do not have physical substance but are identifiable and are controlled by the entity through custody or legal rights' and, as specified in the Companies Act 1985, must be capable of being disposed of separately without disposing of the business of the entity. If it can only be disposed of as part of the revenue-earning activity to which it contributes, it is to be regarded as indistinguishable from goodwill for accounting purposes. FRS 10 stresses that, where expected future benefits are not controlled through legal rights or custody, e.g. in the case of a portfolio of clients or a skilled staff, the entity does not have sufficient control to recognize an intangible asset. Further, software development costs directly attributable to bringing a computer system or computer-operated machinery into working condition for its intended use, should be treated as part of the hardware costs and not as a separate intangible asset. Also, assets that are not fixed assets, such as prepaid expenditure, are not intangible assets. Intangible assets purchased separately from a business and internally generated intangibles (which can only be capitalized if they have a readily ascertainable market value) are considered in United Kingdom – Individual Accounts VI.1.1. Only intangible assets acquired as part of the acquisition of a business are considered here – the third column in Figure VI.2 which shows initial recognition of intangibles.

An intangible asset acquired as part of the acquisition of a business should be capitalized at its fair value, separately from goodwill, only if this value can be measured reliably on initial recognition. If not, such intangibles are to be subsumed within the purchased goodwill amount. FRS 7 requires that fair value should be replacement cost, normally market value, but which can be estimated by other methods. FRS 10 allows that techniques based on, for example, 'indicators of value' such as multiples of turnover, or estimates of the present value of royalties that would be payable to license the asset from a third party, can be used by entities regularly involved in the purchase and sale of unique intangible assets (para. 12).

The fair values of intangible assets acquired as part of the acquisition of a business which do not have readily ascertainable market values are limited to amounts that do not create or increase any negative goodwill at acquisition. The term 'readily ascertainable market value' in this context means that the value is established by reference to an active market for a homogeneous population of assets equivalent in all respects, to which the asset belongs. This excludes unique intangibles such as brands, publishing titles, etc. (para. 2). Further discussed in United Kingdom – Individual Accounts are matters relating to intangibles in a non-group setting, including, for example, that only intangible assets with readily

Figure VI.2 Initial recognition of intangible assets

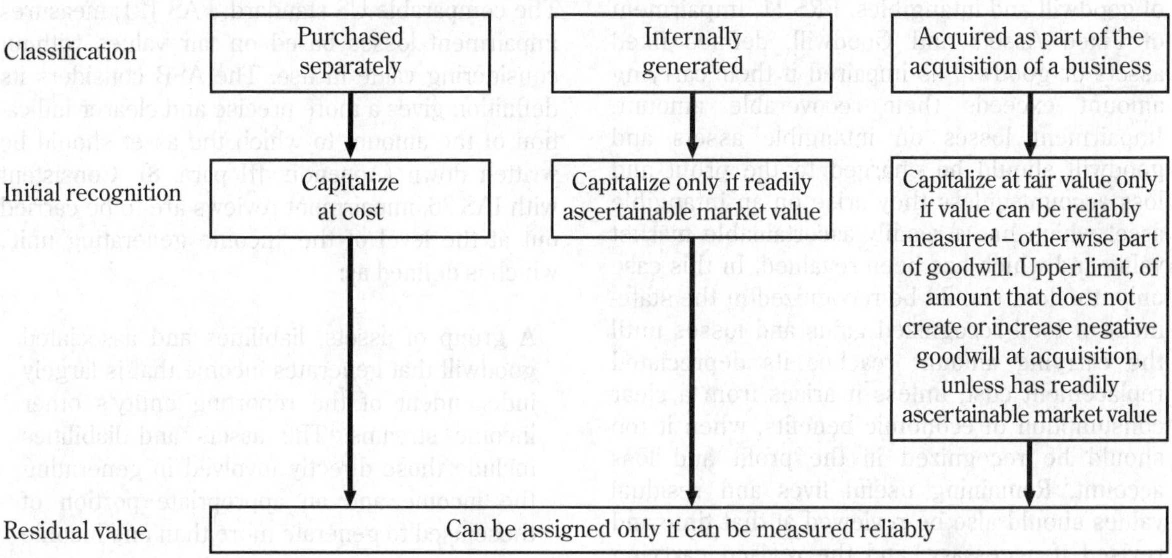

ascertainable market values can subsequently be revalued and must continue to be revalued. This can affect accounting treatment on restoration of past losses and subsequent amortization, which must be based on revalued amounts and remaining economic lives. Note also that internally generated intangibles can only be capitalized if they have a readily ascertainable market value and that this requires an active market for a homogeneous population of assets equivalent in all respects, to which the asset belongs. Such a requirement is likely to prevent the capitalization of internally generated brands and publishing titles.

Accounting requirements relating to useful economic lives, reviews of economic lives and amortization approach are identical to goodwill, discussed earlier. FRS 10 states that the useful economic life of intangible assets must not include additional periods based on assumed renewal of rights, where access to economic benefits is secured through legal rights unless such renewal is assured. Renewal costs must not be included in the asset amount. Unlike goodwill, a residual value may be assigned to an intangible asset only if such a value can be measured reliably, usually when there is a market for the residual asset or there is a legal or contractual right to receive a certain sum at the end of its period of use.

1.4.4 Procedures for impairment reviews for goodwill and intangibles

1.4.4.1 Companies Act requirements

Paragraph 19 of Schedule 4 of the 1985 Act allows provisions for diminutions in value of fixed asset investments, but requires such provisions for any fixed asset if the reduction in value is expected to be permanent. If not shown in the profit and loss account each category of provision must be disclosed separately, or in aggregate in the notes. Where the reasons for making such provisions have ceased, they should be written back to the extent they are no longer necessary, with similar disclosure requirements.

1.4.4.2 FRS 11 – general principles

This matter is only dealt with here in the

context of group accounts, and hence in terms of goodwill and intangibles. FRS 11, Impairment of Fixed Assets and Goodwill, defines fixed assets or goodwill as impaired if their carrying amount exceeds their recoverable amount. Impairment losses on intangible assets and goodwill should be charged to the profit and loss account unless they arise on an intangible asset which has a readily ascertainable market value and which has been revalued. In this case only, the loss should be recognized in the statement of total recognized gains and losses until the carrying amount reaches its depreciated replacement cost, unless it arises from a clear consumption of economic benefits, when it too should be recognized in the profit and loss account. Remaining useful lives and residual values should also be reviewed at that time and revised if necessary, and the revised carrying amount amortized over the revised estimate of remaining economic life. Brown (1998, p. 84), an ASB project director, comments that FRS 11 effectively regards all diminutions in value below carrying amount as permanent, and that the ASB found it 'inherently very difficult to judge whether a fall in a fixed asset's value will be temporary or permanent', which had led to inconsistencies in the way that impairment losses had been measured.

The benchmark for impairment, the recoverable amount, is defined as the higher of the asset's value in use ('the present value of the future cash flows obtainable as a result of the asset's continued use, including those from its ultimate disposal') and its net realizable value ('the amount at which an asset could be disposed of, less any direct selling costs') (para. 2). This is based implicitly on the 'value to the business' concept. If the recoverable amount is below the asset's carrying amount, the asset would not be worth replacing at that amount and would be valued at the value in use if the asset is worth keeping but not replacing, but at its net realizable value if it is neither worth keeping or replacing.

Much of the complexity of FRS 11 is due to the computation of the 'value in use' component. Because assets interact to generate future cash flows it is often impossible to allocate in a non-arbitrary manner such future cash flows to individual assets and liabilities (see Thomas, 1975). The comparable US standard, FAS 121, measures impairment losses based on fair values without considering value in use. The ASB considers its definition gives a more precise and clearer indication of the amount to which the asset should be written down (Appendix III para. 8). Consistent with IAS 36, impairment reviews are to be carried out at the level of the 'income generating unit', which is defined as:

> A group of assets, liabilities and associated goodwill that generates income that is largely independent of the reporting entity's other income streams. The assets and liabilities include those directly involved in generating the income and an appropriate portion of those used to generate more than one income stream.

By doing this the ASB hopes it can overcome criticisms of too much arbitrary allocation of cash flows in determining value in use. However, it cannot avoid some allocation of central cash flows and the ASB claims that earlier field tests demonstrate the feasibility of its income generating unit methodology. The determination of the carrying amount and net realizable value components for impairment testing is less of a problem. IAS 36 contains more detailed guidance on determining net realizable value (net selling price). It follows a similar approach, but using the term 'cash generating unit'.

When impairment losses at income generating unit level have been determined, FRS 11 requires they are allocated to different categories of assets in the unit in the following pragmatic order – categories of assets with most subjective valuations first:

1 to goodwill;
2 next to any capitalized intangible assets (though no intangible asset with a readily ascertainable market value should be written down below its individual net realizable value); and
3 finally to tangible assets in the unit on a pro rata or more appropriate basis (provided individual tangible assets are not written down

below reliably measured net realizable values).

IAS 36 combines the last two categories. The ASB aligns the treatment of intangibles and goodwill, whereas the IASC aligns the treatment of intangibles with tangible assets.

1.4.4.3 *Allocation of impairment losses*

If goodwill cannot be feasibly apportioned its impairment can be assessed as part of an overall impairment test at a higher level of aggregation, by aggregating groups of similar IGUs if such units were acquired as part of the same investment and are involved in similar parts of the business. In such a case capitalized intangible and tangible fixed assets would be tested for impairment at the individual IGU level, and then at the level of appropriate groupings of IGUs, impairment is examined to assess recoverability of goodwill.

Examples of this procedure are shown on the next page. It is easy to see how such procedures might be used in a creative accounting way, by comparing the end results of the two examples.

1.4.4.4 *Acquired businesses merged with existing operations*

In order to 'track' purchased goodwill subsequent to an acquisition when acquired businesses are merged with existing operations, internally generated goodwill on the existing operations has to be estimated and added to the carrying amount of the acquired business for impairment review purposes. The IGU will then comprise both merged businesses. The subsequent notional carrying amount of the internally generated goodwill is assumed to be subject to the same pattern of amortization as the purchased goodwill. At the date of any impairment test, the carrying value of the IGU will include the carrying amount at that date of the purchased goodwill and the notional carrying amount of the original internally generated goodwill at the date of acquisition. The general principle is that at any impairment losses will be allocated pro rata between purchased goodwill and notional internally generated goodwill, but only the former will be recognized in the financial statements. Any amount remaining will then be allocated to capitalized intangibles and tangible assets of the merged business. An exception to this general principle is that any initial impairments arising on first merging the businesses must be allocated solely to the purchased goodwill within the newly acquired business. This is justified by FRS 11 on the grounds that such initial impairment would be solely because of the incremental effect of the acquired business on the combined unit, if internally generated goodwill was correctly estimated and so must, by definition, apply to the acquired business (para. 53). IAS 36 does not contain such requirements for merged operations since, whilst the IASC agrees in theory, it believes 'that it would be costly and difficult, if not impossible, to distinguish items related to internally generated goodwill, especially if businesses are merged' (para. B104).

1.4.4.5 *Reversals of impairment losses*

FRS 11 contains different requirements for the recognition of reversals of past impairment losses for:

(a) Goodwill and intangibles; and
(b) Investments in subsidiaries, associates and joint ventures.

In case (a), for goodwill, reversals of past impairment losses should be recognized in the current period if, and only if, the original recognition of the impairment loss was caused by an external event, and subsequent external events clearly and demonstrably reverse the effects of that event in a way that was not foreseen in the original impairment calculations. Intangibles are covered by the same requirement. Additionally, impairment loss reversals are to be recognized where they arose on an intangible with a readily ascertainable market value and the net realizable value based on that market value has increased to above the intangible asset's impaired carrying amount (para. 60). In both sets of circumstances reversals should be recognized only to the extent that the carrying amount is increased up to the amount

EXAMPLES – ALLOCATION OF IMPARIMENT LOSSES BETWEEN GOODWILL, INTANGIBLES AND NET ASSETS

1. A business comprises two income generating units. Some years after acquisition, the carrying amounts of their net assets, goodwill and intangibles and each IGU's value in use, no reliable net realizable amount for each IGU being available, are as follows:

Income generating unit	1	2	Total
Carrying value (GBP million):			
Goodwill	20	30	50
Capitalized intangibles	10	8	18
Tangible net assets	50	62	112
	80	100	180
Value in use (GBP million)	95	60	155

An impairment loss of GBP40 million will be recognized for the second IGU, to be allocated GBP30 million to goodwill, GBP8 million to intangibles, and GBP2 million to tangible net assets.

2. Suppose both IGUs were acquired as the result of the same investment, and so the carrying amount of the goodwill of GBP50 million has not been allocated to individual units. The above information has been recast as follows:

Income generating unit	1	2	Goodwill	Total
Carrying value (GBP million):				
Goodwill			50	50
Capitalized intangibles	10	8		18
Tangible net assets	50	62		112
	60	70	50	180
Value in use (GBP million)	95	60		155

An impairment loss of GBP10 million would first be recognized in the second IGU, and allocated GBP8 million to intangibles and GBP2 million to tangible net assets. At the next level of aggregation the total impairment loss is GBP25 million (= 180 − 155), so a further GBP15 million will be allocated against goodwill.

It is easy to see how such procedures might be used in a creative accounting way, by comparing the end results of the two examples.

that it would have been if the original impairment had not occurred (para. 61).

In case (b) the reversal of previously recognized impairment losses should be recognized in the profit and loss account when the recoverable amount of the investment increases because of a change in economic conditions or in the use of the asset, but only to the extent that the carrying amount is increased up to the amount if the original impairment had not occurred (para. 56). Increases in value arising simply because of the passage of time (as future cash flows used in discounting become closer) or where the occurrence of forecast cash outflows means that they are no longer included in the calculation of value in use, may not be recognized as reversals of impairment losses. FRS 11 is ambiguous in that it does not make clear whether the term 'investments in associates and joint ventures' refers to such investments in individual entity accounts, in consolidated accounts or both. If one took the view that it applies to consolidated financial statements, this would mean that reversals of impairments in goodwill arising on subsidiaries are to be recognized on a different basis to those on goodwill arising on associates and joint ventures (which are implicitly included in the amount of investments in associates and joint ventures under the equity method). Presumably FRS 10 meant it only to apply to individual entity accounts, since the heading 'investments in subsidiaries' would only appear in consolidated financial statements for subsidiaries excluded from consolidation.

As in the case of allocation of impairment losses, IAS 36 aligns conditions for reversals of impairments for intangibles with its criteria for tangible fixed assets, whereas the ASB, following its reasoning in FRS 10, aligns them with goodwill. Unlike IAS 36, FRS 11 does not contain requirements for income generating units for the order of allocating reversals of impairments between tangible and intangible assets, and goodwill.

1.4.5 Disclosures relating to goodwill and intangibles

Required disclosures under FRS 10 relating to goodwill and intangibles are shown in Table X.1. They include disclosures relating to positive and negative goodwill and to transitional arrangements under FRS 10.

1.4.6 Piecemeal acquisitions

When subsidiaries are acquired in a single transaction, goodwill is calculated as the difference between the fair values of the consideration given and the identifiable net assets acquired; when acquired in a series of transactions different views exist on how goodwill, and hence post-acquisition profits, should be calculated.

1.4.6.1 Conceptual alternatives

Under a single step approach goodwill is calculated as the difference between the aggregate consideration and the fair value of the net assets of the subsidiary at the date control was obtained. In a pure step-by-step approach, total goodwill would be calculated as the aggregate of a series of slices. For each slice of ownership purchased, goodwill would be calculated at that date as the difference between the fair value of the consideration for that slice less the fair value of the relevant proportion of net assets purchased. This approach is termed by the US Discussion Memorandum, Consolidation Policy and Procedures (1992) the parent approach. Post-acquisition profits would also be made up of the sum of a series of slices, each being the relevant proportion of profits from the date that an ownership slice was purchased to the current date. If rigorously followed, the individual assets and liabilities of the subsidiary would be stated at composite fair value amounts, the sum of differently dated fair values based on when the relevant ownership portions in the individual asset were purchased. Minority interests would be based on the original book value of the subsidiary's net assets. This approach can be modified so that, as each slice is purchased, the identifiable assets and liabilities of the subsidiary would be completely revalued rather than just the proportion for the slice purchased.

The credit entry for this extra revaluation would be to consolidated revaluation reserves, the corresponding debit 'topping up' individual assets and liabilities from a composite basis to complete fair values at the date the last slice was purchased. Minority interests would then be based on the fair values of the subsidiary's net assets at that date.

Different proposals have also been made as to how the fair value of the aggregate purchase consideration, the total investment, should be calculated:

- At cost (including all the previous slices at their cost).
- Using equity accounting for all previous slices, from the date of purchase to the date control passes. This alternative treats the subsidiary as though it had previously been an associate.
- Recalculated at fair value at the date control passes, revaluing the previous consideration. The FASB Discussion Memorandum calls this the entity approach. Beyond the point control passes it suggests further revaluations should not take place, since further share acquisitions would be seen as transactions between existing shareholders of the group entity.

The last two alternatives adjust the consideration so that it and the fair values of the net assets acquired relate to the date control passes. Choice between the alternatives may be affected by the route to control and feasibility of obtaining fair values. For example, if a 15 per cent holding were obtained and later a further 45 per cent second stake, the first slice would probably have been accounted for as a passive investment at cost in the consolidated accounts. This makes the application of step-by-step approaches problematical. However, if a 25 per cent initial holding had been obtained and later a further 35 per cent stake, the step-by-step approach would be sensible if the equity approach had been used for the former stake. It is possible that goodwill in a 25 per cent associate would have been calculated on book values if the parent had not been able to obtain fair value information. Thus fair values for the step-by-step approach may not be available.

1.4.6.2 Control newly acquired

The Companies Act 1985 requires that the identifiable assets and liabilities of the new subsidiary undertaking be included in the consolidated financial statements at their fair values at the date it becomes a subsidiary undertaking (Sch 4A, para. 9). This prevents the use of the pure step-by-step approach alternative discussed above (see VI.1.3). FRS 2 takes a pragmatic approach to accounting for previous stakes held prior to control being acquired. The general rule is that goodwill is to be calculated on a single-step basis at the date control passes, and fair values are to be determined at that date even where the group's interest is acquired in stages (para. 50). The cost of the investment used to determine this goodwill would be an aggregate of the costs incurred in gaining control. However, FRS 2 recognizes that using the current fair values of identifiable assets and liabilities together with acquisition costs, some of which may have been established at considerably earlier dates 'may result in accounting that is inconsistent with the way the investment has been treated previously and, for that reason, fail to give a true and fair view'.

So, for example, if the new subsidiary were previously an associated undertaking, then,

> in the rare cases where the Schedule 4A para. 9 calculation of goodwill would be misleading, goodwill should be calculated as the sum of the goodwill arising from each purchase of an interest in the relevant undertaking adjusted as necessary for any subsequent diminution in value. Goodwill ... should be calculated as the difference between the cost of (each) ... purchase and the fair values (at that date) of the (attributable) ... identifiable assets and liabilities. The difference between the goodwill calculated on this method and that calculated on the method provided by the Act is shown in reserves. (para. 89)

This is deemed to be a circumstance where a true and fair override is necessary and the disclosure requirements of S 227(6) showing the particulars, reasons and effects of the departure from the Act's provisions must be followed (para. 89). Presumably the modified step-by-step approach is envisaged in this case so that identifiable assets and liabilities are still valued on a homogeneous basis at their fair values at the date control is obtained. It is not made explicit whether this departure extends to allowing the pure step-by-step approach with its composite asset measurement basis, which would generally be inconsistent with the UK view of a group. However, the requirements below for restatement to fair values of all assets and liabilities on a change in a stake in a subsidiary undertaking after control has been acquired, indicate a pure step-by-step approach is not envisaged by FRS 2 (para. 90). A further example of where departure is allowed is where a previous investment has been substantially restated, for example, by a provision for impairment (para. 89). IAS 22 requires the step-by-step approach and, in terms of the discussion of conceptual alternatives discussed earlier, allows composite [i.e. slice-by-slice] fair values or complete fair values and, if the latter is chosen, the difference between the two bases is to be reflected in consolidated revaluation reserves (paras. 35–36).

1.4.6.3 Increasing a stake in an existing subsidiary

According to FRS 2, 'when a group increases its interest in an undertaking that is already a subsidiary undertaking, the identifiable assets and liabilities should be revalued to fair value and goodwill arising on the increase in interest should be calculated by reference to those fair values (unless the change from previous carrying values is immaterial)' (para. 51). Earlier UK practice had not restated fair values under these circumstances, but FRS 2 argues that calculating incremental goodwill based on carrying values rather than current fair values might have a confounding effect and partially represent the increase in fair values since control was achieved (para. 90).

1.5 Disposals, restructurings, reorganizations and terminations of subsidiary undertakings

FRS 2 requires (paras. 45–47) that:

- The date of disposal is defined as that when control is relinquished.
- Consolidated financial statements in the period of disposal should include the results of the subsidiary up to the date of disposal, and any gain or loss on disposal to the extent it has not already been provided for.

The gain or loss is to be determined by comparing the carrying value of the subsidiary's net assets attributable to the group, including any related goodwill not previously written off through the profit and loss account, with proceeds received, whether the cause of the undertaking ceasing to be a subsidiary is a direct disposal, deemed disposal or other event.

1.5.1 Provisions for restructurings, reorganizations, terminations and sales, and discontinued operations

FRS 12 subsumes provisions for reorganizations, terminations, closures and sales within the overall description of 'restructuring provisions'. Its requirements apply to companies as well as groups, but are discussed here since they arise mainly in a group context. Ernst and Young (FRS 12, 1998, p. 29) express concern that the very wide definition given to 'restructuring' will give users the impression that reorganizations are separate from 'normal' costs of running a business in a dynamic environment and invite management to classify many types of costs under this segregated heading. Reorganizations at acquisition are discussed in VI.1.3.3.4 and dealt with by FRS 7. The general requirements relating to provisions in FRS 12, *Provisions and Contingencies,* issued in September 1998, are examined in United Kingdom – Individual Accounts VI.6.3. It requires provisions to be recognized only when the entity has a legal or constructive obligation as a result of a past event; it is probable that a transfer of economic benefits will be required to settle the obligation; and a reliable estimate can be made of the obligation (para. 14).

FRS 12 states that a constructive obligation to restructure arises only when an entity:

- Has a detailed formal plan for the restructuring identifying at least the business or part of a business concerned, the principal locations affected, the location, function and approximate numbers of employees who will be compensated for terminating their services, the expenditures that will be undertaken, and when the plan will be implemented; and
- Has raised a valid expectation in those affected by it that it will carry out the restructuring by starting to implement the plan or announcing its main features to those affected by it (para. 77).

Implementation must be planned to begin as soon as possible and completed in a timeframe that makes significant changes to the plan unlikely (para. 79). A public announcement or a management board decision alone is not sufficient. However, they may be sufficient if combined with earlier events such as employee negotiations over termination payments or with buyers for an operation, subject only to board approval. Provisions relating to the sale of an operation can only be made where there is a binding sale agreement (para. 83). Some criticize the ASB's rejection of management intentions as the basis for provisions, citing inconsistency with intention-based criteria used in FRS 11 for asset impairments since both may arise from the same company board decision (Ernst and Young, FRS 12, 1998, p. 33).

FRS 12 states that restructuring provisions should only include direct expenditures arising from the restructuring which are both those necessarily entailed by the restructuring and not associated with the ongoing activities of the entity (para. 85). They must not include expenditures relating to the future conduct of the business which are not liabilities at the balance sheet date, such as retraining or relocating existing staff, marketing or investing in new systems and distribution networks. Consistent with its prohibition on provisions for future losses, identifiable future losses up to the date of a restructuring must not be included unless they relate to an onerous contract, nor must any gains on the expected disposal of assets even if they are included in the restructuring plan (paras. 87 and 88). FRS 12 does not give specific examples of types of costs to be included in calculating such provisions. Its exposure draft, FRED 14, had cited costs of making employees redundant, of terminating leases and other contracts whose termination results directly from the reorganization, and also expenditures to be made in the course of the reorganization such as employees' remuneration while they are engaged in such tasks as dismantling plant, disposing of surplus stocks and fulfilling contractual obligations (FRED 14 para. 61).

FRS 12's general disclosure requirements on provisions are not discussed here. FRS 3 requires the setting up of termination and sale provisions and, when they are used up, the costs incurred and the amount previously provided to be disclosed on the face of the profit and loss account (para. 18) (see Table X.1). Prior to both standards common prior practice had been to set up such provisions based on management intentions and to net subsequent costs against the provision.

Discontinued operations: FRS 3 allows provisions for terminations or sales by a reporting entity to be made only to the extent that obligations have been incurred that are not expected to be covered by future profits. It contains similar requirements to FRS 12 regarding the level of commitment necessary and limits the amount of such provisions to direct costs of the sale or termination. However, unlike FRS 12 it specifically includes any operating losses up to the date of sale or termination after taking into account aggregate profit to be recognized in the profit and loss account from the future profits of the operation (para. 18). This seems to be in conflict with FRS 12 since sales and terminations are included within its general category of restructurings. Thus it appears that groups can include operating losses in provisions relating to sales or terminations, but not otherwise.

FRS 3 requires the profit and loss account down to operating profits for the year to be analyzed between continuing and discontinued operations. Discontinued operations are defined as follows:

Operations of the reporting entity that are sold or terminated and that satisfy all of the following conditions:

(i) The sale or termination is completed either in the period or before the earlier of three months after the commencement of the subsequent period and the date on which the financial statements are approved.

(ii) If a termination, the former activities have ceased permanently.

(iii) The sale or termination has a material effect on the nature and focus of the reporting entity's operations and represents a material reduction in its operating facilities resulting either from its withdrawal from a particular market (whether class of business or geographical) or from a material reduction in turnover in the reporting entity's continuing markets.

(iv) The assets, liabilities, results of operations and activities are clearly distinguishable, physically, operationally and for financial reporting purposes.

Operations not satisfying all these conditions are classified as continuing (para. 4).

The setting up of the termination or sale provisions and asset write-downs must be included separately under continuing operations unless the sale or termination has been completed by the qualifying date in (i) above. Only then are they to be classified under discontinued operations. Any such provisions initially set up under continuing operations could then be used in a later period to offset the results of sale or disposal and be reclassified under discontinued operations. Disclosure and analysis of the use of the provision separately between operating losses and losses on sale or termination would be required.

1.5.2 The treatment of goodwill on disposals

As stated in VI.1.5, the gain or loss on disposal of a subsidiary is to be determined by subtracting consolidated carrying values of the subsidiary's net assets at disposal from the consideration received. Problems had arisen whilst immediate write-off of goodwill to reserves had been allowed, in that many groups calculated gains or losses based on carrying values excluding consolidated goodwill. They therefore reported much higher disposal gains or lower losses on disposal than if gradual amortization had been used, when the net book amount of goodwill would have been included in the carrying values of net assets. Immediate write-off would also have reported lower profit and loss charges over the period the subsidiary was controlled. The way that standard setters addressed this problem is discussed below. Though prospectively such a problem should disappear since FRS 10, Goodwill and Intangible Assets, prohibits immediate write-off, its transitional provisions mean that many existing subsidiaries at the date it came into force may continue to have net assets measured under that approach.

1.5.2.1 UITF Abstract 3 and FRS 10

In 1991 Urgent Issues Task Force Abstract 3, Treatment of Goodwill on Disposal of a Business, required the profit or loss on the sale of businesses to be adjusted by writing back any goodwill written off at acquisition to the extent that it had not previously been charged in the profit and loss account. Subsequently this was included in FRS 2's definition of carrying value on disposal (see VI.1.5) and equivalent requirements are now in FRS 10 (para. 71c). Groups which use immediate write-off will show lower profits and higher losses on disposal than similar groups using gradual amortization since the original cost of goodwill, 'the amount not previously charged in profit and loss', must be included, rather than the net book amount of goodwill under the latter approach. Over the period the investment in the business has been held, in principle, both will result in the same overall profit and loss account charges. The lack of annual profit and loss account amortization under immediate write-off are offset in aggregate by an equivalent lower profit or higher loss on disposal. These different effects will still arise because under the transitional goodwill requirements of FRS 10 immediately written off goodwill may continue to exist. If such goodwill is retrospectively restated, *Company Reporting* (March 1998, p. 8) points out that prior year goodwill amortization will be accounted for as a prior year adjustment without ever being charged to any period's profit and loss account. It is concerned that the above profit and loss 'equivalence' between immediate write-off and capitalization and amortization approaches will be lost.

Initially, a number of groups reported immediately written off goodwill brought back to the profit and loss account on disposal separately from the gain or loss on disposal (e.g. see Company Reporting, February 1993, p. 2). UITF information sheet No. 6, issued on 17 December 1992, made clear that such goodwill must be included as part of the profit or loss on disposal, or leading to a single subtotal showing profit or loss on disposals. It also commented that other transfers between profit and loss and reserves should not be used to mask this figure (para. 2). *Company Reporting* (March 1998, p. 3) indicates that 94 per cent of applicable groups now report such a goodwill charge as part of the gain or loss on disposal as opposed to 69 per cent in 1994 (p. 5). Occasionally companies adjust goodwill that had been immediately written off for impairment prior to determining the write-back – such an impairment charge is shown as a separate charge to profit and loss and therefore the 'impaired' value [a pseudo-net book amount] is used to determine profit and loss on disposal (p. 6).

1.5.3 The treatment of foreign currency translation differences on disposals

There is an apparent inconsistency in UK practice on disposals of subsidiaries, in the treatment of goodwill immediately written off directly to reserves (and which may remain there under FRS 10's transitional provisions) and financial statement translation gains and losses under the closing rate approach. Though both are taken direct to reserves, goodwill must be written back to profit and loss in determining the gain or loss on disposal (see VI.1.5.2.1), but closing rate translation differences plays no part in determining such a gain or loss (see Pereira, Paterson and Wilson, 1992, p. 122 and V.4.2.3). The fact that such translation differences are not subsequently written back to profit and loss in determining gains or losses on disposal of foreign subsidiaries is consistent with FRS 3's intention that gains and losses, once reported in the 'Statement of Total Recognized Gains and Losses', should not subsequently be reported in the consolidated profit and loss account – a difference in practice with US financial reporting and international standards (see VI.4). However, immediate write-off of goodwill direct to reserves is not included by FRS 3 in its 'Statement of Recognized Gains and Losses', but in another statement, 'Note on movements in shareholders' funds'. This is consistent with SSAP 22, Accounting for Goodwill, characterizing immediate write-off as a policy decision not a valuation adjustment, whereas, translation differences under the closing rate approach have some similarities with revaluation adjustments of fixed asset investments (see V.4.2). Hence the different treatment – immediate goodwill write-off is a reserve movement which is *not* a recognized gain or loss in FRS 3's terms.

1.5.4 Partial disposals

FRS 2 requires that gain or loss is to be determined as

> [the] difference between the carrying amount of the net assets of the subsidiary attributable to the group . . . before the reduction (in interest) and the carrying amount attributable to the group's interest after the reduction together with any proceeds received. The net assets compared should include any related goodwill not previously written off through the profit and loss account. (para. 52)

Effectively, in ordinary sales, this requirement compares proceeds with the attributable portion sold. So if a 70 per cent holding were reduced to 50 per cent, the carrying amount prior to disposal, i.e. 70 per cent of the net assets of the subsidiary, would be subtracted from the amount derived from the holding as a result of the disposal, i.e. the sum of proceeds plus the remaining 50 per cent of the subsidiary's carrying values. This results in a net deduction from the proceeds of the 20 per cent disposed of. Minority interests would increase by 20 per cent of the carrying values of the net assets disposed of (excluding goodwill). Some argue that under an entity/economic unit view of the group, partial disposal profits are analogous to those on intragroup trading. However, FRS 2 comments that, whereas intragroup trading is wholly between consolidated undertakings under common control, partial disposal involves a direct transaction with third parties (para. 91). Crichton (1990a, p. 27) considers the main criterion for reporting gains and losses on partial disposals is usefulness to the investors in the parent and so should be reported. If by partial disposal a former subsidiary becomes an associate, profits or losses are determined in the same way, but the remaining attributable net assets and goodwill would be aggregated into an equity accounted equitized investment and subsequently reported as discussed below (see VII.2.2).

Deemed disposals can occur through share issues by subsidiary undertakings or rights issues by subsidiaries not taken up by the parent. Applying FRS 2's definition to the rights issue case, the profit or loss on disposal would be determined by subtracting the parent's proportion of the carrying value of the subsidiary's net assets prior to the rights issue, from the amount derived from that holding as a result of the parent not taking up the issue. This would be the proceeds from the rights sold by the parent, plus the parent's new proportionate stake as a result of not taking up the issue, applied to the carrying value of the subsidiary's net assets immediately after the issue (see, for example, Patient, Faris and Holgate, 1993, p. 102).

1.6 Minority interests

1.6.1 Calculation basis

Aggregate minority interests should be placed next to capital and reserves. Profits and losses in subsidiary undertakings should be apportioned between their controlling and minority interests according to their holdings over that relevant period (FRS 2, para. 37), and the minority share in profit or loss on ordinary activities must be shown as an aggregate deduction before extraordinary items. Their aggregate interest in extraordinary profits or losses must also be disclosed separately (see also IX). Though not separately disclosed, the balance sheet aggregate for minority interests is increased by the total minority interest in the profit or loss for the year plus any reserve movements in the subsidiary or applicable consolidation adjustments, and decreased by the amounts of dividends paid to them.

Minority interests under acquisition accounting in the UK include fair value adjustments at acquisition (see VI.1.2) and adjustments to eliminate unrealized profits on intragroup transactions when a less than 100 per cent-owned subsidiary is the originating company (see VI.2.2.1). They also include adjustments to carrying amounts, for example to effect common accounting policies under merger accounting, on the same basis as controlling interests. The United Kingdom follows what Baxter and Spinney (1975, p. 32) term a 'parent extension' approach, which regards minority interests as insiders to the group but, unlike a full-blown economic unit/entity approach, does not attribute goodwill to them (FRS 2, para. 38). This is the allowed alternative treatment for minority interests under IAS 22 (para. 33), unlike (see VI.1.2) the United States and certain continental countries, for example Germany, and IAS 22's benchmark treatment, where the minority share of fair value adjustments at acquisition is not recorded. The difference may reflect generally a more relaxed attitude in the United Kingdom to the inclusion of revaluations in company accounts.

1.6.2 Liability or equity?

The status of minority interests is ambiguous; they do not have the same status as liabilities, nor are they viewed as equal co-owners of the group, as their main information source would be the accounts of the entity in which they had an interest. Their status (see II) depends on the consolidation theory adopted and current practice is ambiguous in this regard. This issue is reflected, for example, in the treatment of minority interests in accumulated losses. FRS 2 requires that such a debit balance should always be recognized, and in addition, a provision made to the extent that the group has a legal obligation to provide finance that is not recoverable in respect of the minority's share of accumulated losses (para. 37). It comments that such a debit balance does not represent a liability by minority interests (Appendix III, para. xi).

FRS 4, Capital Instruments (1994), requires minority interests to be analyzed between the aggregate amounts attributable to equity interests and non-equity interests (para. 50), the latter having curtailed rights because distributions and winding up surpluses are for a limited amount and are not calculated with respect to the subsidiary's assets, profits or equity dividends, or where the shares are redeemable other than at the option of the issuer, for example by the holder (para. 12). Shares issued by subsidiaries should be classified as liabilities, not non-equity minority interests, 'if the group taken as a whole has an obligation to transfer economic benefits in connection with the shares' (para. 49). The example is given of where another group member guarantees payments. Otherwise they should be reported as minority interests.

Non-equity minority instruments are to be accounted for in the same manner as non-equity shares (para. 51). Consequently the amount attributable to non-equity minority interests at the date of issue of the non-equity instrument is the net issue proceeds. From then on they are treated analogously to 'liability type' instruments – increased by the finance costs for the period, equivalent to an interest charge, and reduced by the dividends or other payments during the period, equivalent to loan repayments (para. 41). Finance costs are calculated on the same basis as for debt (para. 42) allocated to periods over their term at a constant rate on the carrying amount (para. 28). Dividends on non-equity instruments which are calculated on a time basis should be accrued for unless the likelihood of ultimate payment is remote (para. 43). Treating non-equity minority interests as if they are liability-type instruments means that the profit and loss charge will be based on the finance cost and not on the dividend due. FRS 4 comments that 'where the finance costs … are not equal to the dividends the difference should be accounted for as an appropriation of profits' (para. 44); i.e. though the charge is analogous to a finance cost, its location is with other appropriations such as normal dividends.

1.7 Multi-level groups

The Companies Act 1985 definition of parent and subsidiary undertakings states that 'a parent undertaking shall be treated as the parent undertaking of undertakings in relation to which any of its subsidiary undertakings are, or are to be treated as, parent undertakings; and references to its subsidiary undertakings shall be construed accordingly' (S 258 (5)). This allows a chain of control in multi-level groups from subsidiaries to sub-subsidiaries and so on.

1.7.1 Simple vertical groups

For capital consolidation purposes, given that the companies meet the criteria for being included in the group accounts, in a simple grandfather–father–son formation, ownership proportions for the lowest level subsidiary can be obtained by multiplying those of the parent and subsidiary. For example, if A owns 70 per cent of B which owns 80 per cent of C, the ownership proportions used in consolidating C will be 70 per cent times 80 per cent = 56 per cent. Then 70 per cent of B's investment in C will be cancelled against 56 per cent of the group's pre-acquisition equity in C; the remaining 30 per cent of the investment belongs to group minority interest. The attribution of profits to the group can be illustrated as follows:

- B and C are acquired on the same date. Consolidated retained earnings would be A's plus 70 per cent of B's, plus 56 per cent of C's since the date of common acquisition.
- C is acquired after A acquired B. Consolidated retained earnings will be A's plus 70 per cent of B's since its acquisition, plus 56 per cent of C's since its later acquisition.
- C is acquired before A acquired B. Company C is deemed to join the group the date A acquired B, so consolidated retained earnings will be A's plus 70 per cent of B's plus 56 per cent of C's, all measured from the date of B's acquisition. A fair value exercise will need to be carried out for C on that date.

In multi-tiered groups ownership percentages may be very low and consequently minority interests very high. For example, four levels of 60 per cent ownership gives a multiplicative ownership stake in the bottom company of 13 per cent! FRS 2 comments that 'despite the title 'minority interests', there is in principle no upper limit to the proportion of shares in a subsidiary undertaking which may be held as minority interests . . .' (para. 80). Some prefer the term 'non-controlling interests', but 'minority interests' is contained in the Companies Act 1985, so has to be lived with. The principles of piecemeal acquisition (VI.1.4.6) apply on the dates that any of the multiplicative ownership proportions in a multi-level group are changed. If a subgroup has already prepared its own consolidated accounts, it may be more efficient to consolidate the parent company's accounts with the subgroup's accounts rather than doing an *ab initio* multiplicative consolidation.

1.7.2 More complex unilateral ownership structures

In addition, the parent or intermediate subsidiaries may hold direct stakes in lower tier subsidiaries. For example, suppose in the above example that A also held a direct 20 per cent stake in C. This is handled by the multiplicative approach, and the ultimate group holding in C will be 76 per cent (i.e. 70 per cent times 80 per cent plus 20 per cent), and so on. In such groups, indirect control can only take place through a chain of subsidiaries, plus any direct holdings, since if there is no intermediate control, there can normally be no further multiplicative control of lower tiers, though there may be significant influence. So multiplicative proportions would not be calculated for any routes of ownership passing through associates, joint ventures, or subsidiaries excluded on the grounds of severe long-term restrictions (i.e. leading to an actual loss of control) or because they are held exclusively for resale. Control could pass through a subsidiary not consolidated on the grounds of too dissimilar activities, but FRS 2's restriction of this category means it is unlikely to have an effect in practice. Post-acquisition profits for each stakeholding must be accrued from the correct date and principles of piecemeal acquisition (VI.1.4.6.2) applied.

1.7.3 Bilateral cross-holdings

In the case of bilateral holdings of subsidiaries in fellow subsidiaries a simultaneous equation can be used to evaluate group ownership proportions (see for example Taylor, 1987, pp. 223–225). However, in the United Kingdom, a subsidiary is prohibited from being a member of its holding company (Companies Act 1985, Sch 4A, para. 23 (1)) except where it is a trustee, a moneylender holding security in the ordinary course of business, or as an authorized market-maker. Because the narrower legal definition of holding company is referred to here rather than that of parent undertaking (see IV.1.1), it would seem legal for a subsidiary undertaking or quasi-subsidiary which is not a subsidiary, to hold shares in its parent undertaking, which would not therefore be its holding company. Even this prohibition for subsidiaries does not apply if the shareholding was acquired prior to the holder becoming a subsidiary (Companies Act 1985, Sch 4A, para. 23(5)). Under such circumstances the subsidiary can remain a member, but cannot vote. Wilkins (1979, p. 187) cites two alternative treatments for such cases:

- To treat the subsidiary's investment as being a repurchase of shares by the group; to deduct their nominal value of the share element from the nominal value of the consolidated share capital of the group, and to write off the difference between nominal and carrying value to consolidated reserves.
- To account for them as a trade investment of the group (under the Companies Act 1985 disclosed under 'Own shares' in 'III Investments' in the consolidated balance sheet).

However, in the case of subsidiary undertakings or quasi-subsidiaries that are not subsidiaries, a simultaneous equations approach could be applied as there are no legal voting restrictions.

2 Consolidation of debt, intragroup transactions and profits

This section covers consolidation adjustments to the financial statements of group companies prior to consolidation and re-measurement of intragroup transactions to a group basis so that consolidated financial statements record transactions between the group and external parties only, and profit is recognized on a group basis.

2.1 Intragroup debt and revenue and expense transactions

The Companies Act 1985 requires (Sch 4A, para. 6 (1)) that if material, 'debts and claims between undertakings included in the consolidation, and income and expenditure relating to transactions between such undertakings, shall be eliminated in preparing the group accounts'. This is effected by aligning transaction treatment in both selling and receiving group companies so that timing differences are removed and balances and flows cancel exactly.

2.1.1 Non-coterminous year ends

Where the subsidiary's financial statement date is different from that of the parent of the reporting group, any changes since the subsidiary's earlier reporting date, representing both transactions with external parties and intragroup transactions which materially affect the view given by the consolidated statements, should be adjusted in preparing those statements (FRS 2, para. 43). Remaining differences between intragroup balances caused by non-coterminous year ends should be immaterial and would normally be included under the appropriate balance sheet or profit and loss heading.

2.1.2 Foreign currency intragroup balances – normal trading transactions

Where there are trading transactions between the UK group and foreign subsidiaries, or between two foreign subsidiaries, transaction exchange differences may arise on inter-company accounts in the individual accounts of one, other or both of the companies involved when their foreign currency transactions and balances are translated (SSAP 20, para. 12). Consequently in the preparation of the consolidated financial statements, when the financial statements of foreign subsidiaries are translated into sterling, any such exchange differences will still be present when the cancellation of the translated inter-company trading balances is considered.

A decision has to be made whether or not to eliminate such translation differences on consolidation. SSAP 20 does not consider this matter. Davies, Paterson and Wilson (1992, p. 406) comment that 'on consolidation, such differences would normally remain in the profit and loss account in the same way as differences on monetary items resulting from transactions with third parties', but do not justify their suggested treatment. Westwick (1986, pp. 96–97) considers that the correct treatment of these transaction exchange gains or losses is to eliminate them if the temporal approach is used, but to leave them in the consolidated profit and loss account if the closing rate approach is used. His proposal is deduced from the underlying rationale for each approach:

- Under the unified group/single functional currency rationale that underpins the temporal approach, internal balances must eliminate exactly and the exchange difference on cancellation should be offset against the overall financial statement translation difference that under this approach appears in the profit and loss account. This is equivalent to using the original transaction exchange rate for translating the foreign currency balance, consistent with the usual principle of no profits or losses being recognized on transactions within a unified group.
- Under the net investment, multiple functional currency rationale for the closing rate approach, he argues that the subsidiary is treated as an autonomous third party and so any exchange difference on cancelling intragroup balances is treated analogously to an exchange gain or loss on a debtor/creditor balance with an independent non-group company and taken to the consolidated profit and loss account.

These proposals reveal the uneasy way in which the conceptualization of foreign subsidiaries under the net investment rationale sits with the overall consolidation concept of the group as a single entity.

2.1.3 Foreign currency intragroup long-term loans and deferred trading balances

An exception to the above is recognized by SSAP 20; the financing of net equity investments in foreign enterprises may be carried out by long-term loans or deferred trading balances rather than by shares, and 'where … (this) is intended to be, for all practical purposes, as permanent as equity, such loans and inter-company balances should be treated as part of the investing company's net investment in the foreign enterprise; hence exchange differences on such loans and inter-company balances should be dealt with as adjustments to reserves' (para. 20). Davies, Paterson and Wilson (1997, pp. 559–560) suggest 'permanent' means not repayable for the foresee-

EXAMPLE

A German subsidiary purchased GBP200,000 of materials from the parent when the exchange rate was GBP1 = €1.60. The sterling balance was outstanding at the group year end, when the exchange rate was GBP1 = €1.75.

The inter-company balances at 30 June will be:	*DR*	*CR*
In the parent's records (debtor)	GBP200,000	
In the subsidiary's records (creditor)		€320,000 (= GBP200,000 × 1.60)
At the group year end they will be:	*DR*	*CR*
In the parent's records (debtor)	GBP200,000	
In the subsidiary's records (creditor)		€350,000 (= GBP200,000 × 1.75)

The subsidiary will record a transaction exchange loss in its individual company profit and loss account of €30,000.

Temporal approach. On translation, the sterling equivalent of the €30,000 will be eliminated and the overall financial statement translation loss adjusted by the same amount. The result will be as if the original transaction rate of €1.60 to the pound sterling had been used to translate the balance.

Closing rate approach. Assume the year end rate is the alternative used under SSAP 20 to translate the German subsidiary's profit and loss account. The transaction exchange loss will be translated as 30,000/1.75 = GBP17,143 and taken to the consolidated profit and loss account. The subsidiary's balance, translated at the closing rate (350,000/1.75 = GBP200,000) will cancel exactly with the parent's balance.

able future and suggest a period of three to five years is often taken as the planning horizon. They also consider short-term loans intended to be rolled over for the foreseeable future can be included in this definition.

Wallace and Ogle (1984, p. 16) conclude that this requirement only applies in consolidated accounts, relating to exchange gains or losses arising from inter-company balances. However, Davies, Paterson and Wilson support the view that where the balance is intended to be as permanent as equity, it should be treated as if equity, and argue that:

> One of the objectives of SSAP 20 is to produce results compatible with the effects of exchange rate changes on future cash flows. As there is no intention for the inter-company account to be repaid until disinvestment or at least in the foreseeable future then there will be no effect on the company's cash flows as no repayments are being made.

Therefore they recommend translating the loan at the rate when the account was first considered as permanent as equity, so that no further exchange differences will accrue after that date (see Davies, Paterson and Wilson, 1997, p. 564). This seems to go beyond SSAP 20 and it seems debatable whether such loans are equity under FRS 4, Accounting for Capital Instruments (1993), since it could be argued that the substance is different, because interest is paid and they have very different liquidation repayment terms. Whether management intention alone is sufficiently strong grounds as a basis for treating as if equity is debatable.

2.2 Removal of intragroup profits

2.2.1 Full elimination

The Companies Act 1985, Sch 4A, para. 6(2) states that 'where profits and losses resulting from transactions between undertakings included in the consolidation are included in the book value of assets, they shall be eliminated in preparing the group accounts'. Proportional elimination relating to the controlling interest only is allowed as an option by the Act, but FRS 2 decides against this option and requires complete elimination 'set against the interests held by the group and the minority interest in respective proportion to their holdings in the undertaking whose financial statements recorded the eliminated profits and losses' (para. 39).

Such an approach is consistent with an entity or extended parent view of consolidation (see II), treating minority interests as insiders and part owners of the group. Thus intragroup transfers not yet resold outside the group are not realized by the group and must be reduced to group cost. So for example, the parent's profits on goods purchased by a 75 per cent-owned subsidiary from it and still held by the subsidiary at the reporting date, termed 'downstream' sales, would be completely eliminated against intragroup stocks, reducing them to group cost. The double entry would eliminate the parent's profit on such goods 100 per cent against consolidated retained profits, as the company recording the profit, the parent, is 100 per cent-owned by the group. If the sale were from the subsidiary to the parent, 'upstream' sales, there would still be complete elimination against intragroup stocks but the double entry would be apportioned 75 per cent against consolidated retained profits and 25 per cent against minority interests, since the subsidiary recorded the original profit. In the consolidated profit and loss account there are two adjustments to cost of goods sold, one restating the opening intragroup stocks element to group cost, and the other adjusting the closing intragroup stocks element. The former is profit-increasing, reflecting profits recognized by individual companies in previous periods now recognized by the group in this period, the latter profit-decreasing, profits recognized by individual companies in the current period, to be recognized by the group in future periods. Where fixed assets are sold between group companies at an amount greater than group cost, it will be necessary to adjust the carrying value to group cost. It will also be necessary to adjust future consolidated depreciation charges so that they will be based on the same group cost.

ARB 51 in the United States requires complete elimination of unrealized profits on intragroup transactions, but allows this profit either to be eliminated 100 per cent against consolidated retained profits or to be apportioned between controlling and minority interests. However, unlike FRS 2, it does not specify that the apportionment must depend on the direction of the sale. IAS 27 also requires complete elimination, but does not specify how such complete elimination should be apportioned between controlling and minority interests (para. 30). If the appropriate options were chosen, both IAS 27 and ARB 51 could be consistent with FRS 2's treatment. Presumably the reason that the United States practices complete profit elimination on unrealized intragroup transactions, but restatement to fair values at acquisition only of the controlling interest's share, is because under its conservatism, write-downs are viewed more favourably than write-ups. The United Kingdom is currently more relaxed about revaluations than many countries. FRS 2 is more consistent with underlying consolidation concepts (see II and particularly Taylor, 1996, Chapter 6).

Subsidiary undertakings excluded from consolidation on the grounds of different activities (Companies Act 1985, S 229(4)) are still controlled, so FRS 2 also requires the same basis for eliminations of intragroup profits if the transactions are between group companies and such excluded subsidiaries (para. 39), but different bases where there are other reasons for exclusion, for example severe restrictions or investments held exclusively for resale.

2.2.2 Foreign currency intragroup transactions

FRS 2 views the group as if a single economic entity, whereas the closing rate–net investment approach to foreign currency translation per SSAP 20 envisages a conceptual separation between the entities in the group. In single currency situations the originating company's profit is deferred, which also reduces intragroup stocks to group cost. However, with foreign currency intragroup transactions both characteristics do not necessarily coincide. Removing the originating company's profit will not necessarily restate goods to group cost.

Under the temporal approach, used for dependent operations, elimination of intragroup profits will be approximately carried out at the transaction rate and so the originating company's translated profit will be removed and also the goods will be restated to translated original group cost. The situation under the closing rate approach, which is used for the vast majority of foreign subsidiaries under SSAP 20, is explored in the example below.

EXAMPLE

1. Sale by parent in Pounds Sterling

The parent sold goods to a German subsidiary for GBP7,500, when the exchange rate was GBP1 = €1.60, which cost the parent GBP5,000. At the group year end the exchange rate was GBP1 = €1.75. All the goods were still in stock at the group year end.

The intragroup stock profits are GBP2,500, and the stocks would be recorded in the subsidiary's records at the transaction date at €12,000 [= GBP7,500 × 1.60]. At the year end this would be translated under the closing rate approach as €12,000/1.75 = GBP6,857, and removing the intragroup profit of GBP2,500 would restate these goods to GBP4,357, which is not group original cost.

2. Sale by foreign subsidiary in euros.

In this case the intragroup profit is in foreign currency and, as SSAP 20 is silent on the matter, it could be translated using either the exchange rate at:

- The transaction date, favoured by SFAS 52 in the United States, arguing that the movement of GBP643 [= 7,500−6,857] results from the effect of changing exchange rates on the subsidiary's stock balance and is not related to the sale; or
- The rate used to translate the subsidiary's profit and loss account – average or year end rate if under the closing rate approach (see Davies, Paterson and Wilson (1997, p. 567)).

Assume the following: the foreign subsidiary sold goods to its parent for €12,000, when the exchange rate was GBP1 = €1.60, which cost the subsidiary €8,000. At the group year end the exchange rate was GBP1 = €1.75. All the goods were still in stock at the group year end. Assume SSAP 20's option of the year end rate is used to translate the subsidiary's profit and loss account, and hence any profits it makes. The intragroup stocks at the parent would be translated at the transaction date at €12,000/1.60 = GBP7,500. The intragroup stock profits at the subsidiary would be €4,000, which could be translated as:

Exchange rate	Sterling intragroup profit	
Transaction date	1/1.60× €4,000	= GBP2,500
Profit and loss – year end rate	1/1.75× €4,000	= GBP2,286

Adjusted stock under transaction date approach would be GBP7,500 − GBP2,500 = GBP5,000, or under profit and loss rate approach, GBP7,500 − GBP2,286 = GBP5,214. The former would reduce the goods to group cost at the transaction date, but Davies, Paterson and Wilson note that the approach would not eliminate the translated profit on the sale in the subsidiary's translated accounts, 4,000/1.75 = GBP2,286 (see Davies, Paterson and Wilson, 1997, p. 567). The second alternative would eliminate the profit, but the goods would not be restated at group cost. This reveals the conflict between the concept of group as a unified entity and the conceptual separation implied by the closing rate approach.

2.3 Foreign currency intragroup dividends

Intragroup dividends from foreign affiliates should be recorded initially by the parent in its individual accounts at the declaration date exchange rate. Unless the profit and loss amount for the foreign affiliate's dividends payable is translated at the declaration date exchange rate, it will not cancel exactly on consolidation with the parent's profit and loss amount for dividends receivable.

SSAP 20 again is silent on which rate to use under each translation approach and UK authorities differ in their recommendations. Wallace and Ogle (1984, p. 15) consider the declaration date rate should be used by the subsidiary and Wainmann (1984, p. 101) suggests as a general organizing principle, where there are parent-subsidiary transactions, these should be segregated in the subsidiary's records and 'retranslated in such a way as to mirror the parent's sterling figures'. However, consistent with his reasoning on the elimination of foreign currency intragroup balances and unrealized intragroup stock profits discussed in VI.2.1.2 and VI.2.2.2, Westwick (1986, p. 101) considers that the original transaction rate should be used under the temporal approach so that any exchange difference is eliminated. Under the closing rate approach, the exchange difference should be 'flowed through into group profit, but as on an inter-company loan if the delay in payment is substantial, included in reserve movements because the unremitted dividend is part of the parent's net investment in the subsidiary'. Davies, Paterson and Wilson (1997, p. 565) on the other hand prefer taking the exchange difference direct to reserves because 'the method will therefore retain the same financial relationships shown in the subsidiary's own financial statements in the consolidated (ones)'. They criticize Westwick's closing rate proposals because they recognize distributed profits of the subsidiary at the transaction rate, but undistributed profits at the rate used to translate the subsidiary's profit and loss account, which may be the year end rate. This displays their strong support for the preservation of foreign currency financial relationships on translation into sterling. The view of the group as a single entity suggests mirror-image elimination, whereas the closing rate approach views the group as a collection of autonomous net investments.

3 Consolidated profit and loss account

3.1 Companies Act requirements

Schedule 4A of the Companies Act 1985 states that 'group accounts shall comply as far as is practicable with the provisions of Schedule 4 as if the undertakings included in the consolidation ('the group') were a single company' (para. 1(1)). Further it states that

> the consolidated (...) profit and loss account shall incorporate in full the information contained in the individual accounts of the undertakings included in the consolidation, subject to adjustments authorized or required by (...) this Schedule (...) and (...) as appropriate in accordance with generally accepted accounting principles or practice. (para. 2(1))

These adjustments include, for example, the determination of goodwill, minority interests, the cancellation of intragroup balances, fair value adjustments at acquisition, the elimination of unrealized intragroup profits and equity accounting adjustments. Table VI.3 gives a brief summary of additional requirements that apply to the consolidated profit and loss account, but not individual company profit and loss accounts. Many are discussed in more detail in the appropriate sections, and disclosures further analysed in Table X.1.

3.2 FRS 3 and other accounting standards

The central definitions within FRS 3, Reporting Financial Performance, issued in 1992 and discussed in United Kingdom – Individual Accounts V.2.1, such as of exceptional items, extraordinary items and prior period adjustments apply to the consolidated profit and loss account in similar ways to individual company accounts. Most of the disclosure requirements of FRS 3, for example acquisitions, discontinued operations and terminations, could, in principle, also be applicable to individual companies (for example, relating to divisions). They are outlined in Table VI.3 because in practice they will crop up a great deal more in groups than in individual companies, and will usually, though not necessarily, apply to subsidiaries. Discontinued operations are defined above (see VI.1.5.1), where matters relating to provisions on termination or sale are discussed. An example of the illustrative layouts of consolidated profit and loss accounts provided by FRS 3 is shown in Figures IX.1 and IX.2.

FRS 14, Earnings Per Share, issued in October 1998, requires basic earnings per share to be calculated by dividing the net profit or loss for the period attributable to ordinary shareholders by the weighted average number of ordinary shares outstanding during the period. It is based on net profit or loss for the period attributable to ordinary shareholders after deducting dividends and

Table VI.3 Profit and loss account: additional consolidation requirements

Description	Summary of requirements
Statutory headings (Co Act 1985 Sch 4A)	'Income from shares in [group undertakings]' would be replaced by the consolidation of the subsidiary's profit and loss account. 'Income from participating interests' is to be replaced by 'Income from associated undertakings' and 'Income from other participating interests' (para. 20(3)). See also Table X.1.
Minority interests (Co Act 1985 Sch 4A / FRS 3 / FRS 4)	The minority share in profit or loss on ordinary activities is shown as an aggregate deduction before extraordinary items and further, the aggregate minority interest in extraordinary profits or losses must also be shown separately (para. 17). FRS 4 requires the charge to be analysed between equity and non-equity interests. Minority interests share in FRS 3's 'super' exceptional items should be given note disclosure.
Analysis of profit and loss items items (FRS 3)	The aggregate results of continuing operations, acquisitions (as a component of continuing operations) and discontinued operations must be separately disclosed from the turnover down to the operating profit level. The minimum disclosure on the face of the consolidated profit and loss account is the analysis of turnover and operating profit (paras. 14–17).
Sales or teminations of operations, fundamental reorganizations or restructurings (FRS 3)	The following, including provisions relating to each item, should be separately disclosed on the face of the profit and loss account *after* operating profit and *before* interest, analyzed appropriately between continuing and discontinued operations: a. profits or losses on the sale or termination of an operation; and b. costs of a fundamental reorganization or restructuring having a material effect on the nature and focus of the reporting entity's operations. Only revenues and expenses directly related to such items must be included. Note disclosures on tax charge effects and minority interests should be given (FRS 3 para. 20).
Associates (FRS 9)	The investor's share of its associates' operating results should be presented immediately after the group operating result (after the investor's share in the results of any joint ventures). Goodwill amortization or write-down should be disclosed and charged at this point. The investor's share of associates' interest and post-operating profit exceptional items should be shown separately from group amounts (according to FRS 9's illustrative example, on the face of the profit & loss account). For items below the level of profit before tax, amounts should relate to the aggregate amount of the group including associates, with the amount relating to associates separately disclosed in the notes. Optionally, the total of the aggregate turnover of the group combined with associates' turnover may be given as a memorandum amount, but associates' turnover must be clearly distinguished from group turnover.
Joint ventures (FRS 9)	The gross equity method requires the same treatment as for associates except that the investor's share of joint ventures' turnover should be shown, separate from group turnover. Any supplemental information in the profit and loss account must be shown clearly separate from group amounts and not included in group totals.

Table VI.3 (contd.)

Description	Summary of requirements
Acquisitions and mergers (FRS 6)	In the year of the combination, under merger accounting – the subsidiary's revenues and expenses, etc. will be consolidated for the whole year, regardless of when the merger took place and comparatives will be restated (para. 17). Under acquisition accounting – revenues and expenses will only be recognized from the date of acquisition and comparatives will not be restated (para. 20). See also X.
Fair values at acquisition (FRS 6)	Exceptional post-acquisition profits and losses that are determined using the fair values recognized at acquisition should be disclosed in accordance with FRS 3's requirements (para. 30).

other appropriations in respect of non-equity shares. Such a profit measure includes all items of income and expense items that are recognized in the period in the determination of such net profit or loss including tax expense, exceptional and extraordinary items and minority interests (paras. 9–11). In material respects FRS 14 is consistent with IAS 33, Earnings Per Share, issued in February 1997. A discussion of FRS 14's requirements is contained in United Kingdom – Individual Accounts IX.6.

4 Consolidated statement of total recognized gains and losses

4.1 Overview of statements required by FRS 3

As discussed in United Kingdom – Individual Accounts V.2.1, in addition to the profit and loss account, FRS 3 requires three other statements of performance, a statement of total recognized gains and losses, a reconciliation of movements in shareholders' funds, and a note of historical cost profits and losses. The first is a primary financial and is discussed further below. The last two are note disclosures, but FRS 3 allows the reconciliation of movements in shareholders' funds to be disclosed as a primary statement provided it is presented separately from the statement of total recognized gains and losses. It is discussed here only in relation to goodwill immediately written off direct to reserves.

The consolidated profit and loss account, the consolidated statement of total recognized gains and losses, and the consolidated note reconciliation of movements in shareholders' funds are a 'nested' system. The first has the narrowest focus. The second includes the single figure, profit for the financial year (before dividends), the end product of the consolidated profit and loss account, together with other reserve movements which FRS 3 classifies as 'recognized gains and losses'. These include, in a group context 'currency translation gains on foreign net investments'. The last includes a summary of total recognized gains and losses, together with capital injections and distributions to shareholders, and other reserve movements not classified as 'recognized' gains and losses, notably goodwill immediately written off direct to reserves. In the United States SFAS 130, Reporting Comprehensive Income, issued in 1997, also requires a primary statement showing total recognized gains and losses for the period (termed comprehensive income), and IAS 1, revised in 1997, requires as a primary statement either one which shows total recognized gains and losses, or one showing movements on shareholder equity in aggregate, which includes as a subtotal gains and losses recognized directly in equity. However, the United Kingdom differs in requiring that once gains and losses have been recognized in the statement of total recognized gains and losses, they should not then be reclassified, for example on disposal of revalued fixed assets.

4.2 Associates and joint ventures, foreign currency translation gains and losses, and goodwill

4.2.1 Associates and joint ventures

FRS 9 requires that the investor's share of associates' and joint ventures' gains and losses should be shown separately under each heading in the statement, or in the notes to the statement, if the amounts are material (para. 28).

4.2.2 Foreign currency translation

The statement of total recognized gains and losses includes 'certain gains and losses ... specifically permitted or required by law or an accounting standard to be taken direct to reserves' (para. 56). Further 'the components should be the gains and losses recognized in the period in so far as they are attributable to shareholders' (para. 27). Whilst 'currency translation differences on foreign currency net investments' (i.e. under the closing rate approach) are not specifically defined as a recognized gain or loss, the illustrative example embodied in FRS 3 makes it clear that they are to be reported in this statement. As discussed in V.4.2, financial statement translation differences can be viewed as somewhat analogous to revaluation adjustments on fixed asset investments and therefore taken direct to reserves. However, in dependent foreign operations, where the temporal approach is used to translate statements, translation differences are viewed as analogous to transaction gains and losses (V.4.3) and taken to the consolidated profit and loss account. They will therefore appear within the 'profit for the financial year' figure in the statement of total recognized gains and losses.

FRS 3 comments that the statement of total recognized gains and losses is a primary statement, so 'the same gains or losses should not be recognized twice' (para. 56) – for example, a gain recognized on fixed asset revaluation should not be recognized as realized when it is sold. The Statement of Principles Exposure Draft (1995) rejects the view that the profit and loss account should show realized gains and losses and the statement of total recognized gains and losses unrealized ones (para. 6.24). Therefore, on disposal of a foreign entity, accumulated statement translation differences under the closing rate approach do not play a part in determining the gain or loss on disposal and will never go through the consolidated profit and loss account (VI.1.5.3). As a result of a change in UK tax law, gains or losses on foreign currency borrowings hedging net investments may be subject to tax. UITF Abstract 19, issued in 1998, requires that the tax consequences of any foreign exchange gains or losses which are reported in the statement of total recognized gains or losses must also be reported there.

4.2.3 Goodwill

Prior to FRS 10, immediate write-off of goodwill direct to reserves was allowed in the United Kingdom. Under FRS 3 this (or credit adjustments to reserves relating to goodwill on disposals of subsidiaries) did not go through either the profit and loss account (unless it is extended to include a full appropriation account) or the statement of total recognized gains and losses (UITF Information Sheet No. 6, 1992, para. 3), but is shown in FRS 3's required note to the accounts, 'Reconciliation of movements in shareholder funds'. See VI.1.5.2 and VI.1.5.3 for a discussion of this treatment, the treatment of goodwill in determining profits and losses on disposals of subsidiaries, and of the apparently contradictory treatments on disposals of goodwill and closing rate translation gains and losses.

Consolidation adjustments may be necessary in the statement of total recognized gains and losses if an asset has been transferred within the group and is then revalued by the receiving company. If the revaluation is in accordance with the Companies Act 1985's requirements, any intragroup profit on sale would not be eliminated, but transferred to the Statement of total recognized gains and losses so the whole increase is reported as a group revaluation gain. Otherwise the profit would be eliminated (see VI.2.2.1).

5 The consolidated cash flow statement

This is another primary statement. In 1996, the revision of FRS 1 changed the UK statement into 'a genuine cash flow statement – the first in the world' (Crichton 1996, p. 97). The revised statement now contains eight categories of cash flows and is based on a narrower definition of 'cash' than in other countries, i.e. cash in hand and deposits repayable on demand less overdrafts. Cash flow statements internationally include some form of cash equivalents in their 'cash' definition. The United States uses a gross concept without deductions of any borrowings, whereas the IASC and Canada use a net concept which allows the deduction of other short-term borrowing in addition to bank borrowing (see Wallace and Collier, 1991, p.49). FRS 1, Cash Flow Statements (revised 1996) requires consolidated cash flow statements in group accounts. Exemptions from producing cash flow statements include small companies under the Companies Act (Table III.2). They also apply to subsidiary undertakings where 90 per cent or more of the voting rights are controlled within the group, provided there are higher 'level' publicly available consolidated financial statements in which the subsidiary's cash flow statement is included (para. 5). Disclosure requirements for consolidated cash flow statements which overlap requirements for individual company cash flow statements are discussed in United Kingdom – Individual Accounts IX.1. Additional requirements of FRS 1 relating only to consolidated cash flow statements are as follows:

- Cash flows internal to the group should be eliminated.
- Dividends to minority interests should be separately disclosed under the heading 'returns on investments and servicing of finance'. Ghosh (1991, p. 58) argues that, where FRS 4 requires minority interests to be classified as liabilities, where their dividends have been guaranteed or similar arrangements made, their dividends paid should be disclosed as if interest paid.
- A note to the cash flow statement should identify the amounts and explain the circumstances where restrictions prevent the transfer of one part of the group or business to another (para. 47).

Acquisitions and disposals of subsidiaries. The amounts of cash and overdrafts acquired or transferred should be shown separately, together with the gross consideration paid or received for the acquisition or disposal (para. 23). The working capital items in the reconciliation of operating profit to cash flow from operations will therefore exclude the effects of the corresponding balances in the acquired or disposed of subsidiaries. A summary of the effects of acquisition and disposal transactions should be included as a note, showing how much of the total consideration comprised cash. Where a subsidiary undertaking joins or leaves the group, cash flows for it should be included for the same period as the group's profit and loss includes its results, i.e.:

- For the whole year under merger accounting (with restatement of comparatives).
- Since acquisition for acquisition accounting.
- To the date of disposal on disposals.

Material effects under each standard heading must be disclosed as a note, as far as is practicable, for subsidiaries acquired or disposed of during the period, and could be given by dividing cash flows between continuing and discontinued operations and acquisitions (para. 45). Companies are encouraged, but not required, to analyze cash flows between continuing and discontinued operations (para. 56).

Associates and joint ventures: FRS 1 requires that cash flows of equity accounted entities should only be included to the extent of cash flows between the group and the entity concerned (paras 37–38). FRS 9 amends FRS 1 by stating that dividends received from joint ventures and associates should be included as separate items between operating activities and returns on investment and servicing of finance (revised para. 12A). It also states that any other cash flows between the investor and its associates and joint ventures (for example loans repaid) should be included under the appropriate cash flow heading giving rise to the cash flow, and that none of the other cash flows of the associates or joint ventures should be included (paras. 21 and 30).

Foreign subsidiaries: cash flows relating to foreign subsidiaries/entities are to be included on the same basis used for translating the results of those activities in the profit and loss account of the reporting entity (whereas IAS 7 and SFAS 52 require the use of actual rates or averages as an approximation). Where there are significant numbers of foreign subsidiaries, this requirement may necessitate translating the individual cash flow statements of the foreign subsidiaries concerned and consolidating directly their cash flow statements with the parent's, rather than derivation from the other consolidated statements (V.4.2.2). FRS 1 also requires that the same basis be used in presenting the movements in stocks, debtors and creditors in the reconciliation between operating profit and cash from operating activities. An exception is in the translation of separately identifiable intragroup cash flows where the actual rate or an approximation may be used to ensure cancellation on consolidation. If the rate used is not an actual rate, exchange differences on cancellation should be included as the effect of exchange rate movements, as part of FRS 1's reconciliation to net debt (para. 41).

VII Proportional consolidation and equity method

1 Overview

The ASB's Draft Statement of Principles, Chapter 7, The Reporting Entity (1995), distinguishes two aspects of control: the ability to deploy economic resources, and the ability to ensure secure the resulting benefits (or suffer exposure to losses) (para. 7.10). By implication, lesser interests may contain the ability to influence but not control deployment. In the United Kingdom the distinction is made between long-term holdings which participate in benefits ('participating interests') and actively exercise influence, from more ephemeral or passive holdings. Table VII.1 shows four levels of influence currently distinguished in the United Kingdom, together with accounting treatments. The differences from international practice are that 'associates' are defined in terms of the actual exercise of influence rather than the ability to exercise influence, and the 'gross equity method' is used for jointly controlled autonomous entities rather than proportional consolidation. The latter treatment is consistent with the alternative treatment in IAS 31, *Financial Reporting of Interests in Joint Ventures.*

The ASB has long held the view that proportional consolidation is not justifiable for genuine joint ventures, and that equity accounting was being incorrectly used for some investment holdings where there was no genuine significant influence. These were set out in an ASB Discussion Paper (DP) issued in June 1994 which tried to unify the treatment of associates and joint ventures using a single concept of 'strategic alliances', under which both would be equity accounted. 'Strategic alliances' were based on the

Table VII.1 Degree of influence and current accounting treatment

Type of influence	Category of relationship	Accounting treatment	Overall profit basis
Non-significant influence	Simple investment	At cost or valuation	Dividends receivable
Significant influence (but not control or joint control)	Associate	Equity method	Attributable profits and gains and losses.
Joint control of an entity carrying on its own trade or business	Joint venture	Gross equity method	Attributable profits and gains and losses
Other joint arrangements	Joint arrangements that are not entities	Include share of assets, liabilities and cash flows in own statements	Attributable profits and gains and losses in investor's own statement
Unilateral control (including dominant influence)	Subsidiaries	Consolidation	Total profits and gains and losses less minority interest

concept of the investor being a partner in the business, in associates with its management, representing the other investors, and in joint ventures with other venturers. The attempt to enforce equity accounting as the sole approach met with considerable opposition. The ASB argued that proportional consolidation misleadingly accounts for the interest as if the investor unilaterally controls a proportional share of the individual assets and liabilities rather than sharing control over the interest and operating and financial policies of the entity as a whole (DP, para. 4.6). Proportional consolidation was widely used in the United Kingdom for joint ventures and arrangements in certain industries, such as the oil and construction industries. The DP proposed size-based disclosures where in aggregate strategic alliances exceeded 15 per cent and 25 per cent of the group's gross assets, gross liabilities, turnover or results and where individual strategic alliances exceeded these limits. It also defined 'significant influence' more rigorously than previously. FRED 11, the subsequent 1996 exposure draft, toned down the size-based disclosure proposals, and reintroduced the idea of dealing with associates and joint ventures separately. It distinguished between two types of joint venture – entities to be equity accounted where the venturers share in common benefits, risks and obligations, and entities to be proportionally consolidated where each venturer has its own separate interest in benefits, risks and obligations. FRED 11 used the term 'joint venture' only for what IAS 31 calls jointly controlled entities.

FRS 9, issued in November 1997, further narrowed the term 'joint venture' to include only jointly controlled entities which carry on a trade or business of their *own*, rather than just being a part of the participants' trades or businesses. It introduces the term 'joint arrangement that is not an entity' for all other joint arrangements, allowing the ASB to require a single treatment for each category. Its 'joint ventures' are to be accounted for using the 'gross equity approach'. 'Joint arrangements that are not entities' are to be accounted for in participants' *own* financial statements by including their own assets, liabilities and cash flows. This allows the ASB to claim conceptual consistency, whilst enabling industries which had formerly used proportional consolidation for 'joint ventures' to account in their own statements for what would now be termed 'joint arrangements that are not entities'. Though not a consolidation procedure, the end product in the consolidated statements may often be very similar to the effects of using proportional consolidation.

IAS 31 defines three types of joint ventures, based on jointly controlled assets, operations and entities. A venturer is required to account for the first two in its separate financial statements. The benchmark treatment for jointly controlled entities is proportionate consolidation either on a line-by-line basis or by including separate line items, and equity accounting is the allowed alternative for jointly controlled entities. It is likely that interests identified as 'jointly controlled assets' or 'jointly controlled operations' would be identified as 'joint arrangement that are not entities' under FRS 9. Therefore they would be similarly treated under both standards. The treatment of 'jointly controlled entities' would depend on whether such entities carried out a trade or business of their own. If they did they would be accounted for under the gross equity method under FRS 9. If not, they would be similarly reflected in the consolidated accounts under both standards, but by being included in the separate statements of the participant under FRS 9.

2 Associates

2.1 Definition

2.1.1 Development

The ASC's first standard SSAP 1, *Accounting for Associated Companies*, issued in 1971, defined the term 'associated companies' to include joint ventures and companies in which more than a 20 per cent long-term stake was held with the ability to exercise a significant influence. Revised twice, in 1974 and 1982, a changed definition made significant influence the guiding principle with the 20 per cent threshold becoming a rebuttable presumption, now including subsidiaries' holdings, though not those of other associates. Holdings in non-corporate joint ventures and consortia were included. The revised Companies Act 1985 embedded definitions and accounting treatments in statute law. Its term 'associated undertakings' additionally included non-corporate entities, slightly changing the relationship between associates and joint ventures. ED 50, Consolidated Accounts, was issued in 1990 to update SSAP 1, but was withdrawn. The ASB's Interim Statement: Consolidated Accounts in 1991 made only the minimum amendments necessary for SSAP 1 to comply with the Act. FRS 9, Associates and Joint Ventures, issued in November 1997 and its preceding exposure draft and Discussion Paper, as discussed above, refined the concept of 'significant influence' to require the actual exercise of such influence rather than just the ability to do so. It appears incongruous that for subsidiaries, as Ernst and Young (FRS 9, 1998, p. 11) point out, FRS 2 defines 'control' as 'the ability to direct the financial and operating policies of another undertaking…' (para. 6). However, the Act's exclusion criteria can effectively modify this where there is not the actual ability or intention to direct such policies.

2.1.2 Companies legislation definition of an associated undertaking

The Companies Act 1985 (Sch 4A, para. 20(1)) defines an associated undertaking as 'an undertaking in which an undertaking included in the consolidation has a participating interest and exercises a significant influence over its operating and financial policy', which is not a subsidiary undertaking or a proportionally consolidated joint venture. It includes partnerships and unincorporated associations.

A participating interest is one in shares held on a long-term basis to secure a contribution to the holding undertaking's activities by the exercise of control or influence arising from or related to that interest. Specifically included are interests:

- Held by subsidiaries.
- Convertible into an interest in shares.
- In the form of options to acquire shares.
- Including equivalent ownership interests in other entities not based on share capital.

The definition rules out solely non-beneficial interests such as those of a trustee or manager, but includes non-dividend means of extracting benefits such as performance-based management fees (DP, para. 3.5). The Companies Act 1985 makes a positive rebuttable presumption that a holding of 20 per cent or more in the shares of an entity is presumed to be a participating interest, and a holding of 20 per cent or more of the voting rights is presumed to exercise a significant influence.

2.1.3 FRS 9 definition of an associate

The definition of an associate in FRS 9, based on the Companies Act 1985 is

> An entity (other than a subsidiary) in which another entity (the investor) has a participating interest and over whose operating and financial policies the investor exercises a significant influence. (para. 4)

The term 'participating interest' also based on the Act is defined as

> An interest in the shares [and other ownership interests which are deemed equivalent] of another entity on a long-term basis for the purpose of securing a contribution to the investor's activities by the exercise of control or influence arising from or related to that interest.… (para. 4).

Consistent with the Act, only beneficial interests count and the term 'interests in shares' includes interests convertible into shares or options to acquire shares, which could arise, for example, in start-up situations where the investor initially has a close involvement in strategic operating and financial policies (para. 13). FRS 9 also restates the Act's 20 per cent of shares rebuttable presumption for a participating interest described above. IAS 28 does not require the existence of a participating influence in its definition of an associate, only significant influence. However, the exception in this standard for the cost method to be used when investments are acquired and held exclusively with a view for disposal in the near future, probably means that such an omission will not create a major difference in practice.

FRS 9 defines the term 'exercising of significant influence' as:

> The investor is actively involved and is influential in the direction of its investee through its participation in policy decisions covering aspects of policy relevant to the investor,

including decisions on strategic issues such as:

(a) the expansion or contraction of the business, participation in other entities or changes in products, markets and activities of its investee; and
(b) determining the balance between dividend and reinvestment.

FRS 9 states that the Act's rebuttable presumption, 'a holding of 20% or more of the voting rights suggests, but does not ensure that the investor exercises significant influence over that entity' (para. 16). It is *not* the operational criterion for 'exercising of significant influence' and FRS 9 requires that the presumption be considered as rebutted if the above conditions for active involvement are not satisfied. FRS 9 stresses the requirement for a formal or informal agreement providing the basis for the exercise of significant influence and the need for direct involvement in the operating and financial policies of the associate, so that the associate is used as a medium for conducting part of the investor's business. The policies implemented by the associate must therefore generally coincide with the strategy and interests of the investor and if this is not the case, significant influence is not deemed to be exercised (para. 14). The investor should have a voice in decisions on strategic issues, usually through board membership or equivalent arrangements. IAS 28, *Accounting for Investment in Associates,* gives further examples of possible indicative factors, such as material transactions between the investor and investee, interchange of management personnel, or provision of essential technical information (para. 5). In terms of the balance between current dividends and reinvestment, the investor's participation in policy decisions is with a view to gaining economic benefits from the associate and also exposes the investor to risks and the possibility of losses. But such benefits may relate to the longer term and FRS 9 comments that demanding high current dividends may not be consistent with the investors' long-term interests, and that such interests are compatible with a policy of reinvestment (para. 15).

FRS 9 stresses a substance over form perspective – that the actual relationship is decisive in identifying associates. It believes that such a relationship is usually clear from the outset, but where an initial assumption is made and the relationship develops differently from the initial assumption, it allows the treatment to be modified. However it is interesting that the criterion for initiating the identification of an associate is different from that of continuing to identify an ongoing associate or terminating its identification as an associate. FRS 9 states that once an actual relationship identified as exercising significant influence has been established, the exercise of such influence should be regarded as continuing unless terminated by a transaction or event removing the investor's *ability* to do this [emphasis added] (para. 17). 'Actual exercise' is the initiating criterion, but the ongoing criterion becomes '*ability* to exercise' since an associate is only de-identified when such '*ability*' is lost. IAS 28 in all cases defines the identification of significant interest in terms of the power to participate (para. 3) not in terms of actual participation. This means that some associates under IAS 28 may not be identified as associates under FRS 9.

2.2 Accounting for associates

2.2.1 The equity method for associates

FRS 9 states (following The Companies Act 1985) that the equity method of accounting must be used for associates in all consolidated primary financial statements. In the investor's individual statements, interests in associates must be accounted for as fixed asset investments at cost less any amounts written off, or at valuation (FRS 9 para. 26). The term 'equity method' is defined in FRS 9 as:

> A method of accounting that brings the investment into its investor's financial statements initially at its cost, identifying any goodwill arising. The carrying amount of the investment is adjusted in each period by the investor's share of the results of its investee less any amortization or write-off for goodwill, the investor's share of any relevant gains or losses, and any other changes in the investee's net assets including distributions to its owners, for example by dividend. The investor's share of its investee's results is recognized in its profit and loss account. The

investor's cash flow statement includes the cash flows between the investor and its investee, for example relating to dividends and loans.

Since the method of accounting for joint ventures under FRS 9 is an expanded version of the equity method used for associates, everything discussed below applies both to associates and joint ventures, unless stated otherwise. The additional requirements relating to joint ventures, but not associates, are discussed in the next section.

The investor's share in an associate or joint venture: this should be based on the aggregate holdings of parent and its subsidiaries, but not of other associates and joint ventures of the group (para. 32). It is normally based on the proportion of equity held. If more complicated sharing arrangements apply, e.g. to dividends, then the investor's share must be based on the substance of the relevant rights. As stated above, FRS 9 requires, in certain circumstances where conditions are appropriate, interests convertible into shares or options to acquire shares to be taken into account in determining the investor's interest. It comments that the costs of exercise or of future payments in the case of non-equity shares should be taken into account in a way that avoids double counting. It uses the example of incorrectly treating options as if exercised for determining the investor's stake and simultaneously valuing the same options at market value (paras. 33 and 34).

The investee's amounts to be included under the equity approach: these are to be based on its consolidated financial statements, including its own associates and joint ventures, and adjusted to give effect to the investor's accounting policies (para. 32). FRS 9 particularly draws attention to the fact that an investor in a non-corporate associate or joint venture should ensure all liabilities are suitably reflected, e.g. because of the effects of joint and several liability in say a partnership. The facts of the case would determine whether this should be treated as an actual or a contingent liability (paras. 46 and 47).

The aggregate amount of investments under the equity approach in the consolidated balance sheet can be analyzed in two complementary ways (see Taylor (1996, pp. 85–87) for a derivation), as:

- Investment at cost plus attributable post-acquisition reserve movements after including the effects of consolidation adjustments.
- The proportionate net assets of the associate after including the effects of consolidation adjustments plus the net book amount of its goodwill.

FRS 9 requires in computing amounts included under the equity method that the same principles be applied as in the consolidation of subsidiaries (para. 31). Figure VII.1 summarizes the effects of consolidation adjustments on the complementary analyses of the investment in associates (see also Ordelheide and Pfaff, 1994, p. 201). Each adjustment is numbered (e.g. item 4) in the discussion below to cross reference to Figure VII.1. The adjustments for the results of the investee and distributions to its owners are included in item 1 in Figure VII.1. One of the changes in net assets is injections of capital, item 4 in the figure. Other items are discussed below.

Adjustments to achieve uniform accounting policies – item 5: according to FRS 9's requirements, the same accounting policies as the investor are to be used for equity accounting. There is no statutory obligation on associated companies or joint ventures to provide necessary information for the parent to calculate fair values at acquisition or, for example, to eliminate unrealized intragroup profits or losses. FRS 9 acknowledges that estimates may need to be used, as access to information may be limited, though if such access is extremely limited there is doubt over whether there is significant influence in the case of associates or joint control in the case of joint ventures (para. 35).

Fair values and goodwill (para. 31(a)): fair value adjustments at acquisition affect proportionate associate asset values with an equal and opposite effect on goodwill. The effects of such adjustments will flow through in higher expenses in post-acquisition periods (item 6). However, revaluations relating to post-acquisition periods will increase the amount of the equity accounted investment (item 7) and would be reflected in the consolidated statement of total recognized gains and losses. At acquisition, fair values for the investee's assets and liabilities are to be determined using the investor's accounting policies in the same way as for a subsidiary, and the require-

Figure VII.1 Alternative breakdowns of aggregate amount of investment in associate under equity approach [left-hand column total = right-hand column total]

Breakdown 1 below = Cost plus attributable post-acquisition reserve movements	Breakdown 2 below = Goodwill plus proportionate net assets
Investment at cost adjusted in consolidated accounts for the items below:	*Parent's goodwill at acquisition adjusted in consolidated accounts for the items below:*
Plus attributable profits less dividends since acquisition (1)	
Less goodwill amortized or written off (2)	Less goodwill amortized or written off (2)
Less any write-down of investment because of impairment (3)	Less any write-down of goodwill because of impairment (3)
Plus any injections/minus any repayments of capital (4)	n/a
	= Consolidated carrying value of goodwill in associate
	Proportionate share of assets and liabilities of associates adjusted in consolidated accounts for items below:
Plus/minus any adjustments to achieve uniform accounting policies (5)	Plus/minus any adjustments to achieve uniform accounting policies (5)
Minus extra expenses arising from flow-though of fair value adjustments at acquisition (6)	Plus carrying amounts of fair value adjustments not yet expensed (6)
Plus investor's share of any unrealized post-acquisition revaluation surpluses or deficits (7)	Plus investor's share of any unrealized post-acquisition revaluation surpluses or deficits (7)
Minus investor's share of adjustments to eliminate unrealized profits on transactions between investing entity and associate (8)	Minus investor's share of adjustments to eliminate unrealized profits on transactions between investing entity and associate (8)
Plus/minus any foreign currency gains or losses arising from closing rate financial statement translation (9)	
= Investment in associates under equity approach	*= Proportional net assets at consolidated carrying amounts*

ments of FRS 7 apply. In determining goodwill at acquisition, any goodwill already in the investee's (consolidated) balance sheet should be ignored in doing the calculation. Subsequent to acquisition, goodwill should be accounted for similarly to other goodwill, in accordance with FRS 10 (items 2 and 3). However, for an associate, the carrying value of such goodwill is part of the fixed asset investment in associates in the consolidated balance sheet, and given separate note disclosure.

Consider the effects of the transitional arrangements for goodwill in FRS 10 on goodwill included in investments of associates and joint ventures. To the extent that goodwill is not reinstated it would result in differences in treatment to IAS 28. Differing treatments for goodwill for associates or joint ventures might include, for those that were acquired:

- Since FRS 10 has been in force or where goodwill immediately written off to reserves prior to FRS 10 has been reinstated – cost at acquisition less amortization for finite life goodwill or cost less impairment for indefinite life goodwill.
- Where goodwill prior to FRS 10 was immediately written off to reserves and not reinstated – at nil amount and the total amount for such goodwill at acquisition must be included in determining the gain or loss on disposal of associates, etc. Where an investment in this category ceases to be an associate, its initial carrying amount will include all the goodwill at cost (see cessation of associates below).

Both categories would be affected by any impairment write-downs. FRS 9 requires that if there has been any impairment in any goodwill attributable to any associate or joint venture it should be written down and the amount written off in the period disclosed (para. 38). It observes that impairments of net assets would usually be reflected at the individual entity level and so would not require further adjustment (para. 39).

Unrealized profits on transactions between investor and associate or joint venture (para. 31(b)): if the carrying amounts of assets of the investing entity or associate/joint venture include profits or losses on transactions between the parties, proportional elimination (i.e. of the investor's share) of such profits and losses is required in the investor's consolidated financial statements (item 8). Compare this with FRS 2's required treatment for group transactions with subsidiary undertakings on the grounds of dissimilar activities, and equity accounted, where complete (100 per cent) elimination is required, reflecting that control is still present under such circumstances. Where the reason for exclusion is severe long-term restrictions, and significant influence is retained, proportional elimination is to be recommended on the grounds that control has been impaired.

Foreign currency translation differences: SSAP 20 comments that 'when preparing group accounts for a company and its foreign enterprises, which includes ... associated companies ... the closing rate/net investment method of translation should normally be used' (para. 52) unless 'the trade of the foreign enterprise is more dependent on the economic environment of the investing company's currency than its own ... (when) the temporal method should be used' (para. 55; see V.4). Under the temporal approach, gains or losses are taken to the consolidated profit and loss account, whereas under the closing rate approach (see V.4.2), attributable financial statement translation differences are part of other attributable reserve movements since acquisition and thus affect the amount of the investment in associate (item 9). If material, such statement translation gains and losses would be shown in the consolidated statement of total recognized gains and losses, or in a note to that statement.

Investment funds: FRS 9 contains an exemption from using the equity or gross equity approach for investment funds, which should include investments held as part of an investment portfolio at cost or market value, even if the fund has significant influence or joint control (para. 48). This accepts industry practice and is to ensure consistency of treatment for investments within the portfolio. However, investments which form the means through which the investor carries out its business are not exempt.

2.2.2 Primary financial statement measurement and disclosure for associates

Consolidated balance sheet: the group's interests in associates should be shown in the consolidated balance sheet as a fixed asset investment including the investor's share of the net assets of its associates and goodwill on associates not yet amortized or written off. Both elements should be shown or disclosed – note disclosure appears to be acceptable, based on FRS 9's illustrative example (para. 29).

Consolidated profit and loss account: the investor's share of its associates' operating results should be presented immediately after the group operating result, after the investor's share in the results of any joint ventures. Goodwill amortization or write-down should be disclosed and charged at this point. The investor's share of associates' interest and post-operating profit exceptional items should be shown separately from group amounts, according to FRS 9's illustrative example, on the face of the profit and loss account. For items below the level of profit before tax, amounts should relate to the aggregate amount of the group including associates, with the amount relating to associates separately disclosed in the notes. Optionally, the total of the aggregate turnover of the group combined with associates' turnover may be given as a memorandum amount, but associates' turnover must be clearly distinguished from group turnover. IAS 28 does not discuss these issues.

Consolidated statement of total recognized gains and losses: FRS 9 requires that the investor's share of associates' and joint ventures' gains and losses should be shown separately under each heading, if the amounts are material, or in the notes to the statement (para. 28).

Consolidated cash flow statement: this should include dividends received from associates separately disclosed between operating activities and returns on investments and servicing of finance. Any other cash flows between investor and associates must be included under the relevant cash flow statement heading. No other cash flows of associates should be included (para. 30).

There is a lack of clarity in FRS 9 whether certain of its disclosures should be shown on the face of the primary financial statements or can be shown in the notes to the accounts. The above interpretations are based mainly on the illustrative examples shown in FRS 9, rather than in the standard itself. For further discussion see Holgate and McCann (1998, p. 94).

2.2.3 Other disclosures for associates and joint ventures

Other disclosures for associates and joint ventures are enumerated in Table X.1. They include disclosure requirements of special circumstances relating to holdings [including details of where and why the 20 per cent presumptions relating to significant influence and participating interests are rebutted], significant statutory, contractual or exchange control restrictions on the ability of an associate or joint venture to distribute its reserves, and notes or matters in the accounts of associates and joint ventures material to understanding the effect on the investor of its investments. Of particular interest are extensive additional size-based disclosures where aggregate associates and separately, aggregate joint ventures exceed 15 per cent of any of gross assets, gross liabilities, turnover or operating results, and further disclosures for individual associates or joint ventures where these exceed 25 per cent of any of these items, which are defined in Table X.1. IAS 28 does not contain requirements for such additional size-based disclosures or of the above circumstances.

2.2.4 Further aspects

Differing year-ends: if an associate or joint venture has a different period-end to the investor, FRS 9 desires that it should be included using financial statements to the same period-end as the investor or, if not practicable, no more than three months prior to it. However, this can be extended to six months if using such financial statements would release restricted, price-sensitive information. In all cases, changes over the intervening period which are material to the investor's financial statements should be taken into account by adjustment (para. 31(d)). IAS 28 does not specify time limits.

Investing entity does not otherwise prepare consolidated financial statements: unless it is exempt from preparing them or would be if it had subsidiaries, such an entity must present the relevant equity method amounts for associates and/or joint ventures either by preparing separate proforma financial statements including them, or by showing such amounts and the effects of including them as additional information to its own statements (para. 48).

Commencement and cessation (associates only): an investment becomes an associate when both criteria for an associate, the holding of a participating influence and the actual exercise of a significant influence begin to be fulfilled. It ceases to be an associate on the date either criterion ceases to hold (para. 40). However one should note paragraph 17 of the standard which states that once the relationship of actually exercising significant influence has been established, the exercise of such influence is to be regarded as continuing unless terminated by a transaction or event removing the investor's *ability* to do this [emphasis added]. IAS 28 also requires the discontinuation of the equity method where the associate operates under severe long-term restrictions that significantly impair its ability to transfer funds to the investor (para. 11). It is not clear that this would necessarily result a loss of associate status under FRS 9.

Carrying amount on cessation of associate or joint venture status: when a company ceases to be an associate or joint venture, its initial carrying amount for subsequent purposes is calculated as the percentage retained of the final carrying amount at the date of cessation. FRS 9 requires this to include including any not as yet amortized goodwill. It should be adjusted for any impairment in value and for any dividends or other distributions to shareholders which may affect the recoverable amount. The investment should continue to be classified as long-term whether the investor intends to dispose of it or not, using the initial carrying amount as its surrogate cost (para. 42). A particular issue arises for associates or joint ventures ceasing in the current period, which were acquired before FRS 10's goodwill accounting requirements came into force. Since FRS 10 allows under its transitional provisions that the accounting goodwill previously immediately written off direct to reserves under SSAP 22 does not have to be reinstated. Therefore, if an investor chose not to reinstate goodwill the carrying value of such an investment would be the carrying amount of attributable net assets plus the *original cost* of the goodwill. As Ernst and Young (1998, pp. 33–34) point out, this would result in a higher carrying value for such associates than under the former SSAP 1, where such goodwill was not reinstated until the date of disposal. It would also tend to be a higher carrying value than for most similar associates acquired after FRS 10 and subject to its requirements. Such a carrying amount would need to be adjusted for any impairment in the current financial statements (para. 40).

Disposal gains and losses on associates and joint ventures: the gain or loss on disposal of an associate or joint venture would be determined by deducting from the proceeds received, the carrying value of the associate's or joint venture's net assets attributable to the group, including any related goodwill not either previously written off through the profit and loss account, or attributed to prior period amortization or impairment on applying the transitional arrangements of FRS 10 (para. 40). Accumulated foreign currency translation differences on associates under the closing rate approach are not in practice transferred from reserves back into profit and loss in the United Kingdom (see V.4.2), and therefore such accumulated translation differences would not usually play a part in determining the gain or loss on disposal of foreign associates or joint ventures (see also VI.1.5.2 and VI.1.5.3).

Piecemeal acquisitions and disposals of associates and joint ventures (para. 41): processes analogous to those required by FRS 2 for subsidiaries should be applied (see VI.1.5.4).

Equity accounted investment becomes a net liability: FRS 9 is stricter than IAS 28 in that it requires an investor in each associate or joint venture to continue to use the equity method even if the result is that the interest is in net liabilities rather than net assets. Such a net liability balance is to be shown as a provision or liability, and not netted off against normal 'positive' balance associates. Equity accounting is only to be discontinued where there is sufficient evidence of an irrevocable change in relationship between investor and investee, marking the investor's irreversible withdrawal from its former associate or joint venture relationship (para. 44). This includes a public statement that it is withdrawing, with a demonstrable commitment to the process of withdrawal or evidence that the direction of the investee's operating and financial policies is to pass from its shareholders to third parties such as its creditors or bankers (para. 34). Under IAS 28 equity accounting is normally discontinued when the equity accounted investment would fall below zero amount if equity accounting were to be used. Losses are then only reported if there are liabilities of the associate or on behalf of the associate that the investor is obliged or committed to support. If the investor subsequently reports profits, then the equity method is resumed only when the subsequent profits exceed the former losses not recognized since discontinuing the use of the equity approach (para. 22). Under FRS 9, once the equity approach has been discontinued, unless the substance of the relationship changes, it would not be reinstated as the investment would no longer be an associate or joint venture.

3 Joint ventures and joint arrangement that are not entities

3.1 Definition

3.1.1 Development

The term 'joint venture' was first used in UK pronouncements by SSAP 1, Accounting for Associated Companies, in 1971, but was not defined. The Companies Act 1985 also does not give an explicit definition, but Schedule 4A, para. 19(1) in the Act implied the criterion of joint management of the undertaking with a non-group undertaking. The Act allows the option of proportional consolidation for unincorporated joint ventures that are not subsidiaries – an option based on legal form and not economic substance and subsequently inconsistent, for example, with FRS 5.

The ASB in its Interim Statement on Consolidated Accounts concluded that the relationship could be more aptly described as joint control, and defined a joint venture as, '... An undertaking by which its participants expect to achieve some common purpose or benefit. It is controlled jointly by two or more venturers. Joint control is the contractually agreed sharing of control' (para. 33). The ASB's Discussion Paper, *Associates and Joint Ventures*, issued in June 1994, focused on a criterion of joint control of an entity (para. 3.12), commenting that 'Jointly controlled operations or jointly controlled assets that do not themselves constitute a business do not amount to a joint venture' (para. 3.16). FRED 11, issued in March 1996, continued in this vein, conceiving a 'joint venture' as a separate entity. It did not explicitly consider jointly controlled assets and operations, which were to be included directly in the investor's individual financial statements (para. 41). It distinguished two types of joint ventures – those where venturers share in common the risks and obligations of their joint venture, and though they were involved in setting its business strategy, the venture had a business strategy in its own right (such joint ventures to be equity accounted), and those where each venturer had its own *separate* interest in the benefits, risks and obligations of the venture (such joint ventures to be proportionately consolidated) (paras. 8 and 9).

FRS 9 using a substance over form argument achieves a similar consolidated accounts solution in a more coherent way, now encompassing all joint arrangements. It narrows the term 'joint venture' to include corporate bodies, partnerships or unincorporated associations which carry on *their own* trade or business [effectively FRS 11's first category]. All other joint arrangements [including FRED 11's second category and also, jointly controlled operations and jointly controlled assets] are termed 'joint arrangements that are not entities'. 'Joint ventures' are equity accounted in an expanded form, the 'gross equity method', including additional information on the face of the financial statements. All 'joint arrangements that are not entities' are to be included in the investor's individual financial statements. This achieves a very similar consolidated financial statement end product for FRED 11's second category, but in addition also affects the individual statements whereas proportional consolidation does not. IAS 31 uses a different classification scheme, distinguishing jointly controlled operations, assets and entities as discussed earlier, but its definition of 'entity' appears not as restrictive as in FRS 9. It is likely therefore that some of its jointly controlled entities and all or nearly all of its jointly controlled operations and assets would be joint arrangements that are not entities – see Figure VII.2.

Figure VII.2 Joint venture accounting – comparison of accounting treatment with international standards

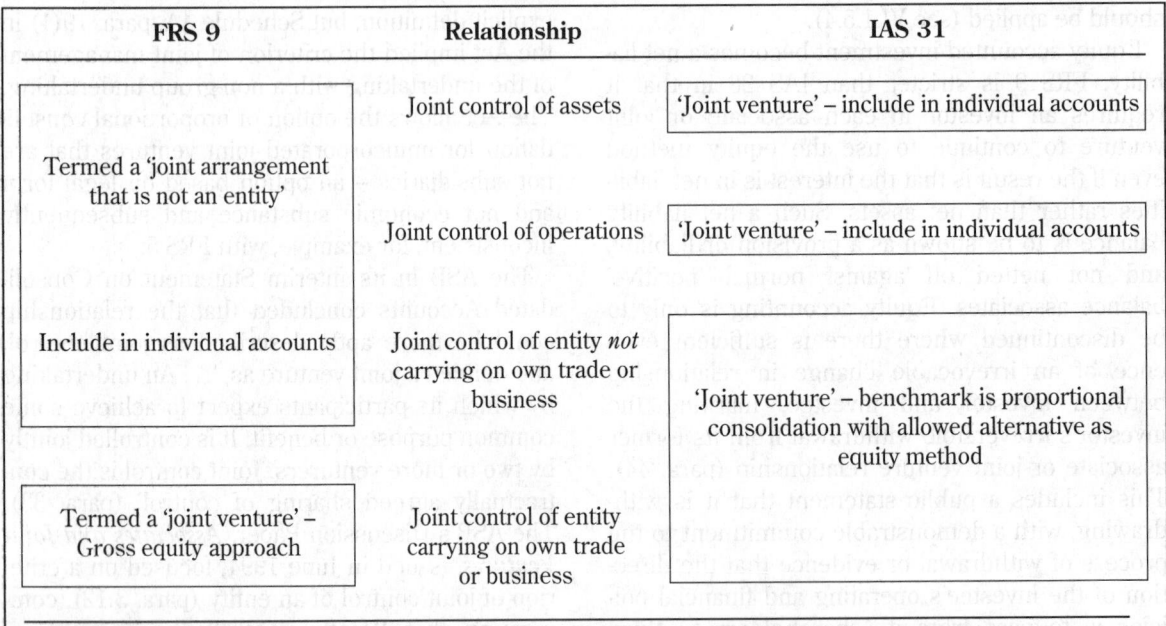

FRS 9	Relationship	IAS 31
Termed a 'joint arrangement that is not an entity'	Joint control of assets	'Joint venture' – include in individual accounts
	Joint control of operations	'Joint venture' – include in individual accounts
Include in individual accounts	Joint control of entity *not* carrying on own trade or business	'Joint venture' – benchmark is proportional consolidation with allowed alternative as equity method
Termed a 'joint venture' – Gross equity approach	Joint control of entity carrying on own trade or business	

Note: Inclusion in individual company accounts results in amounts being reflected in *both* individual accounts and in consolidated accounts. Proportional consolidation affects only consolidated accounts.

3.1.2 FRS 9 definitions of a joint venture and a joint arrangement that is not an entity

The definition of a joint venture in FRS 9 is

> An entity in which the reporting entity holds an interest on a long-term basis and is jointly controlled by the reporting entity and one or more venturers under a contractual arrangement.

Joint control is where 'a reporting entity jointly controls an venture with one or more other entities if none of the entities alone can control that entity but all together can do so and decisions on financial and operating policy essential to the activities, economic performance and financial position of that venture require each venturer's consent'.

On the other hand a joint arrangement that is not an entity is

> A contractual arrangement under which the participants engage in joint activities that do not create an entity because it would not be carrying on a trade of its own. A contractual arrangement where all significant matters of operating and financial policy are predetermined does not create an entity because all policies are those of its participants, not of a separate entity.

The distinguishing feature is that one is an 'entity' and the other is not. FRS 9's definition of an entity is important. It is based on the Companies Act definition of an undertaking, 'a body corporate, partnership, or unincorporated association carrying on a trade with or without a view to a profit', but is more restrictive. It continues, 'the reference to carrying on a trade or business means a trade or business *of its own* and not just a part of the trades or businesses of entities that have interests in it'. (para. 4 emphasis added)

3.1.2.1 Joint ventures – further aspects

FRS 9 specifies that 'joint control' implies taking an active role in setting the operating and financial policies of the venture. This could still be the case even if one venturer manages the joint venture, provided there is collective agreement over the venture's financial and operating policies and the venturers have the power to ensure such policies are followed (para. 11). The standard comments that the power of veto, i.e. to withhold consent on high-level strategic decisions, is what distinguishes a joint venturer from a minority shareholder of shares – the venturer cannot be coerced by majority power. Such requirement for consent can be inferred from the practice of the joint venture and does not have to be set out in the joint venture agreement (para. 12). It is not necessary for all interests in the same entity to be treated the same way – the entity may, for example, be a joint venture for those investors sharing control, and an investment for another investor who does not share control (para. 10).

Subsidiary subject to severe long-term restrictions: The Companies Act 1985 and FRS 2 exclude subsidiary undertakings from consolidation where there are 'severe long-term restrictions' which impair the exercise of unilateral control. Where an entity qualifies as a subsidiary undertaking but there are contractual arrangements with other shareholders that mean in practice there is shared control, FRS 9 states that this amounts to 'severe long-term restrictions' and such a subsidiary should not be consolidated but treated as a joint venture (para. 11). However, Ernst and Young (1998, pp. 13–14) consider that only where the long-term restrictions take the form of requiring mutual consent on matters normally decided by the shareholders in general meeting is it clear that FRS 9's requirement is consistent with FRS 2 and the law. They question whether mere commercial agreements with minority shareholders, for example, to restrict the scope or country of operations in some way would be sufficient. In such circumstances, however, one could argue whether this type of situation is really shared control.

Commencement and cessation (joint ventures only): an investment becomes a joint venture on the date when the investor has both a long-term interest and it begins joint control with other venturers. It ceases on the date the investor ceases to have joint control.

3.1.2.2 Joint arrangement that are not entities – further aspects

FRS 9 points out that participants possessing a long-term interest and having joint control is a necessary but not a sufficient condition for a 'joint venture' to be established. The venture 'entity' must in addition carry on its *own* trade or business. The standard specifies that it must have some independence to pursue its own commercial strategy in its buying and selling within the terms of the agreement establishing the joint arrangement. It must also either have access to the market itself for its main inputs and outputs, or be able to buy from or sell to the participants on genuine market terms. It gives two examples of when this is not the case:

(a) The participants derive their benefits from product or service taken in kind rather than by receiving a share in the results of trading; or
(b) Each participant's share of the output or result of the joint activity is determined by its supply of key inputs to the process producing that output or result (para. 8).

FRS 9 gives the examples of joint arrangements that are not entities such as a joint marketing or distribution network or a shared production facility such as a pipeline providing a service only to the participants. It comments that in most cases a joint arrangement carrying out a single project would not possess the normal signs of an autonomous trade or business of continuing and repetitive activity. The standard allows that, if the substance of the arrangement changes, it could be possible for a joint arrangement that is not an entity to become a joint venture (and presumably vice versa).

3.2 Accounting for joint ventures and joint arrangements that are not entities

The Companies Act 1985 allows, but does not require, proportional consolidation for unincorporated joint ventures which are not subsidiaries (Sch 4A, para. 19(1)), based on the legal structure of the relationship not its substance. FRS 9, on the other hand, takes a substance over form approach, and specifically addresses the issue of joint arrangements which have the form but not the substance of a joint venture. In such circumstances the seeming framework entity is merely an agent for the venturers, who are able to identify their direct share in the assets, liabilities and cash flows of the structure, and requires such a structure to be accounted for as a joint arrangement that is not an entity (para. 24).

3.2.1 Joint ventures

In consolidated financial statements joint ventures are to be accounted for using the gross equity method in all primary financial statements. As with all investments in affiliates, in the individual financial statements of the investor, investments in joint ventures are to be accounted for as fixed asset investments at cost, less amounts written off, or at valuation (para. 20).

The gross equity method is the same as the equity method used for associates, with a few additional disclosures:

- On the face of the consolidated balance sheet the amount for investments in joint ventures under the equity approach must be amplified to show the investor's share of gross assets and liabilities.
- On the face of the consolidated profit and loss account the investor's share of turnover should be shown, but not as a part of group turnover. A similar breakdown of turnover must also be shown in any segmental analysis.

Even though they are short of proportional consolidation, these expanded disclosures would enable analysts to assess in a somewhat crude way, for example, group gearing, profit margins and asset turnover, including joint ventures. Because joint control is a stronger relationship than significant influence, FRS 9 also allows venturers to give additional information about their joint ventures. However, it requires that, except for items below profit before tax in the profit and loss account, such supplementary information must clearly distinguish the amounts for the group from that of joint ventures (para. 23). It permits a columnar presentation that discloses supplementary information about the group, about joint ventures, with an overall total. Such optional presentation, in effect, could reconcile FRS 9's treatment of joint ventures to IAS 28's treatment, proportional consolidation (Appendix 4).

The accounting treatments in VII.2.2.2, Primary financial statement measurement and disclosure for associates, also apply for joint ventures except as modified by the current section, VII.3.2. In the year prior to the implementation of FRS 9, *Company Reporting* (July 1998) found that 10 per cent of UK companies made disclosures about investments in joint ventures (prior to FRS 9 this term would include 'joint arrangements that are not entities'). Forty-two per cent of these used the equity method, 21 per cent the gross equity method, and 28 per cent proportional consolidation. It expected that additional joint ventures and arrangements that had previously been classified with associates would become reclassified under FRS 9.

3.2.2 Joint arrangements that are not entities

These are to be accounted for in the participant's individual financial statements by including the participant's own assets, liabilities and cash flows according to the terms of the agreement covering the arrangement (para. 18). There are no specific disclosures for such arrangements within FRS 9.

4 Other uses for equity accounting

Equity accounting is also used for subsidiary undertakings not consolidated because:

- Their activities are too dissimilar from the rest of the group (FRS 2, para. 30).
- Severe long-term restrictions exist over the parent's rights to control the undertaking where significant influence can still be exercised; but if such influence is not exercisable, the investment must be frozen at its carrying 'equitized' amount at the date the restrictions came into force (FRS 2, para. 27). FRS 9 considers that where significant influence is exercisable the relationship is that of an associate, or where there is joint control of a structure carrying on its own trade or business, a joint venture (VII.3.1.2.1).

Carrying amounts of such investments should be written down for any impairment in value.

VIII Deferred taxation

1 General remarks

In the United Kingdom, SSAP 15, Accounting for Deferred Taxation, issued in 1978 and revised in 1985, requires the liability method for computing taxation effects 'calculated at the rate of tax that it is estimated will be applicable when the timing differences will reverse' (para. 23). Based on this computation, a partial provision approach is used for calculating the deferred taxation provision. Usually the current tax rate is the best estimate of the applicable tax rate unless changes of rates are known in advance (para. 14). The 'partial provision' approach means that 'tax deferred or accelerated by the effects of timing differences should (only) be accounted for to the extent that it is probable that a liability or asset will crystallise' (para. 25) (see United Kingdom – Individual Accounts VII.2 for further discussion of the principles of UK accounting for deferred taxation).

Only particular issues relating to group accounting and consolidation are discussed here. There is very little systematic discussion of the relationship between deferred taxation and group accounting in the published accounting literature in the United Kingdom, and this section attempts to set SSAP 15's requirements within a coherent framework. An ASB Discussion Paper, Accounting for Tax, was issued in March 1995. One of its chapters examines acquisition accounting and taxation. An exposure draft, FRED 19, Deferred Tax, was issued in August 1999. It proposes requiring a full provision incremental liability approach, similar in most respects to international practice. Before examining the group accounting implications of deferred tax, UK tax requirements relating to groups are considered.

2 The taxation of groups

UK taxation is focused on individual companies. However, group relief allows tax group companies to surrender tax reliefs to other tax group companies. Where one company is at least a 75 per cent subsidiary of another or both are at least 75 per cent subsidiaries of the same parent, trading profits and losses computed for tax purposes can be offset within the same accounting period.

Similarly, capital gains and other non-trading income can be set against the same period's trading losses of other group companies, but not the converse, i.e. capital losses against trading profits or capital gains. However, reliefs for tax losses at acquisition of a more than 75 per cent owned subsidiary cannot be surrendered to other tax group companies, as they were made before the subsidiary joined the group, though they can be used by the subsidiary itself in future periods. Advance corporation tax not recoverable by the parent in a period, which is to be abolished with effect from April 1999, can be surrendered to an at least 51 per cent-owned subsidiary to be offset against its mainstream corporation tax charges of that period.

SSAP 15 requires that group companies should in their deferred tax accounting take into account 'any group relief which, on reasonable evidence, is expected to be available and any charge which will be made for such relief'. Some groups require the receiving companies to pay for the use of such reliefs at the going tax rate, whereas others surrender reliefs without charge. Any such payments are intragroup charges and therefore will cancel on consolidation. Where there are no such charges, reported individual company financial statement tax rates will vary from company to company. For accounting purposes, the starting point for the consolidated tax charge is the aggregation of individual company charges and balances. Assumptions made by the group as to the availability of group relief and payments therefore must be stated (para. 43).

3 Group timing differences

3.1 Calculation and definition of timing differences

The calculation of deferred tax in the United Kingdom is a three-stage process:

1 The determination of timing differences between profits and losses as computed for tax purposes and results as stated in financial statements, which arise from the inclusion of items of income and expenditure in tax computations in periods different from those in which they are included in financial statements . . . (They) originate in one period and are capable of reversal in one or more subsequent periods (para. 18).
2 The computation of the full amount of the potential deferred tax liabilities or debit balances for separate categories of timing differences.
3 An assessment, based on forecasts and projections, of the extent that it is probable that a liability or asset will crystallize, resulting in the partial provision computation.

Step 1: the term 'timing differences' in SSAP 15 is based on what IAS 12, Income Taxes, revised in 1996, terms the 'income statement liability method'. 'Timing differences' are narrower in scope than the concept of 'temporary differences' used in IAS under its 'balance sheet liability method', where they are defined as 'differences between the tax basis of an asset or liability and its carrying amount in the balance sheet' (para. 1). Under SSAP 15 deferred taxation liabilities and assets are recognized based on reversal of timing differences rather than taxable temporary differences. The term 'timing differences', used strictly, excludes deferred tax implications of items which affect shareholders' funds or reserves in the balance sheet, but do not pass through the profit and loss account, e.g. fair value adjustments at acquisition and foreign currency financial statement translation differences under the closing rate approach. These are referred to here as quasi-timing differences, and 'temporary differences' comprise therefore both timing and quasi-timing differences. SFAS 109, Accounting for Income Taxes, issued in 1992 in the United States, is based on similar concepts to IAS 12.

Step 2: Table VIII.1 categorizes potential sources of deferred taxation liabilities and assets arising from timing and quasi-timing differences, where consolidated carrying values differ from the tax bases for group assets and liabilities, and such carrying values are based on consolidation adjustments which do not normally arise in individual company accounts. It examines extant UK practice. Because it is not recent, many areas dealt with in IAS 12 and IAS 109 are not covered by SSAP 15.

Step 3: these potential balances would then be assessed to determine the extent to which in net

Table VIII.1 Consolidation adjustments and UK deferred taxation effects

Consolidation adjustment	Deferred tax treatment
Fair values adjustment at acquisition and any subsequent depreciation/expensing	Not discussed in SSAP 15. FRS 7 states that deferred tax assets and liabilities should be determined by considering the enlarged group as a whole (para. 21).
Adapting individual accounts to uniform group accounting policies and UK GAAP and any subsequent depreciation/expensing of such adjustments	As above.
Consolidated goodwill amortization	Not deductible for tax purposes in the United Kingdom, not treated a timing difference.
Elimination of unrealized profits on intragroup transactions	Not discussed in SSAP 15. Some authors suggest treating as if a timing difference.
Foreign currency financial statement translation differences	Not regarded as creating a timing difference. Individual company transaction translation differences may give rise to timing differences depending on whether or not they have a tax effect (SSAP 15 Appendix para. 12 – for guidance only).
Potential additional tax on the remittance of undistributed earnings of subsidiaries	Creates a timing difference and potential deferred tax balance only if (a) there is an intention to remit them; and (b) remittance would result in a tax liability after taking account of any related double taxation relief (SSAP 15 para. 21).
Other temporary differences arising from Investments in Subsidiaries, Branches and Associates, and Interests in Joint Ventures	Not dealt with in UK standards

terms it is probable they will reverse in order to calculate the actual *partial* deferred tax position – the extent to which in future periods it is probable that reversals will not be offset by contemporaneous originating differences. SSAP 15 also enables analysts to determine the full provision amount by requiring note disclosure of any amount unprovided, analysed into its major components (para. 35).

3.2 Fair value adjustments at acquisition

SSAP 15 does not deal with deferred taxation resulting from adjustments to determine fair values at acquisition. Such adjustments adjust company costs to group costs for consolidation purposes. However, it does not allow a 'net of tax' treatment where 'the tax effects of timing differences (are treated as) integral parts of the tax effects of timing differences of the revenues or expenses, assets, provisions or liabilities to which they relate, rather than showing them separately' (para. 16). The ASB's Discussion Paper, *Accounting for Tax*, issued in March 1995, comments that fair value adjustments at acquisition are permanent (i.e. not timing) differences since 'they have the effect that gains and losses that are charged or deducted in post-acquisition tax computations are never recognized in the financial statements of the combined entity' (para. 8.1.5). This is because post-acquisition consolidated accounting profits are based on fair value adjusted group costs, whereas taxable profits are based on original company costs. This can also be characterized by considering such fair valued assets as not fully tax-deductible. The Discussion Paper considers therefore that a strict interpretation of SSAP

15 implies that deferred tax should not be set up relating to such fair value adjustments, 'although it is common – but by no means universal – practice in the United Kingdom to do so' (para. 8.1.4). In the context of its wider discussion on the nature of deferred tax, it comments that those who believe that deferred tax should be characterized as an increase or decrease in a future tax liability would take the view that no deferred tax arises; however, those who think deferred tax is a 'valuation adjustment', the difference in value between a (fair valued) asset which is not fully tax deductible and an asset with the same characteristics which is fully tax-deductible, would provide for deferred tax. The Discussion Paper acknowledged a difference of opinion within the Board, and responses to the paper have not encouraged the ASB to move quickly. Unlike IAS 12, FRED 19, issued in August 1999, requires that deferred taxation should not normally be provided on fair value adjustments at acquisition. Also, no provision is to be recognised in respect of acquired goodwill. IAS 12 regards fair value adjustments at acquisition as causing temporary differences to arise (para. 18).

FRS 7, *Fair Values in Acquisition Accounting,* requires deferred tax to be determined on a group basis. At the end of the accounting period in which an acquisition occurs, the enlarged group's deferred tax provision should be calculated as a single amount, based on assumptions applicable to the group. It argues that, if different assumptions are used at acquisition from those applying to the group as a whole, any post-acquisition profit and loss charge would reflect the effects of changing from one set of assumptions to another rather than any real change in the circumstances of the group (para. 74). Proposals from its previous exposure draft, FRED 7, and a 1993 ASB Discussion Paper on *Fair Values at Acquisition,* that fair value adjustments would only result in potential deferred tax assets and liabilities to the extent that they would be regarded as timing differences (under SSAP 15) if reflected in the acquired entity's individual accounts, were not carried forward. FRS 7 does not include its 1993 precursor Discussion Paper's proposed distinction between the imputed deferred tax position of the acquired company at acquisition, and changes because of group tax planning possibilities and intentions. This is because of the difficulty and feasibility of an acquirer assessing an acquired entity's intentions at acquisition, which is necessary to make such a split.

FRS 7 also discusses group benefits from unrelieved tax losses in an acquired entity, which should be treated as timing differences if they are identified as having a value to the group that can be measured reliably, based on SSAP 15's overall recognition criteria, even if they are not recognized as assets in the acquired company. These should first be deducted from deferred tax liabilities and any remainder treated as deferred tax assets (FRS 7, para. 75). Quantified note disclosures to the profit and loss account of special circumstances affecting the overall tax charge should be given (FRS 3, para. 77). If SSAP 15's criteria applied at acquisition are not met before the ending of FRS 7's investigation period (see VI.1.3.4), FRS 7 requires that such benefits should be deemed post-acquisition.

3.3 Adapting to uniform group accounting policies

The adoption of uniform accounting policies across a group affects the consolidated accounts carrying values of a subsidiary's assets and liabilities, and can be effected either by the subsidiary itself or by consolidation adjustment. This area is not specifically dealt with in SSAP 15, though the Statement by the Accounting Standards Committee on the publication of SSAP No. 15, issued in 1978, commented that 'the need for common accounting policies is considered in SSAP 14 [now in FRS 2]. Clearly, appropriate adjustments may need to be made concerning deferred taxation for consolidation purposes' (para. 32). If uniform accounting policies are effected by consolidation adjustment, the principles of FRS 7 relating to the fair value adjustments at acquisition above would apply.

3.4 Consolidated goodwill amortization

Goodwill amortization is not an allowable deduction for taxation in the United Kingdom, so consolidated goodwill would be regarded as analogous to a permanent difference. General practice is that no deferred tax liability is calculated relating to goodwill at acquisition. This is consistent with IAS 12 which, whilst recognizing that goodwill on consolidation results in the arising of temporary differences, does not permit the recognition of the resulting deferred tax liability.

3.5 Intragroup profits and losses

Such adjustments eliminate unrealized profits on intra-group transactions such as intra-group stocks or fixed assets (see VI.2.2). This area is not dealt with in SSAP 15, but many UK companies currently make deferred tax adjustments for such items. FRED 19 regards them as timing differences and proposes requiring full provision for deferred tax measured using the supplying company's rate of tax. IAS 12 also regards such adjustments as causing temporary differences leading potentially to a deferred tax asset.

3.6 Financial statement foreign currency translation gains and losses

The guidance notes of SSAP 15 state that

> translation of the financial statements of overseas subsidiaries or associated companies is not regarded as creating a timing difference. Gains and losses arising on translation of an enterprise's own overseas assets (including investments in subsidiaries and associated companies) and liabilities may give rise to timing differences depending on whether or not the gains or losses have a tax effect (SSAP 15, Appendix, para. 12).

This guidance seems to apply whether the temporal or closing rate approaches were used. There appears to be no published UK discussion on the basis for this stance. One might argue that even closing rate translation gains and losses affect sterling consolidated carrying amounts of the foreign subsidiary's assets and liabilities in a somewhat similar manner to revaluation adjustments. If the subsidiary is disposed of, such changes in sterling carrying amounts could therefore affect consolidated gains or losses on disposal (see VI.5.3). There is no debate on the similarity to fair value adjustments in the United Kingdom. Such translation gains and losses, unlike fair value adjustments, do not affect local currency carrying amounts and so could not be accounted for as if they had been incorporated in the subsidiary's company financial statements. Also under the partial provision approach, if there is no intention to dispose of a subsidiary in the foreseeable future, it could be argued that, because it is not probable that an asset or liability will crystallize, no deferred tax provision would be necessary.

Unlike IAS 12, FRED 19 proposes not requiring provision on exchange differences on consolidation of non-monetary assets of an entity accounted for under the temporal method. IAS 12 requires recognition of a deferred tax liability for all temporary differences associated with investments in subsidiaries, branches, associates and interests in joint ventures except to the extent that the parent, investor or venturer is able to control the timing of the reversal of the temporary difference, and it is probable that the temporary difference will not reverse in the foreseeable future (para. 39). It seems to imply that where there are dependent foreign operations where the temporal approach is used (in IAS 21's terms 'a foreign operation integral to the enterprise's operations'), temporary differences can arise on non-monetary assets and liabilities where the foreign operation's tax base is determined in foreign currency.

3.7 Potential additional tax on foreign subsidiaries' distributions

Further group deferred tax implications may arise if the group intends to remit the earnings of a foreign subsidiary to the United Kingdom. According to SSAP 15 this creates a timing difference and potential deferred tax balance only if

- There is an intention to remit them; and
- Remittance would result in a tax liability after taking account of any related double taxation relief. (para. 21)

If so, the amount of any additional provision necessary will depend on tax regimes and other group companies' tax positions in the contemplated period of remittance. Where no provision is made on overseas retained earnings, SSAP 15 requires that fact to be noted (para. 44). FRS 2 requires disclosure of the extent to which deferred tax has been accounted for in respect of future remittances of the accumulated reserves of overseas subsidiaries. Also if deferred tax has not been provided in respect of any of the accumulated reserves of overseas subsidiary undertakings, the reasons for not fully providing should be given (FRS 2, para. 54). FRED 19 proposes requiring provision only if dividends payable by subsidiaries, associates or joint ventures have been accrued at the balance sheet date, or a binding agreement to distribute past earnings in the future has been made. IAS 12 requires recognition of a deferred tax liability for such temporary differences except if the parent, investor or venturer is able to control the timing of the reversal of the temporary difference, and it is probable that the temporary difference will not reverse in the foreseeable future (para. 39).

3.8 Other areas

Many areas dealt with in IAS 12 relating to group accounting issues are not covered by SSAP 15. Others are dealt with piecemeal in UK standards as described above

4 Computing partial provisions

The full provision approach, as practised in the United States, broadly determines separate deferred tax assets and liabilities for each type of originating difference, and then considers which of these can be offset against each other. The partial provision approach, as practised in the United Kingdom, starts from the general principle that 'the combined effect of timing differences should be considered when attempting to assess whether a tax liability will crystallise, rather than looking at each timing difference separately' (SSAP 15, Appendix para. 4), and then considers which timing differences should not be offset. SSAP 15 also requires the total amount of any unprovided deferred tax to be disclosed as a note, analysed into its major components (para. 40), so the full provision can in principle be deduced, but in practice in a more netted off form than in the United States.

In determining the partial provision, SSAP 15 is extremely unclear as to its criteria for offset despite its general principle above. For example, Munson (1985, p. 14) reads the standard as requiring timing differences relating to potential deferred tax assets to be separately evaluated from potential deferred tax liabilities, 'just as deferred tax liabilities should not be created unless it is probable that a liability will crystallise, so deferred tax liabilities should not be reduced by deferred tax debit balances which will not crystallise because recovery of the tax is continually deferred' (see Munson, 1985, p. 14). However, Davies, Paterson and Wilson (1997, p. 1227) give an example to show that such an approach can produce anomalous results and consider that the approach required by the standard can be interpreted as leading to timing differences being evaluated in aggregate. These different interpretations of a somewhat ambiguously worded standard have implications for group accounting as discussed below. The key question from a group accounting perspective is how far the effects of timing differences occurring in different group companies can be taken in aggregate in determining the group partial provision. SSAP 15 does not consider this directly and its principles have led to different interpretations. At one end of the spectrum, Wild and Goodhead (1998, p. 498) seem to imply that the group timing differences have to be considered separately in each company, commenting that

> the deferred tax position of each group company should thus be determined separately and the results aggregated to determine the group position. The only adjustments to the group deferred tax liability or asset which may be necessary are those which arise from consolidation adjustments, such as the elimination of intragroup profits.

However, Davies, Paterson and Wilson interpret SSAP 15 as allowing the overall pattern of reversals to be considered on a group basis, unless there are compelling reasons where offset is not possible. This can change the overall amounts of reversals in computing the group partial provision compared with the aggregate of the partial provisions of the separate group companies. They give an example of a two-company group where the parent has an increasing pattern of originating timing differences, and its wholly-owned subsidiary has a reversing pattern of timing differences. On a company basis, the parent makes no partial deferred tax provision since it has no reversals, and the subsidiary's partial provision is based only on its own pattern of reversals. They argue that, for the group, the parent's originating pattern of timing differences is available in each period to set against the reversal pattern of timing differences in the subsidiary, arguing that the deferred tax liability needs only to be calculated on 'the amount which would have been the basis of the provision if all the timing differences had been in a single company. . .'. Such offset reduces the group 'partial' deferred tax provision compared with the aggregate of the parent and subsidiary's deferred tax provisions. The extent of this effect depends on the patterns of originating and reversing differences in each group company and would be unique to each particular case (Davies *et al.*, 1997, p. 1233).

If this general principle of offset is accepted, one must specify instances where it may not apply, and they consider that offset is permissible even where the originating and reversing differences take place in different tax jurisdictions with different tax rates – it would be necessary to convert timing differences in different countries to patterns of potential deferred tax assets and liabilities by applying the different rates. The net aggregate reversal pattern of deferred tax assets and liabilities would then be calculated for the purposes of determining the partial provision. Such offset would be allowable except where 'there is a significant possibility that increased tax charges in one country might not be offset by tax saving in others.' They suggest starting 'by examining the overall position for all the group companies in each tax jurisdiction as an interim stage in the process' Davies *et al.* (1997, p. 1234). They consider offset would not be possible, for example, where the company with the originating differences is tax loss making, and because it is in a different tax jurisdiction its losses cannot be surrendered to other group companies through group relief. They argue tax savings from originating differences in such a company would not then be available to offset the extra tax payment effects of reversals elsewhere in the group. A similar problem in offsetting originating and reversing timing differences on a group basis could arise even if the group were entirely within a single jurisdiction, e.g. if a UK loss-making company with originating timing differences were a 60 per cent owned subsidiary of a UK parent, because the appropriate tax group for surrendering losses would be based on at least 75 per cent-owned subsidiaries (VII.1).

Intermediate between these positions would be the view that offsets in determining net aggregate reversals should only be allowed within each tax jurisdiction separately. SSAP 15 is not explicit on these matters and all seem tenable positions in the United Kingdom.

FRS 7 comments that 'deferred tax has to be determined on a group basis' in which 'the enlarged group's deferred taxation provision will be calculated as a single amount, on assumptions applicable to the group' (para. 74). However, FRS 7 is a discussion of determining the deferred tax position relating to an acquired company and not a discussion of deferred tax and allowable offsets *per se*. Therefore it is dangerous to infer an ASB position on group deferred taxation provisions. Its 1995 Discussion Paper, *Accounting for Tax*, did not address this issue.

Another reason why the group deferred taxation provision will not be a simple aggregation of the group companies' provisions relates to the deferred taxation effects of consolidation adjustments. Because of the requirement for uniform group accounting policies, foreign subsidiaries will need to provide parents with detailed forecasts of probable dates of originating and reversing timing differences and the maturity of any estimated deferred tax balances and estimated tax rates, to be incorporated in the group partial provision calculations (see VIII.3.3). A UK subsidiary of an overseas group conversely will have to prepare individual company accounts complying with SSAP 15, and may need to provide its parent with appropriate information for its group accounting compliance.

To the argument that deferred tax assets not reflected on prudence grounds in individual accounts can be set up and offset at consolidated level because of deferred liabilities in other group companies, Munson (1985, p. 22) comments, 'such a proposal is only justified if the future tax reliefs represented by the debit balances and the future tax liabilities represented by the credit balances will actually be able to be offset (when both arise)'. This would depend on the existence of plausible group tax planning evidence. For example, group relief on trading losses is only available to the extent that gains are available in the tax group in the same trading period.

5 Associates and joint ventures

SSAP 15's guidance section states that foreign currency statement translation gains and losses of overseas associates do not constitute timing differences (para. 12). It is not clear whether SSAP 15's requirements relating to the need to account for deferred taxation on the potential additional taxation effects of the remittance of undistributed overseas earnings, discussed earlier, apply to associated companies. Otherwise SSAP 15 does not address the issue of the deferred tax implications of investments in associates or joint ventures, for example arising from consolidation adjustments.

In the case of associates, the parent only has significant influence over distribution policy. Standard setters in the United States and internationally make the presumption that all earnings will be remitted in determining the deferred tax provision. IAS 12 requires that if there are not specific agreements which give the investor in the associate control over distribution policy, and unless it is probable that temporary differences will not reverse in the foreseeable future, the investor should recognize a deferred tax liability arising from all taxable temporary differences associated with its investment in the associate (para. 42). In cases where investors in joint ventures can, through the joint venture arrangement, control the sharing of profits and it is probable they will not be distributed in the foreseeable future, IAS 12 states that a deferred taxation liability is not to be recognized (para. 43).

IX Formats

1 Companies Act formats

The opening section of Schedule 4A (Form and Content of Group Accounts) of the Companies Act 1985 states 'Group Accounts shall comply so far as is practicable with the provisions of Schedule 4 (i.e. Form and Content of Company Accounts) as if the undertakings included in the consolidation ('the group') were a single company.' There is no requirement, for example, for overseas subsidiaries to follow Companies Act formats, but they must provide the parent with sufficient information for the group to comply with the Schedule 4 formats on a consolidated basis.

1.1 Cancellation of intragroup balances

On consolidation, in balance sheet format 1, amounts in the parent individual company accounts classified under the heading B. III 2. Fixed Asset Investments, 'Loans to group undertakings', C. II 2. Debtors, 'Amounts owed by group undertakings', E. 6. Creditors: amounts falling due within one year, 'Amounts owed to group undertakings', H. 6. Creditors: amounts falling due after more than one year, 'Amounts owed to group undertakings' would be cancelled against the corresponding intra-group balances of subsidiaries, and would not generally appear in the consolidated financial statements. One reason they might continue to appear is if a subsidiary meets the criteria for exclusion from the consolidated financial statements (see IV.3). Similar cancellation is carried out in balance sheet format II, as above except C. 6. Creditors, 'Amounts owed to group undertakings' replaces E. 6. and H. 6. in format 1.

1.2 Consolidated goodwill

Under both balance sheet formats, item B. III 1. Fixed asset investments, 'Shares in group undertakings' would not generally appear in the consolidated financial statements, being part of the consolidation cancellation process in determining goodwill. It would only appear in the case of a subsidiary meeting the criteria for exclusion from consolidation (see IV.3). Consolidated goodwill would be included under heading B. I 3. Intangible assets, 'Goodwill', and intangible assets identified in determining fair values at acquisition of a subsidiary would probably be included under heading B. I 2. Intangible Assets, 'Concessions, patents, licences, trade marks and similar rights and assets'.

1.3 Minority interests

Under both balance sheet formats, non-controlling interests in capital and reserves should be aggregated in the consolidated financial statements under the heading 'minority interests' which is placed next to, but apart from Capital and Reserves. In Format 2 it would be included under the general heading of 'Liabilities', between 'Capital and Reserves' and 'Provisions for Liabilities and Charges' (Sch 4A, para. 17). Under the profit and loss account formats, a new heading for the minority share in profit or loss on ordinary activities is shown as an aggregate deduction before extraordinary items and further, the aggregate minority interest in extraordinary profits or losses must also be shown separately (Sch 4A, para. 17).

1.4 Associated undertakings

Under both balance sheet formats, the heading B. III 3, Fixed asset investments, 'Participating interests' in the individual company balance sheet headings is replaced in the consolidated balance sheet by the two headings, 'Interests in associated undertakings' and 'Other participating interests'. Likewise in the profit and loss account formats 'Income from participating interests' is replaced by 'Income from associated undertakings' and 'Income from other participating interests'.

1.5 Consolidated reserves

The format of consolidated capital and reserves follows that of individual company accounts, adjusted for consolidation matters. Current UK surveys of published accounts do not report the extent to which UK companies take accumulated foreign currency statement translation differences under the closing rate approach to a separate reserve. This is probably rare.

1.6 Information supplementing the consolidated balance sheet

Schedule 4A's requirements that group accounts should comply as far as possible with the individual company accounts requirements in Schedule 4 of the Act would include the adjustments of any note disclosures to a consolidated basis, for example, the necessity to provide a consolidated schedule of fixed asset movements (Sch 4, para. 42 (1)). In preparing such a schedule, unrealized profits on intra-group transactions would be eliminated. Such intra-group transfers of fixed assets would be removed from 'additions' and 'disposals', though not in the case where fixed asset additions for one company are sales of stock-in-trade of another. Note disclosures of, for example, deferred taxation are similar to those of an individual company, but include consolidation adjustments and some suggest partial provision estimation should be carried out on a group basis (VIII.4) where note disclosures relating to deferred taxation on potential remittances from earnings of overseas subsidiaries is discussed.

2 FRS 3, Reporting Financial Performance

In VI.3 and VI.4 the performance statements required by FRS 3, Reporting Financial Performance, issued in 1992, are discussed. Though its two illustrative layouts for the consolidated profit and loss account are not compulsory, they are shown in Figures IX.1 and IX.2, adapted to include an associate per FRS 9. See also Table VI.3. FRS 3's illustrative layout 2 portrays the split between acquisitions, other continuing operations and discontinued operations in a multi-columnar format, which gives more information on the face of the profit and loss account itself. Illustrative layout 1 would give further information in the notes, not shown here (see VI.1.5.1 and VI.3 for further details). Income from investments in associates is gross of tax and the note to the accounts

Figure IX.1 FRS 3 Profit and Loss Account – in GBP million - illustrative layout 1

	1998	1998	1997	1997 restated
Turnover: group and share of joint ventures				
Continuing operations	740		620	
Acquisitions	80			
	820			
Discontinued operations	130		185	
Joint ventures	60		50	
		1,010		855
Less: share of joint ventures' turnover		(60)		(50)
Group turnover		950		805
Cost of sales		(780)		(650)
Gross profit		170		155
Net operating expenses		(95)		(85)
Operating profit				
Continuing operations	80		75	
Acquisitions	10			
	90			
Discontinued operations	(25)		(5)	
Less 1997 provision	10			
		75		70
Provision on sale of properties in continuing operations		12		10
Provision for loss on operations to be discontinued				(35)
Loss on disposal of discontinued operations	(20)			
Less 1997 provision	25			
		5		
Profit on ordinary activities before interest		92		45
Share of operating profit in				
Joint ventures	10		9	
Associates	8		7	
		18		16
Interest payable:				
Group	(15)		(10)	
Joint ventures	(5)		(5)	
Associates	(2)		(3)	
		(22)		(18)
Profit on ordinary activities before taxation		88		43
Tax on profit on ordinary activities*		(25)		(10)
Profit on ordinary activities after taxation		63		33
Minority interests		(5)		(3)
[Profit before extraordinary items]		58		30
[Extraordinary items] (included only to show positioning)		–		–
Profit for the financial year		58		30
Dividends		(12)		(5)
Retained profit for the financial year		46		25

* Note disclosure required of breakdown between parent and subsidiaries, joint ventures and associates.

Figure IX.2 FRS 3 Profit and Loss Account – in GBP million - illustrative layout 2

	Continuing 1998	operations 1998 Acquisitions	Discontinued Operations 1998	Joint ventures 1998	Total 1998	Total 1997 as restated
Turnover: group and share of joint ventures	740	80	130	60	1,010	855
Less: share of joint ventures turnover				(60)	(60)	(50)
Group turnover				–	950	805
Cost of sales	(600)	(63)	(117)		(780)	(650)
Gross profit	140	17	13		170	155
Net operating expenses	(60)	(7)	(38)		(105)	(85)
Less 1997 provision			10		10	
Operating profit	80	10	(15)		75	70
Profit on sale of properties	12				12	10
Provision for loss on operations to be						(35)
Loss on disposal of discontinued			(20)		(20)	
Less 1997 provision			25		25	
Profit on ordinary activities before interest	92	10	(10)		92	45
Share of operating profit in:						
Joint ventures					10	9
Associates					8	7
					110	61
Interest payable:						
Group					(15)	(10)
Joint ventures					(5)	(5)
Associates					(2)	(3)
Profit on ordinary activities before taxation					88	43
Tax on profit on ordinary activities*					(25)	(10)
Profit on ordinary activities after taxation					63	33
Minority interests					(5)	(3)
[Profit before extraordinary items]					58	30
[Extraordinary items]**					–	–
Profit for the financial year					58	30
Dividends					(12)	(5)
Retained profit for the financial year					46	25

* Note disclosure required of breakdown between parent and subsidiaries, joint ventures and associates.
** Included only to show positioning.

accompanying the profit and loss tax charge would separately delineate the taxation charge (VII.2.2.3). Also discussed in VI.4 are two additional statements required by FRS 3: the primary statement, the consolidated statement of total recognized gains and losses, and a note disclosure which FRS 3 allows to be presented as a primary statement provided it is separated from the consolidated statement of total recognized gains and losses, the consolidated reconciliation of movements in shareholders' funds.

Foreign currency financial statement translation gains or losses under the closing rate approach pass through the consolidated statement of total recognized gains and losses (see VI.4.2). Those under the temporal approach go through the consolidated profit and loss account (see V.4.3). The gradual amortization of goodwill is taken to the consolidated profit and loss account, whereas any immediate write-off of goodwill to reserves passes through neither the consolidated profit and loss account nor the consolidated statement of total recognized gains and losses. It is merely shown in the note disclosures of reconciliation of movements in shareholders' funds, and reserve movements. The reasons for these different treatments are discussed in V.4.2, V.4.3, VI.1.4.2.2 and VI.4. For further detail on the elements in these statements and further note disclosures required by FRS 3, and for layouts of consolidated cash flow statements per FRS 1, see United Kingdom – Individual Accounts.

X Notes and additional statements

1 Extra consolidated disclosures checklist

1.1 Overview

Table X.1 gives additional group accounting note disclosures. See also, for example, Hastie (1998).

Discretionary non-disclosure: Companies Act 1985, Schedule 5 disclosures below together with the provisions of FRS 2 marked with an asterisk (*), are subject to the following exemptions:

- If directors consider the number of subsidiaries causes the information requirements to be excessive, disclosures can be confined to principal subsidiaries and excluded subsidiaries, though merger relief disclosures must be given in full. This abridgement must be noted and full information annexed to the next annual return delivered to the registrar.
- They need not be disclosed for a non-UK established undertaking which carries on its business outside the United Kingdom, if its directors consider such disclosure is seriously prejudicial to its business, the business of its parent or of any of its subsidiary undertakings, and approval of the Secretary of State is received. This fact must be stated, and certain disclosures such as the aggregate investment in such subsidiaries under the equity method (Sch 5, para. 5(2)) and the number, description and amount of shares in and debentures of the company held by or on behalf of its subsidiary undertakings, are usually required (Sch 5, paras. 6/20).

Disclosures for exempt groups: where a group is exempt from producing group accounts, disclosures are still onerous (Sch 5, Part I, Companies Act 1985, subject to S 231) but are not dealt with here. They include the fact of exemption; for companies part of a larger group, the name and location of the parent drawing up group accounts; and in all cases, lists of and holdings in subsidiary undertakings and for significant subsidiaries, audit qualifications or savings in their auditors' reports which are material from a group perspective.

1.2 Explanation of specific disclosure items

H Excluded subsidiaries (see Table X.1). Generally, the exclusion disclosures can be aggregated where more appropriate on a sub-unit basis, where such sub-units include only subsidiaries excluded for the same reason. These sub-unit disclosures (a sub-unit may comprise just a single subsidiary) are reportable if, for each, the amount is more than 20 per cent of any of operating profits, turnover or net assets of the group, whose size is to be measured by including all potentially excluded subsidiaries. This implies a consolida-

Table X.1 Additional group accounting note disclosures

Issue	Source
A Shareholdings in other group members and affiliates	
A.1 Subsidiary Undertakings	
A.1.1 For subsidiary undertakings at the end of the financial year:	
a name	
b country of incorporation, registration, or if unincorporated, principal place of business.	
c the fact of inclusion in the consolidation or, if excluded, reasons for exclusion (see VII)	
d reason for it being a subsidiary undertaking if not through majority voting rights with the parent holding the same proportions of shares and voting rights. If the only reason is because its parent has a participating interest and actually exercises a dominant influence, the basis for the dominant influence. An explanation of circumstances if an undertaking becomes or ceases to be a subsidiary undertaking other than by a purchase or exchange of shares is necessary.	Sch 5 para. 15 FRS 2 para. 34* FRS 2 para. 49
e separately for shares held by (a) the parent, and (b) the group, if different: i identity of each class ii proportion of nominal value of the class	Sch 5 para. 16
A.1.2 For each subsidiary undertaking whose results or financial position principally affects the consolidated figures:	
a proportion of *voting rights* held by the parent and its subsidiary undertakings; and b an indication of the nature of its business	FRS 2 para. 33*
A.2 Associates & Joint Ventures	
A.2.1 In financial statements of investing group, for each associate and joint venture	Sch 5 para. 22 FRS 9 para. 52
• name	
• country of incorporation, registration, or principal place of business	
• separately for issued shares held by (a) the parent company, and (b) the group: i identity of each class ii proportion of nominal value of that class (indicating any special rights or constraints attaching to them)	
• nature of business	
A.2.2 Where 20% presumptions relating to significant influence or participating interest are rebutted, why the particular facts rebut these.	FRS 9 para. 56
A.2.3 Certain details are required by The Act for proportionally consolidated joint ventures, including name, location, factors on which joint management is based, proportion of capital held by undertakings included in the consolidation, and non-coterminous year ends. However, this option is prohibited by FRS 9.	Sch 5 para. 21
A.3 Other Significant Shareholdings	
A.3.1 For holdings at the end of its year by the parent and the group in more than:	Sch 5 S23–27 SI 1996/189 SE Ch 5 S 2 para. 21
• 20% of nominal value of any class of any undertaking, or exceeding 20% of its assets: a name b country of incorporation, registration, or principal place of business if unincorporated. c In respect of shares held by the parent company: i identity of each class ii proportion of nominal value of that class	
• 20% of the nominal value of shares in the undertaking, except where holding is less than 50% and the undertaking is under no requirement anywhere to publish its balance sheet:	

Table X.1 (contd.)

Issue		Source
	a the aggregate amount of capital and reserves of the undertaking at the end of its financial year; and b its profit or loss for that year c except for subsidiaries, for group equity holdings of 20%, each entity's principal place of operations These requirements do not apply to subsidiaries [whether consolidated or equity accounted] or associates, or proportionately consolidated joint ventures.	
A.3.2	The number, description and amount of the shares in the company held by or on behalf of its subsidiary undertakings.	Sch 5 para. 6
A.4	**Relationships with Senior Group Companies**	
A.4.1	Number, description and amount of shares and debentures of the reporting company held by subsidiary undertakings, except where held as personal representative, or trustee without beneficial interest	Sch 5 para. 20
A.4.2	For the *ultimate* parent company *and* for the parent of both the largest group for which group accounts are drawn up and the subsidiary undertaking is a member; and the smallest such group, a name, and b if known, country of incorporation or registration. Plus for parent of largest and smallest group only; c if unincorporated, its principal place of business; and d the address from which such group accounts can be obtained.	Sch 5 para. 11 & 12
A.4.3	If applicable, that a subsidiary is a member of a group pension scheme, the nature of the scheme, if appropriate that contributions are based on pension costs across the group as a whole, and in the case of UK and Irish subsidiaries, the name of the holding company in whose financial statements the actuarial valuation of the group scheme are contained.	SSAP 24 para. 90
A.4.4	Related party exemption from disclosure of transactions with the group or related parties of the group, claimed by subsidiaries in their own financial statements where 90% of votes controlled within the group	FRS 8 para. 3c
B	**Balance Sheet Notes**	
B.1	**Associates and Joint Ventures (Section VII.2.2)**	
	In addition to amounts on the face of the primary financial statements discussed in the text, the following additional disclosures are required:	
B.1.1	The group's interest should be shown analyzed between: a share of net assets as a separate item; and b goodwill, less any amortization or write-down as a separate item.	FRS 9 para. 29
B.1.2	For all associates and joint ventures, amounts owing and owed between investor and its associates or its joint ventures, analyzed into loans and trading balances (can be combined with FRS 8 Related Party disclosures).	FRS 9 para. 55
B.1.3	Additional disclosures at 15% and 25% thresholds. Where the relevant thresholds are exceeded: a 15% aggregate threshold for associates: turnover (unless already included as a memorandum item), fixed assets, current assets, liabilities due within one year, and liabilities due after one year or more. b 15% aggregate threshold for joint ventures: fixed assets, current assets, liabilities due within one year, and liabilities due after one year or more.	

Table X.1 (contd.)

Issue		Source
	c 25% aggregate threshold for individual associates or joint ventures: for each, its name and share of the following – turnover, profit before tax, taxation, profit after tax, fixed assets, current assets, liabilities due within one year, and liabilities due after one year or more. If the individual associate or joint venture accounts for nearly all the amounts included for that class of investment, only aggregate information need be given, provided this is explained and the name of the associate or joint venture is given.	
	d Further analysis in addition to a, b, and c above should be given where this is necessary to understand the nature of the total amounts disclosed. It may be important to give an indication of the size and maturity profile of the liabilities held.	
B.2	**Goodwill and Intangible Assets (VI.1.4.2 and VI.1.4.3)** *(see also United Kingdom – Individual Accounts VI.1.1 for Intangible Assets)* In addition to disclosures in 'G Business Combinations',	FRS 10 para. 53
B.2.1	The method used to value intangible assets should be described, and the following information should be disclosed separately for positive goodwill, negative goodwill and each class of intangibles capitalized on the balance sheet: a cost or revalued amount at the beginning of the financial period and at the balance sheet date; b cumulative amount of provisions for amortization or impairment at the beginning of the financial period and at the balance sheet date; c a reconciliation of the movements, separately disclosing additions, disposals, revaluations, transfers, amortization, impairment losses, reversals of past impairment losses and amounts of negative goodwill written back in the financial period; and d the net carrying amount at the balance sheet date.	
B.2.2	Amortization of positive goodwill and intangible assets disclosures: a The methods and period of amortization of goodwill and intangible assets and the reasons for choosing those periods. b Where an amortization period is shortened or extended following a review of remaining useful lives of goodwill and intangible assets, the reason and the effect, if material, in the year of change. c Where there has been a change in the amortization method, the reason and the effect, if material, in the year of change. d Where goodwill or an intangible asset is amortized over a period that exceeds 20 years from the date of acquisition or is not amortized, the grounds for rebutting the 20-year presumption. This must be a reasoned explanation based on the specific factors contributing to the durability of the acquired business or intangible asset. e In cases where goodwill is not amortized, a statement that the financial statements depart from the specific requirement of companies legislation to amortize goodwill over a finite period for the overriding purpose of giving a true and fair view. Particulars of the departure, the reasons for it and its effect should be given in sufficient detail to convey to the reader of the financial statements the circumstances justifying the use of the true and fair override. The reasons for the departure should incorporate the explanation of the specific factors contributing to the durability of the acquired business or intangible asset, required in d. above.	FRS 10 para. 55–59
B.2.3	Negative goodwill disclosures:	FRS 10 paras. 63–64

Table X.1 (contd.)

Issue		Source
	a The period(s) in which negative goodwill is being written back in the profit and loss account.	
	b Where negative goodwill exceeds the fair values of the non-monetary assets, the amount and source of the 'excess' negative goodwill and the period(s) in which it is being written back should be explained.	
B.2.4	Transitional provisions:	
	a If goodwill remains eliminated against reserves, the financial statements should state:	
	i The accounting policy followed in respect of that goodwill;	Sch 4A para. 14
	ii The cumulative amounts of positive goodwill eliminated against reserves and negative goodwill added to reserves, net of any goodwill attributable to businesses disposed of before the balance sheet date *(disclosure of amounts relating to an overseas business need not be given it would be seriously prejudicial and official approval is obtained, and if for acquisitions prior to 23 December 1989 if the information is unavailable or cannot be obtained without unreasonable expense or delay)*	FRS 10 para. 71
	iii The fact that the goodwill had been eliminated as a matter of accounting policy and would be charged or credited in the profit and loss account on subsequent disposal of the business to which it related.	
B.2.5	Statement of non-disclosure relating to B.2.4 a ii above, including the grounds for exclusion.	FRS para. 71 SI (1990) 355
B.2.5	For other aspects of intangible assets including revaluations and impairment of goodwill and intangibles, see UK Individual Accounts section of TRANSACC	
B.3	**Minority interests (VI.1.6)**	
B.3.1	Minority interest should be analyzed between	FRS 4 para. 50
	a equity interests in subsidiaries	
	b non-equity interests in subsidiaries.	
	See also IX 'Formats'. FRS 4 requires additional disclosures relating to non-equity minority interests not discussed here.	FRS 4 paras. 61–63
C	**Profit and Loss (VI.3)**	
C.1	A statement that the exemption provided by Section 230 of the Act from providing parent's profit and loss account applies, giving details of the parent company's profit or loss for the financial year.	Co Act 1985 S230
C.2	Profit and loss components down to operating profits should be analyzed between continuing operations, acquisitions as a component of continuing operations, and discontinued operations, to the extent not reported on the face of the consolidated profit and loss account.	FRS 3
C.3	Income from interests in associated undertakings; and Income from other participating interests.	Sch 4A para. 21(3)
C.4	For associates, the investor's share of its associates' and joint ventures':	FRS 9 para. 27
	a operating result (immediately after group operating result but before its share of joint ventures)	
	b any amortization or write-down arising on acquiring associates	
	c separately disclosed exceptional items	
	d interest	
	e taxation	
	f extraordinary items	
	g retained profits	

Table X.1 (contd.)

Issue		Source
C.5	A total combining the investor's share of its associate's turnover with group turnover may be shown as a memorandum item in the profit and loss account, clearly distinguished from group turnover (a requirement for joint ventures).	FRS 9 para. 27
C.5	Any segmental analysis of turnover and operating profit should clearly distinguish between the group and associates or joint ventures.	FRS 9 para. 27
C.6	Minority interests – see Table VI.3	
C.6	See VI.3. In addition, under the following profit and loss headings, the following amounts relating to group companies must be separately disclosed: a income from other fixed asset investments b other interest receivable and similar income c interest payable and similar charges	Co Act 1985 Profit & Loss account formats
C.7	Where a foreign pension scheme is not dealt with on SSAP 24 basis in consolidated financial statements: a profit and loss charge; b basis for the charge	SSAP 24 para. 91
D	**Cash Flow Statements (VI.5)**	
D.1	See also VI.5. In addition, FRS 1 requires the following note disclosures for acquisitions and disposals of any trade or business, or of an investment in an entity that is or, as a result of the transaction becomes or ceases to be either an associate, joint venture or subsidiary undertaking: • receipts from: a sales of investments in subsidiary undertakings (showing separately cash and overdrafts transferred); b sales of investments in associates, and separately, joint ventures; c sales of trades or businesses • payments: a to acquire investments in subsidiary undertakings (showing separately cash and overdrafts acquired) b to acquire investments in associates, and separately, joint ventures; c to acquire trades or businesses.	FRS 1 paras. 22–24
D.2	Dividends on the parent's equity shares (under equity dividends paid), and dividends to minority interests (under returns on investments and servicing of finance).	FRS 1 paras. 15 and 25
D.3	Dividends received from associates, and from joint ventures as separate lines between operating activities and returns on investments and servicing of finance, and all other cash flows between the investor and its associates, and separately its joint ventures under the appropriate cash flow heading.	FRS 9 para. 30
D.4	Amounts and circumstances where restrictions prevent transfer of one part of business or group to another	FRS 1 para. 47
E	**Statement of Total Recognized Gains and Losses (VI.4)**	
E.1	See also VI.4. Amounts related to associates should be shown separately under each heading, if material, either in the statement or notes to the statement	FRS 9 para. 28
E.2	Where exchange differences on foreign currency borrowings used to finance, or hedge equity investments in foreign enterprises are taken to reserves, and consequently reported in the statement of total recognized gains and losses, tax charges directly and solely attributable to such exchange differences should also be reported separately therein.	UITF Abstract 19 para. 6

Table X.1 (contd.)

Issue		Source
F	**Directors' Report**	
F.1	Separate totals for the aggregate amounts for the company and its subsidiary undertakings (unless it is a wholly-owned subsidiary of a UK-incorporated company) of a) political and b) charitable contributions made during the year	Sch 7 para. 4(2)
G	**Business Combinations**	
G.1	**General**	
G.1.1	In the year of combination: a names of the combining companies other than the combining entity b whether accounted for as an acquisition or merger c the date of the combination d numbers and classes of securities issued and details of other consideration	Sch 4A S13/FRS 6 para. 21
G.2	**Acquisition Accounting (VI.1.2 and 3)**	
G.2.1	For each material acquisition and for other acquisitions in aggregate a composition and fair value of consideration given by acquiring company and its subsidiary undertakings. b nature of any deferred or contingent purchase consideration, and for the latter the range of possible outcomes and principal factors affecting these. c a fair value table showing for each class of assets and liabilities of the acquired entity its recorded book values before acquisition and any fair value adjustments; fair value adjustments analysed between revaluations, adjustments to achieve consistency of accounting policies and other significant adjustments, giving reasons; fair values at the date of acquisition; a statement of the amount of purchased or negative goodwill arising from the acquisition. d In the fair value table provisions for reorganisation and restructuring costs included in the liabilities of the acquired entity and related asset write-downs, made in the 12 months up to the date of acquisition should be separately identified. e if fair values at acquisition are on a provisional basis, that fact and reasons. Subsequent material adjustments to such values and to goodwill to be disclosed and explained. f in the consolidated profit and loss account, acquired entity's post-acquisition results to be shown in continuing operations (FRS 3) unless discontinued in same period. Where it is not possible to determine these an indication of its contribution to turnover and operating profit of continuing operations must be given. If even this cannot be given the fact and reason to be given. g referring to f, where an acquisition has a material impact on a major business segment this should be disclosed and explained. h disclosure per FRS 3 of any exceptional post-acquisition profits or losses determined using fair values at acquisition, identified as relating to the acquisition. i In subsequent periods, disclosure in profit and loss account or notes of costs incurred in those periods relating to reorganizing, restructuring and integrating the acquisition. These are costs which would not have been incurred in the absence of the acquisition, and relate to a project identified and controlled by management as part of a reorganization and integration programme set up at acquisition or as a direct consequence of an immediate post-acquisition review.	FRS 6 paras. 23–34 Sch 4A para. 13

Table X.1 (contd.)

Issue	Source
j movements on provisions relating to an acquisition analyzed between amounts used for the specific purpose they were created for and amounts released unused.	
k for each material acquisition only, profit after tax and minority interests of acquired entity, from the beginning of the acquired entity's financial year to the acquisition date, giving the date of the former.	
G.2.2 For each substantial acquisition in the period of acquisition and in other exceptional cases where the disclosure is necessary for a true and fair view to be given,	
a a summarized profit and loss account of the acquired entity, showing major FRS 3 headings, and its statement of total recognised gains and losses, from the beginning of its financial year to the effective date of acquisition, giving the date on which the period began, and	FRS 6 para. 36
b profit after tax and minority interests for the acquired entity's previous financial year.	
Substantial is defined as where: • for listed companies where the transaction is a Super Class I transaction under the London Stock Exchange Listing Rules and any of the ratios set out in the London Stock Exchange Listing Rules defining Super Class I transactions exceeds 15 per cent [UITF Abstract 15]; or • for other entities, net assets or operating profits of the acquired entity exceed 15 per cent of the acquiring entity, or the fair value of the consideration exceeds 15 per cent of the net assets of the acquiring entity, or in other exceptional cases where the acquisition is of such significance that disclosure is necessary to show a true and fair view.	
Net assets and profits are those shown in the last financial year before acquisition, net assets to be augmented to include purchased goodwill eliminated against reserves as a matter of policy.	
G.3 Mergers (VI.1.1)	
G.3.1 For all material merger accounted combinations in the period, other than group reconstructions:	FRS 6 para. 22
a composition and fair value of the consideration given by issuing company and subsidiary undertakings	
b analysis of principal components of current year's profit and loss account and statement of total recognized gains and losses i for merged party only, after the merger date. ii for each party to the merger, amounts relating to that party for the period up to the merger. At a minimum the profit and loss analysis should show turnover, operating profit and exceptional items, split between continuing operations, discontinued operations and acquisitions; profit before taxation; taxation and minority interests; and extraordinary items.	
c similar analysis of principal components of profit and loss account and statement of total recognized gains and losses between parties to the merger, for the previous financial year.	
d aggregate book values of net assets of each party to the merger at date of merger, and nature and amount of significant adjustments to the net assets of any party to the merger to achieve consistency of accounting policies, and an explanation of other significant adjustments to the net assets of any party to the merger in consequence of it.	
e statement of adjustments to consolidated reserves resulting from the merger.	Sch 4A para. 13(6)*

Table X.1 (contd.)

Issue		Source
G.3.2	Corresponding figures restated in consolidated accounts as if undertaking acquired had been included in the consolidation throughout the year, and at the previous balance sheet date, adjusted so as to achieve uniform accounting policies.	Sch 4A para. 11 FRS 6 para. 17
G.3.3	Group reconstructions using merger accounting are exempt from the disclosure requirements of FRS6 but must still give information required by company legislation.	FRS 6 para. 82
G.4	**Disposals (VI.1.5)**	
G.4.1	In respect of each material disposal: a name of the undertaking, or the parent of the disposed group; and b profit or loss on disposal of each material disposal of a previously acquired business or business segment [FRS 10 para. 54]	Sch 4A para. 15*
G.4.2	For any material undertaking which ceased to be a subsidiary undertaking: • name • any ownership interest retained • the circumstances where its cessation was other than by disposal of at least part of the group interest.	FRS 2 para. 48
G.4.3	Under goodwill transitional arrangements, if goodwill remains eliminated against reserves, in the reporting period in which the business with which the goodwill was acquired is disposed of or closed: a The amount included in the profit and loss account in respect of the profit or loss on disposal or closure should include attributable goodwill to the extent that it has not been previously been charged in the profit and loss account; and b The financial statements should disclose as a component of profit or loss on disposal or closure the attributable amount of goodwill included. c Where it is impractical or impossible to ascertain the goodwill attributable to a business that was acquired before 1 January 1989, this should be stated and the reasons given.	FRS 10 para. 71
H	**Excluded Subsidiaries (IV.3)**	
H.1	**General**	
H.1.1	For each excluded subsidiary: a reasons for exclusion b balances and nature and extent of transactions with the rest of the group; c if not equity accounted, consolidated statement amounts for: i dividends received and receivable from it; and ii any write-down during the period of the investment in or amounts due from it	FRS 2 para. 31*
	d if not equity accounted, unless the group holds less than 50% of the nominal value of its shares and it is not required to deliver or publish its balance sheet anywhere, i its aggregate capital and reserves at the end of the its financial year ending with the parent's or the immediately previous one, and ii its profit or loss for the year	Sch 5 para. 17
H.1.2	Exclusion disclosures can be summarized if they meet certain criteria discussed in X.1.2	
H.2	**Specific Exclusion Reasons**	
H.2.1	When excluded for dissimilar activities, separate financial statements of those undertakings excluded. Summarized information may be provided if the undertaking individually or in combination with undertakings conducting similar operations, do not comprise more than 20% of any one or more of the	FRS 2 para. 31

Table X.1 (contd.)

Issue		Source
	i operating profits ii turnover iii net assets of the group, measured by *including* all excluded subsidiaries	
H.2.2	If excluded on the grounds of severe restrictions, when these cease, the amount of unrecognized profits accrued during the periods of restriction, and any amount previously charged for impairment that needs to be written back must be disclosed in the profit and loss account.	FRS 2 para. 28
J	**Quasi-subsidiaries (IV.2)**	
J.1	For quasi-subsidiaries: a the fact that they are included; and b a summary of the financial statements for each showing separately each major balance sheet, profit and loss account, statement of total recognized gains and losses and cash flow statement heading for which there is a material item, plus comparatives. This may be combined for quasi-subsidiaries of a similar nature if unduly voluminous.	FRS 5 para. 38
J.2	For each excluded quasi-subsidiary: a reasons for exclusion [only allowed if held exclusively with a view to resale and not previously consolidated] b balances and nature and extent of transactions with the rest of the group.	FRS 5 para. 36
J.3	Significant differences between the activities of a quasi-subsidiary's and those of the group that controls it	FRS 5 para. 101 (d)
K	**Additional Size-based Disclosures for Associates and Joint Ventures**	
K.1	Threshold criteria: size-based additional disclosures are necessary if investor's share of any of the following items exceeds the percentage below of any of the following: • Gross assets, • Gross liabilities, • Turnover and • Operating results (on a three year average) as compared with the corresponding amounts for the investor group excluding any amount included by the equity method for associates or the gross equity amount for joint ventures.	
K.2	If the aggregate of the investor's share in associates exceeds 15% of any of the threshold criteria, note disclosure must be made of the investor's share in associates of: • Turnover (unless already disclosed on the face of the profit and loss account). • Fixed assets • Current assets • Liabilities due within one year • Liabilities due after one year or more	
K.3	If the aggregate of the investor's share in joint ventures exceeds 15% of any of the threshold criteria, note disclosure must be made of the investor's share in joint ventures of: • Fixed assets • Current assets • Liabilities due within one year • Liabilities due after one year or more.	

Table X.1 (contd.)

Issue		Source
K.4	If the investor's share in any individual associate or joint venture exceeds 25% of any of the threshold criteria, note disclosure must be given of its name and the investor's share of its: • Turnover • Profit before tax • Taxation • Profit after tax • Fixed assets • Current assets • Liabilities due within one year • Liabilities due after one year or more If that individual associate or joint venture accounts for nearly all of that class of investment, aggregate information for that class will suffice, provided it is explained and associate or joint venture identified.	
K.5	Further analysis should be given where necessary to understand the nature of the total amounts disclosed in size-based disclosures above, for example, a size and maturity profile of liabilities.	FRS 9 para. 58
L	**Miscellaneous Disclosures**	
L.1	For subsidiaries with non-coterminous year ends or accounting periods from the parent: a reasons b date on which last financial year ended before parent's year end. These may be summarized into earliest and latest dates for such subsidiaries c name and accounting date or period	Sch 5, para. 19 FRS 2, para. 44
L.2	Accounting period or date of financial statements used for principal associates and joint ventures if they do not end with that of the group.	FRS 9 para. 52
L.3	Whenever non-uniform accounting policies are used in the consolidated accounts: a particulars of any departure b reasons c effects	Sch 4A, para. 3 Sch 4A S3
L.4	Notes or matters in accounts of associates and joint ventures material to understanding the effect on the investor of its investments, including matters that should have been noted if the investor's accounting policies had been applied, and also particularly relating to investor's share in: • contingent liabilities and • capital commitments.	FRS 9 para. 53
L.5	Foreign currency translation a the methods used to translate foreign currency financial statements; b the net amounts of exchange gains and losses on foreign currency *borrowings less deposits*, identifying separately: i the amounts offset in reserves against foreign currency investments and net investments under the various "cover" provisions disclosing separately gross amounts and tax charges and credits; and ii the net amount charged / credited to the profit and loss account, i.e. not offset, remaining after the offset in i. c the net movement on reserves arising from exchange differences. d the accounting policy adopted in relation to operations in hyper-inflationary (see IV.4.4.4) economies and reasons for any departures from the recommended approaches	SSAP 20 UITF Abstract 9

Table X.1 (contd.)

	Issue	Source
L.6	Differences between the accounting rules adopted in the parent's individual accounts and group accounts should be disclosed with reasons.	Sch 4A para. 4
L.7	Nature and extent of any significant statutory, contractual or exchange control restrictions which materially limit the parent undertaking's access to a subsidiary undertaking's distributable profits.	FRS 2 para. 53
L.8	Similar to L.6 but related to associates and joint ventures.	FRS 9 para. 54
L.9	Assumptions made as to the availability of group relief for tax and payments therefor.	SSAP 15 para. 43
L.10	Extent to which deferred taxation has been accounted for in respect of future remittances of the accumulated reserves of overseas subsidiary undertakings or reasons for not fully providing – see VIII.3.7	SSAP 15 para. 21 FRS 2 para. 54
L.11	Details of guarantees and other financial commitments undertaken on behalf of or for the benefit of any parent undertaking, subsidiary undertaking or fellow subsidiary undertaking must be separately disclosed, including details of charges and amounts secured, unprovided contingent liabilities, contracted capital expenditure unprovided for, authorized but uncontracted capital expenditure, and pension commitments.	Sch 4 para. 50 and 59A
L.12	Certain details of loans, quasi-loans or dealings entered into by the company or its subsidiaries in favour of directors of the company or its holding company, or those connected with such a director.	Sch 6 para. 15
L.13	Disclosures relating to segmental information are dealt with in X.2.3	SSAP 25
L.14	Additional group accounting disclosures relating to related party transactions and balances are discussed in X.3.3 – see also United Kingdom – Individual Accounts IX.11 [Stock Exchange requirements are not covered here].	FRS 8

tion exercise must be carried out internally to test whether aggregation is allowable.

IX Miscellaneous disclosures, note L.12 (see Table X.1). The requirements relating to transactions with directors refers to the legal definitions of holding company and subsidiary, rather than those for accounting purposes of parent undertaking and subsidiary undertaking (see IV.1.1).

2 Segmental reporting

2.1 Scope and exemptions

Only in 1990 did the United Kingdom issue its first standard in this area, SSAP 25, *Segmental Reporting*. Prior to this the United Kingdom had followed minimal Stock Exchange requirements from 1965 for disclosure of turnover and profits by line-of-business, with turnover by geographic segment and profit rates only where these were out of line with the group average. The Companies Act 1967 had extended line-of-business disclosure requirements to all companies. Rennie and Emmanuel (1992) found improved line of business disclosures but not geographical ones for UK companies over the period from 1975 to 1989. In May 1996, the ASB issued a Discussion Paper, *Segmental Reporting* which contrasts changes proposed and now implemented by the FASB and IASC. It is probable that the ASB will supercede SSAP 25 in the near future by a standard similar to the revised IAS 14.

Those of SSAP 25's requirements which are also contained in the Companies Act 1985 apply to all companies. Its other requirements only apply to:

(i) Public limited companies (Plc's) or parents of public limited companies.
(ii) Banking and insurance companies or groups.
(iii) Entities which exceed 10 times the size criteria for a medium-sized company under the Companies Act 1985 (see Table III.2).

A subsidiary that is not a Plc or a banking or insurance company, need not comply if its parent's segmental disclosures comply with SSAP 25. Exemption is given where the directors consider any required information would be 'seriously prejudicial to the interests of the reporting entity' (para. 43), but the fact of non-disclosure must be stated. Such an exemption is not contained in IAS 14. *Company Reporting* (July 1996, p. 4) considers that such an exemption makes the standard 'optional' and finds that seven per cent of UK companies with evidence of segmental information invoke it for some or all of their segmental disclosures. Emmanuel and Garrod (1992) provide a useful discussion of segmental reporting issues and practice in the United Kingdom.

2.2 Identifying segments

SSAP 25 is similar to superceded versions of IAS 14 and FAS 14. Segmental analyses are required by:

- Classes of business: 'a distinguishable component of an entity that provides a separate product or service or a separate group of related products or services'.
- Geographical segments: 'a geographical area comprising an individual country or group of countries in which an entity operates, or to which it supplies products or services'.

Both are director-defined having 'regard to the overall purpose of presenting segmental information and the need of readers of financial statements to be informed if a company carries on operations in different classes of business or in different geographical areas' (para. 8). 'No single factor is universally applicable, nor any single factor is determinative in all cases' (para. 13), but the following may be indicative, where different classes of business or . . . geographical areas:

- Earn a return on investment out of line with the remainder of the business; or
- Are subject to different degrees of risk; or
- Have experienced different rates of growth; or
- Have different potentials for future development. (para. 8)

Directors need annually to review and redefine segments where appropriate, disclosing the nature, reason and effect of any changes, and restating comparatives (para. 39). For classes of business, the nature of the products or services, and the production processes, their markets and distribution channels, organization structure and special legislative frameworks applying to part of the business may be indicative (para. 12); whereas for geographical segments, expansionist or restrictive economic climates, political stability, exchange control regulation and exchange rate fluctuations may be indicative (para. 15). There is no requirement in the United Kingdom to integrate classes of business and geographical disclosures into a matrix approach and indeed Emmanuel and Garrod (1987) in a survey found some preparers questioned the relevance of such an approach where businesses were not organized in that way.

Only identified segments significant to the entity as a whole are reportable, normally those with:

- Third-party turnover 10 per cent or more of the whole entity's third-party turnover (in IAS 14 the analogous requirement is stated in terms of total revenues, external and internal); or
- Segment profit or loss 10 per cent or more of the combined result of all profit-making segments, or of all loss-making segments, whichever is the greater; or
- Net assets 10 per cent or more of the entity's total net assets (para. 9) (IAS 14's requirements refer to total assets).

These are similar to IAS 14's (1997) requirements for reportable segments, but SSAP 25 does not have the requirement that total external turnover of reportable segments must be at least 75 per cent of total consolidated or enterprise revenue, if necessary by identifying additional segments. SSAP 25 does not have any primary and secondary hierarchy of segmentation, and its disclosures are similar for business and geographical segments. Like the revised IAS 14 it bases segment identification on risks and returns. It does not require an enterprise's internal organizational and management structure and internal financial reporting system to be the starting point and contains less guidance on segment identification. Its approach is closer to IAS 14 than to SFAS

131 in the United States, where external segmental reporting is to be based solely on internal operating segmentation.

2.3 Measurement and disclosure

Companies required to report under SSAP 25's additional criteria must disclose turnover, result and net assets by classes of business and geographical segments; analyze significant associates by segment; reconcile segmental totals to consolidated amounts; and provide comparatives. SSAP 25 does not require the basis of inter-segment pricing to be disclosed, IAS 14.

Turnover must be analyzed by segment between inter-segment and external turnover (para. 34). Turnover of geographical segments must be analyzed by origin, and also turnover by destination unless there is no material difference between the two (para. 18). Statutory turnover disclosure exemptions apply here, for example for banking activities.

Segment result is to be measured before tax, minority interests and extraordinary items, normally before interest unless interest earning is part of the business, for example in banking, or interest is central to it, for example in travel or contracting businesses. In these cases it is to be measured after interest. SSAP 25 considers that for most entities financing policy is carried out on a group-wide basis and therefore it is not meaningful to analyze it by segment. Common costs, i.e. costs related to more than one segment, can either be allocated to segments or deducted as a total from segmental results at the directors' discretion. Little guidance is given on which costs should be allocated except that, where common costs are apportioned for internal reporting, it may be reasonable to apportion them for external reporting, and costs should not be apportioned where apportionment would be misleading (para. 23). General corporate expenses and gains or losses on discontinued operations are not specifically excluded and Davies, Paterson and Wilson (1997, p. 1089) give examples of some UK groups which have allocated central expenses to the segmental result.

Segment net assets are defined as normally non-interest-bearing operating assets less non-interest-bearing operating liabilities, unless the after interest alternative is used for measuring the segment result. In this case, for consistency, corresponding interest bearing operating assets and liabilities must be included (para. 24). Similarly, inter-segment balances are not to be included unless the segment result includes interest on such balances. Joint operating assets should be allocated to segments, but not assets or liabilities that are not used in the operations of any segment (para. 25). IAS 14 requires disclosure of segment assets and segment liabilities for the primary segmentation basis, and (gross) segment assets for the secondary one. SFAS 131 requires gross asset disclosure by segment.

If in aggregate, associated undertakings comprise at least 20 per cent of the total result or net assets of the entity as a whole, they are to be analyzed segmentally, disclosing separately the reporting entity's share for associated undertakings of profits or losses before taxation, minority interests and extraordinary items, and net assets (including goodwill to the extent that it has not been written off) stated, where possible, after a fair value exercise at acquisition for each associated undertaking. If such information is not obtainable, reasons for non-disclosure must be given as a note together with a brief description of the omitted businesses (paras. 26–27). FRS 9 states that the segmental analysis of turnover and operating profit (if given) should clearly distinguish between the group and associates. IAS 14 requires the disclosure of share of results of associates and joint ventures, and aggregate investment in these entities for its primary segmentation basis only.

FRS 3 requires that if an acquisition, sale or termination has a material impact on a major business segment, this should be disclosed and explained. SSAP 25 does not require the disclosure of unusual items, depreciation, segment assets, segment liabilities, sales to large customers, basis of inter-segment pricing, non-cash expense or capital expenditure, though Skerratt and Tonkin (1994, p. 244) show that 4 per cent of UK large listed companies disclose capital investment by line of business, and 6 per cent by geographical segment.

The ASB Statement (of best practice), *Operating and Financial Review* (1993), recommends commentary on the operating performance of various business units (para. 19), the principal risks and uncertainties in the main lines of business (para. 12), segmental cash flows which are significantly out of line with segmental profits (para. 31) and the overall level of capital expenditure of the major business segments and geographical areas (para. 14).

3 Related party disclosures

Table X.1 contains a number of disclosures concerning related party disclosures, for example, about shareholdings in related undertakings, guarantees and other financial commitments on behalf of other group undertakings etc. In October 1995 the ASB issued FRS 8, *Related Party Disclosures*, to attempt to bring a more coherent disclosure framework to the area, following ED 46, *Disclosure of Related Party Transactions*, issued by the ASC in April 1989 and FRED 8 in March 1994. FRS 8 decided that all material related party transactions should be disclosed, rather than merely abnormal ones as proposed by ED 46. Only group aspects of such disclosures are discussed here.

3.1 Definition of related parties

FRS 8 spreads its net very widely, outlining general principles, enumerating specific related parties, and specifying further parties which are presumed to be related parties, but which can be rebutted. The lists of specific and presumed related parties are not intended to be exhaustive (para. 2.5). Although the United Kingdom does not require consolidation of entities controlled, for example by private individuals, disclosure of transactions with related parties in group accounts gives limited indication of such information. Because the proposed definitions are tightly defined, they are specified in a somewhat detailed manner below.

General principles – FRS 8 defines a related party transaction as 'the transfer of assets or liabilities or the performance of services by, to or for a related party irrespective of whether a price is charged' (para. 2.6). FRS 8 specifies that two or more parties are related parties when at any time during the financial period:

(i) One party has direct or indirect control of the other party; or
(ii) The parties are subject to common control from the same source; or
(iii) One party has influence over the financial and operating policies of the other party to an extent that the other party might be inhibited from pursuing at all times its own separate interests; or
(iv) The parties, in entering a transaction, are subject to influence from the same source to such an extent that one of the parties to the transaction has subordinated its own separate interests.

'Control' is defined as 'the ability to direct the financial and operating policies of an entity with a view to gaining economic benefits from its activities'. FRS 8 comments that whilst entities subject to common control are included within the definition of a related party because control brings the ability to cause the controlled party to subordinate its separate interests, this is not necessarily the case with common influence. It states that two related parties of a third entity are not necessarily related parties of each other. Normally the relationship between two associates of the same investor is too tenuous to be treated as related parties of each other, as is also the case where one party is subject to control and another influence from the same source, or where two entities simply have a director in common. The decisive issue is whether one or both parties have subordinated their own separate interests in entering transactions. Common control is deemed to exist when both parties are subject to control from boards having a controlling nucleus of directors in common (paras. 13 and 14).

FRS 8 gives further guidance on how to determine the existence of related parties, which are enumerated below either as particular related parties or presumed related parties. These are not exhaustive.

Particular related parties – the following specific relationships are defined as related parties of the reporting entity (para. 2.5(b)):

(i) Its ultimate and intermediate parent undertakings, subsidiary undertakings and fellow subsidiary undertakings;

(ii) Its associates and joint ventures;
(iii) The investor or venturer in respect of which the reporting entity is an associate or a joint venture;
(iv) Directors of the reporting entity and the directors of its ultimate and intermediate parent undertakings (which include shadow directors, defined in Company legislation as persons in accordance with whose directions the directors of the company are accustomed to act); and
(v) Pension funds for the benefit of employees of the reporting entity or of any entity that is a related party of the reporting entity.

Presumed related parties – a rebuttable presumption exists that the following are related parties. Rebuttal must demonstrate that neither party has influenced the financial and operating policies of the other in such a way as to inhibit the pursuit of separate interests (para. 2.5 (c)):

(i) The key management of the reporting entity and the key management of its parent undertaking or undertakings;
(ii) A person owning or able to exercise control over 20 per cent or more of the voting rights of the reporting entity, whether directly or through nominees;
(iii) Each person acting in concert in such a way as to be able to exercise control or influence over the reporting entity, and;
(iv) An entity managing or managed by the reporting entity under a management contract.

'Key management' is defined as 'those persons in senior positions having authority or responsibility for directing or controlling the major activities and resources of the reporting entity'. 'Persons acting in concert' are defined as 'persons who, pursuant to an agreement or understanding (whether formal or informal), actively co-operate, whether by the ownership by any of them of shares in an undertaking or otherwise, to exercise control or influence over that undertaking'. 'Influence' for this purpose is defined as 'influence over the financial and operating policies of the other party to an extent that the other party might be inhibited from pursuing at all times its own separate interests'.

In addition, there is a rebuttable presumption that the following categories are also related parties because of their relationship with parties that are, or are presumed to be, related parties of the reporting entity:

(i) Members of the close family of any individual falling under the parties included in the lists of related parties under the general principles, particular related parties and presumed related party headings above.
(ii) Partnerships, companies, trusts or other entities in which the individual or member of the close family under the general principles, particular related parties and presumed related party headings above has a controlling interest.

'Close family' is defined as 'close members of the family of an individual are those family members, or members of the same household, who may be expected to influence, or be influenced by, that person in their dealings with the reporting entity'.

Wild and Creighton (1996, p. 129) consider that for group accounts, disclosure of transactions with those related to the group as a whole (by implication related parties of the parent) seems to be closer to FRS 8's intentions than aggregation of related party disclosures of group companies. An example given is that a director of a subsidiary is a related party of the subsidiary for its statements, but a related party for group financial statement purposes as part of key management of the group.

3.2 Scope

FRS 8 applies 'to all financial statements intended to give a true and fair view of a reporting entity's financial position and profit and loss (or income and expenditure) for a period' (para. 3) with no exemptions given to small companies, since a majority of respondents to its exposure draft considered that related party transactions were likely to be of greater significance to such companies. The ASB has stated it will review this situation when it has received a report by the Consultative Committee of Accounting Bodies investigating possible bases for exempting small companies from some of the requirements of accounting standards. However, the Financial Reporting Standard for Smaller Entities (FRSSE) only requires that transactions that are material to the reporting entity need to be disclosed in the notes to the financial statements of smaller entities. FRS 8 also requires for larger entities that materiality also be measured with respect to the other related party for certain categories of individuals – see below.

Disclosure exemptions: no disclosure is required:

- In consolidated financial statements, of any transactions or balances between group entities that have been eliminated on consolidation.
- In the parent's own financial statements when these are presented with its consolidated financial statements.
- In the financial statements of subsidiary undertakings, 90 per cent or more of whose voting rights are controlled within the group, of transactions with entities that are part of the group or investees of the group qualifying as related parties, provided that the consolidated financial statements in which the subsidiary is included are publicly available. The fact that this particular exemption has been used must be disclosed.

The ASB feels that the disclosure that the '90 per cent subsidiary' exemption has been invoked is sufficient to alert users of financial statements to the possible existence of related party transactions (Appendix IV para. 13). The corresponding exemption is more restrictive in IAS 24, *Related Party Disclosures*, in which to qualify for the analogous exemption, the subsidiary must be wholly-owned, and its parent must be incorporated and provide consolidated financial statements in the same country. Wild and Creighton (1996, p. 128) note that when an undertaking becomes a subsidiary during the year it is a related party for the whole period. Though transactions on or after the date of acquisition are exempted from disclosure, its material transactions with the rest of the group prior to that date should be disclosed. Conversely, transactions from the date of disposal of a subsidiary to the group year end should also be disclosed.

Non-group accounting exemptions include pension contributions paid to a pension fund, and emoluments in respect of services as an employee of the reporting entity. These are not discussed here. FRS 8 also makes it clear that its provisions do not apply where such compliance conflicts with duties of confidentiality arising through the operation of law, except where confidentiality terms are contractually based. Customer confidentiality for banks is given as an example.

Disclosure of the relationship and transactions between the reporting entity and the following parties is not required simply as a result of their role:

(i) Providers of finance in course of business in that regard.
(ii) Utility companies.
(iii) Government departments and their sponsored bodies

even though any of these three categories may circumscribe the freedom of action of an entity or participate in its decision-making process; and

(iv) A customer, supplier, franchiser, distributor or general agent with whom the entity transacts a significant volume of business.

3.3 Related party disclosures

Transactions and balances: FRS 8 proposes, as discussed above, that all material-related party transactions by a reporting entity with a related party must be disclosed, unless covered by the exemptions above. Disclosure must be made irrespective of whether a price has been charged, and should include (para. 6):

- The names of the transacting related parties.
- A description of the relationship between the parties.
- A description of the transactions.
- The amounts involved.
- Any other elements of the transactions necessary for an understanding of the financial statements (for example an indication that the transfer of a major asset had taken place at an amount materially different from that obtainable on normal commercial terms (para. 22)).
- Amounts due or from related parties at the balance sheet date and provisions for doubtful debts due from such parties at that date.

- Amounts written off in the period in respect of debts due to or from related parties.

Similar transactions by type of related party may be aggregated unless the disclosure of an individual transaction or connected transactions is necessary to understand their financial statement impact or is required by law (para. 8). Aggregation should not be used to obscure or conceal significant transactions or types of transactions (para. 21). Materiality is defined not only in terms of significance to the reporting entity, but also in relation to the other related party when that party is a director, key manager or other individual in a position to influence, or accountable for the stewardship of the reporting entity, or a member of the close family of such individuals, or any entity controlled by such individuals or members of their close family. FRS 8 gives examples of types of related party transactions: donations, purchases or sales of goods, property or other assets, rendering or receiving of services, agency agreements, leasing arrangements, transfer of research and development, licence agreements, provision of finance including loans and equity contributions in cash and kind, guarantees and the provision of collateral security and management contracts (para. 19). London Stock Exchange requirements relating to related parties are not discussed here.

IAS 24's corresponding disclosure requirements are less prescriptive – it contains a more general injunction to 'disclose the nature of the related party relationships as well as the types of the transactions and the elements of the transactions necessary for an understanding of the financial statements' (para. 22). It suggests such elements would normally include an indication of the volume of the transactions either as an amount or appropriate proportion, also amounts or appropriate proportions of outstanding items, and pricing policies (para. 23).

Control: where the reporting entity is controlled by another party, the nature of the related party relationship, the name of that party and, if different, that of the ultimate controlling party should be disclosed. If the ultimate controlling party is not known, this should be stated. Even if there are no transactions by the entity with the controlling parties, these disclosures must be given (para. 5). IAS 24 merely makes a more general requirement that related party relationships should be disclosed where control exists irrespective of whether there have been transactions between the parties (para. 20).

4 Operating and Financial Review

The ASB statement, *Operating and Financial Review (1993)*, resulting from issues raised by the Cadbury Committee on corporate governance, is a statement of best practice. It recommends commentary on reported figures and strategy. Group accounting aspects include in the case of material acquisitions:

- The extent to which the expectations at the time of acquisition have been realized including any unusual effects of seasonal businesses acquired (para. 10).
- The principal risks and uncertainties in the main lines of business (para. 12).
- The overall level of capital expenditure of the major business segments and geographical areas (para. 14).
- Operating performance of the various business units (para. 19).
- The management of exchange rate risk (para. 26).
- Segmental cash flows which are significantly out of line with segmental profits (para. 31).
- Restrictions on the ability to transfer funds from one part of the group to meet another's obligations, where these form or foreseeably could form a significant restraint on the group, e.g. exchange controls (para. 34).
- Strengths and resources of the business not included in the balance sheet, e.g. brands and intangibles (para. 37).

5 Interim Reports and Preliminary Announcements

The ASB issued two statements, one on Interim Reports (September 1997) and the other on Preliminary Announcements (July 1998). These reflect and develop best practice and are persuasive rather than mandatory. The former recommends that interim reports should be prepared using the same measurement basis and accounting principles and practices as used in the annual financial statements, i.e. on a consolidated basis. In them, turnover and operating profit of acquisitions and discontinued operations should be disclosed on the face of the profit and loss account. Segmental turnover (distinguishing inter-segment sales if significant) and segment profit or loss should be disclosed using the same segmentation basis and measurement principles as in the annual report. The ASB statement on Preliminary Announcements contains similar recommendations for preliminary announcements of full year results. Both sets of statements are discussed in United Kingdom – Individual Accounts IX.9. The London Stock Exchange requires listed companies to provide both types of statements – see XII.2.2.

XI Auditing

1 Audit reporting requirements

The eligibility requirement for auditors for UK group accounts is the same as for individual companies. They must be members of a recognized supervisory body and be eligible under the rules of that body (Companies Act 1989, S 25). The Secretary of State authorizes such bodies, which currently include, for example, the Institutes of Chartered Accountants in England and Wales, Ireland, and Scotland, the Chartered Association of Certified Accountants, and the Association of International Accountants, and requires a register of such auditors to be kept. The main regulations are in Part II of the Companies Act 1989 and are discussed in United Kingdom – Individual Accounts X.

The auditor's reporting duties under the Companies Act 1985 includes consolidated accounts (S 235). The report,

> shall state whether in the auditor's opinion the annual accounts have been properly prepared in accordance with the Act, and in particular whether a true and fair view is given . . . in the case of group accounts, of the state of affairs as at the end of the financial year, and the profit and loss for the financial year, of the undertakings included in the consolidation as a whole, so far as concerns members of the company.

2 Rights to information and explanations

The Act imposes the following duties on:

- Subsidiary undertakings incorporated in Great Britain and their auditors, to give to the auditors of any parent company of that undertaking 'such information and explanations as they might reasonably require for (their audit purposes)'. This allows the communication of what might otherwise be regarded as confidential information.
- Parent companies having subsidiary undertakings not incorporated in Great Britain, if required by the parent's auditors, 'to take such steps as are reasonably open to it to obtain from the subsidiary undertaking such information and explanations as they might reasonably require (for the audit of the parent)'.

Default is punishable by a fine, but these duties do not cover associated undertakings (S 389A).

3 Reliance on the work of other auditors

The auditors of the individual subsidiary companies are appointed by these companies, in practice usually by direction of the parent's directors, often the overseas offices or correspondent firms of the parent's auditors. Publication and content of the individual company accounts of overseas subsidiaries will comply with the regulations of their country of registration or incorporation. In addition, auditors of those companies will usually be required by the parent to audit group reporting forms, providing information required for preparing UK group accounts, or the parent's auditors will need to consider what extra audit work is necessary.

Audit Guideline 3.415, *Group Financial Statements – Reliance on the Work of other Auditors*, renders the parent's auditors, the 'primary' auditor, responsible for obtaining sufficient relevant and reliable audit evidence to enable them to draw reasonable conclusions. They must inform group directors that they will communicate with the subsidiary's or associate's auditors, the 'secondary' auditor(s), from time to time and be satisfied as to the latter's independence, the terms, scope and standard of the latter's work for group purposes, and may need to ask them to conduct additional tests, though the primary auditor would not normally need to re-perform audit work and only exceptionally, conduct independent tests themselves.

4 Companies in small groups – exemption from audit

Parent and subsidiary companies in small groups are exempt from audit (SI 1997/936, para. 3). Such groups must have:

(a) A turnover of less than GBP350,000 net of consolidation adjustments or GBP420,000 prior to consolidation adjustments; and
(b) Balance sheet assets of less than GBP1.4 million net of consolidation adjustments or GBP1.68 million prior to consolidation adjustments.

Such groups must also not be ineligible under the Companies Act 1985 S248(2) – i.e. having a member which is a public body, a bank, an insurance company or an authorized person under the Financial Services Act 1986. Oddly but apparently, this exemption does not extend to the consolidated accounts themselves.

XII Filing and publication

1 Filing requirements: Companies Act 1985

The Companies Act 1985 provisions relating to the filing of group accounts with the Registrar of Companies are broadly the same as for individual companies, from the end of the accounting reference period ten months for a private company and seven months for a public company. Where a company carries on business outside the United Kingdom, an extension of three months to the normal filing deadlines may be given after the registrar receives notice (S 244(3)). UK subsidiary undertakings which are companies are obliged to file their own accounts with the registrar, including those excluded from the consolidation.

A parent company excluding a subsidiary from the consolidation on the grounds of dissimilar activities (S 229(4)) must append the subsidiary undertaking's individual accounts, and if the excluded subsidiary is in its own right a parent, its latest group accounts, made up to a date not more than 12 months prior to the financial year end of the parent's accounts. These appended accounts must be audited if required by law. If not prepared for other purposes or otherwise published, the subsidiary's accounts need not be appended, but reasons must be stated (S 243). The parent's record filed with the Registrar includes consolidated financial statements filed since its formation and a microfiche can be inspected or delivered by post.

2 London Stock Exchange requirements

2.1 Annual reports

A listed company must issue a consolidated annual report if the company has subsidiary undertakings, as soon as possible after the accounts have been approved, but certainly within six months of the end of the period to which it relates. In exceptional circumstances the time limit can be extended or other formats can be accepted if the Exchange agrees (but the parent's accounts must also be published if they contain significant additional information) (Ch. 12, para. 42). This annual report and accounts must have been prepared in accordance with the issuer's national law and, in all material respects with UK GAAP, US GAAP or International Accounting Standards. They must be independently audited to UK, US or International Standards on Auditing. The Exchange has the discretion to recognize additional accounting standards (Ch. 17, para. 3). Further information is required as necessary to ensure that the report and accounts give a true and fair view.

2.2 Interim reports and preliminary announcements

A 'half-yearly report' on a consolidated basis on the group's activities and profit or loss during the first six months of each financial year must be sent to holders of listed securities. Unlike in the statutory annual accounts, additional information about the parent's individual company financial statements is not required. Where the figures have been audited or reviewed by auditors according to Auditing Practices Board Bulletins, the auditors report must be reproduced in full. Quoted companies must notify the Exchange of any preliminary statement of annual results as soon as board approval has been given. It must have been agreed with the company's auditors and, if likely to be qualified, they must have given details of the nature of the qualification. . London Stock Exchange requirements relating to accountants' reports, profit forecasts, related parties and pro forma statements are not considered here.

XIII Sanctions

The Financial Reporting Review Panel has been authorized by the Secretary of State for Trade and Industry by Statutory Instrument, based on Companies Act 1985, S 245 B and C to examine departures from the accounting requirements of the Companies Act 1985 and accounting standards, and if the Panel deems it necessary, to seek a court order to remedy them. Usually resort to the courts is only made after attempts to persuade the company to correct the accounts have failed. The Panel may apply to the court for a declaration that the annual accounts do not comply with the Act, which includes applicable accounting standards (Sch 4, para. 36A), and an order requiring the directors of the company to prepare revised accounts. If the court finds non-compliance, it may order the directors who approved the defective accounts to bear the costs of the application, and any reasonable expenses incurred by the company in connection with or consequential on the preparation of revised accounts (S 245B). These powers apply equally to group accounts prepared by the parent company as well as its individual accounts included therein (see Financial Reporting Review Council, 1991, pp. 24–25, 49–50).

The Companies Act 1985 imposes fines for failure to include full information about excluded subsidiaries in the (parent) company's next annual return (S 231(7)), one-fifth of the statutory maximum (GBP5,000), and a daily default fine of one-fiftieth of this maximum (Sch 24). Otherwise defaults are as for single companies.

Bibliography

Archer, S. (1994). *ASB Discussion Paper on Goodwill and Intangible Assets – Comment*, Unpublished paper presented at the Institute of Chartered Accountants in England and Wales Financial Accounting and Auditing Research Conference, July.

Arnold, J., Egginton, D., Kirkham, L., Macve, R. and Peasnell, K. V. (1992). *Goodwill and Other Intangibles* (Institute of Chartered Accountants in England and Wales).

Barwise, P., Higson, C., Likierman, A. and Marsh, P. (1989). *Accounting for Brands* (Institute of Chartered Accountants in England and Wales).

Bailey, G. and Wild K. (1998). *International Accounting Standards: A Guide to Preparing Accounts* (Accountancy Books).

Baxter, G. C. and Spinney, J. C. (1975). A closer look at Consolidated Financial Theory. *CA Magazine*, Part 1, **106** (1), pp. 31–36; Part 2, **106** (2), pp. 31–35.

Bellamy, M. F. and Hastie, S. G (1993). Companies' Accounts Checklist. *Accountants Digest*, **305** (Institute of Chartered Accountants in England and Wales).

Bircher, P. (1988). The Adoption of Consolidated Accounting in Great Britain. *Accounting and Business Research*, Winter, pp. 3–13.

Bircher, P. (1995). Onerous Contracts: Another View, *Accountancy*, p. 82.

Brown, J. (1998). Fixing the Problem of Diminution, *Accountancy*, August, p. 89.

Bryant, R. (1993). Developments in Group Accounts, *Accountants Digest*, **294** (Institute of Chartered Accountants in England and Wales).

Chalmers, J. (1992). *Accounting Guides – Accounting for Subsidiary Undertakings* (Coopers & Lybrand and Gee).

Charity Accounting Review Body (1993). *Accounting by Charities – Statement of Recommended Practice 2 Exposure Draft* (Charities Commission).

Choi, F. D. and Lee, C. (1991). Merger Premia and National Differences in Accounting for Goodwill. *Journal of International Financial Management and Accounting*, **3**, (3), pp. 219–240.

Choi, F. D. and Lee, C. (1992). Effects of Alternative Goodwill Treatments on Merger Premia – Further Empirical Evidence. *Journal of International Financial Management and Accounting*, **4**, (3), pp. 220–236.

Chopping, D. and Skerratt, L. C. L. (1998). *Applying GAAP 1998/9* (Accountancy Books).

Collinson, D., Grinyer, J. and Russell, A. (1993). *Management's Economic Decisions and Financial Reporting*, (Institute of Chartered Accountants in England and Wales).

Company Reporting Limited, *Company Reporting*, Monthly journal.

Coopers & Lybrand (1997). *The Coopers & Lybrand Manual of Accounting* (Accountancy Books).

Crichton, J. (1990a). Consolidation – a Deceptive Simplicity. *Accountancy*, **105**, February, pp. 26–27.

Crichton, J. (1990b). Consolidation – Minority Calculations. *Accountancy*, **105**, June.

Crichton, J. (1996). Cash is King, and has been Re-crowned. *Accountancy*, December, p. 97

Davies, M., Paterson, R. and Wilson, A. (1997). *UK GAAP – Generally Accepted Accounting Practice in the United Kingdom*, Fifth edition (Macmillan, Basingstoke).

Edwards, J. R. and Webb, K. M. (1984). The development of Group Accounting in the UK to 1933. *The Accounting Historians Journal*, **11**, (1), pp. 31–61.

Egginton, D. A. (1990). Towards Some Principles for Intangible Asset Accounting. *Accounting and Business Research*, **20**, Summer, pp. 193–205.

Emmanuel, C. R. and Garrod, N. (1987). On the Segment Identification Issue. *Accounting and Business Research*, **17**, Winter, pp. 235–240.

Emmanuel, C. R. and Garrod, N, (1992). *Segment Reporting – International Issues and Evidence* (Prentice-Hall).

Ernst and Young (1998). *Financial Reporting Developments 1998* (Ernst and Young, London).

Ernst and Young (1998). *FRS: Guides to FRS 9, 10, 11, 12, 13, 14* (Ernst and Young, London).

Financial Accounting Standards Board (1991). *Discussion Memorandum: Consolidation Policy and Procedures* (FASB, Connecticut).

Financial Reporting Review Council (1991). *The State of Financial Reporting – A Review* (The Financial Accounting Council Limited).

Garrod, N. and Emmanuel, C. (1988) The Impact of Company Profile on the Predictive Ability of Disaggregated Data. *Journal of Business Finance and Accounting*, **15**, Summer, pp. 135–154.

Georgiou, G. (1993). Foreign Currency Translation and FRS 1. *Accounting and Business Research*, **23**, Summer, pp. 228–236.

Ghosh, J. (1991). *Accounting Guides – Cash Flow Statements* (Coopers & Lybrand).

Gore, P., Taib, F. and Taylor P. A. (1998). Accounting for Goodwill – What Factors Influence Management Preferences? (University of Lancaster Working Paper).

Gray, S. J., Coenenberg, A. G. and Gordon, P. D. (1993). *International Group Accounting – Issues in European Harmonisation*, 2nd edition (Routledge).

Grinyer, J. R., Russell, A. and Walker, M. (1990). The Rationale for Accounting for Goodwill. *The British Accounting Review*, **22**, pp. 223–235.

Grinyer, J. R., Russell, A. and Walker, M. (1991). Managerial Choices in the Valuation of Acquired Goodwill in the UK. *Accounting and Business Research*, **21**, Winter, pp. 51–55.
Gwilliam, D. and Russell, T. (1991). Polly Peck – Where were the Analysts?, *Accountancy*, **111**, pp. 25–26.
Hastie, S. G. (1998). *Companies' Accounts Checklists*, Accountants Digest issue 399 (Accountancy Books).
Higson, C. (1990). *The Choice of Accounting Method in UK Mergers and Acquisitions* (Institute of Chartered Accountants in England and Wales).
Higson, C. (1994). Goodwill. Unpublished paper presented at the Institute of Chartered Accountants in England and Wales Financial Accounting and Auditing Research Conference, July.
Hodgson, A., Okunev, J. and Willett, R. (1993). Accounting for Intangibles: A Theoretical Perspective. *Accounting and Business Research*, **23**, Spring, pp. 138–150.
Hodgson, E. (1992). *Accounting Guides – Reporting Financial Performance* (Coopers & Lybrand and Gee).
Hodgkinson, R. (1989). Ruling out the unfair advantage. *Accountancy Age*, **20**, July, p. 3.
Holgate, P. A. (1986a). A Guide to Accounting Standards – SSAP 23 Accounting for Acquisitions and Mergers. *Accountants Digest No. 189* (Institute of Chartered Accountants in England and Wales).
Holgate, P. A. (1986b). A Guide to Accounting Standards – Accounting for Goodwill. *Accountants, Digest No. 178* (Institute of Chartered Accountants in England and Wales).
Holgate, P. A. (1993). *Reporting Financial Performance – FRS 3 – 1993 Survey Results* (Coopers & Lybrand).
Holgate, P. and McCann, H. (1998). Accounting solutions. *Accountancy*, June, p. 94.
Johnson, B. and Patient, M. (1990). Tolley's Manual of Accounting, Volume II – *Consolidations and Groups*, and Volume III – *The Accounting Provisions of Specialised Businesses* (Tolley Publishing Company).
Lennard, A. and Peerless, S. (1995). When is a Contract Onerous? *Accountancy*, p. 129.
Ma, R. and Hopkins, R. (1988). Goodwill – an Example of Puzzle-Solving in Accounting. *Abacus*, **24**, (1), pp. 75–85.
Martindale, W. G. (1982). Unfreezing of Pre-acquisition Reserves – it's a Puzzle. *Accountancy*, **90**, September, pp. 129–130.
Mather, P. R. and Peasnell, K. V. (1991). An Examination of the Economic Circumstances Surrounding Decisions to Capitalize Brands. *British Journal of Management*, pp. 151–164.
Munson, R. (1985). A Guide to Accounting Standards – Deferred Tax. *Accountants Digest 174* (Institute of Chartered Accountants in England and Wales).
Nobes, C. W. (1980). A Review of the Translation Debate. *Accounting and Business Research*, **10**, Autumn, pp. 421–431.
Nobes, C. W. (1985). *Some Practical and Theoretical Problems of Group Accounting*, (Deloitte, Haskins and Sells, London).
Nobes, C. W. (1992). A Political History of Goodwill in the U.K.: an Illustration of Cyclical Standard Setting. *Abacus*, **28**, (2), pp. 142–161.
Nobes, C. W. and Parker, R. (1991). *Comparative International Accounting*, 3rd edition (Prentice-Hall International).
Ordelheide, D. and Pfaff, D. (1994). *European Financial Reporting – Germany* (Routledge).
Patient, M., Faris, J. and Holgate, P. (1993). Deemed Disposal. *Accountancy*, **112**, September.
Peasnell, K. V. and Yaansah, R. A. (1988). *Off-Balance Sheet Financing*, Research Report 10 (Chartered Association of Certified Accountants).
Pereira, V., Paterson, R. and Wilson, A. (1992). *UK/US GAAP Comparison*, (Kogan Page).
Pimm, D. (1990). Off Balance Sheet Vehicles Survive Redefinition. *Accountancy*, **105**, June, pp. 89–91.
Power, M. (1992). The Politics of Brand Accounting in the United Kingdom. *The European Accounting Review*, **1**, (1), pp. 39–68.
Rennie, E. D. and Emmanuel, C. R. (1992). Segmental Disclosure Practice: Thirteen Years On. *Accounting and Business Research*, **22**, Spring, pp. 151–159.

Ross, D. (1991). SSAP 20 – Long Overdue for Revision? *Accountancy*, **107**, May, p. 110.
Skerratt, L. C. L. and Tonkin, D. (1991). *Financial Reporting 1990/1* (Institute of Chartered Accountants in England and Wales).
Skerratt, L. C. L. and Tonkin, D. (1993). *Financial Reporting 1992/3* (Institute of Chartered Accountants in England and Wales).
Skerratt, L. C. L. and Tonkin, D. (1994). *Financial Reporting 1993/4* (Institute of Chartered Accountants in England and Wales).
Smith, T. (1992). *Accounting for Growth*, 1st edition (Century Business).
Smith, T. (1996). *Accounting for Growth*, 2nd edition (Century Business).
Swinson, C. (1993). *Group Accounting* (Butterworths).
Taylor, P. A. (1987). *Consolidated Financial Statements – Concepts, Issues and Techniques* (Paul Chapman Publishing).
Taylor, P. A. (1996). *Consolidated Financial Reporting* (Paul Chapman Publishing).
Thomas, A. (1975). The FASB and the allocation fallacy. *Journal of Accountancy*, November, pp. 65–68
Wainmann, D. (1984). *Currency Fluctuation: Accounting and Taxation Implications*, 2nd edition (Woodhead-Faulkner).
Wallace, P. and Ogle, B. D. G. (1984). A Guide to Accounting Standards – Foreign Currency Translation. *Accountants Digest No. 150* (Institute of Chartered Accountants in England and Wales).
Wallace, R. S. O. and Collier, P. A. (1991). The 'Cash' in Cash Flow Statements – A Multi-Country Comparison. *Accounting Horizons*, **91**, December, pp. 44–52.
Weetman, P., Jones, E. A. E., Adams, C. A. and Gray, S. J. (1998). Profit Measurement and UK Accounting Standards: A Case of Increasing Disharmony in Relation to US GAAP and IASs. *Accounting and Business Research*, **28**, Summer, pp. 189–208.
Westwick, C. (1986). *Accounting for Overseas Operations* (Gower).
Wild, K. and Creighton, B. (1996) Implementing FRS 8: Some Practical Aspects. *Accountancy*, October, pp. 128–129.
Wild, K. and Goodhead, C. (1994a). *Financial Reporting and Accounting Manual: Getting Reports Right* (Butterworths).
Wild, K. and Goodhead, C. (1998). *Financial Reporting and Accounting Manual: Getting Reports Right*, 5th edition (Butterworths).
Wild, K. and Goodhead, C. (1994b). Implementing FRS 1. *Accountancy*, **113**, February, pp. 86–87.
Wilkins, R. (1979). *Group Accounts*, 2nd edition (Institute of Chartered Accountants in England and Wales).

Accounting Standards Board (ASB)

Financial Reporting Standards (FRS)

FRS 1	Cash Flow Statements	(1991)
FRS 2	Accounting for Subsidiary Undertakings	(1992)
FRS 3	Reporting Financial Performance	(1993)
FRS 4	Capital Instruments	(1993)
FRS 5	Reporting the Substance of Transactions	(1994)
FRS 6	Acquisitions and Mergers	(1994)
FRS 7	Fair Values in Acquisition Accounting	(1994)
FRS 8	Related Party Disclosures	(1995)
FRS 9	Associates and Joint Ventures	(1997)
FRS 10	Goodwill and Intangible Assets	(1997)
FRS 11	Impairment of Fixed Assets and Goodwill	(1998)
FRS 12	Provisions, Contingent Liabilities and Contingent Assets	(1998)
FRS 13	Financial Instrument Disclosures	(1998)
FRS 14	Earnings Per Share	(1998)
FRSSE	Financial Reporting Standard for Smaller Entities	(1998)

Urgent Issues Task Force (UITF)

Abstract 9	Accounting for Operations in Hyper-inflationary Economies	(1993)
Abstract 15	Disclosure of Substantial Acquisitions	(1996)
Abstract 19	Tax on gains and losses on Foreign Currency Borrowings that Hedge an Investment in a Foreign Enterprise	(1998)
Abstract 21	Accounting Issues arising from the proposed introduction of the euro	(1998)

Financial Reporting Exposure Drafts (FRED)

Statement of Principles:

Chapter 5: Measurement in Financial Statements	(1995)
Chapter 7: The Reporting Entity	(1995)

Statements

Interim Reports	(1997)
Preliminary Announcements	(1998)

Discussion Papers

Accounting for Tax	(1995)
Segmental Reporting	(1996)
Business Combinations	(1998)

Accounting Standards Committee (ASC)

Statements of Standard Accounting Practice (SSAP)

SSAP 20	Foreign Currency Translation	(1983)
SSAP 25	Segmental Reporting	(1990)

Auditing Practices Committee (APC)

Guideline 3.415 Group Financial Statements – Reliance on the Work of other Auditors

USA
INDIVIDUAL ACCOUNTS

Norbert Fischer
Teresa E. Iannaconi
Harald W. Lechner

Acknowledgement

The authors gratefully acknowledge the intense collaboration of Martin Risau and Bernd Wübben, both of KPMG German Practice, New York, which enabled us to undertake and complete this project.

CONTENTS

I Introduction and background information

1 Historical development of financial accounting and reporting standards

In the United States the setting and codification of standards of financial accounting and reporting developed through official pronouncements by authoritative bodies of the private sector, which were empowered to create and select accounting principles. The accounting principles developed when rule-making bodies perceived a need and promulgated standards and opinions or, in their failure to act, when conditions and events arose requiring solutions. A significant event for the establishment of accounting standards in the United States was the stock market crash of 1929, which brought about three major effects.

The first effect was the enactment of the Securities Act in 1933. This law was intended to protect purchasers of securities against manipulation, misrepresentation and other fraudulent practices. The Securities Act required publicly traded securities to be registered with the federal government. The registration process required certain financial statement disclosures to be made before the securities could be sold to the public.

The second significant effect was the passage in 1934 of the Securities Exchange Act, which established the Securities and Exchange Commission (SEC). The Securities Exchange Act required the provision of periodic information to investors including audited financial statements. The SEC was given the authority to establish the accounting and auditing practices employed in the preparation and verification of financial statements issued by publicly traded companies in the United States. However, for various historical reasons, the SEC has decided not to take on the task of establishing Generally Accepted Accounting Principles (GAAP) directly, but rather has delegated the establishment of GAAP to the public accounting profession and the private sector. Therefore, the establishment of GAAP in the United States is characterized by co-operation between the federal government and the public accounting profession.

The third effect was the outcome of a joint effort by the New York Stock Exchange and the American Institute of Certified Public Accountants (AICPA) in 1938 to create a special committee which would establish standards for accounting procedures. The widespread growth in the activities of the Exchange had led to expanded ownership and trading activities by an underinformed public. This lack of information caused the accounting profession to become involved in the concept of GAAP. Although the profession had, through various committees, made previous contributions toward creating uniform accounting standards prior to 1938, the creation of this special committee was different because of the recognition that complexities in business activities and in ownership dispersion required consistency in accounting measurements and in the selection of accounting procedures. This Committee on Accounting Procedure (CAP) created the first authoritative set of statements of accounting principles in the United States. The CAP was followed by the Accounting Principles Board (APB), which was replaced by the Financial Accounting Standards Board (FASB). Each of these bodies has issued pronouncements on accounting issues, whereas the currently predominant type of official pronouncements is the Statement of Financial Accounting Standards (SFASs) by the FASB (see Table I.1). In each case the pronouncements apply to current and future financial statements and are not intended to be applied retroactively unless the pronouncement makes retroactive presentation mandatory.

2 Generally Accepted Accounting Principles

The common set of accounting and reporting standards and procedures adopted by the accounting profession constitute the Generally Accepted Accounting Principles (GAAP). GAAP are not legally compulsory in the United States, since no specific legal regulations require the financial statements of companies to be prepared in accordance with GAAP. But GAAP are binding since a certified public accountant has to express an opinion on financial statements being in conformity with GAAP. Although a statutory audit is not legally mandatory, the SEC requires audited financial statements for public companies. In

Table I.1 Statements of Financial Accounting Standards (SFASs)

No.	Subject
2	Accounting for Research and Development Costs
3	Reporting Accounting Changes in Interim Financial Statements
4	Reporting Gains and Losses from Extinguishment of Debt
5	Accounting for Contingencies
6	Classification of Short-Term Obligations Expected to Be Refinanced
7	Accounting and Reporting by Development Stage Enterprises
10	Extension of 'Grandfather' Provisions for Business Combinations
11	Accounting for Contingencies – Transition Method
13	Accounting for Leases
15	Accounting by Debtors and Creditors for Troubled Debt Restructurings
16	Prior Period Adjustments
19	Financial Accounting and Reporting by Oil and Gas Producing Companies
22	Changes in the Provisions of Lease Agreements Resulting from Refundings of Tax-Exempt Debt
23	Inception of the Lease
25	Suspension of Certain Accounting Requirements for Oil and Gas Producing Companies
27	Classification of Renewals or Extensions of Existing Sales-Type or Direct Financing Leases
28	Accounting for Sales with Leasebacks
29	Determining Contingent Rentals
34	Capitalization of Interest Costs
35	Accounting and Reporting by Defined Benefit Pensions Plans
37	Balance Sheet Classification of Deferred Income Taxes
38	Accounting for Preacquisition Contingencies of Purchased Enterprises
42	Determining Materiality for Capitalization of Interest Costs
43	Accounting for Compensated Absences
44	Accounting for Intangible Assets of Motor Carriers
45	Accounting for Franchise Fee Revenue
47	Disclosure of Long-Term Obligations
48	Revenue Recognition When Right of Return Exists
49	Accounting for Product Financing Arrangements
50	Financial Reporting in the Record and Music Industry
51	Financial Reporting by Cable Television Companies
52	Foreign Currency Translation
53	Financial Reporting by Producers and Distributors of Motion Picture Films
57	Related Party Disclosures
58	Capitalization of Interest Cost in Financial Statements That Include Investments Accounted for by the Equity Method
60	Accounting and Reporting by Insurance Enterprises
61	Accounting for Title Plant
62	Capitalization of Interest Cost in Situations Involving Certain Tax-Exempt Borrowings and Certain Gifts and Grants
63	Financial Reporting by Broadcasters
64	Extinguishment of Debt Made to Satisfy Sinking-Fund Requirements
65	Accounting for Certain Mortgage Banking Activities
66	Accounting for Sales of Real Estate
67	Accounting for Costs and Initial Rental Operations of Real Estate Projects
68	Research and Development Arrangements
69	Disclosures about Oil and Gas Producing Activities
71	Accounting for the Effects of Certain Types of Regulation
72	Accounting for Certain Acquisitions of Banking or Thrift Institutions
73	Reporting a Change in Accounting for Railroad Track Structures
75	Deferral of the Effective Date of Certain Accounting Requirements for Pension Plans of State and Local Governmental Units
76	Extinguishment of Debt
77	Reporting by Transferors for Transfers of Receivables with Recourse
78	Classification of Obligations That Are Callable by the Creditor
79	Elimination of Certain Disclosures for Business Combinations by Nonpublic Enterprises
80	Accounting for Futures Contracts
84	Induced Conversions of Convertible Debt
86	Accounting for the Costs of Computer Software to be Sold, Leased, or Otherwise Marketed

Table I.1 (contd.)

No.	Subject
87	Employers' Accounting for Pensions
88	Employers' Accounting for Settlements and Curtailments of Defined Benefit Pension Plans and for Termination Benefits
89	Financial Reporting and Changing Prices
90	Regulated Enterprises – Accounting for Abandonments and Disallowances of Plant Costs
91	Accounting for Nonrefundable Fees and Costs Associated with Originating or Acquiring Loans and Initial Direct Costs of Leases
92	Regulated Enterprises – Accounting for Phase-in Plans
93	Recognition of Depreciation by Not-for-Profit Organizations
94	Consolidation of All Majority-Owned Subsidiaries
95	Statement of Cash Flows
97	Accounting and Reporting by Insurance Enterprises for Certain Long-Duration Contracts and for Realized Gains and Losses from the Sale of Investments
98	Accounting for Leases: Sale-Leaseback Transactions Involving Real Estate; Sales-Type Leases of Real Estate; Definition of the Lease Term; Initial Direct Costs of Direct Financing Leases
99	Deferral of the Effective Date of Recognition of Depreciation by Not-for-Profit Organizations
101	Regulated Enterprises – Accounting for the Discontinuation of Application of FASB Statement No. 71
102	Statement of Cash Flows – Exemption of Certain Enterprises and Classification of Cash Flows from Certain Securities Acquired for Resale
104	Statement of Cash Flows – Net Reporting of Certain Cash Receipts and Cash Payments and Classification of Cash Flows from Hedging Transactions
105	Disclosure of Information about Financial Instruments with Off-Balance-Sheet Risk and Financial Instruments with Concentrations of Credit Risk
106	Employers' Accounting for Postretirement Benefits Other Than Pensions
107	Disclosures about Fair Value of Financial Instruments
109	Accounting for Income Taxes
110	Reporting by Defined Benefit Pensions Plans of Investment Contracts
111	Rescission of FASB Statement No. 32 and Technical Corrections
112	Employers' Accounting for Postemployment Benefits
113	Accounting and Reporting for Reinsurance of Short-Duration and Long-Duration Contracts
114	Accounting by Creditors for Impairment of a Loan
115	Accounting for Certain Investments in Debt and Equity Securities
116	Accounting for Contributions Received and Contributions Made
117	Financial Statements of Not-for-Profit Organizations
118	Accounting by Creditors for Impairment of a Loan – Income Recognition and Disclosures
119	Disclosure about Derivative Financial Instruments and Fair Value of Financial Instruments
120	Accounting and Reporting by Mutual Life Insurance Enterprises and by Insurance Enterprises for Certain Long-Duration Participating Contracts
121	Accounting for the Impairment of Long-Lived Assets and for Long-Lived Assets to Be Disposed of
123	Accounting for Stock-Based Compensation
124	Accounting for Certain Investments Held by Not-For-Profit Organizations
125	Accounting for Transfers and Servicing of Financial Assets and Extinguishments of Liabilities
126	Exemption from Certain Required Disclosures about Financial Instruments for Certain Nonpublic Entities
127	Deferral of the Effective Date of Certain Provisions of FASB Statement No. 125
128	Earnings per Share
129	Disclosure of Information about Capital Structure
130	Reporting Comprehensive Income
131	Disclosures about Segments of an Enterprise and Related Information
132	Employers' Disclosures about Pensions and Other Postretirement Benefits
133	Accounting for Derivative Instruments and Hedging Activities
134	Accounting for Mortgage-Backed Securities Retained after the Securitization of Mortgage Loans Held for Sale by a Mortgage Banking Enterprise
135	Rescission of FASB Statement No. 75 and Technical Corrections
136	Transfers of Assets to a Not-For-Profit Organization or Charitable Trust That Raises or Holds Contributions for Others
137	Accounting for Derivative Instruments and Hedging Activities – Deferral of the Effective Date of FASB Statement No. 133
138	Accounting for Certain Derivative Instruments and Certain Hedging Activities – An Amendment of FASB Statement No. 133

Statements not included in the table have been superseded. Pronouncements are available from the Financial Accounting Standards Board home page: http://www.rutgers.edu/Accounting/raw/fasb/public/index.html

adopting GAAP for supervising the financial reporting practices of public companies the SEC has established GAAP as the standard for the preparation of financial statements. Financial statements of private companies must only be prepared in accordance with GAAP, if they are audited. However, contractual agreements between creditors and borrowers and other parties may compel such financial statements to be in accordance with GAAP.

The term GAAP evolved through the official accounting pronouncements established by the authoritative standard setting organizations (promulgated principles) and through given accounting practices and procedures which have been accepted over time as appropriate because of their universal application (non-promulgated principles). The determination of the general acceptance of a particular accounting principle is difficult because no single reference source exists for all such principles. To provide guidance for the determination of GAAP in a particular situation, the ASB defined in SAS No. 69, The Meaning of 'Present Fairly in Conformity With Generally Accepted Accounting Principles' in the Independent Auditor's Report, a hierarchy of generally accepted accounting principles, the so-called House of GAAP (see Figure I.1). To prepare financial statements in accordance with GAAP, those principles found in the highest level of that hierarchy must be applied. An alternative accounting principle found in a lower level of the hierarchy cannot override an accounting principle in a higher level (see Figure I.2).

GAAP substitute a codified law of business accounting and constitute a body of case law, based on the system of common law in the United States. Since they are inductively derived from a case-by-case basis, GAAP remain flexible. But the increasing amount and regular amendments of official accounting pronouncements lacks a consistent structure and may lead to confusion in applying GAAP.

3 Organizations for establishing accounting standards

3.1 American Institute of Certified Public Accountants (AICPA)

3.1.1 Committee on Accounting Procedure

The Committee on Accounting Procedure (CAP) was formed in 1938 at the urging of the SEC. This committee had the authority to issue pronounce-

Figure I.1 House of GAAP

'House of GAAP'

Level D (Least Authoritative)	APB Statements	FASB Concept Statement	AICPA Issue Papers	Other Professional Pronouncements
Level C	Prevalent Industry Practices			
Level B	AICPA Industry Guides	AICPA Statements of Position	AICPA Accounting Interpretations	FASB Technical Bulletins
Level A (Most Authoritative)	APB Opinions	FASB Statements and Interpretations	EITF Abstracts	AICPA Accounting Research Bulletins

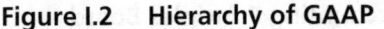

Figure I.2 Hierarchy of GAAP

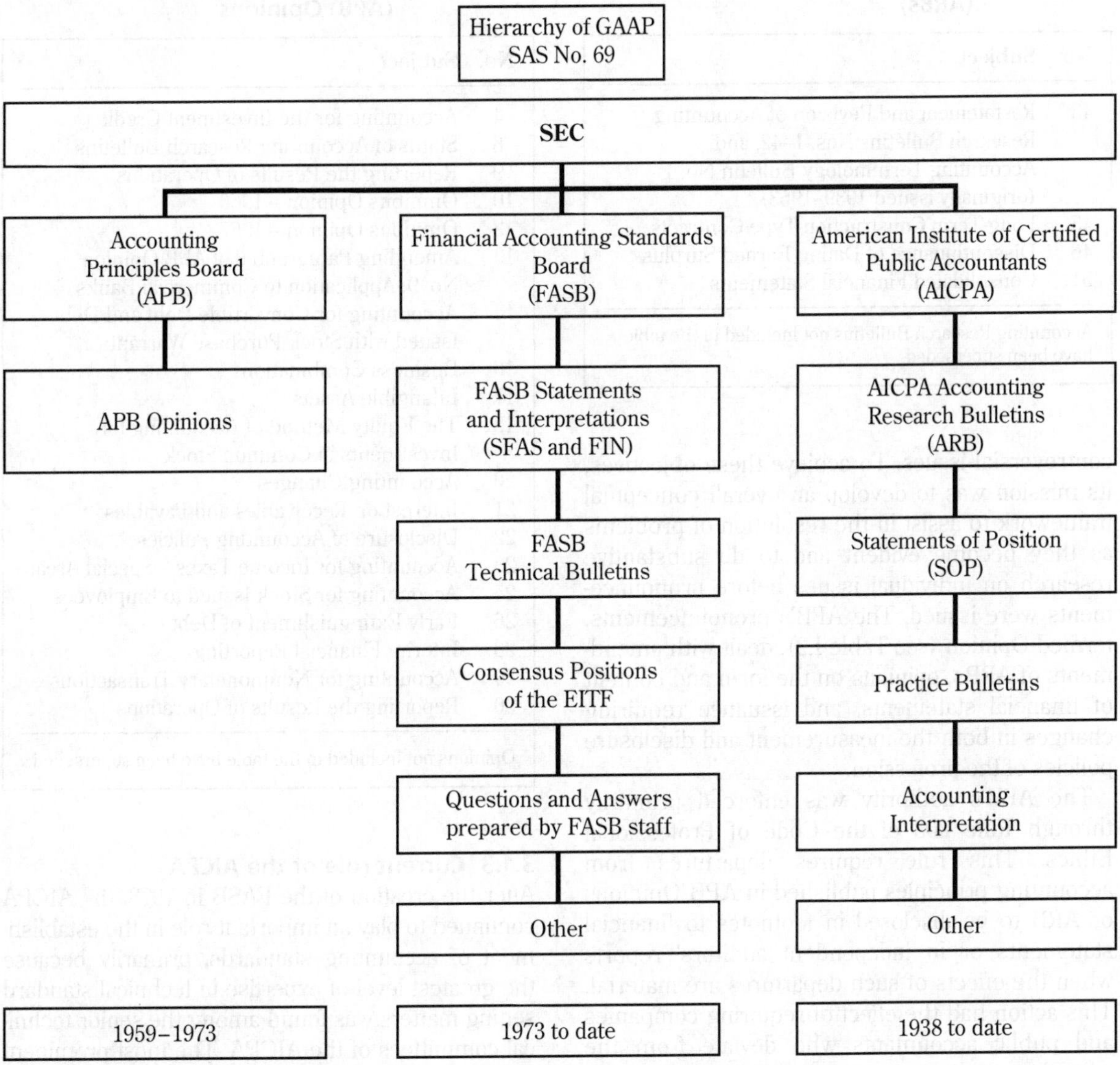

ments which were intended to provide substantial authoritative support for preferred accounting practices. The works of the CAP were published in the form of 51 Accounting Research Bulletins (ARBs, see Table I.2), dealing with a variety of timely accounting problems. These pronouncements were not binding in a legal sense, and received authority only from their general acceptance, whereas the SEC considered bulletins issued by the CAP to be GAAP. However, various alternative accounting principles were allowed to be applied in practice for similar economic transactions and this resulted in confusion and dissatisfaction on the part of some users of financial statements. The problem-by-problem approach failed to provide the structured body of accounting principles that was both needed and desired. In response to this dissatisfaction, in 1959 the AICPA replaced the CAP by the Accounting Principles Board (APB).

3.1.2 Accounting Principles Board (APB)

The APB had as its objectives the advancement of the written expression of GAAP, the narrowing of areas of difference and inconsistency in accounting practice, and the discussion of unsettled and

Table I.2 Accounting Research Bulletins (ARBs)

No.	Subject
43	Restatement and Revision of Accounting Research Bulletins Nos. 1–42, and Accounting Terminology Bulletin No. 1 (originally issued 1939–1953)
45	Long-Term Construction-Type Contracts
46	Discontinuance of Dating Earned Surplus
51	Consolidated Financial Statements

Accounting Research Bulletins not included in the table have been superseded.

Table I.3 Accounting Principles Board (APB) Opinions

No.	Subject
4	Accounting for the Investment Credit
6	Status of Accounting Research Bulletins
9	Reporting the Results of Operations
10	Omnibus Opinion – 1966
12	Omnibus Opinion – 1967
13	Amending Paragraph 6 of APB Opinion No. 9, Application to Commercial Banks
14	Accounting for Convertible Debt and Debt Issued with Stock Purchase Warrants
16	Business Combinations
17	Intangible Assets
18	The Equity Method of Accounting for Investments in Common Stock
20	Accounting Changes
21	Interest on Receivables and Payables
22	Disclosure of Accounting Policies
23	Accounting for Income Taxes – Special Areas
25	Accounting for Stock Issued to Employees
26	Early Extinguishment of Debt
28	Interim Financial Reporting
29	Accounting for Nonmonetary Transactions
30	Reporting the Results of Operations

Opinions not included in the table have been superseded.

controversial issues. To achieve these objectives, its mission was to develop an overall conceptual framework to assist in the resolution of problems as they become evident and to do substantive research on individual issues before pronouncements were issued. The APB's pronouncements, termed Opinion (see Table I.3), dealt with amendments of ARBs, opinions on the form and content of financial statements, and issuance requiring changes in both the measurement and disclosure policies of the profession.

The APB's authority was enforced primarily through Rule 203 of the Code of Professional Ethics. This rule requires departures from accounting principles published in APB Opinions or ARB to be disclosed in footnotes to financial statements or in independent auditors' reports when the effects of such departures are material. This action had the effect of requiring companies and public accountants who deviate from the reporting requirements contained in authoritative pronouncements to justify such departures.

Instead of establishing the principles which would underlie future pronouncements on specific questions, the APB proceeded to deal directly with specific questions. Therefore the impact of research studies on the APB's subsequent decisions was minimal and lead to a growing criticism. As a result, the AICPA appointed a committee to study how financial accounting principles should be established, with the consequence that the APB was abolished and the Financial Accounting Standards Board (FASB) was created in 1973.

3.1.3 Current role of the AICPA

After the creation of the FASB in 1973, the AICPA continued to play an important role in the establishment of accounting standards, primarily because the greatest level of expertise in technical standard setting matters was found among the senior technical committees of the AICPA. The most prominent of these accounting standard setting committees is the Accounting Standards Executive Committee (AcSEC), which was created in 1974.

From time to time AcSEC issues Statements of Position (SOPs) on accounting matters. In addition, AcSEC, in conjunction with the Auditing Standards Board (ASB) of the AICPA, may issue Accounting and Auditing Guides for specific industries. These SOPs and industry audit guides are helpful to practising accountants in that they indicate the preferable accounting practices for specific industries and specialized situations. Thus, while the FASB tends to focus on matters of general application, AcSEC tends to issue guidance on matters which have specific relevance in practice.

3.1.4 Auditing Standards Board

The securities laws in the United States require that the financial statements of all publicly traded companies must be audited by an independent public accountant (i.e. a CPA). The Auditing Standards Board (ASB) of the AICPA is the body currently empowered to establish the auditing standards that must be followed by all CPAs who are members of the AICPA. From time to time the ASB issues Statements of Auditing Standards (SASs) see Table X.1. There are ten specific standards, which are referred to as Generally Accepted Auditing Standards (GAAS), plus various SASs which are interpretations of the ten specific standards. The current sections of the codified SASs are listed in Table X.1.

Most recently, the ASB has issued new auditing standards on internal control, fraud and illegal acts. It has supported and begun to act on the recommendations made by the National Commission on Fraudulent Financial Reporting. As a result, audit committees composed of outside members of the board of directors were required for all public companies and the SEC mandated membership of a professional quality assurance programme, such as the AICPA SEC Practice Section, for all auditors involved in audits of public companies.

3.2 Financial Accounting Standards Board (FASB)

3.2.1 Role and organization

In 1973 the Financial Accounting Standards Board (FASB) became the official body charged with issuing accounting standards. It is composed of seven members appointed by a Board of Trustees. Four of the members are CPAs from public practice. The remaining three members, not necessarily CPAs, are expected to be familiar with the problems of financial reporting. The FASB's mission is to establish and improve standards of financial accounting and reporting for the guidance and education of the public, including issuers, auditors, and the users of financial information.

In addition to the FASB, the Financial Accounting Standards Advisory Council (FASAC) was created. The FASAC has the responsibility for consulting with the FASB on both major policy and technical issues.

3.2.2 Pronouncements

3.2.2.1 Types of pronouncements

The FASB issued four types of pronouncements: Statements of Financial Accounting Standards (SFASs, see Table I.1), Interpretations (FINs, see Table I.4), Technical Bulletins and Statements of Financial Accounting Concepts (SFACs, see Table I.5).

SFACs are intended to establish objectives and concepts that the FASB will use in developing standards of financial accounting and reporting. To date, the FASB has issued six SFACs, as discussed in III. SFACs differ from SFASs in that they are not intended to invoke Rule 203 of the Rules of Conduct of the Code of Professional Ethics.

SFASs indicate required accounting methods and procedures for specific accounting issues. They officially create GAAP.

Interpretations are modifications or extensions of issues related to previously issued SFASs, APB Opinions, or ARBs, with the purpose of clarifying, explaining, or elaborating the issues. The interpretations have the same authority as standards.

Since replacing the APB, the FASB has issued 138 standards and 44 interpretations.

Technical Bulletins are strictly interpretative in nature and do not establish new standards or amend existing standards. They are intended to assist with implementation problems and do not officially create GAAP.

3.2.2.2 Procedure of standard setting

The general approach used in issuing an SFAS by the FASB is as follows:

- An agenda item is selected by the chairman of the FASB on the basis of advice from other FASB members, members of the FASAC and suggestions from interested persons and organizations.

Table I.4 FASB Interpretations (FINs)

No.	Subject
1	Accounting Changes Related to the Cost of Inventory
4	Applicability of FASB Statement No. 2 to Business Combinations Accounted for by the Purchase Method
6	Applicability of FASB Statement No. 2 to Computer Software
7	Applying FASB Statement No. 7 in Financial Statements of Established Operating Enterprises
8	Classification of a Short-Term Obligation Repaid Prior to Being Replaced by a Long-Term Security
9	Applying APB Opinions No. 16 and 17 When a Savings and Loan Association or a Similar Institution is Acquired in a Business Combination Accounted for by the Purchase Method
14	Reasonable Estimation of the Amount of a Loss
18	Accounting for Income Taxes in Interim Periods
19	Lessee Guarantee of the Residual Value of Leased Property
20	Reporting Accounting Changes under AICPA Statements of Position
21	Accounting for Leases in a Business Combination
23	Leases of Certain Property Owned by a Governmental Unit or Authority
24	Leases Involving Only Part of a Building
26	Accounting for Purchase of a Leased Asset by the Lessee during the Term of the Lease
27	Accounting for a Loss on a Sublease
28	Accounting for Stock Appreciation Rights and Other Variable Stock Option or Award Plans
30	Accounting for Involuntary Conversions of Nonmonetary Assets to Monetary Assets
33	Applying FASB Statement No. 34 to Oil and Gas Producing Operations Accounted for by the Full Cost Method
34	Disclosure of Indirect Guarantees of Indebtedness of Others
35	Criteria for Applying the Equity Method of Accounting for Investments in Common Stock
36	Accounting for Exploratory Wells in Progress at the End of a Period
37	Accounting for Translation Adjustments upon Sale of Part of an Investment in a Foreign Entity
38	Determining the Measurement Date for Stock Option, Purchase, and Award Plans involving Junior Stock
39	Offsetting of Amounts Related to Certain Contracts
40	Applicability of Generally Accepted Accounting Principles to Mutual Life Insurance and Other Enterprises
41	Offsetting of Amounts Related to Certain Repurchase and Reverse Repurchase Agreements
42	Accounting for Transfers of Assets in Which a Not-for-Profit Organization is Granted Variance Power
43	Real Estate Sales
44	Accounting for Certain Transactions Involving Stock Compensation

Interpretations not mentioned in the table have been superseded.

Table I.5 Statements of Financial Accounting Concepts (SFACs)

No.	Subject
1	Objectives of Financial Reporting by Business Enterprises
2	Qualitative Characteristics of Accounting Information
4	Objectives of Financial Reporting by Non-business Organizations
5	Recognition and Measurement in Financial Statements of Business Enterprises
6	Elements of Financial Statements of Business Enterprises

SFAC No. 3 has been superseded by SFAC No. 6.

- A task force is then chosen. The chairman of the task force is usually a member of the FASB. The task force members may include FASB and FASAC members and interested persons who are knowledgeable about issues and the needs of financial statement users or who are experts or who have a viewpoint relevant to the project. The task force's responsibilities include refining the definition of the problem and its financial accounting and reporting implications, determining the nature and extent of necessary research, and producing a discussion memorandum.
- The discussion memorandum is then distributed to interested parties for review. Public

hearings are held after the release of the memorandum for the purpose of discussion and receipt of comments from individuals and groups.

- An Exposure Draft of the proposed SFAS is then prepared. Once it receives the support of five of the seven FASB members, it is released to the public for review. The period of public exposure is not less than 30 days. The purpose of the public exposure is to provide the FASB with feedback. Arguments both for and against a position may influence the content of the final statement. The public exposure also allows the public the opportunity to comment on the proposed accounting standard.
- An exposure draft is then revised on the basis of the feedback received from the public. Once five members of FASB agree on a standard, it is issued.

3.2.3 Emerging Issues Task Force

The FASB has been criticized for failing to provide timely guidance on emerging implementation and practice problems. As a response to this criticism the Emerging Issues Task Force (EITF) was established to assist in identifying issues and problems that might require action. The goal of the EITF is to provide timely guidance on new issues while limiting the number of issues whose resolutions require formal pronouncements by the FASB. The purpose of the task force is to reach a consensus on how to account for new and unusual transactions that have the potential for creating differing financial reporting practices.

3.3 Securities and Exchange Commission (SEC)

3.3.1 Tasks and responsibilities

The primary responsibility of the SEC is to regulate the securities markets in the United States, including such markets as the New York Stock Exchange, the American Stock Exchange and the National Association of Securities Dealers' Automated Quotation System (NASDAQ) automated stock exchange. Thus a major portion of the activity of the SEC lies in the regulation and supervision of trading in the securities markets. A necessary component of efficient trading in the securities markets is the availability of reliable information about companies whose securities are being traded. Therefore, the Securities Exchange Act requires all companies with publicly traded securities to file an annual report with the SEC on form. This annual report must include audited financial statements prepared in accordance with GAAP. In addition, these companies must issue an annual report to their shareholders.

A second major task of the SEC is the review of registration statements prior to the issuance of securities, since all companies desiring to issue securities publicly in the United States must file a registration statement with the SEC. A registration statement includes a considerable amount of required information about the company issuing securities, including audited financial statements. The SEC has broad-ranging authority to prevent the issuance of securities which they feel may have incorrect or misleading accounting treatments, and may compel companies who wish to issue securities in the United States to change or correct their accounting treatment to conform to those treatments which the staff of the SEC believe to be preferable.

The SEC is empowered by the Securities Exchange Act to establish accounting standards in the United States. Although the SEC has chosen to finally delegate this task to the FASB in 1973, it nevertheless retains the power to create accounting and reporting regulations and uses it especially in issuing regulations for the standardization of financial reporting for public companies. A summarizing overview of the organizations involved in the development of accounting standards is given in Figure I.1.

3.3.2 Standardization of financial reporting

The most important regulations of the SEC are regulation S-X, which describes the rules for the form and content of financial statements filed with the SEC, and regulation S-K, which covers the required non-financial statement-related disclosures that must be included in the registration statements and the other forms that must be filed with the SEC on form 10-K for domestic registrants and on a form 20-F for foreign filers. From 1933 to 1980 the SEC issued various Accounting

Series Releases (ASR) which were intended to comment on or interpret accounting pronouncements issued by the AICPA and FASB. In addition, some ASRs were issued in response to situations where accountants or auditors were found to have been involved in fraudulent financial reporting activities. In such cases the ASR would describe the situation in detail and the penalties or sanctions would be assessed. In 1980 the SEC codified the various accounting standard-related ASRs into Financial Reporting Release (FRR) No. 1. Therefore matters pertaining to accounting standards are now issued as FRRs and matters pertaining to enforcement action are issued in another series of SEC releases. The SEC also issued Staff Accounting Bulletins (SABs) to assist companies in preparing financial statements included in forms filed with the SEC. In order to reduce the complexity of the registration and reporting process, an Integrated Disclosure System (IDS) was implemented in the early 1980s. The IDS amended regulation S-X for the purpose of standardizing financial statement requirements in SEC filings. Thus financial statements included in annual reports to shareholders of public companies are usually the same as those included in forms filed with the SEC. Under current rules, the content of the annual report to shareholders is the same as that in the 10-K form filed with the SEC. The disclosures required by the SEC for public companies are summarized in Table I.6.

Table I.6 Summary of disclosure required under SEC integrated disclosure system

Article	Subject
Part I	
1	Business
2	Properties
3	Legal Proceedings
4	Submission of Matters to a Vote of Security Holders
Part II	
5	Market for Registrant's Common Equity and Related Stockholder Matters
6	Selected Financial Data
7	Management's Discussion and Analysis of Financial Condition and Results of Operations
7A	Quantitative and Qualitative Disclosures About Market Risk
8	Financial Statements and Supplementary Data
9	Changes in and disagreements with Accountants on Accounting and Financial Disclosure
Part III	
10	Directors and Executive Officers of the Registrant
11	Executive Compensation
12	Security Ownership of Certain Beneficial Owners and Management
13	Certain Relationships and Related Transactions
Part IV	
14	Exhibits, Financial Statement Schedules, and Reports on Form 8-K

3.4 International activities of the SEC, the FASB and the AICPA

3.4.1 International activities of the SEC

3.4.1.1 The IASC core standards project

In December 1998, the International Accounting Standards Committee completed its set of core accounting standards when it approved IAS 39, Financial Instruments – Recognition and Measurement. Accordingly, the International Organization of Securities Commissions and individual securities regulators, including the SEC, will start a process to determine how International Accounting Standards (IASs) could be used in cross-border securities offerings and listings. The SEC acknowledged the relevance of the development of IAS, but repeatedly expressed its demand that a final core set of IASs must be a comprehensive set of high quality standards that result in transparency for the underlying economics of the businesses that are both rigorously interpreted and implemented. A possible result of the concluded assessment of the completed core standards by the SEC could be the amendment of the SEC's current filing requirements for registrants applying IAS (Form 20-F reconciliation).

3.4.1.2 Application of IASs by registrants

At present the SEC is primarily concerned with the question of whether it is appropriate for financial statements prepared in accordance with IASs to be included in SEC filings by foreign registrants without reconciliation to US GAAP. The SEC has become aware of situations where a registrant prepares its financial statements in accordance with local accounting and reporting principles and asserts in footnotes that the financial statements 'comply, in all material respects, with' or 'are consistent with' IASs. In some of these situations, the registrant may have applied only certain IASs or omitted certain information without giving any explanation of why the information was excluded. The SEC has challenged these assertions and will continue to do so. Where the assertion cannot be sustained, the SEC will require either changes to the financial statements to conform with IASs, or removal of the assertion of compliance with IASs. This position is consistent with the prohibition under IAS 1 (Revised).

3.4.2 International activities of the FASB

The FASB issued a report in 1998 which expressed the FASB's vision for the development of international accounting standards in the future. The key points of this outlook are the following:

- The FASB has a leadership role to play in the evolution of the international accounting system and is guided by the belief that, ideally, the ultimate outcome would be the worldwide use of a single set of high-quality accounting standards for both domestic and cross-border financial reporting.
- Until that ideal outcome is achieved, the FASB's objective for participating in the international accounting standard setting process is to increase international comparability while maintaining high-quality accounting standards in the United States. To achieve that objective, the FASB is willing to commit the required resources to the related goals of (i) ensuring that international accounting standards are of high quality; and (ii) increasing the convergence and quality of the accounting standards used in different nations.
- The FASB believes that the establishment of a quality international accounting standard setting structure and process is key to the long-term success and development of international accounting standards. The FASB will participate in establishing that structure and process. The FASB accepts that an increasing and substantial level of resources might be required to support and influence the establishment of that organization.
- The FASB acknowledges that, if a quality international accounting standard setting structure and process emerges, the FASB's commitment and desire to participate in a meaningful way in the operations of that standard setter may ultimately lead to structural and procedural changes to the FASB as well as potential changes in its national role.

3.4.3 International activities of the AICPA

The AICPA is a member body of the International Federation of Accountants (IFAC), which is the worldwide organization of the accounting profession. The IFAC's objective is to develop and establish a core set of International Standards on Auditing (ISAs). In addition to the standardization of accounting standards, the ISAs are a component in the attempt to facilitate cross-border stock offerings and listings by multinational corporations. In implementing and adapting ISA, the AICPA includes in its Statements on Auditing Standards additional remarks for those requirements of ISA exceeding the specifications of the SAS.

II Forms of business organization

1 Introduction

Within this section differences in forms of business organization and different financial reporting requirements are emphasized. In the United States business enterprises are organized primarily in one of three ways: as individual proprietorships, as partnerships or as corporations.

2 Proprietorship

A proprietorship has a single owner, who is also the manager. Proprietorships are usually small stores, restaurants and individual professional businesses. Although from an accounting viewpoint the proprietorship may be distinguished from its owner, from a legal and tax perspective no distinction is made. Every act taken or obligation assumed, legally is an act or an obligation assumed by the business owner as an individual. There are no legal requirements with respect to the accounting records for proprietorships.

3 Partnerships

A partnership joins two or more entities (i.e. individuals, corporations or other partnerships) together as co-owners. Each owner is a partner, with a specified ownership interest. Partnerships may be organized as general partnerships or as limited partnerships.

In a general partnership each partner has the right to make all the management decisions for the business. The partners assume unlimited liability for all the activities of the partnership. When a partner dies or leaves the business, the partnership is dissolved. The interest of a partner is not freely transferable without agreement by the other partners.

A limited partnership has at least one general partner, which may be a corporation. The general partners manage the business and have unlimited liability for the obligations of the limited partnership. The limited partners are liable only to the extent of their investment in the partnership. They have no right to manage the business. If they do manage the business, they may lose their limited liability for the business obligations. A limited partnership may have a life apart from its owners. When a general partner dies or withdraws, the partnership is dissolved in absence of a contrary agreement or admittance of a new partner in case of a transfer.

For certain professions with risk of large personal liabilities for professional malpractice of their partners, several states permitted limited liability partnerships (LLP). An LLP is similar to a partnership, except that a partner's liability for his partners' professional malpractice is limited to the partnership's assets. A partner retains unlimited liability for his own professional malpractice and for all non-professional obligations of the partnership. An LLP is managed from all partners, unless otherwise agreed. The LLP has no life apart from its partners. Ownership interest in an LLP is only transferable if the other partners agree.

A limited liability company (LLC) combines certain features of a limited partnership with those of a corporation. It was created to combine the tax advantages of a partnership with those of a corporation. Similar to a partnership, the LLC is owned by 'members' who may manage the LLC themselves. They also may elect managers, like in a corporation. Members have limited liability for the obligations of the LLC. An LLC is dissolved in case of death, bankruptcy or withdrawal of one member, unless all remaining members vote to continue the business. The transfer of membership interest is prohibited without consent of all members.

The income of a partnership is not taxed under federal law or state law. A partner's share of the partnership's income is taxed as personal income and must be included in the partner's individual income tax return. The partnership laws of the various states generally do not specify how a partnership should keep its accounting records. However, many partnership agreements specify that accounting records shall be maintained in accordance with GAAP.

4 Corporations

A corporation is a business owned by its shareholders who elect a board of directors to manage the business. Ownership and management may be completely separate. No shareholder has the right to manage, and no director or officer needs to be a shareholder. The board of directors has the power and duty to manage the corporation. The directors may manage the corporation only if they act as a board, unless agency power is granted to individual directors. The board usually delegates management responsibility to committees of the board such as an executive committee, to individual board members such as the chairman of the board, and to the officers, especially the chief executive officer. The board has the duty to supervise the activities and the directors have the right to inspect the corporate books and records. A business enterprise becomes a corporation when its owners make an application to one of the states of the United States. A particular state must approve the articles of incorporation of the corporation and then issue a charter allowing the corporation to operate. A corporation is deemed to be a legal entity separate from its shareholders, and may conduct business in its own name, sue and be sued, and generally enjoy the rights of a person under federal and state laws. Shareholders of a corporation have no personal liability with respect to the obligations or activities of the corporation, even if a shareholder is elected as director or officer. Directors and officers are only liable for own misconduct, not for other liabilities committed under the name of the corporation. As a separate legal entity a corporation is directly taxed on its income under federal and many state laws. The laws of the various states dealing with matters of incorporation generally do not specify how the corporation should keep its accounting records other than stating that such records should be kept in accordance with GAAP.

5 Public companies and the SEC

5.1 Filing requirements of public companies

Virtually all business entities with publicly traded securities are organized as corporations (a few large partnerships have publicly traded partnership interests). If a company has publicly traded securities (e.g. shares or bonds) it is required by the federal securities laws to file a registration statement and annual and quarterly reports with the Securities and Exchange Commission (SEC). The SEC was formed to ensure adequate disclosure of financial and other data to keep investors fully informed about publicly held corporations. More than 15,000 companies that are listed on one or more of the organized exchanges or traded in the over-the-counter stock markets are required to file forms with the SEC. Regulations S-X and S-K govern the form and content of financial statements, and relevant accounting principles are further defined and interpreted in Financial Reporting Releases (FRRs); however, for the most part these regulations specify that GAAP should be followed. Regulation S-X is codified in US federal law under title 17, chapter II, part 210 of the Code of Federal Regulations. Part 210 sets forth the form and content of financial statements required to be filed with the SEC. The term 'financial statements' is deemed to include all notes to the statements and all related schedules. Since regulation S-X has the force of federal rule, a company contemplating the issuance of securities publicly should be familiar with the provisions of regulation S-X.

Registration statements are required to be filed with the SEC for offerings or exchanges of securities. These registration statements contain detailed information regarding the extent and the costs of distribution for the securities, copies of audited financial statements and exhibits, and other data concerning the company, its management, any special circumstances, and any risks that investors should recognize. Most public companies must also file an annual form 10-K with detailed information concerning the business, its properties, executive remuneration, legal proceedings, and, in addition to financial statements, other detailed schedules that are of interest to investors. In addition to the annual form 10-K, public companies are required to file quarterly reports on form 10-Q that include unaudited condensed financial statements and form 8-Ks for such events as changes in control of registrant, significant acquisitions or dispositions of assets, bankruptcy, or a change in auditors, etc. (see discussion in IV below).

5.2 Accounting Series Releases (ASRs)

Beginning in 1938, the SEC began to issue opinions relating to accounting matters. These opinions were set forth in approximately 300 Accounting Series Releases (many of which have since been rescinded). Some ASRs were opinions of the Chief Accountant of the SEC. In other cases an ASR initiated new accounting disclosures or changed an existing requirement by amending regulation S-X. In addition to addressing issues on accounting, reporting, and auditing, the ASRs also concerned matters relating to the independence of accountants and disciplinary action against accountants. The ASRs also included amendments to the various SEC filing forms to which financial statements were required to be attached. A particular ASR usually included background information and guidance on the preparation of the specific S-X disclosure promulgated by the ASR. ASRs have been replaced by FRRs, SABs and AAERs.

5.3 Staff Accounting Bulletins (SABs)

In November 1975 the SEC began publishing Staff Accounting Bulletins which were intended to disseminate informal staff interpretations and practices. The SEC felt that the increased availability of these previously unpublished interpretations would reduce repetitive comments and enquiries and thereby save the SEC staff, companies, and their professional advisers time and money in the registration and reporting process.

Unlike regulation S-X, ASRs, and FRRs, the SABs are not official accounting rules of the SEC. Also, SABs do not receive the SEC's official approval. However, since the SEC staff use these positions in reviewing a company's filings with the SEC it is advisable to follow the guidelines and interpretations found in the SABs.

The SABs are usually comprised of a statement of facts, questions, and the staff's interpretation of each item. The first SAB was a comprehensive compilation of the various interpretations that existed at the time the SAB series was initiated.

6 Accounting standards for companies not publicly traded

Although in dollar terms most of the economic activity of the United States is conducted through public corporations, there are a large number of privately held corporations, partnerships and proprietorships which are also very important to the overall economy. All business entities of whatever size are required to file federal, and in most cases state, income tax returns. The federal and state income tax laws do specify that certain accounting practices must be followed with respect to inventory and fixed assets and various other accounting matters such as goodwill amortization. Most business enterprises must also be chartered or licensed by the state in which they operate. However, apart from the tax laws, state laws do not ordinarily specify that any particular accounting practices must be utilized other than GAAP. Hence, for the most part, GAAP, as it has been established by the private sector through the FASB and the AICPA, marks the requirement for accounting standards in the United States for both public and private companies. If a company is audited by a CPA, its auditors will require the company to follow GAAP. If a company is not audited, it may nevertheless be compelled to follow GAAP by provisions contained in contractual agreements with investors, creditors, suppliers or other parties. Thus, even though a company may not be specifically compelled by legal requirements to follow GAAP, it may in effect be required to follow GAAP because to do otherwise would subject the company to questions by parties with whom it would like to have business relationships.

7 Differences between GAAP for public companies and GAAP for private companies

One of the primary mechanisms that has allowed the use of GAAP to be enforced in the United States is the Code of Conduct of the AICPA. This code prevents a CPA from issuing an unqualified audit report on a company's financial statements if they have not been prepared in accordance with GAAP. Nevertheless, it has been argued by CPAs who primarily audit privately held companies that not all accounting pronouncements should be made applicable to all companies, even if they are required to prepare their financial statements in accordance with GAAP. Although there has been a general reluctance on the part of the AICPA to allow a division of GAAP into one part applicable to public companies and another part applicable to private companies, in a few cases exceptions have been made. Some accounting pronouncements have been deemed to be not applicable to privately held companies. The most notable examples of these exceptions are SFAS No. 128 dealing with earnings per share, and SFAS No. 131 dealing with disclosures about segments of an enterprise and related information. These pronouncements are not applicable to private companies. In addition, some accounting pronouncements allow for reduced disclosure for private companies, such as SFAS No. 107 Disclosures about Fair Value of Financial Instruments (applicable only after meeting certain specified criteria mentioned in SFAS No. 126) and SFAS No. 132 Employers' Disclosures about Pensions and Other Postretirement Benefits. Also, the effective date for certain FASB pronouncements has been delayed for non-public and smaller companies in certain cases (e.g. SFAS No. 106, Employers' Accounting for Postretirement Benefits Other than Pensions).

III Objectives, underlying concepts and general principles of financial accounting

1 Conceptual framework

1.1 Development of a conceptual framework

With the intention of establishing a coherent set of financial accounting standards and rules, the FASB has attempted to develop a system of interrelated objectives and concepts that is useful in setting accounting standards and in providing a frame of reference for resolving accounting controversies. The FASB has issued six Statements of Financial Accounting Concepts (see Table I.5). The components of this conceptual framework include objectives, underlying concepts and general principles of financial reporting. Of the six SFACs, SFAC No. 4, Objectives of Financial Reporting by Nonbusiness Organizations, is not presented here due to its specialized nature.

SFAC No. 1, Objectives of Financial Reporting by Business Enterprises, identifies objectives of financial reporting. These were identified to provide useful information for economic decisions, understandable information capable of predicting cash flows, and relevant information about economic resources and the transactions, events, and circumstances that change them.

SFAC No. 2, Qualitative Characteristics of Accounting Information, identifies the qualities which make information useful. Under a cost benefit constraint, the primary qualities of useful information are relevance and reliability. In addition, such information must be comparable, consistent and understandable.

SFAC No. 3, Elements of Financial Statements of Business Enterprises, has been replaced by SFAC 6. This statement has been amended by SFAC 6 to include financial reporting by not-for-profit organizations.

SFAC No. 5, Recognition and Measurement in Financial Statements of Business Enterprises, sets forth recognition criteria which determines what information should be in financial statements and when that information will appear.

SFAC No. 6, Elements of Financial Statements of Business Enterprises, defines ten elements as the basic components of financial statements. Three of these elements (assets, liabilities, and equity) relate to the balance sheet. The remaining elements (comprehensive income, revenue, expenses, gains, losses, investments by owners, and distributions to owners) relate to the performance of an entity over time. These would be displayed on the income statement, statement of changes in financial position, and statement of changes in equity.

Since GAAP may be inconsistent with the principles set forth in the conceptual framework, the FASB expects to reexamine existing accounting standards. Until that time, an SFAC does not require a change in existing GAAP. SFACs do not amend, modify, or interpret existing GAAP, nor do they justify changing GAAP based upon interpretations derived from them.

The following sections present summaries of the SFACs.

1.2 Objectives of financial reporting by business enterprises

In SFAC No. 1, the FASB sets out three basic objectives for financial reporting. In communicating information that relates to an enterprise's resources, obligations, earnings, and other information provided by the accounting system, financial reporting includes not only financial statements but also other means which may take various forms and relate to various matters (SFAC No.1, paragraph 7). The objectives of financial reporting are to provide information that is:

- Useful to those making investment and credit

decisions who have a reasonable understanding of business and economic activities.
- Helpful to present and potential investors, creditors, and other users in assessing the amounts, timing, and uncertainty of future cash flows.
- About the economic resources of an enterprise, the claims to those resources and the changes in them.

The emphasis on 'assessing cash flow prospects' might lead one to suppose that the cash basis is preferred over the accrual basis of accounting. That is not the case. Information should be based on accrual accounting, since it provides a better indication of an enterprise's ability to generate favourable cash flows than information about current cash receipts and payments. Accrual accounting attempts to record the financial effects on an enterprise of transactions and other events and circumstances that have cash consequences for an enterprise in the periods in which those transactions, events and circumstances occur, rather than only in the periods in which cash is received or paid by the enterprise.

The primary focus of financial reporting in the United States is on information about earnings and its components (i.e. the focus is on income measurement, see V.3). The financial reporting is expected to provide information about an enterprise's financial performance during a period of time and about how the management of an enterprise has discharged its stewardship responsibility to the owners (i.e. the periodicity principle). Therefore the objectives enunciated in SFAC No. 1 are directed toward the common interest of many users in the ability of an enterprise to generate favourable cash flows, but are phrased using investment and credit decisions as a reference to give them a focus. Although investment and credit decisions reflect investors' and creditors' expectations about the future performance of the enterprise, those expectations are based at least partly on evaluations of the performance in the past (i.e. the historical cost principle).

The objectives of financial accounting establish the necessary basis in defining qualitative characteristics of accounting information that help in making that information useful. Those underlying characteristics are defined in SFAC No. 2.

1.3 Qualitative characteristics of accounting information

In SFAC No. 2 the FASB presents a hierarchy of qualities of accounting information with usefulness for decision-making as the primary criterion of choice. The hierarchy of those accounting qualities is shown in Figure III.1 and explained in the following.

The primary qualitative characteristics of accounting information are relevance and reliability. For accounting information to be relevant to investors, creditors and other users, it must be capable of making a difference in a decision by improving the decision-makers' capacity to predict (predictive value) or by confirming or correcting earlier expectations (feedback value). To be relevant, it must also be available to decision-makers before it loses its capacity to influence their decisions (timeliness). Information is reliable if it can be verified by agreement among a number of independent observers (verifiability, meaning that several measurers, making independent evaluations, are likely to obtain the same measures), and if it represents what it purports to represent (representational faithfulness). SFAC No. 2, paragraph 65 identifies different degrees of representational faithfulness). Reliability implies completeness and neutrality of information. If either of these primary qualities is missing, the information will not be useful. Though, ideally, the choice of an accounting alternative should produce information that is both more reliable and more relevant, it may be necessary to sacrifice some of one quality for a gain in another.

A secondary quality that interacts with relevance and reliability to contribute to the usefulness of information, is comparability, which includes consistency. Information about an enterprise is more useful if it can be compared with similar information about another enterprise (comparability) and with similar information about the same enterprise at other points in time (consistency).

In addition, the FASB has included two constraints in the hierarchy of qualitative characteristics, cost–benefit and materiality. Unless the benefits to be derived from specific accounting information exceed the costs of providing that

Figure III.1 A Hierarchy of Accounting Qualities

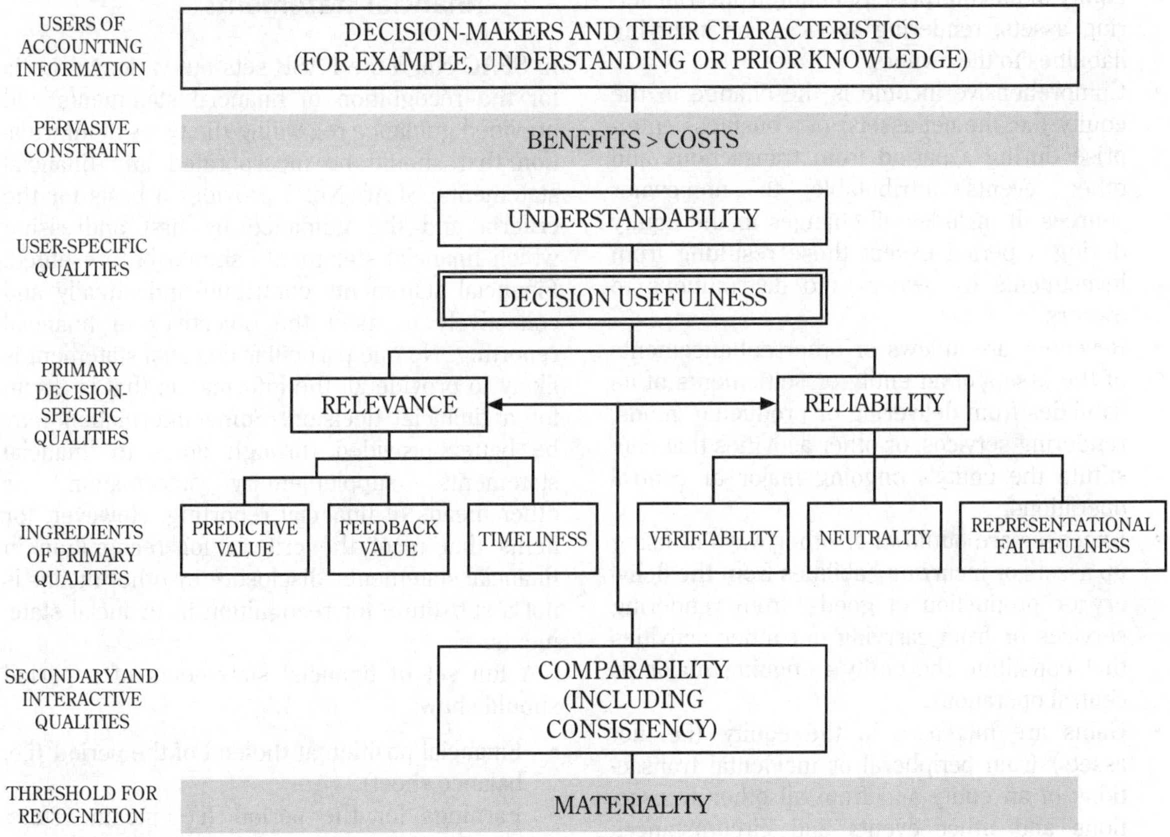

Source: FASB CON2.

information, it should not be provided. Materiality deals with the omission or misstatement of an item on the basis that it is too small to be important. In essence, the omission or misstatement is material if the magnitude of it is such that it is probable that the judgement of a reasonable person relying on the financial information would have been changed or influenced by the inclusion or correction of the item. Whether an item is material will depend on the circumstances surrounding it, and judgements must therefore be made on a case-by-case basis.

1.4 Elements of financial statements

SFAC No. 6 defines ten elements that are most directly related to measuring the performance and status of an enterprise. These elements are defined as follows:

- Assets are probable future economic benefits obtained or controlled by a particular entity as a result of past transactions or events.
- Liabilities are probable future sacrifices of economic benefits arising from present obligations of an entity to transfer assets or provide services to other entities in the future as a result of past transactions or events.
- Equity (i.e. net assets) is the residual interest in the assets of an entity that remains after deducting its liabilities. In a business enterprise the equity is the ownership interest.
- Investments by owners constitute increases in the equity of the business enterprise resulting from transfers to it from other entities of something valuable in order to obtain or increase the ownership interest (i.e. equity). Investments by owners are most commonly received by the enterprise in the form of assets, but the enterprise may also receive services or the satisfaction of liabilities.

- Distributions to owners are decreases in the equity of an enterprise resulting from transferring assets, rendering services, or incurring liabilities to the owners.
- Comprehensive income is the change in the equity (i.e. the net assets) of a business enterprise during a period from transactions and other events attributable to non-owner sources. It includes all changes in the equity during a period except those resulting from investments by owners and distributions to owners.
- Revenues are inflows or other enhancements of the assets of an entity or settlements of its liabilities from delivering or producing goods, rendering services, or other activities that constitute the entity's ongoing major or central operations.
- Expenses are outflows or other ways of using up assets or incurring liabilities from the delivery or production of goods, from rendering services or from carrying out other activities that constitute the entity's ongoing major or central operations.
- Gains are increases in the equity (i.e. net assets) from peripheral or incidental transactions of an entity and from all other transactions and other events and circumstances affecting the entity except those that result from revenues from or investments by owners.
- Losses are decreases in the equity (i.e. net assets) from the peripheral or incidental transactions of an entity and from all other transactions and other events and circumstances affecting the entity except those that result from expenses or distributions to owners.

Since earnings were defined in SFAC No. 5, they were not included in SFAC No. 6 as one of the defined elements of financial statements.

The definition of these elements intends to assure that users of financial statements will receive decision-useful information about an enterprise's resources (assets), claims to those resources (liabilities and equity), and changes therein (the other seven elements). In order to be included in the financial statements, an item must meet the recognition criteria of SFAC No. 5.

1.5 Recognition and measurement in financial statements

In SFAC No. 5 the FASB sets out various criteria for the recognition of financial statements and provided guidance regarding the types of information that should be incorporated into financial statements. SFAC No. 5 provides a basis for the criteria and the guidance by first addressing which financial statements should be presented. Financial statements contribute individually and collectively to meet the objectives of financial reporting. No one particular financial statement is likely to provide all the information that is useful for a financial decision. Some information may be better provided through notes to financial statements, supplementary information, or other means of financial reporting. However, for items that meet the criteria for recognition in financial statements, disclosure by other means is not a substitute for recognition in financial statements.

A full set of financial statements of a period should show:

- Financial position at the end of the period (i.e. balance sheet).
- Earnings for the period (i.e. income statement).
- Comprehensive income for the period (i.e. income statement).
- Cash flows during the period (i.e. cash flow statement).
- Investments by and distributions to owners during the period.

A statement of financial position (i.e. balance sheet) provides information about an entity's assets, liabilities and equity, and their relationship to each other at a given moment in time. The statement delineates the entity's resource structure and its financing structure. A statement of financial position does not purport to show the value of a business enterprise, but, together with other financial statements and other information, should provide information that is useful to those who desire to make their own estimate of the enterprise's value.

Earnings is a measure of enterprise performance during a period and is similar to net income in present practice. It measures the extent to which asset inflows (revenues and gains) associated with cash-to-cash cycles substantially completed during the period exceed asset outflows (expenses and losses) associated with the same cycles.

Comprehensive income is a broad measure of the effects of transactions and other events on an enterprise, comprising all recognized changes in equity of the enterprise during the period from all events and transactions other than those with owners. The statement of earnings and comprehensive income together reflect the extent to which and the ways in which the equity of an enterprise increased or decreased from all sources other than transactions with owners during the period.

A statement of cash flows directly or indirectly reflects an entity's cash receipts classified by major sources and its cash payments classified by major uses during a period, including cash flow information about its operating, financing and investing activities.

A statement of investments by and distributions to owners reflects an entity's capital transactions during a period, and reveals the extent to which, and in what ways, the equity of the entity increased or decreased from transactions with owners in their capacity as owners.

SFAC No. 5 also defines selected purposes and limitations of financial statements as:

- General purpose financial statements, which are directed toward the common interests of various users and are feasible only because groups of users of financial information have similar needs.
- Usefulness of financial statements. Each financial statement provides a different kind of information and, generally, various kinds of information cannot be combined into a smaller number of financial statements without unduly complicating the information.
- Classification and aggregation. Classification in financial statements facilitates analysis by grouping items with essentially similar characteristics and separating items with essentially different characteristics. Financial statements result from processing large amounts of data and involve the need to simplify, condense and aggregate information.
- Articulation of financial statements. Financial statements interrelate, or articulate, because they reflect different aspects of same transactions or other events affecting the entity.

SFAC No. 5, paragraph 63 states that an item must meet four recognition criteria in order to be recognized in the financial statements and should be so recognized whenever the criteria are met, subject to the cost–benefit constraint and the materiality threshold. These criteria are:

- Definition: the item meets the definition criteria of SFAC No. 6 for elements of financial statements .
- Measurability: the item has an attribute that is measurable with sufficient reliability.
- Relevance: the information about the item is capable of making a difference to user decisions.
- Reliability: the information about the item is representationally faithful, verifiable and neutral.

Guidance in applying these recognition criteria is described in VI.2.

2 Basic accounting concepts and principles

Specific measurement and reporting practices followed in the preparation of financial statements are based on several underlying concepts which are often referred to as 'principles'. The following major principles support current accounting practices, whereas a consistent enumeration and sequence of these principles is not provided by the authoritative accounting pronouncements.

2.1 Accrual accounting

Items that qualify under the definitions of elements of financial statements of SFAC No. 6 and that meet criteria for recognition and measurement of SFAC No. 5 are accounted for and included in financial statements by the use of accrual accounting procedures. Accrual accounting attempts to record the financial effects on an entity of transactions and other events and circumstances that have cash consequences for the entity in the periods in which those transactions, events and circumstances occur, rather than only in the periods in which cash is received or paid by the entity.

2.2 Revenue recognition principle

The revenue recognition principle provides that revenue is recognized when it is realized or realizable and when it has been earned (SFAC No. 5). Sales are therefore recorded when a product or service is delivered to the customer, not when cash is received. In accordance with the revenue recognition principle:

- Revenue from selling products is recognized at the date of sale, usually interpreted to mean the date of delivery to customers.
- Revenue from services rendered is recognized when the services have been performed and are billable.
- Revenue from permitting others to use enterprise assets, such as interest, rent and royalties, is recognized as time passes or as the assets are used.
- Revenue from disposing of assets other than products is recognized at the date of sale.

2.3 Matching principle

In recording all transactions under accrual accounting the matching principle is applied. It was put forth in SFAC No. 5 as the principal method by which expenses are to be recognized. The matching principle requires that all expenses incurred in the generating of revenue should be recognized in the same accounting period as the revenues are recognized. Matching is one of the basic processes of income determination; essentially, it is a process of determining the relationships between costs and specific revenues or specific accounting periods (APB Opinion No. 11).

There are inconsistencies in the current application of the matching principle under GAAP. For example, it can be argued that proper matching would require research and development and similar costs to be capitalized over their useful life. However, research and development costs must be expensed as incurred (see VI.5.5).

2.4 Periodic allocation principle

Some expenses cannot readily be matched to the revenues which they produce and must therefore be allocated among reporting periods in some other manner. The most important of these expenditures are governed by the periodic allocation principle, in which costs are distributed over the useful life of an asset in some rational and objective manner. Periodic allocation is used for estimating the depreciation of fixed assets and the amortization of intangible assets. Allocation is necessary because the period in which the asset is acquired and therefore recognized in the financial statements does not coincide with the use of the asset for revenue generation. It is not feasible to measure the exact contribution of a piece of equipment to the revenue generated in any particular accounting period during the asset's useful life. Instead, the cost for each period is estimated in advance, and the capitalized cost is then transferred to expense in accordance with that schedule. This is a well established principle of financial accounting; however, various exceptions exist. The most notable is the required expensing of research and development expenditures (see SFAS No. 2, Accounting for Research and Development Costs). Arguments also continue over the expensing of intangible costs such as goodwill (see APB Opinion No. 17, Intangible Assets).

2.5 Going concern

Most accounting methods proceed on the assumption of a continuance of the business enterprise. This going concern assumption has a strong impact on the valuation of assets, since it provides a strong support for the use of historical cost valuation. Only where liquidation appears imminent is the assumption inapplicable, and the liquidating values should be used to value assets. Under a liquidation situation, asset values are better stated at net realizable value (i.e. sales price less costs of disposal) than at acquisition cost. In these cases a total revaluation of the assets and liabilities can provide information that closely approximates the net realizable value of the entity.

2.6 Conservatism

Conservatism (also prudence) ensures that uncertainty and risks inherent in business situations are adequately considered. Thus, if two estimates of amounts to be paid in the future are equally likely, conservatism dictates using the less favourable estimate. However, if two amounts are not equally likely, conservatism does not necessarily dictate using the more pessimistic amount rather than the more likely one. The rule of conservatism prohibits the recognition of unrealized appreciation in value while at the same time requires reductions in value to be recognized. This has resulted in the 'lower of cost or market' rule (see SFAS No. 5, Accounting for Contingencies; SFAS No. 114, Accounting by Creditors for Impairment of a Loan; SFAS No. 115, Accounting for Certain Investments in Debt and Equity Securities).

2.7 Consistency

Different financial statement outcomes may be created under alternative accounting procedures without necessarily violating GAAP. This lack of precision can be a problem for users of financial statements. For that reason, GAAP generally require businesses to adopt accounting policies that are applied consistently from one period to the next in order to allow comparability. The principle of consistency requires a company to use the same set of accounting procedures from period to period as far as is possible and to disclose clearly the impact of any changes that are made. Consistent use of recognition and valuation methods from one accounting period to another enhances the utility of financial statements to users by facilitating analysis and understanding of comparative accounting data.

The purpose of the required disclosures is for the users of financial statements to be informed as to whether changes in the financial position and results of operations from one period to another are the result of actual transactions or of changes in accounting practices. Thus the consistency concept refers primarily to accounting methods. SFAC No. 2 does not indicate whether the consistency concept also applies to the parameters of a method.

The consistency concept does not necessarily mean that accounting practices will be uniform among affiliated business units. One unit may value inventory on the LIFO basis, whereas another may use the FIFO basis. Similarly, one unit may follow the LIFO inventory method for certain inventory items and the FIFO method for others. Such divergence requires disclosure in the financial statements but it is not a violation of the concept of consistency.

The concept of consistency does not preclude desirable changes in accounting practices, but when a significant change is made, that fact is required to be disclosed, including the dollar effect upon the financial statements and particularly the effect on net income for the period. The certified public accountant is required to report in the opinion paragraph that the statements present fairly the financial position and results of operation, in conformity with GAAP applied on a basis consistent with those of the preceding year. In case of a change in accounting principles, the report is required to be appropriately qualified.

2.8 Substance over form

One principle that is directly linked to the primary objective of financial reporting to provide information that is useful in making business and economic decisions is the substance over form principle. According to this principle, the substance rather than the form of financial information is most important in preparing the financial statements.

In addition, this concept requires to account for the economic substance of events rather than for the legal form (economic substance over form). For example, it is deemed necessary to capitalize assets obtained on finance leases as though they had been bought (SFAS No. 13).

2.9 Accounting changes

APB Opinion No. 20, Accounting Changes, provides guidance when changes in accounting procedures are made. APB Opinion No. 20 identifies three general types of accounting changes: changes in principle, changes in estimate and changes in entity. Even though the correction of an error in previously issued financial statements is not considered an accounting change, it is discussed in the Opinion and will therefore be covered in this section.

For the reporting of accounting changes and the type of change for which each should be used, APB Opinion No. 20 generally allows the retroactive, current and prospective method.

Retroactive treatment requires an adjustment to all current and prior period financial statements for the effect of the accounting change. Prior period financial statements presented currently are to be restated on a basis consistent with the newly adopted principle. Current treatment requires reporting the cumulative effect of the accounting change in the current year's income statement as a special item. Prior period financial statements are not restated. Prospective treatment of accounting changes requires no restatement of prior financial statements and no computing or reporting of the accounting change's cumulative effect in the current period's income statement. Only current and/or future periods' financial report data will reflect the accounting change.

Each of the types of accounting changes and the proper treatment prescribed for them is discussed in detail in the following sections. For more detail on disclosures see V.3.2.3 and IX.1.2.1.

2.9.1 Changes in principle

According to APB Opinion No. 20 the term 'accounting principle' includes the accounting principles and practices used as well as the methods of applying them. SFAS Interpretation No. 1 sets a base for the application of this statement by ruling that a change in the components used to cost a firm's inventory is a change in accounting principle. The Interpretation also stated that the preferability assessment must be made from the perspective of financial reporting basis and not from the income tax perspective.

The Opinion indicates that changes in accounting principles should be accounted for currently using the cumulative effect method with the cumulative net of tax effect reported separately in the income statement after the heading 'Net income' (see also V.3.2.3). The cumulative net of tax effect is the difference between the opening balance in retained earnings and the balance that would have resulted had the changed principle been adopted in all prior periods. Comparative statements from prior periods are to be presented as previously reported when they are included for comparative purposes. The impact on both net income and earnings per share should be disclosed. Although the cumulative effect treatment is deemed to be appropriate for most changes of accounting principle, in some cases retroactive treatment is required. There are five instances which require retroactive treatment:

- Change from LIFO to another inventory method (APB Opinion No. 20, paragraph 27).
- A change in the method of accounting for long-term contracts (APB Opinion No. 20, paragraph 27).
- A change to or from the full cost method of accounting for exploration costs in the extractive industries.
- Any change made by a company first issuing financial statements for the purpose of obtaining additional equity capital, effecting a business combination, or registering securities (APB Opinion No. 20, paragraph 27).
- Any changes mandated by authoritative pronouncements (while this case is not specified by APB Opinion No. 20, the promulgating bodies have in most instances thus far required that the change be made retroactively). FASB Interpretation 20 has also indicated that AICPA SOPs may mandate the treatment given the change.

2.9.2 Changes in estimates

Financial accounting often relies on estimates. Due to new events, changing conditions or additional information, an estimate may prove to be wrong and it should be corrected. This correction is classified as a change in estimate, which is to be accounted for prospectively, without adjustment of previously issued financial statements. For example, if the remaining useful life of a fixed asset is now believed to be five years rather than seven years, the undepreciated balance would be expensed over the next five years. Depreciation expense for prior years would not be restated.

2.9.3 Changes in reporting entity

An accounting change which results in financial statements that are, in effect, the statements of a different reporting entity, should be reported by restating the financial statements of all prior periods presented in order to show financial information for the new reporting entity for all periods. The most important examples of a change in the accounting entity are:

- Consolidated or combined statements in place of individual statements.
- Change in group of subsidiaries for which consolidated statements are prepared.
- Business combination accounted for as a pooling of interests.

Information for non-comparable entities should be restated to a comparable basis.

2.9.4 Error correction

Besides APB Opinion No. 20, SFAS No. 16, Prior Period Adjustments, is promulgated GAAP regarding the accounting for error corrections. Whereas APB Opinion No. 20 identifies examples of some errors and indicates that they are to be treated as prior period adjustments, SFAS No. 16 reiterates this treatment for accounting errors and provides guidance for the disclosing of prior period adjustments.

APB Opinion No. 20, paragraph 13 identifies examples of errors as resulting from mathematical mistakes, mistakes in the application of accounting principles, or the oversight or misuse of facts known to the accountant at the time the financial statements were prepared. It also states that the change from an unacceptable (or incorrect) accounting principle to a correct principle is considered a correction of an error, not a change in accounting principle. According to APB Opinion No. 20, paragraph 37, the disclosure of the correction of an error should include the nature of the error and the effect of its correction on income before extraordinary items, net income, and the related per share amounts in the period in which the error was discovered and corrected. The major criterion for determining whether or not to report the correction of the error is the materiality of the correction.

3 Interrelationships between financial reporting and tax accounting

In determining the taxable income, an enterprise has to follow the requirements and regulations of the Internal Revenue Service. In principal, tax law differs from the recognition and measurement requirements of financial reporting standards. This segregation of accounting for tax purposes and financial accounting reflects the different objectives and general principles of both accounting systems. Despite this general segregation, both methods of income determination influence each other. This interrelation can be derived from § 446 IRC, which requires that taxable income should be computed under the method of accounting on the basis of which the taxpayer regularly computes his income in keeping his books unless the method used does not clearly reflect income. This implies that a transaction should be treated identically for tax and financial reporting purposes, if it clearly reflects income. Deviations from this conformity are based on specific tax requirements with different tax implications for items such as allowances, provisions and goodwill amortization.

Areas in which tax regulations have a direct impact on GAAP are only identifiable for the valuation of inventories. In determining costs for inventories, the LIFO cost flow assumption is only allowed for the computation of taxable income if it is also applied for the financial statements (§ 472 c IRC). An indirect influence of the tax law on financial accounting results from the fact that small companies try to achieve conformity between the two accounting systems in order to minimize their costs for the preparation of financial and taxation statements. Therefore those small and medium-sized firms, which are not required to have an annual audit, usually base their accounting on the common tax law.

IV Bookkeeping and preparation of financial statements

1 Duty to prepare financial statements – overview

Pursuant to the Securities Act of 1934, a company with publicly traded securities in the United States is required to issue periodic financial reports to its shareholders, which include audited financial statements, and to file the financial reports with the SEC. Public companies file an annual form 10-K (including audited financial statements), with detailed information concerning the business, its properties, executive compensation, legal proceedings, and other detail that is of interest to investors. In addition to the annual form 10-K, public companies are required to file quarterly form 10-Qs (including unaudited quarterly condensed financial statements) and form 8-Ks for such events as changes in control of registrant, significant acquisitions or dispositions of assets, bankruptcy, or a change in auditors, etc. Although the basic role of the SEC is to ensure disclosure and to oversee the standard setting process, it has the legal authority to require compliance with GAAP. Regulation S-X specifies the form and content of the financial statements that must be filed with the SEC and also requires that GAAP be followed in preparing the financial statements. Regulation S-X is codified in federal law under title 17, section II, part 210 of the Code of Federal Regulations. Part 210 sets forth the form and content of, and requirements for, financial statements. The term 'financial statements' is deemed to include all notes to the statements and all related schedules. Since regulation S-X has the force of federal law, any company contemplating the issuance of securities publicly in the United States should be familiar with the provisions of regulation S-X.

Companies are subject to the reporting requirements of the 1934 Act under the following circumstances:

- Companies whose securities are listed on a national securities exchange (e.g. the New York Stock Exchange, the American Stock Exchange, or NASDAQ) must register these securities under the 1934 Act, thereby becoming subject to the law's reporting requirements.
- An unlisted company with total assets above a specified minimum (currently USD10 million) or a class of equity securities held by more than a specified minimum number of persons (currently 500) at any fiscal year end must register under the 1934 Act. If the number of equity shareholders falls below the minimum threshold a company may cease reporting if it notifies the SEC.
- Even if a company is otherwise exempt from filing under the 1934 Act it must periodically file reports with the SEC under the 1934 Act, at least for the balance of the year in which a securities registration becomes effective under the Securities Act of 1933. The company's obligation to file with the SEC will continue until such time as the number of holders of the securities registered under the 1933 Act falls below the minimum threshold at the beginning of a fiscal year and it notifies the SEC.

Companies subject to reporting under the 1934 Act must file financial reports with the SEC annually, quarterly, and if a specified reportable event occurs. As discussed above, these reports are filed on forms 10-K, 10-Q, and 8-K, respectively. Most public companies with 1934 Act reporting responsibilities must also conform to the proxy rules when soliciting proxies from their shareholders. If the proxy is being solicited for an annual meeting at which directors are to be elected, an annual report must be furnished to the company's shareholders. Under the current Integrated Disclosure System (IDS) of the SEC, the financial statements contained in the company's form 10-K report to the SEC are essentially the same as the financial statements that are issued to the company's shareholders. The following sections discuss the form and content of each of these filings.

2 Reporting to the SEC

2.1 General instructions for preparing SEC reports

All registrants, with certain exceptions, must file financial statement registration statements and reports using an electronic filing system known as EDGAR (Electronic Data Gathering, Analysis, and Retrieval). Foreign private issuers and foreign governments are exempt from electronic filing requirements unless they file a document jointly with, or as a third-party filer with respect to, a registrant that is subject to the electronic filing requirements.

Regulation S-T provides general rules and regulations for electronic filings. Electronic filings shall be prepared in the manner prescribed by the EDGAR Filer Manual, promulgated by the SEC. It sets out the technical formatting requirements for electronic submissions, such as the maximum number of characters per line, tabular and columnar information. The provisions relating to type, font and its size, paper size and other legibility requirements do not apply to electronically formatted documents. Provisions requiring presentation of information in bold face type shall be satisfied in an electronic format document by presenting such information in capital letters.

Exhibits to an electronic filing that have not previously been filed with the SEC shall be filed in electronic format, absent a hardship exemption. Statements and reports shall be in English. If any exhibit or other paper or document filed with a statement or report is in a foreign language, it shall be accompanied by a translation into English.

Signatures to or within any electronic submission shall be in typed form rather than manual format; however, manual signatures must be obtained before the electronic transmission is made and copies must be maintained by the filer for a period of five years.

For registrants not subject to the EDGAR filing requirements, statements and reports shall be filed on paper.

Statements and reports shall be in English. If any exhibit or other paper or document filed with a statement or report is in a foreign language, it shall be accompanied by a translation into English.

2.2 Form 10-K (annual report)

2.2.1 Content and filing requirements

Form 10-K is the principal annual financial report form used by commercial and industrial companies that are required to file reports with the SEC under the Securities Exchange Act of 1934. In the interests of providing continuing disclosure of material facts concerning the company, the report must include both financial and non-financial information about the company's activities during the year. The financial statements in the form 10-K are usually incorporated by reference from the annual report to shareholders.

Form 10-K must be filed within 90 days of the company's fiscal year end. If a company is unable to file a periodic report when due, it is required to file with the SEC, no later than one day after the due date, a notification that identifies the late report and the reasons for not filing on time. Any report filed with missing information must disclose on the cover the portions omitted. There is also a procedure for a 15-day automatic extension for hardship situations.

The electronically filed form 10-K must include 'electronic' signatures of the company's principal executive officer or officers, its principal financial officer, its controller or principal accounting officer, and by at least a majority of the board of directors. Electronic signatures must be supported by a manually signed document that the company must retain on file for five years.

Form 10-K is not a blank form to be filled in. Therefore the list of required contents is intended as a guide in the preparation of the report. However, the cover page must be produced in accordance with the format specified. The 10-K report comprises both financial and non-financial information. In addition to the requirements set forth in the form itself, a preparer of the report should be aware of the following pronouncements of the SEC, which provide guidance in the preparation of both the financial and the textual portions of form 10-K (see I.4):

- General Rules and Regulations under the Securities Exchange Act of 1934.
- Guides for Preparation and Filing of Reports.
- Regulation S-K.
- Regulation S-X.
- Financial Reporting Releases.
- Staff Accounting Bulletins.

Because the securities regulations are complex, a company that must file a form 10-K should understand that the preparation of the form requires the co-operation of corporate officers, the registrant's attorneys, and independent accountants.

2.2.2 Discussion of items in form 10-K

The following is a discussion of items to be included in the form 10-K that may have relevance to financial matters (see also Table I.6). Certain items are not discussed because they are deemed to have little relevance to financial reporting.

Item 1: Business. The information for this item is covered in item 101 of regulation S-K (see also IV.4.1). General development of part (a) requires a description of the business of the registrant during the previous five years. In describing developments, information should be given as to matters such as the form of organization, bankruptcy, receivership or similar proceedings, merger, consolidation of significant subsidiaries, acquisition or disposition of any material amount of assets, and material changes in the mode of conducting the business. Further requirements apply to the registration statement on Form S-I.

A narrative description of the business by segment is also necessary to provide an understanding of the business. This narrative should contain, *inter alia*:

- The principal products produced and services rendered.
- Sources and availability of raw materials.
- Duration and effect of patents, trademarks, franchises etc.
- The extent to which the business is seasonal.
- Practices relating to working capital items.
- Dependency on a single or few customers.
- The dollar amount of background orders believed to be firm.
- Competitive conditions such as markets, competitors etc.
- Amount spent on research and development.
- Material effects of provisions regulating the discharge of materials into the environment.
- Number of persons employed.

The accountant is concerned principally with part (b) of this item. This calls for financial information about industry segments for each of the previous three fiscal years. In January 1999, the SEC adopted technical amendments to its segment rules, which are required to be disclosed in addition to the disclosure in the basic financial statements required under SFAS No. 131, Disclosures about Segments of an Enterprise and Related Information (see discussion of SFAS No. 131 at IX.1.2.3). The technical amendments were made to ensure that the disclosure requirements conformed generally with the provisions of SFAS No. 131. However, some differences still exist. For example, SFAS No. 131 requires disclosure of revenues from external customers for each product and service, or each group of similar products, unless it is impracticable to do so. On the other hand, item 101 requires the disclosure of product sales within segments for any products (or services) that account for 10 per cent or more of consolidated revenue in any of the previous three years, or 15 per cent or more of consolidated revenue if total revenue did not exceed USD50 million during the previous three fiscal years. In addition, SFAS No. 131 and item 101 both require disclosures about major customers. Item 101 requires registrants to identify these customer by name, while SFAS No. 131 does not.

Item 2: Properties. Requires a list of the principal properties of the company. This is relevant to the preparation of financial statements only as it relates to conformity of the properties described to their accounting in the financial statements.

Item 3: Legal Proceedings. Requires a discussion by management of significant legal proceedings of the company and is deemed to have little or no relevance to the preparation of financial statements unless such proceedings are required to be disclosed in the financial statement footnotes in accordance with GAAP.

Item 4: Submission of Matters to a Vote of Security Holders. Requires a list of securities ownership by larger owners. Not deemed to have relevance to the preparation of financial statements.

Item 5: Market for Registrant's Common Equity and Related Stockholder Matters. Requires disclosure of the stock market listings and the market prices of the company's stock during specified periods of time. Not deemed to have relevance to the preparation of financial statements.

Item 6: Selected Financial Data. Item 301 of regulation S-K calls for a comparative summary of selected consolidated financial data for each of the last five fiscal years. If applicable, the following items are required in the summary:

- Net sales or operating revenues.
- Income/loss from continuing operations.
- Income/loss from continuing operations per common share.
- Total assets.

- Total long-term obligations and redeemable preferred stock (including long-term debt, capital leases, and redeemable preferred stock as defined in ASR 268).
- Cash dividends declared per common share.

Item 7: Management's Discussion and Analysis of Financial Condition and Results of Operations. The requirements for this item are contained in item 303 of regulation S-K. The discussion must, at the minimum, provide information on the following areas:

- Liquidity. Identification of any known trends or other factors likely materially to increase or decrease liquidity. Identification and description of internal and external sources of liquidity such as expected sales of assets or lines of credit.
- Capital resources. Description of material commitments for capital expenditures as of the latest year end, the general purpose of such commitments, and the anticipated source of funds required.
- Results of operations. Description of any unusual or infrequent events, transactions, or economic changes that materially affected income from continuing operations. In addition, information is required regarding any known trends or uncertainties that have had or will have a material effect on sales or income from continuing operations; the impact of inflation and changing prices on sales and on income from continuing operations; the extent to which material increases in sales are attributable to price increases, volume increases, or new products; and any other significant changes in sales, costs, or income from continuing operations. A discussion of the impact of inflation and changing prices is required.

The management's discussion must cover the last three fiscal years on a comparative basis. The instructions for item 11 of regulation S-K contain guidance for preparing the management's discussion.

Item 7a: Quantitative and Qualitative Disclosures About Market Risk. The term market risk refers to the risk of loss arising from adverse changes in: interest rates; foreign currency exchange rates; commodity prices; and other relevant market risks, such as equity price risk. The requirements for this item are contained in item 305 of regulation S-K. The primary objective of item 305 is to provide investors with forward-looking quantitative and qualitative disclosures about a registrant's potential exposures to market risks. The required disclosures are not meant to be precise indicators of expected future losses but, rather, indicators of reasonably possible losses.

Quantitative disclosures are to be presented separately for market risk sensitive instruments that are entered into for trading purposes and those that are entered into for purposes other than trading. Within each of these categories, separate quantitative information is to be presented for each material market risk exposure. The quantitative disclosures are to be presented using one of three disclosure alternatives: tabular presentation, sensitivity analysis, or value at risk. Tabular presentation shall include fair values of the market risk sensitive instruments and contract terms sufficient to determine future cash flows from those instruments, categorized by expected maturity dates. Within each risk exposure category, the market risk sensitive instruments shall be grouped based on common characteristics. Sensitivity analysis shall provide disclosures that express the potential loss in future earnings, fair values or cash flows of market risk sensitive instruments resulting from one or more selected hypothetical changes in interest rates, foreign currency exchange rates, commodity prices or other relevant market rates or prices over a selected period of time. Registrants shall provide a description of the model, assumptions and parameters, which are necessary to understand the disclosures. Value at risk disclosures express the potential loss in future earnings, fair values or cash flows of market risk sensitive instruments over a selected period of time, with a selected likelihood of occurrence, from changes in interest rates, foreign currency exchange rates, commodity prices or other relevant market rates or prices. Registrants shall provide a description of the model, assumptions and parameters, which are necessary to understand the disclosures.

In addition to the above, registrants must provide qualitative disclosures. Such disclosures are necessary to help place the registrant's market risk management activities in context and, therefore, are useful in understanding the quantitative disclosures about market risk. Item 305 requires narrative qualitative disclosure about market risk of:

- The registrant's primary market risk exposures as of the end of latest fiscal year.
- How those exposures are managed.
- Changes in either the primary market risk exposures or how those exposures are managed, when compared to what was in effect during the most recently completed fiscal year and what is known or expected to be in effect in future reporting periods.

Item 8: Financial Statements and Supplementary Data. The consolidated financial statements for a company and its consolidated subsidiaries that are included in its annual report to shareholders may be incorporated by reference into the form 10-K. Since this ability to incorporate by reference substantially reduces the paperwork required of companies, most companies prepare only one set of financial statements for inclusion both in annual reports to shareholders and in the annual report to the SEC. Proxy rules require the following financial statements for the shareholders' annual report (and thus the 10-K):

- Audited consolidated balance sheets as of the end of the two most recent fiscal years.
- Audited consolidated statements of income and consolidated statements of cash flows for each of the three most recent fiscal years.
- Audited consolidated statements of changes in shareholders' equity for each of the three most recent fiscal years.

The financial statements must be prepared in accordance with regulation S-X and GAAP. If required by regulation S-X separate financial statements of the parent, unconsolidated subsidiaries or other significant equity method investees and any other special financial statements, may be filed as 'financial statement schedules' under item 11 of the form 10-K. Item 8 also requires supplementary financial information as specified by item 302 of regulation S-K. This item requires selected quarterly financial data for each of the two most recent fiscal years, as follows:

- Net sales.
- Gross profit.
- Income/loss before extraordinary items and cumulative effect of a change in accounting.
- Income/loss per share (based on the previous amount).
- Net income/loss.

The quarterly information is required only from companies meeting certain tests. Basically, companies will meet the tests if:

- They are registered on a national securities exchange or their shares are quoted on the National Association of Securities Dealers Automated Quotation System (NASDAQ).
- They have had net income, after taxes but before extraordinary items and the cumulative effect of a change in accounting, of greater than a specified amount (currently USD10 million) for each of the previous three fiscal years.
- They had total assets of at least USD200 million at the last fiscal year end.

There are also tests relating to number of shareholders, number of shares held by unrelated shareholders, and the market value of the outstanding shares. If the quarterly data presented vary from the amounts previously reported on form 10-Q for any quarter, the difference must be reconciled and explained. Any unusual transaction that occurred in any quarter should also be described.

Item 9: Changes in and Disagreements with Accountants on Accounting and Financial Disclosure. The required information is described in item 304 of regulation S-K. Item 304 requires companies to disclose the following: whether the former accountants resigned, declined to stand for re-election, or were dismissed; whether the principal accountants' report on the financial statements for either of the past two years contained an adverse opinion, a disclaimer of opinion, or was qualified or modified as to uncertainty, audit scope or accounting principals and what the nature of such opinions were; whether the decision to change auditors was approved by the board of directors, or an audit or similar committee of the board of directors and; whether there were any disagreements with the former accountants during the last two fiscal years and any subsequent interim period preceding the change on any matter of accounting principles or practices, financial statement disclosure, or auditing scope or procedures, which would have caused the accountants to make reference to the subject matter of the disagreement in the audit report if the disagreement had not been resolved to the accountants' satisfaction.

Auditor changes are reported on form 8-K if the principal accountants resign, decline to stand for reelection, or are dismissed, as well as if new auditors are engaged. The form 8-K must be filed within five business days of the change in auditors.

Item 10: Directors and Executive Officers of the Registrant. This item requires the disclosure of the names and ages of the directors and officers of the company. This is not deemed to have relevance to the preparation of financial statements.

Item 11: Executive Compensation. The required information is described in item 402 of regulation S-K, which calls for listing of the remuneration of each of the five most highly compensated executive officers or directors, as well as all officers and directors as a group. The remuneration data must be shown in a table with the following headings:

- Name of individual or number of persons in group.
- Capacities in which served.
- Salaries, fees, directors' fees, commissions, and bonuses.
- Securities or property, insurance benefits or reimbursement, personal benefits.
- Aggregate of contingent forms of remuneration (including pensions and similar contingent benefits).

Remuneration coming from subsidiaries of the reporting company would be included in the above described disclosures.

Item 12: Security Ownership of Certain Beneficial Owners and Management. The required information is described in item 403 of regulation S-K, it requires disclosure of, as of the most recent practicable date, of persons or 'group' who are known to the registrant to be the beneficial owner of more than 5 per cent of any class of the registrant's voting securities. Similar information is required to be provided for the number of shares of each class of registrant's equity securities or any of its parents or subsidiaries owned beneficially by each director and nominees for directorship, naming them, and by each named executive officer (as defined in item 402 of regulation S-K). In addition, disclosure is required for all directors and executive officers as a group. Item 403 provides a specified format for disclosing the information. Also, item 403 requires disclosures regarding any arrangements, known to the registrant, that may at a future date result in a change in control. Pledges of the registrant's securities or its parents are one of the common examples of such arrangements.

Item 13: Certain Relationships and Related Transactions. The required information is described in item 404 of regulation S-K, the required disclosures can be divided into four categories:

- Transactions with management and others: disclosures are required if any of the registrant's directors, officers, nominees for directors, shareholders owning more than 5 per cent of voting stock and their spouses and members of their immediate family have direct or indirect material interest in transactions or series of transactions involving the registrant or any of its subsidiaries, in which the amount involved is USD60,000 or more. The disclosure is required for the transactions entered between the beginning of the registrant's latest fiscal year to a date as current as possible (thus the period covered could be more than 12 months).
- Certain business relationships: item 404(b) requires disclosures for certain relationships involving directors or nominees for directors. No information need to be given with respect to any director who is no longer a director at the time of filing Form 10-K. Item 404(b) describes six situations for which disclosures are required. One of these is when a director or nominee of the registrant is an executive officer or major shareholder (having ownership of 10 per cent or more) of an entity to which the registrant or its subsidiaries made payments for property or services during the registrant's last full fiscal year, or proposes to make during the registrant's current fiscal

year in excess of 5 per cent of (i) the registrant's consolidated gross revenues for its last full fiscal year; or (ii) the other entity's consolidated gross revenues for its last full fiscal year.
- Indebtedness of management: no information is required to be disclosed for amounts due for ordinary travel and expense advances and for transactions entered in the ordinary course of business. Disclosures are required for any indebtedness to the registrant or its subsidiaries at any time since the beginning of the registrant's last fiscal year for amounts in excess of USD60,000. Disclosures should include the name of the person and relationship to the registrant, nature of indebtedness, transaction that resulted in indebtedness, largest aggregate amount outstanding during the period, amount outstanding at the latest practicable date and rate of interest. Such information is required for indebtedness by any director, any nominee for election as a director, any executive officer, any organization in which a director, nominee, or executive officer owns 10 per cent of any class of equity securities or any trust or estate in which a director, nominee, or executive officer has a substantial beneficial interest or serves as a trustee.
- Transactions with promoters (not required in a Form 10-K).

Item 14: Exhibits, Financial Statement Schedules, and Reports on Form 8-K. This item requires a listing of all supplementary financial statements, schedules and exhibits that are filed as part of the 10-K report. Certain supplementary financial statements, such as those of the parent company alone, or those of non-consolidated subsidiaries that are not included in the shareholders' annual report, are included in this item.

2.2.3 Financial statement schedules

Article 12 of regulation S-X prescribes the form and content of financial statement schedules applicable to all other rules of regulation S-X. The financial statement schedules to be used by commercial companies is contained in rule 5-04 of regulation S-X. Other specialized industry rules (Rules 6 to 9) prescribe the schedules for those industries. The financial statement schedules required for commercial and industrial companies are:

- Schedule I: condensed financial information of the registrant. This schedule is required when the restricted net assets of consolidated subsidiaries exceed 25 per cent of consolidated net assets as of the end of the most recently completed fiscal year.
- Schedule II: valuation and qualifying accounts. This schedule requires details for other valuation accounts, such as the allowance for bad debts.
- Schedule III: real estate and accumulated depreciation. Companies in the real estate business must file this schedule.
- Schedule IV: mortgage loans on real estate. This schedule is also required for real estate companies.
- Schedule V: supplemental information concerning property-casualty insurance operations. This schedule is to be filed when the company, its subsidiaries or 50 per cent or less owned equity-based investees, have liabilities for property-casualty insurance claims.

Schedules I and III are filed as of the date of the latest audited balance sheet. Schedule II shall be filed for the periods for which an audited income statement is required. Schedules I and IV shall be filed as of the date and for the periods specified in the schedule. Schedule V shall be presented as of the same dates and the same periods for which the information is reflected in the audited financial statements.

The schedules listed above are required to be filed in support of each set of financial statements filed. Accordingly, if separate parent company financial statements are filed, then applicable schedules are to be filed in support of both consolidated and parent company financial statements. If the financial statements included in the filing are audited, then the schedules must also be audited.

2.3 Form 10-Q (quarterly report)

2.3.1 Filing requirements

Form 10-Q is to be used by companies required to file quarterly reports with the SEC under the

Securities Exchange Act of 1934. Since the initial adoption of the form in 1970 the information required to be included has been expanded. Accordingly, the report has grown from containing only highly condensed information to a comprehensive multi-page report about matters affecting the registrant occurring during the quarter reported upon.

The 10-Q must be filed within 45 days after the end of each of the first three fiscal quarters of each fiscal year; no report need be filed for the fourth quarter. Consistent with the requirements for other filings under the 1934 Act, form 10-Q must be filed electronically using the EDGAR requirements as set forth in regulation S-T and discussed in greater detail above.

The electronically filed form 10-Q must include 'electronic' signatures of a duly authorized officer of the registrant and the principal financial officer or chief accounting officer of the registrant (e.g. the treasurer and controller, respectively). Electronic signatures must be supported by a manually signed document that the company must retain on file for five years. Similar to other reports filed with the SEC, the 10-Q is not a blank to be filled in, but is for use as a guide in the preparation of the report. However, the cover page should be reproduced in accordance with the format specified in the form. As is the case in most SEC filings, the report comprises both financial and non-financial information (text analysis of the quarterly statements, footnotes, management's analysis and exhibits).

2.3.2 Financial statements to be included

The financial statements may be unaudited. The following financial statements for the indicated periods are to be included:

- Income statements for the current quarter and year-to-date for the current year and corresponding periods of the preceding year. These statements may be for the corresponding periods of the cumulative 12-month period ended during the most recent quarter and for the corresponding period of the preceding year in lieu of the year-to-date statements.
- Balance sheets as of the end of the current quarter and as of the end of the most recently completed fiscal year.
- Statements of cash flows for the current year-to-date and for corresponding period of the preceding year.

By way of illustration, and assuming a calendar year company, the following financial statements would be included in the registrant's third-quarter 10-Q report:

- Income statement for the quarter ended 30 September 20X1.
- Income statement for the quarter ended 30 September 20X0.
- Balance sheet as at 30 September 20X1.
- Balance sheet as at 30 September 20X0.
- Income statement for the nine months ended 30 September 20X1.
- Income statement for the nine months ended 30 September 20X0.
- Statement of cash flows for the nine months ended 30 September 20X1.
- Statement of cash flows for the nine months ended 30 September 20X0.

2.3.3 Basis of financial statement preparation

Form 10-Q provides that although the financial statement 'presentation and related disclosures' may be unaudited and on a condensed basis, the statements should be prepared in accordance with the standards of accounting set forth in APB Opinion No. 28, Interim Financial Reporting. APB Opinion No. 28 established guidelines for the preparation of interim financial statements and the application of accounting principles to such statements; the computational guidelines in APB Opinion No. 28 should be adhered to regardless of the extent of optional disclosure condensation opted for by the company.

It is not intended that the form 10-Q financial statements should include all disclosures necessary to comply with GAAP. Accordingly, certain disclosures (e.g. pension and depreciation expense) required in financial statements intended to comply with GAAP, which were previously provided in annual financial statement footnotes, are not required for inclusion in form 10-Q financial statements.

2.3.4 Optional condensation

Although the quarterly financial statements must generally follow the form of presentation specified in regulation S-X, the financial statements may be condensed. Many of the disclosures required to comply with regulation S-X may be omitted; there are no required schedules. Balance sheets and income statements must include major headings, but subheading disclosures are not required. It is not intended that the form 10-Q financial statements should be supplemented by the customary footnotes. As mentioned above, full compliance with the disclosure requirements of regulation S-X and GAAP is not required. The 10-Q form specifies that footnotes summarizing accounting principles and practices as well as other general notes to financial statements are not required. However, disclosures must be adequate to make the information presented not misleading. A company must use judgement in determining which additional disclosures must be included. It may be assumed that normal recurring disclosures, such as those detailing the principles of consolidation and the terms of long-term debt and lease and stock option arrangements, need not be repeated in the quarterly report, since those disclosures would generally be required in the company's annual report.

2.3.5 'Updating' concept

The financial statements and other disclosures in the form 10-Q should be viewed as an updating of that information included in the annual report. Adopting this concept, those significant events modifying previously reported conditions should be considered for disclosure in form 10-Q. However, disclosures concerning litigation or other contingent obligations must be provided even if no changes have occurred.

2.3.6 Specific required disclosures

In addition to the general standards of disclosure discussed above, the instructions to the 10-Q form require certain specific disclosures.

- Business combinations. Where a business combination accounted for as a pooling-of-interests occurs during the quarter covered by the 10-Q, the financial statements for all included periods must reflect the pooling combination as if it had occurred at the beginning of the earliest period presented. Also required is the supplemental disclosure of the separate results of the combined businesses for periods prior to the combination. If the business combination occurs during any period covered by the report and is accounted for as a purchase, condensed pro forma operating results are to be included. These should reflect the results of operations as though the companies had combined at the beginning of the period. Such condensed pro forma information should include at a minimum revenue, income before extraordinary items and cumulative effect of accounting changes, income per share, and net income.
- Discontinued operations. If a significant portion of the company's business has been disposed of during any period covered by the report, disclosure must be made of the effect on revenues and net income, both in total and per share for all periods covered by the report.
- Per-share amounts. If applicable, earnings and dividends per share for each period must be shown on the face of the income statement; the basis on which such amounts are computed and the number of shares used in the computation should also be disclosed. Unless the basis of computation of per-share amounts is very simple (e.g. based on the weighted average number of shares outstanding), or is clearly set out in the footnotes, an exhibit detailing the computation must be included as part of the report.
- Accounting changes. The SEC has expanded on the disclosures relating to changes in accounting principles required by GAAP (see APB Opinion No. 20, Accounting Changes). In addition to those disclosures, the company is required to state the date of the change and the reasons for making the change. A letter from the company's independent accountants must be filed as an exhibit to the first form 10-Q filed subsequent to the date of the accounting change, indicating whether or not the change is to an alternative principle which in the judgement of the auditor is preferable under the circumstances; the accountants'

letter is not required if the change is made in response to a standard adopted by the FASB.

- Statement as to necessary adjustments. The instructions specifically require a statement to be made to the effect that 'all adjustments which are, in the opinion of management, necessary to a fair statement of the results for the interim periods' have been reflected in the preparation of the financial statements. The type of adjustment the SEC is referring to is, for example, quarterly accruals for anticipated year end adjustments (e.g. accruals for bonus and profit sharing arrangements), which should be assigned to the quarter so that the quarter bears a reasonable portion of the anticipated annual amount.
- Prior period adjustments. Material retroactive prior period adjustments are to be disclosed, including the effect on net income, total and per share, and on the balance of retained earnings.
- Management's Discussion and Analysis of Financial Condition and Results of Operations. As is the case in most SEC filings which include profit and loss information, form 10-Q contains a requirement for management's narrative analysis of that data. The format and guidelines for the preparation of the analysis is similar to that to be included in a form 10-K report. Explanations for material changes in financial condition from the end of the preceding fiscal year to the date of the most recent interim balance sheet provided are required. In addition, discussion of material changes in the results of operations are to be provided for the following periods: the most recent quarter compared to the corresponding quarter in the preceding year, and the current year to date compared to the same period in the preceding year.
- Quantitative and qualitative disclosures about market risk. Discussion and analysis shall be provided in interim financial statements so as to enable the reader to assess the sources and effects of material changes in information that would be provided under item 305 of regulation S-K from the end of the preceding fiscal year to the date of the most recent interim balance sheet.
- Other information (Part 11). This part of the form includes the items listed below that call for the disclosure of textual information. Items that are inapplicable or that call for a negative response may be omitted. A copy of the form should be referred to for detailed instructions relating to the responses to the items in part 11. To the extent that responses to certain items lend themselves to legal interpretations, the company should consult its legal counsel. The following items, to the extent applicable, are required in part 11 of form 10-Q:
  - (i) Legal proceedings.
  - (ii) Changes in securities and use of proceeds.
  - (iii) Defaults upon senior securities.
  - (iv) Submission of matters to a vote of security holders.
  - (v) Other materially important events.
  - (vi) Exhibits and reports on form 8-K:
    - (a) List of exhibits.
    - (b) Reports on form 8-K. State whether any reports on form 8-K have been filed during the quarter, listing the items reported, any financial statements filed, and the dates of any such reports.

2.4 Form 8-K (current report)

The third major report required to be filed by companies subject to the periodic reporting requirements of the Securities Exchange Act of 1934 is form 8-K. Form 8-K should be considered an *ad hoc* filing to the extent that it is required to be filed only if one of the eight events identified in the form occurs (see the listing of events below). Actually, only seven specific reportable events are identified in the form; item 5, 'Other materially important events,' is a loosely constructed general item calling for the reporting of events 'not otherwise called for by this form, which the registrant deems of material importance to security holders.' Since the need to file a report under such an item lends itself to a good deal of interpretation, legal counsel should be consulted for guidance as to whether an event should be reported under item 5.

Form 8-K includes the following items:

- Item 1 – Changes in control of the company.

- Item 2 – Acquisition or disposition of assets (including business combinations).
- Item 3 – Bankruptcy or receivership.
- Item 4 – Changes in the company's certifying accountant.
- Item 5 – Other materially important events.
- Item 6 – Resignation of the company's directors.
- Item 7 – Financial statements and exhibits.
- Item 8 – Change in fiscal year.
- Item 9 – Sales of securities pursuant to regulation S.

Most of the events specified in the form are required to be reported on form 8-K within 15 days of the occurrence of the event. If an event is reported under items 4 or 6, the form 8-K is due within five business days of the event. If an event is reported under item 8, the form 8-K is due within 15 calendar days of the date on which the company made the determination to use a different fiscal year. Reports under item 5 are due promptly after the occurrence of the event reported therein. As is the case with forms 10-K and 10-Q, the form 8-K is a guide, not a form to be filled in, except that all the information set forth on the model cover sheet in the form should be furnished. The form 8-K must be filed electronically using the EDGAR requirements as set forth in regulation S-T and discussed in greater detail above. The report must include the 'electronic' signature of a duly authorized officer of the company. Electronic signatures must be supported by a manually signed document that the company must retain on file for five years.

Accounting professionals generally become involved in filings of form 8-K only if a business combination or a change in the certifying accountants is reported. There are no financial statement requirements for form 8-K other than in response to item 2, 'Acquisition or disposition of assets.' If the company acquires a significant amount of assets, financial statements for the acquired business are generally required. 'Acquisition' as used in the form 8-K is a broad term and would include a merger, purchase, consolidation, exchange, succession, or possible other acquisition. An acquisition or disposal of assets is significant if:

- For asset acquisitions, the company's equity in the net book value of the assets or the amount paid or received for such assets exceeded 10 per cent of the total assets of the registrant.
- For businesses acquired, the business acquired or disposed of would be considered a significant subsidiary. Rule 1–02(w) of regulation S-X provides guidance for the determination of a significant subsidiary.
- It involved the acquisition of a business that would meet the test of a significant subsidiary.

If the acquisition involves the acquisition of a business and is determined to be significant, the following financial statements of the acquired business, specified in the form 8-K, are required:

- A balance sheet 'as of a date reasonably close to the date of acquisition.' If that balance sheet is not audited, an additional audited balance sheet as of the close of the acquired business's preceding fiscal year is required to be filed.
- Income statement and statement of cash flows for each of the last full three fiscal years and for the period, if any, between the close of the latest fiscal year and the date of the latest balance sheet filed. Fiscal year requirements may be less than three years depending upon the significance of the acquisition.

The requirements for exhibits to the form 8-K are covered in item 7 of regulation S-K.

2.5 Additional SEC requirements for foreign companies

2.5.1 Overview

A foreign company may choose to enter the US capital markets to raise funds by selling new equity securities or by issuing its securities in an exchange for the securities of a US company. This is called a public offering or an exchange offer. A foreign company may also choose to list its existing securities in the United States without selling any new securities. This is called a public listing. The relevant requirements are forms F-1, F-2, F-3, F-4, and 20-F. Once a foreign issuer is a US reporting company it uses Form 6-K subsequent developments or information. Regardless of the form used, the offering document requires the disclosure of information about the securities offering, such as risk factors, use of the offering proceeds, potential dilution to the new investors, any selling security holders, and the terms of the securities sale transaction.

2.5.2 Public offering with issue of new shares

The form of registration for a public offering frequently used by foreign issuers is the form F-1. It may also be possible to use forms F-2 or F-3. Forms F-1, F-2, F-3, and F-4 are all SEC registration statements (companies choose one) used when a foreign company offers new securities in the United States. All the series F forms have the same requirements relating to such matters as plan of distribution, use of proceeds, description of securities, risk factors and summary. In addition, form F-1 requires all the information and financial statements required in a form 20-F without any incorporation by reference. Form F-2 consists of incorporation by reference of the latest form 20-F financial statements, a copy of which must be delivered with the prospectus and a description of all material developments since the end of the period covered by the form 20-F. A form F-3 also allows incorporation by reference of the latest form 20-F and a description of all material developments since the end of the period covered by the form 20-F. However, the form 20-F is not required to be delivered with the prospectus under an F-3 filing. Form F-4 is used when securities are issued to complete a business combination rather than to raise capital.

The registration statement requires a discussion of the business, which should include a discussion of legal actions and a concise discussion of risk factors, exchange controls, and management and directors.

2.5.3 Public listing without issuing new shares

A public listing occurs when a firm wishes to list its existing shares on the New York Stock Exchange, AMEX or the NASDAQ system, without issuing new equity. The company must file a registration statement on form 20-F (see Table IV.1). The advantage of first listing is that it allows a company that does not need the capital today to establish a presence in the US marketplace and broaden the trading market of its securities.

If the company later decides to issue new equity there are short-form registration statements available (forms F-2 and F-3; see Table IV.2).

2.5.4 Form 20-F

The form most commonly used by the foreign private issuers is Form 20-F. Form 20-F can not only be used as an annual report (similar to Form 10-K) but also as a registration statement (similar to Form 10). It also serves as a transition report for foreign filers that adopt a change in fiscal year (similar to Form 10-K and 10-Q). Foreign registrants eligible to file Form 20-F have several advantages over domestic US filers, the more significant of which are:

- Foreign private issuers using a Form 20-F as an annual report may elect to file financial statements under Item 17. Item 17 does not require supplementing local country footnote disclosures with US GAAP disclosures except as required by Staff Accounting Bulletin No. 88.
- The amount of remuneration paid to individual directors and executive officers need not be disclosed unless such information is disclosed publicly in the home country. However, the total remuneration paid to all directors and executive officers as a group must be disclosed.
- Information on transactions between the registrant and its management is not necessary unless such information is disclosed to its shareholders or made public as required by foreign laws.
- Discussions of the prior business experience and background of the registrant's directors and executive officers need not be given.
- The due date of the foreign private issuer's annual report on Form 20-F is 6 months after fiscal-year end (as opposed to the 90-day deadline of Form 10-K).
- A quarterly report on Form 10-Q is not required to be filed.
- Officers, directors and principal shareholders are exempt from the insider reporting and short-swing profit provisions contained in Section 16 of the Exchange Act.

The information to be included in Form 20-F is divided into four parts. Specific information requirements vary depending on whether the

Table IV.1 SEC requirements for foreign companies

F-1	Form F-1 is the basic registration statement that must be used by foreign issuers initially offering and listing securities in the United States. Alternate forms are permitted, which to some extent lessen the amount of information about the registrant required to be reprinted in the registration statement, once the registrant has been listed with the SEC for a specified period (12 months), and meets certain other requirements.
F-2	Form F-2 is an alternate registration statement form that can be used by foreign issuers to offer and list securities in the United States. Form F-2 provides an option allowing the registrant to incorporate certain information about the issuer by reference to the registrant's latest available annual report filed on Form 20-F (discussed below). If this option is taken, the annual report must be delivered to investors supplementally, along with the prospectus (registration statement). This form may be used by foreign issuers that meet a prescribed worldwide market capitalization *or* the 12-month reporting criteria set out in Form F-3 (discussed below), but not both.
F-3	Form F-3 is available for primary and secondary offerings of securities for cash, and requires prospectus information relating only to the issue being registered. Most other information in the registration statement or Form F-3 is incorporated by reference to the latest annual report of the registrant on Form 20-F. To qualify for the use of this form, a foreign issuer must meet certain criteria which include, but are not limited to the following: the registrant must have an aggregate market value worldwide equal to USD75 million or more of the non-affiliate voting stock, and the registrant must be a *timely* filer under the periodic reporting requirements of the Securities Exchange test of 1934 for at least the most recent 12-month reporting period.
F-4	Form F-4 is issued to register securities (debt or equity) to be exchanged, including shares to be issued in a business combination and the exchange of debt issued in a private placement (e.g rule 144A or Regulation S exempt offering) for debt registered with SEC. Issuers otherwise eligible to use forms F-2 or F-3 may substantially reduce the size of the F-4 document by relying on the incorporation by reference of certain information into the registration statement including for example, financial information of the registrant from its latest annual report filed on form 20-F.
6-K	Used to disclose any material information made public in the issuer's home country or on other exchanges. This information includes interim financial information and events such as major acquisitions, disposals or changes in control.
20-F	Form 20-F serves as the foreign registrant's annual report to the SEC. The primary function of Form 20-F is to provide the SEC and investors with current and continual financial information for securities registered under the provisions of the Securities Act of 1933. Form 20-F can also be used for the initial registration of existing outstanding securities under the Securities Exchange Act of 1934. Securities cannot be offered and issued on Form 20-F.

form is being used for the initial registration of securities under the Exchange Act or as an annual report form. If any item required by the form does not apply to the registrant, a statement to that effect should be made.

Registrants using Form 20-F may respond to information requested by incorporating by reference to documents previously filed with the SEC. Incorporation by reference can also be used instead of presenting certain financial statements and exhibits required by the form's instructions.

A summary of the four parts of Form 20-F is as follows:

- Part I includes items 1–13 – description of business, description of property, legal proceedings, control of registrant, nature of trading market, exchange controls and other limitations affecting security holders, taxation, selected financial data, management's discussion and analysis, directors and officers of registrant, compensation of directors and officers, options to purchase

securities from registrant or subsidiaries, and interest of management in certain transactions.
- Part II includes item 14 – description of securities to be registered.
- Part III includes items 15 and 16 – defaults upon senior securities and changes in securities and changes in security for registered securities.
- Part IV includes items 17 and 18 – financial statements (see further discussion of items 17 and 18 below), as well as item 19 – financial statements and exhibits.

The use of Form 20-F for the initial registration of securities under the Exchange Act requires that Parts I, II and IV be presented. In addition, certain documents must be filed as exhibits to the registration statement.

The prime function of Form 20-F is to provide to the SEC current and continual financial and non-financial information for securities registered under the provisions of the Exchange Act. For this purpose, Parts I, III and IV should be presented. When Form 20-F serves as the foreign registrant's annual report to the SEC, the following exhibits should be filed:

- Copies of all modifications, not previously filed, to all exhibits already filed.
- Copies of new contracts executed during the current fiscal year that would have to be filed if the form were utilized as a registration statement.
- Copies of all documents discussed in response to item 16 of the form.
- Chart· of all the registrant's subsidiaries (if requested by the SEC).

Once registered, the foreign company is obliged under the Securities Exchange Act of 1934 to file an annual report on form 20-F within six months of each year end and to submit a form 6-K to report interim financial results and certain specified events that have been disclosed publicly in its home country.

Registrants filing on form 20-F can elect between item 17 and item 18 financial statements. Electing item 17 financial statements may allow a registrant to utilize the simplified F-2 and F-3 forms in connection with a public offering. The eligibility to use an F form is summarized in Table IV.1. Registrants choosing the item 18 alternative must set forth all other information required by US GAAP and regulation S-X (see IV.4.1). What this means is that these foreign issuers will have to make all the required supplemental disclosures such as appropriate segmental information, pension information and supplemental disclosure for oil and gas producers. The essential difference between item 17 and item 18 is that item 17 calls for footnote disclosure in the local country while item 18 calls for virtually all US GAAP and SEC disclosures.

Foreign private issuers are permitted to prepare their financial statements in accordance with local GAAP (within certain limits) or US GAAP. If prepared in accordance with local GAAP, reconciliation to US GAAP are required. The financial statement requirements for foreign private issuers are: two years of balance sheet and three years of all other statements. However, if a first-time foreign registrant elects to present its financial statements in accordance with US GAAP, the SEC will permit it to provide the audited statements of income and cash flows for only the latest two fiscal years instead of three. This policy is intended to encourage the use of US GAAP.

The financial statements of a foreign registrant may be presented in the currency that the registrant deems to be appropriate. The currency utilized in the financial statements must be disclosed prominently on the face of the financial statements. Item 8 requires disclosure in the forepart of Form 20-F of the exchange rate (as defined) between US dollars and the foreign currency in which the financial statements are denominated, as of the latest practicable date.

Since the consolidated statements need not be made up in accordance with US GAAP, one of the key disclosure requirements within items 17 and 18 is the reconciliation of net income and shareholders' equity with US GAAP. In addition to the numerical US GAAP information, the foreign issuer must describe all significant reconciling items. Generally, the principal reconciling items are:

- Accounting changes.
- Business combinations and goodwill.
- Deferred income taxes.
- Equity method and consolidation.

- Foreign currency translation.
- Leases.
- Pensions.
- Revenue recognition.
- Stock compensation.

Reports of independent auditors for foreign private issuers can be prepared in accordance with the requirements of the accountants' home country, provided the report:

- Does not imply a scope limitation.
- Includes a statement that the audit was conducted in accordance with Generally Accepted Auditing Standards (GAAS), or substantially similar foreign auditing standards.
- Includes a statement identifying the accounting principles followed in its preparation.

A scope limitation is, for example, when the home-country auditing obligations do not require the observation of inventory. Observation of inventory is a mandatory auditing procedure in the United States, and if inventory has not been observed, then the scope of the audit was limited. If inventories are significant a company may have to delay their offering. The audit report must be issued annually (see X).

3 Annual report to shareholders

Companies subject to the SEC's proxy rules are required to send an annual report to their shareholders in connection with proxy solicitations for the election of directors. Revisions of the various security regulations have integrated the requirements of the shareholders' annual report with the requirements of the form 10-K and allow the annual report to be incorporated by reference in the form 10-K.

The financial statement requirements for the shareholders' annual report are explained under item 8 of form 10-K. In addition, the report must include the following information, most of which is also required in form 10-K:

- The supplementary financial information specified in item 12 of regulation S-K (see IV.4.2).
- The selected financial data described in item 10 of regulation S-K (see IV.4.2);
- Management's discussion and analysis in accordance with the provisions of item 11 of regulation S-K (see IV.4.2).
- Quantitative and qualitative disclosures about market risk in accordance with the provisions of item 305 of regulation S-K (see IV.4.2).
- The market price of the company's common stock and related security holder matters in accordance with the provisions of item 9 of regulation S-K (see IV.4.2).

If the annual report is not incorporated into form 10-K, the company must furnish a copy of form 10-K to any shareholder who requests one. In addition to the required financial statements, the shareholders' annual report may contain any other information, graphs, pictures, and so on, that management wishes to provide to the shareholders. The SEC encourages the use of graphics to make the report more readily understandable to a reader, as long as the information is consistent with the required data. Copies of the shareholders' annual report must be mailed to the SEC for its information. If the report has been incorporated by reference in form 10-K, then the required information is considered as 'filed' with the SEC and thus subject to the securities laws.

4 SEC disclosure guides

In addition to the specific guidance and instructions found in the various periodic reporting forms required to be filed under the Securities Exchange Act of 1934, there are a number of general disclosure regulations and guides whose requirements must be adhered to in the preparation of the periodic reports. Principal among these are regulations S-X and S-K, Financial Reporting Releases (FRRs) and Staff Accounting Bulletins (SABs).

4.1 Regulation S-X

4.1.1 Scope and content

Regulation S-X is the principal accounting regulation of the SEC. With regulation S-K and the FRRs it states the requirements applicable to the form and content of the financial statements (both audited and unaudited) required to be filed under the Securities Acts. Regulation S-X also sets out requirements of the accountants' reports and the

Table IV.2 Outline of regulation S-X

Article	Subject
1	Application of regulation S-X
2	Qualifications and reports of accountants
3	General instructions as to financial statements
3A	Consolidated and combined financial statements
4	General rules of form, order and terminology
5	Commercial and industrial companies
6	Investment companies
6A	Employee share purchase
7	Insurance companies
9	Bank holding companies and banks
10	Interim financial statements
11	Pro forma financial information
12	Form and content of schedules

qualifications (e.g. independence) of the accountants issuing the reports included in filings with the SEC. Amendments to the Securities Acts have standardized the financial statement requirements of the shareholders' annual report with the form 10-K and most other registration forms. Financial statements and other items of financial information are now interchangeable among the various reporting and registration forms. Care should be taken to use the most current version of regulation S-X, since it is continually being amended by additions, changes and deletions.

Regulation S-X is divided into a series of articles which are in turn subdivided into rules relating to specific required disclosures. As Table IV.2 indicates, articles 1-4 contain rules of general application. Articles 5-9 prescribe the form and content of financial statements for various types of companies. Article 10 deals with form and content of interim financial statements. Article 11 deals with pro forma financial statements which are required under conditions specified in that article. Article 12 describes the required supplemental S-X schedules. Most commercial and industrial companies need address themselves only to articles 1–5, and 12; articles 6–9 relate to specific types of businesses for which the general disclosures and financial statement presentations may not be appropriate.

4.1.2 Specific disclosure requirements

Most of the specific disclosure requirements for commercial and industrial companies are to be found in articles 4 and 5. These disclosures include those required by GAAP as well as some additional compliance items required to satisfy the increased disclosure requirements of the SEC. The SEC has attempted to make S-X conform with GAAP, but there are disclosure requirements in S-X that exceed the requirements of GAAP. Examples of such disclosures include:

- Allowance for doubtful accounts and notes receivable, and the related provision.
- Aggregate maturities of long-term debt for five years.
- Redeemable preferred stock.
- Interest rates and other details of short-term borrowings.
- Oil and gas reserves.
- Excess of replacement or current cost over the stated LIFO value of inventories.
- Disclosures relating to restrictions on retained earnings, compensating balances, and income tax expense in excess of that required by GAAP.

The SEC has deleted from regulation S-X disclosure requirements that duplicate GAAP, so regulation S-X should be used not as a disclosure checklist but rather as guidance in determining what information is required in excess of GAAP requirements. Certain information may be included in a schedule or in a supplementary financial data section.

4.1.3 Discussion of certain items of disclosure

The following are some of the issues that companies should follow in preparing annual reports for submission to the SEC.

Form, order and terminology. Financial statements should be filed in such form and order, and should use such generally accepted terminology, as will best indicate their significance and character in the light of the provisions applicable thereto. The information required with respect to any statement should be furnished as a minimum requirement, to which should be added such further material information as is necessary to render the required statements, in the light of the circumstances under which they are made, not misleading.

Financial statements not in accordance with GAAP. Financial statements filed with the SEC which are not prepared in accordance with generally accepted accounting principles will be presumed to be misleading or inaccurate, despite footnote or other disclosures, unless the commission has otherwise provided. Regulation S-X provides clarification of certain disclosures which must be included, in any event, in financial statements filed with the commission.

Filings of foreign private issuers. In all filings of foreign private issuers, except as stated otherwise in the applicable form, the financial statements may be prepared according to a comprehensive body of accounting principles other than those generally accepted in the United States, if a reconciliation to US GAAP and the provisions of regulation S-X of the type specified in form 20-F is also filed as part of the financial statements. Alternatively, the financial statements may be prepared according to US GAAP.

Money amounts in whole dollars. All money amounts required to be shown in financial statements may be expressed in whole dollars or multiples thereof, as appropriate, provided that, when they are stated other than in whole dollars, an indication to that effect is inserted immediately beneath the heading of the statement or schedule, at the top of the money columns, or at an appropriate point in narrative material.

Negative amounts. Negative amounts should be shown in a manner which clearly distinguishes the negative attribute. When determining methods of display, consideration should be given to the limitations of reproduction and microfilming processes.

Items not material. If the amount which would otherwise be required to be shown with respect to any item is not material, it need not be separately set forth. The combination of insignificant amounts is permitted.

Inapplicable headings and omission of unrequired or inapplicable financial statements:

- No heading should be shown in any financial statement as to which the items and conditions are not present.
- Financial statements not required or inapplicable because the required matter is not present need not be filed.
- The reasons for the omission of any required financial statements should be indicated.

Omission of substantially identical notes. If a note covering substantially the same subject matter is required with respect to two or more financial statements relating to the same or affiliated persons, for which separate sets of notes are presented, the required information may be shown in a note to only one of such statements, provided that clear and specific reference thereto is made in each of the other statements with respect to which the note is required.

Current assets and current liabilities. If a company's normal operating cycle is longer than one year, generally recognized trade practices should be followed with respect to the inclusion or exclusion of items in current assets or current liabilities. An appropriate explanation of the circumstances should be given, with, if practicable, an estimate of the amount not realizable or payable within one year. The amounts maturing in each year (if practicable) along with the interest rates or range of rates should also be disclosed.

Reacquired evidences of indebtedness. Reacquired evidences of indebtedness should be deducted from the appropriate liability heading. However, reacquired evidences of indebtedness held for pension and other special funds not related to the particular issues may be shown as assets, provided that there be stated the amount of such evidences of indebtedness, the cost thereof, the amount at which they are stated, and the purpose for which they have been acquired.

Discount on shares. Discount on shares, or any unamortized balance thereof, should be shown separately as a deduction from the applicable account(s) as circumstances require.

4.1.4 General notes to financial statements

If applicable to the companies for which the financial statements are filed, the following should be set forth on the face of the appropriate statement or in appropriately headed notes. The information should be provided for each statement required to be filed, except that the information required by the second, third, fourth, fifth and sixth items below should be provided as of the most recent audited balance sheet being filed and for the tenth item (leasing) as specified therein. When specific statements are presented separately, the pertinent notes should accompany such statements unless cross-referencing is appropriate.

- Principles of consolidation or combination. With regard to consolidated or combined financial statements.
- Assets subject to lien. Assets mortgaged, pledged, or otherwise subject to lien, and the approximate amounts thereof, should be designated and the obligations collateralized briefly identified.
- Defaults. The facts and amounts concerning any default in principal, interest, sinking fund, or redemption provisions with respect to any issue of securities or credit agreements, or any breach of covenant of a related indenture or agreement, which default or breach existed at the date of the most recent balance sheet being filed and which has not been subsequently cured, should be stated in the notes to the financial statements. If a default or breach exists but acceleration of the obligation has been waived for a stated period of time beyond the date of the most recent balance sheet being filed, state the amount of the obligation and the period of the waiver.
- Preferred shares.
  - (i) Aggregate preferences on involuntary liquidation, if other than par or stated value, should be shown parenthetically in the equity section of the balance sheet.
  - (ii) Disclosure should be made of any restriction upon retained earnings that arises from the fact that upon involuntary liquidation the aggregate preferences of the preferred shares exceed the par or stated value of such shares.
- Restrictions which limit the payment of dividends by the registrant.
  - (i) Describe the most significant restrictions, other than as reported under (iv) above, on the payment of dividends by the registrant, indicating their sources, their pertinent provisions, and the amount of retained earnings or net income restricted or free of restrictions.
  - (ii) Disclose the amount of consolidated retained earnings which represents undistributed earnings of 50 per cent or less owned persons accounted for by the equity method.
  - (iii) The disclosures (a) and (b) below should be provided when the restricted net assets of consolidated and unconsolidated subsidiaries and the parent's equity in the undistributed earnings of 50 per cent or less owned persons accounted for by the equity method together exceed 25 per cent of consolidated net assets as of the end of the most recently completed fiscal year. For purposes of this test, restricted net assets of subsidiaries mean that amount of the registrant's proportionate share of net assets (after intercompany elimination) reflected in the balance sheets of its consolidated and unconsolidated subsidiaries as of the end of the most recent fiscal year which may not be transferred to the parent company in the form of loans, advances or cash dividends by the subsidiaries without the consent of a third party (i.e. lender, regulatory agency, foreign government, etc.). Not all limitations on the transferability of assets are considered to be restrictions for the purposes of this test, which considers only specific third-party restrictions on the ability of subsidiaries to transfer funds outside the entity. For example, the presence of subsidiary debt which is secured by certain of the subsidiary's assets does not constitute a restriction under this rule. However, if there are any loan provisions prohibiting dividend payments, loans or advances to the parent by a subsidiary, these are considered restrictions for the purposes of computing restricted net assets. When a loan agreement requires a subsidiary to maintain certain working capital, net tangible asset, or net asset levels, or where formal compensating arrangements exist, there is considered to be a restriction under the rule, because the lender's intent is normally to preclude the transfer by dividend or otherwise of funds to the

parent company. Similarly, a provision which requires a subsidiary to reinvest all its earnings is a restriction, since this precludes loans, advances or dividends in the amount of such undistributed earnings by the entity. Where restrictions on the amount of funds which may be loaned or advanced differ from the amount restricted as to transfer in the form of cash dividends, the amount least restrictive to the subsidiary should be used. Redeemable preferred stocks and minority interests should be deducted in computing net assets for the purposes of this test. (a) Describe the nature of any restrictions on the ability of consolidated subsidiaries and unconsolidated subsidiaries to transfer funds to the registrant in the form of cash dividends, loans or advances (e.g. borrowing arrangements, regulatory restraints, foreign governments, etc.). (b) Disclose separately the amounts of such restricted net assets for unconsolidated subsidiaries and consolidated subsidiaries as of the end of the most recently completed fiscal year.

- Significant changes in bonds, mortgages and similar debt. Any significant changes in the authorized or issued amounts of bonds, mortgages and similar debt since the date of the latest balance sheet being filed for a particular person or group should be stated.
- Summarized financial information of subsidiaries not consolidated and 50 per cent or less owned persons.
  - (i) The summarized information as to assets, liabilities and results of operations should be presented in notes to the financial statements on an individual or group basis (v.*a*) for companies not consolidated and (v.*b*) for 50 per cent or less owned persons accounted for by the equity method by the registrant or by a subsidiary of the registrant, if the criteria for a significant subsidiary are met either individually by any subsidiary not consolidated *or* any 50 per cent or less owned person *or* on an aggregate basis by any combination of such subsidiaries and persons.
  - (ii) Summarized financial information should be presented insofar as is practicable as of the same dates and for the same periods as the audited consolidated financial statements submitted and should include the disclosures prescribed. Summarized information of subsidiaries not consolidated should not be combined for disclosure purposes with the summarized information of 50 per cent or less owned persons.
- Income tax expense.
  - (i) Disclosure should be made in the income statement, or in a note thereto, of:
    - (a) The components of income/loss before income tax expense/benefit as either domestic or foreign.
    - (b) The components of income tax expense, including (a) taxes currently payable and (b) the net tax effects, as applicable, of timing differences (indicate separately the amount of the estimated tax effect of each of the various types of timing difference, such as depreciation, warranty costs, etc., where the amount of each such tax effect exceeds 5 per cent of the amount computed by multiplying the income before tax by the applicable statutory federal income tax rate; other differences may be combined).

    Note: Amounts applicable to federal income taxes, to foreign income taxes and to other income taxes should be stated separately for each major component. Amounts applicable to foreign income/loss and amounts applicable to foreign or other income taxes which are less than 5 per cent of the total of income before taxes or the component of tax expense, respectively, need not be separately disclosed. For purposes of this rule, foreign income/loss is defined as income/loss generated from a registrant's foreign operations, i.e. operations that are located outside the registrant's home country.
  - (ii) Provide a reconciliation between the amount of reported total income tax expense/benefit and the amount com-

puted by multiplying the income/loss before tax by the applicable statutory federal income tax rate, showing the estimated dollar amount of each of the underlying causes of the difference. If no individual reconciling item amounts to more than 5 per cent of the amount computed by multiplying the income before tax by the applicable statutory federal income tax rate, and the total difference to be reconciled is less than 5 per cent of such computed amount, no reconciliation need be provided unless it would be significant in appraising the trend of earnings. Reconciling items that are individually less than 5 per cent of the computed amount may be aggregated in the reconciliation. The reconciliation may be presented in percentages rather than in dollar amounts. Where the reporting person is a foreign entity, the income tax rate in that person's country of domicile should normally be used in making the above computation, but different rates should not be used for subsidiaries or other segments of a reporting entity. When the rate used by a reporting person is other than the US federal corporate income tax rate, the rate used and the basis for using such rate should be disclosed.

- Warrants or rights outstanding. Information with respect to warrants or rights outstanding at the date of the related balance sheet should be set out as follows:
  - (i) Title of issue of securities called for by warrants or rights.
  - (ii) Aggregate amount of securities called for by warrants or rights outstanding.
  - (iii) Date from which the warrants or rights are exercisable.
  - (iv) Price at which each warrant or right is exercisable.
- Related-party transactions which affect the financial statements.
  - (i) Related-party transactions should be identified and the amounts stated on the face of the balance sheet, income statement, or statement of changes in the financial position.
  - (ii) In cases where separate financial statements are presented for the registrant, certain investees, or subsidiaries, separate disclosure should be made in such statements of the amounts in the related consolidated financial statements which (a) have been eliminated and (b) have not been eliminated. Also, any inter-company profits or losses resulting from transactions with related parties and not eliminated and the effects thereof should be disclosed.
- Repurchase and reverse repurchase agreements.
  - (i) Repurchase agreements (assets sold under an agreement to repurchase).
    - (a) If, as of the most recent balance sheet date, the carrying amount (or market value, if higher than the carrying amount, or if there is no carrying amount) of the securities or other assets sold under agreements to repurchase (repurchase agreements) exceeds 10 per cent of total assets, disclose separately in the balance sheet the aggregate amount of liabilities incurred pursuant to repurchase agreements, including accrued interest payable thereon.
    - (b) If, as of the most recent balance sheet date, the carrying amount (or market value, if higher than the carrying amount) of securities or other assets sold under repurchase agreements, other than securities or assets specified in (c) below, exceeds 10 per cent of total assets, disclose in an appropriately headed footnote containing a tabular presentation, segregated as to type of such securities or assets sold under agreements to repurchase (e.g. US Treasury obligations, US government agency obligations and loans), the following information as of the balance sheet date for each such agreement or group of agreements (other than agreements involving securities or assets specified in (c) below) maturing: overnight; term up to 30 days; term of 30 to 90 days; term over 90 days and demand:

- The carrying amount and market value of the assets sold under agreement to repurchase, including accrued interest plus any cash or other assets on deposit under the repurchase agreements.
- The repurchase liability associated with such transaction or group of transactions and the interest rate(s) thereon.

(c) For the purposes of (b) above only, do not include securities or other assets for which unrealized changes in market value are reported in current income or which have been obtained under reverse repurchase agreements. (d) If, as of the most recent balance sheet date, the amount at risk under repurchase agreements with any individual counterparty or group of related counterparties exceeds 10 per cent of stockholders' equity (or, in the case of investment companies, net asset value), disclose the name of each such counterparty or group of related counterparties, the amount at risk with each, and the weighted average maturity of the repurchase agreements with each. The amount at risk under repurchase agreements is defined as the excess of carrying amount (or market value, if higher than the carrying amount, or if there is no carrying amount) of the securities or other assets sold under agreement to repurchase, including accrued interest plus any cash or other assets on deposit to secure the repurchase obligation, over the amount of the repurchase liability (adjusted for accrued interest). Cash deposits in connection with repurchase agreements should not be reported as unrestricted cash.

(ii) Reverse repurchase agreements (assets purchased under agreements to resell). If, as of the most recent balance sheet date, the aggregate carrying amount of reverse repurchase agreements (securities or other assets purchased under agreements to resell) exceeds 10 per cent of total assets: (a) disclose separately such amount in the balance sheet; and (b) disclose in an appropriately headed footnote: the registrant's policy with regard to taking possession of securities or other assets purchased under agreements to resell; *and* whether or not there are any provisions to ensure that the market value of the underlying assets remains sufficient to protect the registrant in the event of default by the counterparty *and*, if so, the nature of those provisions.

If, as of the most recent balance sheet date, the amount at risk under reverse repurchase agreements with any individual counterparty or group of related counterparties exceeds 10 per cent of stockholders' equity (or, in the case of investment companies, net asset value), disclose the name of each such counterparty or group of related counterparties, the amount at risk with each, and the weighted average maturity of the reverse repurchase agreements with each. The amount at risk under reverse repurchase agreements is defined as the excess of the carrying amount of the reverse repurchase agreements over the market value of assets delivered pursuant to the agreements by the counterparty to the registrant (or to a third-party agent that has affirmatively agreed to act on behalf of the registrant) and not returned to the counterparty, except in exchange for their approximate market value in a separate transaction.

- Accounting policies for certain derivative instruments. Disclosures regarding accounting policies should include descriptions of the accounting policies used for derivative financial instruments (as defined in SFAS No. 119, paragraphs 5–7) and derivative commodity instruments (commodity instruments with characteristics similar to derivative financial instruments that are permitted by contract or business custom to be settled in cash or with another financial instrument) and the methods of applying those policies that materially affect the determination of financial position, cash flows, or results of operation. This description should include, to the extent material, each of the following items:

(i) A discussion of each method used to account for derivative financial instruments and derivative commodity instruments.
(ii) The types of derivative financial instruments and derivative commodity instruments accounted for under each method.
(iii) The criteria required to be met for each accounting method used, including a discussion of the criteria required to be met for hedge or deferral accounting and accrual or settlement accounting (e.g., whether and how risk reduction, correlation, designation, and effectiveness tests are applied).
(iv) The accounting method used if the criteria specified in paragraph (n) (3) of this section are not met.
(v) The method used to account for terminations of derivatives designated as hedges or derivatives used to affect directly or indirectly the terms, fair values, or cash flows of a designated item.
(vi) The method used to account for derivatives when the designated item matures, is sold, is extinguished, or is terminated. In addition, the method used to account for derivatives designated to an anticipated transaction, when the anticipated transaction is no longer likely to occur.
(vii) Where and when derivative financial instruments and derivative commodity instruments, and their related gains and losses, are reported in the statements of financial position, cash flows, and results of operations.

For purposes of paragraphs 2, 3, 4, and 7, the required disclosures should address separately derivatives entered into for trading purposes and derivatives entered into for purposes other than trading.

For the purposes of paragraph 6, anticipated transactions means transactions (other than transactions involving existing assets or liabilities or transactions necessitated by existing firm commitments) an enterprise expects, but is not obligated, to carry out in the normal course of business.

4.2 Regulation S-K

This regulation reflects the SEC's decision to establish uniform guidelines for disclosures in the various registration and reporting forms to be filed with the SEC under the 1933 and 1934 Acts. It includes guidelines and instructions for disclosures common to more than one SEC filing form. In a sense, it is similar to regulation S-X, the uniform standard for accounting disclosures. Prior to the promulgation of S-K the items in the various forms contained instructions and guidelines for the completion of the items, which varied from form to form. A preparer of an SEC filing is now referred to an item in S-K for guidance in completing the item.

V Balance sheet and profit and loss account formats

1 General classification principles

1.1 Consistency of presentation

The concept of consistency indicates that the procedures used in accounting for a given entity should be appropriate for the measurement of its activities and should be followed consistently from period to period (see also III.2.7). The consistency principle does not mean that a particular method of accounting, once adopted, should not be changed. APB Opinion No. 20, Accounting Changes, states that an accounting principle may be changed only if the enterprise justifies the use of an alternative acceptable accounting principle on the basis that it is preferable (see III.2.9).

1.2 Prior year figures and comparability

Comparative statements are not required by GAAP. However, ARB No. 43 states that the presentation of comparative financial statements in annual and other reports enhances the usefulness of such reports and brings out more clearly the nature and trends of current changes affecting the enterprise. In addition, the SEC requires that all publicly traded companies prepare comparative statements. These statements should include comparative balance sheets, and statements of income and cash flows for each of the three most recent fiscal years. As a result of the SEC requirement comparative statements are the norm in practice.

1.3 Offsetting of assets and liabilities

Assets and liabilities are generally not offset against each other. APB No. 10, Omnibus opinion – 1966, paragraph 7 states that it is a general principle of accounting that the offsetting of assets and liabilities is improper except where a right of set-off exists. The right of set-off exists only when all the following conditions are met:

- Each of the two parties owes the other determinable amounts.
- The entity has the right to set-off against the amount owed by the other party.
- The entity intends to offset.
- The right of set-off is legally enforceable.

In particular cases, state laws or bankruptcy laws may impose restrictions or prohibitions against the right of set-off. Furthermore, when maturities differ, only the party with the nearest maturity can offset because the party with the longer maturity must settle in the manner determined by the earlier maturity party.

The offsetting of cash or other assets against a tax liability or other amounts due to governmental bodies is also not acceptable except under limited circumstances. The only exception is when it is clear that the purchase of securities is in substance an advance payment of taxes payable in the near future and that the securities are acceptable for the payment of taxes. For forward interest rate swaps, currency swaps, options, and other conditional or exchange contracts, the conditions for the right of offset must exist or the fair value of contracts in a loss position cannot be offset against the fair value of contracts in a gain position. Neither can accrued receivable amounts be offset against accrued payable amounts. If, however, there is a master netting arrangement, then fair value amounts recognized for forwards, interest or currency swaps, options, or other such contracts may be offset without respect to the conditions previously specified.

2 Balance sheet

2.1 Prescribed format

Each corporation's balance sheet items and subtotals should be arranged in a manner useful to its various external user groups. Neither the SEC nor GAAP mandate a required format for the balance sheet. Flexibility in classifications is allowed to ensure such usefulness, owing to differences in companies, industries and economic conditions.

There are two commonly accepted arrangements followed in the presentation of the balance sheet: the account form and the report form. The account form lists the assets by section on the left side and the liabilities and stockholders' equity by sections on the right side. The report form lists the liabilities and stockholders' equity directly below the assets and on the same page. Accounting Trends and Techniques 1998 indicates that 461 of 600 companies surveyed use the report form, while 138 companies use the account form, sometimes collectively referred to as the customary form.

The balance sheet contains three major sections: assets, liabilities and shareholders' equity. These are presented in order of decreasing liquidity.

2.1.1 Legal form

There is a difference in the format of the owners' equity section of a balance sheet for different types of business entities. The owners' equity section of a partnership is referred to as partners' equity. A separate capital account is maintained for each partner. The owners' equity section of a corporation's balance sheet is referred to as shareholders' or stockholders' equity. The shareholders' equity section of the balance sheet of a corporation consists of the cumulative net contributions by shareholders in exchange for shares or stock issued. Private partnerships prepare financial statements in order to meet the information needs of the partners. They are not required by law to follow GAAP. However, many partnership agreements will specify the manner in which the accounts should be maintained (see II.3).

2.1.2 Business sector

Depending on the industry, some industries use accounts specific to their industry. The insurance industry, for example, uses the reserve on premiums accounts. The FASB and the AICPA have issued several standards relating to specialized industries. The SEC has also issued various Accounting Series Releases and Financial Reporting Releases dealing with specialized industries.

2.2 Possibilities to summarize items in the balance sheet

Classifications in balance sheets involve aggregations designed to facilitate analysis by grouping items with essentially similar characteristics. Usually the balance sheet is divided into three parts, with separate sections containing: (a) assets; (b) liabilities; and (c) stockholders' equity. The items reported within each of these sections are usually arranged according to liquidity, thereby distinguishing between current and non-current assets and liabilities. Current assets are cash and other assets expected to be converted into cash, sold, or consumed either in one year or in the operating cycle, whichever is longer. The operating cycle refers to the period of time needed to convert cash first into materials and services, then into products, then by sale into receivables, and finally back into cash when collected (ARB No. 43, chapter 3.A). Current assets are presented in order of liquidity and normally include cash, short-term investments, receivables, inventories and prepaid expenses. Current liabilities are defined in ARB No. 43, chapter 3 as those enterprise obligations whose liquidation is reasonably expected to require the use of existing resources properly classifiable as current assets or the creation of other current liabilities. This definition excludes from the current liability classification any currently maturing obligations which will be satisfied by using long-term assets and currently maturing obligations expected to be refinanced.

There is no minimum number of items to be presented in the balance sheet. A representative classification for a corporation would be:

- Assets
  - (i) Current assets
  - (ii) Long-term investments
  - (iii) Property, plant, and equipment
  - (iv) Intangible assets
  - (v) Other assets
  - Total assets
- Liabilities
  - (i) Current liabilities
  - (ii) Long-term liabilities
  - (iii) Other liabilities
  - Total liabilities
- Shareholders' equity
  - (i) Contributed capital
  - (ii) Capital stock
  - (iii) Additional paid-in capital
  - (iv) Retained earnings
  - (v) Other comprehensive income
  - Total shareholders' equity
  - Total liabilities and shareholders' equity.

In addition to 'liquid' and 'separable' sub-classifications, assets may also be classified:

- According to their type or expected function in the central operations or other activities of the entity. For example, assets held for resale (e.g. inventory) may be reported separately from assets held for use in production (e.g. property, plant and equipment).
- According to the implications for the financial flexibility of the entity. For example, assets used in operations, assets held for investment, and assets subject to restrictions (e.g. leased equipment).
- According to the method of measurement used to value the items. For example, assets and liabilities measured at net realizable value versus those measured at current cost.

2.3 Equity

2.3.1 Description of equity items

Stockholders' equity is defined in SFAC No. 6 as the residual interest in the assets of an entity after deducting its liabilities, arising from the investment of owners and the retention of earnings over time. Stockholders' equity is comprised of all capital contributed to the entity plus its accumulated earnings less any distributions that have been made. There are two major categories of stockholders' equity: contributed capital and retained earnings.

Contributed or paid-in capital represents the amount provided by stockholders in the original purchase of shares of stock resulting from subsequent transactions with owners, such as treasury transactions. It is usually separated into two components, capital stock and additional paid-in capital.

2.3.1.1 *Capital stock*

Capital stock or legal capital refers to the amount of capital that must be retained by a corporation for the protection of its creditors. Each state has established a certain amount of legal capital that is required to be retained in the business and must not be paid out as dividends. Generally, legal capital is the aggregate par value of all preferred and common stock issued.

Preferred shares may be participating or non-participating as to the earnings of the corporation, may be cumulative or non-cumulative as to the payment of dividends, and may have a preference claim on assets upon liquidation of the business. Usually, preferred stock does not have voting rights. Common stock usually has the right to vote, the right to share in earnings, a preemptive right to a proportionate share of any additional common stock issued, and the right to share in assets on liquidation.

Generally, stock is issued with a par value. No-par value stock may or may not have a stated value. When shares with no par value are issued, the legal capital may be:

- Total consideration paid in for the shares.
- A minimum amount stated in the applicable state incorporation law.
- An arbitrary amount established by the board of directors at its discretion.

When stock is issued above or below par value, a premium or discount on the stock is recorded, respectively. A premium on stock is often referred to as 'paid-in capital in excess of par value'. The discount is considered a contingent liability of the shareholders to the creditors of the corporation. In the event of a liquidation of corporate assets the creditors may recover unsatisfied obligations by assessing the shareholders for additional contributions up to the amount of the discount.

A corporation's charter contains the types and amounts of stock that it can legally issue – the authorized capital stock. When part or all of the authorized capital stock is issued, it is called issued capital stock. Since a corporation may own issued capital stock in the form of treasury stock, the amount of issued capital stock in the hands of stockholders is called outstanding capital stock.

Redeemable preferred stock with mandatory redemption requirements or whose redemption is outside the control of the issuer must not be included in stockholders' equity (Regulation S-X, rule 5.02.28).

2.3.1.2 *Additional paid-in capital*

Additional paid-in capital represents all capital contributed to a corporation other than that defined as par or stated value. Additional paid-in capital can arise from proceeds received from the sale of common or preferred shares in excess of their par or stated values. It can also arise from transactions relating to the following:

- Sale of shares previously issued and subsequently reacquired by the corporation (treasury stock).
- Retirement of previously outstanding shares.
- Donated assets.
- Transfer of assets or liabilities between entities under common control, i.e. the entities are subject to control by the same parent, investor or common officers, if they are not conducted in an arm's length transaction.
- Payment of stock dividends in a manner which justifies the dividend being recorded at the market value of the shares distributed.
- Conversion of convertible bonds.

Generally, items properly chargeable to current or future years' income accounts may not be charged to contributed capital accounts. An exception to this rule occurs in accounting for a quasi-reorganization, in which case a one-time adjustment to contributed capital is appropriate.

2.3.1.3 Retained earnings

Retained earnings represent the amount of previous income of the corporation that has not been distributed to owners as dividends or transferred to paid-in or contributed capital. If a series of operating losses have been incurred or distributions to shareholders in excess of accumulated earnings have been made and if there is a debit balance in retained earnings, the account is generally referred to as accumulated deficit.

The distributions to shareholders generally take the form of dividend payments but may take other forms as well, such as reacquisition of shares for amounts in excess of the original issuance proceeds. Various forms of dividends are as the following:

- Cash dividends: they are declared on the basis of a specified amount for each share of outstanding stock.
- Property dividends: they are paid in the form of some non-cash asset, such as merchandise, investments, or fixed assets.
- Liquidating dividends: they occur when the corporation uses paid-in capital, rather than retained earnings, as a basis for dividends.
- Stock dividends: they occur when a corporation issues shares of its own stock in the form of a dividend. A stock dividend changes neither the par value of the stock nor the total stockholders' equity. However, the number of shares issued and outstanding is increased.

When a distribution of stock is more than 20 per cent to 25 per cent of the outstanding shares immediately before the distribution, it is considered a stock split. Stock splits involve an increase in the number of shares outstanding and a corresponding decrease in the par or stated value of each share. Stock splits have no impact on retained earnings or total stockholders' equity. No entry is generally required for a stock split.

The legality of dividends can be determined only by reviewing the applicable state law. For most general dividend declarations, the following summary applies:

- Retained earnings, unless legally restricted in some manner, are usually the correct basis for dividend declaration.
- Revaluation capital is seldom the appropriate basis for dividends (except possibly stock dividends).
- In some states, additional paid-in capital may be used for dividends, although such dividends may be limited to preferred stock.
- Deficits in retained earnings and debits in paid-in capital accounts must be restored before payment of any dividend.
- Dividends in most states may not reduce retained earnings below the cost of treasury stock held.

The decision to declare a dividend is made by the board of directors of the corporation.

Appropriation of retained earnings serves disclosure purposes and serves to restrict dividend payments but does nothing to provide any resources for satisfaction of the contingent loss or other underlying purpose for which the appropriation has been made.

2.3.1.4 Treasury stock

Treasury stock consists of a corporation's own stock which has been issued, subsequently reacquired by the firm and not yet reissued or cancelled. Treasury stock does not reduce the number of shares issued but does reduce the number of shares outstanding, as well as total stockholders' equity. These shares are not eligible to receive cash dividends. Treasury stock is not considered to be an asset and is therefore not capitalizable.

Two methods are used to account for transactions involving treasury stock – the cost method and the par value method.

Under the cost method, the gross cost of the shares reacquired is charged to a contra equity account. When the treasury shares are reissued, proceeds in excess of cost are credited to a paid-in capital account. Any deficiency is charged to retained earnings.

Under the par value approach, the treasury stock account is charged only for the aggregate par (or stated) value of the shares reacquired. Other paid-in capital accounts are relived in proportion to the amounts recognized upon the original issuance of the shares. The treasury share acquisition is treated almost as a retirement. However, the common (or preferred) stock account continues at the original amount, thereby preserving the distinction between an actual retirement and a treasury share transaction. When the treasury shares accounted for by the par value method are subsequently resold, the excess of the sale price over par value is credited to paid-in capital. A reissuance for a price below par values does not create a contingent liability for the purchaser.

2.3.1.5 Comprehensive income

Several accounting standards require that certain items be reported as direct charges or credits to equity, without having been recognized in the income statement. Those items qualify as part of comprehensive income.

Comprehensive income is defined in SFAC No. 6 as the change in equity (net assets) of a business enterprise during a period from transactions and other events and circumstances from non-owner sources. It includes all changes in equity during a period, except those resulting from investments by owners and distributions to owners. Thus, comprehensive income includes net income but also includes other components of comprehensive income. Those items of other comprehensive income are defined in SFAS No. 130 as all revenues, expenses, gains, and losses that under generally accepted accounting principles are included in comprehensive income but excluded from net income. Examples of these items include:

- Foreign currency translation adjustments in accordance with SFAS No. 52, Foreign Currency Translation.
- Changes in the market value of a futures contract that qualifies as a hedge of an asset reported at fair value pursuant to SFAS No. 115, Accounting for certain Investments in Debt and Equity Securities.
- Unrealized holding gains and losses on available-for-sale securities in accordance with SFAS No. 115.
- Negative equity adjustments recognized in accordance with SFAS No. 87, Employers' Accounting for Pensions, as an additional pension liability not yet recognized as net periodic pension cost.
- Fair value adjustments for derivatives in a cash flow hedge (SFAS No. 133).

Comprehensive income should be reported in a financial statement that is displayed with the same prominence as other financial statements that are part of a full set of financial statements. Although no specific format is required by the standard, a preference was stated for the components of other comprehensive income and total comprehensive income to be displayed in the income statement or a separate statement of comprehensive income that begins with net income. Another alternative is to show the changes in the components of other comprehensive income in the statement of stockholders' equity. Components of other comprehensive income should also describe the amount of income tax expense or benefit for each component either on the face of the statement in which components are reported or in notes to financial statements.

2.3.2 Disclosures on information about changes in equity and the capital structure

The disclosure of changes in the separate accounts composing shareholders' equity is required to make the financial statements sufficiently informative (see APB Opinion No. 12). Disclosure of such changes may take the form of separate statements (i.e. statement of shareholders' equity or statement of retained earnings) or may be in the basic financial statements or notes thereto. A sample statement of shareholders' equity is presented in Table V.1.

SFAS No. 129, Disclosure on Information about Capital Structure, establishes standards for disclosing information about an entity's capital structure. It requires information about capital structure to be disclosed in three separate categories:

- Information about securities. The entity shall provide within its financial statements a summary explanation of the pertinent rights and privileges of the various securities that are outstanding. In addition, information about the number of shares issued upon conversion, exercise or satisfaction of required conditions should be disclosed.
- Liquidation preference of preferred stock. If an entity issued preferred stock that has a preference in involuntary liquidation considerably in

Table V.1 Sample statement of shareholders' equity

	Common Stock						
	Number of shares outstanding	Par value (in USD)	Additional paid-in capital	Retained earnings	Treasury stock	Other comprehensive income	Total
Balance as of 31 December 1995	12,000,000	120	80,000	20,000		400	100,520
Net earnings				40,000			40,000
Issuance of 11,000,000 common shares in public offering	11,000,000	110	220,000				220,110
Stock option plans, net	200,000	2	4,000				4,002
Net translation adjustment						1,900	1,900
Dividend paid to stockholders			(20,000)	(30,000)			(50,000)
Balance as of 31 December 1996	23,200,000	232	284,000	30,000		2,300	316,532
Net earnings				55,000			55,000
Stock option plans, net	100,000	1	2,000				2,001
Net translation adjustment						(2,500)	(2,500)
Repurchase of common stock					(600)		(600)
Balance as of 31 December 1997	23,300,000	233	286,000	85,000	(600)	(200)	370,433
Net loss				(3,000)			(3,000)
Net translation adjustment						(19,800)	(19,800)
Balance as of 31 December 1998	23,300,000	233	286,000	82,000	(600)	(20,000)	347,633

excess of the par or stated value of the shares the liquidation preference of the stock should be disclosed.

- Redeemable stock. Redeemable stock must be repurchased by the issuing entity. In this situation, the issuing entity is required to disclose the amount of redemption requirements for all issues of stock for which the redemption prices and dates are fixed or determinable.

2.3.3 Partnerships

There are no authoritative pronouncements concerning the accounting for equity in a partnership. In practice a separate capital account is maintained for each partner. Since the partnership is a separate accounting entity, those accounts reflect the assets contributed to and the liabilities assumed by the partnership at fair value. Other changes in the capital accounts relate to withdrawals by the partners and the absorption of their share of the profit or loss of the partnership.

3 Profit and loss account

3.1 Prescribed formats

There are two commonly followed formats in the reporting of revenues, gains, expenses and losses: the single-step income statement and the multiple-step income statement. In a single-step statement total expenses are subtracted from total revenues. The expenses are deducted from the revenues to arrive at the net income or loss; the expression 'single-step' is derived from the single subtraction necessary to arrive at net income. In a multiple-step statement the basic division is between operating and non-operating activities, and revenues and expenses are both subdivided into these two groups. Accounting Trends and Techniques 1998 indicates that 432 of the 600 companies surveyed used the multiple-step form, while 168 companies used the single-step format (see also APB Opinion No. 9, Reporting the Results of Operations).

3.2 Components of net income

3.2.1 Major items

The format of an income statement should show the significant components of net income, thereby allowing financial statements users to assess the quality of earnings and to evaluate the results of normal operations for a prediction of future cash flows. APB Opinion No. 30 requires that extraordinary items, prior period adjustments, discontinued operations and the cumulative effect of an accounting change be separately disclosed net of applicable income taxes following income from continuing operations. Other major items required by APB Opinion No. 30 to be identified in the income statement are:

- Income from continuing operations (i.e. gross margin on sales less operating expenses for the primary activities of the business).
- Other revenues and other expenses (i.e. those related to secondary business activities).
- Income before income taxes and extraordinary items.
- Income tax expense.
- Income before extraordinary items, cumulative effect of accounting changes, and discontinued operations.
- Discontinued operations (i.e. operations of a discontinued business segment must be reported separately from continuing operations, see VII.8).
- Extraordinary items (i.e. events or transactions that are distinguished both by their unusual nature and by the infrequency of their occurrence).
- Cumulative effect of accounting changes (i.e. changes in accounting principle as defined in APB Opinion No. 20).
- Earnings per share of common stock.

In addition SFAS No. 130, Reporting comprehensive income, requires that comprehensive income and its components, as defined in the Statement, be reported in the income statement or alternatively in a separate statement or in a statement of changes in stockholders' equity (see also VI.2.3.1.5).

3.2.2 Extraordinary items

APB Opinion No. 30, paragraph 30 requires the two criteria unusual nature and infrequency of occurrence both to be met in order to classify an event or transaction as an extraordinary item.

The criteria unusual nature implies that the underlying event or transaction should possess a high degree of abnormality and be of a type clearly unrelated to, or only incidentally related to, the ordinary and typical activities of the entity, taking into account the environment in which the entity operates. Special characteristics of the entity include type and scope of operations, lines of business and operating policies.

Infrequency of occurrence implies that the underlying event or transaction should be of a type that would not reasonably be expected to recur in the foreseeable future, taking into account the environment in which the entity operates.

In addition, specific accounting pronouncements require the following items to be disclosed as extraordinary, even though they do not meet the criteria stated above:

- Material gains and losses from the extinguishment of debt (SFAS No. 4, paragraph 8). Does not apply, however, to gains and losses from extinguishments of debt made to satisfy sinking-fund requirements that an enterprise must meet within one year of the date of extinguishment (SFAS No. 64, paragraph 4).
- Profit or loss resulting from the disposal of a significant part of the assets or a separable segment of previously separate companies, provided the profit or loss is material and the disposal is within two years after a pooling of interest (APB Opinion No. 16, paragraph 60).

- Write-off of operating rights of motor carriers (SFAS No. 44, paragraph 6).
- The investor's share of an investee's extraordinary item when the investor uses the equity method of accounting for the investee (APB Opinion No. 18, paragraph 19).

Gains of a debtor related to a troubled debt restructuring (SFAS No. 15, paragraph 21).

Extraordinary items should be segregated from the results of ordinary operations and be shown net of taxes in a separate section of the income statement, following 'discontinued operations' and preceding 'cumulative effect of a change in accounting principle', if any.

The nature, tax effect, effect on earnings per share and principal items entering into the determination of the gain or loss should be disclosed. Significant events that are either unusual or infrequent but not both should be reflected separately as part of income from continuing operations or disclosed in the notes to the financial statements. Such items should not be reported net of taxes, nor should the per-share effects be disclosed (see APB Opinions No. 9 and No. 30).

3.2.3 Cumulative effect of an accounting change

Changes in accounting principle can be mandatory, resulting from changes made to conform with a new FASB pronouncement, or may be discretionary. When a change in accounting principle is adopted, the nature of the change and management's justification for the change must be disclosed in the financial statements of the period in which the change is made. The justification must explain why the new principle is thought to be preferable. In general, the cumulative effect (net of tax) of a change in accounting principle should be recorded in the year of the change and shown on the income statement between extraordinary items and net income, without restating prior years. The effect on net income in the period of the change should also be disclosed, and per-share information for the cumulative effect of the accounting change is required. Though financial statements for prior periods are generally not to be restated for accounting changes, APB Opinion No. 20 and SFAS No. 16 accorded special treatment to four types of change:

- A change from the LIFO method of accounting for inventory to another method.
- A change in the method of accounting for long-term construction contracts.
- A change from or to the full-cost method of accounting used in extractive industries.
- A change from retirement–replacement–betterment accounting to depreciation accounting (see SFAS No. 73).

The APB concluded that for these changes the financial statements for all periods presented should be restated (see IX.1.2.1).

3.2.4 Earnings per share

Earnings per share figures are required to be presented in the income statement of publicly held companies and are presented in a manner consistent with the headings included in the income statement. They are often used in evaluating a firm's stock price and in assessing the firm's future earnings and ability to pay dividends.

In order to assess quality of earnings, SFAS No. 128, Earnings per Share requires presentation of basic earnings per share on the face of the income statement for public-market issued common stock or potential common stock (such as options, warrants, convertible securities or contingent stock agreements). Shares outstanding are determined by the weighted average method. Earnings is defined as both income from continuing operations and net income available to common stockholders requiring the deduction of dividends declared on preferred stock and dividends for the period on cumulative preferred stock, whether or not earned or paid.

Diluted earnings per share is computed similarly to basic earnings per share by changing the number of shares outstanding to include the number of additional common shares that would be issued if the potentially dilutive shares had been issued.

Entities with simple capital structures – that is, only common stock – should display basic earnings per share for both continuing operations and net income on the face of the income statement. All other entities are required to display both basic and diluted per share amounts for both continuing operations and net income. For discontinued operations, extraordinary items, or cumulative effect of an accounting change, basic and diluted per share amounts may be described either on the face of the income statement or in the footnotes.

3.3 Statement of Income and Retained Earnings

An acceptable practice is to combine the income statement and the statement of retained earnings into a single statement called the Statement of Income and Retained Earnings. Net income is computed in the same fashion as in a multiple- or single-step income statement. The beginning balance in retained earnings is added to the net income (loss) figure. Declared dividends are deducted to obtain the retained earnings ending balance.

VI Recognition criteria and valuation

1 Introduction

Through the Conceptual Framework Project a basis of underlying concepts and general principles of financial accounting was built that aided in deriving comprehensive principles for the recognition and valuation of the elements of financial statements. These general principles remain virtually stable, whereas regulations for specific accounting issues and items continue to be promulgated, developed, and evolved and thereby continue to increase the complexity in applying GAAP. Therefore the following discussion of accounting practices for certain assets, liabilities, stockholders' equity and other accounting areas may not represent the authoritative requirements at the moment. Because of the amount of ongoing projects of the FASB and other institutions the latest development of pronouncements for the respective accounting area should be observed (a listing of relevant Internet links is provided in XIII – see also Table I.1).

2 General recognition criteria

2.1 Overview

The definition of accounting income requires certain recognition criteria for assets, liabilities, revenues and expenses. SFAC No. 5 has identified four recognition criteria, which are the following:

- Definition. To be recognized, the item must meet one of the definitions of an element of the financial statements of SFAC No. 6. A resource must meet the definition of an asset; an obligation must meet the definition of a liability; and a change in equity must meet the definition of a revenue, expense, gain, loss, investment by owner, or distribution to owner.
- Measurability. The item must have a relevant attribute that can be quantified in monetary units with sufficient reliability. Measurability must be considered in terms of both relevance and reliability, the two primary qualitative characteristics of accounting information.
- Relevance. An item is relevant if the information about it has the capacity to make a difference in investors', creditors', or other users' decisions.
- Reliability. An item is reliable if the information about it is representationally faithful, verifiable, and neutral. The information must be faithful in its representation, free of error, and unbiased.

All four criteria are subject to the pervasive cost–benefit constraint and the materiality threshold.

2.2 Assets and liabilities

Those general criteria of SFAC No. 5 for the recognition of assets and liabilities are specified in SFAC No. 6.

Assets are defined as probable (likely to occur) future economic benefits obtained or controlled by a particular entity as a result of past transactions or events (SFAC No. 6, paragraph 25).

The following three characteristics must be present for an item to qualify as an asset:

- The asset must provide probable future economic benefit which enables it to provide future net cash inflows.

- The entity is able to receive the benefit and restrict other entities' access to that benefit.
- The event which provides the entity with the right to the benefit has occurred.

Assets remain an economic resource of an enterprise as long as they continue to meet the three requirements identified above. Assets have features that help identify them in that they are exchangeable, legally enforceable, and have future economic benefit (service potential). It is that potential that eventually brings in cash to the entity and that underlies the concept of an asset.

Liabilities are probable (likely to occur, see SFAS No. 5, paragraph 3) future sacrifices of economic benefits arising from present obligations of a particular entity to transfer assets or provide services to other entities in the future as a result of past transactions or events (SFAC No. 6, paragraph 35).

The qualification of an item as a liability requires the following three characteristics to be fulfilled:

- A liability requires that the entity settles a present obligation by the probable future transfer of an asset on demand when a specified event occurs or at a particular date.
- The obligation cannot be avoided.
- The event which obligates the entity has occurred.

Liabilities usually result from transactions which enable entities to obtain resources. Other liabilities may arise from non-reciprocal transfers such as the declaration of dividends to the owners of the entity or the pledge of assets to charitable organizations.

2.3 Revenues and expenses

Revenues are increases in assets or decreases in liabilities during a period from delivering goods, rendering services, or other activities constituting the enterprise's central operations. According to SFAC No. 6, paragraph 79 characteristics of revenues are a culmination of the earning process and an actual or expected cash inflow resulting from central operations. The recognition of revenues depends on their being:

- Realized or realizable. Revenues and gains are generally not recognized as components of earnings until they have been realized or have become realizable.
- Earned. Revenues are not recognized until they have been earned. Revenues are considered to have been earned when the entity has substantially accomplished what it must do to be entitled to the benefits represented by the revenues.

The realization concept stipulates that revenue is only recognized when the earnings process is (virtually) complete and the existence of an exchange transaction, which has taken place, evidences the revenue.

Expenses are decreases in assets or increases in liabilities during a period from delivery of goods, rendering services, or other activities constituting the enterprise's central operations. They are characterized by sacrifices involved in carrying out the earnings process and an actual or expected cash outflow resulting from central operations (see SFAC No. 6, paragraph 81). Guidance for the recognition of expenses is based on:

- Consumption of economic benefit. Expenses are generally recognized when an entity's economic benefits are consumed in revenue-earning activities, or
- Loss or lack of benefit. Expenses are recognized if it becomes evident that the previously recognized future economic benefits of assets have been reduced or eliminated, or that liabilities have been incurred or have increased, without associated economic benefits.

Expenses are expired costs, or items which were assets but which are no longer assets because they have no future value. The matching principle requires that all expenses incurred in the generating of revenue should be recognized in the same accounting period as the revenue is recognized. The matching principle is broken down into three pervasive measurement principles: associating cause and effect, systematic and rational allocation, and immediate recognition.

Costs, such as materials and direct labour consumed in the manufacturing process, are relatively easy to identify with the related revenue elements. These cost elements are included in inventory and expensed as cost of sales when the product is sold and revenue from the sale is recognized. This is associating cause and effect.

Some costs are more closely associated with specific accounting periods. In the absence of a cause and effect relationship, the asset's cost should be allocated to benefiting accounting periods in a systematic and rational manner. This form of expense recognition involves assumptions about the expected length of benefit and the relationship between benefit and cost of each period. Depreciation of fixed assets, amortization of intangibles, and allocation of rent and insurance are examples of costs that would be recognized by the use of a systematic and rational method.

All other costs are normally expensed in the period in which they are incurred. This would include those costs for which no clear-cut future benefits can be identified, costs that were recorded as assets in prior periods but for which no remaining future benefits can be identified, and those other elements of administrative or general expense for which no rational allocation scheme can be devised. The general approach is first to attempt to match costs with the related revenues. Next, a method of systematic and rational allocation should be attempted. If neither of these measurement principles can be applied, the cost should be immediately expensed.

There are inconsistencies in the current application of the matching principle under GAAP. For example, it can be argued that proper matching would require research and development and similar costs to be capitalized over their useful life. However, research and development costs must be expensed as incurred (see VI.5.2).

2.4 Gains and losses

Gains (losses) are increases (decreases) in equity from peripheral transactions of an entity excluding revenues (expenses) and investments by owners (distribution to owners). Characteristics of gains and losses (SFAC 6, paragraph 84–86) include the following:

- Result from peripheral transactions and circumstances which may be beyond entity's control.
- May be classified according to sources or as operating and non-operating.

According to SFAC 5, the recognition of gains and losses should follow the principles stated below:

- Gains often result from transactions, and other events, that involve no 'earnings process'; therefore, in terms of recognition, it is more significant that the gain be realized than earned.
- Losses are recognized when it becomes evident that future economic benefits of a previously recognized asset have been reduced or eliminated, or that a liability has been incurred without associated economic benefits. The main difference between expenses and losses is that expenses result from ongoing major or central operations, whereas losses result from peripheral transactions that may be beyond the entity's control.

3 General valuation principles

The FASB has identified five alternatives for measuring (valuing) elements of the balance sheet. Five different attributes of assets (and liabilities) are used in present practice (SFAC No 5, paragraph 67 – see Table VI.1):

- Historical cost (historical proceeds).
- Current cost.
- Current market value.
- Net realizable value (net realizable settlement value).
- Present (or discounted) value of future cash flows.

Those attributes can be divided into two groups. The first group (historical cost, current cost and current market value) captures the current state of the marketplace. The measurer looks to observable current amounts, transaction prices under current condition. The second group (net realizable value and present value of future cash flows) looks to estimated future amounts and bases today's measurement on an estimate of future benefit or sacrifice.

In recent years, new measurement techniques have emerged from the accounting profession. For example tools such as the Capital Asset Pricing Model and the Contingent Claim Pricing Model use mathematics and statistical theory in an attempt to estimate market values and are

Table VI.1 Attributes of assets and liabilities

Attribute	Assets	Liabilities
1 Historical cost (historical proceeds)	Initially, the amount of cash (or its equivalent) paid to acquire an asset (historical cost); subsequent to acquisition, the historical amount may be adjusted for amortization or other allocations.	Initially, the amount of cash (or its equivalent) received when the obligation was incurred (historical proceeds); subsequent to incurrence, the historical amount may be adjusted for amortization or other allocations.
2 Current cost	Amount of cash (or its equivalent) that would have to be paid if the same or an equivalent asset were acquired currently: 2.1 The 'same asset' may be an identical asset ('current reproduction cost' or 'current cost of replacement in kind'). 2.2 The 'same asset' may be an asset with equivalent productive capacity ('current replacement cost').	Amount of proceeds that would be obtained if the same obligation were incurred currently.
3 Current market value	Amount of cash (or its equivalent) that could be obtained currently by selling the asset in an orderly liquidation (current exit value in orderly liquidation).	Cash outlay that would be required currently to eliminate the liability.
4 Net realizable value (net realizable settlement value)	Non-discounted amount of cash (or its equivalent) into which an asset is expected to be converted in the due course of business, less direct costs necessary to make that conversion.	Non-discounted amount of cash (or its equivalent) expected to be paid to liquidate an obligation in the due course of business, including direct costs necessary to make that payment (non-discounted amount of expected cash outlays).
5 Present (or discounted) value of future cash flows	Present (or discounted) value of future cash inflows into which an asset is expected to be converted in the due course of business, less present values of cash outflows necessary to obtain those inflows. Rate of discount may be: 5.1 Historical rate 5.2 Current rate 5.3 Other rate (for example, average expected rate or weighted average cost of capital).	Present or discounted value of future cash outflows expected to be required to satisfy the liability in due course of business. Rate of discount may be: 5.1 Historical rate 5.2 Current rate 5.3 Other rate (for example, average expected rate or incremental borrowing rate).

used in estimating the market values of derivative financial instruments.

Current cost and current exit value are both market values. Net realizable value differs from current exit value in that it is based upon the expected future sales proceeds of the assets rather than upon the current disposal value of an asset in its existing form. The expected cash flows used to determine present value are similar to those used to determine net realizable value; the difference between the two alternatives is that under the present value approach consideration is given to the time value of money.

In the financial statements of business entities the measurement most often used is historical cost. Historical cost is used extensively as a valuation method because it is based on transactions and provides information that has a high degree of reliability. Some users of financial statements criticize that historical cost is not as relevant as the amounts reported under alternative valuation methods.

Although the historical cost principle is the primary basis of valuation, recording and reporting of fair value information is increasing. SFAC No. 5, paragraph 68 states that although the 'historical cost system' description may be convenient and describes well present practice for some major classes of assets (most inventories, property, plant and equipment, and intangibles), it describes less well present practice for a number of other classes of assets and liabilities, e.g., trade receivables, notes payable, and warranty obligations. The alternative valuation methods of current cost, current market value, net realizable value, and present value are used in certain circumstances for selected elements of the balance sheet.

In recent years the new accounting standards preferred fair value over historical cost, whenever a reliable measurement of the fair market value deemed possible. SFAS No. 115 requires most debt and equity security investments to be reported at fair market value and SFAS No. 133 requires the fair market value of all derivatives. The FASB is trying to require that all financial instruments be accounted and reported at fair market value (see VII.4). Also, the proposed elimination of the pooling-of-interest method of accounting for business combinations in favour of the purchase accounting is another example of the increased acceptance of fair value as a measurement basis over historical cost.

3.1 Historical cost (historical proceeds)

Historical cost is the amount of cash, or its equivalent, paid to acquire an asset, commonly adjusted after acquisition for amortization or other allocations (SFAC No. 5, paragraph 67a). Liabilities that involve obligations to provide goods or services to customers are generally reported at historical proceeds, which is the amount of cash, or its equivalent, received when the obligation was incurred and may be adjusted after acquisition for amortization or other allocations.

GAAP requires that most assets and liabilities be accounted for and reported under the historical cost principle. As a measurement basis, historical exchange prices are the most objectively determinable and capable of being independently verified.

At the date the transaction is recognized, each asset, liability, revenue, expense, gain, or loss arising from the transaction should be measured and recorded in the domestic currency of the recording entity by the use of the exchange rate in effect at that date.

3.1.1 Acquisition cost

3.1.1.1 Definitions

Acquisition cost is the usual basis for valuing tangible assets. Acquisition cost is measured by the cash or cash equivalent price of obtaining the asset and bringing it to the location and condition necessary for its intended use. The net purchase price, freight cost, sales taxes, and installation costs of a productive asset are considered part of the asset's cost. Any related costs incurred after the asset's acquisition, such as additions, improvements, or replacements, are added to the asset's cost if they provide future service potential. Expenditures that simply maintain a given level of service should be expensed.

Cost is the basis used at the date of acquisition because cash or cash equivalent price best measures the value of the asset at that time.

The major exception to the cost principle is acquisition through donation, where the appraisal or fair market value of the asset is used to establish a reasonable basis of asset valuation for the purposes of enterprise accountability. Another approach that is sometimes allowed, and not considered a violation of the historical cost concept, is the prudent cost concept. This concept states that if for some reason one is ignorant about a certain price and paid too much for the asset originally, it is theoretically preferable to charge a loss immediately.

3.1.1.2 Cash discounts

Any cash discount allowed should be recorded whether or not the discount is taken. If payment is not made within the discount period, the discount not taken should be treated as a financing expense. This method is generally preferred. However, the discount not taken should not always be considered a financing expense because the terms may be unfavourable or because it might not be prudent for the company to take the discount.

3.1.1.3 Deferred payment contracts

Assets purchased on long-term credit contracts should be accounted for at the present value of the consideration exchange between the contracting parties at the date of the transaction. When no interest rate is stated, or the specific rate is unreasonable, an appropriate interest rate must be imputed. The objective is to approximate the interest rate that the buyer and seller would negotiate at arm's length in a similar borrowing transaction.

3.1.1.4 Exchanges of non-monetary assets

In general, the accounting for the exchange of non-monetary assets should be based on the fair value. The asset received should be recorded at the fair value of the asset given up or the fair value of the asset received, whichever is clearly more evident (APB Opinion No. 29, Accounting for Nonmonetary Transactions). Fair value of assets in a non-monetary transfer is determined by reference to the estimated realizable value of similar assets that are sold for cash, quoted market prices, independent appraisals, and other available evidence. Thus the difference between the fair value and the book value of the assets surrendered should be recognized as a gain or loss at the time of the exchange. This approach is always employed when the assets are dissimilar in nature. Exceptions to this rule are:

- If the fair value is not determinable within reasonable limits.
- If the earnings process is not completed.

If the fair value is not easily determinable within reasonable limits, the book value of the asset given up is usually taken as the basis for recording a non-monetary exchange. Therefore, no gain or loss is recognized.

In general, when exchanges of similar non-monetary assets would result in a gain and the earnings process is not considered completed, a gain should not be recognized. However, if the exchange transaction involving similar assets would result in a loss, the loss is recognized immediately. When a monetary consideration such as cash is received in addition to the non-monetary asset, it is assumed that a portion of the earnings process is completed and, therefore, a partial gain is recognized. When the non-monetary consideration is significant (25 per cent or more of the fair value of the exchange) the entire transaction is considered monetary and has to be accounted for at fair value (EITF Issue No. 86-29).

In summary, losses on non-monetary transactions are always recognized whether the exchange involves dissimilar or similar assets. Gains on non-monetary transactions are recognized if the exchange involves dissimilar assets; gains are deferred if the exchange involves similar assets, unless cash or some other form of monetary consideration is received, in which case a partial gain is recognized.

3.1.1.5 Individual prices as components of total price

When a group of assets is purchased in a single transaction for a lump-sum price, total cost should be allocated to the individual assets based on the best indicator of their relative market value.

3.1.2 Production cost

3.1.2.1 Definitions

Assets, which are manufactured by the firm as opposed to purchased, are valued at production cost. There is no definition of the term 'production' in GAAP. Typically, the components of production cost or manufacturing cost are divided into direct materials, direct labour and factory overheads. Direct material cost is the cost of materials that become part of the finished product and can be conveniently and economically traced to specific product units. Direct labour cost is the cost of labour services for specific work performed on products that can be conveniently or economically traced to end products. Factory overheads is a varied collection of production-related costs that cannot be practically or conveniently traced to end products. It includes indirect materials, indirect labour, and factory costs such as building and machinery maintenance, depreciation on factory equipment, and factory utilities expense. Direct materials, direct labour and factory overheads are all obligatory components of production cost. They are incurred in making inventoriable products.

General and administrative expenses must be included as period charges, except for the portion of such expenses that may be clearly related to production. Selling expenses constitute no part of production costs (ARB No. 43, chapter 4, paragraph 5).

3.1.2.2 Wages and ancillary costs of employment

Labour costs that can be directly traced to end products under the cost–benefit restraint, are recorded as direct labour, otherwise they are accounted for as factory overhead.

3.1.2.3 Research and development costs as part of production cost

SFAS No. 2, Accounting for Research and Development Costs, requires that all research and development costs be charged to expense when incurred (see VI.5.1). However, if the development of computer software is involved, SFAS No. 86 applies, which requires capitalization of development costs based on the development stage of the software (see VI.5.3).

3.1.2.4 Interest on capital as part of production cost

All assets that require a period of time to get them ready for their intended use should include a capitalized amount of interest cost. According to SFAS No. 34, Capitalization of Interest Cost, interest cost should only be capitalized as a part of the historical cost of the following qualifying assets when such costs are considered to be material:

- Assets constructed for an entity's own use or for which deposit or progress payments are made.
- Assets produced as discrete projects that are intended for lease or sale.
- Equity method investments when the investee is using funds to acquire qualifying assets for its principal operations which have not yet begun.

The capitalization of interest cost does not apply to situations where routine inventories are produced in large quantities on a repetitive basis, effects are not material (compared to the effect of expensing interest), qualifying assets are already in use or ready for use, and principal operations of an investee accounted for under the equity method have already begun.

The amount capitalized in an accounting period shall be determined by applying an interest rate to the average amount of accumulated expenditures for the asset during the period. The capitalization rates used should be based on the rates applicable to borrowings outstanding during the period.

3.1.2.5 Overheads when producing at less than full capacity

Identifying the cost of idle capacity usually requires segregating the overhead into fixed and variable components. Variable costs change in proportion to production volume, and therefore are not generally incurred as a result of excess capacity. Fixed costs (such as depreciation, rent, and property taxes) are incurred regardless of production volume. The fixed components of overheads that relate to excess productive capacity should be excluded from factory overheads and charged to expense in the year incurred.

3.1.2.6 *Treatment of expenditures for repairs, maintenance and modernization*

Maintenance and ordinary repairs to plant assets, factory equipment and tools are charged to factory overheads. Extraordinary repairs, and modernization that extends the life of the asset, are capitalized into property plant and equipment (APB Opinion No. 28).

3.2 Current cost

Current cost is measured by the amount of cash, or its equivalent, that would have to be paid if the same or an equivalent asset were acquired currently (replacement cost: SFAC No. 5, paragraph 67b).

3.3 Current market value

Current market value is the amount of cash, or its equivalent, that could be obtained by selling an asset in orderly liquidation. Current market value is also generally used for assets expected to be sold at prices lower than previous carrying amounts. Some investments in marketable securities are reported at their current market value. Some liabilities that involve marketable commodities and securities, for example, the obligations of writers of options or sellers of common shares who do not own the underlying commodities or securities, are reported at current market value (SFAC No. 5, paragraph 67c).

Fair value is not defined in SFAC No. 5 but used in several SFASs. In the view of the FASB the term fair value has the same meaning as market value. However, the FASB choose fair value to avoid confusion because not all items are actively traded in markets.

Fair value of an asset or liability is the amount at which that asset (liability) could be bought (incurred) or sold (settled) in a current transaction between willing partners, that is, other than in a forced or liquidation sale (SFAS No. 125, paragraph 42).

Quoted market prices are the best evidence of fair value. If no quoted market prices are available the estimate should at first consider market prices of similar assets or otherwise should be based on valuation models appropriate for those circumstances (e.g. discounted cash flow, option pricing models or fundamental analysis).

3.4 Net realizable value (net realizable settlement value)

Net realizable (settlement) value is the non-discounted amount of cash, or its equivalent, into which an asset is expected to be converted in due course of business less direct costs, if any, necessary to make that conversion. Short-term receivables and some inventories are reported at their net realizable value. Liabilities that involve known or estimated amounts of money payable at unknown future dates, for example, trade payables or warranty obligations, generally are reported at their net settlement value, which is the non-discounted amounts of cash, or its equivalent, expected to be paid to liquidate an obligation in the due course of business, including direct costs, if any, necessary to make that payment (SFAC No. 5, paragraph 67d).

3.5 Present (or discounted) value of future cash flows

Present value of future cash flows is the present or discounted value of future cash inflows into which an asset is expected to be converted in due course of business less present values of cash outflows necessary to obtain those inflows. Long-term receivables are reported at their present value (discounted at the implicit or historical rate). The present value (discounted at the implicit or historical rate), which is the present or discounted value of future cash outflows expected to be required to satisfy the liability in due course of business applies for liabilities. Long-term payables are reported at their present value (SFAC No. 5, paragraph 67e).

4 Impairment

Recognition of impairment of assets is generally required when events and circumstances indicate that the carrying amount of assets will not be recovered in the future (SFAS No. 121). For loans, recognition of an impairment loss is required when it is probable that the creditor will not be able to collect all amounts due according to the contractual terms of the loan agreement, including both the contractual interest and the principal receivable (SFAS No. 114 and SFAS No. 118).

4.1 Impairment of long-lived assets

SFAS No. 121, Accounting for the Impairment of Long-Lived Assets and for Long-Lived Assets to Be Disposed Of, requires that long-lived assets and certain identifiable intangible assets and any goodwill related to those assets have to be reviewed for impairment whenever circumstances and situations change such that there is an indication that the carrying amount may not be recoverable. The statement applies to all entities but does not apply to the following types of asset (SFAS No.121, paragraph 3):

- Financial instruments.
- Long-term customer relationships of a financial institution (i.e. core deposit intangibles and credit cardholder intangibles).
- Mortgage and other servicing rights.
- Deferred policy acquisition costs.
- Deferred tax assets.

Certain FASB Statements establish separate standards of accounting for specific long-lived assets in specialized situations. SFAS No. 121 does not change those standards (SFAS No. 121, paragraph 4).

4.1.1 Long-lived assets to be held and used

If, upon review, the undiscounted expected future cash flows (without interest charges) are less than the carrying amount of an asset, an impairment loss has to be recognized. In a second step the carrying amount is reduced to fair value. The fair value is considered the new cost basis, which is not subject to subsequent adjustment except for depreciation and further impairment. For a depreciable asset, the new cost shall be depreciated over the asset's remaining useful life. Restoration of previously recognized impairment losses is prohibited.

In estimating expected future cash flows for determining whether an asset is impaired, assets should be grouped at the lowest level for which there are identifiable cash flows that are largely independent of the cash flows of other groups of assets.

Once the need to recognize an impairment loss is established, the amount of that loss is measured as the excess of the carrying amount of the asset over the fair value of the asset. The fair value of an asset is the amount at which the asset could be bought or sold in a current transaction between willing parties, that is, other than in a forced or liquidation sale. SFAS No. 121, paragraph 7 states that quoted market prices in active markets are the best evidence of fair value and shall be used as the basis for the measurement, if available. If quoted market prices are not available, the estimate of fair value shall be based on the best information available in the circumstances. The estimate of fair value shall consider prices for similar assets and the results of valuation techniques to the extent available in the circumstances.

An impairment loss recognized for assets to be held and used is to be reported as a component of income from continuing operations before income taxes in the income statement. If an entity chooses to report a subtotal such as 'income from operations', that amount must include the impairment loss.

If an impairment loss is recognized the following information must be disclosed:

- Description of the impaired assets and the facts and circumstances leading to the impairment.
- The amount of the impairment loss and how fair value was determined.
- The location of the impairment loss in the income statement in which the impairment loss is aggregated.
- Business segment(s) affected (if applicable).

4.1.2 Long-lived assets to be disposed of

APB Opinion No. 30 (Reporting the Results of Operations, Reporting the Effects of Disposal of a Segment of a Business, and Extraordinary, Unusual and Infrequently Occurring Events and Transactions) specifies accounting for the disposal of a segment of a business and requires certain assets included in that type of transaction to be measured at the lower of carrying amount or net realizable value. SFAS No. 121 requires all long-lived assets and certain identifiable intangibles to be disposed of that are not covered by APB

Opinion No. 30 should be reported at the lower of carrying amount or net realizable value.

Fair value is determined in the same manner as it is determined for long-lived assets to be held and used (see preceding section). The costs to sell includes incremental direct costs to transact the sale of the asset, such as broker commissions, legal and title transfer fees and closing costs. Costs that are generally excluded from the cost to sell are insurance, security services, utility expense, and other costs of protecting or maintaining the asset. If the fair value of the asset is determined by discounting expected future cash flows and if the sale is expected to occur beyond one year, the cost to sell the asset is discounted.

Long-lived assets being held for disposal should not be depreciated while they are being held for disposal.

Subsequent revisions in estimates of fair value less cost to sell should be reported as adjustments to the carrying amount of the asset to be disposed of, provided that the carrying amount of the asset does not exceed the carrying amount of the asset before an adjustment was made to reflect the decision to dispose of the asset (SFAS No. 121, paragraph 17). This may result in an increase or a decrease in the carrying amount of the impaired asset to be disposed of, subsequent to the initial recognition of an impairment loss. Thus an asset held for disposal can be written up or down in future periods, as long as the write-up is never greater than the carrying amount of the asset before the impairment. This is a difference in accounting for assets to be held and used and assets to be disposed of. Once an impairment loss is recognized on assets to be held and used, it cannot be restored.

Losses or gains related to impaired long-lived assets to be disposed of, should be reported in the same manner as a similar adjustment for assets to be held and used, as a component of income from continuing operations before income taxes in the income statement. If an entity chooses to report a subtotal such as 'income from operations', the amount recognized for the impairment of the assets to be disposed of must be disclosed.

The following information must be included:

- Description of assets to be disposed of, including the facts and circumstances leading to the expected disposal, the expected disposal date, and the carrying amount of the assets.
- The business segment(s) in which the assets to be disposed of are held (if applicable).
- The loss, if any, resulting from the application of SFAS No. 121 or APB Opinion No. 30.
- The gain or loss, if any, resulting from changes in the carrying amounts of assets to be disposed of that arise subsequent to the initial recognition of impairment.
- The location in the income statement in which the gains or losses are presented if they are not presented as a separate heading or reported parenthetically on the face of the statement.
- The results of operations for assets to be disposed of to the extent that those results can be identified.

4.2 Impairment of loans

SFAS No. 114, Accounting by Creditors for Impairment of a Loan, specifies that a loan is impaired when, based on current information and events, it is probable that a creditor will be unable to collect all amounts due (both principal and interest) according to the contractual terms of the loan agreement. For purposes of applying SFAS No. 114, a loan is a contractual right to receive money on demand or on fixed or determinable dates that is recognized as an asset in the creditor's statement of financial position. Examples include but are not limited to accounts receivable (with terms exceeding one year) and notes receivable.

SFAS No. 114 applies to all loans that are identified for evaluation, uncollateralized as well as collateralized, except:

- Large groups of smaller-balance homogeneous loans that are collectively evaluated for impairment, such as credit cards, residential mortgages, and consumer instalment loans.
- Loans that are measured at fair value or at the lower of cost or fair value (i.e., in accordance with SFAS No. 65, Accounting for Certain Mortgage Banking Activities, or other specialized industry practice).
- Leases.

- Debt securities as defined in SFAS No. 115, Accounting for Certain Investments in Debt and Equity Securities.

If a loan is considered impaired, the loss due to the impairment should be measured as the difference between the investment in the loan (generally the principal plus accrued interest) and the expected future cash flows discounted at the loan's effective interest rate. If a loan is collateral dependent, as a practical expedient, the creditor may measure impairment based on a loan's observable market price, or the fair value of the collateral.

If the measure of the impaired loan is less than the recorded investment in the loan (including accrued interest, net deferred loan fees or costs, and unamortized premium or discount), the creditor shall recognize an impairment by creating or adjusting a valuation allowance with a corresponding charge to bad-debt expense.

After the initial measurement of impairment, any significant change (increase or decrease) in the amount or timing of an impaired loan's expected or actual future cash flows should be reflected by a recalculation of the impairment and an adjustment to the allowance account. The net carrying amount of the loan shall at no time exceed the recorded investment in the loan.

SFAS No. 118 amends SFAS No. 114 to indicate that guidance is not provided concerning how a creditor should recognize, measure, or display interest income on an impaired loan. Under SFAS No. 118 the creditor may use one of the two alternative interest income recognition methods provided under SFAS No. 114 or may use some other reasonable method.

SFAS No 118 requires that the following information be disclosed, either in the body of the financial statements or in accompanying notes, for loans that meet the definition of an impaired loan:

- As of the date of each statement of financial position, the total recorded investment in the impaired loan at the end of each period and (i) the amount of that recorded investment for which there is a related allowance for credit losses determined in accordance with SFAS No. 118; and (ii) the amount of the recorded investment for which there is no allowance for credit losses in accordance with this SFAS No. 118.
- The creditor's policy for recognizing interest income on impaired loans, including how cash receipts are recorded.
- For each period for which results of operations are presented, the average recorded investment in the impaired loans during each period, the related amount of interest income recognized during the time within that period that the loans were impaired, and, unless not practicable, the amount of interest income recognized using a cash-basis method of accounting during the time within that period that the loans were impaired.

For each period for which results of operations are presented, a creditor shall disclose the activity in the total allowance for credit losses related to loans, including the beginning and end of each period, additions charged to operations, direct write-downs charged against the allowance, and recoveries of amounts previously charged off.

5 Recognition and valuation rules for assets, liabilities and stockholders' equity

5.1 Start-up costs

Guidance on accounting for costs incurred on start-up activities is given in SOP 98-5, Reporting on the Costs of Start-up Activities. The SOP broadly defines start-up activities as those one-time activities related to opening a new facility, introducing a new product or service, conducting business in a new territory, conducting business with a new class of customers or beneficiary, initiating a new process in an existing facility, or commencing some new operation. Start-up activities also include costs related to organizing a new entity, e.g. initial incorporation, legal and accounting fees incurred in connection with establishing the entity. Certain costs incurred in conjunction with start-up activities are not covered by the SOP, such as costs of acquiring or constructing long-lived assets and getting them ready for their intended uses, costs of acquiring or producing inventory, costs of raising capital, costs of advertising and research and development costs. Also, activities related to routine, ongoing effects to refine, enrich, or otherwise improve upon the qualities of an existing product, service, process, or facility are not start-up activities and are not within the scope of SOP 98-5.

The SOP requires costs of start-up activities, including organization costs to be expensed as incurred. This requirement is also applicable for precontract costs in the course of long-term contracts. This treatment runs contrary to IRS guidelines, which dictate that in order for start-up costs to be deductible, they must be amortized.

Special disclosure requirements are not provided, since the AcSEC assumed that the costs of separately identifying start-up costs incurred would likely outweigh the related benefits of disclosing those costs. If an entity submits a disclosure of total start-up costs expected to be incurred, it is usually not included in the financial statements, but may be included in Management's Discussion and Analysis.

5.2 Research and development costs

5.2.1 Recognition

SFAS No. 2, Accounting for Research and Development Costs and SFAS No. 68, Research and Development Agreements, provide guidance on GAAP relating to research and development (R&D) costs. R&D costs incurred in purchasing or internally developing and producing computer software products that are sold, leased, or otherwise marketed by an enterprise, are accounted for in accordance with SFAS No. 86.

Research is defined in SFAS No. 2 as the planned efforts of a company to discover new information that will help create a new product, service, process, or technique or vastly improve one in current use. Development is the translation of research findings or other knowledge into a plan or design for a new product or process or for a significant improvement to an existing product or process whether intended for sale or use. Examples of R&D costs covered by SFAS No. 2 include laboratory research to discover new knowledge, formulation and design of product alternative, preproduction prototypes and models, and engineering activity until the product is ready for manufacture. Engineering during an early phase of commercial production, quality control for commercial production, troubleshooting during a commercial production breakdown and routine, ongoing efforts to improve products are not applicable to the provisions of SFAS No. 2.

SFAS No. 68 covers an enterprise's research and development arrangements that are partially or completely funded by other parties. The actual R&D activities are performed by the enterprise or a related party, usually under a contract with a limited partnership through which the R&D activities are funded. The accounting for the R&D costs depends upon the nature of the obligation that the enterprise incurs in the arrangement.

5.2.2 Valuation

Generally, all R&D costs are charged to expense when incurred (SFAS No. 2, paragraph 12). Exceptions from this rule are given based on the specifications of the elements of costs associated with R&D activities. SFAS No. 2, paragraph 11 identifies five R&D cost categories and requires the following accounting treatment:

- Material, equipment, and facility costs should be expensed, unless the items have alternative future uses in other R&D projects or otherwise. If an alternative future use exists for those expenditures, the costs are capitalizable as a tangible or intangible asset and depreciated or amortized. The periodic depreciation or amortization is identified as R&D expense as long as the asset is used in R&D activity.
- Personnel, salaries, wages, and other related costs of personnel engaged in R&D should be expensed as incurred.
- Cost of purchased intangibles should be expensed, unless the items have alternative future uses in other R&D projects or otherwise.
- Contract services. The cost of services performed by others in connection with the reporting company's R&D should be expensed as incurred.
- Indirect costs. A reasonable allocation of indirect cost should be included in R&D costs, except for general and administrative cost. Indirect costs must be clearly related to be included and expensed.

In the case of an agreement for the funding of R&D activities by others, the nature of the obligation determines the accounting of R&D costs. If the obligation is solely to perform contractual services, all R&D costs are charged to cost of sales. If the obligation represents a liability to repay all of the funds provided by other funds, all R&D costs are charged to expense when incurred. If the nature of the obligation is partly a liability and partly the performance of contractual services, R&D costs are charged partly to expense and partly to cost of sales. The portion charged to expense is related to the funds provided by the other parties that are likely to be repaid by the enterprise. SFAS No. 68, paragraph 9 requires that the enterprise charges its portion of the R&D costs to expense in the same manner as the liability is incurred. An important criterion in determining an enterprise's obligation is whether the financial risk involved in the arrangement has been substantially transferred to other parties. To the extent the financial risk associated with the R&D has been transferred because repayment of any of the funds provided depends solely on the results of the R&D having future economic benefit, the enterprise should account for its obligation as a contract to perform R&D for others.

Costs of the development of computer software are considered to be R&D costs under the provisions of SFAS No. 2, until technological feasibility has been established (SFAS No. 86, paragraph 3). The technological feasibility is established, when the enterprise has completed all the planning, designing, coding, and testing activities that are necessary to meet the design specifications of the product (SFAS No. 86, paragraph 6). Design specifications, in turn, may include such product aspects as functions, features, and technical performance requirements.

5.2.3 Disclosures

SFAS No. 2, paragraph 13, which is also applicable for computer software costs that are classified as R&D costs, requires the disclosure, generally in footnotes, of the total R&D costs charged to expense. For R&D arrangements that are accounted for as contracts to perform R&D service for others, SFAS No. 68, paragraph 14 requires to disclose the terms of the arrangement, including purchase provisions, licence and royalty agreements, and the amount of R&D costs incurred and compensation earned during the period.

5.3 Goodwill

5.3.1 Recognition

In the case of a business combination, that is to be accounted for as a purchase, the assets acquired and liabilities assumed are initially recorded at their fair values. If the assets and liabilities net to an amount other than the total acquisition price, the excess (or deficiency) is generally referred to as goodwill (negative goodwill), which is accounted for in accordance with APB Opinion No. 17, Intangible assets. Goodwill can arise only in the context of a purchase business combination, since in a pooling, the assets and liabilities of the combining entities are carried forward at precombination book values.

When the acquired entity is merged into the acquiring entity or when both entities are consolidated into a new (third) entity, all assets and liabilities are recorded directly on the books of the surviving or the new organization. Depending upon whether the conditions stipulated by APB Opinion No. 16, Business Combinations, are met, this transaction will be treated as either a pooling or a purchase. However, when the acquirer obtains a majority (or all) of the voting common stock of the acquired entity (which maintains a separate legal existence), the assets and liabilities of the acquired company will not be recorded on the acquirer's books. In this case, GAAP normally requires that consolidated financial statements be prepared. In certain cases combined financial statements of entities under common control (but neither of which is owned by the other) are also prepared. This process is very similar to an accounting consolidation using pooling accounting, except that the equity accounts of the combining entities are carried forward intact.

The revaluation of the assets and/or liabilities, including the recognition of goodwill, is reflected in the books of the acquired company when the concept of push-down (or new basis) accounting is applied. Push-down accounting has no impact on the presentation of consolidated financial statements or on the separate financial statements of the parent (investor) company, which would continue to be based on the price paid for the acquisition and not on the acquired entity's book value. However, the use of push-down accounting represents a departure in the way separate financial statements of the acquired entity are presented. For further details see USA – Group Accounts, VI.1.3.5.

Goodwill is a 'going concern' valuation and cannot be separated from the business as a whole. Goodwill generated internally should not be capitalized in the accounts, because measuring the components of goodwill and associating any costs with future benefits is not objectively verifiable.

5.3.2 Valuation

Under the purchase method, the acquiring corporation must allocate the cost of the acquired company to the assets acquired and liabilities assumed. Acquired assets are recorded at their fair market value; likewise, liabilities at the present value of amounts to be paid. The tax basis of an asset or liability shall not be a factor in determining its fair value. However, a deferred tax asset or liability should be recognized for any difference between the fair value and the tax basis of an asset or liability if the difference represents a temporary difference. Contingent assets and liabilities should be included when the fair value of the preacquisition contingency can be determined during the allocation period, or information available prior to the end of the allocation period indicates that it is probable that an asset existed, a liability had been incurred, or an asset had been impaired, and the amount of the asset or liability can be reasonably estimated. The excess of the cost of the acquired company over the sum of the amounts assigned to identifiable net assets should be recorded as goodwill. If the goodwill is tax deductible, an accrual for deferred taxes must be set up.

If the values assigned to the identifiable net assets exceed the cost of the acquired company, such excess should be used to reduce proportionately the assigned values of non-current assets (except long-term investments in marketable securities). If this reduction results in a zero balance in the non-current assets, any remaining excess should be recognized as a deferred credit (negative goodwill) to be amortized systematically to income over the period estimated to be benefited but not in excess of 40 years.

Goodwill is considered an intangible asset and as such is subject to the APB Opinion No. 17 requirements for amortization of cost over useful life or 40 years, whichever comes first. In practice, depending on the type of industry, goodwill is usually amortized over a shorter period than 40 years, thereby often following the tax mandatory 15-year amortization period. A recent Exposure Draft of the FASB indicates a shortening of the accepted maximum period to 20 years, which would correspond to requirements of the SEC. The amortization should be computed using the straight-line method unless another method is deemed more appropriate. Goodwill cannot be written off as a lump sum to paid-in capital or retained earnings, or be reduced to a nominal amount, at or immediately after acquisition. Goodwill is also subject to an impairment test within the provisions of SFAS No. 121. In instances where goodwill is identified with assets that are subject to an impairment loss under SFAS No. 121, the carrying amount of the identified goodwill shall be eliminated before making any reduction of the carrying amounts of impaired long-lived assets and identifiable intangibles (SFAS No. 121, paragraph 12). The reversal of a depreciated amount is prohibited.

APB Opinion No. 16, paragraph 91 requires that the deferred credit resulting from negative goodwill be amortized systematically to income over the period expected to be benefited but not in excess of 40 years. The credit may not be added direct to stockholders' equity at the date of acquisition.

5.3.3 Disclosures

Goodwill is shown in the non-current asset section of the balance sheet after tangible assets such as property, plant and equipment. Accumulated amortization is generally deducted directly from the asset account. However, it is permissible to reflect the accumulated amortization in a separate valuation account that is shown as a deduction from the original cost of the asset on the face of the balance sheet or in the notes.

Write-offs of intangible assets should generally

not be reported as extraordinary items because they are usual in nature or may be expected to recur in the future. However, write-offs of intangible assets may be classified as extraordinary items if they are a direct result of a major casualty, an expropriation, or a prohibition under a newly enacted law or regulation that meets both the unusual nature and infrequency of occurrence tests of APB Opinion No. 30.

APB Opinion No. 22, paragraph 13, Disclosure of Accounting Policies, requires that the policy with respect to amortization of intangible assets be included in an accounting policies note. APB Opinion No. 16, paragraph 95, and APB Opinion No. 7, paragraph 30 require both the method and the period of amortization to be disclosed. The method and period of amortization of a deferred credit should also be disclosed, as required by APB Opinion No. 16, paragraph 91.

5.4 Other intangible fixed assets

5.4.1 Recognition

Intangible assets differ in their characteristics and may be classified on the basis of their identifiability, manner of acquisition, expected period of benefit and separability. A company should record as assets the costs of intangible assets acquired from other enterprises or individuals. Costs of developing, maintaining, or restoring intangible assets which are not specifically identifiable, have indeterminate lives, or are inherent in a continuing business and related to an enterprise as a whole should be deducted from income when incurred (APB Opinion No.17, paragraph 24). Costs for computer software are capitalizable as an intangible asset under certain criteria (SFAS No. 86, SOP 98-1).

Examples for specifically identifiable intangibles are as follows:

- Copyrights. The term of copyright in the United States is the life of the creator plus 50 years.
- Customer and supplier lists. This intangible asset is capitalized based on the assumption of continuing business.
- Franchises. If a franchise agreement covers a specified period of time, the cost of the franchise is capitalized and written off systematically over that period unless the economic life is anticipated to be less.
- Non-compete agreements. These intangibles are capitalized based upon the agreement that the seller of a business will not engage in a competing business in a certain area for a specified period of time.
- Patents. A United States patent has a specified legal life which provides protection for 17 years.
- Royalty and licence agreements. These intangible assets are capitalized based upon the life of the agreement.
- Secret formulas and processes. Costs that can be directly identified with secret formulas and processes are capitalized, except for costs of activities constituting research and development.
- Trademarks and trade names. These intangible assets have an unlimited life as long as they are continually used, although technically the term of registration at the US Patent Office is 20 years, with indefinite renewal for additional 20-year periods.

An example for a non-identifiable intangible asset is goodwill resulting from the acquisition of a company in accordance with APB Opinion No. 17.

5.4.2 Valuation

Intangible assets acquired singly should be recorded at cost at the date of acquisition. Cost is measured by the amount of cash disbursed, the fair value of other assets distributed or the present value of amounts to be paid for liabilities incurred (ABP Opinion No. 17, paragraph 25).

Specific regulations are provided for the accounting for the costs of computer software, whereas it should be distinguished whether the software is to be sold (SFAS No. 86) or developed or obtained for internal use (SOP No. 98-1). In both cases all costs incurred to establish the technological feasibility of a computer software product should be charged to expense when incurred. The technological feasibility is established when the enterprise has completed all planning, designing, coding, and testing activities that are necessary to meet the design specifications of the product (SFAS No. 86, paragraph 4).

For computer software to be sold, leased, or otherwise marketed, capitalization should begin when the technological feasibility is established and shall cease when the product is available for general release to customers. Cost of maintenance and customer support shall be charged to expense when related revenue is recognized or when those costs are incurred, whichever occurs first (SFAS No. 86, paragraph 6).

For computer software developed or obtained for internal use, capitalization of costs should begin when the preliminary project stage is completed and management authorizes and commits to funding the project and it is probable that the project will be completed and the software will be used to perform the intended function. Costs to be capitalized under SOP No. 98-1 are external costs of materials and services consumed, to the extent of the time spent directly on the project payroll and payroll-related costs for employees who are directly associated with and devote time to the internal-use software project. General and administrative costs and overhead costs should not be capitalized.

The cost of each type of intangible asset should be amortized on the basis of the estimated life of that specific asset and should not be written off in the period of acquisition. For intangible assets that have a limited life, the cost of the intangible asset is amortized over the estimated period of benefit of that specific asset up to a maximum period of forty years. Factors which should be considered in estimating the useful lives are as follows:

- Provisions for renewal or extension may alter a specified limit on useful life.
- Effects of obsolescence, demand, competition, and other economic factors may reduce a useful life.
- Expected actions of competitors and others may restrict present competitive advantages.

For intangible assets with indeterminate lives that are likely to exceed 40 years, costs should be amortized over the maximum period of 40 years (APB Opinion No. 17, paragraph 28(f). Write-ups are not allowed for intangible assets.

APB Opinion No. 17, paragraph 30 requires that the straight-line method of amortization be used unless another systematic method appears more appropriate. However, except for specialized industries such as banks and thrifts, methods other than straight-line are rarely encountered. Any change in the estimated remaining life should be treated as a change in accounting estimate and accounted for prospectively. Generally, in computing amortization, no residual value is associated with intangibles. Amortization of acquired intangible assets not deductible in computing income taxes payable does not create a timing difference, and allocation of income taxes is inappropriate.

Intangibles must be assessed periodically under the provisions of SFAS No. 121 to ascertain whether any permanent diminution in value should be recognized in addition to normal amortization. Also, each intangible right must be assessed periodically by management to see if the amortization term continues to be reasonable. If not, adjustment of the term may be necessary. Diminution in value often is recognized through shortening the amortization term. Occasionally, an intangible balance may be completely written off based on the occurrence of a single event.

5.4.3 Disclosures

Intangible assets are shown in the non-current asset section of the balance sheet after tangible assets such as property, plant and equipment. Accumulated amortization is generally deducted directly from the asset account. Unamortized computer software costs are always classified in the balance sheet as an intangible asset. Amortization of intangibles is included in the operating expense category of the income statement. The financial statements should disclose the method and period of amortization (APB Opinion No. 17, paragraph 30).

5.5 Tangible fixed assets

5.5.1 Recognition

Tangible fixed assets are used in a productive capacity, have physical substance, are relatively long-lived, and provide future benefit which is readily measurable. Such assets, referred to as property, plant and equipment, include land, building structures (offices, factories, warehouses) and equipment (machinery, furniture, tools).

The major characteristics of property, plant and equipment are:

- Acquired for use in operations and not for resale.
- Long-term in nature and usually subject to depreciation.
- Possess physical substance.

Only assets used in normal business operations should be classified as property, plant and equipment. They yield services over a number of years and the investment in these assets is assigned to future periods through periodic depreciation charges. Only land is not depreciated, unless a material decrease in value occurs.

Property, plant and equipment differ from intangible assets as they are characterized by physical existence or substance. Unlike raw materials they do not physically become part of a product held for resale.

5.5.2 Valuation

The acquisition of property, plant and equipment should be recorded at historical cost, which includes the costs necessarily incurred to bring them to the condition and location necessary for their intended use.

Any reasonable cost involved in bringing the asset to the buyer and incurred prior to using the asset in actual production is capitalized. Examples include sales taxes, finders' fees, freight costs, installation costs, and set-up costs. These costs are not to be expensed in the period in which they are incurred.

Cost for land typically include:

- Purchase price.
- Costs incurred in 'closing' and obtaining title to the land.
- Cost incurred in preparing the land for its intended use, such as grading, filling, draining and clearing.
- Assumption of any liens or mortgages or encumbrances on the property.
- Any additional land improvements that have an indefinite life.

If a company acquires an option to purchase land and later exercises that option, the cost of the option generally becomes part of the cost of the land. Even if an option elapses without being exercised, its cost can be capitalized if the option is one of a series of options acquired as part of an integrated plan to acquire a site. In that case, if any one of the options is exercised, the cost of all may be capitalized as part of the cost of the site.

Cost for a building typically include:

- Purchase price or materials, labour and overhead costs incurred during construction or excavation costs.
- Capitalized interest (if self-constructed).
- Fees, such as attorneys' and architects'.
- Building permits.

Cost of equipment includes:

- Purchase price.
- Freight and handling charges incurred.
- Insurance on equipment while it is in transit.
- Cost of special foundations if required.
- Assembly and installation costs.
- Cost of conducting trial runs.

Subsequent to acquisition, plant and equipment are reflected at cost less accumulated depreciation. Land is shown at cost.

5.5.2.1 *Costs subsequent to acquisition*

Various expenditures relating to plant and equipment will normally be incurred during the useful lives of those assets. In general, costs incurred to achieve greater future benefits should be capitalized, whereas expenditures that simply maintain a given level of services should be expensed.

In order for costs to be capitalized, one of three conditions must be present:

- The useful life of the asset must be increased.
- The quantity of the units produced from the asset must be increased.
- The quality of the units produced must be enhanced.

Ordinary repairs that represent expenditure to maintain plant assets in operating condition are charged to expense accounts in the period in which they are incurred, on the basis that it is the only period benefiting. Replacement of minor parts, lubricating and adjusting equipment, repainting, and cleaning are examples of the type of maintenance charges that occur regularly and are treated as ordinary operating expenses.

Extraordinary repairs occur infrequently and involve relatively large amounts of money. They tend to increase the usefulness of the asset in the future because of greater efficiency or longer life, or both. As such, the cost is capitalized. Extraordinary repairs are represented by major overhauls, complete reconditioning, and major replacements and betterment. When the quantity or quality of the asset itself has not been improved, but the useful life has been extended, the expenditure may be debited to accumulated depreciation rather than to an asset account.

5.5.2.2 Scheduled depreciation

A depreciation method is deemed to be acceptable if it results in a systematic and rational allocation of the cost of an asset over the asset's useful life. The allowed depreciation methods may be classified as follows:

- Straight-line method.
- Decreasing charge methods:
  - (i) Sum of the years' digits.
  - (ii) Double declining balance.
- Activity method (units of use or production).
- Inventory method.

The straight-line method is used where the asset's economic usefulness is the same each year. This method is applicable to both buildings and machines. The decreasing charge methods are used when the asset is more efficient or suffers the greatest loss of service in the earlier years. This method is normally used if the revenues generated by the asset are higher at the beginning of the life of the asset. It is likewise applicable to both buildings and machines. The activity method is used for assets where the number of units of output or service hours received from the asset can be estimated. This method is to machines used in production applicable.

GAAP does not provide rules recording the useful economic life of assets. Tax regulations do mandate the appropriate useful life; however, conformity is not required. If an asset is deemed to be worthless, then planned depreciation is suspended and the asset is written down to an amount not exceeding its net realizable value.

There are no prerequisites for a change in depreciation methods. APB Opinion No. 20 only requires that the cumulative effect of the change be recognized in the net income of the period of change.

5.5.2.3 Unscheduled depreciation

Specific procedures for identifying assets subject to write-down for impairment, as well as standards for measuring the amount of the write-down, are included in SFAS No. 121 (see VI.4.1).

5.5.2.4 Special problem areas

Minor items are allowed to be expensed immediately, depending on the materiality of the item. The amount capitalized may vary in relation to a certain percentage of net income or total assets.

5.5.3 Disclosures

APB Opinion No. 12, Omnibus Opinion, 1967 requires the following disclosures in the financial statements or in the footnotes for fixed assets:

- Depreciation expense for the period.
- Balances of major classes of depreciable assets, by nature or by function, at the balance sheet date.
- Accumulated depreciation, either by major asset classes or in total, at the balance sheet date.
- A general description of the method or methods used in computing depreciation for the major classes of depreciable assets.

The SEC requires all public companies whose fixed assets (net of depreciation) exceed 25 per cent of net assets to provide additional disclosures (see IV.2.2.2).

5.6 Investments

5.6.1 Equity investments

APB Opinion No. 18 requires the use of the equity method of accounting for investments in common stock when the investor has the ability to exercise significant influence over the operating or financial decisions of the investee, including joint ventures. The equity method is not intended as a substitute for consolidated financial statements when the conditions for consolidation are present.

In order to establish uniform application of the equity method, the presumption in APB Opinion No. 18 is that, absent evidence to the contrary, an investment of 20 per cent or more of the voting stock of the investee indicates that the investor has the ability to influence the investee.

Long-term investments in common stock where the investor lacks the ability to influence the operating or financial decisions of the investee are accounted for following the guidance of SFAS No. 115; see VI.5.6.2.

Under the equity method the investment is originally recorded at acquisition cost but is subsequently adjusted for changes in net assets of the investee. That is, the investment's carrying amount is adjusted for the investor's proportionate share of the income or loss of the investee and decreased by the dividends received by the investor from the investee. For further explanations of the equity method, refer to USA – Group Accounts, VII.2.

5.6.2 Marketable securities and certain financial assets subject to prepayment risk

Marketable securities are equity securities with a readily determinable fair value where the company is not able to control or influence the operating or financial decisions of the investee and debt securities.

At acquisition the securities have to be classified into one of three categories: held to maturity, trading or available for sale. Transfer between these categories is only allowed under certain rare circumstances.

5.6.2.1 *Held to maturity securities*

Debt securities that the enterprise has the intent and ability to hold to maturity. If the enterprise intends to sell the securities (e.g. in cases of liquidity needs or change of interest rates), the security can not be classified as held to maturity. A sale is only allowed under certain circumstances (deterioration of creditworthiness, certain changes in tax law). A security that can be prepaid or otherwise settled in such a way that the holder would not recover substantially all of its investment cannot be classified as held to maturity.

Held to maturity securities are valued at amortized cost. Amortized cost is the acquisition cost adjusted for the amortization of a discount or premium. In case of a not only short-term decline in value they have to be written down to a fair market value. Subsequent recoveries in value do not result in a write-up.

Interest is recorded in the income statement when earned. Gains and losses are recognized in the case of a sale.

5.6.2.2 *Trading securities*

Debt and equity securities that are frequently bought and sold with the objective of generating profit on short-term differences in price are considered trading securities. Trading securities are always presented as current assets. Trading securities are valued at fair value with unrealized gains and losses included in earnings. Premiums or discounts on debt securities are not amortized. Interest from debt securities is recorded in the income statement when earned.

5.6.2.3 *Available for sale*

Securities not classified as held to maturity nor trading have to be classified as available for sale.

They are valued at fair value. Unrealized gains or losses are excluded from earnings and reported in other comprehensive income until realized (net of deferred taxes). Interest earned and gains and losses from sale are recognized in the income statement.

Available for sale securities have to be written down through income in case of a long-term impairment in value. Subsequent recoveries are reported in other comprehensive income.

SFAS No. 125 extends the available for sale or trading approach to non-security financial assets that can contractually be repaid or otherwise settled in such a way that the holder of the asset would not recover substantially all of its recorded investment. Thus, non-security financial assets such as loans, interest-only strips or residual interests in securitization trusts are reported at fair value with the change in fair value accounted for depending on the asset's classification.

5.6.2.4 *Disclosures*

Several disclosures are required for investments in marketable securities.

For available for sale and held to maturity securities only:

- Aggregate fair value by category.
- Gross unrealized holding gains.
- Gross unrealized holding losses.
- Amortized cost basis.
- Contractual maturities of debt securities.

For available for sale securities only:

- Proceeds from sale.
- Gross realized gains and losses.
- Gross gains and losses from transfer to trading securities.
- Change in the net unrealized holding gain and loss that is included in other comprehensive income.

For held to maturity securities only:

- Amortized cost amount of sold or transferred securities and the related realized or unrealized gain or loss.
- Circumstances leading to the decision to sell or transfer.

Further disclosure requirements are:

- Basis on which cost was determined (specific identification, average cost or other method used).
- Change in the net unrealized hedging gain or loss on trading securities.

Financial institutions also have to disclose the types of securities and provide certain groupings for the maturities of debt securities.

5.7 Inventories

5.7.1 Recognition

Definition, valuation and classification of inventories are discussed in ARB No. 43, section 4. The term inventory includes goods awaiting sale (the merchandise of a trading company and the finished goods of a manufacturer), goods in the course of production (work in progress), and goods to be consumed directly or indirectly in production (raw materials and supplies). This definition of inventories excludes long-term assets subject to depreciation accounting, or goods which, when put into use, will be so classified.

The amount presented as inventories in the balance sheet is the aggregate historical cost, determined using selected cost-flow assumptions and valuation conventions, of the physical units on hand at the reporting date, adjusted for certain economic and physical factors such as market conditions and excess, damaged and obsolete stock. In order to assist in the determination of the actual physical quantity of inventory on hand, the proper ownership of inventories has to be determined. In general, a firm should record purchases and sales of inventory when legal title passes. Areas which create a question as to proper ownership are goods in transit and consignment sales. Goods in transit that were shipped free on board (f.o.b.) shipping point at the balance sheet date should be included in inventory of the buyer. Goods in transit shipped f.o.b. destination should be included in the inventory of the seller. Goods out on consignment should be included in the inventory of the consignor. For accounting requirements in the case of long-term contracts, see VII.5.

A special inventory issue is the accounting for product financing arrangements under the provisions of SFAS No. 49, Accounting for Product Financing Arrangements. These arrangements are transactions in which a company sells a portion of its inventory to another entity with the intent to acquire that inventory in the future at a purchase price equal to the original sale price plus carrying and handling costs. Under certain criteria, among them the retaining of the risks and rewards of ownership by the seller, those arrangements are viewed as financing transactions.

5.7.2 Valuation

The primary basis of accounting for inventories is cost, which is the cash equivalent of the expenditures necessarily incurred to bring the items to the condition and location for their intended use. The definition of cost as applied to inventories is understood to mean acquisition and production cost and is commonly referred to as the concept of absorption or full costing. For raw materials and merchandise, inventory cost will include all expenditure incurred in bringing the goods to the point of sale and putting them into a saleable condition. These costs include the purchase price, transportation costs, insurance and handling costs.

Work in progress and finished goods inventory accounts are to include an appropriate portion of direct materials, direct labour, and indirect production costs (both fixed and variable). ARB No. 43, chapter 4 indicates that, under some circumstances, expenses related to an idle facility, excessive spoilage, double freight and rehandling costs may be abnormal and therefore require treatment as period costs. On the other hand, there may be instances in which certain general and administrative expenses are directly related to the production process and therefore should be allocated to the work in progress and finished goods inventory. ARB No. 43 also points out the fact that selling costs do not constitute inventory costs. Interest should not be capitalized for inventories that are routinely manufactured or otherwise produced in large quantities on a repetitive basis.

According to ARB No. 43, chapter 4 cost for inventory purposes shall be determined under any one of several assumptions as to the flow of cost factors. The major objective in selecting a method should be to choose the one which, under the circumstances, most clearly reflects periodic income. The most common cost flow assumptions used are:

- Weighted average method. This method is used where the inventory involved is relatively homogeneous and it is impossible to measure a specific physical flow of inventory, so that it is better to cost items on an average price basis.
- First in, first out (FIFO). This method is based on the assumption that goods are used in the order in which they are purchased; in other words, the first goods are the first used (in a manufacturing concern) or sold (in a trading concern).
- Last in, first out (LIFO). This method is based on the assumption that the cost of the last goods purchased matched against revenue first; in other words, the last goods are the first used (in a manufacturing company) or sold (in a trading company). There are several variations of LIFO (e.g. pooled LIFO; specific goods; dollar value). The objective of pooled LIFO is to group inventory items so as to match most recently incurred costs to current revenues. In the specific goods approach to LIFO, changes in the quantity of individual types of inventory is the basis for determining whether inventory levels have increased or whether a portion of the existing inventory has been liquidated. The purpose of the dollar-value LIFO method is to convert inventory which is priced at end-of-year prices to that same inventory priced at base-year (or applicable LIFO layer) prices. The dollar-value method achieves this result through the use of a conversion price index. The inventory at current year cost is divided by the appropriate index to arrive at the base-year cost. The result is compared with each pool's aggregate base-year cost at the end of the prior year to determine whether or not inventory levels have increased. Thus, the main focus is on the determination of the conversion price index. One approach that can be used in the computation of the LIFO value of a dollar-value pool is the double-extension method. This involves extending the entire quantity of ending inventory for the current year at both base-year prices and end-of-year prices to arrive at a total dollar value for each, hence the title of double extension. The end-of-year dollar total is then divided by the base-year dollar total to arrive at the conversion price index. Other approaches are the link chain, the index method, and the alternative LIFO method for auto dealers.
- Specific identification. This method may be used only in instances where it is practical to separate physically the different purchases made. It can be applied in situations where a relatively small number of costly, easily distinguishable items are handled. In the retail trade this includes some types of jewellery, fur coats, automobiles, and some furniture. In manufacturing, it includes special orders and many products manufactured under a job-cost system.

Standard costs may be used if they are adjusted at reasonable intervals to reflect current conditions. Inventory costs so determined must reasonably approximate costs computed under one of the recognized cost flow methods.

ARB No. 43 requires a departure from the cost basis of pricing the inventory when the utility of the goods is no longer as great as its cost. In such cases, the loss should be recognized in the period in which the decline takes place. This is accomplished by stating inventories at the lower of cost or market value. The market value of inventory is defined as the replacement cost, by purchase or production, except that:

- Market should not exceed the net realizable value (estimated selling price less reasonably predictable costs of completion and disposal).
- Market should not be less than the net realizable value less a normal profit margin.

The upper and lower limits set on replacement cost as an equivalent of market are intended to prevent inventory from being stated at an amount above the net proceeds that could be expected from the disposal of an item or at an amount below the net proceeds less normal profit margin. ARB No. 43 states that the 'lower of cost or market' rule can be applied direct to each item or to the total inventory or in some cases to the total component of each major category. The method should be that which most clearly reflects periodic income.

If the market value of inventory declines from its original cost, for whatever reason (for instance, obsolescence, price-level changes, or damaged goods), the inventory should be written down to reflect this loss. In the accounting practice the assessment of write-downs is often approached by an estimation, e.g. a specific percentage of the total amount. There is no potential or option to write back previous decreases in the value of assets. Increases in the value of assets are recognized only at the point of sale.

Inventory units, which are interchangeable and have an immediate marketability at quoted prices and for which appropriate costs may be difficult to obtain (e.g. agricultural and mineral products), may be stated at sales prices. They should be reduced by expenditures to be incurred in disposal, and the use of such basis should be fully disclosed in the financial statements (ARB No. 43, chapter 4.16).

In contrast to other accounting policy choices for financial statements and for tax purposes, the Internal Revenue Service (IRS) regulations require that the same method of inventory accounting used for tax purposes must also be used for financial reporting purposes.

For product financing arrangements, SFAS No. 49 requires that the merchandise covered by the arrangement be included with the inventory of the company and valued in accordance with the cost-flow assumptions and pricing methods applied by the company, with the proceeds from the sale treated as debt.

5.7.3 Disclosures

GAAP does not address whether the various components of inventory need to be disclosed. Regulation S-X, rule 5–02.6, requires that the major classes of inventory be disclosed in terms of their stage of completion.

Explanatory information in the notes is required for unusual or significant financing arrangements relating to inventories such as related-party transactions, product financing arrangements, firm purchase commitments, involuntary liquidation of LIFO inventories, and pledging of inventory as collateral.

The basis upon which inventory amounts are stated (lower of cost or market) and the method used in determining cost (LIFO, FIFO, average cost, etc.) should also be reported (ARB No. 43, chapter 5.8; APB Opinion No. 22, paragraph 13).

5.8 Trade accounts receivable

5.8.1 Recognition

Trade receivables are amounts owed by customers for goods sold and services rendered as part of normal business operations. The substance of a sale of any asset is that the transaction should transfer the risks and rewards of ownership to the buyer. Before these receivables are recognized in the balance sheet, the conditions of revenue being earned and being realized (or realizable) must be met. SFAC No. 5, paragraph 84 provides the following guidelines for recognizing revenue:

- If sale or cash receipt precedes production and delivery, revenues may be recognized as earned by production and delivery.

- If production is contracted for before production, revenues may be recognized by a percentage-of-completion method as earned – as production takes place – provided reasonable estimates of results at completion and reliable measures of progress are available. For further information, see VII.5.
- If services are rendered or rights to use assets extend continuously over time, reliable measures based on contractual prices established in advance are commonly available, and revenues may be recognized as earned as time passes.

Ordinarily, in the case of product sales a receivable is recognized at the time the purchaser of the product is vested with ownership rights. Therefore, a receivable cannot be recognized for bill and hold-agreements, unless the risk of theft or physical destruction or damage of the product does not rest with the seller.

For sales, in which the buyer has a right to return the product, a receivable from the sale transaction shall be recognized at time of sale only if all of the following criteria are met (SFAS No. 48, Revenue Recognition When Right of Return Exists, paragraph 6):

- The seller's price to the buyer is substantially fixed or determinable at the date of sale.
- The buyer has paid the seller, or the buyer is obligated to pay the seller and the obligation is not contingent on resale of the product.
- The buyer's obligation to the seller would not be changed in the event of theft or physical destruction or damage of the product.
- The buyer acquiring the product for resale has economic substance apart from that provided by the seller.
- The seller does not have significant obligations for future performance to directly bring about resale of the product by the buyer.

The risk of sales returns is usually reflected in a provision for sales returns.

Occasionally, before goods are sold and services are rendered, funds are advanced by the customer. These can be in the form of contractual prepayments or advance billings, or they may be deposits. If the advance was made to secure performance under an agreement and is legally refundable, it should be classified as a deposit within other assets. If the advance is to be applied as payment, it should be classified as unearned (deferred) revenue and credited to revenue when earned within the terms of the agreement.

5.8.2 Valuation

Trade receivables are valued and reported at net realizable value – the net amount expected to be received in cash, which is not necessarily the amount legally receivable. Determining net realizable value requires an estimation of both uncollectible amounts and any returns and allowances to be granted. For accounting requirements in connection with foreign currency translation of receivables, see VII.3.

An allowance is provided by SFAS No. 5, paragraph 23, if it is probable that the enterprise will be unable to collect all amounts due and if the loss can reasonably be estimated. Therefore, GAAP does not permit set up of lump-sum bad-debt provisions. Two general procedures are in use for recording uncollectible amounts: the direct write-off method and the allowance method. The direct write-off method records the bad debt in the year it is determined that a specific receivable cannot be collected. The allowance method enters the expense on an estimated basis in the accounting period in which the sales on account are made. Those estimates are made either on the basis of percentage of sales or on the basis of outstanding receivables. The percentage of receivables may be applied using one composite rate that reflects an estimate of the uncollectible receivables. Another approach sets up an aging schedule and applies a different percentage based on past experience to the various age categories.

To properly match expenses to sales revenues, it is sometimes necessary to establish additional allowance accounts, as allowance for sales returns and for collection expenses.

5.8.3 Disclosures

Trade accounts receivable are classified as either current or non-current assets and have to be shown separately from other receivables, e.g. receivables due from affiliates.

Allowances are considered being part of the

receivables (SFAC No. 6, paragraph 34) and should therefore be deducted from the gross receivables amount. Corporations reporting under the requirements of the SEC have to set forth the amount of the allowances for doubtful accounts in the balance sheet or in a note hereto (Regulation S-X, Rule 5-02.4).

5.9 Notes receivable

5.9.1 Recognition

Notes receivable represent contractual rights to receive money on fixed or determinable dates. They are an unconditional written promise of the maker of the note to pay the payee or holder a specific amount. They may arise from sales, financing, or other transactions. Although notes contain an interest element because of the time value of money, notes are classified as interest-bearing or noninterest-bearing. Interest-bearing notes have a stated rate of interest, whereas zero-interest-bearing notes include interest as part of their face amount instead of stating it explicitly.

5.9.2 Valuation

Notes may be short-term or long-term. Short-term notes are generally recorded at face value less allowances because the interest implicit in the maturity value is immaterial. Long-term notes should be recorded and reported at the present value of the cash expected to be collected. When the interest stated on an interest-bearing note is equal to the effective rate of interest, the note is presented at face value (see APB Opinion No. 21, Interest on Receivables and Payables). When the stated rate is different from the market rate, the cash exchanged (present value) is different from the face value of the note. The difference between the face value and the cash exchanged, either a discount or a premium, is then recorded and amortized over the life of a note to approximate the effective (market) interest rate.

A note receivable is considered impaired when it is probable that the creditor will be unable to collect all amounts due (both principal and interest) according to the contractual terms of the loan. In that case, the present value of the expected future cash is determined by discounting those flows at the effective rate (SFAS No. 114, paragraph 50). This present value amount is deducted from the carrying amount of the receivable to measure the loss.

5.9.3 Disclosures

Notes receivables are usually shown separately in the balance sheet. SEC Regulation S-X, rule 5–02.3b requires a separate disclosure in the balance sheet or in a note, if the aggregate amount of notes receivable exceeds 10 per cent of the aggregate amount of receivables.

5.10 Derecognition of receivables and other financial assets

5.10.1 Derecognition

In most cases receivables and other financial assets are settled by payment from the debtor. For various reasons such as more favourable refinancing or liquidity needs, the owner may transfer receivables or other financial assets to a third party for cash before maturity. SFAS No. 125, Accounting for Transfers and Servicing of Financial Assets and Extinguishments of Liabilities, provides a single approach for all transfers of financial assets. Typical transactions affected are repurchase agreements, securities lending, factoring or securitizations.

The types of contract on the transfer of financial assets extend from outright sales with no continuing involvement by the seller to borrowings collateralized by assets. In between are sales of assets with varying degrees of involvement by the seller.

For the transactions at the end of the spectrum, the accounting as sale or borrowing is straightforward since the substance follows their form. The transaction within in this span have characteristics of both sales and borrowings, and raises the question of whether the transferor should include the assets on its balance sheet by treating the transaction as a financing transaction or record a sale, with recognition of the gain or loss.

A transfer of financial assets is accounted for as a sale if the transferor surrenders control over the assets. Control is surrendered if all of the following three conditions are met:

(i) The transferred assets have been isolated from the transferor, e.g. they are beyond the reach of the creditors of the transferor, even in bankruptcy.
(ii) The transferee obtains the unencumbered right to pledge or exchange the transferred assets or the transferee is a qualified special purpose entity.
(iii) The transferor does not effectively maintain control over the transferred assets through a forward contract or an option to repurchase, in case of transferred assets that are not readily obtainable.

If the three conditions are met, a sale occurs. Otherwise the transferor should record the transfer as a secured borrowing.

If sale accounting is appropriate, it is still necessary to consider assets obtained and liabilities incurred in the transaction. At the time of the transfer the transferred assets are derecognized and broken into separate financial components (financial components approach). Any retained interests are recognized at book value since they are not relinquished or sold. The book value is determined by allocating the previous carrying amount between the assets sold and retained interest, based on their relative fair value at the date of the transfer. Any proceeds for the part of the assets considered sold such as cash or assets obtained and liabilities incurred are also recognized by the transferor at fair value. The subsequent measurement of retained interests, assets obtained and liabilities incurred follows the general rule for this item.

5.10.2 Disclosures

Disclosures required include:

- Policy for requiring collateral or other security under repurchase or securities lending agreements.
- Descriptions of items for which it is not practicable to estimate fair value and reasons why it is not practicable.
- Servicing assets and liabilities.
- Amounts recognized and amortized.
- Fair value recognized, methods and assumptions used to estimate fair value.
- Risk characteristics used to stratify in order to measure impairments.
- Valuation allowance for impairment activity for each period for which results of operations are given.

5.11 Cash and cash equivalents

5.11.1 Recognition

Cash includes money in any form – for example, cash on deposit, cash awaiting deposit, and cash funds available for use. The promulgated GAAP for accounting for cash is ARB 43, chapter 3. To be included as cash in the balance sheet, funds must be represented by actual coins and currency on hand or demand deposits available without restriction.

Cash equivalents are defined in SFAS No. 95, paragraph 7 as short-term, highly liquid investments that are both readily convertible to known amounts of cash and so near maturity that they present insignificant risk of change in value. Generally, only investments with original maturities of three months or less qualify under these definitions. Examples of cash equivalents are treasury bills, commercial paper, and money funds. Cash equivalents must be available upon demand in order to justify inclusion.

5.11.2 Valuation

Cash and cash equivalents are stated at their nominal value. Foreign currency items have to translated with the spot rate of the end-of-the year.

5.11.3 Disclosures

Pursuant to borrowing arrangements with lenders, an entity will often be required to maintain a minimum amount of cash on deposit (compensating balance). The purpose of this balance is to increase the yield on the loan to the lender. The compensating balance is not available for unrestricted use and must be segregated and shown as a non-current asset if the related borrowings are non-current liabilities. If the borrowings are current liabilities, it is acceptable to show the compensating balance as a separately headed current asset.

Separate disclosure shall also be made of the cash and cash items which are restricted as to withdrawal or usage. The provisions of any restrictions shall be described in a note to the financial statements. Restrictions may include legally restricted deposits held as compensating balances against short-term borrowing arrangements, contracts entered into with others, or company statements of intention with regard to particular deposits (Regulation S-X, rule 5–02.1).

When certificates of deposit, time deposits, or savings accounts are included in the cash heading, the amount thereof is customarily given either parenthetically on the balance sheet or in the notes.

5.12 Prepaid expenses and deferred charges

5.12.1 Recognition

Prepaid expenses are expenditures already made for benefits (usually services) to be received within one year or the operating cycle, whichever is longer. A common example is the payment in advance for an insurance policy. It is classified as a prepaid expense at the time of the expenditure because the payment precedes the receipt of the benefit of coverage. Other common prepaid expenses include prepaid rent, advertising, taxes, and office or operating supplies. A discount on a loan is not capitalizable but reduces the loan liability.

Deferred charges is a classification often used to describe a number of different costs which are not prepaids, but which have a future economic benefit. Although the term deferred charges is not defined in the authoritative literature of the profession, the capitalization of deferred charges is considered permissible due to the matching principle in cases of future economic benefits from certain expenses. GAAP presently recognizes such items with a debit balance as property held for sale, segregated cash or securities, cash surrender value of life insurance as eligible for treatment as assets. The possibility of the capitalization of deferred charges was restricted in past years; e.g. start-up costs (see VI.5.1) and advertising costs (see SOP 93-7, Costs of Advertising) are no longer capitalizable.

5.12.2 Valuation

Prepaid expenses are reported at the amount of the unexpired or unconsumed cost and charged to expense in the following financial year. Deferred charges are stated at cost and amortized in the periods in which the attributable benefits occur.

5.12.3 Disclosures

Prepaid expenses are identified as a current asset by ARB 43, chapter 3A. For any significant deferred charges Regulation S-X, rule 5-02.17 requires to state the policy for deferral and amortization.

5.13 Liabilities

5.13.1 Recognition

Liabilities are defined in SFAC No. 6, paragraph 35 as probable (likely to occur) future sacrifices of economic benefits arising from present obligations of a particular entity to transfer assets or provide services to other entities in the future as a result of past transactions or events. A liability has three essential characteristics:

(i) It is a present obligation that entails settlement by probable future transfer or use of cash, goods, or services.
(ii) It is an unavoidable obligation.
(iii) The transaction or other event creating the obligation has already occurred.

As liabilities involve future disbursements of assets or services, liabilities are divided into current liabilities and long-term debt. Current liabilities are obligations whose liquidation is reasonably expected to require use of existing resources properly classified as current assets, or the creation of other current liabilities (ARB 43, chapter 3A, paragraph 7). Long-term debts consist of probable future sacrifices of economic benefits arising from present obligations that are not payable within a year or the operating cycle of the business, whichever is longer.

Liabilities occur when unpaid expenses come into existence as a result of past contractual commitments, past services received, or by the operation of law. The revenue recognition principle dictates that revenue and the corresponding receivables will be recognized only when they are realized (or realizable) and earned. Likewise, expenses and the corresponding payables will be recognized when the items of expense have been incurred.

Instruments with both liability and equity characteristics should be treated as consisting of a liability component and an equity component that should be accounted for separately. Convertible debt, for example, should be viewed as two separate components. The obligation of the issuing corporation to make periodic interest payments and to repay the principal at maturity is a liability as defined in SFAC No. 6. The conversion feature gives the holder an option to convert the bond to common stock. This option is an equity instrument because it has essentially the same economic value as a call on stock represented by a separately traded call option or warrant.

A liability has to be derecognized if and only if either (a) the debtor pays the creditor and is relieved of the obligation for the liability; or (b) the debtor is legally released from being the primary obligor under the liability either judicially or by the creditor. Therefore, a liability is not considered extinguished by an in-substance defeasance.

Gains and losses recognized from extinguishments of debt shall be aggregated and, if material, classified as extraordinary items and additional disclosures shall be made.

Current accounting practice for convertible debt, as set forth in APB Opinion No. 14, assigns no portion of the issuance proceeds to the equity or conversion feature. The most important rationale given for such treatment is the inseparability of the debt and conversion option. Practical estimation difficulties have also been identified as another consideration.

5.13.2 Valuation

Liabilities are recorded at the present value of the goods and services obtained by incurring debt. The book value of long-term debt at the balance sheet date is the present value of all remaining cash payments required, discounted at the market rate of interest at issuance. The rate of interest used for this purpose is not changed during the term of the loan.

As current liabilities are usually recorded at their full maturity value the slight overstatement resulting from the difference between present value and maturity value is accepted as immaterial.

5.13.3 Valuation of specific liabilities

5.13.3.1 Low-interest liabilities and interest imputations

Interest imputation is required in certain circumstances. As a general rule, when one party exchanges goods or services in return for a second party's obligation to pay, in part or in full, cash in the future, and the stated interest rate is not substantially equal to the prevailing interest rates, the stated principal amount of the transaction should be adjusted so that it is substantially equal to the present value of all future payments discounted at the prevailing interest rate. The prevailing interest rate depends largely on the borrower's credit standing and the repayment terms, collateral and other pertinent factors. The general rule has exceptions, among them the 'one year rule', which exempts receivables and payables arising in the normal course of business that are not due in approximately one year.

5.13.3.2 Foreign currency liabilities

Under SFAS No. 52 foreign currency liabilities that do not qualify as hedged items should be measured and recorded in the functional currency of the recording entity by use of the exchange rate in effect at that date. The exposed liabilities must be remeasured each balance sheet date and at the date of settlement. Generally the current exchange rate is the rate that is used to settle a transaction on the date it occurs, or on a subsequent balance sheet date (see VII.3).

Discounts and premiums on foreign currency contracts are separately accounted for and amortized to net income over the life of the contract, unless they are intended as a hedge against an identified foreign currency commitment (see VII.4).

5.13.3.3 Specific features of the valuation of bonds (i.e. convertible bonds, income bonds, payments received)

A bond is valued at the present value of its future cash flow, consisting of interest and principal. The discount rate used to calculate the present value of these cash flows is the interest rate that provides an acceptable return on the bond. This interest rate known as the stated or coupon rate is set by the issuer of the bond. If the effective market rate differs from the stated rate, the present value of the bond will differ from the face value of the bond. The difference between the face value and the present value is either a discount or a premium. If the bonds sell for less than face value, they are sold at a discount. If the bonds sell for more than face value, they are sold at a premium. Such discount or premium is amortized over the lifetime of the bond. For bonds with embedded derivatives, see VII.4.

5.13.3.4 Trade creditors

Accounts payable to trade creditors are recorded at their full maturity amount.

5.13.3.5 Bills payable

Liabilities for certain bills – for example, wages, interest – are recorded at their full maturity amount.

5.13.3.6 Tax liabilities

The tax payable on the corporation's income is recorded at the amount computed per tax return. Payroll tax deductions are recorded at the amount withheld and remittable to the government. A deferred tax liability represents the increase in taxes payable in future years as a result of taxable temporary differences existing at the end of the current year. The measurement of current and deferred tax liabilities is based on provisions of the enacted tax law, effects of future changes in tax laws or rates are not anticipated (see also VII.1).

5.13.3.7 Social security liabilities

Under social security legislation, all employers covered are required to withhold the employee's share of the social security tax and remit it to the government along with the employer's share. The rates used are changed intermittently by Congress. Social security liabilities are recorded at the unremitted employee and employer share of the social security tax.

5.13.4 Disclosures

SFAS No. 47, Disclosure of Long-term Obligations, requires disclosure of information about recorded obligations and unconditional purchase obligations.

An unconditional purchase obligation that meets the following characteristics shall be disclosed:

- Substantially non-cancellable.
- Associated with the financing arrangements for the facilities that will provide the contracted goods or services or for costs related to those goods or services (such as carrying costs).
- Remaining term in excess of one year.

The following information is to be disclosed:

- Description of the nature and term of the obligation(s).
- The total determinable amount of unrecorded unconditional purchase obligations as of the latest balance sheet date and the total determinable amount of unrecorded unconditional purchase obligations for each of the five succeeding fiscal years.
- Description of the nature of any variable components of the unconditional purchase obligations.
- For each income statement presented, the amounts actually purchased under the unconditional purchase obligations.

For recorded obligations the following specific disclosures must be made:

- The aggregate amount of payments for unconditional purchase obligations that meet the criteria to be disclosed and that have been recognized on the purchaser's balance sheet.
- The combined aggregate amount of maturities and sinking fund requirements for all long-term borrowings.
- The amount of redemption requirements for all issues of capital stock that are redeemable at fixed or determinable prices on fixed or determinable dates, separately by issue or combined.

5.14 Accrued liabilities

A provision is (i) an estimated expense that is a charge for diminution in value of an asset or for an estimated liability; or (ii) an estimated amount set up in recognition of a liability whose extent and timing are uncertain.

The generally accepted meaning of the term 'reserve' corresponds to the amount of unidentified or unsegregated assets held or retained for a specific purpose, as in the case of a reserve for betterment or plant expansion, or for the excess cost of replacements of property, or for general contingencies. In this sense, a reserve is frequently referred to as an appropriation of retained earnings.

When a loss contingency exists, the likelihood that the future event or events will confirm the incurrence of a liability can range from probable to remote. The FASB uses the terms probable, reasonably possible and remote to identify three areas within that range and assigns the following meanings (SFAS No. 5, paragraph 3):

- Probable: the future event or events are likely to occur.
- Reasonably possible: the chance of the future event or events occurring is more than remote but less than likely.
- Remote: the chance of the future event or events occurring is slight.

An estimated loss from a loss contingency should be accrued by a charge to expense and a liability recorded only if both of the following conditions are met (SFAS No. 5, paragraph 8):

- It is probable that as of the date of the financial statements a liability has been incurred, based on information available prior to issuance of the financial statements
- The amount of loss can be reasonably estimated.

Depending upon whether it is classified as probable, reasonably probable, or remote, a loss contingency may be:

- Accrued as a charge to income as of the date of the financial statements.
- Disclosed by footnote in the financial statement.
- Neither accrued or disclosed.

When both conditions for the accrual of a loss contingency are met, an accrual for the estimated amount is required. If a loss contingency is classified as probable and only a range of possible loss can be established, then the minimum amount in the range is accrued, unless some other amount within the range appears to be a better estimate (FIN No. 14, paragraph 3). The range of a possible loss must also be disclosed. If one or both conditions for the accrual of a loss contingency are not met and the loss contingency is classified as probable or reasonably possible, financial disclosure of the loss contingency is required. The disclosure shall contain a description of the nature of the loss contingency and the range of the possible loss, or include a statement that no estimate of the loss can be made (SFAS No. 5, paragraph 10).

General risk contingencies that are inherent in business operations, such as the possibility of war, strikes, uninsurable catastrophes, or a business recession, are not reported in the notes to the financial statements.

5.14.1 Pension liabilities

Accrual accounting for pensions is based on the view that cash payments (funding decisions) should not necessarily be used as the basis for accounting recognition of cost. The accounting treatment for pension costs depends on whether they relate to defined benefit pension plans or to defined contribution plans. GAAP accounting for pensions are located in the pronouncements SFAS No. 35, SFAS No. 87, SFAS No. 88, SFAS No. 106, and SFAS No. 132.

Two general types of pension plan can be distinguished, defined benefit plans (SFAS No. 87, paragraph 11) and defined contribution plans (SFAS No. 87, paragraph 63). Under a defined benefit plan the employer provides a benefit at retirement that is based on an agreed-upon formula. As the benefits for the participants are defined, the employer accepts the risk associated with the changes in the variables that determine the amounts needed to meet the obligation to plan participants. Under a defined contribution plan the employer contributes to a pension plan as determined by the provisions of the plan. At retirement the plan participant receives whatever benefits the contribution can provide.

For a pension plan with characteristics of both a defined benefit plan and a defined contribution plan, the accounting and disclosure requirements shall be determined in accordance with the provisions to a defined benefit plan (SFAS No. 87, chapter 66, subject to a materiality threshold).

The resources of a pension plan may be converted into:

- Plan assets that have been segregated and restricted (usually in a trust) to provide for pension benefits.
- Plan assets that are used in the operation.

Assets not segregated in a trust or otherwise effectively restricted so that the company cannot use them for other purposes, are not plan assets under SFAS No. 87 (SFAS No. 87, paragraph 19). In the United States, assets of a pension plan are usually kept in a trust account and are therefore treated as plan assets. Contributions to the pension trust account are made periodically by the company and, if the plan is contributory, by the employees. The company usually cannot withdraw plan assets placed in a trust account. An exception arises, when the plan assets exceed the pension obligation and the plan is terminated.

Pension plan assets that are held as investments to provide pension benefits are measured at fair value (SFAS No. 87, paragraph 23). The plan assets are invested in stocks, bonds, real estate, and other types of investment. Plan assets are increased by earnings and gains on investments and are decreased by losses on investments, payment of pension benefits, and administrative expenses. Pension assets that are used in the operation of the plan are measured at cost, less accumulated depreciation or amortization (SFAS No. 87, paragraph 23).

5.14.1.1 Defined benefit plans

5.14.1.1.1 Recognition

A pension liability (or asset) is recorded if the employer's cumulative contributions to the pension plan (represented by the plan assets) are not equal to the amount of the net periodic pension obligation as determined by the provisions of SFAS No. 87, paragraph 20. The difference is recorded as a liability if the amount of the contribution is less than the amount of net periodic pension cost. If the amount of contribution is more than the amount of net periodic pension cost, an asset (prepaid pension cost) is recognized (SFAS No. 87, paragraph 35).

Under certain circumstances an additional minimum liability must be recognized (see VI.5.14.1.1.9 and V.2.3.1.5).

5.14.1.1.2 Valuation

The projected benefit obligation (PBO) is the actuarial present value of all pension benefits (vested and non-vested) attributed by the pension benefit formula, including consideration of future employee compensation levels. Assumed compensation levels shall reflect changes because of general price levels, productivity, seniority, promotion, and other factors (SFAS No. 87, paragraph 46). Changes resulting from a plan amendment that has become effective, and automatic benefit changes specified by the terms of the pension plan, such as cost-of-living increases, are included in the determination of service cost for a period (SFAS No. 87, paragraph 48). The FASB states that the projected benefit obligation should be used as the basis for determining service cost.

If at year end the projected benefit obligation is in excess of the fair value of pension plan assets, a liability has to be recorded. If the fair value of plan assets are in excess of the projected benefit obligation, an asset has to be recorded.

The FASB refers to delayed recognition as one of the fundamentals of Pension Accounting. Based on actuarial assumptions, the net periodic pension cost is estimated at the beginning of the year. Changes in the pension obligation and changes in the value of pension assets at year end are not recognized as they occur, but systematically and gradually over subsequent years. Only if the accumulated change in the pension obligation and change in the value of pension (unrecognized gains or losses) exceed certain limits must the excess be amortized in future periods (see also VI.5.14.1.1.8). Therefore, the net periodic pension expense is not changed at the year end.

However, if at year end the projected benefit obligation is in excess of the fair value of pension plan assets, a (minimum) liability has to be recorded (see VI.5.14.1.1.9). If the fair value of plan assets are in excess of the projected benefit obligation, an asset has to be recorded. Recognition of the minimum liability and the asset does not affect the net periodic pension expense.

5.14.1.1.3 Net periodic pension cost

The accounting for the plan's costs should be reflected in the financial statements and these amounts should not be discretionary. All pension costs should be charged against income. No amounts should be charged directly to retained earnings. Under SFAS No. 87, paragraph 20 the net periodic pension costs for defined benefit plans include the following components:

- Service cost.
- Interest cost.
- Return on plan assets.
- Amortization of unrecognized prior service cost.
- Recognition of net gain and loss.
- Amortization of unrecognized net obligation or net asset existing at the date of initial application of SFAS 87.

5.14.1.1.4 Service cost

Service cost is the increase in the PBO due to the additional service rendered by the employees during a period. The FASB states that the service cost component recognized in a period should be determined as the actuarial present value of benefits attributed by the pension benefit formula to employee service during a specific period (SFAS No. 87, paragraph 21).

5.14.1.1.5 Interest cost

Interest cost component is the interest for the period on the PBO outstanding during this period. The FASB did not address the question how often to compound the interest cost. The assumed discount rate should reflect the rates at which pension benefits could be effectively settled. That is assumed to be the yield of bonds from virtually risk-free borrowers. In practice this rate is adjusted in case of significant movements in interest rates.

5.14.1.1.6 Return on plan assets

Actual return on the plan assets is the increase in pension funds from interest, dividends, and realized and unrealized changes in the fair market value of the plan assets. Fair value is the amount that a pension plan could reasonably be expected to receive from a current sale of an investment.

Plan assets used in the operation of the pension plan (i.e. buildings and equipment) are valued at cost, less accumulated depreciation (SFAS No. 87, paragraph 51).

SFAS No. 87, paragraph 34 requires the recognition of the actual return or loss on pension plan assets as a component of net periodic pension cost. The actual return is computed by adjusting the change in the plan assets for the effects of contributions during the year and benefits paid out during the year.

As the net periodic pension cost is estimated at the beginning of the year, the expected return on plan assets is to be included as a component of the pension expense, not the actual return in a given year. Differences between the expected return and the actual return, referred to as asset gains and losses, are subject to the corridor amortization (see VI.5.14.1.1.8).

As the net periodic pension cost is estimated at the beginning of the year, the difference between the actual and expected return on plan assets is deferred. This difference between the expected return and the actual return, referred to as asset gains and losses, are subject to the corridor amortization (see VI.5.14.1.1.8). As a consequence, only the expected return on plan assets is to be included as a component of the pension expense.

5.14.1.1.7 Amortization of unrecognized prior service cost

When a defined benefit plan is either initiated or amended, employees may be granted pension benefits for services performed in prior years. The cost of pension benefits that are granted retroactively to employees for services performed in prior periods is referred to as prior service cost. As a result of prior service credits, the projected benefit obligation is usually greater than it was before. Only a portion of the total amount of prior service cost arising in a period, including retroactive benefits that are granted to retirees, is included in net periodic pension cost. SFAS No. 87, paragraph 25 requires that the total prior service cost arising in a period from an adoption or amendment of a plan be amortized in equal amounts over the future service period of active employees who are expected to receive the retroactive benefits. The consistent use of an alternative amortization approach that more rapidly reduces the amount of unrecognized prior service cost is permitted (SFAS No. 87, paragraph 26).

5.14.1.1.8 Recognition of net gain and loss

Gains and losses are changes in the amount of either:

- Projected benefit obligation.
- Pension plan assets.

A gain or loss resulting from a change in the PBO is referred to as an actuarial gain or loss, as it reflects any change in the actuarial assumptions (i.e. mortality rate, retirement rate, turnover rate, disability rate and salary amounts). A gain or loss resulting from a change in the fair value of pension plan assets is referred to as a net asset gain or loss.

The expected return on pension plan assets during the period is computed by multiplying the market-related value of plan assets by the expected long-term rate of return. The current rate of return earned on plan assets and the rates of return expected to be available for plan investments should be considered in estimating the long-term rate of return on plan assets. The expected long-term rate of return on plan assets should reflect the average rate of earnings expected on plan investments (SFAS No. 87, paragraph 45).

The sources of these gains and losses are not distinguished separately, and they include amounts that have been realized as well as amounts that are unrealized. Therefore, the asset gains and losses and the liability gains and losses can offset the accumulated total unrecognized net gain or loss. If the accumulated unrecognized gain or loss does not meet a specific materiality threshold, called the corridor, no gain or loss is recognized in a particular accounting period. The unrecognized net gain or loss balance is considered too large and must be amortized when it exceeds the arbitrarily selected FASB criterion of 10 per cent of the larger of the balances of the projected benefit obligation or the market-related value of the plan assets (SFAS No. 87, paragraph 32). If no unrecognized gain or loss exists at the beginning of the year, no recognition of gains and losses can result in that period.

The minimum recognition at the beginning of the year shall be the excess divided by the average remaining service period of active employees who are expected to receive benefits under the plan. Any systematic method of amortization of unrecognized gains or losses may be used in lieu of the minimum, when the method:

- Is systematic and applied constantly.
- Is applied similarly to both gains and losses.
- Reduces the unamortized balance by an amount greater than the amount that would result from the minimum amortization method provided by SFAS No. 87, paragraph 32.
- Is disclosed in the financial statements.

However, gains and losses that arise from a single occurrence not directly related to the operation of the pension plan and not in the ordinary course of the company's business should be recognized immediately.

5.14.1.1.9 Additional minimum liability

The FASB states that the PBO should be used as the basis for determining service cost. Under SFAS No. 87, paragraph 36, an additional minimum liability must be recognized if an unfunded accumulated benefit obligation (ABO) exists. The unfunded ABO is the amount by which the ABO exceeds the amount of the fair value of plan assets. The amount of the additional minimum liability must at least equal the unfunded ABO.

The ABO is calculated in the same way as the PBO, except that current or past compensation levels instead of future compensation levels are used to determine pension benefits.

If the ABO exceeds the amount of the fair value of plan assets as of a specific date (unfunded ABO), under SFAS No. 87, paragraph 36, an additional minimum liability must be recognized, if:

- An asset has been recognized as prepaid pension cost.
- A liability has been recognized as unfunded accrued pension cost in an amount that is less than the amount of the existing unfunded accumulated benefit obligation.
- No accrued or prepaid pension cost has been recognized.

If an asset has been recognized as prepaid pension cost, the additional minimum liability is the amount of the existing unfunded ABO plus the amount of the prepaid pension cost. If a liability has been recognized as unfunded accrued pension cost in an amount that is less than the amount of the existing unfunded ABO, the additional minimum liability is the amount of the existing unfunded ABO reduced by the amount of the unfunded accrued pension cost. If no accrued or prepaid pension cost has been recognized, the additional minimum liability is the amount of the existing unfunded accumulated benefit obligation (SFAS No. 87, paragraph 36).

If an additional minimum liability is required to be recognized, generally an intangible asset in the same amount as the additional minimum liability is recognized. Therefore an expense is not recorded in the year the unfunded ABO is recognized, but taken into consideration for future contributions. Recognizing the expense in the same year would be a change in the base of accounting.

However, the amount of the intangible asset can not exceed the total amount of any existing unrecognized prior service cost and any unrecognized net obligation. In the event that the intangible asset exceeds the total existing unrecognized prior service cost and unrecognized net obligation, the excess is reported as a separate negative component of stockholders equity, net of related tax benefits (SFAS No. 87, paragraph 37).

5.14.1.1.10 Disclosures

The disclosure requirements of SFAS No. 87 are replaced by the disclosure requirements of SFAS 132, Employers' Disclosures about Pensions and Other Postretirement Benefits. The disclosure requirements of SFAS No. 132 are very detailed. Besides separate disclosures public companies have to disclose primarily:

- Description of the plan.
- Benefit obligation reconciliation.
- Plan assets reconciliation.
- Fund status reconciliation.
- Assumptions.
- Net periodic benefit cost.

Disclosures about plans with accumulated benefits in excess of assets may be combined with other plans, but disclosure is required of the aggregate benefit obligation and aggregate fair value of plans with benefit obligations in excess of assets.

Information about plans outside the United States may be combined with US plans, unless the benefit obligation of the plans outside the United States are significant relative to the total benefit obligation and those plans use significantly different assumptions (SFAS No. 132, paragraph 7).

For non-public companies SFAS No. 132, paragraph 8 permits reduced disclosure requirements.

5.14.1.2 Defined contribution plans

Under a defined contribution plan only a defined contribution has to be made each year based on the provisions established in the plan. As a result, the pension expense is the amount that is obligated to contribute to the pension trust. A liability is reported if the contribution has not been made in full. An asset is reported if more than the required amount has been contributed.

A company with one or more defined contribution pension plans shall disclose the amount of cost recognized for defined contribution pension during the period. The disclosures shall include the nature and effect of any significant changes during the period affecting comparability (SFAS No. 132, paragraph 9).

5.14.2 Post-retirement benefits other than pensions

The FASB requires companies to accrue the costs of providing retirement benefits other than pensions (i.e. post-retirement healthcare) over the working careers of their active workforce (SFAS No. 106, Employers' Accounting for Postretirement Benefits other than Pensions). Prior to SFAS No. 106, companies recorded expenses for these benefits on a pay-as-you-go basis.

SFAS No. 106 requires the accrual of post-retirement benefits in a manner similar to the recognition of net periodic pension cost under SFAS No. 87. The provisions of SFAS No. 106 are similar in most respects to those of SFAS No. 87 and differ only where there are compelling reasons for different treatments.

Under SFAS No. 106 companies can choose immediate recognition or deferred recognition. Under immediate recognition a transition amount (the difference between the accumulated post-retirement benefit obligation and the fair value of the plan assets, plus any accrued obligation or less any prepaid cost) is recognized in the income statement as a change in an accounting principle (net of tax) and also on the balance sheet as a long-term liability. Under the deferred recognition method the transition amount is amortized on a straight-line basis over the average remaining service period to expected retirement.

The disclosure requirements of SFAS No. 106 were replaced by SFAS No. 132 and consolidated with the disclosure of information about pensions (see V.2.3.1.2).

5.14.3 Restructuring costs

A major restructuring of company affairs can include programmes such as closing down, consolidating, relocating, selling, or abandoning operations, providing early retirement, severance, relocation, or transfer benefits to employees, and liquidation of slow-moving inventories and overdue receivables. Restructuring charges include the costs of developing and administrating such programmes, the costs of the programmes themselves, and programme-related asset valuation and liability provisions.

A commitment by an enterprise's management to execute a restructuring plan that will result in the occurrence of costs that have no future economic benefit represents an obligating event for those costs meeting the following criteria (EITF 94-3, Liability Recognition for Certain Employee Termination Benefits and Other Costs to Exit an Activity [Including Certain Costs Incurred in a Restructuring]):

- The restructuring plan specifically identifies all significant actions to be taken to complete the plan, activities that will not be continued, including the method of disposition and location of those activities, and the expected date of completion.
- Actions required by the restructuring plan will begin as soon as possible after the commitment date, and the period of time to complete the plan indicates that significant changes to the restructuring plan are not likely.

Those costs resulting from a restructuring that are not associated with or that do not benefit activities that will be continued should be recognized as a provision at the commitment date.

For involuntary employee termination benefits a provision should be recognized in the period management approves the plan of termination if all of the following conditions exist:

- Prior to the date of the financial statements the benefit arrangement is communicated to employees.
- The plan of termination specifically identifies the number of employees to be terminated, their job classifications and their locations.
- The period of time to complete the plan of termination indicates that significant changes to the plan of termination are not likely.

A restructuring charge should not be reported as an extraordinary item, because these write-offs are considered part of a company's ordinary and typical activities (EITF 86-22, Display of Business Restructuring Provisions in the Income Statement).

5.14.4 Environmental liabilities

An enterprise identifies a loss contingency with respect to environmental matters and accrues an estimated loss by a charge to income in accordance with SFAS No. 5, when available information indicate that the enterprise incurred such a liability whose amount can be estimated reasonably (see EITF 93-5, Accounting for Environmental Liabilities).

Specific guidance on accounting for environmental remediation liabilities is given in SOP 96-1. An entity should accrue environmental remediation liabilities if an assertion has been, or will be, made that the entity is responsible for participating in a remediation process as a result of a past event, and if the outcome of the pending or potential action will probably be unfavourable. In order to consider in evaluating the probability that a reasonable measurable loss has incurred the SOP sets out benchmarks, such as identification and verification of an entity as a Potential Responsible Party, receipt of unilateral administrative order and completion of the feasibility study. If one or both conditions are not met, disclosure of the contingency shall be made when there is at least a reasonable possibility that a loss may have been incurred.

The measurement of the environmental liability

should include the entity's allocable share of the liability for a specific site and the entity's share of amounts related to the site that will not be paid by other potentially responsible parties or the government. The liability should be evaluated independently from any potential claim for recovery. The loss arising from the recognition of an environmental liability should be reduced only when a claim for recovery is probable of realization. Discounting environmental liabilities for a specific clean-up site to reflect the time value of money is allowed, but not required, only if the aggregate amount of the obligation and the amount and timing of the cash payments for that site are fixed or reliably determinable. To be considered reliably determinable the estimate of the expected costs to be incurred should be based on a site-specific plan for the clean-up or remediation of the contamination and the amount and timing of cash payments should be based on objective and verifiable information.

For the valuation of remediation liabilities SOP 96-1 requires recognizing a liability for both incremental direct costs of the remediation effort and costs of compensation and benefits for those employees who are expected to devote a significant amount of time directly to the remediation effort. Incremental direct costs will include such items as the fees to outside law firms for work related to the remediation effort, fees to outside consulting and engineering firms for site investigations and development of remedial action plans and remedial actions, and costs of contractors performing remedial actions. Costs are to be estimated based on existing laws and technologies.

Environmental clean-up costs are not unusual in nature, and thus cannot be shown as extraordinary items in the income statement. If the effect of discounting is material, the financial statements should disclose the undiscounted amounts of the liability and any related recovery, and the discount rate used.

5.14.5 Anticipated losses on firm purchase and selling commitments

Purchase commitments represent contracts and agreements for the future purchase of specified quantities of materials at a specified price. They should be evaluated in the same fashion as inventory on hand for the purpose of determining any lower-of-cost-or-market adjustment.

ARB No. 43, chapter 5 paragraph 17 requires that accrued net losses on firm purchase commitments for goods for inventory, measured in the same way as are inventory losses, should, if material, be recognized in the accounts and the amounts thereof separately disclosed in the income statement. The utility of such commitments is not impaired, and hence there is no loss, when the amounts to be realized from the disposition of the future inventory items are adequately protected by firm sales contracts or when there are other circumstances which reasonably assure continuing sales without price decline.

Similarly, agreements requiring the sale of goods in the future for a specified price should be used in making the lower-of-cost-or-market valuation for existing inventory and should be viewed in the context of whether the cost to acquire and produce the product will exceed the agreed selling price. If firm agreements have been made for future sales, that selling price should usually be used in establishing net realizable value for the product on hand subject to the sales commitment. If the carrying value of the inventory on hand needed to meet the future sales agreements exceeds such net realizable value, losses should be recognized currently in the financial statements.

5.14.6 Anticipated losses on certain construction-type contracts

Under the requirements of SOP 81-1, a provision for anticipated losses on long-term contracts arises when the current estimate of total contract cost exceeds the current estimate of total contract revenue. Provisions for losses should be made in the period in which they become evident under either the percentage-of-completion method or the completed-contract method (SOP 81-1, paragraph 85).

The loss provision should be computed on the basis of the total estimated costs to complete the contract, which would include the contract costs incurred to date plus estimated costs to complete. Factors that should be considered in arriving at the projected loss include target penalties and rewards, non-reimbursable costs on cost-plus contracts, change orders, and potential price redeterminations. In circumstances in which general and administrative expenses are treated as contract costs under the completed-contract method, the estimated loss should include the same types of general and administrative expenses.

The provision should be shown separately as a current liability on the balance sheet. In the income statement the provision should be accounted for as an additional contract cost rather than as a reduction of contract revenue.

5.15 Deferred income

Amounts collected in advance for services to be rendered in later accounting periods do not represent revenue because they have not yet been earned as dictated by the revenue recognition principle. Revenue is reported in the period in which it is earned: therefore, when it is received in advance of its being earned, the amount applicable to future periods is deferred to future periods as deferred income. The balance of the deferred income account is classified as a liability account. Deferred income differs from other liabilities because it is usually settled by rendering services rather than by making cash payments.

Deferred income items include unearned management fees, rent received in advance, interest received in advance on notes receivable, subscriptions and advertising received by publishers in advance, and deposits from customers in advance of the delivery of merchandise.

VII Special accounting areas

1 Deferred taxation

1.1 Recognition

GAAP and tax regulations frequently require different accounting procedures. Accounting for income taxes under SFAS No. 109, Accounting for Income Taxes, is based on the balance sheet focused liability method (asset/liability method) and has two primary objectives (SFAS No 109, paragraph 6):

- To recognize the amount of taxes currently payable or refundable.
- To recognize the deferred tax assets and liabilities for the future tax consequences of temporary differences.

Deferred taxation is required on temporary differences. They arise when items are treated differently in the financial statements and the income tax return. Under SFAS No. 109, paragraph 13, a taxable temporary difference is one that will result in the payment of income taxes in the future when the temporary difference reverses. A deductible temporary difference is one that will result in reduced income taxes in future years when the temporary difference reverses. Taxable temporary differences give rise to deferred tax liabilities, whereas deductible temporary differences give rise to deferred tax assets. No adjustments are permitted for differences that will not result in taxable or deductible amounts in the future (permanent differences).

Under GAAP temporary differences typically arise from:

- Depreciation method: use of modified accelerated cost recovery system (MACRS) for tax purposes and straight-line for accounting purposes.
- Warranty liability: warranty expense and the related liability are recognized on accrual basis for accounting purposes and on cash basis for tax purposes.
- Allowance for doubtful accounts: bad-debt expense and the related allowance are recognized on accrual basis for accounting purposes and deferred for tax purposes until written off (direct-method).
- Inventory reserve: loss from decline and the related allowance are recognized on accrual basis for accounting purposes and deferred for tax purposes until written off (direct-method).
- Estimated liabilities to discontinued operations or restructuring: the estimated liabilities and the related losses are recognized on accrual basis for accounting purposes and deferred for tax purposes.
- Investments: investments accounted for under the equity method for accounting purposes and under the cost method for tax purposes.
- Long-term contracts: contracts accounted for under the percentage-of-completion method (cost method) for accounting purposes and a

portion of related gross profit deferred for tax purposes.
- Instalment sale receivable: instalment sales and the related asset are recognized as a sale at transaction date for accounting purposes and deferred for tax purposes until collected.

Operating loss carryforwards are treated as future tax savings. Therefore there are no different requirements for recognition of a deferred tax asset resulting from deductible temporary differences and operating loss carryforwards.

Under current US tax law a potential tax liability using the regular tax system and an alternative minimum tax system has to be calculated. The company's annual income tax is the greater of the income tax liability calculated of the regular tax or the alternative minimum tax. The paid minimum tax is entitled to a credit against the regular tax liability in subsequent years. Therefore the alternative minimum tax is to be treated like a tax credit carryforward.

On the balance deferred tax liabilities and assets have to be separated into a net current amount and a net non-current amount. All current deferred tax liabilities and assets shall be offset and presented as a single amount and all non-current deferred tax liabilities and assets shall be offset and presented as a single amount. Deferred tax liabilities and assets to different tax jurisdictions shall not be offset. (SFAS No. 109, paragraph 42).

1.2 Valuation

As SFAS No. 109 is based on the liability method the amount of deferred income tax is based on the tax rates expected to be in effect during the periods in which the temporary differences reverse. This tax rate must be legally enacted (SFAS No. 109, paragraph 18). In the United States, the current single flat tax rate is 35 per cent. Deferred tax assets should not be discounted (APB Opinion No. 30, paragraph 6).

When a change in the tax rate is enacted into law, the effect on the existing deferred tax is reported immediately as an adjustment to income tax expense in the period of the change.

The consideration of tax-planning strategies is required by SFAS No. 109, especially for determining the need for, and the amount of, the valuation allowance for deferred tax assets.

A deferred tax asset should be reduced by a valuation allowance if, based on all available evidence, it is more likely than not that some portion or all of the deferred tax asset will not be realized. More likely than not means a level of likelihood that is at least slightly more than 50 per cent.

All available evidence, both positive and negative, should be considered to determine whether, based on the weight of that evidence, a valuation allowance is needed. Examples of negative evidence are listed in SFAS 109, paragraph 23:

- Cumulative losses in recent years.
- History of operating loss or tax credit carryforwards expiring unused.
- Losses expected in early future years.
- Unsettled circumstances that, if unfavourably resolved, would adversely affect future operations and profit levels on a continuing basis in future years.
- Carryback or carryforward period that is so brief that it significantly limits the probability of realizing deferred tax assets.

The FASB states that a cumulative loss in recent years is a significant piece of negative evidence that is difficult to overcome (SFAS No. 109, paragraph 103). There is no definition of the term cumulative losses in recent years. As a practice pointer retroactive three years are often used for determination.

The valuation allowance shall be allocated between current and non-current deferred tax assets on a pro rata basis (SFAS No. 109, paragraph 41).

Not all deferred tax items are immediately recognized in income. The tax effect from certain items follows the treatment of the asset that caused the deferred tax and is charged or credited directly to the related component in shareholders equity. For example, the deferred tax on the fair value adjustment of an available for sale security is recognized in other comprehensive income.

Several specialized applications of SFAS No. 109 exist, i.e. regulated enterprises (SFAS No. 109, paragraph 29); business combinations (SFAS No. 109, paragraph 30); and quasi-reorganizations (SFAS No. 109, paragraph 39).

1.3 Disclosures

Current and non-current classifications of deferred taxes are based on classifications of the asset or liability giving rise to the temporary difference. If the deferred tax does not relate to an underlying asset or liability on the balance sheet, classification is based on the expected timing of reversal (SFAS No. 109, paragraph 41).

There are several additional detailed disclosure requirements specified in SFAS No. 109, paragraph 43–49. In essence the following information have to be disclosed:

- Total of all deferred tax liabilities and deferred tax assets (taxable or deductible temporary differences).
- Components of the net deferred tax liability or asset.
- Total valuation allowance recognized for deferred tax assets and the net change during the year.
- Amount of income tax expense or benefit allocated to continuing operations, discontinued operations, extraordinary items and special items charged or credited directly to related components of shareholders' equity.
- Amounts and expiration dates of operating loss and tax credit carryforward.
- Significant components of income tax expense attributable to continuing operations.
- Additional notes when filing a consolidated tax return.

Additional disclosures are required in Regulation S-X, Rule 4–08(h) by the SEC. In essence, a breakdown of the pretax income or loss in domestic or foreign origin as well as a breakdown of the major US Federal income taxes, foreign income taxes and other income taxes have to be disclosed.

2 Accounting for leases

A lease is a contractual agreement between a lessor and a lessee that gives the lessee the right to use property, plant or equipment (land and/or depreciable assets) usually for a stated period of time in return for stipulated, and generally periodic, cash payments (rents).

GAAP for leases include the largest number of authoritative accounting pronouncements of any single subject. The following pronouncements collectively establish promulgated GAAP for lease accounting:

- SFAS Nos 13, 22, 23, 27, 28, 29, 91, 98.
- FASB Interpretation Nos 19, 21, 23, 24, 26, 27.

The basic lease criteria are established in SFAS No. 13, SFAS No. 28 and SFAS No. 98. Lessees and lessors make their evaluations of leases from their own perspective. For accounting and reporting purposes the lessee has two alternatives in classifying a lease (SFAS No. 13, paragraph 6):

- Capital lease.
- Operating lease.

From the standpoint of the lessor, all leases may be classified for accounting purposes as one of the four alternatives (SFAS No. 13, paragraph 6):

- Sales-type lease.
- Direct financing lease.
- Leveraged lease.
- Operating lease.

The various viewpoints range from no capitalization to capitalization of all leases. The FASB apparently agrees with the capitalization approach when the lease is similar to an instalment purchase, noting that a lease that transfers substantially all of the benefits and risks of property, plant or equipment ownership should be capitalized. Those leases that do not transfer substantially all of the benefits and risks of ownership are operating leases. They should not be capitalized but rather accounted for as rental payments and receipts.

The economic substance of the transaction determines the accounting treatment, even though the legal form of the transaction may indicate a different treatment (substance over form).

2.1 Basic lease capitalization criteria

Four basic capitalization criteria are established in SFAS No. 13, paragraph 7. For a lease to be recorded as a capital lease by the lessee, the lease must be non-cancellable, and meet one or more of the following four criteria:

- The lease transfers ownership of the property to the lessee by the end of the lease term.
- The lease contains an option to purchase the leased property at a bargain price.
- The lease term is equal to or greater than 75 per cent of the estimated economic life of the leased property.
- The present value of rental and other minimum lease payments (excluding portion which represents executory costs) equals or exceeds 90 per cent of the fair value of the leased property, less any investment tax credit retained by the lessor.

The last two criteria are not applicable when the beginning of the lease term falls within the last 25 per cent of the total estimated economic life of the leased property.

If a lease meets at least one of the four criteria indicating that substantially all the benefits and risks of ownership have been transferred to the lessee, and it meets both of the following condition, the lease is classified by the lessor as capital lease (direct financing lease or as a sales-type lease, whichever is appropriate):

- Collectibility of the minimum lease payments is reasonably predictable.
- No important uncertainties surround the amount of unreimbursable costs yet to be incurred by the lessor under the terms of the lease.

Leases that do not meet those criteria are classified and accounted for as operating leases.

2.2 Lessee accounting

A lessee must classify a non-cancellable lease as either a capital lease or an operating lease.

2.2.1 Capital lease

The lessee records a capital lease as an asset and a corresponding liability. The initial recording value of a lease is the lesser of the fair value of the leased property or the present value of the minimum lease payments, excluding any portion representing executory costs and profit thereon to be paid by the lessor. Fair value is determined as of the inception of the lease, and the present value of the minimum lease payments is computed at the beginning of the lease term (SFAS No. 13, paragraph 10).

The asset recorded under a capital lease is amortized in a manner consistent with the lessee's normal depreciation policy for other owned assets. The period for amortization is either the estimated economic life or the lease term, depending on which criterion was used to classify the lease. If the criterion used is either the first two criteria (ownership of the property is transferred to the lessee by the end of the lease term or the lease contains a bargain purchase option), the asset is amortized over its estimated economic life. In all other cases, the asset is amortized over the lease term. Any estimated residual value is deducted from the asset to determine the amortizable base (SFAS No. 13, paragraph 11).

During the lease term, each minimum lease payment shall be allocated between a reduction of the obligation and interest expense so as to produce a constant periodic rate of interest on the remaining balance of the obligation. The effective interest method is used to allocate each lease payment between principal and interest.

Rental expenses amortized over the lease term and interest expenses are recorded in the income statement.

2.2.2 Operating lease

Leases that do not qualify as capital lease in accordance with the provisions of SFAS No. 13, paragraph 7 are classified as operating lease. The lessee does not record the lease as an asset and liability and records lease payments. Expenses are recognized on a straight-line basis unless some other method is deemed to be more appropriate (SFAS No. 13, paragraph 15).

2.3 Lessor accounting

Leases are classified for the lessor as either sales-type, direct financing or operating lease. Both sales-type and direct financing are forms of capital lease. A leveraged lease is a direct financing lease with additional characteristics.

2.3.1 Capital lease

The distinction between a sales-type lease and a direct financing lease is the presence or absence of a profit or loss (difference between fair value and carrying amount at the inception of the lease). Sales-type leases are usually used by sellers of property to increase the marketability of expensive assets. The occurrence of a manufacturer's or dealer's profit or loss generally is present in a sales-type lease. Direct financing leases do not give rise to a manufacturer's or dealer's profit or loss, and the fair value usually is the cost or the carrying amount of the property.

2.3.1.1 *Sales-type lease*

A lease is classified as a sales-type lease when the fair value of the leased property is different from its carrying amount and therefore the lessor (generally a manufacturer or dealer) recognizes a profit or loss on the transaction in addition to interest revenue. Common examples of sales-type leases: (i) when an automobile dealership opts to lease a car to its customers in lieu of making an actual sale; and (ii) the re-lease of equipment coming off an expiring lease.

For a lease classified as a sales-type lease, the present value of the minimum lease payments receivable from the lessee is reported as sales and the carrying amount of the leased property, plus any initial direct costs, less the present value of any unguaranteed residual value, is charged as cost of sales. The lessor reports an asset by recording the gross investment in a sales-type lease in an amount equal to the sum of the minimum lease payments and the unguaranteed residual value, less the unearned interest, calculated using the interest rate implicit in the lease as the discount factor. The difference between the gross investment and the net investment is unearned income, which is amortized over the lease term so as to produce a constant periodic rate of return on the net investment.

2.3.1.2 *Direct financing lease*

A lease is classified as a direct financing lease when, the fair value of the property at the inception of the lease is equal to the cost (carrying value). And therefore the lessor does not realize a profit or loss on the transaction other than interest revenue. This type of lease transaction most often involves entities engaged in financing operations. The lessor (a bank, or other financial institution) purchases the asset and then leases the asset to the lessee.

For a lease classified as a direct financing lease, the lessor reports as an asset in the balance sheet the net investment in a lease consisting of gross investment less unearned income and the unamortized initial direct costs. The gross investment is calculated by adding the minimum lease payments and the unguaranteed residual value. Unearned income is determined by subtracting the sum of the cost or carrying amount of the leased property from the gross investment. Unearned income and the initial direct costs are amortized over the useful life so as to produce a constant periodic rate of return on the lease term net investment. The practice of recognizing a portion of the unearned income at the inception of the lease to offset initial direct costs or other costs of direct financing leases is not acceptable.

2.3.1.3 *Leveraged lease*

A leveraged lease is a direct financing lease that additionally has all of the following characteristics (SFAS No. 13, paragraph 42):

- It involves at least three parties: a lessee, a long-term creditor and a lessor.
- The financing provided by the long-term creditor is substantial to the transaction and is non-recourse to the lessor.
- The lessor's net investment declines during the early years and increases during the later years of the lease term.
- Any investment tax credit retained by the lessor is accounted for as one of the cash flow components of the lease.

The lessor records the investment in a leveraged lease net of the non-recourse debt. Income is recognized only in periods in which the net investment net of related deferred taxes is positive. The total net income over the lease term is calculated by deducting the original investment from total cash receipts. Using projected cash receipts and disbursements, the rate of return on the net investment in the years in which the investment is positive is determined and applied to the net investment to determine the periodic income to be recognized. The assumption that underlies this accounting is that the lessor will earn other income during the years in which the investment is negative (net funds are provided by the lease) and will not expect the lease to provide income during those years.

2.3.2 Operating lease

Leases that do not qualify as capital leases in accordance with the provisions of SFAS No. 13, paragraph 7 are classified as operating leases. The cost of the property leased to the lessee is included in the lessor's balance sheet as property, plant and equipment.

The rent is reported as income over the lease term in a systematic manner, which is usually straight-line; and the leased property is depreciated like other productive assets. Initial direct costs paid to independent third parties such as appraisal fees, finder's fees and costs of credit checks are amortized over the life of the lease.

It is possible that a lessor having not met the criteria for capital lease will classify the lease as an operating lease but the lessee will classify the same lease as a capital lease. In such an event, both the lessor and the lessee will carry the asset on their books, and both will depreciate the capitalized asset.

2.4 Sale and leaseback

The guideline for the determination of sale and leaseback is based upon the classification criteria presented for the lease transaction.

2.4.1 Basic principle

If the lease meets one of the four criteria for treatment as capital lease, the seller-lessee accounts for the transaction as a sale and the lease as a capital lease. Any profit or loss experienced by the seller-lessee from the sale of the asset that is leased back under a capital lease should be deferred and amortized over the lease term in proportion to the amortization of the leased assets. If none of the capital lease criteria are satisfied, the seller-lessee accounts for the transaction as a sale and the lease as an operating lease. Under an operating lease, such profit or loss should be deferred and amortized in proportion to the rental payments over the period of time the assets are expected to be used by the lessee. As a deviation from this general rule SFAS No. 28, paragraph 3 requires the recognition of some profit or loss in the following circumstances:

- When fair value of the asset is less than the carrying amount a loss has to be recognized immediately up to the amount of the difference between the book value and fair value.
- When the seller-lessee retains to use only a minor part of the property or a minor part of the remaining life, the transaction is a sale and a full gain or loss recognition is appropriate.

If the lease meets one of the four criteria and meets both of the additional criteria for treatment as capital lease, the purchaser-lessor records the transaction as a purchase and a direct financing lease. If the lease does not meet the criteria, the purchaser-lessor records the transaction as a purchase and an operating lease.

2.4.2 Sale and leaseback transaction involving real estate

Leases involving real estate are also addressed by SFAS No. 13 and amended by SFAS No. 28, SFAS No. 66 and SFAS No. 98. Three additional requirements are necessary for a sale-leaseback involving real estate (including real estate with equipment) to qualify for sale-leaseback accounting treatment. Those sale-leaseback transactions not meeting the three requirements should be accounted for as a deposit or as financing. The three requirements (SFAS No. 98, paragraph 7) are:

(i) The lease must be a normal leaseback.
(ii) Payment terms and provisions must adequately demonstrate the buyer-lessor's initial and continuing investment in the property.
(iii) Payment terms and provisions must transfer all the risks and rewards of ownership as demonstrated by a lack of continuing involvement by the seller-lessee.

A normal leaseback involves active use of the leased property in the seller-lessee's trade or business during the lease term.

2.5 Disclosures

2.5.1 Lessee presentation

A general description of the lessee's leasing arrangements shall be disclosed in the lessee's financial statements or the footnotes (SFAS No. 13, paragraph 16). For a capital lease, the following information should be disclosed in the lessee's financial statements or the footnotes:

- Gross amount of assets recorded under capital leases as of the date of each balance sheet presented by major classes according to nature or function.
- Future minimum lease payments in total and for each of the next five years.
- Total of minimum sublease rentals to be received in the future under non-cancellable subleases.
- Total contingent rentals actually incurred for each period for which an income statement is presented.

For all operating leases of lessees having initial or remaining non-cancellable lease terms in excess of one year the following information should be disclosed in the lessee's financial statements or the footnotes:

- Future minimum rental payments in total and for each of the next five years.
- Total of minimum rentals to be received in the future under non-cancellable subleases.
- Schedule of total rental expense showing the composition by minimum rentals, contingent rentals, and sublease income (excluding leases with terms of a month or less that were not renewed).

2.5.2 Lessor presentation

When leasing, exclusive of leveraged leasing, is a significant part of the lessor's business activities in terms of revenue, net income, or assets, a general description of the lessor's leasing arrangements should be disclosed in the financial statements or footnotes thereto (SFAS No. 13, paragraph 23).

For sales-type and direct financing leases the following information should be disclosed in the lessor's financial statements or the footnotes:

- Components of the net investment in sales-type and direct financing leases.
- Schedule of the minimum lease payments, in total and for the next five years.

For all operating leases the following information should be disclosed in the lessor's financial statements or the footnotes:

- Schedule of the investment in property on operating leases, and property held for lease, by major categories, less accumulated depreciation.
- Schedule of future minimum rentals on non-cancellable operating leases, in total and for the next five years.
- Amount of contingent rentals.

3 Foreign currency translation

Under SFAS No. 52 the financial statements have to be presented in the functional currency of an entity. The functional currency is the currency of the economic environment in which the entity operates. Usually this is the currency in which the entity primarily generates and expenses cash.

Transactions in another than the functional currency of the entity have to be translated into the functional currency by using the current exchange rate at the day the asset or liability is recognized. At balance sheet date the foreign currency denominated transaction has to be remeasured with the spot rate. Any difference between the translated amount at the balance sheet date and the translated amount that was recorded at the transaction date is reflected as an exchange gain or loss in the income statement.

For the accounting of foreign currency derivatives and hedging of foreign currency transactions see VII.4.

The entity should disclose the aggregate transaction gain or loss included in net income for the period.

4 Financial instruments, derivatives and hedge accounting

4.1 Overview

Due to the increased volume and the high complexity and diversity of innovative financial instruments, the FASB issued several SFAS on recognition and measurement of financial instruments including derivatives, disclosure requirements and finally on hedge accounting. The long-term goal of the FASB is to measure all financial instruments with fair values. The current accounting guidelines do not provide a consistent approach.

The approach of the FASB, to value all financial instruments at fair value, is based on the premise that all, even complex, financial instruments are made up of a few different 'building blocks' (fundamental financial instruments) and that determining how to recognize and measure these fundamental instruments is the key to a consistent solution for the accounting issues.

Hedge accounting requires that gains and losses on hedging instruments be recognized in income at the same time the effects of related changes in the hedged item are recognized.

Under the current accounting rules the hedging instrument follows the accounting for the hedged item, e.g. when an asset is carried at historical cost, the gains or losses on the hedging instrument are deferred. According to SFAS No. 133 all derivatives have to be measured at their market value. Offset is achieved by changing the recognition and measurement of the hedged item.

SFAS No. 133, which became effective for all financial years beginning after 15 June 2000, establishes comprehensive accounting and reporting standards for derivative instruments and hedging activities. The accounting and reporting principles prescribed by the Standard are complex and will significantly change the way entities account for derivative instruments and hedging activities. To avoid a significant volatility in earnings, each company must carefully reconsider its hedging strategy and the instruments used for hedging.

4.2 Definitions

Financial instruments are defined in SFAS No. 105, paragraph 6 as cash, evidence of ownership in an equity or a contract that both:

- Imposes on one entity a contractual obligation to (a) deliver cash or another financial instrument to a second entity; or (b) to exchange financial instruments on potentially unfavorable terms with the second entity
- Conveys to that second entity a contractual right (a) to receive cash or another financial instrument from the first entity; or (b) to exchange other financial instruments on potential favourable terms with the first entity.

The definition covers all fundamental financial instruments like receivables, payables, debt and equity securities, forwards, options and other rights or obligations to receive a financial instrument that are conditional on the occurrence of a specified event outside the control of either party, e.g. a letter of credit.

Due to the rapid development of new derivatives, a general definition of derivatives is introduced by SFAS No. 133. Prior to SFAS No. 133 certain (financial only) derivatives were listed in SFAS No. 119, paragraph 5. Derivatives provide the holder with the ability to participate in changes of market variables depending on one or more underlyings (price, index, rate) and notional amounts. They require no or only a small initial investment, resulting in a leveraged reaction to movements in the underlying. Generally they are settled in cash or there is a market mechanism, such as an exchange, in which it is possible to enter into a closing contract with only cash settlement. Due to the very broad definition several types of contract had to be excluded, like regular sales and purchase of goods or securities, certain insurance contracts, etc.

Embedded derivative instruments are derivatives embedded into a host contract with implicit or explicit terms that affect some or all of the cash flows or the value or other exchanges required by the contract in a manner similar to a derivative instrument.

In trading a party enters into a derivative contract for the purpose of speculating on changes in the underlying (price, rate or index). This type of activity is generally characterized by high volume of transactions as positions are adjusted, closed out or opened as market conditions change. From the accounting perspective not only trading contracts, but also all contracts that do not meet the requirements for hedge accounting are considered trading.

Hedge accounting carries with it the notion of risk reduction. That is, an enterprise enters into a hedging transaction to reduce its risk of unfavourable market prices or cash flows. Therefore it is driven by the identification of risk and the effectiveness of the hedging transaction. To achieve hedge accounting the hedging entity has to meet certain requirements.

4.3 Recognition and measurement

4.3.1 Financial instruments

Under the current accounting regulations there is no comprehensive guidance about recognition and measurement of all financial instruments. The disclosure requirements about the fair value of financial instruments and concentration of credit risk are included in VII.4.4.

4.3.2 Derivatives

4.3.2.1 Trading activities

Under the current accounting practice as well as under SFAS No. 133, derivatives that are considered trading are marked to market with gains or losses recorded in the income statement. The trading activity for derivatives is generally reported in a trading account on the balance sheet. However, some entities report this activity in other assets and other liabilities. Most derivative contracts are based on notional amounts or underlying instruments, but only the initial cash position adjusted for changes in the market value is recorded on the balance sheet. In addition, under FIN No. 39, the assets and liabilities related to trading positions generally must be presented gross, rather than net, unless certain criteria of that interpretation are met.

4.3.2.2 Hedge accounting

4.3.2.2.1 Current hedge accounting

Currently hedge accounting is guided by SFAS No. 52 for foreign currency transactions, SFAS No. 80 for futures transactions, except foreign currency futures, and various EITF statements. Accounting for options and swaps has developed largely through practice and by analogy to SFAS No. 80 and No. 52.

SFAS No. 80 permits hedge accounting for futures (forward) contracts when the futures contract effectively reduces the risk from the item to be hedged. To determine that a futures contract reduces risk, it must be probable that there is a high correlation of 0.8 between the changes in the value of the futures contract and the value of the hedged instrument. Although risk reduction is required on a macro basis, the hedging contract must be designated as a hedge of a specific asset or liability (micro-hedge).

The hedged item can be an existing asset or liability, a firm commitment and an anticipated transaction.

Measurement is as follows:

- If the hedged item is measured at fair value with unrealized changes recognized in income or other comprehensive income, the same accounting is applied to the futures contract.
- A change in the market value of a futures contract that qualifies as a hedge of an existing asset or liability is recognized as an adjustment of the carrying amount of the hedged item. In the case of a firm commitment, the change in the market value of the future is included in the measurement that satisfies the commitment. Premiums or discounts on a hedge contract are deferred and amortized over the life of the contract.
- Changes in the value of a hedge contract of an anticipated transaction are deferred and included in the subsequent measurement of the subsequent transaction.

If a company designates options as hedging instruments, a high correlation must be maintained between the changes in the value of the hedged item and the item underlying the option. The value of the underlying is used rather than the value of the option because option value is affected by the potential volatility of the time value component. The premium paid for an option designated as a hedge generally is amortized over the option period if the option is hedging a recorded balance sheet item carried at cost. If the option is hedging an anticipated purchase of an asset, the premium can be expensed or included in the cost of the acquired asset.

While the concept of hedge accounting implies risk reduction, a swap transaction does not necessarily reduce risk. This issue introduces the concept of 'synthetic alteration', because the swap changes the nature of a financial instrument (e.g., changing a floating-rate debt to a fixed-rate debt obligation). The premise behind synthetic alterations is that two or more financial instruments can be combined to synthetically create another financial instrument with the accounting driven by the nature of the synthetic instrument rather than the individual financial instrument components. Except for cases where the hedged item has to be recorded at fair market value, synthetic alteration results in deferral of changes in value of the hedging instrument like the other types of hedges.

Hedging accounting for exposure from foreign currency transactions is only permitted for hedges of firm commitments and investments in a foreign entity, but not for forecasted transactions. Hedge accounting does not necessarily require a derivative as hedging instrument. Changes in value of a contract hedging a firm commitment are deferred and included in the measurement of the hedged item. In case of the hedge of an investment of a foreign entity the change in value is reported in other comprehensive income. A premium or discount on a forward contract is either amortized over the lifetime of the contract or treated as an adjustment of the transaction price or, in case of a hedge of a net investment in a foreign entity, included in the amount of the equity adjustment from translation.

4.3.2.2.2 Hedge accounting under SFAS No. 133

In developing the Standard, the Board reached four fundamental decisions that became the cornerstones of SFAS No. 133. These are:

- Derivative instruments represent rights or obligations that meet the definitions of assets or liabilities and should be reported in financial statements.
- Fair value is the most relevant measure for financial instruments and the only relevant measure for derivative instruments.
- Only items that represent assets or liabilities should be reported as such in financial statements. The deferral of option premiums and premiums or discounts of other derivative contracts is no longer permitted.
- Special accounting for items designated as being hedged should be provided but should be limited to qualifying transactions. One criterion for qualification should be an assessment of offsetting changes in fair values or cash flows for the risk being hedged.

With these as cornerstones, the Board developed an accounting model for derivative instruments and hedging activities.

SFAS No. 133 requires that all derivative instruments be recorded in the statement of financial position at fair value.

To qualify for hedge accounting certain criteria have to be met by the hedging instrument and the hedged item. Most important is that the hedging instrument must be highly effective (correlation of at least 0.8) in offsetting the changes in value of the hedged item. The time value of an option or a forward premium or discount can be excluded in the effectiveness test. There must be a formal designation of the hedging instrument to the hedged item (micro-hedge). Also a proportion of a derivative can be a hedging instrument, but a component of a derivative representing a specific risk cannot be designated as a hedging instrument. The entity has to document the risk management objective and strategy for this hedge. The hedged item can be an asset or liability, a percentage thereof, a selected cash flow thereof or a portfolio of assets or liabilities that share the same risk exposure and react similar to changes in the underlying. Prior to SFAS No. 133, the derivative could be broken down into components, but now only the hedged item can be broken down into its risk components. An asset or liability that is measured at fair value with changes reported currently in earnings cannot be a hedged item since there is a natural offset in the income statement, if the hedge is effective. In a cash-flow hedge the occurrence of the forecasted transaction must be probable. There are further requirements, e.g. for written options or basis swaps that need to be evaluated carefully in order to achieve hedge accounting.

If the derivative instrument qualifies as a hedge,

the accounting varies based on the type of risk being hedged. The Standard permits hedge accounting for the following three risks:

- Changes in fair values due to fixed rates or prices. For example, the risk of changes in the fair value of a fixed-rate investment security may be reduced or eliminated by a fair value hedge. For fair value hedges, gains or losses on derivative hedging instruments are recorded in income each reporting period. To achieve the offsetting effect in the income statement, gains and losses on the hedged item attributable to the hedged risk are also recognized in income, e.g. the fair value adjustment of an available for sale security related to the hedged risk is included in earnings instead of other comprehensive income. Consequently, if gains or losses on the derivative hedging instrument and the related hedged item do not completely offset, the difference (i.e. ineffective portion of the hedge) is recognized currently in income.
- Changes in cash flows due to variable rates or prices. For example, the risk of changes in cash flows from a variable-rate debt obligation may be reduced or eliminated by a cash flow hedge. For cash flow hedges, the reporting of gains or losses on derivative hedging instruments depends on whether the gains or losses are effective at offsetting the variability in cash flows of the hedged item attributable to the hedged risk. The effective portion of the gain or loss on the derivative hedging instrument is accumulated in other comprehensive income and recognized in income during the period that the hedged forecasted transaction has an impact on earnings. The ineffective portion of the gain or loss from the derivative hedging instrument is recognized in income immediately.
- Changes in fair values, cash flows, or net investment in foreign operations due to variability of foreign currency exchange rates. For example, the risk of changes in the foreign currency exchange rates inherent in the forecasted purchase of equipment to be paid for in a foreign currency may be reduced or eliminated by a foreign currency hedge. For foreign currency hedges, the Standard permits an entity to use the following three accounting models to hedge its exposure to foreign currency risk:
  - (i) The fair value hedge accounting model can be used to account for hedges of the currency exposure inherent in assets or liabilities in a foreign-currency-denominated firm commitment or an available-for-sale security. The accounting requirements for this model are identical to the aforementioned requirements for other fair value hedges.
  - (ii) The cash flow hedge accounting model can be used to account for hedges of the currency exposure inherent in assets or liabilities in a forecasted foreign-currency-denominated transaction or a forecasted foreign-currency-denominated intercompany transaction. The accounting requirements for this model are identical to the aforementioned requirements for other cash flow hedges.
  - (iii) Consistent with SFAS No. 52, Foreign Currency Translation, the Standard permits an entity to hedge the foreign currency risk inherent in a net investment in a foreign operation. The effective portion of the gain or loss on the hedging instrument is reported in other comprehensive income. The ineffective portion of the gain or loss is recognized in income immediately.

As exception to the general rule that only derivatives can be designated as hedging instruments, SFAS No. 133 allows non-derivative financial instruments to be designated as hedging instruments for the hedge of foreign currency exposure of an unrecognized firm commitment and a net investment in a foreign entity.

4.3.2.3 Embedded derivatives

In order to avoid circumvention of the provisions for hedge accounting, SFAS No. 133 stipulated the accounting for embedded derivatives. The embedded derivative has to be separated from the host contract, if the economic characteristics and risks of the embedded derivative are not clearly and closely related to those of the host contract. This is usually the case, but not limited to, if the underlying price, rate or index of the derivative is different from that of the host contract.

The separated host contract has to be accounted for following the GAAP applicable for this type of contract. The derivative is accounted for as a derivative pursuant to SFAS No. 133. If the embedded derivative cannot reliably be identified and measured, the entire contract has to be measured at fair value with gains or losses recognized in income.

4.4 Disclosures

Under the current authoritative guidance disclosures about financial instruments are required by SFAS No. 105, Disclosure of Information about Financial Instruments with Off-Balance-Sheet Risk and Financial Instruments with Concentration of Credit Risk, SFAS No. 107, Disclosures about Fair Value of Financial Instruments, and SFAS No. 119, Disclosure about Derivative Financial Instruments and Fair Values of Financial Instruments. The current disclosure requirements can be summarized as follows.

A company must disclose the fair value and related carrying value of its financial instruments in the body of the financial statements, a single note, or in a summary table in a form that clarifies whether the amounts represent assets or liabilities. The fair value disclosures must distinguish between financial instruments held for trading and held for other than trading purposes. The fair value disclosures have to be presented by different classes of financial instruments. A combination, aggregation or netting is not allowed, even if those instruments are considered to be related. For financial instruments with off-balance-sheet risk and for all financial derivatives, specific disclosures are required, such as the contract principal amount, terms of the instrument and information about their credit or market risk, cash requirements, and related accounting policies.

For financial years beginning after 15 June 2000, SFAS No. 133 supersedes SFAS Nos 105 and 119. The disclosure requirements related to concentrations of credit risk are transferred from SFAS No. 105 to SFAS No. 107. The amended disclosure requirements will have the following main components:

A company must disclose the fair value of its financial instruments, whether recognized or not recognized in the balance sheet, either in the body of the financial statements, or in the accompanying notes. The company must disclose the methods and significant assumptions used to estimate the fair values of financial instruments. A combination, aggregation or netting is not allowed, even if those instruments are considered to be related. Significant concentrations of credit risks from all financial instruments have to be disclosed. For trading derivatives the purpose of entering into such activities shall be described in the context of the overall risk management profile.

For hedging instruments the company must disclose the objectives for holding or issuing the instrument and the strategies to achieve these objectives. The description should distinguish between the hedge categories and include the risk management policy for each type of hedge. For fair value and cash flow hedges disclosure is required regarding the amount of hedges' ineffectiveness, components that are excluded from the assessment of the hedge effectiveness and a description where the gains or losses are reported. Other disclosure requirements are related to the movements in other comprehensive income.

5 Long-term contracts

5.1 General principles

There are two methods of accounting for long-term construction contracts: the percentage-of-completion method and the completed-contract method. Various accounting pronouncements (see ARB No. 45, and SOP No. 81–1) have emphasized that the percentage-of-completion and the completed-contract methods of accounting for long-term construction contracts are not intended to be free-choice alternatives. ARB No. 45 states that the percentage-of-completion method is preferable when estimates of costs of the degree of completion are reliable. SOP No. 81-1 restates the same position.

5.2 Profit recognition

Under the percentage-of-completion method, revenues and gross profit are recognized in each period upon the progress of the construction, that is, the percentage of completion. Construction costs plus gross profit earned to date are accumulated in an inventory account called 'Construction in progress'. The progress is measured using engineering estimates or ratios of costs incurred to expected total costs.

SOP No. 81–1, section 23, requires the percentage-of-completion method to be used when estimates of progress toward completion, revenues, and costs are reasonably dependable and all the following conditions exist:

- The contract clearly specifies the enforceable rights regarding goods or services to be provided and received by the parties, the consideration to be exchanged, and the manner and terms of settlement.
- The buyer can be expected to satisfy all obligations under the contract.
- The contractor can be expected to perform the contractual obligation.

The completed-contract method should be used only when an entity has primarily short-term contracts, or when the conditions for using the percentage-of-completion method can not be met, or when there are inherent hazards in the contract beyond the normal, recurring business risk.

Under the completed-contract method, revenues and profits are recognized only when the contract is completed. Construction costs are accumulated in the inventory account called 'Construction in progress'.

5.3 Measurement of percentage of completion

Estimates of the degree of completion usually are based on one of the following methods (ARB No. 45, paragraph 4):

- Cost-to-cost method.
- Other measures of progress toward completion, such as engineering estimates.

The most popular method to determine the progress toward completion is the cost-to-cost method. Under the cost-to-cost method, the percentage of completion is measured by comparing costs incurred to date with the most recent estimate of the total costs to complete the contract. The cost-to-cost method can result in inaccurate measures, i.e. when some early-stage construction costs do not relate to contract performance. The most common other methods are the efforts-expended method, units-of-delivery method, and units-of-work-performed method.

5.4 Anticipated losses

When current estimates of the total contract costs indicate a loss, a provision for the loss on the entire contract should be made (see VI.5.14.5). When a loss is indicated on a total contract that is part of a related group of contracts, however, the group may be treated as a unit in determining the necessity of providing for losses (ARB No. 45, paragraph 6).

5.5 Disclosures

Construction contractors shall disclose the following information in the footnotes:

- The method used of recognizing revenue.
- The basis used to classify assets and liabilities as current.
- The basis for recording inventory.
- The effects of any revision of estimates.
- The amount of backlog on uncompleted contracts.
- Details about receivables (billed and unbilled, maturity, interest rates, retainage provisions, and significant individual or group concentrations of credit risks).

6 Stock-based compensation

6.1 Introduction

Stock-based compensation plans include all agreements by an employer to issue shares of stock or other equity instruments to employees, or to pay cash to employees in amounts based on the increase in the employer's stock price from a specified level. Common examples of stock-based compensation plans are fixed stock options, restricted stock awards, stock purchase plans and stock appreciation rights (SARs). The FASB decided that, even if a company distributes options in its own equity, the options give rise to compensation to employees that should be included in measuring the company's net income.

The authoritative accounting literature on stock-based compensation is contained primarily in APB Opinion No. 25, Accounting for Stock Issued to Employees, and SFAS No. 123, Accounting for Stock-Based Compensation.

APB Opinion No. 25 sets forth an intrinsic value method of accounting for stock-based compensation. Under this method, compensation cost is measured as the excess, if any, of the quoted market price of the stock at the measurement date over the amount to be paid by the employee.

SFAS No. 123 amended APB Opinion No. 25 by specifying that a fair value method of accounting for stock-based compensation is preferable to the intrinsic value method. SFAS No. 123 allows companies to continue to measure compensation cost in the basic financial statements under the intrinsic value method prescribed by APB Opinion No. 25; however, companies electing to retain the APB Opinion No. 25 approach are required to make pro forma disclosures of net income and earnings per share as if they had applied the fair value method.

In practice, the vast majority of companies have elected to continue to measure compensation cost under APB Opinion No. 25 and provide the disclosures required by SFAS No. 123.

6.2 SFAS No. 123: Accounting for stock-based compensation

6.2.1 Valuation of equity instruments

SFAS No. 123, paragraph 6 applies to all transactions in which an entity acquires goods or services by issuing equity instruments or by incurring liabilities to the supplier in amounts based on the price of the entity's common stock or other equity instruments. Therefore, it applies to all transactions in which an entity grants shares of its common stock, stock options, or other equity instruments to its employees, except for equity instruments held by an employee stock ownership plan.

Frequently, part or all of the consideration received for equity instruments issued to employees is past or future employee services. Equity instruments issued to employees and the cost of the services received as consideration shall be measured and recognized based on the fair value of the equity instruments issued. The portion of the fair value of an equity instrument attributed to employee services is net of the amount, if any, that employees pay for the instrument when it is granted.

The objective of the measurement process is to estimate the fair value, based on the stock price at the grant date, of stock options or other equity instruments to which employees become entitled when they have rendered the requisite service and satisfied any other conditions necessary to earn the right to benefit from the instruments (for example, to exercise stock options or to sell shares of stock). Restrictions that continue in effect after employees have earned the rights to benefit from their instruments, such as the inability to transfer vested employee stock options to third parties, affect the value of the instruments actually issued and therefore are reflected in estimating their fair value. However, restrictions that stem directly from the forfeitability of instruments to which employees have not yet earned the right, such as the inability either to exercise a non-vested option or to sell non-vested stock, do not affect the value of the instruments issued at the vesting date, and their effect therefore is not included in that value. Instead, no value is attributed to instruments that employees forfeit because they fail to satisfy specified service- or performance-related conditions.

The fair value of a share of non-vested stock awarded to an employee should be measured at the market price (or estimated market price, if the stock is not publicly traded) of a share of the same stock as if it were vested and issued on the grant date.

The fair value of a stock option (or its equivalent) granted by a public entity should be estimated using an option-pricing model (for example, the Black–Scholes or a binomial model). The fair value of an option estimated at the grant date should not be subsequently adjusted for changes in the price of the underlying stock or its volatility, the life of the option, dividends on the stock, or the risk-free interest rate.

A non-public entity should estimate the value of its options based on the factors described above, except that a non-public entity need not consider the expected volatility of its stock over the expected life of the option. The result of excluding volatility in estimating an option's value is an amount commonly termed minimum value (see SFAS No. 123, paragraphs 17–20).

6.2.2 Recognition of compensation cost

Under the fair value method, compensation cost is normally measured at the grant date based on the fair value of the award.

Compensation cost is recognized over the service period, which is usually the vesting period, with an offsetting credit to equity (paid-in capital).

6.2.3 Subsequent modifications and settlements

An entity may choose at the grant date to base accruals of compensation cost on the best available estimate of the number of options or other equity instruments that are expected to vest and to revise that estimate, if necessary, if subsequent information indicates that actual forfeitures are likely to differ from initial estimates. Alternatively, an entity may begin accruing compensation cost as if all instruments granted that are subject only to a service requirement are expected to vest. The effect of actual forfeitures would then be recognized as they occur. Initial accruals of compensation cost for an award with a performance condition that will determine the number of options or shares to which all employees receiving the award will be entitled shall be based on the best estimate of the outcome of the performance condition, although forfeitures by individual employees may either be estimated at the grant date or recognized only as they occur (SFAS No. 123, paragraph 28).

Compensation cost estimated at the grant date for the number of instruments that are expected to vest based on performance-related conditions, as well as those in which vesting is contingent only on future service for which the entity chooses to estimate forfeitures at the grant date pursuant to SFAS No. 123, paragraph 28, should be adjusted for subsequent changes in the expected or actual outcome of service- and performance-related conditions until the vesting date. The effect of a change in the estimated number of shares or options expected to vest is a change in an estimate, and the cumulative effect of the change on current and prior periods should be recognized in the period of the change.

6.2.4 Income tax effects

Income tax regulations specify allowable tax deductions for stock-based employee compensation arrangements in determining an entity's income tax liability. Compensation cost recognized under SFAS No. 123 is measured based on the fair value of an award to an employee. Under existing U.S. tax law, allowable tax deductions are generally measured at a specified date as the excess of the market price of the related stock over the amount the employee is required to pay for the stock (that is, at intrinsic value). The time value component of the fair value of an option is not tax deductible. Therefore, tax deductions will generally arise in different amounts and in different periods from compensation cost recognized in financial statements.

The cumulative amount of compensation cost recognized for a stock-based award that ordinarily results in a future tax deduction under existing tax law should be considered to be a deductible temporary difference in applying SFAS No. 109, Accounting for Income Taxes. The deferred tax benefit (or expense) that results from increases (or decreases) in that temporary difference – for example, as additional service is rendered and the related cost is recognized – should be recognized in the income statement. Recognition of compensation cost for an award that ordinarily does not result in tax deductions under existing tax law shall not be considered to result in a deductible temporary difference in applying Statement 109. A future event, such as an employee's disqualifying disposition of stock under existing US tax law, can give rise to a tax deduction for an award that ordinarily does not result in a tax deduction. The tax effects of such an event shall be recognized only when it occurs.

6.3 APB Opinion No. 25: Accounting for stock issued to employees

6.3.1 Valuation of equity instruments

APB Opinion No. 25 classifies plans as either compensatory or non-compensatory. It defines non-compensatory plans as plans having all four of the following characteristics:

- Substantially all full-time employees may participate.

- Stock is offered to eligible employees equally or based on a uniform percentage of salary or wages.
- The time permitted for exercise of an option or purchase right is limited to a reasonable period.
- Any discount from the market price of the stock is no greater than would be reasonable in an offer of stock to shareholders or others.

A compensatory plan is any plan not having all four attributes. Most plans do not meet the above conditions and are, therefore, compensatory plans; however, compensation expense is recognized for compensatory plans only if the application of the measurement principle results in compensation cost. Some existing broad-based employee stock purchase plans (such as Internal Revenue Code Section 423 plans) are non-compensatory under APB Opinion No. 25. Section 423 plans generally provide a 15 per cent discount from the market price of the shares.

Compensation cost is measured as the intrinsic value of the award at the measurement date. The intrinsic value is the amount by which the quoted market price of the stock exceeds the amount, if any, the employee is required to pay. The measurement date is the first date on which both the number of shares that an individual employee is entitled to receive and the option or purchase price are known.

If the measurement date occurs on the grant date, the plan is referred to as a fixed plan because compensation cost is fixed at the grant date. If the measurement date occurs after the grant date, the plan is referred to as a variable plan because compensation cost varies with the quoted market price of the stock at the end of each accounting period until the measurement date.

A requirement that an employee remains in the employ of the company in order to vest in an award does not change an otherwise fixed plan to a variable plan. If an employee forfeits an award because of termination prior to vesting, compensation cost previously recognized is reversed in the period of termination. Compensation cost generally is not reversed in other circumstances involving the lapsing or cancellation of awards (e.g. vested options that expire at the end of their term).

6.3.2 Recognition of compensation cost

Compensation cost is recognized in the periods in which the employee performs services. The method of recognizing compensation cost over the service period depends on whether the plan is fixed or variable. FIN No. 28 provides a specific method of accruing compensation cost for variable plans. Compensation cost for variable plans is recognized over the service period (the vesting period), which is normally the period that ends when all performance conditions have been met. If portions of the award vest at different dates (graded vesting), compensation cost for each portion should be recognized in full by the time that portion vests. Thus, FIN No. 28 requires accelerated expense recognition for variable plans with graded vesting.

Compensation cost for fixed plans may be recognized on a straight-line basis or some other systematic and rational method over the service period. Normally the vesting period is assumed to be the service period.

6.3.3 Subsequent modifications and settlements

The renewal of an option, the extension of its exercise period, a change in the number of shares or the option exercise price or a similar modification generally establishes a new measurement date as if the option were newly granted. Compensation cost previously recognized is not reversed, any unamortized compensation cost measured at the original measurement date continues to be amortized or accrued and any compensation cost associated with the new terms is recognized over the new service period.

Cash settlement of an option or an award within six months of the issuance of the shares requires a new measurement of compensation cost.

6.3.4 FASB Project – Interpretation of APB Opinion No. 25

In March 2000, the FASB issued FIN 44, an Interpretation of APB Opinion No. 25 'Accounting for Certain Transactions Involving Stock Compensation'. This Interpretation clarifies the application of APB Opinion No. 25 in certain situations. It does not address issues related to the application of SFAS No. 123. The issues being addressed were selected after receiving input from members of the FASB's Emerging Issues Task Force and the Task Force on stock compensation that assisted in the development of SFAS No. 123. The Exposure Draft includes 20 questions and answers that cover the following topics:

- Scope of APB Opinion No. 25.
- Non-compensatory plans.
- New measurement date.
- Variable awards.
- Business combinations.
- Grant date.
- Deferred tax assets.
- Cash bonus plan linked to a stock-based plan.

The Interpretation will result in a number of changes to the way APB Opinion No. 25 has been applied in the past. Among those are:

- Independent directors are not considered to be employees and thus option plans involving them are not covered by APB Opinion No. 25.
- Changes to either the number of shares or the option price (repriced options) become variable awards upon the first repricing.
- A new measurement date is required for any modification if the increase in fair value of the option or award is more than de minimus.
- In a purchase business combination, stock options granted by the acquirer in exchange for outstanding vested options are part of the purchase price; those exchanged for non-vested options are accounted for under APB Opinion No. 25 (compensation).

The provisions of this Interpretation are effective from 1 July 2000 and will generally be applied prospectively to new awards. To the extent that certain events covered by this Interpretation occurred after 15 December 1998 but prior to 1 July 2000, the effects of applying this Interpretation will be recognized only on a prospective basis. Accordingly, (a) no adjustments shall be made upon initial application of this Interpretation to financial statements for periods prior to 1 July 2000; and (b) no expense shall be recognized for any additional compensation cost measured that is attributable to periods prior to 1 July 2000. Such events covered by this Interpretation are (a) repricing transactions after 15 December 1998; (b) determination of employee or non-employee status after 15 December 1998; and (c) modifications to add a reload feature apply to modifications made after 12 January 2000.

6.3.5 Accounting for income tax effects

An employer corporation may obtain an income tax benefit related to stock issued to an employee through a stock option, purchase, or award plan. A corporation is usually entitled to a deduction for income tax purposes of the amount that an employee reports as ordinary income, and the deduction is allowable to the corporation in the year in which the amount is includable in the gross income of the employee. Thus, a deduction for income tax purposes may differ from the related compensation expense that the corporation recognizes, and the deduction may be allowable in a period that differs from the one in which the corporation recognizes compensation expense in measuring net income.

An employer corporation should reduce income tax expense for a period by no more than the tax reduction under a stock option, purchase, or award plan than the proportion of the tax reduction that is related to the compensation expense for the period. Compensation expense that are deductible in a tax return in a period different from the one in which they are reported as expense in measuring net income results in temporary differences, and deferred taxes should be recorded. The remainder of the tax reduction, if any, is related to an amount that is deductible for income tax purposes but does not affect net income. The remainder of the tax reduction should not be included in income but should be added to capital in addition to par or stated value of capital stock in the period of the tax reduction. Conversely, a tax reduction may be less than if recorded compensation expenses were deductible for income tax purposes. If so, the corporation may deduct the difference from additional capital in the period of the tax reduction to the extent that tax reductions under the same or similar compensatory stock option, purchase, or award plans have been included in additional capital.

6.4 Special rules for non-public companies and companies going public

6.4.1 Non-public company issues

A private company may have a stock purchase plan for employees who reach a certain level in the company (either by promotion or through direct entry) to acquire shares in the company. Participation in the plan may be either mandatory or optional for employees. The purchase price is a formula price based on book value or earnings. The arrangement requires that the employee sell the shares back to the company at retirement or upon leaving the company; the selling price is determined in the same manner as the purchase price. The issue is whether such a plan for a private company is compensatory pursuant to APB Opinion No. 25.

The EITF addressed these types of plans (referred to as formula or book value purchase plans) in Issue No. 87-23 and concluded that no compensation cost should be recognized for changes in book value during the employment period if the employee makes a substantive investment that will be at risk for a reasonable period of time. The EITF did not provide guidance on the term substantive investment; however, the accounting profession generally has taken the position that a formula equal to book value or a reasonable multiple of earnings would be considered substantive unless the earnings on which the formula is based are abnormally near break-even and future earnings are expected to recover to more normal levels. This portion of the consensus is only applicable to privately-held companies.

Rather than selling shares, a private company may grant options to eligible employees. The exercise price is based on a formula involving book value or earnings, and the employee can resell the options or the shares received upon exercise to the company at fixed or determinable dates or upon retirement or termination at a sales price determined in the same manner as the exercise price. The issue is whether under APB Opinion No. 25, this plan should be accounted for as (i) a fixed plan for which compensation is measured at the date of grant; or (ii) a variable plan for which compensation is measured at a later date.

In Issue No. 87-23, the EITF concluded that this arrangement is a variable plan since the employee has made no investment that is at risk prior to exercise. Accordingly, compensation cost should be recognized for any increase in the formula value from the grant date to the exercise date. Once exercised, no further expense would need to be recognized unless the shares are sold back to the company shortly after exercise. In this situation, additional compensation expense would be recognized as required by APB Opinion No. 25, paragraph 11(g).

6.4.2 Issues for companies going public

In connection with EITF Issue No. 88-6, the SEC Observer stated that book value shares issued within one year of the IPO are presumed to have been issued in contemplation of the IPO and are therefore 'cheap stock'. Compensation cost should be recognized for the difference between the grant price of those shares and the estimated fair value at the date of issuance (considering the IPO price and other evidence of fair value). Subsequent to the IPO, no further compensation expense would be recognized under APB Opinion No. 25 assuming the plan otherwise remains a fixed plan under that Opinion.

The SEC requires that a registration statement filed in conjunction with the IPO include prominent pro forma disclosure of the additional compensation cost related to book value shares issued within one year of the IPO that will be recognized upon the successful completion of the IPO.

6.5 Disclosures

Regardless of the method used to account for stock-based employee compensation arrangements, the financial statements of an entity shall include the following disclosures:

- A description of the plan(s), including the general terms of awards under the plan(s), such as vesting requirements, the maximum term of options granted, and the number of shares authorized for grants of options or other equity instruments.
- The number and weighted-average exercise prices of options for each of the following groups of options: (i) those outstanding at the

beginning of the year, (ii) those outstanding at the end of the year, (iii) those exercisable at the end of the year, and those (iv) granted, (v) exercised, (vi) forfeited, or (vii) expired during the year.

- The weighted-average grant-date fair value of options granted during the year. If the exercise prices of some options differ from the market price of the stock on the grant date, weighted-average exercise prices and weighted-average fair values of options shall be disclosed separately for options whose exercise price (i) equals, (ii) exceeds, or (iii) is less than the market price of the stock on the grant date.
- The number and weighted-average grant-date fair value of equity instruments other than options, for example, shares of non-vested stock, granted during the year.
- A description of the method and significant assumptions used during the year to estimate the fair values of options, including the following weighted-average information: (i) risk-free interest rate, (ii) expected life, (iii) expected volatility, and (iv) expected dividends.
- Total compensation cost recognized in income for stock-based employee compensation awards.
- The terms of significant modifications of outstanding awards.

For options outstanding at the date of the latest statement of financial position presented, the range of exercise prices (as well as the weighted-average exercise price) and the weighted-average remaining contractual life should be disclosed. If the range of exercise prices is wide (for example, the highest exercise price exceeds approximately 150 per cent of the lowest exercise price), the exercise prices should be segregated into ranges that are meaningful for assessing the number and timing of additional shares that may be issued and the cash that may be received as a result of option exercises. The following information should be disclosed for each range:

- The number, weighted-average exercise price, and weighted-average remaining contractual life of options outstanding.
- The number and weighted-average exercise price of options currently exercisable.

In addition, an entity that continues to apply APB Opinion No. 25 shall disclose for each year for which an income statement is provided the pro forma net income and, if earnings per share is presented, pro forma earnings per share, as if the fair value-based accounting method in this Statement had been used to account for stock-based compensation cost. Those pro forma amounts shall reflect the difference between compensation cost, if any, included in net income in accordance with APB Opinion No. 25 and the related cost measured by the fair value-based method, as well as additional tax effects, if any, that would have been recognized in the income statement if the fair value-based method had been used. The required pro forma amounts should reflect no other adjustments to reported net income or earnings per share.

7 Quasi-reorganization

Many state laws preclude the declaration of dividends unless a credit balance exists in the retained earnings account. Corporations with a debit balance (deficit) in retained earnings would be unable to pay dividends even though the corporation might have a promising future if it could attract equity capital. In such cases, under carefully defined circumstances, contributed or paid-in capital generally may be used to restructure a corporation, including the elimination of a deficit in retained earnings. This procedure is called a quasi-reorganization. It is termed 'quasi' since the accumulated deficit is eliminated at a lower cost and with less difficulty than a legal reorganization. The accounting requirements are outlined in ARB No. 43, chapter 7.

The following specific criteria must be fulfilled for a quasi-reorganization to be appropriate:

- Assets are overvalued in the balance sheet.
- The company can reasonably expect to be profitable in the future if the restructuring occurs in such a way that future operations are not burdened with the problems of the past.
- The proposed quasi-reorganization procedure should be submitted to and receive the approval of the corporation's stockholders before it is put into effect.

Although the corporate entity remains unchanged in a quasi-reorganization, a new basis of accountability is established. The assets of the enterprise are restated to their fair values and the liabilities to their present values with the net amount of these adjustments added or deducted from the deficit. Retained earnings (deficit) is raised to a zero balance by charging any deficit accumulated from operations and the asset adjustments to either capital contributed in excess of par of capital contributed other than for capital stock. Contributed capital accounts must be large enough to absorb the deficit in retained earnings, including adjustments made as part of the quasi-reorganization.

The steps in the accounting procedure are as follows:

- Assets are written down to their fair values and liabilities are restated to their present value. The amounts to be written off are charged to retained earnings.
- After all amounts to be written off are recognized and charged to retained earnings, the negative balance is transferred to either capital contributed in excess of par or capital contributed other than for capital stock. Capital contributed in excess of par value may have existed prior to the quasi-reorganization, or it may have been created as a result of a reduction of par value in conjunction with the quasi-reorganization. The excess created by the reduction of par value is credited to 'Paid-in capital from quasi-reorganization'.
- If a deficit in retained earnings is transferred to an allowable capital account, any subsequent balance sheet must disclose, by dating the retained earnings, that the balance in the retained earnings account has accumulated since the date of reorganization. The dating of retained earnings following a quasi-reorganization would rarely, if ever, be of significance after a period of 10 years.
- New or additional shares of stock may be issued or exchanged for other shares or existing indebtedness.
- Corporations with subsidiaries should follow the same procedures, so that no credit balance remains in consolidated retained earnings after a quasi-reorganization in which losses have been charged to allowable capital accounts.
- The effective date of the quasi-reorganization from which income of the corporation is thereafter determined should be as close as possible to the date of formal stockholders' consent and preferably at the start of the new fiscal year.

A simpler procedure, referred to as a deficit reclassification, results solely in eliminating a deficit in retained earnings without restating assets or liabilities. The accounting procedure is limited to a reclassification of a deficit in reported retained earnings as a reduction of paid-in capital.

Careful consideration must be given to the proper accounting for any tax attributes in a quasi-reorganization. Under SFAS No. 109, Accounting for Income Taxes, tax benefits of deductible temporary differences and carryforwards as of the date of the quasi-reorganization ordinarily are reported as a direct addition to capital contributed in excess of par if the tax benefits are recognized in subsequent years.

8 Discontinued operations

Discontinued operations represent separately identifiable segments (components of a major class of business with separately identifiable assets, liabilities, revenues, and expenses) which are being disposed of. After it has been determined that certain discontinued operations constitute the disposal of a segment in accordance with APB Opinion No. 30, Reporting the Results of Operations, results of the discontinued segment must be segregated from normal, recurring operations. The results of discontinued operations should be disclosed separately as a component of income before extraordinary items and the cumulative effect of accounting changes (if applicable).

The discontinued operations section of an income statement consists of two components, the income (loss) from operations and the gain (loss) on the disposal. Income (loss) from operations is disclosed for the current year only if the decision to discontinue operations is made after the beginning of the fiscal year for which the financial statements are being prepared. A gain (loss) on disposal consists of the two elements income (loss) from operations during the phase-out period (the period between the measurement date (the date the decision is made to discontinue a segment's operations, and the disposal date) and the gain (loss) from disposal of segment assets.

When discontinued operations are disclosed in a comparative income statement, the income statement presented for each previous year must be adjusted retroactively to enhance comparability with the current year's income statement. Accordingly, the revenues, cost of goods sold, and operating expenses (including income taxes) for the discontinued segment are removed from the revenues, cost of goods sold, and operating expenses of continuing operations. These items are netted into one figure (i.e. income (loss) from operations), which is disclosed under the discontinued operations section.

The notes to financial statements for the period encompassing the measurement date of the discontinued operations should disclose:

- The identity of the segment of business that has been or will be discontinued.
- The expected disposal date.
- The expected manner of disposal.
- A description of the remaining assets and liabilities of the segment at the balance sheet date.
- The income or loss from operations and any proceeds from disposal of the segment during the period from the measurement date to the balance sheet date.

9 Government grants

An enterprise may receive grants from the government, requiring a determination that the resources are being used in accordance with the grant provisions. A grant is a contribution or gift of cash, or other assets from the government to be used or expended for a specific purpose, activity, or facility. Some grants are restricted by the grantor for the acquisition or construction of fixed assets. These are capital grants. All other grants are operating grants.

Since there are no specific accounting requirements, some grants are recorded as reductions of the cost of the related assets, some as deferred credits to be amortized over the life of the assets, some as income when received and some as additions to contributed capital under present accounting practice. Guidance on accounting is provided in an AICPA Issues Paper, Accounting for grants received from governments. The Issues Paper contains advisory conclusions taking the position that grants related to depreciable fixed assets should be recognized as income over the useful life of the assets, and grants related to land should be amortized into income over the life of the depreciable fixed assets built on the land. In addition, it is concluded that a grant received before the conditions of the grant are met should be recorded as a liability or deferred credit. If it is probable that a grant will be refunded, the grant should be recorded as a liability. If expenses covered by the grant are incurred prior to grant receipt, a receivable should be set up to reflect the grantor's obligation.

10 Special revenue recognition areas

10.1 Introduction

Accrual accounting requires revenue to be recognized when it is realized or realizable and it has been earned. This is the conceptual guideline given by SFAC No. 5 for the recognition of revenue. In some cases the circumstances of the transaction do not allow for the clear application of the revenue realization principle. As a result, specific accounting guidelines on revenue recognition are provided for those special areas, such as long-term construction (see VII.5), instalment sales, sales of real estate, franchising, licence receipts, computer software and mortgage banking as described below. Specific guidance on revenue recognition is also given regarding sales with the right of returns (see VI.5.8), product financing arrangements (see VI.5.7), sales-type lease and sale-leaseback transactions (see VII.2), sale of receivables with recourse (see VI.5.10), and recognition of loan and lease origination fees.

In deviation to the provisions of accrual accounting small companies often recognize revenue in accordance with APB Opinion No. 10, paragraph 12 on a cash collection basis. In addition, the completed contract method in the case of long-term construction type contracts (ARB No. 45) is not in compliance with the accrual principle (see VII.5).

10.2 Instalment sales

An instalment transaction occurs when a seller delivers a product or performs a service and the buyer makes periodic payments over an extended period of time. Under the instalment method, income recognition is deferred until the period(s) of cash collection (see APB Opinion No.10, Omnibus opinion, paragraph 12). The seller recognizes both revenues and cost of sales at the time of the sale. The related gross profit is deferred to those periods in which cash is collected. Under the cost-recovery method, both revenues and cost of sales are recognized at the time of the sale, but none of the related gross profit is recognized until all cost of sales has been recovered. Once the seller has recovered all cost of sales, any additional cash receipts are included in income. APB Opinion No. 10 does not specify when one method is preferred over the other. The cost-recovery method is more conservative than the instalment method because gross profit is deferred until all costs have been recovered. Therefore, it should be reserved for situations of extreme uncertainty.

The instalment method can be used in most sales transactions for which payment is to be made through periodic instalments over an extended period of time and the collectibility of the sales price cannot be reasonably estimated. This method is applicable to the sales of real estate, heavy equipment, home furnishings, and other merchandise sold on an instalment basis. Instalment method revenue recognition is not in accordance with accrual accounting because revenue recognition should not normally be based upon cash collection.

10.3 Sales of real estate

Accounting for sales of real estate is governed by SFAS No. 66, Accounting for Sales of Real Estate. Profit from real estate sales may be recognized in full, provided the following:

- The profit is determinable (i.e. the collectibility of the sales price is reasonably assured or the amount that will not be collectible can be estimated).
- The earnings process is virtually complete, that is, the seller is not obliged to perform significant activities after the sale to earn the profit (SFAS No. 66, paragraph 3).

When both of these conditions are satisfied, the method used to recognize profits on real estate sales is referred to as the full accrual method. If both of these conditions are not satisfied, recognition of all or part of the profit should be postponed.

For real estate sales, the collectibility of the sales price is reasonably assured when the buyer has demonstrated a commitment to pay. This commitment is supported by a substantial initial investment, along with continuing investments that give the buyer a sufficient stake in the property such that the risk of loss through default motivates the buyer to honour its obligations to the seller.

The full accrual method is appropriate and profit shall be recognized in full at the point of sale for real estate transactions when all of the following criteria are met:

- A sale is consummated.
- The buyer's initial and continuing investments are adequate to demonstrate a commitment to pay for the property.
- The seller's receivable is not subject to future subordination.
- The seller has transferred to the buyer the usual risks and rewards of ownership in a transaction that is, in substance, a sale and does not have a substantial continuing involvement in the property.

10.4 Franchise fees

SFAS No. 45, Accounting for Franchise Fee Revenue, requires franchise fee revenue from individual and area franchise sales to be recognized only when all material services or conditions relating to the sale have been substantially performed or satisfied by the franchisor. Substantial performance means that the franchisor has no remaining obligations, virtually all services required of the franchisor have been performed, and no other material obligations of the franchisor remain.

If a portion of the initial franchise fee is designated to compensate the franchisor for continuing services to be rendered during the franchise period, a portion of the fee is deferred and recognized over the franchise period. If the conditions necessary for recognition of franchise fee revenue are not met, direct costs relating to franchise sales are deferred until the related revenue is recognized. Any costs deferred in this manner should not exceed anticipated revenue less estimated additional related costs to be incurred.

10.5 Licence receipts

SFAS No. 50, Financial Reporting in the Record and Music Industry, requires the licensor of a record master or music copyright to recognize the licensing fee as revenue if the licensor:

- Has signed a non-cancellable contract.
- Has agreed to a fixed fee.
- Has delivered the rights to the licensee, who is free to exercise them.
- Has no remaining significant obligations to furnish music or records.
- Is reasonably assured of the collectibility of the full licensing fee.

10.6 Computer software

In issuing SOP 97-2 the same basic criteria for revenue recognition are applied to software arrangements as applied to sales of other products. The SOP sets out four criteria that must be met before revenue should be recognized:

- Persuasive evidence of an arrangement exists.
- Delivery has occurred.
- The vendor's fee is fixed or determinable.
- Collectibility is probable.

Those criteria apply to arrangements to deliver software or a software system which do not require significant production, modification or customization of software. They do not apply to revenue earned on products or services containing software that is incidental to the products or services as a whole.

If an arrangement to deliver software or a software system requires significant production, modification, or customization of software, the entire arrangement should be accounted for in conformity with ARB No. 45, Long-Term Construction-Type Contracts, and SOP 81-1, Accounting for Performance of Construction-Type and Certain Production-Type Contracts. In multiple-element arrangements the fee should be allocated to the various elements based on vendor-specific objective evidence of fair-value, regardless of any separate prices stated within the contract for each element.

11 Prior period adjustments

Prior period adjustments are related to prior accounting periods, and thus should be excluded from the determination of earnings for the current period. They are treated instead as direct adjustments to the opening retained earnings balance. An error in previously issued financial statements and the effect of its correction on income before extraordinary items, net income, and the related earnings per share amounts, both gross and net of applicable income taxes, should be disclosed for all periods presented (see APB Opinion No. 9 and No. 20 and SFAS No. 16).

Prior period adjustments are limited to those material adjustments which:

- Can be specifically identified with and directly related to the business activities of particular prior period.
- Are not attributable to economic events occurring subsequent to the date of the financial statements for the prior period.
- Depend primarily on determinations by persons other than management.
- Were not susceptible of reasonable estimation prior to such determination.

An SEC interpretation of GAAP concluded that management plays a significant role in the settlement of litigation and the settlement of income taxes and that such settlement does not meet the criteria for a prior period adjustment. Prior period adjustments are for the most part limited to the correction of an error in a prior period financial statement or an adjustment for the realization of the tax benefits of the pre-acquisition operating loss carryforward of purchased subsidiaries.

12 Post-balance sheet events (subsequent events)

The balance sheet is dated as of the last day of the fiscal period, but a period of time may elapse before the financial statements are issued. During this period, significant events or transactions may have occurred that materially affect the company's financial position. These events and transactions are called post-balance sheet or subsequent events. Significant events occurring between the balance sheet date and issue date could make the financial statements misleading if not disclosed.

Statement on Auditing Standards No. 1 describes two types of subsequent event. A type-one subsequent event consists of those events providing additional evidence with respect to the conditions that existed at the date of the balance sheet and that affect the estimates inherent in the process of preparing the financial statements; for example, events that affect the realization of assets or the settlement of estimated liabilities. The financial statements should be adjusted for any changes in estimates resulting from the new evidence. A type-two event consists of those events that provide evidence with respect to the conditions that did not exist at the date of the balance sheet but arose subsequent to that date for example, the sale of bonds or capital stock, a stock split or a stock dividend, a pending business combination, loss of inventory from fire or flood. The second type does not require adjustment of the financial statements but may require disclosure in order to keep the financial statements from being misleading. Disclosure can be made in the form of footnotes, supporting schedules, and pro forma statements.

VIII Revaluation accounting

Generally, GAAP do not permit assets to be revalued to a level above acquisition or production cost. However certain financial instruments are recorded at fair value including unrealized gains. For the accounting treatment of these assets or liabilities see VI.5.6, Investments, and VII.4, Financial instruments, derivatives and hedge accounting.

Special rules apply for certain industries like insurance companies, investment companies or broker dealer in securities.

The revaluation of financial statements of entities operating in a highly inflationary environment is described in USA – Group Accounts, V.4.4.

IX Notes and additional statements

1 Notes

1.1 General requirements

Reporting entities must disclose significant accounting policies, which are defined as the specific accounting principles, and the methods of applying those principles that are judged by the management of the entity to be the most appropriate in the circumstances to present fairly financial position, changes in financial position, and results of operations in accordance with GAAP (APB Opinion No. 22, Disclosure of Accounting Policies, paragraph 6). The determination of which accounting policies are significant is a matter of judgement. The preferable presentation of disclosing accounting policies is in the first note to the financial statements, under the statement 'Summary of significant accounting policies'. APB Opinion No. 22 specifically states this preference, but recognizes the need for flexibility in the matter of formats.

1.2 Special areas of additional disclosures

Additional disclosures are often made in respect of the following.

1.2.1 Accounting changes

Required disclosures of mandatory accounting changes are stated in the respective SFASs. The required disclosures for discretionary changes are specified in APB Opinion No. 20. APB Opinion No. 20 requires footnote disclosure of a significant change in accounting principle that has a material effect on the comparability of financial statements. The disclosure should include:

- The nature of the change.
- Justification for the change.
- Pro-forma (as if the new principle had always been used) information on current income before extraordinary operations, net income and earnings per share). The pro-forma statements should be prepared for all periods presented.
- In addition, the dollar effect of the change on current operations.

Discretionary changes in accounting principles by public companies require an independent accountant's comment letter regarding preferability of the change to be filed with the SEC.

1.2.2 Business combinations

In the case of a business combination which was accounted for under the provisions of APB Opinion No. 16, the applied accounting method – whether it is a purchase or a pooling-of-interests – must be disclosed, including the name and a brief description of the companies. Also, the period in which the results of operation are included must be stated. More details on business combinations will be found in USA – Group Accounts.

1.2.3 Segment reporting

SFAS No. 131, Disclosures about Segments of an Enterprise and Related Information, requires public business enterprises to report certain information about operating segments in the annual financial statements and in interim financial reports. It also requires that public business enterprises report certain information about their products and services, the geographic areas in which they operate, and their major customers. Even if the statement encourages non-public business enterprises to disclose the same information, the statement does not apply for (SFAS No. 131, paragraph 9):

- Non-public business enterprises.
- Not-for-profit enterprises.
- The separate financial statements of parent enterprises, subsidiaries, joint ventures, or equity-method investees if those enterprises' statements are consolidated or combined and both the separate company statements and the consolidated or combined statements are included in the same financial report.

The objective of this disaggregation is to report information about the types of activities in which an enterprise is engaged and the economic environment in which those activities take place. The reader of the financial statements is assisted in understanding the enterprise performance and by assessing its prospects for future cash flows.

SFAS No. 131 requires a management approach as the criteria of requiring disclosures about segments is based on the way the management operates the business enterprise. Operating segments are components of an enterprise (SFAS No. 131, paragraph 10):

- That engages in business activities from which revenues may be earned and in which expenses are incurred.
- Whose operating results are regularly reviewed by the top management (enterprise's chief operating decision-maker) for purposes of making decisions with regard to resource allocation and performance evaluation.
- For which discrete financial information is available.

An operating segment is a reportable segment if any of the following quantitative thresholds is met (SFAS No. 131, paragraph 28):

- Revenue test: The total reported revenue (internal and external) is 10 per cent or more of the combined revenue of all reported operating segments.
- Profitability test: The absolute amount of its reported profit or loss of the operating segment is 10 per cent or more of the greater (absolute) amount of the total profit of all operating segments reporting a profit or the total loss of all operating segments reporting a loss.
- Asset test: The assets are 10 per cent or more of the combined assets of all operating segments.

Under certain criteria a reportable segment may combine the activities of two or more operating segments (SFAS No. 131, paragraph 17). If the total revenue of the reportable segments is less than 75 per cent of total consolidated revenue, additional operating segments must be identified as reportable segments until at least 75 per cent of total consolidated revenue is included in reportable segments (even if they do not meet the quantitative criteria).

Information about all operating segments that did not qualify as reportable segments must be combined and disclosed in an all other category. SFAS No. 131, paragraph 24 indicates that 10 reportable segments are probably a reasonable maximum number of reportable segments although a precise limit does not exist.

If an operating segment qualifies in the current period as a reportable segment, but did not qualify as a reportable segment in the prior period(s), prior-period segment data presented for comparative purposes should be restated as if the segment qualified as a reportable segment in the prior period(s).

Segment information is required in four areas:

- Factors used to identify a segment and the types of products from which a segment derives its revenue.
- Information about segment profit or loss and assets.
- Reconciliation of segment information to aggregate enterprise amounts.
- Condensed interim period information.

For any of those four areas additional disclosures are required (SFAS No. 131, paragraphs 26–33). The reporting requirements about the segments profit or loss and assets are often referred to as the heart of the segment reporting. A measure of profit or loss and total assets are required for each reportable segment. In addition, an enterprise must disclose the following about each reportable segment if the specified amounts are included in the segments measure of profit or loss:

- Revenues from external customers.
- Revenues from transactions with other operating segments.
- Interest revenue.
- Interest expense.
- Depreciation, depletion and amortization expense.
- Unusual gains and losses.
- Equity in the net income of equity method investees.
- Income tax expense or benefit.
- Extraordinary items.
- Significant non-cash items other than depreciation, depletion and amortization expense.

In interim financial statements abbreviated segment information is required (SFAS No. 131, paragraph 33). If an enterprise changes the structure of its internal organization previous period information should be presented for comparative purposes including an explanation of the change.

In addition, enterprise-wide information is required in three areas (SFAS No. 131, paragraphs 37–39):

- Information about products and services.
- Information about geographic areas.
- Information about major customers.

1.2.4 Related party disclosures

Under GAAP a related party is one that can exercise control or significant influence over the management and/or policies of another party, to the extent that one of the parties may be prevented from fully pursuing its own interests. As examples SFAS No. 57, Related Party Disclosures, enumerates a parent company and its subsidiaries, subsidiaries of a common parent, an enterprise and trusts for the benefit of employees, such as pension and profit-sharing trusts that are managed by or under the trusteeship of the enterprise's management, an enterprise and its principal owners, management, or members of their immediate families and affiliates.

Transactions involving related parties cannot be presumed to be carried out on an 'arm's length' basis because the requisite conditions of competitive, free-market dealings may not exist.

Substance over form is an important consideration when accounting for transactions involving related parties.

SFAS No. 57, paragraph 2 requires that material related party transactions that are not eliminated in consolidated or combined financial statements have to be disclosed in the financial statements of the reporting entity. Disclosure of related party transactions is required, however, in separate financial statements of a parent company, a subsidiary, a corporate joint venture, or an investee that 50 per cent owned or less.

The following information has to be disclosed:

- The nature and relationship of the related parties.
- A description of the transactions, including amounts and other pertinent information deemed necessary for an understanding of the effects of the related party transactions, for each period in which an income statement is presented (related party transactions of no or nominal amounts must also be disclosed).
- The dollar amount of transactions for each of the period in which an income statements is presented and also the effects of any change in terms between the related parties from terms used in preceding periods.
- Amounts due from or to related parties as of the date of each balance sheet presented and, if not otherwise apparent, the terms and manner of settlement.

Related party transactions involving compensation arrangements, expense allowances, and similar items incurred in the ordinary course of business, however, do not have to be disclosed (SFAS No. 57, paragraph 2).

1.3 SEC required schedules

In addition, all publicly traded companies must prepare the additional disclosures in the form of schedules to the 10-K report.

2 Additional financial statements

2.1 Required statements

The following four financial statements are required:

- Balance sheet (see V.2).
- Income statement (see V.3).
- Statement of changes in shareholders' equity. A statement of changes in the separate accounts of stockholders' equity, including retained earnings, is necessary to conform with GAAP. This disclosure may be presented in the notes or as a separate statement (see V.2.3.2 and V.1).
- Statement of cash flows. SFAS No. 95 requires a statement of cash flows to be included as part of a full set of general-purpose financial statements.

Social accounting and value-added statements are not required. A small number of firms are presenting them.

2.2 Statement of cash flows

According to SFAC No. 5, paragraph 13, a full set of financial statements should show cash flows during the period, in order to help investors and creditors in assessing:

- The enterprise's ability to generate future positive future net cash flows.
- The enterprise's ability to meet its obligations and to pay dividends.
- The reasons for differences between net income and associated cash receipts and payments.
- The effects on an enterprise's financial position of both its cash and non-cash investing and financing transactions during a period.

2.2.1 Classification of cash receipts and cash payments

Under the provisions of SFAS No. 95, a statement of cash flows should clearly specify the amount of net cash provided or used by an enterprise during a period from operating activities, investing activities and financing activities. The statement should clearly indicate the net effects of those cash flows on the enterprise's cash and cash equivalents.

Operating activities generally involve the cash effects of transactions that enter into the determination of net income, e.g. producing and delivering goods and providing services. Investing activities include the acquisition and disposition of long-term productive assets or securities that are not considered cash equivalents, e.g. a net disposal of subsidiaries. They also include the lending of money and collection on loans. Financing activities include obtaining resources from owners and providing them with a return on, and return of, their investment. Also included is obtaining resources from creditors and repaying the amount borrowed.

The statement of cash flows should include only inflows and outflows of cash and cash equivalents. Cash equivalents include any short-term highly liquid investments used as a temporary investment of idle cash. The statement should exclude all transactions that do not directly affect cash receipts and payments. However, the effects of transactions not resulting in receipts or payments of cash, e.g. acquiring an asset through a capital lease, conversion of debt to equity, should be reported in a separate schedule. The reasoning for not including non-cash items in the statement of cash flows and placing them in a separate schedule is that it preserves the statement's primary focus on cash flows from operating, investing and financing activities.

2.2.2 Content and form of statement of cash flows

In reporting cash flows from operating activities, SFAS No. 95, paragraph 27 encourages but does not require an enterprise to use the direct method. Enterprises may also use the indirect method, also referred to as the reconciliation method. There is no difference in reporting the cash flows from investing and financing activities, regardless of whether the direct or indirect method is used to report cash flows from operations.

The direct method shows the items that affected cash flow. Cash received and cash paid are presented, as opposed to converting accrual-basis income to cash flow information. Entities using the direct method are required to report the following classes of operating cash receipts and payments:

- Cash collected from customers.
- Interest and dividends received.
- Cash paid to employees and other suppliers.
- Interest and income taxes paid.
- Other operating cash receipts and payments.

The direct method allows the user to clarify the relationship between the company's net income and its cash flows. In applying the direct method, a separate schedule reconciling net income to net cash flows from operating activities must also be provided.

The indirect method is the most widely used presentation of cash from operating activities. It focuses on the differences between net income and cash flows and begins with net income. Revenue and expense items not affecting cash are added or deducted to arrive at net cash provided by operating activities. The statement of cash flows prepared using the indirect method emphasizes changes in the components of most current asset and current liability accounts. Changes in inventory, accounts receivable, and other current accounts are used to determine the cash flow from operating activities. The change in accounts receivable should be calculated using the balances net of the allowance account in order to assure write-offs of uncollectible accounts are treated properly. However, short-term borrowing used to purchase equipment would not be included since it is not related to operating activities. Instead, the short-term borrowing would be classified as a financing activity. Other adjustments under the indirect method include changes in the account balances of deferred income taxes and the income (loss) from investments under the equity method.

If the indirect method is chosen, then the amount of interest and income tax paid should be included in the related disclosures. Also with the indirect method, SFAS No. 95 permits, but does not require, separate disclosure of cash flows related to extraordinary items and discontinued operations. If an entity chooses to disclose this information, disclosure must be consistent for all periods affected.

2.2.3 Other requirements

The emphasis in the statement of cash flows is on gross cash receipts and cash payments. For instance, reporting the net change in bonds payable would obscure the financing activities of the entity by not disclosing separately cash inflows from issuing bonds and cash outflows from retiring bonds. SFAS No. 95 specifies a few exceptions where netting of cash flows is allowed.

Items having quick turnovers, large amounts, and short maturities may be presented as net cash flows if the cash receipts and payments pertain to investments (other than cash equivalents), loans receivable and debts (original maturity of three months or less).

Other requirements to be considered in preparing a cash flow statement are:

- Foreign operations must prepare a separate statement of cash flows, and translate the statement to the reporting currency using the exchange rate in effect at the time of the cash flow (SFAS No. 95, paragraph 25). A weighted-average exchange rate may be used if the result is substantially the same.
- Non-cash exchange gains and losses recognized on the income statement should be reported as a separate item when reconciling net income and operating activities.
- Per SFAS No. 102, a statement of cash flows is not required for a defined benefit pension plan that presents the financial information under the guidelines of SFAS No. 35.
- Per SFAS No. 102, as amended by SFAS No. 115, cash flows are classified as operating cash flows if originating from the purchase or selling of loans, debt, or equity instruments acquired specifically for resale and carried at market value in a trading account.
- Cash flows from securities held in available-for-sale or held-to-maturity portfolios are disclosed gross in the investing activities section of the cash flow section.
- Banks, savings institutions, and credit unions are allowed to report net cash receipts and payments for within the provisions of SFAS No. 104.

2.3 Directors' report

A directors' report is not required under GAAP. The SEC requires a management discussion and analysis (MD&A) of the financial statements to be provided in the annual reports. The MD&A is expected to include discussion of the results of operations, liquidity and capital resources. In addition the disclosure of future-orientated information is considered desirable (see also IV.2.2.2).

2.4 Interim reporting

The SEC mandates that certain companies file a form 10-Q, which requires a company to disclose quarterly data similar to those disclosed in the annual report (see also IV.2.2.2). It also requires companies to disclose selected quarterly information in the notes to the annual financial statements. In addition to this SEC requirement, the APB has issued Opinion No. 28, Interim Financial Reporting, which attempted to narrow the reporting alternatives relating to interim reports. The profession indicates that the same accounting principles used for annual reports should be employed for interim reports.

The following interim data should be reported as a minimum:

- Sales or gross revenues, provision for income taxes, extraordinary items, the cumulative effect of a change in accounting principles or practices, and net income.
- Primary and fully diluted earnings per share.
- Seasonal revenue, cost or expense.
- Significant changes in estimates or provisions for income taxes.
- Disposal of a segment of the business and extraordinary, unusual, or infrequently occurring items.
- Contingent items.
- Changes in accounting principles or estimates.
- Significant changes in the financial position.

X Auditing

1 The auditor and Statements on Auditing Standards

The federal securities laws require that the financial statements of all companies that have publicly traded securities must be audited by a Certified Public Accountant (CPA). Although CPAs are licensed under the laws of each of the states, the examination process is the same for all CPAs, and the professional control mechanisms for the public accounting profession rest primarily with the AICPA.

In performing audits in accordance with Generally Accepted Auditing Standards (GAAS) a CPA must abide by ten specific standards referred to a GAAS (see I) as well as by the complete body of other standards promulgated through the SASs. Up to June 1999 the ASB had issued 87 SASs (see Table X.1).

2 Types of business organization for which an audit is compulsory

Under the federal securities laws, any company that has securities that are publicly traded must have an annual audit of its financial statements by a CPA. Many privately held business enterprises as well as governmental and not-for-profit organizations are also required to have audits by CPAs, either pursuant to explicit contractual agreements (e.g. loan documents), or by state or local laws (i.e. in the case of governments), or by the organizing charter of a not-for-profit organization such as a hospital or a university.

CPAs are the primary auditors of companies, not-for-profit organizations, and many local governments. Nevertheless, there are other groups of auditors, such as bank auditors who work for the Federal Reserve System or the banking authorities of the various states. Thus banks may be audited not only by CPAs but also by bank auditors, who are civil servants. In addition, many levels of government have their own audit agencies that perform audits of governmental agencies and programmes (e.g. defence contractors). Finally, the federal Internal Revenue Service and the tax authorities of the various states have auditors who perform audits of tax returns.

3 Objectives of the audit

With regard to public companies, the federal securities laws specify that there must be an annual audit of the financial statements by a CPA. What constitutes the financial statements has been discussed previously (see IV.2.2.2). In addition, the required schedules to the financial statements must be audited in many cases. The auditor's opinion must be submitted with the 10-K filing no later than 90 days after the year end of the company. Generally no extensions are granted. For private companies, generally the contractual agreement that calls for the audit will generally also specify when the audit report must be delivered.

4 Auditor's reports

A standard auditor's report for a public company is a public document and it must be included in the 10-K report, which is itself a public document. Furthermore, the standard auditor's report is included in the annual report to shareholders, which is also a public document. A special auditor's report or an auditor's report for a private company is generally not available to those outside the company unless a specific contractual agreement (e.g. a loan document) requires the company to deliver an auditor's report to an outside party. A special auditor's report is one which is prepared pursuant to the order of the board of directors, the management, or other parties interested in a company, and relates to a specific concern of those ordering the report (e.g. a potential take-over of the company; an examination of fraud).

5 Organization of the audit profession

The history and organization of the audit profession were discussed in I above. The most important organization in the public accounting profession is the AICPA, which has over 335,000 members (June 1999). The AICPA prepares and administers the Uniform Certified Public Accountants' Examination, which must be passed satisfactorily in order to become a CPA in each of the states of the United States. In addition to passing the CPA exam, a potential CPA must have a college or university degree with a specialization in accounting and in most states a candidate must work for a CPA firm for a specified period of time. This period varies by state, but is generally two years. There is a growing trend in many states to require approximately one year of further education beyond a first university degree (i.e. a bachelor's degree) as a prerequisite for admission to sitting the CPA exam. However, this is not a uniform requirement in all of the states at this time.

In order to practice as a CPA a person must be

Table X.1 Statements on Auditing Standards (SASs)

No.	Subject
1	Codifications of Auditing Standards and Procedures
8	Other Information in Documents Containing Audited Financial Statements
12	Inquiry of a Client's Lawyer concerning Litigation, Claims, and Assessments
21	Segment Information
22	Planning and Supervision
25	The Relationship of Generally Accepted Auditing Standards to Quality Control Standards
26	Association with Financial Statements
29	Reporting on Information Accompanying the Basic Financial Statements in Auditor-Submitted Documents
31	Evidential Matter
32	Adequacy of Disclosure in Financial Statements
35	Special Reports – Applying Agreed-Upon Procedures to Specified Elements, Accounts, or Items of a Financial Statement
37	Filing under Federal Securities Statutes
39	Audit Sampling
41	Working Papers
42	Reporting on Condensed Financial Statements and Selected Financial Data
43	Omnibus Statement on Auditing Standards
45	Omnibus Statement on Auditing Standards-1983
46	Consideration of Omitted Procedures after the Report Date
47	Audit Risk and Materiality in Conducting an Audit
48	The Effects of Computer Processing on the Examination of Financial Statements
50	Reports on the Application of Accounting Principles
51	Reporting on Financial Statements Prepared for Use in Other Countries
52	Omnibus Statement on Auditing Standards 1987
53	The Auditor's Responsibility to Detect and Report Errors and Irregularities
54	Illegal Acts by Clients
55	Consideration of the Internal Control Structure in a Financial Statement Audit
56	Analytical Procedures
57	Auditing Accounting Estimates
58	Reports on Audited Financial Statements
59	The Auditor's Consideration of an Entity's Ability to continue as a Going Concern
60	Communication of Internal Control Structure Related Matters Noted in an Audit
61	Communication with Audit Committees
62	Special Reports
64	Omnibus Statement on Auditing Standards 1990
65	The Auditor's Consideration of the Internal Audit Function in an Audit of Financial Statements
67	The Confirmation Process
68	Compliance Auditing Applicable to Governmental Entities and Other Recipients of Governmental Financial Assistance
69	The Meaning of 'Present Fairly in Conformity with Generally Accepted Accounting Principles' in the Independent Auditor's Report
70	Reports on the Processing of Transactions by Service Organizations
71	Interim Financial Information
72	Letters for Underwriters and Certain Other Requesting Parties
73	Using the Work of a Specialist
74	Compliance Auditing Considerations in Audits of Governmental Entities and Recipients of Governmental Financial Assistance
75	Engagements to Apply Agreed-Upon Procedures to Specified Elements, Accounts, or Items of a Financial Statement
76	Amendments to Statement on Auditing Standards No. 72, Letters for Underwriters and Certain Other Requesting Parties
77	Amendments to Statement on Auditing Standards No. 22, Planning and Supervision, No. 59, The Auditor's Consideration of an Entity's Ability to Continue as a Going Concern, and No. 62, Special Reports
78	Consideration of Internal Control in a Financial Statement Audit: An Amendment to SAS No. 55
79	Amendment to Statement on Auditing Standards No. 58, Reports on Audited Financial Statements
80	Amendment to Statement on Auditing Standards No. 31, Evidential Matter
81	Auditing Investments
82	Consideration of Fraud in a Financial Statement Audit
83	Establishing an Understanding with the Client
84	Communications Between Predecessor and Successor Auditors
85	Management Prepresentations
86	Amendment to Statement on Auditing Standards No. 72, Letters for Underwriters and Certain Other Requesting Parties
87	Restricting the Use of an Auditor's Report

Statements on Auditing Standards not included in the table have been superseded.

licensed by a particular state, as discussed above. Upon receiving a CPA licence, a person is fully qualified to practice. However, most individuals who are licensed CPAs provide tax preparation and advice and accounting services for privately held and smaller business enterprises, unless they are affiliated to a firm that is capable of performing audits and maintaining the quality control standards that have now become required by the AICPA. Nevertheless, it is not required that any CPA should become a member of the AICPA or of the various state societies of CPAs.

The enforcement and control mechanism of the audit profession is principally the AICPA and the various state societies. However, each state has a board of public accountancy which regulates the CPAs in its state. These boards have the power to punish or rescind the licence of a CPA who has been found guilty of various offences. For example, the preparation of a fraudulent tax return would be cause for punishment or licence removal in many states.

6 Qualifications of auditors

Besides specific education and training requirements for becoming a CPA, it has been generally presumed that any CPA should be qualified to perform an audit in accordance with GAAS; in practice, it is those CPAs who are affiliated to CPA firms that are members of the AICPA or one of the state societies of CPAs, which firms abide by the quality control standards promulgated by the AICPA, that are most qualified to perform audits.

Furthermore, for audits of public companies, it has become necessary for a CPA firm to become a member of the SEC Practice Section of the AICPA, which requires that additional quality control standards be followed, including peer review. Thus it is difficult for an individual CPA or a small firm of CPAs to audit public companies.

7 Characteristics of independence

Independence is the primary requisite and quality of a CPA. Independence is required both in fact and in appearance. Thus a CPA is prevented from performing an audit in accordance with GAAS in any case where he or she is not independent. Independence is defined as not having any type of financial interest in the company under audit. Therefore, there can be no ownership of shares or debt of the company. Furthermore, no close member of the family of the auditor may own shares or debt of the company. This holds true for any partner member of the CPA firm, and not just the primary engagement partner. It also holds true for any non-partner member of the firm who works in the same practice office where the audit is taking place. The AICPA has issued various rules and interpretations concerning independence as part of its Code of Professional Conduct.

XI Filing and publication

1 Public and private companies

All companies with publicly traded securities must file annual reports with the SEC on form 10-K, which include audited financial statements prepared in accordance with GAAP.

Private companies are not required to disclose financial statements publicly, but may be required to deliver audited or unaudited financial statements to third parties pursuant to contractual agreements (e.g. loan documents).

For public companies, the financial statements and other items to be disclosed in reports to the SEC or reports to shareholders have been discussed in previous sections. In addition, most public companies maintain a dialogue with the financial press in order to provide information that may be of use to financial analysts and investors in evaluating the financial position and earnings of the company prior to the official release of financial reports and financial statements. Thus chief financial officers of public companies frequently discuss the prospects for future earnings of their companies with the financial press and the financial analyst community. This is not required by the SEC, nor by any law.

Companies are required to issue their annual reports, including audited financial statements, prior to the annual meeting of shareholders. However, for a calendar year company these meetings are usually held in the spring, so the information is not particularly useful to investors. The annual report to the SEC on form 10-K, including audited financial statements, is required to be filed no later than 90 days after the end of the company's fiscal year, which for a calendar year company would be 31 March. This is still not particularly timely information, therefore most companies will release their annual earnings to the public in a press release some time prior to 31 March. Usually this release of earnings is done at or soon after the point at which the auditor has concluded all audit fieldwork and has proposed any audit adjustments to the financial statements. The announcement of earnings varies substantially. There is no requirement by the SEC or by law to make such an announcement prior to 90 days after the fiscal year end. There is no requirement for private companies to disclose their financial statements to the public or to release information about their earnings to the public, and most do not do so.

2 Research findings regarding resistance to disclosure

Since all public companies are required by law to disclose their audited financial statements by a certain date, there is no resistance to such disclosure. However, various companies are more forthcoming about their financial position and earnings than are other companies. Thus there have been a number of research studies which have investigated the phenomenon of audit delay, or, in other words, the period of time between the end of the fiscal year and the release of the audit report (i.e. often defined as the release of earnings publicly, because of the difficulty of knowing precisely when the audit report was filed). Some of these studies have found that audit reports may be delayed when the earnings announcements are disappointing or below expectations. However, the research findings have nothing to do with legal requirements, because usually the disclosures are made well within the legally specified time frames, except for those companies which have publicly declared that they are not releasing their financial results because of financial difficulties (e.g. owing to bankruptcy or fraud).

3 Access to disclosed information

Access to the audited financial statements of public companies is widely available. Not only are such financial statements available publicly from the SEC via form 10-K, but most companies willingly provide copies of their financial statements free of charge to anyone who asks for them. In addition, most libraries have copies of the publications of financial information services such as Moody's Industrial Manual which include financial statements of public manufacturing and commercial companies in a standardized format. Finally, various on-line electronic information services (e.g. National Automated Accounting Retrieval System, NAARS; Disclosure) provide access to the financial statements of public companies for subscribers to such services.

XII Sanctions

1 Sanctions on the enterprise and its directors

In the United States there is a federal system of law, hence both the federal government and the various state governments have the ability to enact laws which may affect the issuance of securities and other matters related to the governance of business enterprises.

Under the federal securities laws, registration statements and annual reports on form 10-K, which contains audited financial statements, are signed by the chief executive officer, the chief financial officer and the majority of the members of the board of directors. In a registration statement, if any information contained in the audited financial statements is materially misleading, sanctions may be imposed under the federal securities laws. Such sanctions could include fines or possible imprisonment. In the form 10-K annual report, not only would the information have to be misleading to entail sanctions but there would also have to be an intent on the part of the parties involved to deceive. The penalties for violation of the disclosure rules are similar under the annual disclosure requirements to those under the registration statement rules. The Securities and Exchange Commission may itself bring enforcement action and impose penalties, which may be appealed to the federal courts.

In addition to the federal securities laws, there are securities and corporate laws in many of the states. Most of these laws prohibit the issuance of financial statements containing false or misleading information. There has been a significant increase in litigation surrounding securities in recent years.

2 Sanctions on the auditors

In the United States, Certified Public Accountants are licensed by each of the states. As a consequence CPAs may be sanctioned by the laws of the states in which they are licensed. The sanctions may be administrative in nature and may be imposed through the state licensing board or body (e.g. the State Board of Accountancy). Sanctions may include a temporary or permanent suspension of a CPA's licence or some other form of penalty such as mandatory participation in continuing professional education (CPE) programmes. In addition, for more serious acts, criminal penalties may be imposed pursuant to the general laws involving fraudulent activities. Finally, a CPA may be exposed to civil liability and may be sued under various provisions of law, including breach of contract and the law of torts (i.e. fraudulent activity).

CPAs are usually members of professional bodies and therefore may be subject to sanctions imposed by those professional bodies, such as the AICPA or the New York State Society of CPAs. Sanctions imposed on behalf of these bodies may range from such penalties as mandatory participation in CPE programmes to temporary or permanent expulsion from membership of the professional body.

CPAs who are involved in the audit of companies with reporting responsibilities under the federal securities Acts may be subject to significantly greater sanctions and legal liabilities than CPAs who confine their practice to private companies. In the first place, the SEC requires that all CPA firms with SEC clients must expose their practices to mandatory peer review and quality control programmes. This involves a periodic review of the working papers of the firm by another CPA firm. Sanctions may be imposed for failure to maintain adequate quality throughout the practice. Most CPA firms with SEC clients are also members of the SEC Practice Section of the AICPA. Membership in the SEC Practice Section also requires periodic peer review and quality control maintenance programmes, including mandatory continuing professional education for all professional accountants employed by the firm. Failure to abide by the peer review programmes and the quality control programmes can result in sanctions, ranging from temporary or permanent suspension of membership in the SEC Practice Section to temporary or permanent suspension of the right to practice before the SEC. In addition, the SEC can impose penalties on individual CPAs or on firms as a result of specific investigations into violations of federal securities laws. The sanctions can range from temporary or permanent suspension of the right to practice before the SEC to criminal penalties such as fines and even imprisonment. Lesser sanctions and penalties are imposed relatively frequently. More serious penalties involving fines are imposed relatively infrequently and sanctions involving imprisonment are rarely imposed. This conclusion is based on review of publicly disclosed enforcement actions.

Perhaps the most significant sanction facing practising CPAs in the United States is the threat of civil penalties imposed under state common laws and state and federal securities laws. These penalties arise as a result of lawsuits brought on the part of private individuals or groups of individuals who believe that they have been harmed, typically financially, by the company involved and its auditors. Since there is frequently no limit on the size of the penalty that may be imposed, public accounting firms have been subject in recent years to increasingly large damages as a result of claims by creditors or shareholders that the financial statements they relied on to make investment and credit decisions were misleading or fraudulent. Although the management or employees of a company may be the principal parties responsible for misleading or fraudulent financial reporting, it is often the auditors who bear the brunt of the responsibility. This is sometimes the case because, after a significant fraud, there may be no party financially solvent except for the auditors. Thus, under current laws which provide for joint and several liability (i.e. one party may be assessed the whole blame if others are not solvent), the auditor may be held wholly liable. In addition, the legal system in the United States has evolved in such a way that the scale of the monetary penalties imposed is often beyond any measure of the harm actually suffered by the creditors or shareholders (i.e. juries may assess punitive monetary penalties without regard to actual injury). Several proposals have been introduced into the Congress of the United States on behalf of the AICPA and the large public accounting firms which would attempt to limit the size of claims upon the legal liability of auditors. At present the eventual result of these proposed changes to the laws is not predictable.

XIII Databases and Internet links

1 Databases

Below is a list of databases that can be used to access company information. Access to these databases might be on a paid subscription basis but they might also have limited-access free websites:

- Disclosure Global Access – http://www.primark.com/pfid. Contains SEC filings and other company information.
- Dow Jones Interactive – http://djinteractive.com. Contains company information, press releases, articles on all topics.
- Lexis-Nexis – http://www.lexis.com. Contains SEC filings, press releases, company information, articles, SEC releases, etc.

2 Internet links

Annual reports for many companies are available directly from the Annual Report Gallery (http://www.reportgallery.com) which also provides a direct link to the homepages of many companies.

Financial Data Finder (http://www.cob.ohio-state.edu/~fin/osudata.htm) is a collection of references to websites devoted to various areas of finance. It is targeted to the needs of researchers and students.

Hoover's Online (http://www.hoovers.com) provides summary information about individual companies.

The SEC, the FASB and the AICPA have websites which provide important accounting information. In some situations, the website is the only source for the information.

2.1 SEC

The left side of the SEC home page (http://www.sec.gov) contains all of the pages that can be accessed through the SEC home page. Key pages include:

- 'About the SEC': this page contains a search engine where key words can be entered and documents on the SEC website with those key words will be located.
- 'Edgar Database': this page contains search engines that can locate all documents that have been filed, utilizing EDGAR, with the SEC (10-Ks, 10-Qs, S-1s 8-Ks etc.). You can perform a general search or a special purpose search. Within the general search you can either 'Search the EDGAR Archives' or you can do 'Quick Forms Lookup':
  - A 'Search the EDGAR Archives' allows you to search the header information in all the filings in the database by entering key words or phrases. You should follow the instructions contained on the web page to either limit or broaden your search. All filings in the database that match your key word search will be located and you will be able to select the desired filings by clicking on that particular filing.
  - A 'Quick Forms Lookup' allows you to look up common forms for a designated company. This is useful if you already know the company's name and the particular filing you are looking for.
- 'SEC News Digest and Public Statements': this page contains press releases by the SEC and speeches by SEC officials (of primary interest are those by the Commission staff at the AICPA's Annual National Conference on

Current SEC Developments). The speeches can be obtained by clicking on 'SEC Digest & Statements', then clicking on 'Speeches by SEC Officials'.

- 'Current SEC Rulemaking': this page contains proposed and final rules, SEC Concept and Interpretive Releases and selected Staff Accounting Bulletins.

The SEC website also contains pages with information on 'Investor Assistance and Complaints,' 'Enforcement Division' and 'Small Business Information'.

2.2 FASB

The FASB's home page (http://www.rutgers.edu/Accounting/raw/fasb/) is presented on the Rutgers University Accounting website. There is a link (Rutgers Accounting Web) to this website on the FASB home page where there is information on many accounting-related organizations that may be of interest. The left side of the FASB home page contains all the pages that can be accessed through the FASB home page. Key pages include:

- 'Recent Additions': contains any recently posted new or updated information to this web site
- 'Technical Projects': currently contains links to:
  - Project Updates: summaries of selected projects.
  - Technical Inquiry Service: implementation guidance.
  - Technical Plan: quarterly plan for technical projects.
  - Derivatives Implementation Group: activities of the Statement 133 Implementation Group.
  - Emerging Issues Task Force (EITF): general information, current agenda, description of recently discussed issues.
  - FauxCom Inc.: an experiment in business reporting on the Internet.
- 'Publications': currently contains links to:
  - Action Alert: board actions and upcoming meetings.
  - Summaries/Status: summaries and status of all FASB statements.
  - Exposure Drafts: recent FASB exposure drafts.
  - Financial Accounting Research System (FARS): contents, pricing and downloadable demo.
  - Publications: descriptions, order information and pricing.
  - FASB Casebook: FASB cases on recognition and measurement.

The FASB website also contains pages with information on 'FASB Facts', 'News Releases' and 'Membership'.

A new fax-on-demand system has been installed at the FASB, enabling callers to receive information through their fax machine. Information available through this service includes the most frequently requested documents. To use this fax service, call: 1 203 847-0700, press 14, and follow the prompts. Now included on the system are the following:

Code Number	**Document Name**
643	FASB Action Alert
644	Facts about FASB
645	Facts about GASB
646	FASB Subscriptions
647	FASB Publications List
648	GASB Subscriptions
649	GASB Publications List
650	Financial Accounting Research System (FARS)
651	FASB Technical Plan
652	GASB Technical Plan
653	Travel Directions to Standards Boards Offices, Norwalk, Connecticut
654	FASB Summary of the Board's Decisions on Derivatives and Hedging
656	Summary of GASB Exposure Draft: Basic Financial Statements – and Management's Discussion and Analysis – For State and Local Governments

2.3 AICPA

The AICPA's home page (http://www.aicpa.org) is organized into four sections: 'Search', 'What's New', 'Contact AICPA' and 'The Forums'. The left margin of the home page contains various categories of pages ('Home', 'Assurance', 'Members', 'Tax', etc.).

Key accounting information on this website includes exposure drafts and other technical documents, AICPA SEC Regulations Committee highlights, and recent news articles or press releases concerning the accounting profession.

The web page for the Accounting Standards Team contains important information relating to the Accounting Standards Executive Committee (AcSEC). In addition to exposure drafts, the Accounting Standards Team web page contains the following information:

- 'About the Accounting Standards Team': contains information regarding the teams objectives and staff contacts.
- 'General Announcements': contains AcSec meeting dates and agendas, links to the SEC, FASB and IASC websites, changes to rescind and amend certain SOPs and Practice Bulletins affected by recently issued guidance from the FASB and a listing of recently issued SOPs, Practice Bulletins and Accounting and Auditing Guides.
- 'Technical Status Updates': contains highlights of recent AcSEC meetings, the AcSEC newsletter and AcSEC project status.

Bibliography

American Institute of Certified Public Accountants (1998a). *Accounting Trends and Techniques 1998* (American Institute of Certified Public Accountants, New York).

American Institute of Certified Public Accountants (1998b). *AICPA Technical Practice Aids* (American Institute of Certified Public Accountants, New York).

Bailey, L. and Miller, M. (1993). *Comprehensive GAAS Guide* (Harcourt Brace Jovanovich, New York).

Baker, R., Lembke, V. and King, T. (1993). *Advanced Accounting*, 2nd edition (McGraw-Hill, New York).

Beams, F. (1998). *Advanced Accounting*, 5th edition (Prentice-Hall, Englewood Cliffs, NJ).

Bloomenthal, H. (1994). *Securities Law Manual* (Clark, Boardman & Callahan, New York, NY).

Bullen, H. G., Wilkins, R. C. and Woods, C. C. (1987). 'The fundamental financial instruments approach', *Journal of Accountancy.*

Carey, J. (1969–70). *The Rise of the Accounting Profession*, 2 volumes (American Institute of Certified Public Accountants, New York).

Chatfield, M. (1978). *A History of Accounting Thought* (Dryden Press, New York).

Curley, A. (1981). 'Accounting principles', *Corporate Controller's Manual*, Wendell P. (ed.) (Warren Gorham & Lamont, New York).

Davidson, S. and Anderson, G. (1987). 'The development of accounting and auditing standards', *Journal of Accountancy*, May, **163**, pp. 130–135.

Edwards, J. and Miranti, P. (1987). 'The AICPA: a professional institution in a dynamic society', *Journal of Accountancy*, May, **120**, pp. 22–38.

Financial Accounting Standards Board (1976). 'Conceptual framework for financial accounting and reporting: elements of financial statements and their measurement', Discussion Memorandum (FASB, Stamford, C).

Financial Accounting Standards Board (1998a). *Original Pronouncements, Accounting Standards I* (FASB, Norwalk, CT).

Financial Accounting Standards Board (1998b). *Original Pronouncements, Accounting Standards II* (FASB, Norwalk, CT).

Hoffman, W., Raabe, W. and Smith, J. (1996). *West's Federal Taxation: Corporations, Partnerships, Estates, and Trusts* (West Publishing Company, Minneapolis).

Horngren, C. and Harrison, W. (1995). *Accounting*, 3rd edition (Prentice-Hall, Englewood Cliffs, NJ).

Johnson, L. T. and Swieringa, R. J. (1996). 'Derivatives, hedging and comprehensive income', *Accounting Horizons*, December, pp. 109–122.

Kay, R. and Searfoss, G. (1988). *Handbook of Accounting and Auditing*, 2nd edition (Gorham & Lamont, New York).
Kieso, D. and Weygandt, J. (1998). *Intermediate Accounting*, 9th edition (Wiley, New York).
KPMG (1995). *Selling Securities in the United States.*
KPMG (1998). *Derivatives and Hedging Handbook.*
KPMG (1999). *Rechnungslegung nach US-amerikanischen Grundsätzen*, 2nd edition (IdW, Düsseldorf).
Lowe, H. (1987). 'Standards overload: what must be done', *Management Accounting*, June, **68**, pp. 57–61.
May, G. (1934). 'The position of accountants under the Securities Acts', *Journal of Accountancy*, January, **110**, pp. 9–23.
McKinnell, R. (1975). 'The FASB and its role in the development of accounting principles in the United States', *Cost and Management*, May–June, **49**, pp. 51–53.
Metzger, M. B. *et al.* (1995). Business Law and the Regulatory Environment, 9th edition (Irwin, Chicago).
Miller, J. R. (1999). *GAAP Guide* (Harcourt Brace & Co., San Diego).
Miranti, P. (1990). *Accountancy Comes of Age: The Development of an American Profession, 1986–1940* (University of North Carolina Press, Chapel Hill, NC).
Nair, R., Rittenberg, L. and Weygandt, J. (1990). 'Accounting for redeemable preferred stock: unresolved issues', *Accounting Horizons*, June, **4**, no. 2, pp. 33–41.
Paton, W. and Littleton, A. (1940). *An Introduction to Corporate Accounting Standards* (AAA, Sarasota, FL).
Previts, G. (1986). *The Scope of CPA Services: A Study of the Development of Independence and the Profession's Role in Society* (Wiley, New York).
Reckers, P. and Jennings, A. (1984). 'Concepts of materiality and disclosure', *CPA Journal*, December, **54**, no. 12, pp. 20–30.
Schroeder, R. G. and Clark, M. W. (1998). Accounting Theory, 6th edition (Wiley, New York).
Schwartz, B. and Diamond, M. (1980). 'Is dollar value LIFO for you?', *CPA Journal*, July, **50**, no. 7, pp. 33–40.
Securities Exchange Commission (1993). *Survey of Securities Filings* (Securities Exchange Commission, Washington DC).
Solomons, D. (1978). 'The politicization of accounting', *Journal of Accountancy*, November, **162**, no. 5, pp. 65–72.
Weinstein, G. (1989). *Inside Accounting Today* (New American Library, New York).
White, G., Fried, D. and Sondhi, A. (1993) *Financial Statement Analysis* (Wiley, New York).
Williams, J. and Miller, M. (1993). *Comprehensive GAAP Guide* (Harcourt Brace Jovanovich, New York).
Zeff, S. (1972). *Forging Accounting Principles in Five Countries: A History and Analysis of Trends* (University of Illinois Press, Champaign-Urbana, IL).
Zeff, S. (1976). 'Developing accounting principles', *The Accounting Sampler*, Burns, T. J. and Hendrickson, H. (eds), 3rd edition (McGraw-Hill, New York).

USA
GROUP ACCOUNTS

Norbert Fischer
Teresa E. Iannaconi
Martin Klöble
Harald W. Lechner

Acknowledgements

The authors gratefully acknowledge the intense collaboration of Martin Risau and Bernd Wübben, both of KPMG German Practice, New York, which enabled us to undertake and complete this project.

CONTENTS

I Introduction and background information

1 Evolution of group accounting in the United States

1.1 Historic development

Group accounting in the United States evolved in a manner which differed from that in European and other countries. As a matter of fact, there is no such term as 'group' accounting in the United States. The term comparable to group accounting in the US is 'consolidation' accounting. Topics related to consolidation accounting include:

- Accounting for business combinations.
- Accounting for intangible assets.
- Equity method of accounting for investment in common stock.
- Foreign currency translation.

Therefore, when we refer to consolidation accounting and related topics, it should be understood that we are referring to group accounting. Second, there are relatively few official accounting pronouncements dealing with consolidation accounting and related topics in US Generally Accepted Accounting Principles (GAAP) (see Table I.1 for a list of official accounting pronouncements dealing with consolidation accounting and related topics under GAAP). Third, consolidation accounting and related topics has been a fairly settled matter in the United States for some time. Accounting Research Bulletin (ARB) No. 51, which is the primary official pronouncement dealing with consolidation accounting, was issued in 1959. Finally, most of the detailed technical aspects of consolidation accounting and related topics are discussed in accounting textbooks rather than through official accounting pronouncements.

Since there are a relatively small number of official pronouncements dealing with consolidation accounting and related topics under GAAP, the essential aspects of these pronouncements may be summarized as follows:

- The financial statements of all majority-owned subsidiaries must be fully consolidated into the financial statements of the parent company. A subsidiary is defined as a company in which more than 50 per cent of the voting common shares are owned by the parent company (i.e. the parent) (see Statement of Financial Accounting Standards (SFAS) No. 94). There are no exceptions to this rule, except when voting control of the subsidiary does not reside with the parent company (e.g. legal bankruptcy of the subsidiary) (SFAS No. 94).
- In preparing consolidated financial statements, all of the accounts of the subsidiary company must be combined fully (i.e. 100 per cent) with the accounts of the parent company (see ARB No. 51).
- All inter-company transactions must be eliminated fully (i.e. 100 per cent) from the consolidated financial statements (ARB No. 51).
- The claim of the minority shareholders of a subsidiary on the net income of the subsidiary is shown in the consolidated income statement as a deduction in arriving at consolidated net income (ARB No. 51).
- The claim of the minority shareholders on the net assets of the subsidiary company is shown as a credit balance in the consolidated balance sheet (typically between consolidated liabilities and consolidated shareholders' equity) (ARB No. 51).
- A business combination occurs when one company acquires a majority of the voting common shares or substantially all of the assets of another company. If voting common shares are acquired, and if the acquired company remains a separate legal entity, consolidation accounting is required (see Accounting Principles Board (APB) Opinion No. 16).
- If the acquired company is legally dissolved, there is no parent/subsidiary relationship and no consolidated financial statements are required. Likewise, if only assets (other than common shares) are acquired, there is no parent/subsidiary relationship and no consolidated financial statements are required (see ARB No. 51).
- Business combinations must be accounted for either by the purchase method or by the pooling-of-interests method. These are mutually exclusive methods (APB Opinion No. 16).
- The primary requirement for the pooling-of-

interests method is an exchange of voting common shares of the acquiring company for substantially all (i.e. 90 per cent or more) of the voting common shares of the acquired company. If the business combination is effected by any other means, purchase accounting is required (APB Opinion No. 16).

- Goodwill arising from a business combination accounted for by the purchase method must be amortized against income over a period not to exceed 40 years (see APB Opinion No. 17).
- All investments in voting common shares which are not majority-owned must be accounted for either by the equity method (i.e. for investments between 20 per cent and 50 per cent and for which the investor has an ability to exercise significant influence over operating and financial policies) or by the cost method (see APB Opinion No. 18).

1.2 Current developments

In 1982, the Financial Accounting Standards Board (FASB) undertook a broad multiphase project intended to cover all aspects of accounting for affiliations between entities and several potentially related issues. The purpose of the project is to address numerous questions raised by constituents, including the Accounting Standards Executive Committee of the American Institute of Certified Public Accountants (AICPA), the staff of the Securities and Exchange Commission (SEC), and certain issuers and users of financial statements. In 1995, as part of this project, the FASB issued an Exposure Draft that covered both when to include entities (consolidation policy) and procedures on how to combine entities. A revised Exposure Draft, Consolidated Financial Statements – Purpose and Policy, was issued in February 1999 that covers only the consolidation policy guidance. The revised Exposure Draft includes proposed guidance that is very similar to the guidance contained in the first Exposure Draft. Basically, a parent shall consolidate each entity (subsidiary) that it controls unless control is temporary at the date it is acquired or otherwise obtained. Once a subsidiary is consolidated, it shall continue to be included in the consolidated financial statements until the parent ceases to control it. The second Exposure Draft includes the following definition of control and presumptions of control:

- Control is the non-shared decision-making ability to direct the policies and management that guide the ongoing activities of another entity so as to increase its benefits and limit its losses from the other entity's activities.

Table I.1 Accounting pronouncements dealing with consolidation accounting and related topics (asterisked items have been superceded wholly or in part)

Year	Pronouncement	Topic
1953	ARB No. 43*	Restatement and Revision of Accounting Research Bulletins (superseded by various pronouncements)
1959	ARB No. 51*	Consolidated Financial Statements
1966	APB Opinion No. 10*	Pooling-of-Interests – Restatement of Financial Statements (superseded by APB Opinion No. 18)
1967	APB Opinion No. 11*	Accounting for Income Taxes (superseded by SFAS No. 109)
1970	APB Opinion No. 16	Business Combinations
1970	APB Opinion No. 17	Intangible Assets
1971	APB Opinion No. 18	The Equity Method of Accounting for Investments in Common Stock
1972	APB Opinion No. 23*	Accounting for Income Taxes: Special Areas (superseded by SFAS No. 109)
1973	APB Opinion No. 29	Accounting for Non-monetary Transactions
1981	SFAS No. 52	Foreign Currency Translation
1982	SFAS No. 58	Capitalization of Interest Cost in Financial Statements that Include Investments Accounted for by the Equity Method
1982	SFAS No. 71	Accounting for the Effects of Certain Types of Regulation
1987	SFAS No. 94	Consolidation of All Majority-Owned Subsidiaries
1992	SFAS No. 109	Accounting for Income Taxes

- Control is presumed if an entity: (i) has a majority voting interest in the election of a corporation's governing body; (ii) has a large minority voting interest and no other party or organized group of parties has a significant voting interest; (iii) has the unilateral ability to obtain a majority voting interest through the present ownership of convertible securities or other rights that are currently exercisable at the option of the holder and the expected benefits from converting those securities or exercising that right exceeds its expected costs; and (iv) is the only general partner in a limited partnership and no other partner or organized group of partners has the effective current ability to dissolve the limited partnership or otherwise remove the general partner for reasons other than violation of law.

There are ten cases included in the second Exposure Draft that are used to demonstrate how the proposed guidance is intended to work. Although some of the cases involve Special Purpose Entities (SPEs), there is limited guidance directed to SPEs. Transition is proposed to be accomplished by restating prior periods (with certain exceptions provided). As at the time of writing the Board was not able to find a consensus on a model for the consolidation of SPEs. Therefore the timing for the issuance of a final statement on consolidation policy is uncertain, and there is also pressure to abandon the project. This is an Exposure Draft which does not constitute GAAP, but which may lead in the future to a Financial Accounting Standard concerning GAAP. The reader should therefore be alerted to the possibility of changes to the definition of control and subsidiary under GAAP.

In August 1996, the Board added to its agenda a project on business combinations to reconsider APB Opinions No. 16, Business Combinations, and No. 17, Intangible Assets. The objective of the project is to improve the transparency of accounting for business combinations. The project focuses on the accounting for purchased goodwill and other intangible assets and the fundamental issues related to the methods of accounting for business combinations, including whether there is a need for two separate and distinct methods. In September 1999, the Board issued an Exposure Draft of a proposed Statement, Business Combinations and Intangible Assets. The Exposure Draft is divided into two parts: part I addresses the method of accounting for business combinations and part II addresses accounting for intangible assets.

To date the Board's more significant decisions are as follows:

- The purchase method should be the only method used to account for business combinations.
- Purchased goodwill should be amortized on a straight-line basis over its useful life, not to exceed 20 years.
- A review of goodwill for impairment should be performed no later than two years after the acquisition date if certain indicators are present.
- When allocating the cost of an acquired entity to the assets acquired and liabilities assumed, every effort should be made to recognize all of the net assets acquired, including intangible assets, and measure them at their fair values.
- If the fair value of the purchase consideration exceeds the fair value of the net assets acquired, purchased goodwill should be recognized as an asset and measured initially at the amount of that difference.
- Goodwill amortization and impairment charges should be displayed on a net-of-tax basis as a separate line item within income from continuing operations. That line item would be preceded by a subtotal such as 'income before goodwill charges' (which would follow the income tax provision). Optional is a basic diluted per-share disclosure of goodwill amortization expense amounts.
- Unlocated negative goodwill would be recognized immediately in income as an extraordinary gain.
- Intangible assets that are reliably measurable should be recognized separately as assets. Intangible assets should be presumed to have a useful economic life of 20 years or less and generally should be amortized over their useful economic lives.
- Intangible assets that are either exchangeable or control over the future economic benefits is obtained through contractual or legal rights may be amortized over periods longer than 20 years or not amortized at all if they are expected to generate clearly identifiable cash

flows for more than 20 years. Intangible assets with indefinite lives for which an observable market exists would not be subject to amortization but should be reviewed for impairment annually on a fair value basis.
- Costs of internally developed intangible assets are expensed if the asset is not specifically identifiable, has an indeterminate life or is inherent to the business as a whole.

The Board is currently evaluating several models related to goodwill amortization (amortization, non-amortization, or a mixed approach). The Board expects to issue a final statement in the first quarter of 2001.

2 The development of multi-corporate entities

A parent/subsidiary relationship exists when one corporation owns more than 50 per cent of the voting common shares of another corporation. Few companies in the United States with publicly traded shares operate as single legal corporate entities. In a survey of 600 large US companies made by AICPA, virtually all of the companies had at least one subsidiary. Highly diversified companies may have several hundred subsidiaries. In most cases, subsidiaries are created by the parent company for legal and/or economic reasons (e.g. to differentiate between types of products and product lines; to divide manufacturing activities from distribution activities; to operate in a foreign country; to shield the parent company from the liabilities of the subsidiary). Subsidiaries are also created as a result of business combinations. It should be noted that no parent/subsidiary relationship exists unless a subsidiary continues to be a legal entity. If no parent/subsidiary relationship exists, there is no consolidation accounting and all of the principles and rules specified in USA – Individual Accounts are applicable.

Even though most consolidation accounting arises because a parent company has created a subsidiary, the acquisition of subsidiaries through business combinations is an important topic for practicing accountants who deal with major corporations and investment bankers. Business combinations have been a continuing aspect of the business environment in the United States for many years and merger activity in the United States continues to move along at unprecedented levels. Specialty banking firm Houlihan Lokey Howard & Zukin reports 7,700 transactions in 1998 (Heckler and Acciarito, 1999, p. 27) and more than one-third of the 600 companies included in Accounting Trends and Techniques have participated in business combinations (AICPA, 1998, p. 67). Notable examples of business combinations in 1998 have included the merger of Exxon with Mobil Corp.; SBC Communication with Ameritech Corp.; British Petroleum with Amoco Corp.; Bell Atlantic with GTE Corp.; and Daimler-Benz AG with Chrysler Corp.

A number of accounting issues arise when a business combination takes place. One set of issues involves how to account for the business combination at the date of the combination. GAAP allows only two methods of accounting for business combinations: the purchase method and the pooling-of-interests method. These methods are described below (see VI.1.3 and VI.1.4). Another set of issues involves how to account for equity investments in voting common shares in periods following an acquisition of the investment and how to report the results of operations and the financial positions of companies in parent-subsidiary relationships, regardless of whether the parent-subsidiary relationship arises from a business combination or whether the parent company created the subsidiary. Issues related to consolidation accounting will also be discussed in II and III.

3 Types of business combinations and other forms of corporate integration

3.1 Legal methods of business combination

Although there are only two types of business combination specified by GAAP, under the laws of the various states of the US companies may combine or integrate using several legal methods. A 'statutory merger' is a form of business combination defined by law in which only one of the combining companies survives and the other loses its separate identity. The assets and liabilities of the acquired company are transferred to the acquiring company, and the acquired company is dissolved, or liquidated. The operations of the previously separate companies are conducted in a single legal entity thereafter, and, thus, no consolidation accounting is required because there is no parent/subsidiary relationship.

A 'statutory consolidation' (the word consolidation here has an entirely different meaning from its meaning under GAAP) is a second form of business combination defined by law in which both of the combining companies are dissolved and the assets and liabilities of both companies are transferred to a newly created corporation. The operations of the previously separate companies are carried on in a single legal entity. Because neither of the combining companies remains in existence after a statutory consolidation, no consolidation accounting is required because there is no parent/subsidiary relationship.

Business combinations are accomplished through written legal agreements. These legal agreements specify the terms of the business combination, including the form of the combined company, the consideration to be exchanged, the disposition of outstanding securities, and the rights and responsibilities of the participants. Consummation of a legal agreement for a business combination usually requires recognition in the accounting records of one or both of the parties to the combination (see II for a description of the accounting methods to be followed).

3.2 Integration by majority shareholding

Many business combinations are effected by purchasing the voting shares of another company rather than by acquiring assets. In such a situation, the acquired company continues to exist, and the purchasing company records an investment in the shares of the acquired company rather than the individual assets and liabilities. As with the purchase of the assets and liabilities, the cost of the investment is based on the total value of the consideration given in purchasing the shares, together with any additional costs incurred in bringing about the combination. When a business combination is effected through an acquisition of shares, the acquired company may continue to operate as a separate company or it may lose its separate identity and be merged into the acquiring company. If the acquired company is liquidated and its assets and liabilities are transferred to the acquiring company, the result is the same as if the assets and liabilities had been acquired directly.

Integration by majority share holding is the most common form of business combination in the United States (AICPA, 1992). Even if the assets of another company are acquired in a business combination, it is frequently the shares of a subsidiary of the selling company which are acquired rather than tangible assets. Therefore, in the following sections when reference is made to a business combination, it should be understood that the reference is to an integration by majority share holding.

Business combinations can be characterized as either friendly combinations or unfriendly combinations. In a friendly combination, the management of the companies involved reach an agreement on the terms of the combination and recommend approval by the shareholders. Such combinations usually are effected in a single transaction. In an unfriendly combination (hostile takeover), the management of the companies involved are unable to agree on the terms of a combination, and the management of one of the companies makes a tender offer directly to the shareholders of the other company. A tender offer invites the shareholders of the other company to 'tender', or exchange, their shares for securities of the acquiring company or for cash. If sufficient shares are tendered, the combination is consummated.

A business combination effected through a share acquisition does not necessarily involve the acquisition of all of a company's outstanding voting shares. In order for one company to gain control over another through share holding, a majority of the outstanding voting shares is required. An acquisition of less than a majority of the outstanding voting shares is not considered to be a business combination. From an accounting perspective the cost of the shares of the acquired company is recorded in the accounts of the acquiring company as an investment. In subsequent accounting periods, consolidated financial statements are prepared in which all of the inter-company accounts and transactions are eliminated.

3.3 Integration by asset deals

Sometimes one company acquires the assets of another company through direct negotiations with the management of the other company. The acquisition agreement may also specify the assumption of the other company's liabilities. The selling company typically distributes to its shareholders the assets or securities received from the acquiring company and then liquidates, leaving only the acquiring company as the surviving legal entity. The acquiring company accounts for the business combination by recording the fair value of each asset acquired and each liability assumed. The fair value is usually equal to the aggregate consideration given in exchange. As mentioned previously, even though a business combination may involve an acquisition of assets, the acquired assets frequently consist of shares of a subsidiary of the selling company. This makes integration by asset deals essentially the same as integration by majority share holding. There are no particular incentives in the US that would cause companies to prefer asset deals over share deals.

4 Group accounting in the United States

4.1 Development of consolidation accounting and related topics

Consolidation accounting has been required by GAAP since the 1950s (see Table I.1). ARB No. 51, which was issued by the Committee on Accounting Procedures (CAP) of the AICPA, requires the preparation of consolidated financial statements, including the individual accounts of a parent company and all of its majority-owned subsidiaries. Exceptions to the consolidation rule were allowed for majority-owned subsidiaries whose operations were substantially different in nature from that of the parent company. For example, the accounts of a subsidiary involved in leasing, banking or financial services were often not consolidated if the parent company was an industrial or commercial enterprise. In 1987, the FASB issued SFAS No. 94, 'Consolidation of All Majority Owned Subsidiaries', which requires the consolidation of all majority-owned subsidiaries virtually without exception.

USA – Individual Accounts discusses the historical development of GAAP and points out that GAAP developed in the private sector. As a consequence, different accounting practices were created by companies to meet their own needs without any particular regard to uniformity. When the SEC was established by the US Congress in 1934, the SEC was provided with the legal authority to establish GAAP. The SEC delegated this authority to the AICPA. The AICPA began issuing pronouncements on accounting principles in the 1930s. In 1934, six rules dealing with GAAP were adopted by the AICPA. These rules had been recommended in 1933 to the New York Stock Exchange by the AICPA's committee on co-operation with stock exchanges. One of these rules stated that:

> earned surplus (i.e. retained earnings) of a subsidiary company created prior to acquisition does not form a part of the consolidated earned surplus of the parent company and subsidiaries; nor can any dividend declared out of such surplus properly be credited to the income account of the parent company (ARB No. 43, chapter 1, section A 3).

This was the first instance of an official accounting pronouncement dealing with consolidation accounting. What is notable about this rule is that it appears to take for granted the existence of corporate parent/subsidiary relationships and the publication of consolidated financial statements with respect to such relationships, indicating that during the late 1930s consolidation accounting was not a particularly controversial issue among American accounting professionals.

In 1953, the CAP issued ARB No. 43, which was a compilation and revision of all prior ARBs. The rule about the earned surplus of subsidiaries referred to above was included in chapter 1 of ARB No. 43. Other chapters of ARB No. 43 also contained provisions pertaining to consolidation accounting and related topics. For example, chapter 5 of ARB No. 43 dealt with accounting for intangible assets, including accounting for goodwill arising from a business combination (FASB, 1998b, p. 19). Companies were allowed considerable flexibility under ARB No. 43 to decide whether to reflect goodwill in consolidated financial statements at unamortized cost or to amortize the goodwill balance by charges to income over time. However, ARB No. 43, chapter 5 did state that: 'Lump-sum write-offs of intangibles should not be made to earned surplus immediately after acquisition, nor should intangibles be charged against capital surplus' (FASB, 1998b, p. 20). ARB No. 43, chapter 5 was replaced entirely by APB Opinion No. 17 in 1970. APB Opinion No. 17, paragraph 29 requires the amortization of goodwill through an annual charge to income over a period not exceeding 40 years.

Business combinations were addressed in chapter 7, section C of ARB No. 43. Section C had previously been issued as part of ARB No. 40 in 1950. ARB No. 40 was the first official accounting pronouncement to deal with pooling-of-interests accounting for business combinations. Wyatt (1963) has indicated that the first suggestion of pooling-of-interests accounting appeared in unofficial correspondence of the AICPA in 1945. Section C of ARB No. 43, paragraph 2 stated that the primary requisite for pooling-of-interests accounting was that 'all or substantially all of the equity interests in predecessor corporations continue, as such, in a surviving corporation which may be one of the predecessor corporations or a new one created for the purpose'. Furthermore, Section C, paragraph 5 indicated that pooling-of-interests accounting does not give rise to a new basis of accounting, and therefore the recorded balances in the accounts of the combining companies should be carried forward to the new company (FASB, 1998b, p. 26).

Chapter 12 of ARB No. 43 dealt with another topic related to consolidation accounting: foreign operations and foreign exchange. Chapter 12, paragraph 8 stated that:

> in view of the uncertain values and availability of the assets and net income of foreign subsidiaries subject to controls and exchange restrictions and the consequent unrealistic statements of income that may result from the translation of many foreign currencies into dollars, careful consideration should be given to the fundamental question of whether it is proper to consolidate the statements of foreign subsidiaries with the statements of United States companies. Whether consolidation of foreign subsidiaries is decided upon or not, adequate disclosure of foreign operations should be made.

This reluctance to consolidate foreign subsidiaries as was seen in ARB No. 43, chapter 12, reflected the unsettled economic state of the world in the early 1950s and the fact that many countries were still in the process of recovering from World War II or experiencing colonial rebellions. In addition, the United States at that time was not as reliant as it is currently on foreign trade and multinational operations through foreign subsidiaries. The provisions of ARB No. 43, chapter 12 have been superseded by SFAS No. 52, Foreign Currency Translation, issued in 1981.

Although ARB No. 43 took for granted the existence of corporate parent/subsidiary relationships and the publication of consolidated financial statements with respect to these relationships, it was not until the issuance of ARB No. 51 in 1959 that consolidation accounting became the subject of an official accounting pronouncement. Under ARB No. 51 it was presumed that majority-owned subsidiaries would be consolidated unless the subsidiary operated in a substantially different type of business from that of the parent company (e.g. leasing, finance or banking). In addition, foreign subsidiaries located in countries with monetary exchange restrictions were typically not consolidated under ARB No. 51.

ARB No. 51 has been the primary accounting pronouncement dealing with consolidated accounting for many years. ARB No. 51 was amended by SFAS No. 94 in 1987 to require the consolidation of all majority-owned subsidiaries. Otherwise the provisions of ARB No. 51 remain in force. One of the principal reasons expressed by the FASB for requiring the consolidation of all majority-owned subsidiaries was to prevent off-balance sheet financing through the use of finance and leasing subsidiaries (SFAS No. 94, paragraph 7). The consolidation of all majority-owned subsidiaries causes the debt of highly leveraged finance subsidiaries to be reflected in the financial statements of the consolidated group.

4.2 Government regulation

The SEC has the legal power to establish GAAP for companies with publicly traded securities. The SEC has delegated the authority to establish GAAP to the FASB. The SEC has from time to time confirmed the authority of the FASB to establish GAAP. Therefore, it can be presumed that any statement issued by the FASB has the imprimatur of the SEC. In addition, Rule 3A-02 of Regulation S-X of the SEC (discussed in USA – Individual Accounts, IV.4.1) explicitly requires the issuance of consolidated financial statements, which include the financial statements of a parent company and all of its majority-owned subsidiaries. Furthermore, the rule states that:

> In deciding upon consolidation policy, the registrant must consider what financial presentation is most meaningful in the circumstances and should follow in the consolidated financial statements principles of inclusion or exclusion which will clearly exhibit the financial position and results of operations of the registrant. There is a presumption that consolidated statements are more meaningful than separate statements and that they are usually necessary for a fair presentation when one entity directly or indirectly has a controlling financial interest in another entity.

5 Relationship to international accounting and harmonization efforts

As discussed in USA – Individual Accounts, I.3.4, the FASB as well as the SEC are closely monitoring the development of IAS and they will consider accepting them for financial statements to be filed in the United States.

With respect to consolidation the FASB and group of standard setters known as the G4+1 issued an invitation to comment on Methods of Accounting for Business Combinations: Recommendations of the G4+1 for Achieving Convergence (FASB, 1998e). The G4+1 includes representatives from the Accounting Standard Boards of Australia, Canada, New Zealand, the United Kingdom, and the United States. Representatives of the International Accounting Standard Board (IASC) participate as observers. In its position paper the G4+1 consider the methods of accounting for business combinations and address the following fundamental issues:

- Whether a single method of accounting for business combinations is preferable to two (or more) methods.
- If so, which method of accounting should be applied to all business combinations.
- If not, which methods should be applied and to which combinations they should be applied.

In addressing those issues, the position paper considers three broad classes of methods of accounting for business combinations:

- The pooling-of-interests method.
- The purchase method.
- The fresh-start method (FASB, 1998e, p. XV).

The Position Paper concludes that the use of a single method of accounting is preferable and that the purchase method is the appropriate method to use. As a result of this international discussion, the FASB has acknowledged that the widespread use of the pooling-of-interests method is largely an American phenomenon. According to Mr Jenkins, chairman of the FASB, 'the U.S. not only is out of step with other countries on the pooling versus purchase issue, but domestically we have a great deal of diversity in practice as well. Pooling of interests is the exception most everywhere else in the world, and some countries ban the use of it altogether.' Recently, the FASB announced that it would eliminate pooling of interests as a method of accounting for business combinations. In a unanimous vote on 21 April 1999, the Board tentatively decided that using the purchase method is preferable to allowing more than one method to be used when businesses combine. The change will be effective for business combinations initiated after the FASB issues a final standard on the issues, which is expected to be late in 2000 (FASB press release 21 April 1999).

II Objectives of group accounting, underlying concepts and general principles

1 Relevant objectives

In Statement of Financial Accounting Concepts (SFAC) No. 1, Objectives of Financial Reporting by Business Enterprises, the FASB states that the principal objective of financial reporting is to provide information to investors and creditors that is useful in making investment and credit decisions (see USA – Individual Accounts, III.1.2). Because consolidation accounting constitutes part of GAAP, consolidation accounting has the same objective as all other parts of GAAP (i.e. to provide information that is useful to investors and creditors). However, ARB No. 51, paragraph 1 also states:

> The purpose of consolidated statements is to present, primarily for the benefit of the shareholders and creditors of the parent company, the results of operations and the financial position of a parent company and its subsidiaries essentially as if the group were a single company with one or more branches or divisions. There is a presumption that consolidated statements are more meaningful than separate statements and that they are usually necessary for a fair presentation when one of the companies in the group directly or indirectly has a controlling financial interest in the other companies.

2 Underlying concepts and general principles

Although the primary purpose of consolidation accounting is the same as that for GAAP generally, consolidation accounting takes as its focus the information needs of the shareholders of the parent company. From a theoretical perspective, the approach to consolidation accounting followed in the United States is in accordance with the 'parent company' theory. The parent company theory recognizes that even though the parent company does not have direct ownership of the assets of a subsidiary nor direct responsibility for the liabilities of the subsidiary, it has the ability to exercise control over all of the subsidiary's assets and liabilities, not simply a proportionate share. Thus, all the assets and liabilities of the subsidiary are included in the consolidated balance sheet by combining them with the assets and liabilities of the parent. Similarly, all the revenues and expenses of the subsidiary are included in the consolidated income statement. Under the parent company theory, separate recognition is given in the consolidated balance sheet to the non-controlling interest's claim on the assets of the subsidiary, and recognition is given in the consolidated income statement to the earnings assigned to the non-controlling shareholders.

Under the parent company theory, the full amount of the book value of the subsidiary's net assets is included in the consolidated balance sheet. In those cases where the parent company pays more than book value for its share of the subsidiary's net assets, the parent's portion of the fair value increment and any goodwill is included in the consolidated balance sheet. As a result, subsidiary assets would be stated in the consolidated balance sheet at their full fair values only when the parent company purchases complete ownership of the subsidiary. In preparing the consolidated balance sheet, the non-controlling shareholders are assigned a proportionate share of the book value of the subsidiary's net assets. Amounts less than the full fair values of identifiable assets and the full amount of goodwill are reported whenever the parent does not purchase 100 per cent ownership of the subsidiary.

In addition, the full amounts of the revenues and expenses of the subsidiary are included in the consolidated income statement. However, only that portion of subsidiary net income assignable to the parent company is included in consolidated net income. Income assigned to the non-controlling shareholders is treated as a deduction in the consolidated income statement when arriving at consolidated net income.

As it has evolved in the United States, the parent company theory presents a different picture from that of European group accounting. First, no particular distinction is made between the different information interests of shareholders of the parent company and of minority shareholders of the subsidiaries. The primary emphasis is on the information needs of the shareholders of the parent company. Second, no particular distinction is made between the information interest of creditors of the parent company and of creditors of the subsidiaries. Again, the primary emphasis is on the information needs of creditors of the parent company. Under the parent company theory, the equity of the group is the equity of the parent company and the profit of the group is the profit share of the parent company shareholders. The information interests of the minority shareholders of the subsidiaries are less well taken care of. For example, there is no obligation to prepare sub-group accounts even if there are minority shareholders of the parent of the sub-group.

The parent company theory is often thought of as being a blend of the proprietary theory and the entity theory (Baker *et al.*, 1998, pp. 144–145). However, the parent company theory differs from the proprietary theory and entity theory at certain points. For example, when assets of a subsidiary are revalued at the point of consolidation, the revaluation is only to the extent of the majority shareholders' interest in the subsidiary rather than for 100 per cent of the fair value of the assets acquired. This is essentially at variance with the entity theory of consolidation and represents a partial intrusion of the proprietary theory, whereby the consolidated accounts are said to be the accounts that are relevant to the owners of the parent company and not those that would be relevant to both the parent company shareholders and the minority shareholders.

The basic principle of consolidated accounting under GAAP is that consolidated financial statements must be prepared for the parent company and all of its majority-owned subsidiaries. The primary concept underlying this requirement is the notion that the parent company and its subsidiaries are viewed as being part of one economic entity. Prior to the issuance of ARB No. 51 in the 1950s, there was a generally accepted norm that large companies with multiple subsidiaries would present consolidated financial statements. Listing requirements for the New York Stock Exchange also required companies to present consolidated financial statements. The principle of a true and fair view did not enter into discussions of consolidation accounting in the United States, although a parallel concept regarding the problem of off balance sheet financing with unconsolidated finance subsidiaries was a prominent rationale for the requirement to consolidate all majority owned subsidiaries (SFAS No. 94, paragraph 7).

III Group concept and obligation to prepare group accounts

1 Concept of a group

The legal concept of the group is dominant in GAAP. By the term legal, we mean ownership of a majority of the outstanding voting common shares. The rule for the preparation of consolidated financial statements under GAAP is based on the presumption that control rests with the group that owns voting control of the majority of outstanding common shares of a company. There is no concept under GAAP that financial statements of companies which are independent but nevertheless managed on a unified basis should be consolidated.

2 Obligation to prepare group accounts

The obligation to prepare consolidated financial statements arises when one company, either listed or private, has a subsidiary. The primary guidance is found in ARB No. 51 and SFAS No. 94. SFAS No. 94, paragraph 2 requires a parent to consolidate all companies in which it has 'a controlling financial interest through direct or indirect ownership of a majority voting interest', unless control is likely to be temporary or control does not rest with the majority owner (e.g. the subsidiary is in legal reorganization or in bankruptcy or operates under foreign exchange restrictions, controls or other governmentally imposed uncertainties so severe that they cast doubt on the parent's ability to control the subsidiary). Investments in subsidiaries meeting the conditions of either of the exceptions should be accounted for using the cost method.

SFAS No. 94 does not address investments in non-corporate entities such as partnerships. In practice, similar criteria are applied to partnerships, such that the holder of a majority voting interest in a partnership generally will consolidate the partnership. If the parent company has an interest in a joint venture that is not majority owned, there is no requirement to consolidate this interest in the joint venture. Likewise, arrangements with associated companies or strategic alliances are not consolidated. An interest in a joint venture may be included in the individual accounts of the parent company using the equity method (see APB Opinion No. 18).

In consolidation accounting there are no group accounts. In other words, there are no separate accounting records for the consolidated group. Each company (i.e. parent company and subsidiaries) keeps its own accounts and bookkeeping records. Consolidation of accounts and preparation of consolidated financial statements take place as a separate function away from the accounting records. Both private and public companies are required to follow GAAP if the financial statements are audited by a CPA. Subsidiary companies are usually not explicitly audited except as a part of the audit of the financial statements of the parent company. The accounts and bookkeeping records of the subsidiary, if not explicitly audited, may not always be kept in conformity with GAAP. Reconciliations and adjustments may therefore be made during the consolidation process in order to ensure that the consolidated accounts are presented in conformity with GAAP.

3 Additional SEC requirements

Companies with publicly traded securities are subject to SEC regulations. Rule 3A-02 of SEC Regulation S-X, which is the primary SEC rule dealing with consolidation accounting differs in two respects from SFAS No. 94. Rule 3A-02 requires that:

> registrants shall consolidate entities that are majority owned and shall not consolidate entities that are not majority owned. The determination of majority ownership requires a careful analysis of the facts and circumstances of a particular relationship among entities. In rare situations, consolidation of a majority owned subsidiary may not result in a fair presentation, because the registrant, in substance, does not have a controlling financial interest (for example, when the subsidiary is in legal reorganization or in bankruptcy, or when control is likely to be temporary).

The rule also indicates that, in certain rare circumstances, it may be necessary to consolidate an affiliated entity fully, notwithstanding the lack of majority ownership of voting common shares, 'because of the existence of a (*de facto*) parent-subsidiary relationship by means other than recorded ownership of voting stock' (see Rule 3A-02 in Regulation S-X as described in USA – Individual Accounts, IV.4.1).

4 Obligation to prepare subgroup accounts

In principle, there is no requirement for publication of consolidated financial statements at a level other than that of the ultimate parent company of the group. There may be informal consolidations of subgroups for reasons such as management information and control, separate financing arrangements, or internal administration, but these are infrequently published or disclosed outside the consolidated group unless the subgroup itself has publicly issued outstanding securities.

IV Consolidated group

1 Subsidiaries to be consolidated

1.1 Definition of a subsidiary

The theory behind the preparation of consolidated financial statements may be stated quite simply: when one company has the ability to exert control over another company, permitting it to direct the subsidiary's operations and policies, then the assets, liabilities and operations of the two companies should be presented in one set of consolidated financial statements as though they operated as a single company. This general approach was articulated originally by the AICPA in 1929. In 1959, ARB No. 51 was issued, and although modified by SFAS No. 94, it is still the authority for today's practice.

No matter what the form of the association might be, the applicability of the consolidation rules should be based on the substance of the relationship. The underlying consolidation theory requires consolidation of any company substantively controlled by its parent. As stated previously (see III.2) the SEC has occasionally broadened the definition of control to allow for consolidation of another company when *de facto* control exists by means other than ownership of voting common shares. Hereby the SEC considers the diversity of financial and operating arrangements in business environments employed in recent years, which allows the possibility of gaining control with less than majority ownership.

Under GAAP, a subsidiary is defined as a company, the majority of the common shares of which is owned by another company. A corporation becomes a subsidiary when another corporation acquires a majority of its outstanding voting shares. A subsidiary may be defined through indirect ownership as well. For example, if a parent company A owned 40 per cent of company B and 60 per cent of company C and at the same time company C owned 15 per cent of company B, then B would be a subsidiary of company A.

Prior to 15 December 1988, the exclusion of non-homogeneous operations from consolidated financial statements was allowed. However SFAS No. 94 introduced the requirement that even subsidiaries with non-homogeneous operations should be consolidated. SFAS No. 94 also states that subsidiaries should not be consolidated (a) when control is likely to be temporary or (b) when control does not rest with the majority owner. Control does not rest with the majority owner if the subsidiary is in legal reorganization, or in bankruptcy, or operating under severe foreign exchange restrictions or other governmental imposed uncertainties (for more detail see IV.1.4).

1.2 Calculating majority voting rights

ARB No. 51 and SFAS No. 94 state that the usual condition for consolidation is majority ownership of the outstanding voting common shares. It is the formal right to vote the shares that determines the calculation of majority interest. Furthermore, it is the potential to vote the shares and not the actual votes in a general meeting of shareholders that determines majority ownership. It may be possible to avoid consolidation by transferring voting rights. However, this is an unusual practice which may be viewed negatively by the SEC.

1.3 Divergent activities of subsidiaries

SFAS No. 94 amended ARB No. 51 and eliminated non-homogeneity of business operations as a possible reason to exclude a subsidiary from consolidation. Regardless of the nature of their operations, all majority-owned companies must be consolidated, with certain exceptions described below.

1.4 Exclusion of subsidiaries

For the most part, all subsidiaries must be consolidated. However, the ability to exercise control and the permanence of control must be considered. Under certain circumstances, the majority shareholders of a subsidiary may not be able to exercise control even though they hold more than 50 per cent of the outstanding voting common shares. This might occur, for instance, if the subsidiary is in legal reorganization or in bankruptcy. Although the parent might hold majority ownership, control would rest with the courts or a court appointed trustee. Similarly, if the subsidiary is located in a foreign country that places restrictions on the subsidiary that prevent the remittance of profits or assets back to the parent, consolidation of that subsidiary is not appropriate because of the parent's inability to control important aspects of the subsidiary's operations. If the parent's control is expected to be temporary, consolidation of the subsidiary is also not appropriate. This could occur, for example, if the parent intends to sell a recently acquired investment in a subsidiary in the near future. If the political environment is such that the parent expects to lose control shortly, then consolidation is not appropriate because control is temporary.

In addition, the FASB's Emerging Issues Task Force (EITF) agreed in Issue No. 96-16 (FASB, 1998d):

> that the assessment of whether the rights of a minority shareholder should overcome the

presumption of consolidation by the investor with a majority voting interest in its investee is a matter of judgement that depends on facts and circumstances. The Task Force further agreed that the framework in which such facts and circumstances are judged should be based on whether the minority rights, individually or in the aggregate, provide for the minority shareholder to effectively participate in significant decisions that would be expected to be made in the 'ordinary course of business'. Effective participation means the ability to block significant decisions proposed by the investor who has a majority voting interest. That is, control does not rest with the majority owner because the investor with the majority voting interest cannot cause the investee to take an action that is significant in the ordinary course of business if it has been vetoed by the minority shareholder.

The Task Force believes that certain minority rights *would not overcome* the presumption of consolidation by the investor with a majority voting interest since they are considered *protective rights* even if they allow the minority shareholder to block certain corporate actions.

However, the following corporate actions *would overcome* the presumption of consolidation by the investor with a majority voting interest since they are considered *substantive participating rights*:

- Selecting, terminating, and setting the compensation of management responsible for implementing the investee's policies and procedures.
- Establishing operating and capital decisions of the investee, including budgets, in the ordinary course of business.

1.5 Consolidation of special-purpose entities

As expressed in EITF Topic No. D-14:

> the SEC staff is becoming increasingly concerned about certain receivables, leasing, and other transactions involving special-purpose entities (SPEs). Certain characteristics of those transactions raise questions about whether SPEs should be consolidated (notwithstanding the lack of majority ownership) and whether transfers of assets to the SPE should be recognized as sales. Generally, the SEC staff believes that for nonconsolidation and sales recognition by the sponsor or transferor to be appropriate, the majority owner (or owners) of the SPE must be an independent third party who has made a substantive capital investment in the SPE, has control of the SPE, and has substantive risks and rewards of ownership of the assets of the SPE (including residuals). Conversely, the SEC staff believes that nonconsolidation and sales recognition are not appropriate by the sponsor or transferor when the majority owner of the SPE makes only a nominal capital investment, the activities of the SPE are virtually all on the sponsor's or transferor's behalf, and the substantive risks and rewards of the assets or the debt of the SPE rest directly or indirectly with the sponsor or transferor. (FASB, 1998d, p. 4979)

2 Joint ventures

2.1 Definition of a joint venture

Paragraph 3(d) of APB Opinion No. 18, The Equity Method of Accounting for Investments in Common Stock, describes a corporate joint venture as follows:

- A corporation owned and operated by a small group of businesses (the 'joint venturers') as a separate and specific business or project for the mutual benefit of the members of the group. A government may also be a member of the group.
- The purpose of a corporate joint venture frequently is to share risks and rewards in developing a new market, product or technology; to combine complementary technological knowledge; or to pool resources in developing production or other facilities.
- A corporate joint venture also usually provides an arrangement under which each joint venturer may participate, directly or indirectly, in the overall management of the joint venture. Joint venturers thus have an interest or relationship other than as passive investors.

- An entity which is a subsidiary of one of the 'joint venturers' is not a corporate joint venture.
- The ownership of a corporate joint venture seldom changes, and its stock is usually not traded publicly. A minority public ownership, however, does not preclude a corporation from being a corporate joint venture.

Since each venturer commonly participates in the overall management, and significant decisions commonly require the consent of each venturer (regardless of ownership interest), no individual venturer has unilateral control. The venturers will not necessarily have equal ownership interests and a venturers' share could be as low as 5 per cent, or as high as 50 per cent. Joint ventures differ from what is encountered in the normal application of the equity method of consolidation accounting, in that joint venturers typically have special rights and obligations assuring significant influence even when ownership is less than 20 per cent.

A joint venture may be organized as a corporation, a partnership or an undivided interest. If the joint venture is organized as a corporation, APB Opinion No. 18 applies, and the investment in the corporate joint venture would be accounted for using the equity method. If the joint venture is organized as a partnership or in some other unincorporated form, the provisions of APB Opinion No. 18 may be applied in accounting for such investments, or partnership accounting may be used instead.

Since the provisions of APB Opinion No. 18 are applicable primarily to corporate joint ventures, the AICPA Accounting Standards Executive Committee (AcSEC) has prepared an issues paper entitled Joint Venture Accounting (AICPA, 1979), and the AICPA has asked the FASB to consider a project on accounting and reporting for investments in joint ventures. The FASB has added joint ventures to its project on Reporting Entities. In its issue paper, AcSEC reached the conclusion that a joint venture should be defined broadly to include all entities that have joint control of major decisions, regardless of legal form.

2.2 Corporate joint ventures

A corporation owned and operated by a small group of venturers to accomplish a mutually beneficial venture or project should report their investments as equity investments (one-line consolidations) under the provisions of APB Opinion No. 18. The approach for establishing significant influence on corporate joint ventures is quite different from that for most investments in common shares because each venturer usually must consent to each significant venture decision, thus establishing an ability to exercise significant influence regardless of ownership interest. Even so, when a venturer cannot exercise significant influence, for whatever reason, the venturer's investment in the venture should not be accounted for by the equity method but rather by the cost method.

The reporting requirements of APB Opinion No. 18 for corporate joint ventures are described as follows: The equity method best enables investors in corporate joint ventures to reflect the underlying nature of their investment in those ventures. Therefore, investors should account for investments in the common shares of corporate joint ventures by the equity method.

2.3 Accounting for unincorporated joint ventures

Interpretation No. 2 of APB Opinion No. 18 addresses the applicability of APB Opinion No. 18 to investments in partnerships and undivided interests in joint ventures. While the provisions of APB Opinion No. 18 apply only to investments in common shares, and therefore do not cover unincorporated ventures, Interpretation No. 2 explains that many of the provisions of APB Opinion No. 18 are appropriate in accounting for investments in unincorporated entities. For example, partnership profits and losses accrued by investor-partners are generally reflected in the partners' financial statements. Elimination of inter-company profit is also appropriate, as well as providing for deferred income tax liabilities on profits accrued by partner-investors.

Interpretation No. 2 to APB Opinion No. 18 also applies to undivided interests in joint ventures. An undivided interest is a legal term, derived from the law of real property. An undivided interest is defined as a situation in which the investor-venturer is deemed to own a proportionate share of each asset of the venture and is proportionately liable for its share of each liability of the venture. Although the provisions of APB Opinion No. 18 are generally applicable to undivided interests, they do not apply in some industries that have specialized industry practices. For example, the industry practice in oil and gas ventures is for the investor-venturer to account for its pro rata share of the assets, liabilities, revenues and expenses of a joint venture in its own financial statements. This reporting procedure is referred to as pro rata or proportionate consolidation. On the other hand, SOP 78–9 issued by AcSEC recommends that proportionate consolidation should not be used for undivided interests in real estate venturers subject to joint control by investors. A venture is subject to joint control if decisions regarding the financing, development, or sale of property require the approval of two or more owner-venturers. Subsequently, a 1979 AICPA issues paper, Joint Venture Accounting, recommended that a joint venture that is not subject to joint control, and where the venturers are not jointly responsible for the liabilities of the venture, should be required to use the proportionate consolidation method.

2.4 Accounting by a joint venture for businesses received at its formation

Most of the existing accounting guidance on joint ventures focuses on the accounting by the venturers and not on the accounting in the separate, stand-alone financial statements of the joint venture. Current practice generally has been to report the assets that a business contributes to a joint venture at historical cost unless certain conditions are met. In EITF Issue No. 98-4, the Task Force considers how the joint venture should record, in its separate financial statements, the businesses received if the relationship between the joint venture and the venturers has either (a) none of the five attributes of a corporate joint venture described in paragraph 3(d) of APB Opinion No. 18, but there is joint control; or (b) all five of the attributes of a corporate joint venture described in paragraph 3(d) of APB Opinion No. 18, and there is joint control. The issue is if two or more parties contribute businesses to a newly formed entity, whether the characteristics from APB Opinion No. 18, or some other characteristics, must exist in order for the entity to qualify for historical cost accounting. Transactions that are considered business combinations would require acquisition accounting.

The Task Force discussed the fact that the potential for diversity in current practice could be minimized if the Task Force provided guidance on what constitutes 'joint control' and agreed to pursue a definition of joint control. The Task Force was not asked to reach a consensus on this Issue.

The SEC Observer indicated that the SEC staff would object to a conclusion that did not result in the application of APB Opinion No. 16 to transactions in which businesses are contributed to a newly formed, jointly controlled entity if that entity is not a joint venture. The SEC staff also would object to a conclusion that joint control is the only defining characteristic of a joint venture. Further discussion is expected at a future meeting.

V Uniformity of accounts included

1 Uniformity of formats

Although there is a tendency for companies in the same industry to have similar formats for their financial statements, there are no requirements either by the SEC or by the FASB for companies in a given industry to follow a specific format when presenting their financial statements. The member companies of a consolidated group may follow different formats when preparing their financial statements for submission to the parent company of the group. It is the responsibility of the parent company to prepare the consolidated financial statements for the group in whatever format the management of the company feels is appropriate. For example, the consolidated financial statements of General Electric Company are presented with three columns for each year: one column presents the consolidated group numbers; the second column presents the consolidated operations of the manufacturing side of the General Electric Company; and the third column presents the consolidated operations of the financial side of the company. However, it should be noted that this format is entirely voluntary and not required by the SEC or the FASB.

As has been mentioned previously, in consolidation accounting there are no group accounts. There are no accounting records for the consolidated group. Each company (i.e. parent company and subsidiaries) keeps its own accounting and bookkeeping records. Consolidation of accounts and preparation of consolidated financial statements takes place as a separate function away from the accounting records. Consolidated financial statements must be prepared in accordance with GAAP. However, the specific accounts to be included in the consolidated financial statements and the manner of the presentation of these accounts is subject to the discretion of the management of the parent company and the auditors of the consolidated group. Both private and public companies are required to follow GAAP. Public companies must also meet additional SEC requirements regarding publication of accounts of subsidiaries if the subsidiaries have publicly traded securities outstanding.

GAAP require consolidation of worldwide operations of a multi-national company into one set of financial statements stated in USD. When consolidated financial statements are first presented for a group, comparative consolidated financial statements for prior periods should be published, if it is practical to do so. In all years subsequent to the date of consolidation, full comparative statements should be published.

2 Balance sheet date of companies included

The fact that a subsidiary has a balance sheet date which differs from that of the parent company or that of the other companies in the consolidated group does not act as a bar against consolidation. When fiscal periods of the parent company and its subsidiaries are different, consolidated statements are prepared as of the parent company's balance sheet date. If the difference in balance sheet dates is not greater than three months, it is acceptable to use the subsidiaries' statements for consolidation purposes, with the disclosure of the effect of any intervening events which materially affect the financial position or results of operations. Otherwise, the financial statements of the subsidiary should be adjusted so that they correspond as closely as possible to the fiscal period of the parent company (see FASB Interpretation No. 13). This adjustment may be effected through a change in the balance sheet date of the subsidiary, or it may be that an unaudited balance sheet for the subsidiary is prepared as of the balance sheet date of the parent company and that an unaudited income statement is prepared for the subsidiary for the 12 months ending on the balance sheet date of the parent company. This would pose no particular technical problems, and a major portion (i.e. nine months) of the subsidiary's fiscal period could be audited with no difficulties.

It is possible to choose a balance sheet date for the consolidated group which is the same as that of the most important subsidiaries of the group, but it would be more common to change the balance sheet date of the parent company or that of the subsidiaries so that they coincide.

3 Recognition criteria and valuation rules

The recognition criteria and valuation rules applied in consolidation accounting are the same as the rules applied in the accounts of the individual companies which constitute the consolidated group. As long as they fall within the parameters of GAAP, the decision as to which specific recognition criteria and valuation rules to utilize is made by the management of the parent company of the consolidated group. There is no specific requirement to adjust the accounts of the subsidiary companies in order to bring them into congruence with the accounting methods of the parent company. It is permissible for the parent to use one accounting method while the subsidiary uses another accounting method, as long as both methods conform to GAAP. However, since a group has to adopt certain accounting policies within a group, the consistent use of such accounting principles is almost naturally.

It is permissible for the parent company to cause the accounting methods used by the subsidiary to be modified so as to correspond to those of the parent company. Changes in accounting methods are governed by APB Opinion No. 20 (see USA – Individual Accounts, III.2.9). As a general rule, accounting methods should be applied under GAAP on a consistent basis from period to period in order to provide investors and creditors with a consistent view of the financial position and results of operations of the business enterprise. However, if the management of the company feels that a particular accounting method is better suited to portraying the financial position of the company than one that is currently being used, the management may change accounting methods. Typically, this change must receive the concurrence of the company's auditors and be disclosed as a change in accounting principle in the financial statements. If a parent company acquires a subsidiary, the parent company generally causes the subsidiary to adopt the group accounting principles although there is no requirement by the SEC or the FASB to make such changes.

As there are no group accounts under GAAP, consolidated financial statements are prepared through a process of combining the accounts of the individual companies. Thus, the parent company and the subsidiaries may use different accounting methods as long as the methods are acceptable under GAAP. There is no requirement to adjust for differences in accounting methods during the consolidation process. However, it is permissible to make such adjustments if the parent company feels that it is useful to do so.

4 Translation of financial statements reported in a foreign currency

4.1 The objective

SFAS No. 94 requires the consolidation of all subsidiaries, including foreign subsidiaries, unless there are substantial restrictions placed on the remittance of funds from the foreign subsidiary to the parent company. When the financial statements of a foreign subsidiary are consolidated into the financial statements of the US parent, it is necessary to translate the financial statements of the foreign subsidiary from the currency in which they are denominated into the US dollar. SFAS No. 52, Foreign Currency Translation, (FASB, 1998a, pp. 501–31) specifies the GAAP requirements for the translation of foreign currency financial statements. SFAS No. 52 superseded SFAS No. 8, Accounting for the Translation of Foreign Currency Transactions and Foreign Currency Financial Statements, (FASB, 1998a, p. 73). The primary objectives of SFAS No. 52 are stated to be as follows:

- To provide information that is generally compatible with the expected economic effects of a rate change on an enterprise's cash flows and equity.
- To reflect in consolidated statements the financial results and relationships of the individual consolidated entities as measured in their functional currencies in conformity with GAAP.

4.2 The functional currency

4.2.1 Concept

SFAS No. 52, paragraph 5 states that the assets, liabilities, and operations of a foreign entity shall be measured using the functional currency of that entity.

The functional currency of an entity is defined as the currency of the primary economic environment in which the entity generates and expends cash. For example, a German subsidiary or division with relatively self-contained and integrated operations in Germany would have as its functional currency the Deutsche Mark (DM). If the subsidiary also has some transactions and open account balances denominated in currencies other than the DM, those balances first must be restated to the DM, the functional currency, with any gain or loss included in the subsidiary's net income, before the statements are translated into USD in consolidation. The functional currency will not always be the currency of the country in which the foreign entity is located or the currency in which the records are maintained. Such a situation will occur, for example, when the entity is merely an extension of the parent company. In that case, the functional currency would be the reporting currency of the parent company. For example, a German sales branch or subsidiary of a US parent that basically takes order for the parent's merchandise, bills and collects for it, and has a warehouse to facilitate timely delivery probably would have the USD as its functional currency. In such a case, any receivables or payables of the branch or subsidiary denominated in currencies other than the USD would be translated at the current rate, and movements in the exchange rate would result in an exchange gain or loss to be included in current income. Because the functional currency of the operation located in Germany is the same as the reporting currency, technically no translation is needed. However, the remeasurement of the recorded amounts in DM to the USD functional currency is a process essentially like the application of the rules contained in SFAS No. 8. The inventory in the German warehouse, if recorded in the DM by the branch or subsidiary, would have to be restated to the US parent's historical cost in USD, and translation adjustments would be included in income.

4.2.2 Criteria to determine the functional currency

In most cases the functional currency of the subsidiary is the currency of the country in which the subsidiary is located and operates. SFAS No. 52, paragraph 5 provides the following definition of functional currency:

> An entity's functional currency is the currency of the primary economic environment in which the entity operates; normally, that is the currency of the environment in which an entity primarily generates and expends cash.

In complex situations, it may be difficult to determine the functional currency of a foreign operation. For example, a foreign operation may have extensive transactions in many currencies, or its operations may be conducted in different economic environments. Operations that are distinct and separable may have more than one functional currency. The determination of the functional currency in complex situations depends upon management's evaluation of the various economic facts and circumstances pertaining to the operation(s) and the relative weight assigned to each. SFAS No. 52, appendix A provides general guidance on some of the economic factors that should be considered in determining the functional currency. The factors, which should be considered both individually and collectively, include the nature and extent of a foreign operation. The local currency of the foreign subsidiary is considered to be the functional currency when:

- Cash flows are primarily in the foreign currency and do not directly affect the parent's cash flow.
- Sales prices are not responsive to short-term changes in exchange rates but are determined more by local competition or local government regulation.
- There is an active local market for the foreign entity's products or services, although there may be significant exports as well.
- Labour, materials and other costs for the foreign entity's products or services are primarily local costs.
- Financing is primarily denominated in the local currency, and funds generated by the subsidiary's operations are sufficient to service existing and expected debt obligations.
- There is a low volume of inter-company transactions and a limited interrelationship between the subsidiary and its parent. Use of or reliance on the parent's or affiliate's competitive advantages is not a factor in this evaluation (SFAS No. 52, paragraph 42).

In contrast, the parent company's currency is deemed to be the functional currency when:

- The foreign entity's cash flows affect the parent's current cash flows by, for example, frequent remittances.
- Sales prices are responsive to short-term changes in exchange rates with prices determined more by worldwide competition than local conditions.
- Sales are made mostly in the parent's country, or sales contracts written in foreign countries are denominated in the parent's currency.
- Labour, materials and other costs for the

foreign entity's products or services are primarily costs for components obtained from the country in which the parent is located.

- Financing is primarily from the parent or is otherwise arranged in the parent's currency, or the foreign entity cannot service existing or expected debt obligations without the infusion of cash from the parent.
- There is a high volume of inter-company transactions and an extensive interrelationship between the foreign entity and its parent.
- A foreign entity is used as a device or shell corporation for holding non-operating assets and liabilities that could be readily carried on the parent's books (SFAS No. 52, paragraph 42).

Other factors to be considered when determining the functional currency are:

- Separability of operations. If a foreign entity deals in more than one foreign currency, has operations involving both its parent and unrelated entities, or operates in two or more economic environments or countries, each operation may be considered as a separate economic entity for purposes of identifying the functional currency.
- Influence of the parent. The fact that the parent exercises the control or significant influence necessary to permit consolidation or application of the equity method of accounting for foreign investments does not, *per se*, imply that the functional currency of the subsidiary is the parent's currency (SFAS No. 52, paragraphs 43–4).

None of the factors above is considered more important than the others. All factors are to be examined, individually and collectively, to determine the appropriate functional currency. If, after considering all factors, it is not clear whether the foreign currency is the functional currency, it is not unreasonable to use the reporting currency as the functional currency.

4.3 Translation of foreign currency statements

4.3.1 Concept

SFAS No. 52 also states that all elements of financial statements of foreign subsidiaries are to be translated into USD-denominated financial statements by using the 'current exchange rate'. Under the current rate method, all elements of financial statements are translated using the exchange rate in effect as of the dates the elements are reported. Thus:

- Assets and liabilities of a foreign operation are translated from the operation's functional currency using the exchange rate in effect at the foreign operation's balance sheet date.
- Revenues, expenses, gains, and losses are translated in a manner that produces approximately the same result as using the exchange rate in effect at each transaction date. That result typically is achieved by using the weighted average exchange rate for the period (SFAS No. 52, paragraph 12).

Under certain circumstances (e.g. when the operations of the subsidiary are closely tied to the USD or when the subsidiary is located in a country with a highly inflationary economy) the financial statements of the foreign subsidiary must be 'remeasured' before they are translated. The process of remeasuring the financial statements of a foreign subsidiary into USD denominated financial statements or into financial statements denominated in the functional currency of the subsidiary is referred to as the 'temporal method'. The functional currency is defined as the primary currency in which the subsidiary conducts its business (SFAS No. 52, paragraph 5). The use of the current rate method or the temporal method in a particular situation is dependent on the subsidiary's functional currency. If the functional currency of the subsidiary is the local currency of the country in which the subsidiary is domiciled, the current rate method is used. If the functional currency of the subsidiary is not the local currency of the subsidiary, the temporal method is used to remeasure the subsidiary's accounts from the local currency to the functional currency, at which point the current rate method is used to translate the financial statements into the USD. If the functional currency is already the USD, the temporal method is used to remeasure the subsidiary's accounts from local currency to the USD without translation. Accounting for foreign subsidiaries is therefore a two-step process:

- Step 1: remeasure from the local currency into the functional currency using the temporal method.
- Step 2: translate from the functional currency to the USD using the current rate method.

If the local currency of the foreign subsidiary is the functional currency, remeasurement (Step 1) is not necessary.

4.3.2 Current rate method

Under the current rate method, all of the balance sheet accounts except for the shareholders' equity accounts of the subsidiary are translated at the current exchange rate. Income statement accounts are translated at the average rate of exchange for the accounting period. The paid-in capital accounts of the subsidiary are translated at the historical exchange rate on the date when the shares were issued, or the date when the subsidiary was acquired by the parent, if later. Dividends declared by the subsidiary are translated at the rate on the date when the dividends were declared. The retained earnings account is not translated *per se*, but rather is computed as a result of the carried forward translated balance of retained earnings for the prior period plus the translated net income from the income statement accounts minus the translated dividend declaration account.

Upon translation, any translation difference is considered to be a translation adjustment which is carried to a separate component of shareholders' equity and does not enter into the computation of net income for the period. When the translated financial statements of the subsidiary are consolidated with the financial statements of the parent company, the investment in the subsidiary as reflected in the accounts of the parent company is eliminated against the translated shareholders' equity accounts of the subsidiary. Any difference that results from this elimination process, other than that arising from revaluation of assets, reflection of goodwill, or reflection of minority interest, will be carried to the translation adjustment account as a component of consolidated shareholders' equity.

The current rate method is required by GAAP because the FASB felt that other methods of translation reported financial results on an 'as if' basis (i.e. as if the foreign-based operations had been conducted in the parent company's environment). The current rate method preserves the separate relationships, both those existing in the foreign environment and those of the parent company.

4.3.3 Temporal method

Under certain circumstances, prior to translating the financial statements of a foreign subsidiary into USD-denominated financial statements, the financial statements of the subsidiary must be remeasured through the use of the temporal method. In general these circumstances relate to situations in which the functional currency of the subsidiary is already the USD, or situations in which the foreign subsidiary is located in a country with a high rate of inflation. Under the temporal method, monetary assets (i.e. cash, marketable securities and receivables) and liabilities are translated at the current exchange rate. Non-monetary assets and shareholders' equity accounts are translated at historical rates. Income statement accounts are translated at the average rate for the period, except for those expenses that derive from non-monetary assets such as depreciation (i.e. derived from property, plant and equipment). Expenses derived from non-monetary assets are translated at historical rates. Dividends of the subsidiary are translated at the exchange rate at the date of the dividend declaration. Under the temporal method, the remeasurement adjustment is reflected as a component of income from continuing operations for the period.

Appendix B to SFAS No. 52 provides examples of the following non-monetary accounts which should be remeasured using historical exchange rates:

- Marketable securities carried at cost.
- Inventories carried at cost.
- Prepaid expenses such as insurance and rent.
- Property, plant, and equipment.
- Accumulated depreciation on property, plant, and equipment.
- Patents, trademarks, licences and formulas.
- Goodwill.
- Other intangible assets.
- Deferred charges and credits.

- Deferred income.
- Common shares.
- Preferred shares carried at issuance price.
- Cost of goods sold.
- Depreciation on property, plant and equipment.
- Amortization of intangible items such as goodwill.
- Amortization of deferred charges or credits.

4.3.4 Positive and negative translation differences

As discussed above, the treatment of translation differences depends on the functional currency of the foreign subsidiary. If the functional currency is different from that of the parent company and if the foreign subsidiary is not located in a country with a high rate of inflation, translation differences which result from the use of the current rate method are carried to a separate component of shareholders' equity and do not enter into the computation of earnings for the period. On the other hand, if the functional currency of the foreign subsidiary is the USD or if the foreign subsidiary is located in a country with a high rate of inflation, the translation differences that result from the use of the temporal method to remeasure the financial statements of the foreign subsidiary enter into the computation of earnings for the period.

4.3.5 Translation of profit and loss account

Under SFAS No. 52 there is no independent translation of profit. Revenues and expenses are translated at the average rate during the accounting period. The net difference between translated revenues and translated expenses is the profit for the period.

4.3.6 Different methods for different foreign subsidiaries

It is possible for one subsidiary to have a functional currency which is a local currency and another subsidiary to have a functional currency that is the USD. In this case two different methods would be used to translate or remeasure the financial statement for these subsidiaries.

4.3.7 Accounting for changes in the functional currency

Once the functional currency for a foreign entity is determined, that determination shall be used consistently unless significant changes in economic facts and circumstances indicate clearly that the functional currency has changed. Previously issued financial statements shall not be restated for any change in the functional currency (SFAS No. 52, paragraph 9).

No guidance is provided in SFAS No. 52 regarding how to identify the 'significant changes in economic facts and circumstances' that indicate clearly that the functional currency has changed. Presumably, changes in the functional currency should be rare. If such a change does occur, however, paragraph 45 indicates that the change is not a change in accounting principle. Paragraph 46 provides guidance on how to account for a change in the functional currency. It states that: If the functional currency changes from a foreign currency to the reporting currency, translation adjustments for prior periods should not be removed from equity and the translated amounts for nonmonetary assets at the end of the prior period become the accounting basis for those assets in the period of the change and subsequent periods.

That says that when a change in the functional currency from the foreign currency to the reporting currency occurs, the change is not a retroactive one. Rather, the accounting basis of the assets and liabilities affected by the change carries forward at the amounts translated as of the date of the change and is not altered because of the change. The exchange rate on the date of the change becomes the historical rate for subsequent translations.

On the other hand, if the change in functional currency is from the reporting currency to the foreign currency, SFAS No. 52, paragraph 46 states that: the adjustment attributable to current-rate translation of non-monetary assets as of the date of the change should be reported in the cumulative translation adjustments component of equity.

This means that when a change in the functional currency from the reporting currency to the foreign currency occurs, the change is a retroactive one. The accounting basis of the assets and liabilities affected by the change (e.g. fixed assets translated at the historical exchange rate before the change and at the current exchange rate after the change) is adjusted retroactively to reflect the difference between the exchange rate when the asset or liability arose and the exchange rate on the date of the change in functional currency. This effect is just like the effect of the adoption of SFAS No. 52 initially, and the cumulative adjustment is reported as of the date of the change.

4.4 Foreign currency translation for subsidiaries in highly inflationary economies

4.4.1 Concept

The temporal method is used when a subsidiary is located in a country with high rates of inflation. When inflation exceeds 100 per cent over the most recent three-year period, measured cumulatively over the period, the temporal method must be used. The cumulative inflation for three years is a compounded rate, not the sum of the annual rates. If the annual inflation rates for three consecutive years are known, to determine cumulative inflation for those three years, one must: (i) add 1.00 to each year's rate (expressed as a decimal); (ii) multiply the three rates together; and (iii) deduct 1.00 from the product. For example, assume that inflation rates for 1995, 1996 and 1997 were 3.6 per cent, 1.9 per cent and 3.7 per cent, respectively. The cumulative inflation for those three years is 9.5 per cent (1.036 x 1.019 x 1.037) – 1 = .095 or 9.5 per cent). An annual inflation rate of approximately 26 per cent will result in cumulative inflation of 100 per cent over three years.

4.4.2 Criteria to determine a highly inflationary economy

EITF Topic No. D-55, 'Determining a Highly Inflationary Economy under FASB Statement No. 52' states in part:

> The FASB staff believes the determination of a highly inflationary economy must begin by calculating the cumulative inflation rate for the three years that precede the beginning of the reporting period, including interim reporting periods. If that calculation results in a cumulative inflation rate in excess of 100 percent, the economy should be considered highly inflationary in all instances. However, if that calculation results in the cumulative rate being less than 100 percent, the staff believes that historical inflation rate trends (increasing or decreasing) and other pertinent economic factors should be considered to determine whether such information suggests that classification of the economy as highly inflationary is appropriate. The staff believes that projections of future inflation rates were not contemplated by the language in paragraph 109 and thus projections cannot be used to overcome the presumption that an economy is highly inflationary if the 3-year cumulative rate exceeds 100 percent.

4.4.3 Changes from highly inflationary to not highly inflationary

A consistent means of measurement should be applied to the determination of a highly inflationary designation on a year to year basis; that is an economy should not be considered highly inflationary one year but not the next.

Examples:

- Mexico's cumulative inflation rate for the three periods ended November 1998, calculated by reference to the index published by the Mexican Central Bank, is 71.1 per cent. This information continues to support the conclusion that Mexico is not a highly inflationary economy as of 1 January 1999.
- Brazil's inflation statistics are derived from the actual inflation rates from the Brazilian IGP (an inflation index widely used in Brazil). The IGP three-year cumulative rates as of December 31 1997 and 1998 are 35.6 per cent and 19.7 per cent, respectively. Accordingly, entities should have discontinued treating Brazil as a highly inflationary economy for periods after 31 December 1997.

When a designation changes from highly inflationary to not highly inflationary, EITF Issue No. 92-4, 'Accounting for a Change in Functional Currency When an Economy Ceases to Be Considered Highly Inflationary' and EITF Issue No. 92-8, 'Accounting for the Income Tax Effects under FASB Statement No. 109 of a Change in Functional Currency When an Economy Ceases to Be Considered Highly Inflationary' provide guidance for converting functional currency from the U.S. Dollar to the local currency if the local currency is determined to be the appropriate functional currency. The Task Force reached a consensus that paragraph 46 of SFAS No. 52 does not apply to this type of change and that the entity should restate the functional currency accounting bases of non-monetary assets and liabilities at the date of change as follows: the reporting currency amounts at the date of change should be translated into the local currency at current exchange rates and those amounts should become the new functional currency accounting basis for the non-monetary assets and liabilities.

4.5 Special problems

4.5.1 Statement of cash flows

SFAS No. 95, paragraph 25, Statement of Cash Flows, requires companies with foreign subsidiaries to report in the consolidated statement of cash flows the equivalent of foreign currency cash flows using exchange rates in effect at the time of those cash flows or using an appropriately weighted average exchange rate for the period, provided that the results are substantially the same as if actual rates were used. These rules apply whether the functional currency is local currency or the reporting currency. In either case the local foreign currency cash flows are translated to USD using the exchange rate in effect at the time of the cash flows or appropriately weighted exchange rates. Exchange rate changes, whether recognized in the income statement or as a separate component of stockholders' equity, do not give rise to cash flows. Therefore there will be no effect on net cash flow for the period.

4.5.2 Elimination of inter-company profits

At least two alternatives exist for determining the amount of inter-company profit to be eliminated. One method, the one selected by the Board, bases the calculation on the exchange rate as of the date the transaction occurred that gave rise to the inter-company profit (SFAS No. 52, paragraph 25). The other bases the calculation on the exchange rate as of the date the asset (e.g. inventory) or related expense (e.g. cost of sales) that resulted from the inter-company transaction is translated. The FASB decided that inter-company profit arises on the date of the inter-company transaction and that subsequent changes in the exchange rate do not affect that profit.

4.5.3 Sale of foreign subsidiaries

Upon sale or upon complete or substantially complete liquidation of an investment in a foreign entity, the amount attributable to that entity and accumulated in the translation adjustment component of equity shall be removed from the separate component of equity and shall be reported as part of the gain or loss on sale or liquidation of the investment for the period during which the sale or liquidation occurs (SFAS No. 52, paragraph 14).

4.6 Disclosure

SFAS No. 52 requires disclosure of the following items:

- The aggregate transaction gain or loss included in determining net income, including gains or losses related to forward exchange contracts. Certain enterprises, such as banks, that deal in foreign currency may include related transaction gains and losses along with their dealer gains and losses.
- Effects, if significant, of exchange rate changes that occur after the balance sheet date and their effect on unsettled foreign currency balances.
- Cumulative translation adjustment displayed as a separate component of equity together with an analysis of changes during the period, including:
  - Beginning and ending amount of cumulative translation adjustments.
  - Aggregate adjustment for the period resulting from translation adjustments and gains and losses from hedging transactions and inter-company foreign currency transactions which are of a long-term nature.
  - Amount of income taxes for the period allocated to translation adjustments.

- Amounts transferred from the account resulting from the sale or complete or substantially complete liquidation of an investment in a foreign entity.

In addition, the SEC requires disclosures in management's discussion and analysis (Financial Reporting Release (FRR) No. 6, Disclosure Considerations Related to Foreign Operations and Foreign Currency Translation Effects):

- A display of the net investments by major functional currency.
- An analysis of the translation component of equity either by functional currency or on a geographic basis.
- Narrative information and functional currencies used to measure significant foreign operations.
- The degree of exposure to exchange rate risks.
- The nature of the translation component.

VI Full consolidation

1 Consolidation of capital

1.1 Concept

Consolidation of capital is subject to the method of accounting for business combinations. The two methods of accounting for business combinations are the pooling-of-interests method and the purchase method.

Under the purchase method, a business combination is accounted for as the acquisition of one enterprise ('acquired enterprise') by another ('acquiring enterprise'). The acquiring enterprise records the assets acquired and liabilities assumed at its cost. The difference between the cost of an acquired enterprise and the sum of the fair values of tangible and identifiable intangible assets acquired less liabilities assumed is recorded as goodwill. The income of an acquiring enterprise includes the operations of the acquired enterprise subsequent to the acquisition. Intercompany investments in equity securities of subsidiaries are eliminated and shares of a parent held by a subsidiary are not reported as outstanding shares in consolidated financial statements. The non-controlling or minority interest in the shares of the subsidiary is reflected in the consolidated financial statements as a credit balance in the consolidated balance sheet. However, this credit balance is not considered to be part of the capital of the consolidated group. Likewise, the share of retained earnings (i.e. retained surplus or retained capital) of the minority share holding interest is not considered to be part of the consolidated retained earnings and is therefore not available for dividends.

Under the pooling-of-interests method, a business combination is accounted for as the uniting of the ownership interests of two or more enterprises through an exchange of equity interests. No acquisition is recognized because the combination is accomplished without disbursing resources of the combining enterprises. Ownership interests in the combining enterprises continue in the combined enterprise, and the former bases of accounting in the combining enterprises are carried over to the new combined enterprise. The assets and liabilities of the combining enterprises are carried forward to the combined enterprise at their recorded amounts. The income of the combining enterprises for the period of the combination and prior periods is combined and reported as income of the combined enterprise. The capital accounts of the combining companies would be joined together. The paid-in capital or paid-in surplus and the retained earnings accounts of the combining companies would be added together and would be available for the payment of dividends in accordance with the laws of the state in which the companies operate.

1.2 Applicable method

The purchase and pooling-of-interests methods are not alternatives in accounting for the same business combination. Twelve specific conditions must be met for pooling-of-interests accounting to be appropriate. A business combination that meets all 12 conditions must be accounted for as a pooling of interests. A business combination that does not meet all of the specific pooling conditions must be accounted for as a purchase. An entire business combination must be accounted for by a single method. Accounting for a combination as part-pooling, part-purchase is not acceptable.

All business combinations were treated as purchases before pooling-of-interests accounting was created as an acceptable alternative in the 1940s. Although pooling-of-interests accounting was widespread throughout the 1960s, the issuance in 1970 of APB Opinion No. 16 significantly restricted the use of pooling-of-interests. Accounting Trends and Techniques reports that out of 316 business combinations in 1997, 38 were recorded as poolings-of-interests and 278 were recorded as using the purchase method (AICPA, 1998, p. 41).

1.3 Purchase method

1.3.1 Determining the acquiring enterprise

APB Opinion No. 16, paragraph 70 provides the following guidance on determining the acquiring enterprise in a purchase business combination:

> A corporation which distributes cash or other assets or incurs liabilities to obtain the assets or stock of another company is clearly the acquirer. The identities of the acquirer and the acquired company are usually evident in a business combination effected by the issue of stock. The acquiring corporation normally issues the stock and commonly is the larger company. The acquired company may, however, survive as the corporate entity, and the nature of the negotiations sometimes clearly indicates that a smaller corporation acquires a larger company. The Board concludes that presumptive evidence of the acquiring corporation in combinations effected by an exchange of stock is obtained by identifying the former common stockholder interests of a combining company which either retain or receive the larger portion of the voting rights in the combined corporation. That corporation should be treated as the acquirer unless other evidence clearly indicates that another corporation is the acquirer. For example, a substantial investment of one company in common stock of another before the combination may be evidence that the investor is the acquiring corporation.

In addition, one of the existing combining enterprises would be considered the acquiring enterprise on the basis of available evidence when a new enterprise is formed to issue stock to effect a purchase business combination between two or more existing enterprises (APB Opinion No. 16, paragraph 71).

1.3.1.1 Reverse acquisitions

The acquiring enterprise in a purchase business combination that is effected through an exchange of stock generally is determined by identifying the combining enterprise whose stockholders retain or receive the larger portion of the voting rights in the combined enterprise. In a 'reverse acquisition', one enterprise (Company A) purchases another enterprise (Company B) for stock or stock and cash, and the stockholders of the acquired enterprise (Company B) receive the majority of the voting interests in the surviving combined enterprise (Company A). Company B generally would be deemed to be the acquiring enterprise for financial reporting purposes (accounting acquirer). Therefore, all of the assets and liabilities of Company A (legal acquirer) would be recorded at fair value as required by the purchase method and the operations of Company A would be reflected in the operations of the combined enterprise from the date of acquisition. Accordingly, the post-acquisition net assets of the surviving combined enterprise (Company A) would include the historical cost of the net assets of Company B (accounting acquirer) plus the fair value of the net assets of Company A (legal acquirer).

Example:
Company A issues 250 shares of its common stock for all outstanding common shares of Company B.

	Company A	Company B
Number of common shares outstanding	150	800
Book value	USD100,000	USD100,000
Fair value	USD300,000	USD500,000

Since the former stockholders of Company B will own 62.5 per cent of the outstanding shares of Company A after the acquisition (250 shares of the 400 shares outstanding), Company B is considered the acquirer for accounting purposes. Accordingly, the net assets of Company A after the acquisition should consist of the net assets of Company B at historical cost (USD100,000) plus the net assets of Company A at fair value (USD300,000). As such, the cost of the acquisition (USD300,000) should be allocated to the net assets of Company A. There would be a complete (i.e. 100%) change in the accounting basis of Company A's assets and liabilities even though Company B's shareholders hold only 62.5 per cent of the outstanding shares of the combined enterprise after the acquisition.

1.3.1.2 Presentation of reverse acquisitions in the financial statements of the combined enterprise

In a purchase business combination, the assets and liabilities of the acquired enterprise are recorded at fair value by the acquiring enterprise and the results of operations of the acquired enterprise from the date of the acquisition are included in the financial statements of the acquiring enterprise. That guidance also applies to reverse acquisitions in which the acquired enterprise is the surviving enterprise. The historical financial statements of the acquiring enterprise (accounting acquirer) are presented as the historical financial statements of the combined enterprise and the assets and liabilities of the acquired enterprise (legal acquirer) are accounted for as required by the purchase method of accounting without regard to which enterprise is the surviving enterprise. The results of operations of the acquired enterprise (legal acquirer) are included in the financial statements of the combined enterprise only from the date of acquisition even if the acquired enterprise (legal acquirer) is the surviving enterprise. The equity of the accounting acquirer is presented as the equity of the combined enterprise; however, the capital stock account of the accounting acquirer is adjusted to reflect the par value of the outstanding stock of the legal acquirer after giving effect to the number of shares issued in the business combination. The difference between the capital stock account of the accounting acquirer and the capital stock account of the legal acquirer (presented as the capital stock account of the combined enterprise) is recorded as an adjustment to additional paid-in capital of the combined enterprise. For periods prior to the business combination, the equity of the combined enterprise is the historical equity of the accounting acquirer prior to the merger retroactively restated to reflect the number of shares received in the business combination. Retained earnings of the accounting acquirer are carried forward after the acquisition. Earnings per share for periods prior to the business combination are restated to reflect the number of equivalent shares received by the accounting acquirer.

Example:
Company A issues common stock to the stockholders of Company B in exchange for all outstanding common stock of Company B. The business combination does not qualify for pooling-of-interests accounting. Upon consummation of the combination, Company B becomes a wholly-owned subsidiary of Company A and the former stockholders of Company B own 60 per cent of the outstanding common stock of Company A.

Since the former stockholders of Company B receive the larger portion of the voting interests in the combined enterprise, Company B is presumed to be the acquiring enterprise. There is no evidence that would overcome that presumption.

For purposes of the consolidated financial statements of the surviving enterprise (Company A), Company B's historical financial statements should be presented as the historical financial statements of the combined enterprise. The purchase method of accounting should be applied to the assets and liabilities of Company A as of the acquisition date. The results of operations of Company A are included in the financial statements of the combined enterprise only for periods subsequent to the acquisition.

Although the equity of Company B (the acquiring enterprise) is presented as the equity of the combined enterprise, the capital stock account must be adjusted to reflect the outstanding stock of Company A (the surviving enterprise). That is, the retained earnings of Company B would be presented as the retained earnings of the combined enterprise; the capital stock account of the combined enterprise would reflect the par value of Company A's stock; and the additional capital account of Company B would be adjusted for the difference between the capital stock account of Company B and the capital stock account of Company A, and that adjusted amount would be presented as additional capital of the combined enterprise.

1.3.2 Significant dates in a purchase business combination

1.3.2.1 Date to value securities issued ('valuation date')

The date to value securities issued in a purchase business combination generally is significant only when there are significant changes in the value of the acquiring enterprise's securities between the date the terms of the purchase agreement are agreed to and announced and the date that the business combination is consummated. In Issue 95-19, Determination of the Measurement Date of the Market Price of Securities Issued in a Purchase Business Combination, the EITF addressed the issue of what date should be used to value marketable equity securities issued to effect a business combination accounted for by the purchase method:

> The Task Force reached a consensus that the value of marketable equity securities issued to effect a purchase business combination should be determined, pursuant to the guidance in paragraph 74 of Opinion 16, based on the market price of the securities over a reasonable period of time before and after the two companies have reached agreement on the purchase price and the proposed transaction is announced. In other words, the date of measurement of the value of the marketable equity securities should not be influenced by the need to obtain shareholder or regulatory approvals. Task Force members observed that the reasonable period of time referred to in paragraph 74 of Opinion 16 is intended to be very short, such as a few days before and after the acquisition is agreed to and announced. Task Force members also observed that in transactions involving a hostile tender offer, the measurement date for the value of the marketable equity securities occurs when the proposed transaction is announced and sufficient shares have been tendered to make the offer binding or when the proposed acquisition becomes nonhostile, as evidenced by the target company's agreement to the purchase price.

1.3.2.2 Date for allocating the purchase price ('acquisition date' or 'consummation date')

APB Opinion No. 16, paragraph 93 provides the following guidance regarding the acquisition date of a purchase business combination:

> The Board believes that the date of acquisition of a company should ordinarily be the date assets are received and other assets are given or securities are issued. However, the parties may for convenience designate as the effective date the end of an accounting period between the dates a business combination is initiated and consummated. The designated date should ordinarily be the date of acquisition for accounting purposes if a written agreement provides that effective control of the acquired company is transferred to the acquiring corporation on that date without restrictions except those required to protect the stockholders or other owners of the acquired company – for example, restrictions on significant changes in the operations, permission to pay dividends equal to those regularly paid before the effective date, and the like. Designating an effective date other than the date assets or securities are transferred requires adjusting the cost of an acquired company and net income otherwise reported to compensate for recognizing income before consideration is transferred. The cost of an acquired company and net income should therefore be reduced by imputed interest at an appropriate current rate on assets given, liabilities incurred, or preferred stock distributed as of the transfer date to acquire the company.

1.3.2.3 *Date to account for the acquisition ('effective date')*

The acquisition date of a purchase business combination generally is the date that assets are received and securities or other assets are distributed. However, for convenience, the end of an accounting period that falls between initiation and consummation of the combination may be designated as the effective date of the combination. To use a date prior to the acquisition date (unless the effect of doing so is insignificant), the following conditions should be met:

- The parties reach a firm purchase agreement that includes specifying the date of acquisition other than the closing date.
- Effective control of the acquired enterprise, including the risks and rewards of ownership, transfers to the acquiring enterprise as of the designated effective date.
- The time period between the designated effective date and the closing date is relatively short (say, less than 30 days), except if the delay is caused by obtaining required regulatory approval.
- The designated effective date and the closing date fall in the same accounting period (i.e., same interim and annual accounting period).

1.3.3 Determining cost of an acquired enterprise

In accordance with APB Opinion No. 16, paragraph 67, in a purchase business combination the acquiring corporation records at its 'cost' (determined either by the fair value of the consideration given or by the fair value of the property acquired, whichever is more clearly evident) the acquired assets less liabilities assumed.

1.3.3.1 *Direct costs of acquisition*

Only direct costs of a purchase business combination are included in the cost of the acquired enterprise. Costs included in the cost of the acquired enterprise should be limited to direct out-of-pocket costs or incremental costs that are directly related to the purchase business combination. Internal costs should not be included in the cost of the acquired enterprise even if the costs are directly related to a purchase business combination (APB Opinion No. 16, paragraph 76).

Examples of direct costs of a purchase business combination that are included in the cost of the acquired enterprise include finder's fees or other fees paid to outside consultants for accounting, legal, or engineering investigations or for appraisals. Finder's fees and bonuses awarded to employees of the acquiring enterprise would not be included in the cost of the acquired enterprise.

1.3.3.2 *Acquisition costs incurred by acquired enterprise*

While direct acquisition costs incurred by an acquiring enterprise in a purchase business combination are included in the cost of the acquired enterprise, direct acquisition costs incurred by an acquired enterprise, or its major or controlling stockholders, generally would not be included as part of the cost of the acquired enterprise. Acquisition costs incurred by the acquired enterprise are presumed to be considered by the acquiring enterprise in setting the purchase price. If, however, the acquiring enterprise agrees to reimburse the acquired enterprise's major or controlling stockholders for acquisition costs incurred by them, the acquiring enterprise would include those reimbursements in the cost of the acquired enterprise. Direct acquisition costs incurred by an acquired enterprise in a purchase business combination should be charged to expense as incurred without regard to whether negotiations for a combination are successful or unsuccessful.

1.3.3.3 *Equity securities issued to acquire an enterprise*

The quoted market price of an equity security issued to effect a business combination may usually be used to approximate the fair value of an acquired enterprise after recognizing possible effects of price fluctuations, quantities traded, issue costs, and the like. The market price for a reasonable period before and after the date the terms of the acquisition are agreed to and announced shall be considered in determining the fair value of securities issued (APB Opinion No. 16, paragraph 74). The quoted market price of stock issued to effect a purchase business combination may need to be adjusted to determine the fair value of that stock.

For example, quoted market price may need adjustment for the following:

- Stock issued to effect the combination is not actively traded.
- Stock issued to effect the combination contains restrictions that limit its marketability.
- The number of shares of stock issued to effect the combination substantially exceeds the current trading volume of the shares in the marketplace (commonly referred to as 'blockage').

In those situations the advice of an investment banker or broker should be sought in determining the appropriate adjustment and the resultant fair value of the stock issued.

Options issued by the acquiring enterprise in a purchase business combination to acquire common stock of the acquired enterprise are valued at the valuation date and included in the cost of the acquired enterprise. The fair value of the options may be estimated with the assistance of an investment banker or broker or using an option-pricing model.

Example:
Company A acquired all the outstanding stock of Company B in a purchase business combination. In connection with the acquisition, Company A issued common stock and options to acquire shares of Company A's common stock to the shareholders of Company B in exchange for the common stock of Company B. The options are exercisable over the next five years. The exercise price of the options is in excess of the quoted market price of Company A's common stock (i.e. the options are out-of-the-money).

The fair value of these options should be included in Company A's cost of acquiring Company B because the options are issued as consideration for the acquisition of the outstanding stock Company B. The fair value of the options may be estimated with the assistance of an investment banker or broker or using an option-pricing model.

1.3.3.4 Golden parachutes

The acquiring enterprise in a purchase business combination may agree to assume a liability for amounts related to the exercise of 'golden parachutes' by executives of the acquired enterprise. If the liability for the golden parachutes is assumed by the acquiring enterprise as part of the plan of combination, that liability should be included in the cost of the acquired enterprise.

Amounts payable under golden parachutes are obligations of the acquired enterprise because golden parachutes are executory contracts between the executives of the acquired enterprise and the acquired enterprise. Therefore, those costs would be charged to expense in the separate financial statements of the acquired enterprise. Consistent with the SEC staff's position on the settlement of stock options and awards (see EITF Issues No. 84-13 and 85-45), the acquired enterprise should record reimbursements received from the acquiring enterprise for the settlement of the golden parachutes as capital contributions. The acquiring enterprise would treat those reimbursements as part of the cost of the acquired enterprise.

1.3.3.5 Settlement of litigation related to a purchase business combination

Stockholders of an acquired enterprise may bring litigation against the acquiring enterprise in an attempt to obtain additional consideration for their interests in the acquired enterprise. The former shareholders of the acquired enterprise may allege that they were not fully informed of all relevant facts at the consummation date and they did not receive fair consideration in exchange for their interests in the acquired enterprise. To avoid litigation, the acquiring enterprise may offer the former stockholders of the acquired enterprise additional consideration for their interests in the acquired enterprise. Although it appears reasonable that such consideration could be considered additional cost of the acquired enterprise, the SEC staff has taken the position that the additional consideration given to the former shareholders of the acquired enterprise is a payment to avoid litigation and should be charged to expense when the settlement is reached.

1.3.3.6 Contingent consideration

Additional consideration in a purchase business combination that is contingent on maintaining or achieving specified earnings levels in future periods generally is not recorded until the contingency is resolved and the additional consideration is distributable unless the outcome of the contingency is determinable beyond reasonable doubt. When the contingency is resolved and the additional consideration is issued or issuable or the outcome of the contingency is determinable beyond reasonable doubt, the acquiring enterprise should record the current fair value of the additional consideration as additional cost of the acquired enterprise. The additional cost should be allocated to the appropriate assets, usually goodwill, and amortized over the remaining lives of the assets.

Example:
Company A issued 10,000 shares of its common stock to acquire Company B in a purchase business combination in December 1993. The market price of Company A's common stock at the date of the purchase was USD10 per share. The purchase agreement required that Company A issue 1,000 additional shares of its common stock to the former stockholders of Company B if the earnings of Company B were at least USD125,000 in 1994 and USD150,000 in 1995.

Company A recorded the acquisition of Company B in December 1993 at USD100,000 (10,000 shares at USD10 per share) and allocated the cost of USD100,000 to the net assets acquired, including goodwill. The goodwill related to the acquisition is being amortized over 20 years.

Company B's earnings were USD135,000 in 1994 and USD170,000 in 1995. Accordingly, Company A issued an additional 1,000 shares of its common stock to the former stockholders of Company B in early 1996. The market price of Company A's common stock at the time that the additional shares became issuable was USD12 per share. Company A should record the additional consideration of USD12,000 (1,000 shares at USD12 per share) as additional goodwill related to the acquisition. The additional goodwill should be amortized over the remaining life of the goodwill (18 years).

1.3.4 Recording assets acquired and liabilities assumed

1.3.4.1 Concept

APB Opinion No. 16, paragraph 87 provides the following guidance on recording assets acquired and liabilities assumed in a purchase business combination:

> First, all identifiable assets acquired, either individually or by type, and liabilities assumed in a business combination, whether or not shown in the financial statements of the acquired company, should be assigned a portion of the cost of the acquired company, normally equal to their fair values at date of acquisition.
>
> Second, the excess of the cost of the acquired company over the sum of the amounts assigned to identifiable assets acquired less liabilities assumed should be recorded as goodwill. The sum of the market or appraisal values of identifiable assets acquired less liabilities assumed may sometimes exceed the cost of the acquired company. If so, the values otherwise assignable to noncurrent assets acquired (except long-term investments in marketable securities) should be reduced by a proportionate part of the excess to determine the assigned values. A deferred credit for an excess of assigned value of identifiable assets over cost of an acquired company (sometimes called 'negative goodwill') should not be recorded unless those assets are reduced to zero value.
>
> Independent appraisals may be used as an aid in determining the fair values of some assets and liabilities. Subsequent sales of assets may also provide evidence of values.

1.3.4.2 Cost allocation to individual assets and liabilities

The allocation of the cost of an acquired enterprise in a purchase business combination should be based on the fair value of the identifiable assets acquired and liabilities assumed. Current Text Section B50.146 (FASB, 1998c) provides the following guidance on assigning amounts to individual assets acquired and liabilities assumed in a purchase business combination:

- Marketable securities at current net realizable values [fair value].
- Receivables at present values of amounts to be received determined at appropriate current interest rates, less allowances.
- Inventories:
  (i) Finished goods and merchandise at estimated selling prices less the sum of (a) costs of disposal; and (b) a reasonable profit allowance for the selling effort of the acquiring enterprise.
  (ii) Work in process at estimated selling prices of finished goods less the sum of (a) costs to complete; (b) costs of disposal; and (c) a reasonable profit allowance for the completing and selling effort of the acquiring enterprise based on profit for similar finished goods.
  (iii) Raw materials at current replacement costs.
- Plant and equipment:
  (i) To be used, at the current replacement cost for similar capacity unless the expected future use of the assets indicates a lower value to the acquirer, and
  (ii) To be sold, at fair value less cost to sell.
- Intangible assets that can be identified and named, including contracts, patents, franchises, customer and supplier lists, and favorable leases, at appraised values.
- Other assets, including land, natural resources, and non-marketable securities, at appraised values.
- Accounts and notes payable, long-term debt, and other claims payable at present values of amounts to be paid determined at appropriate current interest rates.
- Liabilities and accruals – for example, accruals for warranties, vacation pay, deferred compensation – at present values of amounts to be paid determined at appropriate current interest rates.
- Other liabilities and commitments, including unfavourable leases, contracts, and commitments and plant closing expense incident to the acquisition, at present values of amounts to be paid determined at appropriate current interest rates.
- Pension liabilities:
  (i) defined benefit pension plan, a liability for the projected benefit obligation in excess of plan assets or an asset for plan assets in excess of the projected benefit obligation is recorded,
  (ii) defined benefit post-retirement plan, a liability for the accumulated post-retirement benefit obligation in excess of the fair value of the plan assets or an asset for the fair value of the plan assets in excess of the accumulated post-retirement benefit obligation is recorded.

It should be noted that the above guidance cannot be circumvented by specifying values in the purchase agreement.

1.3.4.3 Present values of receivables and payables

Receivables acquired and payables assumed in purchase business combinations are to be recorded at the present values of amounts to be received and amounts to be paid using appropriate current interest rates. However, discounting may not be necessary when the receivables are to be recovered or the payables are to be settled within a short period of time, such as within several months, provided that the difference between the present values and the gross amounts of the receivables or payables is not significant. When receivables acquired or payables assumed in a purchase business combination are recorded at present values, the acquiring enterprise should recognize interest income or expense related to the receivables or payables in periods subsequent to the acquisition. In addition, allowances for uncollectibility and collection costs should be considered in assigning values to receivables acquired in a purchase business combination.

1.3.4.4 Valuation of inventories

The value added to the inventories by the acquired enterprise during the manufacturing process prior to the acquisition ('manufacturing profit') must be considered in assigning values to inventories acquired in a purchase business combination. Finished goods inventory of the acquired enterprise that remains on hand at balance sheet dates of the acquiring enterprise subsequent to the date of acquisition generally are assigned values in excess of current replacement costs. Current Text Section I78.111 (FASB, 1998c) states that no write-down of the acquired enterprise's inventory by the acquiring enterprise at balance sheet dates following the date of acquisition is necessary if evidence exits that the acquiring enterprise will recover the value assigned to the inventories at the date of acquisition plus a 'normal profit'. If the acquired enterprise's inventories were valued using the LIFO method prior to the acquisition, the allocation of the purchase price to the inventories generally will result in a significant increase in the carrying value of the inventories.

1.3.4.5 Purchased research and development activities of the acquired enterprise

A method that companies use to achieve the objective of minimizing the effect of goodwill in a purchase business combination is through the allocation of the cost of an acquired enterprise to in-process research and development (IPR&D) activities of the acquired enterprise. However, the SEC challenges such accounting if a registrant's valuation of IPR&D is materially misleading and has issued a letter to the AICPA SEC Regulations Committee expressing its view on a number of situations that the SEC has observed with respect to IPR&D.

The applicable accounting rule, FASB Interpretation No. 4, 'Applicability of FASB Statement No. 2 to Business Combinations Accounted for by the Purchase Methods', requires that 'costs assigned to assets to be used in a particular research and development project and that have no alternative future use shall be charged to expense at the date of consummation of the combination' (a purchase business combination). Although it is clear that IPR&D must be expensed in the recording of a purchase business combination, it is becoming increasingly controversial as to whether the correct amount of the purchase cost is being allocated to IPR&D.

1.3.4.6 Identifiable intangible assets

APB Opinion No. 16 requires separate recognition of the value of identifiable intangible assets, such as contracts, patents, franchises, customer and supplier lists, and favourable leases which should not be included in goodwill. Thus, amounts should be assigned to all identifiable intangible assets and allocated as part of the purchase price. It may be necessary to obtain independent appraisals from an investment adviser or broker in determining the fair values of identifiable intangible assets. Other identifiable intangible assets may include trademarks, copyrights, licences, and those resulting from research and development activities. The useful life of identifiable intangibles is determined typically by statute or contract and is usually less than the estimated life of goodwill. As a result, to minimize the annual amortization of acquired intangibles, a bias exists against a thorough and rigorous identification of specific intangible assets in a company's allocation of the purchase price. In addition, recent changes in the tax law allow the tax deductibility of purchased intangibles including goodwill over a single statutory period. Thus, companies in the US will no longer have a tax incentive to recognize specific intangibles separately from goodwill. Despite these factors, the SEC staff continues to expect registrants to comply with the requirements of APB Opinion No. 16 and fully identify acquired intangible assets.

1.3.4.7 Non-compete agreements

The acquiring enterprise in a purchase business combination may enter into non-compete agreements with former stockholders or employees of the acquired enterprise. The acquiring enterprise should recognize a liability in the allocation of the purchase price for the payments to be made under non-compete agreements with former stockholders and employees of the acquired enterprise based on the present value of the amounts to be paid determined using appropriate current interest rates.

The non-compete agreement is an identifiable intangible asset acquired in the combination. Accordingly, the acquiring enterprise should recognize an identifiable intangible asset in the allocation of the purchase price related to the non-compete agreement. As indicated above, identifiable intangible assets acquired in a purchase business combination should be recorded at their appraised values. The present value of the payments under non-compete agreements may approximate the fair value of the agreements; however, the fair value of the agreements must be evaluated based on the specific facts and circumstances involved.

1.3.4.8 Acquired property and equipment

Upon consummation of a purchase business combination, the acquiring enterprise is to assign values to the plant and equipment of the acquired enterprise in accordance with Current Text Section B50.146(d). Those values represent the new cost basis of the plant and equipment. Accordingly, the accumulated depreciation accounts of the acquired enterprise are not carried over to the acquiring enterprise. This guidance also applies to accumulated amortization relating to intangible assets acquired in a purchase business combination.

Footnote 12 to Current Text Section B50.146 indicates that replacement cost for plant and equipment of the acquired enterprise to be used by the acquiring enterprise may be approximated from the replacement cost of new plant and equipment less estimated accumulated depreciation. If that method is used to value the acquired plant and equipment, the acquiring enterprise should reflect only the net amount in the allocation of the purchase price; accumulated depreciation should not be reflected as of the acquisition date.

The allocation of purchase price to assets to be sold is addressed in EITF Issue No. 87-11 and 95-21.

1.3.4.9 Accounting for leases

Current Text Section L10.137 states that the 'classification of a lease [as a capital or operating lease] in accordance with the criteria of [Current Text Section L10, "Leases"] shall not be changed as a result of a business combination unless the provisions of the lease are modified'.

A portion of the purchase price of an acquired enterprise (i.e. the acquired enterprise is the lessee) should be assigned to operating leases of the acquired enterprise for property or equipment that currently could be leased for higher or lower amounts (referred to as 'favourable' and 'unfavourable' leases, respectively). Amounts assigned to favourable leases of the acquired enterprise should be recorded at their appraised values in the allocation of the purchase price. The present value of the excess of the current market rental over the contractual lease payments may provide an acceptable estimate of the fair value of the intangible asset related to a favourable lease. Amounts assigned to unfavourable leases of the acquired enterprise should be based on the present value of amounts to be paid determined at appropriate current interest rates (i.e. a discount rate that is commensurate with the risks involved). Accordingly, unfavourable leases should be recorded at the present value of the excess of the contractual lease payments over the current market rental.

Assets under capital leases acquired in a purchase business combination should be recorded in the allocation of the purchase price at current replacement costs in accordance with the guidance in Current Text Section B50.146(d) regarding plant and equipment. Capital lease obligations assumed in a purchase business combination should be recorded in the allocation of the purchase price at the present values of amounts to be paid using appropriate current interest rates in accordance with the guidance in Current Text Section B50.146(g) regarding long-term debt and other payables.

1.3.4.10 Debt of acquired enterprise

Debt assumed by an acquiring enterprise in a purchase business combination should be recorded at the present value of amounts to be paid, using an appropriate interest rate at the date of acquisition. The following factors are considered in determining the appropriate value to assign to debt of the acquired enterprise:

- Remaining term to maturity of the debt. For example, if the debt assumed has a remaining term to maturity of five years, the market interest rate for new debt with a five-year term to maturity should be considered in determining the appropriate interest rate to use in assigning a value to the debt in the purchase price allocation.
- Conversion features of the debt, if any. For example, if the debt is convertible into preferred or common stock, the market price of

the preferred or common stock into which the debt is convertible should be considered in determining the appropriate value to be assigned to the debt.

- Other terms of the debt instrument, such as prepayment penalties, call provisions, and debt covenants.
- Effective interest rates currently available to the acquiring enterprise for debt obligations with similar terms and features. The effective interest rates used in assigning values to debt assumed in a purchase business combination should be based on rates applicable to the acquiring enterprise rather than rates that would apply to the acquired enterprise.
- If the interest rate on debt assumed in a purchase business combination is dependent on the prime rate or another benchmark rate, but subject to a stated minimum and maximum rate, the limits should be evaluated in light of current market rate conditions at the date of acquisition.
- Estimated debt issuance costs to issue debt with similar terms should be considered in determining an appropriate discount rate to use in assigning a value to the debt. However, no value should be assigned to the unamortized debt issuance costs of the acquired enterprise.

1.3.4.11 Valuing acquired enterprise's pension asset or liability

At the date of acquisition, the acquired enterprise's projected benefit obligation and fair value of plan assets should be remeasured. The difference in those two amounts, whether it represents a net asset or a net liability, should be recorded in the allocation of the purchase price. The recognition of the net pension asset or liability in the allocation of the purchase price eliminates any previously existing unrecognized gain or loss, prior service cost, and transition asset or obligation related to the acquired enterprise's pension plan. Any planned restructuring, such as termination or curtailment, of the acquired enterprise's pension plan should be taken into consideration in remeasuring the acquired enterprise's projected benefit obligation at the date of acquisition.

1.3.4.12 Other liabilities of acquired enterprise

Other liabilities and commitments of the acquired enterprise should be recorded at the present values of amounts to be paid using appropriate current interest rates. Such liabilities and commitments would include accruals for warranties, compensated absences, deferred compensation, self-insurance reserves, and unfavorable contracts and commitments. The acquiring enterprise generally would recognize interest expense subsequent to the acquisition on liabilities that are recorded in the allocation of the purchase price at the present values of amounts to be paid.

1.3.4.13 Recognition of liabilities in connection with a purchase business combination

Direct costs of the acquisition are to be included in the cost of an enterprise acquired in a purchase business combination. Also, commitments and plant closing costs of the acquired enterprise that are incident to an acquisition are to be recorded in the allocation of the purchase price in a purchase business combination. Prior to the consensuses reached on EITF Issue 95-3, Recognition of Liabilities in Connection with a Purchase Business Combination, various costs related to restructuring and integrating an acquired enterprise frequently were included in the cost of an enterprise acquired in a purchase business combination. Issue 95-3 prohibits recording many of these costs as a cost of the acquired enterprise.

In Issue 95-3, the EITF reached a consensus that three types of anticipated costs that meet certain specific criteria would be recognized as liabilities assumed in a purchase business combination and included in the cost of the acquired enterprise:

- Cost of a plan to exit an activity of the acquired enterprise.
- Cost of a plan to involuntarily terminate employees of the acquired enterprise.
- Cost of a plan to relocate employees of the acquired enterprise.

Anticipated costs that do not meet the specific criteria in Issue 95-3 (such as most integration costs) would not be included in the cost of the acquired enterprise. Costs to be incurred by the combined (acquiring) enterprise that are not included in the cost of the acquired enterprise would be expensed or capitalized when incurred based on the nature of the item and the acquiring enterprise's accounting policy for those costs. Costs to be incurred by the combined (acquiring) enterprise that qualify as exit costs would be recognized as a liability assumed as of the consummation date and would be included in the cost of the acquired enterprise. Examples of costs that may qualify as exit costs based on the criteria set out in Issue 95-3 are as follows:

- A lease cancellation penalty paid to terminate a lease on an acquired enterprise's facility that will be closed.
- The cost of a leased acquired enterprise facility that will remain idle.
- The cost of the remaining non-cancellable term of the operating lease after operations of an acquired enterprise plant are ceased.
- Contractual obligations for performance associated with the closing of an acquired enterprise facility.
- The incremental increase in the workers' compensation, health and welfare costs related to claims for injuries and illnesses prior to the date of acquisition as a result of plans to close an acquired enterprise facility.

Examples of costs that would not qualify as exit costs based on the criteria set out in Issue 95-3 are as follows:

- Incremental increase in workers' compensation and health and welfare costs related to claims for injuries and illnesses that arise after the date of acquisition as a result of an exit plan to close a manufacturing plant of the acquired enterprise.
- Unfavourable overhead costs that will result from an exit plan to close a manufacturing plant of the acquired enterprise after outstanding customer orders have been filled.
- Incremental increase in warranty costs related to products sold after the date of acquisition as a result of an exit plan to abandon certain lines of business.

Costs resulting from a plan to involuntarily terminate or relocate employees of an acquired enterprise that meet the criteria would be recognized as a liability assumed as of the consummation date of the purchase business combination and included in the cost of the acquired enterprise. If all or some portion of the termination benefits or relocation costs are payable to employees of the acquired enterprise only if those employees continue to provide services to the combined enterprise until those employees are terminated ('stay bonuses'), those stay bonuses generally should be accounted for as payments for future services rather than recognized as a cost of the acquired enterprise.

1.3.4.14 Subsequent adjustments to liabilities recorded as part of a purchase business combination

Issue 95-3 establishes a one-year period from the date of acquisition for the adjustment of the costs covered by Issue 95-3. Adjustments during the one-year period would result in an adjustment to the cost of the acquired enterprise, usually increasing or decreasing goodwill. Increases to the costs covered by Issue 95-3 after the one-year period would be charged against earnings and decreases to the costs covered by Issue 95-3 would result in an adjustment to the cost of the acquired enterprise (usually decreasing goodwill).

1.3.4.15 Integration costs

Costs incurred to integrate an acquired enterprise into the operations of the acquiring enterprise generally would not meet the criteria in Issue 95-3 for recognition as a liability assumed as of the consummation date of the purchase business combination. Such costs should not be included in the cost of the acquired enterprise. Costs to be incurred by the combined (acquiring) enterprise that are not included in the cost of the acquired enterprise would be capitalized, or expensed as incurred, based on the nature of the item and the acquiring enterprise's accounting policy for these costs. However, plans to incur certain integration costs may impact the determination of the amount (fair value) of the purchase price to be allocated to the assets of the acquired enterprise.

Example:
Company A acquired Company B in a purchase business combination. Subsequent to the acquisition, Company A plans to replace Company B's computer software with the software used by Company A. The fair value of Company B's existing software and thus, the amount of the purchase price allocated to Company B's existing software, would be based on the decision to dispose of Company B's existing software. The costs of the new software purchased would be treated in accordance with Company A's accounting policy for purchased assets.

Examples of planned integration costs that would not be included in the cost of the acquired enterprise but may have an impact on the cost allocated to the acquired assets of the acquired enterprise are as follows:

- Costs to purchase new signs with acquiring enterprise's logo to replace existing acquired enterprise's signs.
- Costs to repaint acquired enterprise's delivery equipment.
- Costs to upgrade acquired enterprise's plant or store locations to meet specifications of the acquiring enterprise.
- Costs to upgrade acquired enterprise's computer software and hardware.
- Costs to purchase new computers for acquired enterprise's location.
- Costs to train acquired enterprise's employees.
- Costs to relocate acquired enterprise's inventory or machinery and equipment.

Examples of integration costs that would not be included in the cost of the acquired enterprise and generally would not have an impact on the cost allocated to the acquired assets of the acquired enterprise are as follows:

- Consulting fees to identify new combining (acquiring) enterprise goals.
- Advertising costs for a programme to announce the acquisition or to promote the new combined enterprise.

1.3.4.16 Reimbursements for operating losses

The former shareholders of an acquired enterprise may agree to reimburse the acquiring enterprise for operating losses generated by the acquired enterprise subsequent to the acquisition. Payments made by the former shareholders of the acquired enterprise to the acquiring enterprise as reimbursements of operating losses of the acquired enterprise in periods subsequent to the acquisition generally should be accounted for as adjustments of the purchase price. The acquiring enterprise generally should not recognize reimbursements for operating losses of the acquired enterprise in income.

1.3.4.17 Preliminary allocation of purchase price

Completion of the allocation process sometimes requires an extended period of time. For example, appraisals might be required to determine replacement cost of plant and equipment acquired, a discovery period may be needed to identify and value intangible assets acquired, and an actuarial determination may be required to determine the pension liability to be accrued. If a business combination is consummated towards the end of the acquiring enterprise's fiscal year or the acquired enterprise is very large or unusually complex, the acquiring enterprise may not be able to obtain some of the data required to complete the allocation of the cost of the acquired enterprise for inclusion in its annual financial statements. In that case, a tentative allocation might be made using the values that have been determined and preliminary estimates of the values that have not yet been determined. The portions of the allocation that relate to the data that were not available subsequently are adjusted to reflect the finally determined amounts, usually by adjusting the preliminary amount with a corresponding adjustment of goodwill.

1.3.4.18 Goodwill

Goodwill is to be recorded in a purchase business combination for an excess of the cost of the acquired enterprise over the total amount assigned to the identifiable assets acquired less liabilities assumed. Goodwill that previously was recorded by an acquired enterprise should not be recorded as a separate asset by the acquiring enterprise. The acquiring enterprise should allocate the purchase price only to the tangible and identifiable intangible assets acquired and liabilities assumed in the purchase business combination without regard to goodwill previously recorded by the acquired enterprise. Any excess of the cost of the acquired enterprise over the values assigned to the identifiable assets acquired less liabilities assumed should be recorded as goodwill of the current acquisition. Within one year from the consummation of a business combination, the goodwill can be adjusted for contingencies existing prior to the acquisition for which an asset, a liability, or an impairment of an asset can not be estimated (see VI.1.3.4.14). Accounting for goodwill arising in separate purchases of stock in a step acquisition is discussed in Current Text Sections I60.503 through .506. For each separate step acquisition of stock of a subsidiary that is consolidated or an investee that is accounted for by the equity method, the acquiring enterprise should determine the cost of each investment, the fair value of the net assets acquired, and the goodwill related to each investment. Goodwill related to each investment should be amortized systematically over the estimated benefit period for that investment. Goodwill arising from investments after the initial investment in a step acquisition is not required to be amortized over the remaining amortization period of goodwill related to the initial investment.

The current authoritative literature does not specifically address the income statement classification of goodwill amortization. However, we believe that amortization of intangible assets including goodwill should be classified as an operating expense in the statement of earnings because footnote 401 to Current Text Section S20, 'Segment of Business Reporting', requires that amortization and depreciation expense be classified as an operating charge in segment reporting.

Goodwill should be amortized by systematic charges to income over the periods estimated to be benefited, not exceeding 40 years (APB 17.29). Current Text Sections B50 and I60 do not specify a minimum period of amortization of goodwill, except that Current Text Section I60.109 states that an intangible asset should not be written off in the period of acquisition. Amortization of goodwill over periods of three to five years is acceptable in some situations, but only after considering all of the facts and circumstances. The SEC staff has recently questioned the period of amortization assigned by registrants to goodwill and other identifiable intangible assets acquired in a purchase business combination. The acquiring enterprise should develop a supportable rationale for determining the period of amortization assigned to goodwill and other identifiable intangibles at the date of acquisition. Determining the appropriate useful life of goodwill is a matter of judgement. The following factors may influence the useful life of an acquired enterprise:

- Position of the acquired enterprise in the market and the extent of barriers to entry into the market for its competitors.
- Age, historical operating performance and quality of earnings of the acquired enterprise.
- Susceptibility of the acquired enterprise's products to technological obsolescence and the level of continued investment required to maintain its technological position.
- Industry practice.

Indications that the goodwill has a limited life include limited operating history, lack of stable earnings history, limited experience on the part of the acquired enterprise's management, a narrow product line, uncertain future product viability, intense product competition, and constantly changing technology. In certain situations, the SEC staff has indicated that goodwill amortization should not exceed three to five years in acquisitions of enterprises in the computer software or other technology industries.

Enterprises often attempt to support a 40-year amortization period for goodwill resulting from acquisitions of entities in the health care industry. A frequent assertion made in support of a 40-year amortization period has been that people always have and always will need health care. None the less, there are other factors in many segments of the industry that raise questions about the use of a 40-year amortization period for goodwill (e.g. non-contractual relationships with customers, significantly increased competition, industry consolidation, changing third party reimbursement requirements, technological innovation, and an uncertain regulatory future). When these factors exist, the SEC staff believes that amortization of goodwill for as few as 5 to 10 years is often appropriate, and may challenge an amortization period of more than 20 years.

1.3.4.19 Goodwill impairment

At the date of acquisition, the purchase price should be allocated (based on relative fair values) to the appropriate underlying operating units or business segments to determine the proper allocation of goodwill. Subsequent evaluations for impairment should be performed on a disaggregated basis at each operating unit or business segment. In any subsequent sale or liquidation, the related goodwill must be included in the costs of assets sold, as discussed in paragraph 32 of APB Opinion No. 17. Accordingly, the allocation of goodwill to the separate acquired business units is important to the future evaluation of the carrying amount of goodwill and to the accounting for any subsequent dispositions of the acquired operations.

The relevant accounting literature regarding the impairment of goodwill is brief. Paragraph 31 of APB Opinion No. 17 states, 'estimation of value and future benefits of an intangible asset may indicate that the unamortized cost should be reduced significantly.' APB Opinion No. 17, however, provides no definitive guidance regarding how to estimate the 'value and future benefits' of intangible assets, such as goodwill.

The SEC staff continues to focus on registrants' accounting for impairment of goodwill. If goodwill is material to the financial statements, the staff expects registrants to disclose, pursuant to APB Opinion No. 22, its accounting policy regarding the periodic evaluation of the appropriateness of the carrying amount of goodwill. The staff believes that such disclosure should describe the methodology used to recognize and measure the possible impairment of goodwill. The staff is concerned that registrants recognize impairments on a timely basis, based on objective, rather than discretionary, factors. The staff emphasized that goodwill write-offs should not be a surprise but rather should be supported by previous disclosure in MD&A regarding known trends and uncertainties that could reasonably be expected to have a material effect on the results of operations. Such write-offs generally should be based on the consistent application of the registrant's goodwill impairment policy.With respect to selection of an accounting policy for assessing and measuring goodwill for impairment, the SEC staff has indicated that alternative methods are available under generally accepted accounting principles. Specifically, the SEC staff has accepted the following four methods to measure goodwill for impairment, but expects registrants to apply the measurement methodology consistently in each accounting period:

- Forecasted undiscounted cash flows (usually this would exclude interest charges).
- Forecasted discounted cash flows (only if interest charges are not included).
- Forecasted operating income (including interest expense and depreciation and amortization charges other than goodwill amortization).
- Fair value.

1.3.4.20 Negative goodwill

Although Current Text Section B50.160 states that negative goodwill should be allocated to reduce proportionately the initial values assigned to non-current assets (except long-term investments in marketable securities), the authors believe that an excess of the fair value of the net assets acquired over the cost of the acquired enterprise ('negative goodwill') should be allocated to reduce proportionately only the values assigned to the acquired enterprise's non-current assets that also are non-monetary ('non-current, non-monetary assets'). As defined in Current Text Section C28.408, monetary assets are 'money or a claim to receive a sum of money the amount of which is fixed or determinable without reference to future prices of specific goods or services'. As defined in Current Text Section N35.403, 'non-monetary assets are assets other than monetary ones'. Examples of non-current, non-monetary assets would include property, plant and equipment, land held for development, patents, copyrights, non-compete agreements, and other identifiable intangible assets.

If the non-current, non-monetary assets have been reduced to zero and unallocated negative goodwill still exists, an issue arises as to whether the unallocated negative goodwill should be allocated to non-current deferred tax assets relating to the acquired enterprise's deductible temporary differences and net operating loss and tax credit carryforwards that exist at the date of acquisition. Non-current deferred tax assets should not be reduced by an allocation of negative goodwill because deferred tax assets are considered monetary items under SFAS No. 109. Therefore, the recognition of deferred tax assets relating to the acquired enterprise's deductible temporary differences and net operating loss and tax credit carryforwards that exist at the date of acquisition may result in an increase in unallocated negative goodwill.

Unallocated negative goodwill should be amortized to income over the period estimated to be benefited, but not in excess of 40 years. In practice, unallocated negative goodwill generally is amortized over periods ranging from 5 to 15 years using the straight-line method. Unallocated negative goodwill should be presented as a deferred credit in the financial statements of the combined (acquiring) enterprise. Negative goodwill should not be netted against positive goodwill related to other acquisitions when reporting those amounts in the consolidated financial statements.

1.3.5 Push-down accounting

Push-down accounting refers to establishing a new basis of accounting in the separate stand-alone financial statements of the acquired enterprise based on a purchase of stock of the acquired enterprise. That is, the purchase adjustments recorded by the acquiring enterprise in a purchase business combination are 'pushed down' to the separate stand-alone financial statements of the acquired enterprise. In Issue 86-9, IRC Section 338 and Push-Down Accounting, the EITF reached a consensus that push-down accounting is not required for enterprises that are not SEC registrants. The EITF did not address circumstances in which push-down accounting may be applied by enterprises that are not SEC registrants. Currently, push-down accounting generally is considered to be acceptable, but not required, for enterprises that are not SEC registrants when 80 percent or more of the enterprise's voting common stock is acquired in a purchase transaction.

The following guidance is provided regarding the application of Staff Accounting Bulletin 54:

- Push-down accounting is required for SEC registrants if more than 95 per cent of the voting securities are acquired in a purchase transaction. Acquisition of between 80 per cent and 95 per cent (inclusive) of the voting securities may be considered 'substantial' depending on the facts and circumstances. The SEC staff generally would not permit push-down accounting if less than 80 per cent of the voting securities are acquired.
- Push-down accounting generally is not required if the acquired company has significant publicly-held debt or preferred stock outstanding prior to the transaction and it will remain outstanding after the transaction.
- The significance and character of the minority interests also should be considered in applying SAB 54. In acquisitions of between 80 per cent and 95 per cent (inclusive) of voting interests, the greater the number of disinterested minority interest holders, the greater the percentage of voting securities that could be acquired before push-down accounting is required. Conversely, if a majority shareholder (or a controlling shareholder group) is the sole remaining minority interest, the SEC staff may require push-down accounting for an 80 per cent acquisition since only one minority interest holder remains.
- Transactions in which the selling shareholder retains a minority interest must be evaluated carefully. If the minority interest holder has a put option and the acquiring enterprise has a call option, the existence of the put and call feature on the minority interest may represent a 100 per cent purchase with delayed settlement (i.e. a synthetic forward contract) if, in substance, the put/call feature ensures that the minority interest will be transferred to the acquiring enterprise in the future. In these circumstances, push-down accounting should be applied.

1.3.6 Step acquisition

A 'step acquisition' is the purchase of shares of stock of another enterprise in a series of two or more transactions in which the accumulation of shares of stock in the other enterprise ultimately results in the acquiring enterprise accounting for its investment in the other enterprise by consolidation or the equity method of accounting. The acquisition of shares of stock of the other enterprise may be planned at the date of initial acquisition of shares of stock of the other enterprise or the accumulation of shares of stock may be the result of independent decisions to acquire shares of the other enterprise. In addition, the accumulation of shares of stock of the other enterprise, whether planned or not, may take place over an extended period of time (i.e. over several years). Accounting procedures to be applied in a step acquisition would be applied in the period in which the accumulation of shares of stock in the other enterprise results in the acquiring enterprise accounting for its investment in the other enterprise by consolidation or the equity method of accounting.

1.3.7 Minority interest

The valuation of assets acquired and liabilities assumed in an acquisition of less than 100 per cent of the outstanding stock of the acquired enterprise should be based on a pro rata allocation of the fair values of the assets acquired and liabilities assumed and the historical financial statement carrying amounts of the assets and liabilities of the acquired enterprise. The values assigned to the assets acquired and liabilities assumed should be based on fair value, determined in accordance with the guidance in Current Text Sections B50.145 and B50.146, to the extent of the ownership interest acquired by the acquiring enterprise. The remaining portion of the values assigned to the assets and liabilities of the acquired enterprise represents the minority interests' ownership in the acquired enterprise and, therefore, should be based on the historical financial statement carrying amounts of the acquired enterprise.

Example:

On 31 December 1995, Company A acquired 60 per cent of the outstanding stock of Company B for cash of USD1,000. The fair value of the identifiable net assets of Company B was USD1,500 and the book value of Company B's net assets was USD1,200.

Company A should record its investment in Company B at its cost of USD1,000. In the consolidated financial statements, Company A should record the identifiable net assets of Company B at USD1,380 (60 per cent of USD1,500 plus 40 per cent of USD1,200), goodwill of USD100 (USD1,000 less 60 per cent of USD1,500), and minority interest in Company B of USD480 (40 per cent of USD1,200).

Company A would record the following entry to reflect Company B in its 31 December 1995, consolidated financial statements:

Identifiable net assets of Company B	USD1,380	
Goodwill	USD100	
Minority interest		USD480
Investment in Company B		USD1,000

The minority interest in the net assets of the subsidiary is generally shown as a single amount in the liability section of the consolidated balance sheet, frequently under the heading of non-current liabilities. The alternative is to include the minority interest in consolidated shareholders' equity or in a separate minority interest section. This is rarely done in practice. The minority shareholders' share of the profit of the subsidiary for the current period is not reflected in the consolidated balance sheet, but rather is reflected as a deduction in the consolidated income statement when arriving at consolidated net income for the current period:

> In the unusual case in which losses applicable to the minority interest in a subsidiary exceed the minority interest in the equity capital of the subsidiary, such excess and any further losses applicable to the minority interest should be charged against the majority interest, as there is no obligation of the minority interest to make good such losses. However, if future earnings do materialize, the majority interest should be credited to the extent of such losses previously absorbed (ARB No. 51, paragraph 15).

1.3.8 Sale of a subsidiary and change from full consolidation

If the investment in the subsidiary is sold, the profit arising in the individual accounts is not reflected in the group accounts. Profit from the sale of the subsidiary would be the difference between the sale price and the net assets of the subsidiary as reflected in the consolidated financial statements, including any related goodwill as well as foreign currency translation adjustments, if any.

If there is a change from full consolidation to use of the equity method, the value of the investment under the equity method would be based on the proportionate share of the net assets of the subsidiary as reflected in the consolidated financial statements.

1.3.9 Multi-level group of companies

When there are multiple levels of consolidation, difficulties may arise in dealing with consolidation differences and minority interests. There are no official accounting pronouncements or regulations dealing with the consolidation of multi-level groups of companies. Baker *et al.* (1998) is one of the leading textbooks dealing with consolidation accounting in the United States. This text discusses accepted practices for the consolidation for multi-level groups (Baker *et al.*, 1998, p. 517). The complexity of the consolidation process increases as additional ownership levels are included. The amount of income and net assets to be assigned to the controlling and non-controlling shareholders, and the amount of unrealized profits and losses to be eliminated, should be determined at each level of ownership.

When there are a number of different levels of ownership, the normal first step is to consolidate the bottom, or most remote, subsidiaries with the companies at the next higher level. This sequence is continued up through the ownership structure until the subsidiaries owned directly by the parent are consolidated with the parent company. Income is apportioned between the controlling and non-controlling shareholders of the companies at each level. For example, in the case of a three-tiered structure involving a parent company, its subsidiary, and the subsidiary's subsidiary, the parent company's equity method net income would be computed by first adding an appropriate portion of the income of the bottom subsidiary to the separate earnings of the parent's subsidiary and then adding an appropriate portion of that total to the separate earnings of the parent company. Consolidated net income is computed in the same manner. Goodwill reflected in the consolidated balance sheet includes the sum of the goodwill balances determined at each of the levels of the consolidation process. The minority interest reflected in the consolidated balance sheet consists of the sum of the minority interest balances as calculated at each level of the consolidation process.

1.3.10 Purchases and reciprocal shareholding

A reciprocal share holding relationship exists when two companies own shares in each other. Reciprocal shareholding relationships are relatively rare in practice, and their accounting impact is often immaterial. There are two approaches to the treatment of reciprocal share holding relationships. The treasury stock method is used most frequently in practice and treats shares of the parent held by a subsidiary as if they had been repurchased by the parent. The entity approach views the parent and subsidiary as a single entity with income shared between the different ownership groups, explicitly taking into consideration the reciprocal relationship. Under the treasury stock method, purchase of a parent's shares by a subsidiary are treated in the same manner as if the parent had repurchased its own shares and is holding them in the treasury. The subsidiary normally accounts for the investment in the parent's shares using the cost method because such investments are usually small and almost never confer the ability to significantly influence the parent. Income assigned to the non-controlling interest in the subsidiary is usually based on the subsidiary's net income, which includes the dividend income arising from the investment in the parent. The parent, however, normally bases its equity method share of the subsidiary's income on the subsidiary's income excluding the dividend income from the parent.

The entity approach to dealing with reciprocal share holdings is often referred to as the traditional or conventional approach. This method is more consistent with the entity theory of consolidation than is the treasury stock method. Under the entity approach, the total of the separate incomes of the consolidating companies is viewed as the total income of the consolidated entity and is apportioned between the controlling and non-controlling shareholders. In assigning income to the shareholder groups, recognition is given to the reciprocal nature of the relationship created when a subsidiary acquires shares of the parent company. A set of simultaneous equations is used to compute a reciprocal income total for each company. The resulting amounts are then used to apportion income to the controlling and non-controlling interests.

There are no official accounting pronouncements under GAAP dealing with the issue of reciprocal share holding. The two methods mentioned above were developed in practice and are only described by accounting textbooks such as Baker *et al.* (1998).

1.3.11 Not completely paid-in shares

In the United States, it is rare for shares to be not completely paid in. If shares are not completely paid in, the capital stock account is not carried at the full value of the subscription price. A separate component of stockholders' equity is created with the heading 'Capital Stock Subscribed'. This component of stockholders' equity would be eliminated in the consolidation process in the same manner as the other components of stockholders' equity.

1.3.12 Financial reporting subsequent to a purchase

Financial statements prepared subsequent to a purchase-type business combination reflect the combined entity only from the date of combination. When a business combination occurs during a fiscal period, income reported by the acquired company before the combination is not reported in the consolidated income statement. If the consolidated group presents comparative financial statements that include statements for periods before the combination, those statements would include only the activities and financial position of the acquiring company and not those of the acquired company.

1.3.13 Disclosure requirements

A number of disclosures are required following a business combination to provide financial statement readers with information about the combination and the expected effects of the combination on operating results. APB Opinion No. 16 requires the following disclosures in the notes to the financial statements when purchase treatment is used. This disclosure is required for each company purchased if the purchase price is material to the financial statements. The disclosure is required to be included in the period of the business acquisition and in any subsequent period for which financial statements are presented that cover the period in which the acquisition took place:

- Name and a brief description of the acquired company.
- A statement that the purchase method has been used.
- The period for which results of operations of the acquired company are included in the income statement of the acquiring company.
- The cost of the acquired company and, if applicable, the number of shares of stock issued or issuable, and the amount assigned to the issued and issuable shares.
- A description of the plan for amortization of acquired goodwill, the amortization method and period.
- Contingent payments, options, or commitments specified in the acquisition agreement and the proposed accounting treatment.

There are no disclosure requirements for minority interests.

Pro forma financial statement data should also be presented to provide statement readers with a better understanding of the potential operating impact of the business combination. At a minimum, supplemental information should be provided to show:

- Operating results as if the acquisition had been made at the start of the period.
- When comparative financial statements are presented, operating results for the preceding period as if the acquisition had occurred at the start of that period.

1.4 Pooling-of-interests method

1.4.1 Concept

Under the pooling-of-interests method, an exchange of stock to effect a business combination is deemed to be a transaction between the combining shareholder groups and not a sale of a corporate entity. Therefore, a new basis of accountability for the assets of the combined corporations is not appropriate under pooling-of-interests accounting. The facts of the situation regarding a business combination dictate whether to use the purchase or pooling method in a specific transaction. As noted previously, the pooling-of-interests method of accounting applies only to business combinations effected by an exchange of outstanding common shares, not those involving cash, other assets, or liabilities. In order to use the pooling-of-interests method, 12 criteria as specified in APB Opinion No. 16 must be met. These criteria are discussed in 1.4.3. If any of the criteria for the application of the pooling-of-interests method are not met, then the purchase method must be used. Under the pooling-of-interests method, the accounts of the acquired entity (i.e. the combined company) are carried forward at their book values at the date of the business combination. The capital accounts of the combined company are replaced by the newly issued capital stock used to effect the combination. The retained earnings account of the combined company is carried forward. No goodwill is recorded. The net income of the consolidated entity includes the separate incomes of all combining entities during the year of the combination.

1.4.2 Reasons for restricted application

The managements of companies involved in business combinations often prefer accounting for business combinations as poolings because of the impact on the financial statements subsequent to the combination. Purchase accounting requires the purchased assets and liabilities to be valued at their fair values. In many cases, the fair values of an acquired company's assets are higher than the previous book values, and if the assets are tangible with limited lives or are intangible, these higher fair values must be amortized. Under pooling-of-interests accounting, the book values of both the acquiring company and the acquired company are carried forward. Therefore, income is often higher subsequent to the combination under pooling accounting than under purchase accounting because of lower depreciation and amortization charges and because of the opportunity to report gains on the sale of the subsidiary's assets with realized market values in excess of cost. Further, the absence of goodwill and subsequent amortization in a pooling results in higher income in later years. Finally, the higher asset values recorded under purchase accounting negatively impact ratios such as return on investment because of the lower income amount in the numerator and the higher asset amount in the denominator. As a result, the pooling method is preferred practice, although it is not always popular with stockholders of the acquired company who may prefer cash rather than the stock of the acquiring company.

Attempts to restrict the use of pooling-of-interests accounting were notably unsuccessful before the issuance of APB Opinion No. 16 in 1970. In particular, Accounting Research Bulletin No. 48, Business Combinations (ARB No. 48), issued by the AICPA's Committee on Accounting Procedure in 1957, established criteria for determining whether a business combination should be treated as a purchase or pooling. However, because some of the provisions of ARB No. 48 were subject to judgement, and because of a lack of general acceptance, the pronouncement had little impact during much of the period from 1957 to 1970. Thus, before APB Opinion No. 16 became effective, companies had rather broad latitude in recording business combinations and there were many abuses of pooling-of-interests accounting. For example, some combinations were recorded as 'part purchase, part pooling' combinations because they seemed to involve elements of both. Others were treated originally as purchases and later changed to 'retroactive poolings' as the standards eroded. Therefore, APB Opinion No. 16 sets very stringent requirements for the use of pooling-of-interests accounting. Many of the rules established in APB Opinion No. 16 are quite complex and exist primarily to eliminate the previous abuses of pooling accounting.

1.4.3 Criteria for application of the pooling-of-interests method

APB Opinion No. 16 allows both purchase and pooling-of-interests as methods of accounting for business combinations, but these methods may not be viewed as alternatives to one another. A series of conditions is set forth in APB Opinion

No. 16 as prerequisite to pooling treatment. If all the criteria are met, the transaction must be recorded as a pooling of interests. If any are not met, the combination must be treated as a purchase. The companies involved in a combination are not permitted a choice of methods except through the structuring of the combination.

The conditions for a pooling of interests are specified in paragraphs 45–48 of Opinion No. 16. The criteria established by APB Opinion No. 16 for determining whether purchase or pooling accounting is appropriate are divided into three categories:

- Attributes of the combining companies.
- Manner of combining interests.
- Absence of planned transactions.

Each of the categories, in turn, contains several specific criteria that must be met in order to use pooling treatment.

1.4.3.1 Attributes of the combining companies

Two of the conditions for a pooling-of-interests are classified as attributes of the combining companies.

The first condition for pooling-of-interests accounting is that the combining enterprises must be autonomous entities. If any of the combining enterprises has been a subsidiary or a division of another entity within two years of the initiation of the business combination, or another entity has had the ability to control any of the combining enterprises during the two-year period prior to initiation of the business combination, pooling-of-interests accounting is not appropriate. The purpose of this condition is to preclude pooling-of-interests accounting for combinations that involve only selected assets, operations, or ownership interests of the combining enterprises. Under the autonomy condition, pooling-of-interests accounting is precluded if one of the combining enterprises has been a subsidiary or a division of another enterprise within two years before the plan of combination is initiated. Current Text Section B50.637 clarifies that the autonomy condition applies to the period beginning two years prior to initiation and ending at consummation of the combination. Although the combining enterprises may have been autonomous during the two years prior to initiation of a combination, pooling-of-interests accounting would not be appropriate if one of the combining enterprises became a subsidiary of another entity between initiation and consummation of the combination.

The second condition for pooling-of-interests accounting is that each of the combining enterprises must be independent of the other combining enterprises. Under this condition, the combining enterprises may hold no more than 10 per cent in total of the outstanding voting common stock of any combining enterprise at the date that the plan of combination is initiated or between initiation and consummation ('intercorporate investment'). The independence condition is not met if the combining enterprises hold more than 10 per cent of the outstanding voting common stock of any combining enterprise at the initiation date, the consummation date, or at any time between the initiation and the consummation dates. Voting common stock that is acquired after the initiation date pursuant to the exchange ratio in exchange for the voting common stock issued to effect the combination is excluded from the computation of the 10 per cent limitation.

1.4.3.2 Manner of combining interests

Seven of the conditions for a pooling-of-interests are classified as manner of combining interest. These conditions are as follows:

- The combination must be effected in a single transaction or be completed in accordance with a specific plan within one year of the plan being initiated. The one-year rule would not apply if there is a delay in completing the combination due to proceedings of a governmental authority or litigation. If the exchange terms of a combination are altered, the new terms result in a new initiation date unless earlier exchanges of common stock are adjusted to the new terms.
- One corporation (the issuing corporation) must offer and issue only common stock in exchange for substantially all (90 per cent) of the outstanding voting stock of another company (combining company) at the date the plan is consummated. The number of shares assumed to be exchanged excludes shares of the combining company held by the issuing

company when the plan is initiated, shares acquired by the issuing company before the plan is consummated, and shares outstanding after the plan is consummated. If the combining company holds shares in the issuing company, these shares must be converted into an equivalent number of shares and deducted from outstanding shares to determine the number of shares assumed to be exchanged. The reason for this adjustment is that the issuing company is issuing its shares to reacquire its own shares. Such shares are not issued to acquire stock of the other combining company. The following example illustrates the calculation of the 90 per cent rule.

Example:
Assume that Company A, Company B and Company C enter into a combination agreement whereby A issues one of its shares for every two shares of B and C. Further assume:

(i) Before the combination, Company A holds 20 shares of Company C and Company C holds 30 shares of Company A.
(ii) Before the combination, B holds 40 shares of C, which it exchanges under the terms of the combination.
(iii) C has 1000 shares outstanding.
(iv) At the completion of the exchange, A holds all the shares of B and 980 shares of C.
(v) All the shares of C are acquired in exchange for A's common stock except those of C already held by A and 10 shares acquired for cash.

For applying the 90 per cent rule, the imputed number of C's shares exchanged is computed as follows:

Number of shares of C held by A after the exchange	980
Deduct:	
Shares of C exchanged by B	(40)
Shares of C acquired for cash	(10)
Shares of C held by A before the exchange	(20)
Shares of A held by C before the exchange, restated into an equivalent number of C's shares (30 *2)	(60)
Imputed number of C's shares acquired in the exchange	850

The combination cannot be treated as a pooling because only 85 per cent of C's shares were viewed as being acquired in the exchange:

- None of the combining companies must change the equity interest of the voting common stock in contemplation of effecting the combination in the two years before initiation of the plan of combination or between the dates of initiation and consummation.
- Each of the combining companies must reacquire shares of voting common stock only for purposes other than a business combination, and no company must reacquire more than a normal number of shares between the dates the plan is initiated and consummated. This restriction on treasury stock transactions generally does not apply to shares purchased for stock option or compensation plans.
- The ratio of the interest of an individual common shareholder to the interests of other common stockholders in a combining company must remain the same as a result of the exchange of stock to effect the combination.
- The voting rights to which the common stock ownership interest in the resulting combined corporation is entitled must be exercisable by the stockholders; the stockholders must be neither deprived of nor restricted in exercising those rights for a period.
- The combination must be resolved at the date the plan is consummated and no provision of the plan relating to the issue of securities or other consideration must be pending.

1.4.3.3 *Absence of planned transactions*

The last group of conditions for a pooling of interests relates to planned transactions of the combined entity. First, the combined corporation must not agree to retire or reacquire stock issued to effect the combination. Second, the combined corporation must not enter into financial arrangements (such as loan guarantees) for the benefit of former stockholders of a combining company. Finally, an intention or a plan to dispose of a significant percentage of the assets of the combining enterprise within two years after the consummation of the business combination would violate the absence-of-planned-transactions condition unless the disposals are (1) in the ordinary course of business, (2) to eliminate duplicate facilities or excess capacity, or (3) the plan of disposition was developed and in process prior to commencement of negotiations with the combining enterprise. The SEC staff has taken the position that all dispositions of a significant percentage of the assets of the combining enterprise (other than disposals to eliminate duplicate facilities or excess capacity) within two years subsequent to the business combination are presumed to have been intended or planned at the consummation of the business combination and, therefore, would violate the absence-of-planned-transactions condition. However, if evidence clearly indicates that the disposition was not intended or planned in contemplation of the business combination, that presumption may be overcome based on the facts and circumstances in each situation. Such evidence must be objectively verifiable and should demonstrate that (i) the disposition was in the ordinary course of business; and (ii) the disposition would have occurred independently and in the absence of the business combination to avoid violating the absence-of-planned-transactions condition. In addition, the closer the disposition is to the consummation of the business combination, the more persuasive the evidence must be to overcome the presumption that the disposition was intended or planned in contemplation of the business combination.

1.4.4 Characteristics of the pooling-of-interests method

Under the pooling-of-interests method:

- Assets and liabilities of the joining companies are reported at the previously existing book values of each. Adjustments may be made to reflect consistent applications of accounting principles. However, fair market values at the time of a combination are not substituted for book values as is the case under the purchase method.
- Goodwill (or differential of any kind) does not result from the combination as would be the case under the purchase method.
- Retained earnings of the combining companies are simply added to determine the retained earnings balance of the combined companies at the date of acquisition.

When one company is merged into another in a pooling of interests, the assets and liabilities of that company are recorded on the books of the continuing company at their book values. The shares issued are recorded at the book values of the net assets received in exchange. In most poolings, the total par value of the shares issued in the combination is different from the total par value of the shares acquired. This occurs because the number of shares issued is usually different from the number of shares acquired and their per share par values are often different. If the total par value of the common shares of the combined company is less than the sum of the total par values of the shares of the combining companies, the total amount assigned to capital stock is more, and the amount assigned to additional paid-in capital decreases. If the total par value of the stock issued in the combination is large enough, the total additional paid-in capital of the combined company is eliminated, and combined retained earnings are reduced.

1.4.5 Disclosure requirements

APB Opinion No. 16 requires the following disclosures in the notes to the financial statements when pooling treatment is used:

- The name and a brief description of the acquired company.
- A statement that pooling treatment has been used.
- Description and number of shares issued in the exchange.
- For the separate companies, revenue, extraordinary items, net income, changes in stockholders' equity, and the amount and handling of inter-company transactions for the portion of the current period before the date of combination.
- A description of any adjustments of net assets or income related to changes in accounting procedures.
- A description of the impact of a change in the fiscal period of the combining company.
- A reconciliation of revenue and income previously reported by the stock-issuing company with the restated amounts reported for those periods for the combined company.

In addition, a material profit or loss from disposing of a significant part of the assets of the combining companies must be reported separately in the income statement as an extraordinary item.

2 Consolidation of debt

In the preparation of consolidated financial statements, any account balances that are reciprocal must be eliminated (see ARB No. 51 and SFAS No. 94). This includes debt of one company within a consolidated group that is owed to another member of the consolidated group. Thus, the consolidated financial statements would not reflect inter-company debt. If the debt of a subsidiary company is owed entirely to third parties outside the consolidated group, and if the parent company also has debt outstanding, both the debt of the parent company and the debt of the subsidiary would be shown together in the consolidated financial statements, even though the debt of the subsidiary company is not a liability of the parent company but rather is a liability of the subsidiary which constitutes a member of the consolidated group.

A complication arises if the debts are reflected in different currencies in the separate financial statements of reciprocal parties. Under SFAS No. 52, the carrying values of payables and receivables denominated in foreign currencies are to be adjusted to the current rate of exchange at the balance sheet date. By this mechanism, the carrying values of the reciprocal debt payables and receivables should be equivalent and should be eliminated against one another. As with any issue of accounting, under GAAP the rules do not need to be applied to immaterial items.

SFAS No. 52 does provide for an exception when inter-company foreign currency transactions will not be settled within the foreseeable future. These inter-company transactions may be considered part of the net investment in the foreign entity. The translation adjustments on these long-term receivables or payables may also be considered part of the net investment in the foreign entity. The translation adjustments on these long-term receivables or payables are deferred and accumulated as part of the cumulative translation adjustment account. For example, a US parent may loan its German subsidiary USD10,000 for which the parent does not expect to be repaid in the foreseeable future. Under the translation method, the USD-denominated loan payable account of the subsidiary would first be adjusted for the effects of any changes in exchange rates during the period. Any exchange gain or loss adjustment relating to this inter-company note should be classified as part of the cumulative translation adjustment account in stockholder's equity.

3 Elimination of inter-company profits

Transactions between affiliated companies involving the sale and purchase of assets create unrealized profits and losses to the consolidated entity (see ARB No. 51 and SFAS No. 94). Such profits and losses, whether resulting from the sale of inventory or of fixed assets, are eliminated fully until such time as the profits are realized by sales or transfers to third parties outside the consolidated group.

The amount of inter-company profit or loss to be eliminated is not affected by the existence of a minority interest. The complete elimination of the inter-company profit or loss is consistent with the underlying assumption that consolidated statements represent the financial position and operating results of a single company. Generally, inter-company profit or loss is eliminated through a proportionate allocation between the majority and minority interests.

4 Elimination of inter-company revenues and expenses

As noted above, all inter-company transactions and balances between affiliated companies shall be eliminated in consolidated financial statements. Such inter-company transactions include for example, sales revenue and cost of sales of the selling affiliate, interest income and interest expense, income taxes, selling, general and administrative expenses.

VII Proportionate consolidation and equity method

1 Proportionate consolidation

The accounting literature has given little attention to accounting issues relating to investments in joint ventures. Generally, investments in common stock are accounted for in accordance with APB Opinion No. 18 (see below) whereas proportionate consolidation may be appropriate for unincorporated joint ventures in the form of undivided interest in industries in which this is common practice. For example, the industry practice in oil and gas ventures is for the investor-venturer to account for its pro rata share of the assets, liabilities, revenues and expenses of a joint venture in its own financial statements (Interpretation No. 2 to APB Opinion No. 18).

However, even if proportionate consolidation would not be in accordance with GAAP, a foreign SEC registrant is permitted to apply the proportionate method of consolidation for investments in joint ventures:

- If the foreign SEC registrant reconciles its net income and stockholder's equity to US GAAP:

  Issuers that prepare financial statements on a basis of accounting other than US generally accepted accounting principles that allows proportionate consolidation for investments in joint ventures that would be accounted for under the equity method pursuant to US GAAP may omit differences in classification or display that result from using proportionate consolidation in the reconciliation to US GAAP specified by paragraphs (c)(2)(i), (c)(2)(ii) and (c)(2)(iii) of this Item; provided, the joint venture is an operating entity, the significant financial operating policies of which are, by contractual arrangement, jointly controlled by all parties having an equity interest in the entity. Financial statements that are presented using proportionate consolidation must provide summarized balance sheet and income statement information using the headings specified in § 210.1–02(aa) of this section and summarized cash flow information resulting from operating, financing and investing activities relating to its pro rata interest in the joint venture (Form 20-F, items 17 (c)(2)(vii) and 18 (c)(2)(vii)).

- If the foreign SEC registrant prepares its financial statements already according to US GAAP (i.e. joint ventures accounted for under the equity method of accounting) a formal request to the SEC is necessary in order to utilize the proportionate method of consolidation. However, the SEC has stated that it would not object to the use of the proportionate method of consolidation as supplemented by the disclosures mentioned above, and if such deviation from US GAAP is explained in the independent auditors' report.

2 Equity method

2.1 The obligation to use the equity method

The equity method of accounting for an investment in common shares should be followed by an investor whose investment in the voting common shares of another company gives it the ability to exercise significant influence over operating and financial policies of the investee even though the investor holds 50 per cent or less of the voting shares (see paragraph 17 of APB Opinion No. 18). APB Opinion No. 18 requires the use of the equity method when a company is deemed to have significant influence over another company, but lacks a majority ownership interest. The general rule has been established that when one company owns more than 20 per cent of the common shares of another company it is deemed to have the ability to exercise significant influence over the investment and is therefore required to use the equity method of accounting for the investment.

APB Opinion No. 18, paragraph 17 states this rule as follows:

> An investment (direct or indirect) of 20 per cent or more of the voting stock of an investee should lead to a presumption that in the absence of evidence to the contrary an

investor has the ability to exercise significant influence over an investee. Conversely, an investment of less than 20 per cent of the voting stock of an investee should lead to a presumption that an investor does not have the ability to exercise significant influence unless such ability can be demonstrated.

2.2 Exemptions from the obligation

If a company owns more than 20 per cent of the voting common shares of another company, but it is unable to exercise significant influence, then the equity method should not be used. For example, if the other 80 per cent of the voting common shares are owned by one shareholder, then the 20 per cent interest would not be able to exercise significant influence.

In addition, the equity method should not be used if the investor's ability to exert significant influence is temporary or if the investee is a foreign company operating under severe exchange restrictions or controls. FASB Interpretation No. 35 cites the following indications that the equity method should not be used:

- Opposition by the investee that challenges the investor's influence.
- Surrender of significant stockholder rights by agreement between investor and investee.
- Concentration of majority ownership.
- Inadequate or untimely information that prevents the equity method from being applied.
- Failure to obtain representation on the investee's board of directors as an indicator of an investor's inability to exercise significant influence.

Application of the equity method should be discontinued when the investor's share of losses reduces the carrying amount to zero.

2.3 Calculation of the equity value

2.3.1 Overview

If a company is using the equity method, the investment balance is not adjusted for changes in the value of the investment. The investment balance is adjusted by increasing the balance for the proportionate share of the income of the investee company and reducing the balance for the proportionate share of dividends declared by the investee. Thus, the investment balance is equal to a proportionate share of the book value of the investee company. To the extent that the book value of the investee company is different from its fair value, there will be a difference between the investment balance and the fair value of that ownership interest.

Under the equity method, the balance in the investor's investment account is calculated by increasing the prior year balance in the investment account by the investor's proportionate share of the net income of the investee and decreasing it by dividends declared by the investee (see Table VII.1). The income from investment recognized by the investor company is calculated as demonstrated in Table VII.2.

2.3.2 Goodwill

Goodwill, or the difference between the purchase price of the investment and the book value or fair value of the percentage ownership interest acquired, is not shown separately (see APB Opinion No. 18). The original cost of the investment (including any goodwill component) is reflected as an investment among the assets of the investor company. The goodwill amount or any other differential implicit in the investment should be amortized against the income recognized by the investor company. No direct offsetting against the shareholders' equity of the investor company is allowed.

Information regarding individual assets and liabilities at the time of purchase is important because subsequent accounting under the equity method entails accounting for any differences between the investment cost and the underlying

Table VII.1 Calculation of the ending equity balance

	Prior year equity value
+	Income recognized by the investor (as calculated above)
–	Investor's share of the dividends declared by the investee
=	End of year equity value

Table VII.2 Calculation of income from investment

The investor's proportionate share of the investee's net income
– Amortization of goodwill
– Amortization of the portion of a positive differential (i.e. the difference between the fair market value of net assets of the investee and the book value of the net assets)
– Amortization of the portion of a positive differential arising from depreciable assets (the portion is dependent on the remaining useful life of the asset)
– Any inter-company profit
+ Amortization of negative differential associated with assets sold
+ Amortization of the portion of a negative differential arising from depreciable assets (the portion is dependent on the remaining useful life of the asset)
+ Any inter-company loss
+ Amortization of any bargain purchase element
= Income recognized by the investor.

equity in the net assets of the investee. Under the equity method of accounting, excess cost over underlying equity is eliminated by periodic charges (debits) and credits to income from the investment and by equal credits or charges to the investment account. Thus, the original difference between investment cost and book value acquired will disappear over the remaining lives of identifiable assets and over a maximum period of 40 years if assigned to goodwill. Unrealized gains and losses in foreign currency transactions are charged directly to the investment when using the one-line consolidation procedure to account for foreign currency.

If the acquisition cost of an investment is lower than the proportionate share of the net assets of the investee company, there is a bargain purchase. In a bargain purchase there is implicit negative goodwill or overstatement of the book value of the net assets of the investee company. This differential should be amortized by a credit to income recognized by the investor. Negative equity value may result when the share of the losses recognized by applying the equity method exceeds the recorded value of the investment. In this case the investment balance is reduced to zero, but not to a negative value.

2.3.3 Elimination of inter-company gains and losses

Inter-company gains and losses should be eliminated as if consolidated financial statements were prepared.

> Paragraph 14 of ARB No. 51 provides for complete elimination of inter-company profits or losses in consolidation. It also states that the elimination of inter-company profit or loss may be allocated proportionately between the majority and minority interests. Whether all or a proportionate part of the inter-company profit or loss should be eliminated under the equity method depends largely upon the relationship between the investor and investee.
>
> When an investor controls an investee through majority voting interest and enters into a transaction with an investee which is not on an 'arm's length' basis, none of the inter-company profit or loss from the transaction should be recognized in income by the investor until it has been realized through transactions with third parties. . . .
>
> In other cases, it would be appropriate for the investor to eliminate inter-company profit in relation to the investor's common stock interest in the investee. In these cases, the percentage of inter-company profit to be eliminated would be the same regardless of whether the transaction is 'downstream' (i.e. a sale by the investor to the investee) or 'upstream' (i.e. a sale by the investee to the investor) (Interpretation No. 2 to APB Opinion No. 18).

2.3.4 Sale of an investment under the equity method and change from the equity method to full consolidation

If the equity investment is sold, the profit on the sale is based upon the difference between the proceeds realized from the sale of the investment and its carrying value. In certain circumstances (e.g. when it is determined that the investor does not have significant influence) it is possible that the investor company could use a method other than the equity method (e.g. the cost method) in accounting for its investment. In this case, the profit on the sale of the investment would still be based upon the difference between the proceeds realized from the sale of the investment and its carrying value.

If the equity method has been correctly applied, there should be no significant problems when changing from the equity method to full consolidation. The income and retained earnings of the parent company ought to be the same as the consolidated net income and consolidated retained earnings (see APB Opinion No. 18).

2.3.5 Foreign currency translation and the equity method

If an investor has an equity investment in a company located in a foreign country, the provisions of SFAS No. 52 should be applied to translate the foreign currency financial statements prior to using the equity method. This means that the current exchange rate should be used to translate the financial statements of a foreign investee company if the functional currency of the foreign investee company is not the US dollar and if the foreign investee company is not located in a highly inflationary economy. If the functional currency of the foreign investee company is the US dollar or if the foreign investee company is located in a country with a high rate of inflation, the financial statements should be remeasured in accordance with the provisions of SFAS No. 52 (see V.4.2) prior to translation and use of the equity method. Subsequent to the application of SFAS No. 52, the foreign currency financial statements will produce a net income figure stated in the currency of the investor company. This net income number should be used in applying the equity method. In addition, the translated value of the dividends declared should be used to adjust the carrying value of the equity investment.

2.3.6 Disclosure requirements

The significance of the investor's financial position and results of operations should be considered when evaluating the extent of disclosures of the financial position and results of operations of an investee. If the investor has more than one investment in common shares, disclosures wholly or partly on a combined basis may be appropriate. The following disclosures are generally applicable to the equity method of accounting for investments in common shares:

- Financial statements of an investor should disclose parenthetically, in notes to financial statements, or in separate statements or schedules (a) the name of each investee and percentage ownership of common stock; (b) the accounting policies of the investor with respect to investments in common stock; and (c) the difference, if any, between the amount at which an investment is carried and the amount of underlying equity in net assets and the accounting treatment of the difference.
- For those investments in common stock for which a quoted market price is available, the aggregate value of each identified investment based on the quoted market price should usually be disclosed. This disclosure is not required for investments in common stock of subsidiaries.
- When investments in common stock of corporate joint ventures or other investments accounted for under the equity method are, in the aggregate, material in relation to the financial position or results of operations of an investor, it may be necessary for summarized information as to assets, liabilities and results of operations of the investees to be presented in the notes or in separate statements, either individually or in groups, as appropriate.
- Conversion of outstanding convertible securities, and the exercise of outstanding options and warrants and other contingent issuances of an investee may have a significant effect on an investor's share of reported earnings or losses. Accordingly, the material effect of possible conversions, and the exercise of contingent issuances should be disclosed in the notes to the financial statements of an investor.

VIII Deferred taxation

1 Overview

1.1 Regulation on deferred taxation

Each member of a consolidated group may have deferred taxes in its individual accounts (see USA – Individual Accounts, VII.1 for a description of deferred tax accounting under GAAP). Under SFAS No. 109, Accounting for Income Taxes, deferred taxes must be provided for temporary differences including tax loss carryforwards arising from recognition of an item in the accounting records and recognition for tax purposes. If a difference between recognition for accounting and recognition for tax purposes is deemed to be permanent, the provision of deferred taxes is not required. There is no concept of a quasi permanent difference under GAAP. In the process of preparing consolidated financial statements, the deferred tax accounts of the individual companies of the consolidated group are combined. In addition, as a result of the consolidation process there are other sources for deferred taxation, including:

- Adaptation of the values from the individual accounts to uniform valuation in the group accounts.
- Foreign currency translation.
- Capital consolidation.
- Elimination of profits from inter-company transactions.
- Application of the equity method.

These sources of deferred taxation will be discussed in the following sections.

1.2 US tax rules

Under US tax law, the sale of assets typically causes the incurrence of a tax on the part of the seller. The buyer of the assets records the assets in its accounting records at the amount paid to acquire the assets. If the buyer purchases common shares of a company (instead of the assets of that company), then the tax is paid by the seller of the shares and the buyer records the investment at the amount paid. In such a case the underlying assets of the purchased company are not revalued. Under certain circumstances, either the selling shareholders or the buyer can elect to pay the tax on the appreciation in the assets and therefore revalue the assets of the purchased company to their fair values. Under US tax law, a parent must own at least 80 per cent of the voting common shares of a subsidiary in order to file a consolidated tax return. In general, the consolidated tax group is treated as if it were one taxpayer. Furthermore, a foreign subsidiary cannot be included in a consolidated tax return.

The equity method is not allowed for tax purposes. Therefore, if a parent company owns less than 80 per cent of the common shares of an investee company, it must use the cost method of accounting for tax purposes. This means that the parent is only taxed on dividends received from the subsidiary or investee company. The parent company is also entitled to a dividends received deduction of 80 per cent (in some cases, 70 per cent) of the dividend. Therefore, the parent or investor company is only taxed on 20 per cent (in some cases, 30 per cent) of the dividend received.

2 Deferred taxes from adaptation to uniform valuation

If assets and liabilities are revalued to comply with uniform valuation principles used in the preparation of consolidated financial statements for accounting purposes, SFAS No. 109, Accounting for Income Taxes, defines these differences as temporary differences and therefore deferred taxes must be provided. The deferred taxes are calculated based upon the difference, if any, between the new valuation for accounting purposes and the valuation for tax purposes.

3 Deferred taxes from foreign currency translation

The provisions of SFAS No. 109 apply to foreign operations that are accounted for by the equity method or that are consolidated in the financial statements of a US parent company. Temporary differences may exist in a foreign tax jurisdiction for differences between the foreign currency financial statement amounts and the foreign currency tax bases of a subsidiary's assets and liabilities due to items such as the use of accelerated depreciation for tax purposes. Deferred taxes should be recognized for the tax consequences of such temporary differences in the foreign tax jurisdiction in accordance with the provisions of SFAS No. 109. SFAS No. 109 includes special requirements regarding recognition of deferred taxes for foreign operations. The requirements of SFAS No. 109 for situations where the local foreign currency is the functional currency differ from the requirements where the reporting currency is the functional currency.

3.1 Local foreign currency is the functional currency

When the local foreign currency is the functional currency for foreign operations under the provisions of SFAS No. 52, Foreign Currency Translation, SFAS No. 109, paragraph 11 requires measurement of foreign deferred tax assets and liabilities for temporary differences between the foreign currency financial statement carrying amounts (using generally accepted accounting principles in the United States but before translation to the reporting currency) and the foreign currency tax bases, including the effects of indexing, if any, of the assets and liabilities, in the foreign tax jurisdiction. The resulting foreign deferred tax assets and liabilities are then translated into the reporting currency with the other foreign assets and liabilities.

Under SFAS No. 52, the assets and liabilities of a foreign subsidiary whose functional currency is the local foreign currency are translated from that foreign currency into the reporting currency using current exchange rates. Translation adjustments that result from this process are not recognized in income but are reported as a separate component of equity (see V.4.3).

The process of translating the assets and liabilities of a foreign subsidiary into the reporting currency after a change in exchange rates gives rise to a temporary difference in an amount equal to the translation adjustment because it changes the parent's financial statement carrying amount of the investment in the foreign subsidiary, but the parent's tax basis in that investment does not change. The recognition or nonrecognition of deferred taxes attributable to a temporary difference resulting from a translation adjustment depends on whether the indefinite reversal criterion of APB Opinion No. 23 applies to the unremitted earnings of the subsidiary. If deferred taxes are not recognized on unremitted earnings of the subsidiary in accordance with the provisions of APB Opinion No. 23, deferred taxes would not be recognized on temporary differences resulting from a translation adjustment relating to that subsidiary.

If deferred taxes are recognized on temporary differences resulting from translation adjustments, the tax effect of the translation adjustments should be charged or credited to the cumulative translation adjustment account in stockholders' equity.

3.2 Reporting currency is the functional currency

When the reporting currency (USD) is the functional currency for foreign operations, SFAS No. 52 requires remeasurement of non-monetary assets and liabilities of the foreign operations in USD using historical exchange rates.

SFAS No. 109, paragraph 9 prohibits the recognition of deferred taxes on temporary differences that result from changes in exchange rates or indexing for tax purposes related to non-monetary assets and liabilities that are remeasured using historical exchange rates when the functional currency is the reporting currency.

However, if the reporting currency is the functional currency and there is a difference between the historical foreign currency financial statement carrying amount of an asset or liability and its historical foreign currency tax basis exclusive of the effects of indexing, if any, foreign deferred taxes are recognized on that temporary difference. For instance, a difference between the historical foreign currency tax basis and the historical foreign currency financial statement carrying amount of an asset may arise due to differences in the method of depreciation or life used for tax and financial reporting purposes. A foreign deferred tax asset or liability should be recognized on that temporary difference in the foreign tax jurisdiction. Under SFAS No. 52, the resulting foreign deferred tax asset or liability is then remeasured into the reporting currency at current exchange rates.

4 Deferred taxes from capital consolidation

APB Opinion No. 16, Business Combinations, requires that amounts assigned to assets acquired and liabilities assumed in a purchase business combination be at fair value. Frequently, fair value will differ from the tax bases of the assets acquired and liabilities assumed, for example, because the transaction is non-taxable. Prior to SFAS No. 96 and SFAS No. 109, the net-of-tax approach was used in assigning values to assets acquired and liabilities assumed in purchase business combinations. Under the net-of-tax approach, the future tax effects of differences between the fair values and tax bases and the timing of those tax effects (discounting) were considered in assigning values to assets acquired and liabilities assumed. Deferred tax assets and liabilities were not recognized in purchase business combinations.

SFAS No. 109 prohibits the net-of-tax approach and requires assets acquired and liabilities assumed to be recorded at their 'gross' fair value. Differences between the assigned values and tax bases of assets acquired and liabilities assumed in purchase business combinations are temporary differences under the provisions of SFAS No. 109. Accordingly, deferred tax assets (assuming that the more likely than not criterion is met) and liabilities are recognized for differences between the assigned values and the tax bases of assets acquired and liabilities assumed in purchase business combinations. However, deferred taxes are not recognized for temporary differences relating to the portion of goodwill for which amortization is not deductible for tax purposes, unallocated negative goodwill, or leveraged leases acquired in purchase business combinations. Also, SFAS No. 109 does not permit discounting future tax effects.

When the purchase price exceeds the fair value of identifiable net assets acquired, the 'gross method' of allocating the purchase price under SFAS No. 109 is accomplished in the following manner:

- Assign fair values (irrespective of the tax bases) to identifiable assets acquired and liabilities assumed.
- Compare the assigned fair values of the identifiable assets and liabilities with their tax bases to determine the temporary differences.
- Recognize deferred tax assets and liabilities for the future tax consequences of the deductible and taxable temporary differences between the assigned values and the tax bases in accordance with the requirements of SFAS No. 109.
- Recognize a deferred tax asset for the tax benefits of operating loss and tax credit carryforwards for tax purposes acquired in the combination.
- Recognize a valuation allowance for all or some portion of the deferred tax asset for acquired deductible temporary differences and acquired carryforwards to reduce that deferred tax asset to the amount that more likely than not will be realized in the future by the consolidated entity.
- Record goodwill for the residual difference between (i) the purchase price; and (ii) the values assigned to identifiable assets and liabilities, including the deferred tax assets (net of any valuation allowance) and deferred tax liabilities.

5 Deferred taxes from elimination of inter-company profits

An inter-company transfer of assets such as the sale of inventory or depreciable assets between tax jurisdictions is a taxable event that establishes a new tax basis for those assets in the buyer's tax jurisdiction. However, no gain on the transfer is recognized in the consolidated financial statements on assets that remain within the consolidated group, based on the provisions of Accounting Research Bulletin No. 51, Consolidated Financial Statements. As a result of the transfers, there will be differences in the buyer's tax jurisdiction between the new tax basis of those assets and the carrying amount of those assets as reported in the consolidated financial statements.

SFAS No. 109, paragraph 9(e) and paragraph 124 retain the provisions of ARB 51 regarding taxes on inter-company profits. Under those provisions, income taxes paid by the seller on inter-company profits on assets that remain within the consolidated group, including the tax effect of any reversing temporary differences in the seller's tax jurisdiction, are deferred. SFAS No. 109 also prohibits the recognition of a deferred tax asset for the excess of the new tax basis of the assets in the buyer's tax jurisdiction over the financial statement carrying amount of the assets in the consolidated financial statements.

6 Deferred taxes from application of the equity method

If an investor company uses the equity method to account for its investment in the common shares of an investee company, SFAS No. 109 requires the provision of deferred taxes for the difference between income from the investment recognized for accounting purposes and dividend income recognized for tax purposes. Since dividend income is subject to the dividends received deduction, which exempts 80 (or 70) per cent of the dividends received from tax, only the portion of the dividend which is subject to tax is considered to be a timing difference. The remainder is considered to be a permanent difference and therefore no deferred taxes are required.

Under current GAAP, if the equity investment is in a corporate joint venture, and if the investor intends that the funds invested in the joint venture will be permanently invested, there is no requirement to provide for deferred taxes (see SFAS No. 109, paragraph 31).

7 Netting of deferred taxes

In accordance with the requirements of SFAS No. 109, deferred taxes must be shown in the balance sheet in four categories depending on the source and the expected duration of the timing difference. These four categories are: current deferred tax asset; non-current deferred tax asset; current deferred tax liability; and non-current deferred tax liability. Netting of the current deferred tax assets and current deferred tax liabilities is allowed as long as these are payable to the same government and owed by the same component of the tax payer. Netting is also allowed in the non-current category on the same basis. If the deferred tax assets and deferred tax liabilities are owed to different governments or by different components of the tax-payer, netting is not allowed.

There is no heading entitled 'deferred taxes' in the consolidated income statement. Consolidated income tax expense as reflected in the consolidated income statement includes both currently payable and deferred taxes. The division of the consolidated income tax expense into the currently payable amount and the deferred tax amount is shown in a footnote to the financial statements. With regard to the presentation of deferred taxes in either the consolidated balance sheet or the consolidated income statement, no distinction is made between deferred taxes which arise in the individual accounts and those which are a result of the consolidation process. In the footnotes to the financial statements disclosures may be made regarding the source of the deferred taxation.

8 Problems relating to applicable tax rate

The objective of SFAS No. 109 is to measure a deferred tax liability or asset using the enacted tax rate expected to apply to taxable income in the periods in which the deferred tax liability or asset is expected to be settled or realized. Under US tax law, if taxable income exceeds a specified amount, all taxable income is taxed at a single tax rate (i.e. the statutory rate). SFAS No. 109 requires most corporations to use the statutory tax rate for the measurement of deferred tax liabilities or assets. SFAS No. 109 requires deferred tax liabilities and assets to be adjusted to take into account the effect of a change in tax laws or a change in tax rates. The effect of the change in laws or rates must be included in income from continuing operations for the period in which the enactment of the law takes place.

9 Disclosure

The provisions of SFAS No. 109 include the following disclosure requirements:

- The following components of the net deferred tax asset or liability must be disclosed (paragraph 43):
  - The total of all deferred tax liabilities.
  - The total of all deferred tax assets (before a valuation allowance).
  - The total valuation allowance recognized for deferred tax assets.
- The net change during the year in the total valuation allowance must also be disclosed (paragraph 43).
- Public companies are required to disclose the tax effect of each type of temporary differences and carryforwards that gives rise to a significant portion of deferred tax assets (before any valuation allowance) and liabilities. Non-public companies may omit disclosures of the tax effects but should disclose the types of temporary differences and carryforwards that result in significant portions of deferred tax assets (before any valuation allowance) and liabilities (paragraph 43).
- The amounts of income tax expense or benefit allocated to the following items must be disclosed (paragraph 46):
  - Continuing operations.
  - Discontinued operations.
  - Extraordinary items.
  - Items charged or credited directly to stockholders' equity.
- Significant components of income tax expense or benefit allocated to continuing operations must be disclosed. Examples of those components include (paragraph 45):
  - Current tax expense or benefit.
  - Deferred tax expense or benefit, excluding the effects of other components.
  - Investment tax credits.
  - Government grants recognized as a reduction of income tax expense.
  - Benefits of operating loss carryforwards.
  - Tax expense that results from allocating certain tax benefits either directly to capital or to reduce goodwill or other noncurrent intangible assets of an acquired entity ('charges equivalent' to recognized tax benefits).
  - Effects on deferred tax assets and liabilities of changes in tax laws or rates or a change in the tax status.
  - Effects of adjustments to the beginning-of-the-year valuation allowance.
- A reconciliation (using percentages or USD amounts) of expected income tax expense attributable to continuing operations (amount of income tax expense that would result from multiplying pre-tax income from continuing operations by the federal statutory tax rate) to the reported amount of income tax expense allocated to continuing operations should be disclosed. Non-public companies are not required to present a numerical reconciliation, but they must disclose the nature of significant reconciling items (paragraph 47).
- The amounts and expiration dates of operating loss and tax credit carryforwards for tax purposes should be disclosed (paragraph 48).
- The portion of the valuation allowance for deferred tax assets for which subsequently recognized tax benefits will be (i) applied to reduce goodwill or other non-current intangible assets, as required by the provisions of the Statement related to purchase business combinations; or (ii) applied directly to contributed capital in accordance with paragraph 36 of SFAS No. 109 should be disclosed (paragraph 48).
- If a deferred tax liability is not recognized for temporary differences that relate to investments in foreign subsidiaries or foreign corporate joint ventures, undistributed earnings of domestic subsidiaries or domestic corporate joint ventures, bad debt reserves of savings and loan associations, or policyholders' surplus of stock life insurance companies under the provisions of APB Opinion No. 23, as amended by SFAS No. 109, or for deposits in statutory reserve funds by US steamship enterprises, the following information should be disclosed (paragraph 44):
  - A description of the types of temporary differences for which a deferred tax liability has not been recognized.
  - The cumulative amount of each such difference.

- A description of the nature of events that would result in those differences becoming taxable.
- The amount of unrecognized deferred tax liability for temporary differences related to investments in foreign subsidiaries or foreign corporate joint ventures that are essentially permanent in duration if determination of that liability is practicable, or a statement that determination is not practicable.
- The amount of unrecognized deferred tax liability for temporary differences other than those related to investments in foreign subsidiaries and foreign corporate joint ventures.

- Public enterprises that are not subject to income taxes directly (such as partnerships) must disclose that fact. The net amount of differences between financial statement carrying amounts of assets and liabilities and their tax bases also must be disclosed (paragraph 43).
- An enterprise that is part of a group that files a consolidated tax return is required to disclose in its separate financial statements the method of allocating taxes to members of the group and the amount of current and deferred tax expense and tax-related receivables from or payables to affiliates. The principal provisions of the method of allocation and any changes in that method also should be disclosed (paragraph 49).
- Additional Disclosures by Public Companies (Rule 4–08(h), 'General notes to financial statements—Income tax expense', of Regulation S-X):
  - The components of pre-tax income (loss) as either domestic or foreign.
  - Current tax expense or benefit and deferred tax expense or benefit applicable to US federal income taxes, foreign income taxes, and other income taxes for the year.

IX Formats

1 Regulation on individual accounts to be applied

GAAP does not prescribe specific formats for the presentation of either individual accounts or group accounts. However, it is traditional for assets to be placed on the left or above liabilities and equities and for liabilities and equities to be combined into one total (see USA – Individual Accounts, V.2 and V.3, for generally accepted formats for the presentation of financial statements). These formats are the same whether the financial statements are for an individual company or for a consolidated group. There is no requirement for a statement of movements in fixed assets.

2 Additional rules for group accounts

2.1 Changes in the consolidated group

Changes in the consolidated group (or changes in reporting entity) are considered an accounting change under US GAAP. An accounting change that results in financial statements that are in effect the statements of a different reporting entity should be reported by restating the financial for the new reporting entity for all periods. The following are changes in reporting entities (APB Opinion No. 20, paragraph 12):

- Presenting consolidated or combined statements in place of individual statements.
- Changing specific subsidiaries comprising the group of enterprises for which consolidated financial statements are presented.
- Changing the enterprises included in combined financial statements.
- A business combination accounted for by the pooling of interests.

Financial statements should describe the nature of the change and the reason for it. In addition, the effect of the change on income before extraordinary items, net income, and related per share amounts should be disclosed for all periods presented. Financial statements of subsequent periods need not repeat the disclosures (APB Opinion No. 20, paragraph 35).

In addition, APB Opinion No. 30 requires special disclosures in the footnotes whenever a segment of the business is disposed of during an accounting period.

2.2 Changes as a result of foreign currency translation

In consolidating the financial statements of subsidiaries that are stated in foreign currencies, US GAAP specifies that the current rate of exchange should be used to translate all balance sheet accounts except for the shareholders' equity accounts of the subsidiary (see discussion of SFAS No. 52 in V.4). The income statement accounts are translated at the average rate of exchange for the accounting period. Upon consolidation there will be a difference between the investment in the subsidiary account on the books of the parent company and the translated shareholders' equity accounts of the subsidiary, apart from asset revaluations and goodwill. This difference is a translation adjustment which is carried directly to a separate component of consolidated shareholders' equity and does not enter into the computation of consolidated net income for the period.

2.3 Changes caused by consolidation procedures

If there are changes caused by consolidation procedures or the application of the equity method, there must be disclosure of these changes and a retroactive restatement of the prior year's consolidated financial statements presented in the same format as the consolidated financial statements of the current period. These types of changes may be listed as follows:

- Change from unconsolidated subsidiary to consolidated subsidiary.
- Change from consolidated subsidiary to unconsolidated subsidiary.
- Change from equity method of accounting to consolidation method.
- Change from consolidation method to equity method of accounting.

2.4 Disclosure related to shareholders' equity accounts

In the consolidated balance sheet the shareholders' equity accounts are essentially those of the parent company. The minority interest reflected in the consolidated balance sheet is not subdivided into components representing paid-in capital or retained earnings, but is treated as a one-line item, normally under the liabilities. There is no requirement to disclose the share of profit or loss which is included in the minority interest. The calculation of the profit or loss to be included in the minority interest is based upon the reported net income of the subsidiary in its individual accounts. If there are inter-company sales which originate with the subsidiary, the minority interest is charged with its proportionate share of the inter-company profit elimination.

Corrections may be made to the consolidated shareholders' equity accounts. These changes are made on a retroactive restatement basis if they involve changes in the consolidated group or movements from or to the equity method. If the corrections involve errors of application of accounting methods, they are also treated in a retroactive manner. If they involve changes in accounting estimates, they are treated prospectively.

It is rare to have appropriations of retained earnings in the United States. If the subsidiary has appropriated retained earnings, these would be eliminated in the consolidation process.

X Notes and additional statements

1 Overview

The footnotes for the consolidated financial statements are the footnotes for the consolidated group. If the parent company is required to issue a complete set of separate financial statements, there must be a separate set of footnotes to accompany these financial statements. However, if the disclosure of the parent company information is contained in a footnote to the consolidated financial statements, no additional footnotes disclosure would be required.

2 Information concerning group structure

If the parent company has unconsolidated majority-owned subsidiaries, there must be disclosure of this fact and in appropriate cases there may be a requirement for condensed financial statement information for the unconsolidated subsidiary. If the parent company has less than majority-owned investments in the common stock of other companies and accounts for these investments under the equity method, this fact must be disclosed. However, information concerning the names of the investments or the precise application of the equity method is not required.

3 Description of consolidation methods

Consolidated statements shall disclose the consolidation policy that is being followed. ABP Opinion No. 22:

- Requires that significant accounting policies be presented as an integral part of the financial statements.
- Sets forth guidelines as to the content and format of disclosures and accounting policies.
- States that the preferable format is to present a separate Summary of Significant Accounting Policies preceding the notes to financial statements or as the initial note.

Accordingly, companies typically disclose their consolidation and foreign currency translation policies within a Summary of Significant Accounting Policies. A short example follows:

> Consolidation: The consolidated financial statements include the accounts of the company and all majority-owned subsidiaries. Investments in affiliates, owned 20 per cent to 50 per cent, are carried at cost plus equity in undistributed earnings since acquisition. The effects of intercompany transactions have been eliminated.
>
> Foreign currencies: The assets and liabilities of foreign subsidiaries are translated at current exchange rates, while revenues and expenses at average rates prevailing during the year. Translation adjustments are reported as a separate component of stockholders' equity.

4 Special items of the group accounts

As no separate group accounts are maintained in the United States, the consolidated financial statements are prepared from the separate accounts and financial statements of the parent company and its subsidiaries. The only special items that would appear in the consolidated financial statements that would not appear in the separate financial statements of the parent or the subsidiary would be minority interest, minority interest income and goodwill arising from the consolidation process (i.e. when the subsidiary is not merged into the parent, but retains a separate legal existence).

5 Additional group-related information

In the form 10-K which must be filed with the SEC on an annual basis, all of the subsidiaries in the consolidated group must be listed according to their legal name. In addition, if any of the subsidiaries have publicly issued securities outstanding, separate financial statements for these subsidiaries may need to be prepared for the benefit of the security holders.

XI Auditing

For the most part the rules regarding auditing for consolidated accounts are exactly the same as the rules for individual accounts. In accordance with the rules of the SEC, the consolidated financial statements of companies with publicly traded securities must be audited by a Certified Public Accountant (CPA) to be filed with the SEC on form 10-K and included in the annual report to the shareholders. Audited consolidated financial statements must be filed with the SEC on form 10-K within 90 days of the date of the consolidated financial statements. The audit report for the consolidated group is typically the responsibility of the public accounting firm which signs the report. In many cases, the public accounting firm will use the work of affiliated or corresponding public accounting firms in other countries which have audited the financial statements of subsidiaries of the US parent company located in those countries. As long as the audit report is signed by a public accounting firm that is licensed to practise within the United States, the SEC has no objection to the fact that audit work is performed by public accountants licensed in other jurisdictions. Moreover, if the audit report and the level of auditing standards are comparable to those of the United States, the report of recognized firms of public accountants, even though not licensed to practise in the United States, is also acceptable to the SEC.

XII Filing and publication

The SEC, which is empowered by the Securities Exchange Act of 1934 to establish accounting standards in the United States, has delegated the establishment of accounting standards to the private sector in the form of the FASB. Nevertheless, the SEC retains the power to establish accounting standards, and it has used this power from time to time when controversies have arisen over the establishment of accounting principles. In addition, the staff of the SEC has broad ranging authority to prevent the issuance of securities which they feel may have incorrect or misleading accounting treatments, and may compel companies who wish to issue securities in the United States to change or correct their accounting treatments to conform to those treatments which the staff of the SEC believe to be preferable.

The responsibility of the SEC is to regulate the securities markets in the United States including such markets as the New York Stock Exchange, the American Stock Exchange and the NASDAQ automated stock exchange. A necessary component of efficient trading in the securities markets is the availability of reliable information about the companies whose securities are being traded. Therefore, the Securities Exchange Act of 1934 requires all companies with publicly traded securities to file an annual report with the SEC on form 10-K. This annual report must include audited consolidated financial statements prepared in accordance with GAAP. In addition, these companies must issue an annual report to their shareholders which includes audited consolidated financial statements.

All companies desiring to issue securities publicly in the United States must file a registration statement with the SEC. A registration statement includes a considerable amount of required information about the company issuing securities, including audited consolidated financial statements. Thus, a second major activity of the SEC lies in the review of registration statements prior to the issuance of securities. Accountants working for the SEC may review the financial statements of companies issuing securities. If the accounting practices are deemed to be in accordance with GAAP, the financial statements portion of the registration statement is approved. If the financial statements contain accounting practices which are not considered by the accounting staff to be in accordance with GAAP, the company would be asked to change or correct these accounting practices prior to allowing the securities to be sold publicly.

The most important regulations of the SEC are Regulation S-X, which describes the rules for the form and content of financial statements filed with the SEC, and Regulation S-K, which covers the required non-financial statement related disclosures that must be included in the registration statements and the other forms that must be filed with the SEC (see USA – Individual Accounts, IV.4.1, for a discussion of these regulations). In order to reduce some of the complexity in the registration and reporting process, an Integrated Disclosure System (IDS) was implemented in the early 1980s. The IDS amended Regulation S-X for the purpose of standardizing financial statement requirements in SEC filings. Thus, financial statements included in annual reports to shareholders of public companies are usually the same as those included in forms filed with the SEC. Under current rules, the content of the annual report to shareholders is the same as that in the 10-K form filed with the SEC. Form 10-K is required to be filed within 90 days of the end of the company's fiscal year. The 10-K must be signed by the chief executive officer of the company, the chief financial officer, the chief accounting officer and a majority of the company's board of directors.

The disclosures required by the SEC for public companies are summarized in Table I.6 in USA – Individual Accounts. As shown in the table, the

SEC divides the disclosures into four groups in order to distinguish the information required to be disclosed in annual reports to shareholders from the complete information package required for filings with the SEC. For example, the information included in Part II of Table I.6 is primarily accounting information that is required for annual reports filed with the SEC as well as the annual reports distributed to the company's shareholders. The disclosure requirements summarized in Parts I, III and IV of the table are only required for filings with the SEC, but they may be included voluntarily in annual reports to shareholders.

XIII Sanctions

The sanctions that may be imposed in the area of consolidation accounting are the same as those for individual accounts. As discussed above, the SEC requires annual audited consolidated financial statements to be filed with the SEC on form 10-K. In the event that there are faults in the consolidation procedures, the auditors should discover these faults and cause them to be corrected by the company. If the auditor does not find the faults in the consolidation procedures, the SEC staff may discover these faults in the process of reviewing the filings that are made with the SEC on an annual basis. If the SEC discovers faults in the consolidation process, it may compel the correction of the faults or it may impose sanctions ranging from temporary suspension of filing and trading of securities to civil fines and even imprisonment for flagrant and intentional violations.

If neither the SEC staff nor the auditors discover the faults in the consolidation procedures, these faults may be subsequently discovered by financial analysts or shareholders. Financial analysts may publish information which they discover concerning the faults and the company may react by performing special investigations in order to determine the facts correctly. Shareholders may also feel that they have been misled by the financial statements, in which case they may bring a lawsuit claiming that the company intended to mislead the shareholders and asking for monetary damages from the court. In recent periods many lawsuits have been brought by shareholders claiming information in financial statements has been misleading. Many of these lawsuits relate to items in the individual accounts, such as inventory items or items of revenue recognition. Lawsuits brought on the basis of faults in consolidation are relatively less common.

Bibliography

American Institute of Certified Public Accountants (1979). *Joint Venture Accounting* (American Institute of Certified Public Accountants, New York).

American Institute of Certified Public Accountants (1989). *Accounting Trends and Techniques* (American Institute of Certified Public Accountants, New York).

American Institute of Certified Public Accountants (1992). *Accounting Trends and Techniques* (American Institute of Certified Public Accountants, New York).

American Institute of Certified Public Accountants (1998). *Accounting Trends and Techniques* (American Institute of Certified Public Accountants, New York).

Baker, R., Lembke, V. and King, T. (1998). *Advanced Financial Accounting*, 4th edition (McGraw-Hill, New York).

Financial Accounting Standards Board (1998a). *Original Pronouncements, Accounting Standards I* (Financial Accounting Standards Board, Norwalk, CT).

Financial Accounting Standards Board (1998b). *Original Pronouncements, Accounting Standards II* (Financial Accounting Standards Board, Norwalk, CT).

Financial Accounting Standards Board (1998c). *Current Text, Accounting Standards Volume I* (Financial Accounting Standards Board, Norwalk, CT).

Financial Accounting Standards Board (1998d). *EITF Abstracts, A Summary of Proceedings of the FASB*

Emerging Issues Task Force as of September 23–24, 1998 (Financial Accounting Standards Board, Norwalk, CT).

Financial Accounting Standards Board (1998e). *Financial Accounting Series No. 192-A, Invitation to Comment, Methods of Accounting for Business Combinations: Recommendations of the G4+1 for Achieving Convergence, 1998* (Financial Accounting Standards Board, Norwalk, CT).

Heckler, B. L. and Acciarito, R. (1999). 'How to Handle the Challenge of Accounting for Leveraged Recapitalizations', *Journal of Corporate Accounting and Finance*, Spring, pp. 22–40.

Wyatt, A. (1963). *A Critical Study of Accounting for Business Combinations*, Accounting Research Study No. 5 (AICPA, New York).

GLOSSARY

Hanne Böckem
Thomas Schröer

Introduction

The *Transnational Accounting* (TRANSACC) glossary aims to provide the reader with a well-structured and comprehensive tool for ease of use in daily practical work. The choice of items included in the glossary derives from close study of other glossaries included in contemporary accounting literature as well as the particular concerns of TRANSACC and the experience gathered in the co-ordination of the TRANSACC project. There is no guarantee of completeness, in part because some terms such as 'true and fair view' have not been included, since their translation into other languages is rather cumbersome and many countries either make use of the original English term or use descriptions of which the equivalence to the original is more than debatable.

The vocabulary relating to finance and accounting is presented in the form of a specialized dictionary in twelve languages: English, Danish, Dutch, Finnish, French, German, Italian, Japanese, Norwegian, Portuguese, Spanish and Swedish. However, other languages, or rather dialects, which are similar to languages already presented in the glossary are presented as well. For example, Dutch and French terms as used in Belgium are included in the respective columns for Dutch (as spoken in the Netherlands) and French (as spoken in France). Any expression used differently in Belgium is followed by '(B)'. Similarly, it has been necessary to supplement the Spanish column for the Argentine terms, followed by '(ARG)', where differences occur. The same applies also to German as different terms are used occasionally in Germany and Switzerland. Swiss terms are followed by '(CH)'. For the Japanese column, the Romanized version as well as the original Japanese characters are provided.

Without the help of native speakers, the task of setting up a glossary would have been insurmountable. We would therefore like to thank the growing number of TRANSACC authors involved in the process for their assistance in choosing and defining the terms, and for providing the respective national terms. However, the responsibility for the correctness of the terms chosen and translated remains that of the authors of this glossary.

From the beginning of this multi-language glossary's development, emphasis was laid not only on linguistic matters, i.e. the dictionary-style ordering of the entries. As languages always derive from a complex cultural base, translation matters can never be regarded as simple transformation processes from one media to another. Complications occur when cultures do not seem to be compatible in the sense that inter-cultural equivalents of nations institutions and customs may not exist. This general problem obviously applies to accounting, being also part of a nation's culture. Consequently, the lack of sufficient cross-cultural or, essentially, *transnational* accounting compatibility became obvious initially in the handling of the purely linguistic aspect of the glossary. Hence, the reader will not find terms such as 'true and fair view' or 'generally accepted accounting principles/practice' in the TRANSACC glossary.

Because of these problems, we have chosen the following approach in setting up the glossary. Having first selected a set of English accounting terms, we added definitions in order to attach meaning to them. These definitions thus served to create some common ground on which to base the corresponding equivalent terms of other languages. In the next step, we asked the TRANSACC authors and editors to comment on the definitions, which were then partly adopted. Finally, we distributed the English terms and the respective definitions to the native speakers of the languages chosen and asked for the national expressions which best fitted the definitions. Thus, the national terms provided are in effect equivalents of the definitions rather than mere translations of the English terms.

From the various terms we were offered by different native speakers of the same language we learned that even terms included in the present glossary sometimes derive from the technical language of an individual. Consequently, the listed terms may not in every case represent the only correct choice, since synonyms do exist and in some cases the given terms may not be the ones most commonly used. At this stage we decided to enlarge the glossary design team to include you, the TRANSACC user, if you wish. The further development of the glossary is thus meant to be an interactive and integrative process between TRANSACC users and TRANSACC authors. We would therefore like to invite readers to provide us with critical comments and proposals on the selection of terms (enlargements or deletions), on national particularities which should be incorporated, and on deficiencies and possible improvements to the definitions and the corresponding national terms.

With your help, the TRANSACC glossary will develop through continuous refinement and enlargement of the chosen terms, thereby becoming more and more what it is intended to be: namely, a valuable instrument for daily practical work.

Definition of terms

account	Any document or device whereby a record is kept of flows of value measured in money terms.
accountability	The responsibility to explain actions involving financial matters to others by rendering an account and submitting to an audit.
accounting	The process of measuring, analyzing and systematically recording operations and transactions, and of summarizing, reporting and interpreting the results thereof.
accounting period	Specified period for which the development of the enterprise's financial situation is accounted for.
accounting policies	Accounting treatment of specific items applied in accordance with accounting principles and followed consistently by an entity.
accounting principles	Rules that guide the measurement, classification and interpretation of economic information, and communication of the results by means of financial statements.
accounting theory	A consistent set of accounting principles.
accruals principle	Revenues and expenses are recognized as they are earned or incurred, not as cash is received or paid.
accrued expenses	An item on the asset side of the balance sheet which represents cash outflows of the current period to be recognized as an expense in the profit and loss account of later periods.
accrued revenue/income	An item on the liability side of the balance sheet which represents cash inflows of the current period to be recognized as revenue/income in the profit and loss account of later periods.
accumulated losses	The sum of losses of the current reporting period and of previous periods.
accumulated profits	The sum of profits of the current reporting period and of previous periods.
acquisition costs	The purchase price of an asset and the costs incurred to get the asset into a usable condition.
amortization	The process of writing off an intangible asset to the profit and loss account over its estimated life.
amount repayable	Amount necessary for settlement of a liability; usually, the amount liabilities are stated in the balance sheet.
annual general meeting	Meeting of a company's shareholders and governing bodies.

annual report	Set of documents, particularly including the primary statements, published annually by a company for external users.
appreciation	Increase in the value of an asset over its cost or book value.
appropriation of profit	The decision as to what uses the profit for the financial year should be put to: e.g. distribution, or transfer to reserves.
articles of association	A document setting out the internal conduct of a company's affairs.
asset	Balance sheet items classified as assets according to national definitions. Usually, expenses are capitalized as an asset if future benefits (which contribute to future cash inflows or to the reduction of future cash outflows) can reasonably be expected.
associate	An enterprise over which another enterprise exercises a significant influence; also termed 'associated company or associated undertaking'.
audit	Examination of the accounts and their underlying records in order to express an opinion on their honesty and veracity.
audit opinion	Usually, a statutory text stating whether or not the accounts that have been audited are in accordance with prescribed principles and present the position of the enterprise fairly.
auditor	The person who examines the accounts of an enterprise in the course of an audit.
auditor's report	A report from the auditor of a company that is required by statute to be annexed to the audited annual accounts, and in which the auditor expresses an opinion as to whether the accounts comply with national accounting regulations.
average cost method	Usually applied where stock units are identical or near identical in order to compute an average unit cost by dividing the total cost by the number of units. Ideally, the average unit cost should be revised with every receipt of stock.
balance sheet	Primary statement of an entity in which all assets, liabilities and equity items are listed.
balance sheet continuity	Identity between the closing balance sheet of the current year and opening balance sheet of the following year.
balance sheet date	The date to which the reported figures in the entity's accounts refer. It is also the deadline for the inclusion of events that influence recognition and valuation of assets and liabilities but occur after the balance sheet date.
balance sheet total	The sum of the values of the stated fixed (tangible and intangible) and current assets, which equals the sum of the values of the stated liabilities plus owners' equity.

base stock cost method A base stock, i.e. a level of stock regarded as the irreducible minimum with which a business can operate, treated as a permanent asset and valued at constant costs. Any excess over the base stock is usually valued by some other method and added to the value of the base stock.

book value The value at which an asset or a liability is stated in the books.

book value method of consolidation A method of capital consolidation in which a subsidiary's identifiable assets, liabilities and goodwill are added on a line-by-line basis to the corresponding amounts for other group undertakings. Goodwill is determined by cancelling the parent's investment against the subsidiary's attributable equity at acquisition, making fair value adjustments to the subsidiary's assets and liabilities at that date for the share of the parent only. Only the post-acquisition profits of subsidiaries are included in the group accounts. Minority interests in the equity of less than wholly-owned subsidiaries are shown at book value of the subsidiaries' equity.

bookkeeping The technical process of maintaining financial records.

brand A proprietary name or label serving as a reputational signal in the marketing process, thus including aspects of goodwill.

business combination A business combination is the bringing together of separate enterprises into one economic entity as a result of one enterprise uniting with or obtaining control over the net assets of another enterprise.

capital Amount invested in a business by its proprietors and finance obtained from external sources, as shown on the liability side of the balance sheet.

capital contribution Increase in equity resulting from investments made by owners in their capacity as owners.

capital grants Money given by an institution (e.g. the government) for specified capital investments.

capital increase Increase of share capital through self-financing or additional capital contributions by existing or new members.

capital maintenance A concept of profit determination which allows the distribution of profits only after the initial value of the capital, as defined by different concepts of capital maintenance, e.g. nominal or real, of the business has been restored.

capital market The capital market is composed of the stock market, the bond market, and the money market. A distinction can be made between organized markets for (a) new issues of equity interests (stocks or shares) and bonds (loan stocks) (primary

	market); and (b) sale of stocks and bonds by the existing holders (secondary market) and non-organized markets for loans, equity interests, and real-estate mortgages.
capital reduction	Reduction of share capital through return of cash to the shareholders or recognising past revenue losses as a permanent capital loss.
capital reserve	Amount set aside by an entity that is not available for distribution, e.g. share premium account or revaluation reserve.
capital stock/share	Total par value of all shares issued by a corporation. In the United States, the issued shares need not have a par value.
capital capitalization	Inclusion of the amount spent for the purchase of an asset into the balance sheet.
cash	Cash comprises cash in hand and demand deposits.
cash equivalents	Cash equivalents are short-term, highly liquid investments that are readily convertible to known amounts of cash, and which are subject to an insignificant risk of change in value.
cash flow	The movement of cash and cash equivalents into and out of an entity within a defined period.
cash flow statement	A financial statement explaining how cash and the cash equivalents held by the business have changed between two balance sheet dates and often classified into cash flow from operating, financing, and investing activities. It is often seen as being synonymous with the 'funds statement'.
chart of accounts	An index to all the accounts in a double entry system.
closing rate	The closing rate is the spot exchange rate at the balance sheet date.
closing rate method	Method of currency translation whereby the balance sheet items are uniformly translated at the exchange rate ruling at the reporting date.
comparability	A qualitative characteristic of financial statements of different entities at the same point in time, or of the same entity at different points in time, which enables comparisons to be made because similar transactions and events have been treated similarly and in a consistent manner. Harmonization of financial reporting in particular has been induced by the need for comparable information.
consistency of valuation	Application of consistent valuation methods across time and within one period for like items.
consistency principle	The principle that similar matters, e.g. concerning valuation methods and the classification of the financial statement, should be treated in a similar way whenever they arise, to

prevent arbitrary variations in accounting methods and to promote inter- and intra-period comparability.

consolidated balance sheet A financial statement which shows the assets and liabilities of a group of companies as if that group were a single entity.

consolidation of capital Offsetting of the investment as stated in the individual accounts of the parent company against the attributable equity of the subsidiary at the date of acquisition as stated in its individual accounts.

consolidation of income and expenses Process of consolidating the group members' profit and loss accounts and eliminating income and expenses resulting from intra-group transactions.

consolidation of inter-company balances Process of offsetting intra-group debts and receivables.

construction contract A construction contract is a contract specifically negotiated for the construction of an asset or a combination of assets that are closely interrelated or interdependent in terms of their design, technology and function, or their ultimate purpose or use.

contingency A contingency is a condition or situation, the ultimate outcome of which, gain or loss, will be confined only on the occurrence, or non-occurrence, of one or more uncertain future events.

contingent liabilities A legal obligation that arises dependent upon a future event, the occurrence of which is uncertain. Contingent liabilities are based, e.g. on the issuance of bills of exchange, on guarantees, or on indemnity agreements.

contingent rent A contingent rent is the portion or the lease payments that is not fixed in amount but is based on a factor other than just the passage of time (e.g. percentage of sales, amount of usage, price indices, market rates of interest).

continuity of classification A concept according to which the formats of the annual accounts should be comparable from one period to the next.

corporation A legal entity which has all the rights and responsibilities of a person except those rights which only a natural person can exercise.

corporation tax A type of income tax that has to be paid by a corporation, itself being a separate legal entity, based on taxable profit (usually not identical to the accounting profit).

cost flow assumptions Systematic bases for the recognition of the cost of sales in order to match costs with related revenues.

cost of sales	Amount of expenses that relate to sales within a reporting period. Cost of sales are matched with the sales in the profit and loss account.
cost of sales method	Method of profit computation where the total amount expended in producing or acquiring the goods sold during a period of time is set off against sales.
cost plus contract	A cost plus contract is a construction contract in which the contractor is reimbursed for allowable or otherwise defined costs, plus a percentage of these costs or a fixed fee.
credit entry	An entry recording the creation of, or an addition to a liability, owners' equity, revenue, as well as the reduction/elimination of an asset or expense.
current asset	An asset which is held temporarily. Assets usually classified as current are stocks, debtors and cash.
current cost	A basis of valuation which values an asset at the amount it would currently cost to acquire.
current cost accounting	The method of accounting which bases valuations on the current cost of assets.
current investment	A current investment is an investment that is by its nature readily realizable and is intended to be held for not more than one year.
current purchasing power accounting	A valuation concept that allows for the effect of inflation on the value of money.
current service costs	Current service costs is the cost to an enterprise under a retirement benefit plan for services rendered in the current period by participating employees.
current tax	Current tax is the amount of income taxes payable (recoverable) in respect of the taxable profit (tax loss) for a period.
current value	As opposed to historical cost, the value of an asset at the reporting date.
de facto control	Control concept for defining a subsidiary based on effective ('after the fact') control; e.g. through the actual exercise of a dominant influence.
de jure control	Control concept for defining a subsidiary according to legal rights; e.g. majority of voting rights, right to appoint or dismiss the majority of members of the administration, management or supervisory body, or the right to exercise a controlling influence based on a contract of domination.
debit entry	An entry recording the creation of, or addition to an asset, the incurring of an expense, or the reduction/elimination of a liability, owners' equity or revenue.

debt	An obligation in money or services owed to a third party.
deferred taxation	Accumulated amounts of notional taxation arising from timing, i.e. reversible differences arising from a difference between profits computed for taxation purposes and profits as reported in the annual accounts. Deferred taxation can be reported as an asset (accumulated tax allocation debit) or as a liability (accumulated tax allocation credit).
defined benefit plan	Defined benefit plans are post-employment benefit plans under which an enterprise pays fixed contributions into a separate entity (a fund) and will have no legal or constructive obligation to pay further contributions if the fund does not hold sufficient assets to pay all employee benefits relating to employee service in the current and prior periods.
defined contribution plan	Defined benefit plans are post-employment plans other than defined benefit plans.
depreciable asset/ amount	The depreciable amount of a depreciable asset is the historical cost or other amount substituted for historical cost in the financial statements, less the estimated residual value.
depreciation	The recognition of the diminution in the value of tangible fixed assets with clearly defined lives.
depreciation method	Any method of systematically charging diminutions in the value of tangible fixed assets to the profit and loss account.
depreciation plan	Systematic documentation of the depreciation process including the depreciation base, period, method and the remaining book values in each period of the useful life of the asset.
development	Development is the application of research findings or other knowledge to a plan or design for the production of new or substantially improved materials, devices, products, processes, systems or services prior to the commencement of commercial production or use.
development costs	Expenses incurred in order to produce new or substantially improved materials, devices, products, processes, systems or services prior to the commencement of commercial production.
directors' report	A statement by the directors of a company containing narrative descriptions of the development and future prospects of the company. The content varies according to national rules.
discontinued operations	A discontinued operation results from the sale or abandonment of an operation that represents a separate, major line of business of an enterprise, and of which the assets, net profit

	or loss and activities can be distinguished physically, operationally and for financial reporting purposes.
distributable reserves	Reserves created from the current and previous years' profit which may be used for dividend payment.
dividends	Payment made by companies to their shareholders out of current profits or distributable reserves.
dominant influence	Criterion constituting the relationship between a parent and a subsidiary resulting from the *de facto* control concept. An influence which one company has over another such that it is able to dictate the commercial and financial policies of the other without regard to the interests of any other.
earnings per share	A company's net profit attributable to ordinary shareholders, divided by the number of shares in issue.
economic ownership	The economic owner can exclude the legal owner from all significant risks and rewards arising from the use of the asset.
elimination of intra-group profits	Elimination of profits on transactions between group members, which have not met realization criteria as far as the group as a whole is concerned. Such realization criteria include, e.g. sale to third parties outside the group, or usage in the case of fixed assets.
enterprise	An organisation undertaking an activity which involves risk and promises a reward commensurate with that risk.
environmental/ green accounting	Accounting for the ecological consequences of the operation of an enterprise.
equity	The excess of total assets over liabilities, encompassing paid-in capital and retained earnings. It is usually increased by profits and reduced by losses, the payment of dividends and the redemption of capital.
equity method	A method of accounting for investments, whereby the investment is initially recorded at cost or at equity and the carrying value adjusted thereafter to show the investor's pro rata share of changes in net assets of the investee since acquisition.
exchange difference	An exchange difference is the difference resulting from reporting the same number of units of a foreign currency in the reporting currency at different exchange rates.
exchange rate	The rate at which one currency can be exchanged for another.
exchange risk	Risks deriving from future changes in the exchange rates of different currencies.
executory/pending contract	Contract whose terms are not fulfilled by all of the contracting parties.
expenditure	Amount of cash disbursement.

expense	A past, current or future cash outflow which reduces profit in the current period's profit and loss account.
extraordinary items	Items in the financial statement which arise from events or transactions that fall outside the ordinary activities of the reporting entity and which are not expected to recur.
fair value	Fair value is the amount for which an asset could be exchanged between knowledgeable, willing parties in an arm's length transaction.
finance lease	A lease which is essentially a method of financing the purchase of economic property.
financial fixed assets	Balance sheet item comprising investments and participations that are held for the long term.
financial instruments	Any contract that gives rise to both a (recognized or unrecognized) financial asset of one enterprise and a (recognized or unrecognized) financial liability or equity instrument of another enterprise.
financial reporting	The process of reporting the financial position and progress of a business to persons outside that business.
financial statement analysis	Process of generating additional information about the company's financial situation by analyzing the structure of the balance sheet items and calculating ratios.
financial year	The period of (usually) one year for which financial statements are regularly prepared.
fixed assets	Assets which are intended for continued use.
fixed-asset movement schedule	A statement detailing changes in the composition of fixed assets through purchases, disposals, depreciation etc.
fixed price contract	A fixed price contract is a construction contract in which the contractor agrees to a fixed contract price or a fixed rate per unit of output, which in some cases is subject to cost escalation clauses.
foreign currency translation	The restatement of items in a foreign currency into the reporting currency.
format	Prescribed form of presenting statements, e.g. the balance sheet and the profit and loss account, the particulars of which are usually dependent on legal form or size criteria.
formation expenses	The expenses required to start up or extend a business's operating activities.
forms of business organization	Legally defined structures of the organization of a business.

forward rate	Exchange rate that is currently determined in order to buy a currency/commodity for future delivery.
funds statement	Statement showing how the activities of the enterprise have been financed, including the extent to which funds have been generated from operations, how the financial resources have been used, and the effects of these activities on the funds of the enterprise. Often used as synonymous with 'cash flow statement'.
gearing	The ratio of loan capital to ordinary share capital; also termed 'leverage'.
geographical segment	Geographical segments are distinguishable components of an enterprise engaged in operations in individual countries or groups of countries within particular geographical areas as may be determined to be appropriate in an enterprise's particular circumstances.
going concern principle	Principle assuming an indefinite life of the business, thereby affecting the valuation of assets and liabilities.
goodwill	The value of a business as a whole in excess of its net assets, as calculated by the separate valuation of its identifiable assets and liabilities. Goodwill arises from the anticipated earning power of an enterprise. It may be self-generated or purchased.
grant	Money given by an institution (e.g. the government) usually for a specified purpose.
group accounts	Financial statements dealing with a group of companies, usually taking the form of consolidated financial statements representing the group of companies as if they were a single entity.
group management report	A statement by the directors of a parent company containing narrative descriptions of the development and future prospects of the group of companies; also termed chairman's/chief executive's report, operating financial review etc. The content varies according to national rules.
group/collective valuation	Similar assets not valued separately but as a group.
guaranteed residual value	Guaranteed residual value is: (a) In the case of the lessee, that part of the residual value which is guaranteed by the lessee or by a party related to the lessee (the amount of the guarantee being the maximum amount that could, in any event, become payable); and

	(b) In the case of the lessor, that part of the residual value which is guaranteed by the lessee or by a third party unrelated to the lessor who is financially capable of discharging the obligations under guarantee.
hedging	Offsetting the risk associated with one item by the benefits of another, the risks and benefits thus being negatively correlated.
hidden reserves	An amount by which the owners' equity in a business has been understated in the annual accounts as a result of an undervaluation or omission of assets or an overstatement of liabilities.
historical cost accounting	Accounting method where assets and liabilities are generally recognised and measured at historical costs.
historical costs	The total expenditure made by the owner to acquire title to an asset and bring it into usable condition.
historical rate	Exchange rate prevailing at the date of first recognition of a transaction in the accounts.
holding company	An entity which owns a controlling interest in the shares of one or more companies. The terms 'parent undertaking' or 'parent company' are used in some jurisdictions.
holding gain	The difference between the current value of an item that is still recorded in the balance sheet, i.e. not sold, and its related book value, sometimes called 'capital gain'.
horizontal/account form	Presentation of the balance sheet and the profit and loss account on facing pages.
imparity principle	Principle governing the anticipation of expected losses, in contrast to the recognition of expected profits which should not be anticipated.
individual accounts	Set of statements, usually comprising the balance sheet, profit and loss account and the notes, that has to be prepared regularly by an individual enterprise.
industry segment	Industry segments are the distinguishable components of an enterprise, each engaged in providing a different product or service, or a different group of related products or services primarily to customers outside the enterprise.
inflation accounting	A technique of revaluation in accordance with the concept of financial capital maintenance (constant purchasing power), either applying a general index or asset specific indices to the book values of items in the annual accounts.
information function	Purpose of setting up annual accounts, being the provision of user-orientated information for decision-making.

initial (or first) consolidation	First-time inclusion of a subsidiary into the group accounts.
intangible assets	An asset that lacks physical substance, e.g. goodwill, patents, copyrights, trademarks, etc.
inter-company profit	A profit realized by one member of a group of companies as a result of transactions with another member of the group.
interim report/ statements	A report or statement on the operational activities of an entity prepared, e.g. quarterly, within the reporting period.
inventory/stocktaking	The physical count of the types and amounts of all assets owned by an entity at a particular date.
investments	An item on the asset side of the balance sheet, classified either as a fixed or a current asset, representing funds invested in other entities, e.g. in shares.
joint venture	Entity managed jointly by one or more entities in order to co-operate in manufacturing, distribution, R&D etc.
lease term	The lease term is the non-cancellable period for which the lessee has contracted to lease the asset together with any further terms for which the lessee has the option to continue to lease an asset, with or without further payment, which option at the inception of the lease it is reasonably certain that the lessee will exercise.
leasing	The conveyance of the right to use an asset by one person (lessor) to another (lessee) for a specified period of time in return for rent.
legal ownership	Title of legal rights to an asset.
legal reserves	Reserves, the setting up of which is required by the national accounting law.
liabilities	An entity's legally enforceable obligation to pay a certain amount of money to another party.
liability	A person or entity may be held liable for its actions, e.g. in case of activities that cause damage to another party.
liquid assets	Class of assets, e.g. cash in hand and in bank, and temporary investments readily convertible into cash, that are instantaneously available for payment of current liabilities.
liquidity	The convertibility of assets into ready cash.
listed companies	An entity that is traded on a stock exchange.
loan	A sum of money advanced by one person to another with the intention that the amount concerned should ultimately be repaid.

long-term contract	A contract entered into for the supply of goods or services such that the activity of meeting the contract extends over more than one accounting period or takes more than twelve months to complete.
loss	The excess of expenses over revenues arising from a single transaction or transactions of an economic agent during a well-defined period of time.
loss brought forward	The sum of losses of accounting periods prior to one the accounts are prepared for, which are brought forward.
loss for the financial year	Amount by which the expenses incurred in an accounting period exceed the revenues and other income of the period.
majority shareholder	A shareholder in a company who holds more than 50 per cent of the total shares.
managing board	Organ of a corporation which plans, organizes, leads and controls in order to achieve the company's objectives.
mandatory reserve	Hidden reserve which is included in the balance sheet as a result of conservative valuation rules, e.g. the acquisition cost principle, etc.
market value	The value of an item determined by the respective market forces.
marketable securities	Securities held by an entity that are readily convertible into cash by a market transaction.
matching principle	Principle according to which expenses are recognized in the period in which their related revenues are recognized as income. Revenues and expenses are matched with one another in so far as their relationship can be established or justifiably assumed.
materiality	The significance of an item or its value that demands recognition in the annual accounts, usually judged in relation to the making of decisions by users, and relative to costs.
measurement	Measurement is the process of determining the monetary amounts at which the elements of the financial statements are to be recognized, and carried in the balance sheet and the income statement.
merger	In general, a situation in which two or more enterprises that were previously separate legal entities agree to cease to be distinct enterprises, thereby becoming one legal entity.
merger surplus	Difference in case of a merger between the net assets of the merged entity and the value of the consideration rendered.

minority interest	The equity of non-group shareholders in a subsidiary company, as shown in the consolidated balance sheet. The minority interests are calculated as a share of capital as defined under the method of capital consolidation.
monetary items	Nominal positions in the balance sheet with a fixed money value, i.e. where the units of value equal the units of quantity.
negative goodwill	Arises where the price paid for an entity or participation is less than the value of the attributable identifiable net assets because of probable or certain future losses anticipated by the buyer.
net assets	The difference between the value of all assets and all liabilities recognized in the balance sheet.
net book value	Value at which a balance sheet item is stated after charging depreciation.
net current assets	Difference between current assets and current liabilities.
net realizable value	Estimated selling price of an asset achievable in the ordinary course of business, less reasonably predictable costs of completion and disposal.
nominal capital	The share capital of a company calculated at its nominal amount, i.e. the number of shares multiplied by their face value.
nominal value	The face value of a security, usually being lower than its traded price.
notes	Explanatory or supplementary information appended to and forming an integral part of the annual accounts.
off-balance sheet transaction	A transaction or an arrangement, the full extent of which is not disclosed in the annual accounts. An example is where 'asset' and 'liability' elements of a transaction are netted in a balance sheet. In an extreme case, such netting may mean that the transaction is not disclosed at all.
operating activities	Operating activities are the principal revenue-producing activities of an enterprise, and other activities that are not investing and financing activities.
operating lease	Any lease which is not a finance lease, i.e. where not all the risks and rewards are transferred to the lessee. It often coincides with the hire period, being substantially less than the life of the leased asset.
operating profit	Profit arising from the usual operating activities of an entity.
parent company	A company heading a group of companies, based on national legal definitions of the parent–subsidiary relationship.

participating interest	Long-term investments in other entities usually assumed to exist when a certain percentage of shares (e.g. 20%) are held.
partnership	Form of business organization in which two or more persons (the partners) engage in common long-term business activities based on a contract between them.
patent	A registered legal right granted to a person or entity to make exclusive use of an invention.
percentage of completion method	An accretion method of recognizing profit according to the degree of completion of goods or services under a long-term contract.
plan assets	Plan assets are assets (other than non-transferable financial instruments issued by the reporting enterprise) held by an entity (a fund) that satisfies all of the following conditions: (a) The entity is legally separate from the reporting enterprise; (b) The assets of the fund are to be used only to settle the employee benefit obligation, are not available to the enterprise's own creditors, and cannot be returned to the enterprise (or can be returned to the enterprise only if the remaining assets of the fund are sufficient to meet the plan's obligations); and (c) To the extent that sufficient assets are in the fund, the enterprise will have no legal or constructive obligation to pay the related employee benefits directly.
pooling-of-interest accounting	A method of capital consolidation in which a subsidiary's identifiable assets, liabilities and goodwill are added on a line-by-line basis to the corresponding amounts of other group undertakings. They are not restated to fair values at acquisition. The parent's purchase consideration is measured at nominal value in the parent's individual accounts, and on consolidation is offset against the nominal amount of the subsidiary's attributable share capital. Any excess or deficit is dealt with in reserves, and goodwill does not arise. Viewed as a change in scope of the accounts, consolidated reserves include both pre- and post-acquisition reserves of the subsidiary. Comparatives are restated. Minority interests in the equity of less than wholly-owned subsidiaries are reported.
post-balance sheet events	Events, both favourable and unfavourable, which become known between the balance sheet date and the date the annual accounts are prepared, and which relate to conditions that existed at the balance sheet date.

pre-payments	Payments in advance.
present value	Discounted future cash flows arising from an asset's use or sale.
principle of completeness	Principle which states that all assets and liabilities of a company must be recorded in the balance sheet.
principle of separate valuation	Principle which states that assets and liabilities shall be valued on an item-by-item basis.
principle of the lower of cost or market	A valuation principle of current assets prescribing the use of either the historical cost or the market value of an asset, whichever is the lower.
production costs	Costs ascribed to an asset that is internally produced. The type of costs to be included in production costs is dependent on national rules.
profit	The excess of revenues over expenses arising from a single transaction or transactions of an economic agent during a well-defined period of time.
profit after tax	The profit of a company after allowing for the corporation or income tax due on the profit.
profit and loss account	Statement in the annual accounts that shows the expenses and revenues of the reporting period.
profit carried forward	Profits of accounting periods prior to one the accounts are prepared for, which are not distributed and thus carried forward to the reporting period.
profit for the financial year	The amount of profit for the reporting period, calculated as the excess of revenues and realized gains over expenses and realized losses in the profit and loss account, which is subject to appropriation. The terms 'gains' and 'losses' are defined slightly differently between jurisdictions.
profit reserves	Reserve in the balance sheet created by the retention of profit.
profit transfer agreement	Agreement under which a corporation assumes the obligation to transfer its entire profits to another entity.
property, plant and equipment	Property, plant and equipment are tangible assets that: (a) Are held by an enterprise for use in the production or supply of goods or services, for rental to others, or for administrative purposes; and (b) Are expected to be used during more than one period.
proportional consolidation	A method of capital consolidation in which a subsidiary's identifiable assets and liabilities are added on a line-by-line basis to the corresponding positions in the group account according to the percentage held. Goodwill is determined by cancelling the parent's investment against the subsidiary's

	attributable proportional equity at acquisition, after making fair value adjustments to the subsidiary's assets and liabilities at that date. Only the post-acquisition profits of subsidiaries are included in the group accounts. Under this approach, minority interests are not reported. Used in many countries where an enterprise is jointly controlled in conjunction with non-group enterprises (see 'joint venture').
provision	An amount set aside by charging it in the profit and loss account to provide for any reasonably expected commitment, the amount of which cannot be determined with complete accuracy.
provision for bad debts	Valuation adjustment for receivables, for which it is reasonable certain that they will never be received.
provision for charges	Provision set up for the purpose of preventing to an extent the distribution of profits in order to be able to meet future expenditure where no obligation to third parties arises, such as general overhauls, etc.
provision for diminution in value	An amount set aside by making charges to the profit and loss account, to cover any anticipated diminution in value of an asset which cannot be determined with complete accuracy.
provision for future losses	Provision set up to cover anticipated losses. There are stringent requirements in certain jurisdictions over the circumstances in which such provisions are permissible.
provision for guarantees/ warranties	Provision set up to cover certain product support commitments resulting from past sales transactions.
provision for pensions	Provisions set up to cover granted future pension rights. The accounting basis for such provisions is usually specified in national accounting regulations.
prudence principle	A principle of dealing with uncertainty in accounting to the effect that accounting statements should be prepared on a cautious basis, i.e. not overestimating asset values and profits, thus avoiding undue optimism. Usually, foreseeable losses are anticipated, but not gains.
publication	The aspect of financial reporting that is concerned with making annual accounts available to outside parties, e.g. via publication in an official gazette.
purchase method of consolidation	A method of capital consolidation in which a subsidiary's identifiable assets, liabilities and goodwill are added on a line-by-line basis to the corresponding amounts for other group undertakings. Goodwill is determined by cancelling the

	parent's investment against the subsidiary's attributable equity at acquisition, after making fair value adjustments to the subsidiary's assets and liabilities at that date. Only the post-acquisition profits of subsidiaries are included in the group accounts. Minority interests in the equity of less than wholly-owned subsidiaries are reported at book value or restated.
realization principle	Principle which states that profits should only be included in the profit and loss account if realized, generally implying a transaction with a third party, thus prohibiting the inclusion of, e.g. unrealized holding gains.
recognition	Recognition is the process of incorporating in the balance sheet or income statement an item that meets the definition of an element and satisfies the prevailing recognition criteria.
recoverable amount	The recoverable amount is the amount which the enterprise expects to recover from the future use of an asset including the residual value on disposal.
reducing balance method	Depreciation method by which the annual depreciation charge is calculated as a fixed percentage of net book value.
reinstatement of value	Reversal to its original value of exceptional diminutions in value of an asset provided for in previous periods.
replacement value	The amount it would cost to replace an asset with another asset that will render similar services.
reportable segment	A reportable segment is an identified business segment or geographical segment based on the foregoing definition, for which segment information is required to be disclosed by the prevailing accounting standard.
reporting currency	The reporting currency is the currency used in presenting the financial statement.
research	Original and planned investigation undertaken with the prospect of gaining new scientific or technical knowledge and understanding.
reserves	Part of the entity's equity. The term 'capital reserves' (comprising, for example, share premiums) indicates that the reserves are regarded as undistributable, either for legal reasons or because of the company's discretion. The term 'revenue reserves' is used in some countries to indicate that such reserves are distributable. Often, though, reserves are classified according to origin, e.g. 'revaluation reserves' and reference must be made to national accounting principles as regards distributability.
retained earnings	Amount of profit retained in the company that is not yet distributed but may be distributed in a future year.

revaluation	The process of placing the fair value or the current market value on an asset which is different from its current recorded value.
revaluation reserve	A reserve to which are credited surpluses arising on the revaluation of assets.
revenue grants	Money given by an institution (e.g. the government) to match specified expenses.
revenues	Proceeds from sales of goods and services as a result of the operating activities of an entity.
sale and lease-back	The process of taking a lease on property previously sold by the lessee to the lessor. The object of this process is to release funds which were previously tied up in fixed assets.
sales	The total value, at selling prices, of the goods or services sold by an entity within one reporting period shown in the profit and loss account.
salvage value	The estimated price of an asset, e.g. a fixed asset, at the date it is expected to be no longer required for use. Also termed 'residual value' or 'scrap value'.
securities	A negotiable claim against a company's assets or activities incorporated in a note such as a share or debenture. Securities in this sense are thus held as income-bearing investments.
segmental reporting	Financial information that relates to geographical areas, major business lines, groups of customers, etc., presented outside the primary financial statement.
selling expenses	Expenses of an organization relating to the selling or marketing of its goods or services, i.e. advertising or salaries and commission paid to travelling sales people, as against expenses incurred for other specialized functions such as administration, financial and manufacturing.
share capital	The ownership interest in a limited company. Issued share capital is the total nominal amount of shares the entity has issued. Authorized share capital is the total nominal amount of the shares the entity is allowed to issue under its constitution.
share premium account	A reserve arising from the issue of new shares at a price in excess of their nominal value.
shareholder	A holder of any class of shares in an entity. Some jurisdictions make a distinction between equity and non-equity shares, whereas others use the terms 'equity' and 'shares' more or less interchangeably.

significant influence	Criterion constituting the relationship between a parent company and an associated company.
spot rate	Exchange rate ruling at the present time.
statutory reserves	Reserves, the setting up of which is prescribed in a company's statutes.
stocks/inventory	An item on the asset side of the balance sheet, usually comprising the value of trading goods, either finished or semi-finished, raw materials, etc.
straight line depreciation	Method of depreciation of an asset by charging to the profit and loss accounts equal amounts over the estimated useful life of the asset.
subordination agreement	Agreement under which a corporation subordinates its management to that of another enterprise.
subsidiary	A member of a group, other than a parent, that is controlled by the parent as defined by national accounting regulations. Usually all subsidiaries (or in some jurisdictions 'subsidiary undertakings') have to be consolidated in preparing group accounts, unless they satisfy exclusion criteria.
sum-of-the-years'-digit method	A depreciation method where the depreciation charge for each year of the estimated useful life decreases from year to year by a fixed amount.
supervisory board	An organ of a company that has a duty to control the activities and decisions of the management board.
tangible assets	Physical items meeting the recognition criteria for assets included in a balance sheet.
tax accounts	Set of financial statements prepared annually according to current tax legislation, forming the basis for the assessment of tax due.
tax base	The tax base of an asset or liability is the amount attributed to that asset or liability for tax purposes.
tax expense (tax income)	Tax expense (tax income) is the aggregate amount included in the determination of net profit or loss for the period in respect of current tax and deferred tax.
taxable profit	Taxable profit (taxable loss) is the profit (loss) for the period, determined in accordance with the rules established by the taxation authorities, upon which income taxes are payable (recoverable).
temporal method	Method of translating foreign currency financial statements, according to which assets and liabilities are translated with the exchange rate ruling at the date relating to the value attributed

	to the item, e.g. for assets recorded at historical cost, the historical rate is applied.
temporary difference	Temporary differences are differences between the carrying amount of an asset or liability in the balance sheet and its tax base. Temporary differences may be either: (a) Taxable temporary differences which are temporary differences that will result in taxable amounts in determining taxable profit (tax loss) of future periods when the carrying amount of the asset or the liability is recovered or settled; or (b) Deductible temporary differences, which are temporary differences that will result in amounts that are deductible in determining taxable profit (tax loss) of futures periods when the carrying amount of the tax asset or liability is recovered or settled.
total costs method	Method of profit computation where the total amount expended in the reporting period is matched with the sales, corrected by the increase or decrease of stocks and own work capitalized within the reporting period.
transfer price	The price charged by one segment of an enterprise for a product or service it provides to another segment of the enterprise.
translation differences	Differences resulting from the process of translating foreign financial statements into the group currency.
uniting/pooling of interest	A uniting of interest is a business combination in which the shareholders of the combining enterprises combine control over the whole, or effectively the whole, of their net assets and operations to achieve a continuing mutual sharing in the risks and benefits attaching to the combined entity such that neither part can be identifies as the acquirer.
unrealized profit	Profits that did not pass the realization test and may thus not be included in the profit and loss account according to the realization principle, e.g. holding gains.
useful economic life	The period during which a fixed asset can efficiently be kept in use and during which the asset is usually being depreciated.
user group	A group of persons having a definable common characteristic who have a legitimate interest in published accounts.
valuation	The process of attributing a monetary value to an item recognized in the financial statement.
value added tax	Tax levied as a certain percentage of the difference between the net selling price and bought-in components used in the

	production process. The total tax change on finished goods is ultimately paid by the customer.
vertical/report form	Presentation of the balance sheet or the profit and loss account in a columnar layout.
voting right	The legal rights which attach to ordinary shares and some other securities, whereby the holder may vote at a meeting of the company.
work in progress	Goods in the course of manufacture usually recorded at production costs.

Language table

ENGLISH	DANISH	DUTCH, BELGIAN (B)	FINNISH	FRENCH, BELGIAN (B)	GERMAN
account	konto	(grootboek) rekening	tili	compte	Konto
accountability	pligt til at aflægge regnskab	verantwoordings-plicht	tilitysvelvollisuus	reddition de comptes	Rechenschaft
accounting	regnskabsvæsen	verslaggeving	laskentatoimi	comptabilité	Rechnungswesen
accounting period	regnskabsperiode	verslagperiode	laskentakausi	exercice comptable (*one year*), periode comptable (*less than one year*)	Rechnungsperiode
accounting policies	regnskabsmetoder, regnskabsprin-cipper	grondslagen van waardering en resultaatbepaling	laskentaperia-atteet	méthodes comptables	Bilanzpolitik
accounting principles	regnskabsprin-cipper	basisprincipes van de verslaggeving	tilinpäätös-periaatteet	principes comptables	Bilanzierungs-grundsätze
accounting theory	regnskabsteori	verslaggevings-theorie	kirjanpitoteoria	théorie comptables	Bilanztheorie
accruals principle	periodiserings-princippet	toerekeningsbe-ginsel	suoriteperuste	comptabilité en créances et dettes, principe de l'annualité de l'exercice (B)	Prinzip der Periodenabgren-zung
accrued expenses	forudbetalinger, periodeafgræns-ningsposter	vooruitbetaalde kosten	siirtosaatava	compte de régularisation actif, charges payées d'avance	aktivischer Rechnungs-abgrenzungs-posten
accrued revenue/income	forudbetalinger, periodeafgræns-ningsposter	vooruitontvangen opbrengsten	siirtovelka	compte de régularisation passif, produits perçus ou comptabilisés d'avance	passivischer Rechnungs-abgrenzungs-posten
accumulated losses	akkumulerede tab, overført resultat	geaccumuleerde verliezen	kertyneet tappiot	pertes cumulées, pertes reportées (B)	Verlustvortrag

ITALIAN	JAPANESE	NORWEGIAN	PORTUGUESE	SPANISH, ARGENTINE (ARG)	SWEDISH
conto	kanjô 勘定	konto	conta	cuenta	konto
responsabilità	kaikei sekinin 会計責任	regnskapsplikt	prestação de contas	responsabilidad financiera	redovisningsskyldighet
contabilizzazione	kaikei 会計	regnskap	contabilidade	contabilidad	redovisning
esercizio	kaikei kikan 会計期間	regnskapsperiode	exercício, período contabilístico	período contable	redovisningsperiod
principi contabili del gruppo	kaikei hôshin 会計方針	regnskapsprinsipper	políticas, práticas contabilísticas	políticas, prácticas contables	redovisningsprinciper
principi contabili	kaikei gensoku 会計原則	regnskapsprinsipper	princípios contabilísticos	principios, normas contables	redovisningsprinciper
principi contabili	kaikei riron 会計理論	regnskapstoeri	teoria da contabilidade	teoría contable	redovisningsteori
principio della competenza temporale ed economica	hassei gensoku 発生原則	periodiserings-prinsippet, (regnskaps-prinsippet)	princípio da especialização (ou do acréscimo)	principio de devengo, principio del devengado (ARG)	bokföringsmässiga grunder
ratei passivi	maebarai hiyô 前払費用	periodiserte utgifter	custos diferidos, despesas antecipadas	gastos anticipados, gastos pagados por adelantado (ARG)	upplupna kostnader
rateo attivo	maeuke shûeki 前受収益	periodiserte inntekter	proveitos diferidos, receitas antecipadas	ingresos anticipados, ingresos cobrados por adelantado (ARG)	upplupna intäkter
perdite portate a nuovo	mishori sonshitsu 未処理損失	akkumulerte tap	prejuízos acumulados	resultados negativos de ejercicios anteriores, pérdidas acumuladas (ARG)	balanserade förluster

ENGLISH	DANISH	DUTCH, BELGIAN (B)	FINNISH	FRENCH, BELGIAN (B)	GERMAN
accumulated profits	akkumuleret overskud, overført resultat	geaccumuleerde winsten	kertyneet voittovarat	bénéfices cumulés, bénéfices reportés (B)	Gewinnvortrag/ Gewinnrücklagen, Gewinnreserven (CH)
acquisition costs	anskaffelsesom-kostninger, anskaffelsespris	kosten van overname	hankintameno	coût d'acquisition, valeur d'acquisition (B)	Anschaffungs-kosten
amortization	afskrivninger på immaterielle anlægsaktiver	aflossing, afschrijving	jaksottaminen	amortissement	Abschreibung
amount repayable	beløb der skal tilbagebetales på gæld	de hoofdsom van een schuld	erääntynyt velka	montant à rembourser, dettes	Rückzahlungs-betrag
annual general meeting	generalforsamling	algemene vergadering van aandeelhouders	yhtiökokous	assemblée générale annuelle	Hauptversamm-lung/Generalver-sammlung, Generalversammlung (CH)
annual report	årsregnskab	jaarrekening	tilinpäätös	rapport annuel, rapport de gestion (B)	Geschäftsbericht
appreciation	værdiforøgelse, opskrivning	herwaardering	arvonkorotus	augmentation de valeur, réévaluation	Zuschreibung, Aufwertung (CH)
appropriation of profit	overskudsforde-ling	winstverdeling	voittovarojen käyttö	affectation des bénéfices	Gewinnverwen-dung
articles of association	vedtægter	statuten	yhtiöjärjestys	statuts	Satzung, Statuten (CH)
asset	aktiv	activa	vastaavaa	actif	Vermögensgegen-stand/Aktivum
associate	associeret virksomhed	geassocieerde onderneming	osakkuusyritys	entreprise associée	assoziiertes Unternehmen, (nicht konsolidierte) Beteiligung (CH)
audit	revision	controle	tilintarkastus	audit, contrôle des comptes	Abschlussprüfung

ITALIAN	JAPANESE	NORWEGIAN	PORTUGUESE	SPANISH, ARGENTINE (ARG)	SWEDISH
utili a nuovo, non distribuiti	mishobun rieki 未処分利益	akkumulerte overskudd	lucros acumulados	beneficios acumulados	balanserade vinster
costo storico, costo di acquisto	shutoku genka 取得原価	anskaffelseskost	custo de aquisição	precio de adquisición, costo de adquisición (ARG)	kostnader i samband med företagsförvärv
ammortamento (di immobilizzazioni immateriali)	shôkyaku 償却	avskrivning	amortização	amortización contable	avskrivning
valore di estinzione	miharai gaku 未払額	beløp som skal tilbakebetales på gjeld	montante em dívida, valor de reembolso	valor de reembolso	förfallen skuld
assemblea degli azionisti, soci	teiji kabunushi sôkai 定時株主総会	generalforsamling	assembleia geral	junta general de accionistas, asamblea ordinaria de accionistas (ARG)	bolagsstämma
bilancio annuale	nenji hôkokusho 年次報告書	årsrapport	relatório e contas, documentos de prestação de contas	cuentas anuales, estados contables anuales (ARG)	årsredovisning
rivalutazione	hyôkazô 評価増	verdiøkning	reavaliação	revalorización	uppskrivning
destnazione dell'utile	rieki shobun 利益処分	disponering av årsoverskudd	(proposta de) aplicação de resultados	acuerdo de distribución del resultado, asignación de resultados	vinstdisposition
atto costitutivo, statuto	teikan 定款	vedtekter	contrato (da sociedade), estatutos, pacto social	códigos de conducta interna	bolagsordning
attività	shisan 資産	eiendel	activo	activo	tillgång
società collegata	kanren kigyô 関連企業 kanren gaisha 関連会社	tilknyttet selskap	empresa associada	sociedad asociada	intressebolag
revisione contabile	kansa 監査	revisjon	auditoria, revisão	auditoría	revision

ENGLISH	DANISH	DUTCH, BELGIAN (B)	FINNISH	FRENCH, BELGIAN (B)	GERMAN
audit opinion	revisionserklæring, revisors konklusion	accountantsverklaring	tilintarkastajan lausuma	opinion d'audit	Bestätigungsvermerk, Prüftestat (CH)
auditor	revisor	controlerend accountant revisor (B)	tilintarkastaja	auditeur, commissaire aux comptes, réviseur d'entreprise (B)	Wirtschaftsprüfer
auditor's report	revisionserklæring	verslag van de accountant, verslag van de revisor (B)	tilintarkastuskertomus	rapport de l'auditeur, rapport de contrôle (B)	Prüfungsbericht
average cost method	varelageropgørelse ved brug af gennemsnitsomkostninger	gemiddelde kostenmethode	Keskimääräiskustannusmenetelmä	méthode du coût moyen, méthode des prix moyens pondérés (B)	Durchschnittskostenmethode
balance sheet	balance	balans	tase	bilan	Bilanz
balance sheet continuity	formel kontinuitet	aansluiting tussen eindbalans en begin balans	tasejatkuvuus	identité du bilan d'ouverture et du bilan de clôture de l'exercice précédent	Bilanzidentität
balance sheet date	regnskabsafslutningstidspunkt	balansdatum	tilinpäätöspäivä	date du bilan, date de clôture	Bilanzstichtag
balance sheet total	balancesum: aktiver ialt passiver ialt	balanstotaal	taseen loppusumma	total du bilan	Bilanzsumme
base stock cost method	basislager metoden, normallager	ijzeren voorraadstelsel	kiinteävarastomenetelmä	méthode du stock de base	Methode des eisernen Bestandes, Eiserne Reserve-Methode (CH)
book value	bogført værdi	boekwaarde	kirjanpitoarvo	valeur nette comptable	Buchwert

ITALIAN	JAPANESE	NORWEGIAN	PORTUGUESE	SPANISH, ARGENTINE (ARG)	SWEDISH
relazione della società di revisione	kansa iken 監査意見	revisors konklusjonsavsnitt	parecer do auditor	opinión del auditor	uttalande i revisionsberättelses
revisore contabile, società di revisione	kansa nin 監査人	revisor	auditor, revisor de contas	auditor	revisor
relazione della società di revisione	kansa hôkokusho 監査報告書	revisjonsberetning	relatório de auditoria, certificação legal de contas	informe de auditoría, informe del auditor (ARG)	revisionsberättelse
metodo del costo medio	heikin genka hô 平均原価法	gjennomsnitts-metoden	método do custo médio ponderado	coste medio ponderado continuo, método de promedios móviles de costos (ARG)	genomsnittskost-nadsmetoden, vid lagervärdering
stato patrimoniale	taishaku taishôhyô 貸借対照表	balanse	balanço	balance de situación, estado de situación patrimonial, hoja de balance (ARG)	balansräkning
corrispondenza delle scritture di riapertura con le scritture di chiusura	taishaku taishôhyô no keizokusei 貸借対照表の継続性	kontinuitets-prinsippet for balansen	correspondência entre o balanço de fecho e o balanço de abertura do exercício seguinte	correspondencia entre balances de cierre y apertura	kontinuitetsprincipen
data di bilancio	taishaku taishôhyô bi 貸借対照表日	balansedato	data (de encerramento) do balanço	fecha de cierre	balansdag
totale attivo	taishaku taishôhyô sôgaku 貸借対照表総額	balansesum	total do balanço	total general del activo, pasivo	balansomslutning
metodo del costo basato sulla scorta base	kijun tanaoroshi hô 基準棚卸法	normallagermetoden	método de custeio básico ou normal	valoración según la existencia base	metod för värdering av varulager där minimilagret värderats till konstant belopp
valore di libro, valore netto contabile	chôbo kagaku 帳簿価額	regnskapsmessig verdi	valor (líquido) contabilístico	valor contable, valor de libros (ARG)	bokfört värde

ENGLISH	DANISH	DUTCH, BELGIAN (B)	FINNISH	FRENCH, BELGIAN (B)	GERMAN
book value method of consolidation	overtagelses-metoden, past equity metoden, kostprismetoden	integrale consolidatie	nimellisarvome-netelmä	consolidation par intégration globale	Buchwert-methode (der Kapitalkonsoli-dierung)
bookkeeping	bogføring	boekhouding	kirjanpito	tenue de(s) comptes	Buchhaltung
brand	varemærke	merk	tavaramerkki	marque	Marke
business combination	virksomhedssam-menslutning	samengaan van bedrijven	yritysyhteen-eiittymä	regroupement d'entreprises	Unternehmens-zusammen-schluss
capital	egenkapital	kapitaal	pääoma	capital	Kapital
capital contribution	kapitalindskud	kapitaalinbreng	pääomansijoitus	apport au capital, contribution de capital (B)	Kapitaleinlage
capital grants	investeringstilskud	investeringspre-mies, kapitaalsub-sidies (B)	yritystuki	subventions d'investissement, subsides en capital (B)	Investitionszu-schuss, Subvention (CH)
capital increase	vækst i egenkapital	kapitaalverhoging	osakepääoman korotus	augmentation de capital	Kapitalerhöhung
capital maintenance	kapitalvedlige-holdelse	instandhoudings-beginsel	pääoman säilyttäminen	maintien du capital	Kapitalerhaltung
capital market	aktiemarked	kapitaalmarkt	pääomamark-kinat	marché des capitaux	Kapitalmarkt
capital reduction	kapitalnedsættelse, nedsættelse af selskabskapitalen	kapitaalvermin-dering	osakepääoman alennus	réduction du capital	Kapitalherabset-zung
capital reserve	bunden reserve	niet uitkeerbare reserve	sidottu varaus	réserve non distribuable	Kapitalrücklage, Kapitalreserven (CH)
capital stock/ share capital	selskabskapital, aktiekapital	uitstaand aandelenkapitaal	osakepääoma	capital social, capital actions (B)	gezeichnetes Kapital, Aktienkapital (CH)

ITALIAN	JAPANESE	NORWEGIAN	PORTUGUESE	SPANISH, ARGENTINE (ARG)	SWEDISH
consolidamento integrale	chôbo kagaku hô renketsu 帳簿価額法連結	*not in use*	método de consolidação integral	método de integración global, consolidación línea por línea (ARG)	parivärdemetoden
tenuta della contabilità	boki 簿記	regnskapsføring	escrituração	llevanza de libros, teneduría de libros (ARG)	bokföring
marchio	shôhyô 商標	varemerke	marca comercial	marca comercial	varumärke
aggregazione di imprese	kigyô ketsugô 企業結合	foretaksintegrasjon	concentração	fusión	koncernbildning
capitale	shihon 資本	egenkapital (sum egenkapital og gjeld)	capital	capital	kapital
versamenti in conto capitale	shihon kyoshutsu 資本拠出	kapitalinnskudd	entradas de capital	aportaciones de socios, aportes de los socios (ARG)	kapitaltillskott (nyemission)
contributi in conto capitale	shihon joseikin 資本助成金	offentlig tilskudd	subsídios ao investimento	subvenciones de capital	statligt stöd
incremento di capitale sociale	zôshi 増資	egenkapitaløkning	aumento de capital	ampliación de capital	kapitalökning
conservazione del capitale	shihon iji 資本維持	kapitalopprettholdelse	manutenção do capital	método del mantenimiento del capital, concepto de mantenimiento de capital (ARG)	skydd för kapital
mercato finanziario, mercato di capitali	shihon shijô 資本市場	kapitalmarked	mercado de capitais	mercado de capitales	kapitalmarknad
riduzione di capitale sociale	genshi 減資	nedskrivning av aksjekapital	redução de capital	reducción de capital	kapitalminskning, nedskrivning av aktiekapital
riserva non distribuibile	shihon jumbikin 資本準備金	*no longer in use* ('bunden egenkapital')	reservas indisponíveis	reservas, reservas de capital (ARG)	bundet kapital
capitale sociale	shihonkin 資本金	nominell aksjekapital	capital social, capital nominal	capital social	aktiekapital

ENGLISH	DANISH	DUTCH, BELGIAN (B)	FINNISH	FRENCH, BELGIAN (B)	GERMAN
capitalization	aktivering	activering	aktivointi	activation, inscription à l'actif, capitaliser (B)	Aktivierung
cash	likvide midler	kasmiddelen	rahat ja pankkisaamiset	banque et caisse, disponible, fonds de caisse (B)	Zahlungsmittel
cash equivalents	likvide midler, mest likvide aktiver	geldmiddelen (excl. kasmiddelen)	rahamääräiset erät	quasi-espèces, valeurs assimilables à des espèces	Zahlungsmittel-äquivalente
cash flow	pengestrøm, cash flow	kasstroom	kassavirta	flux de trésorerie	Cash Flow
cash flow statement	pengestrømsopgørelse, cash flow beskrivelse	kasstroomoverzicht	rahoituslaskelma	tableau des flux de trésorerie, flux de trésorerie (B)	Cash Flow – Rechnung, Geldflussrechnung (CH)
chart of accounts	kontoplan	rekeningen-schema	tilikartta	plan de comptes, plan comptable	Kontenrahmen
closing rate	slutkurs	koers op balansdatum	tilinpäätöspäivän kurssi	taux de clôture	Stichtagskurs
closing rate method	ultimokursmetoden, slutkursmetoden	omrekening tegen eindkoersen	muuntaminen tilinpäätöspäivän kurssiin	méthode du taux de clôture	Stichtagsmethode
comparability	sammenlignelighed	vergelijkbaarheid	vertailukelpoisuus	comparabilité	Vergleichbarkeit
consistency of valuation	kontinuitet i værdiansættelsen	stelselmatigheid van waardering	arvostusperiaatteiden jatkuvuus	permanence des méthodes d'évaluation, identité (B)	Bewertungsstetigkeit
consistency principle	kontinuitetsprincippet	stelselmatigheid-sbeginsel	jatkuvuusperiaate	principe de la permanence des méthodes, principe d'identité des méthodes (B)	Stetigkeitsprinzip

ITALIAN	JAPANESE	NORWEGIAN	PORTUGUESE	SPANISH, ARGENTINE (ARG)	SWEDISH
capitalizzazione	shihonka 資本化 shisan keijô 資産計上	balanseføre	capitalização	activación	aktivering
disponibilità liquide	genkin yokin 現金預金	likvider	disponibilidades (caixa)	tesorería, caja y bancos (ARG)	kassa och bank
disponibilità liquide equivalenti/ valori in cassa	genkin dôtôbutsu 現金同等物	likvide midler	equivalentes de caixa	activos líquidos equivalentes de caja (ARG)	likvida medel andra än kassa och bank
flusso finanziario, cash flow	kyasshu furô キャッシュ・フロー	kontantstrøm	fluxo de caixa, fluxo de tesouraria	flujo de tesorería, flujo de caja (ARG)	kassaflöde
prospetto dei flussi finanziari, rendiconto finanziario	kyasshu furô keisansho キャッシュ・フロー計算書	kontantstrømoppstilling	demonstração dos fluxos de caixa	estado de flujos de tesorería, estado de flujos de fondos (ARG)	kassaflödesanalys
piano dei conti	kanjô shoshiki zu 勘定組織図	kontoplan	quadro de contas, lista de contas	plan de cuentas	kontoplan
cambio di fine esercizio	kessambi rêto 決算日レート	kurs ved regnskapsårets slutt	taxa de encerramento, taxa de fecho	tipo de cambio en la fecha de cierre	balansdagskurs
metodo del cambio corrente	kessambi rêto hô 決算日レート法	dagskursmetoden	método da taxa de fecho	conversión de moneda extranjera según tipo de cambio vigente en fecha de cierre	dagskursmetod
comparabilità, omogeneità	hikaku kanôsei 比較可能性	sammenlignbarhet	comparabilidade	comparabilidad de la información económico-financiera, comparabilidad (ARG)	jämförbarhet
uniformità di applicazione dei criteri di valutazione	hyôka no keizokusei 評価の継続性	konsistens ved vurdering	princípio da consistência, princípio da uniformidade	principio de uniformidad, valuación consistente (ARG)	konsistenta, värderingsprinciper
uniformità di applicazione dei criteri di valutazione e delle classificazioni	keizokusei gensoku 継続性原則	konsistensprinsippet	princípio da consistência, princípio da uniformidade	principio de uniformidad, uniformidad (ARG)	konsistensprincip

ENGLISH	DANISH	DUTCH, BELGIAN (B)	FINNISH	FRENCH, BELGIAN (B)	GERMAN
consolidated balance sheet	koncernbalance	geconsolideerde balans	konsernitase	bilan consolidé	Konzernbilanz
consolidation of capital	kapitaludligning	vermogensmuta-tiemethode	sisäisen omistuksen eliminointi	substitution des capitaux propres aux titres	Kapitalkonsoli-dierung
consolidation of income and expenses	koncernresultat-opgørelse	consolidatie van opbrengsten en kosten	sisäisten tuot-tojen ja kulujen vähentäminen	consolidation des produits et des charges	Aufwands- und Ertragskonsoli-dierung
consolidation of inter-company balances	eliminering af koncerninterne tilgodehavender og gæld	eliminatie van intracompany schulden en vorderingen	keskinäisten saatavien ja velkojen vähentäminen	élimination des soldes intra-groupe	Schuldenkonsoli-dierung
construction contract	entreprisekontrakt	langlopend (bouw)contract	rakentamis-sopimus	contrat de construction	Fertigungsauftrag
contingency	eventualitet	voorwaardelijke baat of last	satunnaisuus	éventualité	Eventualität, Bedingung
contingent liabilities	eventualforplig-telser	voorwaardelijke verplichtingen	vastuut	passif éventuel, dettes conditionnelles (B)	Eventualverbind-lichkeit, Haftungs-verhältnisse
contingent rent	variabel andel af leje, leasingydelse	voorwaardelijke betaling	käytöstä riippuva vuokra	loyer conditionnel	bedingte Miete
continuity of classification	kontinuitet med hensyn til regnskabsposter-nes klassifikation, formel kontinuitet	stelselmatigheid in de tijd	vertailukelpoi-suus	permanence des méthodes de présentation, comparabilité (B)	Gliederungsste-tigkeit

ITALIAN	JAPANESE	NORWEGIAN	PORTUGUESE	SPANISH, ARGENTINE (ARG)	SWEDISH
stato patrimoniale consolidato	renketsu taishaku taishôhyô 連結貸借対照表	konsernbalanse	balanço consolidado	balance de situación consolidado, estado de situación patrimonial consolidado (ARG)	koncernbalansräkning
eliminazione delle partecipazioni consolidate	shihon renketsu 資本連結	*not in use*	compensação de capital, eliminação dos valores contabilísticos das participações de capital	eliminación inversión fondos propios, consolidación de capital (ARG)	eliminering av förvärvat eget kapital
consolidamento dei costi e dei ricavi ed eliminazione di quelli infragruppo	shûeki hiyô renketsu 収益費用連結	konsernresultatregnskap	eliminação de resultados internos	eliminación de gastos e ingresos por operaciones internas, consolidación de estados de resultados (ARG)	konsolidering av intäkter och kostnader
elisione dei crediti e debiti infragruppo	saiken saimu renketsu 債権債務連結	eliminering av intern gjeld og fordringer	eliminação de dívidas nas operações internas	eliminación de créditos y débitos recíprocos	eliminering av interna fordringar och skulder
commesse a lungo termine	kôji keiyaku 工事契約	anleggskontrakt	contrato de construção	contrato de construcción	entreprenaduppdrag
sopravvenienza	gûhatsu jishô 偶発事象	betinget utfall	contingência	contingencia	händelse vars förekomst bekräftas av en eller flera ovissa framtida händelser som helt eller delvis ligger utanför företagets kontroll
passività potenziali	gûhatsu saimu 偶発債務	eventuell forpliktelse	passivo contingente	pasivos contingentes	villkorliga skulder, eventualförpliktelser
canone di locazione variabile	hendô chinshakuryô 変動賃借料	variabel leie	renda contingente	interés variable	variabel avgift
uniformità delle classificazioni negli esercizi	kômoku bunrui no keizokusei 項目分類の継続性	kontinuitet ved klassifisering	uniformidade ou consistência	uniformidad temporal	konsistensprinciper

ENGLISH	DANISH	DUTCH, BELGIAN (B)	FINNISH	FRENCH, BELGIAN (B)	GERMAN
corporation	selskab	vennootschap	yhteisö	personne morale, société anonyme, société (B)	Kapitalgesellschaft
corporation tax	selskabsskat	vennootschapsbelasting	yhteisövero	impôt sur les bénéfices des sociétés	Körperschaftsteuer, Ertragssteuer (CH)
cost flow assumptions	metode for indregning af produktionsomkostninger	veronderstellingen over kostentoerekening	kustannuslaskentaperiaate	conventions de calcul du coût des ventes	Verbrauchsfolgeverfahren
cost of sales	produktionsomkostninger	kost der verkochte goederen	myytyjen tavaroiden hankintameno	coût des ventes	Herstellungskosten der zur Erzielung der Umsatzerlöse erbrachten Leistungen, Herstellungskosten der verkauften Produkte (CH)
cost of sales method	årsagsprincippet	nettoverkoopwaarde	vaihto-omaisuuskulun määrittämismenetelmä	méthode du coût des ventes	Umsatzkostenverfahren
cost plus contract	kontrakt baseret på omkostninger plus (fast) honorar	kostprijs-plus-contract	laskutustyö	contrat à prix coûtant majoré, contrat en régie (B)	Cost-plus-Vertrag
credit entry	kredit postering	creditboeking	hyvitys	écriture créditrice	Habenbuchung
current asset	omsætningsaktiv	vlottende activa	vaihto-omaisuus	actif à court terme, actif circulant	Umlaufvermögen
current cost	omkostninger til genanskaffelsesværdi	actuele waarde	jälleenhankintaarvo	coût de remplacement	Wiederbeschaffungskosten
current cost accounting	genskaffelsesværdibaserede regnskaber	waardering tegen actuele waarde	jälleenhankintahintoihin perustuva kustannuslaskenta	comptabilité au coût de remplacement	Wiederberschaffungswertrechnung, Substanzerhaltungsrechnung

ITALIAN	JAPANESE	NORWEGIAN	PORTUGUESE	SPANISH, ARGENTINE (ARG)	SWEDISH
società di capitali, persona giuridica	hôjin 法人	selskap	sociedade	sociedad	aktiebolag
imposta sul reddito delle persone giuridiche	hôjinzei 法人税	selskappsskatt	imposto sobre rendimento das pessoas colectivas	impuesto sobre sociedades, impuesto a las ganancias (ARG)	bolagsskatt
ipotesi di flusso dei costi (correlazione dei costi e dei ricavi)	genka no nagare no katei 原価の流れの仮定	tilordning	critérios de imputação de custos	criterios de imputación de costes	princip för kostnadsberäkning
costo del venduto	uriage genka 売上原価	kostnad solgte varers	custo das vendas	coste de ventas, costo de mercaderías vendidas, servicios prestados (ARG)	kostnad för sålda varor
metodo del costo del venduto	uriage genka hô 売上原価法	*not in use* (follows from the matching principle)	custo das vendas do período	método del beneficio operativo	*not in use*
contratto a costi maggiorati	genka kasan keiyaku 原価加算契約	kost-pluss-kontrakt	contrato com preço obtido a partir dos custos suportados com acréscimo, cost plus contract	contrato de margen sobre el coste	uppdrag på löpande räkning
avere	kashikata kinyû 貸方記入	kreditpostering	crédito	abono, crédito (ARG)	kreditering
attivo circolante	ryûdô shisan 流動資産	omløpsmidler	activo circulante	activo circulante, activo corriente (ARG)	omsättningstillgångar
costo corrente, costo attuale, costo reale, costo di mercato	genzai genka 現在原価	løpende kost	custo de reposição	coste de reposición	nukostnad
metodo del costo corrente	genzai genka kaikei 現在原価会計	løpende kost regnskap	contabilidade a custos de reposição	contabilidad basada en coste de reposición, contabilidad a costos corrientes (ARG)	nukostnadsredovisning

ENGLISH	DANISH	DUTCH, BELGIAN (B)	FINNISH	FRENCH, BELGIAN (B)	GERMAN
current investment	omsætningsaktiv	korte termijn belegging	lyhytaikainen sijoitus	placement à court terme, placement courant (B)	kurzfristige Finanzinvestition
current purchasing power accounting	inflationsregnskab ved brug af købekraftsprincippet	koopkracht correctiemethode	ostovoimaperusteinen inflaatiolaskenta	comptabilité en pouvoir d'achat de la monnaie	Realkapitalerhaltungs-trechnung
current service costs		pensioenkosten verworven over de prestaties van het huidige dienstjaar	tulevista johdon eläke- ja erorahaeduista yritykselle aiheutuva vuotuinen meno	coût des prestations pour services courants, coût de service rendus au cours de l'exercice (B)	laufender Dienstzeitaufwand
current tax	aktuel skat	verschuldigde of verrekenbare belasting	ennakkovero	impôt courant, impôt exigible (B)	tatsächliche Steuern
current value	genskaffelsesværdi	actuele waarde	jälleenhankintahinta	valeur actuelle, valeur d'inventaire	Zeitwert, Tageswert
de facto control	*de facto* kontrol	feitelijke controle	tosiasiallinen kontrolli	contrôle de fait	faktische Kontrolle
de jure control	*de jure* kontrol	juridische controle	juridinen kontrolli	contrôle de droit	rechtliche Kontrolle
debit entry	debet postering	debetboeking	veloitus	écriture débitrice	Sollbuchung
debt	gæld	schuld	velka	dette	Schuld
deferred taxation	udskudt skat (latent skat, eventualskat)	latente belastingen	laskennalliset verot	impôt différé, fiscalité différée	Steuerabgrenzung
defined benefit plan	pensionsordning	salaris, dienstjarensysteem, vastgelegd doel plan (B)	eläkevakuutus	régime de retraite à prestations définies	leistungsorientierter Plan
defined contribution plan	fratrædelsesforpligtigelse	beschikbaar premiestelsel, vastgelegd middelen plan (B)	eläkerahasto	régimes de retraite à cotisations définies	beitragsorientierter Plan

ITALIAN	JAPANESE	NORWEGIAN	PORTUGUESE	SPANISH, ARGENTINE (ARG)	SWEDISH
investimento a breve termine	tanki tôshi 短期投資	omløps-finansinvestering	investimento corrente	inversión financiera temporal, inversión corriente (ARG)	kortfristig placering
valutazione in base al potere di acquisto attuale	genzai kôbairyoku kaikei 現在購買力会計	prisnivåjustert regnskap	contabilidade a custos constantes	contabilidad en unidades monetarias constantes, contabilidad a precios constantes (ARG)	allmänna penningvärde-principen
costo previdenziale relativo alle prestazioni di lavoro corrente	tôki kimmu hiyô 当期勤務費用	løpende pensjonskostnader	custos dos serviços correntes	coste de los servicios actuales	kostnad för intjänade förmåner under perioden
imposte dell'esercizio	tôki kazei zeikin 当期課税税金	betalbar skatt	imposto por conta	cuota líquida, impuesto efectivo (ARG)	aktuell skatt
valore corrente	genzai kachi 現在価値	løpende verdi	custo actual	coste corriente	nuvärde
controllo di fatto	jisshitsuteki shihai 実質的支配	reell kontroll	controlo de facto	criterio de control efectivo	*de facto* kontroll
controllo giuridico	hôteki shihai 法的支配	formell kontroll	controlo de direito	criterio del tanto de participación, criterio del control legal (ARG)	*de jure* kontroll
dare	karikata kinyû 借方記入	debetpostering	débito	cargo, débito (ARG)	debitering
debito	saimu 債務	gjeld	dívida	deuda	skuld
tassazione differita, tassazione anticipata	kurinobe zeigaku 繰延税額	utsatt skatt	impostos diferidos	método del efecto impositivo, método de impuesto diferido (ARG)	uppskjuten skatt
programmi per il dopo impiego a benefici definiti	kakutei kyûfu seido 確定給付制度	ytelsesplan	plano de benefícios definidos	planes de prestación definida	förmånsbaserad plan
programmi contributivi per il dopo impiego non a benefici definiti	kakutei kyoshutsu seido	tilskuddsplan	plano de contribuição definida	planes de contribución definida	avgiftsbaserad plan

ENGLISH	DANISH	DUTCH, BELGIAN (B)	FINNISH	FRENCH, BELGIAN (B)	GERMAN
depreciable asset/amount		afschrijfbaar bedrag	poistokelpoinen omaisuus	actif amortissable	abschreibungsfähiger Vermögenswert, Abschreibungsvolumen
depreciation	afskrivning	afschrijving	poisto	amortissement, amortissement pour dépréciation	Abschreibung
depreciation method	afskrivningsmetode	afschrijvingsmethode	poistomenetelmä	méthode d'amortissement	Abschreibungsmethode
depreciation plan	afskrivningsplan	afschrijvingsplan	poistosuunnitelma	plan d'amortissement	Abschreibungsplan
development		ontwikkelingsfase	kehittäminen	développement	Entwicklung
development costs	udviklingsomkostninger	ontwikkelingskosten	kehittämismenot	frais de développement	Entwicklungskosten
directors' report	årsberetning	jaarverslag	toimintakertomus	rapport du conseil d'administration	Lagebericht, Bericht des Verwaltungsrates (CH)
discontinued operations	ophør af aktiviteter	beëindiging van bedrijfsactiviteiten	liiketoiminta-alueen lopettaminen, ivestoinnin purkkaminen	secteur (d'activité) abandonné	eingestellte Geschäftsfelder/Bereiche
distributable reserves	frie reserve	uitkeerbare reserves	jakokelpoiset varat	réserves distribuables	(ausschüttungs-) offene Rücklagen, ausschüttbare Reserven (CH)
dividends	dividende	dividend	osingot	dividendes	Dividenden
dominant influence	bestemmende indflydelse	overheersende invloed	määräämisvalta	influence dominante	beherrschender Einfluss

ITALIAN	JAPANESE	NORWEGIAN	PORTUGUESE	SPANISH, ARGENTINE (ARG)	SWEDISH
bene/importo ammortizzabile	genka shôkyaku shisan/gaku 減価償却資産/額	avskrivbar eiendel-avskrivbart beløp	activo amortizável (reintegrável)	base de amortización, base depreciable (ARG)	avskrivningsbart belopp
ammortamento (di immobilizzazioni materiali)	genka shôkyaku 減価償却	avskrivning	depreciação, reintegração	depreciación económica	avskrivning
metodo di ammortamento	genka shôkyaku hôhô 減価償却方法	avskrivningsmetode	método de amortização	métodos de amortización, métodos de depreciación (ARG)	avskrivnings-princip
piano di ammortamento	genka shôkyaku keikaku 減価償却計画	avskrivningsplan	plano de amortização	criterio, plan de amortización, esquema (ARG), plan de depreciación (ARG)	avskrivningsplan
sviluppo	kaihatsu 開発	utviklingskostnader	desenvolvimento	desarrollo	utveckling
costi di sviluppo	kaihatsu hi 開発費	utviklingskostnader	custos de desenvolvimento	costes/gastos de desarrollo	utvecklings-kostnader
relazione sulla gestione	torishimariyaku hôkokusho 取締役報告書	styrets årsberetning	relatório de gestão	informe de gestión, memoria anual (ARG)	verksamhets-berättelse
cessazione di attività	haishi jigyô 廃止事業	ikke videreført driftsaktivitet	operação descontinuada	area de negocio enajenada/abandonada, operaciones discontinuadas (ARG)	avvecklad verksamhetsgren
riserve distribuibili	haitô kanô junbikin 配当可能準備金	fri egenkapital	reservas distribuíveis, reservas livres	reservas libres, reservas distribuibles (ARG)	fria reserver
dividendi	haitôkin 配当金	utbytte	dividendos	dividendos	utdelning
influenza dominante	shihaiteki eikyôryoku 支配的影響力	dominerende innflytelse	influência dominante	relación de dominio y dependencia, criterio del control legal (ARG)	dominerande inflytande

ENGLISH	DANISH	DUTCH, BELGIAN (B)	FINNISH	FRENCH, BELGIAN (B)	GERMAN
earnings per share	resultat pr. aktie	winst per aandeel	tuotto per osake	résultat par action	Gewinn pro Aktie
economic ownership	økonomisk ejerskab	economische eigendom	taloudellinen omistus	propriété économique, droit d'usage (B)	wirtschaftliches Eigentum
elimination of intra-group profits	eliminering af urealiserede interne avancer	eliminatie van intergroepswinst	sisäisten voittojen eliminointi	élimination des profits intragroupe	Zwischengewinn-eliminierung
enterprise	virksomhed	onderneming	yritys	entreprise	Unternehmen
environmental/ green accounting	miljøregnskab/ grønt regnskab	milieuverslagge-ving	ympäristöla-skenta	comptabilité environnementale/ verte	Umweltrech-nungslegung
equity	egenkapital (ialt)	eigen vermogen	oma pääoma	capitaux propres	Eigenkapital
equity method	den indre værdis metode, equity-metoden	vermogensmuta-tiemethode	hankinta-meno-menetelmä	consolidation par mise en équivalence (B)	Equity-Methode
exchange difference	valutakurs difference	omrekenings- of koersverschil	kurssiero	écart de conversion	Umrechnungs-differenz
exchange rate	omregningskurs	wisselkoers	vaihtokurssi	taux de change	Wechselkurs
exchange risk	valutarisiko	valutarisico	kurssiriski	risque de change	Wechselkursrisiko
executory/ pending contract	kontrakt med henblik på senere effektuering	nog niet uitgevoerde contracten	ehdollinen sopimus	contrat en attente d'exécution, contrat conditionnel (B)	schwebender Vertrag
expenditure	udgift	uitgave	meno	dépense	Auszahlung

ITALIAN	JAPANESE	NORWEGIAN	PORTUGUESE	SPANISH, ARGENTINE (ARG)	SWEDISH
utile per azione	hitokabu atari rieki 一株当たり利益	overskudd pr. aksje	lucros por acção	beneficio por acción	vinst per aktie
proprietà economica	keizaiteki shoyûken 経済的所有権	økonomisk eierskap, reelt eierskap	propriedade económica	usufructuario, propiedad económica	äganderätten knyts till de ekonomiska risker och rättigheter som är förenade med tillgångar
elisione di margini infragruppo	shûdan nai rieki no jokyo 集団内利益の除去	eliminering av interngevinster	eliminação de resultados nas operações internas	eliminación de resultados por operaciones internas	eliminering av internvinster
impresa	kigyô 企業	virksomhet	empresa	empresa	företag
contabiltà sociale, ambientale	kankyô/gurîn kaikei 環境/グリーン会計	miljøregnskap	contabilidade ambiental	contabilidad medio-ambiental	miljöredovisning
patrimonio netto	mochibun 持分	egenkapital	capital próprio, situação líquida	fondos propios (neto), patrimonio neto (ARG)	eget kapital
metodo del patrimonio netto	mochibun hô 持分法	egenkapitalmetoden	método da equivalência patrimonial	procedimiento de puesta en equi-valencia, valor patrimonial pro-porcional (ARG)	kapitalandels-metoden
differenza cambio	kawase sagaku 為替差額	valutakursdifferanse	diferença cambial	diferencia de cambio	valutakursdif-ferens
tasso di cambio	kawase rêto 為替レート	valutakurs (omregningskurs)	taxa de câmbio	tipo de cambio, tasa de cambio (ARG)	växelkurs
rischio di cambio	kawase risuku 為替リスク	valutarisiko	risco cambial	riesgo de cambio	valutarisk
contratto esecutivo in data futura, contratto in sospeso	mirikô keiyaku 未履行契約	kontrakt som skal oppfylles senere	contrato com cláusulas pendentes de execução	contrato por claúsulas pendi-entes de ejecución	kontrakt som träder i kraft senare
uscita di cassa	shishutsu 支出	utgift	pagamento	desembolso, pago	utgift

ENGLISH	DANISH	DUTCH, BELGIAN (B)	FINNISH	FRENCH, BELGIAN (B)	GERMAN
expense	omkostning	kosten	kulu	charge	Aufwand
extraordinary items	ekstraordinære poster	buitengewone baten en lasten	satunnaiset kulut	éléments extraordinaires, charges et produits exceptionnelles (B)	außerordentliche (Aufwands- oder Ertrags-) Position
fair value	markedsværdi	reële waarde	käypäarvo	valeur actuelle, juste prix/valeur, valeur du marché	beizulegender Zeitwert, Verkehrswert (CH)
finance lease	finansiel leasing	financiële leasing	rahoitusleasing	crédit bail, location-financement (B)	Finanzierungs-leasing
financial fixed assets	finansielle anlægsaktiver	financiële vaste activa	rahoitusomaisuus	immobilisations financiéres	Finanzanlagevermögen
financial instruments	finansielle instrumenter	financiële instrumenten	rahoitusvälineet	instruments financiers	Finanzinstrumente
financial reporting	regnskabsrapportering	financiële verslaggeving	tilinpäätösraportointi	reporting financier, compte rendu financier (B)	kaufmännische/externe Berichterstattung
financial statement analysis	regnskabsananlyse	financiële analyse	tilinpäätösanalyysi	analyse financière	Bilanzanalyse
financial year	regnskabsår	boekjaar	tilikausi	exercice	Geschäfts-/Berichtsjahr
fixed assets	anlægsaktiver	vaste activa	käyttöomaisuus ja muut pitkävaikutteiset sijoitukset	immobilisations	Anlagevermögen
fixed-asset movement schedule	anlægsnote	mutatieoverzicht vaste activa	käyttöomaisuuskirjanpito	tableau des mouvements des immobilisations	Anlagespiegel

ITALIAN	JAPANESE	NORWEGIAN	PORTUGUESE	SPANISH, ARGENTINE (ARG)	SWEDISH
costo	hiyô 費用	kostnad	custo	gasto	kostnad
partite straordinarie	ijô kômoku 異常項目	ekstraordinære poster	ganhos e perdas, custos e proveitos extraordinários	gastos e ingresos extraordinarios	extraordinära poster
valore normale	kôsei kachi 公正価値	virkelig verdi	valor justo	valor venal, justo valor (ARG)	verkligt värde
locazione finanziaria, leasing finanziario	fainansu rîsu ファイナンス・リース	finansiell leie (lease)	locação financeira	arrendamiento financiero, leasing (ARG)	finansiell lease
immobilizza-zioni finanziarie	kin'yû kotei shisan 金融固定資産	finansielle anleggsmidler	investimentos financeiros	inmovilizaciones financieras, inver-siones financieras no corrientes (ARG)	långfristiga finansiella tillgångar
strumenti finanziari	kin'yû shôhin 金融商品	finansielle instrumenter	instrumentos financeiros	instrumentos financieros	finansiella instrument
rendiconto finanziario	zaimu hôkoku 財務報告	regnskaps-rapportering	relato financeiro, prestação de contas	contabilidad infor-mación financiera externa, contabilidad externa (ARG)	redovisning
analisi finan-ziaria, analisi del bilancio	zaimu shohyô bunseki 財務諸表分析	regnskapsanalyse	análise financeira	análisis de estados financieros, análisis de estados contables (ARG)	räkenskapsanalys
esercizio	kaikei nendo 会計年度	regnskapsår	exercício	ejercicio, ejercicio contable/fiscal (ARG)	räkenskapsår
attività immobilizzate	kotei shisan 固定資産	anleggsmidler	activo fixo, imobilizações	activo fijo, activo no corriente (ARG)	anläggningstill-gångar
contratto a prezzo fisso	kakutei kakaku keiyaku 確定価格契約	fast-pris-kontrakt	contrato de preço pré-fixado	contrato de precio fijo	fastprisuppdrag

ENGLISH	DANISH	DUTCH, BELGIAN (B)	FINNISH	FRENCH, BELGIAN (B)	GERMAN
fixed price contract	fastpriskontrakt	contract met een vaste prijs	kiintehintainen sopimus	contrat à prix ferme, contrat à forfait, contrat à prix fixe	Festpreisvertrag
foreign currency translation	omregning af beløb i udenlands valuta	vertaling van vreemde valuta	ulkomaan rahamääräisten erien muuntaminen	conversion des monnaies étrangères	Währungsumrechnung
format	skema	model	tasekaava	modèle de présentation	Gliederung
formation expenses	stiftelsesomkostninger	oprichtingskosten	perustamiskustannukset	frais d'établissement	Aufwendungen für die Ingangsetzung und Erweiterung des Geschäftsbetriebes
forms of business organization	selskabsform	vennootschapsvormen	yhtiömuodot	structures juridiques d'organisation des entreprises	Gesellschaftsformen
forward rate	forward rate	termijnkoers	sidottu kurssi	taux à terme	Terminkurs
funds statement	pengestrømsopgørelse, cash flow opgørelse	kaastvoom overzicht	rahoituslaskelma	tableau de financement	Kapitalflussrechnung
gearing (leverage)	finansiel gearing, finansiel leverage, gældsandel	hefboomwerking	velkaantumisaste	ratio d'endettement	Verschuldungsgrad
geographical segment	geografisk segment	geografisch segment	maantieteeliset toimialueet	segment géographique, secteur géographique (B)	geographisches Segment

ITALIAN	JAPANESE	NORWEGIAN	PORTUGUESE	SPANISH, ARGENTINE (ARG)	SWEDISH
tabella dei movimenti delle immobilizza-zioni	kotei shisan meisaihyô 固定資産明細表	anleggsmiddeltablå	mapas de variação do activo fixo	estado-análisis de movimientos del activo fijo, anexo de bienes de uso (ARG)	schema som visar förändringen i anläggningstill-gångar
conversione in valuta locale	gaika kansan 外貨換算	valutautenlandsk	conversão	conversión de moneda extranjera	omräkning av utländska valutor
presentazione di bilancio	yôshiki 様式	oppstillingsplan	modelo, formato	modelo de cuentas, formato de pre-sentación (ARG)	schema (för resultat- och balansräkningar)
spese di impianto e ampliamento	sôritsu hi 創立費	stiftelseskostnader	despesas de instalação	gastos de estable-cimiento, gastos de organización (ARG)	bolagsbildnings kostnader
organigramma	kigyô soshiki keitai 企業組織形態	foretaksformer	tipos de sociedade	formas societarias	företagsformer
cambio a termine	sakimono rêto 先物レート	forward rate	taxa a prazo	precio de ejercicio	terminskurs
prospetto fonti e impieghi	shikin keisansho 資金計算書	kontantstrømopp-stilling	mapa de origens e aplicações de fundos	estados de flujos cuadro de financiación, estado de flujo de fondos, estado de origen y aplicación de fondos (ARG)	finansierings-analys
leva, rapporto di indebitamento	giaringu ギアリング	gjeldsgrad	rácio de endividamento	apalancamiento financiero	förhållande mellan främmande och eget kapital
settori geografici	chiikibetsu segumento 地域別セグメント	geografisk segment	segmento regional, segmento geográfico	división por área geográfica	geografisk marknad

ENGLISH	DANISH	DUTCH, BELGIAN (B)	FINNISH	FRENCH, BELGIAN (B)	GERMAN
going concern principle	going concern princippet	continuïteitsprincipe	toiminnan jatkuvuus	principe de continuité d'exploitation	Grundsatz der Unternehmensfortführung
goodwill	goodwill, merværdi ved køb	goodwill	liikearvo	fond commercial, écart d'acquisition	Geschäfts- und Firmenwert
grant	(offentligt) tilskud	subsidie	julkinen tuki	subvention	Zuschuss, Subvention (CH)
group accounts	koncernregnskab	groepsjaarrekening	korsernitilinpäätös	comptes de groupe, comptes consolidés	Konzernabschluss, Konzernrechnung (CH)
group management report	årsberetning	directieverslag, verslag van de raad van bestuur	konsernin toimintakertomus	rapport de gestion du groupe	Konzernlagebericht
group/collective valuation	værdiansættelse ud fra en porteføljesynsvinkel	collectieve wardering	portfolioarvostus, yhteisarvostus	évaluation collective, évaluation sur une base non individuelle, critères d'évaluation de groupe (B)	Gruppenbewertung, Gesamtbewertung (CH)
guaranteed residual value	garanteret scrapværdi	gegarandeerde restwaarde	leasingsopimuksiin liittyvät takuut	valeur résiduelle garantie	garantierter Restwert
hedging	sikring, hedging	hedging	suojaus	couverture	Deckungsgeschäft, Hedging, Absicherungsgeschäft (CH)
hidden reserves	skjulte reserver, hemmelige reserver	geheime/stille reserves	piilovaraus, aliarvostus	réserves occultes, réserves latentes (B)	stille Reserven, stille Rücklagen
historical cost accounting	historisk anskaffelsesværdi princip, historisk kostpris princip	waardering op basis van historische kosten	hankintahintaperiaate	comptabilité au coût historique	Anschaffungs- und Herstellungskostenprinzip
historical costs	historisk anskaffelsesværdi	historische kosten	hankintahinta	coûts historiques	Anschaffungs- und Herstellungskosten

ITALIAN	JAPANESE	NORWEGIAN	PORTUGUESE	SPANISH, ARGENTINE (ARG)	SWEDISH
principio di continuità aziendale	keizoku kigyô no gensoku 継続企業の原則	fortsatt drift prinsipp	princípio da continuidade	principio de empresa en funcionamiento, principio de empresa en marcha (ARG)	going concern principen
avviamento	noren のれん eigyôken 営業権	goodwill	trespasse, diferença de aquisição	fondo de comercio, llave de negocio (ARG)	goodwill
contributo	hojokin 補助金	offentlig tilskudd	subsídio	subvención, subsidio (ARG)	bidrag, statligt stöd
bilancio di gruppo, consolidato	kigyô shûdan zaimu shohyô 企業集団財務諸表	konsernregnskap	contas consolidadas	cuentas anuales consolidadas, estados contables consolidados (ARG)	koncernredovisning
relazione sulla gestione del gruppo	kigyô shûdan jôkyô hôkokusho 企業集団状況報告書	styrets årsberetning for konsernet	relatório consolidado de gestão	informe de gestión consolidado, memoria de estados contables consolidados (ARG)	koncernens förvaltningsberättelse
valutazione di gruppo, collettiva	gurûpu hyôka グループ評価	gruppevurdering av eiendeler	avaliação conjunta, valores agregados	agrupación de partidas	kollektiv värdering
valore residuo garantito	hoshô zanson kachi 保証残存価値	garantert restverdi	valor residual	valor residual garantizado	garanterat restvärde
copertura	hejjingu ヘッジング	sikring	cobertura	cobertura de riesgos, hedging (ARG)	säkring
riserve occulte	himitsu tsumitatekin 秘密積立金	skjulte reserver	reservas ocultas	reservas ocultas	dolda reserver
metodo del costo storico	rekishiteki genka kaikei 歴史的原価会計	historisk kost regnskap	contabilidade a custo histórico	contabilidad histórica, contabilidad a costos históricos (ARG)	anskaffningskostnadsredovisning
costo storico	rekishiteki genka 歴史的原価	historisk anskaffelsekost	custo de aquisição, custo histórico	precio de adquisición, costo histórico (ARG)	anskaffningskostnad

ENGLISH	DANISH	DUTCH, BELGIAN (B)	FINNISH	FRENCH, BELGIAN (B)	GERMAN
historical rate	historisk (valuta omregningskurs)	historische koers	historiallinen kurssi	taux historique	historischer Kurs
holding company	holdingselskab	holding	holdingyhtiö	société mère	Holding (-Gesellschaft)
holding gain	beholdningsgevinst	latente meer-waarde	arvonnousu	gain, plus value latente, plus-value latente (B)	Wertzuwachs am ruhenden Vermögen, Wertsteigerung (CH)
horizontal/ account form	skema i kontoform	scontrovorm	tasemuotoinen	forme 'en tableau', forme de compte (B)	Kontenform
imparity principle	imparitetsprincippet	voorzichtigheids-principe	varovaisuusper-iaate	principe de prudence	Imparitätsprinzip
individual accounts	årsregnskab (for den enkelte regnskabspligtige virksomhed)	enkelvoudige/ vennootschap-pelijke jaarrekening	tilinpäätös	comptes individuels	Einzelabschluss
industry segment	branche segment	bedrijfssegment	toimiala	secteur d'activité, branche d'activité	Geschäftsfeld
inflation accounting	inflationsregnskab	verslaggeving rekeninghoudend met inflatie	inflaatiolaskenta	comptabilité d'inflation	inflationsberei-nigte Rechnungslegung
information function	informationsværdi	informatiefunctie	informointitar-koitus	objectif d'information, fonction d'information (B)	Informationsfunk-tion
initial (or first) consolidation	konsolidering på overtagelsestids-punktet	eerste consolidatie	ensimmäinen konsolidointi hankintahetken jälkeen	première consolidation	Erstkonsolidie-rung

ITALIAN	JAPANESE	NORWEGIAN	PORTUGUESE	SPANISH, ARGENTINE (ARG)	SWEDISH
cambio storico	rekishiteki rêto 歴史的レート	historisk valutakurs	taxa histórica	coste histórico	historisk kurs
società controllante	mochikabu gaisha 持株会社	morselskap (holdingselskap)	empresa-mãe	empresa matriz	moderföretag
plusvalenza inespressa	hoyû ritoku 保有利得	urealisert verdistigning	ganho não efectivado, mais-valia não realizada (In the case of tangible fixed assets)	plusvalía no realizada, ganancias no realizadas (ARG)	orealiserad värdeändring
presentazione di bilancio a sezioni contrapposte	kanjô keishiki 勘定形式	kontoform	formato horizontal, forma de apresentação em conta	formato de cuenta	kontoformat
principio della prudenza	fukintô gensoku 不均等原則	forsiktighetsprin-sippet	princípio da prudência	principio de prudencia	försiktighetsprin-cipen
bilancio d'esercizio	kobetsu zaimu shohyô 個別財務諸表	regnskap for den enkelte regnskapspliktiges	contas individuais	cuentas anuales individuales, estados contables no consolidados (ARG)	årsredovisning
settori di attività, divisioni	jigyôbetsu segumento 事業別セグメント	virksomhetsområde	segmento de negócio	división, segmento de negocio (ARG)	rörelsegren
contabilizza-zione degli effetti inflazionistici	infurêshon kaikei インフレーション会計	inflasjonsregnskap	contabilidade da inflação	contabilidad que considera el efecto de la inflación, contabilidad inflacionista, contabilidad inflacionaria (ARG)	inflationsredovis-ning
funzione informativa	jôhô teikyô kinô 情報提供機能	informasjons-funksjon (verdi)	função informativa	función informativa	informations-funktion
inclusione nell'area di consolidamento di una partecipazione	shutokuji renketsu 取得時連結	konsolidering på oppkjøpstids punktet	primeira consolidação	primera consolida-ción, consolidación inicial (ARG)	konsolidering rid förvärvs-zidpunkten

ENGLISH	DANISH	DUTCH, BELGIAN (B)	FINNISH	FRENCH, BELGIAN (B)	GERMAN
intangible assets	immatrielle anlægsaktiver	immateriële activa	aineettomat hyödykkeet	immobilisations incorporelles	immaterielle Vermögens-gegenstände, immaterielle Anlagen (CH)
inter-company profit	koncern-/intern avance	intercompany-winsten	sisäinen voitto	profit interne, profit intragroupe (B)	Zwischengewinn
interim report/ statements	perioderegnskab	tussentijds verslag/ verslaggeving	osavuotiskatsaus	rapport intérimaire	Zwischenbericht
inventory/ stocktaking	varelager	voorraadopname	inventointi	inventaire, inventorisation (physique) (B)	Inventur
investments	investeringer	beleggingen	investoinnit	titres de participation (fixed), titres de placement (current), immobilisations financières (B), placements de trésorerie (B)	Finanzanlagen und Wertpapiere
joint venture	arbejdsfællesskab, joint venture	joint venture	yhteisyritys	exploitation en commun filiale commune (B)	Gemeinschaftsun-ternehmen, Joint Ventures
lease term	leasingperiode	leasetermijn	leasingaika	durée du bail, durée du contrat de location (B)	Laufzeit des Leasingverhält-nisses
leasing	leasing	leasing	leasing	crédit-bail, location-financement (B)	Leasing
legal ownership	juridisk ejerskab	juridische eigendom	omistusoikeus	droit de propriété, propriété légale (B)	rechtliches Eigentum
legal reserves	lovpligtige reserver	wettelijke reserves	lakisääteiset rahastot	réserves légales	gesetzliche Rücklagen, gesetzliche Reserven (CH)
liabilities	passiver i alt	schulden	velat	dettes	Verbindlichkeiten, Schulden

ITALIAN	JAPANESE	NORWEGIAN	PORTUGUESE	SPANISH, ARGENTINE (ARG)	SWEDISH
immobilizzazioni immateriali	mukei shisan 無形資産	immaterielle eiendeler	imobilizações incorpóreas, activos intangíveis	inmovilizado inmaterial, activos intangibles (ARG)	immateriella tillgångar
utile infragruppo	kaishakan rieki 会社間利益	interngevinst	lucros internos	beneficio intergrupo	internvinst
bilanci interinali	chûkan hôkokusho/zaimu shohyô 中間報告書/財務諸表	delårsrapport/regnskap	informação económico-financeira intercalar	estados financieros intermedios	delårsrapport
inventario fisico	jicchi tanaoroshi 実地棚卸	vareopptelling	inventário	inventario	inventering
investimenti, partecipazioni	tôshi 投資	finansinvesteringer	investimentos financeiros, participações financeiras	inversiones financieras	aktier och andra andelar
joint venture	jointo benchâ ジョイント・ベンチャー	felles kontrollert virksomhet, joint venture	consórcio	asociación/unión temporal de empresas, joint venture	joint venture
durata del leasing	rîsu kikan リース期間	leasingperiode	prazo de locação	vigencia del contrato	leasingavtals giltighetestid
locazione, leasing	rîsu torihiki リース取引	leasing	locação financeira	arrendamiento financiero, leasing (ARG)	leasing
proprietà	hôteki shoyûken 法的所有権	juridisk eierskap	titularidade	propiedad legal	äganderätt
riserve legali	hôtei jumbikin 法定準備金	*no longer in use* ('reservefond')	reservas legais	reservas legales	lagstadgad fond
debiti, passività	fusai 負債	gjeld	passivos, dívidas	obligaciones, pasivo (ARG)	skulder

ENGLISH	DANISH	DUTCH, BELGIAN (B)	FINNISH	FRENCH, BELGIAN (B)	GERMAN
liability	forpligtelse	aansprakelijkheid	vastuu	obligation, responsabilité (B)	Haftung
liquid assets	likvide aktiver	liquide middelen	rahat ja pankkisaamiset	disponibilités	liquide/flüssige Mittel
liquidity	likviditet	liquiditeit	likviditeetti	liquidité	Liquidität
listed conpanies	børsnoterede selskaber	beursgenoteerde ondernemingen	pörssinoteeratut yhtiöt	sociétés cotées, entreprises cotées en bourse (B)	börsennotierte Unternehmen, kotierte Unternehmen (CH)
loan	lån	lening	laina	prêt, emprunt (B)	Kredit, Darlehen
long-term contract	igangværende arbejder for fremmed regning	langlopende overeenkomst	pitkäaikainen sopimus	contrat de longue durée, contrat à long terme (B)	langfristiger Vertrag
loss	tab	verlies	tappio	perte	Verlust
loss brought forward	overført tab	geaccumuleerd verlies	aktivoitu tappio	perte reportée	Verlustvortrag
loss for the financial year	årets tab	verlies van het boekjaar	tilivuoden tappio	perte de l'exercice	Jahresfehlbetrag, Jahresverlust (CH)
majority shareholder	majoritetsaktionær	meerderheids-aandeelhouder	pääosakas	actionnaire majoritaire	Mehrheitsgesell-schafter/-aktionär
managing board	direktion	directie	yhtiön hallitus	conseil d'administration, comité de direction (B)	Vorstand, Geschäftsführung, Geschäftsleitung (CH)
mandatory reserves	lovpligtige-og vedtægtsmæssige reserver	geheime reserves	pakolliset varaukset	réserves latentes, réserves occultes	stille Zwangsrücklagen, Zwangsreserven (CH)
market value	markedsværdi	marktwaarde	markkina-arvo	valeur de marché	Marktwert
marketable securities	omsættelige værdipapirer	verhandelbare effecten	noteeratut arvopaperit	titres négociables, valeurs négociables (B)	marktgängige Wertpapiere, kotierte Wertschriften (CH)

ITALIAN	JAPANESE	NORWEGIAN	PORTUGUESE	SPANISH, ARGENTINE (ARG)	SWEDISH
responsabilità	sekinin 責任	gjeld	responsabilidade	responsabilidad	ansvar
liquidità, disponibilità liquide	temoto ryûdôsei 手元流動性	likvider	disponibilidades	disponible, activos líquidos (ARG)	likvida medel
liquidità, disponibilità	ryûdôsei 流動性	likviditet	liquidez	liquidez	likviditet
società quotate	jôjô gaisha 上場会社	børsnoterte selskap	empresa cotada	sociedades inscritas en bolsa, sociedades cotizadas	börsregistrerade företag
prestito	kashitsukekin 貸付金	lån	empréstimo	préstamo	lån
commesse a lungo termine	chôki keiyaku 長期契約	anleggskontrakter	contrato de longa duração	contrato de suministro a largo plazo	långtidskontrakt
perdita	sonshitsu 損失	tap	prejuízo	pérdida	förlust
perdite portate a nuovo	kurikoshi sonshitsu 繰越損失	fremførbart underskudd	prejuízos de exercícios anteriores, prejuízos reportados	pérdidas de ejercicios anteriores	balanserad förlust
perdita dell'esercizio	tôki sonshitsu 当期損失	årets underskudd	prejuízo do exercício	pérdidas del ejercicio	årets förlust
azionista/socio di maggioranza	shihai kabunushi 支配株主	majoritetsaksjonærer	accionista maioritário	participación mayoritaria, partícipe mayoritario	majoritetsägare
comitato direttivo, esecutivo	torishimariyaku kai 取締役会	ledelsen	conselho de gerência, conselho de administração	comité de dirección, directorio (ARG)	direktion
riserve inespresse	kyôsei himitsu tsumitatekin 強制秘密 積立金	lovbestemte eller vedtektsbestemte reserver	reservas ocultas	plusvalías latentes (seldom used), reservas por ajustes (ARG)	obligatorisken reserver
valore di mercato	shijô kachi 市場価値	markedsverdi	valor de mercado	valor de mercado	marknadsvärde
titoli non immobilizzati	shijôsei aru yûka shôken 市場性ある 有価証券	børsnoterte verdipapirer	títulos negociáveis	valores negociables	marknadsnoterade värdepapper

ENGLISH	DANISH	DUTCH, BELGIAN (B)	FINNISH	FRENCH, BELGIAN (B)	GERMAN
matching principle	matching-princippet	toerekenings-beginsel	kohdistamispe-riaate	principe de rattachement des charges aux produits, principe de rapprochement des produits et des charges (B)	(Aufwands-) Periodisierungs-prinzip
materiality	væsentlighed	materialiteit	oleellisuus	principe d'importance relative	Wesentlichkeit
measurement	måling	waardering	arvostus	mesure, évaluation (B)	Bewertung
merger	fusion	fusie	fuusio	fusion	Verschmelzung, Unternehmenszu-sammenschluss, Fusion
merger surplus	fusionsgevinst	fusiemeerwaarde	fuusiovoitto	prime de fusion, plus-value de fusion (B)	Verschmelzungs-mehrwert, Fusionsgewinn (CH)
minority interest	minoritetsinteresse	minderheids-belang	vähemmistö-osuus	intérêts minoritaires, intérêts de tiers (B)	Minderheitsge-sellschafter/ -aktionär
monetary items	monetære poster	monetaire posten	rahamääräiset erät	actifs ou passifs monétaires	monetäre Positionen
negative goodwill	badwill	badwill	konsernipassiiva	écart d'acquisition negatif, différence de consolidation négative (B)	negativer Geschäfts-/ Firmenwert, Badwill
net assets	nettoaktiver	netto-activa	nettovarallisuus	situation nette, actifs nets (B)	Reinvermögen, Eigenkapital
net book value	bogført værdi (netto)	nettoboekwaarde	jäljelläoleva hankintameno	valeur nette comptable	Restbuchwert
net current assets	arbejdskapital	werkkapitaal	käyttöpääoma	actif circulant net (de dettes)	Nettoumlaufver-mögen

ITALIAN	JAPANESE	NORWEGIAN	PORTUGUESE	SPANISH, ARGENTINE (ARG)	SWEDISH
principio di correlazione costi e ricavi	taiô gensoku 対応原則	sammenstillings-prinsippet	princípio do matching	principio de correlación de ingresos y gastos, principio de asociación de ingresos y gastos (ARG)	matchningsprincipen
rilevanza, materialità	jûyôsei 重要性	vesentlighet	materialidade	principio de importancia relativa principio de materialidad	väsentlighet
misurazione, quantificazione	sokutei 測定	måling	mensuração	valoración, medición (ARG)	värdering
fusione	gappei 合併	fusjon	fusão	fusión	fusion
avanzo di fusione	gappei saeki 合併差益	fusjonsgevinst	ganhos de fusão	plusvalías de fusión, reservas por fusión (ARG)	fusions-vinst
interessenze delle minoranze	shôsû kabunushi mochibun 少数株主持分	minoritetsintresse	interesses minoritários	intereses socios externos, interés minoritario (ARG)	minoritetsintresse
valori monetari	kahei kômoku 貨幣項目	pengeposter	rubricas monetárias	partidas monetarias, cuentas monetarias (ARG)	monetära poster
avviamento negativo (badwill)	shôkyoku noren 消極のれん	negativ goodwill	goodwill negativo, badwill	fondo de comercio negativo, llave de negocio negativa (ARG)	negativ goodwill
attività nette	jun shisan 純資産	*not in use as a presentation* (egenkapital)	situação líquida	neto contable, activos netos (ARG)	eget kapital
valore netto contabile	shômi chôbo kagaku 正味帳簿価額	regnskapsmessig verdi	valor líquido contabilístico	valor neto contable, valor neto de libros (ARG)	bokfört värde
capitale circolante netto	shômi ryûdô shisan 正味流動資産	arbeidskapital	fundo de maneio	activo circulante neto, activos corrientes neto (ARG)	rörelsekapital

ENGLISH	DANISH	DUTCH, BELGIAN (B)	FINNISH	FRENCH, BELGIAN (B)	GERMAN
net realizable value	nettorealisations-værdi	directe opbrengstwaarde	realisointi-arvo	valeur nette réalisable	Nettoveräuße-rungserlös
nominal capital	nominel selskabskapital	nominaal aandelenkapitaal	nimellispääoma	capital, capital nominal (B)	Nominalkapital
nominal value	nominel værdi	nominale waarde	nimellisarvo	valeur nominale	Nominalwert
notes	noter	toelichting	liitetiedot	annexe	Anhang
off-balance sheet transaction	off-balance sheet transaktioner	'off-balance-sheet'-transactie	taseen ulkopuoliset erät	opération hors bilan	bilanzunwirksame Transaktion
operating activities	driftsaktivitet	operationele activiteiten	liiketoimet	activités d'exploitation, activités opérationelles (B)	Geschäftsfelder
operating lease	operationel leasing	operationele leasing	käyttöleasing	location-exploitation, leasing opérationnel (B)	Operating-Leasing
operating profit	ordinært resultat	bedrijfswinst	liikevoitto	bénéfice d'exploitation	Ergebnis der gewöhnlichen Geschäftstätigkeit
parent company	moderselskab	moedermaat-schappij	emoyhtiö	société mère	Mutterunterneh-men, Holdingge-sellschaft (CH)
participating interest	kapitalinteresse	deelneming	omistusyhteys-yritys	participation	Beteiligung
partnership	interessentskab	joint-venture	henkilöyhtiö	société en nom collectif, société de personnes (B)	Personengesell-schaft, Kollektiv-gesellschaft (CH)
patent	patent	patent	patentti	brevet	Patent

ITALIAN	JAPANESE	NORWEGIAN	PORTUGUESE	SPANISH, ARGENTINE (ARG)	SWEDISH
valore di netto realizzo	shômi jitsugen kanô kagaku 正味実現可能価額	salgsverdi	valor realizável líquido	valor neto de realización, valor realizable neto	nettoförsäljning-svärdet
capital sociale al valore nominale	meimoku shihon 名目資本	pålydende aksjekapital	capital nominal, capital social	capital nominal	aktiekapital
valore nominale	gakumen kagaku 額面価額	nominell verdi	valor nominal, valor facial	valor nominal	nominellt värde
nota integrativa, note illustrative, note esplicative	chûki 注記	noteopplysninger	anexo	memoria, notas a los estados contables (ARG)	noter
compensazione di partite	ofubaransu torihiki オフバランス取引	transaktioner utenfor balansen	operações fora do balanço	operación fuera de balance	off-balance sheet transaktioner
attività operative	eigyô katsudô 営業活動	driftsaktiviteter	actividades de exploração, actividades operacionais	actividades de explotación, actividades operativas (ARG)	löpande verksamhet
locazione, leasing operativo	operêitingu rîsu オペレーティング・リース	operasjonell leie	locação financeira operacional	leasing operativo	operationell lease
risultato operativo	eigyô rieki 営業利益	driftsresultat	lucro de exploração	beneficio de explotación, resultado operativo (ARG)	rörelseresultat
società controllante	oya gaisha 親会社	morselskap	empresa-mãe	empresa matriz	moderföretag
società controllate, società collegate	shihon sanka 資本参加	tilknyttet selskap	participação	participación	andel i intresseföretag
società di persona, associazione	kumiai 組合	partnerskap, interesselskap	sociedade de pessoas	sociedad comanditaria, sociedad de responsabilidad limitada – SRL (ARG)	konsortium
brevetto	tokkyoken 特許権	patent	patente	patente	patent

ENGLISH	DANISH	DUTCH, BELGIAN (B)	FINNISH	FRENCH, BELGIAN (B)	GERMAN
percentage of completion method	produktionsmetoden	'percentage of completion'-methode, methode met toegerekende winst	valmistusasteen mukainen tuloutus	comptabilisation bénéfice à l'avancement, calculation compte tenu du degré d'avancement (B)	Produktionsfort-schrittmethode
plan assets		fondsbeleggingen	sijoitukset eläkerahastoissa	actif du régime	Planvermögen
pooling-of-interest accounting	fusionsmetoden, pooling-of-interest metoden	pooling of interest methode	yhdistelmäme-netelmä	mise en commun d'intérêts	Pooling of interest-Methode, Interessenzusa-mmenführungs-methode
post-balance sheet events	begivenheder efter balancedagen	gebeurtenissen na balansdatum	tilikauden päättymisen jälkeiset tapahtumat	événements postérieurs à la clôture de l'exercice, événements après bilan (B)	wertaufhellende Tatsachen, Ereignisse nach dem Bilanzstichtag (CH)
pre-payments	forudbetalinger	vooruitbetalingen	ennakkomaksu	avance, acomptes (B)	Anzahlung
present value	nutidsværdi	contante waarde	nykyarvo	valeur actualisée, valeur actuelle (B)	Barwert
principle of completeness	fuldstændigheds-princippet	principe van volledigheid (B)	täydellisyyden periaate	l'obligation d'exhaustivité, principe d'universalité (B)	Vollständigkeits-prinzip
principle of separate valuation	individuel værdiansættelse	individuele waardering (B)	erillisarvostus	évaluation individuelle (B)	Einzelbewertungs-grundsatz
principle of the lower of cost or market	laveste værdis princip	minimumwaar-deringsregel	alimman arvon periaate	principe de prudence, évaluation à la valeur inférieure de marché (B)	Niederwerts-prinzip
production costs	produktionsom-kostninger	vervaardigings-prijs	valmistuskust-annukset	coûts de production, coût de revient (B)	Herstellungskos-ten

ITALIAN	JAPANESE	NORWEGIAN	PORTUGUESE	SPANISH, ARGENTINE (ARG)	SWEDISH
metodo della percentuale di completamento	kôji shinkô kijun 工事進行基準	løpende avregningsmetode	método de percentagem de acabamento	método del grado de cumplimiento, método del porcentaje de obra realizada, método del grado de avance (ARG)	successiv vinstavräkning
patrimonio dedicato al fondo pensione	nenkin shisan 年金資産	*not in use*	plano de contribuições (fundos)	activos del fondo de pensiones	förvaltningstill-gångar
metodo dell'aggrega-zione dei valori (pooling-of-interest)	mochibun pûringu kaikei 持分プーリング会計	kontinuitetsmetoden	método da comunhão de interesses	método de la fusión de intereses	poolnings-metoden
eventi successivi	kôhatsu jishô 後発事象	hendelser etter balansedagen	acontecimentos posteriores ao fecho	acontecimientos posteriores al cierre, hechos posteriores al cierre (ARG)	viktiga händelser efter räkenskaps-årets utgång
pagamenti in anticipo	maebaraikin 前払金	forskuddsbetaling	pagamentos antecipados	anticipos	förutbetalda kostnader
valore attualizzato	genzai kachi 現在価値	nåverdi	valor actual, valor presente	valor actual, valor presente (ARG)	nuvärde
principio della completezza	kanzensei no gensoku 完全性の原則	fullstendighets-prinsippet	princípio da não compensação	principio del registro total de derechos y obligaciones	*not in use*
principio della valutazione separata	kobetsu hyôka no gensoku 個別評価の原則	individuell vurdering	princípio da avaliação parcela a parcela	principio de no compensación	individuell värdering
principio del minore fra costo e mercato	teika kijun 低価基準	laveste verdis prinsipp	princípio da avaliação dos 'dois o mais baixo'	valoración por el precio de adquisición o el de mercado, si éste fuese inferior, costo o mercado el menor (ARG)	lägsta värdets princip
costo di produzione	seizô genka 製造原価	tilvirknings-kostnader	custo de produção	costes de producción	tillverknings-kostnader

ENGLISH	DANISH	DUTCH, BELGIAN (B)	FINNISH	FRENCH, BELGIAN (B)	GERMAN
profit	overskud	winst	voitto	profit, bénéfice, bénéfice (B)	Gewinn
profit after tax	overskud efter skat	nettowinst	voitto verojen jälkeen	profit net (de taxe), bénéfice après impôts (B)	Gewinn nach Steuern
profit and loss account	resultatopgørelse	winst- en verliesrekening	tuloslaskelma	compte de résultat	Gewinn- und Verlustrechnung, Erfolgsrechnung (CH)
profit carried forward	overført overskud (fra tidligere år)	ingehouden winst	edellisten tilikausien voitto	report à nouveau, bénéfice reporté (B)	Gewinnvortrag
profit for the financial year	årets overskud, årets resultat	winst boekjaar	tilikauden voitto	bénéfice de l'exercice, bénéfice à affecter (B)	Jahresüberschuss, Jahresergebnis, Jahresgewinn (CH)
profit reserves	reserver, overført resultat	reserve ingehouden winst	rahastoitu voitto	réserves	Gewinnrücklagen, Gewinnreserven (CH)
profit transfer agreement	udlodningsaftale	overeenkomst tot uitkering van winsten	voitonsiirto-sopimus	clause de transfert de bénéfice, accord de tramsfert de bénéfice (B)	Gewinnabführungsvertrag
property, plant and equipment	ejendom, tekniske anlæg og maskiner	bedrijfsgebouwen en -terreinen, machines en installaties	aineelliset hyödykkeet	immobilisations corporelles	Sachanlagen
proportional consolidation	pro rata konsolidering	proportionele/ partiële consolidatie	omistusosuuden mukainen konsolidointi	intégration proportionnelle, consolidation proportionnelle (B)	Quotenkonsolidierung
provision	betinget forpligtigelse, hensættelse	voorziening	varaus	provision	Rückstellung
provision for bad debts	hensættelse til tab på debitorer	voorziening voor debiteuren, waardevermindering voor dubieuze debiteuren (B)	luottotappio-varaus	provision pour créances douteuses	Wertberichtigung auf uneinbringliche Forderungen

ITALIAN	JAPANESE	NORWEGIAN	PORTUGUESE	SPANISH, ARGENTINE (ARG)	SWEDISH
utile	rieki 利益	overskudd	lucro	plusvalía, beneficio (ARG)	vinst
utile netto	zeibiki go rieki 税引後利益	overskudd etter skatt	lucro pós-impostos	beneficio después de impuestos	vinst efter skatt
conto economico	son'eki keisansho 損益計算書	resultatregnskap	demonstração dos resultados	cuenta de pérdidas y ganancias, estado de resultados (ARG)	resultaträkning
utili portati a nuovo	kurikoshi sonshitsu 繰越利益	*not in use*	resultados transitados	remanente, resultados de ejercicios anteriores (ARG)	balanserade vinstmedel
utile d'esercizio	tôki rieki 当期利益	årsoverskudd	lucro do exercício	beneficio del ejercicio, resultado del ejercicio (ARG)	årets vinst
riserve di utili indivisi	rieki jumbikin 利益準備金	opptjent egenkapital, tilbakeholdt overskudd	reservas de lucros autofinanciación)	reservas (de vinstmedel	balanserade
accordo per il trasferimento degli utili	rieki iten kyôtei 利益移転協定	overskuddsavtale	convenção de atribuição de lucros	acuerdo de traspaso de beneficios	
immobili, impianti e macchinari	yûkei kotei shisan 有形固定資産	varige driftsmidler	terrenos, edifícios e outras construções	activo inmovilizado material terminado, bienes de uso (ARG), edificios (ARG)	materiella anläggningstill-gångar
consolidamento proporzionale	hirei renketsu 比例連結	bruttometoden	consolidação proporcional	método de integración proporcional, método de consolidación proporcional (ARG)	klyvningsmetoden
(accantona-mento al) fondo	hikiatekin 引当金	aavsetning	provisão	provisión, previsión (ARG)	reservering, avsättning
(accantona-mento al) fondo svalutazione crediti	kashidaore hikiatekin 貸倒引当金	eavsetning usikre fordringer	provisão para cobranças duvidosas	provisión para insolvencias, previsión por deudores incobrables (ARG)	avsättning för osäkra fordringar

ENGLISH	DANISH	DUTCH, BELGIAN (B)	FINNISH	FRENCH, BELGIAN (B)	GERMAN
provision for charges	hensættelse	voorziening voor kosten (B)	varaus vastaisia menoja ja menetyksiä varten	provision pour charges	Aufwandsrückstellung
provision for diminution in value	nedskrivning	voorziening (B)	varaus vastaisia arvonalentumisia varten	provision pour dépréciation	Wertberichtigung
provision for future losses	hensættelse til imødegåelse affremtidige tab	voorziening voor risico's en kosten (B)	varaus vastaisten tappioden varalle	provision pour pertes probables, provision pour risques et charges (B)	Drohverlustrückstellung, Rückstellungen für zukünftige Verluste (CH)
provision for guarantees/ warranties	hensættelse vedrørende garantier, warrants	garantievoorziening	takuuvaraus	provision pour garantie, provision pour les charges découlant de garanties techniques (B)	Garantierückstellung
provision for pensions	hensættelse til pensionsforgligtelse, provision pour retraites	pensioenvoorziening	eläkevaraus	provision pour retraites obligations du personnel, provision pour pensions (B)	Pensionsrückstellung
prudence principle	forsigtighedsprincippet	voorzichtigheidsprincipe	varovaisuusperiaate	principe de prudence	Vorsichtsprinzip
publication	offentliggørelse	publicatie	julkistaminen	publicité (des comptes), dépôt (B), publication (B)	Offenlegung, Veröffentlichung (CH)
purchase method of consolidation	overtagelsesmetoden, past equity metoden, kostprismetoden	integrale consolidatie	hankinta-menomenetelmä	consolidation globale ou intégration	Erwerbsmethode
realization principle	realisationsprincippet	realisatieprincipe	realisointiperiaate	principe de réalisation	Realisationsprinzip

ITALIAN	JAPANESE	NORWEGIAN	PORTUGUESE	SPANISH, ARGENTINE (ARG)	SWEDISH
(accantonamento al) fondo rischi e oneri	hiyôsei hikiatekin 費用性引当金	fremtidige utgifter	provisões para riscos e encargos	provisión para riesgos y gastos	avsättning för framtida utgifter
svalutazione, fondo per perdite durevoli di valore	kagaku teika hikiatekin 価額低下引当金	*not in use* ('avsetning for forventet verdifall')	provisões para depreciação de activos, provisões para correcção de activos	provisión por depreciación, previsión por pérdida de valor (ARG)	värdeminsknings-reserv
accantonamento per perdite future	shôrai sonshitsu hikiatekin 将来損失引当金	avsetning for forventet tap	provisão para riscos e encargos	provisión para riesgos, previsión para futuras pérdidas (ARG)	reservering för framtida förluster
accantonamento per garanzia prodotti	seihin hoshô hikiatekin 製品保証引当金	avsetning for garantiansvar	provisões para garantias	provisión para garantías	garantiavsättning
fondo per trattamento di quiescenza ed oneri simili	taishoku kyûfu hikiatekin 退職給付引当金	pensionsforpligtelse	provisões para pensões	provisiones para pensiones	avsatt till pensioner
principio della prudenza	shinchôsei no gensoku 慎重性の原則	forsiktighetsprinsippet	princípio da prudência	principio de prudencia	försiktighets-principen
pubblicazione del bilancio	kôkoku 公告	offentliggjørelse	publicação, depósito	publicidad de las cuentas, publicación de los estados contables (ARG)	offentliggörande (av årsredovisning)
consolidamento delle partecipazioni sulla base del costo d'acquisto	baishû hô renketsu 買収法連結	oppkjøpsmetoden	método da compra	método de compra	förvärvsmetoden
prevalenza del principio di prudenza su quello di competenza (obbligo di considerare le perdite presunte e divieto di considerare gli utili non realizzati)	jitsugen gensoku 実現原則	realisasjonsprinsippet	princípio da realização	principio de realización (ARG)	realisationsprincipen

ENGLISH	DANISH	DUTCH, BELGIAN (B)	FINNISH	FRENCH, BELGIAN (B)	GERMAN
recognition	indregning	verwerking	tilinpäätökseen merkitseminen	principe de reconnaissance	Ansatz
recoverable amount	genindvindnings-værdi	netto opbrengstwaarde	investoinnin tuottoarvo	montant recouvrable, valeur recouvrable (B)	erzielbarer Betrag
reducing balance method	saldoafskrivning-smetoden	afschrijven in procenten van boekwaarde	jäännösarvo-poisto	amortissement dégressif, amortissements selon la valeur comptable (B)	geometrisch-degressive Abschreibung
reinstatement of value	korrektion af bogført værdi	teruneming van waardevermin-deringen	arvonkorjaus	reprise de provision, reprise de réductions de valeur (B)	Wertaufholung, Aufwertung auf Anschaffungs-kosten (CH)
replacement value	genanskaffelsesværdi	vervangings-waarde	jälleenhankin-tahinta	coût de remplacement, valeur de remplacement (B)	Wiederbeschaf-fungskosten
reportable segment	segment	segment waarover wordt gerapporteerd	eirllisesti raportoitavissa oleva liiketoiminta-alue	secteur isolable, (publication d') information(s) sectorielle(s), secteur pour lequel une information sectorielle est imposée (B)	berichtspflichtiges Segment
reporting currency	rapporteringsvaluta	rapporterings-valuta	raportointi-valuutta	monnaie de presentation (des états financiers), monnaie de publication (des comptes)	Berichtswährung
research	forskning	onderzoeksfase	tutkimus	recherche	Forschung
reserves	reserver	reserves	rahastot	réserves	Rücklagen, Reserven (CH)

ITALIAN	JAPANESE	NORWEGIAN	PORTUGUESE	SPANISH, ARGENTINE (ARG)	SWEDISH
riconoscimento, imputazione	ninshiki 認識	regnskapsføring	reconhecimento	reconocimiento	när skall en post redovisas i balans- eller resultaträkningen
valore realizzabile	kaishû kanô kagaku 回収可能価額	bruksverdi	montante recuperável	valor actual neto, valor recuperable (ARG)	nyttovärde
ammortamento a quote decrescenti	teiritsu hô 定率法	saldometoden	método de amortização degressiva	amortización degresira, método del saldo decreciente (ARG)	*not in use*
ripristino di valore	genson no modoshiire 減損の戻入れ	reversering av nedskrivning	reposição de valores	corrección valorativa, reexpresión (ARG)	uppskrivning, reversering av nedskrivning
valore di sostituzione	torikae kachi 取替価値	gjenanskaffelsesverdi	custo de reposição, custo de substituição	valor de reposición	återanskaffnings-värde
settore rappresentato separatamente	hôkoku segumento 報告セグメント	virksomhetsrappor-tering	segmento reportável	segmento, información por segmentos (ARG)	geografisk marknad, rörel-segren för vilken information skall lämnas
valuta di redazione del bilancio	hôkoku tsûka 報告通貨	Tapporterings-valuta	moeda de relato, moeda de referência	unidad monetaria	rapportvaluta
ricerca	kenkyû 研究	forskning	investigação, pesquisa	investigación	forskning
riserve	jumbikin 準備金	*no longer in use; equity is divided into paid-in* (innbetalt kapital) *and retained earnings* (tilbakeholdt overskudd)	reservas	reservas	bundna reserver, fria reserver

ENGLISH	DANISH	DUTCH, BELGIAN (B)	FINNISH	FRENCH, BELGIAN (B)	GERMAN
retained earnings	henlagt af årets overskud	ingehouden winst	jakamatta jätetty voitto	réserves, bénéfice reporté (B)	einbehaltene/ thesaurierte Gewinne, Gewinnreserven (CH)
revaluation	opskrivning	herwaardering	arvonkorotus	réévaluation	Neubewertung, Aufwertung (CH)
revaluation reserve	opskrivninghenlæggelse	herwaarderingsreserve	arvonkorotusrahasto	réserve de réévaluation, plus-value de réévaluation (B)	Neubewertungsrücklage, Aufwertungsreserve (CH)
revenue grants	(offentligt) tilskud, omkostningsrefusion	subsidies	investointiavustus	subventions d'exploitation	Aufwandszuschüsse
revenues	indtægter	omzet	myyntitulot	produits	Erträge
sale and lease back	sale and leaseback	sale and leaseback	myynti ja takaisinvuokraus	cession-bail, vente et reprise en location-financement (B)	Sale und leaseback
sales	nettomomsætning	verkopen, omzet	liikevaihto	chiffre d'affaires	Umsatzerlöse
salvage value	scrap værdi	restwaarde	jäännösarvo	valeur-résiduelle	Rest-/ Schrottwert
securities	omsætningsværdipapirer og kapitalandele	effecten	arvopaperit	titres, valeurs mobilières (B)	Wertpapier, Wertschriften (CH)
segmental reporting	segmentopgørelse, segment oplysninger	segmentatie	toimialakohtainen raportointi	information sectorielle, rapport segmentaire (B)	Segmentberichterstattung
selling expenses	distributionsomkostninger, salgsomkostninger	verkoopkosten	myyntikustannukset	charges commerciales, frais de vente (B)	Vertriebskosten
share capital	aktiekapital, selskabskapital	aandelenkapitaal	osakepääoma	capital	gezeichnetes Kapital, Aktienkapital (CH)

ITALIAN	JAPANESE	NORWEGIAN	PORTUGUESE	SPANISH, ARGENTINE (ARG)	SWEDISH
utili portati a nuovo	ryûho rieki 留保利益	opptjent egenkapital, tilbakeholdt overskudd	lucros retidos resultados no	remanente, vinstmedel	balanserade asignados (ARG)
rivalutazione	sai hyôka 再評価	*no longer permitted* ('oppskrivning')	reavaliação	revalorización, revaluación (ARG)	omvärdering till marknadsvärde
riserva di rivalutazione	sai hyôka tsumitatekin 再評価積立金	*no longer in use* (oppskrivningsfond)	reserva de reavaliação	reserva de revalorización, reserva por revalúos (ARG)	omvärderings-reserv
contributi in conto esercizio	shûeki joseikin 収益助成金	offentlig tilskudd	subsídios à exploração	subvención, subsidios recibidos (ARG)	bidrag, anslag
ricavi	shûeki 収益	inntekter	proveitos	ingresos, ingresos por ventas (ARG)	intäkter
sale and lease back	rîsu modoshi jôken tsuki baikyaku リース戻し条件付き売却	salg og tilbakeleie	venda seguida de locação sale and lease back	venta y rearrendamiento, venta y lease back	sale and lease back
vendite	uriagedaka 売上高	salgsinntekt	vendas, volume de negócios	ventas	fakturering
valore di realizzo/di mercato/di recupero	zanson kagaku 残存価額	utrangeringsverdi, restverdi	valor residual	valor residual	restvärde
titoli	yûka shôken 有価証券	verdipapirer	títulos/valores mobiliários	títulos/valores negociables	värdepapper
informazioni di settore	segumento betsu hôkoku セグメント別報告	virksomhetsområ-derapportering	informação por segmentos	contabilidad segmentada, información por segmentos de negocios (ARG)	redovisning av rörelsegrenar
costi di vendita	hambai hi 販売費	salgskostnader	despesas de venda	gastos de venta, gastos de comercialización (ARG)	försäljningskost-nader
capitale sociale	kabushiki shihonkin 株式資本金	aksjekapital	capital social, capital nominal	capital social	aktiekapital

ENGLISH	DANISH	DUTCH, BELGIAN (B)	FINNISH	FRENCH, BELGIAN (B)	GERMAN
share premium account	overkursfond	agioreserve, uitgiftepremies (B)	ylikurssirahasto	prime d'émission	(Agio-) Kapitalrücklage, Agioreserve (CH)
shareholder	aktionær, selskabsdeltager	aandeelhouder	osakkeenomistaja	actionnaire (SA), porteur de parts (SARL), associé (SNC)	Aktionär
significant influence	betydelig indflydelse	invloed van betekenis	merkittävä vaikutusvalta	influence notable, influence significative (B)	maßgeblicher Einfluss
spot rate	spot rate	actuele koers	päivän kurssi	taux (de charge) courant, taus de charge comptant (B)	Kassakurs, Tageskurs (CH)
statutory reserves	vedtægtsbestemte reserver	statutaire reserves	sidottu rahasto	réserves statutaires	satzungsmäßige Rücklagen, statutarische Reserven (CH)
stocks/inventory	varelager	voorraad	varasto	stocks	Vorräte
straight line depreciation	lineære afskrivninger	lineaire afschrijving	tasapoisto	amortissement linéaire	lineare Abschreibung
subordination agreement	aftale om fælles ledelse	bestuursovere-enkomst	päätäntävaltaan alistaminen	contrat de subordination	Beherrschungs-vertrag
subsidiary	dattervirksomhed	dochteronder-neming	tytäryhtiö	filiale	Tochtergesell-schaft
sum-of-the-years'-digit method	faldende brøks metode	methode som van de jaarnummers	aleneva poisto	amortissement dégressif	arithmetisch-degressive Abschreibung
supervisory board	respræsentantskab	raad van commissarissen	hallintoneuvosto	conseil de surveillance	Aufsichtsrat, Verwaltungsrat (CH)
tangible assets	materielle anlægsaktiver	materiële activa	kiinteä omaisuus	immobilisations corporelles, actif corporel (B)	materielle Vermö-gensgegenstände (Sachanlagen und Vorräte)

ITALIAN	JAPANESE	NORWEGIAN	PORTUGUESE	SPANISH, ARGENTINE (ARG)	SWEDISH
riserva sovrapprezzo azioni	kabushiki haraikomi jôyokin 株式払込剰余金	overkursfond	prémio de emissão	prima de emisión	reservfond, överkursfond
azionista	kabunushi 株主	aksjonær	accionista	accionista	aktieägare
influenza significativa	juyô na eikyô 重要な影響	betydelig innflytelse	influência significativa	influencia significativa	väsentligt inflytande
cambio a pronti	jikimono rêto 直物レート	spot rate	taxa de câmbio corrente	tipo de cambio vigente	dagskurs
riserve statutarie	teikan jumbikin 定款準備金	vedtektsbestemt reserveavsetning	reservas estatutárias	reservas estatutarias	bundna fonder
rimanenze di magazzino	tanaoroshi shisan 棚卸資産	varer	existências	existencias, inventario (ARG)	varulager
ammortamento a quote costanti	teigaku hô 定額法	lineær avskrivning	método das quotas constantes, depreciação linear	amortización lineal, depreciación en línea recta (ARG)	linjär avskrivning
accordo di subordinazione	jûzoku kyôtei 従属協定	*not in use*	contrato de subordinação	acuerdo de subordinación	avtal som ger ett företag rätt att utöva kontroll över annat företag
società controllata	ko gaisha 子会社	datterselskap	(empresa) subsidiária, filial	empresa subsidiaria, filial	dotterföretag
ammortamento a quote decrescenti	kyûsû hô 級数法	årssiffermetoden	método dos dígitos, método das quotas decrescentes em progressão aritmética	amortización degresiva, método de la suma de dígitos (ARG)	*not in use*
collegio sindacale	kansayaku kai 監査役会	representantskap/ styre	órgão de fiscalização, conselho fiscal	órgano supervisor	styrelse
immobilizza-zioni materiali	yûkei shisan 有形資産	materielle eiendeler, varige driftsmidler	activos tangíveis, activos corpóreos	inmovilizado material, activos fijos (ARG)	materiella tillgångar

ENGLISH	DANISH	DUTCH, BELGIAN (B)	FINNISH	FRENCH, BELGIAN (B)	GERMAN
tax accounts	selvangivelse, skatteopgørelse	aangifte belastingen	tilinpäätös ja veroilmoitus	liasse fiscale	Steuerbilanz
tax base	skatte (beregnings) grundlag	fiscale grondslag	verotusarvo	assiette de l'impôt, assiette fiscale, base fiscale d'un actif ou d'un passif (B)	Steuerwert
tax expense (tax income)	skatteomkostning	belastinglast (belastingop-brengst)	jaksotettu tulovero	charge fiscale, charge d'impôts, produit d'impôts	Steueraufwand (-ertrag)
taxable profit	skattepligtig indkomst	belastbare winst	verotettava tulo	bénéfice imposable	zu versteuerndes Einkommen, steuerlicher Verlust
temporal method	temporal metoden	temporale methode, tijdstipmethode (B)	temporaalime-netelmä	méthode temporelle	Zeitbezugsme-thode
temporary difference	midlertidig afvigelse	tijdelijk verschil	jaksotuseroista johtuvat verovelat ja verosaamiset	écart/différence temporaire, différence temporelle (B)	zeitliche Differenz
total costs method	matching princippet	integrale kostenmethode	täyskatteellinen laskenta	méthode du coût complet	Gesamtkosten-verfahren
transfer price	intern afregningspris	transferprijs	siirtohinta	prix de transfert	Verrechnungspreis
translation differences	(valuta) omreg-ningsdifference	omrekeningsver-schillen	muuntoerot	écarts de conversion	Umrechnungs-differenz
uniting/pooling of interest	fusionsmetoden	samensmelting van belangen	liiketoimintojen yhdistäminen	groupement d'intérêts communs, mise en commun d'intérêts (B)	Interessenzusam-menführung
unrealized profit	urealiseret avance	ongerealiseerde winst	realisoitumaton voitto	profit (bénéfice) non réalisé	unrealisierter Gewinn
useful economic life	brugstid, økonomisk levetid	economische levensduur	taloudellinen käyttöaika	durée de vie utile	Nutzungsdauer

ITALIAN	JAPANESE	NORWEGIAN	PORTUGUESE	SPANISH, ARGENTINE (ARG)	SWEDISH
bilancio fiscale	zeimu zaimu shohyô 税務財務諸表	skatteregnskap	contas (para efeitos) fiscais	balance impositivo (ARG)	skatte-bokslut
valore fiscale	kazei kijun gaku 課税基準額	skattemessig verdi	base fiscal	valor fiscal	skattemässigt värde
onere fiscale (provento fiscale)	sonkin (ekikin) 損金（益金）	skattekostnad (skattemessig inntekt)		gasto por impuesto de sociedades, gasto por impuesto a las ganancias (ARG)	periodens skattekostnad (skatteintäkt)
reddito imponibile	kazei shotoku 課税所得	skattepliktig overskudd	lucro tributável	base imponible	skattepliktigt resultat
metodo temporale	temporaru hô テンポラル法	temporalmetoden, penge- ikke pengeposter	método temporal	método monetario/ no monetario	monetära metoden
differenza temporanea	ichiji sai 一時差異	midlertidige forskjeller	diferença temporária	diferencias entre el valor contable de un activo o pasivo y su valor fiscal	temporär skillnad
metodo del costo totale	sô genka hô 総原価法	full tilvirkningskost	(método de) custeio total, (método de) custeio completo	método del costo total (ARG)	*not in use*
transfer price, prezzo di trasferimento	iten/furikae kakaku 移転/振替価格	internpris	preço de transferência	precio de transferencia	internpris
differenze di cambio	kansan sagaku 換算差額	omregningsdifferanse valuta	diferenças de tradução	diferencias de cambio	omräkningsdifferens
unificazione di impresa	mochibun pûringu 持分プーリング	kontinuitetsmetoden ved sammenslutning	unificação de interesses	fusión	samgående
utile non realizzato	mijitsugen rieki 未実現利益	urealisert gevinst	lucros não realizados	ganancias no realizadas	orealiserade vinster
vita utile	taiyô kikan/nensû 耐用期間/年数	økonomisk levetid	vida útil económica	vida económica útil	ekonomisk livslängd

ENGLISH	DANISH	DUTCH, BELGIAN (B)	FINNISH	FRENCH, BELGIAN (B)	GERMAN
user group	regnskabsbruger gruppe	gebruikersgroep	sidosryhmä	groupe d'usagers, groupe d'utilisateurs	Jahresab-schlussadressaten
valuation	værdiansættelse	waardering	arvostus	évaluation	Bewertung
value added tax	moms, merværdiafgift	belasting op de toegevoegde waarde	arvonlisävero	taxe à la valeur ajoutée	Mehrwert-/ Umsatzsteuer
vertical/report form	skema i beretningsform	verticale vorm	vertikaalinen muoto	forme (présentation) en liste	Staffelform, Berichtsform (CH)
voting right	stemmeret	stemrecht	äänioikeus	droit de vote	Stimmrecht
work in progress	varer under fremstilling	onderhanden werk	keskeneräiset suoritteet	produits en cours (stocks on travaux), travaux en cours (B)	unfertige Erzeugnisse, angefangene Arbeiten (CH)

ITALIAN	JAPANESE	NORWEGIAN	PORTUGUESE	SPANISH, ARGENTINE (ARG)	SWEDISH
destinatari	riyôsha shûdan 利用者集団	brukergrupper	(grupo de) utilizadores	grupo de usuarios	intressentgrupp
valutazione	hyôka 評価	vurdering	avaliação, valorização	valoración, valuación (ARG)	värdering
imposta sul valore aggiunto	fuka kachi zei 付加価値税	merverdiavgift, moms	imposto sobre o valor acrescentado	impuesto sobre el valor añadido, impuesto al valor agregado (ARG)	mervärdeskatt
bilancio in forma scalare	hôkoku keishiki 報告形式	rapportform	formato vertical, modelo de apresentação em lista	formato de lista, presentación vertical (ARG)	rapportform
diritto di voto	giketsuken 議決権	stemmerett	direito de voto	derecho de voto	rösträtt
prodotti in corso di lavorazione	shikakarihin 仕掛品	varer under tilvirkning	produtos e trabalhos em curso	productos en curso, producción en proceso (ARG)	pågående arbeten

INDEX

Introduction

Apart from the standardized contents structure and the Reference Matrix the index will provide another medium to allow easy access to and quick comparisons between the individual contributions.
In setting up the index the following abbreviations have been used:

A	Austria	**FIN**	Finland
ARG	Argentina	**I**	Italy
AUS	Australia	**IASC**	International Accounting Standards Committee
B	Belgium	**J**	Japan
CDN	Canada	**N**	Norway
CH	Switzerland	**NL**	Netherlands
D	Germany	**P**	Portugal
DK	Denmark	**S**	Sweden
E	Spain	**UK**	United Kingdom
EU	European Union	**USA**	United States of America
F	France		

Non-English terms used in the contributions will be included in the index, in order to allow quick access to the descriptions in case the English term the reader would like to refer to is unknown.

The following example with explanations will help to understand the technical presentation of the index and the page references.

prudence *see also* conservatism, scepticism

A 274, 287, 291, 295, AUS 137, 151, B 381, 393, 426, 438, CH 2472, 2483, 2538, D 1223, 1232, 1242 . . ., UK 2644, 2748

prudencia *see* prudence

E 2232, 2248, 2261, 2262

In the example above, the term ‘prudence’ is used as a headword, followed by cross-references to ‘conservatism’ and ‘scepticism’. The relevant pages where the term is used or described in more detail are given in the indented paragraph below, e.g. ‘prudence’ is to be found in the contributions to Austria on pages 274, 287 etc., as well as, for example in Germany, page 1223. Although the page numbering would suffice to find the respective page in the TRANSACC volumes (the numbering is continuous for all volumes), the inclusion of the country-abbreviations will help to identify immediately the description within a specific country contribution, thus making it very easy from the very beginning to look up those countries the reader is interested in.

Under the next headword ‘*prudencia*’, a Spanish expression, there is a cross-reference to the English term, under which, as shown before, all references to ‘prudence’ are included, as well as the page reference to Spain, page 2232.